STUDENT SOLUTIONS MANUAL
TO ACCOMPANY

CALCULUS FIFTH EDITION

by

STANLEY I. GROSSMAN

RICHARD B. LANE

University of Montana

SAUNDERS COLLEGE PUBLISHING

Harcourt Brace Jovanovich College Publishers

Fort Worth • Philadelphia • San Diego • New York • Orlando • Austin • San Antonio
Toronto • Montreal • London • Sydney • Tokyo

Grossman: Student Solutions Manual for <u>CALCULUS</u>, 5/E

ISBN 0–03–096968–9

8 9 0 1 2 3 4   021   12 11 10 9 8

Preface

This volume contains a solution for each odd-numbered problem in *Calculus*, 5th edition, by Stanley Grossman. In each set of problems, the solutions for early problems use techniques shown in the text; solutions to later problems, or alternative solutions to early ones, may demonstrate other useful techniques. Figures, tables, and graphs used in a solution are included. Related examples or problems and relevant theorems or equations are often cited.

Doing this text's calculus problems exercises a learner's algebraic skills. Solutions of most problems in the early chapters show algebraic steps in detail. In later chapters, routine algebra may be spelled out only for the first few problems of a section so that attention can focus on those parts of a solution that are genuinely new. Nonroutine algebraic manipulations are always shown.

In each section, solutions are written to be read by students whose knowledge of calculus is based on study of prior (nonoptional) sections of *Calculus*. (Note that "study" includes working a moderate number of problems!) Students with more advanced skills or knowledge will often be able to construct shorter, more clever solutions.

Please report any errors, pass along suggestions for improvements, or send other comments to me:

Richard B. Lane
Department of Mathematical Sciences
University of Montana
Missoula, MT 59812–1032

Note to Students

These solutions are written for you to use as you study *Calculus*. I recommend you defer reading my solution for a problem until you have completed your own work or have become stuck.

If you can not finish working a problem, read just enough of my solution to get past a sticking point; then try to complete your solution. For a difficult problem, you may need to go back and forth several times. (You can also use text examples this way.)

When you complete the writing of your solution, compare and contrast your work with the solution(s) shown here. Even if there is quite a difference, don't panic; there are several ways for both of us to be right. If an alternative solution is given, compare your work with it. After using standard tools on ten similar drill problems in a section, I may have chosen to demonstrate an alternative technique on the eleventh. You may prefer to keep using familiar tools and thus construct a solution that differs from mine. The Remark preceding the solutions for Chapter 2 discusses other legitimate reasons why you may get, and prefer, answers that look different from the ones shown here.

Each individual must develop her or his own ability to work routine problems; most students will benefit from spending part of their studying effort working on harder problems with fellow students. One way to check whether you're learning with others, or just letting them do your work, is to see if you can rework problems done in a group session; another is to see if you can work new problems that are similar. (Do the same type of self-test to monitor your use of these solutions.)

As you progress with your studies, you'll evolve a personal balance between reading mathematics and doing mathematics, a personal set of strategies for tackling worthwhile problems, a personal style for presenting your solutions to those problems. If you'd like to think more about general procedures for solving problems, I highly recommend *How To Solve It*, 2nd edition, by G. Polya (Princeton University Press, 1975).

Note to Instructors

These solutions are written to be read by average students as they progress through a course based upon Grossman's *Calculus*. For each problem, the solution will suggest to you the level of effort required from an average student whose knowledge of calculus is based solely on the material discussed in earlier sections (and, occasionally, in preceding problems) of that text.

On the other hand, because these solutions are complete, your students (and paper graders) will profit from your discussing the standard of work you ask of them — a brief outline of computations may suffice in many instances. Some problems have subtleties that many students will overlook, e.g., 1.7.65, 2.6.25, 3.3.31; you may want to add simplifying conditions to such problems before assigning them. Computational problems often allow several forms for a correct answer, e.g., 2.6.11 or 7.5.13; the Remark preceding the solutions for Chapter 2 discusses such ambiguities in one context.

I have occasionally cited articles published in journals such as *The American Mathematical Monthly*, *Mathematics Magazine*, or *The Mathematical Gazette*. Some of these are reprinted in *Selected Papers on Calculus* published by The Mathematical Association of America, 1969. The National Council of Teachers of Mathematics published a similar effort in 1977, *Calculus: Readings from The Mathematics Teacher*. Both have annotated bibliographies.

While preparing these solutions I made extensive use of REDUCE, muMATH, and Maple; all are systems for doing symbolic computations on a computer. They do rapid and accurate substitution, factoring, simplification, differentiation, and integration. They let me check key parts of a solution that had been done first by hand — they also helped me explore the consequences of different ways of tackling a particular problem.

In preparing this edition of these solutions I have used a variety of computer tools to do computation (numeric and symbolic), plotting, note-taking, and text-processing. I have prepared a memo discussing my experiences, good and bad. Single copies of that memo are available upon request — send a self-addressed stamped envelope (an SASE) to me at the address shown in the Preface.

Acknowledgments

Carol Johnson prepared annotated answers for problems in the first edition of Grossman's *Calculus*; I consulted her work frequently while writing my solutions for the odd-numbered problems of the second edition. Leon Gerber wrote solutions for some chapters of the 2nd–4th editions; his work and aid is much appreciated. Josef Crepeau helped with the checking of answers and solutions for this edition.

Finally, I want to express my abiding gratitude that Kenneth O. May and John Dyer-Bennet introduced me to the excitement of working with problems and to "the pleasure of taking pains" with their solution.

Dick Lane
June 1992

Table of Contents

0
Review of Some Topics in Algebra

Solutions 0.1   *Review of real numbers and absolute value*   (pages 8-10)

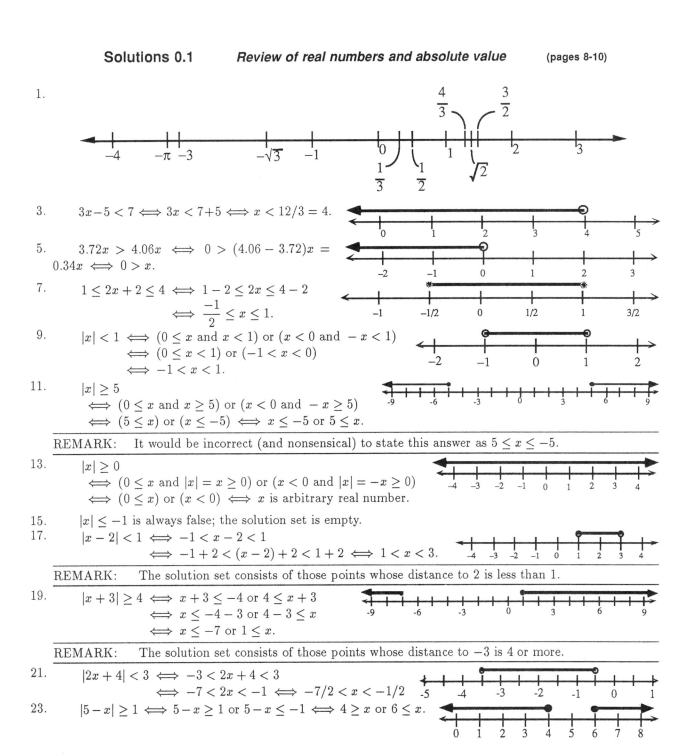

1.

3. $3x - 5 < 7 \iff 3x < 7 + 5 \iff x < 12/3 = 4.$

5. $3.72x > 4.06x \iff 0 > (4.06 - 3.72)x = 0.34x \iff 0 > x.$

7. $1 \leq 2x + 2 \leq 4 \iff 1 - 2 \leq 2x \leq 4 - 2$
     $\iff \dfrac{-1}{2} \leq x \leq 1.$

9. $|x| < 1 \iff (0 \leq x \text{ and } x < 1) \text{ or } (x < 0 \text{ and } -x < 1)$
     $\iff (0 \leq x < 1) \text{ or } (-1 < x < 0)$
     $\iff -1 < x < 1.$

11. $|x| \geq 5$
     $\iff (0 \leq x \text{ and } x \geq 5) \text{ or } (x < 0 \text{ and } -x \geq 5)$
     $\iff (5 \leq x) \text{ or } (x \leq -5) \iff x \leq -5 \text{ or } 5 \leq x.$

REMARK: It would be incorrect (and nonsensical) to state this answer as $5 \leq x \leq -5$.

13. $|x| \geq 0$
     $\iff (0 \leq x \text{ and } |x| = x \geq 0) \text{ or } (x < 0 \text{ and } |x| = -x \geq 0)$
     $\iff (0 \leq x) \text{ or } (x < 0) \iff x \text{ is arbitrary real number.}$

15. $|x| \leq -1$ is always false; the solution set is empty.

17. $|x - 2| < 1 \iff -1 < x - 2 < 1$
     $\iff -1 + 2 < (x - 2) + 2 < 1 + 2 \iff 1 < x < 3.$

REMARK: The solution set consists of those points whose distance to 2 is less than 1.

19. $|x + 3| \geq 4 \iff x + 3 \leq -4 \text{ or } 4 \leq x + 3$
     $\iff x \leq -4 - 3 \text{ or } 4 - 3 \leq x$
     $\iff x \leq -7 \text{ or } 1 \leq x.$

REMARK: The solution set consists of those points whose distance to -3 is 4 or more.

21. $|2x + 4| < 3 \iff -3 < 2x + 4 < 3$
     $\iff -7 < 2x < -1 \iff -7/2 < x < -1/2$

23. $|5 - x| \geq 1 \iff 5 - x \geq 1 \text{ or } 5 - x \leq -1 \iff 4 \geq x \text{ or } 6 \leq x.$

25. $$-4 + |-3x - 4| > 2$$
$$\iff |-3x - 4| > 2 + 4 = 6$$
$$\iff -3x - 4 > 6 \text{ or } -3x - 4 < -6$$
$$\iff -3x > 10 \text{ or } -3x < -2$$
$$\iff x < -10/3 \text{ or } x > 2/3.$$

27. $6 - 4x$ changes sign at $6/4 = 3/2$ and $x - 2$ changes sign at 2. These two points split the number line into three intervals and split our analysis into three cases.

$x < 3/2$: $|6 - 4x| \geq |x - 2| \iff 6 - 4x \geq -(x - 2) \iff 6 - 2 \geq -x + 4x \iff 4/3 \geq x$; this part of the solution set is $(-\infty, 3/2) \cap (-\infty, 4/3] = (-\infty, 4/3]$.

$3/2 \leq x < 2$: $|6 - 4x| \geq |x - 2| \iff -(6 - 4x) \geq -(x - 2) \iff 4x + x \geq 2 + 6 \iff x \geq 8/5$; this part of the solution set is $[3/2, 2) \cap [8/5, \infty) = [8/5, 2)$.

$2 \leq x$: $|6 - 4x| \geq |x - 2| \iff -(6 - 4x) \geq x - 2 \iff 4x - x \geq -2 + 6 \iff x \geq 4/3$; this part of the solution set is $[2, \infty) \cap [4/3, \infty) = [2, \infty)$.

The full solution set is $(-\infty, 4/3] \cup [8/5, 2) \cup [2, \infty) = (-\infty, 4/3] \cup [8/5, \infty)$.

REMARK: $|6 - 4x| \geq |x - 2| \iff |6 - 4x|^2 \geq |x - 2|^2 \iff 0 \leq |6 - 4x|^2 - |x - 2|^2 = (3x - 4)(5x - 8)$; this is true if and only if $3x - 4$ and $5x - 8$ have the same sign or one is zero.

29.
$$\left| \frac{3x + 17}{4} \right| > 9 \iff \frac{3x + 17}{4} > 9 \text{ or } \frac{3x + 17}{4} < -9$$
$$\iff 3x + 17 > 36 \text{ or } 3x + 17 < -36$$
$$\iff 3x > 19 \text{ or } 3x < -53$$
$$\iff x > 19/3 \text{ or } x < -53/3.$$

31.   If $c > 0$, then $|ax + b| \geq c \iff ax + b \geq c \text{ or } ax + b \leq -c \iff ax \geq c - b \text{ or } ax \leq -c - b$. If we also know that $a < 0$, then we may further infer $x \leq (c - b)/a \text{ or } x \geq (-c - b)/a$.

33. $1/x$ is defined if and only if $x \neq 0$. If $x < 0$, then $1/x < 0$ (because x times $1/x$ equals 1 which is greater than 0); therefore

$1/x > 3 \iff 1/x > 3 \text{ and } x > 0 \iff 1/3 > x \text{ and } x > 0$ which implies the solution set is $(0, 1/3)$.

35. $1/x$ is undefined if $x = 0$ so we can exclude that value of x. The following equivalences use the fact that a product of two terms is positive if and only if both terms are positive (and the fact that such a product is negative if and only if the terms have opposite sign).

$$x > \frac{1}{x} \iff (x < 0 \text{ and } x^2 < 1) \text{ or } (x > 0 \text{ and } x^2 > 1)$$
$$\iff (x < 0 \text{ and } x^2 - 1 < 0) \text{ or } (x > 0 \text{ and } (x + 1)(x - 1) > 0)$$
$$\iff \begin{cases} x + 1 < 0 < x - 1 \text{ or } x - 1 < 0 < x + 1 & \text{if } x < 0 \\ x - 1 < x + 1 < 0 \text{ or } 0 < x - 1 < x + 1 & \text{if } 0 < x \end{cases} \iff \begin{cases} -1 < x < 1 & \text{if } x < 0 \\ x < -1 \text{ or } 1 < x & \text{if } 0 < x \end{cases}$$
$$\iff -1 < x < 0 \text{ or } 1 < x.$$

REMARK: If $x \neq 0$, then $x^2 > 0$ and $x > 1/x \iff x^3 > x \iff x^3 - x = (x + 1) \cdot x \cdot (x - 1) > 0$. The product $(x + 1) \cdot x \cdot (x - 1)$ changes sign at -1, 0, and 1; the product is positive if and only if an even number (0 or 2) of terms are negative.

37.
$$-1 < x(x - 2) \iff 0 < x(x - 2) + 1 = (x - 1)^2$$
$$\iff 0 \neq x - 1 \iff 1 \neq x;$$
the solution set is $\mathbb{R} - \{1\} = (-\infty, 1) \cup (1, \infty)$.

39.   a. $x \notin (-3, 3) \iff x \leq -3 \text{ or } 3 \leq x \iff 3 \leq |x|$.
    b. $-2 < x < 4 \iff -3 < x - 1 < 3 \iff |x - 1| < 3$.
    c. $-4 < x < 10 \iff -7 < x - 3 < 7 \iff |x - 3| < 7$.
    d.   8 is midway between 5 and 11; the distance from 8 to either 5 or 11 is 3. Therefore $x \notin [5, 11] \iff (x < 5) \text{ or } (11 < x) \iff (x - 8 < -3) \text{ or } (3 < x - 8) \iff 3 < |x - 8|$.
    e.   11/2 is midway between 2 and 9; the distance from 11/2 to either 2 or 9 is 7/2. Therefore $x \in (-\infty, 2] \cup [9, \infty) \iff |x - 11/2| \geq 7/2$.
    f. $|x - 5| < |x - 0| \iff |x - 5| < |x|$.

41. $(s+t)/2$ is the number midway between s and t while $|s-t|$ is the distance between s and t. If you start at the midpoint of an interval and then move to the right by a distance equal to half the interval's length, you arrive at the right end of the interval. Hence $(s+t)/2 + |s-t|/2$ computes the right end of the interval whose endpoints are s and t . ∎

43.   Suppose p_0, q_0, p_1, q_1 are all positive, then

$$\frac{p_0}{q_0} < \frac{p_1}{q_1} \implies p_0 q_1 < p_1 q_0 \implies p_0 q_0 + p_0 q_1 < p_0 q_0 + p_1 q_0 \text{ and } p_0 q_1 + p_1 q_1 < p_1 q_0 + p_1 q_1$$

$$\implies p_0 (q_0 + q_1) < (p_0 + p_1) q_0 \text{ and } (p_0 + p_1) q_1 < p_1 (q_0 + q_1)$$

$$\implies \frac{p_0}{q_0} < \frac{p_0 + p_1}{q_0 + q_1} \text{ and } \frac{p_0 + p_1}{q_0 + q_1} < \frac{p_1}{q_1}. \text{∎}$$

45.   a.   If $0 \leq x, y$, then $0 \leq x + y$ and $|x+y| = x+y = |x| + |y|$; if $x, y \leq 0$, then $x + y \leq 0$ and $|x+y| = -(x+y) = (-x) + (-y) = |x| + |y|$.

   b.   If $x < 0 < y$, then $|x| + |y| = -x + y$. $x < 0 < y$ also implies $x < -x$ and $-y < y$; these imply $x + y < -x + y$ and $-x - y < -x + y$. Since $|x+y| = \max\{x+y, -(x+y)\}$, we conclude that $x < 0 < y \implies |x+y| < |x| + |y|$. Finally, because the statement is symmetric in x and y, we have also proved it true in case $y < 0 < x$. ∎

47. $|x| = |(x-y) + y| \leq |x-y| + |y|$, thus $|x| - |y| \leq |x-y|$; similarly, $|y| = |(y-x) + x| \leq |y-x| + |x| \implies |y| - |x| \leq |y-x|$. Because $|y-x| = |-(x-y)| = |x-y|$ and because $|y| - |x| = -(|x| - |y|)$, we have shown that $|x| - |y| \leq |x-y|$ and $-(|x| - |y|) \leq |x-y|$ Therefore $\big||x| - |y|\big| \leq |x-y|$. ∎

49. $|w - 1.3| \leq 0.05 \implies 1.25 \leq w \leq 1.35 \implies 1.25^2 = 1.5625 \leq w^2 \leq 1.8225 = 1.35^2 \implies -0.1375 \leq w^2 - 1.7 \leq 0.1225 \implies |x^2 - 1.7|$ may be as large as 0.1375. Therefore 1.7 does not estimate w^2 with one decimal place accuracy.

51.   a. $0 \leq \left(\sqrt{a} - \sqrt{b}\right)^2 = a - 2\sqrt{ab} + b \implies \sqrt{ab} \leq (a+b)/2.$ ∎   (Equality holds if and only if $a = b$.)

   b.   If the rectangle has sides h and w, then $15 = \sqrt{225} = \sqrt{hw} \leq (h+w)/2$; therefore the perimeter $= 2(h+w) \geq 60$ with "equality holding" only if $h = w = 15$.

   c. $300/4 = (h+w)/2 \geq \sqrt{hw} = \sqrt{\text{area}}$, thus $75^2 >$ area unless $h = w = 75$.

53.   *Proof, by contradiction, that $\sqrt{2}$ is irrational*: Suppose $\sqrt{2}$ were rational; then there would be integers M and N, $N \neq 0$, such that $\sqrt{2} = M/N$. Let C be the largest common factor of M and N; then there are integers m and n such that $M = mC$, $N = nC$, $n \neq 0$, and 1 is the only positive integer dividing both m and n. Therefore $\sqrt{2} = (mC)/(nC) = m/n$. This implies $2 = (m/n)^2 = m^2/n^2$ and $2n^2 = m^2$. Hence m^2 is divisible by 2, so m too must be divisible by 2. Thus there is an integer K such that $m = 2K$. Hence $2n^2 = m^2 = (2K)^2 = 4K^2$. But $n^2 = 2K^2$ implies that n^2 and n are divisible by 2. Therefore $\sqrt{2} = m/n$ implies 2 divides both m and n, which is impossible since 1 is the only positive integer dividing both. Because a contradiction follows from the assumption that $\sqrt{2}$ is rational, that assumption must be false — we conclude that $\sqrt{2}$ must be irrational. ∎

55. $-1 = a^3 \iff 0 = a^3 + 1 = (a+1)(a^2 - a + 1) \iff 0 = a+1 \text{ or } 0 = a^2 - a + 1 \iff a = -1 \text{ or } a^2 - a + 1 = 0$. Although $a^2 - a + 1 = 0 \implies a^3 = -1$, the implication does not go the other way.

1.

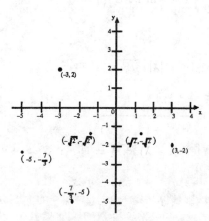

3. $\sqrt{(4-(-7))^2+(3-2)^2} = \sqrt{11^2+1^2} = \sqrt{122}$.

5. $\sqrt{(b-a)^2+(a-b)^2} = \sqrt{2(a-b)^2} = \sqrt{2}\,|a-b|$.

7.   The circle of radius $\sqrt{2}$ centered at $(1,1)$ satisfies the equation $(x-1)^2+(y-1)^2 = 2$.

7.

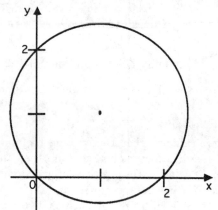

9.

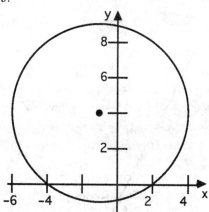

9.   The circle of radius 5 centered at $(-1,4)$ satisfies the equation $(x+1)^2+(y-4)^2 = 5^2$.

11. $(x-7)^2+(y-3)^2 = (-1-7)^2+(7-3)^2 = 64+16 = 80$.

13.   The distance from $(3,3)$ to $(7,-5)$ is $\sqrt{(7-3)^2+(-5-3)^2} = \sqrt{80}$; because that distance is less than the circle's radius $(80 < 81 = 9^2)$, the point $(3,3)$ is inside the circle.

15.   a. $((2+5)/2,(5+12)/2) = (7/2,17/2)$.

     b. $((-3+4)/2,(7+(-2))/2) = (1/2,5/2)$.

17.   The center of a circle is the midpoint of each diameter; the midpoint of the line segment between $(8,-4)$ and $(2,6)$ is $((8+2)/2,(-4+6)/2) = (5,1)$. The radius of the circle is one-half the length of a diameter: $\frac{1}{2}\sqrt{(8-2)^2+(-4-6)^2} = \frac{1}{2}\sqrt{136} = \sqrt{34}$. Points on this circle satisfy the equation $(x-5)^2+(y-1)^2 = 34$.

19. $\sqrt{(a-a)^2+(y_1-y_2)^2} = \sqrt{(y_1-y_2)^2} = |y_1-y_2|$.

21. $x^2-6x+y^2+4y-12 = 0 \iff (x^2-6x+9)+(y^2+4y+4) = 12+9+4 \iff (x-3)^2+(y+2)^2 = 5^2$; this is an equation for the circle of radius 5 centered at $(3,-2)$. ∎

23.   A rectangle can be rotated so that its sides are parallel to the coordinate axes, it can then be moved up-down and left-right so that its lower-left corner is at the origin. If the bottom side has length W and the left side has length H, the corners of the rectangle are then at the points $(0,0)$, $(W,0)$, (W,H), and $(0,H)$. The diagonal from lower-left to upper-right has length $\sqrt{(W-0)^2+(H-0)^2} = \sqrt{W^2+H^2}$; the diagonal from upper-left to lower-right has length $\sqrt{(W-0)^2+(0-H)^2} = \sqrt{W^2+H^2}$. ∎

25.   Let P_λ be the point $([1 - \lambda]x_0 + \lambda x_1, [1 - \lambda]y_0 + \lambda y_1)$. If $0 < \lambda < 1$, P_λ is on the line segment between $P_0 = (x_0, y_0)$ and $P_1 = (x_1, y_1)$ because $\overline{P_0 P_\lambda} + \overline{P_\lambda P_1} = \overline{P_0 P_1}$. The proof will be completed by doing the computations which verify that last equation:

$$\overline{P_0 P_\lambda} = \sqrt{([1 - \lambda]x_0 + \lambda x_1 - x_0)^2 + ([1 - \lambda]y_0 + \lambda y_1 - y_0)^2}$$
$$= \sqrt{[\lambda(x_1 - x_0)]^2 + [\lambda(y_1 - y_0)]^2} = |\lambda|\sqrt{(x_1 - x_0)^2 + (y_1 - y_0)^2} = |\lambda|\,\overline{P_0 P_1}$$

and

$$\overline{P_\lambda P_1} = \sqrt{(x_1 - [1 - \lambda]x_0 - \lambda x_1)^2 + (y_1 - [1 - \lambda]y_0 - \lambda y_1)^2}$$
$$= \sqrt{[(1 - \lambda)(x_1 - x_0)]^2 + [(1 - \lambda)(y_1 - y_0)]^2} = |1 - \lambda|\sqrt{(x_1 - x_0)^2 + (y_1 - y_0)^2} = |1 - \lambda|\,\overline{P_0 P_1}\,.$$

If $0 < \lambda < 1$, then $|\lambda| = \lambda$, $|1 - \lambda| = 1 - \lambda$, and $\overline{P_0 P_\lambda} + \overline{P_\lambda P_1} = \lambda\,\overline{P_0 P_1} + (1 - \lambda)\,\overline{P_0 P_1} = \overline{P_0 P_1}$. ∎

Solutions 0.3   *Lines*   (pages 21-3)

1. $m = \dfrac{4 - 6}{2 - 1} = \dfrac{-2}{1} = -2.$

1.

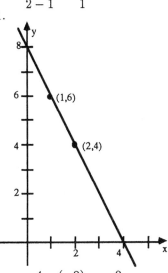

3.

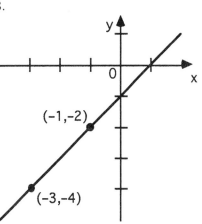

3. $m = \dfrac{-4 - (-2)}{-3 - (-1)} = \dfrac{-2}{-2} = 1.$

5. $m = \dfrac{7 - 7}{-4 - 1} = \dfrac{0}{-5} = 0.$

5.

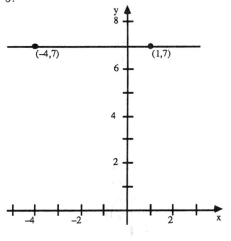

7.

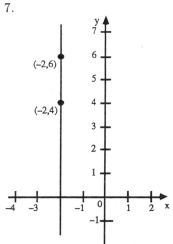

7. m is undefined, the line is vertical, because $x_1 - x_0 = (-2) - (-2) = 0$.

9. $\dfrac{y - (-7)}{x - 4} = 0$ implies $y = -7$.

9.

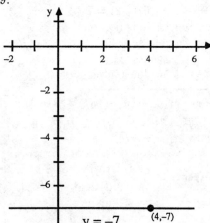

1.

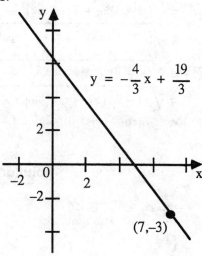

11. $\dfrac{y - (-3)}{x - 7} = -\dfrac{4}{3} \implies y + 3 = -\dfrac{4}{3}(x - 7) \implies y - -\dfrac{4}{3}x + \dfrac{19}{3}$.

13. $\dfrac{y - 2}{x - 1} = \dfrac{6 - 2}{3 - 1} = 2 \implies y - 2 = 2(x - 1) \implies y = 2x$.

13.   15.

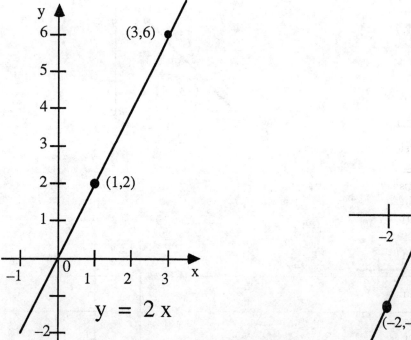

$$y = 2x$$

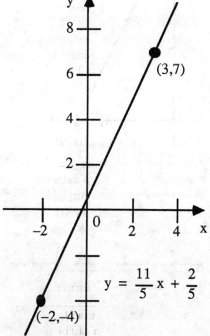

15. $\dfrac{y - (-4)}{x - (-2)} = \dfrac{7 - (-4)}{3 - (-2)} = \dfrac{11}{5} \implies y + 4 = \dfrac{11}{5}(x + 2) \implies y = \dfrac{11}{5}x + \dfrac{2}{5}$.

17.   The line through (a, b) with slope $m = 0$ satisfies the equation $\dfrac{y - b}{x - a} = 0$; this simplifies to $y = b$.

19.   The line $2x + 5y = 6$ has slope $-2/5$; any line parallel to it will also have slope equal to $-2/5$. Because $y = (-2/5)x + b \iff 5y = -2x + 5b \iff 2x + 5y = 5b$, the line through $(-1, 1)$ that is parallel to $2x + 5y = 6$ has standard equation $2x + 5y = 2(-1) + 5(1) = 3$.

21. If $x + 3y = 7$, then $3y = -x + 7$ and $y = \frac{-1}{3} \cdot x + \frac{7}{3}$. Any line perpendicular to this one will have slope $-1/\left(\frac{-1}{3}\right) = 3$. The line with slope 3 which is on $(0,1)$ has point-slope equation $(y-1)/(x-0) = 3$; this implies $y = 3x + 1$ and $-3x + y = 1$.

23. If $y - 2x = 4$, then $y = 2x + 4$; if $4x - 2y = 6$, then $y = [(-4)/(-2)]x + 6/(-2) = 2x - 3$. The two lines are parallel because they have the same slope but different y-intercepts; there is no point of intersection.

25. Suppose (a,b) is on both lines, then $3a + 4b = 5$ and $6a - 7b = 8$. Thus $3a = 5 - 4b$ and $8 + 7b = 6a = 2 \cdot (3a) = 2 \cdot (5 - 4b) = 10 - 8b$; the equation involving only b is equivalent to $15b = 7b + 8b = 10 - 8 = 2$, hence $b = 2/15$. Therefore $a = (5 - 4b)/3 = (5 - 8/15)/3 = 67/45$. The two lines intersect at $(67/45, 2/15)$.

27. $m = \tan 30° = 1/\sqrt{3}$.

29. $m = \tan 45° = 1$.

31. $m = \tan 120° = -\sqrt{3}$.

33. If $0° \le \theta < 180°$ and $\tan\theta = 1$, then $\theta = 45°$. (Measured in radians, $\theta = \pi/4$.)

35. If $0° \le \theta < 180°$ and $\tan\theta = -1/\sqrt{3}$, then $\theta = 150°$. (Measured in radians, $\theta = 5\pi/6$.)

37. If $0° \le \theta < 180°$ and $\tan\theta = 2$, then $\theta \approx 63.43°$. (Measured in radians, $\theta \approx 1.107$.)

39. The line on $(0,2)$ and $(1,1)$ has slope $(1-2)/(1-0) = -1$; the line on $(0,2)$ and $(5,-3)$ has slope $(-3-2)/(5-0) = -1$. These lines have the same slope and pass through a common point, $(0,2)$, hence they are the same line; therefore $(0,2)$, $(1,1)$, $(5,-3)$ are collinear. (The line has equation $x + y = 2$.)

41. The line on $(2,1)$ and $(3,3)$ has slope $(3-1)/(3-2) = 2$; the line on $(2,1)$ and $(-5,-13)$ has slope $(-13-1)/(-5-2) = 2$. These lines have the same slope and pass through a common point, $(2,1)$, hence they are the same line; therefore $(2,1)$, $(3,3)$, $(-5,-13)$ are collinear. (The line has equation $y = 2x - 3$.)

43. The line on $(1,8)$ and $(2,9)$ has slope $(9-8)/(2-1) = 1$; the line on $(1,2)$ and $(0,1)$ has slope $(1-2)/(0-1) = 1$. The two lines are parallel because they have the same slope. (The lines satisfy $y = x + 7$, $y = x + 1$ respectively.)

45. The line on $(0,2)$ and $(-2,0)$ has slope $(0-2)/(-2-0) = 1$; the line on $(0,3)$ and $(3,0)$ has slope $(0-3)/(3-0) = -1$. These lines are perpendicular because the product of their slopes is $1 \cdot (-1) = -1$.

47. The line on $(5,2)$ and $(1,7)$ has slope $(7-2)/(1-5) = -5/4$; the line on $(2,5)$ and $(7,1)$ has slope $(1-5)/(7-2) = -4/5$. Because $-5/4 \ne -4/5$, the lines are not parallel; because $(-5/4) \cdot (-4/5) = 1 \ne -1$, the lines are not perpendicular.

49. The circle $x^2 + y^2 = 1$ is centered at the origin, the radial line through $\left(1/\sqrt{2}, 1/\sqrt{2}\right)$ has slope 1 (the radial line is $y = x$). The tangent at that point has slope -1 and equation $y = (-1)(x - 1/\sqrt{2}) + 1/\sqrt{2} = -x + \sqrt{2}$.

51. $(3,-2)$ is the center of the circle, the radial line through $(4,0)$ has slope $(0-(-2))/(4-3) = 2$. Hence the line tangent at $(4,0)$ has slope $-1/2$ and equation $y = -\frac{1}{2}(x-4) + 0 = -\frac{1}{2}x + 2$.

53. If $2x + 3y = -1$, then $y = -\frac{2}{3}x - \frac{1}{3}$. Hence L_P has slope $(-1)/(-\frac{2}{3}) = \frac{3}{2}$; since L_P goes through $(0,0)$, its y-intercept is 0 and L_P satisfies the equation $y = \frac{3}{2}x$. If (a,b) is the point where L and L_P intersect, then $-\frac{2}{3}a - \frac{1}{3} = b = \frac{3}{2}a$ which implies $-\frac{1}{3} = \left(\frac{3}{2} + \frac{2}{3}\right)a = \frac{13}{6}a$. Therefore $a = -\frac{1}{3} \cdot \frac{6}{13} = -\frac{2}{13}$ and $b = \frac{3}{2} \cdot \frac{-2}{13} = -\frac{3}{13}$. The distance between L and $(0,0)$ is the distance between $\left(-\frac{2}{13}, -\frac{3}{13}\right)$ and $(0,0)$ which equals
$$\sqrt{\left(\frac{-2}{13} - 0\right)^2 + \left(\frac{-3}{13} - 0\right)^2} = \sqrt{\frac{4+9}{13^2}} = \frac{1}{\sqrt{13}}.$$

REMARK: Using the formula in Problem 63 we compute $\dfrac{|2 \cdot 0 + 3 \cdot 0 + 1|}{\sqrt{2^2 + 3^2}}$.

55. Any line perpendicular to $L: x + 2y = 1$ (divide the text expressions by 3 and write the x-term first) has a standard equation of the form $2x - y = C$; since L_P passes through $(8,-1)$ we know that L_P satisfies $2x - y = 2 \cdot 8 - (-1) = 17$. Therefore

(a,b) is on L and $L_P \iff \begin{array}{l} a + 2b = 1 \\ 2a - b = 17 \end{array} \iff \begin{array}{l} [1 + 2(2)]a \quad = 1 + 2(17) \\ [-1 - 2(2)]b = 17 - 2(1) \end{array} \iff \begin{array}{l} a \quad = 35/5 = 7 \\ b = 15/(-5) = -3 \end{array}$

The distance between L and $(8,-1)$ is $\sqrt{(8-7)^2 + (-1-(-3))^2} = \sqrt{1^2 + 2^2} = \sqrt{5}$.

REMARK: Using the formula in Problem 63 we compute $\dfrac{|3 \cdot 8 + 6 \cdot (-1) + (-3)|}{\sqrt{3^2 + 6^2}}$.

57. The line on $(a, \sqrt{a})$ and $(a+h, \sqrt{a+h})$ has slope

$$\frac{\sqrt{a+h} - \sqrt{a}}{(a+h) - a} = \frac{\sqrt{a+h} - \sqrt{a}}{h} \cdot \frac{\sqrt{a+h} + \sqrt{a}}{\sqrt{a+h} + \sqrt{a}} = \frac{\left(\sqrt{a+h}\right)^2 - \left(\sqrt{a}\right)^2}{h \cdot \left(\sqrt{a+h} + \sqrt{a}\right)}$$

$$= \frac{(a+h) - a}{h \cdot \left(\sqrt{a+h} + \sqrt{a}\right)} = \frac{h}{h \cdot \left(\sqrt{a+h} + \sqrt{a}\right)} = \frac{1}{\left(\sqrt{a+h} + \sqrt{a}\right)}. \blacksquare$$

59. The circle $x^2 + y^2 = r^2$ is centered at $(0,0)$. The radial line from $(0,0)$ to (a,b) satisfies the equation $y = \frac{b}{a}x$. As discussed prior to Problems 49-52, the line tangent to the circle at (a,b) will pass through that point and will be perpendicular to this radial line; therefore it will satisfy the equation $y - b = -\frac{a}{b}(x - a)$. This equation is equivalent to $ax + by = a^2 + b^2 = r^2$ (recall that (a,b) satisfies the equation $x^2 + y^2 = r^2$).

61. Suppose line L_1 has slope m_1 and line L_2 has slope m_2. If L_1 is parallel to L_2, then $m_1 = \tan\theta = m_2$, where θ is the common angle of inclination of the two lines. If $\tan\theta_1 = m_1 = m_2 = \tan\theta_2$, then $\theta_1 = \theta_2$ and the lines have the same inclination. $\blacksquare$

63. The line perpendicular to $ax + by + c = 0$ and passing through (x_0, y_0) has standard equation $bx - ay = bx_0 - ay_0$. (See Problem 22.) The two lines intersect at a point whose x-coordinate satisfies

$$\frac{b}{a}(x - x_0) + y_0 = y = -\frac{a}{b}x - \frac{c}{b}.$$

Solving for that x-coordinate we find

$$\left(\frac{a}{b} + \frac{b}{a}\right)x = \frac{b}{a}x_0 - y_0 - \frac{c}{b}, \qquad \text{thus} \qquad x = \frac{\frac{b}{a}x_0 - y_0 - \frac{c}{b}}{\frac{a}{b} + \frac{b}{a}} = \frac{b(bx_0 - ay_0) - ac}{a^2 + b^2}.$$

Therefore the y-coordinate of the intersection point is

$$y = -\frac{a}{b}\left(\frac{b(bx_0 - ay_0) - ac}{a^2 + b^2}\right) - \frac{c}{b} = \frac{a(ay_0 - bx_0) - bc}{a^2 + b^2}.$$

(Although the computations shown require $a \neq 0 \neq b$, the formulas for the intersection point are correct even if a or b is 0.) The distance between (x_0, y_0) and the line $ax + by + c = 0$ is

$$\sqrt{\left(x_0 - \frac{b(bx_0 - ay_0) - ac}{a^2 + b^2}\right)^2 + \left(y_0 - \frac{a(ay_0 - bx_0) - bc}{a^2 + b^2}\right)^2}$$

$$= \sqrt{\left(\frac{a^2 x_0 + aby_0 + ac}{a^2 + b^2}\right)^2 + \left(\frac{b^2 y_0 + abx_0 + bc}{a^2 + b^2}\right)^2}$$

$$= \sqrt{\frac{(a^2 + b^2)(ax_0 + by_0 + c)^2}{(a^2 + b^2)^2}} = \frac{|ax_0 + by_0 + c|}{\sqrt{a^2 + b^2}}. \blacksquare$$

REMARK: Once you understand vectors, you'll be able to design a briefer proof.

65.
$$(x - 3)^2 + (y - 10)^2 = (x + 1)^2 + (y - 5)^2$$
$$\Longleftrightarrow x^2 - 6x + 9 + y^2 - 20y + 100 = x^2 + 2x + 1 + y^2 - 10y + 25$$
$$\Longleftrightarrow -8x - 10y = -83 \qquad \Longleftrightarrow \qquad y = -\frac{4}{5}x + \frac{83}{10}.$$

This is the equation for a straight line; the line passes through $(1, 15/2)$, the midpoint of the segment between $(3, 10)$ and $(-1, 5)$; the slope of this line, $-4/5$, is the negative reciprocal of slope of the line $y = \frac{5}{4}(x - 3) + 10$ determined by $(3, 10)$ and $(-1, 5)$ — this line is the perpendicular bisector of the segment joining $(3, 10)$ and $(-1, 5)$. $\blacksquare$

REMARK: A point is on the perpendicular bisector of a line segment if and only if it is equidistant from the endpoints of the segment.

67. The area of a triangle equals one-half the product of the length of one side times the length of the altitude from the opposite vertex.

The side between (x_1, y_1) and (x_2, y_2) has length $\sqrt{(x_2 - x_1)^2 + (y_1 - y_2)^2}$. That side is on the line with standard equation $(y_1 - y_2)x + (x_2 - x_1)y + (x_1 y_2 - x_2 y_1) = 0$ and we compute the length of the altitude from (x_3, y_3) to this line by using the result of Problem 63. Therefore the triangle's area is

$$\frac{1}{2} \cdot \sqrt{(x_2 - x_1)^2 + (y_2 - y_1)^2} \cdot \frac{|(y_1 - y_2)\,x_3 + (x_2 - x_1)\,y_3 + (x_1 y_2 - x_2 y_1)|}{\sqrt{(y_1 - y_2)^2 + (x_2 - x_1)^2}}$$

$$= \frac{1}{2} \cdot |(y_1 - y_2)\,x_3 + (x_2 - x_1)\,y_3 + (x_1 y_2 - x_2 y_1)| = \frac{1}{2} \cdot |x_1 y_2 + x_2 y_3 + x_3 y_1 - x_1 y_3 - x_2 y_1 - x_3 y_2|. \blacksquare$$

REMARK: I remember this formula for a triangle's area more easily in the following form:

$$\frac{1}{2} \cdot |(x_1 y_2 - x_2 y_1) + (x_2 y_3 - x_3 y_2) + (x_3 y_1 - x_1 y_3)|.$$

Solutions 0.4   *Functions and their graphs*   (pages 35-8)

1.   If $f(x) = x^2$, then $f(0) = 0^2 = 0$, $f(5) = 5^2 = 25$, $f(v) = v^2$, and $f(2 + h) = (2 + h)^2 = 4 + 4h + h^2$.

3.   If $f(x) = \sqrt{x}$, then $f(0) = \sqrt{0} = 0$, $f(1) = \sqrt{1} = 1$, $f(9) = \sqrt{9} = 3$, $f(v^2) = \sqrt{v^2} = |v|$, $f(z^4) = \sqrt{z^4} = z^2$, $f(4 + h) = \sqrt{4 + h}$.

REMARK: Let $v = -3$ to see that it is not correct to state $\sqrt{v^2} = v$.

5.   If $f(x) = x^4$, then $f(0) = 0^4 = 0$, $f(2) = 2^4 = 16$, $f(-2) = (-2)^4 = 16$, $f(\sqrt{5}) = (\sqrt{5})^4 = 5^2 = 25$, $f(s^{1/5}) = (s^{1/5})^4 = s^{4/5}$, and $f(s - 1) = (s - 1)^4$.

7.   If $g(t) = \sqrt{t + 1}$, then $g(0) = \sqrt{0 + 1} = \sqrt{1} = 1$, $g(-1) = \sqrt{-1 + 1} = \sqrt{0} = 0$, $g(3) = \sqrt{3 + 1} = \sqrt{4} = 2$, $g(7) = \sqrt{7 + 1} = \sqrt{8} = 2\sqrt{2}$, $g(n^3 - 1) = \sqrt{n^3 - 1 + 1} = n^{3/2}$, and $g(1/w) = \sqrt{1/w + 1}$.

9.   If $h(z) = 1 + z + z^2$, then $h(0) = 1 + 0 + 0^2 = 1 + 0 + 0 = 1$, $h(2) = 1 + 2 + 2^2 = 1 + 2 + 4 = 7$, $h(1/3) = 1 + 1/3 + 1/3^2 = 13/9$, $h(-1/2) = 1 + (-1/2) + (-1/2)^2 = 1 - 1/2 + 1/4 = 3/4$, $h(z^5) = 1 + z^5 + (z^5)^2 = 1 + z^5 + z^{10}$, and $h(1/u^2) = 1 + 1/u^2 + (1/u^2)^2 = 1 + 1/u^2 + 1/u^4$.

11. f is a function.

13. $g(a)$ does not have a unique value, hence g is not a function.

15. f is a function.

17. $f(w)$ is undefined, hence F is not a function.

19. $x - 6 = 2(y - 3) \iff y = \frac{1}{2}x$; y is a function of x with domain $= (-\infty, \infty)$.

21. $x^2 - 3y = 4 \iff y = \frac{1}{3}x^2 - \frac{4}{3}$; y is a function of x with domain $= (-\infty, \infty)$.

23.   If $y = |x| - 4$, then y is a function of x with domain $= (-\infty, \infty)$.

25. $(1, \sqrt{3})$ and $(1, -\sqrt{3})$ are both on the graph of $x^2 + y^2 = 4$. Since we can't get a unique value of y corresponding to $x = 1$, y is not a function of x. (Note that the discussion is similar for any other choice of x such that $-2 < x < 2$.)

27.   If $y^3 - x = 0$, then $y = \sqrt[3]{x}$; y is a function of x with domain $(-\infty, \infty)$.

29. $g(t) = 4t - 5$ is defined for each choice of t so the domain of g is $\mathbb{R}$. For any choice of s there is some t so that $s = g(t)$ (choose t to be $(s + 5)/4$); therefore the range of g is $\mathbb{R}$.

31. $h(u) = 1/(u + 1)$ is defined provided $u \neq -1$ so the domain of h is $\mathbb{R} - \{-1\} = (-\infty, -1) \cup (-1, \infty)$. $1/(u + 1)$ is never equal to zero, but for any nonzero choice of v there is some u such that $v = h(u)$ (choose $u = -1 + 1/v$); hence the range of h is $\mathbb{R} - \{0\} = (-\infty, 0) \cup (0, \infty)$.

33. $1/(x^2 + 1)$ is defined for all x and so is $g(x) = 3 + 1/(x^2 + 1)$, the domain of g is $\mathbb{R}$. Because $1 \leq x^2 + 1$, we infer that $0 < 1/(x^2 + 1) \leq 1$ and $3 < 3 + 1/(x^2 + 1) \leq 4$. In fact the range of this function is all of $(3, 4]$ since if $y \in (3, 4]$, then letting $x = \sqrt{(4 - y)/(y - 3)}$ yields $y = g(x)$.

35. $\sqrt{t^3 - 1}$ is defined $\iff t^3 \geq 1 \iff t \geq 1$, the domain of this function is $[1, \infty)$.
Since the radical, $\sqrt{}$, indicates the positive square root, the range of this function is contained in $[0, \infty)$; any non-negative y is the value of this function at $x = \sqrt[3]{1 + y^2}$, hence the range is $[0, \infty)$.

37.   The function defined by $s = \begin{cases} -t & \text{if } t < 0 \\ t & \text{if } 0 \leq t \end{cases}$ has a value for any choice of t, thus its domain is $\mathbb{R}$. The range is contained in $[0, \infty)$ because the two cases always yield zero or a positive value; in fact, choosing $x = y$ for any $y \geq 0$, we see the range is exactly $[0, \infty)$.

REMARK: This function is the absolute value function, $|t|$.

39. $(2x+3)/(3x+4)$ is defined provided $x \neq \frac{-4}{3}$, the domain is $\mathbb{R} - \{\frac{-4}{3}\} = (-\infty, \frac{-4}{3}) \cup (\frac{-4}{3}, \infty)$. Any y except $\frac{2}{3}$ is the value of this function at $x = (3 - 4y)/(3y - 2)$ so the range is $\mathbb{R} - \{\frac{2}{3}\} = (-\infty, \frac{2}{3}) \cup (\frac{2}{3}, \infty)$.

41. $(f+g)(x) = (2x - 5) + (-4x) = -2x - 5$, $\text{dom}(f+g) = \mathbb{R}$; $(f-g)(x) = (2x-5) - (-4x) = 6x - 5$, $\text{dom}(f-g) = \mathbb{R}$; $(f \cdot g)(x) = (2x-5) \cdot (-4x) = -8x^2 + 20x$, $\text{dom}(f \cdot g) = \mathbb{R}$; $(f/g)(x) = (2x-5)/(-4x) = (-1/2) + (5/(4x))$, $\text{dom}(f/g) = \mathbb{R} - \{0\}$. The domain of f/g excludes 0 so we don't divide by zero.

43. $(f+g)(x) = \sqrt{x+2} + \sqrt{2-x}$, $\text{dom}(f+g) = [-2,2]$. The domain of f is $[-2, \infty)$ because we need $x + 2 \geq 0$ for $\sqrt{x+2}$ to be defined. The domain of g is $(-\infty, 2]$ since we need $2 - x \geq 0$ for $\sqrt{2-x}$ to be defined. The domain of $f + g$ is the intersection of those: $[-2, \infty) \cup (-\infty, 2] = [-2, 2]$.

$(f-g)(x) = \sqrt{x+2} - \sqrt{2-x}$, $\text{dom}(f-g) = [-2,2]$; $(f \cdot g)(x) = \sqrt{x+2} \cdot \sqrt{2-x} = \sqrt{4 - x^2}$, $\text{dom}(f \cdot g) = [-2, 2]$; $(f/g)(x) = \sqrt{x+2}/\sqrt{2-x} = \sqrt{(x+2)/(2-x)}$, $\text{dom}(f/g) = [-2, 2)$. The domain of f/g excludes 2 so we don't divide by zero.

45. $(f \circ g)(x) = (2x) + 1 = 2x + 1$, $\text{dom}(f \circ g) = \mathbb{R}$; $(g \circ f)(x) = 2(x+1) = 2x + 2$, $\text{dom}(g \circ f) = \mathbb{R}$.

47. $(f \circ g)(x) = 2 - \sqrt{(x+1)^2} = 2 - |x + 1|$, $\text{dom}(f \circ g) = \mathbb{R}$; $(g \circ f)(x) = (2 - \sqrt{x} + 1)^2 = 9 - 6\sqrt{x} + x$, $\text{dom}(g \circ f) = [0, \infty)$.

49. $(f \circ g)(x) = \dfrac{\frac{x-1}{x}}{\frac{x-1}{x} + 2} = \dfrac{x-1}{3x-1}$ and $\text{dom}(f \circ g) = \mathbb{R} - \{0, \frac{1}{3}\}$; 0 is excluded from the domain of $f \circ g$ because $g(0)$ is undefined; $\frac{1}{3}$ is excluded because $g\left(\frac{1}{3}\right) = -2$ and $f(-2)$ is undefined.

$(g \circ f)(x) = \dfrac{\frac{x}{x+2} - 1}{\frac{x}{x+2}} = \dfrac{-2}{x}$, and $\text{dom}(g \circ f) = \mathbb{R} - \{-2, 0\}$. -2 is not in the domain of $g \circ f$ because $f(-2)$ is undefined, 0 is excluded because $f(0) = 0$ which is not in the domain of g.

51. $(f \circ g)(x) = |x^2| / x^2 = x^2 / x^2 = 1$, $\text{dom}(f \circ g) = \mathbb{R} - \{0\}$; $(g \circ f)(x) = (|x|/x)^2 = |x|^2/x^2 = x^2/x^2 = 1$, $\text{dom}(g \circ f) = \mathbb{R} - \{0\}$.

53. Let $g(s) = s - \frac{1}{s}$ and $f(t) = t^2$, then $(f \circ g)(x) = f(g(x)) = f(x - \frac{1}{x}) = (x - \frac{1}{x})^2 = h(x)$ for all x; therefore $f \circ g = h$.

55. Let $g(u) = (u-3)(u+2)$ and $f(w) = \sqrt{w}$, then $(f \circ g)(x) = f(g(x)) = f((x-3)(x+2)) = \sqrt{(x-3)(x+2)} = h(x)$.

57. Let $g(x) = x^2$ and $f(x) = (1-x)/(1+x)$, then $(f \circ g)(x) = f(x^2) = (1 - x^2)/(1 + x^2) = h(x)$.

59. Let $g(x) = x - \frac{1}{x} = f(x)$, then $(f \circ g)(x) = f\left(x - \frac{1}{x}\right) = \left(x - \frac{1}{x}\right) - 1/\left(x - \frac{1}{x}\right) = h(x)$.

61. If $f(x) = 1.25x^2 - 3.74x + 14.38$, then $f(2.34) = 12.4729$, $f(-1.89) = 25.913725$, and $f(10.6) = 115.186$.

63. If $h(z) = (z+3)/(z^2 - 4)$, then $h(38.2) = 41.2/1455.24 \approx 0.0283114813$, $h(57.9) = 60.9/3348.41 \approx 0.0181877369$, and $h(238.4) = 241.4/56830.56 \approx 0.0042477146$.

65. Consider a rectangle with 50 cm perimeter. Let W be its width and H its height, both measured in cm. Therefore $2W + 2H = $ perimeter of rectangle $= 50$ cm, hence $W + H = 25$ and $H = 25 - W$. In order that we describe an actual rectangle with $W > 0$ and $H > 0$, we restrict W to the domain $(0, 25)$. The area of the rectangle is $A(W) = W \cdot (25 - W) = \left(\frac{25}{2}\right)^2 - \left(\frac{25}{2} - W\right)^2$. The range of $A(W)$ is $(0, A\left(\frac{25}{2}\right)] = (0, 12.5^2] = (0, 156.25]$.

67. To make this problem manageable, let's make some assumptions about Casey's running:

a. he runs in straight lines, directly from one base to the next;

b. he makes right angle turns at each base and immediately resumes running at 30 ft/sec;

c. we forget about Casey after he crosses Home Plate (this means the domain of our function is just $[0 \text{ sec}, 12 \text{ sec}]$).

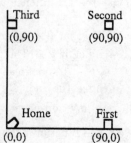

To help describe Casey's position, superimpose a coordinate system on the diamond. The following table summarizes Casey's position and line-of-sight distance from second base for various intervals of time.

Time	Baseline	Position	Line of sight distance to Second
$0 \le t \le 3$	Home to First	$(30t, 0)$	$\sqrt{(30t - 90)^2 + (0 - 90)^2}$
$3 < t \le 6$	First to Second	$(90, 30(t - 3))$	$\sqrt{(90 - 90)^2 + (30t - 90 - 90)^2}$
$6 < t \le 9$	Second to Third	$(90 - 30(t - 6), 90)$	$\sqrt{(270 - 30t - 90)^2 + (90 - 90)^2}$
$9 < t \le 12$	Third to Home	$(0, 90 - 30(t - 9))$	$\sqrt{(0 - 90)^2 + (360 - 30t - 90)^2}$

After some algebraic simplification, the result is

$$d(t) = \begin{cases} 30\sqrt{(3 - t)^2 + 3^2} & \text{if } 0 \le t \le 3, \\ 30\,|6 - t| & \text{if } 3 < t \le 9, \\ 30\sqrt{3^2 + (t - 9)^2} & \text{if } 9 < t \le 12. \end{cases}$$

69.   Both $f(x) = 2\sqrt{x}$ and $g(x) = 3\sqrt{x}$ have domains $[0, \infty)$, that interval is also the domain of $f \cdot g$; on the other hand, $h(x) = 6x$ had domain $(-\infty, \infty)$. It is true that $(f \cdot g)(x) = 6x = h(x)$ for x in the common part of their domains, but the functions are different because their domains are not the same.

71.   If $f(x) = 2x + 14$ and $g(x) = 0.5\,x - 7$, then $(f \circ g)(x) = f(g(x)) = f(0.5x - 7) = 2(0.5x - 7) + 14 = (x - 14) + 14 = x$ and $(g \circ f)(x) = g(f(x)) = g(2x + 14) = 0.5(2x + 14) - 7 = (x + 7) - 7 = x$. ∎

73.   Suppose $f(x) = ax + b$ where $a \ne 0$. Then $(f \circ g)(x) = x \iff a \cdot g(x) + b = x \iff g(x) = (x - b)/a$. For that choice of g, we also see that $(g \circ f)(x) = (f(x) - b)/a = ((ax + b) - b)/a = (ax)/a = x$.

75.   Because the function f is strictly increasing, it takes on each of its values only once. (That is, because $w < x \implies f(w) < f(x)$, we also know that $f(w) = f(x) \implies w = x$.) Therefore, for each s in the range of f, there is a unique (!) t in the domain of f such that $s = f(t)$. Define the function g by setting $g(s) = t$. Then the domain of g equals the range of f, the range of g equals the domain of f, and $s = f(t) \iff g(s) = t$.

If $s \in \text{dom}(g)$, then $f(g(s)) = f(t)$ where $s = f(t)$; therefore, $f(g(s)) = f(t) = s$. On the other hand, if $t \in \text{dom}(f)$, then $g(f(t)) = g(s)$ where $g(s) = t$; therefore $g(f(t)) = g(s) = t$. ∎

REMARK: The only part of this proof which required careful attention was showing that $G = \{(f(x), x) : x \in \text{dom}(f)\}$ is the graph of a function.

77.   Begin by slightly extending standard notation: if T is a subset of $\text{dom}(f)$, let $f(T) = \{f(t) : t \in T\}$.

Lemma. $S \subseteq T$ implies $f(S) \subseteq f(T)$.

If $w \in f(S)$, then there is some $s \in S$ such that $w = f(s)$. But $s \in T$ because $S \subseteq T$; therefore $w \in f(T)$. ∎

Suppose A and B are subsets of $\text{dom}(f)$.

a. $f(A) \cup f(B) \subseteq f(A \cup B)$ because the Lemma implies $f(A) \subseteq f(A \cup B)$ and $f(B) \subseteq f(A \cup B)$. If $w \in f(A \cup B)$, then either $w = f(a)$ for some $a \in A$ or $w = f(b)$ for some $b \in B$; in either case, $w \in f(A) \cup f(B)$; therefore $f(A \cup B) \subseteq f(A) \cup f(B)$. Because each set $f(A \cup B)$ and $f(A) \cup f(B)$ is contained in the other, they must be the same set: $f(A \cup B) = f(A) \cup f(B)$. ∎

b.   The Lemma implies $f(A \cap B) \subseteq f(A)$ and $f(A \cap B) \subseteq f(B)$ because $A \cap B \subseteq A$ and $A \cap B \subseteq B$; therefore $f(A \cap B) \subseteq f(A) \cap f(B)$. ∎

An example shows that $f(A \cap B)$ may be a proper subset of $f(A) \cap f(B)$. Let $f(x) = |x|$, $A = [-5, -2)$, $B = [0, 3]$. Then $f(A) = (2, 5]$, $f(B) = [0, 3]$, $f(A) \cap f(B) = (2, 3]$, $A \cap B = \emptyset$, and $f(A \cap B) = f(\emptyset) = \emptyset$.

This type of example can not be found if the function takes each of its values only once. In fact, it is possible to prove the following result.

Theorem. f is a one-to-one function if and only if $f(A \cap B) = f(A) \cap f(B)$ for all subsets A, B of $\text{dom}(f)$.

79.
$$(R \circ R)(x) = \frac{1}{R(x)} = \frac{1}{1/x} = x = \mathcal{J}(x),$$
$$(S \circ S)(x) = 1 - S(x) = 1 - (1 - x) = x = \mathcal{J}(x);$$
therefore $R \circ R = \mathcal{J} = S \circ S$.

NOTE: If f, g, h are any three functions such that $f \circ (g \circ h)$ is defined, then $(f \circ g) \circ h$ is also defined and $f \circ (g \circ h) = (f \circ g) \circ h$; thus it is unambiguous to write just $f \circ g \circ h$.

a. $\quad T(x) = R(S(x)) = \dfrac{1}{S(x)} = \dfrac{1}{1-x}.$

$$(T \circ T)(x) = \frac{1}{1 - T(x)} = \frac{1}{1 - \left(\frac{1}{1-x}\right)} = \frac{1-x}{(1-x)-1} = \frac{1-x}{-x} = \frac{-1}{x} + 1 = 1 - R(x) = (S \circ R)(x). \blacksquare$$

$$(T \circ T \circ T)(x) = (T \circ T)(T(x)) = 1 - \frac{1}{T(x)} = 1 - \frac{1}{\frac{1}{1-x}} = 1 - (1-x) = x = \mathcal{J}(x). \blacksquare$$

Alternative proof: $\quad T \circ T \circ T = T \circ (T \circ T) = (R \circ S) \circ (S \circ R) = R \circ (S \circ S) \circ R = R \circ \mathcal{J} \circ R = R \circ R = \mathcal{J}.$
(In addition to using the relations $T = R \circ S$, $T \circ T = S \circ R$, $S \circ S = \mathcal{J}$, $R \circ R = \mathcal{J}$, this proof uses the fact that $\mathcal{J} \circ f = f = f \circ \mathcal{J}$ for any function f.) $\blacksquare$

b. $\quad T = R \circ S$ implies $U = R \circ S \circ R = T \circ R$; therefore

$$U(x) = T(R(x)) = T\left(\frac{1}{x}\right) = \frac{1}{1 - \frac{1}{x}} = \frac{x}{x-1},$$

$$(S \circ R \circ S)(x) = (S \circ T)(x) = 1 - T(x) = 1 - \frac{1}{1-x} = \frac{(1-x)-1}{1-x} = \frac{-x}{1-x} = \frac{x}{x-1} = U(x), \blacksquare$$

$$(U \circ U)(x) = \frac{U(x)}{U(x) - 1} = \frac{\frac{x}{x-1}}{\left(\frac{x}{x-1}\right) - 1} = \frac{x}{x - (x-1)} = \frac{x}{1} = x = \mathcal{J}(x). \blacksquare$$

Alternative proof: $\quad U \circ U = (R \circ S \circ R) \circ (S \circ R \circ S) = (R \circ S) \circ (R \circ S) \circ (R \circ S) = T \circ T \circ T = \mathcal{J}. \blacksquare$

c. $\quad T \circ R = (R \circ S) \circ R = R \circ (S \circ R) = R \circ (T \circ T)$, $S \circ T = S \circ (R \circ S) = (S \circ R) \circ S = (T \circ T) \circ S$; the first result of part (b) is that each of these expression equals U. $\blacksquare$

d. $\quad$ Each of the functions $\mathcal{J}, R, S, T, T \circ T, U$ is composed from R and S in some way; hence any composition of two of them can also be expressed as a sequence of R's and S's. Because $R \circ R = \mathcal{J} = S \circ S$ and because $\mathcal{J}$ is the identity function, any such sequence can be simplified to consist of alternating R's and S's, e.g., $U \circ S = (S \circ R \circ S) \circ S = S \circ R$. Because $R \circ S \circ R = S \circ R \circ S$, alternating R and S sequences can be further reduced to have length at most three, e.g., $U \circ T = (S \circ R \circ S) \circ (R \circ S) = (R \circ S \circ R) \circ (R \circ S) = R$. The result of composing any pair of functions chosen from $\{\mathcal{J}, R, S, T, T \circ T, U\}$ is summarized in the following table:

		$\mathcal{J}$	R	S	U	T	$T \circ T$
	$\mathcal{J}$	$\mathcal{J}$	R	S	U	T	$T \circ T$
	R	R	$\mathcal{J}$	T	$T \circ T$	S	U
f	S	S	$T \circ T$	$\mathcal{J}$	T	U	R
	U	U	T	$T \circ T$	$\mathcal{J}$	R	S
	T	T	U	R	S	$T \circ T$	$\mathcal{J}$
	$T \circ T$	$T \circ T$	S	U	R	$\mathcal{J}$	T

where the top row is g and the column labels are $f \circ g$.

e. $\quad$ Consult the table; each column contains $\mathcal{J}$ and each row contains $\mathcal{J}$.

REMARK: For more information about the interesting collections of functions in Problems 78 and 79, see *On functions which form a group* by F.J. Budden, *The Mathematical Gazette*, 54(1970), pages 9-18.

Solutions 0.5 *Shifting the graphs of functions* (pages 44-7)

1. a.

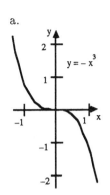

b.

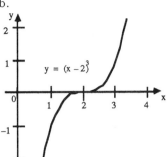

c.

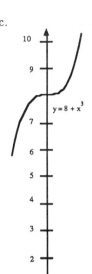

d.

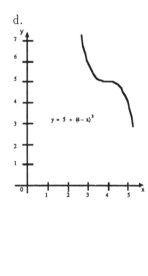

3.

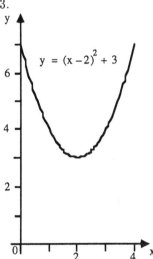

5.

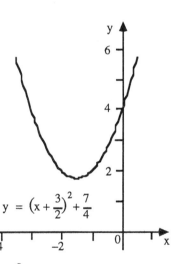

7.

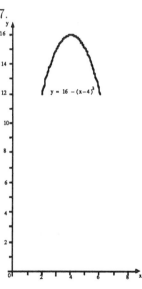

9.

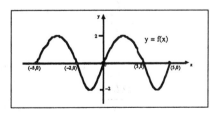

a.

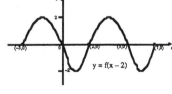

b.

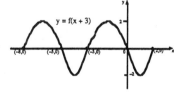

c.

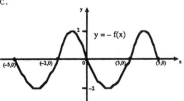

d.

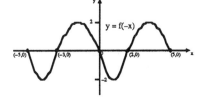

e.

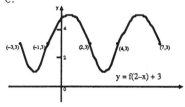

0.5 — Shifting the graphs of functions

11.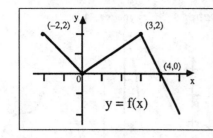

a.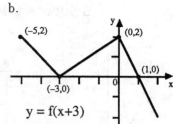

b.

c.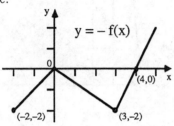

d.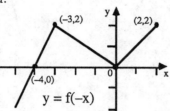

e.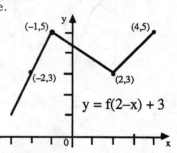

13.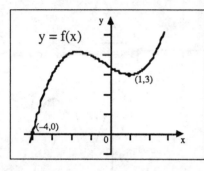

a.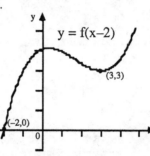

b.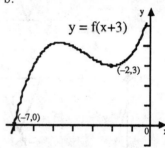

c.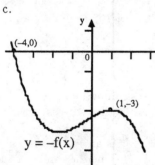

d.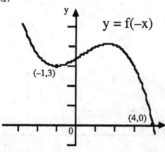

e.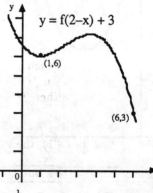

15.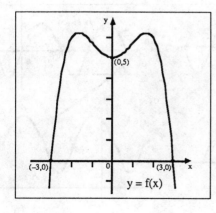

a.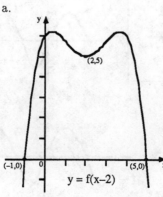

b.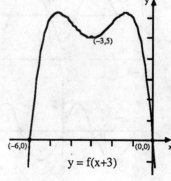

14

c.

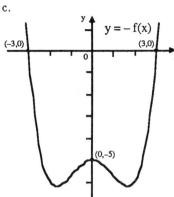

d.

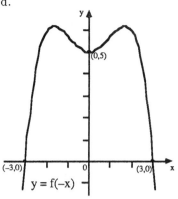

e.

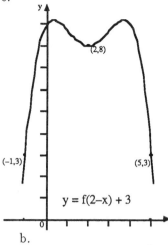

17.

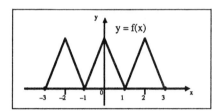

a.

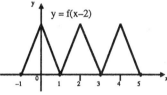

b.

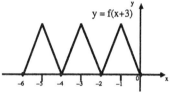

c.

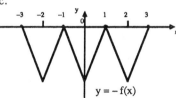

d.

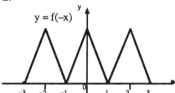

e.

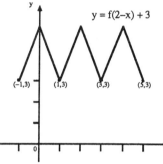

19. If $f(x) = x^4 + x^2$, then $f(-x) = (-x)^4 + (-x)^2 = x^4 + x^2 = f(x)$; therefore f is an even function.

21. If $f(x) = 1/x^2$, then $f(-x) = 1/(-x)^2 = 1/x^2 = f(x)$; therefore f is an even function.

23. If $f(x) = x^2 + x$, then $f(1) = 2$ and $f(-1) = 0$; since $f(-1) \neq \pm f(1)$, f is neither an even nor an odd function.

25. If $f(x) = 1/x$, then $f(-x) = 1/(-x) = -1/x = -(1/x) = -f(x)$; therefore f is an odd function.

27. If $f(x) = x/(x^3 + 1)$, then $f(1) = 2$ and $f(-1) = 0$; since $f(-1) \neq \pm f(1)$, f is neither an even nor an odd function.

REMARK: $f(1) = 1/2$ but $f(-1)$ is undefined, the domain of f is not symmetric with respect to 0.

29.

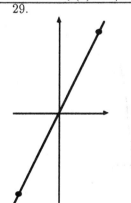

31.

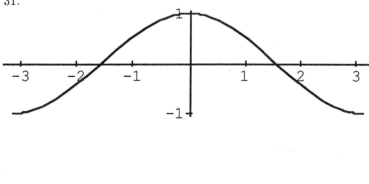

0.5 — Shifting the graphs of functions

33.

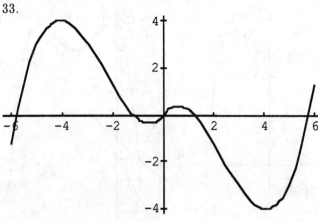

35.

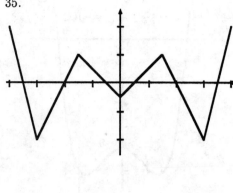

Solutions 0.C   *Computer Exercises*   (pages 47-8)

1. a. $f(x) = 2x^3 - 7x^2 - 17x + 10 = (x+2)(2x-1)(x-5)$; $\{x : 2x^3 - 7x^2 + 10 \le 17x\} = \{x : f(x) \le 0\} = (-\infty, -2] \cup [1/2, 5]$.

$y = 2x^3 - 7x^2 - 17x + 10$

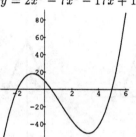

$y_1 = 2x^3 - 7x^2 + 10, \quad y_2 = 17x$

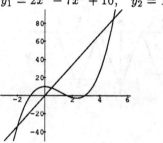

b. $f(x) = 2x^3 - 77x^2 - 122x + 80 = (x+2)(2x-1)(x-40)$; $\{x : 2x^3 - 77x^2 + 80 \le 122x\} = \{x : f(x) \le 0\} = (-\infty, -2] \cup [1/2, 40]$.

$y = 2x^3 - 77x^2 - 122x + 80$

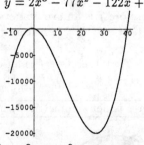

$y_1 = 2x^3 - 77x^2 + 80, \quad y_2 = 122x$

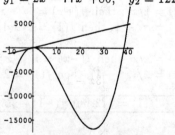

c. $f(x) = 2x^4 - x^3 - 24x^2 - 20x + 16 = (x+2)^2(2x-1)(x-4)$; $\{x : 2x^4 + 16 > x^3 + 24x^2 + 20x\} = \{x : f(x) > 0\} = (-\infty, -2) \cup (-2, 1/2) \cup (4, \infty)$.

$y = 2x^4 - x^3 - 24x^2 - 20x + 16$

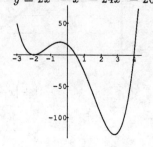

$y_1 = 2x^4 + 16, \quad y_2 = x^3 + 24x^2 + 20x$

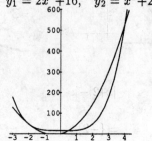

d. $\quad f(x) = |x| + 4 - |2x-1| = \begin{cases} x+3 & \text{if } x \le 0, \\ 3x+3 & \text{if } 0 < x \le 1/2, \\ -x+5 & \text{if } 1/2 < x. \end{cases}$ Thus $\{x : |x| + 4 > |2x-1|\} = \{x : f(x) > 0\} =$

$(-3, 5)$.

$y = |x| + 4 - |2x-1|$

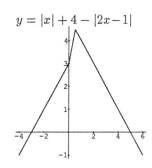

$y_1 = |x| + 4, \quad y_2 = |2x-1|$

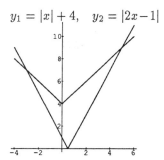

e. $\quad f(x) = |x+3| + |x| - 9 = \begin{cases} -4x - 12 & \text{if } x \le -3, \\ -2x - 6 & \text{if } -3 < x \le 0, \\ 4x - 6 & \text{if } 0 < x. \end{cases}$ Thus $\{x : |x+3| + |x| \le 9\} = \{x : f(x) \le 0\} =$

$[-3, 3/2]$.

$y = |x+3| + |x| - 9$

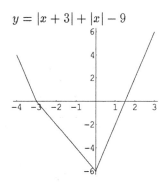

$y_1 = |x+3| + |x|, \quad y_2 = 9$

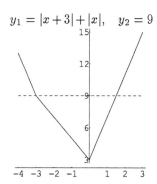

f. $\quad f(x) = |3x - 10| + |x| - 2 = \begin{cases} -4x + 8 & \text{if } x \le 0, \\ -2x + 8 & \text{if } 0 < x \le 10/3, \\ 4x - 12 & \text{if } 10/3 < x. \end{cases}$ Therefore $\{x : |3x - 10| + |x| < 2\} =$

$\{x : f(x) < 0\} = \emptyset$.

$y = |3x - 10| + |x| - 2$

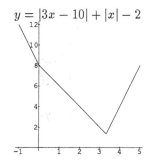

$y_1 = |3x - 10| + |x|, \quad y_2 = 2$

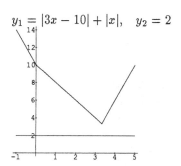

g. $\quad f(x) = |x-1| + |x+2| - |x+1| - 2 = \begin{cases} -x - 2 & \text{if } x \le -2, \\ x + 2 & \text{if } -2 < x \le -1, \\ -x & \text{if } -1 < x \le 1, \\ x - 2 & \text{if } 1 < x. \end{cases}$ Thus $\{x : |x-1| + |x+2| \le |x+1| + 2\}$

$= \{x : f(x) \le 0\} = \{-2\} \cup [0, 2]$.

17

0.C — Computer Exercises

$$y = |x{-}1| + |x{+}2| - |x{+}1| - 2$$

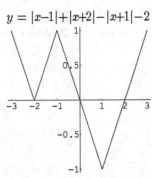

$$y_1 = |x{-}1| + |x{+}2|, \quad y_2 = |x{+}1| + 2$$

3. $x^4 + 4x^3 + x^2 - 8x - 3 = \left(x^2 + 3x + 1\right)\left(x^2 + x - 3\right)$ changes sign at $\left\{ \dfrac{-3 \pm \sqrt{5}}{2}, \dfrac{-1 \pm \sqrt{13}}{2} \right\} \approx$
$\{-2.618, -.382, -2.303, 1.303\}$ and is non-negative on

$$\left(-\infty, \frac{-3-\sqrt{5}}{2}\right] \cup \left[\frac{-1-\sqrt{13}}{2}, \frac{-3+\sqrt{5}}{2}\right] \cup \left[\frac{-1+\sqrt{13}}{2}, \infty\right) = (-\infty, -2.618] \cup [-2.303, -.382] \cup [1.303, \infty)$$

which is the domain of $\sqrt{x^4 + 4x^3 + x^2 - 8x - 3}$.

$$y = x^4 + 4x^3 + x^2 - 8x - 3$$

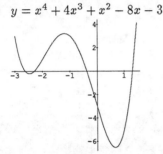

$$y = \sqrt{x^4 + 4x^3 + x^2 - 8x - 3}$$

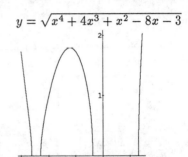

5. The circle centered at $(1, 2)$ with radius 3 satisfies the equation $(x-1)^2 + (y-2)^2 = 3^2$; the point on the circle that is closest to $(15, 1)$ is on the right-hand semicircle which satisfies the equation $x = 1 + \sqrt{5 + 4y - y^2}$. The distance from (x, y) to $(15, 1)$ is $\sqrt{(x-15)^2 + (y-1)^2}$; if (x, y) is on the right-hand semicircle, this distance is $f(y) = \sqrt{202 + 2y - 28\sqrt{5 + 4y - y^2}}$. Plotting this function and zooming on its lowest portion shows the minimum corresponds to $y \approx 1.786$ which then yields $x \approx 3.992$.

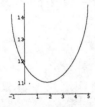 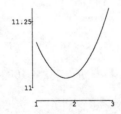 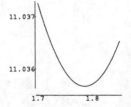 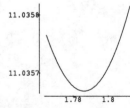

REMARK: The line through $(1, 2)$ and (15.1) satisfies equation $x + 14y = 29$; this line meets the right-hand semicircle at the point $\left(1 + \dfrac{42}{\sqrt{197}}, 2 - \dfrac{3}{\sqrt{197}}\right) \approx (3.992376, 1.786259)$.

7. A sphere 30 inches in diameter has radius 15 inches and volume $(4/3)\pi(15/12)^3 = (125/48)\pi$ ft^3; if the material has specific gravity .104 and we use 62.4 pounds per cubic foot as the density of water, then the sphere has weight $(.104)(62.4)(125/48)\pi = 16.9\pi$ pounds. If the sphere is placed in water and sinks to a depth d where d is less than the radius (i.e., $0 \leq d \leq 15/12 = 5/4$), then the displaced water has volume $\pi d^2(5/4 - d/3)$ and weight $(62.4)\pi d^2(5/4 - d/3)$. Archimedes' principle implies $(.104)(62.4)(125/48)\pi = (62.4)\pi d^2(5/4 - d/3)$ which is equivalent to $13/48 = d^2(5/4 - d/3)$. This last equation can be rewritten and then factored: $0 = d^2(5/4 - d/3) - 13/48 = (2d - 1)\left(8d^2 - 26d - 13\right)/48$. The zeros of $8d^2 - 26d - 13$ are $\{-.44, 3.69\}$ which are not in $[0, 1.25]$, so the conclusion is that the sphere sinks to a depth of $d = 1/2$ feet.

18

9.   A sphere of radius r feet has volume $(4/3)\pi r^3$; if the sphere is hollow with a one inch thick shell wall, then the volume of that shell wall is $(4/3)\pi \left(r^3 - (r-1)^3\right) = (4/3)\pi \left(3r^2 - 3r + 1\right)$. If this hollow sphere is made of a material with specific gravity γ, then its weight is $\gamma(62.4)(4/3)\pi \left(3r^2 - 3r + 1\right)$ pounds. This hollow sphere will float provided that water of the same volume has greater weight, i.e., if $(62.4)(4/3)\pi r^3 > \gamma(62.4)(4/3)\pi \left(3r^2 - 3r + 1\right)$. This inequality is equivalent to $r^3 > \gamma \left(3r^2 - 3r + 1\right)$. The specific gravity of aluminum is 2.70 at 20°C (consult a dictionary or CRC Handbook); let $f(r) = r^3 - (2.70) \left(3r^2 - 3r + 1\right)$. Examination of plots of f on the interval $(1, \infty)$ shows that $f(r) > 0$, and the hollow sphere will float, provided $r > 6.9976$ feet. (Since the problem asks about diameter, this answer can be rephrased: "provided $d \geq 14$ feet".)

$$y = r^3 - (2.70) \left(3r^2 - 3r + 1\right)$$

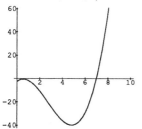
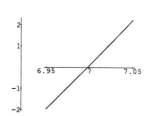

1
Limits and Derivatives

Solutions 1.2   *The calculation of limits*   (pages 65-9)

1. a.

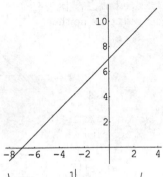

b.

x	$x+7$
1	8
1.5	8.5
1.9	8.9
1.99	8.99

c.

x	$x+7$
3	10
2.5	9.5
2.1	9.1
2.01	9.01

d. $\lim_{x\to2}(x+7) = 9$.

3. a.

b.

x	x^2-4
0	-4
0.5	-3.75
0.9	-3.19
0.99	-3.0199

c.

x	x^2-4
2	0
1.5	-1.75
1.1	-2.79
1.01	-2.9799

d. $\lim_{x\to1}\left(x^2-4\right) = -3$.

5. a. $f(1)$ is undefined because no fraction with a zero denominator makes sense. Provided you understand why division by zero is forbidden, that's a sufficient answer; otherwise, read on.

   An explanation of why no rational person can divide by zero: Division is related to multiplication by the equivalence $N/D = F$ if and only if $N = D\cdot F$. Because $A = A+0$ and $A\cdot F = (A+0)\cdot F = A\cdot F+0\cdot F$, we know $0 = 0\cdot F$ for any number F. Suppose $D = 0$. If $N\neq 0$, there is no choice of F such that $0\cdot F$ equals N; if $N = 0$, then $0\cdot F$ equals N for any and all choices for F. In either case, there is not a unique choice of F such that $N = 0\cdot F$; thus

   b. $f(x) = (x-1)(x-2)/(x-1) = x-2$ whenever $f(x)$ is defined, i.e., when $x\neq 1$. Let $g(x) = x-2$; then $f(x) = g(x)$ if $x\neq 1$ is defined and $g(1)$ is defined.

   c. $\lim_{x\to1}\left[(x-1)(x-2)/(x-1)\right] = \lim_{x\to1}(x-2) = 1-2 = -1$.

7.   If $x < -1$, then $x+1 < 0$ and $\sqrt{x+1}$ is undefined; there is no open interval containing -1 for which that function is defined.

9. $\lim_{x\to5}\left(x^2-6\right) = 5^2-6 = 25-6 = 19$.

11. $\displaystyle\lim_{x\to0}\frac{1}{x^5+6x+2} = \frac{1}{0^5+6(0)+2} = \frac{1}{2}$.

13. $\displaystyle\lim_{x\to-1}\frac{(x+1)^2}{x+1} = \lim_{x\to-1}(x+1) = (-1)+1 = 0$.

15. $\lim_{x\to4}\sqrt{25-x^2} = \sqrt{25-4^2} = \sqrt{9} = 3$.

17. $\lim_{x\to-4}\sqrt{x+4}$ does not exist because $\sqrt{x+4}$ is undefined if $x < -4$.

19. $\lim_{x\to2}\sqrt[3]{x^3-8} = \sqrt[3]{2^3-8} = \sqrt[3]{0} = 0$.

21. $\displaystyle\lim_{x\to3}\frac{x^2-4x+3}{x-3} = \lim_{x\to3}\frac{(x-3)(x-1)}{x-3} = \lim_{x\to3}(x-1) = 3-1 = 2$.

23.   If $0 < x$, then $x = \left(\sqrt{x}\right)^2$ and $x - 1 = \left(\sqrt{x} - 1\right)\left(\sqrt{x} + 1\right)$. Therefore

$$\lim_{x \to 1} \frac{\sqrt{x} - 1}{x - 1} = \lim_{x \to 1} \frac{\sqrt{x} - 1}{\left(\sqrt{x} - 1\right)\left(\sqrt{x} + 1\right)} = \lim_{x \to 1} \frac{1}{\sqrt{x} + 1} = \frac{1}{1 + 1} = \frac{1}{2}.$$

25.   If $x > 2$, then $x - 2 > 0$ and $\sqrt{x - 2}$ is defined; $\lim_{x \to 2+}(x - 2) = 2 - 2 = 0$ and $\lim_{x \to 2+} \sqrt{x - 2} = 0$.

27.   If $x < -1$, then $0 < -1 - x$ and $\sqrt{-1 - x}$ is defined; $\lim_{x \to -1-} \sqrt{-1 - x} = \sqrt{0} = 0$.

29.   If $x > -2$, then $x + 2 > 0$, $|x + 2| = x + 2$, and $3x|x + 2|/(x + 2) = 3x$; thus $\lim_{x \to -2+} 3x|x + 2|/(x + 2) = \lim_{x \to -2+} 3x = 3(-2) = -6$.

31.   If $3/2 < x < 2$, then $[x] = 1$; therefore $\lim_{x \to \frac{3}{2}+} [x] = 1$.

33. $\lim_{x \to -2+} \sqrt{x + 2} = \sqrt{0} = 0$.

35.   If $x > 1$, then $x - 1 > 0$, $|x - 1| = x - 1$ and $|x - 1|/(x - 1) = 1$; hence $\lim_{x \to 1+} |x - 1|/(x - 1) = 1$.

37.   If $1 < x < 2$, then $(x - 1)(x - 2) < 0$ and $\sqrt{(x - 1)(x - 2)}$ is undefined; thus neither $\lim_{x \to 1+} \sqrt{(x - 1)(x - 2)}$ nor $\lim_{x \to 1} \sqrt{(x - 1)(x - 2)}$ exist.

39. $\lim_{x \to 3-} f(x) = \lim_{x \to 3-} x = 3$.

41. $\lim_{x \to 3} f(x) = \lim_{x \to 3} x = 3$.

43. $\lim_{x \to 1+} g(x) = \lim_{x \to 1+} x^2 = 1^2 = 1$.

45. $\lim_{x \to -1-} h(x) = \lim_{x \to -1-}(2x + 3) = -2 + 3 = 1$.

47. $\lim_{x \to -1-} h(x) = \lim_{x \to -1-}(2x + 3) = -2 + 3 = 1$ and $\lim_{x \to -1+} h(x) = \lim_{x \to -1+}(3 - x) = 3 + 1 = 4$; both one-sided limits exist but they are unequal, hence $\lim_{x \to -1} g(x)$ does not exist.

49. $\lim_{x \to 3} f(x)/g(x) = \lim_{x \to 3} x/x^2 = \lim_{x \to 3} 1/x = 1/3$.

51.   a.

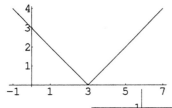

    b. $\lim_{x \to 3+} |x - 3| = \lim_{x \to 3+}(x - 3) = 3 - 3 = 0$ and $\lim_{x \to 3-} |x - 3| = \lim_{x \to 3-}(3 - x) = 0$; therefore $\lim_{x \to 3} |x - 3|$ exists and also equals zero. [See Example 5.]

53.   a.

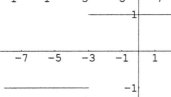

    b. $$\lim_{x \to -3+} \frac{|x + 3|}{x + 3} = \lim_{x \to -3+} \frac{x + 3}{x + 3} = \lim_{x \to -3+} 1 = 1; \qquad \lim_{x \to -3-} \frac{|x + 3|}{x + 3} = \lim_{x \to -3-} \frac{-(x + 3)}{x + 3} = -1.$$

$\lim_{x \to -3} \frac{|x + 3|}{x + 3}$ does not exist because the one-sided limits do not have the same value,

    c.   Because $x + 3 > 0$ when x is near 5, $\lim_{x \to 5} \frac{|x + 3|}{x + 3} = \lim_{x \to 5} \frac{x + 3}{x + 3} = 1$. On the other hand, $x + 3 < 0$ if x is close to -4; therefore $\lim_{x \to -4} \frac{|x + 3|}{x + 3} = \lim_{x \to -4} \frac{-(x + 3)}{x + 3} = -1$.

55.   a.   b.   c.

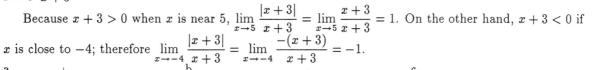

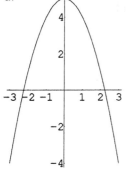

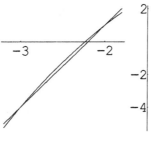

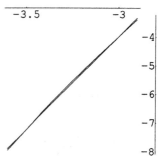

1.2 — The calculation of limits

d. $\dfrac{5 - (-3 + h)^2) - (-4)}{(-3 + h) - (-3)} = \dfrac{5 - (-3 + h)^2) + 4}{h}, \qquad h \neq 0$

is the slope of the secant line through $(-3 + h, 5 - (-3 + h)^2)$ and $(-3, -4)$ of the curve $y = 5 - x^2$.

e. If $h \neq 0$,

$$\frac{\left[5 - (-3 + h)^2\right] + 4}{h} = \frac{6h - h^2}{h} = 6 - h\,;$$

thus

$$\lim_{h \to 0} \frac{\left[5 - (-3 + h)^2\right] + 4}{h} = \lim_{h \to 0}(6 - h) = 6 - 0 = 6.$$

f. The line tangent at $(-3, -4)$ to the curve $y = 5 - x^2$ has slope 6.

57.

a & b.

x	$\sqrt{x^3 + 12}/(x + 8)$	x	$\sqrt{x^3 + 12}/(x + 8)$
-1	0.473804	-3	undefined
-1.5	0.451821	-2.5	undefined
-1.9	0.371701	-2.1	0.280507
-1.99	0.337709	-2.01	0.328818
-1.999	0.333777	-2.001	0.332888

c. $0.3333326343 \approx \dfrac{1}{2}\left[f(-1.999) + f(-2.001)\right]$ is an estimate of $\lim_{x \to -2} f(x)$.

d. $f(-2) = \frac{1}{3}$; the estimate of the limit in part (c) is very close to $f(-2)$ (they are less than 7×10^{-7} apart). In fact, for the function in this problem, $\lim_{x \to -2} f(x) = f(-2)$.

59.

x	$(x^2 - 1)/\sqrt{x - 1}$
2	3
1.5	1.767766953
1.1	0.6640783086
1.01	0.201
1.0036	0.120216
1.000025	0.010000125
1.00000016	0.000800000064

$\lim_{x \to 1+} (x^2 - 1)/\sqrt{x - 1} = 0.$

61.

x	$(4 - x)/(4 - 2\sqrt{x})$	x	$(4 - x)/(4 - 2\sqrt{x})$
0	1	9	2.5
3	1.866025404	5	2.118033989
3.5	1.935414347	4.5	2.060660172
3.9	1.987420883	4.1	2.012422837
3.99	1.998749218	4.01	2.001249220
3.999	1.999874992	4.001	2.000124992
3.9999	1.999987500	4.0001	2.000012500

$\lim_{x \to 4}(4 - x)/(4 - 2\sqrt{x}) = 2.$

63.

h	$(\sqrt{0.25 - h} - 0.5)/h$	h	$(\sqrt{0.25 - h} - 0.5)/h$
0.5	undefined	-0.5	-0.7320508076
0.1	-1.127016654	-0.1	-0.9160797831
0.01	-1.010205144	-0.01	-0.9901951359
0.001	-1.001002005	-0.001	-0.9990019950
0.0001	-1.000100020	-0.0001	-0.9999000200
0.00001	-1.000010000	-0.00001	-0.9999900002

$\lim_{h \to 0} (\sqrt{0.25 - h} - 0.5)/h = -1.$

65.

x	$(\sqrt{x}-2)/(x-4)$	x	$(\sqrt{x}-2)/(x-4)$
4.5	0.2426406871	3.5	0.2583426132
4.1	0.2484567313	3.9	0.2515823419
4.01	0.2498439450	3.99	0.2501564456
4.001	0.2499843770	3.999	0.2500156270
4.0001	0.2499984375	3.9999	0.2500015625
4.00001	0.2499998438	3.99999	0.2500001562

$\lim_{x\to4}(\sqrt{x}-2)/(x-4)=1/4$.

67. This problem is essentially identical to problem 63; read solution 63 with Δx substituted for h.

69.

Δx	$(\sqrt[3]{8+\Delta x}-2)/\Delta x$	Δx	$(\sqrt[3]{8+\Delta x}-2)/\Delta x$
0.5	0.08165510192	−0.5	0.08513235883
0.1	0.08298850247	−0.1	0.08368298710
0.01	0.08329863520	−0.01	0.08336807969
0.001	0.08332986135	−0.001	0.08333680580
0.0001	0.08333298611	−0.0001	0.08333368056
0.00001	0.08333329862	−0.00001	0.08333336804

$\lim_{\Delta x\to0}(\sqrt[3]{8+\Delta x}-2)/\Delta x=1/12$.

71.

x	$(\sqrt{x}-5)/(x-25)$	x	$(\sqrt{x}-5)/(x-25)$
25.5	0.09950493836	24.5	0.1005050634
25.1	0.09990019950	24.9	0.1001002005
25.01	0.09999000200	24.99	0.1000100020
25.001	0.09999900002	24.999	0.1000010000
25.0001	0.09999990001	24.9999	0.1000001000
25.00001	0.09999999008	24.99999	0.1000000099

$\lim_{x\to25}(\sqrt{x}-5)/(x-25)=1/10$.

73.

x	$(x^2-250^2)/(x-250)$	x	$(x^2-250^2)/(x-250)$
251	501	249	499
250.5	500.5	249.5	499.5
250.1	500.1	249.9	499.9
250.01	500.01	249.99	499.99
250.001	500.001	249.999	499.999
250.0001	500.0001	249.9999	499.9999
250.00001	500.00001	249.99999	499.99999

$\lim_{x\to250}(x^2-250^2)/(x-250)=500$.

75.

x	$(x^3-125)/(x-5)$	x	$(x^3-125)/(x-5)$
6	91	4	61
5.5	82.75	4.5	67.75
5.1	76.51	4.9	73.51
5.01	75.1501	4.99	74.8501
5.001	75.015001	4.999	74.985001
5.0001	75.00150001	4.9999	74.99850001
5.00001	75.0001500001	4.99999	74.9998500001

$\lim_{x\to5}(x^3-125)/(x-5)=75$.

77.

x	$(x^5-32)/(x-2)$	x	$(x^5-32)/(x-2)$
3	211	1	31
2.5	131.3125	1.5	48.8125
2.1	88.4101	1.9	72.3901
2.01	80.80401001	1.99	79.20399001
2.001	80.080040010001	1.999	79.920039990001
2.0001	80.008000400010	1.9999	79.992000399990
2.00001	80.00080000400001	1.99999	79.99920000399999

1.2 — The calculation of limits

$\lim_{x \to 2} \left(x^5 - 32\right)/(x - 2) = 80$.

79.

x	$(x^5 - 1)/(x^3 - 1)$	x	$(x^5 - 1)/(x^3 - 1)$
2	4.428571429	0	1.000000000
1.5	2.776315790	0.5	1.107142857
1.1	1.844441088	0.9	1.511107011
1.01	1.683444444	0.99	1.650111111
1.001	1.668334444	0.999	1.665001111
1.0001	1.666833344	0.9999	1.666500011
1.00001	1.666683333	0.99999	1.666650000

$\lim_{x \to 1}(x^5 - 1)/(x^3 - 1) = 5/3 \approx 1.666667$.

81.

x	$(x^5 + 32)/(x^2 - 4)$	x	$(x^5 + 32)/(x^2 - 4)$
-1	-10.33333333	-3	-42.20000000
-1.5	-13.94642857	-2.5	-29.18055556
-1.9	-18.56156410	-2.1	-21.56343902
-1.99	-19.85062406	-2.01	-20.15062594
-1.999	-19.98500625	-2.001	-20.01500625
-1.9999	-19.99850006	-2.0001	-20.00150006
-1.99999	-19.99985000	-2.00001	-20.00015000

$\lim_{x \to -2}(x^5 + 32)/(x^2 - 4) = -20$.

83.

x	$\sin(2x)/x$	x	$\sin(2x)/x$
1	0.9092974268	-1	0.9092974268
0.5	1.6829419696	-0.5	1.6829419696
0.1	1.9866933080	-0.1	1.9866933080
0.01	1.9998666693	-0.01	1.9998666693
0.001	1.9999986667	-0.001	1.9999986667
0.0001	1.9999999867	-0.0001	1.9999999867
0.00001	1.9999999999	-0.00001	1.9999999999

$\lim_{x \to 0} \sin(2x)/x = 2$.

85.

x	$(1 - \cos x)/x$	x	$(1 - \cos x)/x$
1	0.4596976941	-1	-0.4596976941
0.5	0.2448348762	$-1/2$	-0.2448348762
0.1	0.04995834722	$-1/4$	-0.1243503132
0.01	0.004999958333	$-1/8$	-0.06241866217
0.001	0.0004999999583	$-1/16$	-0.03123982880
0.0001	0.00004999999942	$-1/32$	-0.01562372848
0.00001	0.000004999990699	$-1/64$	-0.007812341056

$\lim_{x \to 0}(1 - \cos x)/x = 0$.

87.

x	$\tan(3x)/(5x)$	x	$\tan(3x)/(5x)$
1	-0.02850930861	-1	-0.02850930861
0.5	5.64056797887	$-1/2$	5.64056797887
0.1	0.61867249922	$-1/4$	0.74527716796
0.01	0.60018006482	$-1/8$	0.62980252148
0.001	0.60000180001	$-1/16$	0.60713155430
0.0001	0.60000001800	$-1/32$	0.60176401437
0.00001	0.60000000018	$-1/64$	0.60043983971

$\lim_{x \to 0} \tan(3x)/(5x) = 3/5$.

89.

x	$(1+3x)^{1/x}$	x	$(1+3x)^{1/x}$
1	4	-1	-0.5
0.5	6.25	-0.5	4
0.1	13.78584918	-0.1	35.40133175
0.01	19.21863198	-0.01	21.02938507
0.001	19.99553462	-0.001	20.17630751
0.0001	20.07650227	-0.0001	20.09457926
0.00001	20.08463311	-0.00001	20.08644081

$\lim_{x\to 0}(1+3x)^{1/x} = e^3 \approx 20.08554.$

91.

x	$(3^x-1)/(2^x-1)$	x	$(3^x-1)/(2^x-1)$
1	2	-1	1.333333333
0.5	1.767326988	-0.5	1.443016443
0.1	1.617912386	-0.1	1.553623830
0.01	1.588183802	-0.01	1.581757309
0.001	1.585283905	-0.001	1.584641258
0.0001	1.584994634	-0.0001	1.584930369
0.00001	1.584965714	-0.00001	1.584959288

$\lim_{x\to 0}(3^x-1)/(2^x-1) = (\ln 3)/(\ln 2) \approx 1.584963.$

93.

x	$\ln(1+x)/x$	x	$\ln(1+x)/x$
1	0.6931471806	-1	undefined
0.5	0.8109302162	-0.5	1.386294361
0.1	0.9531017980	-0.1	1.053605157
0.01	0.9950330853	-0.01	1.005033585
0.001	0.9995003331	-0.001	1.000500334
0.0001	0.9999500033	-0.0001	1.000050003
0.00001	0.9999950000	-0.00001	1.000005000

$\lim_{x\to 0}\ln(1+x)/x = 1.$

95. $\lim_{x\to 1-} f(x) = \lim_{x\to 1-} x^2 = 1^2 = 1$ and $\lim_{x\to 1+} f(x) = \lim_{x\to 1+} x^3 = 1^3 = 1$; since $\lim_{x\to 1-} f(x) = \lim_{x\to 1+} f(x)$, $\lim_{x\to 1} f(x)$ exists and it equals the common value, 1, of those one-sided limits. [See Theorem 1.]

97. $\lim_{x\to 2-} f(x) = \lim_{x\to 2-}(0) = 0$, $\lim_{x\to 2+} f(x) = \lim_{x\to 2+}(x-2) = 2 - 2 = 0$. (Since $\lim_{x\to 2-} f(x) = \lim_{x\to 2+} f(x)$, $\lim_{x\to 2} f(x)$ exists and it equals the common value, 0, of those one-sided limits.)

99. $\lim_{x\to 2-} f(x) = f(2) = 1.$

101. $\lim_{x\to -1-} g(x) = 5.$

103. $\lim_{x\to -1} g(x) = 5.$

105. $\lim_{x\to 0} h(x) = h(0) = 3/2.$

107. $\lim_{x\to 5+} h(x) = h(5) = -2.$

109. $\lim_{x\to 2+} h(x) = 1.$

111.   If $-1 \le x \le 1$ but $x \neq 0$, then $-1 \le x^2 < 0$ and $[-x^2] = -1$; therefore $\lim_{x\to 0}[-x^2] = -1.$

113. $A(x)$ is the area of a triangle with base 1 and altitude x^2, thus $A(x) = x^2/2$. $B(x)$ is the area of the triangle with base 1 and altitude x, thus $B(x) = x/2$. Therefore,

$$\lim_{x\to 0}\frac{A(x)}{B(x)} = \lim_{x\to 0}\frac{x^2/2}{x/2} = \lim_{x\to 0} x = 0.$$

Solutions 1.3   *Some limit theorems*   (pages 73-4)

1.   Using Theorem 1, $\lim_{x\to 3}(x^2 - 2x - 1) = 3^2 - 2(3) - 1 = 2.$

3. $\lim_{x\to 1} 3 = 3.$ (Use Theorem 1; also look at Example 2.)

5. $\lim_{x\to 1} 8(x^{101} + 2) = 8(1^{101} + 2) = 8(1 + 2) = 24.$

7. $\lim_{x\to 2}(x^2 - 1)^5 = (2^2 - 1)^5 = 3^5 = 243.$

1.3 — Some limit theorems

9. $\lim_{x \to 3} 1/\left(x^3 - 8\right) = 1/\left(3^3 - 8\right) = 1/19.$

11. $\lim_{x \to 1} \dfrac{x^8 + x^6 + x^4 + x^2 + 1}{x^7 + x^5 + x^3 + x} = \dfrac{1^8 + 1^6 + 1^4 + 1^2 + 1}{1^7 + 1^5 + 1^3 + 1} = \dfrac{5}{4}.$

13. $\lim_{x \to 2} \dfrac{x^2 - x - 12}{x^2 - 5x + 4} = \dfrac{4 - 2 - 12}{4 - 10 + 4} = 5.$

15. $\lim_{x \to 0} \dfrac{x^{28} - 17x^{14} + x^2 - 3}{x^{51} + x^{31} - 23x^2 + 2} = \dfrac{(0)^{28} - 17(0)^{14} + (0)^2 - 3}{(0)^{51} + (0)^{31} - 23(0)^2 + 2} = \dfrac{-3}{2}.$

17. $\lim_{x \to 5} 3\sqrt{x - 1} = 3 \cdot \lim_{x \to 5} \sqrt{x - 1} = 3 \cdot \sqrt{\lim_{x \to 5}(x - 1)} = 3 \cdot \sqrt{4} = 3 \cdot 2 = 6.$

REMARK: The following computations show that we don't need the full power of the result stated in Problem 28. Suppose $\lim_{x \to 5} 3\sqrt{x - 1}$ exists, call it L. Then, using the Corollary to Theorem 4,

$$L^2 = \left(\lim_{x \to 5} 3\sqrt{x - 1}\right)^2 = \lim_{x \to 5} \left(3\sqrt{x - 1}\right)^2 = \lim_{x \to 5} 9(x - 1) = 9(5 - 1) = 36.$$

Because $3\sqrt{x - 1}$ is positive for x near 5, we know $L \geq 0$. Therefore $L = \sqrt{36} = 6$.

19. $\lim_{x \to -2} \left(-4\sqrt{x + 3}\right) = -4\sqrt{-2 + 3} = -4.$

21. $\lim_{x \to 5} \left(\sqrt{x - 1} + \sqrt{x^2 - 9}\right) = \lim_{x \to 5} \sqrt{x - 1} + \lim_{x \to 5} \sqrt{x^2 - 9} = \sqrt{5 - 1} + \sqrt{25 - 9} = 2 + 4 = 6$ (using Theorems 3 and 1).

23. $\lim_{x \to 0} \dfrac{\sqrt{x + 1}}{\sqrt{x^2 - 3x + 4}} = \dfrac{\lim_{x \to 0} \sqrt{x + 1}}{\lim_{x \to 0} \sqrt{x^2 - 3x + 4}} = \dfrac{\sqrt{1}}{\sqrt{4}} = \dfrac{1}{2}.$

25. $\lim_{x \to 5} \sqrt{\dfrac{x + 1}{x^2 - 9}} = \sqrt{\dfrac{\lim_{x \to 5}(x - 1)}{\lim_{x \to 5}(x^2 - 9)}} = \sqrt{\dfrac{-1}{-4}} = \sqrt{\dfrac{1}{4}} = \dfrac{1}{2}.$

27. Let $f(x) = x - 7$ and $g(x) = 5/f(x)$; then $\lim_{x \to 7} f(x) = \lim_{x \to 7}(x - 7) = 7 - 7 = 0$ but $\lim_{x \to 7} \left[f(x) \cdot g(x)\right] = \lim_{x \to 7} \left[5 \cdot \left(\dfrac{f(x)}{f(x)}\right)\right] = \lim_{x \to 7} 5 = 5.$

REMARK: Obviously there are many other correct pairs of functions f and g. (For a more complicated example, let $f(x) = 49 - x^2$ and $g(x) = (x + 5)/(x^2 + x - 56)$.)

29. Let $f(x) = \begin{cases} 7 - |x - 3| & \text{if } x \neq 3 \\ 5 & \text{if } x = 3 \end{cases}$,   Then $f(x) < 7$ for all x but $\lim_{x \to 3} f(x) = 7$. This single counterexample is a disproof of the problem statement.

REMARK: There are many such functions that could serve as a counterexample; however it can be proven that if $f(x) < 7$ for all x and if $\lim_{x \to 3} f(x)$ exists, then $\lim_{x \to 3} f(x) \leq 7$.

Solutions 1.4   *Infinite limits and Limits at infinity*   (pages 81-2)

1.   The text discussion preceding Definition 1 shows that $1/x^2$ increases without bound as x tends to 0. If $0 < |x| < 1$, then $0 < x^4 < x^2$ and $1/x^2 < 1/x^4$. Therefore $1/x^4$ also increases without bound as x tends to 0 and $\lim_{x \to 0}(1/x^4) = \infty$.

3.   Consider the following table of values $1/(x - 5)^3$ for several values of x near 5.

x	$1/(x - 5)^3$	x	$1/(x - 5)^3$
4.5	−8	5.08	1953
4.8	−125	5.04	15625
4.9	−1000	5.02	125000
4.95	−8000	5.01	1000000

If $x < 5$, then $1/(x - 5)^3$ decreases without bound as x tends to 5; however, if $x > 5$, then $1/(x - 5)^3$ increases without bound as x approaches 5. Therefore $\lim_{x \to 5} 1/(x - 5)^3$ does not exist. [See Example 3.]

5.   If $|x - \pi| < 1$ and $x \neq \pi$, then $0 < (x-\pi)^6 < (x-\pi)^4 < (x-\pi)^2$ and $1/(x-\pi)^2 < 1/(x-\pi)^6 < \pi/(x-\pi)^6$. Therefore $\lim_{x \to \pi} \left(\pi/(x-\pi)^6 \right) = \infty$ because $\lim_{x \to \pi} \left(1/(x-\pi)^6 \right) = \lim_{w \to 0} \left(1/w^2 \right) = \infty$.

7. $\dfrac{x + x^2}{x^2 + x^3} = \dfrac{x(1+x)}{x^2(1+x)} = \dfrac{1}{x}$, $x \neq -1$.

Because $\lim_{x \to 0} \dfrac{1}{x}$ does not exist (Example 3), neither does $\lim_{x \to 0} \dfrac{x + x^2}{x^2 + x^3}$.

9. $\lim_{x \to -\infty} \dfrac{x}{x + 1} = \lim_{x \to -\infty} \dfrac{1}{1 + (1/x)} = \dfrac{1}{1 + \lim_{x \to -\infty} (1/x)} = \dfrac{1}{1 + 0} = 1$.   [See Table 4.]

11. $\lim_{x \to \infty} \dfrac{2x}{3x^3 + 4} = \lim_{x \to \infty} \dfrac{(2/x^2)}{3 + (4/x^3)} = \dfrac{0}{3 + 0} = 0$.

13. $\lim_{x \to \infty} \dfrac{5x - x^2}{3x + x^2} = \lim_{x \to \infty} \dfrac{(5/x) - 1}{(3/x) + 1} = \dfrac{0 - 1}{0 + 1} = -1$.

15. $\lim_{x \to \infty} \dfrac{2x^2 + 3x + 5}{3x^2 - x + 2} = \lim_{x \to \infty} \dfrac{2 + (3/x) + (5/x^2)}{3 - (1/x) + (2/x^2)} = \dfrac{2 + 0 + 0}{3 - 0 + 0} = \dfrac{2}{3}$.

17. $\lim_{x \to \infty} \dfrac{x^5 - 3x + 4}{7x^6 + 8x^4 + 2} = \lim_{x \to \infty} \dfrac{(1/x) - (3/x^5) + (4/x^6)}{7 + (8/x^2) + (2/x^6)} = \dfrac{0 - 0 + 0}{7 + 0 + 0} = 0$.

19. $\lim_{x \to \infty} \dfrac{3x^{5/3} + 2\sqrt{x} - 3}{7x^{5/3} - 3x + 6} = \lim_{x \to \infty} \dfrac{3 + (2/x^{7/6}) - (3/x^{5/3})}{7 - (3/x^{2/3}) + (6/x^{5/3})} = \dfrac{3 + 0 - 0}{7 - 0 + 0} = \dfrac{3}{7}$.

21.   If $-1 < x < 1$, then $x^2 - 1 = (x+1)(x-1) < 0$; therefore $\dfrac{1}{x^2 - 1} = \left(\dfrac{1}{x+1} \right) \cdot \left(\dfrac{1}{x-1} \right) < 0$. Because

$\lim_{x \to 1^-} \dfrac{1}{x+1} = \dfrac{1}{2}$ and $\lim_{x \to 1^-} \dfrac{1}{x-1} = -\infty$, we conclude that $\lim_{x \to 1^-} \dfrac{1}{x^2 - 1} = -\infty$.

23.   The solutions of Problems 21 and 22 show that

$$\lim_{x \to 1^-} \dfrac{1}{x^2 - 1} = -\infty \quad \neq \quad +\infty = \lim_{x \to 1^+} \dfrac{1}{x^2 - 1};$$

since the one-sided limits do not have the same value, $\lim_{x \to 1} \dfrac{1}{x^2 - 1}$ does not exist.

25.

$$\lim_{x \to \infty} \left(x - \sqrt{x^2 + 2x} \right) = \lim_{x \to \infty} \left(x - \sqrt{x^2 + 2x} \right) \left(\dfrac{x + \sqrt{x^2 + 2x}}{x + \sqrt{x^2 + 2x}} \right)$$

$$= \lim_{x \to \infty} \dfrac{-2x}{x + \sqrt{x^2 + 2x}} = \lim_{x \to \infty} \dfrac{-2}{1 + \sqrt{1 + (2/x)}} = \dfrac{-2}{1 + \sqrt{1 + 0}} = -1.$$

REMARK: Since $x^2 + 2x = (x+1)^2 - 1$ is approximately $(x+1)^2$ for large x, this result is not surprising. See Problem 30 for a generalization.

27.   a. $\lim_{x \to -\infty} \dfrac{3x^5 - 3x^3 + 17x + 2}{x^5 + 4x^2 + 16} = \lim_{x \to -\infty} \dfrac{3 - 3/x^2 + 17/x^4 + 2/x^5}{1 + 4/x^3 + 16/x^5} = \dfrac{3 - 0 + 0 + 0}{1 + 0 + 0} = 3$.

        b.   If $x < -2$, then $-1/2 < 4/x^3 < 0$ and $-1/2 < 16/x^5 < 0$; therefore the denominator, $1 + 4/x^3 + 16/x^5$, (of the second fraction above) is positive, less than 1, and increases to 1 as $x \to -\infty$. Hence, if $x < -2$, the quotient is greater than its numerator, $3 - 3/x^2 + 17/x^4 + 2/x^5$ which turns out to be greater than $3 - 3/x^2 + 0 + 2(-1/x^2) = 3 - 5/x^2$ because $1/|x|^5 < 1/|x|^2$. $3 - 5/x^2 > 2.999$ when $0.001 > 5/x^2$; if $x < -2$, it suffices to require $x < -\sqrt{5/0.001} = -\sqrt{5000} \approx -70.71$. If $x < -70$, then (the denominator) $1 + 4/x^3 + 16/x^5 > 0.9999$ so the quotient is less than $1/0.9999 < 1.0002$ times the numerator. That, in turn, is less than 2.0001 provided $x < -\sqrt[4]{17/(3.001/1.002 - 3)} \approx -14.36$. Therefore, if $x < -71$, then

$$2.999 < \dfrac{3x^5 - 3x^3 + 17x + 2}{x^5 + 4x^2 + 16} < 3.001.$$

REMARK: An alternative analysis shows the fraction is always less than 3 (if $x < -2$) and it exceeds 2.999 provided $x < -52.54$.

        c.   The denominator is zero when $x = -2$ but is nonzero elsewhere; the stipulation $x < -2$ thus guarantees the quotient is actually defined.

29.

x	$(1-1/x)^x$	x	$(1-1/x)^x$
$2 = 2^1$	0.25	$10 = 10^1$	0.3486784401
$4 = 2^2$	0.31640625	$100 = 10^2$	0.366032341273
$8 = 2^3$	0.343608915806	$1000 = 10^3$	0.367695424771
$16 = 2^4$	0.356074130452	$10,000 = 10^4$	0.367861046433
$32 = 2^5$	0.362055289256	$100,000 = 10^5$	0.367877601767
$64 = 2^6$	0.364986524244	$1,000,000 = 10^6$	0.367879257231
$128 = 2^7$	0.366437715922	$10,000,000 = 10^7$	0.367879422767
$256 = 2^8$	0.367159754892	$100,000,000 = 10^8$	0.367879439526
$512 = 2^9$	0.367519891255	$1,000,000,000 = 10^9$	0.367879441181
$1024 = 2^{10}$	0.367699739411	$10,000,000,000 = 10^{10}$	0.367879461768

$\lim_{x \to \infty}(1-1/x)^x = e^{-1} \approx 0.367879441171.$

31.

x	$(1+x)^{1/x}$	x	$(1+x)^{1/x}$
$2 = 2^1$	1.73205080757	$10 = 10^1$	1.27098161521
$4 = 2^2$	1.49534878122	$100 = 10^2$	1.04723274599
$8 = 2^3$	1.31607401295	$1000 = 10^3$	1.00693267528
$16 = 2^4$	1.19372161438	$10,000 = 10^4$	1.00092146833
$32 = 2^5$	1.11545886826	$100,000 = 10^5$	1.00011513598
$64 = 2^6$	1.06739894987	$1,000,000 = 10^6$	1.00001381561
$128 = 2^7$	1.03869725073	$10,000,000 = 10^7$	1.00000161181
$256 = 2^8$	1.02191271131	$100,000,000 = 10^8$	1.00000018421
$512 = 2^9$	1.01226261558	$1,000,000,000 = 10^9$	1.00000002072
$1024 = 2^{10}$	1.00679293668	$10,000,000,000 = 10^{10}$	1.00000000230

$\lim_{x \to \infty}(1+x)^{1/x} = e^0 = 1.$

33.

x	$e^{x/10}/x^4$	x	$e^{x/10}/x^4$
$2 = 2^1$	0.076337672385	$10 = 10^1$	0.000271828183
$4 = 2^2$	0.005827440225	$100 = 10^2$	0.000220264658
$8 = 2^3$	0.000543344953	$1000 = 10^3$	$2.69 \cdot 10^{31}$
$16 = 2^4$	0.000075577277	$10,000 = 10^4$	$1.97 \cdot 10^{418}$
$32 = 2^5$	0.000023396044	$100,000 = 10^5$	$8.81 \cdot 10^{4322}$
$64 = 2^6$	0.000035872760		
$128 = 2^7$	0.001349365151		
$256 = 2^8$	30.547725223675		
$512 = 2^9$	$2.50 \cdot 10^{11}$		
$1024 = 2^{10}$	$2.69 \cdot 10^{32}$		

$\lim_{x \to \infty} e^{x/10}/x^4 = \infty.$

35.

x	$x^2 e^x$	x	$x^2 e^x$
$-2 = -2^1$	$5.41 \cdot 10^{-1}$	$-10 = -10^1$	$4.54 \cdot 10^{-3}$
$-4 = -2^2$	$2.93 \cdot 10^{-1}$	$-100 = -10^2$	$3.72 \cdot 10^{-40}$
$-8 = -2^3$	$2.15 \cdot 10^{-2}$	$-1000 = -10^3$	$5.08 \cdot 10^{-429}$
$-16 = -2^4$	$2.88 \cdot 10^{-5}$	$-10,000 = -10^4$	$1.14 \cdot 10^{-4335}$
$-32 = -2^5$	$1.30 \cdot 10^{-11}$		
$-64 = -2^6$	$6.57 \cdot 10^{-25}$		
$-128 = -2^7$	$4.21 \cdot 10^{-52}$		
$-256 = -2^8$	$4.34 \cdot 10^{-107}$		
$-512 = -2^9$	$1.15 \cdot 10^{-217}$		
$-1024 = -2^{10}$	$2.01 \cdot 10^{-439}$		

$\lim_{x \to -\infty} x^2 e^x = 0.$

37. If $x > 0$, then $0 < \dfrac{1}{1+x} < 1$ and $\dfrac{-1}{x+x^2} = \left(\dfrac{-1}{x}\right) \cdot \left(\dfrac{1}{1+x}\right) < \left(\dfrac{-1}{x}\right)$; this implies $\lim\limits_{x \to 0^+} \dfrac{-1}{x+x^2} = -\infty.$

On the other hand, if $x < -1$, then $1 > \dfrac{-1}{x} > 0$ and $\dfrac{-1}{x+x^2} = \left(\dfrac{1}{1+x}\right) \cdot \left(\dfrac{-1}{x}\right) > \left(\dfrac{1}{1+x}\right)$; this implies

$\lim\limits_{x\to 0^-} \dfrac{-1}{x+x^2} = \infty$. Because the one-sided limits do not have the same value, the two-sided limit does not exist. ∎

39. An easy counterexample disproves the statement of the problem. Let $f(x) = x$, then $\lim_{x\to\infty} x = \infty$ but $\lim_{t\to 0}(1/t)$ does not exist (since the one-sided limits do not agree).

REMARK: The statement would be true if it referred to $\lim_{t\to 0^+} f(1/t)$; Problem 1.8.22 asks you to prove this revised assertion.

41. The solutions to problems 32 and 34 suggest that $\lim_{x\to\infty} xe^{-x} = 0$ and $\lim_{x\to\infty} x^5 e^{-x} = 0$; the solution to problem 33 implies $\lim_{x\to\infty} x^4 e^{-x} = 0$. Since $x < x^2 < x^3 < x^4 < x^5$ if $1 < x$, we can infer $\lim_{x\to\infty} x^2 e^{-x} = 0 = \lim_{x\to\infty} x^3 e^{-x}$ (use the Squeezing Theorem). Then we can apply Theorem 1.3.2 to infer $\lim_{x\to\infty} \left(a_0 + a_1 x + a_2 x^2 + a_3 x^3 + a_4 x^4 + a_5 x^5\right) e^{-x} = 0$ for any choice of coefficients $a_0, \ldots, a_5$.

Solutions 1.5   *Tangent lines and derivatives*   (pages 95-7)

1. $f'(x) = \lim\limits_{\Delta x\to 0} \dfrac{4-4}{\Delta x} = \lim\limits_{\Delta x\to 0} \dfrac{0}{\Delta x} = 0.$
The tangent at $(3,4)$ has slope 0 and equation $y = 0\cdot(x-3) + 4 = 4$.

3. $f'(x) = \lim\limits_{\Delta x\to 0} \dfrac{(-4(x+\Delta x)+6) - (-4x+6)}{\Delta x} = \lim\limits_{\Delta x\to 0} \dfrac{-4\,\Delta x}{\Delta x} = \lim\limits_{\Delta x\to 0}(-4) = -4.$
The tangent at $(3,-6)$ has slope -4 and equation $y = -4(x-3) + (-6) = -4x + 6$.

5. $f'(x) = \lim\limits_{\Delta x\to 0} \dfrac{2(x+\Delta x-3)^2 - 2(x-3)^2}{\Delta x}$
$ = 2\lim\limits_{\Delta x\to 0} \dfrac{[(x+\Delta x-3)-(x-3)][(x+\Delta x-3)+(x-3)]}{\Delta x}$ $\left(\text{because } a^2 - b^2 = (a-b)(a+b)\right)$
$ = 2\lim\limits_{\Delta x\to 0} \dfrac{\Delta x(2x-6+\Delta x)}{\Delta x} = 2\lim\limits_{\Delta x\to 0}(2x-6+\Delta x) = 2(2x-6) = 4x - 12.$
The tangent at $(4,2)$ has slope $f'(4) = 16 - 12 = 4$ and equation $y = 4(x-4) + 2 = 4x - 14$.

7. $f'(x) = \lim\limits_{\Delta x\to 0} \dfrac{(-(x+\Delta x)^2 + 3(x+\Delta x) + 5) - (-x^2 + 3x + 5)}{\Delta x}$
$ = \lim\limits_{\Delta x\to 0} \dfrac{-x^2 - 2x(\Delta x) - (\Delta x)^2 + 3x + 3\Delta x + x^2 - 3x}{\Delta x}$
$ = \lim\limits_{\Delta x\to 0} \dfrac{-2x(\Delta x) - (\Delta x)^2 + 3\Delta x}{\Delta x} = \lim\limits_{\Delta x\to 0}(-2x - \Delta x + 3) = -2x + 3.$
The tangent at $(0,5)$ has slope $f'(0) = 0 + 3 = 3$ and equation $y = 3(x-0) + 5 = 3x + 5$.

9. $f'(x) = \lim\limits_{\Delta x\to 0} \dfrac{(x+\Delta x)^3 - x^3}{\Delta x} = \lim\limits_{\Delta x\to 0} \dfrac{(x^3 + 3x^2\,\Delta x + 3x\,\Delta x^2 + \Delta x^3) - x^3}{\Delta x}$
$ = \lim\limits_{\Delta x\to 0} \dfrac{3x^2\,\Delta x + 3x\,\Delta x^2 + \Delta x^3}{\Delta x} = \lim\limits_{\Delta x\to 0}\left(3x^2 + 3x\,\Delta x + \Delta x^2\right) = 3x^2.$
The tangent at $(2,8)$ has slope $f'(2) = 3\cdot 2^2 = 12$ and equation $y = 12(x-2) + 8 = 12x - 16$.

11. $f'(x) = \lim\limits_{\Delta x\to 0} \dfrac{-\sqrt{x+\Delta x+3} - (-\sqrt{x+3})}{\Delta x}$
$ = \lim\limits_{\Delta x\to 0} \left(\dfrac{\sqrt{x+3} - \sqrt{x+\Delta x+3}}{\Delta x}\right)\left(\dfrac{\sqrt{x+3} + \sqrt{x+\Delta x+3}}{\sqrt{x+3} + \sqrt{x+\Delta x+3}}\right)$
$ = \lim\limits_{\Delta x\to 0} \dfrac{(x+3) - (x+\Delta x+3)}{\Delta x\left(\sqrt{x+3} + \sqrt{x+\Delta x+3}\right)} = \lim\limits_{\Delta x\to 0} \dfrac{-1}{\sqrt{x+3} + \sqrt{x+\Delta x+3}} = \dfrac{-1}{2\sqrt{x+3}}.$
The tangent at $(6,-3)$ has slope $f'(6) = -1/\left(2\sqrt{9}\right) = -1/6$ and equation $y = (-1/6)(x-6) + (-3) = (-1/6)\,x - 2$.

13. $f'(x) = \lim\limits_{\Delta x\to 0} \dfrac{(x+\Delta x)^4 - x^4}{\Delta x} = \lim\limits_{\Delta x\to 0} \dfrac{x^4 + 4x^3\Delta x + 6x^2\Delta x^2 + 4x\Delta x^3 + \Delta x^4 - x^4}{\Delta x}$
$ = \lim\limits_{\Delta x\to 0} \left(4x^3 + 6x^2\Delta x + 4x\Delta x^2 + \Delta x^3\right) = 4x^3.$
The tangent at $(2,16)$ has slope $f'(2) = 4(2)^3 = 32$ and equation $y = 32(x-2) + 16 = 32x - 48$.

1.5 — Tangent lines and derivatives

15.
$$\frac{dy}{dx} = \lim_{\Delta x \to 0} \frac{(m(x + \Delta x) + b) - (mx + b)}{\Delta x} = \lim_{\Delta x \to 0} \left(m \frac{(x + \Delta x) - (x)}{\Delta x} + \frac{b - b}{\Delta x} \right)$$

$$= \lim_{\Delta x \to 0} \left(m \frac{\Delta x}{\Delta x} + 0 \right) = \lim_{\Delta x \to 0} m = m. \ \blacksquare$$

17.   Because $P^3 - Q^3 = (P - Q) \cdot (P^2 + PQ + Q^2)$,

$$\frac{d(\alpha x^3)}{dx} = \lim_{\Delta x \to 0} \frac{\alpha(x + \Delta x)^3 - \alpha x^3}{\Delta x} = \alpha \lim_{\Delta x \to 0} \frac{(x + \Delta x)^3 - x^3}{\Delta x}$$

$$= \alpha \lim_{\Delta x \to 0} \frac{[(x + \Delta x) - x] \cdot [(x + \Delta x)^2 + (x + \Delta x)x + x^2]}{\Delta x}$$

$$= \alpha \lim_{\Delta x \to 0} \left[\frac{\Delta x}{\Delta x} \right] \cdot [(x + \Delta x)^2 + (x + \Delta x)x + x^2]$$

$$= \alpha (1) \left[x^2 + x^2 + x^2 \right] = 3 \alpha x^2. \ \blacksquare$$

Alternatively,
$$\alpha \lim_{\Delta x \to 0} \frac{(x + \Delta x)^3 - x^3}{\Delta x} = \alpha \lim_{\Delta x \to 0} \frac{x^3 + 3x^2 \Delta x + 3x \Delta x^2 + \Delta x^3 - x^3}{\Delta x}$$

$$= \alpha \lim_{\Delta x \to 0} (3x^2 + 3x \Delta x + \Delta x^2) = \alpha 3x^2. \ \blacksquare$$

---

19.   The tangents at $(x, f(x))$ and $(x, g(x))$ are parallel if and only if they have the same slope, i.e., if and only if $f'(x) = g'(x)$. If $f(x) = x$, then $f'(x) = 1$ (see Problem 15); if $g(x) = x^2$, then $g'(x) = 2x$ (see Example 2). These derivatives are equal if and only if $x = \frac{1}{2}$.

21.   If $g(x) = 6x - 5$, then $g'(x) = 6$ (see Problem 15); if $f(x) = ax^2$, then $f'(x) = 2ax$ (see Problem 16). The tangent to g at $(2, 7)$ has slope $g'(2) = 6$; the tangent to f at $(2, 4a)$ is parallel if and only if its slope, $f'(2) = 4a$, is also equal to 6. This requires that $a = 6/4 = 3/2$.

23.   Remember that $\lfloor x \rfloor = N$ if N is the unique integer such that $N \leq x < N + 1$; therefore $\lfloor x \rfloor = -1$ if $-1 \leq x < 0$. Let $f(x) = x \cdot \lfloor -x^2 \rfloor$. Then $f(0) = 0 \cdot \lfloor 0 \rfloor = 0$ and

$$f'(0) = \lim_{\Delta x \to 0} \frac{f(\Delta x) - f(0)}{\Delta x} = \lim_{\Delta x \to 0} \frac{\Delta x \cdot \lfloor -(\Delta x)^2 \rfloor - 0}{\Delta x} = \lim_{\Delta x \to 0} \lfloor -(\Delta x)^2 \rfloor = -1.$$

(The final equality follows from the fact that $-1 \leq -(\Delta x)^2 < 0$ if $-1 \leq \Delta x \leq 1$ and $\Delta x \neq 0$.)

25.

Δx	$2 + \Delta x$	a. $-3(2 + \Delta x)^2$	b. $\dfrac{-3(2 + \Delta x)^2 - (-3(2)^2)}{\Delta x}$
-0.5	1.5	-6.75	-10.5
0.1	2.1	-13.23	-12.3
-0.01	1.99	-11.8803	-11.97
0.0001	2.001	-12.012003	-12.003

Looking at the last column of this table, I guess $f'(2)$ is close to -12.

c.
$$f'(x) = \lim_{\Delta x \to 0} \frac{-3(x + \Delta x)^2 - (-3x^2)}{\Delta x} = \lim_{\Delta x \to 0} \frac{-3 \left(x^2 + 2x \Delta x + (\Delta x)^2 \right) + 3x^2}{\Delta x}$$

$$= \lim_{\Delta x \to 0} \frac{-6x \Delta x - 3(\Delta x)^2}{\Delta x} = \lim_{\Delta x \to 0} (-6x - 3 \Delta x) = -6x.$$

Therefore, $f'(2) = -6 \cdot 2 = -12$.

d.   Since the tangent at $(2, -12)$ has slope -12, the tangent satisfies $y = -12(x - 2) - 12 = -12x + 12$.

e. $f'(x) = -6x$ implies tangent lines have positive slope if $x < 0$ and they have negative slope if $x > 0$; hence $f(x)$ increases on $(-\infty, 0)$ and decreases on $(0, \infty)$.

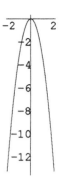

27.

Δx	$\cos\left(\frac{\pi}{4} + \Delta x\right)$	$\frac{\cos\left(\frac{\pi}{4}+\Delta x\right)-\cos\left(\frac{\pi}{4}\right)}{\Delta x}$	Δx	$\cos\left(\frac{\pi}{4} + \Delta x\right)$	$\frac{\cos\left(\frac{\pi}{4}+\Delta x\right)-\cos\left(\frac{\pi}{4}\right)}{\Delta x}$
0.1	0.63298131	−0.74125475	−0.1	0.77416708	−0.67060297
0.01	0.70000048	−0.71063050	−0.01	0.71414238	−0.70355949
0.001	0.70639932	−0.70746022	−0.001	0.70781353	−0.70675311
0.0001	0.70703607	−0.70714214	−0.0001	0.70717749	−0.70707142
0.00001	0.70709971	−0.70711032	−0.00001	0.70711385	−0.70710325
0.000001	0.70710607	−0.70710713	−0.000001	0.70710749	−0.70710643

If $f = \cos$, then $f'(\pi/4) = -1/\sqrt{2} \approx -0.70710678$.

29.

Δx	$\tan\left(\frac{\pi}{4} + \Delta x\right)$	$\frac{\tan\left(\frac{\pi}{4}+\Delta x\right)-\tan\left(\frac{\pi}{4}\right)}{\Delta x}$	Δx	$\tan\left(\frac{\pi}{4} + \Delta x\right)$	$\frac{\tan\left(\frac{\pi}{4}+\Delta x\right)-\tan\left(\frac{\pi}{4}\right)}{\Delta x}$
0.1	1.2230489	2.2304888	−0.1	0.81762881	1.8237119
0.01	1.0202027	2.0202700	−0.01	0.98019737	1.9802634
0.001	1.0020020	2.0020027	−0.001	0.99800200	1.9980027
0.0001	1.0002000	2.0002000	−0.0001	0.99980002	1.9998000
0.00001	1.0000200	2.0000200	−0.00001	0.99998000	1.9999800
0.000001	1.0000020	2.0000020	−0.000001	0.99999800	1.9999980

If $f = \tan$, then $f'(\pi/4) = 2$.

31.

Δx	$\ln\left(1 + \Delta x\right)$	$\frac{\ln(1+\Delta x)-\ln(1)}{\Delta x}$	Δx	$\ln\left(1 + \Delta x\right)$	$\frac{\ln(1+\Delta x)-\ln(1)}{\Delta x}$
0.1	0.09531018	0.95310180	−0.1	−0.10536052	1.0536052
0.01	0.00995033	0.99503309	−0.01	−0.01005034	1.0050336
0.001	0.00099950	0.99950033	−0.001	−0.00100050	1.0005003
0.0001	0.00010000	0.99995000	−0.0001	−0.00010001	1.0000500
0.00001	0.00001000	0.99999500	−0.00001	−0.00001000	1.0000050
0.000001	0.00000100	0.99999950	−0.000001	−0.00000100	1.0000005

If $f = \ln$, then $f'(1) = 1$.

33.

Δx	$\ln\left(10 + \Delta x\right)$	$\frac{\ln(10+\Delta x)-\ln(10)}{\Delta x}$	Δx	$\ln\left(10 + \Delta x\right)$	$\frac{\ln(10+\Delta x)-\ln(10)}{\Delta x}$
0.1	2.31253542	0.099503309	−0.1	2.29253476	0.100503359
0.01	2.30358459	0.099950033	−0.01	2.30158459	0.100050033
0.001	2.30268509	0.099995000	−0.001	2.30248509	0.100005000
0.0001	2.30259509	0.099999500	−0.0001	2.30257509	0.100000500
0.00001	2.30258609	0.099999950	−0.00001	2.30258409	0.100000050
0.000001	2.30258519	0.099999995	−0.000001	2.30258499	0.100000005

If $f = \ln$, then $f'(10) = 1/10 = 0.1$.

35.

Δx	$e^{0 + \Delta x}$	$\frac{e^{0+\Delta x}-e^0}{\Delta x}$	Δx	$e^{0 + \Delta x}$	$\frac{e^{0+\Delta x}-e^0}{\Delta x}$
0.1	1.10517091808	1.05170918076	−0.1	0.904837418036	0.951625819640
0.01	1.01005016708	1.00501670842	−0.01	0.990049833749	0.995016625083
0.001	1.00100050017	1.00050016671	−0.001	0.999000499833	0.999500166625
0.0001	1.00010000500	1.00005000167	−0.0001	0.999900005000	0.999950001666
0.00001	1.00001000005	1.00000500002	−0.00001	0.999990000050	0.999995000014
0.000001	1.00000100000	1.00000050004	−0.000001	0.999999000001	0.999999499984

If $f(x) = e^x$, then $f'(0) = 1$.

37.

Δx	$e^{-1+\Delta x}$	$\frac{e^{-1+\Delta x}-e^{-1}}{\Delta x}$	Δx	$e^{-1+\Delta x}$	$\frac{e^{-1+\Delta x}-e^{-1}}{\Delta x}$
0.1	0.40656966	0.38690219	−0.1	0.33287108	0.35008357
0.01	0.37157669	0.36972499	−0.01	0.36421898	0.36604616
0.001	0.36824750	0.36806344	−0.001	0.36751175	0.36769556
0.0001	0.36791623	0.36789784	−0.0001	0.36784266	0.36786105
0.00001	0.36788312	0.36788128	−0.00001	0.36787576	0.36787760
0.000001	0.36787981	0.36787963	−0.000001	0.36787907	0.36787926

If $f(x) = e^x$, then $f'(-1) = 1/e \approx 0.367879441171$.

39.   Let $f(x) = x\,|x|$. If $x > 0$, then $f(x) = x \cdot x = x^2$ and Example 4 shows that $f'(x) = 2x$. If $x < 0$, then $f(x) = x \cdot (-x) = -x^2$ and Problem 16 says that $f'(x) = -2x$. So we are left to find whether $f'(0)$ is defined. Consider both one-sided limits:

$$\lim_{\Delta x \to 0^+} \frac{f(0 + \Delta x) - f(0)}{\Delta x} = \lim_{\Delta x \to 0^+} \frac{(\Delta x)^2 - 0 \cdot |0|}{\Delta x} = \lim_{\Delta x \to 0^+} \Delta x = 0,$$

$$\lim_{\Delta x \to 0^-} \frac{f(0 + \Delta x) - f(0)}{\Delta x} = \lim_{\Delta x \to 0^-} \frac{-(\Delta x)^2 - 0 \cdot |0|}{\Delta x} = \lim_{\Delta x \to 0^-} -\Delta x = 0.$$

Because the right and left limits exist and agree, we conclude that

$$f'(0) = \lim_{\Delta x \to 0} \frac{f(0 + \Delta x) - f(0)}{\Delta x}$$

exists and equals 0. We can summarize these results as $f'(x) = 2\,|x|$. ∎

41.   a.   Suppose $f(x) = x^n$. Then $f(-x) = (-x)^n = (-1)^n\, x^n = (-1)^n\, f(x)$. If n is an even integer, then $(-1)^n = 1$ and f is an even function; if n is an odd integer, then $(-1)^n = -1$ and f is an odd function. ∎

   b.   If $g(x) = |x|$, then $g(-x) = |-x| = |x| = g(x)$; $|x|$ is an even function. ∎

   c.   If $h(x) = x \cdot |x|$, then $h(-x) = (-x) \cdot |-x| = -x \cdot |x| = -h(x)$; $x \cdot |x|$ is an odd function. ∎

43.   Suppose $f(-x) = f(x)$ for all x and suppose that $f'(x)$ exists for all x. Then

$$f'(-x) = \lim_{\Delta x \to 0} \frac{f(-x + \Delta x) - f(-x)}{\Delta x} = \lim_{\Delta x \to 0} \frac{f(x - \Delta x) - f(x)}{\Delta x}$$

$$= (-1) \cdot \lim_{-\Delta x \to 0} \frac{f(x + (-\Delta x)) - f(x)}{-\Delta x} = (-1) \cdot f'(x). \quad ∎$$

(Note that if Δx tends to 0, then so does $-\Delta x$.)

45.   Suppose f is a periodic differentiable function such that $f(x + \omega) = f(x)$ for all x. Then

$$f'(x + \omega) = \lim_{\Delta x \to 0} \frac{f(x + \omega + \Delta x) - f(x + \omega)}{\Delta x} = \lim_{\Delta x \to 0} \frac{f(x + \Delta x) - f(x)}{\Delta x} = f'(x)$$

so f' is also a periodic function. ∎

REMARK: We have only shown f' is periodic; showing that f and f' have the same period will follow from the Fundamental Theorem of Calculus which you will study in Chapter 4.

---

47.   If $f(x) = \ln x$, then $f'(x) = 1/x$.

49.   Suppose f is differentiable at x, then

$$\lim_{\Delta x \to 0^+} \frac{f(x + \Delta x) - f(x - \Delta x)}{2\,\Delta x} = \frac{1}{2} \lim_{\Delta x \to 0^+} \frac{f(x + \Delta x) - f(x)}{\Delta x} + \frac{1}{2} \lim_{\Delta x \to 0^+} \frac{f(x) - f(x - \Delta x)}{\Delta x}$$

$$= \frac{1}{2} \lim_{\Delta x \to 0^+} \frac{f(x + \Delta x) - f(x)}{\Delta x} + \frac{1}{2} \lim_{\Delta x \to 0^+} \frac{f(x + (-\Delta x)) - f(x)}{-\Delta x}$$

$$= \frac{1}{2} f'(x) + \frac{1}{2} f'(x) = f'(x).$$

On the other hand, consider the absolute value function, $|x|$, which we already know is not differentiable at $x = 0$ (see Example 7). Nevertheless,

$$\lim_{\Delta x \to 0^+} \frac{|0 + \Delta x| - |0 - \Delta x|}{2\,\Delta x} = \lim_{\Delta x \to 0^+} \frac{|\Delta x| - |\Delta x|}{2\,\Delta x} = \lim_{\Delta x \to 0^+} \frac{0}{2\,\Delta x} = 0,$$

so the proposed "alternative definition" is not equivalent to the standard definition.

REMARK: For many "nice" functions, this double-difference quotient not only converges to the derivative but does so more quickly than the regular difference quotient used in defining the derivative. The discussion of the Mean Value Theorem (3.7.2) may suggest what the graphs of such "nice" functions will look like; Section 3.3 and Problems 3.6.15-8 may suggest situations where the double-difference quotient is not preferred; Taylor's Theorem (9.4.1) and Section 9.5 will provide a concrete computational procedure. Problem 10.10.36, posed after more tools have been developed, will offer you the opportunity to compare these two difference quotients from a broader perspective.

Solutions 1.6 *The derivative as a rate of change* (pages 111-2)

1.
$$V(t) = s'(t) = \lim_{\Delta t \to 0} \frac{\left(1 + (t + \Delta t) + (t + \Delta t)^2\right) - \left(1 + t + t^2\right)}{\Delta t}$$
$$= \lim_{\Delta t \to 0} \frac{\Delta t + t^2 + 2t(\Delta t) + (\Delta t)^2 - t^2}{\Delta t} = \lim_{\Delta t \to 0} (1 + 2t + \Delta t) = 1 + 2t;$$

therefore, $V(4) = 1 + 2(4) = 9$.

REMARK: Problem 1.5.16 painlessly yields $s'(t) = 2t + 1$ because $s(t) = t^2 + t + 1$.

3.
$$V(t) = s'(t) = \lim_{T \to t} \frac{(T^3 - T^2 + 3) - (t^3 - t^2 + 3)}{T - t}$$
$$= \lim_{T \to t} \left[\frac{T^3 - t^3}{T - t} - \frac{T^2 - t^2}{T - t} \right] = \lim_{T \to t} \left[(T^2 + Tt + t^2) - (T + t) \right] = 3t^2 - 2t;$$

therefore, $V(5) = 75 - 10 = 65$.

REMARK: The result of Problem 1.5.18 could be applied here.

5.
$$V(t) = s'(t) = \lim_{T \to t} \frac{\left(1 + \sqrt{2T}\right) - \left(1 + \sqrt{2t}\right)}{T - t} = \lim_{T \to t} \left(\frac{\sqrt{2T} - \sqrt{2t}}{T - t} \right) \left(\frac{\sqrt{2T} + \sqrt{2t}}{\sqrt{2T} + \sqrt{2t}} \right)$$
$$= \lim_{T \to t} \frac{2T - 2t}{(T - t)\left(\sqrt{2T} + \sqrt{2t}\right)} = \lim_{T \to t} \frac{2}{\sqrt{2T} + \sqrt{2t}} = \frac{2}{2\sqrt{2t}} = \frac{1}{\sqrt{2t}};$$
$$V(8) = \frac{1}{\sqrt{2(8)}} = \frac{1}{\sqrt{16}} = \frac{1}{4}.$$

REMARK: Compare this with the solution for Problem 1.5.12.

7. $h(t) = 70t^2$ ft for $0 \le t \le 210$. Thus
$$V(t) = h'(t) = \lim_{\Delta t \to 0} \frac{70(t + \Delta t)^2 - 70t^2}{\Delta t} = 70 \lim_{\Delta t \to 0} \frac{t^2 + 2t\Delta t + (\Delta t)^2 - t^2}{\Delta t}$$
$$= 70 \lim_{\Delta t \to 0} (2t + \Delta t)\left(\frac{\Delta t}{\Delta t}\right) = 70(2t) = 140t.$$
$$V(3) = 140(3) = 420 \text{ ft/sec},$$
$$V(10) = 140(10) = 1400 \text{ ft/sec}.$$

9. $P'(5 \text{ hr}) = 4(5^3) = 500$ individuals/ hr, $P'(24 \text{ hr}) = 4(24^3) = 55,296$ individuals/ hr. The function considered in Example 3, $P(t) = 100 + t^4$, is not realistic beyond some finite interval of time — for instance, because the function is unbounded as t increases, it eventually has values greater than the number of atoms in planet Earth!

11.
$$V'(r) = \frac{d}{dr}\left(\frac{4}{3}\pi r^3\right) = \lim_{\Delta r \to 0} \frac{(4/3)\pi(r + \Delta r)^3 - (4/3)\pi r^3}{\Delta r} = \frac{4}{3}\pi \lim_{\Delta r \to 0} \frac{(r + \Delta r)^3 - r^3}{\Delta r}$$
$$= \frac{4}{3}\pi \lim_{\Delta r \to 0} \left(3r^2 + 3r\Delta r + \Delta r^2\right) = \frac{4}{3}\pi \left(3r^2\right) = 4\pi r^2;$$
$$V'(10 \,\mu\text{m}) = 4\pi 10^2 = 400\pi \,\mu\text{m}^3/\mu\text{m}.$$

1.6 — The derivative as a rate of change

13. a.
$$V = \frac{4}{3}\pi r^3 = \frac{4}{3}\pi \left(\left(\frac{S}{4\pi}\right)^{1/2}\right)^3 = \frac{S^{3/2}}{6\sqrt{\pi}}.$$

b. It is shown in Example 4 that $\frac{d}{dx}x^{3/2} = \frac{3}{2}x^{1/2}$, therefore
$$\frac{dV}{dS} = \frac{d}{dS}\frac{S^{3/2}}{6\sqrt{\pi}} = \frac{1}{6\sqrt{\pi}}\cdot\frac{3}{2}\cdot S^{1/2} = \frac{1}{4}\sqrt{\frac{S}{\pi}}$$
$$V'(100\,\mu\,\mathrm{m}^2) = \frac{5}{2\sqrt{\pi}}\ \mu\,\mathrm{m}^3\ \text{per}\ \mu\,\mathrm{m}^2.$$

15.
$$\frac{dC}{dq} = \lim_{Q\to q}\frac{[200 + 6Q - 0.01Q^2 + 0.01Q^3] - [200 + 6q - 0.01q^2 + 0.01q^3]}{Q - q}$$
a.
$$= \lim_{Q\to q}\left[6\frac{Q-q}{Q-q} - 0.01\frac{Q^2-q^2}{Q-q} + 0.01\frac{Q^3-q^3}{Q-q}\right]$$
$$= \lim_{Q\to q}\left[6 - 0.01(Q+q) + 0.01(Q^2+Qq+q^2)\right] = 6 - 0.02q + 0.03q^2.$$

b. The marginal cost function is increasing when $q > 1/3$, the manufacturer should not buy in large quantities.

Solutions 1.7 *Continuity* (pages 120-3)

1. If $f(x) = x^{17} - 3x^{15} + 2$, then f is continuous at every real number (see Example 2); therefore f is continuous on $(-\infty,\infty)$.

3.
$$f(x) = \frac{|x+2|}{x+2} = \begin{cases} -1 & \text{if } x < -2 \\ \text{undefined} & \text{if } x = -2 \\ 1 & \text{if } x > -2 \end{cases}$$
is continuous except at -2; therefore f is continuous on $(-\infty,-2)\cup(-2,\infty)$.

5. If $f(x) = -17x/(x^2-1)$, then f is continuous at every real number except -1 and 1 (see Example 3); therefore f is continuous on $(-\infty,-1)\cup(-1,1)\cup(1,\infty)$.

7. $f(x) = x^{1/3}$ is defined for all x and is continuous everywhere; f is continuous on $(-\infty,\infty)$.

9. The greatest integer function, $\lfloor x\rfloor$, is constant on each interval $(N,N+1)$ where N is an integer; on each such interval the function $f(x) = x - \lfloor x\rfloor$ increases continuously with its values lying in the interval $(0,1)$. This function is discontinuous at each integer N because
$$\lim_{x\to N+}(x - \lfloor x\rfloor) = \lim_{x\to N+}(x - N) = N - N = 0 = f(N),$$
but
$$\lim_{x\to N-}(x - \lfloor x\rfloor) = \lim_{x\to N-}(x - (N-1)) = N - (N-1) = 1 \neq f(N).$$

REMARK: This greatest integer function is also known as the "floor" function; the notation $\lfloor x\rfloor$ is becoming preferred to $[x]$.

11. The graph of f is a straight line. Therefore, $\lim_{x\to x_0}f(x) = f(x_0)$ and f is continuous at x_0. Furthermore, f' is constant, so it too is continuous.

13. $f(x_0)$ is undefined, thus f is not continuous at x_0; moreover, $f'(x_0)$ does not exist (because $f(x_0)$ is undefined) so f' also is not continuous at x_0.

15. The graph of f is two half-lines which do not meet. Examining the graph of f, we see that $\lim_{x\to x_0^-}f(x) = f(x_0) > \lim_{x\to x_0^+}f(x)$ and f is not continuous at x_0. We also see that $f'(x)$ equals a positive constant to the left of x_0 and a negative constant to the right of x_0, hence f' is not continuous at x_0.

REMARKS: If $f'(x_0)$ existed, then we could invoke Theorem 3 to infer that f was continuous at x_0 — we already know that is not true, so it must also be false to say $f'(x_0)$ exists.

That is a proof-by-contradiction. We used a theorem of the form "$P \implies Q$" restated in the form "not $Q \implies$ not P". Logicians refer to the second form as the *contrapositive* of the first.

17.   The graph of f consists of two arcs which meet at the point (x_0, y_0). Therefore, $\lim_{x \to x_0^-} f(x) = y_0 = f(x_0) = \lim_{x \to x_0^+} f(x)$ so f is continuous at x_0. There is a sharp corner at that point — it appears that $\lim_{x \to x_0^-} f'(x) < 0 < \lim_{x \to x_0^+} f'(x)$ so f' is not continuous at x_0.

19.   The graph of f is two arcs which do not meet. Examining the graph of f, we see that $\lim_{x \to x_0^-} f(x) = f(x_0) < \lim_{x \to x_0^+} f(x)$ and f is not continuous at x_0. Although it appears that $\lim_{x \to x_0^-} f'(x) = 0 = \lim_{x \to x_0^+} f'(x)$, f' is not continuous at x_0 because $f'(x_0)$ does not exist.

REMARK: Since differentiability implies continuity (Theorem 3), $f'(x_0)$ does not exist because f is not continuous at x_0. (contrapositive of Theorem 3).

21.   The speed of the car is a continuous function of time, hence (Theorem 5) it takes on each value (in particular, 50) between 0 and 80 (i.e., between its starting speed and its speed after 30 seconds).

23. $f(x) = (x^3 - 1)/(x - 1) = x^2 + x + 1$ when $x \neq 1$; that is a polynomial function and is obviously continuous. $\lim_{x \to 1} f(x) = \lim_{x \to 1} (x^2 + x + 1) = 3 = f(1)$ so f is also continuous at 1 too.

25.   Using Example 2 we can state that f is continuous on the interval $(-\infty, 0)$ because $f(x) = 0$ there, f is continuous on $(0, 2)$ because $f(x) = x^2$ there, and f is continuous on $(2, \infty)$ because $f(x) = 4$ there. [Note that the constant functions 0 and 4 are both polynomial functions.] Now the only job left is to verify that f is continuous at 0 and 2.

$\lim_{x \to 0^-} f(x) = \lim_{x \to 0^-} 0 = 0 = 0^2 = \lim_{x \to 0^+} x^2 = \lim_{x \to 0^+} f(x)$, thus $\lim_{x \to 0} f(x) = 0 = 0^2 = f(0)$ and f is continuous at 0.

$\lim_{x \to 2^-} f(x) = \lim_{x \to 2^-} x^2 = 2^2 = 4 = \lim_{x \to 2^+} 4 = \lim_{x \to 2^+} f(x)$, thus $\lim_{x \to 2} f(x) = 4 = f(2)$ and f is continuous at 2. ∎

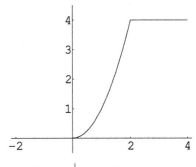

27. f is continuous at 0 and 1, it is discontinous at all other integers.

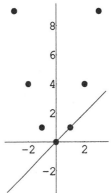

29.   Let $f(A) = \sqrt{A}$ and $g(B) = B - 1$. $\lim_{x \to 2} g(x) = \lim_{x \to 2} (x - 1) = 2 - 1 = 1$. f is continuous at 1; thus, using theorem 2, $\lim_{x \to 2} \sqrt{x - 1} = \sqrt{\lim_{x \to 2} (x - 1)} = \sqrt{1} = 1$. Since $\lim_{x \to 2} (x + 1) = 3$, we invoke part (iv) of Theorem 1 to conclude

$$\lim_{x \to 2} \frac{\sqrt{x - 1}}{x + 1} = \frac{\sqrt{2 - 1}}{2 + 1} = \frac{1}{3}.$$

31.   If $f(x) = x^3 + 4x + 7$, then $f(-1.25538) \approx 2.755 \cdot 10^{-5}$.

a	$f(a)$	$b = a + 0.5$	$f(b)$	$f(c) = 0$ for some c in (a, b)
-1.8	-6.032	-1.3	-0.397	no
-1.7	-4.713	-1.2	0.472	Yes
-1.6	-3.496	-1.1	1.269	Yes
-1.5	-2.375	-1.0	2.000	Yes
-1.4	-1.344	-0.9	2.671	Yes
-1.3	-0.397	-0.8	3.288	Yes
-1.2	0.472	-0.7	3.857	no

33.   If $f(x) = 2x^5 + 5x^3 + 8x + 9$,
then $f(-0.77078228516) = 8.422152 \cdot 10^{-13}$.

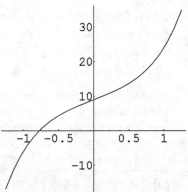

a	$f(a)$	$b = a + 0.5$	$f(b)$	$f(c) = 0$ for some c in (a, b)
-1.2707823	-18.05516172	-0.7707823	-0.00000030	no
-1.2707822	-18.05515589	-0.7707822	0.00000174	Yes
-0.7707823	-0.00000030	-0.2707823	6.73155708	Yes
-0.7707822	0.00000174	-0.2707822	6.73155799	no

REMARK: The other zeros of f are the complex numbers $-0.496765 \pm 1.53758\, i$ and $0.882156 \pm 1.20743\, i$.

35.   If $f(x) = \left(3x^3 + 2x - 20\right) / \sqrt{3x + 8}$,
then $f(1.76416344944834) = 3.654166 \cdot 10^{-16}$.

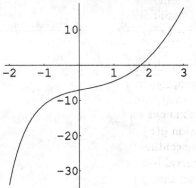

a	$f(a)$	$b = a + 0.5$	$f(b)$	$f(c) = 0$ for some c in (a, b)
1.6641634	-0.78936451	1.7641634	-0.00000041	no
1.6641635	-0.78936375	1.7641635	0.00000042	Yes
1.7641634	-0.00000041	1.8641634	0.85787646	Yes
1.7641635	0.00000042	1.8641635	0.85787736	no

REMARK: The other zeros of f are the complex numbers $-0.882082 \pm 1.7323\, i$.

37.   If $f(x) = \sin x - x/2$,
then $f(1.895494267034) = -1.5557 \cdot 10^{-14}$.

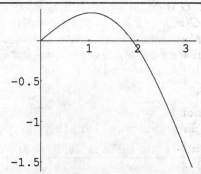

a	$f(a)$	$b = a + 0.1$	$f(b)$	$f(c) = 0$ for some c in (a, b)
1.7954942	0.07711437	1.8954942	0.00000005	no
1.7954943	0.07711430	1.8954943	-0.00000003	Yes
1.8954942	0.00000005	1.9954942	-0.08658384	Yes
1.8954943	-0.00000003	1.9954943	-0.08658393	no

39.   If $f(x) = x - \tan x$,
then $f(4.493409457909) = 1.293854 \cdot 10^{-12}$.

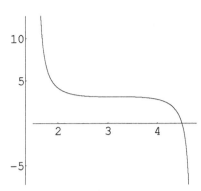

a	$f(a)$	$b = a + 0.1$	$f(b)$	$f(c) = 0$ for some c in (a, b)
4.3934094	1.36546731	4.4934094	0.00000117	no
4.3934095	1.36546639	4.4934095	-0.00000085	Yes
4.4934094	0.00000117	4.5934094	-3.77169681	Yes
4.4934095	-0.00000085	4.5934095	-3.77170381	no

41.   a. $f(-1)$ is not defined. (See condition (i) of the definition.)
    b. $\lim_{x \to 1} f(x) = \lim_{x \to 1}(x^2 - 1)/(x - 1) = \lim_{x \to 1}(x + 1) = 2$.
    c. f has a removable discontinuity at $x_0 = 1$ because $f(x)$ has a finite limit as $x \to 1$. Let $g(x) = x + 1$; then g is continuous everywhere and $g(x) = f(x)$ provided $x \neq 1$.

43. $$f(x) = \frac{x^3 - 6x^2 + 11x - 6}{x - \alpha} = \frac{(x - 1)(x - 2)(x - 3)}{x - \alpha}.$$
f has a removable discontinuity at $x = \alpha$ if and only if α is one of the values 1, 2, or 3.

45.   Because $\lim_{x \to 2^-}(x + 1)/(x - 2) = -\infty$ while $\lim_{x \to 2^+}(x + 1)/(x - 2) = \infty$, the discontinuity at 2 is not a jump discontinuity; therefore the function is not piecewise continuous on $[-3, 3]$. On the other hand, the function is continuous on $[-1, 1]$, so it is certainly piecewise continuous there.

47.   The function $g(x) = x - \lfloor x \rfloor$ is continuous on each interval $(N, N + 1)$ where N is an integer. g does have a jump discontinuity at each integer point because $\lim_{x \to N^+} g(x) = 0 = g(N)$ while $\lim_{x \to N^-} g(x) = 1$.
    In the interval $[-2, 2]$, g has a jump discontinuity only at the four points $-1, 0, 1, 2$ and g is continuous everywhere else; thus g is piecewise continuous on $[-2, 2]$. ∎

49. f is continuous on each of the intervals $[-2, -1)$ and $(-1, 0)$ but it has a jump discontinuity at -1 because $\lim_{x \to -1^-} f(x) = \lim_{x \to -1^-}(x + x^2) = -1 + (-1)^2 = 0$ but $\lim_{x \to -1^+} f(x) = \lim_{x \to -1^+}(x^3) = (-1)^3 = -1$.

51. f is continuous throughout the entire interval $[-5, 5]$. It is obviously continuous on the subintervals $[-5, 0)$ and $(0, 5]$ because $f(x)$ equals a polynomial in each case. f is also continuous at 0 because $\lim_{x \to 0^-} f(x) = \lim_{x \to 0^-}(2x^2 + 4) = 2(0)^2 + 4 = 4$ and $\lim_{x \to 0^+} f(x) = \lim_{x \to 0^+}(x - 2)^2 = (-2)^2 = 4$.

53. $G(x) = \lfloor x \rfloor + (x - \lfloor x \rfloor)^2$ is continuous on every interval of the form $(N, N + 1)$, N an integer, because $\lfloor x \rfloor$ is. In fact, G is continuous everywhere because $\lim_{x \to N^+} G(x) = \lim_{x \to N^+}(x - N)^2 = N + 0^2 = G(N)$ and $\lim_{x \to N^-} G(x) = (N - 1) + \lim_{x \to N^-}(x - (N - 1))^2 = (N - 1) + 1^2 = N = G(N)$.

55.   Suppose N is an integer and $N < x < N + 1$, then $N = \lfloor x \rfloor$, $-N > -x > -N - 1$, $\lfloor -x \rfloor = -N - 1$, and $\lfloor x \rfloor + \lfloor -x \rfloor + 1 = N + (-N - 1) + 1 = 0$. If $x = N$, then $\lfloor x \rfloor + \lfloor -x \rfloor + 1 = N + (-N) + 1 = 1$. Therefore

$$\lfloor x \rfloor + \lfloor -x \rfloor + 1 = \begin{cases} 0 & \text{if } x \text{ is not an integer,} \\ 1 & \text{if } x \text{ is an integer.} \end{cases}$$

    The function $\lfloor x \rfloor + \lfloor -x \rfloor + 1$ is continuous on each interval of the form $(N, N + 1)$, N an integer, and it has a jump discontinuity at each integer.

57.   Let $f(x) = \lfloor -x \rfloor$ for any x in the interval $[0, 1]$, then $f(x) = \begin{cases} -1 & \text{if } 0 < x \leq 1, \\ 0 & \text{if } 0 = x. \end{cases}$

59.   a.   If $x \leq 1$, then $f(x) = x + 3$, $g(x) = x^2 + 6$, and $f(x)g(x) = (x + 3)(x^2 + 6)$. If $x > 1$, then $f(x) = 2x + 5$, $g(x) = 5x^3 - 1$, and $f(x)g(x) = (2x + 5)(5x^3 - 1)$.
    b. $\lim_{x \to 1^-} f(x) = \lim_{x \to 1^-}(x + 3) = 1 + 3 = 4$ but $\lim_{x \to 1^+} f(x) = \lim_{x \to 1^+}(2x + 5) = 2 + 5 = 7$. Because the left and right limits differ, $\lim_{x \to 1} f(x)$ does not exist.
    c. $\lim_{x \to 1^-} g(x) = \lim_{x \to 1^-}(x^2 + 6) = 1 + 6 = 7$ but $\lim_{x \to 1^+} g(x) = \lim_{x \to 1^+}(5x^3 - 1) = 5 - 1 = 4$. Because the right and left limits differ, $\lim_{x \to 1} g(x)$ does not exist.

d. $\lim_{x\to 1-} f(x)g(x) = \lim_{x\to 1-}(x+3)(x^2+6) = (4)(7) = 28$ and $\lim_{x\to 1+} = \lim_{x\to 1+}(2x+5)(5x^3-1) = (7)(4) = 28$. Because both one-sided limits exist and because they agree, $\lim_{x\to 1} f(x)g(x)$ exists (and equals 28).

61.   Let $P(x) = x^3 + cx + d$ and suppose $c > 0$. Since $\lim_{x\to -\infty} P(x) = -\infty$ and $\lim_{x\to\infty} P(x) = \infty$, there must be points x_0 and x_1 such that $P(x_0) < 0$ and $0 < P(x_1)$. The Intermediate Value Theorem (theorem 5) then guarantees the existence of at least one point $\overline{x}$ such that $P(\overline{x}) = 0$. On the other hand, $P(x)$ takes on any particular value at most once because P is an increasing function (since both x^3 and $cx + d$ are increasing functions). Hence there is a unique x such that $P(x) = 0$. ∎

REMARK: With a bit more algebra, we can sidestep the use of the limit argument to infer that $P(x)$ is sometimes negative and sometimes positive. First notice that $P(0) = d$. If $d = 0$, then $P(-1) < 0 < P(1)$. If $d < 0$, then picking $x > -d/c$ will yield $P(x) > 0$:

$$x > \frac{-d}{c} \implies x > \frac{-d}{x^2 + c} \implies P(x) = x\left(x^2 + c\right) + d > 0\,.$$

On the other hand, if $d > 0$, then picking $x < -d/c$ will yield $P(x) < 0$:

$$x < \frac{-d}{c} \implies x < \frac{-d}{x^2 + c} \implies P(x) = x\left(x^2 + c\right) + d < 0\,.$$

63.   Suppose f is continuous at x_0, then $f(x_0) = \lim_{x\to x_0} f(x)$. Pick a small interval, $(f(x_0) - \epsilon, f(x_0) + \epsilon)$, centered at $f(x_0)$ and consider the meaning of the limit statement. It would not be correct to say that $f(x)$ gets close to $f(x_0)$ as x tends to x_0 if $f(x)$ keeps getting outside that interval. Eventually, when x is close enough to x_0, $f(x)$ must get in and stay in that interval. This means there is some positive number δ small enough that x in $(x_0 - \delta, x_0 + \delta)$ implies $f(x)$ in $(f(x_0) - \epsilon, f(x_0) + \epsilon)$.

Now we're ready to tackle Problem 63. If we know $f(x_0) > 0$, then pick $\epsilon = \frac{1}{2}|f(x_0)| = \frac{1}{2}f(x_0)$. The interval $\left(\frac{1}{2}f(x_0), \frac{3}{2}f(x_0)\right)$ is centered at $f(x_0)$ and excludes 0. The preceding discussion shows that if f is continuous at x_0, there is some δ, positive and small, so that $x_0 - \delta < x < x_0 + \delta$ implies $0 < \frac{1}{2}f(x_0) < f(x) < \frac{3}{2}f(x_0)$. ∎

REMARK: If we use the definition of a limit in Section 1.8 rather than the one in Section 1.2, then this proof becomes shorter.

65.   **Lemma.** *If f is continuous and if $f'(c) < 0$, then there is an interval containing c in which $x < c$ implies $f(x) > f(c)$ and $c < x$ implies $f(c) > f(x)$.*

Proof of Lemma: Remember that $f'(c)$ is defined in terms of a limit: $\frac{f(c+\Delta x)-f(c)}{\Delta x}$ is close to $f'(c)$ provided Δx is close enough to 0. Because $f'(c)$ is negative, we can decide that "close to $f'(c)$" excludes 0 and all positive numbers (they're all at least $|f'(c)|/2$ away from $f'(c)$). Therefore $(f(c + \Delta x) - f(c))/\Delta x$ is negative if Δx is close to 0. Suppose $\Delta x > 0$ and Δx is small. Then $\frac{f(c+\Delta x)-f(c)}{\Delta x} < 0$ implies $f(c + \Delta x) - f(c) < 0 \cdot \Delta x = 0$, while $\frac{f(c-\Delta x)-f(c)}{-\Delta x} < 0$ implies $f(c - \Delta x) - f(c) > 0 \cdot (-\Delta x) = 0$. Therefore, $f(c - \Delta x) > f(c) > f(c + \Delta x)$. ∎

If f is continuous on $[a, b]$, there is some c in $[a, b]$ such that $f(c) = m$ is the greatest lower bound for f on $[a, b]$ (see Theorem 4). If $f'(a) < 0$, the Lemma implies there is some $\Delta x_1 > 0$ such that $f(a) > f(a+\Delta x_1) \geq m$; if $f'(b) > 0$, the Lemma (applied to $-f$) implies there is some $\Delta x_2 > 0$ such that $m \leq f(b-\Delta x_2) < f(b)$. Therefore $a < c < b$. If $f'(c)$ does not exist, then we're done. If $f'(c) < 0$, the Lemma says there are points in (c, b) where $f(x)$ is smaller than $m = f(c)$; impossible (because m is the smallest value of f on the whole interval $[a, b]$). If $f'(c) > 0$, there are points in (a, c) where $f(x)$ is smaller than m; also impossible. The only remaining possibility, $f'(c) = 0$, must be true. ∎

REMARK: The Lemma does not assert that $f'(c) < 0$ implies f is a decreasing function throughout a neighborhood of c; there exist functions for which this stronger claim is false. (See Theorem 3.2.1 and the Remark following the solution of Problem 3.3.31.)

Solutions 1.8   *The theory of limits*   (pages 129-30)

1. Note that $|7x - 7| = 7|x - 1|$. Therefore $|7x - 7| < \epsilon$ if and only if $|x - 1| < \epsilon/7$. If a positive ϵ is given, choose $\delta = \epsilon/7$. Then $0 < |x - 1| < \delta \implies |x - 1| < \epsilon/7 \implies |7x - 7| = 7|x - 1| < \epsilon$. Therefore, $\lim_{x \to 1} 7x = 7$. If $\epsilon = 1/10$, then δ may not exceed $1/70$; $\delta = 1/70$ works. If $\epsilon = 1/100$, then δ must be $1/700$ or less; $\delta = 1/700$ works.

3. Note that $|(5x + 1) - (-9)| = |5x + 10| = 5|x - (-2)|$. Therefore $|(5x + 1) - (-9)| < \epsilon$ if and only if $|x - (-2)| < \epsilon/5$. If a positive ϵ is given, choose $\delta = \epsilon/5$. Then $0 < |x - (-2)| < \delta \implies |x + 2| < \epsilon/5 \implies |(5x + 1) - (-9)| = 5|x + 2| < \epsilon$. Therefore, $\lim_{x \to -2} 5x + 1 = 9$. If $\epsilon = 1/10$, then δ must be chosen from $(0, 1/50]$; if $\epsilon = 1/100$, then δ must be positive and no larger than $1/500$.

5. Note that $|(x^2 - 6) - 3| = |x^2 - 9| = |x + 3| \cdot |x - 3|$. If we require $\delta \le 1$, then
$$|x - 3| < \delta \implies -1 \le -\delta < x - 3 < \delta \le 1 \implies 5 < x + 3 < 7 \implies |x + 3| < 7$$
$$\implies |(x^2 - 6) - 3| = |x + 3| \cdot |x - 3| < 7\delta.$$
If a positive ϵ is given, choose δ to be the smaller of 1 and $\epsilon/7$. Then
$$0 < |x - 3| < \delta \implies |x + 3| < 7 \text{ and } 0 < |x - 3| < \delta \le \epsilon/7 \implies |(x^2 - 6) - 3| < 7(\epsilon/7) = \epsilon.$$
Therefore, $\lim_{x \to 3} (x^2 - 6) = 3$. If $\epsilon = 1/10$, then choose δ from $(0, 1/70]$; if $\epsilon = 1/100$, then choose δ from $(0, 1/700]$.

REMARK: Alternative choices for $\delta(\epsilon)$ include $\min\{2, \epsilon/8\}$ and $\min\{0.25, \epsilon/6.25\}$. We can find a usable δ without making an auxiliary assumption such as $\delta \le 1$ but the work is harder.
$$|(x^2 - 6) - 3| < \epsilon \iff |x^2 - 9| < \epsilon \iff -\epsilon < x^2 - 9 < \epsilon \iff 9 - \epsilon < x^2 < 9 + \epsilon$$
$$\iff \sqrt{9 - \epsilon} < x < \sqrt{9 + \epsilon} \qquad \text{if} \quad x \ge 0 \text{ and } \epsilon \le 9$$
$$\iff -3 + \sqrt{9 - \epsilon} < x - 3 < -3 + \sqrt{9 + \epsilon}.$$
With some extra algebra, it can be shown that $\left|-3 + \sqrt{9 + \epsilon}\right| < \left|-3 + \sqrt{9 - \epsilon}\right|$. After tidying a few details, such as the case with $\epsilon > 9$, it follows that if $0 < |x - 3| < -3 + \sqrt{9 + \epsilon}$, then $|(x^2 - 6) - 3| < \epsilon$ **and** any larger choice of δ will not work.
   Moral: Make the auxiliary assumption that $\delta \le 1$ and simplify your life.

7. Note that $|(1 + x + x^2) - 1| = |x + x^2| = |x| \cdot |x - (-1)|$. If we require $\delta \le 1$, then $|x - (-1)| < \delta \le 1$ implies $-1 < x + 1 < 1$, $-2 < x < 0$, $|x| < 2$, and $|(1 + x + x^2) - 1| < 2\delta$. If a positive ϵ is given, choose δ to be the smaller of 1 and $\epsilon/2$. Then $|x - (-1)| < \delta \le 1$ implies $|x| < 2$ and $0 < |x - (-1)| < \delta \le \epsilon/2$ implies $|(1 + x + x^2) - 1| = |x||x - (-1)| < 2(\epsilon/2) = \epsilon$. Therefore $\lim_{x \to -1} (1 + x + x^2) = 1$. If $\epsilon = 0.1$, then choose δ from $(0, 0.05]$; if $\epsilon = 0.01$, then choose δ from $(0, 0.005]$.

REMARK: If $0 < |x - (-1)| < \sqrt{\frac{1}{4} + \epsilon} - \frac{1}{2}$, then $|(1 + x + x^2) - 1| < \epsilon$ **and** any larger choice of δ fails.

9. a. $\left|\dfrac{1}{\sqrt{x/10}} - 0\right| < \epsilon \iff \sqrt{\dfrac{10}{x}} < \epsilon \iff 0 < x \text{ and } \dfrac{10}{x} < \epsilon^2 \iff \dfrac{10}{\epsilon^2} < x$. If a positive ϵ is given, let $N = 10/\epsilon^2$; then $x > N$ implies $\left|1/\sqrt{x/10}\right| < \epsilon$; therefore $\lim_{x \to \infty} 1/\sqrt{x/10} = 0$.

   b. If $\epsilon = 0.01$, then let $N = 100,000$.

11. a. Since we are considering a limit as $x \to -\infty$, assume $x < -3$; in this case, $\left||1/(x + 3)| - 0\right| < \epsilon \iff 0 < -1/(x + 3) < \epsilon \iff 1/\epsilon < -(x + 3) \iff x < -(3 + 1/\epsilon)$. Choose $N = 3 + 1/\epsilon$, then $x < -N$ implies $\left||1/(x + 3)| - 0\right| < \epsilon$; therefore $\lim_{x \to -\infty} |1/(x + 3)| = 0$.

   b. If $\epsilon = 0.01$, then choose $N = 103$.

13. If $x < 4$, then $x + 4 < 8$. For any positive ϵ, let $\delta = \epsilon/8$; then
$$4 - \delta < x < 4 \implies |x - 4| < \epsilon/8 \text{ and } x + 4 < 8 \implies |f(x) - 16| = |x^2 - 16| = |x + 4| \cdot |x - 4| < 8 \cdot \delta = \epsilon.$$
Therefore $\lim_{x \to 4^-} f(x) = 16$
   If $4 < x$, then $2 < \sqrt{x}$, $4 < 2 + \sqrt{x}$, and $1/(\sqrt{x} + 2) < 1/4$. For any positive ϵ, let $\delta = \epsilon/(1/4) = 4\epsilon$; then
$$4 < x < 4 + \delta \implies |x - 4| < 4\epsilon \text{ and } \frac{1}{\sqrt{x} + 2} < \frac{1}{4}$$
$$\implies |f(x) - 2| = |\sqrt{x} - 2| = |\sqrt{x} - 2| \cdot \left|\frac{\sqrt{x} + 2}{\sqrt{x} + 2}\right| = |x - 4| \cdot \left|\frac{1}{\sqrt{x} + 2}\right| < 4\epsilon \cdot \frac{1}{4} = \epsilon.$$

Therefore $\lim_{x\to 4+} f(x) = 2$. ∎

15.   Let $f(x) = \begin{cases} 1 & \text{if } x \text{ is rational,} \\ 0 & \text{if } x \text{ is irrational.} \end{cases}$ Pick any real number x_0 and consider the interval $(x_0 - \delta, x_0 + \delta)$ where $\delta > 0$. This interval contains at least one rational number, pick one and call it R; the interval also contains at least one irrational number, pick one and call it I. (See (2) in Section 1.1.) Then one of $|f(R) - f(x_0)|$, $|f(I) - f(x_0)|$ is 1 and the other is 0. It is impossible to pick a $\delta > 0$ so that $0 < |x - x_0| < \delta \implies |f(x) - f(x_0)| < 1/2$. Therefore $\lim_{x\to x_0} f(x)$ never exists. ∎

17. $\lim_{x\to 2-} f(x) = \lim_{x\to 2-}(3x + 2) = 3(2) + 2 = 8$. $\lim_{x\to 2+} f(x) = \lim_{x\to 2+} 2x^2 = 2(2)^2 = 8$. Both one-sided limits exist; since they have the same value, $\lim_{x\to 2} f(x)$ exists and equals that value, 8. ∎

19.   If x_0 is not an integer, then let δ be the smaller of $x_0 - \lfloor x_0 \rfloor$ and $\lfloor x_0 \rfloor + 1 - x_0$. Because $\lfloor w \rfloor \leq w < \lfloor w \rfloor + 1$ is always true, if $0 < |x - x_0| < \delta$, then $\lfloor x_0 \rfloor < x < \lfloor x_0 \rfloor + 1$ and $\lfloor x \rfloor = \lfloor x_0 \rfloor$, therefore, $|\lfloor x \rfloor - \lfloor x_0 \rfloor| = 0$ which is less than any $\epsilon > 0$. Thus $\lim_{x\to x_0} \lfloor x \rfloor = \lfloor x_0 \rfloor$. If x_0 is an integer, then let $\delta = 2/3$. Then $x_0 < x < x_0 + \delta \implies x_0 = \lfloor x \rfloor \implies \lim_{x\to x_0^+} \lfloor x \rfloor = x_0$, but $x_0 - \delta < x < x_0 \implies x_0 - 1 = \lfloor x \rfloor \implies \lim_{x\to x-} x = x_0 = 1$. $\lim_{x\to x_0} \lfloor x \rfloor$ does not exist because the one-sided limits do not agree in value. ∎

21.   Suppose $\lim_{x\to a} f(x) = \infty$. Then for every $N > 0$, there is a $\delta = \delta(N)$ such that $0 < |x - a| < \delta \implies N < f(x)$. Hence

$$a - \delta < x < a \implies -\delta < x - a < 0 \implies 0 < |x - a| < \delta \implies N < f(x)$$

and

$$a < x < a + \delta \implies 0 < x - a < \delta \implies 0 < |x - a| < \delta \implies N < f(x);$$

therefore $\lim_{x\to a-} f(x) = \infty$ and $\lim_{x\to a+} f(x) = \infty$.

Suppose $\lim_{x\to a-} f(x) = \infty$ and $\lim_{x\to a+} f(x) = \infty$ with corresponding functions $\delta_-(N)$ and $\delta_+(N)$. Let $\delta = \min\{\delta_-(N), \delta_+(N)\}$.

$$0 < |x - a| < \delta \implies -\delta < x - a < 0 \text{ or } 0 < x - a < \delta \implies -\delta_- < x - a < 0 \text{ or } 0 < x - a < \delta_+$$
$$\implies a - \delta_- < x < a \text{ or } a < x < a + \delta_+ \implies N < f(x) \text{ or } N < f(x) \implies N < f(x);$$

therefore $\lim_{x\to a} f(x) = \infty$. ∎

23. $|f(x) - f(1)| = |(2x + 3) - (5)| = 2 \cdot |x - 1| < \epsilon \iff |x - 1| < \epsilon/2$; choose $\delta = 0.005$ when $\epsilon = 0.01$.

25. $|x^2 - 0^2| = |x^2| < \epsilon \iff |x - 0| = |x| < \sqrt{\epsilon}$; choose $\delta = \sqrt{0.1} \approx 0.3162$ when $\epsilon = 0.1$.

27.   It is convenient to impose an auxiliary assumption about δ to keep x close to 2 and far enough from 0 that $|1/x|$ is not too big, let's try $\delta \leq 1$. Then

$$|x - 2| < \delta \implies -1 < x - 2 < 1 \implies 1 < x < 3 \implies 2 < 2x = |2x|.$$

Because we always have

$$\left|\frac{1}{x} - \frac{1}{2}\right| < \epsilon \iff \left|\frac{x - 2}{2x}\right| < \epsilon \iff |x - 2| < |2x| \cdot \epsilon,$$

if we choose $\delta = \min\{1, 2\epsilon\}$, then

$$|x - 2| < \delta \implies |x - 2| < 2\epsilon \leq |2x| \cdot \epsilon \implies \left|\frac{x - 2}{2x}\right| < \epsilon.$$

Choose $\delta = 0.2$ when $\epsilon = 0.1$.

REMARK: If $a > 0$, then $|x - a| < \dfrac{\epsilon a^2}{1 + \epsilon a} \implies \left|\dfrac{1}{x} - \dfrac{1}{a}\right| < \epsilon$.

29.   If a positive ϵ is given, let $\delta = \min\{1/2, \epsilon/2\}$. Then

$$|x - 1| < \delta \implies \frac{1}{2} \leq 1 - \delta < x < 1 + \delta \implies 0 < \frac{1}{x} < 2 \implies \left|\frac{1}{x} - \frac{1}{1}\right| = \left|\frac{1}{x}\right| \cdot |x - 1| < 2 \cdot \frac{\epsilon}{2} = \epsilon.$$

Choose $\delta = 0.05$ when $\epsilon = 0.1$.

31. f is continuous at x_0 according to Definition 1.7.1 if and only if $\lim_{x\to x_0} f(x) = f(x_0)$. (The three conditions shown in Section 1.7 are merely for emphasis.)

Now consider Definition 2.8.1 of a limit with L replaced by $f(x_0)$. Because $f(x_0)$ exists, we can relax the condition $0 < |x - x_0| < \delta$ to require merely $|x - x_0| < \delta$.

Our statement that $\lim_{x\to x_0} f(x) = f(x_0)$ now corresponds to asserting conditions (i) and (ii) of the alternative definition of continuity. ∎

Solutions 1.R   *Review*   (pages 131-2)

1.

x	$x^2 - 3x + 6$	x	$x^2 - 3x + 6$
1	4	3	6
1.5	3.75	2.5	4.75
1.9	3.91	2.1	4.11
1.99	3.9901	2.01	4.0101

The numerical data suggest $\lim_{x \to 2} \left(x^2 - 3x + 6 \right)$ is 4.

3.   a. $\lim_{x \to 1} \left(x^3 - 3x + 2 \right) = 1^3 - 3 + 2 = 0.$

     b. $\lim_{x \to 5} \left(-x^3 + 17 \right) = -5^3 + 17 = -125 + 17 = -108.$

     c. $\lim_{x \to 3} \dfrac{x^4 - 2x + 1}{x^3 + 3x - 5} = \dfrac{81 - 6 + 1}{27 + 9 - 5} = \dfrac{76}{31}.$

     d. $\lim_{x \to -1} \dfrac{x^3 + x^2 + x + 1}{x^4 + x^3 + x^2 + x + 1} = \dfrac{-1 + 1 - 1 + 1}{1 - 1 + 1 - 1 + 1} = \dfrac{0}{1} = 0.$

5.   a. $\lim_{x \to 3} \dfrac{(x-3)(x-4)}{x-3} = \lim_{x \to 3}(x-4) = 3 - 4 = -1.$

     b. $\lim_{x \to 5} \dfrac{x^2 - 6x + 5}{x - 5} = \lim_{x \to 5} \dfrac{(x-5)(x-1)}{x-5} = \lim_{x \to 5}(x-1) = 5 - 1 = 4.$

7.   a. $\lim_{x \to 1} 23\sqrt{x - 17}$ does not exist because $\sqrt{x - 17}$ is defined only on $[17, \infty)$.

     b. $\lim_{x \to -1} \left(1 - x + x^2 - x^3 + x^4 \right) = 1 - (-1) + 1 - (-1) + 1 = 5.$

     c. $\lim_{x \to 4} \dfrac{x^2 + 9}{x^2 - 9} = \dfrac{16 + 9}{16 - 9} = \dfrac{25}{7}.$

     d. $\lim_{x \to -1} 5x^{250} = 5(-1)^{250} = 5(1) = 5.$

     e. $\lim_{x \to -1} 6x^{251} = 6(-1)^{251} = 6(-1) = -6.$

     f. $\lim_{x \to 3} \left(x^2 + x - 8 \right)^5 = (9 + 3 - 8)^5 = 4^5 = 1024.$

     g. $\lim_{x \to 0} \dfrac{x^8 - 7x^5 + x^3 - x^2 + 3}{x^{23} - 2x + 9} = \dfrac{0 - 0 + 0 - 0 + 3}{0 - 0 + 9} = \dfrac{3}{9} = \dfrac{1}{3}.$

     h. $\lim_{x \to 0} \dfrac{ax^2 + bx + c}{dx^2 + ex + f} = \dfrac{0 + 0 + c}{0 + 0 + f} = \dfrac{c}{f}.$

9.   If $x < 0$, then $1/x^3 < 0$ and $\lim_{x \to 0^-} \left(1/x^3 \right) = -\infty$ while if $x > 0$, then $1/x^3 > 0$ and $\lim_{x \to 0^+} \left(1/x^3 \right) = \infty$; because the one-sided limits are different, $\lim_{x \to 0} \left(1/x^3 \right)$ does not exist. On the other hand if $x \ne 0$, then $x^4 > 0$ and $\lim_{x \to 0} \left(1/x^4 \right) = \infty$.

11.   a. $\lim_{x \to \infty} \left(1/x^3 \right) = 0.$

     b. $\lim_{x \to \infty} \left(1/\sqrt{x + 2} \right) = 0.$

     c. $\lim_{x \to \infty} \dfrac{\sqrt{x}}{x^2 + 3} = \lim_{x \to \infty} \dfrac{1/\left(x\sqrt{x} \right)}{1 + \left(3/x^2 \right)} = \dfrac{0}{1 + 0} = 0.$

     d. $\lim_{x \to \infty} \dfrac{x^3 + 6x^2 + 4x + 2}{3x^3 - 9x^2 + 11} = \lim_{x \to \infty} \dfrac{1 + (6/x) + \left(4/x^2 \right) + \left(2/x^3 \right)}{3 - (9/x) + \left(11/x^3 \right)} = \dfrac{1 + 0 + 0 + 0}{3 - 0 + 0} = \dfrac{1}{3}.$

     e. $\lim_{x \to \infty} \dfrac{3x^5 - 6x^2 + 3}{x^7 - 2} = \lim_{x \to \infty} \dfrac{\left(3/x^2 \right) - \left(6/x^5 \right) + \left(3/x^7 \right)}{1 - \left(2/x^7 \right)} = \dfrac{0 - 0 + 0}{1 - 0} = 0.$

     f. $\lim_{x \to \infty} \dfrac{x^7 - 9}{30x^5 + x^4 + 161} = \lim_{x \to \infty} \left(x^2 \right) \left(\dfrac{1 - \left(9/x^7 \right)}{30 + (1/x) + \left(161/x^5 \right)} \right) = \infty$

     because $\lim_{x \to \infty} \dfrac{1 - \left(9/x^7 \right)}{30 + (1/x) + \left(161/x^5 \right)} = \dfrac{1}{30}$ and $\lim_{x \to \infty} x^2 = \infty.$

13.   a. f' is positive on the intervals $(-\infty, x_3)$ and (x_5, x_7); is negative on (x_3, x_5) and (x_7, ∞); and is zero at x_3, x_5, and x_7.

     b. f' is positive on $(-\infty, x_3)$, (x_4, x_5), (x_7, x_8), (x_9, ∞); negative on (x_3, x_4), (x_5, x_7), (x_8, x_9); and is zero at x_3, x_4, x_5, x_7, x_8, x_9.

15.   Let $f(x) = |x + 1|$; then

$$\lim_{\Delta x \to 0^-} \frac{f(-1 + \Delta x) - f(-1)}{\Delta x} = \lim_{\Delta x \to 0^-} \frac{|(-1 + \Delta x) + 1| - |-1 + 1|}{\Delta x} = \lim_{\Delta x \to 0^-} \frac{|\Delta x|}{\Delta x} = \lim_{\Delta x \to 0^-} \frac{-\Delta x}{\Delta x} = -1,$$

$$\lim_{\Delta x \to 0^+} \frac{f(-1 + \Delta x) - f(-1)}{\Delta x} = \lim_{\Delta x \to 0^+} \frac{|\Delta x|}{\Delta x} = \lim_{\Delta x \to 0^+} \frac{\Delta x}{\Delta x} = 1.$$

$\lim_{\Delta x \to 0} \dfrac{f(-1 + \Delta x) - f(-1)}{\Delta x}$ does not exist because the one-sided limits have different values, therefore $f'(-1)$ does not exist. Thinking geometrically, the graph of f has a sharp corner at $(-1, 0)$; it does not have a tangent there. You can't find the slope of a nonexistent tangent line, so the derivative of f is undefined at -1.

17. $\dfrac{dV}{dt} = \dfrac{d}{dt}\left(\dfrac{4}{3}\pi r^3\right) = \dfrac{4}{3}\pi 3r^2 = 4\pi r^2$; thus $V'(2 \text{ ft}) = 16\pi \text{ ft}^3/\text{ft}$.

19.   The sharp corner at $x = -3$ means the derivative does not exist there; the discontinuity at $x = 2$ means the derivative can not exist there either.

21. $\lim_{x \to 3^-} f(x) = \lim_{x \to 3^-} (x^2 + 3) = 3^2 + 3 = 12$ and $\lim_{x \to 3^+} f(x) = \lim_{x \to 3^+} (x^3 + \alpha) = 3^3 + \alpha = 27 + \alpha$; these one-sided limits have the same value (12) if and only if $\alpha = -15$.

23.   If $f(x) = 3\sqrt[3]{x}$, then f is continuous on $(-\infty, \infty)$.

25.   If $f(x) = 1/(x^2 - 6)$, then f is continuous on $(-\infty, -\sqrt{6}) \cup (-\sqrt{6}, \sqrt{6}) \cup (\sqrt{6}, \infty)$; the discontinuities at $\pm\sqrt{6}$ are not removable.

27. $f(x) = |x + 2|$ is continuous on $(-\infty, \infty)$.

29.   If $f(x) = (x^2 - 9)/(x + 3)$, then f is continuous on $(-\infty, -3) \cup (-3, \infty)$; because $f(x) = x - 3$ for $x \neq -3$ the discontinuity at -3 is removable since $\lim_{x \to -3} f(x)$ exists.

31.   If $f(x) = (x^3 + 1)/(x + 1)$, then f is continuous on $(-\infty, -1) \cup (-1, \infty)$; because $f(x) = x^2 + x + 1$ for $x \neq -1$ the discontinuity at -1 is removable since $\lim_{x \to -1} f(x)$ exists.

33.   Make the auxiliary assumption that $\delta \leq 1$; then $|x - 3| < \delta$ implies $|x + 3| < 7$. Therefore, if we let $\delta = \min\{1, \epsilon/7\}$, we see that

$$|x - 3| < \delta \implies |x + 3| < 7 \text{ and } |x - 3| < \epsilon 7 \implies |(x^2 + 2) - 11| = |x^2 - 9| = |x + 3| \cdot |x - 3| < 7 \cdot \left(\frac{\epsilon}{7}\right) = \epsilon;$$

we have shown that $\lim_{x \to 3}(x^2 + 2) = 11$. ∎

35.   For any ϵ, let $N = N(\epsilon) = 1/\epsilon^3$. Then

$$0 < \frac{1}{\epsilon^3} = N < x \implies 0 < \frac{1}{\epsilon^3} < 3 + \frac{1}{\epsilon^3} < 3 + x \implies 0 < \frac{1}{\epsilon} < \sqrt[3]{3 + x} \implies 0 < \frac{1}{\sqrt[3]{3 + x}} < \epsilon;$$

we have shown $\lim_{x \to \infty} 1/\sqrt[3]{x + 3} = 0$. ∎

37.   No matter how large we set N, there are choices of integers $x > N$ for which $f(x) = 1$ and there are nonintegers $w > N$ for which $f(w) = 2$. Let $\epsilon = 1/3$; there is no value L such that $|1 - L| < \epsilon$ and $|2 - L| < \epsilon$. Thus $\lim_{x \to \infty} f(x)$ does not exist. ∎

Solutions 1.C   *Computer Exercises*   (pages 132-3)

1.   A simple first choice of points within $[-.05, .05]$ produces a table

x	$-.05$	$-.04$	$-.03$	$-.02$	$-.01$	$.01$	$.02$	$.03$	$.04$	$.05$
$(1 + x)^{1/x}$	2.7895	2.7747	2.7602	2.7460	2.7320	2.7048	2.6916	2.6786	2.6658	2.6533

sufficient to estimate $\lim_{x \to 0}(1 + x)^{1/x} \approx 2.7$ to just one decimal place. If we narrow our focus to compute

x	$-.005$	$-.004$	$-.003$	$-.002$	$-.001$	$.001$	$.002$	$.003$	$.004$	$.005$
$(1 + x)^{1/x}$	2.7251	2.7237	2.7224	2.7210	2.7196	2.7169	2.7156	2.7142	2.7129	2.7115

then we can estimate $\lim_{x \to 0}(1 + x)^{1/x} \approx 2.72$ with two decimal place accuracy.

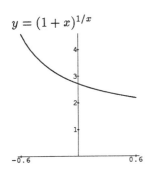
$y = (1+x)^{1/x}$

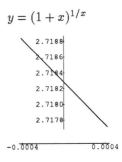
$y = (1+x)^{1/x}$

$y = (1+x)^{1/x}$

3.   I chose $b = 7.491$ which corresponds to July 4, 1991. A quick scan of some graphs

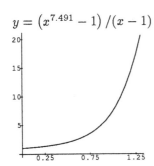
$y = \left(x^{7.491} - 1\right) / (x - 1)$

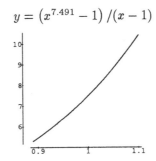
$y = \left(x^{7.491} - 1\right) / (x - 1)$

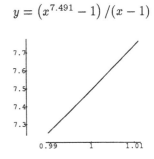
$y = \left(x^{7.491} - 1\right) / (x - 1)$

suggests we will need to tabulate our function at points fairly close to 1.

x	.9995	.9996	.9997	.9998	.9999	1.0001	1.0002	1.0003	1.0004	1.0005
$\left(x^{7.491} - 1\right) / (x - 1)$	7.4789	7.4813	7.4837	7.4861	7.4886	7.4934	7.4959	7.4983	7.5007	7.5031

provides sufficient evidence to estimate that $\lim_{x \to 1} \left(x^{7.491} - 1\right) / (x - 1) \approx 7.49$ with two decimal place accuracy.

You can check your final estimate by using the fact that $b > 0$ implies $\lim\limits_{x \to 1} \dfrac{x^b - 1}{x - 1} = b$

5.

x	.95	.96	.97	.98	.99	1.01	1.02	1.03	1.04	1.05
$(2^x - 2) / (x - 1)$	1.3625	1.3673	1.3720	1.3767	1.3815	1.3911	1.3959	1.4008	1.4057	1.4106

gives evidence sufficient to infer $\lim_{x \to 1} (2^x - 2) / (x - 1) \approx 1.4$ to one decimal place.

x	.995	.996	.997	.998	.999	1.001	1.002	1.003	1.004	1.005
$(2^x - 2) / (x - 1)$	1.3839	1.3844	1.3849	1.3853	1.3858	1.3868	1.3873	1.3877	1.3882	1.3887

implies $\lim_{x \to 1} (2^x - 2) / (x - 1) \approx 1.39$ with two decimal place accuracy.

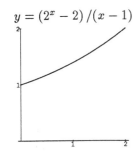
$y = (2^x - 2) / (x - 1)$

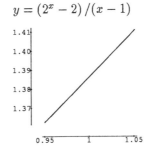
$y = (2^x - 2) / (x - 1)$

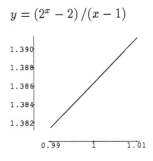
$y = (2^x - 2) / (x - 1)$

7.   Plots of $d(x) = \dfrac{\ln(x + .00001) - \ln x}{.00001}$ and $\dfrac{1}{x}$ look very similar; a plot of $\dfrac{1}{x} - d(x)$ confirms the impression that their difference is small. Hence we might guess the derivative (with respect to x) of $\ln x$ is $1/x$.

1.C — Computer Exercises

$$y = \frac{\ln(x+.00001) - \ln x}{.00001}$$

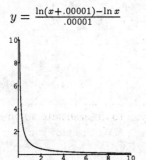

$$y = 1/x$$

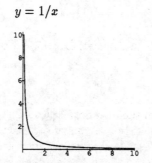

$$y = \frac{1}{x} - \frac{\ln(x+.00001) - \ln x}{.00001}$$

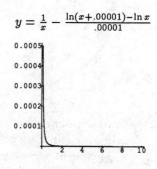

2
More about Derivatives

Note about "simplification": Many of the drill problems in Chapter 2 ask you to differentiate a non-simple function. Their main purpose is served once you've finished differentiating each piece of the function.

The result at that stage may be an algebraic nightmare. In an applied problem (see Chapter 3), what you do next will depend on the application. If you only want to find the slope of the tangent line at a single point, there is little reason to simplify the general expression before you evaluate it (especially if you're using a pocket calculator). On the other hand, if you need to know where the derivative is positive (i.e., where the original function is increasing), then it is worth spending effort to obtain a formula for the derivative which can be analyzed easily.

In the absence of an application, it is a very personal decision when to be satisfied with an algebraic simplification. The answers in the back of *Calculus*, 5th edition, indicate where I would quit. However, because reasonable people can disagree about such matters, I have also provided simplified answers here. (Also note that reasonable people can differ on what "simplified" means.)

Solutions 2.1   *Some differentiation formulas*   (pages 137-9)

1. $f(x) = 3^4 = 81$, a constant; therefore $f'(x) = 0$.

3.   If $h(x) = x^2 + (a+b)x + ab$, then $h'(x) = \dfrac{d}{dx}\left(x^2 + (a+b)x + ab\right) = \dfrac{d}{dx}\left(x^2\right) + \dfrac{d}{dx}\left((a+b)x\right) + \dfrac{d}{dx}\left(ab\right) = \dfrac{d}{dx}\left(x^2\right) + (a+b)\left(\dfrac{d}{dx}x\right) + \dfrac{d}{dx}\left(ab\right) = 2x + (a+b)(1) + 0 = 2x + (a+b)$.

5. $\dfrac{d}{dt}\left(1 - t + t^4 - t^7\right) = \left(\dfrac{d}{dt}1\right) - \left(\dfrac{d}{dt}t\right) + \left(\dfrac{d}{dt}t^4\right) - \left(\dfrac{d}{dt}t^7\right) = 0 - 1 + 4t^{4-1} - 7t^{7-1} = -1 + 4t^3 - 7t^6$.

7. $\dfrac{d}{dw}\left(27w^6 - (3w^5 - 4w)\right) = 27\dfrac{d}{dw}w^6 - \left(3\dfrac{d}{dw}w^5 - 4\dfrac{d}{dw}w\right) = 27(6w^5) - \left(3(5w^4) - 4(1)\right) = 162w^5 - 15w^4 + 4$.

9. $\dfrac{d}{dz}\left(-3z^{12} + 12z^3\right) = (-3)\left(12z^{12-1}\right) + (12)\left(3z^{3-1}\right) = -36z^{11} + 36z^2$.

11. $\dfrac{d}{ds}\left(3s^8 - 8s^6 - 7s^4 + 2s^2 + 3\right) = 3\left(\dfrac{d}{ds}s^8\right) - 8\left(\dfrac{d}{ds}s^6\right) - 7\left(\dfrac{d}{ds}s^4\right) + 2\left(\dfrac{d}{ds}s^2\right) + 3\left(\dfrac{d}{ds}1\right)$
$$= 3\left(8s^7\right) - 8\left(6s^5\right) - 7\left(4s^3\right) + 2\left(2s^1\right) + 3(0)$$
$$= 24s^7 - 48s^5 - 28s^3 + 4s.$$

13. $\dfrac{d}{dx}x^4 = 4x^3$; when $x = 1$, $\dfrac{dy}{dx} = 4\left(1^3\right) = 4$; the tangent at $(1,1)$ has slope 4 and equation $y = 4(x-1) + 1 = 4x - 3$.

15. $\dfrac{d}{dx}\left(2x^7 - x^6 - x^3\right) = 2\left(\dfrac{d}{dx}x^7\right) - \left(\dfrac{d}{dx}x^6\right) - \left(\dfrac{d}{dx}x^3\right)$
$$= 2\left(7x^6\right) - \left(6x^5\right) - \left(3x^2\right) = 14x^6 - 6x^5 - 3x^2\,;$$
when $x = 1$, $\dfrac{dy}{dx} = 14 - 6 - 3 = 5$; the tangent at $(1,0)$ has slope 5 and equation $y = 5(x-1) + 0 = 5x - 5$.

17. $\dfrac{d}{dx}\left(1 + x + x^2 + x^3 + x^4 + x^5\right) = 0 + 1 + 2x^{2-1} + 3x^{3-1} + 4x^{4-1} + 5x^{5-1} = 1 + 2x + 3x^2 + 4x^3 + 5x^4$;
when $x = 0$, $\dfrac{dy}{dx} = 1 + 0 + 0 + 0 + 0 = 1$; the tangent at $(0,1)$ has slope 1 and equation $y = (1)(x-0) + 1 = x + 1$.

2.1 — Some differentiation formulas

19.
$$\frac{d}{dx}\left(x^6 - 6\sqrt{x}\right) = \frac{d}{dx}x^6 - 6\frac{d}{dx}\sqrt{x} = 6x^5 - 6\left(\frac{1}{2\sqrt{x}}\right) = 6x^5 - \frac{3}{\sqrt{x}};$$

when $x = 1$, $\frac{dy}{dx} = 6\left(1^5\right) - \frac{3}{\sqrt{1}} = 3$; the tangent at $(1, -5)$ has slope 3 and equation $y = 3(x - 1) + (-5) = 3x - 8$.

21.
$\frac{d}{dx}x^3 = 3x^2$; when $x = 1$, $\frac{dy}{dx} = 3$; the tangent at $(1, 1)$ has slope 3, therefore the normal at $(1, 1)$ has slope $-1/3$ and equation $y = (1/3)(x - 1) + 1 = (x/3) + (4/3)$.

23.
$\frac{d}{dx}\left(x^6 - 6\sqrt{x}\right) = 6x^5 - \frac{3}{\sqrt{x}}$; when $x = 1$, $\frac{dy}{dx} = 6 - 3 = 3$; the tangent at $(1, -5)$ has slope 3, therefore the normal at $(1, -5)$ has slope $-1/3$ and equation $y = (-1/3)(x - 1) - 5 = (-1/3)x - (14/3)$.

25.
$\frac{d}{dx}\left(2x^3 + 3x^2 - 6x + 1\right) = 2\left(3x^2\right) + 3\left(2x^1\right) - 6(1) = 6x^2 + 6x - 6$; the curve has a horizontal tangent if and only if its derivative is 0, locate these by solving the equation $\frac{dy}{dx} = 0$ for x-values:

$$\frac{dy}{dx} = 0 \iff x^2 + x - 1 = 0 \iff x = \frac{-1 \pm \sqrt{5}}{2}.$$

The points are $\left((-1 - \sqrt{5})/2, (9 + 5\sqrt{5})/2\right)$ and $\left((-1 + \sqrt{5})/2, (9 - 5\sqrt{5})/2\right)$.

27. a. $P_T(t) = P_1(t) + P_2(t) = 3000 - 20\sqrt{t} - 30t + 30t^2$.

   b.
$$P_T'(t) = \frac{d}{dt}\left(3000 - 20\sqrt{t} - 30t + 30t^2\right)$$
$$= 0 - 20\left(\frac{1}{2\sqrt{t}}\right) - 30(1) + 30(2t) = \frac{-10}{\sqrt{t}} - 30 + 60t \text{ organisms per hour.}$$

   c. $P_T'(4\,\text{hr}) = -10/2 - 30 + 240 = 205$ organisms/hr; $P_T'(6\,\text{hr}) = -10/\sqrt{6} - 30 + 360 \approx 325.9$ org/hr.

29.
A circle of radius r has area πr^2, $\frac{d}{dr}\left(\text{Area}(r)\right) = \frac{d}{dr}\left(\pi r^2\right) = \pi\frac{d}{dr}r^2 = \pi(2r) = 2\pi r$, a circle of radius r has circumference $2\pi r$.

REMARK: This result is not just a coincidence; hang on until Chapter 4 (The Integral) and you'll understand it forwards and backwards.

31.
The equation $(1/8)x - 8y = 1$ is equivalent to $y = (1/64)x - (1/8)$; it describes a line whose slope is always $1/64$. The graph of $y = \sqrt{x}$ has a tangent with slope $1/64$ when $y' = 1/\left(2\sqrt{x}\right) = 1/64$, i.e., when $x = \left(64/2\right)^2 = 1024$ (and nowhere else).

33.
The line through the points $(a, a^2 + 3)$ and $(1, 0)$ has slope $((a^2 + 3) - 0)/(a - 1)$. If this line is also tangent to the graph of $y = x^2 + 3$ at $(a, a^2 + 3)$, then its slope equals the derivative of $x^2 + 3$ evaluated for $x = a$. Because $\frac{d}{dx}\left(x^2 + 3\right) = 2x$, this means our line is tangent if and only if $(a^2 + 3)/(a - 1) = 2a \iff a^2 + 3 = 2a(a - 1) = 2a^2 - 2a \iff 0 = a^2 - 2a - 3 = (a + 1)(a - 3) \iff a = -1$ or $a = 3$.
   CASE $a = -1$: $y = -2x + 2$ is tangent at $(-1, 4)$ and passes through $(1, 0)$,
   CASE $a = 3$: $y = 6x - 6$ is tangent at $(3, 12)$ and passes through $(1, 0)$.

35.
Some thinking about the geometric relation between the graphs of f and g will simplify our algebraic task. The graph of g is the reflection through the origin of the graph f, i.e., if (a, b) is on the graph of f, then $(-a, -b)$ is on the graph of g. This symmetry about the origin implies any line tangent to both parabolas must pass through the origin. The line tangent to $y = f(x) = x^2 - 2x + 4$ at $(p, f(p))$ passes through the origin if and only if $f(p)/p = f'(p)$, i.e., $p^2 = 4 \iff p \in \{2, -2\}$. The tangent to $y = f(x)$ at $(2, 4)$ has equation $y = 2x$; it is also tangent to $y = g(x)$ at $(-2, -4)$. The tangent to $y = f(x)$ at $(-2, 12)$ has equation $y = -6x$; it is also tangent to $y = g(x)$ at $(2, -12)$.

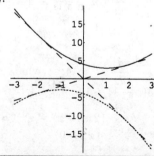

REMARK: Here is an outline of a less geometric approach. The line tangent to $y = f(x)$ at $(a, f(a))$ has equation $y = 2(a - 1)x + 4 - a^2$, that line meets the graph of $y = g(x)$ at points with x-coordinates $-a \pm \sqrt{2\left(a^2 - 4\right)}$. The line is tangent there if and only if $\sqrt{2\left(a^2 - 4\right)} = 0$, i.e., $a^2 = 4$. (Note also that if $a^2 < 4$, the tangent at $(a, f(a))$ does not meet the graph of g).

37.   Suppose f is a differentiable function and c is an arbitrary real number. Let $g(x) = c \cdot f(x)$; then

$$g'(x) = \lim_{w \to x} \frac{g(w) - g(x)}{w - x} = \lim_{w \to x} \frac{c \cdot f(x) - c \cdot f(x)}{w - x} = c \cdot \lim_{w \to x} \frac{f(w) - f(x)}{w - x} = c \cdot f'(x). \ \blacksquare$$

39.   a.   Suppose f_1, f_2, and f_3 are differentiable functions. Then $f_1 + f_2$ is differentiable (Theorem 4) and so is $(f_1 + f_2) + f_3$ (Theorem 4 again). Furthermore

$$\begin{aligned}
\frac{d}{dx}\left(f_1 + f_2 + f_3\right) &= \frac{d}{dx}\left((f_1 + f_2) + f_3\right) \\
&= \frac{d}{dx}(f_1 + f_2) + \frac{d}{dx}f_3 && \text{(Theorem 4)} \\
&= \left(\frac{d}{dx}f_1 + \frac{d}{dx}f_2\right) + \frac{d}{dx}f_3 && \text{(Theorem 4)} \\
&= \frac{d}{dx}f_1 + \frac{d}{dx}f_2 + \frac{d}{dx}f_3. \ \blacksquare
\end{aligned}$$

Suppose f_1, f_2, f_3, f_4 are differentiable functions. Then $f_1 + f_2 + f_3$ is differentiable (just proved) and so is $(f_1 + f_2 + f_3) + f_4$ (Theorem 4). Furthermore

$$\begin{aligned}
\frac{d}{dx}\left(f_1 + f_2 + f_3 + f_4\right) &= \frac{d}{dx}\left((f_1 + f_2 + f_3) + f_4\right) \\
&= \frac{d}{dx}(f_1 + f_2 + f_3) + \frac{d}{dx}f_4 && \text{(Theorem 4)} \\
&= \left(\frac{d}{dx}f_1 + \frac{d}{dx}f_2 + \frac{d}{dx}f_3\right) + \frac{d}{dx}f_4 && \text{(case } n = 3) \\
&= \frac{d}{dx}f_1 + \frac{d}{dx}f_2 + \frac{d}{dx}f_3 + \frac{d}{dx}f_4. \ \blacksquare
\end{aligned}$$

b.   Theorem 5 has been proven for the cases $n = 2$, $n = 3$, and $n = 4$. Suppose it is true for the case $n = k$, k an integer 2 or larger. If f_1, f_2, ..., f_k, f_{k+1} are differentiable functions, then so are $[f_1 + f_2 + \ldots + f_k]$ (case $n = k$) and $[f_1 + f_2 + \cdots + f_k] + f_{k+1}$ (Theorem 4, i.e., case $n = 2$ of Theorem 5). Furthermore

$$\begin{aligned}
\frac{d}{dx}\left(f_1 + f_2 + \cdots + f_k + f_{k+1}\right) &= \frac{d}{dx}\left((f_1 + f_2 + \cdots + f_k) + f_{k+1}\right) \\
&= \frac{d}{dx}(f_1 + f_2 + \cdots + f_k) + \frac{d}{dx}f_{k+1} && \text{(Theorem 4)} \\
&= \left(\frac{d}{dx}f_1 + \frac{d}{dx}f_2 + \cdots + \frac{d}{dx}f_k\right) + \frac{d}{dx}f_{k+1} && \text{(case } n = k) \\
&= \frac{d}{dx}f_1 + \frac{d}{dx}f_2 + \cdots + \frac{d}{dx}f_k + \frac{d}{dx}f_{k+1}.
\end{aligned}$$

Because **(true for case $n = k$)** implies **(true for case $n = k + 1$)** and because case $n = 2$ is true, Theorem 5 is true for all integers n, $n \geq 2$. $\ \blacksquare$

Solutions 2.2   *The product and quotient rules*   (pages 144-6)

1. $$\frac{d}{dx}\left((x^2 - 9) \cdot (x - 5)\right) = \left(\frac{d}{dx}(x^2 - 9)\right) \cdot (x - 5) + (x^2 - 9) \cdot \left(\frac{d}{dx}(x - 5)\right)$$
$$= (2x) \cdot (x - 5) + (x^2 - 9) \cdot 1 = 3x^2 - 10x - 9.$$

3. $$\frac{d}{dt}\left(\frac{1}{t^6}\right) = \frac{d}{dt}t^{-6} = (-6)t^{-6-1} = -6t^{-7} = \frac{-6}{t^7}.$$

5. $$\frac{d}{dx}\left(2x(x^2 + 1)\right) = \left(\frac{d}{dx}(2x)\right)(x^2 + 1) + (2x)\left(\frac{d}{dx}(x^2 + 1)\right) = (2)(x^2 + 1) + (2x)(2x) = 6x^2 + 2.$$

2.2 — The product and quotient rules

7. $\dfrac{d}{dt}\, t^{-100} = (-100)\cdot t^{-100-1} = -100\cdot t^{-101}$.

9. $\dfrac{d}{dx}\left((1+x+x^5)\cdot(2-x+x^6)\right) = \dfrac{d}{dx}\left(1+x+x^5\right)\cdot\left(2-x+x^6\right) + \left(1+x+x^5\right)\cdot\dfrac{d}{dx}\left(2-x+x^6\right)$

$\qquad\qquad = \left(1+5x^4\right)\cdot\left(2-x+x^6\right) + \left(1+x+x^5\right)\cdot\left(-1+6x^5\right)$

$\qquad\qquad = 1 - 2x + 10x^4 + 7x^6 + 11x^{10}$.

11.
$$\dfrac{d}{dx}\left(\dfrac{1}{\sqrt{x}}\right) = \dfrac{(\sqrt{x})\dfrac{d}{dx}1 - (1)\dfrac{d}{dx}\sqrt{x}}{(\sqrt{x})^2} = \dfrac{0-\left(\dfrac{1}{2\sqrt{x}}\right)}{x} = \dfrac{-1}{2x\sqrt{x}} = \dfrac{-1}{2x^{3/2}} = \left(-\dfrac{1}{2}\right)x^{-3/2}.$$

REMARK: Compare this work with the proof of Theorem 3.

13.
$$\dfrac{d}{dt}\left(t^3\cdot\left(1+\sqrt{t}\right)\right) = \left(\dfrac{d}{dt}t^3\right)\cdot\left(1+\sqrt{t}\right) + t^3\cdot\dfrac{d}{dt}\left(1+\sqrt{t}\right)$$

$$= \left(3t^2\right)\cdot\left(1+\sqrt{t}\right) + (t^3)\cdot\left(\dfrac{1}{2\sqrt{x}}\right) = 3t^2 + \dfrac{7}{2}x^{5/2}.$$

15.
$$\dfrac{d}{dt}\left(\dfrac{1+\sqrt{t}}{1-\sqrt{t}}\right) = \dfrac{\left(1-\sqrt{t}\right)\cdot\dfrac{d}{dt}\left(1+\sqrt{t}\right) - \left(1+\sqrt{t}\right)\cdot\dfrac{d}{dt}\left(1-\sqrt{t}\right)}{\left(1-\sqrt{t}\right)^2}$$

$$= \dfrac{\left(1-\sqrt{t}\right)\left(\dfrac{1}{2\sqrt{t}}\right) - \left(1+\sqrt{t}\right)\left(\dfrac{-1}{2\sqrt{t}}\right)}{\left(1-\sqrt{t}\right)^2} = \dfrac{1}{\sqrt{t}\left(1-\sqrt{t}\right)^2}.$$

17.
$$\dfrac{d}{dv}\left((v^3-\sqrt{v})(v^2+2\sqrt{v})\right) = \left(3v^2 - \dfrac{1}{2\sqrt{v}}\right)(v^2+2\sqrt{v}) + (v^3-\sqrt{v})\left(2v + \dfrac{2}{2\sqrt{v}}\right)$$

$$= 5v^4 + 7v^{5/2} - \dfrac{5}{2}v^{3/2} - 2.$$

19.
$$\dfrac{d}{dx}x^{5/2} = \dfrac{d}{dx}\left(x^2\cdot x^{1/2}\right) = \dfrac{d}{dx}\left(x^2\cdot\sqrt{x}\right)$$

$$= x^2\dfrac{d}{dx}\sqrt{x} + \sqrt{x}\dfrac{d}{dx}x^2 = x^2\left(\dfrac{1}{2\sqrt{x}}\right) + \sqrt{x}\,(2x) = \dfrac{1}{2}x^{3/2} + 2x^{3/2} = \dfrac{5}{2}x^{3/2}.$$

REMARK: If you have discovered $\dfrac{d}{dx}x^{3/2} = \dfrac{3}{2}\sqrt{x}$ (Problem 12), then

$$\dfrac{d}{dx}x^{5/2} = \dfrac{d}{dx}\left(x\cdot x^{3/2}\right) = x\cdot\left(\dfrac{3}{2}\sqrt{x}\right) + x^{3/2}\cdot(1) = \dfrac{5}{2}x^{3/2}.$$

21.
$$\dfrac{d}{dt}\left(\dfrac{1}{t^4+2}\right) = \dfrac{(t^4+2)\dfrac{d}{dt}1 - (1)\dfrac{d}{dt}(t^4+2)}{(t^4+2)^2} = \dfrac{-4t^3}{(t^4+2)^2},$$

therefore, using the product rule and the result of Problem 11,

$$\dfrac{d}{dt}\left(\left(\dfrac{1}{\sqrt{t}}\right)\left(\dfrac{1}{t^4+2}\right)\right) = \left(\dfrac{1}{\sqrt{t}}\right)\left(\dfrac{-4t^3}{(t^4+2)^2}\right) + \left(\dfrac{1}{t^4+2}\right)\left(\dfrac{-1}{2t\sqrt{t}}\right) = \dfrac{-(9t^4+2)}{2t\sqrt{t}\,(t^4+2)^2}.$$

23. $\dfrac{d}{dx}\left((1-x^{-2})\cdot(1+x)\cdot(1-x)\right)$

$= (-(-2)x^{-3}\cdot(1+x)\cdot(1-x) + (1-x^{-2})\cdot(1)\cdot(1-x) + (1-x^{-2})\cdot(1+x)\cdot(-1) = 2x^{-3} - 2x$.

25. $\dfrac{d}{dx}\left((x^2+1)(x^3+2)(x^4+3)\right)$

$= (2x)(x^3+2)(x^4+3) + (x^2+1)(3x^2)(x^4+3) + (x^2+1)(x^3+2)(4x^3)$.

27. $f'(x) = \dfrac{d}{dx}\left(4x(x^5+1)\right) = 4x(5x^4) + (x^5+1)(4) = 24x^5 + 4$; the tangent line at $(1,8)$ has slope
$f'(1) = 24 + 4 = 28$ and equation $y = 28(x-1) + 8 = 28x - 20$.

29. $h'(u) = \dfrac{d}{du}\left(\dfrac{1+\sqrt{u}}{u^2}\right) = \dfrac{\left(\frac{1}{2\sqrt{u}}\right)u^2 - (2u)(1+\sqrt{u})}{(u^2)^4};$ the tangent line at $(1,2)$ has slope

$h'(1) = \left(\dfrac{1}{2} - (2)(2)\right)\Big/ 1^8 = \dfrac{-7}{2};$ and equation $y = \left(-\dfrac{7}{2}\right)(u-1) + 2 = \left(-\dfrac{7}{2}\right)u + \left(\dfrac{11}{2}\right).$

REMARK: (i)   You might prefer to describe the tangent line with the equation $y = \left(-\dfrac{7}{2}\right)x + \left(\dfrac{11}{2}\right).$

(ii)   We're interested in the value of the derivative at just one point, it seems a waste of effort to simplify algebraically before evaluating.

---

31. $f'(x) = \dfrac{d}{dx}\left(\dfrac{1}{\sqrt{x}}\right) = \dfrac{-1}{2x\sqrt{x}}$ (see the solution for Problem 11); the tangent at $(1,1)$ has slope $f'(1) =$

$-\dfrac{1}{2}$, hence the normal there has slope $-1/(-1/2) = 2$ and equation $y = 2(x-1) + 1 = 2x - 1.$

33. $h'(x) = \dfrac{d}{dx}\left(\dfrac{1-\sqrt{x}}{1+3\sqrt{x}}\right) = \dfrac{\left(\frac{-1}{2\sqrt{x}}\right)(1+3\sqrt{x}) - (1-\sqrt{x})\left(\frac{3}{2\sqrt{x}}\right)}{(1+3\sqrt{x})^2};$ the tangent at $(1,0)$ has slope

$h'(1) = \dfrac{\left(\frac{-1}{2}\right)(1+3) - (1-1)\left(\frac{3}{2}\right)}{(1+3)^2} = \dfrac{-1}{8}$, hence the normal line there has slope $-1/(-1/8) = 8$ and

equation $y = 8(x-1) + 0 = 8x - 8.$

35. $\dfrac{d}{dr}R = \dfrac{d}{dr}\left(\alpha L r^{-4}\right) = \alpha L(-4)r^{-5} = \dfrac{-4\alpha L}{r^5};\quad R'(0.2\,\text{mm}) = \dfrac{-4\alpha L}{(1/5)^5} = -4\left(5^5\right)\alpha L = -12500\alpha L.$

37. $\left(\dfrac{fg}{f+g}\right)' = \dfrac{(f+g)\cdot(fg)' - (fg)\cdot(f+g)'}{(f+g)^2} = \dfrac{(f+g)(f'g+fg') - (fg)(f'+g')}{(f+g)^2} = \dfrac{f'g^2 + f^2g'}{(f+g)^2}.$

39.   Since $\dfrac{d}{dx}\left(f(x)\cdot g(x)\right) = f(x)\cdot\dfrac{dg}{dx} + g(x)\cdot\dfrac{df}{dx}$ if f and g are any differentiable functions, we get a special case of the Product Rule when $g = f$:

$$\dfrac{d}{dx}\left(f(x)\cdot f(x)\right) = f(x)\cdot\dfrac{df}{dx} + f(x)\cdot\dfrac{df}{dx} = 2f(x)\dfrac{df}{dx} = 2f(x)\cdot f'(x).$$

Therefore

$$\dfrac{d}{dx}\left((f(x))^2\right) = 2f(x)\cdot f'(x). \ \blacksquare$$

41.   Suppose $H(x) = F(x)/x^n$, then $x^n\cdot H(x) = F(x)$. Therefore

$$F'(x) = (x^n\cdot H(x))' = (x^n)'\cdot H(x) + x^n\cdot H'(x)$$
$$= \left(nx^{n-1}\right)\cdot(F(x)/x^n) + x^n\cdot H'(x) = nF(x)/x + x^n\cdot H'(x),$$

this implies $x\cdot F'(x) = nF(x) + x^{n+1}\cdot H'(x)$ and $H'(x) = \dfrac{x\cdot F'(x) - n\,F(x)}{x^{n+1}}. \ \blacksquare$

43. $f(x+\Delta x)g(x+\Delta x) - f(x)g(x) = f(x)\left[g(x+\Delta x) - g(x)\right] + \left[f(x+\Delta x) - f(x)\right]g(x+\Delta x)$
$\approx f(x)\left[g(x+\Delta x) - g(x)\right] + \left[f(x+\Delta x) - f(x)\right]g(x)$
$\approx f(x)\left[g'(x)\cdot\Delta x\right] + \left[f'(x)\cdot\Delta x\right]g(x).$

2.2 — The product and quotient rules

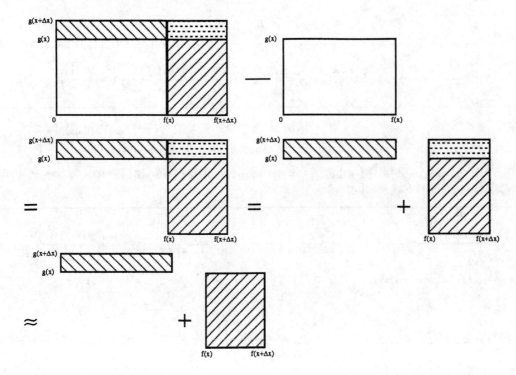

Solutions 2.3   *The derivative of composite functions: the chain rule*   (pages 153-4)

1. Let $u = x + 1$, then $\dfrac{du}{dx} = 1$ and $\dfrac{d}{dx}(x+1)^3 = \dfrac{d}{dx}u^3 = \left(\dfrac{d}{du}u^3\right)\dfrac{du}{dx} = (3u^2)(1) = 3(x+1)^2$.

3. Let $u = 1 + x^6$, then $\dfrac{du}{dx} = 6x^5$ and

$$\frac{d}{dx}\left(1+x^6\right)^6 = \frac{d}{dx}u^6 = \left(\frac{d}{du}u^6\right)\frac{du}{dx} = \left(6u^5\right)\left(6x^5\right) = 36x^5\left(1+x^6\right)^5.$$

5. Let $u = x^2 - x^3$, then

$$\frac{d}{dx}\left(x^2-x^3\right)^4 = \left(\frac{d}{du}u^4\right)\frac{du}{dx} = \left(4u^3\right)\left(2x-3x^2\right) = 4\left(x^2-x^3\right)^3\left(2x-3x^2\right) = 4x^7\left(1-x\right)^3\left(2-3x\right).$$

7. Let $u = 1 - x^2 + x^5$, then

$$\frac{d}{dx}\left(1-x^2+x^5\right)^3 = \left(\frac{d}{du}u^3\right)\frac{du}{dx} = \left(3u^2\right)\left(-2x+5x^4\right) = 3\left(1-x^2+x^5\right)^2\left(-2x+5x^4\right).$$

9. Let $u = \sqrt{x} - x$, then $\dfrac{d}{dx}\left(\sqrt{x}-x\right)^3 = \left(\dfrac{d}{du}u^3\right)\dfrac{du}{dx} = \left(3u^2\right)\left(\dfrac{1}{2\sqrt{x}}-1\right) = 3\left(\sqrt{x}-x\right)^2\left(\dfrac{1}{2\sqrt{x}}-1\right).$

11. Let $u = y^2 + 3$, then $\dfrac{d}{dy}\left(y^2+3\right)^{-4} = \left(\dfrac{d}{du}u^{-4}\right)\dfrac{du}{dy} = \left(-4u^{-5}\right)(2y) = \dfrac{-8y}{\left(y^2+3\right)^5}.$

13. Let $u = \dfrac{t+1}{t-1}$, then $\dfrac{du}{dt} = \dfrac{(t-1)(1)-(t+1)(1)}{(t-1)^2} = \dfrac{-2}{(t-1)^2}$ and

$$\frac{d}{dt}\left(\frac{t+1}{t-1}\right)^3 = \left(\frac{d}{du}u^3\right)\frac{du}{dt} = \left(3u^2\right)\left(\frac{-2}{(t-1)^2}\right) = (-6)\left(\frac{t+1}{t-1}\right)^2\frac{1}{(t-1)^2} = \frac{-6(t+1)^2}{(t-1)^4}.$$

REMARK: If you notice $u = \dfrac{t+1}{t-1} = \dfrac{t-1+2}{t-1} = 1 + \dfrac{2}{t-1} = 1 + 2(t-1)^{-1}$ and then apply the Power

Rule (Theorem 2), you find again that $\dfrac{du}{dt} = -2(t-1)^{-2}$.

15.
$$\frac{d}{dx}\left[\left(x^2+2\right)^5\left(x^4+3\right)^3\right] = \left[\frac{d}{dx}\left(x^2+2\right)^5\right]\left(x^4+3\right)^3 + \left(x^2+2\right)^5\left[\frac{d}{dx}\left(x^4+3\right)^3\right]$$
$$= \left[5\left(x^2+2\right)^4(2x)\right]\left(x^4+3\right)^3 + \left(x^2+2\right)^5\left[3\left(x^4+3\right)^2\left(4x^3\right)\right]$$
$$= 10x\left(x^2+2\right)^4\left(x^4+3\right)^3 + 12x^3\left(x^2+2\right)^5\left(x^4+3\right)^2$$
$$= 2x\left(x^2+2\right)^4\left(x^4+3\right)^2\left[5\left(x^2+2\right) + 6x^2\left(x^4+3\right)\right]$$
$$= 2x\left(x^2+2\right)^4\left(x^4+3\right)^2\left[11x^4+12x^2+15\right].$$

REMARK: The first step uses the Product Rule in the form $(f \cdot g)' = f'g + fg'$, the second step involves two applications of the Power Rule, the remaining steps are optional "simplifications" of the result.

17.
$$\frac{d}{dt}\left(\frac{\sqrt{t^2+1}}{(t+2)^4}\right) = \frac{(t+2)^4\frac{d}{dt}\sqrt{t^2+1} - \sqrt{t^2+1}\frac{d}{dt}(t+2)^4}{\left((t+2)^4\right)^2}$$
$$= \frac{(t+2)^4\left(1/2\sqrt{t^2+1}\right)\left[\frac{d}{dt}(t^2+1)\right] - \sqrt{t^2+1}\left((4)(t+2)^3\right)\left[\frac{d}{dt}(t+2)\right]}{(t+2)^8}$$
$$= \frac{(t+2)^4\left(1/2\sqrt{t^2+1}\right)[2t] - \sqrt{t^2+1}(4)(t+2)^3[1]}{(t+2)^8}$$
$$= (t+2)^3\left[\frac{(t+2)t/\sqrt{t^2+1} - 4\sqrt{t^2+1}}{(t+2)^8}\right] = \frac{(t^2+2t) - 4(t^2+1)}{(t+2)^5\sqrt{t^2+1}} = \frac{-3t^2+2t-4}{(t+2)^5\sqrt{t^2+1}}.$$

REMARK: If the original function is written as $(t^2+1)^{-1/2}(t+2)^{-4}$ and differentiated using the Product and Power Rules, then the fifth power of $(t+2)$ in the denominator of the result may be less surprising.

19.
$$\frac{d}{dx}\left[(x^2+1)^2(x^3+2)^3(x^4+3)^{-1/2}\right]$$
$$= \begin{aligned}&\left[2(x^2+1)(2x)\right](x^3+2)^3(x^4+3)^{-1/2} + (x^2+1)^2\left[3(x^3+2)^2(3x^2)\right](x^4+3)^{-1/2}\\&\qquad\qquad + (x^2+1)^2(x^3+2)^3\left[-\frac{1}{2}(x^4+3)^{-3/2}(4x^3)\right]\end{aligned}$$
$$= \frac{(x^2+1)(x^3+2)^2}{(x^4+3)^{3/2}}\left[(4x)(x^3+2)(x^4+3) + (x^2+1)(9x^2)(x^4+3) + (x^2+1)(x^3+2)(-2x^3)\right]$$
$$= \frac{(x^2+1)(x^3+2)^2}{(x^4+3)^{3/2}}\left[x\left(11x^7+7x^5+4x^4+39x^3-4x^2+27x+24\right)\right].$$

21.
$$\frac{d}{dx}\sqrt{x^2+\sqrt{1+x^2}} = \frac{1}{2\sqrt{x^2+\sqrt{1+x^2}}}\frac{d}{dx}\left(x^2+\sqrt{1+x^2}\right)$$
$$= \frac{1}{2\sqrt{x^2+\sqrt{1+x^2}}}\left(2x + \frac{1}{2\sqrt{1+x^2}}\frac{d}{dx}\left(1+x^2\right)\right)$$
$$= \frac{1}{2\sqrt{x^2+\sqrt{1+x^2}}}\left(2x + \frac{1}{2\sqrt{1+x^2}}(2x)\right).$$

23.
$$\frac{d}{dx}\left[(1+x)^{-1} + (1-x)^{-1}\right]^{-1} = (-1)\left[(1+x)^{-1} + (1-x)^{-1}\right]^{-2}\frac{d}{dx}\left[(1+x)^{-1}+(1-x)^{-1}\right]$$
$$= (-1)\left[(1+x)^{-1}+(1-x)^{-1}\right]^{-2}\left[(-1)(1+x)^{-2}+(-1)(1-x)^{-2}(-1)\right]$$
$$= \frac{(1+x)^{-2} - (1-x)^{-2}}{\left[(1+x)^{-1}+(1-x)^{-1}\right]^2} = \frac{(1-x)^2-(1+x)^2}{\left[(1-x)+(1+x)\right]^2} = \frac{-4x}{4} = -x.$$

REMARK: $\left[(1+x)^{-1}+(1-x)^{-1}\right]^{-1} = \dfrac{1}{2}\left(1-x^2\right)$ if $x \neq \pm 1$.

25. $\frac{d}{dx}\text{Homer}(2x+1) = \text{Virgil}(2x+1)\cdot\frac{d}{dx}(2x+1) = \text{Virgil}(2x+1)\cdot 2.$

2.3 — The derivative of composite functions: the chain rule

27. $\frac{d}{dx}\text{Homer}^6(x) = 6\text{Homer}^5(x) \cdot \frac{d}{dx}\text{Homer}(x) = 6\text{Homer}^5(x) \cdot \text{Virgil}(x).$

29. $\frac{d}{dx}L(3x) = L'(3x) \cdot \frac{d}{dx}(3x) = \frac{1}{3x} \cdot 3 = \frac{1}{x}.$

31. $\frac{d}{dx}\sqrt{L(7x)} = \frac{1}{2\sqrt{L(7x)}} \cdot \frac{d}{dx}L(7x) = \frac{1}{2\sqrt{L(7x)}} \cdot \frac{1}{7x} \cdot \frac{d}{dx}(7x) = \frac{1}{2\sqrt{L(7x)}} \cdot \frac{1}{7x} \cdot 7 = \frac{1}{2x\sqrt{L(7x)}}.$

33.
$$\frac{d}{dx}L^{12}\left(\frac{x+1}{x-1}\right) = 12L^{11}\left(\frac{x+1}{x-1}\right) \cdot \frac{d}{dx}L\left(\frac{x+1}{x-1}\right) = 12L^{11}\left(\frac{x+1}{x-1}\right) \cdot \frac{1}{(x+1)/(x-1)} \cdot \frac{d}{dx}\left(\frac{x+1}{x-1}\right)$$
$$= 12L^{11}\left(\frac{x+1}{x-1}\right) \cdot \frac{x-1}{x+1} \cdot \frac{-2}{(x-1)^2}$$
$$= \frac{-24}{x^2-1} \cdot L^{11}\left(\frac{x+1}{x-1}\right)$$

REMARK: $\dfrac{x+1}{x-1} = \dfrac{(x-1)+2}{x-1} = 1 + \dfrac{2}{x-1} = 1 + 2(x-1)^{-1}$, thus $\dfrac{d}{dx}\left(\dfrac{x+1}{x-1}\right) = 0 + 2(-1)(x-1)^{-2}.$

---

35. $\frac{d}{dx}\frac{L(x)}{x} = \frac{x \cdot L'(x) - L(x) \cdot \frac{d}{dx}x}{x^2} = \frac{x \cdot (1/x) - L(x) \cdot 1}{x^2} = \frac{1 - L(x)}{x^2}.$

37. $\frac{d}{dx}E(4x) = E'(4x) \cdot \frac{d}{dx}(4x) = E(4x) \cdot 4 = 4\,E(4x).$

39. $\frac{d}{dx}E\left(x^3 + 2x + 3\right) = E'\left(x^3 + 2x + 3\right) \cdot \frac{d}{dx}\left(x^3 + 2x + 3\right) = E\left(x^3 + 2x + 3\right) \cdot \left(3x^2 + 2\right).$

41.
$$\frac{d}{dx}\left(x\,E^{1/2}(x+1)\right) = \left(\frac{d}{dx}x\right) \cdot E^{1/2}(x+1) + x \cdot \left(\frac{d}{dx}E^{1/2}(x+1)\right)$$
$$= E^{1/2}(x+1) + x \cdot (1/2)E^{-1/2}(x+1) \cdot E'(x+1)$$
$$= E^{1/2}(x+1) + x \cdot (1/2)E^{-1/2}(x+1) \cdot E(x+1)$$
$$= \left(1 + \frac{x}{2}\right)E^{1/2}(x+1).$$

43. $\frac{d}{dx}E(L(x)) = E'(L(x)) \cdot L'(x) = E(L(x)) \cdot \frac{1}{x} = \frac{E(L(x))}{x}.$

45. If $f(x) = 6(x-3)^3 + 3x$, then $f(1) = -45$, $f'(x) = 18(x-3)^2 + 3$, $f'(1) = 75$. The line tangent to the path at $(1, -45)$ has equation $y = 75(x-1) - 45 = 75x - 120$, this line passes through the point $(4, 180)$.

47. a. Suppose $g(x) = 5x$ and $f(x) = (1/5)x$. Then $f(g(x)) = f(5x) = (1/5)(5x) = x$; $g'(x) = 5$, $f'(x) = 1/5$, and $f'(g(x)) = 1/5 = 1/g'(x)$. ∎

   b. Suppose $g(x) = 17x - 8$ and $f(x) = (1/17)x + (8/17)$. Then $f(g(x)) = (1/17)(17x - 8) + (8/17) = x - (8/17) + (8/17) = x$; $g'(x) = 17$, $f'(x) = 1/17$ and $f'(g(x)) = 1/17 = 1/g'(x)$. ∎

   c. Suppose $g(x) = \sqrt{x}$ and $f(x) = x^2$. Then $f(g(x)) = \left(\sqrt{x}\right)^2 = x$; $g'(x) = 1/\left(2\sqrt{x}\right)$, $f'(x) = 2x$, and $f'(g(x)) = 2g(x) = 2\sqrt{x} = 1/\left(1/\left(2\sqrt{x}\right)\right) = 1/g'(x)$. ∎

   d. Suppose $g(x) = x^2$ with domain $(-\infty, 0]$ and $f(x) = -\sqrt{x}$. Then the domain of $f \circ g$ is $(-\infty, 0]$. If $x \leq 0$, then
$$(f \circ g)(x) = f(g(x)) = -\sqrt{x^2} = -|x| = -(-x) = x\,;$$
$$g'(x) = 2x \quad \text{and} \quad f'(x) = \frac{-1}{2\sqrt{x}}\,,$$
$$f'(g(x)) = \frac{-1}{2\sqrt{g(x)}} = \frac{-1}{2\sqrt{x^2}} = \frac{-1}{2|x|} = \frac{-1}{2(-x)} = \frac{1}{2x} = \frac{1}{g'(x)}. \text{ ∎}$$

49. Suppose $(x-17)^2$ divides $p(x)$. Then there is a (unique) polynomial q such that $p(x) = (x-17)^2 \cdot q(x)$. Therefore
$$p'(x) = \left(\frac{d}{dx}(x-17)^2\right) \cdot q(x) + (x-17)^2 \cdot \left(\frac{d}{dx}q(x)\right)$$
$$= 2(x-17)(1) \cdot q(x) + (x-17)^2 \cdot q'(x) = (x-17) \cdot \left[2q(x) + (x-17)q'(x)\right],$$
hence $x - 17$ divides $p'(x)$ with quotient $2q(x) + (x-17)q'(x)$. ∎

51. $x = f(g(x)) \implies 1 = \frac{dx}{dx} = \frac{d}{dx}f(g(x)) = f'(g(x)) \cdot g'(x) \implies g'(x) \neq 0$ and $f'(g(x)) = \frac{1}{g'(x)}$. ∎

53.   Let $G(x) = f(cx)$ where c is fixed and positive. Then

$$G'(x) = \left[\frac{d}{d(cx)} f(cx)\right] \cdot \frac{d}{dx}(cx) \qquad \text{(Chain Rule)}$$
$$= [f'(cx)] \cdot (c) = c \cdot f'(cx).$$

If f has the special property that $f(A \cdot B) = f(A) + f(B)$ when A and B are positive, then $G(x) = f(cx) = f(c) + f(x)$ and $G'(x) = \frac{d}{dx} f(c) + \frac{d}{dx} f(x) = 0 + f'(x) = f'(x)$. Therefore $cf'(cx) = G'(x) = f'(x)$, $f'(cx) = f'(x)/c$, and $f'(c) = f'(1)/c$ (letting $x = 1$). But c was arbitrary, therefore $f'(c) = f'(1)/c$ for all $c > 0$; this is what we were to show! ∎

Solutions 2.4   *The derivative of a power function*   **(pages 159-60)**

1. $\dfrac{d}{dx}\left(x^{2/5} + 2x^{1/3}\right) = \left(\dfrac{2}{5}\right) x^{2/5-1} + 2\left(\dfrac{1}{3}\right) x^{1/3-1} = \left(\dfrac{2}{5}\right) x^{-3/5} + \left(\dfrac{2}{3}\right) x^{-2/3}$.

3.   Let $u = x^2 + 1$. Then $\dfrac{du}{dx} = 2x$ and

$$\frac{d}{dx}\left(x^2 + 1\right)^{5/3} = \frac{d}{dx} u^{5/3} = \left[\frac{d}{du} u^{5/3}\right]\left[\frac{du}{dx}\right] = \left[\left(\frac{5}{3}\right) u^{2/3}\right](2x)$$
$$= \left(\frac{5}{3}\right)\left(x^2 + 1\right)^{2/3}(2x) = \left(\frac{10}{3}\right) x \left(x^2 + 1\right)^{2/3}.$$

5. $\dfrac{d}{dx}\left(x^3 + 3\right)^{1/10} = \dfrac{1}{10}\left(x^3 + 3\right)^{-9/10} \dfrac{d}{dx}\left(x^3 + 3\right) = \dfrac{1}{10}\left(x^3 + 3\right)^{-9/10}\left(3x^2\right) = \dfrac{3x^2}{10\left(x^3 + 3\right)^{9/10}}$.

7. $\dfrac{d}{dt}\left(t^{10} - 2\right)^{4/7} = \dfrac{4}{7}\left(t^{10} - 2\right)^{-3/7} \dfrac{d}{dt}\left(t^{10} - 2\right) = \dfrac{4}{7}\left(t^{10} - 2\right)^{-3/7}\left(10t^9\right) = \dfrac{40t^9}{7\left(t^{10} - 2\right)^{3/7}}$.

9. $\dfrac{d}{du}\sqrt[3]{u^3 + 3u + 1} = \dfrac{d}{du}\left(u^3 + 3u + 1\right)^{1/3} = \dfrac{1}{3}\left(u^3 + 3u + 1\right)^{-2/3}\dfrac{d}{du}\left(u^3 + 3u + 1\right)$
$$= \frac{1}{3}\left(u^3 + 3u + 1\right)^{-2/3}\left(3u^2 + 3\right) = \frac{u^2 + 1}{\left(u^3 + 3u + 1\right)^{2/3}}.$$

11. $\dfrac{d}{dz}\left(\dfrac{z^2 - 1}{z^2 + 1}\right)^{-7/17}$
$$= \frac{-7}{17}\left(\frac{z^2 - 1}{z^2 + 1}\right)^{-24/17}\frac{d}{dz}\left(\frac{z^2 - 1}{z^2 + 1}\right) = \frac{-7}{17}\left(\frac{z^2 + 1}{z^2 - 1}\right)^{24/17}\frac{d}{dz}\left(1 - \frac{2}{z^2 + 1}\right)$$
$$= \frac{-7}{17}\left(\frac{z^2 + 1}{z^2 - 1}\right)^{24/17}\frac{4z}{(z^2 + 1)^2} = \frac{-28\,z}{17\,(z^2 + 1)^{10/17}(z^2 - 1)^{24/17}}.$$

13. $\dfrac{d}{dr}\left(\left(r^2 + 1\right)^{5/6}\left(r - 1\right)^{1/2}\right) = \left[\dfrac{5}{6}\left(r^2 + 1\right)^{-1/6}(2r)\right](r - 1)^{1/2} + \left(r^2 + 1\right)^{5/6}\left[\dfrac{1}{2}(r - 1)^{-1/2}(1)\right]$
$$= \frac{13r^2 - 10r + 3}{6\left(r^2 + 1\right)^{1/6}(r - 1)^{1/2}}.$$

15. $\dfrac{d}{dx}\left(3x^{\sqrt{2}}\right) = 3\dfrac{d}{dx}\left(x^{\sqrt{2}}\right) = 3\sqrt{2}\, x^{\sqrt{2}-1}$.

17. $\dfrac{d}{dt}\left(t^2 + 1\right)^{\sqrt{3}} = \sqrt{3}\left(t^2 + 1\right)^{\sqrt{3}-1}\left[\dfrac{d}{dt}\left(t^2 + 1\right)\right] = \sqrt{3}\left(t^2 + 1\right)^{\sqrt{3}-1}(2t) = 2\sqrt{3}t\left(t^2 + 1\right)^{\sqrt{3}-1}$.

19. $\dfrac{d}{du}\left(u^{\sqrt{2}} - u^{\sqrt{3}}\right)^4 = 4\left(u^{\sqrt{2}} - u^{\sqrt{3}}\right)^3\left[\dfrac{d}{du}\left(u^{\sqrt{2}} - u^{\sqrt{3}}\right)\right] = 4\left(u^{\sqrt{2}} - u^{\sqrt{3}}\right)^3\left(\sqrt{2}u^{\sqrt{2}-1} - \sqrt{3}u^{\sqrt{3}-1}\right)$.

21. $\dfrac{d}{dx}\left(x^{1/2} + x^{3/2}\right) = \dfrac{1}{2}x^{-1/2} + \dfrac{3}{2}x^{1/2}$. When $x = 1$, $\dfrac{dy}{dx} = \dfrac{1}{2\sqrt{1}} + \dfrac{3\sqrt{1}}{2} = 2$; the tangent at $(1, 2)$ has slope 2 and equation $y = 2(x - 1) + 2 = 2x$; the normal at $(1, 2)$ has slope $-1/2$ and equation $y = (-1/2)(x - 1) + 2 = (-1/2)(x - 5)$.

23. $\dfrac{d}{dx}(x - 1)^{\sqrt{2}} = \sqrt{2}(x - 1)^{\sqrt{2}-1}$. When $x = 1$, $\dfrac{dy}{dx} = 0$; the tangent at $(1, 0)$ has slope 0 and equation $y = 0(x - 1) + 0 = 0$; the normal is vertical because the tangent is horizontal, its equation is $x = 1$.

2.4 — The derivative of a power function

25. a. $v_{rms} = \sqrt{\dfrac{3(1)}{0.0899}} \approx \sqrt{33.37} \approx 5.777$.

    b. $\dfrac{d}{d\rho}\, v_{rms} = \dfrac{d}{d\rho}\left(\sqrt{3P}\cdot\rho^{-1/2}\right) = \sqrt{3P}\left(\dfrac{-1}{2}\right)\rho^{-3/2} = \left(\dfrac{-1}{2}\right)\sqrt{\dfrac{3P}{\rho^3}}$.

    c. $v'_{rms}(0.0899) = \left(\dfrac{-1}{2}\right)\sqrt{\dfrac{3(1)}{0.0899^3}} \approx -32.1285$.

27.   The circle centered at the origin and with radius r contains exactly those points whose distance to the center equals r, i.e., the circle is specified by the equation $r = \sqrt{(x-0)^2 + (y-0)^2} = \sqrt{x^2 + y^2}$. The upper half of the circle described by $r^2 = x^2 + y^2$ satisfies the equation $y = \sqrt{r^2 - x^2}$. This is differentiable on the open interval $(-r, r)$ where we find $y' = -x/\sqrt{r^2 - x^2} = -x/y$. Therefore, the line tangent to the semicircle at point $(x_0, y_0) = (x_0, \sqrt{r^2 - x_0^2})$ has slope $-x_0/\sqrt{r^2 - x_0^2} = -x_0/y_0$ and satisfies the equation $y - y_0 = (-x_0/y_0)(x - x_0)$. On the other hand, the line through $(0,0)$, the circle's center, and (x_0, y_0) is the radial line that passes through (x_0, y_0); this line has slope $(y_0 - 0)/(x_0 - 0) = y_0/x_0$ and satisfies $y - y_0 = (y_0/x_0)(x - x_0)$ [which simplifies to $y = (y_0/x_0)\,x$]. The tangent line is perpendicular to the radial line because the product of their slopes, $(-x_0/y_0)\cdot(y_0/x_0)$, equals -1 [see Theorem 1.4.3]. Symmetry, geometric and algebraic, implies that tangents are perpendicular to radial lines on the lower semicircle too; it only remains to show that points $(-r, 0)$ and $(0, r)$ are not exceptions. Since the limit of $|y'|$ as $x \to -r^+$ and as $x \to r^-$ is ∞, the tangent lines at those points are vertical; the radial line is the x-axis which is obviously perpendicular to those vertical tangents. ∎

REMARK: Any circle and its family of tangent lines can be rigidly translated (moved) to be centered at the origin. The above proof assures us that such origin-centered circles have tangents perpendicular to radial lines; hence, all circles do.

Solutions 2.5   *The derivatives of the trigonometric functions*   (pages 167-8)

1.   Let $u = 3x$. Then $\dfrac{du}{dx} = 3$ and $\dfrac{d}{dx}\sin 3x = \left(\dfrac{d}{du}\sin u\right)\left(\dfrac{du}{dx}\right) = (\cos u)(3) = (\cos 3x)(3) = 3\cos 3x$.

3. $\dfrac{d}{dx}(x\cdot\cos x) = \left(\dfrac{d}{dx}x\right)\cdot\cos x + x\cdot\left(\dfrac{d}{dx}\cos x\right) = (1)\cdot\cos x + x\cdot(-\sin x) = \cos x - x\cdot\sin x$.

5. $\dfrac{d}{dx}\sqrt{\sin x} = \dfrac{d}{dx}(\sin x)^{1/2} = \left(\dfrac{1}{2}\right)(\sin x)^{-1/2}\cdot\left(\dfrac{d}{dx}\sin x\right) = \left(\dfrac{1}{2\sqrt{\sin x}}\right)\cdot(\cos x) = \dfrac{\cos x}{2\sqrt{\sin x}}$.

7. $\dfrac{d}{dx}\sin^2 x = \dfrac{d}{dx}(\sin x)^2 = 2(\sin x)^1\left(\dfrac{d}{dx}\sin x\right) = 2(\sin x)(\cos x)$.

REMARK: Explore the chain rule a bit further after noting $\sin^2 x = (1 - \cos 2x)/2$ and $2\sin x\cos x = \sin 2x$.

9. $\dfrac{d}{dx}(2\sin x\,\cos x) = 2\left(\dfrac{d}{dx}\sin x\right)(\cos x) + 2(\sin x)\left(\dfrac{d}{dx}\cos x\right) = 2(\cos x)(\cos x) + 2(\sin x)(-\sin x) = 2\left(\cos^2 x - \sin^2 x\right)$.

REMARK: $2\sin x\,\cos x = \sin(2x)$ and $\cos^2 x - \sin^2 x = \cos(2x)$.

11. $\dfrac{d}{dx}\left(\sin^2 x + \cos^2 x\right) = 2\sin x\left(\dfrac{d}{dx}\sin x\right) + 2\cos x\left(\dfrac{d}{dx}\cos x\right) = 2\sin x\,\cos x + 2\cos x(-\sin x) = 0$.

REMARK: The result should not surprise because $\sin^2 x + \cos^2 x$ is constant.

13. $\dfrac{d}{dx}\left(\dfrac{\sin x}{\tan x}\right) = \dfrac{(\tan x)\left(\frac{d}{dx}\sin x\right) - (\sin x)\left(\frac{d}{dx}\tan x\right)}{\tan^2 x} = \dfrac{(\tan x)(\cos x) - (\sin x)(\sec^2 x)}{\tan^2 x}$

$= (\sin x)\left(\dfrac{1 - \sec^2 x}{\tan^2 x}\right) = -\sin x$.

REMARK: $(\sin x)/(\tan x) = \cos x$.

15. $\dfrac{d}{dx}\left(\dfrac{1+\sec x}{1-\sec x}\right) = \dfrac{(1-\sec x)(\sec x \tan x) - (1+\sec x)(-\sec x \tan x)}{(1-\sec x)^2} = \dfrac{2 \sec x \tan x}{(1-\sec x)^2}.$

REMARK: $\dfrac{d}{dx}\left(\dfrac{1+\sec x}{1-\sec x}\right) = \dfrac{d}{dx}\left(\dfrac{2}{1-\sec x}-1\right) = 2\,(-1)\,(1-\sec x)^{-2}\,(-\sec x \tan x).$

17. $\dfrac{d}{dx}\left(\sin\left(x^3 - 2x + 6\right)\right) = \left(3x^2 - 2\right)\cos\left(x^3 - 2x + 6\right).$

19. $\dfrac{d}{dx}\tan^2 x = \dfrac{d}{dx}\left(\tan x\right)^2 = 2 \tan x \sec^2 x.$

21. $\dfrac{d}{dx}\sec^3 x = 3 \sec^2 x\,(\sec x \tan x) = 3 \sec^3 x \tan x.$

23. $\dfrac{d}{dx}\csc x^2 = (-\csc x^2 \cot x^2)\cdot(2x) = -2x \csc x^2 \cot x^2.$

25. $\displaystyle\lim_{x\to 0}\dfrac{\sin(x/2)}{x} = \left(\dfrac{1}{2}\right)\lim_{x\to 0}\dfrac{\sin(x/2)}{(x/2)} = \left(\dfrac{1}{2}\right)\lim_{(x/2)\to 0}\dfrac{\sin(x/2)}{(x/2)} = \left(\dfrac{1}{2}\right)(1) = \dfrac{1}{2}.$ (See Example 1.)

27. $\displaystyle\lim_{x\to 0}\dfrac{\sin^2 x}{x} = \left(\lim_{x\to 0}\dfrac{\sin x}{x}\right)\left(\lim_{x\to 0}\sin x\right) = (1)(0) = 0.$

29. $\displaystyle\lim_{x\to 0}\dfrac{\sin^2 4x}{x^2} = \lim_{x\to 0}\left(\dfrac{\sin 4x}{x}\right)^2 = \left(4\lim_{x\to 0}\dfrac{\sin 4x}{4x}\right)^2 = 4^2\left(\lim_{(4x)\to 0}\dfrac{\sin 4x}{4x}\right)^2 = 16(1)^2 = 16.$

31. $\displaystyle\lim_{x\to 0}\dfrac{\sin 3x}{\sin 4x} = \lim_{x\to 0}\left(\dfrac{3}{4}\right)\left(\dfrac{\left(\dfrac{\sin 3x}{3x}\right)}{\left(\dfrac{\sin 4x}{4x}\right)}\right) = \left(\dfrac{3}{4}\right)\dfrac{\lim_{x\to 0}\left(\dfrac{\sin 3x}{3x}\right)}{\lim_{x\to 0}\left(\dfrac{\sin 4x}{4x}\right)} = \left(\dfrac{3}{4}\right)\dfrac{\lim_{3x\to 0}\left(\dfrac{\sin 3x}{3x}\right)}{\lim_{4x\to 0}\left(\dfrac{\sin 4x}{4x}\right)} = \left(\dfrac{3}{4}\right)\left(\dfrac{1}{1}\right) = \dfrac{3}{4}.$

33. $\displaystyle\lim_{x\to 0}\dfrac{\cos 2x - 1}{x} = 2\lim_{x\to 0}\dfrac{\cos 2x - 1}{2x} = 2\lim_{2x\to 0}\dfrac{\cos 2x - 1}{2x} = 2\,(0) = 0.$ (See Example 2.)

35. $\displaystyle\lim_{x\to 0}\dfrac{\sin^2 2x}{(7x)^2} = \lim_{x\to 0}\left(\dfrac{2}{7}\right)^2\left(\dfrac{\sin 2x}{2x}\right)^2 = \left(\dfrac{2}{7}\right)^2\left(\lim_{x\to 0}\dfrac{\sin 2x}{2x}\right)^2 = \left(\dfrac{4}{49}\right)(1)^2 = \dfrac{4}{49}.$

37. $\dfrac{d}{dx}\cot x = \dfrac{d}{dx}\left(\dfrac{\cos x}{\sin x}\right) = \dfrac{(\sin x)\left(\frac{d}{dx}\cos x\right) - (\cos x)\left(\frac{d}{dx}\sin x\right)}{(\sin x)^2}$

$= \dfrac{(\sin x)(-\sin x) - (\cos x)(\cos x)}{\sin^2 x} = \dfrac{-\left(\sin^2 x + \cos^2 x\right)}{\sin^2 x} = \dfrac{-1}{\sin^2 x} = -\csc^2 x$

because $\sin^2\theta + \cos^2\theta = 1$ for all real numbers θ. ∎

39. Let $\theta = \left(\dfrac{\pi}{180}\right)x$ where x is measured in degrees and θ in radians, thus sine $x = \sin\theta$ and cosine $x = \cos\theta$;

$$\lim_{x\to 0}\dfrac{\text{sine }x}{x} = \lim_{\theta\to 0}\dfrac{\sin\theta}{180\,\theta/\pi} = \left(\dfrac{\pi}{180}\right)\left(\lim_{\theta\to 0}\dfrac{\sin\theta}{\theta}\right) = \left(\dfrac{\pi}{180}\right)(1) = \dfrac{\pi}{180}.$$

41. a. $\displaystyle\lim_{x\to 0}\dfrac{\sin x}{x} = 1$ and $\displaystyle\lim_{x\to 0}\cos x = 1$ imply $\displaystyle\lim_{x\to 0}\dfrac{\tan x}{x} = \lim_{x\to 0}\dfrac{\left(\dfrac{\sin x}{x}\right)}{\cos x} = \dfrac{\lim_{x\to 0}\left(\dfrac{\sin x}{x}\right)}{\lim_{x\to 0}\cos x} = \dfrac{1}{1} = 1.$

b. $\dfrac{d}{dx}\tan x = \displaystyle\lim_{\Delta x\to 0}\dfrac{\tan(x+\Delta x) - \tan x}{\Delta x} = \lim_{\Delta x\to 0}\dfrac{\dfrac{\tan x + \tan\Delta x}{1 - \tan x \tan\Delta x} - \tan x}{\Delta x}$

$= \displaystyle\lim_{\Delta x\to 0}\left(\dfrac{\tan\Delta x}{\Delta x}\right)\left(\dfrac{1 + \tan^2 x}{1 - \tan x \tan\Delta x}\right) = (1)\left(\dfrac{1 + \tan^2 x}{1 - \tan x\cdot 0}\right)$

$= 1 + \tan^2 x = 1 + \dfrac{\sin^2 x}{\cos^2 x} = \dfrac{\cos^2 x + \sin^2 x}{\cos^2 x} = \dfrac{1}{\cos^2 x} = \sec^2 x.$ ∎

REMARK: The identity $\tan A - \tan B = \dfrac{\sin(A-B)}{\cos A\,\cos B}$ can be used to construct a briefer proof.

2.6 — Implicit differentiation

Solutions 2.6 *Implicit differentiation* (pages 173-4)

1. If $3 = x^3 + y^3$, then $0 = \frac{d}{dx}(3) = \frac{d}{dx}(x^3 + y^3) = \left(\frac{d}{dx}x^3\right) + \left(\frac{d}{dy}y^3\right)\left(\frac{dy}{dx}\right) = 3x^2 + 3y^2\left(\frac{dy}{dx}\right)$;

   therefore $3y^2\left(\frac{dy}{dx}\right) = -3x^2$ and, provided $y \neq 0$, $\frac{dy}{dx} = \frac{-3x^2}{3y^2} = \frac{-x^2}{y^2} = -\left(\frac{x}{y}\right)^2$.

3. If $2 = \sqrt{x} + \sqrt{y}$, then $0 = \frac{d}{dx}(2) = \frac{d}{dx}(\sqrt{x} + \sqrt{y}) = \frac{1}{2\sqrt{x}} + \frac{1}{2\sqrt{y}}\frac{dy}{dx}$; thus $\frac{dy}{dx} = \frac{-\sqrt{y}}{\sqrt{x}} = -\sqrt{\frac{y}{x}}$.

5. If $\frac{7}{8} = x^{-7/8} + y^{-7/8}$, then $0 = \frac{d}{dx}\left(\frac{7}{8}\right) = \frac{d}{dx}\left(x^{-7/8} + y^{-7/8}\right) = \left(\frac{-7}{8\,x^{15/8}}\right) + \left(\frac{-7}{8\,y^{15/8}}\right)\frac{dy}{dx}$;

   thus $\left(\frac{7}{8\,y^{15/8}}\right)\frac{dy}{dx} = \frac{-7}{8\,x^{15/8}}$ and $\frac{dy}{dx} = -\left(\frac{y}{x}\right)^{15/8}$.

7. If $x^2 = (3xy + 1)^5$, then $2x = \frac{d}{dx}x^2 = \frac{d}{dx}(3xy + 1)^5 = 5(3xy + 1)^4\left(3y + 3x\frac{dy}{dx}\right)$;

   therefore $3y + 3x\frac{dy}{dx} = \frac{2x}{5(3xy + 1)^4}$ and $\frac{dy}{dx} = \frac{2}{15}(3xy + 1)^{-4} - \frac{y}{x}$ provided $x \neq 0$ and $3xy + 1 \neq 0$.

9. If $1 = (4x^2y^2)^{1/5}$, then $0 = \frac{d}{dx}(1) = \frac{d}{dx}(4x^2y^2)^{1/5} = \frac{1}{5}(4x^2y^2)^{-4/5}(4)\left((2x)(y^2) + (x^2)(2y)\frac{dy}{dx}\right)$;

   therefore $(2x)(y^2) + (x^2)(2y)\frac{dy}{dx} = 0$ and, provided $xy \neq 0$, $\frac{dy}{dx} = \frac{-(2x)(y^2)}{(x^2)(2y)} = \frac{-y}{x}$.

   REMARK: $(4x^2y^2)^{1/5} = 1 \iff 4x^2y^2 = 1^5 = 1 \iff |xy| = \sqrt{1/4} \implies y + xy' = 0 \implies y' = -y/x$.

11. $(x + y)^{1/2} = (x^2 + y)^{1/3} \implies (x + y)^3 = (x^2 + y)^2$

   $\implies 3(x + y)^2\left(1 + \frac{dy}{dx}\right) = 2(x^2 + y)\left(2x + \frac{dy}{dx}\right)$

   $\implies \left(3(x + y)^2 - 2(x^2 + y)\right)\frac{dy}{dx} = 2(x^2 + y)(2x) - 3(x + y)^2$

   $\implies \frac{dy}{dx} = \frac{4x(x^2 + y) - 3(x + y)^2}{3(x + y)^2 - 2(x^2 + y)} = \frac{4x^3 - 3x^2 - 2xy - 3y^2}{x^2 + 6xy + 3y^2 - 2y}$.

   REMARK: "Straightforward" differentiation of the original relation will yield

   $$\frac{dy}{dx} = \frac{\left(\frac{1}{3}\right)(x^2 + y)^{-2/3}(2x) - \left(\frac{1}{2}\right)(x + y)^{-1/2}}{\left(\frac{1}{2}\right)(x + y)^{-1/2} - \left(\frac{1}{3}\right)(x^2 + y)^{-2/3}}.$$

13. $7 = (x + y)(x - y) = x^2 - y^2 \implies 0 = \frac{d}{dx}(x^2 - y^2) = 2x - 2y\frac{dy}{dx} \implies \frac{dy}{dx} = \frac{2x}{2y} = \frac{x}{y}$.

15. The sine function has range $[-1, 1]$, it never takes on 2 as a value; hence there is no choice of y as a function of x for which $\sin xy = 2$ is true.

17. $y = (\sin x)(\cos y) \implies \frac{dy}{dx} = (\cos x)(\cos y) + (\sin x)(-\sin y)\frac{dy}{dx}$

   $\implies (1 + (\sin x)(\sin y))\frac{dy}{dx} = (\cos x)(\cos y) \implies \frac{dy}{dx} = \frac{(\cos x)(\cos y)}{1 + (\sin x)(\sin y)}$.

19. $x^5 = xy + x^2y^2 = (xy) + (xy)^2 \implies 5x^4 = \frac{d}{d(xy)}\left((xy) + (xy)^2\right)\frac{d}{dx}(xy) = (1 + 2(xy))\left(y + x\frac{dy}{dx}\right)$

   $\implies y + x\frac{dy}{dx} = \frac{5x^4}{1 + 2xy}$

   $\implies \frac{dy}{dx} = \frac{5x^3}{1 + 2xy} - \frac{y}{x}$.

REMARK: $x^5 = xy + x^2y^2 \implies x^4 = y + xy^2 \implies 4x^3 = y^2 + (1+2xy)\dfrac{dy}{dx} \implies \dfrac{dy}{dx} = \dfrac{4x^3 - y^2}{1+2xy}.$

21. $\dfrac{1}{\sqrt{x^2+y^2}} = 4 \implies \dfrac{1^2}{4} = x^2 + y^2 \implies 0 = 2x + 2y\dfrac{dy}{dx} \implies \dfrac{dy}{dx} = \dfrac{-x}{y}.$

23. $\dfrac{x^2+y^2}{x^2-y^2} = 4 \implies x^2 + y^2 = 4\left(x^2 - y^2\right) \implies 5y^2 = 3x^2 \implies 10y\dfrac{dy}{dx} = 6x \implies \dfrac{dy}{dx} = \dfrac{6x}{10y} = \dfrac{3x}{5y}.$

REMARK: Since $5y^2 = 3x^2$ we also find that $\dfrac{dy}{dx} = \left(\dfrac{3x}{5y}\right)\left(\dfrac{y}{y}\right) = \dfrac{3xy}{5y^2} = \dfrac{3xy}{3x^2} = \dfrac{y}{x}.$

25. $0 = \dfrac{d}{dy}(1) = \dfrac{d}{dy}\left(\sqrt{x} + \sqrt{y}\right) = \dfrac{1}{2\sqrt{x}}\dfrac{dx}{dy} + \dfrac{1}{\sqrt{y}}$ therefore $\dfrac{dx}{dy} = \dfrac{-\sqrt{x}}{\sqrt{y}} = -\sqrt{\dfrac{x}{y}}$; this equals 0 when $x = 0$;

the curve has a vertical tangent at $(0,1)$. Similarly, $0 = \dfrac{1}{2\sqrt{x}} + \dfrac{1}{2\sqrt{y}}\dfrac{dy}{dx}$ implies $\dfrac{dy}{dx} = -\sqrt{\dfrac{y}{x}}$ which is 0 when

$y = 0$; the curve has a horizontal tangent at $(1,0)$.

REMARK for A-students and instructors: There are some subtle difficulties here. For instance, $\frac{1}{2\sqrt{x}}\frac{dx}{dy} + \frac{1}{2\sqrt{y}} = 0$ does not make sense if $x = 0$ (mustn't divide by 0). One tactic would be to differentiate an explicit expression for $x(y)$, e.g., $x = 1 - 2\sqrt{y} + y$; another would be to consider the limit of dx/dy as x tends to 0. In fact, both tactics require considering one-sided limits because $(0,1)$ is an endpoint of the curve (we need $0 \le x$ and $0 \le y$ for $\sqrt{x}$ and $\sqrt{y}$ to be defined, we need $x \le 1$ and $y \le 1$ because $\sqrt{x} + \sqrt{y} = 1$). Careful analysis will confirm the correctness of the locations declared above for the various tangents.

27. $0 = \dfrac{d}{dy}(1) = \dfrac{d}{dy}(xy) = \left(\dfrac{dx}{dy}\right)y + x \implies \dfrac{dx}{dy} = \dfrac{-x}{y}.$ The expression $\dfrac{-x}{y}$ is 0 when $x = 0$ but no point on the curve has an x-coordinate which is 0 (because $0 \cdot y = 1$ is never true), hence there is no tangent which is vertical. Similarly, $dy/dx = -y/x$, this is never 0 if $xy = 1$, hence the curve has no horizontal tangents.

REMARK: These conclusions are obvious from explicit expressions: $x'(y) = -1/y^2$ and $y'(x) = -1/x^2$.

29. $0 = \dfrac{d}{dy}(1) = \dfrac{d}{dy}\left(\dfrac{x^2}{a^2} - \dfrac{y^2}{b^2}\right) = \left(\dfrac{2x}{a^2}\right)\dfrac{dx}{dy} - \dfrac{2y}{b^2}$; thus $\dfrac{dx}{dy} = \left(\dfrac{a^2}{b^2}\right)\left(\dfrac{y}{x}\right)$, this is 0 if and only if $y = 0$. If $y = 0$, then $x^2/a^2 - 0 = 1$, and thus $x = \pm a$; the curve has vertical tangents at $(-a, 0)$ and $(a, 0)$. Similar computation shows $\dfrac{dy}{dx} = \left(\dfrac{b^2}{a^2}\right)\left(\dfrac{x}{y}\right)$, this equals 0 only if $x = 0$; however $|x| \ge |a|$ on the curve; hence the curve has no horizontal tangents.

31. If $y = \csc x = (\sin x)^{-1}$, then $\dfrac{dy}{dx} = \dfrac{d}{dx}\left((\sin x)^{-1}\right) = (-1)(\sin x)^{-2}\cos x$; this is 0 if and only if $\cos x = 0$, this means $x = \pi/2 + k\pi$, k an integer. Because $\sin(\pi/2 + k\pi) = (-1)^k \ne 0$, the curve has horizontal tangents at each of the points $\left(\pi/2 + k\pi, (-1)^k\right)$.

$\dfrac{dx}{dy} = \left(\dfrac{dy}{dx}\right)^{-1} = \dfrac{-\sin^2 x}{\cos x}$ could be zero only if $\sin x = 0$; that is impossible because $y = 1/\sin x$ would be undefined there. The curve has no vertical tangents.

33. $2 = x^5 + y^5 \implies 0 = 5x^4 + 5y^4\dfrac{dy}{dx} \implies \dfrac{dy}{dx} = \dfrac{-x^4}{y^4}$; thus $y'(1) = -1/1 = -1$; the line $y = (-1)(x-1) + 1 = -x + 2$ is tangent to the curve at $(1,1)$.

35. Suppose p and q are integers, $q \ne 0$.

$$y = x^{p/q} \implies y^q = x^p \implies qy^{q-1}\dfrac{dy}{dx} = px^{p-1}$$
$$\implies \dfrac{dy}{dx} = \left(\dfrac{p}{q}\right)\left(\dfrac{x^{p-1}}{y^{q-1}}\right) = \left(\dfrac{p}{q}\right)\left(\dfrac{x^p}{y^q}\right)(x^{-1}y) = \left(\dfrac{p}{q}\right)\left(\dfrac{x^p}{x^p}\right)\left(x^{-1}x^{p/q}\right) = \left(\dfrac{p}{q}\right)x^{p/q-1}. \ \blacksquare$$

37.   If $0 = 3x - 2y + x^3 - x^2 y$, then $0 = 3 - 2\dfrac{dy}{dx} + 3x^2 - \left(2xy + x^2\dfrac{dy}{dx}\right)$ and $\dfrac{dy}{dx} = \dfrac{3 + 3x^2 - 2xy}{2 + x^2}$; on this

curve $\dfrac{dy}{dx}\bigg|_{(0,0)} = \dfrac{3}{2}$. If $0 = x^2 - 2x + y^2 - 3y$, then $0 = 2x - 2 + (2y - 3)\dfrac{dy}{dx}$ and $\dfrac{dy}{dx} = \dfrac{2 - 2x}{2y - 3}$; on this second

curve $\dfrac{dy}{dx}\bigg|_{(0,0)} = \dfrac{-2}{3}$. Both curves pass through $(0,0)$ and the product of the slopes of their tangents is -1,

therefore those tangents are perpendicular. (In such a case, we will say the curves are perpendicular.)

39.
$$1 = \frac{d}{dy}(y) = \frac{d}{dy}\left((L \circ E)(y)\right) = \frac{d}{dy}L(E(y)) = L'(E(y)) \cdot E'(y) \qquad \text{[Chain Rule]}$$
$$= \frac{1}{E(y)} \cdot E'(y) \qquad \left[L'(x) = \frac{1}{x}\right].$$

Therefore $E(y) = E'(y)$ for all real y, this means $E = E'$. ∎

REMARK: You'll meet such functions, logarithm and exponential, in Chapter 6.

41. $y^2 = \dfrac{137}{\sqrt{\pi}} - x^2 + 2ax \iff x^2 - 2ax + y^2 = \dfrac{137}{\sqrt{\pi}} \iff (x - a)^2 + y^2 = a^2 + \dfrac{137}{\sqrt{\pi}}$;

the curve is a circle centered at $(a, 0)$. The normal at any point on a circle is a radial line and passes through the center of the circle. ∎

This solution uses special information about circles; the solution to Problem 40 uses a systematic procedure which works for any differentiable curve.

Although the term "Calculus" is now considered synonymous with the use of derivatives (and integrals), it was originally applied to any general system for doing calculations which routinely obtains results without specialized knowledge or tricks. For instance, we could construct a tangent or a normal to a parabola, an ellipse, or a hyperbola by knowing special (but different) features of each of those curves; given our present knowledge of derivatives, we would probably find it easier to do the same basic task (compute and evaluate an appropriate derivative) for each curve.

Solutions 2.7   *Higher-order derivatives*   (pages 177-8)

1.   If $y = 3$, then $y' = 0$, $y'' = 0$, and $y''' = 0$.

3.   If $y = 4x^2$, then $y' = 8x$, $y'' = 8$, and $y''' = 0$.

5.   If $y = \sqrt{x} = x^{1/2}$, then $y' = \left(\dfrac{1}{2}\right)x^{-1/2}$, $y'' = \left(\dfrac{1}{2}\right)\left(\dfrac{-1}{2}\right)x^{-3/2} = \left(\dfrac{-1}{4}\right)x^{-3/2}$, and

$y''' = \left(\dfrac{-1}{4}\right)\left(\dfrac{-3}{2}\right)x^{-5/2} = \left(\dfrac{3}{8}\right)x^{-5/2}$.

7.   If $y = \dfrac{1}{x^2} = x^{-2}$, then $y' = (-2)x^{-3}$, $y'' = (-2)(-3)x^{-4} = 6x^{-4}$, and $y''' = 6(-4)x^{-5} = (-24)x^{-5}$.

9.   If $y = ax^2 + bx + c$, then $y' = 2ax + b$, $y'' = 2a$, and $y''' = 0$.

11.   If $y = \dfrac{1}{(x + 1)^5} = (x + 1)^{-5}$, then $y' = (-5)(x + 1)^{-6}$, $y'' = (-5)(-6)(x + 1)^{-7} = 30(x + 1)^{-7}$, and

$y''' = (30)(-7)(x + 1)^{-8} = (-210)(x + 1)^{-8}$.

13.   If $y = \sqrt{1 - x^2} = \left(1 - x^2\right)^{1/2}$, then
$$y' = \left(\frac{1}{2}\right)\left(1 - x^2\right)^{-1/2}(-2x) = (-x)\left(1 - x^2\right)^{-1/2},$$
$$y'' = (-1)\left(1 - x^2\right)^{-1/2} + (-x)\left(\frac{-1}{2}\right)\left(1 - x^2\right)^{-3/2}(-2x) = -\left(1 - x^2\right)^{-3/2},$$
$$y''' = (-1)\left(\frac{-3}{2}\right)\left(1 - x^2\right)^{-5/2}(-2x) = (-3x)\left(1 - x^2\right)^{-5/2}.$$

15.   If $y = \sin x^2$, then
$$y' = \cos x^2 \cdot (2x) = 2x\cos x^2,$$
$$y'' = 2\cos x^2 + 2x\left(-\sin x^2 \cdot (2x)\right) = 2\cos x^2 - 4x^2\sin x^2,$$
$$y''' = 2\left(-\sin x^2 \cdot (2x)\right) - 4\left(2x\sin x^2 + x^2\cos x^2 \cdot (2x)\right) = -8x^3\cos x^2 - 12x\sin x^2.$$

17. If $y = \tan x$, then $y' = \sec^2 x$, $y'' = 2 \sec x \cdot (\sec x \tan x) = 2 \sec^2 x \tan x$, and
$y''' = 2 \left[2 \sec x \cdot (\sec x \tan x) \right] \tan x + 2 \sec^2 x \sec^2 x = 4 \sec^2 x \tan^2 x + 2 \sec^4 x$.

$$
\begin{aligned}
y = \tan x &\implies y' = \sec^2 x = 1 + \tan^2 x, \\
\text{REMARK:} \qquad &\implies y'' = 2 \tan x \, \sec^2 x = 2 \left(\tan x + \tan^3 x \right), \\
&\implies y''' = 2 \left(1 + 3 \tan^2 x \right) \sec^2 x = 2 \left(1 + 4 \tan^2 x + 3 \tan^4 x \right).
\end{aligned}
$$

19. $$
\begin{aligned}
y = \csc x &\implies y' = -\csc x \cot x, \\
&\implies y'' = -(-\csc x \cot x) \cot x - \csc x \left(-\csc^2 x \right) = \csc x \cot^2 x + \csc^3 x, \\
&\implies y''' = (-\csc x \cot x) \cot^2 x + \csc x \left(2 \cot x \left(-\csc^2 x \right) \right) + 3 \csc^2 x \left(-\csc x \cot x \right) \\
&\qquad = -\csc x \cot^3 x - 5 csc^3 x \cot x.
\end{aligned}
$$

21. $a(t) = \text{acceleration}(t) = \dfrac{d}{dt} \text{velocity}(t) = \dfrac{d}{dt} v(t) = \dfrac{d}{dt} 50t = 50$.

23. $x(t) = (t-4)(t^2 - 11t + 4) = t^3 - 15t^2 + 48t - 16$ with domain $[0, \infty)$.
   a. $v(t) = x'(t) = 3t^2 - 30t + 48 = 3(t-2)(t-8)$; the particle moves to the right when $v > 0$, this occurs when $0 \le t < 2$ or $8 < t$.
   b. $a(t) = x''(t) = 6t - 30 = 6(t-5)$; the particle is speeding up when $a > 0$, this occurs when $5 < t$.
   c. The particle is moving to the right and is speeding up when both v and a are positive, i.e., when $8 < t$.
   d. The particle is moving to the left and slowing down when both v and a are negative, i.e., when t is in $(2, 8) \cap [0, 5) = (2, 5)$.

25. $$
\begin{aligned}
f(x) &= \frac{1}{x^2 - 49} = \frac{1}{(x+7)(x-7)} = \frac{1}{14} \left[\frac{1}{x-7} - \frac{1}{x+7} \right] = \frac{1}{14} \left[(x-7)^{-1} - (x+7)^{-1} \right] \\
f'(x) &= \frac{1}{14} \left[\frac{-1}{(x-7)^2} + \frac{1}{(x+7)^2} \right] = \frac{-2x}{\left(x^2 - 49 \right)^2} \\
f''(x) &= \frac{1}{14} \left[\frac{2}{(x-7)^3} - \frac{2}{(x+7)^3} \right] = \frac{2 \left(3x^2 + 49 \right)}{\left(x^2 - 49 \right)^3} \\
f'''(x) &= \frac{1}{14} \left[\frac{-6}{(x-7)^4} + \frac{6}{(x+7)^4} \right] = \frac{-24x \left(x^2 + 49 \right)}{\left(x^2 - 49 \right)^4}
\end{aligned}
$$

27. $(uV)' = u'V + uV'$.
$(uV)'' = (u'V + uV')' = (u'V)' + (uV')' = (u''V + u'V') + (u'V' + uV'') = u''V + 2u'V' + uV''$.
$(uV)''' = (u''V + 2u'V' + uV'')' = (u''V)' + 2(u'V')' + (uV'')'$
$\qquad = (u'''V + u''V') + 2(u''V' + u'V'') + (u'V'' + uV''') = u'''V + 3u''V' + 3u'V'' + uV'''$.

29. Let $\mathcal{P}(n)$ represent the following assertion: $\dfrac{d^n}{dx^n} x^n = n!$.

$\mathcal{P}(1)$ is obviously true because $\dfrac{d^1}{dx^1} x^1 = \dfrac{d}{dx} x = 1 = 1!$. Suppose $\mathcal{P}(k)$ were true.
$$
\begin{aligned}
\frac{d^{k+1}}{dx^{k+1}} x^{k+1} = \frac{d^k}{dx^k} \left(\frac{d}{dx} x^{k+1} \right) = \frac{d^k}{dx^k} \left((k+1)x^k \right) = (k+1) \left(\frac{d^k}{dx^k} x^k \right) \\
= (k+1)(k!) \qquad\qquad \mathcal{P}(k) \\
= (k+1)!\,.
\end{aligned}
$$
Hence $\mathcal{P}(k{+}1)$ would also be true.
$\mathcal{P}(1)$ is true and $\mathcal{P}(k)$ implies $\mathcal{P}(k{+}1)$, therefore $\mathcal{P}(n)$ is true for each positive integer n. ∎

31. $\dfrac{d}{dx} \sin x = \cos x$, $\quad \dfrac{d^2}{dx^2} \sin x = \dfrac{d}{dx} (\cos x) = -\sin x$, $\quad \dfrac{d^3}{dx^3} \sin x = \dfrac{d}{dx} (-\sin x) = -\cos x$.

If we use the convention $\dfrac{d^0}{dx^0} f(x) = f(x)$, then we have just proved case $k = 0$ of the statement

(∗) $$
\frac{d^n}{dx^n} \sin x = \begin{cases}
\sin x & \text{if } n = 4k, \\
\cos x & \text{if } n = 4k + 1, \\
-\sin x & \text{if } n = 4k + 2, \\
-\cos x & \text{if } n = 4k + 3,
\end{cases}
$$

where k is a nonnegative integer. Observe that $\dfrac{d^4}{dx^4}\sin x = \dfrac{d}{dx}\left(\dfrac{d^3}{dx^3}\sin x\right) = \dfrac{d}{dx}(-\cos x) = \sin x$ and

$\dfrac{d^{n+4}}{dx^{n+4}}\sin x = \dfrac{d^n}{dx^n}\left(\dfrac{d^4}{dx^4}\sin x\right) = \dfrac{d^n}{dx^n}\sin x$. Therefore, if the statement is true for some value of k, it is also true for $k+1$ because $4(k+1)+r = (4k+r)+4$ for each of the choices $r \in \{0,1,2,3\}$. The principle of mathematical induction guarantees us that statement (*) is true for each positive integer k and each n. ∎

REMARK: $\quad \dfrac{d^n}{dx^n}\sin x = \sin\left(x + n\dfrac{\pi}{2}\right)$.

$\boxed{\text{Proof}}$ $\sin\left((x+a)+\dfrac{\pi}{2}\right) = \sin(x+a)\cos\dfrac{\pi}{2} + \cos(x+a)\sin\dfrac{\pi}{2} = \cos(x+a) = \dfrac{d}{dx}\sin(x+a)$.

If $\dfrac{d^k}{dx^k}\sin x = \sin\left(x + k\dfrac{\pi}{2}\right)$, then

$$\dfrac{d^{k+1}}{dx^{k+1}}\sin x = \dfrac{d}{dx}\left(\dfrac{d^k}{dx^k}\sin x\right) = \dfrac{d}{dx}\left(\sin\left(x + k\dfrac{\pi}{2}\right)\right) = \sin\left(x + k\dfrac{\pi}{2} + \dfrac{\pi}{2}\right)\sin\left(x + (k+1)\dfrac{\pi}{2}\right),$$

therefore the statement is true for each positive integer. ∎

33.   Let $\mathcal{P}(n)$ represent the following assertion: $\quad \dfrac{d^n}{dx^n}\left(\dfrac{1}{x}\right) = \dfrac{(-1)^n\, n!}{x^{n+1}}$.

$\mathcal{P}(1)$ is obviously true because $\dfrac{d^1}{dx^1}\left(\dfrac{1}{x}\right) = \dfrac{d}{dx}\left(\dfrac{1}{x}\right) = \dfrac{d}{dx}(x^{-1}) = (-1)x^{-2} = \dfrac{(-1)^1\, 1!}{x^{1+1}}$.

Suppose $\mathcal{P}(k)$ were true.

$$\begin{aligned}
\dfrac{d^{k+1}}{dx^{k+1}}\left(\dfrac{1}{x}\right) &= \dfrac{d}{dx}\left(\dfrac{d^k}{dx^k}\left(\dfrac{1}{x}\right)\right) \\[4pt]
&= \dfrac{d}{dx}\left(\dfrac{(-1)^k\, k!}{x^{k+1}}\right) \\[4pt]
&= (-1)^k\, k!\,\dfrac{d}{dx}\left(x^{-(k+1)}\right) = (-1)^k\, k!\left(-(k+1)\,x^{-(k+1)-1}\right) = \dfrac{(-1)^{k+1}(k+1)!}{x^{(k+1)+1}}.
\end{aligned}$$ $\mathcal{P}(k)$

Hence $\mathcal{P}(k+1)$ would also be true.

$\mathcal{P}(1)$ is true and $\mathcal{P}(k)$ implies $\mathcal{P}(k+1)$, therefore $\mathcal{P}(n)$ is true for each positive integer n. ∎

35.   Suppose f and g are smooth functions and n is a positive integer.

$$(f\cdot g)^{(n)} = \sum_{k=0}^{n}\binom{n}{k}f^{(n-k)}\cdot g^{(k)} \qquad\qquad \mathcal{P}(n)$$

where $\dbinom{n}{k} = \dfrac{n!}{k!\,(n-k)!}$ and $H^{(0)} \equiv H$ for every function H. (This result is known as Leibniz's rule.)

$\boxed{\text{Proof}}$ For small n we have the following familiar facts:

$$\begin{aligned}
(f\cdot g)' &= f'g + fg' & \mathcal{P}(1) \\
(f\cdot g)'' &= f''g + 2f'g' + fg'' & \mathcal{P}(2) \\
(f\cdot g)''' &= f'''g + 3f''g' + 3f'g'' + fg''' & \mathcal{P}(3) \\
(f\cdot g)^{(4)} &= f^{(4)}g + 4f'''g' + 6f''g'' + 4f'g''' + fg^{(4)} & \mathcal{P}(4)
\end{aligned}$$

Suppose statement $\mathcal{P}(n)$ were true for a particular choice of n.

$$\begin{aligned}
(f\cdot g)^{(n+1)} &= \left[(f\cdot g)^{(n)}\right]' & \dfrac{d^{n+1}}{dx^{n+1}} = \dfrac{d}{dx}\dfrac{d^n}{dx^n} \\[4pt]
&= \left[\sum_{k=0}^{n}\binom{n}{k}f^{(n-k)}\cdot g^{(k)}\right]' & \mathcal{P}(n) \\[4pt]
&= \sum_{k=0}^{n}\binom{n}{k}\left[f^{(n-k)}\cdot g^{(k)}\right]' & (u+v)' = u' + v' \\[4pt]
&= \sum_{k=0}^{n}\binom{n}{k}\left[\left(f^{(n-k)}\right)'\cdot g^{(k)} + f^{(n-k)}\cdot\left(g^{(k)}\right)'\right] & (uv)' = u'v + uv' \\[4pt]
&= \sum_{k=0}^{n}\binom{n}{k}\left[f^{(n-k+1)}\cdot g^{(k)} + f^{(n-k)}\cdot g^{(k+1)}\right] & \dfrac{d}{dx}\dfrac{d^i}{dx^i} = \dfrac{d^{i+1}}{dx^{i+1}}
\end{aligned}$$

$$= \sum_{k=0}^{n} \binom{n}{k} f^{(n-k+1)} \cdot g^{(k)} + \sum_{k=0}^{n} \binom{n}{k} f^{(n-k)} \cdot g^{(k+1)} \qquad \sum_k (a_k + b_k) = \sum_k a_k + \sum_k b_k$$

$$= \sum_{k=0}^{n} \binom{n}{k} f^{(n+1-k)} \cdot g^{(k)} + \sum_{j=1}^{n+1} \binom{n}{j-1} f^{(n+1-j)} \cdot g^{(j)} \qquad j = k+1$$

$$= \binom{n}{0} f^{(n+1)} g^{(0)} + \sum_{i=1}^{n} \left[\binom{n}{i} + \binom{n}{i-1} \right] f^{(n+1-i)} g^{(i)} + \binom{n}{n} f^{(0)} g^{(n+1)}$$

$$= \binom{n+1}{0} f^{(n+1)} g^{(0)} + \sum_{i=1}^{n} \binom{n+1}{i} f^{(n+1-i)} g^{(i)} + \binom{n+1}{n+1} f^{(0)} g^{(n+1)} \qquad \binom{n+1}{i} = \binom{n}{i} + \binom{n}{i-1}$$

$$= \sum_{i=0}^{n+1} \binom{n+1}{i} f^{(n+1-i)} g^{(i)}.$$

Hence $\mathcal{P}(n+1)$ would also be true.

$\quad$ $\mathcal{P}(1)$ is true and $\mathcal{P}(n)$ implies $\mathcal{P}(n+1)$, therefore $\mathcal{P}(n)$ is true for each positive integer n. ∎

Solutions 2.R *Review* (page 178-9)

1. $\quad \dfrac{dy}{dx} = \dfrac{d}{dx}(3x+4) = 3$

3. $\quad 0 = \dfrac{d}{dx} 3 = \dfrac{d}{dx}(x+y+xy) = 1 + \dfrac{dy}{dx} + y + x\dfrac{dy}{dx}$, therefore $(1+x)\dfrac{dy}{dx} = -(1+y)$ and, provided $x \neq 1$,

$\dfrac{dy}{dx} = \dfrac{-(1+y)}{1+x}$.

REMARK: $\quad 3 = x + y + xy$ if and only if $3 - x = (1+x)y$, solving explicitly for y yields $y = \dfrac{3-x}{1+x}$ and

$\dfrac{dy}{dx} = -4(1+x)^{-2}$.

5. $\quad \dfrac{d}{dx}\left((3x^2+1)\sqrt{x+1}\right) = 3(2x)\sqrt{x+1} + (3x^2+1)\dfrac{1}{2\sqrt{x+1}} = \dfrac{15x^2 + 12x + 1}{2\sqrt{x+1}}.$

7. $\quad 0 = \dfrac{d}{dx} 4 = \dfrac{d}{dx}\left(x^{3/4} + y^{3/4}\right) = \dfrac{3}{4}x^{-1/4} + \dfrac{3}{4}y^{-1/4}\dfrac{dy}{dx}$, thus $\dfrac{dy}{dx} = \dfrac{-(3/4)x^{-1/4}}{(3/4)y^{-1/4}} = -\dfrac{y^{1/4}}{x^{1/4}} = -\left(\dfrac{y}{x}\right)^{1/4}.$

9. $\quad \dfrac{dy}{dx} = \dfrac{d}{dx}\left(\dfrac{x+2}{x-3}\right)^{1/3} = \dfrac{1}{3}\left(\dfrac{x+2}{x-3}\right)^{-2/3}\left(\dfrac{(x-3)(1) - (x+2)(1)}{(x-3)^2}\right) = \left(\dfrac{-5}{3}\right)(x+2)^{-2/3}(x-3)^{-4/3}.$

11. $\quad \dfrac{d}{dx}\left(\dfrac{x^2 - 2x + 3}{x^3 + 5x - 6}\right) = \dfrac{(x^3 + 5x - 6)(2x - 2) - (x^2 - 2x + 3)(3x^2 + 5)}{(x^3 + 5x - 6)^2} = \dfrac{-(x+1)\left(x^3 - 5x^2 + 9x + 3\right)}{(x-1)^2 \left(x^2 + x + 6\right)^2}.$

13. $\quad \dfrac{d}{dx}\left((x^3 - 3)^{5/6} \cdot (x^3 + 4)^{6/7}\right)$

$\quad = \left(\dfrac{5}{6}\right)(x^3 - 3)^{-1/6}(3x^2) \cdot (x^3 + 4)^{6/7} + (x^3 - 3)^{5/6} \cdot \left(\dfrac{6}{7}\right)(x^3 + 4)^{-1/7}(3x^2)$

$\quad = \dfrac{x^2 \left(71x^3 + 32\right)}{14 \left(x^3 - 3\right)^{1/6} \left(x^3 + 4\right)^{1/7}}.$

15. $\quad 1 + \left(\dfrac{1}{2}\right)(x+y)^{-1/2}\left(1 + \dfrac{dy}{dx}\right) = \dfrac{d}{dx}(x + \sqrt{x+y}) = \dfrac{d}{dx}y^{1/3} = \left(\dfrac{1}{3}\right)y^{-2/3}\dfrac{dy}{dx};$

multiply by $6y^{2/3}(x+y)^{1/2}$ to get the relation $6y^{2/3}\sqrt{x+y} + 3y^{2/3}\left(1 + \dfrac{dy}{dx}\right) = 2\sqrt{x+y}\dfrac{dy}{dx}.$

Thus $6y^{2/3}\sqrt{x+y} + 3y^{2/3} = \left(2\sqrt{x+y} - 3y^{2/3}\right)\dfrac{dy}{dx}$ and $\dfrac{dy}{dx} = \dfrac{3y^{2/3}\left(2\sqrt{x+y} + 1\right)}{2\sqrt{x+y} - 3y^{2/3}}.$

17. $\dfrac{d}{dx}\left(1+x^3\right)^{\sqrt{2}} = \sqrt{2}\left(1+x^3\right)^{\sqrt{2}-1}\left(3x^2\right).$

19. $\dfrac{d}{dx}\left(\dfrac{\left(x^3-1\right)^{2/3}}{\left(x^5+3\right)^{3/4}}\right) = \dfrac{\left(x^5+3\right)^{3/4}\cdot(2/3)\left(x^3-1\right)^{-1/3}\left(3x^2\right) - \left(x^3-1\right)^{2/3}\cdot(3/4)\left(x^5+3\right)^{-1/4}\left(5x^4\right)}{\left(x^5+3\right)^{3/2}}$

$$= \dfrac{x^2\left(-7x^5+15x^2+24\right)}{4\left(x^3-1\right)^{1/3}\left(x^5+3\right)^{7/4}}.$$

21. $\dfrac{d}{dx}\sin x^3 = \left(\cos x^3\right)\left(3x^2\right) = 3x^2\cos x^3.$

23. $\dfrac{d}{dx}\left(\tan x\cdot\sec x\right) = \sec^2 x\cdot\sec x + \tan x\cdot\sec x\tan x = \sec^3 x + \sec x\tan^2 x.$

25. $\dfrac{d}{dx}\left(x^3-x+4\right) = 3x^2-1$; hence $y'(1) = 3-1 = 2$; the tangent at $(1,4)$ has slope 2 and equation $y = 2(x-1)+4 = 2x+2$; the normal at $(1,4)$ has slope $-1/2$ and equation $y = (-1/2)(x-1)+4 = (-1/2)(x-9)$.

27. $\dfrac{d}{dx}\left(\left(x^2-4\right)^2\left(\sqrt{x}+3\right)^{1/2}\right) = 2\left(x^2-4\right)(2x)\left(\sqrt{x}+3\right)^{1/2} + \left(x^2-4\right)^2\left(\dfrac{1}{2}\right)\left(\sqrt{x}+3\right)^{-1/2}\dfrac{1}{2\sqrt{x}};$

thus $y'(1) = 2(-3)(2)(4)^{1/2} + (-3)^2(1/2)(4)^{-1/2}(1/2) = -24+(9/8) = -183/8$; the tangent at $(1,18)$ has slope $-183/8$ and equation $y = (-183/8)(x-1)+18 = (-183/8)x+(327/8)$; the normal at $(1,18)$ has slope $-1/(-183/8) = 8/183$ and equation $y = (8/183)(x-1)+18 = (8x+3286)/183.$

29. $\dfrac{dy}{dx} = \dfrac{d}{dx}\cos\sqrt{x} = \dfrac{-\sin\sqrt{x}}{2\sqrt{x}}$; therefore $y'\left(\dfrac{\pi^2}{9}\right) = \dfrac{-\sin(\pi/3)}{2\pi/3} = \left(\dfrac{-\sqrt{3}}{2}\right)\left(\dfrac{3}{2\pi}\right) = \dfrac{-3\sqrt{3}}{4\pi}$; the tangent

at $\left(\pi^2/9, 1/2\right)$ has slope $-3\sqrt{3}/4\pi$ and equation $y = \left(-3\sqrt{3}/4\pi\right)\left(x-\pi^2/9\right)+(1/2)$; the normal has slope $-1/(-3\sqrt{3}/4\pi) = 4\pi/3\sqrt{3}$ and equation $y = \left(4\pi/3\sqrt{3}\right)\left(x-\pi^2/9\right)+(1/2).$

31. If $y = x^7-7x^6+x^3+3$, then $y' = 7x^6-42x^5+3x^2$, $y'' = 42x^5-210x^4+6x$, and $y''' = 210x^4-840x^3+6.$

33. If $y = 1/(1+x) = (1+x)^{-1}$, then $y' = (-1)(1+x)^{-2}$, $y'' = (-1)(-2)(1+x)^{-3} = 2(1+x)^{-3}$, and $y''' = 2(-3)(1+x)^{-4} = (-6)(1+x)^{-4}.$

35. $y = \dfrac{x^2-3}{x^2+5} = 1-\dfrac{8}{x^2+5} = 1-8(x^2+5)^{-1}$

$y' = (-8)(-1)(x^2+5)^{-2}(2x) = 16x\left(x^2+5\right)^{-2}$

$y'' = 16(1)(x^2+5)^{-2}+16x(-2)(x^2+5)^{-3}(2x) = 16\left(5-3x^2\right)\left(x^2+5\right)^{-3}$

$y''' = 16\left((-6x)(x^2+5)^{-3}+\left(5-3x^2\right)(-3)(x^2+5)^{-4}(2x)\right) = 192x\left(x^2-5\right)\left(x^2+5\right)^{-4}.$

37. $y = \dfrac{\cos x}{x} = x^{-1}\cos x$

$y' = -x^{-2}\cos x + x^{-1}(-\sin x) = -x^{-2}\cos x - x^{-1}\sin x$

$y'' = -(-2)x^{-3}\cos x - x^{-2}(-\sin x) - (-1)x^{-2}\sin x - x^{-1}(\cos x) = \left(2x^{-3}-x^{-1}\right)\cos x + 2x^{-2}\sin x$

$y''' = \left(-6x^{-4}+x^{-2}\right)\cos x + \left(2x^{-3}-x^{-1}\right)(-\sin x) + (-4)x^{-3}\sin x + 2x^{-2}(\cos x)$

$= \left(-6x^{-4}+3x^{-2}\right)\cos x + \left(-6x^{-3}+x^{-1}\right)\sin x.$

Solutions 2.C   *Computer Exercises*   (page 179)

1. One computer algebra system may display the various derivatives of $\tan x$ as sums of terms consisting of powers of $\sin x$ divided by powers of $\cos x$; another system may display them as sums of products of powers of $\sec x$ and $\tan x$. Closer examination (factoring and substitution may help at this stage) will show that each derivative can be written as the product of $\sec^2 x$ and a polynomial in $\tan x$; more precisely, for the n-th derivative of $\tan x$, the polynomial is of degree $n-1$ and the various powers have the same parity, i.e., all powers are even or all are odd. (Since $\sec^2 x = 1+\tan^2 x$, the previous sentence also implies the k-th derivative of $\tan$ can be expressed as a polynomial function of $\tan$ with degree $2+(k-1) = k+1$.)

Let $y = \tan x$ and examine several derivatives of tan.

$$y' = \sec^2 x = 1 + \tan^2 x = 1 + y^2,$$

$$y'' = 2\sec x(\sec x \tan x) = 2\tan x \sec^2 x = 2yy' = 2y\left(1 + y^2\right) = 2\left(y + y^3\right),$$

$$y''' = 2\left(1 + 3y^2\right)y' = 2\left(1 + 3\tan^2 x\right)\sec^2 x = 2\left(1 + 3y^2\right)\left(1 + y^2\right) = 2\left(1 + 4\tan^2 x + 3\tan^4 x\right),$$

$$y^{(4)} = 2\left(8\tan x + 12\tan^3 x\right)\sec^2 x = 8\left(2\tan x + 3\tan^3 x\right)\left(1 + \tan^2 x\right) = 8\left(2\tan x + 5\tan^3 x + 3\tan^5 x\right),$$

$$y^{(5)} = 8\left(2 + 15\tan^2 x + 15\tan^4 x\right)\sec^2 x = \ldots$$

If $y^{(k)}$ has the form $\sec^2 x \, P_{k-1}(\tan x) = P_{k+1}(\tan x)$, then $y^{(k+1)}$ has form $P_k(\tan x)\sec^2 x = P_{k+2}(\tan x)$. ∎

3.   If $f(x) = \dfrac{1}{1+x^2}$, then $f^{(n)}(0) = \begin{cases} (-1)^{n/2}\, n! & \text{if } n \text{ is even,} \\ 0 & \text{if } n \text{ is odd.} \end{cases}$

5.   If $f(x) = x^2 + \sin x$, then $f'(x) = 2x + \cos x$. Because $|\cos x| \le 1$, all solutions to $0 = f'(x)$ must lie in the interval $[-1/2, 1/2]$. A numerical search begun with initial guess $x_0 = -.5$ leads to approximate solution .450184.

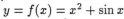

$y = f(x) = x^2 + \sin x$ $y = f'(x) = 2x + \cos x$ $y = f'(x) = 2x + \cos x$

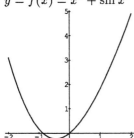 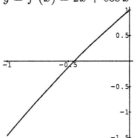 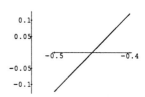

3

Applications of the Derivative

1. $6 = xy$ implies $0 = \dfrac{d}{dt}(6) = \dfrac{d}{dt}(xy) = \left(\dfrac{dx}{dt}\right)y + x\left(\dfrac{dy}{dt}\right)$. If $x = 3$, then $y = \dfrac{6}{x} = \dfrac{6}{3} = 2$; if also

$\dfrac{dx}{dt} = 5$, then $0 = (5)(2) + (3)\dfrac{dy}{dt}$; thus $\dfrac{dy}{dt} = \dfrac{-10}{3}$.

REMARK: $6 = xy$ implies $y = 6x^{-1}$ and $\frac{dy}{dt} = -6x^{-2}\frac{dx}{dt} = -6(3)^{-2}(5) = -10/3$.

3.   The Pythagorean theorem guarantees $10^2 = x^2 + y^2$; differentiate this relation with respect to t: $0 = 2x\dfrac{dx}{dt} + 2y\dfrac{dy}{dt}$. We're given $\frac{dx}{dt} = 2$ ft/sec and asked to find $\frac{dy}{dt}$ when $x = 8$. Note that $y = \sqrt{100 - x^2} = \sqrt{100 - 8^2} = \sqrt{36} = 6$ when $x = 8$. Thus $0 = 2(8)(2) + 2(6)\dfrac{dy}{dt}$ and $\dfrac{dy}{dt} = \dfrac{-32}{12} = \dfrac{-8}{3}$ ft/sec.

REMARK: $\frac{dy}{dt} < 0$ indicates the top of the ladder is moving down (at speed 8/3 ft/sec.)

5.   Let $x(t)$ be the distance of the man from the lamp-post and $s(t)$ be the length of his shadow. By similar triangles, $\dfrac{x + s}{12} = \dfrac{s}{6}$; thus $6x + 6s = 12s$, and $x = s$. Therefore $\dfrac{dx}{dt} = \dfrac{ds}{dt}$. If the man walks away from the lamp-post at 5 ft/sec, then $\dfrac{ds}{dt} = \dfrac{dx}{dt} = 5$ ft/sec. (Note that $\dfrac{ds}{dt}$ does not vary with x.)

7.   Let $R(t)$ be the distance of the runner from third base, $B(t)$ be the distance of the ball from third base, and $s(t)$ be the distance between the ball and the runner. Assuming both runner and ball are directly on the base paths, we use the Pythagorean theorem to infer $s^2 = B^2 + R^2$. Therefore $2s\dfrac{ds}{dt} = 2B\dfrac{dB}{dt} + 2R\dfrac{dR}{dt}$ and $\dfrac{ds}{dt} = \dfrac{BB' + RR'}{s}$. If the runner approaches third at a speed of 25 ft/sec, then $R' = -25$ ft/sec; similarly, if the catcher throws the ball at a speed of 120 ft/sec, then $B' = -120$ ft/sec. Therefore, when $R = 30$ ft and $B = 90$ ft,

$$\dfrac{ds}{dt} = \dfrac{90(-120) + 30(-25)}{\sqrt{90^2 + 30^2}} = \dfrac{-11550}{\sqrt{9000}} \approx -121.75 \text{ ft/sec.}$$

9.   Let $s(t)$ be the distance between the boats t hours after 2 PM; then $s(t)^2 = (15t)^2 + (100 - 20t)^2$. Therefore $\dfrac{ds}{dt} = \dfrac{(15t)(15) + (100 - 20t)(-20)}{s(t)}$ and $s'(3\text{ hr}) = \dfrac{-125}{\sqrt{3625}} \approx -2.076$ km/hr; the distance is decreasing.

11.   Because a cylinder of radius r and height h has volume $V = \pi r^2 h$, we have $h = V/(\pi r^2)$ and $\dfrac{dh}{dt} = \dfrac{dV}{dt}\Big/(\pi r^2) = (10/1000)\Big/(\pi\, 2^2) \approx 0.000796$ m/min.

13.   If the circular wave has radius r, its circumference is $C = 2\pi r$. Thus $\dfrac{dC}{dt} = 2\pi\dfrac{dr}{dt}$. If we're given $\dfrac{dr}{dt} = 25$ cm/sec, then $\dfrac{dC}{dt} = 2\pi(25) = 50\pi \approx 157.08$ cm/sec. (Note that $\frac{dC}{dt}$ does not vary with r.)

15. $\dfrac{dV}{dt} = \dfrac{d}{dr}\left(\dfrac{4}{3}\pi r^3\right) = 4\pi r^2\dfrac{dr}{dt}$; therefore when $r = 3$ cm, $\dfrac{dr}{dt} = \dfrac{-8\pi}{4\pi 3^2} = -\dfrac{2}{9}\dfrac{\text{cm}}{\text{hr}}$

17.   If $PV^{5/3}$ is constant, then

$$\left(\frac{dP}{dt}\right)V^{5/3} + P\left(\frac{5}{3}\right)V^{2/3}\left(\frac{dV}{dt}\right) = 0$$

and

$$\frac{dV}{dt} = \frac{-V^{5/3}}{P(5/3)V^{2/3}}\left(\frac{dP}{dt}\right) = \frac{-V}{(5/3)P}\left(\frac{dP}{dt}\right).$$

Given $V = 18$ m^3, $P = 0.3$ kg/m^2, and $\frac{dP}{dt} = 0.01$ kg/m^2 sec, we compute $\frac{dV}{dt} = \frac{-18}{(5/3)(0.3)}(0.01) = -0.36$ m^3/ sec. Because $\frac{dv}{dt} < 0$, we infer that V is decreasing.

19.   Suppose (x, y) is a point on the rim of the wheel and θ is the angle between the radius to (x, y) and the positive x-axis. Then $x = 13\cos\theta$ and $y = 13\sin\theta$. If the point revolves counterclockwise at 10 rpm, then $\frac{d\theta}{dt} = 10(2\pi)$ radians per minute. $\frac{dx}{dt} = 13(-\sin\theta)\frac{d\theta}{dt} = (-20\pi)(13\sin\theta) = -20\pi y$ and $\frac{dy}{dt} = (13\cos\theta)\frac{d\theta}{dt} = (20\pi)(13\cos\theta) = 20\pi x$. Thus, when $x = 12$ and $y = 5$, $\frac{dx}{dt} = -100\pi$ cm/min and $\frac{dy}{dt} = 240\pi$ cm/min.

21. a.   If $z = 2$ ft, $r = 4$ ft and $P = 10,000$ lb, then $\sigma_z = \frac{3(10000)}{2\pi(2^2)}\left[1 + \left(\frac{4}{2}\right)^2\right]^{2.5} \approx 66,727.7\frac{\text{lb}}{\text{ft}^2}$.

   b. $\frac{d\sigma_z}{dt} = \frac{(2.5)(3750)}{\pi}\left[1 + \left(\frac{r}{2}\right)^2\right]^{1.5}\left(\frac{r}{2}\right)\frac{dr}{dt} = -26,226.3$ lb/ft^2/hr when $r = 3$ and $\frac{dr}{dt} = -1$.

23.   The distance from $(2, 5)$ to (x, x^2) is $s = \sqrt{(x-2)^2 + (x^2-5)^2} = \sqrt{29 - 4x - 9x^2 + x^4}$. Given the information $\frac{dx}{dt} = 3$,

$$\frac{ds}{dt} = \frac{-4 - 18x + 4x^3}{2\sqrt{29 - 4x - 9x^2 + x^4}}\left(\frac{dx}{dt}\right) = \left(\frac{3}{2}\right)\frac{(-4 - 18x + 4x^3)}{\sqrt{29 - 4x - 9x^2 + x^4}}.$$

At $(1, 1)$,

$$\frac{ds}{dt} = \left(\frac{3}{2}\right)\frac{-4 + 18 - 4}{\sqrt{29 + 4 - 9 + 1}} = \left(\frac{3}{2}\right)\frac{10}{\sqrt{25}} = 3.$$

At $(3, 9)$,

$$\frac{ds}{dt} = \left(\frac{3}{2}\right)\frac{-4 - 54 + 108}{\sqrt{29 - 12 - 81 + 81}} = \left(\frac{3}{2}\right)\frac{50}{\sqrt{17}} = \frac{75}{\sqrt{17}}.$$

25.   Establish a coordinate system with the light at the origin, the screen on the line $x = 25$, and the center of the sphere on the x-axis, x m from light. The line through the light and the top of the shadow is tangent to the sphere. Since tangents to a sphere are perpendicular to a radius, this line has slope equal to $\tan\theta = 1/\sqrt{x^2 - 1}$. Therefore $H/2 = (1/\sqrt{x^2 - 1})25$ or the height of the shadow is $H = 50/\sqrt{x^2 - 1} = 50(x^2 - 1)^{-1/2}$. If $\frac{dx}{dt} = 3$ m/ sec, then $\frac{dH}{dt} = 50\left(-\frac{1}{2}\right)(x^2 - 1)^{-3/2}(2x)\frac{dx}{dt} = -150x(x^2 - 1)^{-3/2}$ When the center of the sphere is 15 m from the screen, $x = 25 - 15 = 10$ m and $\frac{dH}{dt} = -1500(99)^{-3/2} \approx -1.52$ m/ sec.

Solutions 3.2   *Curve sketching I: Increasing & decreasing, first-deriv test*   (pages 196-7)

1.   If $f(x) = x^2 + x - 30$, then $f'(x) = 2x + 1$. $f'(x) > 0$ if $x > -1/2$, so f is increasing on $(-1/2, \infty)$; $f'(x) < 0$ if $x < -1/2$, thus f decreases on $(-\infty, -1/2)$; $f'(x)$ is always defined and $f'(-1/2) = 0$, hence $-1/2$ is the only critical point of f. The y-intercept is $f(0) = -30$; because $x^2 + x - 30 = (x + 6)(x - 5)$, -6 and 5 are the x-intercepts. Use part (ii) of Theorem 3 to infer that f has a local minimum at $-1/2$, it is $-121/4$.

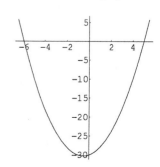

3.   If $f(x) = 4x^2 - 8$, then $f'(x) = 8x$. $f'(x) < 0$ if $x < 0$ so f is decreasing on $(-\infty, 0)$; $f'(x) > 0$ if $x > 0$ so f is increasing on $(0, \infty)$; $f'(x)$ is always defined and is 0 only at $x = 0$, the unique critical point of f. The y-intercept is $f(0) = -8$; the x-intercepts are $\pm\sqrt{8/4} = \pm\sqrt{2}$. Part (ii) of Theorem 3 implies f has a local minimum of -8 at $x = 0$; indeed, that is the global minimum.

3: $y = 4x^2 - 8$

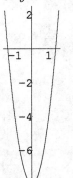

5: $y = 2x^2 + 4x + 6$

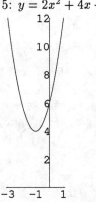

5.   If $f(x) = 2x^2 + 4x + 6 = 2\left((x + 1)^2 + 2\right)$, then $f'(x) = 4x + 4 = 4(x + 1)$. We see that f' is negative on $(-\infty, -1)$, zero at -1, and positive on $(-1, \infty)$. Therefore $f(x)$ decreases on $(-\infty, -1)$, reaches its global minimum value of 4 at $x = -1$, and then increases on $(-1, \infty)$. The y-intercept is $f(0) = 6$, there are no x-intercepts.

7.   If $f(x) = x^3 + x$, then $f'(x) = 3x^2 + 1$. Because $f'(x) \geq 1$, f is always increasing, it has no critical points and no local extrema. The y-intercept is $f(0) = 0$; because $f(x) = x\left(x^2 + 1\right)$, the unique x-intercept is 0.

7: $y = x^3 + x$

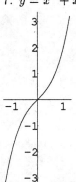

9: $y = x^3 - x^2$

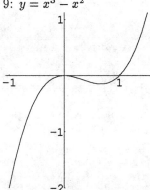

9.   Let $g(x) = x^3 - x^2 = x^2(x - 1)$, then $g'(x) = 3x^2 - 2x = x(3x - 2)$. g' is zero at 0 and $2/3$, negative between 0 and $2/3$, and positive everywhere else; g increases on $(-\infty, 0)$, has a local maximum of 0 at 0, decreases on $(0, 2/3)$, has a local minimum of $-4/27$ at $2/3$, and increases on $(2/3, \infty)$. The y-intercept is $g(0) = 0$; the x-intercepts are 0 and 1.

11.   If $f(x) = x^4 - 18x^2$, then $f'(x) = 4x^3 - 36x = 4(x + 3)x(x - 3)$. The following table summarizes our analysis of f' and f: ($-$ means a term is negative, $+$ means it is positive)

interval or point	$x + 3$	x	$x - 3$	f'	f
$(-\infty, -3)$	$-$	$-$	$-$	$-$	decreasing
-3	0	$-$	$-$	0	local min
$(-3, 0)$	$+$	$-$	$-$	$+$	increasing
0	$+$	0	$-$	0	local max
$(0, 3)$	$+$	$+$	$-$	$-$	decreasing
3	$+$	$+$	0	0	local min
$(3, \infty)$	$+$	$+$	$+$	$+$	increasing

The locally minimum value of f at -3 and at 3 is -81; the locally maximum value at 0 is 0. The y-intercept is 0; the x-intercepts are 0 and $\pm 3\sqrt{2}$.

11: $y = x^4 - 18x^2$

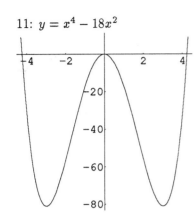

13: $y = |2x + 7|$

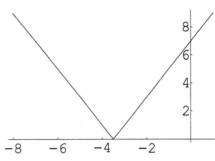

13. If $f(x) = |2x + 7|$, then $f'(x) = \begin{cases} -2 & \text{if } x < -7/2, \\ \text{undefined} & \text{if } x = -7/2, \\ 2 & \text{if } x > -7/2. \end{cases}$ f increases on $(-7/2, \infty)$ and decreases on $(-\infty, -7/2)$. $-7/2$, where f' is undefined, is the only critical point for f; the minimum value of f is $f(-7/2) = 0$. The y-intercept is 7, the x-intercept is $-7/2$.

15. If $g(x) = (x + 1)^{1/3}$, then $g'(x) = \frac{1}{3}(x + 1)^{-2/3}$. If $x \neq -1$, then $g'(x) = \frac{1}{3}\left[(x + 1)^{-1/3}\right]^2 > 0$. Thus g is increasing on $(-\infty, -1) \cup (-1, \infty)$. In fact g is increasing on $(-\infty, \infty)$ because g is continuous at -1 (although not differentiable there). Because $g'(x)$ is never zero, -1 is the only critical point of g. There are no local extrema; $1 = g(0)$ is the y-intercept; $-1 = 0^3 - 1$ is the x-intercept.

15: $y = (x + 1)^{1/3}$

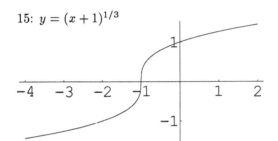

17: $y = \sin 4x$

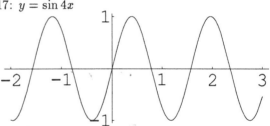

17. If $f(x) = \sin 4x$, then $f'(x) = 4 \cos 4x$. $f' > 0$ and f increases on $\left(-\frac{\pi}{8}, \frac{\pi}{8}\right)$ and on each interval $\left(-\frac{\pi}{8} + k\frac{\pi}{2}, \frac{\pi}{8} + k\frac{\pi}{2}\right)$ where k is an integer. $f' < 0$ and f decreases on $\left(\frac{\pi}{8}, \frac{3\pi}{8}\right)$ and on each interval $\left(\frac{\pi}{8} + k\frac{\pi}{2}, \frac{3\pi}{8} + k\frac{\pi}{2}\right)$. $f'(x) = 0$ if and only if $x = \frac{\pi}{8} + k\frac{\pi}{4}$; these are the critical points for f. The maximum value, 1, of $f(x)$ is obtained if $x = \frac{\pi}{8} + k\frac{\pi}{2}$; the minimum value, -1, is obtained if $x = -\frac{\pi}{8} + k\frac{\pi}{2}$. The y-intercept is 0; the x-intercepts are $k\frac{\pi}{4}$.

19. Let $f(x) = \sqrt{x + 1}$, then $f'(x) = \frac{1}{2\sqrt{x + 1}}$.

a. $f'(x) > 0$ whenever $x > -1$ because $\sqrt{x + 1} > 0$ if $x > -1$, therefore f is increasing on $(-1, \infty)$. ∎

b. $f'(c) = 0$ implies $\frac{1}{2\sqrt{c + 1}} = 0$ and $1 = 0 \cdot 2\sqrt{c + 1} = 0$. Because $1 \neq 0$, there can be no c such that $f'(c) = 0$. ∎

c. The y-intercept is $1 = \sqrt{0 + 1}$ and the x-intercept is $-1 = 0^2 - 1$.

d. $\lim_{x \to -1+} f(x) = 0$ but $\lim_{x \to -1+} f'(x) = \infty$. The tangent at $(-1, 0)$ to the graph of f is vertical.

19 (d): $y = \sqrt{x+1}$

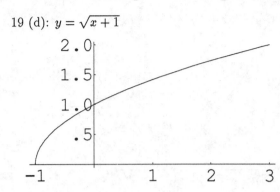

21 (f): $y = x/(1+x)$

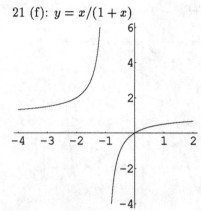

21.   a.   If $f(x) = \dfrac{x}{1+x} = 1 - (1+x)^{-1}$, then $f'(x) = (1+x)^{-2} = \left(\dfrac{1}{1+x}\right)^2$. f is increasing because $f'(x) > 0$. ∎

      b. $\lim\limits_{x \to -\infty} f(x) = \lim\limits_{x \to -\infty} \dfrac{1}{1+(1/x)} = \dfrac{1}{1+0} = 1$; similarly, $\lim_{x \to \infty} f(x) = 1$.

      c. $\lim\limits_{x \to -1^-} f(x) = \lim\limits_{x \to -1^-} \left(\dfrac{x}{1+x}\right) = \infty$, but $\lim\limits_{x \to -1^+} f(x) = -\infty$.

      d.   If $x < -1$, then $1 + x < 0$, $(1+x)^{-1} < 0$ and $-(1+x)^{-1} > 0$; therefore $f(x) = 1 - (1+x)^{-1} > 1$. ∎

      e.   If $x > -1$, then $1 + x > 0$, $(1+x)^{-1} > 0$ and $-(1+x)^{-1} < 0$; therefore $f(x) = 1 - (1+x)^{-1} < 1$. ∎

23.   Begin by observing that $S = D^2$ and $D(x) \geq 0$ for all x. Therefore, $D(c) \leq D(x) \implies D^2(c) \leq D^2(x) \iff S(c) \leq S(x)$. Since the square-root function is increasing, $S(c) \leq S(x) \iff D^2(c) \leq D^2(x) \implies D(c) \leq D(x)$. ∎

25.   Let $P(x) = ax^2 + bx + c$, $a \neq 0$. Since P is unbounded on $(-\infty, \infty)$, if we assume $P(x) \geq 0$, then a must be positive. The minimum value of P is attained at the solution to $P'(x) = 0$; that solution is $x = -b/(2a)$. Therefore, $0 \leq \min P = P(-b/2a) = c - b^2/(4a)$, $b^2/(4a) \leq c$, and $b^2 \leq 4ac$ because $a > 0$. If $a = 0$, then $b = 0$ is necessary if $P(x) \geq 0$; in this case, the conclusion $b^2 \leq 4ac$ reduces to $0 \leq 0$, which is clearly true. ∎

27.   a.   Let $f(x) = (x+1)^p (x-1)^q$ where p and q are integers ≥ 2. Then

$$f'(x) = p(x+1)^{p-1}(x-1)^q + (x+1)^p q(x-1)^{q-1}$$
$$= [p(x-1) + q(x+1)] \, (x+1)^{p-1}(x-1)^{q-1} = (p+q)\left[x - \frac{p-q}{p+q}\right](x+1)^{p-1}(x-1)^{q-1}$$

Clearly, the only solutions to $f'(x) = 0$ are $(p-q)/(p+q)$, -1, and 1. To simplify the discussion in part (b), let $\lambda = (p-q)/(p+q)$ and note that $-1 < \lambda < 1$. ∎

      b. p and q odd is the easiest case because the sign of f' depends only on the sign of $x - \lambda$ because $p-1$ and $q-1$ are even. $f' \leq 0$ if $x < \lambda$ and $f' \geq 0$ if $x > \lambda$. Therefore f has a global minimum at λ.

         If p is even and q is odd, then $f' > 0$ on $(-\infty, -1)$, $f' < 0$ on $(-1, \lambda)$, and $f' \geq 0$ on (λ, ∞). Hence f has a local maximum at -1 and a local minimum at λ.

         If p is odd and q is even, then $f' \geq 0$ on $(-\infty, \lambda)$, $f' < 0$ on $(\lambda, 1)$, and $f' > 0$ on $(1, \infty)$. Therefore f has a local maximum at λ and a local minimum at 1.

         If p and q are even, then f' changes sign at each critical point. Since $f' < 0$ on $(-\infty, -1)$ and $f' > 0$ on $(1, \infty)$, we must have local minima (= global minimum value of 0) at -1 and 1; f has a local maximum at λ. ∎

Solutions 3.3 *Curve sketching II: Concavity & second derivative test* (pages 203-5)

1. $f(x) = x^2 - x + 30$ has domain $\mathbb{R}$. $f'(x) = 2x - 1$ is negative if $x < 1/2$, zero if $x = 1/2$, and positive if $x > 1/2$; thus f decreases on $(-\infty, 1/2)$, has a global minimum value of $119/4$ at $1/2$, and increases on $(1/2, \infty)$. The graph of f is always concave up because $f''(x) = 2 > 0$. The y-intercept is 30, there are no x-intercepts.

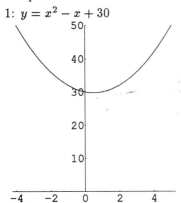

1: $y = x^2 - x + 30$

3: $y = x^3 - 3x^2 - 45x + 25$

3. If $g(x) = x^3 - 3x^2 - 45x + 25$, then $g'(x) = 3x^2 - 6x - 45 = 3(x+3)(x-5)$ and $g''(x) = 6x - 6 = 6(x-1)$. $g'(x) > 0$ if $x < -3$ or $x > 5$ and $g'(x) < 0$ if $-3 < x < 5$; hence g is increasing on $(-\infty, -3) \cup (5, \infty)$ and decreasing on $(-3, 5)$. $g''(x)$ is negative if $x < 1$, is zero when $x = 1$, and is positive if $x > 1$; thus the graph of g is concave down on $(-\infty, 1)$, has an inflection point at $(1, -22)$, and is concave up on $(1, \infty)$. The point $(-3, 106)$ is a local maximum and $(5, -150)$ is a local minimum. The y-intercept is 25; the x-intercepts are approximately -5.6865, 0.5396, and 8.1469.

5. $f(x) = 2x^3 - 9x^2 + 12x - 3$ has domain $\mathbb{R}$. $f'(x) = 6x^2 - 18x + 12 = 6(x-1)(x-2)$ is positive if $x < 1$ or $2 < x$, negative if $1 < x < 2$, zero if $x = 1$ or $x = 2$. Hence f increases on $(-\infty, 1) \cup (2, \infty)$, decreases on $(1, 2)$, has a local maximum of 2 at 1 and a local minimum of 1 at 2. $f''(x) = 12x - 18 = 12(x - 3/2)$; thus the graph of f is concave down on $(-\infty, 3/2)$, concave up on $(3/2, \infty)$, and has $(3/2, 3/2)$ as a point of inflection. The y-intercept is -3; the only x-intercept is approximately 0.3223.

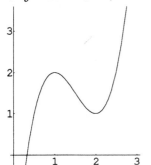

5: $y = 2x^3 - 9x^2 + 12x - 3$

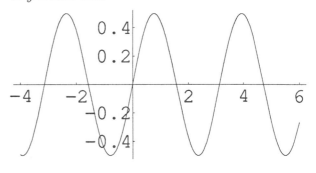

7: $y = \sin x \cos x$

7. Let $f(x) = \sin x \cos x$. $f'(x) = \cos^2 x - \sin^2 x = 2\cos^2 x - 1$. The critical points of f are at $\dfrac{\pi}{4} + k\dfrac{\pi}{2}$ where $\cos x = \pm\dfrac{1}{\sqrt{2}}$ and $f'(x) = 0$; f increases when $\dfrac{1}{\sqrt{2}} < \cos x$ or $\cos x < -\dfrac{1}{\sqrt{2}}$, i.e., when $-\dfrac{\pi}{4} < x - 2k\pi <$ $\dfrac{\pi}{4}$ or $3\dfrac{\pi}{4} < x - 2k\pi < 5\dfrac{\pi}{4}$. $f''(x) = -4\sin x \cos x = -4f(x)$; therefore the graph is concave down when it is above the x-axis, concave up when it is below, and has an inflection point wherever it crosses the x-axis. The minimum value is $-\dfrac{1}{2} = f\left(-\dfrac{\pi}{4} + k\pi\right)$ and the maximum value is $\dfrac{1}{2} = f\left(\dfrac{\pi}{4} + k\pi\right)$.

This function has period π: $f(x+\pi) = \sin(x+\pi)\cos(x+\pi) = (-\sin x)(-\cos x) = \sin x \cos x = f(x)$; it is also an odd function, symmetric about the origin: $f(-x) = \sin(-x)\cos(-x) = (-\sin x)(\cos x) = -f(x)$.

3.3 — Curve sketching II: Concavity & second derivative test

	f'	f
$-\pi/4 + k\pi$	0	local min
$(-\pi/4, \pi/4) + k\pi$	+	increasing
$\pi/4 + k\pi$	0	local max
$(\pi/4, 3\pi/4) + k\pi$	−	decreasing

	f	f''	f
$(-\pi/2, 0) + k\pi$		−	+	concave up
$0 + k\pi$	0	0	inflection point
$(0, \pi/2) + k\pi$	+	−	concave down
$\pi/2 + k\pi$	0	0	inflection point

REMARK: $\sin x \, \cos x = \dfrac{1}{2} \sin 2x.$

---

9. Let $f(x) = (x+2)^2(x-2)^2$; then $f'(x) = 2(x+2)(x-2)^2 + (x+2)^2(2)(x-2)] = 4(x+2)(x)(x-2)$ and $f''(x) = 4(3x^2 - 4)$.

	$x+2$	x	$x-2$	f'	f
$(-\infty, -2)$	−	−	−	−	decreasing
-2	0	−	−	0	global min
$(-2, 0)$	+	−	−	+	increasing
0	+	0	−	0	local max
$(0, 2)$	+	+	−	−	decreasing
2	+	+	0	0	global min
$(2, \infty)$	+	+	+	+	increasing

	f''	f
$(-\infty, -2/\sqrt{3})$	+	concave up
$-2/\sqrt{3}$	0	inflection point
$(-2/\sqrt{3}, 2/\sqrt{3})$	−	concave down
$2/\sqrt{3}$	0	inflection point
$(2/\sqrt{3}, \infty)$	+	concave up

The global minimum value of f is $0 = f(-2) = f(2)$, f has a local maximum of $16 = f(0)$; the graph has inflection points $\left(\pm \dfrac{2}{\sqrt{3}}, \dfrac{64}{9} \right)$.

9: $y = (x+2)^2(x-2)^2 = (x^2-4)^2$

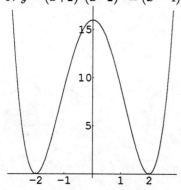

11: $y = |x| + |x - 1| + |2 - x|$

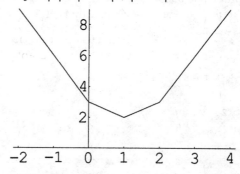

11.

$$|x| + |x - 1| + |2 - x| = \begin{cases} (-x) + (1 - x) + (2 - x) & \text{if } x < 0 \\ (x) + (1 - x) + (2 - x) & \text{if } 0 \le x < 1 \\ (x) + (x - 1) + (2 - x) & \text{if } 1 \le x < 2 \\ (x) + (x - 1) + (x - 2) & \text{if } 2 \le x \end{cases} = \begin{cases} 3 - 3x & \text{if } x < 0 \\ 3 - x & \text{if } 0 \le x < 1 \\ 1 + x & \text{if } 1 \le x < 2 \\ 3x - 3 & \text{if } 2 \le x. \end{cases}$$

13. If $f(x) = \dfrac{x^2 - 1}{x^2 + 1} = 1 - \dfrac{2}{x^2 + 1}$, then $f'(x) = \dfrac{4x}{(x^2 + 1)^2}$ and $f''(x) = 4 \dfrac{1 - 3x^2}{(x^2 + 1)^3}$. f decreases on $(-\infty, 0)$, has a global minimum of -1 at 0, and increases on $(0, \infty)$. The graph is concave down on $(-\infty, -1/\sqrt{3}) \cup (1/\sqrt{3}, \infty)$ and concave up on $(-1/\sqrt{3}, 1/\sqrt{3})$; its points of inflection are $(\pm 1/\sqrt{3}, -1/2)$. The horizontal line $y = 1$ is an asymptote because $\lim_{x \to -\infty} f(x) = 1 = \lim_{x \to \infty} f(x)$ and $\lim_{x \to -\infty} f'(x) = 0 = \lim_{x \to \infty} f'(x)$.

13: $y = (x^2 - 1)/(x^2 + 1) = 1 - 2/(x^2 + 1)$

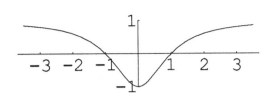

15: $y = 3x^5 - 5x^3 = x^3(3x^2 - 5)$

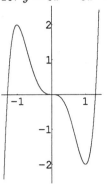

15. If $f(x) = 3x^5 - 5x^3$, then $f'(x) = 15x^4 - 15x^2 = 15(x+1)x^2(x-1)$ and $f''(x) = 60x^3 - 30x = 60(x + 1/\sqrt{2})x(x - 1/\sqrt{2})$.

	$x+1$	x^2	$x-1$	f'	f
$(-\infty, -1)$	$-$	$+$	$-$	$+$	increasing
-1	0	$+$	$-$	0	local max
$(-1, 0)$	$+$	$+$	$-$	$-$	decreasing
0	$+$	0	$-$	0	(inflection pt)
$(0, 1)$	$+$	$+$	$-$	$-$	decreasing
1	$+$	$+$	0	0	local min
$(1, \infty)$	$+$	$+$	$+$	$+$	increasing

	f''	f
$(-\infty, -1/\sqrt{2})$	$-$	concave down
$(-1/\sqrt{2}, 0)$	$+$	concave up
$(0, 1/\sqrt{2})$	$-$	concave down
$(1/\sqrt{2}, \infty)$	$+$	concave up

f has a local maximum of $2 = f(-1)$ and a local minimum of $-2 = f(1)$; the graph has inflection points $\left(\pm\dfrac{1}{\sqrt{2}}, \mp\dfrac{7}{4\sqrt{2}}\right)$ and $(0, 0)$.

17. If $g(x) = x\sqrt{1+x} = x(1+x)^{1/2}$, then $g'(x) = \sqrt{1+x} + \dfrac{x}{2\sqrt{1+x}} = \left(1 + \dfrac{3}{2}x\right)(1+x)^{-1/2}$ and $g''(x) = \dfrac{3}{2}(1+x)^{-1/2} + \left(1 + \dfrac{3}{2}x\right)\left(-\dfrac{1}{2}\right)(1+x)^{-3/2} = \left(1 + \dfrac{3}{4}x\right)(1+x)^{-3/2}$. The domain of g is $[-1, \infty)$; g decreases on $[-1, -2/3)$, increases on $(-2/3, \infty)$, and has its global minimum of $-2/(3\sqrt{3})$ at $-2/3$. Because $1 + 3x/4$ is positive throughout the domain of g, g'' is always positive and the graph of g is concave up. The vertical line $x = -1$ is tangent to the graph of g at $(-1, 0)$ because $\lim_{x \to -1+} g'(x) = -\infty$.

17: $y = x\sqrt{1+x}$

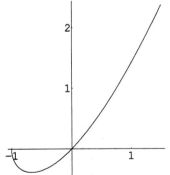

19: $y = x - \sqrt[3]{x}$

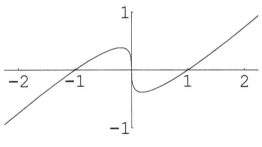

19. If $f(x) = x - x^{1/3}$, then $f'(x) = 1 - \dfrac{1}{3}x^{-2/3}$. Unless $x = 0$, $x^{-2/3}$ is positive, hence f increases on $\left(-\infty, -3^{-3/2}\right) \cup \left(3^{-3/2}, -\infty\right)$ and decreases on $\left(-3^{-3/2}, 3^{-3/2}\right)$, f has a local maximum of $2/3^{3/2}$ at $-3^{-3/2}$ and a local minimum of $-2/3^{3/2}$ at $3^{3/2}$. Because $f(0) = 0$ and $\lim_{x \to 0} f'(x) = -\infty$, the tangent at $(0, 0)$ to the graph of f is vertical. $f''(x) = \dfrac{2}{9}x^{-5/3}$; f is concave down on $(-\infty, 0)$ and concave up on $(0, \infty)$; although $f''(0)$ does not exist, $(0, 0)$ is a point of inflection for f because concavity changes there.

3.3 — Curve sketching II: Concavity & second derivative test

21.   If $g(x) = x^{2/3}(x-3)$, then $g'(x) = (1/3)(5x-6)x^{-1/3}$. g increases on $(-\infty, 0) \cup (6/5, \infty)$, decreases on $(0, 6/5)$, has a local maximum of 0 at 0 and a local minimum of $(-9/5)(6/5)^{2/3}$ at $6/5$. Note that $\lim_{x \to 0} |g'(x)| = \infty$ implies the tangent at $(0,0)$ is vertical. $g''(x) = (2/9)(5x + 3)x^{-4/3}$; g is concave down on $(-\infty, -3/5)$, has inflection point $\left(-3/5, (-18/5)(-3/5)^{2/3}\right)$, and is concave up on $(-3/5, 0) \cup (0, \infty)$.

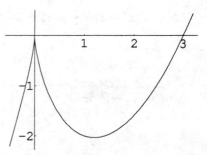

23.   Graphs of three typical solutions appear in the text's answers; the curves are parallel.

25.   Let $p(x) = x^3 + x^2 + 5x - 15$. Because $p'(x) = 3x^2 + 2x + 5$ is always positive (its minimum value is 14/3), p is always increasing so the equation $p(x) = 0$ has at most one real root. Since $p(0) = -15 < 0 < 7 = p(2)$ and p is continuous, $p(x) = 0$ has a root somewhere in $(0, 2)$. (It's close to 1.6215.) ∎

27.   Let $q(x) = x^{11} + x^3 + x + 3$. $q'(x) = 11x^{10} + 3x^2 + 1 \geq 1$, thus q always increases. Because $q(-2) = -2055 < 0 < 3 = q(0)$, the unique real solution to $q(x) = 0$ lies in $(-2, 0)$. (The root is -1.)

29. $f(x) = x^3 + cx^2 + bx + a = x^3\left(1 + \dfrac{c}{x} + \dfrac{b}{x^2} + \dfrac{a}{x^3}\right)$ and $\displaystyle\lim_{x \to \pm\infty}\left(1 + \dfrac{c}{x} + \dfrac{b}{x^2} + \dfrac{a}{x^3}\right) = 1$, therefore $\lim_{x \to -\infty} f(x) = -\infty$ and $\lim_{x \to \infty} f(x) = \infty$. In particular, there are choices of x_0 and x_1 such that $f(x_0) < -1$ and $1 < f(x_1)$. By the Intermediate Value Theorem, there must be some x^* between x_0 and x_1 where $f(x^*) = 0$. ∎

31.   Suppose f is differentiable on some open interval containing x_0, $f'(x_0) = 0$, $f''(x_0)$ exists and is negative. Because $f''(x_0)$ is computed as a limit and because $f''(x_0) < 0$, there is some interval $[c, d]$ containing x_0 in its interior such that $\dfrac{f'(x) - f'(x_0)}{x - x_0} < 0$ if $c \leq x \leq d$ but $x \neq x_0$. Hence $f'(x) - f'(x_0) > 0$ if $c - x_0 \leq x - x_0 < 0$ and $f'(x) - f'(x_0) < 0$ if $0 < x - x_0 \leq d - x_0$. We are supposing that $f'(x_0) = 0$, hence $f'(x) > 0$ if $c \leq x < x_0$ and $f'(x) < 0$ if $x_0 < x \leq d$. Therefore f increases on (c, x_0) and decreases on (x_0, d); since f is continuous on $[c, d]$, $f(x_0)$ must be the maximum value of f on $[c, d]$. (More formally, apply the First Derivative Test, Theorem 3.3.3(i).) ∎

REMARK: $f''(x_0) < 0$ does not imply f' is a decreasing function on some interval containing x_0.

(Such a function might wiggle a lot near x_0. You can verify that $f'(x) = \begin{cases} -x - 3x^2 \sin(1/x) & \text{if } x \neq 0 \\ 0 & \text{if } x = 0 \end{cases}$

behaves that way near $x_0 = 0$.)

---

33.   Suppose $f''(x_0) = 0$ and $f'''(x_0) \neq 0$. The solutions to Problems 31-32 show that there must be an interval (c, d) containing x_0 such that f'' changes sign at x_0. This implies f is concave one way on (c, x_0) and concave the other way on (x_0, d), hence $(x_0, f(x_0))$ is an inflection point for f. ∎

35.   Suppose p and q satisfy the relations $p > 1$, $q > 1$, $1/p + 1/q = 1$. Pick (& fix) a positive V and consider $f(x) = x^p/p - Vx$ on $[0, \infty)$. $f'(x) = x^{p-1} - V$ and $f''(x) = (p-1)x^{p-2}$. Since $p > 1$, we infer $f''(x) > 0$ on $(0, \infty)$. Therefore f has its global minimum at $x = V^{1/(p-1)}$, the unique critical point of f. $1/p + 1/q = 1$ implies $q = 1 \left/ \left(1 - \dfrac{1}{p}\right)\right. = \left(\dfrac{p}{p-1}\right)$. Therefore

$$\frac{x^p}{p} - xV = f(x) \geq f\left(V^{1/(p-1)}\right) = \frac{1}{p}V^{p/(p-1)} - V^{1+1/(p-1)} = V^{p/(p-1)}\left(\frac{1}{p} - 1\right) = V^q\left(\frac{-1}{q}\right)$$

and $$\frac{x^p}{p} + \frac{V^q}{q} \geq xV. \ ∎$$

Solutions 3.4   *Curve sketching III: Asymptotes*   (page 212)

1.   Let $f(x) = (x-3)^{-1}$. The x-axis is a horizontal asymptote because $\lim_{x \to \pm\infty} f(x) = 0$, the line $x = 3$ is a vertical asymptote because $\lim_{x \to 3-} f(x) = -\infty$ and $\lim_{x \to 3+} f(x) = +\infty$. The function always decreases because $f'(x) = -(x-3)^{-2} < 0$. The graph is concave down on $(-\infty, 3)$ and concave up on $(3, \infty)$ because $f''(x) = 2(x-3)^{-3}$ has the same sign as $x - 3$.

1: $y = 1/(x-3)$

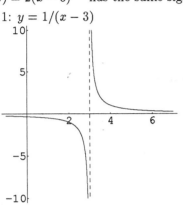

3: $y = 1 + 1/x^2$

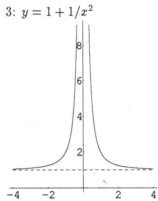

3. $g(x) = 1 + x^{-2}$ has domain $\mathbb{R} - \{0\}$. $g'(x) = (-2)x^{-3}$; hence g increases on $(-\infty, 0)$ and decreases on $(0, \infty)$. $g''(x) = 6x^{-4} > 0$, hence g is always concave up. $\lim_{x \to 0} g(x) = \infty$ so the y-axis is a vertical asymptote; $\lim_{x \to \pm\infty} g(x) = 1$ so the line $y = 1$ is a horizontal asymptote.

5. $f(x) = \dfrac{1}{(x-1)(x-2)} = \dfrac{1}{x-2} - \dfrac{1}{x-1}$ has domain $\mathbb{R} - \{1, 2\}$. $f'(x) = (-1)(x-2)^{-2} + (x-1)^{-2} = (3-2x)(x-1)^{-2}(x-2)^{-2}$; thus f is increasing on $(-\infty, 3/2)$ and decreasing on $(3/2, \infty)$. [Since 1 and 2 are outside the domain of f, it is more correct to say f increases on $(\infty, 1) \cup (1, 3/2)$ and decreases on $(3/2, 2) \cup (2, \infty)$.] f does have a local maximum of -4 at $3/2$. $f''(x) = 2(x-2)^{-3} - 2(x-1)^{-3} = 2(3x^2 + 9x - 7)(x-1)^{-3}(x-2)^{-3}$; because the numerator, $3x^2 - 9x + 7$, is always positive (with its minimum being $1/4$), the graph of f is concave up on $(-\infty, 1) \cup (2, \infty)$ and concave down on $(1, 2)$, it has no points of inflection. $\lim_{x \to 1} f'(x) = \infty = \lim_{x \to 1-} f(x)$ and $\lim_{x \to 1+} f(x) = -\infty$, hence the vertical line $x = 1$ is an asymptote; $\lim_{x \to 2} f'(x) = -\infty = \lim_{x \to 2-} f(x)$ and $\lim_{x \to 2+} f(x) = \infty$, the vertical line $x = 2$ is also an asymptote. The x-axis is an asymptote because $\lim_{x \to \pm\infty} f(x) = 0 = \lim_{x \to \pm\infty} f'(x)$.

5: $y = (x-1)^{-1}(x-2)^{-1} = (x-2)^{-1}-(x-1)^{-1}$

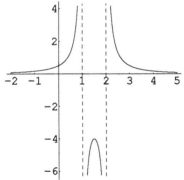

7: $y = (x^2-1)^{-1} = (1/2)\left((x-1)^{-1} - (x+1)^{-1}\right)$

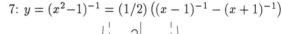
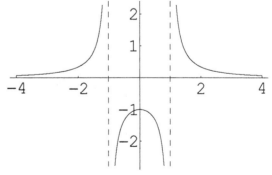

7.   Let $g(x) = 1/(x^2 - 1) = (1/2)\left((x-1)^{-1} - (x+1)^{-1}\right)$, the domain of g is $\mathbb{R} - \{-1, 1\}$. g increases on $(-\infty, -1) \cup (-1, 0)$, decreases on $(0, 1) \cup (1, \infty)$, and has a local maximum of -1 at 0 because $g'(x) = (1/2)\left(-(x-1)^{-2} + (x+1)^{-2}\right) = -2x(x^2 - 1)^{-2}$. The graph of g is concave up on $(-\infty, 1) \cup (1, \infty)$ and concave down on $(-1, 1)$ with no inflection points because $g''(x) = \left((x-1)^{-3} - (x+1)^{-3}\right) = 2(3x^2 + 1)(x^2 - 1)^{-3}$. The vertical line $x = -1$ is an asymptote since $\lim_{x \to -1} g'(x) = \infty = \lim_{x \to -1-} g(x)$ and $\lim_{x \to -1+} g(x) = -\infty$; similarly, the vertical line $x = 1$ is an asymptote since $\lim_{x \to 1} g'(x) = -\infty = \lim_{x \to 1-} g(x)$ and $\lim_{x \to 1+} g(x) = \infty$. The x-axis is an asymptote since $\lim_{x \to \pm\infty} g(x) = 0 = \lim_{x \to \pm\infty} g'(x)$.

9. $f(x) = (x^2 + 1)/(x^2 - 1) = 1 + ((x-1)^{-1} - (x+1)^{-1}) = 1 + 2g(x)$ where g is the function analyzed in the solution of Problem 7. The intervals of increase and decrease are the same as are the concavities and vertical asymptotes, the horizontal asymptote is $y = 1$.

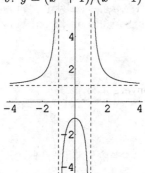

9: $y = (x^2 + 1)/(x^2 - 1)$

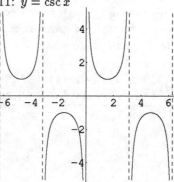

11: $y = \csc x$

11.   If $f(x) = \csc x = (\sin x)^{-1}$, then $f'(x) = -(\cos x)(\sin x)^{-2}$ and $f''(x) = (\sin x)^{-1} + 2(\cos x)^2 (\sin x)^{-3} = \left(1 + (\cos x)^2\right)(\sin x)^{-3}$. We can focus on the interval $[0, 2\pi]$ because f is a periodic function with period 2π. $f(x)$ is undefined wherever $\sin x = 0$, therefore the domain of f excludes all multiples of π. The sign of $f'(x)$ is opposite to that of $\cos x$, thus f increases on $(\pi/2, \pi) \cup (\pi, 3\pi/2)$ and decreases on $(0, \pi/2) \cup (3\pi/2, 2\pi)$; $f(\pi/2) = 1$ is a local minimum, $f(3\pi/2) = -1$ is a local maximum. The sign of $f''(x)$ is the same as that of $\sin x$, hence the graph is concave up on $(0, \pi)$ and concave down on $(\pi, 2\pi)$.

13. $g(x) = (x + 1)/\sqrt{x - 1} = (x + 1)(x - 1)^{-1/2}$ implies $g'(x) = (1/2)(x - 3)(x - 1)^{-3/2}$ and $g''(x) = (1/4)(7 - x)(x - 1)^{-5/2}$. The domain of g is $(1, \infty)$; g decreases on $(1, 3)$, reaches its (global) minimum value of $2\sqrt{2} = g(3)$, and increases on $(3, \infty)$. The graph is concave up on $(1, 7)$ and concave down on $(7, \infty)$; the point $(7, 8/\sqrt{6})$ is an inflection point. The line $x = 1$ is a vertical asymptote because $\lim_{x \to 1+} g(x) = \infty$ and $\lim_{x \to 1+} g'(x) = -\infty$.

REMARK: Notice that $g(x) = \sqrt{x - 1} + 2/\sqrt{x - 1}$ and the graph of g is asymptotic to the graph of $y = \sqrt{x - 1}$ as $x \to \infty$.

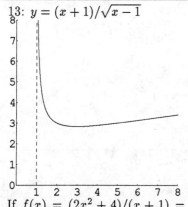

13: $y = (x + 1)/\sqrt{x - 1}$

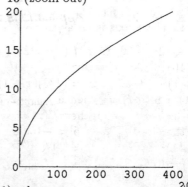

13 (zoom out)

15.   If $f(x) = (2x^2 + 4)/(x + 1) = 2(x - 1) + 6/(x + 1)$, then $f'(x) = 2 - 6/(x+1)^2$ and $f''(x) = 12/(x+1)^3$. The domain of f is $\mathbb{R} - \{1\} = (-\infty, -1) \cup (-1, \infty)$; the graph of f is concave down on the left portion, $(-\infty, -1)$, and concave up on the right part, $(-1, \infty)$. f increases on $(-\infty, -1 - \sqrt{3})$, reaches a local maximum of $-4 - 4\sqrt{3}$ at $-1 - \sqrt{3}$, and then decreases on $(-1 - \sqrt{3}, -1)$; f decreases on $(-1, -1 + \sqrt{3})$, reaches a local minimum of $-4 + 4\sqrt{3}$ at $-1 + \sqrt{3}$, and then increases on $(-1 + \sqrt{3}, \infty)$. The line $x = -1$ is a vertical asymptote and the line $y = 2(x - 1)$ is an oblique asymptote.

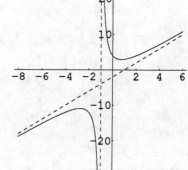

17.   Since $\tan x$ increases from $-\infty$ to ∞ in each interval $I_k = \left(\left(k - \frac{1}{2} \right) \pi, \left(k + \frac{1}{2} \right) \pi \right)$, it must meet the graph of any continuous function f at least once in that interval. (*Proof:* f is bounded on I_k so the continuous function $f - \tan$ takes negative and positive values; the Intermediate Value Theorem then assures us it must also take the value zero.) To apply that result to solutions of $x = -\tan x$, we just let $f(x) = -x$. Since this f is monotonic, there is a unique solution in each I_k; since the intervals do not overlap, the set of solutions to $x = -\tan x$ has infinitely many members. ∎

REMARK: If x is a solution, then so is $-x$. The first six positive solutions are approximately 2.02876, 4.91318, 7.97867, 11.08554, 14.20744, and 17.33638.

    The sum of the reciprocals of the squares of the positive solutions converges to $1/3$.

17: $y = -\tan x$ & $y = x$

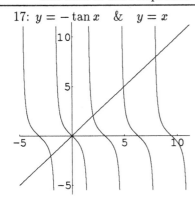

19: $y = x \sin x$

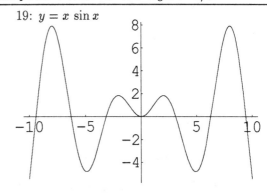

19.   Let $f(x) = x \sin x$. f is an even function because $f(-x) = (-x) \sin(-x) = (-x)(-\sin x) = x \sin x = f(x)$, hence the graph of f is symmetric with respect to the y-axis.

21.   Let $f(x) = x$. It is certainly true that $\lim_{x \to \infty} |f(x) - x| = 0$; therefore, according to the text's definition, the curve $y = x$ is an asymptote to the graph of f.

    If that counterexample seems overly simple, then consider $f(x) = x + (\sin x)/x^2$. $f(x_0) = x_0$ provided x_0 is a multiple of π; the straight line is an asymptote because $\lim_{x \to \infty} (\sin x)/x^2 = 0$.

Solutions 3.5   *Applications of maxima and minima*   (pages 223-7)

1.   If $f(x) = x^3 - 12x + 10$, then $f'(x) = 3x^2 - 12 = 3(x + 2)(x - 2)$. $f(-10) = -870$, $f(-2) = 26$, $f(2) = -6$, and $f(10) = 890$. Therefore $f_{\min}[-10, 10] = -870$ and $f_{\max}[-10, 10] = 890$.

3.   If $f(x) = (x + 1)^{1/3}$, then $f'(x) = (1/3)(x + 1)^{-2/3} > 0$; $f'(x)$ is never 0 but it is undefined at $x = -1$. $f(-2) = -1$, $f(-1) = 0$, $f(7) = 2$; hence $f_{\min}[-2, 7] = -1$ and $f_{\max}[-2, 7] = 2$.

5.   If $f(x) = x^3 - 3x^2 - 45x + 25$, then $f'(x) = 3x^2 - 6x - 45 = 3(x + 3)(x - 5)$. $f(-5) = 50$, $f(-3) = 106$, and $f(5) = -150$; therefore $f_{\min}[-5, 5] = -150$ and $f_{\max}[-5, 5] = 106$.

7.   If $f(x) = (x - 1)^{1/5}$, then $f'(x) = (1/5)(x - 1)^{-4/3} > 0$; $f'(x)$ is never 0 but it is undefined at $x = 1$. $f(-31) = -2$, $f(1) = 0$, $f(33) = 2$; hence $f_{\min}[-31, 33] = -2$ and $f_{\max}[-31, 33] = 2$.

9.   If $f(x) = (x - 1)/(x - 2) = 1 + (x - 2)^{-1}$, then $f'(x) = -(x - 2)^{-2} < 0$ and f is always decreasing, therefore $f_{\min}[-3, 1] = f(1) = 0$ and $f_{\max}[-3, 1] = f(-3) = 4/5$.

11.   If $f(x) = |x^2 - 2x|$, then $f'(x) = \begin{cases} 2(x - 1) & \text{if } x < 0 \text{ or } 2 < x, \\ \text{undefined} & \text{if } x = 0 \text{ or } x = 2, \\ 2(1 - x) & \text{if } 0 < x < 2. \end{cases}$ We have three critical points, $\{0, 1, 2\}$, and two endpoints, $\{-2, 3\}$, to consider. $f(-2) = 8$, $f(0) = 0$, $f(1) = 1$, $f(2) = 0$, $f(3) = 3$; therefore $f_{\min}[-2, 3] = 0$ and $f_{\max}[-2, 3] = 8$.

13.   If $f(x) = \sin x + \cos x$, then $f'(x) = \cos x - \sin x$; on the interval $[0, 2\pi]$, the only critical points are $\pi/4$ and $5\pi/4$. $f(0) = 1$, $f(\pi/4) = \sqrt{2}$, $f(5\pi/4) = -\sqrt{2}$, and $f(2\pi) = 1$; therefore $f_{\min}[0, 2\pi] = -\sqrt{2}$ and $f_{\max}[0, 2\pi] = \sqrt{2}$.

REMARK: $\sin x + \cos x = \sqrt{2} \sin(x + \pi/4)$.

3.5 — Applications of maxima and minima

15. If $f(x) = \cos\left(\dfrac{1-2x}{3}\right)$, then $f'(x) = \dfrac{2}{3}\sin\left(\dfrac{1-2x}{3}\right)$. The only critical point in $[-2,2]$ is $1/2$ because $f'(x) = 0$ if and only if $(1-2x)/3$ is a multiple of π, i.e., $x = (1 - 3k\pi)/2$, and the only such choice of x is with $k = 0$. $f(-2) = \cos(5/3) \approx -0.0957$, $f(1/2) = \cos 0 = 1$, $f(2) = \cos(-1) \approx 0.5403$; thus $f_{\min}[-2,2] = \cos(5/3)$ and $f_{\max}[-2,2] = 1$.

17. If $f(x) = \begin{cases} x^2 & \text{if } -4 \le x \le 1, \\ x^3 & \text{if } 1 < x \le 2, \end{cases}$ then $f'(x) = \begin{cases} 2x & \text{if } -4 < x < 1, \\ \text{undefined} & \text{if } x = 1, \\ 3x^2 & \text{if } 1 < x < 2. \end{cases}$ At the critical points we have $f(0) = 0^2 = 0$ and $f(1) = 1^2 = 1$, at the endpoints we have $f(-4) = (-4)^2 = 16$ and $f(2) = 2^3 = 8$; therefore $f_{\min}[-4,2] = 0$ and $f_{\max}[-4,2] = 16$.

19. Suppose $2x + 2y = 1000$ and $A = xy$. Then, differentiating implicitly with respect to x, $2 + 2y' = 0$ and $A' = y + xy'$. These imply $y' = (-2)/2 = -1$ and $A' = y + x(-1) = y - x$. Therefore $A' = 0$ if and only if $x = y$; the constraint then yields $x = 1000/4 = 250$. Furthermore, $A'' = (y - x)' = y' - 1 = (-1) - 1 < 0$ so $A(250) = 250^2$ is a local maximum for A. We also have the constraints that $x \ge 0$ and $y \ge 0$; because $A = 0$ at the endpoint of either one, we conclude that $A(250)$ is the global maximum.

21. A rectangle of length x and width y has perimeter $P = 2(x + y)$ and area $A = xy$. If we restrict ourselves to rectangles with $P = 300\,\text{m}$, then $300 = 2(x+y)$. Thus, differentiating implicitly with respect to x, $0 = 300' = P' = 2(x+y)' = 2(1 + y')$ and $A' = y + xy'$. These imply $y' = -1$ and $A' = y + x(-1) = y - x$. Hence $A' = 0$ if and only if $x = y$; the constraint then yields $x = 300/4 = 75$. Furthermore, $A'' = (y - x)' = y' - 1 = (-1) - 1 < 0$ so $A(75) = 75^2$ is a local maximum for A. We also have the constraints that $x \ge 0$ and $y \ge 0$; since $A = 0$ at the endpoint of either one, we conclude that $A(75)$ is the global maximum.

23. Suppose $3x - 4y - 12 = 0$ and $D = (x-1)^2 + (y-2)^2$. ($\sqrt{D}$ is the distance between (x,y) and $(1,2)$; D and $\sqrt{D}$ reach their respective minimums simultaneously.) Implicit differentiation with respect to x yields $3 - 4y' = 0$ and $D' = 2(x-1) + 2(y-2)y'$. Therefore $y' = 3/4$ and $D' = 2(x-1) + (3/2)(y-2)$; furthermore, $D'' = 2 + (3/2)y' = 2 + (3/2)(3/4) > 0$, so D is always concave up and assumes its global minimum at the unique critical point which the following work will identify. If $D' = 0$, then $y = (10 - 4x)/3$; substituting this expression for y into the constraint, we get $12 = 3x - 4(10 - 4x)/3$ which implies $x = 76/25$ and $y = -18/25$. Therefore $(76/25, -18/25)$ is the point on the straight line $3x - 4y - 12 = 0$ that is closest to the point $(1,2)$.

25. a. If $h = 48t - 16t^2$, then $h' = 48 - 32t$ and $h'(0) = 48\,\text{ft/sec}$ is the initial vertical velocity.
   b. $h'(t) = 0 \iff t = 48/32 = 1.5\,\text{sec}$.
   c. $h(1.5) = 36\,\text{ft}$.

27. Let x be the length of the fence which is parallel to the river, let y be the length of one of the 7 fences perpendicular to the river. The area enclosed by the fencing is $A = 20\,\text{acres} = xy$ and the total length of all fences is $P = x + 7y$. Implicit differentiation with respect to x yields $0 = y + xy'$ and $P' = 1 + 7y'$; these imply $y' = -y/x$ and $P' = 1 - 7y/x$. $P' = 0$ if and only if $7y = x$, i.e., the sum of the lengths of the 7 cross fences equals the length of the one fence parallel to the river. Since $P'' = -7(y/x)' = -7(y'/x - y/x^2) = 14y/x^2 > 0$, the total amount of fencing, P, is minimized at the critical point we've found. ∎

29. A cylinder of radius r and height h has volume $V = \pi r^2 h$ and area $A = 2(\pi r^2) + (2\pi r)h$. Implicit differentiation with respect to r yields $\begin{cases} V' = \pi\left(2rh + r^2 h'\right) \\ A' = 2\pi\left(2r + h + rh'\right). \end{cases}$ If $V = 50$, then $V' = 0$, $h' = -2h/r$, and $A' = 2\pi(2r - h)$. We have a critical point of A if and only if $h = 2r$. Furthermore, $A'' = 2\pi(2 - h') = 2\pi(2 + 2h/r) > 0$ so that critical point corresponds to the global minimum of A. The dimensions of the optimal cylinder are $r = \sqrt[3]{25/\pi}$ and $h = 2\sqrt[3]{25/\pi}$.

31. Cut x cm from a 35 cm wire to be bent into a circle, bend the remaining $35 - x$ cm into a square. A circle with circumference x has radius $r = x/(2\pi)$ and area $\pi r^2 = x^2/(4\pi)$; a square with perimeter $35 - x$ has sides of length $(35 - x)/4$ and area $(35 - x)^2/16$. The total area enclosed by the circle and the square is $A = \dfrac{x^2}{4\pi} + \dfrac{(35-x)^2}{16}$, $0 \le x \le 35$. Therefore $\dfrac{dA}{dx} = \dfrac{x}{2\pi} - \dfrac{35 - x}{8} = \dfrac{(4+\pi)x - 35\pi}{8\pi}$ and $\dfrac{d^2A}{dx^2} = \dfrac{4+\pi}{8\pi} > 0$. A has a local minimum of $1225/(16 + 4\pi) \approx 42.88$ when $x = 35\pi/(4 + \pi)$; $A(0) = 1225/16 = 76.5625$ and $A(35) = 1225/(4\pi) \approx 97.48$. The total area is maximized by using all wire for the circle, i.e., let $x = 35$ cm. The total area is minimized by letting $x = 35\pi/(4 + \pi)$, this produces a circle with radius $r = 35/(8 + 2\pi)$ and a square with sides $35/(4 + \pi)$.

33.   Suppose the semicircle has radius r and the rectangle has height h (and width $2r$). The perimeter is $\pi r + (2h + 2r) = 30$ ft, thus the total area is $A = (1/2)\left(\pi r^2\right) + 2rh = (1/2)\left(\pi r^2\right) + r(30 - (2 + \pi)r) = 30r - (2 + \pi/2)r^2$. Thus $dA/dr = 30 - (4 + \pi)r$ and $d^2A/dr^2 = -(4 + \pi) < 0$. A has its maximum value of $450/(4 + \pi)$ ft^2 when $r = 30/(4 + \pi)$ ft and $h = 30/(4 + \pi)$ ft.

35.   Let x be the distance between points P and R. Using the fact that TIME equals DISTANCE divided by RATE, the total travel time for canoeing and walking is $T = \sqrt{1 + x^2}/3 + (10 - x)/5$, $0 \le x \le 10$. Thus $\dfrac{dT}{dx} = \dfrac{x}{3\sqrt{1 + x^2}} - \dfrac{1}{5}$; this derivative is zero if and only if $x = 3/4$. Evaluating T at this critical point and at the two endpoints we find $T(3/4) = 34/15$ hr $= 136$ min, $T(0) = 7/3$ hr $= 140$ min, $T(10) = \sqrt{101}/3$ hr ≈ 201 min; The quickest route requires canoeing to a point 0.75 km from point P and walking from there.

37.   Let x be the distance from the base of the taller tree to the spot where the rope is pegged to the ground. The total length of the rope is $s = \sqrt{x^2 + 8^2} + \sqrt{(10 - x)^2 + 6^2}$. Then $\dfrac{ds}{dx} = \dfrac{x}{\sqrt{x^2 + 64}} - \dfrac{10 - x}{\sqrt{x^2 - 20x + 136}}$. $ds/dx = 0$ if and only if $x^2\left(x^2 - 20x + 136\right) = (10 - x)^2\left(x^2 + 64\right)$ which simplifies to $0 = 7x^2 - 320x + 1600 = (7x - 40)(x - 40)$. Since the domain of s is $[0, 10]$, $40/7$ is the only critical point for s. $s(0) = 8 + 2\sqrt{34} \approx 19.66$. $s(40/7) = 2\sqrt{74} \approx 17.20$, and $s(10) = 2\sqrt{41} + 6 \approx 18.81$ feet; the total length of the rope is minimized by pegging it $40/7$ ft from the taller tree (and $30/7$ ft from the shorter).

39.   Let θ be the angle between the two sides after the sheet metal is bent. The volume of the vee-shaped trough is the area of the triangular cross section times the length: $V = (1/2)(1)(\sin\theta) \cdot (6) = 3\sin\theta$ where $0 \le \theta \le \pi$. It is a well-known result that $\sin\theta$ is maximized by choosing $\theta = \pi/2$.

41.   Example 6 shows that the branching angle θ which minimizes blood resistance satisfies the relation $\cos\theta = \left(r_2/r_1\right)^4$.

    a.   If $r_1 = r_2$, then $\cos\theta = 1^4 = 1$; since it is implicit that $0 \le \theta \le \pi$, this implies $\theta = 0$.

    b.   If $r_1 = 2r_2$, then $\cos\theta = (1/2)^4 = 0.0625$ and $\theta \approx 1.508$ radians $\approx 86.4°$.

43.   Using Snell's Law (equation 11), we have $\sin\theta_1/\sin 38° = 220400/300000$; therefore $\theta_1 = \sin^{-1}\left((2204/3000) \cdot \sin 38°\right) = \sin^{-1} 0.452306 = 26.8917°$.

45.   If $V = \alpha\left(R_0 - R\right)R^2$, then $dV/dR = \alpha\left(2R_0 - 3R\right)R$. Because $V(0) = 0$, $V\left(2R_0/3\right) = 4\alpha R_0^3/27$ and $dV/dR < 0$ on $\left(2R_0/3, \infty\right)$, we infer $V(R)$ is maximized on $[0, R_0]$ at $R = 2R_0/3$. ∎

47.   Suppose $p(x) = a_nx^n + a_{n-1}x^{n-1} + \cdots a_1x + a_0$, $a_n \ne 0$, has the property that $p(x) \ge 0$ for all x. Then n must be even and $a_n > 0$. Let $F(x) = p(x) + p'(x) + p''(x) + \cdots + p^{(n)}(x)$. Then $F(x)$ is also a polynomial of degree n and $F(x) = a_nx^n + \cdots$. Because F is a polynomial of even degree with positive leading coefficient, F has a finite global minimum. Suppose that minimum is attained at x_0, then $F'(x_0) = 0$. However $F'(x) = p'(x) + p''(x) + p'''(x) + \cdots + p^{(n+1)}(x) = F(x) - p(x)$ because $p^{(n+1)}(x) \equiv 0$. Therefore, for all x, $F(x) \ge F\left(x_0\right) = \left(F' + p\right)\left(x_0\right) = F'\left(x_0\right) + p\left(x_0\right) = 0 + p\left(x_0\right) \ge 0$. ∎

REMARK: This problem appears in *Mathematical Spectrum*, volume 1 (1968), page 29.

49.   Theorem: *Let $\{a_1, a_2, \ldots, a_n\}$ be an arbitrary collection of real numbers and let $D(x) = \displaystyle\sum_{k=1}^{n}\left(a_k - x\right)^2$.*

The choice of x that minimizes D is $x = \dfrac{1}{n}\displaystyle\sum_{k=1}^{n} a_k$, the arithmetic mean of the a_k's.

$\boxed{\text{Proof}}$ $D'(x) = \displaystyle\sum_{k=1}^{n} 2\left(a_k - x\right)(-1) = (-2)\left(\sum_{k=1}^{n} a_k - \sum_{k=1}^{n} x\right) = (-2)\left(\sum_{k=1}^{n} a_k - nx\right) = 2n\left(x - \frac{1}{n}\sum_{k=1}^{n} a_k\right)$ and $D''(x) = 2n > 0$. The graph of D is concave up, hence the function takes its minimum value at its unique critical point: $D'(x) = 0 \iff x = \dfrac{1}{n}\displaystyle\sum_{k=1}^{n} a_k$. ∎

REMARK: One simple task of data analysis is to characterize a set of values by a few summary statistics; the sample mean is an estimate for the "center" of a set of data. This result says the sample mean has the desirable property of minimizing the aggregate measure of deviation computed by D.

    You might find it interesting to think about the distinction between "approximation" and "estimation". Consult several dictionaries, encyclopedias, and teachers.

1. Let $f(x) = x^2 - 90$. Any interval $[a, b]$ such that $0 < a < b$ and $a^2 < 90 < b^2$ satisfies the conditions of Theorem 2.

    Proof $f(a) = a^2 - 90 < 0 < b^2 - 90 = f(b)$ because $a^2 < 90 < b^2$. Since $0 < a < b$, $f'(x) = 2x > 0$ for every x in $[a, b]$; $f''(x) = 2$ never changes sign. Because f' is an increasing function, the fourth condition to be verified is $|f(x)/f'(a)| \le b - a \iff (90 - a^2)/2a \le b - a \iff 90 - a^2 \le 2ab - 2a^2 \iff 90 + a^2 \le 2ab$. For any choice of a such that $0 < a$ and $a^2 < 90$, choose any number B such that $b \le B$ and $(90 + a^2)/2a \le B$; Theorem 2 guarantees any initial choice of x_0 in $[a, B]$ will yield a sequence which converges to a solution of $f(x)$. Because $[a, b]$ is a subset of $[a, B]$, $[a, b]$ inherits this property. ∎

3. If we begin with $x_0 = 10$, our computations progress as follows:

n	x_n	$x_n^2 - 90$	$-\left(x_n^2 - 90\right)/\left(2x_n\right)$
0	10.0	10.0	−0.5
1	9.5	0.25	−0.01315789
2	9.48684211	0.00017313	−0.00000912
3	9.48683298	0.00000000	0.00000000

Because $f(x_3)$ and $x_4 - x_3 = -f(x_3)/f'(x_3)$ are "essentially zero", $x_3 = 9.486833$ seems to be a satisfactory approximation to $\sqrt{90}$. If $x_0 = 9$, the resulting table has the same numerical entries (to eight decimal places) as the one above with $x_0 = 10$. The following table summarizes the computations which begin with $x_0 = 8$.

n	x_n	$x_n^2 - 90$	$-\left(x_n^2 - 90\right)/\left(2x_n\right)$
0	8.0	−26.0	1.625
1	9.625	2.640625	−0.13717533
2	9.48782468	0.01881707	−0.00099164
3	9.48683303	0.00000098	−0.00000005
4	9.48683298	0.00000000	0.00000000

(With $x_0 = 2$, it took six iterations to get a good approximation for $\sqrt{90}$; with $x_0 = 0.001$, it took seventeen.)

5. Let $f(x) = x^4 - 25$.

n	x_n	$x^4 - 25$	$-\left(x^4 - 25\right)/\left(4x_n^3\right)$
0	2.0	−9.0	0.28125
1	2.28125	2.08267307	−0.04385736
2	2.23739264	0.05929346	−0.00132349
3	2.23606915	0.00005259	−0.00000118
4	2.23606798	0.00000000	−0.00000000

7. Let $f(x) = x^2 - 7x + 5$, then $x_{n+1} - x_n = -\left(x_n^2 - 7x_n + 5\right)/\left(2x_n - 7\right)$.

n	x_n	$x_n^2 - 7x_n + 5$	$x_{n+1} - x_n$	n	x_n	$x_n^2 - 7x_n + 5$	$x_{n+1} - x_n$
0	0.0	5.0	0.71428571	0	7.0	5.0	−0.71428571
1	0.71428571	0.51020408	0.09157509	1	6.28571429	0.51020408	−0.09157509
2	0.80586081	0.00838600	0.00155634	2	6.19413919	0.00838600	−0.00155634
3	0.80741715	0.00000242	0.00000045	3	6.19258285	0.00000242	−0.00000045
4	0.80741760	0.00000000	0.00000000	4	6.19258240	0.00000000	−0.00000000

7: $y = x^2 - 7x + 5$

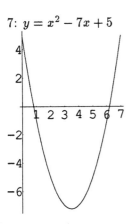

9: $y = x^3 - 8x^2 + 2x - 15$

9. Let $f(x) = x^3 - 8x^2 + 2x - 15$, then $f'(x) = 3x^2 - 16x + 2$ has zeroes $\left(8 \pm \sqrt{58}\right)/3$. At $\left(8 - \sqrt{58}\right)/3$, f has a local maximum ≈ -14.87; at $\left(8 + \sqrt{58}\right)/3$, f has a local minimum at ≈ -80.31. Therefore f has at most one real zero. Because $f(7) = -50 < 0 < 1 = f(8)$, the unique real zero of f is in the interval $[7, 8]$.

n	x_n	$f(x_n)$	$x_{n+1} - x_n$
0	7.0	-50.0	1.35135135
1	8.35135135	26.20772708	-0.33766932
2	8.01368203	1.90601172	-0.02868841
3	7.98499362	0.01317857	-0.00020114
4	7.98479248	0.00000065	-0.00000001
5	7.98479247	-0.00000000	0.00000000

n	x_n	$f(x_n)$	$x_{n+1} - x_n$
0	8.0	1.0	-0.01515152
1	7.98484848	0.00366962	-0.00005601
2	7.98479247	0.00000005	-0.00000000

11. a. See the graph below.

b. Let $f(x) = x - \cos x$, then $f'(x) = 1 + \sin x$ and the Newton sequence is $x_{n+1} = x_n - \left(\dfrac{x - \cos x}{1 + \sin x}\right)$.

n	x_n	$f(x_n)$	$x_{n+1} - x_n$
0	0.0	-1.0	1.0
1	1.0	0.45969769	-0.24963613
2	0.75036387	0.01892307	-0.01125098
3	0.73911289	0.00004646	-0.00002776
4	0.73908513	0.00000000	-0.00000000

n	x_n	$f(x_n)$	$x_{n+1} - x_n$
0	1.57079633	1.57079633	-0.78539816
1	0.78539816	0.07829138	-0.04586203
2	0.73953613	0.00075487	-0.00045096
3	0.73908518	0.00000008	-0.00000004
4	0.73908513	0.00000000	0.00000000

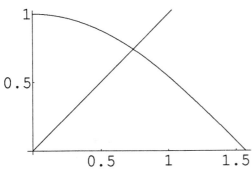

11 (a): $y = x$ & $y = \cos x$

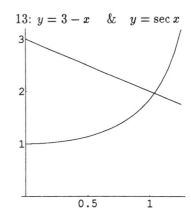

13: $y = 3 - x$ & $y = \sec x$

13. Let $f(x) = \sec x - (3 - x)$, then $f'(x) = \sec x \tan x + 1$ and the Newton sequence is $x_{n+1} = x_n - (\sec x - 3 + x)/(\sec x \tan x + 1)$.

n	x_n	$f(x_n)$	$x_{n+1} - x_n$
0	0.5	-1.36050607	0.83852022
1	1.33852022	2.68269906	-0.13853278
2	1.19998744	0.95960186	-0.11850034
3	1.08148710	0.20907294	-0.04185261
4	1.03963449	0.01382999	-0.00317169
5	1.03646280	0.00006707	-0.00001553
6	1.03644727	0.00000000	0.00000000

n	x_n	$f(x_n)$	$x_{n+1} - x_n$
0	1.0	-0.14918428	0.03842505
1	1.03842505	0.00856609	-0.00197181
2	1.03645324	0.00002581	-0.00000598
3	1.03644727	0.00000000	0.00000000

15. $f(x) = x^2 + 5x + 7 = \left(x + \frac{5}{2}\right)^2 + \frac{3}{4} \geq \frac{3}{4} > 0$; there is no real solution to $f(x) = 0$.

If we persist and start Newton's method with $x_0 = 0$, we obtain a nonconvergent sequence that begins
$0, -1.4, -2.29, -4.19, -3.12, -2.21, -3.64, -2.74, -1.08, -2.05, -3.11, -2.19, -3.56, -2.67, -0.422$.

15: $y = x^2 + 5x + 7$

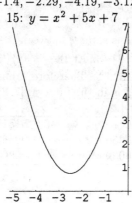

17: $y = x^4 - 8x^2 - 17$

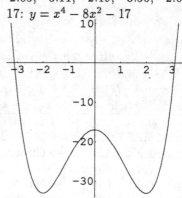

17. If $f(x) = x^4 - 8x^2 - 17$, then $f'(x) = 4x^3 - 16x = 4(x+4)x(x-4)$. There are no zeros of f in the interval $[-2, 2]$ because f has local minima of -33 at ± 2 and a local maximum of -17 at 0. Hence it is impossible to find an interval $[a, b]$ containing 1 and simultaneously satisfying conditions (i) and (ii) of Theorem 1 or Theorem 2. (Keeping f' bounded away from zero would force the interval to be a subinterval of $(0, 2)$ but finding an interval such that f changes sign would force the interval to extend left of 0 or to extend right of 2.)

Observe that $f(\pm 1) = -24$ and $f'(\pm 1) = \mp 12$; a Newton's method iteration started with $x_0 = 1$ will continue $-1, 1, -1, 1, \ldots, (-1)^k, \ldots$; this sequence oscillates between 1 and -1, it does not converge.

REMARK: The two real zeros of f are $\pm\sqrt{4 + \sqrt{33}} \approx \pm 3.122$.

Solutions 3.7   *The mean value theorem*   (pages 245-7)

1. $f'(c) = 4c = \dfrac{f(1) - f(-1)}{1 - (-1)} = \dfrac{(1 + 2(1)^2) - (1 + 2(-1)^2)}{1 - (-1)} = \dfrac{3 - 3}{2} = 0$ if and only if $c = 0$ (which is in $(-1, 1)$).

3. If $1 < c < 4$, then $f'(c) = \dfrac{1}{2\sqrt{c}} = \dfrac{f(4) - f(1)}{4 - 1} = \dfrac{\sqrt{4} - \sqrt{1}}{4 - 1} = \dfrac{1}{3}$ if and only if $c = \left(\dfrac{3}{2}\right)^2 = \left(\dfrac{9}{4}\right) = 2.25$.

5. If $1 < c < 2$, then $f'(c) = 3c^2 = \dfrac{2^3 - 1^3}{2 - 1} = 7$ if and only if $c = \sqrt{7/3} \approx 1.53$.

7. $f'(c) = (c - 1)(c - 5) + (c + 3)(c - 5) + (c + 3)(c - 1) = 3c^2 - 6c - 13$
$$= \dfrac{f(1) - f(-3)}{1 - (-3)} = \dfrac{(1 + 3)(1 - 1)(1 - 5) - (-3 + 3)(-3 - 1)(-3 - 5)}{1 - (-3)} = \dfrac{0 - 0}{4} = 0$$

if and only if $c = \dfrac{-(-6) \pm \sqrt{(-6)^2 - 4(3)(-13)}}{2(3)} = \dfrac{6 \pm \sqrt{192}}{6} = 1 \pm \dfrac{4}{\sqrt{3}}$. To get c in the interval $(-3, 1)$, we must choose $c = 1 - \dfrac{4}{\sqrt{3}} \approx -1.309$.

9. $\dfrac{f(\pi/2) - f(0)}{(\pi/2) - 0} = \dfrac{\cos(\pi/2) - \cos(0)}{(\pi/2) - 0} = \dfrac{0 - 1}{\pi/2} = -\dfrac{2}{\pi}$; in the interval $\left(0, \dfrac{\pi}{2}\right)$, $c = \sin^{-1}\left(\dfrac{2}{\pi}\right) \approx 0.6901$ radians is the unique solution to $f'(c) = -\sin c = -2/\pi$.

11. If $0 < c < \dfrac{\pi}{12}$, then $f'(c) = 2\cos 2c = \dfrac{\sin 2(\pi/12) - \sin 2(0)}{(\pi/12) - 0} = \dfrac{6}{\pi}$ if and only if $c = \dfrac{1}{2}\cos^{-1}\left(\dfrac{3}{\pi}\right) \approx$ 0.1507 radians.

13. If $r = p^n$, then $dr/dp = np^{n-1}$. The ESC (defined in Example 7) is the value of p such that dr/dp equals 1; if $np^{n-1} = 1$, then $p^{n-1} = 1/n$ and
$$\epsilon_n = p = \left(\frac{1}{n}\right)^{1/(n-1)} = \frac{1}{^{n-1}\!\sqrt{n}}.$$

15. Let t be the number of minutes elapsed from the moment when the first police officer "clocks" the car and let $f(t)$ be the distance traveled by the car from that point. Thus, $f(0) = 0$ and $f(6) = 7$. It is physically plausible that f is differentiable; therefore, the Mean Value Theorem implies there is some c between 0 and 6 such that $f'(c) = (7-0)/(6-0) = 7/6$. Since f' measures the car's speed and since $7/6$ miles per minute equals 70 miles per hour, there must have been at least one instant when the car was going that fast. ∎

REMARK: We did not use the information that $f'(0) = 65/60$ and $f'(6) = 55/60$.

17. If $p(x) = 3x^4 - 20x^3 + 36x^2 - 1$, then $p'(x) = 12x^3 - 60x^2 + 72x = 12x(x-2)(x-3)$. We see that $p'(0) = 0 = p'(2)$ and $p'(x) \neq 0$ if $0 < x < 2$; the result of Problem 23 then implies there is at most one r such that $0 < r < 2$ and $p(r) = 0$.

19. a. $f(3) = |3 - 1| - 2 = 2 - 2 = 0$; $f(-1) = |-1 - 1| - 2 = 2 - 2 = 0$.
    b. $$f'(x) = \begin{cases} -1 & \text{if } x < 1 \\ \text{undefined} & \text{if } x = 1 \\ 1 & \text{if } x > 1 \end{cases}$$
    c. f is not differentiable on $(-1, 3)$ because $f'(1)$ does not exist; f does not satisfy the hypotheses of Rolle's theorem on $(-1, 3)$.

21. If $f(a) = f(b)$, then $f(b) - f(a) = 0$; if in addition the Mean Value Theorem applies to f on (a, b), there is some c in (a, b) such that $f'(c) = \dfrac{f(b) - f(a)}{b - a} = \dfrac{0}{b - a} = 0$. ∎

23. Let $g(x) = f(x) - cx$, then $g'(x) = f'(x) - c = 0$. Therefore, by Theorem 3, there is some constant d such that $g(x) = d$; thus $f(x) = cx + g(x) = cx + d$. ∎

25. Suppose the cubic polynomial has four distinct roots $r_1 < r_2 < r_3 < r_4$. By Rolle's theorem, its derivative has a root in each of the intervals (r_1, r_2), (r_2, r_3), and (r_3, r_4). On the other hand, that derivative is a quadratic polynomial which, according to Problem 30, can have at most two roots. This contradiction implies the cubic can have at most three roots. ∎

27. If p were zero at two or more points within (a, b), then there would be some c between those two points such that $p'(c) = 0$. We are told that p' is never zero in (a, b), therefore p can be zero at no more than one point. ∎

REMARK: The assumption that p is a polynomial is used merely to obtain the benefits of Rolle's theorem; the assumption $p'(a) = 0 = p'(b)$ was not used.

29. Consider the statement that each polynomial of degree n has the value 0 at no more than n places. It is true if $n = 1$ because every polynomial of degree one, $a_0 + a_1 x$ with $a_1 \neq 0$, has exactly one real zero, $x = -a_0/a_1$. Suppose the statement is true if $n = k$ and also suppose $p(x) = a_0 + a_1 x + \cdots + a_{k+1} x^{k+1}$ with $a_{k+1} \neq 0$ is a polynomial of degree $k + 1$. Rolle's theorem implies there is a zero of P' between each pair of zeroes of P; $P'(x) = a_1 + \cdots + (k+1)a_{k+1}x^k$ is a polynomial of degree k and therefore has at most k zeroes; P has at most one more zero than P'; hence P has at most $k + 1$ zeroes. Therefore the statement is also true if $n = k + 1$. The Principle of Mathematical Induction lets us conclude that the statement is true for each positive integer choice of n. ∎

31. The Mean Value theorem says w_1 is strictly between 0 and w. (In fact, $w_1 = w/2$.)

33. Suppose $\alpha > 1$. We can prove $(1+x)^\alpha > 1 + \alpha x$ for all $x > 0$ by letting $f(x) = (1+x)^\alpha - (1+\alpha x)$ and then showing $f(x) > 0$ on the interval $(0, \infty)$. f is differentiable on $(-1, \infty)$ so it certainly satisfies the hypotheses of the Mean Value Theorem on any interval $[0, b]$. Therefore there is at least one c such that $0 < c < b$ and $f'(c) = \dfrac{f(b) - f(0)}{b - 0}$. Since $f(0) = 1^\alpha - 1 = 0$, this is equivalent to $f(b) = f'(c)b$. It is easy to show $f'(x) >$

0 on $(0, \infty)$ [since $x > 0 \implies 1 + x > 1 \implies (1+x)^{\alpha - 1} > 1^{\alpha - 1} = 1 \implies f'(x) = \alpha(1+x)^{\alpha - 1} - \alpha > 0$]; therefore $f'(c)b$ is the product of two positive values. We conclude that $f(b)$ must be positive. Since b was arbitrary in $(0, \infty)$, the proof is complete. ∎

REMARK: Problem 3.6.46 generalizes this result.

35.   Suppose $1 < a$. If x and y are in $[a, \infty)$, then $0 < 1/x, 1/y \le 1/a < 1$ and

$$\left| \frac{1}{x} - \frac{1}{y} \right| = \frac{|y - x|}{|xy|} \le \frac{|y - x|}{a^2} = \left(\frac{1}{a^2} \right) |x - y|.$$

Because $1/a^2 < 1$, this means $f(x) = 1/x$ is a contraction on $[a, \infty)$. ∎

REMARK: $f(x) = 1/x$ is not a contraction on $[1, \infty)$.

37. $\left| (1 + x^2) - (1 + y^2) \right| = |x + y| \cdot |x - y|$; if x and y are in $[0, b)$, the best bound for $|x + y|$ is $2b$; $2b < 1$ if and only if $b < 1/2$. Therefore $f(x) = 1 + x^2$ is a contraction on $[0, b)$ if and only if $0 < b < 1/2$.

39.   a. $(C^2 + S^2)' = 2CC' + 2SS' = 2C(-S) + 2S(C) = 0$; Theorem 3 implies $C^2 + S^2$ is constant. ∎

   b.   Let $A(x) = C(x) - \cos x$ and $B(x) = S(x) - \sin x$; then $A' = -B$ and $B' = A$. Therefore, $(A^2 + B^2)' = 2A(-B) + 2B(A) = 0$ and $(C(x) - \cos x)^2 + (S(x) - \sin x)^2$ is constant. ∎

   c.   If $C(0) = 1$ and $S(0) = 0$, then $A(x)^2 + B(x)^2 = A(0)^2 + B(0)^2 = (1 - 1)^2 + (0 - 0)^2 = 0$. Therefore $A(x) = C(x) - \cos x = 0$ and $B(x) = S(x) - \sin x = 0$. ∎

41.   Suppose there exist $a > b \ge 1$ such that $f(a) = a^2$ and $f(b) = b^2$. Then there is a c between a and b such that $f'(c) = \dfrac{b^2 - a^2}{b - a} = b + a > 2$; this contradicts the assumption that $f'(x) \le 2$. ∎

43.   If $f'(a) < 0 < f'(b)$, then Problem 1.7.65 says there is some c in (a, b) such that $f'(c) = 0$ or $f'(c)$ does not exist. The second case cannot occur if f is differentiable. If $f'(a) > 0 > f'(b)$, then $-f'(a) < 0 < -f'(b)$ and the preceding case implies $-f'(c) = 0$ for some c in (a, b), thus $f'(c) = 0$. ∎

REMARK: The last part of Solution 1.7.65 may be rephrased as a direct proof using tactics demonstrated in the text's proof of Rolle's theorem. Ideas in Section 3.3 may help you shorten that solution further.

Solutions 3.8   *Application of mean-value th'm: linearization of a function*   (pages 254-5)

1.   The linearization of $f(x) = x^6$ at $x_0 = 1$ is $P_1(x) = f(1) + f'(1)(x - 1) = 1^6 + 6(1^5)(x - 1) = 1 + 6(x - 1)$. We approximate $f(1.1)$ by $P_1(1.1) = 1 + 6(1.1 - 1) = 1 + 0.6 = 1.6$. Because $|f''(x)| = 30x^4 \le 30(1.1^4)$ on the interval $(1, 1.1)$, the corollary to Theorem 4 implies $|f(1.1) - P_1(1.1)| \le 30(1.1^4)(1.1 - 1)^2/2 = 0.219615$.

REMARK: $1.1^6 = 1.771561$ and the approximation error is 0.171561.

3.   The linearization of $f(x) = \sqrt{x}$ at $x_0 = 4$ is $P_1(x) = f(4) + f'(4)(x - 4) = \sqrt{4} + (1/2\sqrt{4})(x - 4) = 2 + (1/4)(x - 4)$. We approximate $f(4.2)$ by $P_1(4.2) = 2 + (1/4)(4.2 - 4) = 2 + 0.05 = 2.05$. On the interval $(4, 4.2)$ $|f''(x)| = (1/4)x^{-3/2} \le (1/4)(4^{-3/2}) = 1/32$, thus $|f(4.2) - P_1(4.2)| \le (1/32)(4.2 - 4)^2/2 = 0.000625$.

REMARK: $\sqrt{4.2} \approx 2.0493902$ and the approximation error is about -0.0006098.

5.   The linearization of $f(x) = (1+x)/(1-x) = -1 + 2(1-x)^{-1}$ at $x_0 = 0$ is $P_1(x) = 1 + 2(1^{-2})(x - 0) = 1 + 2x$. We approximate $f(0.05)$ by $P_1(0.05) = 1 + 2(0.05) = 1.1$. On the interval $(0, 0.05)$ $|f''(x)| = 4|1 - x|^{-3} \le 4(0.95^{-3}) \approx 4.665$, thus $|f(0.05) - P_1(0.05)| \le 4(0.95^{-3})(0.05 - 0)^2/2 \approx 0.00583$.

REMARK: $1.05/0.95 \approx 1.10526$ and the approximation error is about 0.00526.

7.   If $f(x) = \cos x$, then $f'(x) = -\sin x$ and $f'(0) = 0$. Therefore the linearization of $\cos$ at $x_0 = 0$ is $P_1(x) = \cos 0 = 1$ and we approximate $\cos 0.1$ by 1. $|f''(x)| = \cos x \le \cos 0 = 1$, thus $|\cos 0.1 - P_1(0.1)| \le (1/2)(0.1 - 0)^2 = 0.005$.

REMARK: $\cos 0.1 \approx 0.995004$ and the approximation error is about -0.004996.

9.   The linearization of $\sin$ at $x_0 = \pi$ is $P_1(x) = \sin \pi + (\cos \pi)(x - \pi) = -(x - \pi)$ and we approximate $\sin(\pi - 0.05)$ by $-((\pi - 0.05) - \pi) = 0.05$. On the interval $(\pi - 0.05, \pi)$, $\left| \dfrac{d^2}{dx^2} \sin x \right| = |-\sin x| \le \sin \dfrac{5\pi}{6} = 0.5$, thus $|\sin(\pi - 0.05) - P_1(\pi - 0.05)| \le (0.5/2)((\pi - 0.05) - \pi)^2 = 0.000625$.

REMARK: $\sin(\pi - 0.05) \approx 0.049979$ and the approximation error is about -0.0000208.

11.   If $f(x) = ax + b$, then $f'(x) = a$ and the linearization at $(x_0, f(x_0))$ is $y = f(x_0) + f'(x_0) \cdot (x - x_0) = (ax_0 + b) + a \cdot (x - x_0) = ax + b = f(x)$. ∎

13.   Let A be a fixed number such that $a < A < b$ and define
$$Q(x, A) = f(A) + f'(A)(x - A) + \frac{f''(A)(x - A)^2}{2}.$$

(Note that setting $A = x_0$ gives $P_2(x) = Q(x, x_0)$, our quadratic approximation to f at x_0.) Next, we let B also be a fixed number such that $a < B < b$ and $B \neq A$, then let
$$K = \frac{f(B) - Q(B, A)}{(B - A)^3}.$$

Now we define the auxiliary function
$$g(t) = f(B) - Q(B, t) - K \cdot (B - t)^3 = f(B) - \left(f(t) + f'(t)(B - t) + \frac{f''(t)(B - t)^2}{2} \right) - K \cdot (B - t)^3.$$

Thus $g(B) = f(B) - (f(B) + 0 + 0) - K \cdot 0 = f(B) - f(B) = 0$ and $g(A) = f(B) - Q(B, A) - K \cdot (B - A)^3 = f(B) - Q(B, A) - (f(B) - Q(B, A)) = 0$. Rolle's theorem now implies there is some c between A and B such that $g'(c) = 0$. Our next step is get a formula for g':
$$g'(t) = 0 - \left(f'(t) + f''(t)(B - t) + f'(t)(-1) + \frac{f'''(t)(B - t)^2 + f''(t)2(B - t)(-1)}{2} \right) - K \cdot 3(B - t)^2(-1)$$
$$= -\frac{f'''(t)(B - t)^2}{2} + 3K \cdot (B - t)^2 = \left(3K - \frac{f'''(t)}{2} \right)(B - t)^2.$$

The c guaranteed by Rolle's theorem is different from B, thus we can solve $g'(c) = 0$ to find $K = \frac{f'''(c)}{6}$. Therefore, for every choice of $A = x_0$ and $B = x$ in $[a, b]$, there is some c between x_0 and x such that
$$f(x) = Q(x, x_0) + K \cdot (x - x_0)^3 = \left(f(x_0) + f'(x_0)(x - x_0) + \frac{f''(x_0)(x - x_0)^2}{2} \right) + \frac{f'''(c)(x - x_0)^3}{6}. ∎$$

15. a.   If $f(x) = x^6$ and $x_0 = 1$, then $P_2(x) = 1 + 6(x - 1) + 15(x - 1)^2$.

 b. $f(1.1) \approx P_2(1.1) = 1 + 0.6 + 0.15 = 1.75$.

 c. $f'''(x) = 120x^3$, let $K = 120 \cdot 1.1^3 = 159.72$; then $|f(1.1) - P_2(1.1)| \leq (K/6) \cdot 0.1^3 = 0.02662$.

 d. $P_1(1.1) = 1.6 < P_2(1.1) = 1.75 < 1.1^6 = 1.771561$.

17. a.   If $f(x) = \sqrt{x}$ and $x_0 = 4$, then $P_2(x) = 2 + (1/4)(x - 4) - (1/64)(x - 4)^2$.

 b. $f(4.2) \approx P_2(4.2) = 2 + 0.05 - 0.000625 = 2.049375$.

 c. $f'''(x) = (3/8)x^{-5/2}$, let $K = (3/8) \cdot 4^{-5/2} = 3/256$; then $|f(4.2) - P_2(4.2)| \leq (K/6) \cdot 0.2^3 = 0.000015625$.

 d. $P_2(4.2) = 2.049375 < \sqrt{4.2} \approx 2.049390153 < P_1(4.2) = 2.05$.

19. a.   If $f(x) = (1 + x)/(1 - x) = -1 + 2/(1 - x)$ and $x_0 = 0$, then $P_2(x) = 1 + 2x + 2x^2$.

 b. $1.05/0.95 = f(0.05) \approx P_2(0.05) = 1 + 0.10 + 0.005 = 1.105$.

 c. $f'''(x) = 12(1 - x)^{-4}$, let $K = 12 \cdot 0.95^{-4} \approx 14.73$; then $|f(0.05) - P_2(0.05)| \leq (K/6) \cdot 0.05^3 \approx 0.000307$.

 d. $P_1(0.05) = 1.1 < P_2(0.05) = 1.105 < f(0.05) = 1.05/0.95 \approx 1.105263158$.

21. a.   If $f(x) = \cos x$ and $x_0 = 0$, then $P_2(x) = 1 - (1/2)x^2$.

 b. $\cos 0.1 \approx P_2(0.1) = 1 - 0.005 = 0.995$.

 c. $f'''(x) = \sin x$, let $K = \sin 0.1 \approx 0.0998$; then $|\cos(0.1) - P_2(0.1)| \leq (K/6) \cdot 0.1^3 = 0.000017$.

 d. $P_2(0.1) = 0.995 < \cos 0.1 \approx 0.995004 < P_1(0.1) = 1$.

23. a.   If $f(x) = \sin x$ and $x_0 = \pi$, then $P_2(x) = -(x - \pi)$.

 b. $\sin(\pi - 0.05) \approx P_2(\pi - 0.05) = -((\pi - 0.05) - \pi) = 0.05$.

 c. $f'''(x) = -\cos x$, let $K = -\cos \pi = 1$; then $|\sin(\pi - 0.05) - P_2(\pi - 0.05)| \leq (K/6) \cdot 0.05^3 = 0.0000208$.

 d. $\sin(\pi - 0.05) \approx 0.049979 < P_1(\pi - 0.05) = P_2(\pi - 0.05) = 0.05$.

Solutions 3.9   *Newton's method and Chaos*   (pages 260-1)

1. a. Let $g(x) = 2x^3/(3x^2 - 1)$, then the Newton iterates for $f(x) = x^3 - x$ are given by $x_{n+1} = g(x_n)$. $g'(x) = 6x^2(x^2 - 1)/(3x^2 - 1)^2$. If $x > 1/\sqrt{3}$, then $g(x) > 0$; if $1/\sqrt{3} < x < 1$, then $g'(x) < 0$ and if $1 < x$, then $g'(x) > 0$. Since $g'(1) = 0$, this implies $g(1) = 1$ is the minimum value of g on the interval $(1/\sqrt{3}, \infty)$. ∎

    b. If $f(x) = x^3 - x$, then $f'(x) = 3x^2 - 1$ and $f''(x) = 6x$. Both f' and f'' are positive and increasing on $(1/\sqrt{3}, \infty)$. Therefore, if $1/\sqrt{3} < a < b$, then the minimum of $|f'|$ on $[a, b]$ is $f'(a)$ and the maximum of $|f''|$ on $[a, b]$ is $f''(b)$. If $a = 0.99$ and $b = 1.2$, then $f(a) = -0.019701 < 0 < 0.528 = f(b)$. Furthermore the m and M of Theorem 2 are $m = f'(a) = 1.9403$ and $M = f''(b) = 7.2$; condition (iv) of Theorem 2 is satisfied because $M/m \approx 3.71$. Hence any choice of x_0 in $[a, b]$ yields a sequence of Newton iterates which converge to 1, the unique solution of $f(x) = 0$ in that interval. ∎

    c. Let $h(x) = g(x)/x = 2x/(3x^2 - 1)$, then $h'(x) = -3(2x^2 + 1)/(3x^2 - 1)^2 < 0$ if $x > 1/\sqrt{3}$; hence h decreases throughout $(1/\sqrt{3}, \infty)$. Since $h(3/\sqrt{7}) = 3\sqrt{7}/10 \approx 0.794 < 0.8$, we see that $x_n > 3/\sqrt{7}$ implies $x_{n+1} = g(x_n) = h(x_n) \cdot x_n < h(3/\sqrt{7}) \cdot x_n < 0.8 \cdot x_n$. ∎

    d. The sequence $\{0.8^n\}$ decreases to zero; thus, if $x_0 > 0$, there is an $N = N(x_0)$ such that $n \geq N$ implies $0.8^n < 1.2/x_0$. Therefore, if $x_0 > 1.2 > 3/\sqrt{7}$ and $n \geq N$, then $x_{n+1} < 0.8^n \cdot x_0 < 1.2$. ∎

    e. Part (d) implies that a sequence of Newton iterates begun at some $x_0 > 1.2$ eventually yields a term x_N less than 1.2. The sequence from then on coincides with a sequence begun at $x_0^* = x_N$ and part (b) implies that will converge to 1. ∎

3. If $x_0 = -0.5$, then $x_1 = 2 \cdot (-0.5)^3 / \left(3(-0.5)^2 - 1\right) = 2 \cdot (-0.125)/(3 \cdot 0.25 - 1) = -0.25/(0.75 - 1) = -0.25/(-.25) = 1$.

5. If $f(x) = x^3 - x$, then $f(-w) = (-w)^3 - (-w) = -w^3 + w = -(w^3 - w) = -f(w)$; we also can see that $f'(-w) = 3(-w)^2 - 1 = 3w^2 - 1 = f'(w)$. Therefore, if $g(x) = x - f(x)/f'(x)$, then $g(-w) = (-w) - f(-w)/f'(-w) = -w - (-f(w))/f'(w) = -w + f(w)/f'(w) = -g(w)$. ∎

REMARK: Solution 2 says this in a more compact way.

7. If $x_0 = 2.5$, then $x_1 = 2.5 - p(2.5)/p'(2.5) = 2.5 - 0.375/0.25 = 1.0$ and $x_2 = 1 - p(1)/p'(1) = 1 - 0/2 = 1 = x_1$.

9.

n	x_n	$f(x_n)$	$x_{n+1} - x_n$
0	2.445	−0.35687888	−0.87917442
1	1.56582558	0.35232932	0.81092607
2	2.37675165	−0.32327484	−0.56302535
3	1.81372631	0.17981039	0.20070222
4	2.01442853	−0.01442552	−0.01443454
5	1.99999399	0.00000601	0.00000601
6	2.00000000	0.00000000	0.00000000

11. $y = 3x^4 - 4x^3 - 12x^2 + 3$

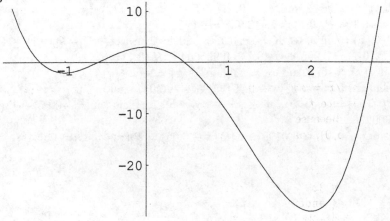

13.

n	w_n	W_n	x_n	X_n
0	-4.000000	-0.1640000	0.0730000	7.000000
1	-3.033565	-0.9215343	1.6931700	5.438690
2	-2.340384	-0.1733025	0.1349586	4.311094
3	-1.859610	-0.8860536	0.9437495	3.530389
4	-1.547578	-0.3742832	0.5708496	3.041982
5	-1.373115	-0.6122055	0.4839787	2.806206
6	-1.305842	-0.5909310	0.4763493	2.746725
7	-1.295258	-0.5910426	0.4762877	2.743106
8	-1.295005	-0.5910426	0.4762877	2.743093
9	-1.295005	-0.5910426	0.4762877	2.743093

15. If $p(x) = (x-1)(x-2)(x-3)$ and $x_0 = 3+t$, then $p(x_0) = p(3+t) = t^3 + 3t^2 + 2t = P(t)$ and $p'(x_0) = p'(3+t) = 3t^2 + 6t + 2 = P'(t)$. If $x_0 > 3$, then $t > 0$ and

$$x_1 = x_0 - \frac{p(x_0)}{p'(x_0)} = (3+t) - \frac{P(t)}{P'(t)} = 3 + \frac{t \cdot P'(t) - P(t)}{P'(t)} = 3 + \frac{2t^3 + 3t^2}{3t^2 + 6t + 2} = 3 + t \cdot \frac{2t^2 + 3t}{3t^2 + 6t + 2}$$
$$< 3 + t \cdot \frac{2}{3}.$$

Therefore $3 < x_k < 3 + t \cdot (2/3)^k$ and $\lim_{k \to \infty} x_k = 3$. ∎

Solutions 3.R   *Review*   (pages 261-3)

1. When the mass is x meters from the base of the 5 m tower, the straight line distance from the mass to the top of the tower is $s = \sqrt{5^2 + x^2}$. Assuming s measures the length of the rope between mass and tower and assuming $\dfrac{ds}{dt} = 1.5$ m/sec, then $1.5 = \dfrac{ds}{dt} = \left(\dfrac{x}{\sqrt{25 + x^2}}\right)\dfrac{dx}{dt}$ or $\dfrac{dx}{dt} = 1.5\dfrac{\sqrt{25 + x^2}}{x}$. When $x = 3$ m, $dx/dt = 1.5\sqrt{34}/3 = \sqrt{17/2} \approx 2.915$ m/sec.

3. $f(x) = x^2 - 3x - 4 = (x+1)(x-4)$ has domain $\mathbb{R}$, y-intercept -4, and x-intercepts -1 and 4. $f'(x) = 2x - 3$ so f decreases on $(-\infty, 3/2)$, has its global minimum of $-25/4$ at $3/2$, and increases on $(3/2, \infty)$. Because $f''(x) = 2$ is always positive, f is always concave up.

3: $y = x^2 - 3x - 4$

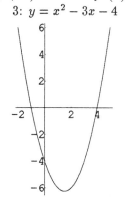

5: $y = \sqrt[3]{x}$

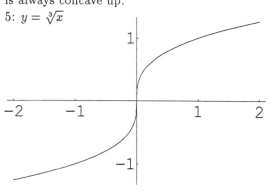

5. If $f(x) = \sqrt[3]{x} = x^{1/3}$, then $f'(x) = (1/3)x^{-2/3}$ and $f''(x) = -(2/9)x^{-5/3}$. Since $f'(x) > 0$ for all nonzero x and since f is continuous everywhere, the function is always increasing. The y-axis is a vertical tangent at $(0,0)$ because $\lim_{x \to 0} f'(x) = \infty$. Since $x^{-5/3}$ has the same sign as x, the graph is concave up on the interval $(-\infty, 0)$, concave down on $(0, \infty)$, and has inflection point $(0,0)$.

7. If $f(x) = |x^2 - 4| = \begin{cases} x^2 - 4 & \text{if } x \le -2 \text{ or } 2 \le x \\ 4 - x^2 & \text{if } -2 < x < 2 \end{cases}$, then

$$f'(x) = \begin{cases} 2x & \text{if } x < -2 \text{ or } 2 < x \\ \text{undefined} & \text{if } x = -2 \text{ or } x = 2 \\ -2x & \text{if } -2 < x < 2 \end{cases} \quad \text{and} \quad f''(x) = \begin{cases} 2 & \text{if } x < -2 \text{ or } 2 < x \\ \text{undefined} & \text{if } x = -2 \text{ or } x = 2 \\ -2 & \text{if } -2 < x < 2. \end{cases}$$

The function decreases on $(-\infty, -2) \cup (0, 2)$, reaches its minimum value of 0 at ± 2, and increases on $(-2, 0) \cup (2, \infty)$. The graph is concave down on $(-2, 2)$, has $(\pm 2, 0)$ as inflection points, and is concave up elsewhere.

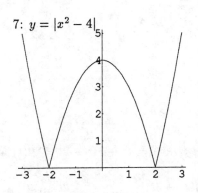

7: $y = |x^2 - 4|$

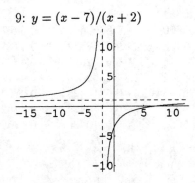

9: $y = (x - 7)/(x + 2)$

9.   If $f(x) = (x - 7)/(x + 2) = 1 - 9(x + 2)^{-1}$, then $f'(x) = 9(x + 1)^{-2}$ and $f'(x) = -18(x + 1)^{-3}$. The function is always increasing since $f'(x) > 0$ whenever $f(x)$ is defined; the graph is concave up on $(-\infty, -2)$ and concave down on $(-2, \infty)$. The line $x = -2$ is a vertical asymptote and the line $y = 1$ is an horizontal asymptote.

11. $f(x) = x(x - 1)(x - 2)(x - 3)$ has domain $\mathbb{R}$, y-intercept 0, and x-intercepts 0, 1, 2, 3. $f'(x) = 2(2x - 3) \cdot (x^2 - 3x + 1)$ and $f''(x) = 2(6x^2 - 18x + 11)$. f decreases on $(-\infty, (3 - \sqrt{5})/2) \cup (3/2, (3 + \sqrt{5})/2)$, increases on $((3 - \sqrt{5})/2, 3/2) \cup ((3 + \sqrt{5})/2, \infty)$, reaches its global minimum of -1 at $(3 \pm \sqrt{5})/2$ and a local maximum of $9/16$ at $3/2$. The graph of f is concave up on $(-\infty, (9 - \sqrt{15})/6) \cup ((9 + \sqrt{15})/6, \infty)$, concave down on $((9 - \sqrt{15})/6.(9 + \sqrt{15})/6)$, and has inflection points $((9 \pm \sqrt{15})/6, -11/36)$.

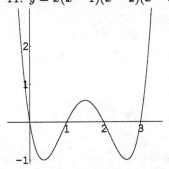

11: $y = x(x - 1)(x - 2)(x - 3)$

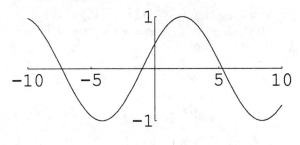

13: $y = \sin((x + 1)/2)$

13. $f(x) = \sin\left(\dfrac{x + 1}{2}\right)$ has domain $\mathbb{R}$, y-intercept $\sin 0.5 \approx 0.479$ and x-intercepts $-1 + 2k\pi$, k an integer. Because $f(x + 4\pi) = f(x)$, it is sufficient to analyze f on an interval of length 4π, say $[-1, 4\pi - 1]$. $f'(x) = \dfrac{1}{2} \cos\left(\dfrac{x + 1}{2}\right)$ so f increases on $(-1, \pi - 1) \cup (3\pi - 1, 4\pi - 1)$, decreases on $(\pi - 1, 3\pi - 1)$, reaches its global maximum of 1 at $\pi - 1$ and its global minimum of -1 at $3\pi - 1$. $f''(x) = -\dfrac{1}{4} \sin\left(\dfrac{x + 1}{2}\right)$ so f is concave down on $(-1, 2\pi - 1)$, concave up on $(2\pi - 1, 4\pi - 1)$, and has inflection points at $(-1, 0)$, $(2\pi - 1, 0)$, and $(4\pi - 1, 0)$.

15. $f(x) = \cot(\pi - x)$ has domain $\mathbb{R} - \{k\pi\}$; no y-intercept and x-intercepts $(2k + 1)(\pi/2)$. Because $f'(x) = \csc^2(\pi - x) > 0$, f is always increasing (on the intervals where f is defined). $f''(x) = 2\csc^2(\pi - x)\cot(\pi - x) = 2\csc^2(\pi - x)f(x)$, so f is concave up where f is positive, concave down where it's negative, and has an inflection point where it crosses the x-axis. The lines $x = k\pi$ are vertical asymptotes because $\lim_{x \to k\pi}|f(x)| = \infty = \lim_{x \to k\pi}|f'(x)|$.

15: $y = \cot(\pi - x)$

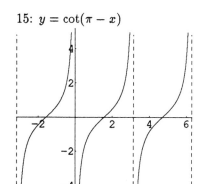

17: $y = (x+2)(x-2)/x = x-4/x$

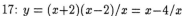
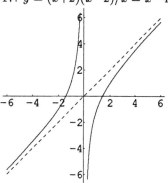

17.   If $f(x) = \dfrac{(x+2)(x-2)}{x} = x - \dfrac{4}{x}$, then $f'(x) = 1 + 4x^{-2} > 0$ and $f''(x) = -8x^{-3}$. The function is always increasing; the graph is concave up on $(-\infty, 0)$ and concave down on $(0, \infty)$, there are no inflection points. The y-axis is an asymptote since $\lim_{x \to 0\pm} f(x) = \mp\infty$, the line $y = x$ is an asymptote because $\lim_{x \to \pm\infty} (f(x) - x) = \lim_{x \to \pm\infty} (-4/x) = 0$.

19.   a. $f(x) = 2x^3 + 9x^2 - 24x + 3$ implies $f'(x) = 6(x+4)(x-1)$ and $f''(x) = 6(2x+3)$. The only critical point for f which is in $[-2, 5]$ is 1. $f(-2) = 71$, $f(1) = -10$, $f(5) = 358$; hence $f_{\max}[-2, 5] = 358$ and $f_{\min}[-2, 5] = -10$.

19 (a): $y = 2x^3 + 9x^2 - 24x + 3$

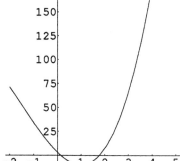

19 (b): $y = (x-2)^{1/3}$

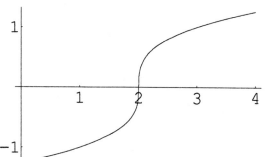

   b. $g(x) = (x-2)^{1/3}$ implies $g'(x) = \frac{1}{3}(x-2)^{-2/3}$. Because $g'(2)$ is not defined, 2 is a critical point. $g(0) = (-2)^{1/3} = -2^{1/3} \approx -1.26$, $g(2) = 0$, $g(4) = 2^{1/3} \approx 1.26$; thus $g_{\max}[0, 4] = 2^{1/3}$ and $g_{\min}[0, 4] = -2^{1/3}$.

   c. $h(x) = \dfrac{x+1}{x^2 - 4} = \dfrac{1}{4}\left(\dfrac{1}{x+2} + \dfrac{3}{x-2}\right)$ implies $h'(x) = -\dfrac{1}{4}\left((x+2)^{-2} + 3(x-2)^{-2}\right) = -\dfrac{x^2 + 2x + 4}{\left(x^2 - 4\right)^2}$

   which is always negative because $x^2 + 2x + 4 = (x+1)^2 + 3 \geq 3$, thus h has no critical points and is always decreasing. (h and h' are undefined at the same places, those places are outside the interval $[-1, 1]$.) Hence $h_{\max}[-1, 1] = h(-1) = 0$ and $h_{\min}[-1, 1] = h(1) = -2/3$.

19 (c): $y = (x+1)/(x^2 - 4)$

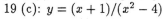
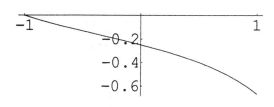

19 (d): $y = (x-1)^2 + (2x+1)^2$

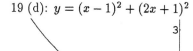
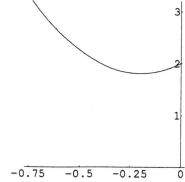

d. $F(x) = (x-1)^2 + (2x+1)^2 = 5x^2 + 2x + 2 = 5\left(x + \dfrac{1}{5}\right)^2 + \dfrac{9}{5}$ implies $F'(x) = 10x + 2$. $F_{\min}(-\infty, 0] = f(-1/5) = 9/5$ but $F_{\max}(-\infty, 0]$ does not exist because F is unbounded.

21.   A rectangle of length x and width y has perimeter $P = 2(x + y)$ and area $A = xy$. We're told P is constant, we find critical points of A by analyzing the equation $A' = 0$; differentiate implicitly with respect to x: $\begin{cases} P = 2(x+y) \\ A = xy \end{cases} \implies \begin{cases} 0 = P' = 2(1 + y') \\ 0 = A' = y + xy' \end{cases} \implies \begin{cases} y' = -1 \\ y = -xy' = x. \end{cases}$ Therefore $x = P/4 = 800/4 = 200\,\text{m} = y$ yields the only critical point of A. Furthermore, $A'' = (y + xy')' = (y - x)' = y' - 1 = (-1) - 1 < 0$ so $A(200\,\text{m}) = 200^2\,\text{m}^2$ is a global maximum for A.

23.   Suppose the barrel has radius r and height h. Volume V is constant, function C is proportional to cost.
$\begin{cases} 128\pi = V = \pi r^2 h \\ \quad C = (1)2\left(\pi r^2\right) + (3)(2\pi rh) = 2\pi\left(r^2 + 3rh\right) \end{cases} \overset{' \equiv \frac{d}{dr}}{\implies} \begin{cases} 0 = V' = \pi\left(2rh + r^2 h'\right) \\ \quad C' = 2\pi\left(2r + 3\left(h + rh'\right)\right) \end{cases}$
$\implies \begin{cases} h' = -2h/r \\ C' = 2\pi(2r + 3(h - 2h)) = 2\pi(2r - 3h) \\ C'' = 2\pi\left(2 - 3h'\right) = 2\pi(2 + 6h/r) > 0. \end{cases}$

If $C' = 0$, then $h = 2r/3$, $128\pi = \pi r^2(2r/3) = (2\pi/3)r^3$, $r = \sqrt[3]{192} = 4\sqrt[3]{3}$ and $h = (8/3)\sqrt[3]{3}$.

25.   Let $f(x) = x^3 - 2x^2 + 5x - 8$, then $f'(x) = 3x^2 - 4x + 5 = 3\left(x - \dfrac{2}{3}\right)^2 + \dfrac{11}{3} \geq \dfrac{11}{3} > 0$; therefore, f has at most one real zero. Because $f(1) = -4 < 0 < 16 = f(3)$, the unique real zero of f is in the interval $[1, 3]$. The Newton iteration is $x_{n+1} = x_n - \left(\dfrac{x^3 - 2x^2 + 5x - 8}{3x^2 - 4x + 5}\right)$.

n	x_n	$f\left(x_n\right)$	$x_{n+1} - x_n$	n	x_n	$f\left(x_n\right)$	$x_{n+1} - x_n$
0	1.0	−4.0	1.0	0	3.0	16.0	−0.8
1	2.0	2.0	−0.22222222	1	2.2	3.968	−0.37014925
2	1.77777778	0.18655693	−0.02531174	2	1.82985075	0.57953384	−0.07501417
3	1.75246603	0.00211940	−0.00029422	3	1.75483658	0.01921403	−0.00266159
4	1.75217182	0.00000028	−0.00000004	4	1.75217499	0.00002311	−0.00000321
5	1.75217178	0.00000000	−0.00000000	5	1.75217178	0.00000000	−0.00000000

27. $5c^4 = f'(c) = \dfrac{f(1) - f(0)}{1 - 0} = 1$ provided $c = 5^{-1/4} \approx 0.6687$.

29. $\dfrac{1}{5}c^{-4/5} = f'(c) = \dfrac{f(-1) - f(-32)}{-1 - (-32)} = \dfrac{-1 - (-2)}{31} = \dfrac{1}{31}$ provided $c = -\left(\dfrac{31}{5}\right)^{5/4} \approx -9.7834$.

31.   a.   If $f(x) = \sin x$ and $x_0 = 0$, then $P_1(x) = x$.
    b. $\sin(0.15) \approx P_1(0.15) = 0.15$.
    c. $f''(x) = -\sin x$, let $K = |-\sin(0.15)| \approx 0.149$; then $|\sin(0.15) - P_1(0.15)| \leq (K/2) \cdot 0.1^2 = 0.00075$.

33.   a.   If $f(x) = 1/x$ and $x_0 = 2$, then $P_1(x) = (1/2) - (1/4)(x - 2)$.
    b. $1/2.05 \approx P_1(2.05) = 0.5 - 0.0125 = 0.4875$.
    c. $f''(x) = 2x^{-3}$, let $K = 2 \cdot 2^{-3} = 0.25$; then $|1/2.05 - P_1(2.05)| \leq (K/2) \cdot 0.1^2 = 0.00125$.

Solutions 3.C   *Computer Exercises*   (pages 263-4)

1.   a. $\sqrt{2} \approx 1.41421\ 35623\ 73095\ 04880\ 16887\ 24210$
    b.   If we apply Newton's method to $f(x) = x^2 - 2$ with $x_0 = 1$ and if we do our intermediate calculations to high precision, then $|-f(x_5)/f'(x_5)| \approx 9.0 \cdot 10^{-25}$ and $|-f(x_6)/f'(x_6)| \approx 2.9 \cdot 10^{-49}$; we obtain the approximation stated in part (a) after 6 iterations.

REMARK: I used a stopping condition which looks at $|x_{k+1} - x_k|$ because the text does not give us an estimate for $|\sqrt{2} - x_k|$. The partial proof that Newton's method works will also justify this heuristic.

3.   a. $\sqrt{703} \approx 26.51414\ 71671\ 25704\ 10183\ 28635\ 74050$
    b.   If we apply Newton's method to $f(x) = x^2 - 703$ with $x_0 = 20$ and if we do our intermediate calculations to high precision, then $|-f(x_5)/f'(x_5)| \approx 2.5 \cdot 10^{-26}$ and $|-f(x_6)/f'(x_6)| \approx 1.2 \cdot 10^{-53}$; we obtain the approximation stated in part (a) after 6 iterations.

5.   If $x > 0$ and $a > 0$, then $\left|x-\sqrt{a}\right|\left|x+\sqrt{a}\right| = \left|x^2 - a\right| < \epsilon \implies \left|x-\sqrt{a}\right| < \dfrac{\epsilon}{\left|x+\sqrt{a}\right|} < \dfrac{\epsilon}{\sqrt{a}}$. If $a \geq 1$, then $\sqrt{a} \geq 1$ and $\epsilon/\sqrt{a} \leq \epsilon$; on the other hand, if $0 < a < 1$, then $a < \sqrt{a} < 1$ and $\epsilon < \epsilon/\sqrt{a} < \epsilon/a$. Hence, if $a \geq 1$ and x^2 is a "good" approximation to a, then x is a "better" approximation to $\sqrt{a}$; if $0 < a < 1$, then our bound for $\left|x - \sqrt{a}\right|$ is a constant multiple of our bound for $\left|x^2 - a\right|$ so we can obtain a "good" approximation to $\sqrt{a}$ by requiring x^2 to be a "very good" approximation to a.

An alternative approach is to examine "sharp" bounds for $x - \sqrt{a}$ (remember that a and x are positive):

$$\left|x^2 - a\right| < \epsilon \iff -\epsilon < x^2 - a < \epsilon \iff a - \epsilon < x^2 < a + \epsilon \iff \sqrt{a-\epsilon} < x < \sqrt{a+\epsilon}$$
$$\iff \sqrt{a-\epsilon} - \sqrt{a} < x - \sqrt{a} < \sqrt{a+\epsilon} - \sqrt{a}.$$

The following plots display both sets of bounds.

$$\sqrt{2 \pm \epsilon} - \sqrt{2}, \quad \pm\epsilon/\sqrt{2} \qquad\qquad \sqrt{.034 \pm \epsilon} - \sqrt{.034}, \quad \pm\epsilon/\sqrt{.034}$$

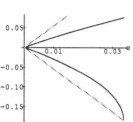

7.   The distance from a point (x, y) on the curve $y = 1/x$ to the point $(1, 0)$ is $d(x) = \sqrt{(x - 1)^2 + (y - 0)^2} = \sqrt{(x - 1)^2 + x^{-2}}$. If we were working this problem "by hand", we would probably observe that d is minimized if and only if $f(x) = d(x)^2 = (x - 1)^2 + x^{-2}$ is minimized; we would then proceed to analyze the simpler function f. Since we have a computer algebra system to keep track of the details, we will proceed to plot and analyze d. First note that d has two critical points because $d'(x) = \left((x - 1) - x^{-3}\right)/\sqrt{(x - 1)^2 + x^{-2}}$ has two real zeros. Starting Newton's method for $d' = 0$ with $x_0 = -1$, we approximate one zero to be $z_1 \approx -.819173$ with $d(z_1) \approx 2.1908$; starting Newton's method with $x_0 = 1$, we approximate the other zero of d' to be $z_2 \approx 1.38028$ with $d(z_2) \approx .81823$. The point $(z_2, 1/z_2) \approx (1.38028, .724492)$ is the point on $y = 1/x$ closest to $(1, 0)$.

$$d(x) = \sqrt{(x - 1)^2 + x^{-2}} \qquad\qquad d'(x) = \frac{(x - 1) - x^{-3}}{\sqrt{(x - 1)^2 + x^{-2}}} \qquad\qquad y = 1/x$$

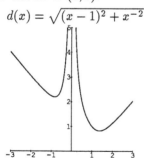 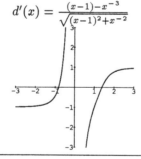 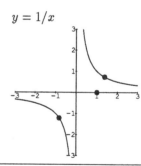

REMARK: The line through $(1, 0)$ and $(z_i, 1/z_i)$ is perpendicular to the curve's tangent at $(z_i, 1/z_i)$.

9.   A rectangular box with square base of size x and height y has volume $x^2 y$; if the box must have volume 1029 cubic feet, then $y = 1029x^{-2}$. If the base costs \$6 per square foot and the sides cost \$3 per square foot, then the cost of materials is $(x^2)(6) + (4xy)(3)$. If welding costs \$4.50 per foot, then the assembly cost is $(4x + 4y)(4.50)$. Adding these two costs and using our earlier expression for y, we have total cost $C(x) = 6x^2 + 18x + 12348x^{-1} + 18522x^{-2}$. This unit cost is minimized at the function's unique critical point on $(0, \infty)$ which is located by solving $0 = C'(x) = 12x + 18 - 12348x^{-2} - 37044x^{-3} = 12(x - 10.5)\left(1 + 12x^{-1} + 126x^{-2} + 294x^{-3}\right)$. The solution $x = 10.5$ implies $y = 28/3$ and $C(10.5) = \$2194.50$.

There are two interpretations of "the contractor wants to make a profit of 17%". If the profit is to be 17% of the unit cost, then the bid price should be $(1 + .17)(2194.50) = \$2567.565$ per box; on the other hand, if the profit is to be 17% of the final bid price, then that bid price should be $(2194.50)/(1 - .17) = \$2643.9759$ per box.

3.C — Computer Exercises

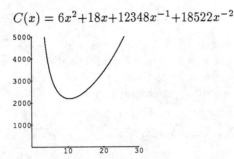

$$C(x) = 6x^2 + 18x + 12348x^{-1} + 18522x^{-2}$$

$$C'(x) = 12x + 18 - 12348x^{-2} - 37044x^{-3}$$

11. Establish a coordinate system with origin at the SouthWest (lower-left) corner of the lake, town A at $(1,0)$ and town B at $(-1,3)$. Let the underwater section of the pipeline meet the Western edge of the lake at $(0,t)$, $0 \le t \le 3$.

a. The cost of the pipeline is proportional to $c(t) = K\sqrt{1+t^2} + \sqrt{1+(3-t)^2}$. Let $t_m(K)$ denote the value of t which minimizes the cost for a fixed K and let $c_m(K)$ denote that minimal cost. Then $\lim_{K \to \infty} t_m(K) = 0$ and $\lim_{K \to \infty}(c_m(K) - K) = \sqrt{10}$.

$$\frac{d}{dt}c = \frac{kt}{\sqrt{1+t^2}} - \frac{3-t}{\sqrt{1+(3-t)^2}}.$$ Solving the equation $c' = 0$ with a computer algebra system will probably, at an intermediate and behind-the-scenes stage, involve a quartic polynomial which has two complex roots, a negative root, and a positive root. The full expressions for these roots are very messy; numerical exploration will identify the one which has positive values for $t_m(K)$ for $K > 1$. If we consider the positive root as a function of K and also evaluate the minimal cost at that value of t_m, we can obtain the following graphs.

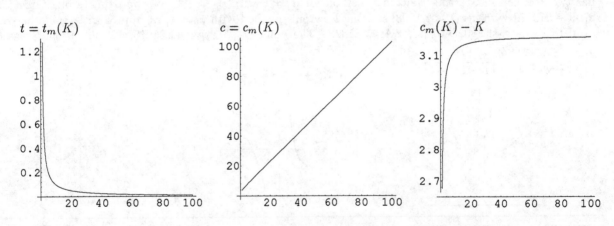

b. $c'(0) = -3 < 0$ and $c'(3) = 3K/\sqrt{10} > 0$; hence $c(t)$ is strictly decreasing at the left endpoint of $[0,3]$ and increasing at the right endpoint, therefore, the minimum of c occurs in the interior of the interval $[0,3]$ This means the minimum cost is never achieved by laying the underwater segment in the due-West direction. ∎

13. $p(x) = x^5 - 4x^4 - 6x^3 + 8x^2 - 5x + 3$ has three real zeros; they are approximately -1.96839, $.77208$, 4.92519. The following table summarizes some computations using Newton's method.

x_0	n	x_n	$|p(x_n)/p'(x_n)|$
-2	4	$-1.96838\ 52369\ 13186\ 01327$	$8.5 \cdot 10^{-23}$
1	6	$.77207\ 79274\ 39673\ 25093$	$6.7 \cdot 10^{-33}$
4	5	$4.92518\ 50206\ 02793\ 22476$	$2.5 \cdot 10^{-40}$

$p(x) = x^5 - 4x^4 - 6x^3 + 8x^2 - 5x + 3$

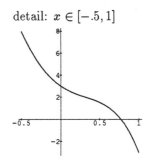

detail: $x \in [-.5, 1]$

4
The Integral

Solutions 4.2   *Antiderivatives*   (pages 274-5)

1. $\int(-5)\,dx = -5x + C.$

3. $\int 10x\,dx = 10\int x\,dx = 10\left(\dfrac{x^2}{2}+C\right) = 5x^2 + C_*.$

5. $\int\left(1 + 2x + 3x^2 + 4x^3\right)dx = \int dx + \int 2x\,dx + \int 3x^2\,dx + \int 4x^3\,dx = x + x^2 + x^3 + x^4 + C.$

7. $\int x^{-5}\,dx = \dfrac{x^{-5+1}}{-5+1} + C = \dfrac{x^{-4}}{-4} + C.$

9. $\int\left(7x^{-1/2}\right)dx = 7\left(\dfrac{x^{-1/2+1}}{-1/2+1}\right) + C = 14x^{1/2} + C = 14\sqrt{x} + C.$

11. $\int\left(\dfrac{-17}{x^{13/17}} + \dfrac{3}{x^{4/9}}\right)dx = -17\int x^{-13/17}\,dx + 3\int x^{-4/9}\,dx$

$$= (-17)\dfrac{x^{4/17}}{4/17} + 3\dfrac{x^{5/9}}{5/9} + C = \dfrac{-289}{4}x^{4/17} + \dfrac{27}{5}x^{5/9} + C.$$

13. $\int\left(2\sin x - 3\cos x\right)dx = 2\int\sin x\,dx - 3\int\cos x\,dx = 2(-\cos x) - 3(\sin x) + C = -2\cos x - 3\sin x + C.$

15. $\int\left(4x + 7\sin x\right)dx = 4\int x\,dx + 7\int\sin x\,dx = 4\left(\dfrac{x^2}{2}\right) + 7(-\cos x) + C = 2x^2 - 7\cos x + C.$

17. If $\dfrac{dy}{dx} = 2x(x+1) = 2x^2 + 2x,$ then

$$y = \int\left(2x^2 + 2x\right)dx = 2\int x^2\,dx + 2\int x\,dx = 2\left(\dfrac{x^3}{3}\right) + 2\left(\dfrac{x^2}{2}\right) + C = \dfrac{2}{3}x^3 + x^2 + C.$$

If $y(2) = 0,$ then $0 = \left(\dfrac{2}{3}\right)(2^3) + 2^2 + C;$ this implies $C = \dfrac{-28}{3}$ and $y = \dfrac{2}{3}x^3 + x^2 - \dfrac{28}{3}.$

19. If $\dfrac{dy}{dx} = \sqrt[3]{x} + x - \dfrac{1}{3\sqrt[3]{x}},$ then $y = \int\left(x^{1/3} + x - \dfrac{1}{3}x^{-1/3}\right)dx = \dfrac{3}{4}x^{4/3} + \dfrac{1}{2}x^2 - \dfrac{1}{2}x^{2/3} + C.$

If $-8 = y(-1) = \dfrac{3}{4} + \dfrac{1}{2} - \dfrac{1}{2} + C,$ then $C = \dfrac{-35}{4}$ and $y = \dfrac{3}{4}x^{4/3} + \dfrac{1}{2}x^2 - \dfrac{1}{2}x^{2/3} - \dfrac{35}{4}.$

21. If $y' = \cos x,$ then $y = \int\cos x\,dx = \sin x + C.$ If $4 = y(\pi/6) = \sin(\pi/6) + C = 1/2 + C,$ then $C = 7/2$ and $y = \sin x + 7/2.$

23. See Example 6. Because acceleration is the derivative of velocity with respect to time and because $a(t) = 5.8$ m/sec^2, we infer that $V(t) = \int a(t)\,dt = \int 5.8\,dt = 5.8t + C.$ Therefore 0.2 m/sec $= V(0) = (5.8)(0) + C = C$ implies $V(t) = 5.8t + 0.2.$ Because velocity is the derivative of position with respect to time, $s(t) = \int V(t)\,dt = \int(5.8t + 0.2)\,dt = 5.8\int t\,dt + 0.2\int dt = (5.8)\left(\dfrac{t^2}{2}\right) + (0.2)t + C_* = 2.9t^2 + 0.2t + C_*.$ Therefore 25 m $= s(0) = (2.9)(0^2) + (0.2)(0) + C_* = C_*$ implies $s(t) = 2.9t^2 + 0.2t + 25$ m.

25. $h(t) = \int(2000 - 32t)\,dt = 2000t - 16t^2 + C;$ therefore $0 = h(0) = C$ implies $h(t) = 2000t - 16t^2$ ft.

27. See Example 7. If $\mu(x)$ is the total mass up to position $x,$ then

$$\mu(x) = \int\rho(x)\,dx = \int\left(3x - x^{3/2} + 2x^2\right)dx = 3\int x\,dx - \int x^{3/2}\,dx + 2\int x^2\,dx = \dfrac{3}{2}x^2 - \dfrac{2}{5}x^{5/2} + \dfrac{2}{3}x^3 + C.$$

Hence $0 = \mu(0) = C$ implies $\mu(x) = \dfrac{3}{2}x^2 - \dfrac{2}{5}x^{5/2} + \dfrac{2}{3}x^3;$ therefore $\mu(10\,\text{m}) = \dfrac{3}{2}(100) - \dfrac{2}{5}\left(100\sqrt{10}\right) +$

$\dfrac{2}{3}(1000) = \dfrac{450}{3} - 40\sqrt{10} \approx 690.2$ kg is the total mass of the 10 m beam.

92

29. If $C(q)$ is the total cost of producing q units, then $C(q) = \int MC \, dq = \int (20 + 25q - 0.02q^2) \, dq = 20q + 25\dfrac{q^2}{2} - 0.02\dfrac{q^3}{3} + K$ where K is some constant. Hence $2000 = C(10) = 25\left(\dfrac{100}{2}\right) - 0.02\left(\dfrac{1000}{3}\right) + 20(10) + K$ implies $K = \dfrac{1670}{3}$ and $C(q) = \dfrac{1670}{3} + 20q + \dfrac{25}{2}q^2 - \dfrac{1}{150}q^3$. The cost of producing 100 units is $C(100) = \$120,890$; the cost of producing 500 units is $C(500) \approx \$2,302,223.33$.

31. Total revenue increases while marginal revenue is positive and decreases when marginal revenue is negative. Because $MR = 100 - 0.03q$, the point of diminishing returns is the solution to $MR = 0$, i.e., $q = 100/0.03 \approx 3333.33$. (If q is required to be an integer, then revenue drops with items produced after the $3,333^{\text{rd}}$ one.)

33. If $F(x) = \int \dfrac{1}{x} \, dx$, then $F'(x) = \dfrac{1}{x}$. Therefore F increases on $(0, \infty)$ because $F'(x) > 0$. The graph is concave down because $F''(x) = \dfrac{-1}{x^2} < 0$. The y-axis is an asymptote because $\lim_{x \to 0+} F'(x) = \infty$.

Solutions 4.3   *The $\sum$ notation*   (pages 278-9)

1. $\displaystyle\sum_{j=1}^{5} \dfrac{j}{2} = \dfrac{1}{2} + \dfrac{2}{2} + \dfrac{3}{2} + \dfrac{4}{2} + \dfrac{5}{2} = \dfrac{15}{2}$.

3. $\displaystyle\sum_{k=0}^{5} (-2)^k = (-2)^0 + (-2)^1 + (-2)^2 + (-2)^3 + (-2)^4 + (-2)^5 = 1 - 2 + 4 - 8 + 16 - 32 = -21$.

5. $\displaystyle\sum_{j=5}^{9} \dfrac{j}{j+1} = \dfrac{5}{6} + \dfrac{6}{7} + \dfrac{7}{8} + \dfrac{8}{9} + \dfrac{9}{10} = \dfrac{10973}{2520}$.

7. $1 + 2 + 4 + 8 + 16 = 2^0 + 2^1 + 2^2 + 2^3 + 2^4 = \displaystyle\sum_{k=0}^{4} 2^k$.

9. $\dfrac{2}{3} + \dfrac{3}{4} + \dfrac{4}{5} + \dfrac{5}{6} + \dfrac{6}{7} + \dfrac{7}{8} = \displaystyle\sum_{j=2}^{7} \dfrac{j}{j+1} = \displaystyle\sum_{k=3}^{8} \dfrac{k-1}{k}$.

11. $1 \cdot 3 + 3 \cdot 5 + 5 \cdot 7 + 7 \cdot 9 + 9 \cdot 11 + 11 \cdot 13 + 13 \cdot 15 + 15 \cdot 17$

$$= (2-1)(2+1) + (4-1)(4+1) + (6-1)(6+1) + \cdots + (16-1)(16+1) = \sum_{k=1}^{8} (2k-1)(2k+1).$$

13. $1 - 2x + 4x^2 - 8x^3 + 16x^4 - 32x^5 = (-2x)^0 + (-2x)^1 + (-2x)^2 + (-2x)^3 + (-2x)^4 + (-2x)^5 = \displaystyle\sum_{k=0}^{5} (-2x)^k$.

15. $1 + x^3 + x^6 + x^9 + x^{12} + x^{15} + x^{18} + x^{21}$

$$= (x^3)^0 + (x^3)^1 + (x^3)^2 + (x^3)^3 + (x^3)^4 + (x^3)^5 + (x^3)^6 + (x^3)^7 = \sum_{k=0}^{7} (x^3)^k = \sum_{k=0}^{7} x^{3k}.$$

17. $\left(\dfrac{1}{5}\right)^2 + \left(\dfrac{2}{5}\right)^2 + \left(\dfrac{3}{5}\right)^2 + \left(\dfrac{4}{5}\right)^2 + \left(\dfrac{5}{5}\right)^2 = \displaystyle\sum_{k=1}^{5} \left(\dfrac{k}{5}\right)^2 = \left(\dfrac{1}{5}\right)^2 \displaystyle\sum_{k=1}^{5} k^2$.

19. $0.2 \sin(0.1) + 0.2 \sin(0.3) + 0.2 \sin(0.5) + 0.2 \sin(0.7) + 0.2 \sin(0.9) = 0.2 \displaystyle\sum_{k=0}^{4} \sin\left(\dfrac{1+2k}{10}\right)$.

21.

$$
\begin{array}{ccccccccccccc}
i & \ 0 & 1 & 2 & 3 & 4 & & k & \ 5 & 6 & 7 & 8 & 9 & 10 \\
  & \downarrow & \downarrow & \downarrow & \downarrow & \downarrow & & & \downarrow & \downarrow & \downarrow & \downarrow & \downarrow & \downarrow
\end{array}
$$

$$\sum_{i=0}^{4} 1 = 1+1+1+1+1 = \sum_{j=1}^{5} 1 \neq \sum_{k=5}^{10} 1 = 1+1+1+1+1+1$$

$$
\begin{array}{ccccc}
\uparrow & \uparrow & \uparrow & \uparrow & \uparrow \\
1 & 2 & 3 & 4 & 5 \quad j
\end{array}
$$

4.3 — The $\sum$ notation

23.
$$\sum_{k=1}^{7} k^2 = 1^2 + 2^2 + 3^2 + 4^2 + 5^2 + 6^2 + 7^2 = 7^2 + 6^2 + 5^2 + 4^2 + 3^2 + 2^2 + 1^2 = \sum_{j=0}^{6}(7-j)^2$$
$$\neq 6^2 + 5^2 + 4^2 + 3^2 + 2^2 + 1^2 + 0^2 = \sum_{i=1}^{7}(7-i)^2$$

25. a.
$$\sum_{k=1}^{n}\left[g(k) - g(k-1)\right] = \left[\sum_{k=1}^{n} g(k)\right] - \left[\sum_{k=1}^{n} g(k-1)\right]$$
$$= \sum_{k=1}^{n} g(k) - \sum_{i=0}^{n-1} g(i) = \left[\sum_{k=1}^{n-1} g(k) + g(n)\right] - \left[g(0) + \sum_{i=1}^{n-1} g(i)\right]$$
$$= [g(n) - g(0)] + \left[\sum_{k=1}^{n-1} g(k) - \sum_{i=1}^{n-1} g(i)\right] = [g(n) - g(0)] + [0] = g(n) - g(0). \ \blacksquare$$

b. Let $g(k) = \dfrac{k(k+1)}{2}$, then $g(k) - g(k-1) = \dfrac{k(k+1)}{2} - \dfrac{(k-1)k}{2} = [(k+1) - (k-1)]\left(\dfrac{k}{2}\right) = $

$[2]\left(\dfrac{k}{2}\right) = k.$ Therefore $\displaystyle\sum_{k=1}^{n} k = \sum_{k=1}^{n}[g(k) - g(k-1)] = g(n) - g(0) = \dfrac{n(n+1)}{2} - \dfrac{0(1)}{2} = \dfrac{n(n+1)}{2}. \ \blacksquare$

27. If $g(k) = \dfrac{k(k+1)(k+2)(k+3)}{4}$, then

$$g(k) - g(k-1) = \frac{k(k+1)(k+2)(k+3)}{4} - \frac{(k-1)(k)(k+1)(k+2)}{4}$$
$$= [(k+3) - (k-1)]\left(\frac{k(k+1)(k+2)}{4}\right) = [4]\left(\frac{k(k+1)(k+2)}{4}\right) = k(k+1)(k+2).$$

Thus
$$\sum_{k=1}^{n} k(k+1)(k+2) = g(n) - g(0) = \frac{k(k+1)(k+2)(k+3)}{4} - \frac{(0)(1)(2)(3)}{4} = \frac{n(n+1)(n+2)(n+3)}{4};$$

on the other hand, $k(k+1)(k+2) = k^3 + 3k^2 + 2k$ implies $k^3 = k(k+1)(k+2) - 3k^2 - 2k$ and

$$\sum_{k=1}^{n} k^3 = \sum_{k=1}^{n}\left[k(k+1)(k+2) - 3k^2 - 2k\right] = \sum_{k=1}^{n} k(k+1)(k+2) - 3\sum_{k=1}^{n} k^2 - 2\sum_{k=1}^{n} k$$
$$= \frac{n(n+1)(n+2)(n+3)}{4} - 3\left[\frac{n(n+1)(2n+1)}{6}\right] - 2\left[\frac{n(n+1)}{2}\right]$$
$$= \frac{n(n+1)}{4}\left[(n+2)(n+3) - 2(2n+1) - 4\right]$$
$$= \frac{n(n+1)}{4}\left[n^2 + n\right] = \left[\frac{n(n+1)}{2}\right]^2. \ \blacksquare$$

REMARK: It is more elegant to note $k^3 = k(k+1)(k+2) - 3k(k+1) + k$; therefore
$$\sum_{k=1}^{n} k^3 = \frac{n(n+1)(n+2)(n+3)}{4} - 3\frac{n(n+1)(n+2)}{3} + \frac{n(n+1)}{2}.$$

29.
$$\frac{1}{3} + \frac{1}{8} + \frac{1}{15} + \cdots + \frac{1}{n^2 - 1} = \frac{1}{1\cdot 3} + \frac{1}{2\cdot 4} + \frac{1}{3\cdot 5} + \cdots + \frac{1}{(n-1)(n+1)}$$
$$= \sum_{k=2}^{n}\frac{1}{(k-1)(k+1)} = \sum_{k=2}^{n}\frac{1}{2}\left(\frac{1}{k-1} - \frac{1}{k+1}\right)$$
$$= \frac{1}{2}\left[\sum_{k=2}^{n}\frac{1}{k-1} - \sum_{k=2}^{n}\frac{1}{k+1}\right] = \frac{1}{2}\left[\sum_{i=1}^{n-1}\frac{1}{i} - \sum_{j=3}^{n+1}\frac{1}{j}\right]$$
$$= \frac{1}{2}\left[\left(\frac{1}{1} + \frac{1}{2}\right) - \left(\frac{1}{n} + \frac{1}{n+1}\right)\right] = \frac{(n-1)(3n+2)}{4n(n+1)} = \frac{3}{4} - \frac{n+\frac{1}{2}}{n(n+1)}.$$

REMARK: $\sum_{k=2}^{n}[g(k+1) - g(k-1)] = g(n+1) + g(n) - g(2) - g(1);$ choose $g(k) = -1/(2k).$

---

31. $\sum_{i=1}^{2}(k \cdot a_i) = (k \cdot a_1) + (k \cdot a_2) = k \cdot (a_1 + a_2) = k \cdot \sum_{i=1}^{2} a_i.$ If $n > 1$, then $\sum_{i=1}^{n+1} b_i = \sum_{i=1}^{n} b_i + b_{n+1}.$ Suppose k is a constant such that $b_i = k \cdot a_i$ for all $i \geq 1.$ Therefore, if the statement we are to prove is true for some positive integer n, then it is also true for the subsequent integer:

$$\sum_{i=1}^{n+1}(k \cdot a_i) = \sum_{i=1}^{n}(k \cdot a_i) + (k \cdot a_{n+1}) = k \cdot \sum_{i=1}^{n} a_i + k \cdot a_{n+1} = k \cdot \left(\sum_{i=1}^{n} a_i + a_{n+1}\right) = k \cdot \sum_{i=1}^{n+1} a_i.$$

Hence the result is true for any constant k, any sequence $\{a_i\}$, and any positive integer n. ∎

Solutions 4.4   *Approximations to area*   (pages 290-3)

1.
$$s = \left(\frac{0}{16}\right)^2 \cdot \frac{1}{16} + \left(\frac{1}{16}\right)^2 \cdot \frac{1}{16} + \left(\frac{2}{16}\right)^2 \cdot \frac{1}{16} + \cdots + \left(\frac{14}{16}\right)^2 \cdot \frac{1}{16} + \left(\frac{15}{16}\right)^2 \cdot \frac{1}{16}$$

$$= \frac{0^2 + 1^2 + 2^2 + \cdots + 14^2 + 15^2}{16^3} = \frac{1240}{4096} = \frac{155}{512} \approx 0.303,$$

$$S = \left(\frac{1}{16}\right)^2 \cdot \frac{1}{16} + \left(\frac{2}{16}\right)^2 \cdot \frac{1}{16} + \cdots + \left(\frac{14}{16}\right)^2 \cdot \frac{1}{16} + \left(\frac{15}{16}\right)^2 \cdot \frac{1}{16} + \left(\frac{16}{16}\right)^2 \cdot \frac{1}{16}$$

$$= \frac{1^2 + 2^2 + \cdots + 14^2 + 15^2 + 16^2}{16^3} = \frac{1496}{4096} = \frac{187}{512} \approx 0.365.$$

3. Since $y = x^2$ is strictly increasing on the interval $[0, 1]$, the under-approximation s is computed using rectangles whose upper left corner is on the curve. On the kth interval, $\left[\frac{k-1}{n}, \frac{k}{n}\right]$, the small rectangle has height $\left(\frac{k-1}{n}\right)^2$ and area $\left(\frac{k-1}{n}\right)^2 \frac{1}{n}$. Therefore,

$$s = \sum_{k=1}^{n} \left(\frac{k-1}{n}\right)^2 \frac{1}{n} = \sum_{j=0}^{n-1} \left(\frac{j}{n}\right)^2 \frac{1}{n} = \left(\frac{1}{n^3}\right) \sum_{j=0}^{n-1} j^2 = \left(\frac{1}{n^3}\right) \left(\frac{(n-1)(n)(2n-1)}{6}\right) = \frac{1}{3} - \frac{1}{2n} + \frac{1}{6n^2}.$$

The over-approximation S is computed using rectangles whose upper right corner is on the graph of $y = x^2$; on $\left[\frac{k-1}{n}, \frac{k}{n}\right]$, the large rectangle has height $\left(\frac{k}{n}\right)^2$ and area $\left(\frac{k}{n}\right)^2 \frac{1}{n}$. Therefore,

$$S = \sum_{k=1}^{n} \left(\frac{k}{n}\right)^2 \frac{1}{n} = \left(\frac{1}{n^3}\right) \sum_{k=1}^{n} k^2 = \left(\frac{1}{n^3}\right) \left(\frac{n(n+1)(2n+1)}{6}\right) = \frac{1}{3} + \frac{1}{2n} + \frac{1}{6n^2}.$$

5. Let $f(x) = 7x$. Let A be the area of the triangular region bounded by the line $y = 7x$, the x-axis, and satisfying $0 \leq x \leq 4$. The exact value of A is $\frac{1}{2}(4)(28) = 56$. In each of the following approximations we compute various sums of areas of rectangles of height $f(x_i^*)$. Those sums denoted s_n have each x_i^* chosen to be the left endpoint of the ith interval, those denoted S_n will have every x_i^* chosen as the right endpoint. Because f is an increasing function, $s_n < A < S_n$.

If $n = 2$, then $(b-a)/n = (4-0)/2 = 2$, the partition points are 0, 2, 4. Using $x_1^* = 0$ and $x_2^* = 2$, $s_2 = f(0) \cdot 2 + f(2) \cdot 2 = (0) \cdot 2 + (14) \cdot 2 = 28 < A$; on the other hand, $A < S_2 = f(2) \cdot 2 + f(4) \cdot 2 = (14) \cdot 2 + (28) \cdot 2 = 84$ if $x_1^* = 2$ and $x_2^* = 4$ are chosen.

If $n = 4$, then $(b-a)/n = 1$ and the partition points are 0, 1, 2, 3, 4. Thus $s_4 = f(0) \cdot 1 + f(1) \cdot 1 + f(2) \cdot 1 + f(3) \cdot 1 = 0 + 7 + 14 + 21 = 42 < A$ and $A < S_4 = f(1) \cdot 1 + f(2) \cdot 1 + f(3) \cdot 1 + f(4) \cdot 1 = 7 + 14 + 21 + 28 = 70$.

If $n = 8$, then $\Delta x = (4-0)/8 = 1/2$ and the partition points are 0, 1/2, 1, 3/2, 2, 5/2, 3, 7/2, 4. Therefore

$$s_8 = \left[7\left(\frac{0}{2}\right) + 7\left(\frac{1}{2}\right) + 7\left(\frac{2}{2}\right) + 7\left(\frac{3}{2}\right) + 7\left(\frac{4}{2}\right) + 7\left(\frac{5}{2}\right) + 7\left(\frac{6}{2}\right) + 7\left(\frac{7}{2}\right)\right] \cdot \frac{1}{2}$$

$$= (7) \cdot [0 + 1 + 2 + 3 + 4 + 5 + 6 + 7] \cdot \left(\frac{1}{2}\right)^2 = \left[\frac{(7)(8)}{2}\right] \cdot \left(\frac{7}{4}\right) = 49 < A$$

and

4.4 — Approximations to area

$$A < S_8 = \left[7\left(\frac{1}{2}\right) + 7\left(\frac{2}{2}\right) + 7\left(\frac{3}{2}\right) + 7\left(\frac{4}{2}\right) + 7\left(\frac{5}{2}\right) + 7\left(\frac{6}{2}\right) + 7\left(\frac{7}{2}\right) + 7\left(\frac{8}{2}\right) \right] \cdot \frac{1}{2}$$

$$= 7\left(\frac{1}{2}\right)^2 \cdot [1 + 2 + 3 + 4 + 5 + 6 + 7 + 8] = \left(\frac{7}{4}\right) \cdot \left[\frac{(8)(9)}{2}\right] = 63.$$

REMARK: In each of the cases $n = 2, 4,$ or 8, I have computed two approximations for A. Because the function $f(x) = 7x$ is increasing, however you choose the various x_i^* points, your approximations must fall in the interval $[s_n, S_n]$.

7.   Let $g(x) = 3x + 2$. Let A be the area of the trapezoidal region bounded by the line $y = 3x + 2$, the x-axis, and satisfying $0 \le x \le 3$. A is the sum of the area of a 3 by 2 rectangle and the area of a triangle with base 3 and height $11 - 2 = 9$; thus $A = (3)(2) + (1/2)(3)(9) = 19.5$.

If $n = 2$, then $\Delta x = (3 - 0)/2 = 3/2$ and $0, 3/2, 3$ are the partition points. Therefore

$$s_2 = g(0) \cdot \left[\frac{3}{2}\right] + g\left(\frac{3}{2}\right) \cdot \left[\frac{3}{2}\right] = 2 \cdot \left[\frac{3}{2}\right] + \frac{13}{2} \cdot \left[\frac{3}{2}\right] = 12.75 < A,$$

$$S_2 = g\left(\frac{3}{2}\right) \cdot \left[\frac{3}{2}\right] + g(3) \cdot \left[\frac{3}{2}\right] = \frac{13}{2} \cdot \left[\frac{3}{2}\right] + 11 \cdot \left[\frac{3}{2}\right] = 26.25 > A.$$

If $n = 4$, then $\Delta x = 3/4$ and $0, 3/4, 3/2, 9/4, 3$ are the partition points. If the left end of each subinterval is chosen to be the point x_i^*, then

$$s_4 = g(0) \cdot \left[\frac{3}{4}\right] + g\left(\frac{3}{4}\right) \cdot \left[\frac{3}{4}\right] + g\left(\frac{3}{2}\right) \cdot \left[\frac{3}{4}\right] + g\left(\frac{9}{4}\right) \cdot \left[\frac{3}{4}\right] = \left[2 + \frac{17}{4} + \frac{13}{2} + \frac{35}{4}\right] \cdot \left[\frac{3}{4}\right] = 16.125 < A$$

while choosing x_i^* to be the right end of a subinterval produces

$$S_4 = g\left(\frac{3}{4}\right) \cdot \left[\frac{3}{4}\right] + g\left(\frac{3}{2}\right) \cdot \left[\frac{3}{4}\right] + g\left(\frac{9}{4}\right) \cdot \left[\frac{3}{4}\right] + g(3) \cdot \left[\frac{3}{4}\right] = \left[\frac{17}{4} + \frac{13}{2} + \frac{35}{4} + 11\right] \cdot \left[\frac{3}{4}\right] = 22.875 > A.$$

If $n = 8$, then $\Delta x = 3/8$ and the partition points are $k(3/8)$ for $k = 0, 1, 2, \ldots, 7, 8$. Letting x_i^* be left ends of subintervals,

$$s_8 = \sum_{k=0}^{7} \left[3\left(\frac{3k}{8}\right) + 2\right] \cdot \left[\frac{3}{8}\right] = \left[2 + \frac{25}{8} + \frac{34}{8} + \frac{43}{8} + \frac{52}{8} + \frac{61}{8} + \frac{70}{8} + \frac{79}{8}\right] \cdot \left[\frac{3}{8}\right] = \frac{285}{16} = 17.8125 < A$$

and if right ends of subintervals are chosen for x_i^*, then

$$S_8 = \sum_{k=1}^{8} \left[3\left(\frac{3k}{8}\right) + 2\right] \cdot \left[\frac{3}{8}\right] = \left[\frac{25}{8} + \frac{34}{8} + \frac{43}{8} + \frac{52}{8} + \frac{61}{8} + \frac{70}{8} + \frac{79}{8} + 11\right] \cdot \left[\frac{3}{8}\right] = \frac{339}{16} = 21.1875 > A.$$

REMARK: In each of the cases $n = 2, 4, 8$, I have done double the work requested by the directions for the problem. Because g is an increasing function, any approximation you compute for a particular n must fall between my values s_n and S_n.

NOTE: In the solutions to Problems 9-20, the right-hand endpoint of each subinterval has been chosen to be x_i^*; any sum based on n subintervals will be denoted S_n even if $f(x_i^*)$ is not the maximum of f in the ith subinterval. By using left-hand endpoints to compute the text answers and right-hand endpoints here, I hope most people will be able to check their work.

9.   a. $f(x) = x^2/2$; $\Delta x = (2 - 0)/4 = 1/2$; the partition points are $0, 1/2, 2/2, 3/2, 4/2$.

$$S_4 = f\left(\frac{1}{2}\right) \cdot \frac{1}{2} + f\left(\frac{2}{2}\right) \cdot \frac{1}{2} + f\left(\frac{3}{2}\right) \cdot \frac{1}{2} + f\left(\frac{4}{2}\right) \cdot \frac{1}{2} = \left[\frac{1}{8} + \frac{4}{8} + \frac{9}{8} + \frac{16}{8}\right] \cdot \frac{1}{2} = \frac{15}{8} = 1.875.$$

   b. $\Delta x = (2 - 0)/8 = 1/4$; the partition points are $0, 1/4, 2/4, 3/4, 4/4, 5/4, 6/4, 7/4, 8/4$; then

$$S_8 = \left[f\left(\frac{1}{4}\right) + f\left(\frac{2}{4}\right) + f\left(\frac{3}{4}\right) + f\left(\frac{4}{4}\right) + f\left(\frac{5}{4}\right) + f\left(\frac{6}{4}\right) + f\left(\frac{7}{4}\right) + f\left(\frac{8}{4}\right)\right] \cdot \frac{1}{4}$$

$$= \left[\frac{1}{32} + \frac{4}{32} + \frac{9}{32} + \frac{16}{32} + \frac{25}{32} + \frac{36}{32} + \frac{49}{32} + \frac{64}{32}\right] \cdot \frac{1}{4} = \frac{51}{32} = 1.59375.$$

   c.   If $\Delta x = 2/n$, the partition points are $k(2/n)$ for $k = 0, 1, \ldots, n$.

$$S_n = \sum_{k=1}^{n} f\left(x_i^*\right) \Delta x = \sum_{k=1}^{n} f\left(\frac{2k}{n}\right) \cdot \frac{2}{n} = \sum_{k=1}^{n} \left[\left(\frac{2k}{n}\right)^2 \cdot \frac{1}{2}\right] \cdot \frac{2}{n} = \sum_{k=1}^{n} \frac{2k^2}{n^2} \cdot \frac{2}{n} = \frac{4}{n^3} \sum_{k=1}^{n} k^2$$

$$= \frac{4}{n^3} \cdot \frac{n(n+1)(2n+1)}{6} = \frac{2(n+1)(2n+1)}{3n^2} = \frac{4}{3} + \frac{2}{n} + \frac{2}{3n^2}.$$

d. $\quad \lim\limits_{n \to \infty} S_n = \dfrac{2}{3}\left(\lim\limits_{n \to \infty} \dfrac{n+1}{n}\right)\left(\lim\limits_{n \to \infty} \dfrac{2n+1}{n}\right) = \dfrac{2}{3}(1)(2) = \dfrac{4}{3}.$

11.   Let $g(x) = 1 - x^2$; note that $g(x) \geq 0$ on $[0, 1]$.

a. $\Delta x = 1/8$; the partition points are 0, $1/8$, $1/4$, $3/8$, $1/2$, $5/8$, $3/4$, $7/8$, 1.

$$S_8 = \left[g\left(\dfrac{1}{8}\right) \cdot \dfrac{1}{8} + g\left(\dfrac{1}{4}\right) \cdot \dfrac{1}{8} + g\left(\dfrac{3}{8}\right) \cdot \dfrac{1}{8} + g\left(\dfrac{1}{2}\right) \cdot \dfrac{1}{8} + g\left(\dfrac{5}{8}\right) \cdot \dfrac{1}{8} + g\left(\dfrac{3}{4}\right) \cdot \dfrac{1}{8} + g\left(\dfrac{7}{8}\right) \cdot \dfrac{1}{8} + g(1) \cdot \dfrac{1}{8} \right]$$

$$= [63 + 60 + 55 + 48 + 39 + 28 + 15 + 0] \cdot \left(\dfrac{1}{8}\right)^3 = \dfrac{77}{128} \approx 0.602.$$

b. $\Delta x = 1/16$; the partition points are $k/16$ for $k = 0, 1, 2, 3, \ldots, 15, 16$.

$$S_{16} = \sum_{k=1}^{16} \left[1 - \left(\dfrac{k}{16}\right)^2 \right] \cdot \dfrac{1}{16} = \sum_{k=1}^{16} \dfrac{1}{16} - \left(\dfrac{1}{16}\right)^3 \sum_{k=1}^{16} k^2 = 1 - \dfrac{1}{16^3} \cdot \dfrac{(16)(17)(33)}{6} = \dfrac{325}{512} \approx 0.635.$$

c. $\Delta x = 1/n$; the partition points are k/n for $k = 0, 1, 2, \ldots, n$.

$$S_n = \sum_{k=1}^{n} \left[1 - \left(\dfrac{k}{n}\right)^2 \right] \dfrac{1}{n} = \left(\dfrac{1}{n}\right) \sum_{k=1}^{n} 1 - \left(\dfrac{1}{n}\right)^3 \sum_{k=1}^{n} k^2 = \left(\dfrac{1}{n}\right) n - \dfrac{1}{n^3} \cdot \dfrac{n(n+1)(2n+1)}{6} = \dfrac{2}{3} - \dfrac{1}{2n} - \dfrac{1}{6n^2}.$$

d. $\lim\limits_{n \to \infty} S_n = \lim\limits_{n \to \infty} \left(\dfrac{2}{3} - \dfrac{1}{2n} - \dfrac{1}{6n^2}\right) = \dfrac{2}{3} - 0 - 0 = \dfrac{2}{3}.$

13.   a. $\Delta x = (2 - 1)/8 = 1/8$; the partition points are $1 = 8/8$, $9/8$, $10/8 = 5/4$, $11/8$, $12/8 = 3/2$, $13/8$, $14/8 = 7/4$, $15/8$, and $16/8 = 2$.

$$S_8 = \left(\dfrac{9}{8}\right)^2 \cdot \dfrac{1}{8} + \left(\dfrac{10}{8}\right)^2 \cdot \dfrac{1}{8} + \left(\dfrac{11}{8}\right)^2 \cdot \dfrac{1}{8} + \left(\dfrac{12}{8}\right)^2 \cdot \dfrac{1}{8} + \left(\dfrac{13}{8}\right)^2 \cdot \dfrac{1}{8} + \left(\dfrac{14}{8}\right)^2 \cdot \dfrac{1}{8} + \left(\dfrac{15}{8}\right)^2 \cdot \dfrac{1}{8} + \left(\dfrac{16}{8}\right)^2 \cdot \dfrac{1}{8}$$

$$= \dfrac{81 + 100 + 121 + 144 + 169 + 196 + 225 + 256}{8^3} = \dfrac{323}{128} \approx 2.523.$$

b. $\Delta x = 1/16$; the partition points are $k/16$ for $k = 16, 17, 18, \ldots, 32$.

$$S_{16} = \sum_{k=17}^{32} \left(\dfrac{k}{16}\right)^2 \cdot \left(\dfrac{1}{16}\right) = \dfrac{1}{16^3} \sum_{k=17}^{32} k^2 = \dfrac{1243}{512} \approx 2.428.$$

c. $\Delta x = 1/n$; the partition points are k/n for $k = n, n+1, \ldots, 2n$.

$$S_n = \sum_{k=n+1}^{2n} \left(\dfrac{k}{n}\right)^2 \left(\dfrac{1}{n}\right) = \dfrac{1}{n^3} \sum_{k=n+1}^{2n} k^2 = \dfrac{1}{n^3}\left[\sum_{k=1}^{2n} k^2 - \sum_{k=1}^{n} k^2 \right]$$

$$= \dfrac{1}{n^3}\left[\dfrac{2n(2n+1)(4n+1)}{6} - \dfrac{n(n+1)(2n+1)}{6} \right] = \dfrac{7}{3} + \dfrac{3}{2n} + \dfrac{1}{6n^2}.$$

d. $\lim\limits_{n \to \infty} S_n = \lim\limits_{n \to \infty} \left(\dfrac{7}{3} + \dfrac{3}{2n} + \dfrac{1}{6n^2}\right) = \dfrac{7}{3} + 0 + 0 = \dfrac{7}{3}.$

15.   a. $\Delta x = (5 - 0)/4 = 5/4 = 1.25$; the partition points are 0, 1.25, 2.5, 3.75, 5.

$$S_4 = (1.25)^3 \cdot 1.25 + (2.5)^3 \cdot 1.25 + (3.75)^3 \cdot 1.25 + (5)^3 \cdot 1.25 = \dfrac{625}{256} + \dfrac{625}{32} + \dfrac{16875}{256} + \dfrac{625}{4} = \dfrac{15625}{64} \approx 244.141.$$

b. $\Delta x = 5/8$; the partition points are $k \cdot 5/8$ for $k = 0, 1, 2, \ldots, 7, 8$.

$$S_8 = \sum_{k=1}^{8} \left(k \cdot \dfrac{5}{8}\right)^3 \cdot \left(\dfrac{5}{8}\right) = \left(\dfrac{5}{8}\right)^4 \sum_{k=1}^{8} k^3 = \dfrac{50625}{256} \approx 197.754.$$

c. $\Delta x = 5/n$; the partition points are $k5/n$ for $k = 0, 1, \ldots, n$.

$$S_n = \sum_{k=1}^{n} \left(k \dfrac{5}{n}\right)^3 \left(\dfrac{5}{n}\right) = \left(\dfrac{5}{n}\right)^4 \sum_{k=1}^{n} k^3 = \dfrac{5^4}{n^4} \cdot \left(\dfrac{n(n+1)}{2}\right)^2 = 5^4 \left(\dfrac{1}{4} + \dfrac{1}{2n} + \dfrac{1}{4n^2}\right).$$

d. $\lim\limits_{n \to \infty} S_n = \lim\limits_{n \to \infty} 5^4 \left(\dfrac{1}{4} + \dfrac{1}{2n} + \dfrac{1}{4n^2}\right) = 5^4 \cdot \left(\dfrac{1}{4} + 0 + 0\right) = \dfrac{625}{4} = 156.25.$

4.4 — Approximations to area

17.   Let $g(x) = x + x^2 = x(1+x)$.

    a. $\Delta x = (1-0)/6 = 1/6$; the partition points are 0, 1/6, 1/3, 1/2, 2/3, 5/6, 1.

$$S_6 = \left[\left(\frac{1}{6}\right)\left(\frac{7}{6}\right) + \left(\frac{1}{3}\right)\left(\frac{4}{3}\right) + \left(\frac{1}{2}\right)\left(\frac{3}{2}\right) + \left(\frac{2}{3}\right)\left(\frac{5}{3}\right) + \left(\frac{5}{6}\right)\left(\frac{11}{6}\right) + 1 \cdot 2\right]\frac{1}{6}$$

$$= \left[\frac{7}{36} + \frac{4}{9} + \frac{3}{4} + \frac{10}{9} + \frac{55}{36} + 2\right]\frac{1}{6} = \frac{217}{216} \approx 1.0046.$$

    b. $\Delta x = 1/12$; the partition points are $k/12$ for $k = 0, 1, 2, ..., 12$.

$$S_{12} = \sum_{k=1}^{12}\left[\frac{k}{12} + \left(\frac{k}{12}\right)^2\right] \cdot \frac{1}{12}$$

$$= \frac{13 + 28 + 45 + 64 + 85 + 108 + 133 + 160 + 189 + 220 + 253 + 288}{12^3} = \frac{793}{864} \approx 0.9178.$$

    c. $\Delta x = 1/n$; the partition points are k/n for $k = 0, 1, 2, ..., n$.

$$S_n = \sum_{k=1}^{n}\left[\frac{k}{n} + \left(\frac{k}{n}\right)^2\right] \cdot \frac{1}{n} = \frac{1}{n^2}\sum_{k=1}^{n}k + \frac{1}{n^3}\sum_{k=1}^{n}k^2 = \frac{1}{n^2}\cdot\frac{n(n+1)}{2} + \frac{1}{n^3}\cdot\frac{n(n+1)(2n+1)}{6} = \frac{5}{6} + \frac{1}{n} + \frac{1}{6n^2}.$$

    d. $\displaystyle\lim_{n\to\infty} S_n = \lim_{n\to\infty}\left(\frac{5}{6} + \frac{1}{n} + \frac{1}{6n^2}\right) = \frac{5}{6} + 0 + 0 = \frac{5}{6}.$

19.   a. $\Delta x = (1-0)/4 = 0.25$; the partition points are 0, 0.25, 0.5, 0.75, 1.

$$S_4 = \left(\left[1 + 2(0.25) + 3(0.25)^2\right] + \left[1 + 2(0.5) + 3(0.5)^2\right] + \left[1 + 2(0.75) + 3(0.75)^2\right] + \left[1 + 2(1) + 3(1)^2\right]\right) \cdot 0.25$$

$$= \left(\frac{27}{16} + \frac{11}{4} + \frac{67}{16} + 6\right)\frac{1}{4} = \frac{117}{32} \approx 3.656.$$

    b. $\Delta x = 1/8$; the partition points are $k/8$ for $k = 0, 1, 2, ..., 7, 8$.

$$S_8 = \sum_{k=1}^{8}\left[1 + 2\left(\frac{k}{8}\right) + 3\left(\frac{k}{8}\right)^2\right]\cdot\frac{1}{8} = \frac{425}{128} \approx 3.320.$$

    c. $\Delta x = 1/n$; the partition points are k/n for $k = 0, 1, 2, ..., n$.

$$S_n = \sum_{k=1}^{n}\left[1 + 2\left(\frac{k}{n}\right) + 3\left(\frac{k}{n}\right)^2\right]\cdot\frac{1}{n} = \frac{1}{n}\sum_{k=1}^{n}1 + \frac{2}{n^2}\sum_{k=1}^{n}k + \frac{3}{n^3}\sum_{k=1}^{n}k^2$$

$$= \frac{1}{n}\cdot n + \frac{2}{n^2}\cdot\frac{n(n+1)}{2} + \frac{3}{n^3}\cdot\frac{n(n+1)(2n+1)}{6} = 3 + \frac{5}{2n} + \frac{1}{2n^2}.$$

    d. $\displaystyle\lim_{n\to\infty} S_n = \lim_{n\to\infty}\left(3 + \frac{5}{2n} + \frac{1}{2n^2}\right) = 3 + 0 + 0 = 3.$

21.   a.   The first subinterval is $[a, a + \Delta x] = \left[a, a + \dfrac{b-a}{n}\right]$, its endpoints are $x_0 = a$ and $x_1 = a + (b-a)/n$.

        The ith subinterval is $\left[a + (i-1)\Delta x, a + i\Delta x\right]$, its endpoints are x_{i-1} and x_i where, in general,

$$x_k = a + k\Delta x = a + k\left(\frac{b-a}{n}\right) = \left(\frac{n-k}{n}\right)a + \left(\frac{k}{n}\right)b.$$

    b.   If x_i^* is the right-hand endpoint of the ith subinterval, then $x_i^* = x_i = a + i(b-a)/n$ and

$$f(x_i^*) = \left(a + i\frac{b-a}{n}\right)^2 = a^2 + \frac{2i}{n}a(b-a) + \frac{i^2(b-a)^2}{n^2}.$$

c.
$$S_n = \sum_{i=1}^{n} f\left(x_i^*\right) \Delta x = \sum_{i=1}^{n} \left(a + i\frac{b-a}{n}\right)^2 \Delta x$$

$$= \left[\sum_{i=1}^{n} a^2 + \frac{2a(b-a)}{n}\sum_{i=1}^{n} i + \frac{(b-a)^2}{n^2}\sum_{i=1}^{n} i^2\right]\Delta x$$

$$= \left[na^2 + \frac{2a(b-a)}{n}\cdot\frac{n(n+1)}{2} + \frac{(b-a)^2}{n^2}\cdot\frac{n(n+1)(2n+1)}{6}\right]\Delta x$$

$$= \left(\frac{b-a}{n}\right)\left[\frac{na^2 + (n+1)a(b-a) + (n+1)(2n+1)(b-a)^2}{6n}\right]$$

$$= \left(\frac{b-a}{n}\right)\left[\frac{\left(b^2 + ab + a^2\right)n}{3} + \frac{b^2 - a^2}{2} + \frac{(b-a)^2}{6n}\right]$$

$$= \frac{b^3 - a^3}{3} + \frac{(b-a)^2(b+a)}{2n} + \frac{(b-a)^3}{6n^2}.$$

$$S_4 = \frac{(b-a)\left(15b^2 + 10ab + 7a^2\right)}{32}, \qquad S_8 = \frac{(b-a)\left(51b^2 + 42ab + 35a^2\right)}{128},$$

$$S_{16} = \frac{(b-a)\left(187b^2 + 170ab + 155a^2\right)}{512}, \qquad S_{32} = \frac{(b-a)\left(715b^2 + 682ab + 651a^2\right)}{2048}.$$

d. $\quad A_a^b = \lim_{n\to\infty} S_n = \lim_{n\to\infty}\left[\dfrac{b^3 - a^3}{3} + \dfrac{(b-a)^2(b+a)}{2n} + \dfrac{(b-a)^3}{6n^2}\right] = \dfrac{b^3 - a^3}{3} + 0 + 0 = \dfrac{b^3 - a^3}{3}.$

23. $s_{10} = 8.10656,\ s_{20} = 7.22666$

25. $s_{10} = 5.97,\ s_{20} = 5.8175$

27. $s_{10} = 18.1157464834,\ s_{20} = 17.7289017602$

29. $s_{10} = 0.7865659209,\ s_{20} = 0.7391767435$

31.   Let $f(x) = x^4$ and $F(x) = x^5/5$, then $f = F'$. The mean value theorem now implies that for any a and b, there is some c between them such that $F(b) - F(a) = f(c)\cdot(b-a)$. Therefore, for any partition of $[0,1]$, there are choices $x_i^* = c_i$ such that the sum in (18) telescopes down to $F(1) - F(0) = 1/5 - 0 = 1/5$. ∎

REMARK: $c = \left(b^4 + b^3 a + b^2 a^2 + ba^3 + a^4\right)/5.$

33.   Let $f(x) = \sqrt{x}$ and $F(x) = (2/3)x^{3/2}$, then $f = F'$. The mean value theorem now implies that for any a and b, there is some c between them such that $F(b) - F(a) = f(c)\cdot(b-a)$. Therefore, for any partition of $[0,1]$, there are choices $x_i^* = c_i$ such that the sum in (18) telescopes down to $F(1) - F(0) = 2/3 - 0 = 2/3$. ∎

REMARK: $c = (2/3)\left(b + \sqrt{ab} + a\right)/\left(\sqrt{b} + \sqrt{a}\right).$

35.   Let $f(x) = \sin x$ and $F(x) = -\cos x$, then $f = F'$. The mean value theorem implies that for any a and b, there is some c between them such that $F(b) - F(a) = f(c)\cdot(b-a)$. Thus, for any partition of $[\pi/6, \pi/3]$, there are choices $x_i^* = c_i$ such that the sum in (18) telescopes down to $F(\pi/3) - F(\pi/6) = (-\cos(\pi/3)) - (-\cos(\pi/6)) = (-1/2) - (-\sqrt{3}/2) = \left(\sqrt{3} - 1\right)/2$. ∎

37.   Let $f(x) = 1/x^2$ and $F(x) = -1/x$, then $f = F'$. The mean value theorem implies that for any a and b having the same sign, there is some c between them such that $F(b) - F(a) = f(c)\cdot(b-a)$. Thus, for any partition of $[1,2]$, there are choices $x_i^* = c_i$ such that the sum in (18) telescopes down to $F(2) - F(1) = (-1/2) - (-1) = 1/2$. ∎

REMARK: $c = \sqrt{ab}$; f is discontinuous at 0 so the "same sign" condition is necessary.

39.   Let $f(x) = x^5$ and $F(x) = (1/6)x^6$, then $f = F'$. The mean value theorem implies that for any a and b, there is some c between them such that $F(b) - F(a) = f(c)\cdot(b-a)$. Thus, for any partition of $[0,1]$, there are choices $x_i^* = c_i$ such that the sum in (18) telescopes down to $F(1) - F(0) = 1/6 - 0 = 1/6$.

41.   Let $f(x) = x^3 - x^2$ and $F(x) = (1/4)x^4 - (1/3)x^3$, then $f = F'$. The mean value theorem implies that for any a and b, there is some c between them such that $F(b) - F(a) = f(c)\cdot(b-a)$. Thus, for any partition of $[1,2]$, there are choices $x_i^* = c_i$ such that the sum in (18) telescopes down to $F(2) - F(1) = (4 - 8/3) - (1/4 - 1/3) = 17/12$.

43.   Let $f(x) = 1/x^2$ and $F(x) = -1/x$, then $f = F'$. The mean value theorem implies that for any a and b, there is some c between them such that $F(b) - F(a) = f(c)\cdot(b-a)$. Thus, for any partition of $[1,3]$, there are choices $x_i^* = c_i$ such that the sum in (18) telescopes down to $F(3) - F(1) = (-1/3) - (-1) = 2/3$.

4.4 — Approximations to area

45.   Let $f(x) = 2\sin x$ and $F(x) = -2\cos x$, then $f = F'$. The mean value theorem implies that for any a and b, there is some c between them such that $F(b) - F(a) = f(c) \cdot (b-a)$. Thus, for any partition of $[0, \pi/4]$, there are choices $x_i^* = c_i$ such that the sum in (18) telescopes down to $-2\cos(\pi/4) + 2\cos(0) = -2/\sqrt{2} + 2 = 2 - \sqrt{2}$.

47.   a. $S_n = \dfrac{1}{n} \sum_{k=1}^{n} \sqrt{1 - \left(\dfrac{k}{n}\right)^2}$ is a sum of the form $\Delta x \sum_{k=1}^{n} f(x_k^*) = \sum_{k=1}^{n} f(x_k^*) \, \Delta x$ where $\Delta x = \dfrac{1}{n}$, $x_k^* = k/n$, and $f(x) = \sqrt{1 - x^2}$. The various x_k^* are between 0 and 1; the graph of $y = f(x)$ on the interval $[0, 1]$ is that part of the circle $x^2 + y^2 = 1$ which is in the first quadrant.

   b.   Because the area of a circle of radius r is πr^2, we see that $\lim\limits_{n \to \infty} S_n = \frac{1}{4} \cdot \pi \cdot 1^2 = \frac{\pi}{4}$.

Solutions 4.5   *The definite integral*   (pages 307-8)

1. $4\sqrt{x}$ increases on the interval $[1, 4]$, thus $4 = 4\sqrt{1} \le 4\sqrt{x} \le 4\sqrt{4} = 8$ if $1 \le x \le 4$. Hence, using Theorems 7 and 2 (or Theorem 8), we infer $12 = \int_1^4 4\,dx \le \int_1^4 \sqrt{x}\,dx \le \int_1^4 8\,dx = 24$.

3. $0 \le x^{10} \le x^2$ on $[-1, 1] \implies 0 = \int_{-1}^{1} 0\,dx \le \int_{-1}^{1} x^{10}\,dx \le \int_{-1}^{1} x^2\,dx = \dfrac{1}{3}\left(1^3 - (-1)^3\right) = \dfrac{2}{3}$.

5. $1/x$ decreases on the interval $[1, 100]$, thus $1/100 \le 1/x \le 1/1 = 1$ if $1 \le x \le 100$. Therefore
$$\frac{99}{100} = \int_1^{100} \frac{1}{100}\,dx \le \int_1^{100} \frac{1}{x}\,dx \le \int_1^{100} 1\,dx = 99.$$

REMARK: We can improve this result by combining easy bounds for $\int_1^2 \dfrac{1}{x}\,dx$ and $\int_2^{100} \dfrac{1}{x}\,dx$:
$$1.48 = \frac{1}{2} + \frac{98}{100} \le \int_1^{100} \frac{1}{x}\,dx \le 1 + \frac{98}{2} = 50.$$

---

7. $0 \le \sin x \le 1$ on $\left[0, \dfrac{\pi}{2}\right] \implies 0 = \int_0^{\pi/2} 0\,dx \le \int_0^{\pi/2} \sin x\,dx \le \int_0^{\pi/2} 1\,dx = \dfrac{\pi}{2}$.
We can improve the lower bound by observing that on the interval $[0, \pi/2]$ the graph of $y = \sin x$ lies on or above the straight line $y = 2x/\pi$ which connects $(0, 0) = (0, \sin 0)$ and $(\pi/2, 1) = (\pi/2, \sin \pi/2)$. because $\sin$ is concave down on this interval; hence $2x/\pi \le \sin x$ which, in turn, implies $\dfrac{\pi}{4} = \int_0^{\pi/2} \dfrac{2x}{\pi}\,dx \le \int_0^{\pi/2} \sin x\,dx$.

9. x^3 increases on $[-5, -4]$, hence
$$-125 = (-5)^3 = \int_{-5}^{-4} (-5)^3\,dx \le \int_{-5}^{-4} x^3\,dx \le \int_{-5}^{-4} (-4)^3\,dx = (-4)^3 = -64.$$

11. $\int_0^4 7x\,dx = \lim\limits_{n \to \infty} \left[\sum_{k=1}^{n} 7\left(\dfrac{4k}{n}\right) \cdot \left(\dfrac{4}{n}\right)\right] = \lim\limits_{n \to \infty} \left[\dfrac{112}{n^2} \sum_{k=1}^{n} k\right] = \left[\lim\limits_{n \to \infty} \dfrac{112}{n^2} \cdot \dfrac{n(n+1)}{2}\right] = 112\left(\dfrac{1}{2}\right) = 56$.

13. $\int_2^5 (3x + 2)\,dx = \int_2^5 3x\,dx + \int_2^5 2\,dx = 3\int_2^5 x\,dx + \int_2^5 2\,dx$ by Theorems 5 and 4.
$$\int_2^5 x\,dx = \lim_{n \to \infty} \sum_{k=1}^{n} \left[2 + k\left(\frac{3}{n}\right)\right] \cdot \left(\frac{3}{n}\right) = \lim_{n \to \infty} \left[\left(\frac{6}{n}\right) \sum_{k=1}^{n} 1 + \left(\frac{9}{n^2}\right) \sum_{k=1}^{n} k\right]$$
$$= \lim_{n \to \infty} \left[\left(\frac{6}{n}\right) \cdot n + \frac{9}{n^2} \cdot \frac{n(n+1)}{2}\right] = 6 + \frac{9}{2} = \frac{21}{2}.$$
Theorem 2 implies $\int_2^5 2\,dx = 2(5 - 2) = 6$. Therefore $\displaystyle\int_2^5 (3x + 2)\,dx = 3\left(\dfrac{21}{2}\right) + 6 = \dfrac{75}{2}$.

15. $\displaystyle\int_0^5 [x]\,dx = \int_0^1 [x]\,dx + \int_1^2 [x]\,dx + \int_2^3 [x]\,dx + \int_3^4 [x]\,dx + \int_4^5 [x]\,dx$
$$= \int_0^1 0\,dx + \int_1^2 1\,dx + \int_2^3 2\,dx + \int_3^4 3\,dx + \int_4^5 4\,dx = 0 + 1 + 2 + 3 + 4 = 10.$$

17. $\int_0^2 x^2\, dx = \lim_{n\to\infty} \sum_{k=1}^n \left(\frac{2k}{n}\right)^2 \cdot \left(\frac{2}{n}\right) = \lim_{n\to\infty} \left(\frac{8}{n^3}\right) \sum_{k=1}^n k^2 = \lim_{n\to\infty} \left(\frac{8}{n^3}\right) \cdot \frac{n(n+1)(2n+1)}{6} = \frac{8}{3}.$

19. $\int_{-1}^1 x^3\, dx = \lim_{n\to\infty} \sum_{k=1}^n \left[-1 + k\left(\frac{2}{n}\right)\right]^3 \cdot \left(\frac{1-(-1)}{n}\right) = \lim_{n\to\infty} \left(\frac{2}{n}\right) \sum_{k=1}^n \left[-1 + \frac{6k}{n} - \frac{12k^2}{n^2} + \frac{8k^3}{n^3}\right]$

$= \lim_{n\to\infty} \left[\frac{2}{n}\sum_{k=1}^n (-1) + \frac{12}{n^2}\sum_{k=1}^n k - \frac{24}{n^3}\sum_{k=1}^n k^2 + \frac{16}{n^4}\sum_{k=1}^n k^3\right]$

$= \lim_{n\to\infty} \left[\frac{2}{n}\cdot(-n) + \frac{12}{n^2}\cdot\frac{n(n+1)}{2} - \frac{24}{n^3}\cdot\frac{n(n+1)(2n+1)}{6} + \frac{16}{n^4}\cdot\frac{n^2(n+1)^2}{4}\right] = -2 + 6 - 8 + 4 = 0.$

REMARK: $(-x)^3 = -x^3$ implies $\int_{-1}^0 x^3\, dx = -\int_0^1 x^3\, dx$; hence $\int_{-1}^1 x^3\, dx = \int_{-1}^0 x^3\, dx + \int_0^1 x^3\, dx = 0.$

21. $\int_0^1 (1-t^2)\, dt = \int_0^1 1\, dt - \int_0^1 t^2\, dt$ (use Theorems 5 and 4). Theorem 2 implies $\int_0^1 1\, dt = 1(1-0) = 1$; Example 4.4.1 implies $\int_0^1 t^2\, dt = 1/3$. Therefore $\int_0^1 (1-t^2)\, dt = 1 - 1/3 = 2/3.$

23. $\int_0^1 (1 + x + x^2 + x^3)\, dx = \lim_{n\to\infty} \sum_{k=1}^n \left[1 + \frac{k}{n} + \left(\frac{k}{n}\right)^2 + \left(\frac{k}{n}\right)^3\right] \cdot \frac{1}{n}$

$= \lim_{n\to\infty} \left[\frac{1}{n}\sum_{k=1}^n 1 + \frac{1}{n^2}\sum_{k=1}^n k + \frac{1}{n^3}\sum_{k=1}^n k^2 + \frac{1}{n^4}\sum_{k=1}^n k^3\right]$

$= \lim_{n\to\infty} \left[\frac{n}{n} + \frac{n(n+1)}{2n^2} + \frac{n(n+1)(2n+1)}{6n^3} + \frac{n^2(n+1)^2}{4n^4}\right]$

$= 1 + \frac{1}{2} + \frac{1}{3} + \frac{1}{4} = \frac{25}{12}.$

25. Equation 13 of the text implies $\int_1^0 t\, dt = -\int_0^1 t\, dt = -\lim_{n\to\infty} \sum_{k=1}^n \frac{k}{n}\cdot\frac{1}{n} = -\lim_{n\to\infty} \frac{n(n+1)}{2n^2} = -\frac{1}{2}.$

27. If we apply the definition of A_a^b and the subsequent discussion, we can interpret $\int_{-1}^1 |x|\, dx$ as the sum of the areas of the two shaded triangles shown in either of the following pictures. Each triangle has base 1, height 1, and area $\frac{1}{2}$; thus $\int_{-1}^1 |x|\, dx = 2\left(\frac{1}{2}\right) = 1.$

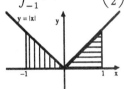
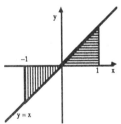

REMARK: $\int_{-1}^1 |x|\, dx = \int_{-1}^0 |x|\, dx + \int_0^1 |x|\, dx = \int_{-1}^0 (-x)\, dx + \int_0^1 x\, dx = 2\int_0^1 x\, dx = 1.$

29. The solution to Problem 4.4.21, especially parts (c) and (d), shows that $\int_a^b x^2\, dx = \frac{b^3 - a^3}{3}.$

31. If $1 < x$, then $1 < x^{4/3}$, $1 < x^{17}$, and $x^{17} = x^{17}\cdot 1 < x^{17}\cdot x^{4/3} = x^{55/3}$. Thus, applying Theorem 7, $\int_{23}^{47} x^{17}\, dx < \int_{23}^{47} x^{55/3}\, dx.$

33. $x > 1 \implies x > \sqrt{x}$; hence, using Theorem 7, $\int_1^2 x\, dx > \int_1^2 \sqrt{x}\, dx.$

35. $0 < x \implies 1 < 1 + x^3 \implies 1 = \sqrt{1} < \sqrt{1 + x^3}$; therefore $1 = \int_0^1 1\, dx < \int_0^1 \sqrt{1 + x^3}\, dx.$

The same sort of quick estimate shows that $0 < x < 1 \implies 0 < x^3 < 1 \implies 1 < 1 + x^3 < 1 + 1 = 2 \implies 1 < \sqrt{1 + x^3} < \sqrt{2}$; thus $\int_0^1 \sqrt{1 + x^3}\, dx < \int_0^1 \sqrt{2}\, dx = \sqrt{2}$. Unfortunately, this does not wrap up the problem because $\sqrt{2}$ is bigger than 5/4. The following discussion uses Theorem 3 and the fact that $\sqrt{1 + x^3}$ does not increase quickly on $[0, 1]$.

$1 + x^3$ and $\sqrt{1 + x^3}$ increase on $[0, \infty)$, thus $0 \le x \le \frac{1}{2} \implies \sqrt{1 + x^3} \le \sqrt{1 + \left(\frac{1}{2}\right)^3} = \frac{3}{2\sqrt{2}}$; similarly, $\frac{1}{2} \le x \le 1 \implies \sqrt{1 + x^3} \le \sqrt{1 + 1^3} = \sqrt{2}$. Therefore

$$\int_0^1 \sqrt{1 + x^3}\, dx = \int_0^{1/2} \sqrt{1 + x^3}\, dx + \int_{1/2}^1 \sqrt{1 + x^3}\, dx$$

$$\le \int_0^{1/2} \frac{3}{2\sqrt{2}}\, dx + \int_{1/2}^1 \sqrt{2}\, dx = \left(\frac{3}{2\sqrt{2}}\right)\left(\frac{1}{2}\right) + \left(\sqrt{2}\right)\left(\frac{1}{2}\right) = \frac{7\sqrt{2}}{8}.$$

Now we're done: $7\sqrt{2}/8 < 5/4$ because $(7\sqrt{2}/8)^2 = 98/64 < 100/64 = 25/16 = (5/4)^2$, hence

$$1 < \int_0^1 \sqrt{1 + x^3}\, dx \le \frac{7\sqrt{2}}{8} < \frac{5}{4}. \quad \blacksquare$$

REMARK: $\quad -1 \le u \implies 0 \le 1 + u \le 1 + u + \frac{u^2}{4} = \left(1 + \frac{u}{2}\right)^2 \implies \sqrt{1 + u} \le \left|1 + \frac{u}{2}\right| = 1 + \frac{u}{2}$, thus

$$\int_0^1 \sqrt{1 + x^3}\, dx \le \int_0^1 \left(1 + \frac{x^3}{2}\right) dx = 1 + \frac{1}{8}.$$

---

37.   a. $v(4 \sec) = 32 \cdot 4 = 128 \text{ ft/sec.}$

    b. $s(4 \sec) = \int_0^4 32t\, dt = 32 \int_0^4 t\, dt = 32 \lim_{n \to \infty} \sum_{k=1}^n \left(k\frac{4}{n}\right) \cdot \frac{4}{n} = 32 \lim_{n \to \infty} \frac{16n(n+1)}{2n^2} = 256 \text{ ft.}$

    c. $s(T) = \int_0^T v(t)\, dt = 32 \int_0^T t\, dt = 32 \lim_{n \to \infty} \sum_{k=1}^n \left(k\frac{T}{n}\right) \cdot \frac{T}{n} = 32 \lim_{n \to \infty} \frac{T^2 n(n+1)}{2n^2} = 16\, T^2;$

      $400 = s(T) = 16T^2 \iff 25 = T^2$, thus the ball hits the ground at $T = 5 \sec.$

39.   Suppose the function f is constant on $[a, b]$, i.e., there is some c such that $f(x) = c$ for all x satisfying $a \le x \le b$. Then for any partition of $[a, b]$ and for any choice of the various x_k^* points in the subintervals, we obtain a telescoping sum: $\sum_{k=1}^n f(x_k^*) \cdot \Delta x_k = \sum_{k=1}^n c \cdot (x_k - x_{k-1}) = c \sum_{k=1}^n (x_k - x_{k-1}) = c(x_n - x_0) = c \cdot (b - a).$

Therefore, $\int_a^b c\, dx = \lim_{n \to \infty} \sum_{k=1}^n f(x_k^*) \cdot \Delta x_k = \lim_{n \to \infty} c \cdot (b - a) = c \cdot (b - a). \quad \blacksquare$

41.   If f is integrable on an interval which contains a, b, and c, then each of the following integrals exist

$$\int_a^b f, \quad \int_b^a f, \quad \int_b^c f, \quad \int_c^b f, \quad \int_c^a f, \quad \int_a^c f.$$

Furthermore, using the text's equation (13), and possibly multiplying by -1,

$$\int_a^b f = -\int_b^a f, \quad \int_b^c f = -\int_c^b f, \quad \int_c^a f = -\int_a^c f.$$

Note: In this problem, f is the only function being integrated, so $\int_a^b f$ will be abbreviated to $\int_a^b$, etc. We may assume a, b, c are three distinct numbers, otherwise the conclusion follows directly from equation (12).

| Case 1, $a < c < b$ | Theorem 3 states that $\int_a^b = \int_a^c + \int_c^b$.

| Case 2, $b < c < a$ | Theorem 3 implies that $\int_b^a = \int_b^c + \int_c^a$ which then implies, using equation (13), that $-\int_a^b = \left(-\int_c^b\right) + \left(-\int_a^c\right)$; hence $\int_a^b = \int_c^b + \int_a^c = \int_a^c + \int_c^b$.

| Case 3, $a < b < c$ | $\int_a^c = \int_a^b + \int_b^c$ by Theorem 3, thus $\int_a^b = \int_a^c - \int_b^c = \int_a^c - \left(-\int_c^b\right) = \int_a^c + \int_c^b$.

| Case 4, $c < a < b$ | Theorem 3 implies $\int_c^b = \int_c^a + \int_a^b$, thus $\int_a^b = -\int_c^a + \int_c^b = -\left(-\int_a^c\right) + \int_c^b = \int_a^c + \int_c^b$.

| Case 5, $b < a < c$ | $\int_b^c = \int_b^a + \int_a^c$ by Theorem 3, thus $-\int_b^a = \int_a^c - \int_a^c$; using equation (13), this implies $\int_a^b = -\left(-\int_b^a\right) = \int_a^c - \left(-\int_c^b\right) = \int_a^c + \int_c^b$.

| Case 6, $c < b < a$ | By Theorem 3, $\int_c^a = \int_c^b + \int_b^a$; rearranging and using equation (13), we obtain $-\int_b^a = -\int_c^a + \int_c^b$ and $\int_a^b = \int_a^c + \int_c^b$. $\blacksquare$

43. a. $\int_{-1}^{2} \lfloor x \rfloor \, dx = \int_{-1}^{0} \lfloor x \rfloor \, dx + \int_{0}^{1} \lfloor x \rfloor \, dx + \int_{1}^{2} \lfloor x \rfloor \, dx = \int_{-1}^{0} (-1) \, dx + \int_{0}^{1} 0 \, dx + \int_{1}^{2} 1 \, dx = (-1) + 0 + 1 =$
$0 = (-1) \cdot 0 = \int_{2}^{-1} \lfloor x \rfloor \, dx$, but $\lfloor x \rfloor$ is not equal to 0 throughout the interval $[-1, 2]$. This counterexample disproves the assertion.

  b. $\int_{-1}^{1} x \, dx = 0 = \int_{1}^{-1} x \, dx$, but x is not equal to 0 throughout the interval $-1, 1]$. This counterexample disproves the assertion.

  c. $\int_{a}^{b} |f| = \int_{b}^{a} |f| = -\int_{a}^{b} |f|$ implies $\int_{a}^{b} |f| = 0$. Suppose there were some $c \in (a, b)$ such that $f(c) \neq 0$. If f is continuous on $[a, b]$, then must be a $\delta > 0$ such that $a < c - \delta < x < c + \delta < b \implies 0 < |f(c)|/2 < |f(x)|$; therefore $\int_{a}^{b} |f| \geq \int_{c-\delta}^{c+\delta} |f| \geq \int_{c-\delta}^{c+\delta} |f(c)|/2 = |f(c)| \cdot \delta > 0$.

Therefore, f continuous and $\int_{a}^{b} |f| = \int_{b}^{a} |f|$ implies $f \equiv 0$ throughout $[a, b]$. ∎

45.   Suppose $0 \leq a \leq b$.

$$\int_{a}^{b} \lfloor x \rfloor \, dx = \int_{a}^{\lfloor a \rfloor} \lfloor x \rfloor \, dx + \int_{\lfloor a \rfloor}^{\lfloor b \rfloor} \lfloor x \rfloor \, dx + \int_{\lfloor b \rfloor}^{b} \lfloor x \rfloor \, dx = -\int_{\lfloor a \rfloor}^{a} \lfloor x \rfloor \, dx + \sum_{k=\lfloor a \rfloor}^{\lfloor b \rfloor - 1} \int_{k}^{k+1} \lfloor x \rfloor \, dx + \int_{\lfloor b \rfloor}^{b} \lfloor x \rfloor \, dx$$

$$= -\int_{\lfloor a \rfloor}^{a} \lfloor a \rfloor \, dx + \sum_{k=\lfloor a \rfloor}^{\lfloor b \rfloor - 1} \int_{k}^{k+1} k \, dx + \int_{\lfloor b \rfloor}^{b} \lfloor b \rfloor \, dx = -\lfloor a \rfloor (a - \lfloor a \rfloor) + \sum_{k=\lfloor a \rfloor}^{\lfloor b \rfloor - 1} k + \lfloor b \rfloor (b - \lfloor b \rfloor)$$

$$= -\lfloor a \rfloor (a - \lfloor a \rfloor) + \left(\frac{(\lfloor b \rfloor - 1) \lfloor b \rfloor}{2} - \frac{(\lfloor a \rfloor - 1) \lfloor a \rfloor}{2} \right) + \lfloor b \rfloor (b - \lfloor b \rfloor)$$

$$= \frac{\lfloor b \rfloor (2b - \lfloor b \rfloor - 1)}{2} - \frac{\lfloor a \rfloor (2a - \lfloor a \rfloor - 1)}{2}.$$

47. a.   Every nonempty interval contains a rational number, the value of f there is 1.
  b.   Every nonempty interval contains an irrational number, the value of f there is 0.
  c.   For any partition we can choose x_k^* to be rational, then

$$\lim_{\max \Delta x_i \to 0} \sum_{k=1}^{n} f(x_i^*) \cdot \Delta x_i = \lim_{\max \Delta x_i \to 0} \sum_{k=1}^{n} 1 \cdot \Delta x_i = \lim_{\max \Delta x_i \to 0} \sum_{k=1}^{n} \Delta x_i = \lim_{\max \Delta x_i \to 0} (x_n - x_0) = b - a;$$

on the other hand, we could also force each x_k^* to be irrational, then

$$\lim_{\max \Delta x_i \to 0} \sum_{k=1}^{n} f(x_i^*) \cdot \Delta x_i = \lim_{\max \Delta x_i \to 0} \sum_{k=1}^{n} 0 \cdot \Delta x_i = \lim_{\max \Delta x_i \to 0} \sum_{k=1}^{n} 0 = 0.$$

  d.   If $a \neq b$, then the preceding part shows that $\lim_{\max \Delta x_i \to 0} \sum_{k=1}^{n} f(x_i^*) \cdot \Delta x_i$ does not have a unique value independent of the choice of x_k^*; therefore $\int_{a}^{b} f$ does not exist.

Solutions 4.6   *The fundamental theorem of calculus*   (pages 318-20)

1. $\int_{-1}^{2} x^4 \, dx = \left. \frac{x^5}{5} \right|_{-1}^{2} = \frac{2^5}{5} - \frac{(-1)^5}{5} = \frac{32}{5} - \left(\frac{-1}{5} \right) = \frac{33}{5}.$

3. $\int_{1}^{9} \frac{\sqrt{t}}{2} \, dt = \frac{1}{2} \int_{1}^{9} t^{1/2} \, dt = \frac{1}{2} \cdot \left. \frac{t^{3/2}}{3/2} \right|_{1}^{9} = \left. \frac{t^{3/2}}{3} \right|_{1}^{9} = \frac{9^{3/2}}{3} - \frac{1^{3/2}}{3} = 9 - \frac{1}{3} = \frac{26}{3}.$

5. $\int_{1}^{8} \left(\frac{1}{\sqrt[3]{x}} + 7\sqrt[3]{x} \right) dx = \int_{1}^{8} \left(x^{-1/3} + 7x^{1/3} \right) dx = \left. \left(\frac{x^{2/3}}{2/3} + 7 \cdot \frac{x^{4/3}}{4/3} \right) \right|_{1}^{8}$

$$= \left(\frac{8^{2/3}}{2/3} + 7 \cdot \frac{8^{4/3}}{4/3} \right) - \left(\frac{1^{2/3}}{2/3} + 7 \cdot \frac{1^{4/3}}{4/3} \right) = (6 + 84) - \left(\frac{3}{2} + \frac{21}{4} \right) = \frac{333}{4} = 83.25.$$

7. $\int_{0}^{\pi/4} 2 \sin x \, dx = \left. -2 \cos x \right|_{0}^{\pi/4} = \left(-2 \cos \left(\frac{\pi}{4} \right) \right) - (-2 \cos 0) = \left(-2 \cdot \frac{1}{\sqrt{2}} \right) - (-2 \cdot 1) = 2 - \sqrt{2}.$

9. $\int_{-1}^{1} \left(p^9 - p^{17} \right) dp = \left. \left(\frac{p^{10}}{10} - \frac{p^{18}}{18} \right) \right|_{-1}^{1} = \left(\frac{1^{10}}{10} - \frac{1^{18}}{18} \right) - \left(\frac{(-1)^{10}}{10} - \frac{(-1)^{18}}{18} \right)$

$$= \left(\frac{1}{10} - \frac{1}{18} \right) - \left(\frac{1}{10} - \frac{1}{18} \right) = 0.$$

4.6 — The fundamental theorem of calculus

11.
$$\int_2^3 (s-1)(s+2)\,ds = \int_2^3 \left(s^2 + s - 2\right) ds = \left(\frac{s^3}{3} + \frac{s^2}{2} - 2\cdot s\right)\Big|_2^3$$
$$= \left(\frac{3^3}{3} + \frac{3^2}{2} - 2\cdot 3\right) - \left(\frac{2^3}{3} + \frac{2^2}{2} - 2\cdot 2\right) = \frac{15}{2} - \frac{2}{3} = \frac{41}{6}.$$

13.
$$\int_0^1 \left(\sqrt{x} - x\right)^2 dx = \int_0^1 \left(x - 2x^{3/2} + x^2\right) dx = \left(\frac{1}{2}\cdot x^2 - \frac{4}{5}\cdot x^{5/2} + \frac{1}{3}\cdot x^3\right)\Big|_0^1$$
$$= \left(\frac{1}{2}\cdot 1^2 - \frac{4}{5}\cdot 1^{5/2} + \frac{1}{3}\cdot 1^3\right) - \left(\frac{1}{2}\cdot 0^2 - \frac{4}{5}\cdot 0^{5/2} + \frac{1}{3}\cdot 0^3\right) = \frac{1}{2} - \frac{4}{5} + \frac{1}{3} = \frac{1}{30}.$$

15.
$$\int_2^4 \frac{6 + 7x + x^2}{1 + x}\,dx = \int_2^4 \frac{(6+x)(1+x)}{1+x}\,dx = \int_2^4 (6 + x)\,dx = \left(6\cdot x + \frac{1}{2}\cdot x^2\right)\Big|_2^4$$
$$= \left(6\cdot 4 + \frac{1}{2}\cdot 4^2\right) - \left(6\cdot 2 + \frac{1}{2}\cdot 2^2\right) = 32 - 14 = 18.$$

17.
$$\int_9^{16} \frac{s + 1}{\sqrt{s}}\,ds = \int_9^{16}\left(s^{1/2} + s^{-1/2}\right) ds = \left(\frac{2}{3}\cdot s^{3/2} + 2\cdot s^{1/2}\right)\Big|_9^{16} = \left(\frac{128}{3} + 8\right) - (18 + 6) = \frac{80}{3}.$$

19.
$$\int_{-5}^3 |x|\,dx = \int_{-5}^0 (-x)\,dx + \int_0^3 x\,dx = (-1)\cdot\left(\frac{x^2}{2}\right)\Big|_{-5}^0 + \frac{x^2}{2}\Big|_0^3 = (-1)\left(\frac{0}{2} - \frac{25}{2}\right) + \left(\frac{9}{2} - \frac{0}{2}\right) = 17.$$

REMARK: $\quad \displaystyle\int_{-5}^3 |x|\,dx = \left(\frac{x\cdot|x|}{2}\right)\Big|_{-5}^3 = \left(\frac{3\cdot|3|}{2}\right) - \left(\frac{-5\cdot|-5|}{2}\right) = \frac{9}{2} - \frac{-25}{2} = 17.$

(The solution to Problem 4.2.32 obtains this indefinite integral of $|x|$.)

---

21. Area $= \displaystyle\int_{-3}^3 |9 - x^2|\,dx = \int_{-3}^3 (9 - x^2)\,dx = \left(9x - \frac{x^3}{3}\right)\Big|_{-3}^3 = \left(27 - \frac{27}{3}\right) - \left(-27 - \frac{-27}{3}\right) = 36.$

23. Area $= \displaystyle\int_1^5 |x^2 - 6x + 5|\,dx = \int_1^5 \left(-x^2 + 6x - 5\right) dx$
$$= \left(\frac{-x^3}{3} + 3x^2 - 5x\right)\Big|_1^5 = \left(\frac{-125}{3} + 75 - 25\right) - \left(\frac{-1}{3} + 3 - 5\right) = \frac{32}{3}.$$

25. Area $= \displaystyle\int_a^b |(x - a)(x - b)|\,dx = \int_a^b (x - a)(b - x)\,dx = \int_a^b \left(-ab + (a + b)x - x^2\right) dx$
$$= \left(-abx + \frac{(a+b)x^2}{2} - \frac{x^3}{3}\right)\Big|_a^b$$
$$= \left(-ab^2 + \frac{(a+b)b^2}{2} - \frac{b^3}{3}\right) - \left(-a^2b + \frac{(a+b)a^2}{2} - \frac{a^3}{3}\right)$$
$$= \frac{-6ab^2 + 3ab^2 + 3b^3 - 2b^3 + 6a^2b - 3a^3 - 3a^2b + 2a^3}{6}$$
$$= \frac{b^3 - 3ab^2 + 3a^2b - a^3}{6} = \frac{(b - a)^3}{6}.$$

27. $x^3 + 2x^2 - x - 2 = (x + 2)(x + 1)(x - 1)$ is positive on $(-2, -1)$ and negative on $(-1, 1)$.

Area $= \displaystyle\int_{-2}^1 |x^3 + 2x^2 - x - 2|\,dx = \int_{-2}^{-1} \left(x^3 + 2x^2 - x - 2\right) dx - \int_{-1}^1 \left(x^3 + 2x^2 - x - 2\right) dx$
$$= \left(\frac{x^4}{4} + \frac{2x^3}{3} - \frac{x^2}{2} - 2x\right)\Big|_{-2}^{-1} - \left(\frac{x^4}{4} + \frac{2x^3}{3} - \frac{x^2}{2} - 2x\right)\Big|_{-1}^1$$
$$= \left[\left(\frac{1}{4} + \frac{-2}{3} - \frac{1}{2} + 2\right) - \left(4 + \frac{-16}{3} - 2 + 4\right)\right] - \left[\left(\frac{1}{4} + \frac{2}{3} - \frac{1}{2} - 2\right) - \left(\frac{1}{4} + \frac{-2}{3} - \frac{1}{2} + 2\right)\right]$$
$$= \left[\frac{13}{12} - \frac{2}{3}\right] - \left[\frac{-19}{12} - \frac{13}{12}\right] = \frac{5}{12} - \frac{-8}{3} = \frac{37}{12}.$$

29. Area $= \displaystyle\int_{-2}^4 |x^4|\,dx = \int_{-2}^4 x^4\,dx = \frac{x^5}{5}\Big|_{-2}^4 = \frac{1024}{5} - \frac{-32}{5} = \frac{1056}{5} = 211.2.$

31.

$$\text{Area} = \int_0^\pi |\cos x - \sin x|\, dx = \int_0^{\pi/4} (\cos x - \sin x)\, dx + \int_{\pi/4}^\pi (\sin x - \cos x)\, dx$$

$$= (\sin x + \cos x)\Big|_0^{\pi/4} + (-\cos x - \sin x)\Big|_{\pi/4}^\pi$$

$$= \left[\left(\frac{1}{\sqrt{2}} + \frac{1}{\sqrt{2}}\right) - (0+1)\right] + \left[(-(-1) - 0) - \left(-\frac{1}{\sqrt{2}} - \frac{1}{\sqrt{2}}\right)\right]$$

$$= \left[\sqrt{2} - 1\right] + \left[1 + \sqrt{2}\right] = 2\sqrt{2}.$$

33. If $F(x) = \int_3^x \left(\frac{1}{1+t^3}\right) dt$, then $F'(x) = \frac{1}{1+x^3}$ by Theorem 2; thus $F'(2) = \frac{1}{1+8} = \frac{1}{9}$.

35. If $F(x) = \int_0^x \sqrt{\frac{u-1}{u+t}}\, du$, then $F'(x) = \sqrt{\frac{x-1}{x+1}}$ and $F'(1) = \sqrt{\frac{0}{1+1}} = \sqrt{0} = 0$.

37. If $F(x) = \int_1^{1+3x} \left(\frac{1}{t}\right) dt$, then $F'(x) = \left(\frac{1}{1+3x}\right) \cdot \frac{d}{dx}(1+3x) = \frac{3}{1+3x}$ and $F'(1) = \frac{3}{4}$.

39. $\int |x|\, dx = \frac{x \cdot |x|}{2} + C$ (see the solution to Problem 4.2.32).

   a. $\int_{-1}^1 |x|\, dx = \frac{x \cdot |x|}{2}\Big|_{-1}^1 = \frac{(1)(1)}{2} - \frac{(-1)(1)}{2} = \frac{1}{2} - \frac{-1}{2} = 1.$

   b. $\int_0^2 |t-1|\, dt = \frac{(t-1)\cdot|t-1|}{2}\Big|_0^2 = \frac{(1)(1)}{2} - \frac{(-1)(1)}{2} = 1.$

   c. $\int_1^3 |u-2|\, du = \frac{(u-2)\cdot|u-2|}{2}\Big|_1^3 = \frac{(1)(1)}{2} - \frac{(-1)(1)}{2} = 1.$

   d. $\int_{-5}^{-3} |s+4|\, ds = \frac{(s+4)\cdot|s+4|}{2}\Big|_{-5}^{-3} = \frac{(1)(1)}{2} - \frac{(-1)(1)}{2} = 1.$

41.
   a. If $a(t) = k$, then $V(t) = \int a(t)\, dt = \int k\, dt = kt + V(0)$. If $V(0) = 500$ and $V(1/100) = 10500$, then $10500 = k(1/100) + 500$, $10000 = k/100$, $k = (10000)(100) = 10^4 10^2 = 10^6$. Therefore acceleration $a(t) = 10^6$ m/sec^2.

   b. $V(t) = 10^6 t + 500$. The distance traveled in the first $1/100$ sec after entering the linear accelerator is $\int_0^{1/100} v(t)\, dt = ((5)10^5 t^2 + 500t)\Big|_0^{10^{-2}} = (5)10^5 10^{-4} + (5)10^2 10^{-2} = 50 + 5 = 55$ m.

43. The number of insects added between the beginning of the ninth week (the end of the eighth week) and the end of the twenty-fifth week is $\int_8^{25} 3000\, t^{-1/2}\, dt = 6000\, t^{1/2}\Big|_8^{25} = 30,000 - 12,000\sqrt{2} \approx 13,029$.

   The number of insects added between the end of the ninth week and the beginning of the twenty-fifth week is $\int_9^{24} 3000\, t^{-1/2}\, dt = 6000\, t^{1/2}\Big|_9^{24} = 12,000\sqrt{6} - 18,000 \approx 11,394$.

   It is numerically simple to compute $\int_9^{25} 3000\, t^{-1/2}\, dt = 6000\, t^{1/2}\Big|_9^{25} = 30,000 - 18,000 = 12,000$, the number of insects added from the tenth week through the twenty-fifth week.

45. $R(100) - R(50) = \int_{50}^{100} MR(q)\, dq = \int_{50}^{100} (2 - 0.02q + 0.003q^2)\, dq = (2q - 0.01q^2 + 0.001q^3)\Big|_{50}^{100} = \$900$.

47.

$$f' \text{ exists} \quad\Longrightarrow\quad f \text{ is continuous} \quad\Longrightarrow\quad \frac{d}{dx}\int_a^x f = f(x),$$

$$f' \text{ is continuous} \quad\Longrightarrow\quad \int_a^x f' = f(x) - f(a),$$

$$\frac{d}{dx}\int_a^x f = \int_a^x f' \quad\Longleftrightarrow\quad f(a) = 0.$$

49. The average value of $f(x) = C$ over $[a,b]$ is $\left(\frac{1}{b-a}\right)\int_a^b C\, dx = \left(\frac{1}{b-a}\right)C(b-a) = C$.

51. $\left(\frac{1}{3-1}\right)\int_1^3 \frac{1}{x^2}\, dx = \frac{1}{2} \cdot \frac{-1}{x}\Big|_1^3 = \frac{1}{2} \cdot \left(\frac{-1}{3} - \frac{-1}{1}\right) = \frac{1}{2} \cdot \frac{2}{3} = \frac{1}{3}$.

4.6 — The fundamental theorem of calculus

53. $\left(\dfrac{1}{2-0}\right)\displaystyle\int_0^2 x^3\, dx = \dfrac{1}{2}\cdot\dfrac{x^4}{4}\Big|_0^2 = \dfrac{1}{2}\cdot\left(\dfrac{16}{4}-\dfrac{0}{4}\right) = 2.$

55. $\left(\dfrac{1}{2-(-2)}\right)\displaystyle\int_{-2}^2 (x^2-2x+5)\, dx = \left(\dfrac{1}{4}\right)\cdot\left(\dfrac{x^3}{3}-x^2+5x\right)\Big|_{-2}^2 = \left(\dfrac{1}{4}\right)\cdot\left(\dfrac{26}{3}-\dfrac{-50}{3}\right) = \dfrac{19}{3}.$

57. Let $s(t)$ measure the distance of the ball above the ground at time t, $v(t) = s'(t)$ is the velocity and $a(t) = v'(t)$ is the acceleration. $s(0) = 400$ and $v(0) = 0$ because the ball starts 400 feet above the ground and is dropped from rest, $a(t) = -32\,\text{ft/sec}^2$ because the ball moves toward the ground and the magnitude of gravitational acceleration is $32\,\text{ft/sec}^2$. Therefore $v(T) = v(0) + \displaystyle\int_0^T a(t)\, dt = 0 + \int_0^T -32\, dt = -32T$

and $s(T) = s(0) + \displaystyle\int_0^T v(t)\, dt = 400 + \int_0^T -32t\, dt = 400 - 16T^2$. Because $s(5) = 0$ and $s > 0$ on $[0,5)$, the ball hits the ground 5 seconds after it is dropped. The average value of $v(t) = -32t$ on $[0,5]$ is

$\dfrac{1}{5-0}\displaystyle\int_0^5 (-32t)\, dt = \dfrac{1}{5}(-16t^2)\Big|_0^5 = -80\,\text{ft/sec}.$

REMARK: Example 4.2.6 shows a different point of view. Suppose $S(t)$ is the distance traveled by the ball after t seconds. Then $S(0) = 0$; $V(t) = S'(t)$ and $V(0) = 0$; $A(t) = V'(t)$ and $A(t) = 32\,\text{ft/sec}^2$. Because the ball hits the ground 400 feet from its starting position, that occurs when $400 = S(t) = 16t^2$, i.e., when $t = 5$ sec. The average velocity during the time interval $[0,5]$ is $\dfrac{1}{5-0}\displaystyle\int_0^5 32t\, dt = \dfrac{1}{5}(16t^2)\Big|_0^5 = \dfrac{1}{5}(400) = 80\,\text{ft/sec}$. Note that s and S measure distance in opposite directions; in fact $s(t) + S(t) = 400$ for all t, $0 \le t \le 5$.

---

59. Assuming the total cost of producing $q = 0$ units is \$0, the average cost per unit if $q = 200$ is

$\left(\dfrac{1}{200-0}\right)\displaystyle\int_0^{200} MC\, dq = \dfrac{1}{200}\int_0^{200}\left(50-\dfrac{q}{20}\right) dq = \dfrac{1}{200}\left(50q-\dfrac{q^2}{40}\right)\Big|_0^{200} = \dfrac{1}{200}(9000) = \$45/\text{unit}.$

61. *Lemma:* If f is integrable, then $\displaystyle\int_{-a}^0 f(-x)\, dx = \int_0^a f(x)\, dx$.

If we could assume f were continuous, then f would have an antiderivative F and $\dfrac{d}{dx}\left(-F(-x)\right) = -F'(-x)\cdot(-1) = F'(-x) = f(-x)$ would imply $-F(-x)$ is an antiderivative for $f(-x)$. Hence

$\displaystyle\int_0^a f(x)\, dx = F(x)\Big|_0^a = F(a) - F(0) = -F(-0) - \left(-F(-(-a))\right) = -F(-x)\Big|_{-a}^0 = \int_{-a}^0 f(-x)\, dx.$

If we only know f is integrable, we must work a bit harder. Any partition P of $[-a,0]$ reflects through the origin into a partition $\hat{P}$ of $[0,a]$; the ith subinterval $[x_{i-1}, x_i]$ reflects into $[-x_i, -x_{i-1}] = [w_{n-i}, w_{n-i+1}]$, the $n+1-i^{\text{th}}$ subinterval of $\hat{P}$; similarly, $-x_i^* = w_{n+1-i}^*$. Therefore

$\displaystyle\int_{-a}^0 f(-x)\, dx = \lim_{\max \Delta x_i \to 0} \sum_{i=1}^n f(-x_i^*)\Delta x_i$

$= \displaystyle\lim_{\max \Delta w_j \to 0} \sum_{j=n}^1 f(w_{n+1-j}^*)\Delta w_{n+1-j} = \lim_{\max \Delta w_k \to 0} \sum_{k=1}^n f(w_k^*)\Delta w_k = \int_0^a f(w)\, dw.\ \blacksquare$

Proof: If f is an even function which is integrable, then

$$\int_{-a}^a f(x)\, dx = \int_{-a}^0 f(x)\, dx + \int_0^a f(x)\, dx$$

$$= \int_{-a}^0 f(-x)\, dx + \int_0^a f(x)\, dx \qquad\qquad f(x) = f(-x)$$

$$= \int_0^a f(x)\, dx + \int_0^a f(x)\, dx \qquad\qquad \text{Lemma}$$

$$= 2\int_0^a f(x)\, dx.\ \blacksquare$$

The geometric interpretation is that the regions to the left and right of the y-axis are congruent.

63. $0 < p$ and $0 < x \implies (1 + x^p) \cdot (1 - x^p) = 1 - x^{2p} < 1 < 1 + x^p$

$$\implies 1 - x^p < \frac{1}{1 + x^p} < 1$$

$$\implies \frac{p}{p+1} = \left(x - \frac{x^{p+1}}{p+1} \right) \Big|_0^1 = \int_0^1 (1 - x^p) \, dx < \int_0^1 \left(\frac{1}{1+x^p} \right) dx < \int_0^1 1 \, dx = 1. \; \blacksquare$$

Solutions 4.7   *Integration by substitution and Differentials*   (page 327)

1.   If $y = 3x + 2$, then $dy = 3\,dx$.

3.   If $y = x^4$, then $dy = 4x^3\,dx$.

5.   If $y = (1 + x^2)^4$, then $dy = 4(1 + x^2)^3(2x)\,dx = 8x(1 + x^2)^3\,dx$.

7.   If $y = \sqrt{1 + x^2}$, then $dy = x\left(1 + x^2\right)^{-1/2}\,dx$.

9.   If $y = \cos\sqrt{x}$, then $dy = \dfrac{-\sin\sqrt{x}}{2\sqrt{x}}\,dx$.

11.   Let $u = 1 + x^2$, then $du = 2x\,dx$ and $\int \left(1 + x^2\right)^5 2x\,dx = \int u^5\,du = (1/6)u^6 + C = (1/6)\left(1 + x^2\right)^6 + C$.

13.   Let $u = 3 - x$, then $du = -\,dx$ and $\int (3 - x)^3\,dx = \int u^3(-du) = (-1/4)u^4 + C = (-1/4)(3 - x)^4 + C$.

Therefore $\int_0^2 (3 - x)^3\,dx = (-1/4)(3 - x)^4 \Big|_0^2 = (-1/4)(1)^4 - (-1/4)(3)^4 = (-1/4)(1 - 81) = 20$.

15.   Let $u = 2 - x^4$, then $du = -4x^3\,dx$ and $\displaystyle\int \left(2 - x^4\right) 4x^3\,dx = \int u(-du) = \frac{-u^2}{2} + C = \frac{-\left(2 - x^4\right)^2}{2} + C$.

Therefore $\displaystyle\int_0^1 \left(2 - x^4\right) 4x^3\,dx = \frac{-\left(2 - x^4\right)^2}{2} \Big|_0^1 = \frac{-1}{2} - \frac{-4}{2} = \frac{3}{2}$.

17.   Let $u = 9 + x$, then $du = dx$, and $\displaystyle\int (9 + x)^{1/2}\,dx = \int u^{1/2}\,du = \frac{2}{3}u^{2/3} + C = \frac{2}{3}(9 + x)^{3/2} + C$. Therefore

$$\int_0^7 \sqrt{9 + x}\,dx = \frac{2}{3}(9 + x)^{3/2} \Big|_0^7 = \frac{2}{3}\left(16^{3/2}\right) - \frac{2}{3}\left(9^{3/2}\right) = \frac{128}{3} - 18 = \frac{74}{3}.$$

19.   Let $u = 10 - 9x$, then $du = -9\,dx$, $dx = \dfrac{-1}{9}du$, and

$$\int (10 - 9x)^{1/2}\,dx = \int u^{1/2}\left(\frac{-1}{9}\right)du = \frac{-1}{9}\int u^{1/2}\,du = \left(\frac{-1}{9}\right)\left(\frac{2}{3}\right)u^{2/3} + C = \left(\frac{-2}{27}\right)(10 - 9x)^{3/2} + C.$$

Thus $\displaystyle\int_0^1 \sqrt{10 - 9x}\,dx = \left(\frac{-2}{27}\right)(10 - 9x)^{3/2} \Big|_0^1 = \left(\frac{-2}{27}\right)\left(1^{3/2}\right) - \left(\frac{-2}{27}\right)\left(10^{3/2}\right) = \frac{20\sqrt{10} - 2}{27}$.

21.   Let $u = 1 + 3x^4$, then $du = 12x^3\,dx$, and $\displaystyle\int x^3 \sqrt[5]{1 + 3x^4}\,dx = \int (1 + 3x^4)^{1/5}\left(\frac{1}{12}\right)(12x^3\,dx) =$

$\displaystyle\int u^{1/5}\left(\frac{1}{12}\right)du = \left(\frac{1}{12}\right)\left(\frac{5}{6}\right)u^{6/5} + C = \left(\frac{5}{72}\right)(1 + 3x^4)^{6/5} + C.$

23.   Let $u = s^5 + 5s$, then $du = (5s^4 + 5)ds = 5(s^4 + 1)ds$ and

$$\int (s^4 + 1)\sqrt{s^5 + 5s}\,dx = \int (s^5 + 5s)^{1/2}\left(\frac{1}{5}\right)(5(s^4 + 1)ds)$$

$$= \int u^{1/2}\left(\frac{1}{5}\right)du = \left(\frac{1}{5}\right)\left(\frac{2}{3}\right)u^{2/3} + C = \left(\frac{2}{15}\right)(s^5 + 5s)^{3/2} + C.$$

25.   Let $u = t^2 + 2t^3$, then $du = (2t + 6t^2)dt$, $(t + 3t^2)\,dt = (1/2)\,du$, and $\displaystyle\int \frac{1}{\sqrt{t^2 + 2t^3}}(t + 3t^2)\,dt =$

$\displaystyle\int u^{-1/2}\left(\frac{1}{2}\,du\right) = u^{1/2} + C = \sqrt{t^2 + 2t^3} + C.$ Therefore

$$\int_1^2 \frac{t + 3t^2}{\sqrt{t^2 + 2t^3}}\,dt = \sqrt{t^2 + 2t^3} \Big|_1^2 = \sqrt{4 + 16} - \sqrt{1 + 2} = \sqrt{20} - \sqrt{3}.$$

27.   Let $u = 1 + \sqrt{x}$, then $du = \dfrac{dx}{2\sqrt{x}}$ and

$$\int \frac{dx}{\sqrt{x}\,(1 + \sqrt{x})^3} = \int (1 + \sqrt{x})^{-3}\,(2)\left(\frac{dx}{2\sqrt{x}}\right) = 2\int u^{-3}\,du = 2\left(\frac{1}{-2}\right)u^{-2} + C = -\,(1 + \sqrt{x})^{-2} + C.$$

29.   Let $u = 1 + \dfrac{1}{v^2} = 1 + v^{-2}$, then $du = (-2)v^{-3}\,dv$ and

$$\int \frac{\left(1 + \frac{1}{v^2}\right)^{5/3}}{v^3}\,dv = -\frac{1}{2}\int \left(1 + \frac{1}{v^2}\right)^{5/3}(-2v^{-3}\,dv) = -\frac{1}{2}\int u^{5/3}\,du$$

$$= \left(\frac{-1}{2}\right)\left(\frac{3}{8}\right)u^{8/3} + C = \left(\frac{-3}{16}\right)\left(1 + \frac{1}{v^2}\right)^{8/3} + C.$$

31.   Let $u = ax^2 + 2bx + c$, then $du = (2ax + 2b)\,dx$, $\dfrac{1}{2}du = (ax + b)\,dx$, and $\displaystyle\int (ax + b)\sqrt{ax^2 + 2bx + c}\,dx =$

$$\int (ax^2 + 2bx + c)^{1/2}\,(ax + b)\,dx = \int u^{1/2}\left(\frac{1}{2}du\right) = \frac{1}{2}\cdot\frac{2}{3}u^{3/2} + C = \frac{1}{3}(ax^2 + 2bx + c)^{3/2} + C.$$

33.   Let $u = t^2 + \alpha^2$, then $\displaystyle\int t\sqrt{t^2 + \alpha^2}\,dt = \int u^{1/2}\left(\frac{du}{2}\right) = \frac{1}{2}\cdot\frac{2}{3}u^{3/2} + C = \frac{1}{3}(t^2 + \alpha^2)^{3/2} + C$; therefore

$$\int_0^\alpha t\sqrt{t^2 + \alpha^2}\,dt = \frac{1}{3}(t^2 + \alpha^2)^{3/2}\Big|_0^\alpha = \frac{2\sqrt{2} - 1}{3}\cdot|\alpha|^3 = \frac{2\sqrt{2} - 1}{3}\cdot\alpha^3.$$ (Remember the text states $\alpha > 0$.)

35.   Let $u = a + bs^n$, then $\displaystyle\int \frac{s^{n-1}}{\sqrt{a + bs^n}}\,ds = \int u^{-1/2}\left(\frac{du}{nb}\right) = \frac{1}{nb}\cdot 2\,u^{1/2} + C = \frac{2\sqrt{a + bs^n}}{nb} + C.$

37.   Let $u = \alpha^6 - p^6$, then $\displaystyle\int p^5\sqrt{\alpha^6 - p^6}\,dp = \int u^{1/2}\left(\frac{du}{-6}\right) = \frac{-1}{6}\cdot\frac{2}{3}u^{3/2} + C = \frac{-1}{9}(\alpha^6 - p^6)^{3/2} + C;$

therefore $\displaystyle\int_{-\alpha}^\alpha p^5\sqrt{\alpha^6 - p^6}\,dp = \frac{-1}{9}(\alpha^6 - p^6)^{3/2}\Big|_{-\alpha}^\alpha = \left(\frac{-2}{9}\right)(0 - 0) = 0.$

REMARK: The integrand is an odd function and the interval is symmetric about 0; the result of Problem 4.6.62 implies the integral must be zero.

39.   Let $u = 2x$, then $\displaystyle\int \sin 2x\,dx = \int \sin u\left(\frac{du}{2}\right) = \frac{1}{2}(-\cos u) + C = \frac{-\cos 2x}{2} + C;$

therefore $\displaystyle\int_0^{\pi/4} \sin(2x)\,dx = \left(\frac{-1}{2}\right)\cos(2x)\Big|_0^{\pi/4} = \left(\frac{-1}{2}\right)\left(\cos\frac{\pi}{2} - \cos 0\right) = \left(\frac{-1}{2}\right)(0 - 1) = \frac{1}{2}.$

41.   Let $u = \sqrt{x}$, then $\displaystyle\int \frac{\sin\sqrt{x}}{\sqrt{x}}\,dx = \int \sin\sqrt{x}\left(\frac{dx}{\sqrt{x}}\right) = \int \sin u\,(2\,du) = 2\,(-\cos u) + C = -2\,\cos\sqrt{x} + C.$

43.   Let $u = x + 2$; then Area $= \displaystyle\int_{x=0}^7 \sqrt{x + 2}\,dx = \int_{u=2}^9 \sqrt{u}\,du = \frac{2}{3}u^{3/2}\Big|_2^9 = \left(\frac{2}{3}\right)(27 - 2\sqrt{2}).$

45.   Let $u = \dfrac{x^3}{3} + 1$; then Area $= \displaystyle\int_{x=0}^3 x^2\sqrt{\frac{x^3}{3} + 1}\,dx = \int_{u=1}^{10} u^{1/2}\,du = \left(\frac{2}{3}\right)u^{3/2}\Big|_1^{10} = \left(\frac{2}{3}\right)(10\sqrt{10} - 1).$

47.   Let $u = \dfrac{x}{3}$; then

Area $= \displaystyle\int_{x=0}^\pi \sin\left(\frac{x}{3}\right)dx = \int_{u=0}^{\pi/3} \sin u\,(3\,du) = -3\cos u\Big|_0^{\pi/3} = (-3)\left(\cos\frac{\pi}{3} - \cos 0\right) = (-3)\left(\frac{1}{2} - 1\right) = \frac{3}{2}.$

49.   Distance travelled $= \displaystyle\int_{t=2}^7 \frac{1}{3\sqrt{2 + t}}\,dt \overset{u \equiv 2 + t}{=} \int_{u=4}^9 \frac{1}{3}u^{-1/2}\,du = \frac{2}{3}u^{1/2}\Big|_4^9 = \left(\frac{2}{3}\right)(9^{1/2} - 4^{1/2}) = \frac{2}{3}$ m.

51. $\displaystyle\int_{q=60}^{77} MC(q)\,dq = \int_{q=60}^{77} \frac{4}{\sqrt{q + 4}}\,dq \overset{u \equiv q + 4}{=} \int_{u=64}^{81} 4u^{-1/2}\,du = 8u^{1/2}\Big|_{64}^{81} = 8(\sqrt{81} - \sqrt{64}) = 8(9 - 8) = \$8.$

53.   Average density $= \dfrac{\text{mass}}{\text{length}} = \dfrac{5/2\text{ kg}}{19\text{ m}} = \dfrac{5}{38}$ kg/m.

Solutions 4.8   *Additional integration theory*   (pages 331-2)

1. $\dfrac{d}{dx} \displaystyle\int_0^{5x} \cos t \, dt = (\cos 5x) \cdot \dfrac{d}{dx}(5x) = (\cos 5x) \cdot (5) = 5 \cdot \cos 5x.$

3. $\dfrac{d}{dx} \displaystyle\int_{1+2x}^{5} \dfrac{t}{\sqrt{1+t^2}} \, dt = -\left(\dfrac{1+2x}{\sqrt{1+(1+2x)^2}} \right) \cdot \dfrac{d}{dx}(1+2x) = -\left(\dfrac{1+2x}{\sqrt{1+(1+2x)^2}} \right) \cdot (2).$

5. $\dfrac{d}{dx} \displaystyle\int_{\sin x}^{\cos x} (t)(5-t)\, dt = (\cos x)(5-\cos x)\left(\dfrac{d}{dx}\cos x \right) - (\sin x)(5-\sin x)\left(\dfrac{d}{dx}\sin x \right)$

$\qquad = (\cos x)(5-\cos x)(-\sin x) - (\sin x)(5-\sin x)(\cos x)$

$\qquad = (\sin x)(\cos x)\left[-(5-\cos x) - (5-\sin x)\right] = (\sin x)(\cos x)\left[\sin x + \cos x - 10\right].$

7. $\displaystyle\int_0^{2\pi} \sin t^3 \, dt = \left(\sin c^3\right) \cdot (2\pi - 0) \le 1 \cdot 2\pi = 2\pi.$ ∎

9. $\displaystyle\int_0^{\pi} f(t) \cdot \sin t \, dt = \sin c \cdot \displaystyle\int_0^{\pi} f(t) \, dt \le 1 \cdot \displaystyle\int_0^{\pi} f(t) \, dt = \displaystyle\int_0^{\pi} f(t) \, dt.$ ∎

11. Let $g(t) = 1$ for all t. Theorem 2 implies there is some c in (a, b) such that

$$\int_a^b f(t) \, dt = \int_a^b f(t) g(t) \, dt = f(c) \int_a^b g(t) \, dt = f(c) \int_a^b dt = f(c)(b-a). \quad ∎$$

REMARK: If f is continuous on $[a, b]$ then $G(x) = \int_a^x f$ is continuous on $[a, b]$ and differentiable on (a, b). The Mean Value Theorem for Derivatives (Theorem 3.7.2) implies there is some c in (a, b) such that $G'(c) = \dfrac{G(b) - G(a)}{b - a}$. Because $G' = f$ and $G(a) = 0$, this implies $f(c) = \left(\dfrac{1}{b-a} \right) \int_a^b f.$ ∎

(You may find it interesting to interpret this last equation in terms of the average value of f on $[a, b]$.)

---

13. a.   If $f \ge 0$ and f is not identically equal to zero, then there is some $x_1 \in (a, b)$ such that $f(x_1) > 0$. If f is also continuous, then there exists some $\delta > 0$ such that $\delta < \min\{x_1 - a, b - x_1\}$ and $\left|x - x_1\right| \le \delta \implies f(x) \ge \dfrac{1}{2} \cdot f(x_1)$. Therefore

$$\int_a^b f = \int_a^{x_1-\delta} f + \int_{x_1-\delta}^{x_1+\delta} f + \int_{x_1+\delta}^b f \ge 0 + \int_{x_1-\delta}^{x_1+\delta} f + 0 \ge \int_{x_1-\delta}^{x_1+\delta} \dfrac{f(x_1)}{2} = \dfrac{f(x_1)}{2} \cdot 2\delta = f(x_1) \cdot \delta > 0. ∎$$

   b.   Suppose $g \ge 0$. If f is not constant, then $(f - m) \cdot g$ is nonnegative and is not identically zero. Thus $0 < \int_a^b (f - m) \cdot g = \int_a^b (f \cdot g) - m \int_a^b g$ which, in turn, implies $m \int_a^b g < \int_a^b (f \cdot g)$. Similarly, $(M - f) \cdot g$ is nonnegative and not identically zero; that implies $\int_a^b (f \cdot g) < M \int_a^b g.$ ∎

15. $\dfrac{d}{dx}\left(\displaystyle\int_a^u f \right) = \dfrac{d}{du}\left(\displaystyle\int_a^u f \right) \cdot \dfrac{d}{dx} u = f(u) \cdot \dfrac{du}{dx}.$ ∎

17. $\dfrac{d}{dx}\left(\displaystyle\int_{a(x)}^b f \right) = \dfrac{d}{dx}\left(-\displaystyle\int_b^{a(x)} f \right) = -\dfrac{d}{dx}\left(\displaystyle\int_b^{a(x)} f \right) = -f(a(x)) \cdot a'(x).$ ∎

Solutions 4.R   *Review*   (page 332)

1. $f'(x) = \dfrac{d}{dx}\left(\displaystyle\int_0^x \dfrac{t^3}{\sqrt{t^2+17}} \, dt \right) = \dfrac{x^3}{\sqrt{x^2+17}}$ by Theorem 4.6.2; thus $f'(8) = \dfrac{8^3}{\sqrt{8^2+17}} = \dfrac{512}{\sqrt{81}} = \dfrac{512}{9}.$

3. $F'(s) = \dfrac{d}{ds}\left(\displaystyle\int_{-1}^{2s} \left(1+u^2\right)^{2001} du \right) = \left(1+(2s)^2\right)^{2001} \cdot \dfrac{d}{ds}(2s) = 2\left(1+4s^2\right)^{2001};$

therefore $F'(0) = 2\left(1 + 4 \cdot 0^2\right)^{2001} = 2.$

5. $y = \dfrac{x}{3} \implies dy = \dfrac{1}{3}\, dx.$

7. $y = x^2 + 1 \implies dy = 2x \, dx.$

9. $\int x^5\, dx = \frac{x^{5+1}}{5+1} + C = \frac{x^6}{6} + C.$

11.
$$\int_{-1}^{1} (x^2 - 3)(x^5 + 2)\, dx = \int_{-1}^{1} (x^7 - 3x^5 + 2x^2 - 6)\, dx = \left(\frac{x^8}{8} - \frac{x^6}{2} + \frac{2x^3}{3} - 6x \right)\Big|_{-1}^{1}$$
$$= \left(\frac{1}{8} - \frac{1}{2} + \frac{2}{3} - 6 \right) - \left(\frac{1}{8} - \frac{1}{2} + \frac{-2}{3} + 6 \right) = \frac{4}{3} - 12 = \frac{-32}{3}.$$

REMARK: x^7 and x^5 are odd functions, so $\int_{-1}^{1} x^7\, dx = 0 = \int_{-1}^{1} x^5\, dx$. 6 and x^2 are even functions, so $\int_{-1}^{1} 6\, dx = 2\int_{0}^{1} 6\, dx$ and $\int_{-1}^{1} x^2\, dx = 2\int_{0}^{1} x^2\, dx$. See Problems 4.6.61-2.

---

13. $\int_{0}^{\pi/3} (\sin 2x + \cos 3x)\, dx = \left(\frac{-\cos 2x}{2} + \frac{\sin 3x}{3} \right)\Big|_{0}^{\pi/3} = \left(\frac{1/2}{2} + \frac{0}{3} \right) - \left(\frac{-1}{2} + \frac{0}{3} \right) = \frac{3}{4}.$

15.   Let $u = t^3 + 7$, then $\int t^2 \left(t^3 + 7 \right)^{-3/4} dt = \int u^{-3/4}\, \frac{du}{3} = \frac{1}{3} \cdot 4 \cdot u^{1/4} + C = \frac{4}{3} \cdot \left(t^3 + 7 \right)^{1/4} + C.$

17.   Let $u = s + 2$, then $\int_{s=-3}^{4} |s + 2|\, ds = \int_{u=-1}^{6} |u|\, du = \frac{u \cdot |u|}{2}\Big|_{-1}^{6} = \frac{36}{2} - \frac{-1}{2} = \frac{37}{2}.$

19.
$$\text{Area} = \int_{-2}^{5} |3x - 7|\, dx = \int_{-2}^{7/3} (7 - 3x)\, dx + \int_{7/3}^{5} (3x - 7)\, dx$$
$$= \left(7x - \frac{3x^2}{2} \right)\Big|_{-2}^{7/3} + \left(\frac{3x^2}{2} - 7x \right)\Big|_{7/3}^{5}$$
$$= \left[\frac{49}{6} - (-20) \right] + \left[\frac{5}{2} - \frac{-49}{6} \right] = \frac{169}{6} + \frac{32}{3} = \frac{233}{6}.$$

21. $y = 2 - x - x^2 = (2 + x)(1 - x)$ is positive on the interval $(-2, 1)$. The area of the region bounded by it and the x-axis is $\int_{-2}^{1} (2 - x - x^2)\, dx = \left(2x - \frac{x^2}{2} - \frac{x^3}{3} \right)\Big|_{-2}^{1} = \frac{7}{6} - \frac{-10}{3} = \frac{9}{2}.$

23. $y = (x + 2)^{-2}$ is positive on $[0, 3]$; $\text{Area} = \int_{0}^{3} (x + 2)^{-2}\, dx = -(x + 2)^{-1}\Big|_{0}^{3} = \frac{-1}{5} - \frac{-1}{2} = \frac{3}{10}.$

25.   a. $\int_{0}^{15} v(t)\, dt = \int_{0}^{15} \left(t + (t + 1)^{-1/2} \right) dt = \left(\frac{t^2}{2} + 2(t + 1)^{1/2} \right)\Big|_{0}^{15} = \frac{241}{2} - 2 = \frac{237}{2}$ m.

      b. $\left(\frac{1}{15 - 0} \right) \int_{0}^{15} v(t)\, dt = \frac{1}{15} \cdot \frac{237}{2} = \frac{237}{30}$ m/sec.

27.   The solution to Problem 1.5.39 shows that the function $F(x) = x|x|$ is differentiable everywhere and $F'(x) = 2|x|$. Therefore $\int |x|\, dx = \frac{1}{2} x\, |x| + C.$

REMARK: If you do not know the result of Problem 1.5.39, you might begin with the two-part formula:
$$\int |x|\, dx = \begin{cases} \int -x\, dx \\ \int x\, dx \end{cases} = \begin{cases} -\frac{1}{2}x^2 + C_- & \text{if } x < 0, \\ \frac{1}{2}x^2 + C_+ & \text{if } 0 \leq x. \end{cases}$$

Next, note that $C_- = C_+$ is necessary in order that the antiderivative of $|x|$ be continuous at 0. The chore remains of showing the resulting function is differentiable at 0; when you've done that, you might present your answer in the form $\int |x|\, dx = \begin{cases} C - \frac{1}{2}x^2 & \text{if } x < 0, \\ C + \frac{1}{2}x^2 & \text{if } 0 \leq x. \end{cases}$

Solutions 4.C   *Computer Exercises*   (page 333)

1. a.   If $[0,1]$ is divided into n subintervals of equal length, then the ith subinterval is $[(i-1)/n, i/n]$.

    b.   Let $x_i^* = \dfrac{i-1}{n}$, left endpoint of the ith subinterval, and let $L(n) = \sum\limits_{i=1}^{n} \sin\sqrt{x_i^*}\,\Delta x_i = \dfrac{1}{n}\sum\limits_{i=1}^{n}\sin\sqrt{\dfrac{i-1}{n}}$.
    Then $L(10) \approx 0.5539283868$, $L(50) \approx 0.5933439023$, and $L(100) \approx 0.5979244105$.

    c.   Let $x_i^{**} = \dfrac{i}{n}$, right endpoint of the ith subinterval, and let $R(n) = \sum\limits_{i=1}^{n}\sin\sqrt{x_i^{**}}\,\Delta x_i = \dfrac{1}{n}\sum\limits_{i=1}^{n}\sin\sqrt{\dfrac{i}{n}}$.
    Then $R(10) \approx 0.6380754853$, $R(50) \approx 0.610173322$, and $R(100) \approx 0.6063391203$.

    d.   Observe that $x_i^{**} = x_{i-1}^*$ for $1 < i < n$, hence most terms in the difference of Riemann sums will cancel and we have $R(n) - L(n) = \left(\sin\sqrt{n/n} - \sin\sqrt{0/n}\right)(1/n) = (\sin 1)/n$.

    e. $\sqrt{x}$ increases on $(0,\infty)$ and $\sin\sqrt{x}$ increases on $\left[0, (\pi/2)^2\right] \supset [0,1]$. Thus $L(n) \leq \int_0^1 \sin\sqrt{x}\,dx \leq R(n)$ for every positive integer n. Since $R(n) - L(n) = (\sin 1)/n$ and $L(n) \leq (L(n)+R(n))/2 \leq R(n)$, we can infer $\left|\displaystyle\int_0^1 \sin\sqrt{x}\,dx - \dfrac{L(n)+R(n)}{2}\right| \leq \dfrac{\sin 1}{n}$. Hence $\left(L(100)+R(100)\right)/2 \approx 0.6021317654$ has error bounded by $(\sin 1)/100 \approx 0.00841$. (The actual error is about 0.000206.)

3. $f(x) = x^5 - 3x^3 - 2x^2 + 6 = \left(x^2 - 3\right)\left(x^3 - 2\right)$ has zeros $\pm\sqrt{3}$ and $\sqrt[3]{2}$. If $F'(x) = f(x)$, then $F(x) = x^6/6 - 3x^4/4 - 2x^3/3 + 6x + C$; the area of the region bounded by the x-axis and the graph of $y = f(x)$ is

$$\int_{-\sqrt{3}}^{\sqrt{3}} |f(x)|\,dx = \int_{-\sqrt{3}}^{\sqrt[3]{2}} f(x)\,dx + \int_{\sqrt[3]{2}}^{\sqrt{3}} -f(x)\,dx = F(x)\big|_{-\sqrt{3}}^{\sqrt[3]{2}} - F(x)\big|_{\sqrt[3]{2}}^{\sqrt{3}} = 2F\left(\sqrt[3]{2}\right) - F\left(-\sqrt{3}\right) - F\left(\sqrt{3}\right)$$

$$= 2\left(\frac{9}{2}\sqrt[3]{2} - \frac{2}{3}\right) - \left(-4\sqrt{3} - \frac{9}{4}\right) - \left(4\sqrt{3} - \frac{9}{4}\right) = 9\sqrt[3]{2} + \frac{19}{6} \approx 14.51.$$

$y = \left(x^2 - 3\right)\left(x^3 - 2\right)$

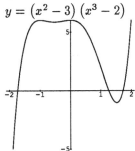

$y = |f(x)|, \quad -\sqrt{3} \leq x \leq \sqrt{3}$

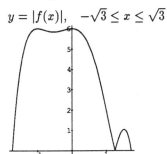

5.
$$\int \cos^2 x\,dx = \frac{x}{2} + \frac{\cos x \sin x}{2} + C,$$

$$\int \cos^4 x\,dx = \frac{3x}{8} + \frac{3\cos x \sin x}{8} + \frac{\cos^3 x \sin x}{4} + C,$$

$$\int \cos^6 x\,dx = \frac{5x}{16} + \frac{5\cos x \sin x}{16} + \frac{5\cos^3 x \sin x}{24} + \frac{\cos^5 x \sin x}{6} + C,$$

$$\int \cos^8 x\,dx = \frac{35x}{128} + \frac{35\cos x \sin x}{128} + \frac{35\cos^3 x \sin x}{192} + \frac{7\cos^5 x \sin x}{48} + \frac{\cos^7 x \sin x}{8} + C,$$

$$\int \cos^{10} x\,dx = \frac{63x}{256} + \frac{63\cos x \sin x}{256} + \frac{21\cos^3 x \sin x}{128} + \frac{21\cos^5 x \sin x}{160} + \frac{9\cos^7 x \sin x}{80} + \frac{\cos^9 x \sin x}{10} + C.$$

If k is a positive integer, then the general pattern is that $\displaystyle\int \cos^{2k} x\,dx = ax + (\cos x \sin x) \cdot P\left(\cos^2 x\right) + C$ where $P(t)$ is a polynomial of degree $k-1$ which has $1/(2k)$ as the coefficient of t^{k-1}.

5
Applications of the definite Integral

Solutions 5.1   *The area between two curves*   (page 339)

1.   The graphs of $y = x$ and $y = x^2$ intersect at $(0,0)$ and $(1,1)$. If $0 \le x \le 1$, then $x^2 \le x$. The area of

the region bounded by the two graphs is $\displaystyle\int_0^1 (x - x^2)\,dx = \left(\dfrac{x^2}{2} - \dfrac{x^3}{3}\right)\Big|_0^1 = \left(\dfrac{1}{2} - \dfrac{1}{3}\right) - 0 = \dfrac{1}{6}$.

Because $0 \le y \le 1$ implies $y \le \sqrt{y}$, we may also compute the region's area as

$$\int_0^1 (\sqrt{y} - y)\,dy = \left(\frac{2}{3}y^{3/2} - \frac{1}{2}y^2\right)\Big|_0^1 = \left(\frac{2}{3} - \frac{1}{2}\right) - 0 = \frac{1}{6}.$$

3.   The graphs of $f(x) = 3x^2 + 6x + 8$ and $g(x) = 2x^2 + 9x + 18$ meet at $(-2,8)$ and $(5,113)$. If $-2 \le x \le 5$,
then $g(x) - f(x) = 10 + 3x - x^2 = (2+x)(5-x) \ge 0$. The area of the region bounded by the two graphs is

$$\int_{-2}^5 |g(x) - f(x)|\,dx = \int_{-2}^5 (10 + 3x - x^2)\,dx = \left(10x + \frac{3x^2}{2} - \frac{x^3}{3}\right)\Big|_{-2}^5 = \frac{275}{6} - \left(\frac{-34}{3}\right) = \frac{343}{6}.$$

5.   The graphs of $f(x) = 3 - 7x + x^2$ and $g(x) = 5 - 4x - x^2$ meet at $(-1/2, 27/4)$ and $(2,-7)$. If
$-1/2 \le x \le 2$, then $f(x) - g(x) = 2x^2 - 3x - 2 = (2x+1)(x-2) \le 0$. The area of the region bounded by
the two graphs is

$$\int_{-1/2}^2 |g(x) - f(x)|\,dx = \int_{-1/2}^2 (-2x^2 + 3x + 2)\,dx = \left(\frac{-2x^3}{3} + \frac{3x^2}{2} + 2x\right)\Big|_{-1/2}^2 = \frac{14}{3} - \left(\frac{-13}{24}\right) = \frac{125}{24}.$$

7.   The graphs of $f(x) = x^2$ and $g(x) = x^3$ over the interval $[0,3]$ cross when $x = 1$. The region's area is

$$\int_0^3 |f(x) - g(x)|\,dx = \int_0^1 (x^2 - x^3)\,dx + \int_1^3 (x^3 - x^2)\,dx = \left(\frac{x^3}{3} - \frac{x^4}{4}\right)\Big|_0^1 + \left(\frac{x^4}{4} - \frac{x^3}{3}\right)\Big|_1^3$$
$$= \left[\left(\frac{1}{3} - \frac{1}{4}\right) - 0\right] + \left[\left(\frac{81}{4} - 9\right) - \left(\frac{1}{4} - \frac{1}{3}\right)\right] = \frac{1}{12} + \frac{34}{3} = \frac{137}{12}.$$

We may also integrate with respect to y but then there are three intervals to be considered:

$$\text{Area} = \int_0^1 \left(y^{1/3} - y^{1/2}\right)dy + \int_1^9 \left(y^{1/2} - y^{1/3}\right)dy + \int_9^{27} \left(3 - y^{1/3}\right)dy$$
$$= \frac{1}{12} + \left[\frac{217}{12} - \left(\frac{27}{4}\right)9^{1/3}\right] + \left[\left(\frac{27}{4}\right)9^{1/3} - \frac{27}{4}\right] = \frac{218 - 81}{12} = \frac{137}{12}.$$

9.   The graphs of $f(x) = x^3$ and $g(x) = x^3 + x^2 + 6x + 5$ meet at $(-5,-125)$ and $(-1,-1)$. If $-5 \le x \le -1$,
then $g(x) - f(x) = x^2 + 6x + 5 = (x+5)(x+1) \le 0$. The area of the region bounded by the two graphs is

$$\int_{-5}^{-1} |f(x) - g(x)|\,dx = \int_{-5}^{-1} (-x^2 - 6x - 5)\,dx = \left(-\frac{x^3}{3} - 3x^2 - 5x\right)\Big|_{-5}^{-1} = \frac{7}{3} - \left(\frac{-25}{3}\right) = \frac{32}{3}.$$

11.   The finite region bounded by $xy^2 = 1$, $x = 5$, $y = 5$ is $\{(x,y) : 1/25 \le x \le 5, x^{-1/2} \le y \le 5\}$; its area is

$$\int_{1/25}^5 \left(5 - x^{-1/2}\right)dx = \left(5x - 2x^{1/2}\right)\Big|_{1/25}^5 = \left(25 - 2\sqrt{5}\right) - \left(\frac{-1}{5}\right) = \frac{126}{5} - 2\sqrt{5}.$$

The region can also be characterized as $\{(x,y) : 1/\sqrt{5} \le y \le 5, y^{-2} \le x \le 5\}$ and its area computed as

$$\int_{1/\sqrt{5}}^5 \left(5 - y^{-2}\right)dy = \left(5y + y^{-1}\right)\Big|_{1/\sqrt{5}}^5 = \left(25 + \frac{1}{5}\right) - \left(\frac{5}{\sqrt{5}} + \sqrt{5}\right) = \frac{126}{5} - 2\sqrt{5}.$$

13. The graphs of $y = x^{1/3}$ and $y = x$ meet at $(-1,-1)$, $(0,0)$, and $(1,1)$. Observe that $x^{1/3} \leq x$ on the interval $[-1,0]$ and $x \leq x^{1/3}$ on $[0,1]$. Therefore

$$\text{Area} = \int_{-1}^{1} \left| x - x^{1/3} \right| dx = \int_{-1}^{0} \left(x - x^{1/3} \right) dx + \int_{0}^{1} \left(x^{1/3} - x \right) dx = \left(\frac{x^2}{2} - \frac{3x^{4/3}}{4} \right) \Big|_{-1}^{0} + \left(\frac{3x^{4/3}}{4} - \frac{x^2}{2} \right) \Big|_{0}^{1}$$

$$= \left[0 - \left(\frac{1}{2} - \frac{3}{4} \right) \right] + \left[\left(\frac{3}{4} - \frac{1}{2} \right) - 0 \right] = \frac{1}{4} + \frac{1}{4} = \frac{1}{2}.$$

Alternatively,

$$\text{Area} = \int_{-1}^{0} (y^3 - y) dy + \int_{0}^{1} (y - y^3) dy = 2 \int_{0}^{1} (y - y^3) dy = 2 \cdot \left(\frac{1}{2} - \frac{1}{4} \right) = \frac{1}{2}.$$

15.
$$\text{Area} = \int_{0}^{1} \left[\sin \left(\frac{\pi x}{2} \right) - x \right] dx + \int_{1}^{2} \left[x - \sin \left(\frac{\pi x}{2} \right) \right] dx$$

$$= \left[\left(\frac{-2}{\pi} \right) \cos \left(\frac{\pi x}{2} \right) - \frac{x^2}{2} \right] \Big|_{0}^{1} + \left[\frac{x^2}{2} - \left(\frac{-2}{\pi} \right) \cos \left(\frac{\pi x}{2} \right) \right] \Big|_{1}^{2} = \left[\frac{-1}{2} - \frac{-2}{\pi} \right] + \left[\left(2 - \frac{2}{\pi} \right) - \frac{1}{2} \right] = 1.$$

17. The graphs of $x = f(y) = 8 - y^2$ and $x = g(y) = 2 - y$ meet at $(4,-2)$ and $(-1,3)$. If $-2 \leq y \leq 3$, then $f(y) - g(y) = 6 + y - y^2 = (2+y)(3-y) \geq 0$. The area of the region bounded by the two graphs is

$$\text{Area} = \int_{-2}^{3} \left(6 + y - y^2 \right) dy = \left(6y + \frac{y^2}{2} - \frac{y^3}{3} \right) \Big|_{-2}^{3} = \frac{27}{2} - \left(\frac{-22}{3} \right) = \frac{125}{6}.$$

Alternatively,

$$\text{Area} = \int_{-1}^{4} \left(\sqrt{8-x} - (2-x) \right) dx + \int_{4}^{8} \left(\sqrt{8-x} - (-\sqrt{8-x}) \right) dx$$

$$= \left(\left(\frac{-2}{3} \right) (8-x)^{3/2} - 2x + \frac{x^2}{2} \right) \Big|_{-1}^{4} + \left(\frac{-4}{3} \right) (8-x)^{3/2} \Big|_{4}^{8} = \left[\frac{-16}{3} - \frac{-31}{2} \right] + \left[0 - \frac{-32}{3} \right] = \frac{125}{6}.$$

19. The graphs of $x = y^2$ and $x^2 = 6 - 5y$ meet at $(1,1)$ and $(4,-2)$; the region they bound is $\{(x,y) : -2 \leq y \leq 1, y^2 \leq x \leq \sqrt{6-5y}\}$ with

$$\text{Area} = \int_{-2}^{1} \left(\sqrt{6-5y} - y^2 \right) dy = \left(\frac{-2}{15} (6-5y)^{3/2} - \frac{1}{3} y^3 \right) \Big|_{-2}^{1} = \left(\frac{-2}{15} - \frac{1}{3} \right) - \left(\frac{-128}{15} + \frac{8}{3} \right) = \frac{27}{5}.$$

Alternatively,

$$\text{Area} = \int_{0}^{1} \left(x^{1/2} - (-x^{1/2}) \right) dx + \int_{1}^{4} \left(\frac{1}{5} (6 - x^2) - (-x^{1/2}) \right) dx = \frac{4}{3} + \frac{61}{15} = \frac{27}{5} = 5.4.$$

21. If $\sqrt{x} + \sqrt{y} = 4$, then $y = (4 - \sqrt{x})^2 = 16 - 8\sqrt{x} + x$; this curve meets $x = 0$ at $(0,16)$ and $y = 0$ at $(16,0)$. $\text{Area} = \int_{0}^{16} \left(16 - 8\sqrt{x} + x \right) dx = \left(16x - \frac{16}{3} x^{3/2} + \frac{1}{2} x^2 \right) \Big|_{0}^{16} = \left(256 - \frac{1024}{3} + 128 \right) = \frac{128}{3}.$ Because the region is symmetric with respect to the line $y = x$, it is just as easy to integrate with respect to y: $\text{Area} = \int_{0}^{16} (4 - \sqrt{y})^2 dy = \int_{0}^{16} (16 - 8\sqrt{y} + y) dy = \ldots .$

23. The graphs of $y = f(x) = ax + b$ and $y = g(x) = ax^2$ meet when $x = x_{\pm} = \frac{1}{2} \pm \frac{\sqrt{4ab + a^2}}{2a}$. Evaluating the following integrals can be simplified by observing that since $x_{\pm}$ solve the equation $ax^2 = g(x) = f(x) = ax + b$, we know $ax_{\pm}^2 = ax_{\pm} + b$ and $ax_{\pm}^3 = (ax_{\pm}^2) x_{\pm} = (ax_{\pm} + b) x_{\pm} = ax_{\pm}^2 + bx_{\pm} = (a+b)x_{\pm} + b$. The region in the first quadrant between these curves has

$$\text{Area}_1 = \int_{0}^{x_+} \left(b + ax - ax^2 \right) dx$$

$$= \left(bx + \frac{ax^2}{3} - \frac{ax^3}{3} \right) \Big|_{0}^{x_+} = bx_+ + \frac{ax_+^2}{3} - \frac{ax_+^3}{3} = bx_+ + \frac{ax_+ + b}{2} - \frac{(a+b)x_+ + b}{3}$$

$$= \left(\frac{a}{6} + \frac{2b}{3} \right) x_+ + \frac{b}{6}$$

$$= \left(\frac{a}{6} + \frac{2b}{3} \right) \left(\frac{1}{2} + \frac{\sqrt{4ab + a^2}}{2a} \right) + \frac{b}{6} = \left(\frac{a}{12} + \frac{b}{2} \right) + \left(\frac{1}{12} + \frac{b}{3a} \right) \sqrt{4ab + a^2}.$$

The region in the second quadrant between these curves has

$$\text{Area}_2 = \int_{x_-}^{0} (b + ax - ax^2)\, dx = -\left[\left(\frac{a}{6} + \frac{2b}{3}\right)x_- + \frac{b}{6}\right]$$

$$= -\left(\frac{a}{6} + \frac{2b}{3}\right)\left(\frac{1}{2} - \sqrt{\frac{1}{4} + \frac{b}{a}}\right) - \frac{b}{6} = -\left(\frac{a}{12} + \frac{b}{2}\right) + \left(\frac{a}{6} + \frac{2b}{3}\right)\sqrt{\frac{1}{4} + \frac{b}{a}}.$$

25.   The "bottom" of the triangle is the line through $(-3, 7)$ and $(2, 4)$, which has equation

$$y = \left(\frac{4-7}{2-(-3)}\right)(x - (-3)) + 7 = \frac{-3x + 26}{5}.$$

The "top left" side of the triangle is the line through $(1, 6)$ and $(-3, 7)$, which has equation

$$y = \left(\frac{7-6}{-3-1}\right)(x - 1) + 6 = \frac{25 - x}{4}.$$

The "right" side of the triangle is the line through $(2, 4)$ and $(1, 6)$ which has equation

$$y = \left(\frac{6-4}{1-2}\right)(x - 2) + 4 = -2x + 8.$$

The area of the triangle is

$$\int_{-3}^{1}\left(\frac{-3x + 26}{5} - \frac{25 - x}{4}\right)dx + \int_{1}^{2}\left(-2x + 8 - \frac{25 - x}{4}\right)dx = \int_{-3}^{1}\left(\frac{7x + 21}{20}\right)dx + \int_{1}^{2}\left(\frac{-7x + 14}{4}\right)dx$$

$$= \left(\frac{49}{40} - \frac{-63}{40}\right) + \left(\frac{14}{5} - \frac{21}{10}\right) = \frac{7}{2}.$$

Alternatively, the area can be computed as

$$\int_{4}^{6}\left(\frac{8 - y}{2} - \frac{26 - 5y}{3}\right)dy + \int_{6}^{7}\left((25 - 4y) - \frac{26 - 5y}{3}\right)dy = \int_{4}^{6}\left(\frac{7y}{6} - \frac{14}{3}\right)dy + \int_{6}^{7}\left(\frac{49}{3} - \frac{7y}{3}\right)dy$$

$$= \frac{7}{3} + \frac{7}{6} = \frac{7}{2}.$$

REMARK: This problem provides "drill" in integration; the solutions shown above are not necessarily the simplest ways to compute a triangle's area. The formula given in Problem 0.3.67 yields

$$\frac{|(1)(4) + (2)(7) + (-3)(6) - (1)(7) - (2)(6) - (-3)(4)|}{2} = \frac{|4 + 14 - 18 - 7 - 12 + 12|}{2} = \frac{|-7|}{2} = \frac{7}{2}.$$

---

27.   The lines with equations $x - y + 6 = 0$ and $2x - y - 2 = 0$ meet at $(8, 14)$ because

$$\begin{cases} x - y + 6 = 0 \\ 2x - y - 2 = 0 \end{cases} \implies \begin{cases} y = x + 6 \\ y = 2x - 2 \end{cases} \implies \begin{cases} y = x + 6 \\ x + 6 = 2x - 2 \end{cases} \implies \begin{cases} y = x + 6 \\ 8 = x \end{cases} \implies \begin{cases} y = 8 + 6 = 14 \\ x = 8 \end{cases}$$

The lines with equations $x - y + 6 = 0$ and $3x - y + 4 = 0$ meet at $(1, 7)$ because

$$\begin{cases} x - y + 6 = 0 \\ 3x - y + 4 = 0 \end{cases} \implies \begin{cases} y = x + 6 \\ y = 3x + 4 \end{cases} \implies \begin{cases} 3x + 4 = x + 6 \\ y = 3x + 4 \end{cases} \implies \begin{cases} 2x = 2 \\ y = 3x + 4 \end{cases} \implies \begin{cases} x = 1 \\ y = 3(1) + 4 = 7 \end{cases}$$

The lines with equations $2x - y - 2 = 0$ and $3x - y + 4 = 0$ meet at $(-6, -14)$ because

$$\begin{cases} y = 2x - 2 \\ y = 3x + 4 \end{cases} \implies \begin{cases} 0 = y - y = (2x - 2) - (3x + 4) = -x - 6 \\ y = 3x + 4 \end{cases} \implies \begin{cases} x = -6 \\ y = 3(-6) + 4 = -14 \end{cases}$$

The area of the triangle is

$$\int_{-6}^{1}\left([3x + 4] - [2x - 2]\right)dx + \int_{1}^{8}\left([x + 6] - [2x - 2]\right)dx = \int_{-6}^{1}(x + 6)\,dx + \int_{1}^{8}(-x + 8)\,dx$$

$$= \left(\frac{x^2}{2} + 6x\right)\Big|_{-6}^{1} + \left(\frac{-x^2}{2} + 8x\right)\Big|_{1}^{8}$$

$$= \left(\frac{13}{2} - (-18)\right) + \left(32 - \frac{15}{2}\right) = \frac{49}{2} + \frac{49}{2} = 49.$$

REMARK: According to the formula in the Remark following Solution 0.3.67, the triangle with vertices at $(-6, -14)$, $(1, 7)$, $(8, 14)$ has area

$$\frac{\left|(-6)(7) - (1)(-14) + (1)(14) - (8)(7) + (8)(-14) - (-6)(14)\right|}{2} = \frac{\left|-42 + 14 + 14 - 56 - 112 + 84\right|}{2}$$

$$= \frac{\left|-98\right|}{2} = 49.$$

29.   The region bounded by the x-axis and $y = a^2 - x^2$ has area $\displaystyle\int_{-a}^{a} \left(a^2 - x^2\right) dx = \frac{4a^3}{3} = \int_0^{a^2} 2\left(a^2 - y\right)^{1/2} dy.$

The upper portion of this region lying above the horizontal line $y = b$ has area $\displaystyle\int_b^{a^2} 2\left(a^2 - y\right)^{1/2} dy =$ $\dfrac{4\left(a^2 - b\right)^{3/2}}{3}$. This portion has area equal to half of the whole provided

$$\frac{4\left(a^2 - b\right)^{3/2}}{3} = \frac{1}{2} \cdot \frac{4a^3}{3} \iff a^2 - b = \left(\frac{a^3}{2}\right)^{2/3} = 2^{-2/3} a^2 \iff b = \left(1 - 2^{-2/3}\right) a^2.$$

31.   The region $\left\{(x, y) : -|a| \le x \le |a|, x^2 - a^2 \le y \le a^2 - x^2\right\}$ is symmetric with respect to the origin, i.e., if (X, Y) is inside, then so is $(-X, -Y)$. Consider a pair of points (X, Y) and $(-X, -Y)$ that lie inside the region. If one falls on a fixed line through the origin, then so does the other point; otherwise, one point is above the line and the other is below. Therefore any line through the origin splits the region into two parts of the same shape and size (and, thus, with the same area). ∎

REMARK: If you really feel you must compute some integrals for this problem, then consider the following. If $m \ge 0$, the line $y = mx$ meets the upper curve when $x = x_+ = \dfrac{-m + \sqrt{m^2 + 4a^2}}{2}$ and it meets the lower curve when $x = x_- = \dfrac{m - \sqrt{m^2 + 4a^2}}{2} = -x_+$. Therefore, using the substitution $u = -x$,

$$\text{Area}_+ = \int_{x = x_+}^{x_-} \left(\left[a^2 - x^2\right] - mx\right) dx + \int_{x = x_-}^{|a|} \left(\left[a^2 - x^2\right] - \left[x^2 - a^2\right]\right) dx$$

$$= \int_{u = |a|}^{-x_-} \left(\left[a^2 - (-u)^2\right] - \left[(-u)^2 - a^2\right]\right) (-du) + \int_{u = -x_-}^{-x_+} \left(\left[a^2 - (-u)^2\right] - m(-u)\right) (-du)$$

$$= \int_{u = -x_-}^{|a|} \left(\left[a^2 - u^2\right] - \left[u^2 - a^2\right]\right) du + \int_{u = -x_+}^{-x_-} \left(\left[a^2 - u^2\right] + mu\right) du$$

$$= \int_{u = x_+}^{|a|} \left(\left[a^2 - u^2\right] - \left[u^2 - a^2\right]\right) du + \int_{u = x_-}^{x_+} \left(mu - \left[u^2 - a^2\right]\right) du \qquad -x_- = x_+$$

$$= \int_{u = x_-}^{x_+} \left(mu - \left[u^2 - a^2\right]\right) du + \int_{u = x_+}^{|a|} \left(\left[a^2 - u^2\right] - \left[u^2 - a^2\right]\right) du = \text{Area}_-.$$

If $m < 0$, then $x_+ = \dfrac{-m - \sqrt{m^2 + 4a^2}}{2}$, $x_- = \dfrac{m + \sqrt{m^2 + 4a^2}}{2} = -x_+$, and

$$\text{Area}_+ = \int_{x = x_+}^{x_-} \left(\left[a^2 - x^2\right] - mx\right) dx + \int_{x = x_-}^{|a|} \left(\left[a^2 - x^2\right] - \left[x^2 - a^2\right]\right) dx = \cdots = \text{Area}_-. ∎$$

Solutions 5.2 *Volumes* (pages 348-9)

1. See Example 1. $\boxed{\text{Disk method}}$ $\int_0^2 \pi \left[y(x)\right]^2 \, dx = \int_0^2 \pi x^2 \, dx = \pi \left.\frac{x^3}{3}\right|_0^2 = \frac{8\pi}{3}.$

$\boxed{\text{Cylindrical shell method}}$

$$\int_0^2 2\pi y \left[2 - x(y)\right] \, dy = \int_0^2 2\pi y (2-y) \, dy = \int_0^2 2\pi \left(2y - y^2\right) \, dy = 2\pi \left.\left(y^2 - \frac{y^3}{3}\right)\right|_0^2 = 2\pi \left(\frac{4}{3}\right) = \frac{8\pi}{3}.$$

3. $\boxed{\text{Disk method}}$ $\int_0^1 \pi y^2 \, dx = \int_0^1 \pi \left(2x^2\right)^2 \, dx = 4\pi \int_0^1 x^4 \, dx = \frac{4\pi}{5}.$

$\boxed{\text{Cylindrical shell method}}$

$$\int_0^2 2\pi y \left(1 - x(y)\right) \, dy = \int_0^2 2\pi y \left(1 - \sqrt{\frac{y}{2}}\right) \, dy = \pi \int_0^2 \left(2y - \sqrt{2} y^{3/2}\right) \, dy$$

$$= \pi \left.\left(y^2 - \frac{2\sqrt{2}}{5} y^{5/2}\right)\right|_0^2 = \pi \left(4 - \frac{16}{5}\right) = \frac{4\pi}{5}.$$

5. $\boxed{\text{Disk method}}$ $\int_1^5 \pi y^2 \, dx = \int_1^5 \pi \left(\sqrt{x+1}\right)^2 \, dx = \pi \int_1^5 (x+1) \, dx = \pi \left.\frac{(x+1)^2}{2}\right|_1^5 = \pi (18 - 2) = 16\pi.$

7. $\boxed{\text{Disk method}}$ $\int_2^5 \pi \left(y_2^2 - y_1^2\right) \, dx = \int_2^5 \pi \left((2x)^2 - x^2\right) \, dx = \int_2^5 \pi 3x^2 \, dx = \pi x^3 \big|_2^5 = \pi(125 - 8) = 117\,\pi.$

9. $x = 2 - |y - 2| \iff x = \begin{cases} 2 - (y-2) = 4 - y & \text{if } 2 \le y \\ 2 - (2-y) = y & \text{if } y < 2 \end{cases} \iff y = \begin{cases} y_+ = 4 - x & \text{if } 2 \le y \\ y_- = x & \text{if } y < 2 \end{cases}.$

$\boxed{\text{Disk method}}$ $\int_0^2 \pi \left(y_+^2 - y_-^2\right) \, dx = \int_0^2 \pi \left((4-x)^2 - x^2\right) \, dx = \int_0^2 \pi (16 - 8x) \, dx = \pi \left(16x - 4x^2\right)\big|_0^2 = 16\,\pi.$

11. $\boxed{\text{Disk method}}$ $\int_0^3 \pi \left(x(y)\right)^2 \, dy = \int_0^3 \pi (3-y)^2 \, dy = \frac{-\pi}{3} (3-y)^3 \big|_0^3 = 9\pi.$

$\boxed{\text{Cylindrical shell method}}$ $\int_0^3 2\pi x(3-x) \, dx = \int_0^3 \pi \left(6x - 2x^2\right) \, dx = \pi \left.\left(3x^2 - \frac{2x^3}{3}\right)\right|_0^3 = 9\pi.$

13. $\boxed{\text{Cylindrical shell method}}$ $\int_0^1 2\pi x \cdot \frac{x^2}{2} \, dx = \int_0^1 \pi x^3 \, dx = \left.\frac{\pi x^4}{4}\right|_0^1 = \frac{\pi}{4}.$

$\boxed{\text{Disk method}}$ $\int_0^{1/2} \pi \left(1^2 - \left(\sqrt{2y}\right)^2\right) \, dy = \int_0^{1/2} \pi (1 - 2y) \, dy = \pi \left(y - y^2\right)\big|_0^{1/2} = \frac{\pi}{4}.$

15. $\boxed{\text{Cylindrical shell method}}$ $\int_0^1 2\pi x \left(\sqrt{x} - x^2\right) \, dx = \int_0^1 \pi \left(2x^{3/2} - 2x^3\right) \, dx = \pi \left.\left(\frac{4x^{5/2}}{5} - \frac{x^4}{2}\right)\right|_0^1 = \frac{3\pi}{10}.$

$\boxed{\text{Disk method (adapt formula (7))}}$ $\int_0^1 \pi \left[\left(\sqrt{y}\right)^2 - \left(y^2\right)^2\right] \, dy = \pi \int_0^1 \left[y - y^4\right] \, dy = \pi \left.\left[\frac{y^2}{2} - \frac{y^5}{5}\right]\right|_0^1 = \frac{3\pi}{10}.$

17. $\boxed{\text{Disk method (adapt formula (7))}}$

$$\int_{-1}^1 \pi \left(2^2 - \left(1 + y^2\right)^2\right) \, dy = \pi \int_{-1}^1 \left(3 - 2y^2 - y^4\right) \, dy = \pi \left.\left(3y - \frac{2y^3}{3} - \frac{y^5}{5}\right)\right|_{-1}^1 = \frac{64\pi}{15}.$$

$\boxed{\text{Cylindrical shell method}}$ $\int_{x=1}^2 2\pi x \cdot 2\sqrt{x-1} \, dx \overset{x-1 \equiv u}{=} \int_{u=0}^1 2\pi (u+1) \cdot 2u^{1/2} \, du = 4\pi \left.\left(\frac{2u^{5/2}}{5} + \frac{2u^{3/2}}{3}\right)\right|_0^1.$

19. $\boxed{\text{Disk method}}$

$$\int_{-r}^r \pi \left(\sqrt{r^2 - y^2}\right)^2 \, dy = \int_{-r}^r \pi \left(r^2 - y^2\right) \, dy = \pi \left.\left(r^2 y - \frac{y^3}{3}\right)\right|_{-r}^r = \pi \left[\left(r^3 - \frac{r^3}{3}\right) - \left(-r^3 - \frac{-r^3}{3}\right)\right] = \frac{4\pi r^3}{3}.$$

$\boxed{\text{Cylindrical shell method}}$

$$\int_0^r 2\pi x \left(2\sqrt{r^2 - x^2}\right) \, dx = \int_0^r 4\pi x \sqrt{r^2 - x^2} \, dx = \left(\frac{-4\pi}{3}\right) \left(r^2 - x^2\right)^{3/2} \big|_0^r = \frac{4\pi r^3}{3}.$$

21. The circle of radius r centered at $(a, 0)$ has equation $(x - a)^2 + y^2 = r^2$. If $r < a$, the torus generated by rotating this circle about the y-axis has volume

$$\int_{x=a-r}^{a+r} 2\pi x \cdot 2\sqrt{r^2 - (x-a)^2}\, dx = 4\pi \int_{u=-r}^{r} (u+a)\sqrt{r^2 - u^2}\, du \qquad u = x - a$$

$$= 4\pi \int_{-r}^{r} u\left(r^2 - u^2\right)^{1/2} du + 4\pi a \int_{-r}^{r} \sqrt{r^2 - u^2}\, du$$

$$= \frac{-4\pi}{3}\left(r^2 - u^2\right)^{3/2}\Big|_{-r}^{r} + 4\pi a \int_{-r}^{r} \sqrt{r^2 - u^2}\, du$$

$$= 0 + 4\pi a \int_{-r}^{r} \sqrt{r^2 - u^2}\, du$$

$$= 4\pi a \cdot (\text{area of semicircle of radius } r)$$

$$= 4\pi a \cdot \left(\frac{\pi r^2}{2}\right) = 2\pi^2 a\, r^2.$$

23. Rotating the region $\{(x, y) : 0 \leq x \leq 1, 0 \leq y \leq \sqrt{x}\} = \{(x, y) : 0 \leq y \leq 1, y^2 \leq x \leq 1\}$ about the line $x = 4$ generates a solid with volume

$$\int_0^1 2\pi(4 - x) \cdot \sqrt{x}\, dx = \pi\left(\frac{16x^{3/2}}{3} - \frac{4x^{5/2}}{5}\right)\Big|_0^1 = \frac{68\pi}{15} = \int_0^1 \pi\left((4 - y^2)^2 - (4 - 1)^2\right) dy.$$

Rotating the region about the line $y = 2$ yields a solid with volume

$$\int_0^1 2\pi(2 - y) \cdot \left(1 - y^2\right) dy = \frac{13\pi}{6} = \int_0^1 \pi\left(2^2 - \left(2 - \sqrt{x}\right)^2\right) dx.$$

25. In a sphere of radius 10 m, let x be the distance of a horizontal section from the sphere's center; then the cross-section has area $\pi r^2 = \pi\left(\sqrt{10^2 - x^2}\right)^2 = \pi\left(100 - x^2\right)$. A fluid that partially fills the sphere to a depth of 2 m will occupy a volume of

$$\int_8^{10} \pi\left(100 - x^2\right) dx = \pi\left(100x - \frac{x^3}{3}\right)\Big|_8^{10} = \frac{112\,\pi}{3}\ \text{m}^3.$$

If the fluid is water with density $\approx 1000\ \text{kg/m}^3$, the mass of the 2 m's of water in the tank is $\dfrac{112000\,\pi}{3}$ kg.

27. (Compare with Example 5.) Suppose the base of the solid is the disk $x^2 + y^2 \leq 4$ in the xy-plane and suppose cross sections perpendicular to the x-axis are square. Then, for any x in $[-2, 2]$, the square cross-section has corners $\left(x, \pm\sqrt{4 - x^2}, 2\sqrt{4 - x^2}\right)$ and area $\left(2\sqrt{4 - x^2}\right)^2 = 16 - 4x^2$. The volume of the solid is

$$\int_{-2}^{2} (16 - 4x^2)\, dx = \left(16x - \frac{4x^3}{3}\right)\Big|_{-2}^{2} = \frac{64}{3} - \frac{-64}{3} = \frac{128}{3}.$$

REMARK: This solid is the region inside the circular cylinder $x^2 + y^2 = 4$ centered on the z=axis, inside the elliptical cylinder $x^2 + (z/2)^2 = 4$ centered on the y-axis, and above the xy-plane.

29. Suppose (a_1, b_1) and (a_2, b_2) are two points in the first quadrant that are on the line $y = mx + p$. If $a_1 < a_2$, the solid generated by rotating the trapezoid with vertices (a_1, b_1), (a_2, b_2), $(a_2, 0)$, $(a_1, 0)$ about the x-axis has volume

$$V_{12} = \int_{a_1}^{a_2} \pi(mx + p)^2\, dx = \frac{\pi(mx + p)^3}{3m}\Big|_{a_1}^{a_2} = \frac{\pi}{3m}\left(b_2^3 - b_1^3\right) = \frac{\pi}{3}\left(a_2 - a_1\right)\left(b_1^2 + b_1 b_2 + b_2^2\right).$$

(Remember that $b_k = ma_k + p$ for $k = 1, 2$ and $\dfrac{1}{m} = \dfrac{a_2 - a_1}{b_2 - b_1}$.) Observe that the final formula for V_{12} is correct also in case $a_1 = a_2$. If the triangle is oriented so that $a_1 \leq a_2 \leq a_3$, $b_1 \leq b_2$, and $b_3 \leq b_2$, then rotating it about the x-axis generates a solid with volume equal to $V_{12} + V_{23} - V_{13}$ where $V_{23} = \dfrac{\pi}{3}\left(a_3 - a_2\right)\left(b_2^2 + b_2 b_3 + b_3^2\right)$ and $V_{13} = \dfrac{\pi}{3}\left(a_3 - a_1\right)\left(b_1^2 + b_1 b_3 + b_3^2\right)$ are easy adaptations of V_{12}. A moderate bit of algebra show that the total volume $V_{12} + V_{23} - V_{13}$ equals

$$\frac{\pi}{3}\left(b_1 + b_2 + b_3\right)\left((a_2 b_1 - a_1 b_2) + (a_3 b_2 - a_2 b_3) + (a_1 b_3 - a_3 b_1)\right).$$

Any other arrangement of the three points results in this expression or its negative computing the total volume. Therefore, the general answer just replaces the brackets with absolute value bars.

1. $f(x) = 2 + x^2/2 \implies f'(x) = x \implies 1 + \left(f'(x)\right)^2 = 1 + x^2 \implies s = \int_0^1 \sqrt{1 + x^2}\,dx.$

3. $f(x) = \cos x \implies f'(x) = -\sin x \implies 1 + \left(f'(x)\right)^2 = 1 + \sin^2 x \implies s = \int_0^\pi \sqrt{1 + \sin^2 x}\,dx.$

5. $f(x) = 3x + 5 \implies f'(x) = 3 \implies s = \int_1^5 \sqrt{1 + 3^2}\,dx = \sqrt{1+9} \cdot (5 - 1) = \sqrt{10} \cdot 4.$

REMARK: The arc is the straight line segment between $(1, 8)$ and $(5, 20)$ whose length can also be computed as $\sqrt{(5-1)^2 + (20-8)^2} = \sqrt{16 + 144} = \sqrt{160} = 4\sqrt{10}.$

7. If $f(x) = x^{3/2}$, then $f'(x) = \dfrac{3x^{1/2}}{2}$ and

$$s = \int_0^1 \sqrt{1 + \left(\frac{3x^{1/2}}{2}\right)^2}\,dx = \int_0^1 \sqrt{1 + \frac{9x}{4}}\,dx = \int_0^1 \frac{1}{2}\sqrt{4 + 9x}\,dx = \frac{1}{2} \cdot \frac{2}{3} \cdot \frac{1}{9} \cdot (4 + 9x)^{3/2}\Big|_0^1 = \frac{13\sqrt{13} - 8}{27}.$$

9. If $f(x) = \dfrac{1}{6}\left(x^3 + \dfrac{3}{x}\right)$, then $f'(x) = \dfrac{1}{2}\left(x^2 - x^{-2}\right)$ and

$$s = \int_1^3 \sqrt{1 + \frac{1}{4}\left(x^4 - 2 + x^{-4}\right)}\,dx = \int_1^3 \sqrt{\frac{x^4 + 2 + x^{-4}}{4}}\,dx = \int_1^3 \left(\frac{x^2 + x^{-2}}{2}\right)dx = \frac{1}{2}\left(\frac{x^3}{3} - \frac{1}{x}\right)\Big|_1^3 = \frac{14}{3}.$$

11. $$f(x) = \left(\frac{1}{3}\right)\left(x^{3/2} - 3\sqrt{x}\right) \implies f'(x) = \left(\frac{1}{2}\right)\left(x^{1/2} - x^{-1/2}\right)$$

$$\implies 1 + \left(f'(x)\right)^2 = 1 + \left(\frac{1}{4}\right)\left(x - 2 + x^{-1}\right) = \left(\frac{1}{4}\right)\left(x + 2 + x^{-1}\right)$$

$$\implies s = \int_1^4 \frac{1}{2}\sqrt{x + 2 + x^{-1}}\,dx = \int_1^4 \frac{1}{2}\left(x^{1/2} + x^{-1/2}\right)dx = \frac{10}{3}.$$

13. If $f(x) = \dfrac{2}{3}\left(a + \dfrac{x^2}{a^2}\right)^{3/2}$, then $f'(x) = \dfrac{2x}{a^2}\left(a + \dfrac{x^2}{a^2}\right)^{1/2}$; thus $1 + \left(f'(x)\right)^2 = 1 + \dfrac{4x^2}{a^4}\left(a + \dfrac{x^2}{a^2}\right) =$

$1 + \dfrac{4x^2}{a^3} + \dfrac{4x^4}{a^6} = \left(1 + \dfrac{2x^2}{a^3}\right)^2$ and $s = \int_0^1 \left(1 + \dfrac{2x^2}{a^3}\right)dx = \left(x + \dfrac{2x^3}{3a^3}\right)\Big|_0^1 = 1 + \dfrac{2}{3a^3}.$

15. The graph of $x^{2/3} + y^{2/3} = 1$ is symmetric with respect to the x-axis and the y-axis; its total length is 4 times the length of that part lying in the first quadrant where $y = \left(1 - x^{2/3}\right)^{3/2}$. On that arc, we have

$y' = \left(\dfrac{3}{2}\right)\left(1 - x^{2/3}\right)^{1/2} \cdot \left(\dfrac{-2}{3}\right)x^{-1/3} = \dfrac{-\sqrt{1 - x^{2/3}}}{x^{1/3}}$ and $1 + (y')^2 = 1 + \dfrac{1 - x^{2/3}}{x^{2/3}} = x^{-2/3}$. Therefore the

total arc length is $s = 4\int_0^1 \sqrt{1 + (y')^2}\,dx = 4\int_0^1 x^{-1/3}\,dx = 4\left(\dfrac{3x^{2/3}}{2}\right)\Big|_0^1 = 6.$

17. If $y = 2 + x^2/2$, then $y' = x$ and $ds = \sqrt{1 + x^2}\,dx$; the arc length over the interval $[0, 1]$ is $\int_0^1 \sqrt{1 + x^2}\,dx$. The following table shows approximations using regular partitions of various sizes and function evaluations at left and right endpoints of subintervals.

n	left	right
10	1.127672	1.169094
20	1.137586	1.158296
30	1.140955	1.154763
40	1.142653	1.153008
50	1.143675	1.151959
60	1.144358	1.151262
80	1.145214	1.150392
100	1.145728	1.149871

REMARK: The arc length is $\left(\sqrt{2} + \ln\left|1 + \sqrt{2}\right|\right)/2 \approx 1.147794.$

19. If $y = \cos x$, then $y' = -\sin x$; the arc length over the interval $[0, \pi]$ is $\int_0^\pi \sqrt{1 + \sin^2 x}\, dx$. The following table shows approximations using regular partitions of various sizes and function evaluations at left and right endpoints of subintervals.

n	left	right
2	3.792238	3.792238
4	3.819944	3.819944
6	3.820194	3.820194
8	3.820198	3.820198
10	3.820198	3.820198

REMARK: The arc length is approximately 3.8201978.

21. a.   On the unit circle in the first quadrant $y = \sqrt{1 - x^2}$, thus $y' = \dfrac{1}{2}\left(1 - x^2\right)^{-1/2} \cdot (-2x) = \dfrac{-x}{\sqrt{1 - x^2}}$.
The upper half of the arc in this quadrant has length

$$\int_0^{1/\sqrt{2}} \sqrt{1 + \left(\frac{-x}{\sqrt{1 - x^2}}\right)^2}\, dx = \int_{x=0}^{1/\sqrt{2}} \frac{dx}{\sqrt{1 - x^2}}$$
$$= \int_{\theta=\pi/2}^{\pi/4} \frac{-\sin\theta\, d\theta}{\sqrt{1 - \cos^2\theta}} \qquad\qquad \cos\theta \equiv x$$
$$= \int_{\pi/2}^{\pi/4} \frac{-\sin\theta\, d\theta}{\sin\theta} = \int_{\pi/2}^{\pi/4} -d\theta = -\left(\frac{\pi}{4} - \frac{\pi}{2}\right) = \frac{\pi}{4}.$$

Therefore the whole circle has arc length equal to $8\left(\dfrac{\pi}{4}\right) = 2\pi$.

   b.   The cosine and sine functions are defined in terms of radian measure, but radian measure θ involves the arc length on the unit circle from $(1, 0)$ to the point $(\cos\theta, \sin\theta)$. [See sections A1.1 and 3.5 in the text's appendices.]
One might say that part (a) uses circular reasoning.

23. a.   If $x \geq 0$ and $y \geq 0$, then $y = x^2/4 \iff 4y = x^2 \iff 2\sqrt{y} = x$; reflecting this curve through the line $y = x$ yields the curve $2\sqrt{x} = y$, the arc of $y = x^2/4$ between $(2, 1)$ and $(4, 4)$ reflects into the arc of $y = 2\sqrt{x}$ between $(1, 2)$ and $(4, 4)$.

   b.   If $u = \dfrac{x^2}{4}$, then $x = 2\sqrt{u}$ and $dx = 2 \cdot \dfrac{1}{2\sqrt{u}} \cdot du = \dfrac{du}{\sqrt{u}}$. The substitution $u = \dfrac{x^2}{4}$ transforms integral (IV.d) into integral (V.a):

$$\int_{x=2}^{4} \sqrt{1 + \frac{x^2}{4}}\, dx = \int_{u=1}^{4} \sqrt{1 + u} \cdot \frac{du}{\sqrt{u}} = \int_{u=1}^{4} \sqrt{\frac{1 + u}{u}}\, du = \int_{u=1}^{4} \sqrt{\frac{1}{u} + 1}\, du.$$

Solutions 5.4   *Work, Power, and Energy*   (pages 364-5)

1.   (See Example 2.) $120 = 20 \cdot 6$ foot-pounds of work are done when a 20 lb stone is lifted 6 ft.

3.   (See Example 3.) The spring constant is $k = (3\,\text{nt})/(0.5\,\text{m}) = 6\,\text{nt/m}$; the work to stretch the spring 3 m is $\int_0^3 F(x)\, dx = \int_0^3 kx\, dx = \int_0^3 6x\, dx = 3x^2\big|_0^3 = 27$ J.

5. $8\,\text{nt} = F(x) = 6x \implies x = \dfrac{8}{6} = \dfrac{4}{3}$ m.

7.   (See Example 4; we too assume the pump outlet is at the top.) The volume of a cylinder of radius r and height h is $\pi r^2 h$. (A cylinder with 10 m diameter has a radius of 5 m.) Let the x-axis be oriented vertically, origin at top of cylinder, positive direction going down. Consider a horizontal slice with thickness Δx and distance x_i^* from the top:

$$
\begin{aligned}
V_i &= \pi(5^2)\Delta x = 25\pi\Delta x \text{ m}^3 && \text{Volume}\\
\mu_i &= \left(1000\,\text{kg/m}^3\right)\left(25\pi\Delta x\ \text{m}^3\right) = 25{,}000\pi\Delta x\ \text{kg} && \text{Mass}\\
F_i &= \left(9.81\,\text{m/sec}^2\right)\left(25{,}000\pi\Delta x\ \text{kg}\right) = 245{,}250\pi\Delta x\ \text{nt} && \text{Force to overcome gravity}\\
W_i &= (245{,}250\pi\Delta x\ \text{nt})(x_i^*\ \text{m}) = 245{,}250\pi x_i^*\Delta x\ \text{J}. && \text{Work to raise slice of water } x_i^*\ \text{m}
\end{aligned}
$$

5.4 — Work, Power, and Energy

The limit of the sum of the various W_i's is the definite integral

$$\int_0^{10} (245,250\pi)x\,dx = (245,250\pi)\left(\frac{x^2}{2}\right)\Big|_0^{10} = (245,250\pi)(50) = 12,262,500\pi \text{ nt} \approx 3.852 \cdot 10^7 \text{ J}.$$

9.   See the solution to Problem 7. The total work to pump water out of the top 5 m of the cylinder is

$$\int_0^5 (245,250\pi)\,x\,dx = (245,250\pi)\left(\frac{x^2}{2}\right)\Big|_0^5 = 3,065,625\pi \text{ J, one-quarter of the answer to Problem 7.}$$

11.   Put the origin of the x-axis at the base of the cone and have the positive direction be upwards, then a horizontal slice at location x has radius $x/3$ and it has been lifted $8 + x$ to its position. The work to fill the conical tank is

$$\int_0^9 (9810\pi)\left(\frac{x}{3}\right)^2 (8+x)\,dx = (9810\pi)\left(\frac{8x^3}{27} + \frac{x^4}{36}\right)\Big|_0^9 = (9810\pi)\left(\frac{1593}{4}\right) = 3,906,832.5\,\pi \text{ J} \approx 12,273,676 \text{ J}.$$

13.   When the top of the chain is x ft above the ground, there is $\min(x, 10)$ ft of chain off the ground; the work to lift this flexible chain from one end until that end is 15 ft above the ground is

$$\int_0^{15} 1.5 \cdot \min(x, 10)\,dx = 1.5\left(\int_0^{10} x\,dx + \int_{10}^{15} 10\,dx\right) = 1.5(50 + 50) = 150 \text{ ft-lb}.$$

REMARK: Focus, instead, on the variable distance each piece of chain is lifted: the small portion of chain s ft from the top end is lifted $(15 - s)$ ft and the total work is

$$\int_0^{10} 1.5 \cdot (15 - s)\,ds = 1.5\left(\frac{-(15-s)^2}{2}\right)\Big|_0^{10} = 1.5\left(\frac{-25}{2} - \frac{-225}{2}\right) = 150 \text{ ft-lb}.$$

---

15.   A horizontal slice with thickness Δx weighs $(62.4 \text{ lb/ft}^3) \cdot (20^2\,\Delta x \text{ ft}^3) = 24,960\Delta x$ lb. If that slice is x ft above the bottom, the work done by it as it flows out of the pool is $(x \text{ ft}) \cdot (24,960\,\Delta x \text{ lb}) = 24,960\,x\,\Delta x$ ft-lb.

The total work in emptying a 6 ft deep pool is $\int_0^6 24,960\,x\,dx = 12,480\,x^2\big|_0^6 = 449,280$ ft-lb.

17.   The work to fill the pool to a 2 ft depth is $\int_0^2 (62.4)\left(20^2\right)x\,dx = 12,480\,x^2\big|_0^2 = 49,920$ ft-lb. Since power $P = (\text{work } W)/(\text{time } T)$, if $W = 49,920$ ft-lb and $P = 2$ horsepower $= 2 \cdot 550 = 1100$ ft-lb/sec, then $T = \dfrac{W}{P} = \dfrac{49920}{1100} \approx 45.38$ seconds.

19.   If $30 + 4t - \dfrac{t^2}{2} = P(t) = \dfrac{dW}{dt}$, the work done in the first 15 seconds is

$$\int_0^{15} \frac{dW}{dt}\,dt = \int_0^{15}\left(30 + 4t - \frac{t^2}{2}\right)dt = \left(30t + 2t^2 - \frac{t^3}{6}\right)\Big|_0^{15} = 450 + 450 - 562.5 = 337.5 \text{ J}.$$

21.   If the force of the bow is proportional to the draw, then $F(x) = kx$ and $35 \text{ lb} = k(16 \text{ in}) = k(4/3 \text{ ft})$ imply $F(x) = 26.25x$ for a draw of x feet. The work done by the bow when released from a draw of $4/3$ ft is $\int_0^{4/3} 25.25x\,dx = (26.25)\left(\frac{1}{2}\right)\left(\frac{4}{3}\right)^2 = \frac{70}{3}$ ft-lb. This work equals the change in kinetic energy of the arrow; because the velocity of the arrow is 0 prior to release, $\dfrac{70}{3} = \dfrac{mv^2}{2}$ where v is the velocity of the arrow as it leaves the bowstring and $m = 0.0322/32.2 = 0.001$ slug is the mass of the arrow. Therefore

$$v = \sqrt{\left(\frac{1}{0.001}\right) \cdot 2 \cdot \left(\frac{70}{3}\right)} = \sqrt{\frac{140000}{3}} = 100\sqrt{\frac{14}{3}} \approx 216.02 \text{ ft/sec} \approx 147.29 \text{ mi/hr}.$$

REMARK: If you use 32 ft/sec^2, instead of 32.2 ft/sec^2, as the value of the acceleration due to gravity, then you should compute the velocity to be ≈ 215.35 ft/sec.

---

23.   See Example 8, especially equation (14). For Mars, $\delta = (6860/2)$ km $= 3,430,000$ m and $g = 3.92$ m/sec^2. The escape velocity from Mars is $\sqrt{2g\delta} = \sqrt{2(3.92)(3430000)} \approx 5185.67$ m/sec ≈ 18668 km/hr.

25. Since "dropping" implies an initial velocity of zero, the kinetic energy of the object when it hits the earth equals the work needed to lift it from the earth's surface to a distance of x_0 m above the surface: $K = \int_R^{R+x_0} mg \, dx = mg \, x_0$. (Remember that we are assuming x_0 is small enough that we can also assume g is constant.)

   If v is the final velocity, then $\frac{1}{2}mv^2 = K + \frac{1}{2}mv_0^2 = mgx_0 + 0 = mgx_0 \implies v = \sqrt{2gx_0}$.

Solutions 5.5   *Center of mass and the first moment*   (pages 368-9)

1. The first moment of the system about the origin (i.e., about the y-axis) is $4 \cdot 3 + 6 \cdot 5 = 12 + 30 = 42$ gm·cm. The center of mass is $\overline{x} = 42/(4+6) = 4.2$ cm.

3. The first moment about the origin is $2 \cdot (-6) + 8 \cdot 2 + 5 \cdot 20 + 3 \cdot (-1) = -12 + 16 + 100 - 3 = 101$ gm·cm; the center of mass is $\overline{x} = 101/(2+8+5+3) = 101/18$ cm.

5. a. The mass is $\int_0^1 \rho(x) \, dx = \int_0^1 x \, dx = 1/2$.

   b. The first moment about the y-axis (the line $x = 0$) is $\int_0^1 x \cdot \rho(x) \, dx = \int_0^1 x^2 \, dx = 1/3$.

   c. The center of mass is $\overline{x} = (1/3)/(1/2) = 2/3$.

7. a. The mass is $\int_{-1}^1 \rho(x) \, dx = \int_{-1}^1 |x| \, dx = 1$.

   b. The first moment about the y-axis is $\int_{-1}^1 x \cdot \rho(x) \, dx = \int_{-1}^1 x \cdot |x| \, dx = 0$.

   c. The center of mass is $\overline{x} = 0/1 = 0$.

9. a. The mass is $\int_1^8 \rho(x) \, dx = \int_1^8 x^{1/3} \, dx = (3/4)x^{4/3}\Big|_1^8 = (3/4)(16 - 1) = 45/4$.

   b. The first moment about the y-axis is $\int_1^8 x \cdot \rho(x) \, dx = \int_1^8 x \cdot x^{1/3} \, dx = \int_1^8 x^{4/3} \, dx = (3/7)x^{7/3}\Big|_1^8 = 381/7$.

   c. The center of mass is $\overline{x} = (381/7)/(45/4) = 508/105$.

11. Formula (8) may be rearranged to say that the first moment around the y-axis equals $m \cdot \overline{x}$ where m is the total mass and $\overline{x}$ is the x-coordinate of the center of mass. This relation is also true for each piece of the rod; thus

$$m \cdot \overline{x} = \int_a^b x \cdot \rho(x) \, dx = \int_a^c x \cdot \rho(x) \, dx + \int_c^b x \cdot \rho(x) \, dx = m_1 \cdot \overline{x}_1 + m_2 \cdot \overline{x}_2.$$

Since $m = m_1 + m_2$, we have proved

$$\overline{x} = \frac{m_1 \cdot \overline{x}_1 + m_2 \cdot \overline{x}_2}{m_1 + m_2}. \blacksquare$$

Solutions 5.6   *The centroid of a plane region*   (pages 376-7)

1. $\mu = 4 + 6 = 10$ kg; $M_y = 4 \cdot 3 + 6 \cdot (-5) = 12 - 30 = -18$ kg·m; $M_x = 4 \cdot 4 + 6 \cdot 3 = 16 + 18 = 34$ kg·m; thus $(\overline{x}, \overline{y}) = (M_y/\mu, M_x/\mu) = (-18/10, 34/10) = (-1.8$ m$, 3.4$ m$)$.

3. $\mu = 2 + 8 + 5 + 3 = 18$ kg; $M_y = 2 \cdot (-6) + 8 \cdot 2 + 5 \cdot 20 + 3 \cdot (-1) = -12 + 16 + 100 - 3 = 101$ kg·m; $M_x = 2 \cdot 2 + 8 \cdot 3 + 5 \cdot (-5) + 3 \cdot 8 = 4 + 24 - 25 + 24 = 27$ kg·m; thus $(\overline{x}, \overline{y}) = (M_y/\mu, M_x/\mu) = (101/18, 27/18) = (101/18$ m$, 3/2$ m$)$.

5. $\mu = A = \int_0^1 (2x + 3) \, dx = (x^2 + 3x)\Big|_0^1 = 4$ kg; $M_y = \int_0^1 x(2x + 3) \, dx = ((2/3)x^3 + (3/2)x^2)\Big|_0^1 = $
   $13/6$ kg·m; $M_x = \int_0^1 (1/2)(2x + 3)^2 \, dx = (1/12)(2x + 3)^3\Big|_0^1 = 49/6$ kg·m; $(\overline{x}, \overline{y}) = (M_y/\mu, M_x/\mu) = $
   $((13/6)/4, (49/6)/4) = (13/24$ m$, 49/24$ m$)$.

5.6 — The centroid of a plane region

7. $\mu = \int_0^4 (4-x)\, dx = (-1/2)(4-x)^2 \Big|_0^4 = 8$ kg; $M_y = \int_0^4 x(4-x)\, dx = \left(2x^2 - (1/3)x^3\right)\Big|_0^4 = 32/3$ kg $\cdot$ m;

$M_x = \int_0^4 (1/2)(4-x)^2\, dx = (-1/6)(4-x)^3 \Big|_0^4 = 32/3$ kg$\cdot$m; $(\bar{x}, \bar{y}) = (M_y/\mu, M_x/\mu) = ((32/3)/8, (32/3)/8) = (4/3$ m, $4/3$ m$)$.

REMARK: The triangular region is symmetric about the line $y = x$ and the centroid lies on that line.

9.   Let $f(x) = 4x$ and $g(x) = x^2$; then $g(x) \le f(x)$ for all $x \in [0,1]$.

$$\mu = A = \int_0^1 \left(4x - x^2\right)\, dx = \left(2x^2 - (1/3)x^3\right)\Big|_0^1 = 5/3 \text{ kg};$$

$$M_y = \int_0^1 x\left(4x - x^2\right)\, dx = \left((4/3)x^3 - (1/4)x^4\right)\Big|_0^1 = 13/12 \text{ kg} \cdot \text{m};$$

$$M_x = \int_0^1 (1/2)\left((4x)^2 - \left(x^2\right)^2\right)\, dx = \left((8/3)x^3 - (1/10)x^5\right)\Big|_0^1 = 77/30 \text{ kg} \cdot \text{m};$$

Therefore $(\bar{x}, \bar{y}) = (M_y/\mu, M_x/\mu) = ((13/12)/(5/3), (77/30)/(5/3)) = (13/20$ m, $77/50$ m$)$.

11.   The graphs of $x + y^2 = 8$ and $x + y = 2$ meet at the points $(-1, 3)$ and $(4, -2)$. The region is bounded by the graphs of $f_1(x) = \sqrt{8-x}$ and $g_1(x) = 2 - x$ for $-1 \le x \le 4$ and by the graphs of $f_2(x) = \sqrt{8-x}$ and $g_2(x) = -\sqrt{8-x}$ for $4 \le x \le 8$.

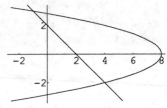

REMARK: We integrate $x\sqrt{8-x}$ by rewriting it as $(8 - 8 + x)\sqrt{8-x} = 8\sqrt{8-x} - (8-x)^{3/2}$.

$$\mu = \int_{-1}^4 \left(\sqrt{8-x} - (2-x)\right)\, dx + \int_4^8 \left(\sqrt{8-x} - (-\sqrt{8-x})\right)\, dx$$

$$= \left((-2/3)(8-x)^{3/2} + (-1/2)(2-x)^2\right)\Big|_{-1}^4 + (-4/3)(8-x)^{3/2}\Big|_4^8 = 61/6 + 32/3 = 125/6 \text{ kg};$$

$$M_y = \int_{-1}^4 x\left(\sqrt{8-x} - (2-x)\right)\, dx + \int_4^8 x\left(\sqrt{8-x} - (-\sqrt{8-x})\right)\, dx$$

$$= \left((2/5)(8-x)^{5/2} - (16/3)(8-x)^{3/2} - x^2 + (1/3)x^3\right)\Big|_{-1}^4 + \left((4/5)(8-x)^{5/2} - (32/3)(8-x)^{3/2}\right)\Big|_4^8$$

$$= 118/5 + 896/15 = 250/3 \text{ kg} \cdot \text{m};$$

$$M_x = (1/2)\int_{-1}^4 \left((8-x) - (2-x)^2\right)\, dx + (1/2)\int_4^8 \left((8-x) - (8-x)\right)\, dx$$

$$= \left(4x - (1/4)x^2 + (1/6)(2-x)^3\right)\Big|_{-1}^4 + 0\Big|_4^8 = 125/12 + 0 = 125/12 \text{ kg} \cdot \text{m};$$

Therefore $(\bar{x}, \bar{y}) = (M_y/\mu, M_x/\mu) = ((250/3)/(125/6), (125/12)/(125/6)) = (4$ m, $1/2$ m$)$.

REMARK: It is simpler to integrate with respect to y over this region; adapt formulas (13)–(16):

$$\mu = \int_{-2}^3 \left((8 - y^2) - (2 - y)\right)\, dy = \left(6y + y^2/2 - y^3/3\right)\Big|_{-2}^3 = 125/6 \text{ kg};$$

$$M_y = (1/2)\int_{-2}^3 \left((8 - y^2)^2 - (2 - y)^2\right)\, dy = \left(30y + y^2 - 17y^3/6 + y^5/10\right)\Big|_{-2}^3 = 250/3 \text{ kg} \cdot \text{m};$$

$$M_x = \int_{-2}^3 y\left((8 - y^2) - (2 - y)\right)\, dy = \left(3y^2 + y^3/3 - y^4/4\right)\Big|_{-2}^3 = 125/12 \text{ kg} \cdot \text{m}.$$

13. $f(x) = 2 - x^2$ and $g(x) = x^2$ are both even functions; therefore the region they bound must be symmetric about the y-axis and $\bar{x} = 0$. Furthermore, that region is symmetric about the line $y = 1$ because $f(x) - 1 = 1 - x^2 = 1 - g(x)$; hence $\bar{y} = 1$.

122

15.   The unit circle, radius 1 and center $(0,0)$, is symmetric about the y-axis and the x-axis, thus its centroid is $(0,0)$ Using the result of Problem 0.3.63, we compute the distance from that centroid to the line $y = 4 - x$ (i.e., $x + y - 4 = 0$) to be

$$\frac{|1 \cdot 0 + 1 \cdot 0 + (-4)|}{\sqrt{1^2 + 1^2}} = \frac{4}{\sqrt{2}} = 2\sqrt{2}.$$

The unit circle does not intersect the line $y = 4 - x$ because $1 < 2\sqrt{2}$. The area of the unit circle is $\pi \cdot 1^2 = \pi$; the circumference traced by rotating its centroid $(0,0)$ about the line $y = 4 - x$ is $2\pi \cdot 2\sqrt{2} = 4\sqrt{2}\,\pi$. The first theorem of Pappus implies the torus generated by rotating the unit circle about the line $y = 4 - x$ has volume $\pi \cdot 4\sqrt{2}\,\pi = 4\sqrt{2}\,\pi^2$.

17.   For the following discussion, assume plane region R is of the form $\{(x, y) : a \leq x \leq b, 0 \leq y \leq f(x)\}$ for some continuous non-negative function f. R will be symmetric about the line $x = c$ provided $c = (a + b)/2$ (i.e., the midpoint of $[a, b]$) and $f(c - w) = f(c + w)$ for all w between 0 and $d = (b - a)/2$ (i.e., half the length of $[a, b]$).

$\boxed{\text{Informal discussion using the hint}}$ The moment of the region about the line $x = c$ is $\int_a^b (x - c)f(x)\,dx$. Partition $[a, b]$ into $2n$ subintervals, each of length $\Delta x = d/n = (b - a)/(2n)$. The definite integral is approximated by sums of the form $\sum_{k=1}^{2n} (x_k^* - c) f(x_k^*) \Delta x$ where x_k^* is the midpoint of its subinterval. If x_i^* is in $[a, c]$, there is an x_j^* in $[c, b]$ such that $c - x_i^* = x_j^* - c$; then $f(x_i^*) = f(x_j^*)$, $(x_i^* - c) f(x_i^*) \Delta x + (x_j^* - c) f(x_j^*) \Delta x = 0$, and the whole sum of $2n$ terms collapses to zero. Because $(x - c)f(x)$ is continuous, Theorem 4.5.1 implies the limit of this particular sequence of approximating sums (all equal to 0) is the value of definite integral. Therefore $\overline{x} = c$ since $\int_a^b (x - c)f(x)\,dx = 0$. $\blacksquare$

$\boxed{\text{Proof using substitution}}$

$$\int_a^b x f(x)\,dx = \int_{-d}^{d} (c + u)f(c + u)\,du \qquad\qquad \text{let } u = c - x$$

$$= \int_{-d}^{d} (c + u)f(c - u)\,du \qquad\qquad \text{symmetry of } f \text{ about } c$$

$$= \int_{b}^{a} (2c - v)f(v)(-dv) \qquad\qquad \text{let } v = c - u$$

$$= 2c \int_a^b f(v)\,dv - \int_a^b v f(v)\,dv.$$

Since $\int_a^b x f(x)\,dx = M_y = \int_a^b v f(v)\,dv$, we obtain the relations $2M_y = 2c \int_a^b f(v)\,dv = 2cA$ and $\overline{x} = M_y/A = c = (a + b)/2$. $\blacksquare$

REMARK: If $I = \int_a^b (x - c)f(x)\,dx$, these substitutions show $I = -I$ which has unique solution $I = 0$.
         If the lower boundary of the region is the graph of $y = g(x)$ instead of the x-axis, the second proof shows $\int_a^b x g(x)\,dx = c \int_a^b g(x)\,dx$ provided g is symmetric about c. Then

$$M_y = \int_a^b x \cdot \big(f(x) - g(x)\big)\,dx = \int_a^b x f(x)\,dx - \int_a^b x g(x)\,dx$$

$$= c \int_a^b f(x)\,dx - c \int_a^b g(x)\,dx = c \int_a^b \big(f(x) - g(x)\big)\,dx = c \cdot A.$$

5.6 — The centroid of a plane region

18. a,c. If $(\overline{x}, \overline{y})$ is the centroid of the plane region R and if L is any line through $(\overline{x}, \overline{y})$, then the first moment of R about L is equal to zero. The converse is also true: if R has zero moment about L, then L passes through the centroid of R. The computational procedure we have used so far works with vertical or horizontal lines, but this problem is simplified if we use median lines (i.e., connect a vertex to the midpoint of the opposite side).

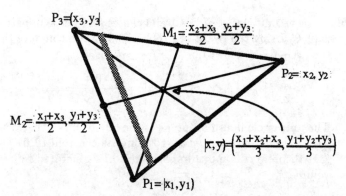

Let M_2 be the midpoint of $P_1 P_3$ and let L be the line $P_2 M_2$. Now consider a thin rectangular strip which is parallel to $P_1 P_3$: its centroid is its center and that point lies on the median line L. Each such strip has zero moment about L; hence, so does the whole triangle. Therefore the triangle's centroid is on the median line $P_2 M_2$.

A similar argument proves the centroid also lies on the median lines $P_1 M_1$ and $P_3 M_3$. All three median lines intersect at the point $((x_1 + x_2 + x_3)/3, (y_1 + y_2 + y_3)/3)$ so that must be the centroid of the triangle. ∎

REMARK: The main idea of this proof is due to Archimedes.

b. If equal-size weights, each of size w, are placed at the vertices of an otherwise weightless triangle, then $\mu = w + w + w = 3w$, $M_y = w \cdot x_1 + w \cdot x_2 + w \cdot x_3$, and $M_x = w \cdot y_1 + w \cdot y_2 + w \cdot y_3$. Therefore $\overline{x} = M_y/\mu = (x_1 + x_2 + x_3)/3$ and $\overline{y} = M_x/\mu = (y_1 + y_2 + y_3)/3$. ∎

Solutions 5.7   *Fluid pressure*   (pages 382-3)

1. Pressure $P = \rho \cdot d \cdot g = (1000\ \text{kg/m}^3) \cdot (2\ \text{m}) \cdot (9.81\ \text{m/sec}^2) = 19620\ \text{nt/m}^2$.
Force $F = P \cdot A = (19620\ \text{nt/m}^2) \cdot \pi(6\ \text{m})^2 = 706320\pi\ \text{nt}$.

3. a. $P_A = P_D = P_C = \rho \cdot h \cdot g = (13.6\ \text{gm/cm}^3) \cdot h \cdot (9.81\ \text{m/sec}^2) = (13600\ \text{kg/m}^3) \cdot h \cdot (9.81\ \text{m/sec}^2) = 133416\ h\ \text{nt/m}^2$ if h is measured in meters.

b. If $h = 76\ \text{cm} = 0.76\ \text{m}$, then $P_A = 133416 \cdot 0.76 = 101396.16\ \text{nt/m}^2$.

5. A layer at height y above the bottom of the trough is $5\sqrt{3}/2 - y$ below the surface of the water.

$$F = \int_0^{5\sqrt{3}/2} \left(\frac{5\sqrt{3}}{2} - y \right) \left(\frac{2y}{\sqrt{3}} \right) (\rho\, dy) = (62.4\ \text{lb/ft}^3) \int_0^{5\sqrt{3}/2} \left(5y - \frac{2y^2}{\sqrt{3}} \right) dy = (62.4\ \text{lb/ft}^3)\frac{125}{8}\ \text{ft}^3 = 975\ \text{lb}.$$

7. $F = \int_0^1 \rho g (1 - y) 2\sqrt{1 - (y-1)^2}\, dy = (9.81\ \text{m/sec}^2)(810\ \text{kg/m}^2)(2/3) \left(1 - (y-1)^2 \right)^{3/2} \Big|_0^1 = 5297.4\ \text{nt}$.

9. $F = \rho g \int_0^{6.25} (6.25 - y) 2\sqrt{25y}\, dy = (1000\ \text{kg/m}^3)(9.81\ \text{m/sec}^2) \left((125/3)y^{3/2} - 4y^{5/2} \right) \Big|_0^{6.25} = (9810) \cdot$
$(3125/12) \approx 2.55 \cdot 10^6\ \text{nt}$.

11. The vertical cross section of the dam has the shape of the parabola $y = x^2/25$; if the river is 25 m wide at the top of the dam, then its depth is $(25/2)^2/25 = 6.25$ m. The width at distance y from the bottom is $2x = 10\sqrt{y}$. Suppose the upstream face of the dam slopes at an angle θ with the horizontal. If L is the slant length along the face from the bottom to a position y directly above the bottom, then $L = (\csc \theta)y$; that implies $dL = \csc \theta\, dy$ and

$$F = \int_0^{6.25} \rho g \cdot \text{depth} \cdot dA = \rho g \int_0^{6.25} (6.25 - y) \cdot 10\sqrt{y}(\csc \theta)\, dy$$
$$= 98100\, \csc \theta \int_0^{6.25} \left(6.25y^{1/2} - y^{3/2} \right) dy \approx 2,554,687.5\ \csc \theta\ \text{nt}.$$

If the angle is 30 degrees, then $\csc \theta = 2$ and the force is approximately $5,109,378$ newtons.

13. $F = \rho g \int_0^{3/2} (3/2 - y) 2(2y/3 + 1)\, dy = \rho g \int_0^{3/2} (3 - 4y^2/3)\, dy = 1030 \cdot 9.81 \cdot 3 = 30312.9$ newtons.

124

15.   Because the side of the trough is sloping, the area of a horizontal strip is $3\sqrt{(1/2)^2 + (3/2)^2}\,\Delta x = 3\sqrt{5/2}\,\Delta x$ if it appears Δx wide when viewed directly above. $F = \rho g \int_0^1 \left(0 - (3x/2 - 3/2)\right) 3\sqrt{5/2}\, dx = \rho g (-9/4)\sqrt{5/2}(x-1)^2 \Big|_0^1 \approx 34899.7$ nt.

17. $A = \int_0^1 x^{1/3}\, dx = (3/4) x^{4/3} \Big|_0^1 = (3/4)\ \text{m}^2$; therefore $F = 1000 \cdot 9.81 \cdot 15 \cdot 3/4 = 110362.5$ nt.

19.   Suppose the region R is $\{(x, y) : a \le y \le b, g(y) \le x \le f(y)\}$ and the surface is the x-axis. The total force on one side of the region is

$$F = \rho \int_a^b |y| \cdot \left(f(y) - g(y)\right) dy = -\rho \int_a^b y \cdot \left(f(y) - g(y)\right) dy = -\rho \cdot M_x = -\rho \cdot (\bar{y}A) = (\rho(-\bar{y})) \cdot A = \rho\,|\bar{y}| \cdot A$$

but $\rho\,|\bar{y}|$ is the pressure at the centroid of R. ∎

Solutions 5.R   *Review*   (pages 383-4)

1.
$$\int_{-2}^5 |3x - 7|\, dx = \int_{-2}^{7/3} (7 - 3x)\, dx + \int_{7/3}^5 (3x - 7)\, dx = \left(7x - \frac{3x^2}{2}\right)\Big|_{-2}^{7/3} + \left(\frac{3x^2}{2} - 7x\right)\Big|_{7/3}^5$$
$$= \left(\frac{49}{6} - (-20)\right) + \left(\frac{5}{2} - \frac{-49}{6}\right) = \frac{169}{6} + \frac{32}{3} = \frac{233}{6}.$$

3. $y = 2 - x - x^2 = (2 + x)(1 - x)$ is positive on the interval $(-2, 1)$. The area of the region bounded by the curve and the x-axis is $\int_{-2}^1 \left(2 - x - x^2\right) dx = \left(2x - \frac{x^2}{2} - \frac{x^3}{3}\right)\Big|_{-2}^1 = \frac{7}{6} - \frac{-10}{3} = \frac{9}{2}.$

5.   The straight line $y = 2x + 1$ meets the parabola $x = -y^2$ at the points $(-1, -1)$ and $(-1/4, 1/2)$. If $-1 \le y \le 1/2$, the straight line is to the left of the parabola so the area is
$$\int_{-1}^{1/2} \left((-y^2) - \frac{y - 1}{2}\right) dy = \left(\frac{-y^3}{3} - \frac{(y-1)^2}{4}\right)\Big|_{-1}^{1/2} = \frac{7}{48} - \frac{-5}{12} = \frac{9}{16}.$$

7.
$$\int_0^\pi |\cos x - \sin x|\, dx = \int_0^{\pi/4} (\cos x - \sin x)\, dx + \int_{\pi/4}^\pi (\sin x - \cos x)\, dx$$
$$= (\sin x + \cos x)\Big|_0^{\pi/4} + (-\cos x - \sin x)\Big|_{\pi/4}^\pi$$
$$= \left[\left(\frac{1}{\sqrt 2} + \frac{1}{\sqrt 2}\right) - (0 + 1)\right] + \left[(-(-1) - 0) - \left(\frac{-1}{\sqrt 2} - \frac{1}{\sqrt 2}\right)\right]$$
$$= \left[\sqrt 2 - 1\right] + \left[\sqrt 2 + 1\right] = 2\sqrt 2.$$

9. $\boxed{\text{Disk method}}$ $V = \int_0^2 \pi\, [y(x)]^2\, dx = \int_0^2 \pi (3x + 2)^2\, dx = \pi\, \frac{(3x + 2)^3}{9}\Big|_0^2 = \pi\, \frac{512 - 8}{9} = 56\,\pi.$

11. $\boxed{\text{Disk method}}$ $V = \int_0^4 \pi\, [y(x)]^2\, dx = \int_0^4 \pi \left(\sqrt{2x + 1}\right)^2\, dx = \int_0^4 \pi(2x + 1)\, dx = \pi\left(x^2 + x\right)\Big|_0^4 = 20\pi.$

13. $\boxed{\text{Disk method}}$ $V = \int_0^1 \pi \left(x^3\right)^2\, dx = \pi \int_0^1 x^6 dx = \pi\, \frac{x^7}{7}\Big|_0^1 = \frac{\pi}{7}.$

$\boxed{\text{Cylindrical shell method}}$ $V = \int_0^1 2\pi y \left(1 - y^{1/3}\right) dy = \pi\left(y^2 - \frac{6y^{7/3}}{7}\right)\Big|_0^1 = \frac{\pi}{7}.$

15.   Rotating the region $\{(x, y) : 0 \le x \le 4, \sqrt{2x + 1} \le y \le 5\}$ about the x-axis generates a solid with volume

$\boxed{\text{Disk method}}$ $\int_0^4 \pi \left(5^2 - \left(\sqrt{2x + 1}\right)^2\right) dx = \int_0^4 \pi\, (25 - (2x + 1))\, dx = \pi\left(24x - x^2\right)\Big|_0^4 = 80\pi.$

17.   Suppose the base of the solid is the circle $x^2 + y^2 = 3^2$ and vertical cross sections perpendicular to the x-axis are square. The square through the point (x, y) has area $(2y)^2 = 4y^2 = 4\left(9 - x^2\right)$ and the volume of the solid is $\int_{-3}^3 4\left(9 - x^2\right) dx = 4\left(9x - \frac{x^3}{3}\right)\Big|_{-3}^3 = 4 \cdot 18 - 4 \cdot (-18) = 144.$

19. If $y = 2x + 3$, then $s = \int_2^7 \sqrt{1 + (y')^2}\, dx = \int_2^7 \sqrt{1 + 2^2}\, dx = \sqrt{5}(7 - 2) = 5\sqrt{5}$.

21. If $y = \left(\dfrac{2}{3}\right) x^{3/2}$, then $y' = x^{1/2}$ and

$$s = \int_0^1 \sqrt{1 + (y')^2}\, dx = \int_0^1 \sqrt{1 + x}\, dx = \left(\frac{2}{3}\right)(1 + x)^{3/2}\Big|_0^1 = \left(\frac{2^{5/2} - 2}{3}\right).$$

23. If $y = x \sin x$, then $y' = \sin x + x \cos x$ and $s = \int_0^{\pi/2} \sqrt{1 + (\sin x + x \cos x)^2}\, dx$.

25. If $F(x) = kx$ and $F(30\,\text{cm}) = 8\,\text{N}$, then the spring constant is $k = (8\,\text{N})/(0.30\,\text{m}) = (80/3)\,\text{N/m}$.

Stretching the spring 60 cm does $\displaystyle\int_0^{0.6} \left(\frac{80}{3}\right) x\, dx = \left(\frac{40}{3}\right) x^2 \Big|_0^{0.6} = \frac{24}{5} = 4.8\,\text{J}$ of work.

27. $$W = \int_0^{25} \frac{dW}{dt}\, dt = \int_0^{25} P(t)\, dt = \int_0^{25} \left(45\sqrt{t} + \frac{t}{10} + \frac{t^2}{100}\right) dt = \left(30 t^{3/2} + \frac{t^2}{20} + \frac{t^3}{300}\right)\Big|_0^{25} = \frac{11500}{3}\,\text{J}.$$

29. $\mu = 2 + 5 + 8 = 15\,\text{kg}$; $M_y = 2 \cdot 4 + 5 \cdot (-9) + 8 \cdot 2 = 8 - 45 + 16 = -21\,\text{kg} \cdot \text{m}$; therefore $\overline{x} = M_y/\mu = (-21)/15 = -7/5\,\text{m}$.

31.
$$\overline{x} = \frac{\int_0^3 x\rho(x)\, dx}{\int_0^3 \rho(x)\, dx} = \frac{\int_0^3 x \cdot x^2\, dx}{\int_0^3 x^2\, dx} = \frac{x^4/4\big|_0^3}{x^3/3\big|_0^3} = \frac{3^4/4}{3^3/3} = \frac{9}{4}.$$

33.
$$\mu = \int_1^3 x^2\, dx = \frac{x^3}{3}\Big|_1^3 = \frac{26}{3},$$

$$M_y = \int_1^3 x \cdot x^2\, dx = \frac{x^4}{4}\Big|_1^3 = 20,$$

$$M_x = \frac{1}{2} \int_1^3 \left(x^2\right)^2 dx = \frac{x^5}{10}\Big|_1^3 = \frac{121}{5};$$

$$(\overline{x}, \overline{y}) = \left(\frac{M_y}{\mu}, \frac{M_x}{\mu}\right) = \left(\frac{20}{26/3}, \frac{121/5}{26/3}\right) = \left(\frac{30}{13}, \frac{363}{130}\right).$$

35. If $f(x) = x + 3$ and $g(x) = x^2 + 5x + 6 = (x + 2)(x + 3)$, then $f(x) = g(x)$ if x is -3 or -1 and $g(x) < f(x)$ if $x \in (-3, -1)$.

$$\mu = \int_{-3}^{-1} \left((x + 3) - (x^2 + 5x + 6)\right) dx = \left(\frac{-x^3}{3} - 2x^2 - 3x\right)\Big|_{-3}^{-1} = \frac{4}{3} - 0 = \frac{4}{3},$$

$$M_y = \int_{-3}^{-1} x \cdot \left((x + 3) - (x^2 + 5x + 6)\right) dx = \left(\frac{-x^4}{4} - \frac{4x^3}{3} - \frac{3x^2}{2}\right)\Big|_{-3}^{-1} = \frac{-5}{12} - \frac{9}{4} = \frac{-8}{3},$$

$$M_x = \frac{1}{2} \int_{-3}^{-1} \left((x + 3)^2 - (x^2 + 5x + 6)^2\right) dx = \frac{1}{2} \int_{-3}^{-1} \left(-x^4 - 10x^3 - 36x^2 - 54x - 27\right) dx = \frac{97}{20} - \frac{81}{20} = \frac{4}{5};$$

$$(\overline{x}, \overline{y}) = \left(\frac{M_y}{\mu}, \frac{M_x}{\mu}\right) = \left(\frac{-8/3}{4/3}, \frac{4/5}{4/3}\right) = \left(-2, \frac{3}{5}\right).$$

37. Force $=$ Pressure $\cdot$ Area $= (\rho d g) \cdot A = (1000 \cdot 9.81 \cdot 2) \cdot (4 \cdot 5) = 392400\,\text{N}$.

39. $F = \int_{-2}^0 \rho g |y| 2 \left(-3y/8 + 3/4\right) dy = \rho g \int_{-2}^0 \left(3y^2/4 - 3y/2\right) dy = 49050\,\text{N}$.

41. a. $F = \int_{-2}^{-3/2} \rho g \left(|y| - 3/2\right) \left(2(-3y/8 + 3/4)\right) dy = \rho g (135/64 - 7/4) \approx 3525\,\text{N}$.

    b. $F = \int_{-2}^{-1} \rho g \left(|y| - 1\right) \left(-3y/4 + 3/2\right) dy = \rho g(7/8 - (-1/2)) \approx 13489\,\text{N}$.

    c. $F = \int_{-2}^{-1/2} \rho g \left(|y| - 1/2\right) \left(-3y/4 + 3/2\right) dy = \rho g(13/64 - (-11/4)) \approx 28970\,\text{N}$.

6
Transcendental Functions and Their Inverses

1. If $f(x) = 3x + 5$, then $f'(x) = 3 > 0$ and f is 1–1 on $\mathbb{R}$. Because $y = 3x + 5 \iff y - 5 = 3x \iff$ $(y - 5)/3 = x$; we conclude that $f^{-1}(y) = (y - 5)/3$. Theorem 4 says that $\dfrac{d}{dy} f^{-1}(y) = \dfrac{1}{f'(x)} = \dfrac{1}{3}$.

3. If $f(x) = 1/x$, then $f'(x) = -1/x^2 < 0$ and f is 1–1 on $\mathbb{R} - \{0\}$. Since $y = 1/x \iff xy = 1 \iff x = 1/y$, we infer that $f^{-1}(y) = 1/y$. (Indeed, f is its own inverse.) $\dfrac{d}{dy} f^{-1}(y) = \dfrac{1}{\frac{d}{dx} f(x)} = \dfrac{1}{-1/x^2} = -x^2 = \dfrac{-1}{y^2}$.

5. If $f(x) = \sqrt{4x + 3}$, then $f'(x) = 2/\sqrt{4x + 3} > 0$; f is 1–1 on $[-3/4, \infty)$ and the range of f is $[0, \infty)$. If $-3/4 \le x$, then $y = \sqrt{4x + 3} \iff y^2 = 4x + 3 \iff (y^2 - 3)/4 = x$; therefore $f^{-1}(y) = (y^2 - 3)/4$ with domain $[0, \infty)$. Theorem 4 says $\dfrac{d}{dy} f^{-1}(y) = \dfrac{1}{f'(x)} = \dfrac{\sqrt{4x + 3}}{2} = \dfrac{y}{2}$.

7. If $f(x) = 1 - x^3$, then $f'(x) = -3x^2 \le 0$. Because f is continuous and because $f'(x) < 0$ if $x \ne 0$, f is strictly decreasing and thus is 1–1 on all of $\mathbb{R}$. Because $y = 1 - x^3 \iff x^3 = 1 - y \iff x = (1 - y)^{1/3}$; we conclude that $f^{-1}(y) = (1 - y)^{1/3}$. Theorem 4 says that $\dfrac{d}{dy} f^{-1}(y) = \dfrac{1}{f'(x)} = \dfrac{1}{-3x^2} = \left(\dfrac{-1}{3} \right)(1 - y)^{-2/3}$.

9. If $f(x) = \dfrac{x + 1}{x} = 1 + \dfrac{1}{x} = 1 + x^{-1}$, then $f'(x) = -x^{-2} < 0$ and f is 1–1 on $\mathbb{R} - \{0\}$. Since $y = \dfrac{x + 1}{x} \iff xy = x + 1 \iff x(y - 1) = xy - x = 1 \iff x = \dfrac{1}{y - 1}$, we infer $f^{-1}(y) = \dfrac{1}{y - 1} = (y - 1)^{-1}$. Theorem 4 says $\dfrac{d}{dy} f^{-1}(y) = \dfrac{1}{f'(x)} = \dfrac{1}{-x^{-2}} = -x^2 = -\left(\dfrac{1}{y - 1} \right)^2 = -(y - 1)^{-2}$.

11. If $f(x) = (x - 2)(x - 3) = x^2 - 5x + 6$, then $f'(x) = 2x - 5$. f is 1–1 and decreasing on $(-\infty, 5/2]$ while it is 1–1 and increasing on $[5/2, \infty)$.

$$y = x^2 - 5x + 6 = \left(x - \frac{5}{2} \right)^2 - \frac{1}{4} \iff \sqrt{y + \frac{1}{4}} = \left| x - \frac{5}{2} \right| = \begin{cases} -\left(x - \dfrac{5}{2} \right) & \text{if } x \le \dfrac{5}{2} \\ x - \dfrac{5}{2} & \text{if } \dfrac{5}{2} \le x \end{cases}$$

$$\iff \begin{cases} x = f_-^{-1}(y) = \dfrac{5}{2} - \sqrt{y + \dfrac{1}{4}} = \dfrac{5 - \sqrt{4y + 1}}{2} \\ x = f_+^{-1}(y) = \dfrac{5}{2} + \sqrt{y + \dfrac{1}{4}} = \dfrac{5 + \sqrt{4y + 1}}{2} \end{cases}$$

$$\implies \begin{cases} \dfrac{d}{dy} f_-^{-1}(y) = \dfrac{1}{2x - 5} = \dfrac{1}{-\sqrt{4y + 1}} = \dfrac{-1}{\sqrt{4y + 1}} \\ \dfrac{d}{dy} f_+^{-1}(y) = \dfrac{1}{2x - 5} = \dfrac{1}{2(x - 5/2)} = \dfrac{1}{2\sqrt{y + 1/4}} = \dfrac{1}{\sqrt{4y + 1}} \end{cases}$$

13. If $x = (\alpha + \beta)^{1/3} + (\alpha - \beta)^{1/3}$, then
$$x^3 = (\alpha + \beta) + 3(\alpha + \beta)^{2/3}(\alpha - \beta)^{1/3} + 3(\alpha + \beta)^{1/3}(\alpha - \beta)^{2/3} + (\alpha - \beta)$$
$$= 2\alpha + 3(\alpha + \beta)^{1/3}(\alpha - \beta)^{1/3}\left((\alpha + \beta)^{1/3} + (\alpha - \beta)^{1/3} \right) = 2\alpha + 3(\alpha^2 - \beta^2)^{1/3} x.$$

If $\alpha = \dfrac{y}{2}$ and $\beta = \sqrt{\alpha^2 - \left(\dfrac{-1}{3} \right)^3} = \sqrt{\dfrac{y^2}{4} + \dfrac{1}{27}}$, then $2\alpha = y$, $3(\alpha^2 - \beta^2)^{1/3} = -1$, and $x^3 = y - x$. On the other hand, if $f(x) = x^3 + x$, then $f'(x) = 3x^2 + 1 > 0$ and f is 1–1 on $\mathbb{R}$. Hence f has a unique inverse and $f^{-1}(y) = \sqrt[3]{\dfrac{y}{2} + \sqrt{\dfrac{y^2}{4} + \dfrac{1}{27}}} + \sqrt[3]{\dfrac{y}{2} - \sqrt{\dfrac{y^2}{4} + \dfrac{1}{27}}}$. Theorem 4 says $\dfrac{d}{dy} f^{-1}(y) = \dfrac{1}{f'(x)} = \dfrac{1}{3x^2 + 1}$.

REMARK: The algebra shown here also yields a solution to the equation $0 = a + bx + x^3$, the substitution $w = x - C/3$ transforms the general cubic $A + Bw + Cw^2 + w^3$ into one with zero coefficient for the quadratic term; therefore we can obtain a "closed-form" expression for roots of a general cubic equation.

15.   If $f(x) = \cos x$, then $f'(x) = -\sin x < 0$ on $(0, \pi)$; f decreases and is 1–1 on $[0, \pi]$. The inverse function, the ArcCosine function, is discussed in section 6.7; for now we can refer to Theorem 4 and say
$$\frac{d}{dy} f^{-1}(y) = \frac{1}{f'(x)} = \frac{1}{-\sin x} = \frac{-1}{\sin x} = \frac{-1}{\sqrt{1 - \cos^2 x}} = \frac{-1}{\sqrt{1 - y^2}}.$$

17.   Let $f(x) = \tan x$. $f(x)$ is defined only if $x \neq (k + 1/2)\pi$ for any integer k. In fact, f is differentiable on each interval $(-\pi/2 + k\pi, \pi/2 + k\pi)$. Since $f'(x) = \sec^2 x$ which is strictly positive, f must be strictly increasing on each of those intervals; hence it is 1–1 there. On each interval there is an inverse function; it satisfies $\dfrac{d}{dy} f^{-1}(y) = \dfrac{1}{f'(x)} = \dfrac{1}{\sec^2 x} = \dfrac{1}{1 + \tan^2 x} = \dfrac{1}{1 + y^2}$.

19. $y = f(x)$ $y = f^{-1}(x)$

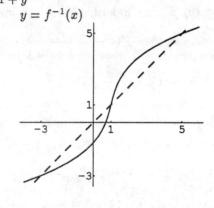

21. $y = f(x)$ $y = f^{-1}(x)$

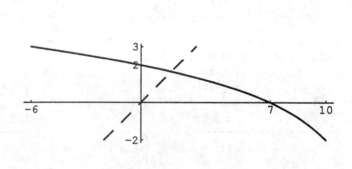

23.   Suppose n is a positive integer; consider the functions $f(t) = t^n$ and $g(t) = t^{1/n}$, each with domain and range equal to $[0, \infty)$. Since $g = f^{-1}$ and $f'(t) = nt^{n-1}$, if $y = g(x) = x^{1/n}$, then $x = f(y) = y^n$ and
$$\frac{d}{dx} x^{1/n} = g'(x) = \frac{1}{f'(y)} = \frac{1}{ny^{n-1}} = \frac{1}{n} \cdot y^{1-n} = \frac{1}{n} \cdot \left(x^{1/n}\right)^{1-n} = \frac{1}{n} \cdot x^{(1-n)/n} = \frac{1}{n} \cdot x^{(1/n)-1}. \ \blacksquare$$

25.   Suppose the function g is continuous at x_0 and $g(x_0) \neq 0$, then there is a neighborhood of x_0 in which g does not change sign. (See Problem 1.7.63.) Therefore, if f' exists in a neighborhood of x_0, is continuous at x_0, and $f'(x_0) \neq 0$, then there is a (possibly smaller) neighborhood of x_0 in which f' does not change sign. Theorem 3.2.1 then guarantees that f is either strictly increasing or strictly decreasing throughout that neighborhood. In either case, f is 1–1 on the neighborhood (see Theorem 1). $\blacksquare$

Solutions 6.2   *Review of Exponential and Logarithm functions*   (pages 406-7)

1. $4 = 2^2$ implies $\log_2 4 = \log_2 \left(2^2\right) = 2$.

3. $27 = 3^3 = (1/3)^{-3}$ implies $\log_{1/3} 27 = \log_{1/3}\left((1/3)^{-3}\right) = -3$.

5. $1/\pi^4 = \pi^{-4}$ implies $\log_\pi(1/\pi^4) = \log_\pi\left(\pi^{-4}\right) = -4$; therefore $\pi \log_\pi(1/\pi^4) = -4\pi$.

7. $\ln\left(e^5\right) = \log_e\left(e^5\right) = 5$.

9. $2 = \sqrt{4} = 4^{1/2} = (1/4)^{-1/2}$ implies $\log_{1/4} 2 = \log_{1/4}(1/4)^{-1/2} = -1/2$.

REMARK: If $y = \log_{1/4} 2$, then $(1/4)^y = 2$. Because $1/4 = 4^{-1} = 2^{-2}$, this implies $2^1 = 2 = (1/4)^y = (2^{-2})^y = 2^{-2y}$. Therefore $1 = -2y$ which then implies $-1/2 = y$.

---

11. $36 = 6^2$ and $1/5 = 1/\sqrt{25} = 25^{-1/2}$; thus $\log_6 36 \cdot \log_{25}(1/5) = 2 \cdot (-1/2) = -1$.

13.   If $2^{x^2} = 64 = 2^6$, then $x^2 = \log_2\left(2^{x^2}\right) = \log_2 2^6 = 6$. This implies $x = \pm\sqrt{6}$.

15.   If $b > 0$ and $x > 0$, then $b^{\log_b x} = x$; thus $y = e^{\ln\sqrt{2}} = e^{\log_e\sqrt{2}} = \sqrt{2}$.

17. $e^{\ln e^\pi} = e^{\log_e e^\pi} = e^\pi$.

REMARK: Suppose $b > 0$; either of two rules justifies the above result:

a. $b^{\log_b w} = w$ for any positive w,

b. $\log_b b^x = x$ for any real x.

---

19. $\log_x 64 = 3$ implies $64 = x^3$ which then implies $x = \sqrt[3]{64} = \sqrt[3]{2^6} = 2^{6/3} = 2^2 = 4$.

21. $\log_x 32 = -5$ implies $32 = x^{-5}$ which then implies $x = \sqrt[5]{1/32} = 1 \left/ \sqrt[5]{2^5}\right. = 1/2$.

23. $10^{-23} = \log x = \log_{10} x$ implies $10^{10^{-23}} = x$.

25. $\log x = 4\log(1/2) - 3\log(1/3) = \log(1/2)^4 + \log(1/3)^{-3} = \log\left((1/2)^4 \cdot (1/3)^{-3}\right)$ implies $x = (1/2)^4 \cdot (1/3)^{-3} = (1/16) \cdot 27 = 27/16$.

27. $\log x = \log 1 + \log 2 + \log 3 + \log 4 = \log(1 \cdot 2 \cdot 3 \cdot 4)$ implies $x = 1 \cdot 2 \cdot 3 \cdot 4 = 4! = 24$.

29. $\log x^3 = 2\log 5 - 3\log 2 = \log 5^2 - \log 2^3 = \log\left(5^2/2^3\right)$ implies $x^3 = 5^2/2^3 = 25/8$ which then implies $x = \sqrt[3]{25}/2$.

31. $\ln x^2 - \ln x = \ln(x^2/x) = \ln x$ and $\ln 18 - \ln 6 = \ln(18/6) = \ln 3$; thus $\ln x^2 - \ln x = \ln 18 - \ln 6$ implies $\ln x = \ln 3$ which then implies $x = 3$.

33. $\log\left(x^3 \cdot \sqrt{x}\right) = \log\left(x^3 \cdot x^{1/2}\right) = \log x^{3+1/2} = \log x^{7/2} = (7/2)\log x$.

REMARK: $\log\left(x^3 \cdot \sqrt{x}\right) = \log x^3 + \log\sqrt{x} = 3\log x + (1/2)\log x = (3 + 1/2)\log x$.

---

35. $\log\left(x^5/(1+x)^{18}\right) = \log x^5 - \log(1+x)^{18} = 5\log x - 18\log(1+x)$.

37. $y = -e^x$

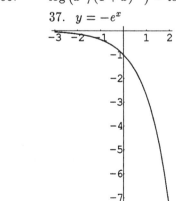

39. $y = e^{x-2}$

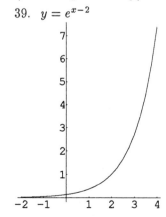

41. $y = 3 - e^{2-x}$

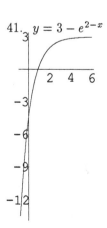

43. $y = \ln(x+2)$

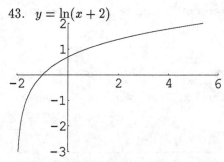

45. $y = 1 + \ln(3-x)$

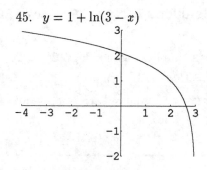

47. a. $\text{pH} = -\log\left(4.2 \cdot 10^{-6}\right) = 6 - \log 4.2 \approx 5.37675$.

    b. $\text{pH} = -\log\left(8 \cdot 10^{-6}\right) = 6 - \log 8 \approx 5.09691$.

    c. $\text{pH} = -\log\left(0.6 \cdot 10^{-7}\right) = 7 - \log 0.6 \approx 7.22185$.

49. a. Suppose the intensity I_2 of one sound is twice the intensity I_1 of another, then $I_2 = 2I_1$. Because the perceived loudness L equals $10\log\left(I/I_0\right)$, we compute

$$L_2 = 10\log\left(\frac{I_2}{I_0}\right) = 10\log\left(\frac{2I_1}{I_0}\right) = 10\left(\log 2 + \log\left(\frac{I_1}{I_0}\right)\right) = 10\log 2 + L_1.$$

Therefore $\dfrac{L_2}{L_1} = \dfrac{10\log 2}{L_1} + 1$.

    b. If $L_2 = 2L_1$, then $10\log\left(\dfrac{I_2}{I_0}\right) = 2 \cdot 10\log\left(\dfrac{I_1}{I_0}\right) = 10\log\left(\dfrac{I_1}{I_0}\right)^2$. Thus $\dfrac{I_2}{I_0} = \left(\dfrac{I_1}{I_0}\right)^2$ which implies

$I_2 = \dfrac{I_1^2}{I_0}$, and $\dfrac{I_2}{I_1} = \dfrac{I_1}{I_0} = 10^{12}I_1$ because $I_0 = 10^{-12}$.

    c. $L_c = 6L_w$ implies $\log\left(\dfrac{I_c}{I_0}\right) = 6\log\left(\dfrac{I_w}{I_0}\right) = \log\left(\dfrac{I_w}{I_0}\right)^6$ and $\dfrac{I_c}{I_0} = \left(\dfrac{I_w}{I_0}\right)^6$. Therefore $I_c = \dfrac{I_w^6}{I_0^5}$ and

$\dfrac{I_c}{I_w} = \dfrac{I_w^5}{I_0^5} = 10^{60}I_w^5$.

51. Suppose $b > 1$. Let $f(x) = (1+x)^b - (1+bx)$. f is continuous on $[-1,\infty)$ and differentiable on $(-1,\infty)$; therefore, if $A > -1$, and $A \neq 0$ then f is differentiable on the interval with endpoints A and 0, $[A,0]$ or $[0,A]$. The Mean Value Theorem for Derivatives (Theorem 3.7.2) guarantees there is some c in that interval such that $f'(c) = \dfrac{f(A) - f(0)}{A - 0}$. Because $f(0) = 0$ and $f'(x) = b(1+x)^{b-1} - b = b[(1+x)^{b-1} - 1]$, this means $f(A) = Af'(c) = Ab\left((1+c)^{b-1} - 1\right)$. If $A > 0$, then $c > 0$, $1+c > 1$, $(1+c)^{b-1} > 1$, and $Ab\left((1+c)^{b-1} - 1\right)$ is the product of three positive terms; hence, $A > 0$ implies $f(A) > 0$. On the other hand, if $-1 < A < 0$, then $-1 < A < c < 0$, $0 < (1+c)^{b-1} < 1$, $(1+c)^{b-1} - 1 < 0$, and $A\left((1+c)^{b-1} - 1\right) > 0$ because it is the product of two negative terms; hence, $-1 < A < 0$ implies $f(A) > 0$. In summary, $x > -1$ and $x \neq 0$ implies $f(x) > 0$; $x > -1$ implies $f(x) \geq 0$. ∎

53. If $n > 0$, then $\dfrac{1}{n} > 0$ and $T_n = \left(1 + \dfrac{1}{n}\right)^{n+1} = \left(1 + \dfrac{1}{n}\right) \cdot \left(1 + \dfrac{1}{n}\right)^n = \left(1 + \dfrac{1}{n}\right) \cdot S_n > 1 \cdot S_n = S_n$.

Furthermore, $\displaystyle\lim_{n\to\infty} \dfrac{T_n}{S_n} = \lim_{n\to\infty}\left(1 + \dfrac{1}{n}\right) = 1 + 0 = 1$. ∎

55.
$$T_n = \left(1 + \frac{1}{n}\right)^{n+1} = \left(\frac{n+1}{n}\right)^{n+1} \implies \frac{1}{T_n} = \left(\frac{n}{n+1}\right)^{n+1} = \left(1 - \frac{1}{n+1}\right)^{n+1}$$

$$T_{n-1} = \left(1 + \frac{1}{n-1}\right)^{(n-1)+1} = \left(\frac{n}{n-1}\right)^n \implies \frac{1}{T_{n-1}} = \left(\frac{n-1}{n}\right)^n = \left(1 - \frac{1}{n}\right)^n.$$

Apply Bernoulli's inequality (Problem 52) with $b = \dfrac{n+1}{n}$ and $x = \dfrac{-1}{n+1}$ to infer

$$\left(\frac{1}{T_n}\right)^{1/n} = \left(1 - \frac{1}{n+1}\right)^{(n+1)/n} > 1 + \left(\frac{n+1}{n}\right)\left(\frac{-1}{n+1}\right) = 1 - \frac{1}{n} = \left(\frac{1}{T_{n-1}}\right)^{1/n}.$$

Since T_k is always positive, we see that

$$\left(\frac{1}{T_n}\right)^{1/n} > \left(\frac{1}{T_{n-1}}\right)^{1/n} > 0 \implies \frac{1}{T_n} > \frac{1}{T_{n-1}} > 0 \implies T_{n-1} > T_n > 0.$$

Therefore T_n is a decreasing function of n. ∎

57.   The transition from step (a) to step (b) would be okay if we could cite a theorem of the form:

If $0 < A < B$ and $\alpha < \beta$, then $A\alpha < B\beta$.

No such luck. It's easily shown to be false with a single counterexample: let $A = 1$, $B = 2$, $\alpha = -5$, $\beta = -3$; then $A\alpha = -5 > -6 = B\beta$. If $B < 1$, then $\log B < 0$ and $\log A < \log B$ does not imply $A \log A < B \log B$.

Solutions 6.3   *The natural logarithm function*   (pages 418-9)

1. $\ln|x^7| - 2\ln|x| = \ln|x|^7 - 2\ln|x| = 7\ln|x| - 2\ln|x| = 5\ln|x|.$

3. $\ln\left(x^3 \cdot \sqrt{x}\right) = \ln\left(x^{3+(1/2)}\right) = (3 + (1/2))\ln x = (7/2)\ln x.$

REMARK: The domain of this function must be contained in $[0, \infty)$ because $\sqrt{x}$ is defined only if $x \geq 0$, therefore we do not need to write the answer as $(7/2)\ln|x|$. (The domain actually is $(0, \infty)$.)

5. $\ln\left(\sqrt[3]{x^2+3} \cdot \left(x^5 - 9\right)^{1/8}\right) = \ln\left(x^2+3\right)^{1/3} + \ln\left(x^5 - 9\right)^{1/8} = (1/3)\ln\left(x^2 + 3\right) + (1/8)\ln\left(x^5 - 9\right).$

7.   The domain of $\ln x$ is $(0, \infty)$; $\dfrac{d}{dx}\ln x = \dfrac{1}{x}.$

9. $1 + 2x > 0 \iff x > -1/2$, the domain of $\ln(1+2x)$ is $(-1/2, \infty)$;
$\dfrac{d}{dx}\ln(1+2x) = \dfrac{1}{1+2x} \cdot \dfrac{d}{dx}(1 + 2x) = \dfrac{2}{1+2x}.$

11. $(x+3)(x-2) > 0 \iff x < -3$ or $x > 2$, the domain of $\ln((x+3)(x-2))$ is $(-\infty, -3) \cup (2, \infty)$;
$\dfrac{d}{dx}\ln((x+3)(x-2)) = \dfrac{d}{dx}(\ln(x+3) + \ln(x-2)) = \dfrac{1}{x+3} + \dfrac{1}{x-2}.$

13. $\dfrac{x+1}{x}$ is defined provided $x \neq 0$, $0 < \left|\dfrac{x+1}{x}\right| = |x+1| \cdot \left|\dfrac{1}{x}\right| \implies 0 \neq x+1 \implies -1 \neq x$, the domain of
$\ln\left|\dfrac{x+1}{x}\right|$ is $\mathbb{R} - \{-1, 0\}$; $\dfrac{d}{dx}\ln\left|\dfrac{x+1}{x}\right| = \dfrac{d}{dx}\ln\dfrac{|x+1|}{|x|} = \dfrac{d}{dx}(\ln|x+1| - \ln|x|) = \dfrac{1}{x+1} - \dfrac{1}{x}.$

15. $\ln(\sin x)$ is defined if and only if $\sin x > 0$, the domain of $\ln(\sin x)$ is the union of the open intervals
$(2k\pi, (2k+1)\pi)$ where k is an integer. $\dfrac{d}{dx}\ln(\sin x) = \dfrac{1}{\sin x} \cdot \dfrac{d}{dx}\sin x = \dfrac{\cos x}{\sin x} = \cot x.$

17. $x \cdot \ln x$ and $\ln x$ have the same domain, $(0, \infty)$;
$\dfrac{d}{dx}(x \cdot \ln x) = \left(\dfrac{d}{dx}x\right) \cdot (\ln x) + (x) \cdot \left(\dfrac{d}{dx}\ln x\right) = 1 \cdot \ln x + x \cdot \dfrac{1}{x} = \ln x + 1 = 1 + \ln x.$

19. $\dfrac{x+1}{x-1}$ is defined and nonzero provided $x \notin \{1, -1\}$, the domain of $\ln\left|\dfrac{x+1}{x-1}\right|$ is $\mathbb{R} - \{1, -1\}$;
$\dfrac{d}{dx}\ln\left|\dfrac{x+1}{x-1}\right| = \dfrac{d}{dx}\ln\dfrac{|x+1|}{|x-1|} = \dfrac{d}{dx}(\ln|x+1| - \ln|x-1|) = \dfrac{1}{x+1} - \dfrac{1}{x-1}.$

21.
$$y = \sqrt[5]{\frac{x^3-3}{x^2+1}} \implies \ln y = \ln\left(\frac{x^3-3}{x^2+1}\right)^{1/5} = \frac{1}{5}\ln\left(\frac{x^3-3}{x^2+1}\right) = \frac{1}{5}\left(\ln\left(x^3-3\right) - \ln\left(x^2+1\right)\right)$$
$$\implies \frac{1}{y} \cdot \frac{dy}{dx} = \frac{1}{5} \cdot \left(\frac{3x^2}{x^3-3} - \frac{2x}{x^2+1}\right)$$
$$\implies \frac{dy}{dx} = \left(\frac{1}{5}\right) \cdot \left(\frac{3x^2}{x^3-3} - \frac{2x}{x^2+1}\right) \cdot y = \left(\frac{1}{5}\right) \cdot \left(\frac{3x^2}{x^3-3} - \frac{2x}{x^2+1}\right) \cdot \left(\frac{x^3-3}{x^2+1}\right)^{1/5}$$
$$= \left(\frac{1}{5}\right)\left(x^4 + 3x^2 + 6x\right)\left(x^3-3\right)^{-4/5}\left(x^2+1\right)^{-6/5}.$$

23. $y = \sqrt{x} \cdot \sqrt[3]{x+2} \cdot \sqrt[5]{x-1} = x^{1/2} \cdot (x+2)^{1/3} \cdot (x-1)^{1/5}$

$\implies \ln y = \frac{1}{2} \ln x + \frac{1}{3} \ln(x+2) + \frac{1}{5} \ln(x-1)$

$\implies \frac{1}{y} \cdot \frac{dy}{dx} = \frac{1}{2x} + \frac{1}{3(x+2)} + \frac{1}{5(x-1)}$

$\implies \frac{dy}{dx} = \left(\frac{1}{2x} + \frac{1}{3(x+2)} + \frac{1}{5(x-1)} \right) \cdot y = \left(\frac{31x^2 + 17x - 30}{30} \right) \cdot x^{-1/2} \cdot (x+2)^{-2/3} \cdot (x-1)^{-4/5}.$

25. Let $u = 1 + x^2$, then $du = 2x \, dx$ and $\int \frac{2x}{1+x^2} \, dx = \int \frac{1}{u} \, du = \ln|u| + C = \ln(1+x^2) + C.$

27. Let $u = 7 + x^3$, then $du = 3x^2 \, dx$ and $\int \frac{x^2}{7+x^3} \, dx = \frac{1}{3} \int \frac{1}{u} \, du = \frac{1}{3} \ln|u| + C = \frac{1}{3} \ln|7+x^3| + C.$

29. $\int \left(\frac{1}{x-1} + \frac{1}{x+1} \right) dx = \int \frac{1}{x-1} \, dx + \int \frac{1}{x+1} \, dx = \ln|x-1| + \ln|x+1| + C = \ln|x^2 - 1| + C.$

REMARK: Let $u = x^2 - 1$, then $du = 2x \, dx$ and

$$\int \left(\frac{1}{x-1} + \frac{1}{x+1} \right) dx = \int \frac{2x \, dx}{x^2 - 1} = \int \frac{du}{u} = \ln|u| + C = \ln|x^2 - 1| + C.$$

---

31. We set $u(x) = x^4$ in (8); if $x > 0$, then $\frac{d}{dx} \int_1^{x^4} \frac{1}{t} \, dt = \frac{u'(x)}{u(x)} = \frac{4x^3}{x^4} = \frac{4}{x}.$

REMARK: $\frac{d}{dx} \int_1^{x^4} \frac{1}{t} \, dt = \frac{d}{dx} \ln x^4 = \frac{d}{dx} 4 \ln x = 4 \cdot \frac{1}{x}.$

---

33. We set $u(x) = \cos^2 x$ in (8); if $0 < x < \pi/2$, then $u(x) > 0$ and we compute

$$\frac{d}{dx} \int_1^{\cos^2 x} \frac{1}{t} \, dt = \frac{u'(x)}{u(x)} = \frac{-2 \cos x \sin x}{\cos^2 x} = \frac{-2 \sin x}{\cos x} = -2 \tan x.$$

REMARK: $\frac{d}{dx} \int_1^{\cos^2 x} \frac{1}{t} \, dt = \frac{d}{dx} \ln \cos^2 x = \frac{d}{dx} 2 \ln \cos x = 2 \cdot \frac{1}{\cos x} \cdot (-\sin x).$

---

35. If $x = 0.8$, then $A = \frac{0.8 - 1}{0.8 + 1} = \frac{-1}{9}$ and

$$\ln 0.8 \approx \left[\left(\frac{3(-1/9)^2}{5} + 1 \right) \cdot \frac{(-1/9)^2}{3} + 1 \right] \cdot 2(-1/9) = \frac{-65882}{295245} \approx -0.223143491.$$

If $x = 1.2$, then $A = \frac{1.2 - 1}{1.2 + 1} = \frac{1}{11}$ and

$$\ln 1.2 \approx \left[\left(\frac{3(1/11)^2}{5} + 1 \right) \cdot \frac{(1/11)^2}{3} + 1 \right] \cdot 2(1/11) = \frac{440446}{2415765} \approx 0.182321542.$$

REMARK: A pocket calculator with a $\boxed{\ln}$ key yields $\ln 0.8 \approx -0.223143551$ and $\ln 1.2 \approx 0.182321557.$

---

37. $f(x) = \ln|x| \implies f'(x) = \left(\frac{1}{|x|} \right) \cdot \frac{x}{|x|} = \frac{1}{x} \implies f''(x) = \frac{-1}{x^2}$; f decreases on $(-\infty, 0)$ and increases on $(0, \infty)$, the graph is always concave down, there are no extrema or points of inflection.

37. $y = \ln|x|$

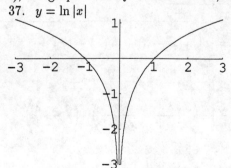

39. $y = x \ln x - x$

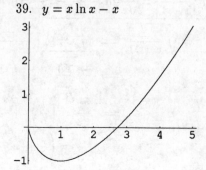

39. $f(x) = x \ln x - x \implies f'(x) = \ln x \implies f''(x) = \dfrac{1}{x}$; f decreases on $(0,1)$ and increases on $(1, \infty)$, $f(1) = -1$ is the global minimum, there is no maximum nor are there points of inflection.

41. $\displaystyle\int_4^{16} \dfrac{3}{x-1}\, dx = 3 \ln|x-1| \Big|_4^{16} = 3(\ln 15 - \ln 3) = 3 \ln\left(\dfrac{15}{3}\right) = 3 \ln 5 = \ln 5^3 = \ln 125.$

43. $\displaystyle\int_2^4 \dfrac{\ln x}{x}\, dx = \dfrac{1}{2}(\ln x)^2 \Big|_2^4 = \dfrac{1}{2}\left((\ln 4)^2 - (\ln 2)^2\right) = \dfrac{1}{2}\left((2\ln 2)^2 - (\ln 2)^2\right) = \dfrac{3}{2}(\ln 2)^2.$

45.   If $x > 0$, then $\displaystyle\int_{-1}^{-x} \dfrac{1}{t}\, dt = \ln|t| \Big|_{-1}^{-x} = \ln|-x| - \ln|-1| = \ln x - \ln 1 = \ln x - 0 = \ln x.$

47. $\displaystyle\sum_{k=nq+1}^{np} \dfrac{1}{k} = \dfrac{1}{nq+1} + \dfrac{1}{nq+2} + \cdots + \dfrac{1}{np-1} + \dfrac{1}{np} = \left(\dfrac{1}{q+1/n} + \dfrac{1}{q+2/n} + \cdots + \dfrac{1}{p-1/n} + \dfrac{1}{p}\right)\left(\dfrac{1}{n}\right)$

is a Riemann sum to approximate $\int_q^p x^{-1}\, dx$ using a partition with $\Delta x = 1/n$. Therefore

$$\lim_{n \to \infty} \sum_{k=nq+1}^{np} \dfrac{1}{k} = \int_q^p \dfrac{1}{x}\, dx = \ln p - \ln q = \ln\left(\dfrac{p}{q}\right). \ \blacksquare$$

49.   Suppose there exist polynomial functions p and q such that $\ln x = \dfrac{p(x)}{q(x)}$. If so, then for any integer k,

we have the identity $k \cdot \dfrac{p(x)}{q(x)} = k \cdot \ln x = \ln x^k = \dfrac{p\left(x^k\right)}{q\left(x^k\right)}$ which implies $k \cdot p(x) \cdot q\left(x^k\right) = p\left(x^k\right) \cdot q(x)$. By examining the coefficients of the highest-order terms in this last equation, we obtain a contradiction.

If $p(x) = \sum_{i=0}^m a_i x^i$ with $a_m \neq 0$ and $q(x) = \sum_{j=0}^n b_j x^j$ with $b_n \neq 0$, then $k a_m b_n x^{m+kn}$ is the highest-order term of $k \cdot p(x) \cdot q\left(x^k\right)$ while $a_m b_n x^{km+n}$ is the highest-order term of $p\left(x^k\right) \cdot q(x)$. The requirement that these have the same degree implies $m + kn = km + n$ which holds for all k if and only if $m = n$. However, even if we have $m = n$, it is impossible for $k a_m b_n = a_m b_n$ to be true for all k. This contradiction to the assumption of a rational-function (i.e., quotient of polynomials) expression for the logarithm function proves that it can not exist. $\blacksquare$

Solutions 6.4   *The exponential function e^x*   (pages 430-2)

1. $\ln e^5 = 5 \cdot \ln e = 5 \cdot 1 = 5.$

3. $\ln\left(\dfrac{1}{e^{2.718}}\right) = \ln e^{-2.718} = -2.718.$

5. $\ln x = \ln 1 + \ln 2 + \ln 3 + \ln 4 = \ln(1 \cdot 2 \cdot 3 \cdot 4) = \ln 24 \iff x = e^{\ln x} = e^{\ln 24} = 24.$

7. $\ln(x^2) - \ln x = \ln 18 - \ln 6 \iff \ln\left(\dfrac{x^2}{x}\right) = \ln\left(\dfrac{18}{6}\right) \iff \ln x = \ln 3 \iff x = e^{\ln x} = e^{\ln 3} = 3.$

9. $e^x = e^2 \cdot e^{-\pi} = e^{2-\pi} \iff x = \ln e^x = \ln e^{2-\pi} = 2 - \pi.$

11. $\dfrac{d}{dx}\left(e^x \cdot e^{x+1}\right) = \left(\dfrac{d}{dx} e^x\right) \cdot e^{x+1} + e^x \cdot \left(\dfrac{d}{dx} e^{x+1}\right) = e^x \cdot x^{x+1} + e^x \cdot e^{x+1} = 2 \cdot e^x \cdot e^{x+1} = 2\, e^{2x+1}.$

13. $\dfrac{d}{dx} \sqrt{e^x} = \dfrac{1}{2\sqrt{e^x}} \cdot \dfrac{d}{dx} e^x = \dfrac{e^x}{2\sqrt{e^x}} = \dfrac{\sqrt{e^x}}{2}.$

REMARK: $\dfrac{d}{dx} \sqrt{e^x} = \dfrac{d}{dx}(e^x)^{1/2} = \dfrac{d}{dx} e^{x/2} = e^{x/2} \dfrac{d}{dx} \dfrac{x}{2} = \dfrac{1}{2} e^{x/2}.$

15. $\dfrac{d}{dx}\left(e^{-x} \cdot \cos x\right) = (-e^{-x}) \cdot \cos x + e^{-x} \cdot (-\sin x) = -e^{-x}(\cos x + \sin x).$

17. $\dfrac{d}{dx}\left(xe^x\right) = 1 \cdot e^x + x \cdot e^x = (1+x)e^x.$

19. $\dfrac{d}{dx}\left((x-1)e^x\right) = 1 \cdot e^x + (x-1) \cdot e^x = xe^x.$

21. $\dfrac{d}{dx} \ln(xe^x) = \dfrac{d}{dx}(\ln x + \ln e^x) = \dfrac{d}{dx}(\ln x + x) = \dfrac{1}{x} + 1.$

REMARK: $\dfrac{d}{dx} \ln(xe^x) = \dfrac{1}{xe^x} \cdot \dfrac{d}{dx}(xe^x) = \dfrac{1}{xe^x} \cdot (e^x + xe^x) = \dfrac{1+x}{x}.$

23. $\displaystyle\int_1^{\ln 4} e^x \, dx = e^x \Big|_1^{\ln 4} = e^{\ln 4} - e^1 = 4 - e.$

25. $\displaystyle\int e^{3x} \, dx = \frac{1}{3} e^{3x} + C.$

27. $\displaystyle\int e^x \cdot e^{x+1} \, dx = \int e^{2x+1} \, dx = \frac{1}{2} e^{2x+1} + C.$

29. $\displaystyle\int (\sin e^x)(e^x \, dx) \overset{e^x \equiv u}{=} \int \sin u \, du = -\cos u + C = -\cos e^x + C.$

31. $\displaystyle\int \left(\frac{e^{1/x}}{x^2}\right) dx = \int \left(-e^{1/x}\right)\left(\frac{-1}{x^2} \cdot dx\right) \overset{1/x \equiv u}{=} \int -e^u \, du = -e^u + C = -e^{1/x} + C.$

33. $\displaystyle\int_1^4 \left(\frac{e^{\sqrt{x}}}{\sqrt{x}}\right) dx = \int_{x=1}^4 2e^{\sqrt{x}}\left(\frac{1}{2\sqrt{x}} \, dx\right) \overset{\sqrt{x} \equiv u}{=} \int_{u=1}^2 2e^u \, du = 2e^u \Big|_1^2 = 2\left(e^2 - e\right).$

35. $\displaystyle\int xe^{x^2}\left(5 + e^{x^2}\right)^3 dx = \int \frac{1}{2}\left(5 + e^{x^2}\right)^3 \left(2xe^{x^2} \, dx\right) \overset{5+e^{x^2} \equiv u}{=} \int \frac{u^3}{2} \, du = \frac{u^4}{8} + C = \frac{1}{8}\left(5 + e^{x^2}\right)^4 + C.$

37. The expression written in the text is a fifth-degree polynomial which is factored in such a way that computation can be done without using special keys or memory registers. The basic computational step is to multiply the result of a previous step by x, divide by a particular number, add 1; begin by entering 1. The following table shows the results of successive steps for two different values of x. (Read top to bottom, then think about the numbering scheme S_6, S_5, S_4, etc.)

Step	Computation	(a) $x = 0.13$	(b) $x = -0.37$
S_6	enter 1	1	1
S_5	$S_6 \cdot x/5 + 1$	1.026	0.926
S_4	$S_5 \cdot x/4 + 1$	1.033345	0.914345
S_3	$S_4 \cdot x/3 + 1$	1.044778283	0.8872307833
S_2	$S_3 \cdot x/2 + 1$	1.067910588	0.8358623051
S_1	$S_2 \cdot x/1 + 1$	1.138828376	0.6907309471

Thus $e^{0.13} \approx 1.138828$ and $e^{-0.37} \approx 0.690731.$

REMARK: This procedure of nesting a polynomial for easier computation is commonly called Horner's Rule (although it was used earlier by Newton and others).

c. $e^{0.13} - 1.138828376 \approx 0.000000007$; $e^{-0.37} - 0.6907309471 \approx 0.000003384$

39. $f(x) = e^{2x+3} \implies f'(x) = 2e^{2x+3} \implies f''(x) = 4e^{2x+3}$; f always increases and the graph is always concave up, there are no extrema or points of inflection.

39. $y = e^{2x+3}$

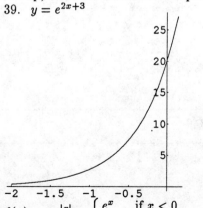

41. $y = e^{-|x|}$

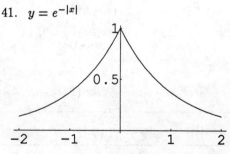

41. $f(x) = e^{-|x|} = \begin{cases} e^x & \text{if } x < 0 \\ e^{-x} & \text{if } 0 \le x \end{cases} \implies f'(x) = -e^{-|x|} \cdot \frac{x}{|x|} = \begin{cases} e^x & \text{if } x < 0 \\ -e^{-x} & \text{if } 0 \le x \end{cases} \implies f''(x) = e^{-|x|};$

f increases on $(-\infty, 0)$ and decreases on $(0, \infty)$, $f'(0)$ is undefined and $f(0) = 1$ is the global maximum, there is no minimum, the graph is always concave up.

43. If $f(x) = e^{-x}\cos x$, then $f'(x) = -e^{-x}(\cos x + \sin x)$ which is 0 if $x = \dfrac{-\pi}{4} + k\pi$. Furthermore, $f''(x) = 2e^{-x}\sin x$ which is 0 if $x = k\pi$. Since $f''(x) < 0$ if $x = \dfrac{-\pi}{4} + 2k\pi$, f has local maxima for those values of x; similarly, f has local minima when $x = \dfrac{3\pi}{4} + 2k\pi$ because $f'' > 0$ there. The points $\left(k\pi, (-1)^k e^{-k\pi}\right)$ are inflection points for f because $\sin$, and hence f'', changes sign there.

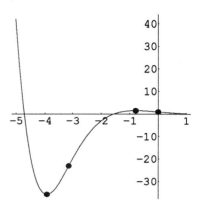

REMARK: The extrema are not where the graph of f touches the graph of $y = \pm e^{-x}$.

45. $\displaystyle\int_0^1 (5 - e^x)\,dx = (5x - e^x)\big|_0^1 = (5 - e) - (0 - 1) = 6 - e.$

47. $\displaystyle\int_{-1}^1 (e^{2x} - 2x)\,dx = \left(\tfrac{1}{2}e^{2x} - x^2\right)\Big|_{-1}^1 = \left(\dfrac{e^2}{2} - 1\right) - \left(\dfrac{e^{-2}}{2} - 1\right) = \dfrac{e^2 - e^{-2}}{2}.$

49. $s(t) = 30 + 3t + 0.01t^2 + \ln t + e^{-2t^2} \implies v(t) = \dfrac{d}{dt}\,s(t) = 3 + 0.02t + \dfrac{1}{t} - 4te^{-2t^2} \implies a(t) = \dfrac{d}{dt}\,v(t) = 0.02 - \dfrac{1}{t^2} - 4\left(1 - 4t^2\right)e^{-2t^2}$; therefore $a(10) = 0.02 - 0.01 + 1596e^{-200} \approx 0.01 + 2.2 \cdot 10^{-84} \approx 0.01$ m/min^2.

51. $t > 0 \implies e^{-t} < e^{-0} = 1$ since e^{-x} is a decreasing function. If f is a positive function, then $t > 0 \implies e^{-t} \cdot f(t) < 1 \cdot f(t) = f(t)$. If $0 < A$, we apply the comparison theorem for integrals (Theorem 4.5.7) to infer $\int_0^A e^{-t}f(t)\,dt < \int_0^A f(t)\,dt$. ∎

53. Suppose $b > 1$. Let $f(x) = (1+x)^b - (1+bx)$. f is continuous on $[-1, \infty)$ and differentiable on $(-1, \infty)$; therefore, if $A > -1$, and $A \neq 0$ then f is differentiable on the interval with endpoints A and 0, $[A, 0]$ or $[0, A]$. The Mean Value Theorem for Derivatives (Theorem 3.7.2) guarantees there is some c in that interval such that $f'(c) = \dfrac{f(A) - f(0)}{A - 0}$. Because $f(0) = 0$ and $f'(x) = b(1+x)^{b-1} - b = b[(1+x)^{b-1} - 1]$, this means $f(A) = Af'(c) = Ab\left((1+c)^{b-1} - 1\right)$. If $A > 0$, then $c > 0$, $1 + c > 1$, $(1+c)^{b-1} > 1$, and $Ab\left((1+c)^{b-1} - 1\right)$ is the product of three positive terms; hence, $A > 0$ implies $f(A) > 0$. On the other hand, if $-1 < A < 0$, then $-1 < A < c < 0$, $0 < (1+c)^{b-1} < 1$, $(1+c)^{b-1} - 1 < 0$, and $A\left((1+c)^{b-1} - 1\right) > 0$ because it is the product of two negative terms; hence, $-1 < A < 0$ implies $f(A) > 0$. In summary, $x > -1$ and $x \neq 0$ implies $f(x) > 0$; $x > -1$ implies $f(x) \geq 0$. ∎

55. If $n > 0$, then $\dfrac{1}{n} > 0$ and $T_n = \left(1 + \dfrac{1}{n}\right)^{n+1} = \left(1 + \dfrac{1}{n}\right) \cdot \left(1 + \dfrac{1}{n}\right)^n = \left(1 + \dfrac{1}{n}\right) \cdot S_n > 1 \cdot S_n = S_n$. Furthermore, $\displaystyle\lim_{n\to\infty} \dfrac{T_n}{S_n} = \lim_{n\to\infty}\left(1 + \dfrac{1}{n}\right) = 1 + 0 = 1$. ∎

57.
$$T_n = \left(1 + \dfrac{1}{n}\right)^{n+1} = \left(\dfrac{n+1}{n}\right)^{n+1} \implies \dfrac{1}{T_n} = \left(\dfrac{n}{n+1}\right)^{n+1} = \left(1 - \dfrac{1}{n+1}\right)^{n+1}$$
$$T_{n-1} = \left(1 + \dfrac{1}{n-1}\right)^{(n-1)+1} = \left(\dfrac{n}{n-1}\right)^n \implies \dfrac{1}{T_{n-1}} = \left(\dfrac{n-1}{n}\right)^n = \left(1 - \dfrac{1}{n}\right)^n.$$

Apply Bernoulli's inequality (Problem 54) with $b = \dfrac{n+1}{n}$ and $x = \dfrac{-1}{n+1}$ to infer

$$\left(\dfrac{1}{T_n}\right)^{1/n} = \left(1 - \dfrac{1}{n+1}\right)^{(n+1)/n} > 1 + \left(\dfrac{n+1}{n}\right)\left(\dfrac{-1}{n+1}\right) = 1 - \dfrac{1}{n} = \left(\dfrac{1}{T_{n-1}}\right)^{1/n}.$$

Since T_k is always positive, we see that

$$\left(\dfrac{1}{T_n}\right)^{1/n} > \left(\dfrac{1}{T_{n-1}}\right)^{1/n} > 0 \implies \dfrac{1}{T_n} > \dfrac{1}{T_{n-1}} > 0 \implies T_{n-1} > T_n > 0.$$

Therefore T_n is a decreasing function of n. ∎

59. Using the functional equation $E(U + V) = E(U)E(V)$ in the special case $U = O = V$, we obtain $E(0) = E(0+0) = E(0)E(0) = E(0)^2$; therefore $E(0) = 0$ or $E(0) = 1$. If $E(0) = 0$, then $E(x) = E(x+0) =$

6.4 — The exponential function e^x

$E(x)E(0) = E(x)0 = 0$ for all x, i.e., E is the constant function always equal to 0. In the remaining cases, if there really are any, $E(0) = 1$. Now use the functional equation in the definition of the derivative:

$$E'(x) = \lim_{\Delta x \to 0} \frac{E(x + \Delta x) - E(x)}{\Delta x} = \lim_{\Delta x \to 0} \frac{E(x)E(\Delta x) - E(x)}{\Delta x}$$

$$= \lim_{\Delta x \to 0} \left[\frac{E(\Delta x) - 1}{\Delta x} \cdot E(x) \right] = \left[\lim_{\Delta x \to 0} \frac{E(\Delta x) - E(0)}{\Delta x} \right] E(x) = E'(0) \cdot E(x).$$

If we knew $E(x) \neq 0$, we could then get $E'(0) = \dfrac{E'(x)}{E(x)} = \dfrac{d}{dx} \ln |E(x)|$; using a trick, we can bypass that assumption. Rewrite the differential equation as $0 = E'(x) - E'(0)E(x)$ and multiply by the function $e^{-\alpha x}$ where $\alpha = E'(0)$. Thus $0 = E'(x)e^{-\alpha x} + E(x)(-\alpha e^{-\alpha x}) = \frac{d}{dx}\left(E(x)e^{-\alpha x}\right)$ which implies $E(x)e^{-\alpha x} = C$ for some constant C, thus $E(x) = Ce^{\alpha x} = E(0)e^{\alpha x} = e^{\alpha x} = e^{E'(0)x}$. ∎

REMARK: This type of argument exploiting the functional equation $E(s+t) = E(s) \cdot E(t)$ is part of the theory of continuous semigroups.

Solutions 6.5   *The functions a^x and $\log_a x$*   (pages 437-8)

1. $\log 10000 = \log_{10} 10^4 = 4$.

3. $\log_7 \left(\frac{1}{7}\right) = \log_7 7^{-1} = -1$.

5. $\log_{1/2} 32 = \log_{1/2} 2^5 = \log_{1/2} \left(\frac{1}{2}\right)^{-5} = -5$.

7. $\log_{25} 5 = \log_{25} \sqrt{25} = \log_{25} 25^{1/2} = \frac{1}{2}$.

9. $\log_a a \cdot \log_b b^2 \cdot \log_c c^3 = 1 \cdot 2 \cdot 3 = 6$.

11. $10^{-23} = \log x = \log_{10} x \implies x = 10^{\log_{10} x} = 10^{10^{-23}}$.

REMARK: 10^{-23} is close to zero, therefore $10^{10^{-23}}$ is close to $10^0 = 1$.

13. $3 = \log_x 64 \iff x^3 = x^{\log_x 64} = 64 = 4^3 \iff x = 4$.

15. $\log x = 3\log 2 + 2\log 3 = \log 2^3 + \log 3^2 = \log\left(2^3 \cdot 3^2\right) = \log 72 \iff x = 72$.

17. $\log x^3 = 2\log 5 - 3\log 2 = \log 5^2 + \log 2^{-3} = \log\left(5^2 \cdot 2^{-3}\right) \iff x^3 = 5^2 \cdot 2^{-3} \iff x = 5^{2/3} \cdot 2^{-3/3} = \dfrac{5^{2/3}}{2}$.

19. $\dfrac{d}{dx} 5^x = 5^x \cdot \ln 5$.

21. $\dfrac{d}{dx} 2^{3x} = \left(2^{3x} \cdot \ln 2\right) \cdot 3 = 8^x \cdot \ln 8$.

23. $\dfrac{d}{dx} \log(100\,x) = \dfrac{1}{100x\,\ln 10} \cdot 100 = \dfrac{1}{x\,\ln 10} = \dfrac{\log e}{x}$.

25. $\dfrac{d}{dx} \log_\pi(3 - 2x) = \dfrac{1}{(3-2x)\ln \pi} \cdot (-2) = \dfrac{-2}{(3-2x)\ln \pi} = \dfrac{-2\log_\pi e}{3 - 2x}$.

27. $y = x^{\sqrt{x}} \implies \ln y = \sqrt{x} \cdot \ln x \implies \dfrac{1}{y} \cdot \dfrac{dy}{dx} = \dfrac{1}{2\sqrt{x}} \cdot \ln x + \sqrt{x} \cdot \dfrac{1}{x} = \dfrac{\ln x + 2}{2\sqrt{x}}$

$$\implies \dfrac{dy}{dx} = \dfrac{\ln x + 2}{2\sqrt{x}} \cdot y = \dfrac{\ln x + 2}{2\sqrt{x}} \cdot x^{\sqrt{x}} = \dfrac{1}{2}\left(\ln x + 2\right) x^{\sqrt{x} - 1/2}.$$

29. $\displaystyle\int \left(\frac{1}{2}\right)^x dx = \dfrac{\left(\frac{1}{2}\right)^x}{\ln\left(\frac{1}{2}\right)} + C = \dfrac{2^{-x}}{-\ln 2} + C = \dfrac{-2^{-x}}{\ln 2} + C$.

31. $\int \pi^x \, dx = \dfrac{\pi^x}{\ln \pi} + C$.

33. $\log_5 40 = (\ln 40)/(\ln 5) \approx 2.2920296742$.

35. $\log_2 0.1524 = (\ln 0.1524)/(\ln 2) \approx -2.7140651921$.

37. $\log_\pi 2 = (\ln 2)/(\ln \pi) \approx 0.6055115614$.

39. $\log_\pi 1.285 = (\ln 1.285)/(\ln \pi) \approx 0.2190549242$.

41. $\log_{1/3} 26 = (\ln 26)/(\ln 1/3) \approx -2.965647273$.

43. $\log 999^{1000} = 1000 \log 99 = 1000 \log \left((10^2)(9.99) \right) = 1000 \left(\log 10^2 + \log 9.99 \right) \approx 1000(2 + 0.999565) = 2999.565$ but $\log 1000^{999} = 999 \log 1000 = 999 \log 10^3 = 999(3) = 2997$. Because $\log x$ and its inverse, 10^x, are increasing functions and because $\log 999^{1000} > \log 1000^{999}$, we infer $999^{1000} > 1000^{999}$.

45. $y = \pi^{x-1} = \left(e^{\ln \pi} \right)^{x-1} = \dfrac{e^{(\ln \pi)x}}{\pi}$.

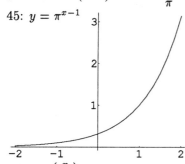

45: $y = \pi^{x-1}$

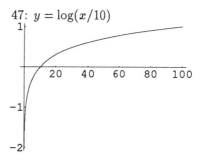

47: $y = \log(x/10)$

47. $y = \log \left(\dfrac{x}{10} \right) = \log x - \log 10 = \dfrac{\ln x}{\ln 10} - 1$.

49.   a.   Suppose the intensity I_2 of one sound is twice the intensity I_1 of another, then $I_2 = 2I_1$. Because the perceived loudness L equals $10 \log (I/I_0)$, we compute

$$L_2 = 10 \log \left(\frac{I_2}{I_0} \right) = 10 \log \left(\frac{2I_1}{I_0} \right) = 10 \left(\log 2 + \log \left(\frac{I_1}{I_0} \right) \right) = 10 \log 2 + L_1.$$

Therefore $\dfrac{L_2}{L_1} = \dfrac{10 \log 2}{L_1} + 1$.

   b.   If $L_2 = 2L_1$, then $10 \log \left(\dfrac{I_2}{I_0} \right) = 2 \cdot 10 \log \left(\dfrac{I_1}{I_0} \right) = 10 \log \left(\dfrac{I_1}{I_0} \right)^2$. Thus $\dfrac{I_2}{I_0} = \left(\dfrac{I_1}{I_0} \right)^2$ which implies $I_2 = \dfrac{I_1^2}{I_0}$, and $\dfrac{I_2}{I_1} = \dfrac{I_1}{I_0} = 10^{12} I_1$ because $I_0 = 10^{-12}$.

   c. $L_c = 6L_w$ implies $\log \left(\dfrac{I_c}{I_0} \right) = 6 \log \left(\dfrac{I_w}{I_0} \right) = \log \left(\dfrac{I_w}{I_0} \right)^6$ and $\dfrac{I_c}{I_0} = \left(\dfrac{I_w}{I_0} \right)^6$. Therefore $I_c = \dfrac{I_w^6}{I_0^5}$ and $\dfrac{I_c}{I_w} = \dfrac{I_w^5}{I_0^5} = 10^{60} I_w^5$.

51.   Let $f(x) = 2^x$. Then $f'(x) = (\ln 2)f(x)$ and $f''(x) = (\ln 2)^2 f(x$. The graph of f is concave up because $f'' > 0$ everywhere. Therefore the graph $y = f(x)$ lies below the chord connecting any two points $(a, f(a))$ and $(b, f(b))$. The chord between $(0, 2^0) = (0, 1)$ and $(1, 2^1) = (1, 2)$ has equation $y = 1 + x$. Thus $2^x = f(x) < 1 + x$ if $0 < x < 1$. ∎

REMARK: An alternative tactic is to modify Bernoulli's inequality (see Problem 6.4.54). Suppose $0 < \alpha < 1$. Let $G(x) = (1 + \alpha x) - (1 + x)^\alpha$ with domain $[-1, \infty)$. $G'(x) = \alpha \left(1 - (1 + x)^{\alpha-1} \right)$ and $G''(x) = \alpha(1 - \alpha)(1 + x)^{\alpha-2}$. Because $G'' > 0$ on $(-1, \infty)$, $G'(0) = 0$, and $G(-1) = 1 - \alpha > 0 = G(0)$, we infer $G(0)$ is the global minimum of G. Also since $G'' > 0$, $G'(x) < G'(0) = 0$ if $x < 0$ and $0 = G'(0) < G'(x)$ if $x > 0$; thus $G(x) > G(0)$ if $x \neq 0$. Therefore $1 + \alpha x > (1 + x)^\alpha$ if $x \neq 0$. Now choose $x = 1$ to infer $1 + \alpha > 2^\alpha$; this is true for any α in $(0, 1)$. ∎

53. $$A'(x) = \lim_{\Delta x \to 0} \frac{A(x + \Delta x) - A(x)}{\Delta x} = \lim_{\Delta x \to 0} \frac{a^{x+\Delta x} - a^x}{\Delta x} = \lim_{\Delta x \to 0} \frac{a^x a^{\Delta x} - a^x}{\Delta x} = \lim_{\Delta x \to 0} \left[\frac{a^{\Delta x} - 1}{\Delta x} \right] a^x$$

$$= \left[\lim_{\Delta x \to 0} \frac{a^{\Delta x} - a^0}{\Delta x} \right] a^x = \left[\lim_{\Delta x \to 0} \frac{A(\Delta x) - A(0)}{\Delta x - 0} \right] A(x) = A'(0) \cdot A(x). ∎$$

REMARK: A is differentiable everywhere provided it is differentiable at 0. This sort of stuff is part of the theory of continuous semigroups. (Also see the solution to Problem 6.3.59.)

55.   Let $f(x) = \dfrac{e^x}{x^e}$.

   a. $f(1) = e^1/1^e = e/1 = e > 1$; $f(e) = e^e/e^e = 1$. The fact that $e > 2$ implies $e^{10} > 2^{10} = 1024$, the fact that $e < 3$ implies $10^e < 10^3 = 1000$; therefore $f(10) = e^{10}/10^e > 1024/1000 > 1$. ∎

b. f is continuous on $[1, 10]$, therefore it does reach its minimum value somewhere on that closed interval. Part (a) shows that minimum is not attained at one of the ends of the interval. ∎

c. $f'(x) = \dfrac{d}{dx}\left(e^x \cdot x^{-e}\right) = e^x \cdot x^{-e} + e^x \cdot (-e)x^{-e-1} = f(x)\cdot\left(1 - \dfrac{e}{x}\right).$ ∎

d. $f'(x)$ is negative on $(0, e)$, zero at $x = e$, positive on (e, ∞); $f''(x) = f(x)\cdot\left(\left(1 - \dfrac{e}{x}\right)^2 + \dfrac{e}{x^2}\right) > 0$;

therefore $f(e) = 1$ is the global minimum value of f. ∎

e. $1 = f(e) < f(\pi) = \dfrac{e^\pi}{\pi^e} \implies \pi^e < e^\pi.$ ∎

f.   Since f is strictly decreasing on $(0, e)$ and strictly increasing on (e, ∞), we see that $0 < x$ and $x \neq$

    $e \implies 1 = f(e) < f(x) = \dfrac{e^x}{x^e} \implies x^e < e^x.$ ∎

57.   Suppose that x and y are positive real numbers such that $x + y = 1$. Let $z = x^x + y^y$. Then

$$\frac{dz}{dx} = x^x(1 + \ln x) + y^y(1 + \ln y)\frac{dy}{dx} = x^x(1 + \ln x) - y^y(1 + \ln y)$$

because $x + y = 1$ implies $\dfrac{dy}{dx} = -1$. Although it is clear that $\dfrac{dz}{dx} = 0$ when $x = y = \dfrac{1}{2}$, it's not immediately obvious whether other critical points exist. Onward — let's examine the second derivative:

$$\frac{d^2 z}{dx^2} = x^x(1 + \ln x)^2 + x^x\left(0 + \frac{1}{x}\right) - y^y(1 + \ln y)^2\frac{dy}{dx} - y^y\left(0 + \frac{1}{y}\right)\frac{dy}{dx}$$

$$= x^x(1 + \ln x)^2 + x^{x-1} + y^y(1 + \ln y)^2 + y^{y-1}.$$

$\dfrac{d^2 z}{dx^2} > 0$ because each of the terms is always positive. Therefore $\dfrac{dz}{dx}$ is strictly increasing and has at most one zero, the graph of z is concave up and has its global minimum when $x = 1/2 = y$. That minimum is $\sqrt{1/2} + \sqrt{1/2} = \sqrt{2}$; furthermore, $x^x + y^y > \sqrt{2}$ unless $x = 1/2 = y$. ∎

REMARK: This is problem 803 from *Mathematics Magazine*, Vol. 46 (1973), p 238.

Solutions 6.6   *Differential equations of exponential growth and decay*   (pages 450-2)

1.   See Example 2. $\frac{dP}{dt} = \alpha P$ implies $P(t) = Ce^{\alpha t} = P(0)e^{\alpha t}$. We're told that $P(0) = 10000$ and $P(10) = 25000$. This implies $25000 = P(10) = 10000e^{10\alpha}$, $2.5 = P(10)/P(0) = e^{10\alpha}$, and $\alpha = (\ln 2.5)/10 \approx 0.091629$. Therefore $P(20) = 10000e^{20\alpha} = 10000\left(e^{10\alpha}\right)^2 = 10000(2.5)^2 = 62500$ and $P(30) = 10000e^{30\alpha} = 10000\left(e^{10\alpha}\right)^3 = 10000(2.5)^3 = 156250$.

REMARK: It was sufficient to find $e^{10\alpha} = 2.5$, computing an explicit value for α was not required.

3.   Let $P(t)$ be the population size after t hours, $P(t) = P(0)e^{\alpha t}$. If the mathematical model is valid for all t, then the "best" numerical estimate for α uses the longest time interval:

$$\alpha \approx \left(\frac{1}{20 - 5}\right)\ln\left(\frac{P(20)}{P(5)}\right) = \frac{1}{15}\ln\left(\frac{11986}{936}\right) \approx 0.16999.$$

a. $P(0) \approx P(5)/e^{5\alpha} \approx 936/e^{0.84996} \approx 400.076 \approx 400$ individuals.

b. $P(60\text{days}) = P(20)e^{40\alpha} \approx 11986e^{6.79968} \approx 10758128.79 \approx 10,758,129$ individuals.

REMARK: Using the 5 and 10 day data to estimate the model parameter yields $\alpha \approx 0.17001$, $P(0) \approx 400.044$, and $P(60) \approx 10,767,795$; using the 10 and 20 day data to fit the model yields $\alpha \approx 0.16998$, $P(0) \approx 400.142$, and $P(60) \approx 10,754,616$.

5.   See Examples 3 and 4. Let $T(t)$ be the temperature of the object at time t. Then $T(t) = 70 + (170 - 70)e^{-\alpha t}$. We are told that $T(0.5) = 140$, therefore $e^{-\alpha/2} = (140 - 70)/(170 - 70) = 0.7$ and $\alpha = -2\ln\left(\dfrac{140 - 70}{170 - 70}\right) = -2\ln 0.7 \approx 0.7133$.

a. $T(1\,\text{hr}) = 70 + 100\,e^{-\alpha} = 70 + 100\left(e^{-\alpha/2}\right)^2 = 70 + 100\,(0.7)^2 = 70 + 49 = 119°\text{F}.$

b. $90° = T(t) = 70 + 100\,e^{-\alpha t} \iff e^{-\alpha t} = \dfrac{90 - 70}{100} = 0.2 \iff t = \dfrac{\ln 0.2}{-\alpha} \approx 2.256\,\text{hr}.$

REMARK: $e^{-\alpha/2} = 0.7$ implies $T(t) = 70 + 100\,(0.7)^{2t}$.

7. See Example 5. Let $M(t)$ be the amount of ^{14}C in the fossil t years after death. Then $M(t) = M(0)e^{-\alpha t}$ where $\alpha = (\ln 2)/5580$. If $M(t) = 0.7\,M(0)$, then $0.7 = e^{-\alpha t}$ and $t = \dfrac{\ln 0.7}{-\alpha} = -5580\dfrac{\ln 0.7}{\ln 2} \approx 2871.3$ years.

9. See Examples 6-9.

   a. 8 years of simple 7% interest on \$5000 yields total interest of $(\$5000)(8)(0.07) = \$2800$.

   b. After 8 years of 7% interest on \$5000 compounded annually,
   the investment is worth $(\$5000)\,(1 + 0.07)^8 \approx (\$5000)(1.718) \approx \$8590.93$.

   c. If 7% per year interest on \$5000 is compounded monthly for 8 years,
   the investment is worth $(\$5000)\left(1 + \dfrac{0.07}{12}\right)^{(12)(8)} \approx (\$5000)(1.748) \approx \$8739.13$.

   d. If 7% per year interest on \$5000 is compounded continuously for 8 years,
   the investment is worth $(\$5000)\,e^{(8)(0.07)} \approx (\$5000)(1.751) \approx \$8753.36$.

11. See Example 12.

   a. $3A_0 = A_0 e^{15i} \iff i = \dfrac{\ln 3}{15} \approx 0.07324 \approx 7.324\%$.

   b. $2A_0 = A_0 e^{10i} \iff i = \dfrac{\ln 2}{10} \approx 0.06931 \approx 6.931\%$.

13. Let a be the altitude, measured in meters, above sea level and $P(a)$ be the atmospheric pressure, measured in millibars. $P(0) = 1013.25$ mbar and $\dfrac{dP}{da} = \beta P$ implies $P(a) = 1013.25\,e^{\beta a}$. If $P(1500) = 845.6$, then $\beta = \left(\dfrac{1}{1500}\right)\ln\left(\dfrac{845.6}{1013.25}\right) \approx -1.2058 \times 10^{-4}$.

   a. $P(4000) = 1013.25\,e^{4000\beta} \approx 1013.25\,e^{-0.4823} \approx 625.53$ mbar.

   b. $P(10\text{ km}) = P(10000\text{ m}) \approx 1013.25\,e^{-1.2058} \approx 303.42$ mbar.

   c. $P(\text{MtWhitney}) - P(\text{DeathValley}) = P(4418) - P(-86) \approx 1013.25\left(e^{-0.5327} - e^{0.0104}\right) \approx -429.03$ mbar

   d. $P(8848) \approx 1013.25\,e^{-1.0669} \approx 348.63$ mbar.

   e. $1\text{ mbar} = P(a) = 1013.25\,e^{\beta a}$ if and only if $a = \dfrac{1}{\beta}\ln\left(\dfrac{1}{1013.25}\right) = -1500\dfrac{\ln 1013.25}{\ln(845.6/1013.25)} \approx 57396.3\text{ m} \approx 57.4$ km.

15. Let $M(t)$ be the mass of the salt t hours after it is put into water. $\dfrac{dM}{dt} = -\alpha M$ implies $M(t) = M(0)e^{-\alpha t}$. We're told that $M(0) = 25$ kg and $M(10) = 15$ kg, thus $e^{-10\alpha} = \dfrac{15}{25}$ and $\alpha = \dfrac{-1}{10}\ln\left(\dfrac{15}{25}\right) = \dfrac{1}{10}\ln\left(\dfrac{5}{3}\right) \approx 0.05108$.

   a. $M(24) = 25\exp\left(\dfrac{-24}{10}\ln\left(\dfrac{5}{3}\right)\right) = 25\left(\dfrac{5}{3}\right)^{-24/10} = 25(0.6)^{2.4} \approx 7.3367$ kg.

   b. $\dfrac{1}{2} > M(t) = 25e^{-\alpha t} \iff e^{\alpha t} > 50 \iff \alpha t > \ln 50 \iff t > \dfrac{\ln 50}{\alpha} = 10\dfrac{\ln 50}{\ln(5/3)} \approx 76.58$ hours.

17. The estimated population in the year t is $P(t) = 4.845 \times 10^9 \times 1.01^{t-1986}$. Hence $P(t) \geq 8 \times 10^9 \iff 1.01^{t-1986} \geq \dfrac{8}{4.845} \iff t \geq 1986 + \dfrac{\ln(8/4.845)}{\ln 1.01} \approx 1986 + 50.4 \approx 2036$.

19. If we fit the model $P(t) = P_{1970}\,(1+r)^{t-1970}$ to the data $P_{1970} = 203302031$ and $P(1980) = 226549448$, then $(1+r)^{1980-1970} = \dfrac{226549448}{203302031} \iff r = \left(\dfrac{226549448}{203302031}\right)^{1/10} - 1 \approx 0.0108859$.

   a. $P(2000) = P_{1970}\,(1+r)^{30} = P_{1970}\left(\dfrac{P(1980)}{P_{1970}}\right)^3 \approx P_{1970} \cdot 1.384 \approx 281,323,227$.

   b. $P(t) \geq 5 \times 10^8 \iff (1+r)^{t-1970} \geq \dfrac{5 \times 10^8}{P_{1970}} \iff t \geq 1970 + \dfrac{\ln\left(5 \times 10^8/P_{1970}\right)}{\ln(1+r)} \approx 1970 + 83.12 \approx 2053$.

21. $213.4 = 100e^{(1920-1913)r} = 100e^{7r} \iff r = \dfrac{\ln 2.134}{7} \approx 0.108285 \approx 10.83\%$ per year.

6.6 — Differential equations of exponential growth and decay

23. $116.3 = 88.7e^{(1970-1960)r} = 88.7e^{10r} \iff r = \dfrac{\ln(116.3/88.7)}{10} \approx 0.027091 \approx 2.71\%$ per year.

25. Choose a thermometer scale where the room temperature is 0 and the temperature of the coffee is 1. Let the unit of time be 10 minutes and the unit of volume be the coffee cup, let $-c$ be the temperature of the cream and v its volume. The temperatures are as follows:

time	0	0^+	1^-	1
President	1	$\dfrac{1-cv}{1+v}$	$\dfrac{1-cv}{1+v}e^{-\alpha}$	$\dfrac{1-cv}{1+v}e^{-\alpha} = T_1$
Vice-President	1	1	$e^{-\alpha}$	$\dfrac{e^{-\alpha}-cv}{1+v} = T_2$

$(T_1 - T_2)(1+v)e^\alpha = (1-cv) - (1-cve^\alpha) = cv(e^\alpha - 1) > 0 \implies T_1 > T_2$; the President's coffee is hotter.

Solutions 6.7   *Integration of trigonometric functions*   (pages 454-6)

1. Let $u = 2x$, then $du = 2\,dx$ and
$$\int \cos 2x\,dx = \frac{1}{2}\int (\cos 2x)(2\,dx) = \frac{1}{2}\int \cos u\,du = \frac{1}{2}(\sin u) + C = \frac{1}{2}\sin 2x + C.$$

3. Let $u = 2x$, then $du = 2\,dx$ and $\displaystyle\int 3\cos 2x\,dx = \frac{3}{2}\int \cos u\,du = \frac{3}{2}\sin u + C = \frac{3}{2}\sin 2x + C.$ Therefore
$$\int_0^{\pi/12} 3\cos 2x\,dx = \frac{3}{2}\sin 2x\Big|_0^{\pi/12} = \frac{3}{2}\sin\left(\frac{\pi}{6}\right) - \frac{3}{2}\sin 0 = \frac{3}{2}\cdot\frac{1}{2} - 0 = \frac{3}{4}.$$

5. $\displaystyle\int e^{\cos x}\sin x\,dx = -\int e^{\cos x}\,d(\cos x) = -e^{\cos x} + C.$

7. Let $u = 1 + \sin x$, then $du = \cos x\,dx$ and $\displaystyle\int \frac{\cos x}{1 + \sin x}\,dx = \int \frac{1}{u}\,du = \ln|u| + C = \ln|1 + \sin x| + C.$

9. $\displaystyle\int_{x=0}^{3\pi/4} \tan\left(\frac{x}{3}\right)dx \overset{x/3\equiv u}{=} \int_{u=0}^{\pi/4} \tan u\,(3\,du) = -3\ln|\cos u|\big|_0^{\pi/4} = -3\left(\ln\frac{1}{\sqrt{2}} - \ln 1\right) = \frac{3}{2}\ln 2.$

11. $\displaystyle\int \cot(2-x)\,dx = \int \frac{\cos(2-x)\,dx}{\sin(2-x)} = \int \frac{(-1)\,d(\sin(2-x))}{\sin(2-x)} = (-1)\ln|\sin(2-x)| + C.$

13. Let $u = \sin 2x$, then $du = 2\cos 2x\,dx$ and
$$\int_0^{\pi/3} \sin 2x\,\cos 2x\,dx = \frac{1}{2}\int_{x=0}^{\pi/3} (\sin 2x)(2\cos 2x\,dx) = \frac{1}{2}\int_{u=0}^{\sqrt{3}/2} u\,du = \frac{u^2}{4}\Big|_0^{\sqrt{3}/2} = \frac{3/4}{4} - 0 = \frac{3}{16}.$$

$\boxed{\text{First alternative computation}}$ Let $V = \cos 2x$, then $dV = -2\sin 2x\,dx$ and
$$\int_{x=0}^{\pi/3} \sin 2x\,\cos 2x\,dx = \frac{-1}{2}\int_{V=1}^{-1/2} V\,dV = \frac{-V^2}{4}\Big|_1^{-1/2} = \frac{-1/4}{4} - \frac{-1}{4} = \frac{3}{16}.$$

$\boxed{\text{Second alternative computation}}$ Use the identity $2\sin\theta\cos\theta = \sin 2\theta$:
$$\int_0^{\pi/3} \sin 2x\,\cos 2x\,dx = \int_0^{\pi/3} \frac{1}{2}\sin 4x\,dx = \frac{1}{8}\int_0^{\pi/3} (\sin 4x)\,d(4x) = \frac{-\cos 4x}{8}\Big|_0^{\pi/3} = \frac{1/2}{8} - \frac{-1}{8} = \frac{3}{16}.$$

15. $\displaystyle\int \sin^2 x\,\cos x\,dx = \int \sin^2 x\,d(\sin x) = \frac{1}{3}\sin^3 x + C.$

17. $\displaystyle\int_0^{\pi/6} \sqrt{\sin x}\,\cos x\,dx = \int_0^{\pi/6} (\sin x)^{1/2}\,d(\sin x) = \frac{2}{3}(\sin x)^{3/2}\Big|_0^{\pi/6} = \frac{2}{3}\left(\frac{1}{2}\right)^{3/2} - 0 = \frac{1}{3\sqrt{2}}.$

19. $\displaystyle\int (\sin 2x)^{-1/3}\cos 2x\,dx = \frac{1}{2}\int (\sin 2x)^{-1/3}\,d(\sin 2x) = \frac{3}{4}(\sin 2x)^{2/3} + C.$

21. $\displaystyle\int \sec 4x\,\tan 4x\,dx \overset{4x\equiv u}{=} \frac{1}{4}\int \sec u\,\tan u\,du = \frac{1}{4}\sec u + C = \frac{\sec 4x}{4} + C.$

23. $\displaystyle\int_{x=0}^{\pi/4} 2\sec x\,(\sec x\,\tan x\,dx) \overset{\sec x \equiv u}{=} \int_{u=1}^{\sqrt{2}} 2u\,du = u^2\Big|_1^{\sqrt{2}} = 2 - 1 = 1.$

REMARK: $\displaystyle\int_{x=0}^{\pi/4} 2\tan x\,(\sec^2 x\,dx) \overset{\tan x \equiv V}{=} \int_{V=0}^{1} 2V\,dV = V^2\Big|_0^1 = 1 - 0 = 1.$

25. Let $u = x^2$, then $du = 2x\,dx$ and

$$\int x\cos x^2\,\csc x^2\,dx = \frac{1}{2}\int \csc u\,\cot u\,du = \frac{1}{2}(-\csc x) + C = \frac{-\csc x^2}{2} + C.$$

27. Let $u = \csc x$, then $du = -\csc x\,\cot x\,dx$ and

$$\int \left(\sqrt{\csc x}\,\tan x\right)^{-1} dx = \int (\csc x)^{-3/2}\,(\csc x\,\cot x\,dx) = -\int u^{-3/2}\,du = 2u^{-1/2} + C = \frac{2}{\sqrt{\csc x}} + C.$$

29. $\displaystyle\int_0^{\pi/3} 2\tan(x+\pi)\,(\sec^2(x+\pi)\,dx) = \int_0^{\pi/3} 2\tan(x+\pi)\,d\big(\tan(x+\pi)\big) = \tan^2(x+\pi)\Big|_0^{\pi/3} = 3 - 0 = 3.$

31. Let $u = \tan x$, then $du = \sec^2 x\,dx$ and

$$\int \sec^2 x\,\cot x\,dx = \int \left(\frac{1}{\tan x}\right)(\sec^2 x\,dx) = \int \left(\frac{1}{u}\right) du = \ln|u| + C = \ln|\tan x| + C.$$

33. Identity (xxi) of Appendix 1.2 is useful here.

$$\int \sin^2(x/2)\,dx = \int \frac{1 - \cos 2(x/2)}{2}\,dx = \int \frac{1 - \cos x}{2}\,dx = \frac{x - \sin x}{2} + C.$$

35. $\sin^2 x + \cos^2 x = 1 \implies 1 + \cot^2 x = \csc^2 x$, therefore $\displaystyle\int_{\pi/4}^{\pi/2} \frac{1 + \cot^2 x}{\csc^2 x}\,dx = \int_{\pi/4}^{\pi/2} 1\,dx = \frac{\pi}{2} - \frac{\pi}{4} = \frac{\pi}{4}.$

37. Let $u = \sec x + \tan x$, then $du = (\sec x\,\tan x + \sec^2 x)\,dx = \sec x(\sec x + \tan x)\,dx.$ Thus

$$\int \sec x\,dx = \int \sec x\left(\frac{\sec x + \tan x}{\sec x + \tan x}\right) dx = \int \left(\frac{1}{\sec x + \tan x}\right)(\sec x\,\tan x + \sec^2 x)\,dx$$

$$= \int \left(\frac{1}{u}\right) du = \ln|u| + C = \ln|\sec x + \tan x| + C.$$

39. a. Let $u = 1 + \sin\theta$, then $du = \cos\theta\,d\theta$ and $\displaystyle\int \frac{\cos\theta}{1 + \sin\theta}\,d\theta = \int \frac{1}{u}\,du = \ln|1 + \sin\theta| + C_1.$

b. Let $V = \cos\theta$, then $dV = -\sin\theta\,d\theta$ and $\displaystyle\int \frac{\sin\theta}{\cos\theta}\,d\theta = \int \frac{-1}{V}\,dV = -\ln|V| + C_2 = -\ln|\cos\theta| + C_2.$

c. $\sin^2\theta + \cos^2\theta = 1$ implies $\cos^2\theta = 1 - \sin^2\theta = (1 - \sin\theta)(1 + \sin\theta)$ and $\dfrac{\cos\theta}{1 + \sin\theta} = \dfrac{1 - \sin\theta}{\cos\theta}$ (provided we're not dividing by zero). ∎

$$\sec\theta = \frac{1}{\cos\theta} = \frac{1 - \sin\theta + \sin\theta}{\cos\theta} = \frac{1 - \sin\theta}{\cos\theta} + \frac{\sin\theta}{\cos\theta} = \frac{\cos\theta}{1 + \sin\theta} + \frac{\sin\theta}{\cos\theta},\ \text{therefore}$$

$$\int \sec\theta\,d\theta = \int \frac{\cos\theta}{1 + \sin\theta}\,d\theta + \int \frac{\sin\theta}{\cos\theta}\,d\theta = \ln|1 + \sin\theta| - \ln|\cos\theta| + C_3$$

$$= \ln\left|\frac{1 + \sin\theta}{\cos\theta}\right| + C_3 = \ln\left|\frac{1}{\cos\theta} + \frac{\sin\theta}{\cos\theta}\right| + C_3 = \ln|\sec\theta + \tan\theta| + C_3.$$

REMARK: Compare this work with the solution for Problem 37.

41. $\displaystyle\text{Area} = \int_{\pi/4}^{\pi/3} \cot x\,dx = \ln|\sin x|\,\Big|_{\pi/4}^{\pi/3} = \ln\left(\frac{\sqrt{3}}{2}\right) - \ln\left(\frac{1}{\sqrt{2}}\right) = \ln\sqrt{\frac{3}{2}}.$

43. $\displaystyle\text{Area} = \int_{\pi/4}^{\pi/2} \csc^2 x\,dx = -\cot x\,\Big|_{\pi/4}^{\pi/2} = 0 - (-1) = 1.$

45. $\displaystyle\text{Area} = \int_{-3\pi/2}^{3\pi/2} \cos^2(x/3)\,dx = \int_{-3\pi/2}^{3\pi/2} \frac{1 + \cos(2x/3)}{2}\,dx = \left(\frac{x}{2} + \frac{3}{4}\sin\left(\frac{2x}{3}\right)\right)\Big|_{-3\pi/2}^{3\pi/2}$

$$= \left(\frac{3\pi/2}{2} - 0\right) - \left(\frac{-3\pi/2}{2} - 0\right) = \frac{3\pi}{2}.$$

6.7 — Integration of trigonometric functions

47.   a. $\dfrac{d}{dx} e^x \sin x = e^x \sin x + e^x \cos x = e^x (\sin x + \cos x).$

     b. $\dfrac{d}{dx} e^x \cos x = e^x \cos x + e^x (-\sin x) = e^x (-\sin x + \cos x).$

     c. $\dfrac{d}{dx} e^x (\sin x + \cos x) = \dfrac{d}{dx} e^x \sin x + \dfrac{d}{dx} e^x \cos x = e^x (\sin x + \cos x) + e^x (-\sin x + \cos x) = 2e^x \cos x.$

     d. $\dfrac{d}{dx} e^x (\sin x - \cos x) = \dfrac{d}{dx} e^x \sin x - \dfrac{d}{dx} e^x \cos x = e^x (\sin x + \cos x) - e^x (-\sin x + \cos x) = 2e^x \sin x.$

     e. $\displaystyle\int e^x \sin x \, dx = \dfrac{1}{2} \int 2e^x \sin x \, dx = \dfrac{1}{2} e^x (\sin x - \cos x) + C$

     f. $\displaystyle\int e^x \cos x \, dx = \dfrac{1}{2} \int 2e^x \cos x \, dx = \dfrac{1}{2} e^x (\sin x + \cos x) + C.$

49.   Because $\sin^2 x + \cos^2 x = 1$ for all x, $\left(\dfrac{1}{2}\right) \sin^2 x + C = \left(\dfrac{1}{2}\right) (1 - \cos^2 x) + C = \left(\dfrac{-1}{2}\right) \cos^2 x + \left(C + \dfrac{1}{2}\right).$
$C + 1/2$ is an arbitrary constant because C is.

REMARK: Although the C does signify an arbitrary constant, the usual procedure in situation of this sort is to write C in one antiderivative and C_1 in the other. For example,

$$\int 3(x-1)^2 \, dx = (x-1)^3 + C \quad \text{and} \quad \int 3(x-1)^2 \, dx = \int (3x^2 - 6x + 3) \, dx = x^3 - 3x^2 + 3x + C_1.$$

---

51. $\dfrac{d}{dx}\left(e^{ax} \cdot \sin bx\right) = (ae^{ax}) \sin bx + e^{ax}(b \cos bx) = e^{ax} (a \sin bx + b \cos bx),$

     $\dfrac{d}{dx}\left(e^{ax} \cdot \cos bx\right) = (ae^{ax}) \cos bx + e^{ax}(-b \sin bx) = e^{ax} (a \cos bx - b \sin bx).$

Therefore

$\dfrac{d}{dx}\left(be^{ax} \sin bx + ae^{ax} \cos bx\right) = e^{ax}\left[(ab \sin bx + b^2 \cos bx) + (a^2 \cos bx - ab \sin bx)\right] = e^{ax}\left(a^2 + b^2\right) \cos bx,$

$\dfrac{d}{dx}\left(ae^{ax} \sin bx - be^{ax} \cos bx\right) = e^{ax}\left[(a^2 \sin bx + ab \cos bx) - (ab \cos bx - b^2 \sin bx)\right] = e^{ax}\left(a^2 + b^2\right) \sin bx.$

If we let $r = \sqrt{a^2 + b^2}$, then $\left(\dfrac{a}{r}, \dfrac{b}{r}\right)$ is a point on the unit circle; thus there is a unique θ in $[0, 2\pi)$ such that $\cos \theta = \dfrac{a}{r}$ and $\sin \theta = \dfrac{b}{r}$. Thus

$\displaystyle\int e^{ax} \cos bx \, dx = \dfrac{be^{ax} \sin bx + ae^{ax} \cos bx}{r^2} + C = \dfrac{e^{ax}}{r} (\sin bx \sin \theta + \cos bx \cos \theta) + C = \dfrac{e^{ax}}{r} \cos(bx - \theta) + C,$

$\displaystyle\int e^{ax} \sin bx \, dx = \dfrac{ae^{ax} \sin bx - be^{ax} \cos bx}{r^2} + C = \dfrac{e^{ax}}{r} (\sin bx \cos \theta - \cos bx \sin \theta) + C = \dfrac{e^{ax}}{r} \sin(bx - \theta) + C. \blacksquare$

Solutions 6.8   *The inverse trigonometric functions*   (pages 463-6)

1. $\sin(\pi/3) = \sqrt{3}/2$ and $\pi/3$ is in $[-\pi/2, \pi/2]$ implies $\pi/3 = \sin^{-1}\left(\sqrt{3}/2\right).$

3. $\sin(-\pi/6) = -1/2$ and $-\pi/6$ is in $[-\pi/2, \pi/2]$ implies $-\pi/6 = \sin^{-1}(-1/2).$

5. $\tan(\pi/6) = 1/\sqrt{3}$ and $\pi/6$ is in $(-\pi/2, \pi/2)$ implies $\pi/6 = \tan^{-1}\left(1/\sqrt{3}\right).$

7. $\tan(-\pi/6) = -1/\sqrt{3}$ and $-\pi/6$ is in $(-\pi/2, \pi/2)$ implies $-\pi/6 = \tan^{-1}\left(-1/\sqrt{3}\right).$

9. $\theta = \cos^{-1}(3/5) \implies \cos \theta = 3/5 \implies \cos^2 \theta = (3/5)^2 \implies \sin^2 \theta = 1 - \cos^2 \theta = 1 - (3/5)^2 = 16/25;$ since $\sin$ is non-negative on the range of $\cos^{-1}$, we see that $\sin\left(\cos^{-1}(3/5)\right) = \sin \theta = \sqrt{16/25} = 4/5.$

11.

$$\theta = \sin^{-1}\left(\frac{3}{5}\right) \implies \sin \theta = \frac{3}{5} \text{ and } \theta \in \left[\frac{-\pi}{2}, \frac{\pi}{2}\right] \implies \cos \theta = \sqrt{1 - \sin^2 \theta} = \sqrt{1 - \left(\frac{3}{5}\right)^2} = \sqrt{\frac{16}{25}} = \frac{4}{5}$$

$$\implies \tan\left(\sin^{-1}\left(\frac{3}{5}\right)\right) = \tan \theta = \frac{\sin \theta}{\cos \theta} = \frac{3/5}{4/5} = \frac{3}{4}.$$

13. $$\theta = \tan^{-1}(-5) \implies \tan\theta = -5 \text{ and } \theta \in \left(\frac{-\pi}{2}, \frac{\pi}{2}\right) \implies \sec\theta = \sqrt{1+\tan^2\theta} = \sqrt{1+(-5)^2} = \sqrt{26}$$

$$\implies \sin\left(\tan^{-1}(-5)\right) = \sin\theta = \frac{\tan\theta}{\sec\theta} = \frac{-5}{\sqrt{26}}.$$

15. $\theta = \cos^{-1}x \iff \cos\theta = x \text{ and } \theta \in [0,\pi] \implies \sin^2\theta = 1 - \cos^2\theta = 1 - x^2$. The sine function is non-negative on $[0,\pi]$, the range of $\cos^{-1}$, thus $\sin\left(\cos^{-1}x\right) = \sin\theta = \sqrt{\sin^2\theta} = \sqrt{1-x^2}$ for all $x \in [-1,1]$.

17. $$x \in (-1,1) \text{ and } \theta = \sin^{-1}x \iff \theta \in \left(\frac{-\pi}{2}, \frac{\pi}{2}\right) \text{ and } \sin\theta = x$$

$$\implies \cos\theta > 0 \text{ and } \cos\theta = \sqrt{1-\sin^2\theta} = \sqrt{1-x^2}$$

$$\implies \tan\left(\sin^{-1}x\right) = \tan\theta = \frac{\sin\theta}{\cos\theta} = \frac{x}{\sqrt{1-x^2}}.$$

REMARK: Because the tangent function is undefined at $\pm\pi/2 = \sin^{-1}(\pm 1)$, the preceding result applies only if $-1 < x < 1$.

---

19. $$\frac{d}{dx}\sin^{-1}(3x) = \frac{1}{\sqrt{1-(3x)^2}} \cdot \frac{d}{dx}(3x) = \frac{3}{\sqrt{1-9x^2}}.$$

21. $$\frac{d}{dx}\cos^{-1}(x-5) = \frac{-1}{\sqrt{1-(x-5)^2}} \cdot \frac{d}{dx}(x-5) = \frac{-1}{\sqrt{1-(x-5)^2}} = \frac{-1}{\sqrt{-24+10x-x^2}}.$$

23. $$\frac{d}{dx}\tan^{-1}\left(\frac{x}{2}\right) = \frac{1}{1+(x/2)^2} \cdot \frac{d}{dx}\left(\frac{x}{2}\right) = \frac{1/2}{1+(x/2)^2} = \frac{2}{4+x^2}.$$

25. $$\frac{d}{dx}\sin^{-1}\left(\sqrt{x}\right) = \frac{1}{\sqrt{1-\left(\sqrt{x}\right)^2}} \cdot \frac{d}{dx}\left(\sqrt{x}\right) = \frac{1}{\sqrt{1-x}} \cdot \frac{1}{2\sqrt{x}} = \frac{1}{2\sqrt{x-x^2}}.$$

27. $x \neq 0 \implies 1+x^2 > 1 \implies \sin^{-1}\left(1+x^2\right)$ is undefined, thus the domain of $\sin^{-1}\left(1+x^2\right)$ is the singleton set $\{0\}$ and the function is not differentiable.

29. $x^3 + 1 \in \text{dom}\cos^{-1} = [-1,1] \iff -1 \leq x^3 + 1 \leq 1 \iff -2 \leq x^3 \leq 0 \iff -2^{1/3} \leq x \leq 0$; on the interval $\left(-2^{1/3}, 0\right]$ we have

$$\frac{d}{dx}\cos^{-1}\left(x^3+1\right) = \frac{-1}{\sqrt{1-(x^3+1)^2}} \cdot \frac{d}{dx}\left(x^3+1\right) = \frac{-3x^2}{\sqrt{-x^6-2x^3}} = -3\sqrt{\frac{x^4}{-x^6-2x^3}} = -3\sqrt{\frac{-x}{x^3+2}}.$$

31. $$\frac{d}{dx}\left(\left(\sin^{-1}x\right) \cdot (\sin x)\right) = \left(\frac{1}{\sqrt{1-x^2}}\right) \cdot (\sin x) + \left(\sin^{-1}x\right) \cdot (\cos x).$$

33. $$\frac{d}{dx}\left(x\sqrt{1-x^2} + \sin^{-1}x\right) = 1 \cdot \sqrt{1-x^2} + x \cdot \frac{-x}{\sqrt{1-x^2}} + \frac{1}{\sqrt{1-x^2}} = \sqrt{1-x^2} + \frac{1-x^2}{\sqrt{1-x^2}} = 2\sqrt{1-x^2}.$$

35. $$\int_0^2 \frac{1}{1+x^2}\,dx = \tan^{-1}x\Big|_0^2 = \tan^{-1}2.$$

37. $$\int \frac{1}{x^2-2x+2}\,dx = \int \frac{1}{(x-1)^2+1}\,d(x-1) = \tan^{-1}(x-1) + C.$$

39. Let $u = e^{-x}/2$, then $du = -(e^{-x}/2)\,dx$ and

$$\int \frac{e^{-x}\,dx}{4+e^{-2x}} = \frac{1}{4}\int \frac{e^{-x}\,dx}{1+(e^{-x}/2)^2} = \frac{-1}{2}\int \frac{du}{1+u^2} = \left(\frac{-1}{2}\right)\tan^{-1}u + C = \left(\frac{-1}{2}\right)\tan^{-1}\left(\frac{e^{-x}}{2}\right) + C.$$

REMARK: $$\int \frac{e^{-x}\,dx}{4+e^{-2x}} = \int \frac{e^x\,dx}{4e^{2x}+1} = \frac{1}{2}\int \frac{d\left(2e^x\right)}{\left(2e^x\right)^2+1} = \frac{1}{2}\tan^{-1}\left(2e^x\right) + C_1.$$

(Problem 72 will help you reconcile these two answers.)

---

41. $$\int \frac{\cos x\,dx}{1+\sin^2 x} = \int \frac{d(\sin x)}{1+(\sin x)^2} = \tan^{-1}(\sin x) + C.$$

43. $$\int_0^{1/2} \frac{dx}{\sqrt{1-x^2}} = \sin^{-1}x\Big|_0^{1/2} = \frac{\pi}{6} - 0 = \frac{\pi}{6}.$$

45. $$\int \frac{dx}{\sqrt{1-9x^2}} = \frac{1}{3}\int \frac{d(3x)}{\sqrt{1-(3x)^2}} = \frac{1}{3}\sin^{-1}(3x) + C.$$

47. $$\int \frac{\sin x\, dx}{\sqrt{4 - \cos^2 x}} \overset{\cos x \equiv 2u}{=} \int \frac{-2\, du}{\sqrt{4 - 4u^2}} = \int \frac{-\, du}{\sqrt{1 - u^2}} = \cos^{-1} u + C = \cos^{-1}\left(\frac{\cos x}{2}\right) + C.$$

49. $$\int \frac{\cos^{-1} x}{\sqrt{1 - x^2}}\, dx = \int \cos^{-1} x \cdot \frac{dx}{\sqrt{1 - x^2}} \overset{\cos^{-1} x \equiv u}{=} \int u\,(-du) = \left(\frac{-1}{2}\right) u^2 + C = \left(\frac{-1}{2}\right)\left(\cos^{-1} x\right)^2 + C.$$

REMARK: It is incorrect to write $\cos^{-2} x$ instead of $\left(\cos^{-1} x\right)^2$.

---

51. $$\text{Area} = \int_0^1 \frac{1}{1 + x^2}\, dx = \tan^{-1} x \Big|_0^1 = \frac{\pi}{4} - 0 = \frac{\pi}{4}.$$

53. $$\text{Area} = \int_0^{1/2} \frac{1}{\sqrt{1 - x^2}}\, dx = \sin^{-1} x \Big|_0^{1/2} = \frac{\pi}{6} - 0 = \frac{\pi}{6}.$$

55. Let x be the distance from the boat to the base of the cliff, it is related to θ, the angle of depression of the viewer's line of sight from the horizontal, by the equation $\tan \theta = \dfrac{100}{x}$. Therefore $\theta = \tan^{-1}\left(\dfrac{100}{x}\right)$, and $\dfrac{d\theta}{dt} = \left(\dfrac{1}{1 + (100/x)^2}\right)\left(\dfrac{-100}{x^2}\right)\dfrac{dx}{dt} = \left(\dfrac{-100}{x^2 + 100}\right)\dfrac{dx}{dt}$. We're told $\dfrac{dx}{dt} = 30$ m/min, thus

$$\frac{d\theta}{dt}\Big|_{x=60} = \left(\frac{-100}{60^2 + 100^2}\right) \cdot 30 = \frac{-3000}{13600} \approx -0.22 \text{ radians/min} \approx -12.64 \text{ degrees/min}.$$

57. Let θ be the angle between the observer's line of sight to the rocket and a horizontal, let y be the altitude of the rocket. Then $y = 3 \tan \theta$. When $\theta = 45° = \pi/4$ radians, $\dfrac{d\theta}{dt} = 20$ degrees/sec $= \dfrac{\pi}{9}$ radians/sec. At that moment, the vertical speed of the rocket is $\dfrac{dy}{dt}\Big|_{\theta = \pi/4} = \left(3 \sec^2 \theta\, \dfrac{d\theta}{dt}\right)\Big|_{\theta = \pi/4} = 3\left(\sqrt{2}\right)^2\left(\dfrac{\pi}{9}\right) = \dfrac{2\pi}{3}$ km/sec.

59. $\sin^{-1} 0.78 \approx 0.8946658172$.

61. $\sin^{-1}\left(-\sqrt{3}/2\right) \approx -1.0471975512$.

63. $\cos^{-1}(\pi/4) \approx 0.667457216$

65. $\cos^{-1}(-0.93) \approx 2.7652091713$.

67. $\tan^{-1} 10^5 \approx 1.5707863268$.

69. If $\Psi = \sin^{-1} y$, then $\sin \Psi = y$ and $\Psi \in [-\pi/2, \pi/2]$. Therefore $\cos(\sin^{-1} y) = \cos \Psi = \sqrt{1 - \sin^2 \Psi}$ since $\cos \geq 0$ on $[-\pi/2, \pi/2]$ and $\cos(\sin^{-1} y) = \sqrt{1 - y^2}$ for all $y \in [-1, 1]$. The solution to Problem 15 shows that $\sin\left(\cos^{-1} x\right) = \sqrt{1 - x^2}$ for all $x \in [-1, 1]$. Therefore $\sin(\cos^{-1} w) = \sqrt{1 - w^2} = \cos(\sin^{-1} w)$ for all $w \in [-1, 1]$. ∎

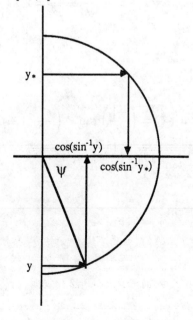

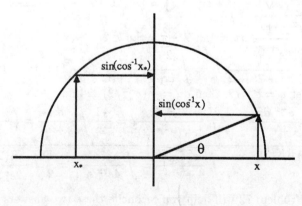

71.   Each of the functions $\sin^{-1} x$, $\cos^{-1} x$ is differentiable on $(-1,1)$; therefore so is their sum.

$$\frac{d}{dx}\left(\sin^{-1} x + \cos^{-1} x\right) = \frac{1}{\sqrt{1-x^2}} + \frac{-1}{\sqrt{1-x^2}} = 0.$$

The function $\sin^{-1} x + \cos^{-1} x$ must be constant on $(-1,1)$ because its derivative is zero throughout the interval. Pick a convenient x to compute that constant value:

$$\sin^{-1} 0 + \cos^{-1} 0 = 0 + \frac{\pi}{2}$$

$$\sin^{-1}\left(\frac{-1}{\sqrt{2}}\right) + \cos^{-1}\left(\frac{-1}{\sqrt{2}}\right) = \frac{-\pi}{4} + \frac{3\pi}{4}$$

$$\sin^{-1}\left(\frac{1}{2}\right) + \cos^{-1}\left(\frac{1}{2}\right) = \frac{\pi}{6} + \frac{\pi}{3}$$

$$\sin^{-1}\left(\frac{-\sqrt{3}}{2}\right) + \cos^{-1}\left(\frac{-\sqrt{3}}{2}\right) = \frac{-\pi}{3} + \frac{5\pi}{6}$$

(Since we know the function is constant, one of these computations is enough.) ∎

73.   If $x \neq 0$, then $\displaystyle\int \frac{dx}{\sqrt{a^2 - x^2}} = \int \frac{d(x/|a|)}{\sqrt{1 - (x/|a|)^2}} = \sin^{-1}\left(\frac{x}{|a|}\right) + C.$ ∎

75. $$\frac{d}{dx}\left(x \cdot \cos^{-1} x - \sqrt{1-x^2}\right) = 1 \cdot \cos^{-1} x + x \cdot \left(\frac{-1}{\sqrt{1-x^2}}\right) - \left(\frac{1}{2\sqrt{1-x^2}}\right)(-2x)$$

$$= \cos^{-1} x + \frac{-x}{\sqrt{1-x^2}} + \frac{x}{\sqrt{1-x^2}} = \cos^{-1} x. ∎$$

77. $f(x) = \sin^{-1}\left(\dfrac{x^2 - 8}{8}\right)$ and $g(x) = 2\sin^{-1}\left(\dfrac{x}{4}\right) - \dfrac{\pi}{2}.$

$f(x) = \sin^{-1}((x^2 - 8)/8)$

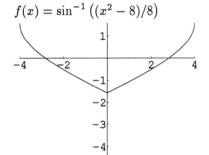

$g(x) = 2\sin^{-1}(x/4) - \pi/2$

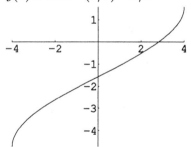

a. $f(0) = \sin^{-1}(-1) = \dfrac{-\pi}{2}$, $g(0) = 2\sin^{-1} 0 - \dfrac{\pi}{2} = 0 - \dfrac{\pi}{2} = \dfrac{-\pi}{2}.$

b. $f(x)$ is defined $\Longleftrightarrow -1 \leq \dfrac{x^2 - 8}{8} \leq 1 \Longleftrightarrow 0 \leq x^2 \leq 16 \Longleftrightarrow -4 \leq x \leq 4.$

       $g(x)$ is defined $\Longleftrightarrow -1 \leq \dfrac{x}{4} \leq 1 \Longleftrightarrow -4 \leq x \leq 4.$ Therefore $\operatorname{dom} f = [-4, 4] = \operatorname{dom} g.$

c. $f'(x) = \left(\dfrac{1}{\sqrt{1 - \left(\frac{x^2-8}{8}\right)^2}}\right)\left(\dfrac{2x}{8}\right) = \dfrac{2x}{\sqrt{8^2 - (x^2 - 8)^2}} = \dfrac{2x}{\sqrt{x^2(16 - x^2)}} = \left(\dfrac{x}{|x|}\right)\dfrac{2}{\sqrt{16 - x^2}}.$

       $g'(x) = \left(\dfrac{2}{\sqrt{1 - (x/4)^2}}\right)\left(\dfrac{1}{4}\right) = \dfrac{2}{\sqrt{16 - x^2}}.$

       $f'(x) = g'(x)$ on $(0, 4)$, $f'(x) = -g'(x)$ on $(-4, 0)$, $f'(0)$ does not exist, $g'(0) = 1/2 = \lim_{x \to 0+} f'(x).$

d. $f(x) = g(x)$ on $[0, 4]$ because $f(0) = -\pi/2 = g(0)$, $f(4) = \pi/2 = g(4)$, and $f'(x) = g'(x)$ on $(0, 4)$.
       $f(x) \neq g(x)$ on $[-4, 0)$ because f is an even function while g is a vertical translation of an odd function.

REMARK: This is an adaptation of problem Q464 from *Mathematics Magazine*, vol. 42 (1969), pp. 277
       and 243-244.

79.   Pick a real number r. The line through the origin and the point $(r, 1)$ meets the positive x-axis in an angle θ such that $0 < \theta < \pi$ and $\cot \theta = r$. θ is a function of r because θ is uniquely determined by the line from the origin to $(r, 1)$, and we may write $\theta = \cot^{-1} r$. Because r was arbitrary, the domain of $\cot^{-1}$ is $(-\infty, \infty)$; by definition the range of $\cot^{-1}$ is $(0, \pi)$.

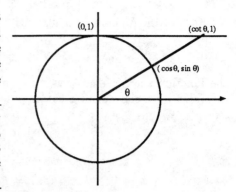

REMARK: Because $\frac{d}{dx} \cot x = -\csc^2 x \leq -1$, cotangent is always decreasing. We also know $\cot$ is continuous on $(0, \pi)$, $\lim_{x \to 0^+} \cot x = \infty$, and $\lim_{x \to \pi^-} \cot x = -\infty$. Therefore $\cot$ is 1–1 on $(0, \pi)$ with range $(-\infty, \infty)$; hence there is an inverse function with domain $(-\infty, \infty)$ and range $(0, \pi)$.

81.   Suppose $\cot^{-1} x = y$, then $x = \cot y$ and y is in $(0, \pi)$. Implicit differentiation produces

$$1 = \frac{d}{dx} x = \frac{d}{dx} \cot y = -\csc^2 y \frac{dy}{dx} = -\left(1 + \cot^2 y\right) \frac{dy}{dx} = -\left(1 + x^2\right) \frac{dy}{dx}.$$

Therefore $\dfrac{d}{dx} \cot^{-1} x = \dfrac{dy}{dx} = \dfrac{-1}{1 + x^2}$ for all real x. ∎

83. $\dfrac{d}{dx} \left(\tan^{-1} x + \cot^{-1} x\right) = \dfrac{1}{1 + x^2} + \dfrac{-1}{1 + x^2} = 0$; hence $\tan^{-1} x + \cot^{-1} x$ is constant throughout its domain, $\mathbb{R}$; that constant value is $\tan^{-1} 0 + \cot^{-1} 0 = 0 + \dfrac{\pi}{2} = \dfrac{\pi}{4} + \dfrac{\pi}{4} = \tan^{-1} 1 + \cot^{-1} 1 = \cdots$.

85.   a. $A(1) = \cot^{-1} 1 = \pi/4$ because $1 = \cot \pi/4$ and $\pi/4$ is in $(0, \pi)$. $B(1) = \tan^{-1}(1/1) = \tan^{-1} 1 = \pi/4$ because $1 = \tan \pi/4$ and $\pi/4$ is in $(-\pi/2, \pi/2)$. Therefore $A(1) = B(1)$.
        $B(-1) = \tan^{-1}(-1) = -\pi/4$, $A(-1) = \cot^{-1}(-1) = 3\pi/4 = (-\pi/4) + \pi = B(-1) + \pi$.

    b.   Problem 81 implies $A'(x) = \dfrac{-1}{1 + x^2}$. Theorem 3 and the chain rule imply $B'(x) = \dfrac{d}{dx} \tan^{-1} \left(\dfrac{1}{x}\right) = \left(\dfrac{1}{1 + (1/x)^2}\right) \left(\dfrac{-1}{x^2}\right) = \dfrac{-1}{x^2 + 1}$. B is discontinuous at $x = 0$ and $B'(0)$ is not defined; if $x \neq 0$, $B'(x)$ is defined and equals $A'(x)$.

    c. $(A - B)' = A' - B' = 0$ on $(0, \infty)$, thus $A - B$ is constant there, part (a) implies $(A - B)(x) = (A - B)(1) = 0$ on $(0, \infty)$. Similarly, $A - B$ is constant on $(-\infty, 0)$, the constant value of the function on this interval is $(A - B)(-1) = \pi$ (using result of the second half of part (a)).

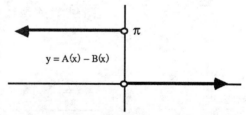

$y = A(x) - B(x)$

87. $\sec x$ is continuous on $[0, \pi/2) \cup (\pi/2, \pi]$. $\sec x$ increases throughout that domain because $\dfrac{d}{dx} \sec x = \sec x \tan x = \dfrac{\sin x}{\cos^2 x}$ is positive on $(0, \pi/2) \cup (\pi/2, \pi)$ and zero at 0 and π. The range includes $[1, \infty)$ because $\sec 0 = 1$, $\lim_{x \to \pi/2^-} \sec x = \infty$, and $\sec x$ is continuous on $[0, \pi/2)$; similarly, the range includes $(-\infty, -1]$ because $\lim_{x \to \pi/2^+} \sec x = -\infty$, $\sec \pi = -1$ and $\sec x$ is continuous on $(\pi/2, \pi]$. Therefore $\sec x$ is 1–1 from the domain $[0, \pi/2) \cup (\pi/2, \pi]$ onto the range $(-\infty, -1] \cup [1, \infty)$; it must have an inverse function, $\sec^{-1}$, with domain $(-\infty, -1] \cup [1, \infty)$ and range $[0, \pi/2) \cup (\pi/2, \pi]$. ∎

REMARK: $\sec x = 1/\cos x$ is undefined when $\cos x$ is 0; if $x = \pi/2$, then $\cos x = 0$ so $\sec(\pi/2)$ is not defined. $|\cos x| \leq 1$ implies $|\sec x| \geq 1$ and it is not possible to describe an inverse to secant which has a larger domain than $(-\infty, -1] \cup [1, \infty)$.

89.   a. $2 = \sec(\pi/3)$ and $\pi/3 \in [0, \pi/2) \implies \sec^{-1} 2 = \pi/3$.
    b. $-2 = \sec(2\pi/3)$ and $2\pi/3 \in (\pi/2, \pi] \implies \sec^{-1}(-2) = 2\pi/3$.
    c. $\sqrt{2} = \sec(\pi/4)$ and $\pi/4 \in [0, \pi/2) \implies \sec^{-1} \sqrt{2} = \pi/4$.
    d. $2/\sqrt{3} = \sec(\pi/6)$ and $\pi/6 \in [0, \pi/2) \implies \sec^{-1} \left(2/\sqrt{3}\right) = \pi/6$.
    e. $1 = \sec 0$ and $0 \in [0, \pi/2) \implies \sec^{-1} 1 = 0$.
    f. $-1 = \sec \pi$ and $\pi \in (\pi/2, \pi] \implies \sec^{-1}(-1) = \pi$.

91. $y = \sec^{-1} x$

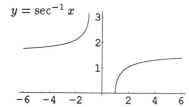

93. $\csc x$ is continuous on $[-\pi/2, 0) \cup (0, \pi/2]$. $\csc x$ decreases throughout that domain because $\dfrac{d}{dx} \csc x = -\csc x \cot x = \dfrac{-\cos x}{\sin^2 x}$ is negative on $(\pi/2, 0) \cup (0, \pi/2)$ and zero at $-\pi/2$ and $\pi/2$. The range includes $[1, \infty)$ because $\csc \pi/2 = 1$, $\lim_{x \to 0+} \csc x = \infty$, and $\csc x$ is continuous on $(0, \pi/2]$; similarly, the range includes $(-\infty, -1]$ because $\lim_{x \to 0-} \csc x = -\infty$, $\csc(-\pi/2) = -1$ and $\csc x$ is continuous on $(0, \pi/2]$. Therefore $\csc x$ is 1–1 from the domain $[-\pi/2, 0) \cup (0, \pi/2]$ onto the range $(-\infty, -1] \cup [1, \infty)$; it must have an inverse function, $\csc^{-1}$, with domain $(-\infty, -1] \cup [1, \infty)$ and range $[-\pi/2, 0) \cup (0, \pi/2]$. ∎

REMARK: $\csc x = 1/\sin x$ is undefined when $\sin x$ is 0; if $x = 0$, then $\sin x = 0$ so $\csc 0$ is not defined. $|\sin x| \leq 1$ implies $|\csc x| \geq 1$ and it is not possible to describe an inverse to cosecant which has a larger domain than $(-\infty, -1] \cup [1, \infty)$.

95. a. $2 = \csc(\pi/6)$ and $\pi/6 \in (0, \pi/2] \implies \csc^{-1} 2 = \pi/6$.

b. $-2 = \csc(-\pi/6)$ and $-\pi/6 \in [-\pi/2, 0) \implies \csc^{-1}(-2) = -\pi/6$.

c. $\sqrt{2} = \csc(\pi/4)$ and $\pi/4 \in (0, \pi/2] \implies \csc^{-1}\left(\sqrt{2}\right) = \pi/4$.

d. $2/\sqrt{3} = \csc(\pi/3)$ and $\pi/3 \in (0, \pi/2] \implies \csc^{-1}\left(2/\sqrt{3}\right) = \pi/3$.

e. $1 = \csc(\pi/2)$ and $\pi/2 \in (0, \pi/2] \implies \csc^{-1} 1 = \pi/2$.

f. $-1 = \csc(-\pi/2)$ and $-\pi/2 \in [-\pi/2, 0) \implies \csc^{-1}(-1) = -\pi/2$.

97. The results of Problems 90 and 96, part (c), imply

$$\frac{d}{dx}\left(\sec^{-1} x + \csc^{-1} x\right) = \frac{1}{|x|\sqrt{x^2-1}} + \frac{-1}{|x|\sqrt{x^2-1}} = \frac{1-1}{|x|\sqrt{x^2-1}} = 0;$$

thus the function is constant on each half of its domain.

$$x > 1 \implies \sec^{-1} x + \csc^{-1} x = \sec^{-1} 2 + \csc^{-1} 2 = \frac{\pi}{3} + \frac{\pi}{6} = \frac{\pi}{2},$$

$$x < 1 \implies \sec^{-1} x + \csc^{-1} x = \sec^{-1}(-2) + \csc^{-1}(-2) = \frac{2\pi}{3} + \frac{-\pi}{6} = \frac{\pi}{2};$$

therefore $\sec^{-1} x + \csc^{-1} x = \pi/2$ for all x in the function's domain. ∎

REMARK: This result can also be inferred from the information in Problems 88, 94, and 71.

99. $\dfrac{d}{dx} \cot^{-1}(e^x) = \dfrac{-1}{1+(e^x)^2} \cdot e^x = \dfrac{-e^x}{1+e^{2x}} = \dfrac{-1}{e^x + e^{-x}}.$

101. $\dfrac{d}{dx} \sec^{-1}(4x+2) = \dfrac{1}{|4x+2|\sqrt{(4x+2)^2-1}} \cdot 4 = \dfrac{2}{|2x+1|\sqrt{16x^2+16x+3}}.$

103. $\dfrac{d}{dx} \csc^{-1}\left(\sqrt{x}\right) = \dfrac{-1}{|\sqrt{x}|\sqrt{(\sqrt{x})^2-1}} \cdot \dfrac{1}{2\sqrt{x}} = \dfrac{-1}{2x\sqrt{x-1}}.$

105. $\dfrac{d}{dx}\left(\sec^{-1}(\cot x)\right) = \dfrac{1}{|\cot x|\sqrt{(\cot x)^2-1}} \cdot (-\csc^2 x) = \dfrac{-\csc^2 x}{|\cot x|\sqrt{\cot^2 x - 1}}.$

REMARK: Problem 88 implies

$$\frac{d}{dx}\left(\sec^{-1}(\cot x)\right) = \frac{d}{dx}\left(\cos^{-1}\left(\frac{1}{\cot x}\right)\right) = \frac{d}{dx}\left(\cos^{-1}(\tan x)\right)$$

$$= \frac{-1}{\sqrt{1-(\tan x)^2}} \cdot \sec^2 x = \frac{-1}{|\cos x|\sqrt{\cos^2 x - \sin^2 x}}.$$

107. $\displaystyle\int \frac{dx}{1+(3-x)^2} = \int \frac{-d(3-x)}{1+(3-x)^2} = \cot^{-1}(3-x) + C.$

6.8 — The inverse trigonometric functions

REMARK: $\displaystyle\int \frac{dx}{1+(3-x)^2} = \int \frac{dx}{1+(x-3)^2} = \int \frac{d\,(x-3)}{1+(x-3)^2} = \tan^{-1}(x-3) + C_1.$

109. $\displaystyle\int_{2/\sqrt{3}}^{2} \frac{dx}{x\sqrt{x^2-1}} = \sec^{-1} x\Big|_{2/\sqrt{3}}^{2} = \sec^{-1} 2 - \sec^{-1}\left(\frac{2}{\sqrt{3}}\right) = \frac{\pi}{3} - \frac{\pi}{6} = \frac{\pi}{6}.$

111. The result of Problem 92 is that $\displaystyle\int \frac{dx}{x\sqrt{x^2-a^2}} = \frac{1}{|a|}\sec^{-1}\left|\frac{x}{a}\right| + C = \frac{1}{|a|}\cos^{-1}\left|\frac{a}{x}\right| + C\,.$

Solutions 6.9   *The hyperbolic functions*   (pages 470-1)

1. $\cosh^2 x - \sinh^2 x = 1$ and $\cosh x \ge 1$ for all x, thus $\cosh x = \sqrt{1+\sinh^2 x}$. If $\sinh x = -2/5$, then
$\cosh x = \sqrt{1 + (-2/5)^2} = \sqrt{29/25} = \sqrt{29}/5,$

$\tanh x = \dfrac{\sinh x}{\cosh x} = \dfrac{-2/5}{\sqrt{29}/5} = \dfrac{-2}{\sqrt{29}},$

$\coth x = \dfrac{1}{\tanh x} = \dfrac{1}{-2/\sqrt{29}} = \dfrac{-\sqrt{29}}{2},$

$\operatorname{sech} x = \dfrac{1}{\cosh x} = \dfrac{1}{\sqrt{29}/5} = \dfrac{5}{\sqrt{29}},$

$\operatorname{csch} x = \dfrac{1}{\operatorname{sech} x} = \dfrac{1}{-2/5} = \dfrac{-5}{2}.$

3. $\dfrac{d}{dx}\sinh(4x+2) = \big(\cosh(4x+2)\big) \cdot \dfrac{d}{dx}(4x+2) = \big(\cosh(4x+2)\big)\cdot(4) = 4\cdot\cosh(4x+2).$

5. $\dfrac{d}{dx}\sin(\sinh x) = \big(\cos(\sinh x)\big)\cdot\dfrac{d}{dx}(\sinh x) = \big(\cos(\sinh x)\big)\cdot(\cosh x) = (\cosh x)\cdot\cos(\sinh x).$

7. $\dfrac{d}{dx}\tanh\left(\dfrac{1}{x}\right) = \left(\operatorname{sech}^2\left(\dfrac{1}{x}\right)\right)\cdot\dfrac{d}{dx}\left(\dfrac{1}{x}\right) = \left(\operatorname{sech}^2\left(\dfrac{1}{x}\right)\right)\cdot\left(\dfrac{-1}{x^2}\right) = \dfrac{-\operatorname{sech}^2(1/x)}{x^2}.$

9. $\dfrac{d}{dx}\tanh\left(\tan^{-1} x\right) = \big(\operatorname{sech}^2\left(\tan^{-1} x\right)\big)\cdot\dfrac{d}{dx}\left(\tan^{-1} x\right) = \dfrac{\operatorname{sech}^2\left(\tan^{-1} x\right)}{1+x^2}.$

11. The range of $\cosh$ is $[1,\infty)$ and the domain of $\sin^{-1}$ is $[-1,1]$, therefore $\sin^{-1}(\cosh x)$ is defined only for $x = 0$. A function whose domain is just $\{0\}$ is not differentiable, thus $\frac{d}{dx}\sin^{-1}(\cosh x)$ is undefined.

13. $\displaystyle\int \sinh 2x\,dx = \frac{1}{2}\int \sinh(2x)\,d(2x) = \frac{1}{2}\cosh(2x) + C.$

15. $\displaystyle\int \frac{\cosh 2x}{1+\sinh 2x}\,dx = \frac{1}{2}\int \frac{d(1+\sinh 2x)}{1+\sinh 2x} = \frac{1}{2}\ln|1+\sinh 2x| + C = \ln\sqrt{1+\sinh 2x} + C.$

17. $\displaystyle\int \frac{\operatorname{sech}^2\sqrt{x}}{\sqrt{x}}\,dx \overset{x\equiv u^2}{=} \int \frac{\operatorname{sech}^2 u}{u}\,(2u\,du) = 2\int \operatorname{sech}^2 u\,du = 2\tanh u + C = 2\tanh\sqrt{x} + C.$

19. $\displaystyle\int \coth x\,dx = \int \frac{\cosh x}{\sinh x}\,dx = \int \frac{1}{\sinh x}\,d(\sinh x) = \ln|\sinh x| + C.$

21. $\displaystyle\int \frac{\sinh(1/x)}{x^2}\,dx = \int \sinh\left(\frac{1}{x}\right)\cdot\frac{dx}{x^2} \overset{1/x\equiv u}{=} -\int \sinh u\,du = -\cosh u + C = -\cosh\left(\frac{1}{x}\right) + C.$

23. $\displaystyle\int_{-a}^{a} a\cosh\left(\frac{x}{a}\right)dx = a^2\int_{-a}^{a}\cosh\left(\frac{x}{a}\right)d\left(\frac{x}{a}\right) = a^2\sinh\left(\frac{x}{a}\right)\Big|_{-a}^{a}$
$$= a^2\sinh 1 - a^2\sinh(-1) = 2a^2\sinh 1 = a^2\left(e - \frac{1}{e}\right).$$

25. $\cosh^2 x - \sinh^2 x = (\cosh x + \sinh x)(\cosh x - \sinh x)$
$$= \left(\frac{e^x + e^{-x}}{2} + \frac{e^x - e^{-x}}{2}\right)\left(\frac{e^x + e^{-x}}{2} - \frac{e^x - e^{-x}}{x}\right) = e^x\,e^{-x} = e^{x-x} = e^0 = 1.\ \blacksquare$$

148

27. $\sinh(-x) = \dfrac{e^{(-x)} - e^{-(-x)}}{2} = \dfrac{e^{-x} - e^{x}}{2} = \dfrac{-(e^{x} - e^{-x})}{2} = -\dfrac{e^{x} - e^{-x}}{2} = -\sinh x.$ ∎

29.

$$\frac{d}{dx}\tanh x = \frac{d}{dx}\left(\frac{\sinh x}{\cosh x}\right) = \frac{(\cosh x)\left(\dfrac{d}{dx}\sinh x\right) - (\sinh x)\left(\dfrac{d}{dx}\cosh x\right)}{(\cosh x)^2}$$

$$= \frac{\cosh^2 x - \sinh^2 x}{\cosh^2 x} = \frac{1}{\cosh^2 x} = \left(\frac{1}{\cosh x}\right)^2 = \operatorname{sech}^2 x.$$ ∎

31. $\dfrac{d}{dx}\operatorname{sech} x = \dfrac{d}{dx}\left(\dfrac{1}{\cosh x}\right) = \dfrac{d}{dx}(\cosh x)^{-1} = (-1)(\cosh x)^{-2}\left(\dfrac{d}{dx}\cosh x\right) = (-1)(\cosh x)^{-2}(\sinh x)$

$$= (-1)\left(\frac{1}{\cosh x}\right)\left(\frac{\sinh x}{\cosh x}\right) = -\operatorname{sech} x\,\tanh x.$$ ∎

33. $(\cosh x + \sinh x)^n = (e^x)^n = e^{nx} = \cosh nx + \sinh nx.$ ∎

35.
$$\sinh(x + y) = \frac{e^{x+y} - e^{-(x+y)}}{2} = \frac{e^x e^y - e^{-x}e^{-y}}{2}$$

$$= \frac{(\cosh x + \sinh x)(\cosh y + \sinh y) - (\cosh x - \sinh x)(\cosh y - \sinh y)}{2}$$

$$= \frac{0 \cdot \cosh x \cdot \cosh y + 2 \cdot \sinh x \cdot \cosh y + 2 \cdot \cosh x \cdot \sinh y + 0 \cdot \sinh x \cdot \sinh y}{2}$$

$$= \sinh x \cdot \cosh y + \cosh x \cdot \sinh y.$$ ∎

REMARK:
$$\sinh x \cdot \cosh y + \cosh x \cdot \sinh y = \left(\frac{e^x - e^{-x}}{2}\right)\left(\frac{e^y + e^{-y}}{2}\right) + \left(\frac{e^x + e^{-x}}{2}\right)\left(\frac{e^y - e^{-y}}{2}\right)$$

$$= \frac{e^{x+y} + e^{x-y} - e^{-x+y} - e^{-x-y}}{4} + \frac{e^{x+y} - e^{x-y} + e^{-x+y} - e^{-x-y}}{4}$$

$$= \frac{2e^{x+y} - 2e^{-x-y}}{4} = \frac{e^{x+y} - e^{-(x+y)}}{2} = \sinh(x + y).$$ ∎

37. $2\sinh x \cosh x = 2\left(\dfrac{e^x - e^{-x}}{2}\right)\left(\dfrac{e^x + e^{-x}}{2}\right) = \dfrac{(e^x)^2 - (e^{-x})^2}{2} = \dfrac{e^{2x} - e^{-2x}}{2} = \sinh 2x.$ ∎

REMARK: Problem 35 implies $\sinh 2x = \sinh(x + x) = \sinh x \cdot \cosh x + \cosh x \cdot \sinh x = 2\sinh x \cosh x$.

39. a. $\cosh^2 x - \sinh^2 x = 1 \implies \cosh^2 x = 1 + \sinh^2 x \implies 1 = \dfrac{1}{\cosh^2 x} + \dfrac{\sinh^2 x}{\cosh^2 x} = \operatorname{sech}^2 x + \tanh^2 x.$ ∎

   b. $\cosh^2 x - \sinh^2 x = 1 \implies \sinh^2 x = \cosh^2 x - 1 \implies 1 = \dfrac{\cosh^2 x}{\sinh^2 x} - \dfrac{1}{\sinh^2 x} = \coth^2 x - \operatorname{csch}^2 x.$ ∎

Solutions 6.10   *The inverse hyperbolic functions*   (page 474)

1.   The domain of $\sinh^{-1}(3x + 2)$ is $\mathbb{R}$.

$$\frac{d}{dx}\sinh^{-1}(3x + 2) = \frac{1}{\sqrt{(3x + 2)^2 + 1}} \cdot \frac{d}{dx}(3x + 2) = \frac{3}{\sqrt{9x^2 + 12 + 5}}.$$

3. $\tanh^{-1}(\ln x)$ is defined if and only if $\ln x \in (-1, 1)$, i.e., $x \in (e^{-1}, e) = \operatorname{dom}\left(\tanh^{-1}(\ln x)\right)$.

$$\frac{d}{dx}\tanh^{-1}(\ln x) = \left(\frac{1}{1 - (\ln x)^2}\right) \cdot \frac{d}{dx}(\ln x) = \left(\frac{1}{1 - (\ln x)^2}\right) \cdot \left(\frac{1}{x}\right).$$

5. $\sqrt{\sinh^{-1} x}$ is defined $\iff \sinh^{-1} x \in [0, \infty) \iff x \in [0, \infty) = \operatorname{dom}\left(\sqrt{\sinh^{-1} x}\right)$.

$$\frac{d}{dx}\sqrt{\sinh^{-1} x} = \left(\frac{1}{2\sqrt{\sinh^{-1} x}}\right) \cdot \frac{d}{dx}\left(\sinh^{-1} x\right) = \frac{1}{2\sqrt{\sinh^{-1} x}} \cdot \frac{1}{\sqrt{x^2 + 1}}.$$

Note that the domain of the derivative is $(0, \infty)$.

6.10 — The inverse hyperbolic functions

7. $\text{I\!R} = \text{dom}\left(\sinh^{-1}\right) = \text{dom}\left(\cosh\right)$, therefore $\text{I\!R} = \text{dom}\left(\cosh \circ \sinh^{-1}\right)$.

$$\frac{d}{dx}\left(\cosh\left(\sinh^{-1}x\right)\right) = \left(\sinh\left(\sinh^{-1}x\right)\right) \cdot \frac{d}{dx}\left(\sinh^{-1}x\right) = (x) \cdot \frac{1}{\sqrt{x^2+1}} = \frac{x}{\sqrt{x^2+1}}.$$

9. $\displaystyle\int \frac{x\,dx}{\sqrt{x^4-4}} \overset{x^2\equiv 2u}{=} \int \frac{du}{\sqrt{(2u)^2-4}} = \frac{1}{2}\int \frac{du}{\sqrt{u^2-1}} = \frac{1}{2}\cosh^{-1}u + C = \frac{1}{2}\cosh^{-1}\left(\frac{x^2}{2}\right) + C.$

REMARK: Problem 24 implies this integral can also be expressed as $\frac{1}{2}\ln\left(\left(x^2 + \sqrt{x^4-4}\right)\right) + C_1$.

---

11. $\displaystyle\int \frac{x\,dx}{4-x^4} \overset{x^2\equiv 2u}{=} \int \frac{du}{4-(2u)^2} = \frac{1}{4}\int \frac{du}{1-u^2} = \frac{1}{4}\tanh^{-1}u + C = \frac{1}{4}\tanh^{-1}\left(\frac{x^2}{2}\right) + C.$

REMARK:

$$\int \frac{x}{4-x^4}\,dx = \int \frac{x}{(2+x^2)(2-x^2)}\,dx = \frac{1}{4}\int \left(\frac{x}{2+x^2} + \frac{x}{2-x^2}\right)dx$$

$$= \frac{1}{4}\left(\frac{1}{2}\ln\left|2+x^2\right| - \frac{1}{2}\ln\left|2-x^2\right|\right) + C_1 = \frac{1}{8}\ln\left|\frac{2+x^2}{2-x^2}\right| + C_1.$$

The antiderivative involving $\tanh^{-1}$ is valid only on $\left(-\sqrt{2}, \sqrt{2}\right)$; the one involving $\ln$ is valid on any interval not containing $\pm\sqrt{2}$. Note that $C_1 = C$. (See Problem 25.)

---

13. $\displaystyle\int \frac{e^x\,dx}{\sqrt{1+e^{2x}}} \overset{e^x\equiv u}{=} \int \frac{du}{\sqrt{1+u^2}} = \sinh^{-1}u + C = \sinh^{-1}\left(e^x\right) + C.$

REMARK: Problem 23 implies this result can also be expressed as $\ln\left(e^x + \sqrt{1+e^{2x}}\right) + C.$

---

15. $\displaystyle\int \frac{\cos x\,dx}{\sqrt{1+\sin^2 x}} = \int \frac{d(\sin x)}{\sqrt{1+(\sin x)^2}} = \sinh^{-1}(\sin x) + C.$

REMARK: Problem 23 implies this result can also be expressed as $\ln\left(\sin x + \sqrt{1+\sin^2 x}\right) + C.$

---

17. $\displaystyle\int \frac{dx}{x\sqrt{\ln^2 x - 25}} \overset{\ln x\equiv 5u}{=} \int \frac{5\,du}{\sqrt{(5u)^2-5^2}} = \int \frac{du}{\sqrt{u^2-1}} = \cosh^{-1}u + C = \cosh^{-1}\left(\frac{\ln x}{5}\right) + C.$

REMARK: Problem 24 implies this result can also be expressed as $\ln\left|\ln x + \sqrt{\ln^2 x - 25}\right| + C_1.$

---

19. $\dfrac{d}{dx}\sinh^{-1}\left(\dfrac{x}{a}\right) = \left(\dfrac{1}{\sqrt{(x/a)^2+1}}\right)\left(\dfrac{1}{a}\right) = \left(\dfrac{\sqrt{a^2}}{\sqrt{x^2+a^2}}\right)\left(\dfrac{1}{a}\right) = \dfrac{1}{\sqrt{x^2+a^2}} \cdot \dfrac{|a|}{a}. \ \blacksquare$

REMARK: If $a > 0$, then $\dfrac{d}{dx}\sinh^{-1}\left(\dfrac{x}{a}\right) = \dfrac{1}{\sqrt{x^2+a^2}}.$

---

21. $\dfrac{d}{dx}\tanh^{-1}\left(\dfrac{x}{a}\right) = \left(\dfrac{1}{1-(x/a)^2}\right)\left(\dfrac{1}{a}\right) = \left(\dfrac{a^2}{a^2-x^2}\right)\left(\dfrac{1}{a}\right) = \dfrac{a}{a^2-x^2}. \ \blacksquare$

REMARK: $\tanh^{-1}(x/a)$ is defined if and only if x/a is in $(-1, 1)$; this means $(x/a)^2 < 1$, hence $x^2 < a^2$.

---

23. $$\sinh^{-1}x = y \iff x = \sinh y = \frac{e^y - e^{-y}}{2} \iff e^y - 2x - e^{-y} = 0 \iff \left(e^y\right)^2 - 2x\left(e^y\right) - 1 = 0$$

$$\implies e^y = \frac{-(-2x) \pm \sqrt{(-2x)^2 - 4(1)(-1)}}{2} = x \pm \sqrt{x^2+1}.$$

Since $x - \sqrt{x^2+1} < x - \sqrt{x^2} = x - |x| \le 0 < e^y$, we infer $e^y = x + \sqrt{x^2+1}$ which is equivalent to $\sinh^{-1}x = y = \ln\left(x + \sqrt{x^2+1}\right). \ \blacksquare$

150

REMARK: $x + \sqrt{x^2 + 1} > x + \sqrt{x^2} = x + |x| \geq 0$, thus $\ln\left(x + \sqrt{x^2 + 1}\right)$ is defined for all x;

$$\frac{d}{dx} \ln\left(x + \sqrt{x^2 + 1}\right) = \left(x + \sqrt{x^2 + 1}\right)^{-1} \cdot \frac{d}{dx}\left(x + \sqrt{x^2 + 1}\right) = \left(\frac{1}{x + \sqrt{x^2 + 1}}\right)\left(1 + \frac{x}{\sqrt{x^2 + 1}}\right)$$

$$= \left(\frac{1}{x + \sqrt{x^2 + 1}}\right)\left(\frac{\sqrt{x^2 + 1} + x}{\sqrt{x^2 + 1}}\right) = \frac{1}{\sqrt{x^2 + 1}} = \frac{d}{dx} \sinh^{-1} x;$$

$\sinh^{-1} 0 = 0 = \ln 1 = \ln\left(0 + \sqrt{0^2 + 1}\right)$. The functions $\sinh^{-1} x$ and $\ln\left(x + \sqrt{x^2 + 1}\right)$ have $\mathbb{R}$ as their common domain, they differ by a constant throughout that domain because their derivatives are identical, that constant is 0 because they have the same function value at a particular point; therefore $\sinh^{-1} x = \ln\left(x + \sqrt{x^2 + 1}\right)$. ∎

25. $\dfrac{1 + x}{1 - x}$ is positive on $(-1, 1)$, zero at -1, undefined at 1, and negative otherwise; thus $(-1, 1)$ is the domain of $\ln\left(\dfrac{1 + x}{1 - x}\right)$; it is also the domain of $\tanh^{-1} x$.

$$\tanh^{-1} x = y \iff x = \tanh y = \frac{e^y - e^{-y}}{e^y + e^{-y}} = \frac{e^{2y} - 1}{e^{2y} + 1} \iff e^{2y} = \frac{1 + x}{1 - x} \iff \tanh^{-1} x = y = \frac{1}{2} \ln\left(\frac{1 + x}{1 - x}\right). \ ∎$$

27. $\displaystyle\int \frac{du}{\sqrt{u^2 - a^2}} = \cosh^{-1}\left(\frac{u}{|a|}\right) + C$ ⠀⠀⠀⠀⠀⠀⠀⠀⠀⠀⠀⠀⠀⠀⠀⠀⠀⠀ Problem 20

$$= \ln\left(\frac{u}{|a|} + \sqrt{\left(\frac{u}{|a|}\right)^2 - 1}\right) + C$$ ⠀⠀⠀⠀⠀⠀⠀⠀⠀⠀⠀ Problem 24

$$= \ln\left(\frac{u + \sqrt{u^2 - a^2}}{|a|}\right) + C = \ln\left(u + \sqrt{u^2 - a^2}\right) - \ln|a| + C = \ln\left(u + \sqrt{u^2 - a^2}\right) + C_1. \ ∎$$

Solutions 6.R ⠀⠀⠀ *Review* ⠀⠀⠀ (pages 475–6)

1. ⠀ $\log_3 9 = \log_3 3^2 = 2 \log_3 3 = 2(1) = 2.$

3. ⠀ $4 = \log_x\left(\dfrac{1}{16}\right) \iff x^4 = \dfrac{1}{16} = \dfrac{1}{2^4} = \left(\dfrac{1}{2}\right)^4$ and $x > 0 \implies x = \dfrac{1}{2}.$

5. ⠀ $\log x = 10^{-2} \implies x = 10^{\log x} = 10^{10^{-2}} = 10^{0.01} \approx 1.02329.$

7. ⠀ $\sin\left(\cos^{-1}\left(\dfrac{7}{10}\right)\right) = \sqrt{1 - \left(\dfrac{7}{10}\right)^2} = \sqrt{1 - \dfrac{49}{100}} = \dfrac{\sqrt{51}}{10}.$

REMARK: ⠀ Solution 6.7.59 shows $\sin\left(\cos^{-1} x\right) = \sqrt{1 - x^2}$.

9. ⠀ $\theta = \cot^{-1}(2/3) \implies \cot\theta = 2/3 \implies \csc^2\theta = 1 + \cot^2\theta = 1 + (2/3)^2 = 13/9$
⠀⠀⠀⠀⠀⠀⠀⠀⠀⠀⠀⠀ $\implies \csc\left(\cot^{-1}(2/3)\right) = \csc\theta = \sqrt{13}/3.$
(Recall that $\csc$ is positive on $(0, \pi)$, the range of $\cot^{-1}$.)

REMARK: ⠀ $\csc\left(\cot^{-1} x\right) = \sqrt{1 + x^2}$.

11. ⠀ $t = \tanh^{-1}(3/7) \implies \tanh t = 3/7 \implies \operatorname{sech}^2 t = 1 - \tanh^2 t = 1 - (3/7)^2 = 40/49$
⠀⠀⠀⠀⠀⠀⠀⠀⠀⠀⠀⠀ $\implies \operatorname{sech}\left(\tanh^{-1}(3/7)\right) = \operatorname{sech} t = \sqrt{40}/7.$

REMARK: ⠀ $\operatorname{sech}\left(\tanh^{-1} x\right) = \sqrt{1 - x^2}$.

13. ⠀ If $y = f(x) = 3\ln x$, then $f(2x) = 3\ln(2x) = 3\left(\ln 2 + \ln x\right) = 3\ln 2 + 3\ln x = 3\ln 2 + f(x)$. If x is doubled, then y increases by $3\ln 2 = \ln 8$.

15. ⠀ $\dfrac{d}{dx} \ln\left(1 + x^2\right) = \left(\dfrac{1}{1 + x^2}\right) \cdot \dfrac{d}{dx}\left(1 + x^2\right) = \left(\dfrac{1}{1 + x^2}\right) \cdot (2x) = \dfrac{2x}{1 + x^2}.$

6.R — Review

17. $\dfrac{d}{dx} e^{x^2+2x+1} = e^{x^2+2x+1} \cdot \dfrac{d}{dx}\left(x^2+2x+1\right) = e^{x^2+2x+1} \cdot (2x+2) = 2(x+1)e^{x^2+2x+1} = 2(x+1)e^{(x+1)^2}.$

19. $\dfrac{d}{dx} \ln(x+\ln(x+3)) = \left(\dfrac{1}{x+\ln(x+3)}\right) \cdot \dfrac{d}{dx}\left(x+\ln(x+3)\right) = \left(\dfrac{1}{x+\ln(x+3)}\right) \cdot \left(1+\dfrac{1}{x+3}\right)$
$$= \left(\dfrac{1}{x+\ln(x+3)}\right) \cdot \left(\dfrac{x+4}{x+3}\right).$$

21. $y = 2^{x+5} \implies \ln y = \ln 2^{x+5} = (x+5)\cdot\ln 2 \implies \dfrac{1}{y}\dfrac{dy}{dx} = \ln 2 \implies \dfrac{d}{dx} 2^{x+5} = y\cdot\ln 2 = (\ln 2)\cdot 2^{x+5}.$

23. $\dfrac{d}{dx}\tan^{-1}\left(\dfrac{1}{x^2}\right) = \left(\dfrac{1}{1+(1/x^2)^2}\right)\cdot\dfrac{d}{dx}\left(\dfrac{1}{x^2}\right) = \left(\dfrac{x^4}{x^4+1}\right)\cdot\left(\dfrac{-2}{x^3}\right) = \dfrac{-2x}{x^4+1}.$

25. $\dfrac{d}{dx}\csc\sqrt{x} = (-\csc\sqrt{x}\,\cot\sqrt{x})\cdot\dfrac{d}{dx}\sqrt{x} = (-\csc\sqrt{x}\,\cot\sqrt{x})\cdot\left(\dfrac{1}{2\sqrt{x}}\right) = \dfrac{-\csc\sqrt{x}\,\cot\sqrt{x}}{2\sqrt{x}}.$

27. $\dfrac{d}{dx}\ln|\sec x-\tan x| = \left(\dfrac{1}{\sec x-\tan x}\right)\cdot\dfrac{d}{dx}\left(\sec x-\tan x\right) = \left(\dfrac{1}{\sec x-\tan x}\right)\cdot\left(\sec x\,\tan x-\sec^2 x\right)$
$$= -\sec x.$$

29. $\dfrac{d}{dx}\tan^{-1}\sqrt[3]{x} = \left(\dfrac{1}{1+(\sqrt[3]{x})^2}\right)\cdot\dfrac{d}{dx}\sqrt[3]{x} = \dfrac{1}{1+x^{2/3}}\cdot\dfrac{1}{3x^{2/3}} = \dfrac{1}{3\,x^{2/3}\left(1+x^{2/3}\right)}.$

31. $\dfrac{d}{dx}\sin^{-1}\left(e^{-x}\right) = \left(\dfrac{1}{\sqrt{1-(e^{-x})^2}}\right)\cdot\dfrac{d}{dx}\left(e^{-x}\right) = \dfrac{1}{\sqrt{1-e^{-2x}}}\cdot(-e^{-x}) = \dfrac{-e^{-x}}{\sqrt{1-e^{-2x}}} = \dfrac{-1}{\sqrt{e^{2x}-1}}.$

33. $\dfrac{d}{dx}\sinh\left(x^2+3x\right) = \left(\cosh\left(x^2+3x\right)\right)\cdot\dfrac{d}{dx}\left(x^2+3x\right) = (2x+3)\,\cosh\left(x^2+3x\right).$

35. $\dfrac{d}{dx}\tanh^{-1}\sqrt{1+x} = \left(\dfrac{1}{1-(\sqrt{1+x})^2}\right)\cdot\dfrac{d}{dx}\sqrt{1+x} = \left(\dfrac{1}{1-(1+x)}\right)\cdot\left(\dfrac{1}{2\sqrt{1+x}}\right) = \dfrac{-1}{2x\sqrt{1+x}}.$

37. $\displaystyle\int \dfrac{3x\,dx}{1+x^2} = \dfrac{3}{2}\int\dfrac{d\left(1+x^2\right)}{1+x^2} = \dfrac{3}{2}\,\ln\left(1+x^2\right)+C.$

39. $\displaystyle\int \dfrac{e^{-1/x^2}}{x^3}\,dx \overset{-1/x^2\equiv u}{=} \dfrac{1}{2}\int e^u\,du = \dfrac{1}{2}e^u+C = \dfrac{1}{2}e^{-1/x^2}+C.$

41. $\displaystyle\int \dfrac{4^{\ln x}}{x}\,dx = \int\left(e^{\ln 4}\right)^{\ln x}\left(\dfrac{dx}{x}\right) \overset{(\ln 4)(\ln x)\equiv u}{=} \dfrac{1}{\ln 4}\int e^u\,du = \dfrac{e^u}{\ln 4}+C = \dfrac{e^{(\ln 4)(\ln x)}}{\ln 4}+C = \dfrac{4^{\ln x}}{\ln 4}+C.$

REMARK: $4^{\ln x} = e^{(\ln 4)(\ln x)} = x^{\ln 4}$, therefore $\displaystyle\int\dfrac{4^{\ln x}}{x}\,dx = \int\dfrac{x^{\ln 4}}{x}\,dx = \int x^{\ln 4-1}\,dx = \dfrac{x^{\ln 4}}{\ln 4}+C.$

43. $\displaystyle\int_{x=0}^{5\pi/6}\tan\left(\dfrac{x}{5}\right)dx \overset{x\equiv 5u}{=} \int_{u=0}^{\pi/6}\tan u\,(5\,du) = 5\int_{u=0}^{\pi/6}\dfrac{\sin u\,du}{\cos u} = -5\,\ln|\cos u|\Big|_0^{\pi/6}$
$$= -5\,\ln\cos\left(\dfrac{\pi}{6}\right)+5\,\ln\cos 0 = -5\,\ln\left(\dfrac{\sqrt{3}}{2}\right) \approx 0.7192.$$

45. $\displaystyle\int\dfrac{3x\,dx}{1+x^4} = \dfrac{3}{2}\int\dfrac{2x\,dx}{1+(x^2)^2} \overset{x^2\equiv u}{=} \dfrac{3}{2}\int\dfrac{du}{1+u^2} = \dfrac{3}{2}\tan^{-1}u+C = \dfrac{3}{2}\tan^{-1}x^2+C.$

47. $\int\sin 3x\,\cos 3x\,dx = \dfrac{1}{3}\int\sin 3x\,d(\sin 3x) = \dfrac{1}{6}\cdot\sin^2 3x+C = \dfrac{-1}{6}\cdot\cos^2 3x+C_1 = \dfrac{-1}{12}\cdot\cos 6x+C_2.$

49. $\displaystyle\int\coth x\,dx = \int\dfrac{\cosh x\,dx}{\sinh x} = \int\dfrac{d(\sinh x)}{\sinh x} = \ln|\sinh x|+C.$

51. $\displaystyle\int\dfrac{\text{sech}^2(\ln x)}{x}\,dx = \int\text{sech}^2(\ln x)\,d(\ln x) = \tanh(\ln x)+C.$

REMARK: $\cosh(\ln x) = \dfrac{e^{\ln x}+e^{-\ln x}}{2} = \dfrac{x+x^{-1}}{2} = \dfrac{x^2+1}{2x}$, therefore $\text{sech}(\ln x) = \dfrac{2x}{x^2+1}$ and
$$\int\dfrac{\text{sech}^2(\ln x)}{x}\,dx = \int\left(\dfrac{2x}{x^2+1}\right)^2\left(\dfrac{1}{x}\right)dx = \int\left(x^2+1\right)^{-2}\,d\left(x^2+1\right) = -\left(x^2+1\right)^{-1}+C_1.$$

53. $\displaystyle\int \frac{\sec^2 x\, dx}{\sqrt{\tan^2 x - 4}} \overset{\tan x \equiv 2u}{=} \int \frac{du}{\sqrt{u^2 - 1}} = \cosh^{-1} u + C = \cosh^{-1}\left(\frac{\tan x}{2}\right) + C.$

55. If $f(x) = \sqrt{2x - 5}$, then f is 1–1 on its domain, $[5/2, \infty)$, because $f'(x) = 1/\sqrt{2x - 5}$ is positive on $(5/2, \infty)$. If $y = \sqrt{2x - 5}$, then $y^2 = 2x - 5$ and $(y^2 + 5)/2 = f^{-1}(y) = x$.

$$\frac{dx}{dy} = \frac{d}{dy} f^{-1}(y) = \frac{1}{f'(x)} = \frac{1}{1/\sqrt{2x - 5}} = \sqrt{2x - 5} = y.$$

57. The domain of $g(x) = \ln(3 - 4x)$ is $(-\infty, 3/4)$ because the domain of $\ln$ is $(0, \infty)$ and $0 < 3 - 4x \iff 4x < 3 \iff x < 3/4$. This function g decreases on its domain so it is 1–1 throughout. $y = \ln(3 - 4x) \iff e^y = 3 - 4x \iff 4x = 3 - e^y \iff x = \dfrac{3 - e^y}{4} = g^{-1}(y)$; therefore

$$\frac{dx}{dy} = \frac{d}{dy} g^{-1}(y) = \frac{1}{g'(x)} = \frac{1}{-4/(3 - 4x)} = \frac{3 - 4x}{-4} = \frac{e^y}{-4} = \frac{-e^y}{4}.$$

59. Let $G(x) = \tan(x/2)$; $G(x)$ is undefined at odd multiples of π. Since $G'(x) = (1/2)\sec^2(x/2) > 0$, G is strictly increasing between successive odd multiples of π; i.e., G is 1–1 on each interval of the form $((2k - 1)\pi, (2k + 1)\pi)$ where k is an integer. Let G_k denote the restriction of G to this k^{th} interval. If $x \in \text{dom}\,(G_k)$, then $(x - 2k\pi)/2 \in (-\pi/2, \pi/2)$ and

$$\frac{x - 2k\pi}{2} = \tan^{-1}\left(\tan\left(\frac{x - 2k\pi}{2}\right)\right) = \tan^{-1}\left(\tan\left(\frac{x}{2} - k\pi\right)\right) = \tan^{-1}\left(\tan\left(\frac{x}{2}\right)\right) = \tan^{-1}\left(G_k(x)\right).$$

Hence $G_k(x) = y \iff (x - 2k\pi)/2 = \tan^{-1} y \iff x = 2k\pi + 2\tan^{-1} y = G_k^{-1}(y)$. Furthermore,

$$\frac{dx}{dy} = \frac{d}{dy} G_k^{-1}(y) = \frac{1}{G_k'(x)} = \frac{2}{\sec^2(x/2)} = \frac{2}{1 + \tan^2(x/2)} = \frac{2}{1 + y^2}.$$

61. $\dfrac{dP/dt}{P} = 6\% \implies \dfrac{dP}{dt} = 0.06\,P \implies P(t) = P(0)e^{0.06\,t}$ where t measures years elapsed from a reference time. Let 1970 correspond to $t = 0$, then $P(0) = 250{,}000$. This continuous growth model estimates the 1980 population to be $P(10) = 250000e^{0.6} \approx 455{,}530$ and the population in 2000 is predicted to be $P(30) = 250000e^{1.8} \approx 1{,}512{,}412$.

63. Newton's law of cooling yields $T(t) - 23 = D(t) = (125 - 23)e^{-\alpha t} = 102\,e^{-\alpha t}$. Therefore $T(10) = 80 \implies 57 = 80 - 23 = D(10) = 102\,e^{-10\alpha} \implies \alpha = (1/10)\ln(102/57) \approx 0.05819$.

    a. $T(20) = 23 + 102\,e^{-20\alpha} \approx 54.85°\text{C}$.

    b. $25 = T(t) = 23 + 102\,e^{-\alpha t} \iff t = (1/\alpha)\ln(102/2) \approx 67.57$ min.

65. Suppose $M(t) = M_0 e^{-\alpha t}$ where t is measured in days. If 20% of the mass is lost in one week (7 days), then $0.8 M_0 = M(7) = M_0 e^{-7\alpha}$ and $\alpha = (-1/7)\ln 0.8$. The half life is $t = \dfrac{\ln 0.5}{-\alpha} = 7\left(\dfrac{\ln 0.5}{\ln 0.8}\right) \approx 21.74$ days ≈ 3.11 weeks. 75% of the mass will be lost in $t = \dfrac{\ln 0.25}{-\alpha} = 7\left(\dfrac{\ln 0.25}{\ln 0.8}\right) \approx 43.49$ days ≈ 6.21 weeks. 95% of the mass will be lost in $t = \dfrac{\ln 0.05}{-\alpha} = 7\left(\dfrac{\ln 0.05}{\ln 0.8}\right) \approx 93.98$ days ≈ 13.43 weeks.

67. Compounding 6% quarterly, $A(8) = 10000 \times (1 + 0.06/4)^{4 \times 8} = 10000 \times 1.015^{32} \approx \$16103.24$. On the other hand, 6% compounded continuously yields $A_*(8) = 10000 \times e^{0.06 \times 8} = 10000 \times e^{0.48} \approx \$16160.74$.

69. $y = \sinh(x - 1)$ 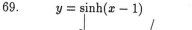 $\qquad\qquad\qquad y = 1 + \sinh^{-1} x$

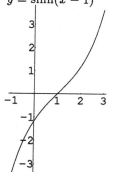 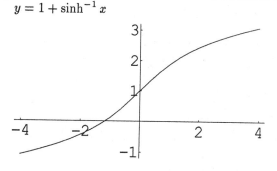

Solutions 6.C   *Computer Exercises*   (page 476)

1. a. $\displaystyle\int \frac{dx}{(x-1)(x-2)} = \ln|x-2| - \ln|x-1| + C = \ln\left|\frac{x-2}{x-1}\right| + C.$

   b. $\displaystyle\int \frac{dx}{(x+3)(x-4)} = \frac{\ln|x-4|}{7} - \frac{\ln|x+3|}{7} + C = \frac{1}{7}\ln\left|\frac{x-4}{x+3}\right| + C.$

   c. $\displaystyle\int \frac{dx}{(x-2)(x-5)} = \frac{\ln|x-5|}{3} - \frac{\ln|x-2|}{3} + C = \frac{1}{3}\ln\left|\frac{x-5}{x-2}\right| + C.$

More generally, if $p \neq q$, then $\displaystyle\int \frac{dx}{(x-p)(x-q)} = \frac{\ln|x-p|}{p-q} - \frac{\ln|x-q|}{p-q} + C = \left(\frac{1}{p-q}\right)\ln\left|\frac{x-p}{x-q}\right| + C.$

Thus, if $ax^2 + bx + c$ has zeros p and q with $p \neq q$, then $ax^2 + bx + c = a(x-p)(x-q)$ and

$$\int \frac{dx}{ax^2 + bx + c} = \left(\frac{1}{a(p-q)}\right)\ln\left|\frac{x-p}{x-q}\right| + C.$$

3.   If $f(x) = \ln^3 x - x$, then $f'(x) = 3x^{-1}\ln^2 x - 1$. Quick examination of graphs of $y = f'(x)$ indicates f' has one zero near .75 a second zero near 2.5, and a third zero near 40. Newton's method applied to f' with $x_0 = .75$ converges to .6319392204, this implies $f(.6319392204) \approx -.7286178215$ is a local maximum. Another use of Newton's method starting this time with $x_0 = 2.5$ converges to 2.4843415229, this implies $f(2.4843415229) \approx -1.7307515335$ is a local minimum. A third use of Newton begun with $x_0 = 40$ converges to 41.8076392317, this implies $f(41.8076392317) \approx 10.2161006654$ is a local maximum. $f''(x) = 3x^{-2}(\ln x)(2 - \ln x)$ does not change sign on $(0,1)$, $(1, e^2) \approx (1, 7.39)$, and (e^2, ∞); this implies $f'(x)$ has at most one zero on each interval. (We have already located a zero of f' in each interval.)

$y = f(x) = \ln^3 x - x$

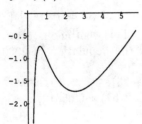

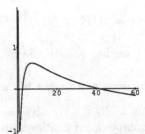

$y = f'(x) = 3x^{-1}\ln^2 x - 1$

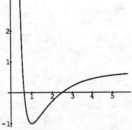

5.   Equation (6.6.13) implies that 10% yearly interest compounded x times per year will yield $A_0\left(1 + .1/x\right)^{7x}$ after 7 years for initial investment A_0. Let $f(x) = \left(1 + .1/x\right)^{7x}$, then $f(1) \approx 1.9487$, $f(2) \approx 1.9799$, $f(3) \approx 1.9909$, $f(4) \approx 1.9965$, $f(5) \approx 1.9999$, $f(6) \approx 2.0022$. Compounding 5 times per year has a result closer to doubling after 7 years than does compounding 6 times per year $(.0001 < .0022)$; on the other hand, compounding 6 times per year has a simple colloquial translation ("compounded bimonthly") but compounding 5 times per year does not.

REMARK: Logarithmic differentiation implies $f'(x) = f(x)\left(7\ln(1 + .1/x) - .7(x + .1)^{-1}\right)$. Newton's method for $f(x) - 2 = 0$ starting with $x_0 = 5$ or $x_0 = 6$ converges to 5.041.

7
Techniques of Integration

1.   Let $u = x$ and $dv = e^{3x}\, dx$, then $du = dx$ and $v = \dfrac{1}{3}e^{3x}$. Therefore

$$\int xe^{3x}\, dx = x\left(\frac{1}{3}e^{3x}\right) - \int \frac{1}{3}e^{3x}\, dx = \frac{xe^{3x}}{3} - \frac{e^{3x}}{9} + C = \left(\frac{3x-1}{9}\right)e^{3x} + C.$$

3.   We need to get rid of the x^2 term. Let $u = x^2$ and $dv = e^{-x}\, dx$. Then $du = 2x\, dx$ and $v = -e^{-x}$, so that

$\displaystyle\int x^2 e^{-x}\, dx = x^2\left(-e^{-x}\right) - \int \left(-e^{-x}\right) 2x\, dx = -x^2 e^{-x} + 2\int xe^{-x}\, dx.$ We see that $\int xe^{-x}\, dx$ is simpler than

$\int x^2 e^{-x}\, dx$, but it is still necessary to integrate by parts once more. This time setting $u_* = x$ and $dv = e^{-x} dx$,

we have $du_* = dx$ and $v = -e^{-x}$, so that $\displaystyle\int xe^{-x}\, dx = x\left(-e^{-x}\right) - \int \left(-e^{-x}\right)\, dx = -xe^{-x} - e^{-x} + C.$

Therefore $\displaystyle\int x^2 e^{-x}\, dx = -x^2 e^{-x} + 2\left(-(x+1)e^{-x}\right) + C = -\left(x^2 + 2x + 2\right)e^{-x} + C.$

5. $\displaystyle\int_1^2 x\ln x\, dx = \int_1^2 (\ln x)(x\, dx) = \int_1^2 (\ln x)\, d\left(\frac{x^2}{2}\right) = (\ln x)\left(\frac{x^2}{2}\right)\Big|_1^2 - \int_1^2 \left(\frac{x^2}{2}\right)\left(\frac{dx}{x}\right)$

$\displaystyle = \frac{x^2 \ln x}{2}\Big|_1^2 - \frac{1}{2}\int_1^2 x\, dx = x^2\left(\frac{\ln x}{2} - \frac{1}{4}\right)\Big|_1^2 = (2\ln 2 - 1) - \frac{-1}{4} = 2\ln 2 - \frac{3}{4}.$

7. $\displaystyle\int x\sinh x\, dx = \int x\, d(\cosh x) = x(\cosh x) - \int \cosh x\, dx = x\cosh x - \sinh x + C.$

9.   Let $u = x$ and $dv = \sqrt{1 - \dfrac{x}{2}}\, dx = \left(1 - \dfrac{x}{2}\right)^{1/2} dx$, then $du = dx$ and $v = \left(\dfrac{2}{3}\right)(-2)\left(1 - \dfrac{x}{2}\right)^{3/2}$. Thus

$\displaystyle\int_0^2 x\sqrt{1 - \frac{x}{2}}\, dx = \int_0^2 x\left(1 - \frac{x}{2}\right)^{1/2} dx = x\left(\frac{2}{3}\right)(-2)\left(1 - \frac{x}{2}\right)^{3/2}\Big|_0^2 - \int_0^2 \left(\frac{-4}{3}\right)\left(1 - \frac{x}{2}\right)^{3/2} dx$

$\displaystyle = (0 - 0) + \left(\frac{4}{3}\right)\int_0^2 \left(1 - \frac{x}{2}\right)^{3/2} dx = \left(\frac{4}{3}\right)\left(\frac{2}{5}\right)(-2)\left(1 - \frac{x}{2}\right)^{5/2}\Big|_0^2 = 0 - \frac{-16}{15} = \frac{16}{15}.$

REMARK: $\displaystyle\int x\sqrt{1 - \frac{x}{2}}\, dx \overset{2-x\equiv 2u}{=} \int 2(1-u)\sqrt{u}(-2\, du) = 4\left[\left(\frac{2}{5}\right)\left(1 - \frac{x}{2}\right)^{5/2} - \left(\frac{2}{3}\right)\left(1 - \frac{x}{2}\right)^{3/2}\right] + C.$

11. $\displaystyle\int x^2 \cosh 2x\, dx = \frac{1}{2}\int x^2\, d(\sinh 2x) = \frac{1}{2}\left[x^2 \sinh 2x - \int \sinh 2x\, d\left(x^2\right)\right] = \frac{x^2 \sinh 2x}{2} - \int x\sinh 2x\, dx$

$\displaystyle = \frac{x^2 \sinh 2x}{2} - \frac{1}{2}\int x\, d(\cosh 2x) = \frac{x^2 \sinh 2x}{2} - \frac{1}{2}\left[x\cosh 2x - \int \cosh 2x\, dx\right]$

$\displaystyle = \frac{x^2 \sinh 2x}{2} - \frac{1}{2}\left[x\cosh 2x - \frac{1}{2}\sinh 2x\right] + C.$

Therefore $\displaystyle\int_0^1 x^2 \cosh 2x\, dx = \left(\frac{2x^2 + 1}{4}\sinh 2x - \frac{x}{2}\cosh 2x\right)\Big|_0^1 = \frac{3}{4}\sinh 2 - \frac{1}{2}\cosh 2.$

13.   Let $I = \int \cos(\ln x)\, dx$. Two integrations by parts yields an equation of the form $I = f(x) - I$ which can be

solved to get a formula for I (see Example 4). First let $u_1 = \cos(\ln x)$ and $dv = dx$, then $du_1 = \dfrac{-\sin(\ln x)\, dx}{x}$,

$v = x$, and

$$I = \int \cos(\ln x)\, dx = x\cos(\ln x) - \int x\left(\frac{-\sin(\ln x)\, dx}{x}\right) = x\cos(\ln x) + \int \sin(\ln x)\, dx.$$

7.2 — Integration by parts

Now let $u_2 = \sin(\ln x)$ and $dv = dx$, then $du_2 = \dfrac{\cos(\ln x)\,dx}{x}$, $v = x$ and

$$I = x\cos(\ln x) + \left[x\sin(\ln x) - \int x\left(\frac{\cos(\ln x)\,dx}{x}\right)\right] = x\cos(\ln x) + x\sin(\ln x) - \int \cos(\ln x)\,dx$$

$$= x\left[\cos(\ln x) + \sin(\ln x)\right] - I.$$

Therefore $2I = x\left[\cos(\ln x) + \sin(\ln x)\right]$ and $\displaystyle\int \cos(\ln x)\,dx = I = \frac{x}{2}\left[\cos(\ln x) + \sin(\ln x)\right] + C.$

15. Let $u = x^3$ and $dv = x^2 e^{x^3}\,dx$, then $du = 3x^2\,dx$ and $v = e^{x^3}/3$. Therefore

$$\int_0^1 x^5 e^{x^3}\,dx = \int x^3\left(x^2 e^{x^3}\,dx\right) = x^3\left(\frac{1}{3}e^{x^3}\right)\Big|_0^1 - \int_0^1 \frac{1}{3}e^{x^3}3x^2\,dx$$

$$= \frac{e}{3} - \int_0^1 x^2 e^{x^3}\,dx = \frac{e}{3} - \left(\frac{1}{3}e^{x^3}\right)\Big|_0^1 = \frac{e}{3} - \left(\frac{e}{3} - \frac{1}{3}\right) = \frac{1}{3}.$$

REMARK: $\displaystyle\int x^5 e^{x^3}\,dx \overset{x^3 \equiv u}{=} \frac{1}{3}\int u e^u\,du = \frac{1}{3}(u-1)e^u + C = \left(\frac{x^3-1}{3}\right)e^{x^3} + C.$

(See Example 1 for computation of $\int u e^u\,du$.)

17. $$I = \int e^x \cos x\,dx = \int e^x\,d(\sin x) = e^x\sin x - \int \sin x\,(e^x\,dx) = e^x\sin x + \int e^x\,d(\cos x)$$

$$= e^x\sin x + e^x\cos x - \int \cos x\,(e^x\,dx) = e^x(\sin x + \cos x) - I$$

implies $\displaystyle\int e^x \cos x\,dx = I = \frac{1}{2}e^x(\sin x + \cos x) + C.$

19. Use the results of Examples 7 and 8.

$$\int \sin^5 x\,dx = \frac{-\sin^4 x\cos x}{5} + \frac{4}{5}\int \sin^3 x\,dx = \frac{-\sin^4 x\cos x}{5} + \frac{4}{5}\left(\frac{-\sin^2 x\cos x}{3} + \frac{-2\cos x}{3}\right) + C$$

$$= \left(\frac{-\cos x}{15}\right)\left(3\sin^4 x + 4\sin^2 x + 8\right) + C.$$

21. Use formula (6) of the text's highlighted *A Perspective: On Integrating an Inverse Function* (page 484) plus the facts that $\cos^{-1} 0 = \dfrac{\pi}{2}$ and $\cos^{-1}\left(\dfrac{1}{\sqrt{2}}\right) = \dfrac{\pi}{4}.$

$$\int_0^{1/\sqrt{2}} \cos^{-1} y\,dy = \left(\frac{1}{\sqrt{2}}\right)\cdot\cos^{-1}\left(\frac{1}{\sqrt{2}}\right) - 0\cdot\cos^{-1} 0 - \int_{\cos^{-1} 0}^{\cos^{-1}(1/\sqrt{2})} \cos x\,dx$$

$$= \frac{1}{\sqrt{2}}\cdot\frac{\pi}{4} - \left(\sin x\big|_{\pi/2}^{\pi/4}\right) = \frac{\pi}{4\sqrt{2}} - \left(\frac{1}{\sqrt{2}} - 1\right) = 1 + \frac{\pi-4}{4\sqrt{2}}.$$

REMARK: $\displaystyle\int \cos^{-1} t\,dt = t\cdot\cos^{-1} t - \int t\cdot\frac{-dt}{\sqrt{1-t^2}} = t\cdot\cos^{-1} t - \sqrt{1-t^2} + C.$

23. $$\int_0^1 \tan^{-1} y\,dy = 1\cdot\tan^{-1} 1 - 0\cdot\tan^{-1} 0 - \int_{\tan^{-1} 0}^{\tan^{-1} 1} \tan x\,dx = \frac{\pi}{4} - \int_0^{\pi/4} \frac{-d(\cos x)}{\cos x}$$

$$= \frac{\pi}{4} + \ln|\cos x|\big|_0^{\pi/4} = \frac{\pi}{4} + \ln\left(\frac{1}{\sqrt{2}}\right) - \ln 1 = \frac{\pi}{4} - \frac{1}{2}\ln 2.$$

REMARK: $\displaystyle\int \tan^{-1} t\cdot dt = t\cdot\tan^{-1} t - \int t\cdot\frac{dt}{1+t^2} = t\cdot\tan^{-1} t - \frac{1}{2}\ln\left(1+t^2\right) + C.$

25. $$\int x\tan^{-1} x\,dx = \int \tan^{-1} x\,(x\,dx) = \int \tan^{-1} x\,d\left(\frac{x^2}{2}\right) = (\tan^{-1} x)\left(\frac{x^2}{2}\right) - \int\left(\frac{x^2}{2}\right)\left(\frac{dx}{1+x^2}\right)$$

$$= \frac{x^2\tan^{-1} x}{2} - \frac{1}{2}\int\left(1 - \frac{1}{1+x^2}\right)dx = \frac{x^2\tan^{-1} x}{2} - \frac{1}{2}\left(x - \tan^{-1} x\right) + C.$$

REMARK: $\displaystyle\int \tan^{-1} x\,(x\,dx) = (\tan^{-1} x)\left(\frac{1+x^2}{2}\right) - \int \left(\frac{1+x^2}{2}\right)\left(\frac{dx}{1+x^2}\right) = \left(\frac{1+x^2}{2}\right)\tan^{-1} x - \frac{x}{2} + C.$

27.
$$I = \int e^{ax}\cos bx\,dx = \int \cos bx\,(e^{ax}\,dx) = \int \cos bx\,d\left(\frac{e^{ax}}{a}\right)$$
$$= (\cos bx)\left(\frac{e^{ax}}{a}\right) - \int \left(\frac{e^{ax}}{a}\right)d(\cos bx) = (\cos bx)\left(\frac{e^{ax}}{a}\right) + \frac{b}{a}\int \sin bx\,(e^{ax}\,dx)$$
$$= (\cos bx)\left(\frac{e^{ax}}{a}\right) + \left(\frac{b}{a}\right)(\sin bx)\left(\frac{e^{ax}}{a}\right) - \left(\frac{b}{a}\right)\int \left(\frac{e^{ax}}{a}\right)d(\sin bx)$$
$$= \left[\frac{a\cos bx + b\sin bx}{a^2}\right]e^{ax} - \frac{b^2}{a^2}\int e^{ax}\cos bx\,dx = \left[\frac{a\cos bx + b\sin bx}{a^2}\right]e^{ax} - \left[\frac{b^2}{a^2}\right]I$$

implies $\displaystyle\int e^{ax}\cos bx\,dx = I = \left(1 + \frac{b^2}{a^2}\right)^{-1}\left(\frac{a\cos bx + b\sin bx}{a^2}\right)e^{ax} + C = \left(\frac{a\cos bx + b\sin bx}{a^2 + b^2}\right)e^{ax} + C.$

29.
$$I = \int (\sinh 2x)(\cosh 3x)\,dx = \int (\sinh 2x)(\cosh 3x\,dx)$$
$$= (\sinh 2x)\left(\frac{\sinh 3x}{3}\right) - \int \left(\frac{\sinh 3x}{3}\right)(2\cosh 2x\,dx)$$
$$= \frac{(\sinh 2x)(\sinh 3x)}{3} - \frac{2}{3}\int (\cosh 2x)(\sinh 3x\,dx)$$
$$= \frac{(\sinh 2x)(\sinh 3x)}{3} - \frac{2}{3}\left[(\cosh 2x)\left(\frac{\cosh 3x}{3}\right) - \int \left(\frac{\cosh 3x}{3}\right)(2\sinh 2x\,dx)\right]$$
$$= \frac{(\sinh 2x)(\sinh 3x)}{3} - \frac{2(\cosh 2x)(\cosh 3x)}{9} + \frac{4}{9}\int (\sinh 2x)(\cosh 3x)\,dx$$
$$= \frac{3(\sinh 2x)(\sinh 3x) - 2(\cosh 2x)(\cosh 3x)}{9} + \frac{4}{9}I$$

implies
$$\int (\sinh 2x)(\cosh 3x)\,dx = I = \left(1 - \frac{4}{9}\right)^{-1}\left(\frac{3(\sinh 2x)(\sinh 3x) - 2(\cosh 2x)(\cosh 3x)}{9}\right) + C$$
$$= \frac{3(\sinh 2x)(\sinh 3x) - 2(\cosh 2x)(\cosh 3x)}{5} + C.$$

REMARK: It is also true that $I = (\cosh 3x)\left(\dfrac{\cosh 2x}{2}\right) - (3\sinh 3x)\left(\dfrac{\sinh 2x}{2^2}\right) + \dfrac{3^2}{2^2}\cdot I.$ Alternatively,
$$\int (\sinh 2x)(\cosh 3x)\,dx = \int \left(\frac{e^{2x} - e^{-2x}}{2}\right)\left(\frac{e^{3x} + e^{-3x}}{2}\right)dx = \frac{1}{4}\int (e^{5x} + e^{-x} - e^{x} - e^{-5x})\,dx$$
$$= \frac{1}{2}\int (\sinh 5x - \sinh x)\,dx = \frac{\cosh 5x}{10} - \frac{\cosh x}{2} + C_1.$$

31.
$$\int_{-1}^{x} \cos^{-1} t\,dt = x\cdot\cos^{-1} x - (-1)\cdot\cos^{-1}(-1) - \int_{\cos^{-1}(-1)}^{\cos^{-1} x}\cos t\,dt = x\cdot\cos^{-1} x + \pi - \left(\sin t\Big|_{\pi}^{\cos^{-1} x}\right)$$
$$= x\cdot\cos^{-1} x + \pi - \left(\sin(\cos^{-1} x) - \sin \pi\right) = x\cdot\cos^{-1} x + \pi - \sqrt{1 - x^2}.$$

33. Replacing $\cos^2 x$ with $1 - \sin^2 x$ and using reduction formula (5) we find
$$\int \cos^2 x\,dx = \int (1 - \sin^2 x)\,dx = x + \frac{1}{2}\sin x\cos x - \frac{1}{2}\int dx = x + \frac{1}{2}\sin x\cos x - \frac{1}{2}x + C = \frac{x + \sin x\cos x}{2} + C.$$
Using the disk method, we compute the volume generated to be
$$\int_{0}^{\pi/2}\pi(\cos x)^2\,dx = \pi\left(\frac{x + \sin x\cos x}{2}\Big|_0^{\pi/2}\right) = \pi\left(\frac{\pi/2 + 1\cdot 0}{2} - \frac{0 + 0\cdot 1}{2}\right) = \frac{\pi^2}{4}.$$

7.2 — Integration by parts

35. $\displaystyle\int \ln^2 x \, dx = x \ln^2 x - \int x \left[\frac{2 \ln x \, dx}{x}\right] = x \ln^2 x - 2\left(x \ln x - \int x \left[\frac{dx}{x}\right]\right) = x \left(\ln^2 x - 2 \ln x + 2\right) + C;$
therefore the volume generated is

$$\int_1^e \pi(\ln x)^2 \, dx = \pi x \left(\ln^2 x - 2 \ln x + 2\right)\Big|_1^e = \pi \cdot e \cdot (1^2 - 2 \cdot 1 + 2) - \pi \cdot 1 \cdot (0^2 - 2 \cdot 0 + 2) = \pi(e - 2).$$

37.

$$\int_0^{\pi/4} \pi \left(\cos^2 x - \sin^2 x\right) dx = \pi \int_0^{\pi/4} (\cos x + \sin x)(\cos x - \sin x) \, dx$$

$$= \pi \int_0^{\pi/4} (\cos x + \sin x)(-\sin x + \cos x) \, dx$$

$$= \pi \int_0^{\pi/4} (\cos x + \sin x) \, d(\cos x + \sin x)$$

$$= \frac{\pi}{2}(\cos x + \sin x)^2 \Big|_0^{\pi/4} = \frac{\pi}{2}\left(\left(\frac{1}{\sqrt{2}} + \frac{1}{\sqrt{2}}\right)^2 - (1 + 0)^2\right) = \frac{\pi}{2}(2 - 1) = \frac{\pi}{2}.$$

REMARK: The reduction formula (5) yields

$$\int \left(\cos^2 x - \sin^2 x\right) dx = \int \left(1 - 2\sin^2 x\right) dx = x - 2\left(\frac{-1}{2}\sin x \cos x + \frac{1}{2}x\right) + C = \sin x \, \cos x + C.$$

39. The cylindrical shell method and the computations in Solution 5 yield

$$\int_1^3 2\pi x \, \ln x \, dx = \pi x^2 \left(\ln x - \frac{1}{2}\right)\Big|_1^3 = \pi \left[9\left(\ln 3 - \frac{1}{2}\right) - \pi\left(\frac{-1}{2}\right)\right] = \pi(9 \ln 3 - 4).$$

REMARK: The disk method yields $\displaystyle\int_0^{\ln 3} \pi \left(3^2 - (e^y)^2\right) dy = \pi \left(9y - \frac{e^{2y}}{2}\right)\Big|_0^{\ln 3} = \pi\left(9 \ln 3 - \frac{9}{2}\right) - \pi\left(\frac{-1}{2}\right).$

41. The cylindrical shell method yields

$$\int_0^{\pi/4} 2\pi x(\cos x - \sin x) \, dx = 2\pi x(\sin x + \cos x)\Big|_0^{\pi/4} - 2\pi \int_0^{\pi/4} (\sin x + \cos x) \, dx$$

$$= 2\pi\left(\frac{\pi}{4}\right)\left(\frac{2}{\sqrt{2}}\right) - 2\pi(-\cos x + \sin x)\Big|_0^{\pi/4} = \frac{\pi^2}{\sqrt{2}} - 2\pi.$$

43. a. Let $f(x) = xe^{-x}$. Then $f'(x) = (1 - x)e^{-x}$, so f increases on $(-\infty, 1)$, has its global maximum value of e^{-1} at $x = 1$, and decreases on $(1, \infty)$. $f''(x) = (x - 2)e^{-x}$; thus f is concave down on $(-\infty, 2)$, concave up on $(2, \infty)$, and has point of inflection $(2, 2/e^2)$. The positive x-axis is an asymptote because $\lim_{x \to \infty} f(x) = 0 = \lim_{x \to \infty} f'(x)$.

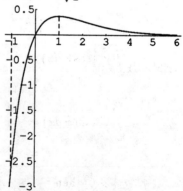

b.

$$\int_{-1}^1 \left|xe^{-x}\right| dx = \int_{-1}^0 -xe^{-x} \, dx + \int_0^1 xe^{-x} \, dx = (x + 1)e^{-x}\Big|_{-1}^0 + \left(-(x + 1)e^{-x}\right)\Big|_0^1$$

$$= (1 - 0) + \left(-2e^{-1} - (-1)\right) = 2 - \frac{2}{e}.$$

45. $\displaystyle\int x^4 e^x \, dx = e^x \left(x^4 - \frac{4 \cdot x^{4-1}}{1} + \frac{4 \cdot 3 \cdot x^{4-2}}{1^2} - \frac{4 \cdot 3 \cdot 2 \cdot x^{4-1}}{1^3} + \frac{(-1)^4 \, 4!}{1^4}\right) + C$

$$= e^x \left(x^4 - 4x^3 + 12x^2 - 24x + 24\right) + C.$$

47. $\displaystyle\int \sec^5 x \, dx = \frac{1}{4} \tan x \sec^3 x + \frac{3}{4} \int \sec^3 x \, dx = \frac{1}{4} \tan x \sec^3 x + \frac{3}{4}\left(\frac{1}{2} \tan x \sec x + \frac{1}{2} \int \sec x \, dx\right)$

$$= \frac{1}{4} \tan x \sec^3 x + \frac{3}{8} \tan x \sec x + \frac{3}{8} \ln |\sec x + \tan x| + C.$$

158

49.
$$\int x^3 \sin ax \, dx = x^3 \left(\frac{-\cos ax}{a}\right) - \int \left(\frac{-\cos ax}{a}\right)(3x^2 \, dx) = \frac{-x^3 \cos ax}{a} + \frac{3}{a} \int x^2 \cos ax \, dx$$

$$= \frac{-x^3 \cos ax}{a} + \frac{3}{a}\left[x^2\left(\frac{\sin ax}{a}\right) - \int\left(\frac{\sin ax}{a}\right)(2x \, dx)\right]$$

$$= \frac{-x^3 \cos ax}{a} + \frac{3x^2 \sin ax}{a^2} - \frac{6}{a^2} \int x \sin ax \, dx$$

$$= \frac{-x^3 \cos ax}{a} + \frac{3x^2 \sin ax}{a^2} - \frac{6}{a^2}\left[x\left(\frac{-\cos ax}{a}\right) - \int\left(\frac{-\cos ax}{a}\right) dx\right]$$

$$= \frac{-x^3 \cos ax}{a} + \frac{3x^2 \sin ax}{a^2} + \frac{6x \cos ax}{a^3} + \frac{-6 \sin ax}{a^4} + C$$

$$= \left(\frac{-x^3}{a} + \frac{6x}{a^3}\right) \cos ax + \left(\frac{3x^2}{a^2} + \frac{-6}{a^4}\right) \sin ax + C. \quad \blacksquare$$

51. The two integrals involve constants of integration which are different.

53.
$$I = \int \cos^n x \, dx = \int \cos^{n-1} x \, (\cos x \, dx) = \int \cos^{n-1} x \, d(\sin x)$$

$$= \cos^{n-1} x \cdot \sin x - \int \sin x \, d\left(\cos^{n-1} x\right)$$

$$= \cos^{n-1} x \cdot \sin x - \int \sin x \cdot (n-1) \cos^{n-2} x (-\sin x) \, dx$$

$$= \cos^{n-1} x \cdot \sin x + (n-1) \int \sin^2 x \cdot \cos^{n-2} x \, dx$$

$$= \cos^{n-1} x \cdot \sin x + (n-1) \int \left(1 - \cos^2 x\right) \cdot \cos^{n-2} x \, dx$$

$$= \cos^{n-1} x \cdot \sin x + (n-1) \int \cos^{n-2} x \, dx - (n-1) \int \cos^n x \, dx$$

$$= \cos^{n-1} x \cdot \sin x + (n-1) \int \cos^{n-2} x \, dx - (n-1)I$$

implies
$$\int \cos^n x \, dx = I = \left(\frac{1}{1 + (n-1)}\right)\left(\cos^{n-1} x \cdot \sin x + (n-1) \int \cos^{n-2} x \, dx\right)$$

$$= \frac{1}{n} \cos^{n-1} x \cdot \sin x + \frac{n-1}{n} \int \cos^{n-2} x \, dx. \quad \blacksquare$$

55. a.
$$\int x^n (\cos x \, dx) = \int x^n \, d(\sin x) = x^n (\sin x) - \int (\sin x) \, d\left(x^n\right)$$

$$= x^n \sin x - \int \sin x \cdot nx^{n-1} \, dx = x^n \sin x - n \int x^{n-1} \sin x \, dx. \quad \blacksquare$$

b.
$$\int x^n (\sin x \, dx) = \int x^n \, d(-\cos x) = x^n (-\cos x) - \int (-\cos x) \, d\left(x^n\right)$$

$$= -x^n \cos x + \int \cos x \cdot nx^{n-1} \, dx = -x^n \cos x + n \int x^{n-1} \cos x \, dx. \quad \blacksquare$$

57. a. If $n > 0$, then
$$G_{n+1}(x) = \int_0^x e^{-t} t^{(n+1)-1} \, dt = \int_0^x t^n \, d\left(-e^{-t}\right) = t^n \left(-e^{-t}\right)\Big|_0^x - \int \left(-e^{-t}\right) \, d\left(t^n\right)$$

$$= x^n \left(-e^{-x}\right) - 0 + n \int e^{-t} t^{n-1} \, dt = -x^n e^{-x} + nG_n(x).$$

b.
$$G_1(x) = \int_0^x e^{-t} \, dt = -e^{-t}\Big|_0^x = -e^{-x} - (-1) = 1 - e^{-x}.$$
$$G_2(x) = -x^1 e^{-x} + 1 \cdot \left(1 - e^{-x}\right) = 1 - (1 + x)e^{-x}.$$
$$G_3(x) = -x^2 e^{-x} + 2 \cdot \left(1 - e^{-x} - xe^{-x}\right) = 2 - \left(2 + 2x + x^2\right) e^{-x}.$$

59. For $0 < x < \pi$, let $C_n(x) = \displaystyle\int_{\pi/2}^x \cot^n \theta \, d\theta.$

7.2 — Integration by parts

a.
$$C_0(x) = \int_{\pi/2}^{x} 1 \, d\theta = \theta\big|_{\pi/2}^{x} = x - \frac{\pi}{2}.$$

$$C_1(x) = \int_{\pi/2}^{x} \cot\theta \, d\theta = \ln|\sin\theta|\big|_{\pi/2}^{x} = \ln|\sin x| - \ln|1| = \ln|\sin x|.$$

b. Suppose $n \geq 2$.

$$\int \cot^n \theta \, d\theta = \int \cot^{n-2}\theta \cdot \cot^2\theta \, d\theta = \int \cot^{n-2}\theta \left(\csc^2\theta - 1\right) d\theta = \int \cot^{n-2}\theta \sec^2\theta \, d\theta - \int \cot^{n-2}\theta \, d\theta$$

$$= \int (\cot\theta)^{n-2} d(-\cot\theta) - \int \cot^{n-2}\theta \, d\theta = \frac{-(\cot\theta)^{n-1}}{n-1} - \int \cot^{n-2}\theta \, d\theta.$$

Since $\dfrac{-(\cot\theta)^{n-1}}{n-1}\bigg|_{\pi/2}^{x} = \dfrac{-\cot^{n-1}x}{n-1}$, we have shown $C_n(x) = \dfrac{-\cot^{n-1}x}{n-1} - C_{n-2}(x)$.

c.
$$C_2(x) = -\cot x - C_0(x) = -\cot x - \left(x - \frac{\pi}{2}\right).$$

$$C_3(x) = \frac{-\cot^2 x}{2} - C_1(x) = \frac{-\cot^2 x}{2} - \ln|\sin x|.$$

61. Let $J_n(x) = \int \dfrac{x^n}{\sqrt{x^k+1}} \, dx$, where k is a fixed integer. (Problem 60 discusses the case $k=2$.)

$$J_n(x) = \int \frac{x^n}{\sqrt{x^k+1}} \, dx = \int x^{n-k+1} \left(\frac{2}{k}\right) \left(\frac{kx^{k-1}\,dx}{2\sqrt{x^k+1}}\right)$$

$$= \frac{2}{k} \int x^{n-k+1} \, d\left(\sqrt{x^k+1}\right)$$

$$= \frac{2}{k} \left(x^{n-k+1}\sqrt{x^k+1} - \int \sqrt{x^k+1}(n-k+1)x^{n-k} \, dx\right)$$

$$= \left(\frac{2}{k}\right) x^{n-k+1}\sqrt{x^k+1} - \frac{2(n-k+1)}{k} \int \frac{x^k+1}{\sqrt{x^k+1}} x^{n-k} \, dx$$

$$= \left(\frac{2}{k}\right) x^{n-k+1}\sqrt{x^k+1} - \frac{2(n-k+1)}{k} \left(J_n(x) + J_{n-k}(x)\right).$$

Therefore

$$J_n(x) = \left[1 + \frac{2(n-k+1)}{k}\right]^{-1} \cdot \left[\left(\frac{2}{k}\right) x^{n-k+1}\sqrt{x^k+1} - \left(\frac{2(n-k+1)}{k}\right) J_{n-k}(x)\right]$$

$$= \left(\frac{2}{2n-k+2}\right) x^{n-k+1}\sqrt{x^k+1} - \left(\frac{2(n-k+1)}{2n-k+2}\right) J_{n-k}(x).$$

Solutions 7.3   *Integrals of certain trigonometric functions*   **(pages 490-2)**

1.
$$\int_0^{\pi} \sin^3 x \cos^2 x \, dx = \int_0^{\pi} \left(1 - \cos^2 x\right) \sin x \cos^2 x \, dx = \int_0^{\pi} \left(\cos^2 x \sin x - \cos^4 x \sin x\right) dx$$

$$= \left[\left(\frac{-1}{3}\right) \cos^3 x + \left(\frac{1}{5}\right) \cos^5 x\right]\Bigg|_0^{\pi}$$

$$= \left[\left(\frac{-1}{3}\right)(-1)^3 + \left(\frac{1}{5}\right)(-1)^5\right] - \left[\left(\frac{-1}{3}\right)(1)^3 + \left(\frac{1}{4}\right)(1)^5\right] = \frac{4}{15}.$$

3.
$$\int \cos^3 x \sqrt{\sin x} \, dx = \int \cos^2 x \sqrt{\sin x} \cos x \, dx = \int \left(1 - \sin^2 x\right)(\sin x)^{1/2} \, d(\sin x)$$

$$= \int \left((\sin x)^{1/2} - (\sin x)^{5/2}\right) d(\sin x) = \frac{2}{3}(\sin x)^{3/2} - \frac{2}{7}(\sin x)^{7/2} + C.$$

5.
$$\int \sin^5 x \cos^2 x \, dx = \int \left(1 - \cos^2 x\right)^2 \cos^2 x(\sin x \, dx) = \int -\left(1 - 2\cos^2 x + \cos^4 x\right)\cos^2 x \, d(-\cos x)$$

$$= \int \left(-\cos^2 x + 2\cos^4 x - \cos^6 x\right) d(\cos x) = \frac{-\cos^3 x}{3} + \frac{2\cos^5 x}{5} + \frac{-\cos^7 x}{7} + C.$$

7.
$$\int \sin^2 x \cos^5 x \, dx = \int \sin^2 x \, (\cos^2 x)^2 \cos x \, dx = \int \sin^2 x \, (1 - \sin^2 x)^2 \, (\cos x \, dx)$$
$$= \int (\sin^2 x - 2\sin^4 x + \sin^6 x) \, d(\sin x) = \frac{\sin^3 x}{3} - \frac{2\sin^5 x}{5} + \frac{\sin^7 x}{7} + C.$$

9.
$$\int_0^{\pi/4} \cos^2 x \, dx = \int_0^{\pi/4} \left(\frac{1 + \cos 2x}{2} \right) dx = \left(\frac{x}{2} + \frac{\sin 2x}{4} \right) \Big|_0^{\pi/4} = \left(\frac{\pi}{8} + \frac{1}{4} \right) - (0 + 0) = \frac{\pi}{8} + \frac{1}{4}.$$

REMARK: Problem 7.2.53 implies $\int \cos^2 x \, dx = \frac{1}{2}(\cos x \sin x + x) + C_*.$

11.
$$\int \cos^4 x \, dx = \int (\cos^2 x)^2 \, dx = \int \left(\frac{1 + \cos 4x}{2} \right)^2 dx = \frac{1}{4} \int (1 + 2\cos 4x + \cos^2 4x) \, dx$$
$$= \frac{1}{4} \int \left(1 + 2\cos 4x + \left(\frac{1 + \cos 8x}{2} \right) \right) dx$$
$$= \int \left(\frac{3}{8} + \frac{\cos 4x}{2} + \frac{\cos 8x}{8} \right) dx = \frac{3x}{8} + \frac{\sin 4x}{8} + \frac{\sin 8x}{64} + C.$$

REMARK: Problem 7.2.53 implies $\int \cos^4 2x \, dx = \frac{\cos^3 2x \sin 2x}{8} + \frac{3\cos 2x \sin 2x}{16} + \frac{3x}{8} + C_*.$

13. See Example 2.
$$\int \sin^4 x \cos^2 x \, dx = \int \sin^2 x \, (\sin x \cos x)^2 \, dx = \int \left(\frac{1 - \cos 2x}{2} \right) \left(\frac{\sin 2x}{2} \right)^2 dx$$
$$= \int \left(\frac{1 - \cos 4x}{16} - \frac{\sin^2 2x \cos 2x}{8} \right) dx = \frac{x}{16} - \frac{\sin 4x}{64} - \frac{\sin^3 2x}{48} + C.$$

REMARK: Using reduction formula (7.2.5) we find
$$\int \sin^4 x \cos^2 x \, dx = \int \sin^4 x \, (1 - \sin^2 x) \, dx = \frac{\sin^5 x \cos x}{6} - \frac{\sin^3 x \cos x}{24} - \frac{\sin x \cos x}{16} + \frac{x}{16}.$$

15.
$$\int_0^{\pi/8} \tan^2 2x \, dx = \int_0^{\pi/8} (\sec^2 2x - 1) \, dx = \left(\frac{\tan 2x}{2} - x \right) \Big|_0^{\pi/8} = \left(\frac{1}{2} - \frac{\pi}{8} \right) - (0 - 0) = \frac{1}{2} - \frac{\pi}{8}.$$

17.
$$\int_0^{\pi/4} \sec^4 x \, dx = \int_0^{\pi/4} \sec^2 x \cdot \sec^2 x \, dx = \int_0^{\pi/4} (1 + \tan^2 x) \cdot \sec^2 x \, dx$$
$$= \int_0^{\pi/4} (1 + \tan^2 x) \, d(\tan x) = \left(\tan x + \frac{\tan^3 x}{3} \right) \Big|_0^{\pi/4} = \left(1 + \frac{1}{3} \right) - (0 + 0) = \frac{4}{3}.$$

REMARK: The reduction formula of Problem 7.2.54 yields $\int \sec^4 x \, dx = \frac{\tan x \sec^2 x}{3} + \frac{2\tan x}{3} + C.$

19. See Example 4.
$$\int \tan^3 x \sec^4 x \, dx = \int \tan^3 x \, (1 + \tan^2 x) \sec^2 x \, dx = \int (\tan^3 x + \tan^5 x) \, d(\tan x) = \frac{\tan^4 x}{4} + \frac{\tan^6 x}{6} + C.$$

REMARK:
$$\int \sec^4 x \tan^3 x \, dx = \int \sec^3 x \, (\sec^2 x - 1) \, (\sec x \tan x \, dx)$$
$$= \int (\sec^5 x - \sec^3 x) \, d(\sec x) = \frac{\sec^6 x}{6} - \frac{\sec^4 x}{4} + C_1.$$

21.
$$\int_0^{\pi/6} \sec^3 2x \tan 2x \, dx = \int_0^{\pi/6} \sec^2 2x \, (\sec 2x \tan 2x \, dx) = \frac{1}{2} \int_0^{\pi/6} \sec^2 2x \, d(\sec 2x)$$
$$= \frac{\sec^3 2x}{6} \Big|_0^{\pi/6} = \frac{8}{6} - \frac{1}{6} = \frac{7}{6}.$$

23.
$$\int \sin 2x \cos 3x \, dx = \int \frac{1}{2} (\sin(2x + 3x) + \sin(2x - 3x)) \, dx = \frac{1}{2} \int (\sin 5x - \sin x) \, dx$$
$$= \frac{1}{2} \left(\frac{-\cos 5x}{5} + \cos x \right) + C = \frac{\cos x}{2} - \frac{\cos 5x}{10} + C.$$

25.
$$\int \sin\left(\sqrt{2}\,x\right)\cos\left(\frac{x}{\sqrt{2}}\right)dx = \int \frac{1}{2}\left(\sin\left(\sqrt{2}\,x + \frac{x}{\sqrt{2}}\right) + \sin\left(\sqrt{2}\,x - \frac{x}{\sqrt{2}}\right)\right)dx$$
$$= \frac{1}{2}\int\left(\sin\left(\frac{3x}{\sqrt{2}}\right) + \sin\left(\frac{x}{\sqrt{2}}\right)\right)dx$$
$$= \frac{1}{2}\left(\left(\frac{-\sqrt{2}}{3}\right)\cos\left(\frac{3x}{\sqrt{2}}\right) - \sqrt{2}\cos\left(\frac{x}{\sqrt{2}}\right)\right) + C.$$

27.
$$\int \cos(10x)\cos(100x)\,dx = \int \frac{1}{2}\left(\cos(10x - 100x) + \cos(10x + 100x)\right)dx = \frac{\sin 90x}{180} + \frac{\sin 110x}{220} + C.$$

29.
$$\int \cot^2 x\,dx = \int\left(\csc^2 x - 1\right)dx = -\cot x - x + C.$$

31.
$$\int_{\pi/4}^{\pi/2}\csc^4 x\,dx = \int_{\pi/4}^{\pi/2}\left(1 + \cot^2 x\right)\left(\csc^2 x\,dx\right) = \left(-\cot x - \frac{\cot^3 x}{3}\right)\Bigg|_{\pi/4}^{\pi/2} = (0 - 0) - \left(-1 - \frac{1}{3}\right) = \frac{4}{3}.$$

33.
$$\int_{\pi/6}^{\pi/2}\cot^7 x\,\csc^4 x\,dx = \int_{\pi/6}^{\pi/2}\cot^7 x\left(1 + \cot^2 x\right)\left(\csc^2 x\,dx\right) = \int_{\pi/6}^{\pi/2} -\left(\cot^7 x + \cot^9 x\right)\,d(\cot x)$$
$$= -\left(\frac{\cot^8 x}{8} + \frac{\cot^{10} x}{10}\right)\Bigg|_{\pi/6}^{\pi/2} = -(0 + 0) + \left(\frac{3^{8/2}}{8} + \frac{3^{10/2}}{10}\right) = \frac{1377}{40}.$$

35.
Lemma:
$$\int \csc^n x\,dx = \frac{-1}{n - 1}\cot x\,\csc^{n-2} x + \frac{n - 2}{n - 1}\int \csc^{n-2} x\,dx.$$

$\boxed{\text{Proof}}$ Since $\sin\left(\frac{\pi}{2} - t\right) = \cos t$ and $\cos\left(\frac{\pi}{2} - t\right) = \sin t$, the result of Problem 7.2.54 implies

$$\int \csc^n x\,dx \overset{x \equiv \frac{\pi}{2} - t}{=} \int \sec^n t\,(-dt) = -\left(\frac{1}{n - 1}\tan t\,\sec^{n-2} t + \frac{n - 2}{n - 1}\int \sec^{n-2} t\,dt\right)$$
$$= \frac{-1}{n - 1}\tan t\,\sec^{n-2} t + \frac{n - 2}{n - 1}\int \sec^{n-2} t\,(-dt)$$
$$= \frac{-1}{n - 1}\cot x\,\csc^{n-2} x + \frac{n - 2}{n - 1}\int \csc^{n-2} x\,dx.\ \blacksquare$$

Therefore, using this Lemma and integration formula 7.1.14, we discover
$$\int \cot^2 x\,\csc x\,dx = \int\left(\csc^2 x - 1\right)\csc x\,dx = \int\left(\csc^3 x - \csc x\right)dx$$
$$= \frac{-1}{2}\cot x\,\csc x + \left(\frac{1}{2} - 1\right)\int \csc x\,dx = \frac{-1}{2}\cot x\,\csc x + \frac{1}{2}\ln|\csc x + \cot x| + C.$$

REMARK: Because $1 = \csc^2 x - \cot^2 x = (\csc x + \cot x)(\csc x - \cot x)$, we can also express this answer as
$$\int \cot^2 x\,\csc x\,dx = \frac{-1}{2}\cot x\,\csc x - \frac{1}{2}\ln|\csc x - \cot x| + C.$$

37. a.
$$\int\left(\cos^2 x - \sin^2 x\right)dx = \int(\cos x + \sin x)(\cos x - \sin x)\,dx = \int(\sin x + \cos x)(\cos x - \sin x)\,dx$$
$$= \int(\sin x + \cos x)\,d(\sin x + \cos x)$$
$$= \frac{(\sin x + \cos x)^2}{2} + C_* = \sin x \cos x + C \qquad \text{where} \qquad C = C_* + \frac{1}{2}.$$

REMARK: $\int \cos^2 x\,dx = \int \cos x\,d(\sin x) = \cos x \sin x - \int \sin x\,d(\cos x) = \cos x \sin x + \int \sin^2 x\,dx$; therefore $\int\left(\cos^2 x - \sin^2 x\right)dx = \cos x \sin x + C_1.$

b.
$$\int \cos^2 x\,dx = \int\left(\frac{(\cos^2 x + \sin^2 x) + (\cos^2 x - \sin^2 x)}{2}\right)dx = \int\left(\frac{1 + (\cos^2 x - \sin^2 x)}{2}\right)dx$$
$$= \int \frac{1}{2}\,dx + \frac{1}{2}\int\left(\cos^2 x - \sin^2 x\right)dx = \frac{x + \sin x \cos x}{2} + C;$$
$$\int \sin^2 x\,dx = \int\left(\frac{(\cos^2 x + \sin^2 x) - (\cos^2 x - \sin^2 x)}{2}\right)dx = \int\left(\frac{1 - (\cos^2 x - \sin^2 x)}{2}\right)dx$$
$$= \frac{1}{2}\int dx - \frac{1}{2}\int\left(\cos^2 x - \sin^2 x\right)dx = \frac{x - \sin x \cos x}{2} + C.$$

39.
$$\sin(A + B) + \sin(A - B) = (\sin A \cos B + \cos A \sin B) + (\sin A \cos B - \cos A \sin B)$$
$$= (1 + 1)\sin A \cos B + (1 - 1)\cos A \sin B = 2\sin A \cos B,$$

therefore $\quad \sin mx \cos nx = \dfrac{1}{2}\left(\sin(mx + nx) + \sin(mx - nx)\right) = \dfrac{1}{2}\left(\sin(m + n)x + \sin(m - n)x\right).$ ∎

41.
$$\cos(A - B) + \cos(A + B) = (\cos A \cos B + \sin A \sin B) + (\cos A \cos B - \sin A \sin B) = 2\cos A \cos B,$$

therefore $\quad \cos mx \cos nx = \dfrac{1}{2}\left(\cos(mx - nx) + \cos(mx + nx)\right) = \dfrac{1}{2}\left(\cos(m - n)x + \cos(m + n)x\right).$ ∎

43.
$$\int \cos mx \cos nx\, dx = \int \frac{1}{2}\left(\cos(m - n)x + \cos(m + n)x\right)\, dx$$
$$= C + \begin{cases} \dfrac{\sin(m - n)x}{2(m - n)} + \dfrac{\sin(m + n)x}{2(m + n)} & \text{if } m \neq \pm n, \ \blacksquare \\[2mm] \dfrac{x}{2} + \dfrac{\sin 2mx}{4m} & \text{if } m = \pm n \neq 0, \\[2mm] x & \text{if } m = n = 0. \end{cases}$$

45.   Problem 43 displays an antiderivative of $\cos mx \cos nx$ when $m \neq \pm n$; that antiderivative is of the form $a \sin(jx) + b\sin(kx)$ where j and k are integers. Since $\sin x$ is zero at each integer multiple of π, it is easy to see that $\int_0^{2\pi} \cos mx \cos nx\, dx = 0 - 0 = 0.$ ∎

47.
$$\int_0^{\pi/2} \sin^2 x\, dx = \int_0^{\pi/2} \frac{1}{2}(1 - \cos 2x)\, dx = \frac{1}{2}\left(x - \frac{\sin 2x}{2}\right)\Big|_0^{\pi/2} = \frac{1}{2}\left(\frac{\pi}{2} - 0\right) - \frac{1}{2}(0 - 0) = \frac{1}{2}\cdot\frac{\pi}{2} = \frac{\pi}{4}.$$ ∎

49.   Use the reduction formula of Example 7.2.7 and the result of the preceding problem.

$$\int_0^{\pi/2} \sin^6 x\, dx = \frac{-\sin^5 x \cos x}{6}\Big|_0^{\pi/2} + \frac{5}{6}\int_0^{\pi/2} \sin^4 x\, dx = \frac{-1^5 \cdot 0}{6} - \frac{-0^5 \cdot 1}{6} + \frac{5}{6}\cdot\frac{3}{4}\cdot\frac{1}{2}\cdot\frac{\pi}{2} = \frac{5}{6}\cdot\frac{3}{4}\cdot\frac{1}{2}\cdot\frac{\pi}{2}.$$ ∎

51.   Solution 49 shows that $\displaystyle\int_0^{\pi/2} \sin^{2n} x\, dx = \left(\frac{2n - 1}{2n}\right)\left(\frac{2n - 3}{2n - 2}\right)\cdots\left(\frac{1}{2}\right)\left(\frac{\pi}{2}\right)$ is true if $n = 3$. With the understanding that $(2n - 1)(2n - 3)\cdots(1)$ means $(3)(1)$ when $n = 2$ and means (1) when $n = 1$, etc., then Solutions 47 and 48 show the equation also holds if $n = 1$ or $n = 2$. Suppose the equation is correct if $n = k$. Use Problem 50 to infer

$$\int_0^{\pi/2} \sin^{2(k+1)} x\, dx = \left(\frac{2k + 1}{2k + 2}\right)\int_0^{\pi/2} \sin^{2k} x\, dx = \left(\frac{2k + 1}{2k + 2}\right)\left[\left(\frac{2k - 1}{2k}\right)\left(\frac{2k - 3}{2k - 2}\right)\cdots\left(\frac{3}{4}\right)\left(\frac{1}{2}\right)\left(\frac{\pi}{2}\right)\right]$$

which means the equation is also correct for $n = k + 1$. The principal of mathematical induction guarantees the equation is correct for all positive integers n. ∎

53. $-1 \leq \sin x \leq 1$ implies $|\sin x| \leq 1$ and $\left|\sin^{2n} x\right| \leq 1$ for all real x and all positive integers n. Therefore,

$$\left|\int_0^{\pi/2} \sin^{2n} x\, dx\right| \leq \int_0^{\pi/2} \left|\sin^{2x} x\right|\, dx \leq \int_0^{\pi/2} 1\, dx = \frac{\pi}{2}.$$ ∎

REMARK: Problem 4.5.42 states the comparison theorem used above. Because $(2k - 1)/2k < 1$ for all k, another proof can be based on the result of Problem 51.

---

55.   Problems 51 and 52 imply

$$\frac{\pi}{2} = \left(\frac{2}{1}\right)\left(\frac{4}{3}\right)\cdots\left(\frac{2n}{2n - 1}\right)\int_0^{\pi/2} \sin^{2n} x\, dx$$

$$1 = \left(\frac{3}{2}\right)\left(\frac{5}{4}\right)\cdots\left(\frac{2n + 1}{2n}\right)\int_0^{\pi/2} \sin^{2n+1} x\, dx$$

Therefore,

$$\frac{\pi}{2} = \frac{\pi/2}{1} = \frac{(2/1)}{(3/2)} \cdot \frac{(4/3)}{(5/4)} \cdots \frac{\frac{2n}{2n-1} \int_0^{\pi/2} \sin^{2n} x \, dx}{\frac{2n+1}{2n} \int_0^{\pi/2} \sin^{2n+1} x \, dx}$$

$$= \left(\frac{2}{1}\right) \left(\frac{2}{3}\right) \cdot \left(\frac{4}{3}\right) \left(\frac{4}{5}\right) \cdots \left(\frac{2n}{2n-1}\right) \left(\frac{2n}{2n+1}\right) \frac{\int_0^{\pi/2} \sin^{2n} x \, dx}{\int_0^{\pi/2} \sin^{2n+1} x \, dx}$$

Problem 54 assures us the quotient of integrals tends to 1 as $n \to \infty$; therefore

$$\lim_{n\to\infty} \left(\frac{2}{1}\right) \left(\frac{2}{3}\right) \cdot \left(\frac{4}{3}\right) \left(\frac{4}{5}\right) \cdots \left(\frac{2n}{2n-1}\right) \left(\frac{2n}{2n+1}\right) = \frac{\pi}{2} \cdot \lim_{n\to\infty} \frac{\int_0^{\pi/2} \sin^{2n+1} x \, dx}{\int_0^{\pi/2} \sin^{2n} x \, dx} = \frac{\pi}{2} \cdot 1 = \frac{\pi}{2}. \ \blacksquare$$

Solutions 7.4   *Trigonometric substitutions*   (pages 501-2)

1. $\displaystyle\int_0^1 \frac{dx}{\sqrt{4-x^2}} \overset{x \equiv 2\sin\theta}{=} \int_0^{\pi/6} \frac{2\cos\theta \, d\theta}{\sqrt{4 - 4\sin^2\theta}} = \int_0^{\pi/6} \frac{2\cos\theta \, d\theta}{|2\cos\theta|} = \int_0^{\pi/6} d\theta = \theta\big|_0^{\pi/6} = \frac{\pi}{6}.$

3. $\displaystyle\int \frac{x \, dx}{\sqrt{9-x^2}} \overset{x \equiv 3\sin\theta}{=} \int \frac{(3\sin\theta)(3\cos\theta \, d\theta)}{\sqrt{9-9\sin^2\theta}} = \int \frac{9\sin\theta\cos\theta \, d\theta}{|3\cos\theta|} = 3\int \sin\theta \, d\theta = -3\cos\theta + C$

$$= -3\cos\left(\sin^{-1}\frac{x}{3}\right) + C$$

$$= -3\sqrt{1 - \left(\frac{x}{3}\right)^2} + C = -\sqrt{9-x^2} + C.$$

REMARK: θ in range of $\sin^{-1}$ implies $\cos\theta \geq 0$; Solution 6.8.69 shows that $\cos\left(\sin^{-1} w\right) = \sqrt{1-w^2}$.

5. $\displaystyle\int_0^1 \sqrt{4-x^2} \, dx \overset{x \equiv 2\sin\theta}{=} \int_0^{\pi/6} \sqrt{4-4\sin^2\theta} \, (2\cos\theta \, d\theta) = \int_0^{\pi/6} 4\cos^2\theta \, d\theta = \int_0^{\pi/6} 2(1 + \cos 2\theta) \, d\theta$

$$= (2\theta + \sin 2\theta)\big|_0^{\pi/6} = \frac{\pi}{3} + \frac{\sqrt{3}}{2}.$$

7. $\displaystyle\int_0^5 \frac{x \, dx}{\sqrt{25+x^2}} \overset{x \equiv 5\tan\theta}{=} \int_0^{\pi/4} \frac{(5\tan\theta)(5\sec^2\theta \, d\theta)}{|5\sec\theta|} = 5\int_0^{\pi/4} \sec\theta\tan\theta \, d\theta = 5\sec\theta\big|_0^{\pi/4} = 5\left(\sqrt{2} - 1\right).$

REMARK: $\displaystyle\int_0^5 \frac{x \, dx}{\sqrt{25+x^2}} = \sqrt{25+x^2}\,\Big|_0^5.$

9. Use the substitution $x = 2\sec\theta$, let $S = \mathrm{sgn}(x) = x/|x| = \mathrm{sgn}(\tan\theta)$.

$$\int \frac{x \, dx}{\sqrt{x^2-4}} = \int \frac{(2\sec\theta)(2\sec\theta\tan\theta \, d\theta)}{|2\tan\theta|} = 2S\int \sec^2\theta \, d\theta = 2S\tan\theta + C = 2\,|\tan\theta| + C = \sqrt{x^2-4} + C.$$

11. $\displaystyle\int \frac{dx}{(x^2+36)^{3/2}} \overset{x \equiv 6\tan\theta}{=} \int \frac{6\sec^2\theta \, d\theta}{|6^3\sec^3\theta|} = \frac{1}{36}\int \frac{d\theta}{\sec\theta} = \frac{1}{36}\int \cos\theta \, d\theta$

$$= \frac{1}{36}\sin\theta + C = \frac{1}{36}\frac{x/6}{\sqrt{1+(x/6)^2}} + C = \frac{x}{36\sqrt{36+x^2}} + C.$$

REMARK: $\theta \in (-\pi/2, \pi/2) = $ range $\tan^{-1} \implies \sec\theta > 0.$

13. $\displaystyle\int \frac{x \, dx}{36+x^2} \overset{x \equiv 6\tan\theta}{=} \int \frac{(6\tan\theta)(6\sec^2\theta \, d\theta)}{36\sec^2\theta} = \int \tan\theta \, d\theta = \ln|\sec\theta| + C_*$

$$= \ln\sqrt{1 + \left(\frac{x}{6}\right)^2} + C_* = \frac{\ln(36+x^2)}{2} + C.$$

15. After the substitution $x = 2\sin\theta$, apply the reduction formula of Problem 7.2.53.

$$\int_0^1 (4-x^2)^{3/2}\,dx = \int_0^{\pi/6} (4\cos^2\theta)^{3/2}(2\cos\theta\,d\theta) = \int_0^{\pi/6} 16\cos^4\theta\,d\theta$$

$$= 4\cos^3\theta\sin\theta\Big|_0^{\pi/6} + 12\int_0^{\pi/6}\cos^2\theta\,d\theta = \frac{3\sqrt{3}}{4} + 6\cos\theta\sin\theta\Big|_0^{\pi/6} + 6\int_0^{\pi/6}d\theta$$

$$= \frac{3\sqrt{3}}{4} + \frac{3\sqrt{3}}{2} + \pi = \frac{9\sqrt{3}}{4} + \pi.$$

17. After the substitution $x = 2\tan\theta$, use the result of Example 7.3.5.

$$\int_0^2 \frac{x^2\,dx}{\sqrt{x^2+4}} = \int_0^{\pi/4} \frac{(4\tan^2\theta)(2\sec^2\theta\,d\theta)}{|2\sec\theta|} = 4\int_0^{\pi/4}\tan^2\theta\sec\theta\,d\theta = 2\left(\sec\theta\tan\theta - \ln|\sec\theta+\tan\theta|\right)\Big|_0^{\pi/4}$$

$$= 2\left(\sqrt{2} - \ln\left|\sqrt{2}+1\right|\right) - 2(0-0) = 2\sqrt{2} - \ln\left(3+2\sqrt{2}\right).$$

19. Use the substitution $x = 6\sec\theta$, let $S = \text{sgn}(x) = x/|x| = \text{sgn}(\tan\theta)$, use the result of Example 7.2.6.

$$\int \frac{x^2\,dx}{\sqrt{x^2-36}} = \int \frac{(36\sec^2\theta)(6\sec\theta\tan\theta\,d\theta)}{|6\tan\theta|} = 36S\int\sec^3\theta\,d\theta = 18S\left(\sec\theta\tan\theta + \ln|\sec\theta+\tan\theta|\right) + C_*$$

$$= 18S\left(\left(\frac{x}{6}\right)\left(S\sqrt{\left(\frac{x}{6}\right)^2-1}\right) + \ln\left|\frac{x}{6} + S\sqrt{\left(\frac{x}{6}\right)^2-1}\right|\right) + C_*$$

$$= \frac{x\sqrt{x^2-36}}{2} + 18\ln\left|x+\sqrt{x^2-36}\right| + C.$$

21.
$$\int_4^5 \frac{dx}{(x^2-9)^{3/2}} \overset{x\equiv 3\sec\theta}{=} \int_{\sec^{-1}(4/3)}^{\sec^{-1}(5/3)} \frac{3\sec\theta\tan\theta\,d\theta}{|27\tan^3\theta|} = \frac{1}{9}\int_{\sec^{-1}(4/3)}^{\sec^{-1}(5/3)} \frac{\cos\theta\,d\theta}{\sin^2\theta} = \frac{-\csc\theta}{9}\Big|_{\sec^{-1}(4/3)}^{\sec^{-1}(5/3)}$$

$$= \left(\frac{-1}{9}\right)\left(\frac{5/3}{\sqrt{(5/3)^2-1}} - \frac{4/3}{\sqrt{(4/3)^2-1}}\right) = \left(\frac{-1}{9}\right)\left(\frac{5}{4} - \frac{4}{\sqrt{7}}\right).$$

23.
$$\int \frac{dx}{x\sqrt{x^2+a^2}} \overset{x\equiv|a|\tan\theta}{=} \int \frac{|a|\sec^2\theta\,d\theta}{(|a|\tan\theta)\,|a\sec\theta|} = \frac{1}{|a|}\int\csc\theta\,d\theta = \frac{-1}{|a|}\ln|\csc\theta+\cot\theta| + C$$

$$= \frac{-1}{|a|}\ln\left|\frac{\sqrt{(x/|a|)^2+1}}{x/|a|} + \frac{1}{x/a}\right| + C = \frac{-1}{|a|}\ln\left|\frac{\sqrt{x^2+a^2}+|a|}{x}\right| + C.$$

We can replace $|a|$ with a in this final expression because

$$\ln\left|\frac{\sqrt{x^2+a^2}-a}{x}\right| = \ln\left|\left(\frac{\sqrt{x^2+a^2}-a}{x}\right)\left(\frac{\sqrt{x^2+a^2}+a}{\sqrt{x^2+a^2}+a}\right)\right| = \ln\left|\frac{x}{\sqrt{x^2+a^2}+a}\right| = -\ln\left|\frac{\sqrt{x^2+a^2}+a}{x}\right|.$$

25. Use the substitution $x = |a|\tan\theta$ and the result of Example 7.2.6.

$$\int \sqrt{a^2+x^2}\,dx = \int |a\sec\theta|(|a|\sec^2\theta\,d\theta) = a^2\int\sec^3\theta\,d\theta = \frac{a^2}{2}\left(\sec\theta\tan\theta + \ln|\sec\theta+\tan\theta|\right) + C_*$$

$$= \frac{a^2}{2}\left(\sqrt{1+\left(\frac{x}{|a|}\right)^2}\left(\frac{x}{|a|}\right) + \ln\left|\sqrt{1+\left(\frac{x}{|a|}\right)^2} + \left(\frac{x}{|a|}\right)\right|\right) + C_*$$

$$= \frac{1}{2}\left(x\sqrt{a^2+x^2} + a^2\ln\left(x+\sqrt{a^2+x^2}\right)\right) + C.$$

27.
$$\int \frac{x^2\,dx}{x^2+a^2} \overset{x\equiv a\tan\theta}{=} \int \frac{(a^2\tan^2\theta)(a\sec^2\theta\,d\theta)}{a^2\sec^2\theta} = a\int\tan^2\theta\,d\theta = a\int(\sec^2\theta-1)\,d\theta$$

$$= a(\tan\theta-\theta) + C = x - a\tan^{-1}\left(\frac{x}{a}\right) + C.$$

7.4 — Trigonometric substitutions

29.

$$\int \frac{dx}{x^3\,(x^2+a^2)} \stackrel{x=a\tan\theta}{=} \int \frac{a\sec^2\theta\,d\theta}{(a^3\tan^3\theta)\,(a^2\sec^2\theta)} = a^{-4}\int \cot^3\theta\,d\theta = a^{-4}\int \left(\cot\theta\csc^2\theta - \cot\theta\right)\,d\theta$$

$$= a^{-4}\left(\frac{-\cot^2\theta}{2} - \ln|\sin\theta|\right) + C$$

$$= a^{-4}\left(\frac{-(a/x)^2}{2} - \ln\left|\frac{x/a}{\sqrt{(x/a)^2+1}}\right|\right) + C = \frac{-1}{2a^2x^2} - \frac{1}{a^4}\ln\left|\frac{x}{\sqrt{x^2+a^2}}\right| + C.$$

31. Use the substitution $x = |a|\sin\theta$ and the result of Example 7.2.8.

$$\int \frac{x^3\,dx}{\sqrt{a^2-x^2}} = \int \frac{(|a|^3\sin^3\theta)\,(|a|\cos\theta\,d\theta)}{|a\cos\theta|} = |a|^3\int \sin^3\theta\,d\theta = |a|^3\left(\frac{-\sin^2\theta\cos\theta - 2\cos\theta}{3}\right) + C$$

$$= \frac{-|a|^3}{3}\left(\sin^2\theta + 2\right)\cos\theta + C = \frac{-|a|^3}{3}\left(\left(\frac{x}{|a|}\right)^2 + 2\right)\sqrt{1 - \left(\frac{x}{|a|}\right)^2} + C$$

$$= \frac{-1}{3}\left(x^2 + 2a^2\right)\sqrt{a^2-x^2} + C.$$

33.

$$\int \frac{\sqrt{5x^2+9}}{x}\,dx \stackrel{\sqrt{5}x=3\tan\theta}{=} \int \frac{|3\sec\theta|}{(3\tan\theta)/\sqrt{5}} \cdot \frac{3\sec^2\theta\,d\theta}{\sqrt{5}} = 3\int \frac{\sec^3\theta\,d\theta}{\tan\theta} = 3\int \frac{d\theta}{\sin\theta\,\cos^2\theta}$$

$$= 3\int \frac{(\sin^2\theta + \cos^2\theta)\,d\theta}{\sin\theta\,\cos^2\theta} = 3\int \left(\sec\theta\tan\theta + \csc\theta\right)\,d\theta$$

$$= 3\left(\sec\theta - \ln|\csc\theta + \cot\theta|\right) + C$$

$$= 3\left(\sqrt{\left(\frac{\sqrt{5}x}{3}\right)^2 + 1} - \ln\left|\frac{\sqrt{(\sqrt{5}x/3)^2+1}}{\sqrt{5}x/3} + \frac{1}{\sqrt{5}x/3}\right|\right) + C$$

$$= \sqrt{5x^2+9} - 3\ln\left|\frac{\sqrt{5x^2+9}+3}{\sqrt{5}\,x}\right| + C.$$

35. Use the substitution $x = |a|\cos\theta$ and the result of Example 7.2.6.

$$\int \frac{dx}{x^3\sqrt{a^2-x^2}} = \int \frac{-|a|\sin\theta\,d\theta}{(|a|^3\cos^3\theta)\,|a\sin\theta|} = \frac{-1}{|a|^3}\int \sec^3\theta\,d\theta = \frac{-1}{2|a|^3}\left(\sec\theta\tan\theta + \ln|\sec\theta + \tan\theta|\right) + C$$

$$= \frac{-1}{2|a|^3}\left(\left(\frac{|a|}{x}\right)\left(\frac{\sqrt{a^2-x^2}}{x}\right) + \ln\left|\frac{|a|}{x} + \frac{\sqrt{a^2-x^2}}{x}\right|\right) + C$$

$$= \frac{-\sqrt{a^2-x^2}}{2a^2x^2} - \frac{1}{2|a|^3}\ln\left|\frac{|a| + \sqrt{a^2-x^2}}{x}\right| + C.$$

37.

$$\int \frac{\sqrt{9-5x^2}}{x}\,dx \stackrel{\sqrt{5}x=3\sin\theta}{=} \int \frac{|3\cos\theta|}{(3/\sqrt{5})\sin\theta} \cdot \frac{3\cos\theta\,d\theta}{\sqrt{5}} = 3\int \frac{\cos^2\theta}{\sin\theta}\,d\theta = 3\int \frac{1-\sin^2\theta}{\sin\theta}\,d\theta$$

$$= 3\int \left(\csc\theta - \sin\theta\right)\,d\theta = 3\left(-\ln|\csc\theta + \cot\theta| + \cos\theta\right) + C_*$$

$$= 3\left(-\ln\left|\frac{3}{\sqrt{5}x} + \frac{\sqrt{9-5x^2}}{\sqrt{5}x}\right| + \frac{\sqrt{9-5x^2}}{3}\right) + C_*$$

$$= 3\ln\left|\frac{x}{3+\sqrt{9-5x^2}}\right| + \sqrt{9-5x^2} + C.$$

39. Use the substitution $x = (3/\sqrt{5})\sec\theta$, and let $S = \operatorname{sgn}(x) = x/|x| = \operatorname{sgn}(\tan\theta)$.

$$\int \frac{\sqrt{5x^2 - 9}}{x}\,dx = \int \frac{|3\tan\theta|}{(3/\sqrt{5})\sec\theta}\cdot \frac{3\sec\theta\tan\theta\,d\theta}{\sqrt{5}} = 3S\int \tan^2\theta\,d\theta = 3S\int (\sec^2\theta - 1)\,d\theta$$

$$= 3S(\tan\theta - \theta) + C = 3\left(|\tan\theta| - S\theta\right) + C = 3\left(\frac{\sqrt{5x^2 - 9}}{3} - S\cos^{-1}\left(\frac{3}{\sqrt{5}\,x}\right)\right) + C$$

$$= \sqrt{5x^2 - 9} - 3S\sec^{-1}\left(\frac{\sqrt{5}\,x}{3}\right) + C.$$

REMARK: Because $\sec^{-1}(t) + \sec^{-1}(-t) = \pi$ for all t such that $|t| \geq 1$, these two cases, inherent in the use of S above, can be unified in the form $\sqrt{5x^2 - 9} - 3\sec^{-1}\left(\dfrac{\sqrt{5}\,|x|}{3}\right) + C_*$.

41.
$$\int \frac{dx}{(a^2 - x^2)^{3/2}} \overset{x \equiv |a|\sin\theta}{=} \int \frac{|a|\cos\theta\,d\theta}{|a\cos\theta|^3} = \frac{1}{a^2}\int \sec^2\theta\,d\theta = \frac{1}{a^2}\tan\theta + C$$

$$= \frac{1}{a^2}\cdot \frac{x/|a|}{\sqrt{1 - (x/|a|)^2}} + C = \frac{x}{a^2\sqrt{a^2 - x^2}} + C.$$

43. Use the substitution $x = |a|\sec\theta$, and let $S = \operatorname{sgn}(x) = x/|x| = \operatorname{sgn}(\tan\theta)$.

$$\int \frac{(x^2 - a^2)^{3/2}}{x^3}\,dx = \int \frac{|a\tan\theta|^3}{|a|^3\sec^3\theta}\left(|a|\sec\theta\tan\theta\,d\theta\right) = |a|S\int \frac{\tan^4\theta}{\sec^2\theta}\,d\theta = |a|S\int \frac{(\sec^2\theta - 1)^2}{\sec^2\theta}\,d\theta$$

$$= |a|S\int (\sec^2\theta - 2 + \cos^2\theta)\,d\theta = |a|S\left(\tan\theta - 2\theta + \frac{1}{2}(\cos\theta\sin\theta + \theta)\right) + C_*$$

$$= |a|S\left(S\sqrt{\left(\frac{x}{|a|}\right)^2 - 1} - \frac{3}{2}\sec^{-1}\frac{x}{|a|} + \frac{1}{2}\frac{1}{x/|a|}\cdot \frac{\sqrt{(x/|a|)^2 - 1}}{|x/a|}\right) + C_*$$

$$= \left(1 + \frac{a^2}{2x^2}\right)\sqrt{x^2 - a^2} - \frac{3|a|}{2}\sec^{-1}\left|\frac{x}{a}\right| + C.$$

45. Use the substitution $x = 7\sec\theta$, and let $S = \operatorname{sgn}(x) = x/|x| = \operatorname{sgn}(\tan\theta)$.

$$\int x^3\sqrt{x^2 - 49}\,dx = \int (7^3\sec^3\theta)\,|7\tan\theta|\,(7\sec\theta\tan\theta\,d\theta) = 7^5 S\int \tan^2\theta\,\sec^4\theta\,d\theta$$

$$= 7^5 S\int \tan^2\theta\,(\tan^2\theta + 1)\,(\sec^2\theta\,d\theta) = 7^5 S\int (\tan^4\theta + \tan^2\theta)\,d(\tan\theta)$$

$$= 7^5 S\left(\frac{1}{5}\tan^5\theta + \frac{1}{3}\tan^3\theta\right) + C$$

$$= 7^5 S\left(\frac{1}{5}\left[S\sqrt{[\tfrac{x}{7}]^2 - 1}\right]^5 + \frac{1}{3}\left[S\sqrt{[\tfrac{x}{7}]^2 - 1}\right]^3\right) + C = \left[\frac{3x^2 + 98}{15}\right][x^2 - 49]^{3/2} + C.$$

47.
$$\int_0^{\pi/2} \pi(\cos x)^2\,dx = \int_0^{\pi/2} \frac{\pi}{2}(1 + \cos 2x)\,dx = \frac{\pi}{2}\left(x + \frac{1}{2}\sin 2x\right)\Big|_0^{\pi/2} = \left(\frac{\pi}{2}\right)^2 = \frac{\pi^2}{4}.$$

49.
$$\int_0^{\pi/4} \pi(\cos^2 x - \sin^2 x)\,dx = \int_0^{\pi/4} \pi\cos 2x\,dx = \frac{\pi}{2}\sin 2x\Big|_0^{\pi/4} = \frac{\pi}{2}.$$

51. $\boxed{\text{Cylindrical shell method}}$ Use the substitution $x^2 = \tan\theta$ and the result of Example 7.2.6.

$$\int_0^1 2\pi x\sqrt{x^4 + 1}\,dx = \pi\int_0^1 \sqrt{x^4 + 1}\,(2x\,dx) = \pi\int_0^{\pi/4} \sqrt{\tan^2\theta + 1}\,(\sec^2\theta\,d\theta) = \pi\int_0^{\pi/4} \sec^3\theta\,d\theta$$

$$= \frac{\pi}{2}\left(\sec\theta\tan\theta + \ln|\sec\theta + \tan\theta|\right)\Big|_0^{\pi/4} = \frac{\pi}{2}\left(\sqrt{2} + \ln\left(\sqrt{2} + 1\right)\right).$$

7.4 — Trigonometric substitutions

Use the substitution $y = \sec\theta$ and the result of Example 7.3.5.

$$\int_0^{\sqrt 2} \pi\left(1^2 - x^2\right) dy = \pi \int_0^{\sqrt 2} \left(1 - \sqrt{y^2 - 1}\right) dy = \pi\sqrt 2 - \pi \int_0^{\pi/4} \tan\theta \; (\sec\theta \; \tan\theta \; d\theta)$$

$$= \pi\sqrt 2 - \frac{\pi}{2} \left(\sec\theta\;\tan\theta - \ln|\sec\theta + \tan\theta|\right)\Big|_0^{\pi/2} = \pi\sqrt 2 - \frac{\pi}{2}\left(\sqrt 2 - \ln\left(\sqrt 2 + 1\right)\right).$$

53. a. The curve $\{(x,y) : 0 \le x \le 1, y = \sqrt{1 - x^2}\}$ is a quarter circle.

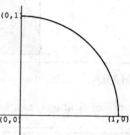

b.
$$\int_0^1 \sqrt{1 - x^2}\, dx \overset{x \equiv \sin\theta}{=} \int_0^{\pi/2} \sqrt{1 - \sin^2\theta}\; d(\sin\theta) = \int_0^{\pi/2} \cos^2\theta\, d\theta = \frac{1}{2}\left(\theta + \sin\theta\cos\theta\right)\Big|_0^{\pi/2}$$

$$= \frac{1}{2}\left(\left(\frac{\pi}{2} + 0\right) - (0 + 0)\right) = \frac{\pi}{4}.$$

REMARK: If you prefer that your work for part (b) not mimic Example 1, then consider

$$\int_0^1 \sqrt{1 - x^2}\, dx \overset{x \equiv \cos\theta}{=} \int_{\pi/2}^0 \sqrt{1 - \cos^2\theta}\; d(\cos\theta) = \int_{\pi/2}^0 -\sin^2\theta\; d\theta = \int_0^{\pi/2} \frac{1}{2}\left(1 - \cos 2\theta\right) d\theta = \dots.$$

55. If $x < -a < 0$ and $\theta = \sec^{-1}(x/a)$, then θ is in $(\pi/2, \pi)$ and $\tan\theta$ is negative on that interval. Since $a > 0$, this implies $a\tan\theta < 0$. Therefore $\sqrt{x^2 - a^2} = \sqrt{(a\sec\theta)^2 - a^2} = |a|\sqrt{\sec^2\theta - 1} = a\sqrt{\tan^2\theta} = a\,|\tan\theta| = -a\tan\theta$, thus (also using the result of Example 7.3.5),

$$\int \sqrt{x^2 - a^2}\, dx = \int (-a\tan\theta)\,(a\sec\theta\,\tan\theta\, d\theta) = -a^2 \int \tan^2\theta\,\sec\theta\, d\theta$$

$$= \frac{-a^2}{2}\left(\sec\theta\,\tan\theta - \ln|\sec\theta + \tan\theta|\right) + C_*$$

$$= \frac{-a^2}{2}\left[\left(\frac{x}{a}\right)\left(\frac{\sqrt{x^2 - a^2}}{-a}\right) - \ln\left|\frac{x}{a} + \frac{\sqrt{x^2 - a^2}}{-a}\right|\right] + C_*$$

$$= \left(\frac{1}{2}\right) x\sqrt{x^2 - a^2} + \left(\frac{a^2}{2}\right)\ln\left|x - \sqrt{x^2 - a^2}\right| + C. \;\blacksquare$$

57. a. Because $\theta \in \left(\frac{-\pi}{2}, \frac{\pi}{2}\right) = \text{range } \tan^{-1} \implies \sec\theta > 0$, we can infer

$$\int \frac{dx}{\sqrt{1 + x^2}} \overset{x \equiv \tan\theta}{=} \int \frac{\sec^2\theta\, d\theta}{|\sec\theta|} = \int \sec\theta\, d\theta = \ln|\tan\theta + \sec\theta| + C = \ln\left|x + \sqrt{x^2 + 1}\right| + C. \;\blacksquare$$

b. $\sinh^{-1} x$ and $\ln\left|x + \sqrt{x^2 + 1}\right|$ have domain $\mathbb{R}$, each is an antiderivative of $\left(1 + x^2\right)^{-1/2}$; therefore they differ by a constant. Because $\sinh^{-1} 0 = 0 = \ln 1 = \ln\left|0 + \sqrt{0^2 + 1}\right|$, that constant difference is zero and the functions are always equal. $\blacksquare$

59. $y^2 = 2x - x^2 = 1 - (x - 1)^2$ is the equation of a circle of radius 1 centered at $(1,0)$; thus

$$\int_0^1 \sqrt{2x - x^2}\, dx = \text{area of quarter circle with radius } 1 = \left(\frac{1}{4}\right)\pi \cdot 1^2 = \frac{\pi}{4}.$$

$y^2 = 2 - x^2$ is the equation of a circle of radius $\sqrt 2$ centered at $(0,0)$. The region whose area is computed by $\int_0^1 \sqrt{2 - x^2}\, dx$ can be decomposed into a right triangle with vertices $(0,0)$, $(1,0)$, $(1,1)$ and a circular sector with arc from $(1,1)$ to $(0,\sqrt 2)$. The triangle has area $1/2$, the sector is one-eighth of the circle and its area is $\left(\frac{1}{8}\right)\pi\left(\sqrt 2\right)^2 = \frac{\pi}{4}$; thus $\int_0^1 \sqrt{2 - x^2}\, dx = \frac{1}{2} + \frac{\pi}{4}$.

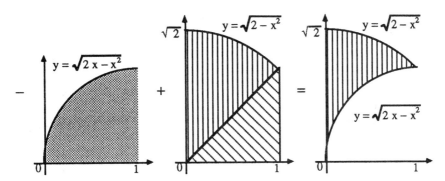

Therefore

$$\int_0^1 \left(\sqrt{2-x^2} - \sqrt{2x-x^2}\right) dx = \left(\frac{1}{2} + \frac{\pi}{4}\right) - \frac{\pi}{4} = \frac{1}{2}.$$

REMARK: This is Problem Q370 from *Mathematics Magazine*, volume 38 (1965), pages 326 and 302.

Solutions 7.5   *Other substitutions*   (pages 504-5)

1. $\displaystyle \int \frac{x}{(x-1)^4} \, dx \stackrel{x-1\equiv u}{=} \int \frac{u+1}{u^4} \, du = \int \left(u^{-3} + u^{-4}\right) du = \frac{-u^{-2}}{2} + \frac{-u^{-3}}{3} + C$

$$= \frac{-1}{2(x-1)^2} + \frac{-1}{3(x-1)^3} + C = \frac{-3x+1}{6(x-1)^3} + C.$$

3. $\displaystyle \int_0^{13} x(1+2x)^{2/3} \, dx \stackrel{1+2x\equiv u}{=} \int_1^{27} \left(\frac{u-1}{2}\right) u^{2/3} \left(\frac{du}{2}\right)$

$$= \frac{1}{4} \int_1^{27} \left(u^{5/3} - u^{2/3}\right) du = \left(\frac{1}{4}\right) \left(\frac{3}{8} u^{8/3} - \frac{3}{5} u^{5/3}\right)\Bigg|_1^{27}$$

$$= \left(\frac{1}{4}\right) \left(\frac{3}{8} 3^8 - \frac{3}{5} 3^5\right) - \left(\frac{1}{4}\right) \left(\frac{3}{8} - \frac{3}{5}\right) = \frac{92583}{160} + \frac{9}{160} = 578.7.$$

5. $\displaystyle \int_0^1 \left(1-x^3\right)^{1/3} x^8 \, dx = \int_0^1 \left(1-x^3\right)^{1/3} \left(x^3\right)^2 \left(x^2 \, dx\right) \stackrel{1-x^3\equiv u^3}{=} \int_1^0 u \cdot \left(1-u^3\right)^2 \left(-u^2 \, du\right)$

$$= \int_0^1 \left(u^3 - 2u^6 + u^9\right) du$$

$$= \left(\frac{u^4}{4} - \frac{2u^7}{7} + \frac{u^{10}}{10}\right)\Bigg|_0^1 = \frac{1}{4} - \frac{2}{7} + \frac{1}{10} = \frac{9}{140}.$$

7. $\displaystyle \int x\sqrt{1+x} \, dx \stackrel{1+x\equiv u}{=} \int (u-1)\sqrt{u} \, du = \int \left(u^{3/2} - u^{1/2}\right) du = \frac{2}{5} u^{5/2} - \frac{2}{3} u^{3/2} + C$

$$= \frac{2}{5}(1+x)^{5/2} - \frac{2}{3}(1+x)^{3/2} + C = \frac{2}{15}(3x-2)(x+1)^{3/2} + C.$$

9. $\displaystyle \int \frac{3x^2 - x}{\sqrt{1+x}} \, dx \stackrel{1+x\equiv V^2}{=} \int \frac{3\left(V^2-1\right)^2 - \left(V^2-1\right)}{V} (2V \, dV) = 2 \int \left(3V^4 - 7V^2 + 4\right) dV$

$$= 2 \left(\frac{3V^5}{5} - \frac{7V^3}{3} + 4V\right) + C = 2 \left(\frac{3}{5}(1+x)^{5/2} - \frac{7}{3}(1+x)^{3/2} + 4(1+x)^{1/2}\right) + C$$

$$= \frac{2}{15} \left(9x^2 - 17x + 34\right) (1+x)^{1/2} + C.$$

11. $\displaystyle \int \frac{dx}{\sqrt{e^{2x} - 1}} = \int \left(\frac{e^{-x}}{e^{-x}}\right) \frac{dx}{\sqrt{e^{2x} - 1}} = \int \frac{e^{-x} \, dx}{\sqrt{1 - (e^{-x})^2}} \stackrel{e^{-x}\equiv u}{=} \int \frac{-du}{\sqrt{1-u^2}} = -\sin^{-1} u + C$

$$= -\sin^{-1}\left(e^{-x}\right) + C.$$

7.5 — Other substitutions

$u = \tan\left(\dfrac{x}{2}\right)$: Some consequences of this substitution (see Example 2) are gathered here for reference.

$$x = 2\tan^{-1} u$$

$$dx = \frac{2\,du}{1 + u^2}$$

$$u = \tan\left(\frac{x}{2}\right) \quad \text{and} \quad \frac{-\pi}{2} < \frac{x}{2} < \frac{\pi}{2}$$

$$\sec\left(\frac{x}{2}\right) = \sqrt{1 + \tan^2\left(\frac{x}{2}\right)} = \sqrt{1 + u^2}$$

$$\cos\left(\frac{x}{2}\right) = \frac{1}{\sec(x/2)} = \frac{1}{\sqrt{1 + u^2}}$$

$$\sin\left(\frac{x}{2}\right) = \tan\left(\frac{x}{2}\right) \cdot \cos\left(\frac{x}{2}\right) = \frac{u}{\sqrt{1 + u^2}}$$

$$\sin x = 2\sin\left(\frac{x}{2}\right) \cdot \cos\left(\frac{x}{2}\right) = \frac{2u}{1 + u^2}$$

$$1 + \sin x = \frac{(1 + u^2) + 2u}{1 + u^2} = \frac{(1 + u)^2}{1 + u^2}$$

$$1 - \sin x = \frac{(1 + u^2) - 2u}{1 + u^2} = \frac{(1 - u)^2}{1 + u^2}$$

$$\cos x = \cos^2\left(\frac{x}{2}\right) - \sin^2\left(\frac{x}{2}\right) = 1 - 2\sin^2\left(\frac{x}{2}\right) = 2\cos^2\left(\frac{x}{2}\right) - 1 = \frac{2}{1 + u^2} - 1 = \frac{1 - u^2}{1 + u^2}$$

$$1 + \cos x = 2\cos^2\left(\frac{x}{2}\right) = \frac{2}{1 + u^2}$$

$$1 - \cos x = 2\sin^2\left(\frac{x}{2}\right) = \frac{2u^2}{1 + u^2}$$

$$\frac{1 - \cos x}{1 + \cos x} = \frac{2\sin^2(x/2)}{2\cos^2(x/2)} = \tan^2\left(\frac{x}{2}\right) = u^2$$

$$\frac{\sin x}{1 + \cos x} = \frac{2\sin(x/2)\cos(x/2)}{1 + (2\cos^2(x/2) - 1)} = \frac{\sin(x/2)}{\cos(x/2)} = \tan\left(\frac{x}{2}\right) = u$$

$$\frac{1 - \cos x}{\sin x} = \frac{1 - (1 - 2\sin^2(x/2))}{2\sin(x/2)\cos(x/2)} = \frac{\sin(x/2)}{\cos(x/2)} = \tan\left(\frac{x}{2}\right) = u$$

$$\tan x = \frac{\sin x}{\cos x} = \frac{2u}{1 - u^2} = \frac{1}{1 - u} - \frac{1}{1 + u}$$

13. Read Example 2 and refer to the table above with information about the substitution $u = \tan(x/2)$.

$$\int \frac{dx}{1 + \sin x} = \int (1 + \sin x)^{-1}\,dx = \int \left(\frac{(1 + u)^2}{1 + u^2}\right)^{-1}\left(\frac{2\,du}{1 + u^2}\right) = \int \frac{2\,du}{(1 + u)^2}$$

$$= \frac{-2}{1 + u} + C = \frac{-2}{1 + (1 - \cos x)/\sin x} + C = \frac{-2\sin x}{\sin x + 1 - \cos x} + C.$$

15. $$\int_0^{\pi/3} \frac{dx}{3 + 2\cos x} \overset{\tan(x/2)\equiv u}{=} \int_0^{1/\sqrt{3}} \left(3 + 2\left(\frac{1 - u^2}{1 + u^2}\right)\right)^{-1}\left(\frac{2\,du}{1 + u^2}\right)$$

$$= 2\int_0^{1/\sqrt{3}} \frac{du}{5 + u^2} = \frac{2}{\sqrt{5}}\tan^{-1}\left(\frac{u}{\sqrt{5}}\right)\Big|_0^{1/\sqrt{3}} = \frac{2}{\sqrt{5}}\tan^{-1}\left(\frac{1}{\sqrt{15}}\right).$$

17. $$\int \frac{x\,dx}{\sqrt{9 - x^2}} \overset{9 - x^2 \equiv u}{=} \int \frac{(-1/2)\,du}{\sqrt{u}} = \frac{-1}{2}\int u^{-1/2}\,du = -\sqrt{u} + C = -\sqrt{9 - x^2} + C.$$

19. $$\int \frac{x\,dx}{36 + x^2} \overset{36 + x^2 \equiv u}{=} \int \frac{(1/2)\,du}{u} = \frac{1}{2}\int u^{-1}\,du = \frac{1}{2}\ln|u| + C = \frac{1}{2}\ln(36 + x^2) + C.$$

21. $$\int_0^1 x^3\sqrt{4-x^2}\,dx = \int_0^1 x^2\sqrt{4-x^2}\,(x\,dx) \overset{4-x^2\equiv u}{=} \int_4^3 (4-u)\sqrt{u}\left(\frac{du}{-2}\right) = \frac{-1}{2}\int_4^3\left(4u^{1/2}-u^{3/2}\right)du$$

$$= \frac{-1}{2}\left(\frac{8}{3}u^{3/2}-\frac{2}{5}u^{5/2}\right)\Big|_4^3 = \frac{-11\sqrt{3}}{5}-\frac{-64}{15} = \frac{64-33\sqrt{3}}{15}.$$

23. $$\int\frac{x^3\,dx}{\sqrt{a^2-x^2}} = \int\frac{x^2}{\sqrt{a^2-x^2}}(x\,dx) \overset{a^2-x^2\equiv V}{=} \int\frac{a^2-V}{\sqrt{V}}\left(\frac{dV}{-2}\right) = \int\left(\frac{-a^2}{2}V^{-1/2}+\frac{1}{2}V^{1/2}\right)dV$$

$$= -a^2V^{1/2}+\frac{1}{3}V^{3/2}+C$$

$$= -a^2\left(a^2-x^2\right)^{1/2}+\frac{1}{3}\left(a^2-x^2\right)^{3/2}+C$$

$$= \frac{-1}{3}\left(x^2+2a^2\right)\sqrt{a^2-x^2}+C.$$

25. $$\int x^3\sqrt{x^2-49}\,dx = \int x^2\sqrt{x^2-49}\,(x\,dx) \overset{x^2-49\equiv V}{=} \int(V+49)\sqrt{V}\left(\frac{dV}{2}\right)$$

$$= \frac{1}{2}\int\left(V^{3/2}+49V^{1/2}\right)dV = \frac{1}{5}V^{5/2}+\frac{49}{3}V^{3/2}+C$$

$$= \frac{1}{5}\left(x^2-49\right)^{5/2}+\frac{49}{3}\left(x^2-49\right)^{3/2}+C.$$

27. $\boxed{\text{Disk method}}$ $\displaystyle\int_1^e \pi(\ln x)^2\,dx \overset{x\equiv e^u}{=} \int_0^1 \pi u^2 e^u\,du = \pi\left(u^2-2u+2\right)e^u\Big|_0^1 = \pi e - \pi 2 = \pi(e-2).$

REMARK: $\boxed{\text{Cylindrical shell method}}$ $\displaystyle\int_0^1 2\pi y\left(e-e^y\right)dy = \pi\left(ey^2-2(y-1)e^y\right)\Big|_0^1 = \pi(e-2)$

29. $\boxed{\text{Disk method}}$ $\displaystyle\int_0^{\ln 3}\pi\left(3^2-(e^y)^2\right)dy = \pi\left(\left(9y-\frac{e^{2y}}{2)}\right)\right)\Big|_0^{\ln 3} = \pi\left(9\ln 3-\frac{9}{2}\right)-\pi\left(\frac{-1}{2}\right) = \pi(9\ln 3-4).$

$\boxed{\text{Cylindrical shell method}}$ $\displaystyle\int_1^3 2\pi x\ln x\,dx = \pi x^2\left[\ln x-\frac{1}{2}\right]\Big|_1^3 = 9\pi\left[\ln 3-\frac{1}{2}\right]-\pi\left[\frac{-1}{2}\right] = \pi(9\ln 3-4).$

31. The writing of this solution is simplified by noting that $\cosh w+\sinh w = \dfrac{e^w+e^{-w}}{2}+\dfrac{e^w-e^{-w}}{2} = e^w$.

If $x = \tan^{-1}(\sinh(e^u))$, then

$$dx = \frac{1}{1+(\sinh(e^u))^2}\cdot(\cosh(e^u))\cdot e^u\cdot du = \frac{e^u}{\cosh e^u}\,du,$$

$$\sec x = \sqrt{1+\tan^2 x} = \sqrt{1+(\sinh(e^u))^2} = \cosh(e^u),$$

$$\int\sec x\,dx = \int e^u\,du = e^u+C = \ln e^{(e^u)}+C = \ln|\cosh(e^u)+\sinh(e^u)|+C = \ln|\sec x+\tan x|+C.$$

REMARK: See *The American Mathematical Monthly*, volume 62 (1955), pages 363-364.

33. If $t = x+\sqrt{x^2+a}$, then $(t-x)^2 = x^2+a$, $t^2-2xt = a$, and

$$x = \frac{1}{2}\left(\frac{t^2-a}{t}\right) = \frac{1}{2}\left(t-\frac{a}{t}\right)$$

$$dx = \frac{1}{2}\left(1+\frac{a}{t^2}\right)dt = \frac{(t^2+a)\,dt}{2t^2}$$

$$\sqrt{x^2+a} = t-x = t-\frac{1}{2}\left(t-\frac{a}{t}\right) = \frac{t^2+a}{2t}$$

$$\int\frac{1}{\sqrt{x^2+5}}\,dx = \int\left(\frac{t^2+5}{2t}\right)^{-1}\cdot\frac{(t^2+5)\,dt}{2t^2} = \int\frac{1}{t}\,dt = \ln|t|+C = \ln\left|x+\sqrt{x^2+5}\right|+C.$$

35. $\displaystyle\int_a^b f(x)\,dx \overset{x\equiv(a+b)-u}{=} \int_{(a+b)-a}^{(a+b)-b} f(a+b-u)\,d(a+b-u) = \int_b^a f(a+b-u)\,(-du) = \int_a^b f((a+b-u)\,du.$ ∎

37.   The relation $x^2 = t$ does not uniquely determine t as a function for x for all $x \in [-1, 7]$. The function $x = -\sqrt{t}$ does map $[0, 1]$ onto $[-1, 0]$ and

$$\int_{-1}^{0} \left(3 + 5x^2\right) x^2 \, dx = \int_{1}^{0} (3 + 5t)t \left(\frac{-dt}{2\sqrt{t}}\right) = \int_{0}^{1} \left(\frac{3}{2}t^{1/2} + \frac{5}{2}t^{3/2}\right) dt = \left(t^{3/2} + t^{5/2}\right)\Big|_{0}^{1} = 2;$$

on the other hand, $x = \sqrt{t}$ maps $[0, 49]$ onto $[0, 7]$ and

$$\int_{0}^{7} \left(3 + 5x^2\right) x^2 \, dx = \int_{0}^{49} (3 + 5t)t \left(\frac{dt}{2\sqrt{t}}\right) = \left(t^{3/2} + t^{5/2}\right)\Big|_{0}^{49} = 7^3 + 7^5 = 17150;$$

hence $\int_{-1}^{7} \left(3 + 5x^2\right) x^2 \, dx = \int_{-1}^{0} + \int_{0}^{7} = 2 + 17150 = 17152.$

Solutions 7.6   *Integration of rational fcns I: linear & quadratic denom*   (pages 510-1)

1. $\displaystyle \int \frac{dx}{2x - 5} = \frac{1}{2} \int \frac{d(2x - 5)}{2x - 5} = \frac{1}{2} \ln |2x - 5| + C.$

3. $\displaystyle \int \frac{dx}{3x + 11} = \frac{1}{3} \int \frac{d(3x + 11)}{3x + 11} = \frac{1}{3} \ln |3x + 11| + C.$

5. $\displaystyle \int \frac{dx}{9 + x^2} = \frac{1}{9} \int \frac{dx}{1 + (x^2/9)} = \frac{1}{3} \int \frac{d(x/3)}{1 + (x/3)^2} = \frac{1}{3} \tan^{-1}\left(\frac{x}{3}\right) + C.$

REMARK: See equation (4) in this section or Example 6.8.5.

7. $\displaystyle \int \frac{dx}{x^2 - 16} = \frac{1}{8} \int \left(\frac{1}{x - 4} - \frac{1}{x + 4}\right) dx = \frac{1}{8}\left(\ln|x - 4| - \ln|x + 4|\right) + C = \frac{1}{8} \ln\left|\frac{x - 4}{x + 4}\right| + C.$

REMARK: See equation (6) or Example 7.4.4.

9. $\displaystyle \int \left(\frac{x - 1}{x + 1}\right) dx = \int \left(1 - \frac{2}{x + 1}\right) dx = x - 2 \ln|x + 1| + C.$

11. $\displaystyle \int \frac{2x + 3}{x - 4} \, dx = \int \left(2 + \frac{11}{x - 4}\right) dx = 2x + 11 \ln|x - 4| + C.$

13. $\displaystyle \int \left(\frac{x^3 - x}{2x + 6}\right) dx = \int \left(\frac{x^2}{2} - \frac{3x}{2} + 4 - \frac{12}{x + 3}\right) dx = \frac{x^3}{6} - \frac{3x^2}{4} + 4x - 12 \ln|x + 3| + C.$

15.   Use equation (4) with $a = \sqrt{12} = 2\sqrt{3}$; thus $\displaystyle \int \frac{dx}{x^2 + 12} = \frac{1}{2\sqrt{3}} \tan^{-1}\left(\frac{x}{2\sqrt{3}}\right) + C.$

17. $\displaystyle \int \left(\frac{3x + 5}{x^2 + 9}\right) dx = \int \left(\frac{3}{2} \cdot \frac{2x}{x^2 + 9} + 5 \cdot \frac{1}{x^2 + 3^2}\right) dx = \frac{3}{2} \ln\left(x^2 + 9\right) + \frac{5}{3} \tan^{-1}\left(\frac{x}{3}\right) + C.$

19. $\displaystyle \int \left(\frac{2x - 3}{x^2 - 4}\right) dx = \int \left(\frac{2x}{x^2 - 4} + (-3) \cdot \frac{1}{x^2 - 2^2}\right) dx = \ln|x^2 - 4| + \left(\frac{-3}{4}\right) \ln\left|\frac{x - 2}{x + 2}\right| + C$

  $\displaystyle = \left(1 + \frac{-3}{4}\right) \ln|x - 2| + \left(1 - \frac{-3}{4}\right) \ln|x + 2| + C = \frac{1}{4} \ln|x - 2| + \frac{7}{4} \ln|x + 2| + C.$

21. $\displaystyle \int \left(\frac{x - 1}{x^2 - 6x + 16}\right) dx = \int \left(\frac{(x - 3) + 2}{x^2 - 6x + 16}\right) dx = \int \left(\frac{1}{2} \cdot \frac{2x - 6}{x^2 - 6x + 16} + 2 \cdot \frac{1}{(x - 3)^2 + 7}\right) dx$

  $\displaystyle = \frac{1}{2} \ln|x^2 - 6x + 16| + \frac{2}{\sqrt{7}} \tan^{-1}\left(\frac{x - 3}{\sqrt{7}}\right) + C.$

23.
$$\int \left(\frac{x^3 - x^2 + 2x + 5}{x^2 + x + 1}\right) dx = \int \left(x - 2 + \frac{3x + 7}{x^2 + x + 1}\right) dx$$

$$= \frac{x^2}{2} - 2x + \int \left(\frac{3}{2}\right) \frac{2x + 1}{x^2 + x + 1} dx + \int \left(\frac{11}{2}\right) \frac{1}{x^2 + x + 1} dx$$

$$= \frac{x^2}{2} - 2x + \left(\frac{3}{2}\right) \ln |x^2 + x + 1| + \left(\frac{11}{2}\right) \int \frac{d(x + 1/2)}{(x + 1/2)^2 + 3/4}$$

$$= \frac{x^2}{2} - 2x + \left(\frac{3}{2}\right) \ln |x^2 + x + 1| + \left(\frac{11}{2}\right) \left(\frac{1}{\sqrt{3/4}}\right) \tan^{-1}\left(\frac{x + 1/2}{\sqrt{3/4}}\right) + C$$

$$= \frac{x^2}{2} - 2x + \left(\frac{3}{2}\right) \ln |x^2 + x + 1| + \left(\frac{11}{\sqrt{3}}\right) \tan^{-1}\left(\frac{2x + 1}{\sqrt{3}}\right) + C.$$

25.
$$\int_0^1 \left(\frac{1 + x^2}{(1 + x)^2}\right) dx = \int_0^1 \left(1 - \frac{2}{1 + x} + \frac{2}{(1 + x)^2}\right) dx = \left(x - 2 \ln |1 + x| - \frac{2}{1 + x}\right)\Big|_0^1$$

$$= \left(1 - 2 \ln 2 - \frac{2}{2}\right) - \left(0 - 2 \ln 1 - \frac{2}{1}\right) = 2\,(1 - \ln 2).$$

27.
$$\int \left(\frac{\sqrt{x + 2} - 1}{\sqrt{x + 2} + 1}\right) dx \overset{x + 2 \equiv u^2}{=} \int \left(\frac{u - 1}{u + 1}\right) (2u\,du) = 2 \int \left(\frac{u^2 - u}{u + 1}\right) du = 2 \int \left(u - 2 + \frac{2}{u + 1}\right) du$$

$$= u^2 - 4u + 4 \ln |u + 1| + C = (x + 2) - 4\sqrt{x + 2} + 4 \ln \left(\sqrt{x + 2} + 1\right) + C.$$

29.
$$\int \frac{\sin x\,dx}{\cos^2 x + 4 \cos x + 4} \overset{\cos x \equiv u}{=} \int \frac{-du}{u^2 + 4u + 4} = \int \frac{-du}{(u + 2)^2} = \frac{1}{u + 2} + C = \frac{1}{\cos x + 2} + C.$$

31.
$$\int_0^{1/\sqrt{2}} \frac{x^3\,dx}{2 + x^2} = \int_0^{1/\sqrt{2}} \left(x - \frac{2x}{2 + x^2}\right) dx = \left(\frac{x^2}{2} - \ln |2 + x^2|\right)\Big|_0^{1/\sqrt{2}} = \frac{1}{4} - \ln \frac{5}{4}.$$

33.
$$\int \frac{dx}{x^{1/3} + x^{1/4}} \overset{x \equiv u^{12}}{=} \int \frac{12u^{11}\,du}{u^4 + u^3} = 12 \int \left(\frac{u^8}{u + 1}\right) du$$

$$= 12 \int \left(u^7 - u^6 + u^5 - u^4 + u^3 - u^2 + u - 1 + \frac{1}{u + 1}\right) du$$

$$= 12 \left(\frac{u^8}{8} - \frac{u^7}{7} + \frac{u^6}{6} - \frac{u^5}{5} + \frac{u^4}{4} - \frac{u^3}{3} + \frac{u^2}{2} - u + \ln |u + 1|\right) + C$$

$$= 12 \left(\frac{x^{2/3}}{8} - \frac{x^{7/12}}{7} + \frac{x^{1/2}}{6} - \frac{x^{5/12}}{5} + \frac{x^{1/3}}{4} - \frac{x^{1/4}}{3} + \frac{x^{1/6}}{2} - x^{1/12} + \ln \left(x^{1/12} + 1\right)\right) + C$$

35.
$$\int \frac{(x - 4)\,dx}{\sqrt{x}\,(1 + \sqrt[3]{x})} \overset{x \equiv u^6}{=} \int \frac{(u^6 - 4)\,(6u^5\,du)}{u^3\,(1 + u^2)} = 6 \int \left(\frac{u^8 - 4u^2}{1 + u^2}\right) du$$

$$= 6 \int \left(u^6 - u^4 + u^2 - 5 + \frac{5}{1 + u^2}\right) du = 6 \left(\frac{u^7}{7} - \frac{u^5}{5} + \frac{u^3}{3} - 5u + 5 \tan^{-1} u\right) + C$$

$$= 6 \left(\frac{x^{7/6}}{7} - \frac{x^{5/6}}{5} + \frac{x^{3/6}}{3} - 5x^{1/6} + 5 \tan^{-1} x^{1/6}\right) + C.$$

37.
$$\int \frac{\csc x\,dx}{3 \csc x + 2 \cot x + 2} = \int \frac{dx}{3 + 2(\cos x + \sin x)}$$

$$\overset{\tan(x/2) \equiv u}{=} \int \frac{\dfrac{2\,du}{1 + u^2}}{3 + 2\left(\dfrac{1 - u^2}{1 + u^2} + \dfrac{2u}{1 + u^2}\right)} = \int \frac{2\,du}{u^2 + 4u + 5}$$

$$= 2 \int \frac{du}{(u + 2)^2 + 1} = 2 \tan^{-1}(u + 2) + C = 2 \tan^{-1}\left(\tan\left(\frac{x}{2}\right) + 2\right) + C.$$

39. If $r > q^2/4$, then $r - q^2/4 > 0$. Therefore $\sqrt{r - q^2/4}$ exists and is strictly positive, hence $1 \Big/ \sqrt{r - q^2/4} =$

$2 \big/ \sqrt{4r - q^2}$ exists. Completing the square for the quadratic denominator and using Example 6.7.4, we find

$$\int \frac{dx}{x^2 + qx + r} = \int \frac{d(x + q/2)}{(x + q/2)^2 + (r - q^2/4)} = \frac{1}{\sqrt{r - q^2/4}} \, \tan^{-1}\left(\frac{x + q/2}{\sqrt{r - q^2/4}} \right) + C$$

$$= \frac{2}{\sqrt{4r - q^2}} \, \tan^{-1}\left(\frac{2x + q}{\sqrt{4r - q^2}} \right) + C. \ \blacksquare$$

REMARK: The above solution uses the ideas of Section 7.6; however, we may opt to just differentiate the specific expression asserted by the problem and show $\left(x^2 + qx + 4\right)^{-1}$ results.

Solutions 7.7   *Integration of rational functions II: partial fractions*   (pages 517-8)

1. $\dfrac{A}{x - 1} + \dfrac{B}{x - 4} = \dfrac{1}{(x - 1)(x - 4)} \implies A(x - 4) + B(x - 1) = 1$

$\implies \begin{cases} -3A = A(1 - 4) + B(1 - 1) = 1 & \text{(let } x = 1) \\ 3B = A(4 - 4) + B(4 - 1) = 1 & \text{(let } x = 4) \end{cases}$

$\implies A = \dfrac{-1}{3} \quad \text{and} \quad B = \dfrac{1}{3}.$

Therefore $\displaystyle \int \frac{dx}{(x - 1)(x - 4)} = \frac{1}{3} \int \left(\frac{1}{x - 4} - \frac{1}{x - 1} \right) dx = \frac{1}{3} \left(\ln|x - 4| - \ln|x - 1| \right) + C = \frac{1}{3} \ln\left| \frac{x - 4}{x - 1} \right| + C.$

3. $\dfrac{A}{x + 3} + \dfrac{B}{x - 7} = \dfrac{3x - 5}{(x + 3)(x - 7)} \implies \begin{cases} A + B \cdot \left(\dfrac{x + 3}{x - 7} \right) = \dfrac{3x - 5}{x - 7} \\ A \cdot \left(\dfrac{x - 7}{x + 3} \right) + B = \dfrac{3x - 5}{x + 3} \end{cases}$

$\implies \begin{cases} A = A + B \cdot \left(\dfrac{-3 + 3}{-3 - 7} \right) = \dfrac{3(-3) - 5}{-3 - 7} = \dfrac{-14}{-10} = \dfrac{7}{5} & \text{(let } x = -3) \\ B = A \cdot \left(\dfrac{7 - 7}{7 + 3} \right) + B = \dfrac{3(7) - 5}{7 + 3} = \dfrac{16}{10} = \dfrac{8}{5} & \text{(let } x = 7). \end{cases}$

Therefore $\displaystyle \int \frac{(3x - 5)\, dx}{(x + 3)(x - 7)} = \int \left(\frac{7/5}{x + 3} + \frac{8/5}{x - 7} \right) dx = \frac{7}{5} \ln|x + 3| + \frac{8}{5} \ln|x - 7| + C.$

5. $\dfrac{A}{x - a} + \dfrac{B}{x - b} = \dfrac{1}{(x - a)(x - b)} \implies A(x - b) + B(x - a) = 1$

$\implies \begin{cases} A(a - b) = A(a - b) + B(a - a) = 1 & \text{(let } x = a) \\ B(b - a) = A(b - b) + B(b - a) = 1 & \text{(let } x = b) \end{cases}$

$\implies A = \dfrac{1}{a - b} = \dfrac{-1}{b - a} \quad \text{and} \quad B = \dfrac{1}{b - a}.$

Therefore

$$\int \frac{dx}{(x - a)(x - b)} = \frac{1}{b - a} \int \left(\frac{1}{x - b} - \frac{1}{x - a} \right) dx = \frac{1}{b - a} \left(\ln|x - b| - \ln|x - a| \right) + C = \frac{1}{b - a} \ln\left| \frac{x - b}{x - a} \right| + C.$$

7. $\dfrac{x^2 + 3x + 4}{x^2 + x} = 1 + \dfrac{2x + 4}{x^2 + x} = 1 + \dfrac{A}{x} + \dfrac{B}{x + 1} \implies 2x + 4 = A(x + 1) + Bx \implies \begin{cases} 4 = A & \text{(let } x = 0) \\ 2 = -B & \text{(let } x = -1). \end{cases}$

Therefore $\displaystyle \int \left(\frac{x^2 + 3x + 4}{x^2 + x} \right) dx = \int \left(1 + \frac{4}{x} - \frac{2}{x + 1} \right) dx = x + 4 \ln|x| - 2 \ln|x + 1| + C.$

9. $\displaystyle \int \frac{x^5}{x^3 - x} \, dx = \int \frac{x^4}{x^2 - 1} \, dx = \int \left(x^2 + 1 + \frac{1}{x^2 - 1} \right) dx = \int \left(x^2 + 1 + \frac{1}{2} \left(\frac{1}{x - 1} - \frac{1}{x + 1} \right) \right) dx$

$$= \frac{x^3}{3} + x + \frac{1}{2} \ln\left| \frac{x - 1}{x + 1} \right| + C.$$

11.   Use the antiderivative found in Example 5.

$$\int_{-1}^{0} \frac{dx}{(x-1)(x-2)(x-3)} = \ln \frac{\sqrt{|x-1| \cdot |x-3|}}{|x-2|} \Bigg|_{-1}^{0} = \ln \frac{\sqrt{|-1| \cdot |-3|}}{|-2|} - \ln \frac{\sqrt{|-2| \cdot |-4|}}{|-3|}$$

$$= \ln \frac{\sqrt{3}}{2} - \ln \frac{\sqrt{8}}{3} = \frac{3}{2} \ln 3 - \frac{5}{2} \ln 2 = \ln \sqrt{\frac{27}{32}}.$$

13. $\dfrac{A}{x-1} + \dfrac{B}{x-2} + \dfrac{C}{x-3} = \dfrac{x^2+3x+4}{(x-1)(x-2)(x-3)}$

$\implies A(x-2)(x-3) + B(x-1)(x-3) + C(x-1)(x-2) = x^2 + 3x + 4$

$$\implies \begin{cases} 2A = A(-1)(-2) + B(0)(-2) + C(0)(-1) = 8 & \text{(let } x=1) \\ -B = A(0)(-1) + B(1)(-1) + C(1)(0) = 14 & \text{(let } x=2) \\ 2C = A(1)(0) + B(2)(0) + C(2)(1) = 22 & \text{(let } x=3) \end{cases} \implies \begin{cases} A = 4 \\ B = -14 \\ C = 11. \end{cases}$$

Therefore

$$\int \frac{x^2+3x+4}{(x-1)(x-2)(x-3)} \, dx = \int \left(\frac{4}{x-1} + \frac{-14}{x-2} + \frac{11}{x-3} \right) dx = 4 \ln|x-1| - 14 \ln|x-2| + 11 \ln|x-3| + C.$$

15. $\dfrac{x^2+2}{x(x-1)^2(x+1)} = \dfrac{A}{x} + \dfrac{B_1}{x-1} + \dfrac{B_2}{(x-1)^2} + \dfrac{C}{x+1}$

$$\implies A = \frac{x^2+2}{(x-1)^2(x+1)} \bigg|_{x=0} = 2, \qquad B_2 = \frac{x^2+2}{x(x+1)} \bigg|_{x=1} = \frac{3}{2}, \qquad C = \frac{x^2+2}{x(x-1)^2} \bigg|_{x=-1} = \frac{-3}{4}.$$

The value of B_1 can be found by differentiating

$$B_1(x-1) + B_2 + \left(\frac{A}{x} + \frac{C}{x+1} \right)(x-1)^2 = \frac{x^2+2}{x(x+1)} = 1 + \frac{2}{x} - \frac{3}{x+1}$$

and then evaluating at $x=1$: $B_1 = B_1 + 0 + 0 = \left(\dfrac{-2}{x^2} + \dfrac{3}{(x+1)^2} \right) \bigg|_{x=1} = -2 + \dfrac{3}{4} = \dfrac{-5}{4}.$ Thus

$$\int \frac{x^2+2}{x(x-1)^2(x+1)} \, dx = \int \left(\frac{2}{x} + \frac{-5/4}{x-1} + \frac{3/2}{(x-1)^2} + \frac{-3/4}{x+1} \right) dx$$

$$= 2 \ln|x| - \frac{5}{4} \ln|x-1| - \frac{3}{2}(x-1)^{-1} - \frac{3}{4} \ln|x+1| + K.$$

17. $\dfrac{1}{x^2(x-1)} = \dfrac{A_1}{x} + \dfrac{A_2}{x^2} + \dfrac{B}{x-1}$

$\implies 1 = A_1 x(x-1) + A_2(x-1) + Bx^2$

$$\implies \begin{cases} 1 = -A_2 & \text{(let } x=0) \\ 1 = B & \text{(let } x=1) \\ 1 = 2A_1 + A_2 + 4B & \text{(let } x=2) \end{cases} \implies \begin{cases} A_2 = -1 \\ B = 1 \\ A_1 = (1 - A_2 - 4B)/2 = (1 + 1 - 4)/2 = -1. \end{cases}$$

Therefore $\displaystyle \int \frac{dx}{x^2(x-1)} = \int \left(\frac{-1}{x} + \frac{-1}{x^2} + \frac{1}{x-1} \right) dx = -\ln|x| + \frac{1}{x} + \ln|x-1| + C = \frac{1}{x} + \ln \left| \frac{x-1}{x} \right| + C.$

19. $\displaystyle \int \frac{dx}{x^2+x^4} = \int \left(\frac{1}{x^2} - \frac{1}{x^2+1} \right) dx = \frac{-1}{x} - \tan^{-1} x + C.$

21. $\dfrac{A_1}{x} + \dfrac{A_2}{x^2} + \dfrac{B_1}{x+2} + \dfrac{B_2}{(x+2)^2} = \dfrac{1}{x^2(x+2)^2}$

$$\implies (A_1 x + A_2)(x+2)^2 + (B_1(x+2) + B_2)x^2 = 1 \implies \begin{cases} 4A_2 = 1 & \text{(let } x=0) \\ 4B_2 = 1 & \text{(let } x=-2) \end{cases} \implies \begin{cases} A_2 = 1/4 \\ B_2 = 1/4 \end{cases}$$

$$\implies \frac{A_1}{x} + \frac{B_1}{x+2} = \frac{1}{x^2(x+2)^2} - \frac{1/4}{x^2} - \frac{1/4}{(x+2)^2} = \frac{-2x(x+2)}{4x^2(x+2)^2} = \frac{-1}{2x(x+2)}$$

$$\implies A_1(x+2) + B_1 x = \frac{-1}{2} \implies \begin{cases} 2A_1 = -1/2 & \text{(let } x=0) \\ -2B_1 = -1/2 & \text{(let } x=-2) \end{cases} \implies \begin{cases} A_1 = -1/4 \\ B_1 = 1/4. \end{cases}$$

Thus $\displaystyle \int \frac{dx}{x^2(x+2)^2} = \frac{1}{4} \int \left[\frac{-1}{x} + \frac{1}{x^2} + \frac{1}{x+2} + \frac{1}{(x+2)^2} \right] dx = \frac{1}{4} \left[-\ln|x| + \frac{-1}{x} + \ln|x+2| + \frac{-1}{x+2} \right] + C.$

23.

$$\frac{A}{x+1} + \frac{B}{x-1} + \frac{Cx+D}{x^2+1} = \frac{x}{(x+1)(x-1)(x^2+1)} = \frac{x}{(x^2-1)(x^2+1)} = \frac{x}{x^4-1}$$

$$\implies A = \left.\frac{x}{(x-1)(x^2+1)}\right|_{x=-1} = \frac{1}{4} \quad \text{and} \quad B = \left.\frac{x}{(x+1)(x^2+1)}\right|_{x=1} = \frac{1}{4}$$

$$\implies Cx+D = \frac{x}{x^2-1} - \left(\frac{1/4}{x+1} + \frac{1/4}{x-1}\right)(x^2+1) = \frac{x-x^3}{2(x^2-1)} = \frac{-x}{2}.$$

$$\int \frac{x\,dx}{x^4-1} = \frac{1}{4}\int\left(\frac{1}{x+1} + \frac{1}{x-1} + \frac{-2x}{x^2+1}\right)dx = \frac{1}{4}\left(\ln|x+1| + \ln|x-1| - \ln|x^2+1|\right) + K$$

$$= \frac{1}{4}\ln\left|\frac{(x+1)(x-1)}{x^2+1}\right| + K = \frac{1}{4}\ln\left|\frac{x^2-1}{x^2+1}\right| + K.$$

25.

$$\frac{A}{x+2} + \frac{Bx+C}{x^2+1} = \frac{1}{(x+2)(x^2+1)}$$

$$\implies A = \left.\left(A + \frac{(Bx+C)(x+2)}{x^2+1}\right)\right|_{x=-2} = \left.\frac{1}{x^2+1}\right|_{x=-2} = \frac{1}{5}$$

$$\implies 1 = \frac{1}{5}(x^2+1) + (Bx+C)(x+2) = \left(\frac{1}{5}+B\right)x^2 + (2B+C)x + \left(\frac{1}{5}+2C\right).$$

The coefficients of x and x^2 must be 0, hence $B = \frac{-1}{5}$ and $C = -2B = \frac{2}{5}$; with these values for B and C, it is true that $1 = \frac{1}{5} + 2C = \frac{1}{5} + \frac{4}{5}$. Therefore

$$\int \frac{dx}{(x+2)(x^2+1)} = \frac{1}{5}\int\left(\frac{1}{x+2} + \frac{-x+2}{x^2+1}\right)dx = \frac{1}{5}\int\left(\frac{1}{x+2} - \frac{x}{x^2+1} + \frac{2}{x^2+1}\right)dx$$

$$= \frac{1}{5}\left(\ln|x+2| - \frac{1}{2}\ln|x^2+1| + 2\tan^{-1}x\right) + K.$$

27.

$$\frac{3+4x}{(x^2+1)(x^2+2)} = \frac{A_0+A_1x}{x^2+1} + \frac{B_0+B_1x}{x^2+2}$$

$$\implies \quad 3+4x = (A_0+A_1x)(x^2+2) + (B_0+B_1x)(x^2+1)$$

$$= (2A_0+B_0) + (2A_1+B_1)x + (A_0+B_0)x^2 + (A_1+B_1)x^3.$$

We proceed by comparing the coefficients of $3+4x$ with those of our last polynomial. The constant terms are equal, hence $3 = 2A_0+B_0$; the coefficients of x^2 are equal, hence $0 = A_0+B_0$; these two equations imply $A_0 = (2A_0+B_0)-(A_0+B_0) = 3-0 = 3$ and $B_0 = 0-A_0 = -3$. Similarly, examining the coefficients of x and x^3 yields $4 = 2A_1+B_1$ and $0 = A_1+B_1$ which imply $A_1 = 4$ and $B_1 = -4$. Therefore

$$\int\left(\frac{3+4x}{(x^2+1)(x^2+2)}\right)dx = \int\left(\frac{3+4x}{x^2+1} - \frac{3+4x}{x^2+2}\right)dx$$

$$= \int\left(\frac{3}{x^2+1} + \frac{2(2x)}{x^2+1} - \frac{3}{x^2+2} - \frac{2(2x)}{x^2+2}\right)dx$$

$$= 3\tan^{-1}x + 2\ln(x^2+1) - \frac{3}{\sqrt{2}}\tan^{-1}\left(\frac{x}{\sqrt{2}}\right) - 2\ln(x^2+2) + C.$$

29. $\displaystyle\int_1^{32}\frac{dx}{x(1+x^{1/5})} \stackrel{x \equiv u^5}{=} \int_1^2 \frac{5u^4\,du}{u^4(1+u)} = 5\int_1^2 \frac{du}{u(1+u)} = 5\int_1^2\left(\frac{1}{u} - \frac{1}{1+u}\right)du = 5\ln\left|\frac{u}{1+u}\right|\Big|_1^2 = 5\ln\frac{4}{3}.$

31. $\displaystyle\int_0^{\pi/2}\frac{\cos x\,dx}{\sin^2 x + 2\sin x + 2} \stackrel{\sin x \equiv u}{=} \int_0^1 \frac{du}{u^2+2u+2} = \int_0^1 \frac{du}{(u+1)^2+1} = \tan^{-1}(u+1)\Big|_0^1 = \tan^{-1}2 - \frac{\pi}{4}.$

33. Let $u = \tan\left(\frac{2x}{2}\right) = \tan x$, then $x = \tan^{-1}u$, $dx = \frac{du}{1+u^2}$, and $\sin 2x + \cos 2x = \frac{2u}{1+u^2} + \frac{1-u^2}{1+u^2} =$

$\dfrac{1+2u-u^2}{1+u^2}$. Therefore

$$\int \frac{dx}{\sin 2x + \cos 2x} = \int \frac{du}{1+2u-u^2} = \frac{1}{2\sqrt{2}} \int \left(\frac{1}{u-1+\sqrt{2}} - \frac{1}{u-1-\sqrt{2}} \right) du$$

$$= \frac{1}{2\sqrt{2}} \ln \left| \frac{u-1+\sqrt{2}}{u-1-\sqrt{2}} \right| + C = \frac{1}{2\sqrt{2}} \ln \left| \frac{\tan x - 1 + \sqrt{2}}{\tan x - 1 - \sqrt{2}} \right| + C.$$

REMARK: $\sin 2x + \cos 2x = \sqrt{2} \sin(2x + \pi/4)$, thus

$$\int \frac{dx}{\sin 2x + \cos 2x} = \frac{1}{2\sqrt{2}} \int \csc(2x + \pi/4)\, d(2x + \pi/4) = \frac{-1}{2\sqrt{2}} \ln |\csc(2x + \pi/4) + \cot(2x + \pi/4)| + C_1$$

$$= \frac{1}{2\sqrt{2}} \ln \left| \frac{\sin(2x + \pi/4)}{1 + \cos(2x + \pi/4)} \right| + C_1 = \frac{1}{2\sqrt{2}} \ln \left| \frac{\sin 2x + \cos 2x}{\sqrt{2} + \cos 2x - \sin 2x} \right| + C_1.$$

35.   Applying the result of Problem 39 with $A = 1$, $B = -1$, $C = 1$, $D = -2$, $N = 2$ yields

$$\int \frac{x^2 - 1}{x^3 - 2x}\, dx = \int \frac{x^2 - 1}{x(x^2 - 2)}\, dx = \frac{-1}{-2} \ln |x| + \frac{1}{2} \left(\frac{1}{1} - \frac{-1}{-2} \right) \ln |x^2 - 2| + C = \frac{1}{2} \ln |x| + \frac{1}{4} \ln |x^2 - 2| + C.$$

REMARK: $\displaystyle \int \frac{x^2 - 1}{x(x^2 - 2)}\, dx = \frac{1}{2} \int \left(\frac{1}{x} + \frac{x}{x^2 - 2} \right) dx = \frac{1}{2} \ln |x| + \frac{1}{4} \ln |x^2 - 2| + C.$

37.   Applying the result of Problem 39 with $A = 3$, $B = 4$, $C = 2$, $D = 5$, $N = 4$ yields

$$\int \frac{3x^4 + 4}{2x^5 + 5x}\, dx = \int \frac{3x^4 + 4}{x(2x^4 + 5)}\, dx = \frac{4}{5} \ln |x| + \frac{1}{4} \left(\frac{3}{2} - \frac{4}{5} \right) \ln |2x^4 + 5| + C = \frac{4}{5} \ln |x| + \frac{7}{40} \ln |2x^4 + 5| + C.$$

REMARK: $\dfrac{3x^4 + 4}{x(2x^4 + 5)} = \dfrac{1}{5} \left(\dfrac{4}{x} + \dfrac{7x^3}{2x^4 + 5} \right).$

39. a.   Suppose $D \neq 0$.

$$\frac{Ax^N + B}{x(Cx^N + D)} = \frac{\alpha}{x} + \frac{\beta x^{N-1}}{Cx^N + D} \iff Ax^N + B = \alpha(Cx^N + D) + \beta x^N = (\alpha C + \beta)x^N + \alpha D$$

$$\iff \begin{cases} B = \alpha D \\ A = \alpha C + \beta \end{cases} \iff \begin{cases} \alpha = B/D \\ \beta = A - \alpha C = A - BC/D. \end{cases} \blacksquare$$

   b.   Suppose $C \neq 0$ and $D \neq 0$.

$$\int \frac{Ax^N + B}{x(Cx^N + D)}\, dx = \int \left\{ \frac{B}{D} \cdot \frac{1}{x} + \left(A - \frac{BC}{D} \right) \cdot \frac{x^{N-1}}{Cx^N + D} \right\} dx$$

$$= \frac{B}{D} \ln |x| + \left(A - \frac{BC}{D} \right) \cdot \frac{1}{NC} \ln |Cx^N + D| + k$$

$$= \frac{B}{D} \ln |x| + \frac{1}{N} \left(\frac{A}{C} - \frac{B}{D} \right) \ln |Cx^N + D| + k. \blacksquare$$

Solutions 7.8   *Using the integral tables*   (page 521)

1.   Let $u = \sqrt{3}\, x$ and $a = 4$ in entry 68: $\displaystyle \int \frac{x\, dx}{\sqrt{3x^2 + 16}} = \frac{1}{3} \int \frac{(\sqrt{3}\, x)\, d(\sqrt{3}\, x)}{\sqrt{(\sqrt{3}\, x)^2 + 4^2}} = \frac{1}{3} \sqrt{3x^2 + 16} + C.$

3.   Let $u = 2x$ and $a = 3$ in entry 61 (or entry 19):

$$\int \frac{dx}{9 - 4x^2} = \frac{1}{2} \int \frac{d(2x)}{3^2 - (2x)^2} = \left(\frac{1}{2} \right) \left(\frac{1}{2 \cdot 3} \right) \ln \left| \frac{3 + 2x}{3 - 2x} \right| + C = \frac{1}{12} \ln \left| \frac{3 + 2x}{3 - 2x} \right| + C.$$

5.   Let $u = x$, $a = 2$, $b = 3$ in entry 28: $\displaystyle \int \frac{x\, dx}{2x + 3} = \frac{x}{2} - \frac{3}{2^2} \ln |2x + 3| + C.$

7.   Let $u = x$, $a = 2$, $b = 3$ in entry 30: $\displaystyle \int \frac{dx}{x(2x + 3)} = \frac{1}{3} \ln \left| \frac{x}{2x + 3} \right| + C.$

9.   Let $u = x$, $a = 2$, $b = 3$ in entry 42:

$$\int x\sqrt{2x+3}\,dx = \frac{2\,(3\cdot 2\cdot x - 2\cdot 3)}{15\cdot 2^2}\sqrt{(2x+3)^3} + C = \frac{1}{5}\,(x-1)\,(2x+3)^{3/2} + C.$$

11.   Let $u = x$ and $a = 3$ in entry 165: $\displaystyle\int x^2 e^{3x}\,dx = \frac{e^{3x}}{3}\left(x^2 - \frac{2x}{3} + \frac{2}{3^2}\right) + C = \frac{1}{27}\left(9x^2 - 6x + 2\right)e^{3x} + C.$

13.   Let $u = x$, $a = 1$, $m = 3$, $n = 4$ in the top line of entry 136; then use entry 126.

$$\int \sin^3 x \cos^4 x\,dx = \frac{-\sin^2 x \cos^5 x}{(3+4)} + \left(\frac{3-1}{3+4}\right)\int \sin x \cos^4 x\,dx$$

$$= \frac{-1}{7}\sin^2 x \cos^5 x + \frac{2}{7}\cdot\frac{-\cos^5 x}{4+1} + C = \frac{-1}{7}\sin^2 x \cos^5 x + \frac{-2}{35}\cos^5 x + C.$$

REMARK: Use $\sin^2 x = 1 - \cos^2 x$ and entry 126 twice.

$$\int \sin^3 x \cos^4 x\,dx = \int \sin x\,(1 - \cos^2 x)\cos^4 x\,dx = \int \sin x \cos^4 x\,dx - \int \sin x \cos^6 x\,dx$$

$$= \frac{-1}{5}\cos^5 x + \frac{1}{7}\cos^7 x + C_1.$$

15.   Let $u = x$, $p = 3$, $q = 4$ in entry 124:

$$\int \sin 3x \cos 4x\,dx = \frac{-\cos(3-4)x}{2(3-4)} - \frac{\cos(3+4)x}{2(3+4)} + C = \frac{1}{2}\cos x - \frac{1}{14}\cos 7x + C.$$

17.   Let $u = x$, $a = 1/3$, $n = 4$ in entry 140; then use entry 138.

$$\int \tan^4\left(\frac{x}{3}\right)dx = \frac{\tan^3(x/3)}{3(1/3)} - \int \tan^2\left(\frac{x}{3}\right)dx = \tan^3\left(\frac{x}{3}\right) - \left[\frac{\tan(x/3)}{1/3} - x\right] + C = \tan^3\left(\frac{x}{3}\right) - 3\tan\left(\frac{x}{3}\right) + x + C.$$

19.   Let $u = x^2$ and $a = 1$ in entry 158:

$$\int x \tan^{-1} x^2\,dx = \frac{1}{2}\int \tan^{-1} x^2\,d\left(x^2\right) = \frac{1}{2}\left(x^2 \tan^{-1} x^2 - \frac{1}{2}\ln\left(\left(x^2\right)^2 + 1^2\right)\right) + C$$

$$= \frac{1}{2}x^2 \tan^{-1} x^2 - \frac{1}{4}\ln\left(x^4 + 1\right) + C.$$

21.   Let $u = x$ and $a = 1/2$ in entry 157:

$$\int x \cos^{-1} 2x\,dx = \left(\frac{x^2}{2} - \frac{(1/2)^2}{4}\right)\cos^{-1} 2x - \frac{x\sqrt{(1/2)^2 - x^2}}{4} + C$$

$$= \frac{1}{16}\left(8x^2 - 1\right)\cos^{-1} 2x - \frac{1}{8}x\sqrt{1 - 4x^2} + C.$$

23.   Let $u = \sqrt{3}\,x$ and $a = 4$ in entry 91.

$$\int \sqrt{16 - 3x^2}\,dx = \frac{1}{\sqrt{3}}\int \sqrt{4^2 - \left(\sqrt{3}\,x\right)^2}\,d\left(\sqrt{3}\,x\right) = \frac{1}{\sqrt{3}}\left(\frac{(\sqrt{3}\,x)\sqrt{16 - 3x^2}}{2} + \frac{16}{2}\sin^{-1}\left(\frac{\sqrt{3}\,x}{4}\right)\right) + C$$

$$= \frac{1}{2}x\sqrt{16 - 3x^2} + \frac{8}{\sqrt{3}}\sin^{-1}\left(\frac{\sqrt{3}\,x}{4}\right) + C.$$

25.   Let $u = \sqrt{2}\,x$ and $a = \sqrt{10}$ in entry 74.

$$\int \frac{\sqrt{2x^2 + 10}}{x}\,dx = \int \frac{\sqrt{\left(\sqrt{2}\,x\right)^2 + 10}}{\sqrt{2}\,x}\,d\left(\sqrt{2}\,x\right) = \sqrt{2x^2 + 10} - \sqrt{10}\,\ln\left|\frac{\sqrt{10} + \sqrt{2x^2 + 10}}{\sqrt{2}\,x}\right| + C.$$

REMARK:

$$\int \frac{\sqrt{2x^2 + 10}}{x}\,dx = \sqrt{2}\int \frac{\sqrt{x^2 + 5}}{x}\,dx = \sqrt{2}\left(\sqrt{x^2 + 5} - \sqrt{5}\,\ln\left|\frac{\sqrt{5} + \sqrt{x^2 + 5}}{x}\right|\right) + C.$$

27.   Let $u = \sqrt{2}\,x$ and $a = 3$ in entry 79: $\displaystyle\int \frac{dx}{x\sqrt{2x^2 - 9}} = \int \frac{d\left(\sqrt{2}\,x\right)}{\left(\sqrt{2}\,x\right)\sqrt{\left(\sqrt{2}\,x\right)^2 - 3^2}} = \frac{1}{3}\sec^{-1}\left|\frac{\sqrt{2}\,x}{3}\right| + C.$

29.   Let $u = x$ and $a = 2$ in entry 85: $\displaystyle\int \frac{dx}{(4x^2 - 16)^{3/2}} = \frac{1}{4^{3/2}}\int \frac{dx}{(x^2 - 4)^{3/2}} = \left(\frac{1}{8}\right)\frac{-x}{4\sqrt{x^2 - 4}} + C.$

31.   Let $u = x$ and $a^2 = 5/2$ in entry 78:

$$\int \frac{x^2\,dx}{\sqrt{2x^2 - 5}} = \frac{1}{\sqrt{2}} \int \frac{x^2\,dx}{\sqrt{x^2 - 5/2}} = \frac{1}{\sqrt{2}}\left(\frac{x\sqrt{x^2 - 5/2}}{2} + \frac{5/2}{2}\ln\left| x + \sqrt{x^2 - 5/2}\right|\right) + C_*$$

$$= \frac{1}{4}x\sqrt{2x^2 - 5} + \frac{5}{4\sqrt{2}}\ln\left|\sqrt{2}x + \sqrt{2x^2 - 5}\right| + C.$$

33.   Let $u = x^3$, $a = 4$, and $b = 7$ in entry 169:

$$\int x^2 e^{4x^3} \cos 7x^3\,dx = \frac{1}{3}\int e^{4u}\cos 7u\,du = \left(\frac{1}{3}\right)\frac{e^{4u}(4\cos 7u + 7\sin 7u)}{4^2 + 7^2} + C$$

$$= \frac{1}{195}e^{4x^3}\left(4\cos 7x^3 + 7\sin 7x^3\right) + C.$$

Solutions 7.9   *Numerical integration*   (pages 532-4)

1. $\Delta x = (1 - 0)/4 = 1/4$; $f(x_k) = x_k = 0 + k \cdot \Delta x = k/4$ for $k = 0, 1, 2, 3, 4$.

a. $\left(\frac{1}{2}\right)\left(\frac{1}{4}\right)\left[\left(\frac{0}{4}\right) + 2\left(\frac{1}{4}\right) + 2\left(\frac{2}{4}\right) + 2\left(\frac{3}{4}\right) + \left(\frac{4}{4}\right)\right] = \left(\frac{1}{8}\right)[4] = \frac{1}{2}$.

b. $\left(\frac{1}{3}\right)\left(\frac{1}{4}\right)\left[\left(\frac{0}{4}\right) + 4\left(\frac{1}{4}\right) + 2\left(\frac{2}{4}\right) + 4\left(\frac{3}{4}\right) + \left(\frac{4}{4}\right)\right] = \left(\frac{1}{12}\right)[6] = \frac{1}{2}$.

c.   If $f(x) = x$, then $f''(x) = 0$. Thus we can choose $M = 0$ and infer that $\left|\epsilon_4^T\right| \le 0$, i.e., $\epsilon_4^T = 0$; this means the trapezoidal rule approximation is exactly correct for the function $f(x) = x$.

d. $f^{(4)}(x) = 0$; choose $M = 0$ to infer $\epsilon_4^S = 0$, Simpson's rule produces an approximation equal to the actual value of the integral.

e. $\int_0^1 x\,dx = \left.\frac{x^2}{2}\right|_0^1 = \frac{1}{2} - 0 = \frac{1}{2}$.

f.   As noted in parts (c) and (d), the maximum possible error is 0 for both trapezoidal and Simpson's rule applied to this function (on any interval).

3. a. $\left(\frac{1}{2}\right)\left(\frac{1}{4}\right)\left\{\left(\frac{0}{16}\right) + 2\left(\frac{1}{16}\right) + 2\left(\frac{4}{16}\right) + 2\left(\frac{9}{16}\right) + 2\left(\frac{16}{16}\right)\right\} = \frac{1}{8}\left\{\frac{44}{16}\right\} = \frac{11}{32}$.

b. $\left(\frac{1}{3}\right)\left(\frac{1}{4}\right)\left\{\left(\frac{0}{16}\right) + 4\left(\frac{1}{16}\right) + 2\left(\frac{4}{16}\right) + 4\left(\frac{9}{16}\right) + 2\left(\frac{16}{16}\right)\right\} = \frac{1}{12}\left\{\frac{64}{16}\right\} = \frac{1}{3}$.

c.   If $f(x) = x^2$, then $f''(x) = 2$; therefore $\left|\epsilon_4^T\right| \le 2\frac{(1 - 0)^3}{12(4)^3} = \frac{1}{96}$.

d. $f^{(4)}(x) = 0$, therefore $\epsilon_4^S = 0$.

e. $\int_0^1 x^2\,dx = \left.\frac{x^3}{3}\right|_{-2}^2 = \frac{1}{3} - 0 = \frac{1}{3}$.

f.   The actual error in the trapezoidal rule approximation is $1/3 - 11/32 = -1/96$; the upper bound in part (c) cannot be improved. The Simpson's rule approximation is exactly correct; this agrees with the result in part (d).

5. a. $\left(\frac{1}{2}\right)\left(\frac{1 - 0}{4}\right)\left[e^0 + 2e^{1/4} + 2e^{1/2} + 2e^{3/4} + e^1\right] \approx \left(\frac{1}{8}\right)(13.8177752364) \approx 1.7272219045$.

b. $\left(\frac{1}{3}\right)\left(\frac{1 - 0}{4}\right)\left[e^0 + 4e^{1/4} + 2e^{1/2} + 4e^{3/4} + e^1\right] \approx \left(\frac{1}{12}\right)(20.6198261031) \approx 1.7183188419$.

c.   If $f(x) = e^x$, then $f''(x) = e^x$ has maximum value e on $[0, 1]$; thus $\left|\epsilon_4^T\right| \le \frac{e(1 - 0)^3}{12 \cdot 4^2} = \frac{e}{192} \approx 0.0141577$.

d. $f^{(4)}(x) = e^x$; thus $\left|\epsilon_4^S\right| \le \frac{e(1 - 0)^5}{180 \cdot 4^4} = \frac{e}{46080} \approx 0.00005899$.

e. $\int_0^1 e^x\,dx = \left. e^x\right|_0^1 = e - 1 \approx 1.718281828459045$.

f. $\epsilon_4^T \approx -0.008940076$ and $\epsilon_4^S \approx -0.000037013$; each bound is about 1.5 times larger than the actual error.

REMARK: My bound for $\left|\epsilon_4^S\right|$ is $\dfrac{e(1-0)^5}{180\cdot4^4}=\dfrac{e(1-0)^5}{2880\cdot2^4}$; if you computed $\dfrac{e(1-0)^5}{2880\cdot4^4}$, then reread Theorem 3 and Corollary 2 paying especial attention to the notation: $2n$ is the total number of intervals considered.

7. a. $\left(\dfrac{1}{2}\right)\left(\dfrac{\pi/2}{4}\right)\left[\cos\left(0\right)+2\cos\left(\dfrac{\pi}{8}\right)+2\cos\left(\dfrac{\pi}{4}\right)+2\cos\left(\dfrac{3\pi}{8}\right)+\cos\left(\dfrac{\pi}{2}\right)\right]\approx0.9871158010.$

   b. $\left(\dfrac{1}{3}\right)\left(\dfrac{\pi/2}{4}\right)\left[\cos\left(0\right)+4\cos\left(\dfrac{\pi}{8}\right)+2\cos\left(\dfrac{\pi}{4}\right)+4\cos\left(\dfrac{3\pi}{8}\right)+\cos\left(\dfrac{\pi}{2}\right)\right]\approx1.0001345850.$

   c. If $f(x)=\cos x$, then $|f''(x)|=|-\cos x|\le1$; thus $\left|\epsilon_4^T\right|\le\dfrac{(\pi/2)^3}{12\cdot4^2}=\dfrac{\pi^3}{1536}\approx0.0201863780.$

   d. $f^{(4)}(x)=\cos x$; thus $\left|\epsilon_4^S\right|\le\dfrac{(\pi/2)^5}{180\cdot4^4}=\dfrac{\pi^5}{1474560}\approx0.0002075329.$

   e. $\displaystyle\int_0^{\pi/2}\cos x\,dx=\sin x\Big|_0^{\pi/2}=1.$

   f. $\epsilon_4^T\approx0.0128841990$ and $\epsilon_4^S\approx-0.0001345850$; each bound is about 1.5 times larger than the actual error.

9. a. $\dfrac{1}{2}\left[\dfrac{e-1}{6}\right]\left[\ln1+2\ln\dfrac{e+5}{6}+2\ln\dfrac{e+2}{3}+2\ln\dfrac{e+1}{2}+2\ln\dfrac{2e+2}{3}+2\ln\dfrac{5e+1}{6}+\ln e\right]\approx0.99569713.$

   b. $\dfrac{1}{3}\left[\dfrac{e-1}{6}\right]\left[\ln1+4\ln\dfrac{e+5}{6}+2\ln\dfrac{e+2}{3}+4\ln\dfrac{e+1}{2}+2\ln\dfrac{2e+2}{3}+4\ln\dfrac{5e+1}{6}+\ln e\right]\approx0.99993607.$

   c. If $f(x)=\ln x$, then $f''(x)=-x^{-2}$; on $[1,e]$, $|f''(x)|\le|f''(1)|=1$; thus $\left|\epsilon_6^T\right|\le\dfrac{(e-1)^3}{12\cdot6^2}\approx0.01174355.$

   d. $f^{(4)}(x)=-6x^{-4}$; on $[1,e]$ we have $\left|f^{(4)}(x)\right|\le\left|f^{(4)}(1)\right|=6$ which implies $\left|\epsilon_6^S\right|\le\dfrac{6(e-1)^5}{180\cdot6^4}\approx0.00038525.$

   e. $\displaystyle\int_1^e\ln x\,dx=x\left(\ln x-1\right)\Big|_1^e=e(1-1)-1(0-1)=1.$

   f. $\epsilon_6^T\approx0.00430287$ and $\epsilon_6^S\approx0.00006393$; the bound on the trapezoidal error is about 2.7 times the size of the actual error while the bound on the Simpson's rule error is about 6 times oversize.

11. $\displaystyle\int_0^1\sqrt{x+x^2}\,dx\underset{\text{trapezoidal}}{\approx}0.8194482758\underset{\text{Simpson's}}{\approx}0.8304090035.$

REMARK: $\displaystyle\int_0^1\sqrt{x+x^2}\,dx=\dfrac{3\sqrt{2}}{4}-\dfrac{1}{8}\ln\left(3+2^{3/2}\right)\approx0.8403167750.$

13. $\displaystyle\int_1^2\dfrac{\sin x}{x}\,dx\underset{\text{trapezoidal}}{\approx}0.6590191268\underset{\text{Simpson's}}{\approx}0.6593301635.$

15. $\displaystyle\int_0^1 e^{\sqrt{x}}\,dx\underset{\text{trapezoidal}}{\approx}1.9877954995\underset{\text{Simpson's}}{\approx}1.9945037403.$

17. $\displaystyle\int_{-1}^1 e^{-x^2}\,dx\underset{\text{trapezoidal}}{\approx}1.4887366795\underset{\text{Simpson's}}{\approx}1.4936741098.$

19. $\displaystyle\int_0^1\sqrt{1+x^3}\,dx\underset{\text{trapezoidal}}{\approx}1.1123323908\underset{\text{Simpson's}}{\approx}1.1114458169.$

21. $\displaystyle\int_0^1\ln\left(1+e^x\right)\,dx\underset{\text{trapezoidal}}{\approx}0.9841199254\underset{\text{Simpson's}}{\approx}0.9838189135.$

23. If $g(x)=e^{-x^2}$, then $g''(x)=2\left(2x^2-1\right)e^{-x^2}$. The maximum value of $2\left|2x^2-1\right|$ on $[-1,1]$ is 2 and the maximum value of $\left|e^{-x^2}\right|$ is 1, therefore $|g''(x)|\le|g''(0)|=2$ on $[-1,1]$ and $\left|\epsilon_{10}^T\right|\le\dfrac{2\left(1-(-1)\right)^3}{12\cdot10^2}=\dfrac{1}{75}\approx0.0133333333.$

$\left|g^{(4)}(x)\right|=\left|4\left(4x^4-12x^2+3\right)e^{-x^2}\right|\le\left|g^{(4)}(0)\right|=12$, thus $\left|\epsilon_{10}^S\right|\le\dfrac{12\cdot2^5}{180\cdot10^4}=\dfrac{2}{9375}\approx0.0002133333.$

25.   If $f(x) = \sqrt{1 + x^3}$, then $f''(x) = 3x \left(1 + \frac{x^3}{4}\right) (1 + x^3)^{\frac{-3}{2}}$ and $f'''(x) = \left(\frac{-3}{8}\right) (x^6 + 20x^3 - 8) (1 + x^3)^{\frac{-5}{2}}$.

Locating the only critical point of f''' in $[0, 1]$ implies $|f''(x)| \leq \left|f''\left(\sqrt[3]{-10 + \sqrt{108}}\right)\right| \approx 1.468$; thus

$|\epsilon_{10}^T| \leq \dfrac{1.468 \cdot 1^3}{12 \cdot 10^2} \approx 0.001223$. $|f^{(4)}(x)| = \left|(9/16)x^2 (x^6 + 56x^3 - 80) (1 + x^3)^{-7/2}\right| \leq |f^{(4)}(0.56)| \approx 7.09$,

thus $|\epsilon_{10}^S| \leq \dfrac{7.09 \cdot 1^5}{180 \cdot 10^4} \approx 0.0000039389$.

27.   If $f(x) = \ln(1 + e^x)$, then $f''(x) = e^x (1 + e^x)^{-2}$ and $f'''(x) = e^x (1 - e^x)(1 + e^x)^{-3}$. The global

maximum of f'' is $f''(0) = \dfrac{1}{4}$, therefore $|\epsilon_8^T| \leq \dfrac{(1/4)(1 - 0)^3}{12 \cdot 8^2} = \dfrac{1}{3072} \approx 0.00032552$.

$f^{(4)}(x) = e^x (e^{2x} - 4e^x + 1) (1 + e^x)^{-4}$ and $f^{(5)}(x) = e^x (1 - e^x) (e^{2x} - 10e^x + 1) (1 + e^x)^{-5}$; $f^{(4)}$ is

negative but increasing on $[0, 1]$, thus $\left|f^{(4)}(x)\right| \leq \left|f^{(4)}(0)\right| = \dfrac{1}{8}$ and $|\epsilon_8^S| \leq \dfrac{0.125 \cdot 1^5}{180 \cdot 8^4} \approx 0.0000001695$.

29.   Let $f(x) = e^{-x^2/2}$. Because $f(-x) = f(x)$, Problem 4.6.61 guarantees us that $\int_{-a}^{a} f = 2 \int_0^a f$ for every

real number a. $|f^{(4)}(x)| = \left|(x^4 - 6x^2 + 3) e^{-x^2/2}\right| \leq |f^{(4)}(0)| = 3$. To approximate $(1/\sqrt{2\pi}) \int_0^1 f$ with error

less than $0.01/2 = 0.005$, pick integer n such that $\left(\dfrac{1}{\sqrt{2\pi}}\right) \dfrac{3(1 - 0)^5}{180 \cdot n^4} \leq 0.005$, i.e., $n \geq \left(\dfrac{3}{180\sqrt{2\pi}} \cdot \dfrac{1}{0.005}\right)^{1/4}$

$\approx (1.3298)^{1/4} \approx 1.0739$; apply Simpson's rule with $n = 2$ subintervals.

$$\frac{1}{\sqrt{2\pi}} \int_{-1}^{1} e^{-x^2/2}\, dx = \frac{2}{\sqrt{2\pi}} \int_0^1 e^{-x^2/2}\, dx \approx \frac{2}{\sqrt{2\pi}} \cdot \frac{1}{3} \cdot \frac{1}{2} (f(0) + 4f(0.5) + f(1))$$

$$= \frac{1}{3\sqrt{2\pi}} \left(1 + 4e^{-1/8} + e^{-1/2}\right) \approx 0.6830581040.$$

31.   a.   Consult the solution for Problem 29.

$$\frac{1}{\sqrt{2\pi}} \int_{-50}^{50} e^{-x^2/2}\, dx = \frac{2}{\sqrt{2\pi}} \int_0^{50} e^{-x^2/2}\, dx = \frac{2}{\sqrt{2\pi}} \int_0^{10} e^{-x^2/2}\, dx + \frac{2}{\sqrt{2\pi}} \int_{10}^{50} e^{-x^2/2}\, dx.$$

On the interval $[10, 50]$, $0 \leq e^{-x^2/2} \leq e^{-50} \approx 1.9 \cdot 10^{-22}$; thus $\dfrac{2}{\sqrt{2\pi}} \displaystyle\int_{10}^{50} e^{-x^2/2}\, dx$ is a number between

0 and $(2/\sqrt{2\pi}) \left(1.9 \cdot 10^{-22}\right) (50 - 10) \approx 6.2 \cdot 10^{-21}$. Let's use 0 as an easy approximation for $\int_{10}^{50}$ and

approximate $\dfrac{2}{\sqrt{2\pi}} \displaystyle\int_0^{10} e^{-x^2/2}\, dx$ with error less than 0.099 (i.e., use a "quick and dirty" approximation

for $\int_{10}^{50}$ and be a bit more careful with $\int_0^{10}$). How many intervals will Simpson's rule need for $\int_0^{10}$? Well

— the global maximum of $f^{(4)}(x) = \left(x^4 - 6x^2 + 3\right) e^{-x^2/2}$ is $f^{(4)}(0) = 3$ and the global minimum is

$f^{(4)} \left(\pm\sqrt{5 - \sqrt{10}}\right) \approx -1.85$, thus $|f^{(4)}(x)| \leq |f^{(4)}(0)| = 3$. Pick an integer n such that

$$\left(3/\sqrt{2\pi}\right) (10 - 0)^5 / 2880 n^4 \leq 0.099/2;$$

this inequality implies $n \geq 5.383$, so choose $n = 6$ and apply Simpson's rule with $2n = 12$ subintervals.

$$2 \left(\frac{1}{\sqrt{2\pi}}\right) \int_0^{10} e^{-x^2/2}\, dx \approx 2(0.4997266418) \approx 0.9994532836$$

and

$$\frac{1}{\sqrt{2\pi}} \int_{-50}^{50} e^{-x^2/2}\, dx \approx 0.99945328365 + 0 = 0.9994532836.$$

(The true value is approximately $1 - 10^{-545}$.)

REMARK: If we wanted to use Simpson's rule for the full interval, $[0, 50]$, then the technique of Example 5 (and Problem 29) shows 82 subintervals would suffice. However, on most pocket calculators, an attempt to evaluate $e^{-x^2/2}$ for $|x| > 21.36$ produces an error condition; on many computers, a similar condition results if $|x| > 13.2$. In some cases, the error condition causes prior work to be lost; in others, calculations can be continued.

b.   Example 14.4.5 offers evidence that $\displaystyle\lim_{N \to \infty} \frac{1}{\sqrt{2\pi}} \int_{-N}^{N} e^{-x^2/2}\, dx = 1$.

33. If $f(x) = x^{-1}$, then $|f''(x)| = |2x^{-3}| \leq |f''(1)| = 2$ on $[1, 2]$. Choose integer n such that $\dfrac{2(2-1)^3}{12n^2} <$

10^{-10}, i.e., $n > \sqrt{\dfrac{10^{10} \cdot 2(2-1)^3}{12}} \approx 40824.8$. Pick $n = 40825$.

35. If $J_{1/2}(x) = \sqrt{\dfrac{2}{\pi x}} \sin x$, then $J_{1/2}''(x) = \sqrt{\dfrac{2}{\pi}} \left(-x^2 \sin x - x \cos x + \dfrac{3}{4} \sin x \right) x^{-5/2}$.

$$\left| -x^2 \sin x - x \cos x + \dfrac{3}{4} \sin x \right| \leq \left| x^2 \sin x \right| + \left| x \cos x \right| + \left| \dfrac{3}{4} \sin x \right| \leq 1 + 1 + \dfrac{3}{4} = \dfrac{11}{4}$$

and $\left| x^{-5/2} \right| \leq (1/2)^{-5/2} = 4\sqrt{2}$ on $[1/2, 1]$. Therefore $\left| J_{1/2}''(x) \right| \leq \sqrt{\dfrac{2}{\pi}} \left(\dfrac{11}{4} \right) \left(4\sqrt{2} \right) = \dfrac{22}{\sqrt{\pi}} \approx 12.4$. With

the trapezoidal rule, use n intervals where $\left(\dfrac{22}{\sqrt{\pi}} \right) \left(1 - \dfrac{1}{2} \right)^3 \dfrac{1}{12} n^{-2} < 0.01$, i.e., $n > \sqrt{12.929} \approx 3.6$. Choose $n = 4$ and compute approximation 0.3099246654.

REMARK:

$$\left| J_{1/2}''(x) \right| \leq \sqrt{\dfrac{2}{\pi}} \left(\left| -x^{1/2} + \dfrac{3}{4} x^{-5/2} \right| \cdot |\sin x| + \left| x^{-3/2} \right| \cdot |\cos x| \right) \leq \sqrt{\dfrac{2}{\pi}} \left((\sqrt{2} + 3\sqrt{2}) \sin 1 + 2\sqrt{2} \cos 0 \right)$$

$$\leq \sqrt{\dfrac{2}{\pi}} \left(\left(6\sqrt{2} \right) \right) = \dfrac{12}{\sqrt{\pi}} \approx 6.77.$$

Use trapezoidal rule with $n > \sqrt{7.05} \approx 2.7$; pick $n = 3$ and compute 0.3095670957.
(The true value of $\int_{1/2}^{1} J_{1/2}$ is approximately 0.3103850549.)

37. a. For any three points (x_0, y_0), (x_1, y_1), (x_2, y_2) with distinct x-coordinates there is a quadratic polynomial curve passing through them:

$$P(x) = \dfrac{(x - x_2)(x - x_3)}{(x_1 - x_2)(x_1 - x_3)} \cdot y_1 + \dfrac{(x - x_3)(x - x_1)}{(x_2 - x_3)(x_2 - x_1)} \cdot y_2 + \dfrac{(x - x_1)(x - x_2)}{(x_3 - x_1)(x_3 - x_2)} \cdot y_3.$$

For our purposes it is sufficient to observe that this expression can be rearranged into the form $P(x) = ax^2 + bx + c$. ∎ (Explicit expressions for a, b, c as functions of the coordinates are a bit messy and are not required for the discussion below.)

b. Suppose $\beta = \alpha + 2\Delta x$.

$$\int_{\alpha}^{\beta} 1 \, dx = x \Big|_{\alpha}^{\beta} = \beta - \alpha = 2\Delta x,$$

$$\int_{\alpha}^{\beta} x \, dx = \dfrac{x^2}{2} \Big|_{\alpha}^{\beta} = \dfrac{1}{2} \left(\beta^2 - \alpha^2 \right) = \dfrac{1}{2}(\beta - \alpha)(\beta + \alpha) = \dfrac{1}{2}(2\Delta x)((\alpha + 2\Delta x) + \alpha) = 2\Delta x(\alpha + \Delta x),$$

$$\int_{\alpha}^{\beta} x^2 \, dx = \dfrac{x^3}{3} \Big|_{\alpha}^{\beta} = \dfrac{1}{3} \left(\beta^3 - \alpha^3 \right) = \dfrac{1}{3}(\beta - \alpha)\left(\beta^2 + \beta\alpha + \alpha^2 \right) = \dfrac{2}{3}\Delta x \left((\alpha + 2\Delta x)^2 + (\alpha + 2\Delta x)\alpha + \alpha^2 \right)$$

$$= \dfrac{2}{3}\Delta x \left(3\alpha^2 + 6\alpha\Delta x + 4\Delta x^2 \right).$$

Now let $\alpha = x_i$ and $\beta = x_{i+2} = x_i + 2\Delta x$.

$$A_{i+2} = \int_{\alpha}^{\beta} \left(ax^2 + bx + c \right) dx = a \int_{\alpha}^{\beta} x^2 \, dx + b \int_{\alpha}^{\beta} x \, dx + c \int_{\alpha}^{\beta} 1 \, dx$$

$$= a \dfrac{2}{3}\Delta x \left(3x_i^2 + 6x_i\Delta x + 4\Delta x^2 \right) + b2\Delta x(x_i + \Delta x) + c2\Delta x$$

$$= \dfrac{\Delta x}{3} \left[a \left(6x_i^2 + 12x_i\Delta x + 8\Delta x^2 \right) + b \left(6x_i + 6\Delta x \right) + c(6) \right]. \blacksquare$$

c. If $P(x) = ax^2 + bx + c$, then

$$P(x_i) = ax_i^2 + bx_i + c,$$

$$P(x_{i+1}) = ax_{i+1}^2 + bx_{i+1} + c = a\left(x_i + \Delta x \right)^2 + b\left(x_i + \Delta x \right) + c$$

$$= a\left(a^2 + 2x_i\Delta x + \Delta x^2 \right) + b\left(x_i + \Delta x \right) + c,$$

$$P(x_{i+2}) = ax_{i+2}^2 + bx_{i+2} + c = a\left(x_i + 2\Delta x \right)^2 + b\left(x_i + 2\Delta x \right) + c$$

$$= a\left(a^2 + 4x_i\Delta x + 4\Delta x^2 \right) + b\left(x_i + 2\Delta x \right) + c. \blacksquare$$

d.
$$\frac{\Delta x}{3}\left[P(x_i) + 4P(x_{i+1}) + P(x_{i+2})\right]$$

$$= \frac{\Delta x}{3}\left[a\left(x_i^2 + 4x_{i+1}^2 + x_{i+2}^2\right) + b\left(x_i + 4x_{i+1} + x_{i+2}\right) + c\left(1 + 4 + 1\right)\right]$$

$$= \frac{\Delta x}{3}\left[a\left(x_i^2 + 4\left(x_i + \Delta x\right)^2 + \left(x_i + 2\Delta x\right)^2\right) + b\left(x_i + 4\left(x_i + \Delta x\right) + \left(x_i + 2\Delta x\right)\right) + c\left(1+4+1\right)\right]$$

$$= \frac{\Delta x}{3}\left[a\left([1+4+1]x_i^2 + [8+4]x_i\Delta x + [4+4]\Delta x^2\right) + b\left([1+4+1]x_i + [4+2]\Delta x\right) + c[1+4+1]\right]$$

$$= \frac{\Delta x}{3}\left[a\left(6x_i^2 + 12x_i\Delta x + 8\Delta x^2\right) + b\left(6x_i + 6\Delta x\right) + c(6)\right]$$

$$= A_{i+2}. \ \blacksquare$$

Solutions 7.R   *Review*   (pages 535-7)

1. Let $x = 2\tan\theta$, then $dx = 2\sec^2\theta\,d\theta$, $\theta = \tan^{-1}(x/2)$, $\theta = \tan^{-1}0 = 0$ when $x = 0$, $\theta = \tan^{-1}(1/2)$ when $x = 1$, $\sqrt{4+x^2} = 2|\sec\theta| = 2\sec\theta$ if $0 \le \theta \le \tan^{-1}(1/2)$. Therefore, using Example 7.3.5 and $\sec\left(\tan^{-1}(1/2)\right) = \sqrt{1 + (1/2)^2} = \sqrt{5}/2$,

$$\int_0^1 \frac{x^2\,dx}{\sqrt{4+x^2}} = 4\int_0^{\tan^{-1}(1/2)} \tan^2\theta\sec\theta\,d\theta = 2\left(\sec\theta\,\tan\theta - \ln|\sec\theta + \tan\theta|\right)\Big|_0^{\tan^{-1}(1/2)}$$

$$= 2\left(\frac{\sqrt{5}}{4} - \ln\left(\frac{\sqrt{5}+1}{2}\right)\right) - 2(0 - \ln 1) = \frac{\sqrt{5}}{2} + 2\ln\left(\frac{2}{\sqrt{5}+1}\right).$$

3. $$\int \frac{x^2 + 3}{(x^2+1)^2}\,dx = \int \frac{(x^2+1) + 2}{(x^2+1)^2}\,dx = \int \frac{dx}{x^2+1} + 2\int \frac{dx}{(x^2+1)^2} = \tan^{-1}x + 2\int \frac{dx}{(x^2+1)^2}.$$

Handle the remaining integral with the substitution $x = \tan\theta$.

$$\int \frac{2\,dx}{(x^2+1)^2} = \int \frac{2\sec^2\theta\,d\theta}{(\sec^2\theta)^2} = \int 2\cos^2\theta\,d\theta = \int (1 + \cos 2\theta)\,d\theta = \theta + \frac{1}{2}\sin 2\theta + C$$

$$= \tan^{-1}x + \sin\theta\cos\theta + C = \tan^{-1}x + \left(\frac{x}{\sqrt{x^2+1}}\right)\left(\frac{1}{\sqrt{x^2+1}}\right) + C = \tan^{-1}x + \frac{x}{x^2+1} + C.$$

Therefore $\displaystyle\int \frac{x^2+3}{(x^2+1)^2}\,dx = 2\tan^{-1}x + \frac{x}{x^2+1} + C.$

REMARK: Alternative computation (using integration by parts):

$$\int \frac{x^2+3}{(x^2+1)^2}\,dx = \int \frac{3\,dx}{x^2+1} - \int \frac{2x^2\,dx}{(x^2+1)^2} = 3\tan^{-1}x + \int x\,d\left(\frac{1}{x^2+1}\right)$$

$$= 3\tan^{-1}x + \frac{x}{x^2+1} - \int \frac{1}{x^2+1}\,dx = 3\tan^{-1}x + \frac{x}{x^2+1} - \tan^{-1}x + C.$$

5. $$\int_0^{\pi/2} \cos^3 x \sin^2 x\,dx = \int_0^{\pi/2} \cos x\left(1 - \sin^2 x\right)\sin^2 x\,dx = \int_0^{\pi/2} \sin^2 x \cos x\,dx - \int_0^{\pi/2} \sin^4 x \cos x\,dx$$

$$= \left(\frac{1}{3}\sin^3 x - \frac{1}{5}\sin^5 x\right)\Big|_0^{\pi/2} = \left(\frac{1}{3} - \frac{1}{5}\right) - (0 - 0) = \frac{2}{15}.$$

7. $$\int xe^{-2x}\,dx = \frac{1}{-2}\int x\,d\left(e^{-2x}\right) = \frac{-1}{2}\left(xe^{-2x} - \int e^{-2x}\,dx\right) = \frac{-1}{2}\left(xe^{-2x} - \frac{e^{-2x}}{-2}\right) + C$$

$$= \frac{-1}{4}(2x+1)e^{-2x} + C.$$

9. $$\int \tan^5 x\,dx = \int (\sec^2 x - 1)^2 \tan x\,dx = \int (\sec^3 x - 2\sec x)(\sec x \tan x\,dx) + \int \tan x\,dx$$

$$= \left(\frac{1}{4}\sec^4 x - \sec^2 x\right) - \ln|\cos x| + C = \frac{1}{4}\left(\tan^2 x - 1\right)^2 + \ln|\sec x| + C_1.$$

11.
$$\int_0^1 \frac{dx}{1+\sqrt[3]{x}} \overset{x=u^3}{=} \int_0^1 \frac{3u^2\,du}{1+u} = \int_0^1 3\left(u-1+\frac{1}{1+u}\right)du = \left(\frac{3}{2}u^2 - 3u + 3\ln|1+u|\right)\Big|_0^1$$
$$= \left(\frac{3}{2} - 3 + 3\ln 2\right) - 0 = \frac{-3}{2} + 3\ln 2.$$

13.
$$\int_{\sqrt 2}^2 \frac{x^2\,dx}{(x^2-1)^{3/2}} \overset{x=\sec\theta}{=} \int_{\pi/4}^{\pi/3} \frac{\sec^2\theta(\sec\theta\tan\theta\,d\theta)}{(\tan^2\theta)^{3/2}} = \int_{\pi/4}^{\pi/3} \frac{d\theta}{\sin^2\theta\cos\theta} = \int_{\pi/4}^{\pi/3} (\sec\theta + \csc\theta\cot\theta)\,d\theta$$
$$= (\ln|\sec\theta + \tan\theta| - \csc\theta)\big|_{\pi/4}^{\pi/3} = \left(\ln\left(2+\sqrt 3\right) - \frac{2}{\sqrt 3}\right) - \left(\ln\left(\sqrt 2 + 1\right) - \sqrt 2\right).$$

15.   Let $I = \int e^{2x}\sin 2x\,dx$ and use integration by parts twice.
$$I = \int \sin 2x\left(e^{2x}\,dx\right) = \frac{1}{2}\int \sin 2x\,d\left(e^{2x}\right) = \frac{1}{2}\left(e^{2x}\sin 2x - \int e^{2x}\,d(\sin 2x)\right)$$
$$= \frac{1}{2}e^{2x}\sin 2x - \int \cos 2x\left(e^{2x}\,dx\right) = \frac{1}{2}e^{2x}\sin 2x - \left(\frac{1}{2}e^{2x}\cos 2x - \int \frac{1}{2}e^{2x}\,d(\cos 2x)\right)$$
$$= \frac{1}{2}e^{2x}(\sin 2x - \cos 2x) - \int e^{2x}\sin 2x\,dx = \frac{1}{2}e^{2x}(\sin 2x - \cos 2x) - I + C_*,$$
therefore $2I = (1/2)e^{2x}(\sin 2x - \cos 2x) + C_*$ and $\int e^{2x}\sin 2x\,dx = I = (1/4)e^{2x}(\sin 2x - \cos 2x) + C.$

REMARK: The substitution $u = 2x$ would let us use the result of Example 7.2.4.
            Hint: Get ready for Problems 47, 49, 63, 65 and 69 by comparing and contrasting this solution
            with the work for Example 7.2.4.

17.
$$\int \sin^3 x\,\cos^3 x\,dx = \int \sin^3 x\left(1 - \sin^2 x\right)\cos x\,dx = \int \sin^3 x\cos x\,dx - \int \sin^5 x\cos x\,dx$$
$$= \frac{1}{4}\sin^4 x - \frac{1}{6}\sin^6 x + C.$$

REMARK:
$$\int \sin^3 x\cos^3 x\,dx = \int \sin x\left(1 - \cos^2 x\right)\cos^3 x\,dx = \int \cos^3 x\sin x\,dx - \int \cos^5 x\sin x\,dx$$
$$= \frac{-1}{4}\cos^4 x + \frac{1}{6}\cos^6 x + C_1.$$

19.
$$\int \csc^3 x\,\cot^3 x\,dx = \int \csc^3 x\left(\csc^2 x - 1\right)\cot x\,dx = \int \left(-\csc^4 x + \csc^2 x\right)(-\csc x\cot x\,dx)$$
$$= \frac{-1}{5}\csc^5 x + \frac{1}{3}\csc^3 x + C.$$

21.
$$\int \frac{x^5-1}{x^3-x}\,dx = \int \left(x^2 + 1 + \frac{x-1}{x^3-x}\right)dx = \int \left(x^2 + 1 + \frac{1}{x} - \frac{1}{x+1}\right)dx$$
$$= \frac{x^3}{3} + x + \ln|x| - \ln|x+1| + C = \frac{x^3}{3} + x + \ln\left|\frac{x}{x+1}\right| + C.$$

23.
$$\int x\sinh\,dx = \int x\,d(\cosh x) = x\cosh x - \int \cosh x\,dx = x\cosh x - \sinh x + C.$$

25.
$$\int \frac{dx}{x^2 - 4x + 4} = \int (x-2)^{-2}\,dx = -(x-2)^{-1} + C = \frac{-1}{x-2} + C.$$

27.   Use identity (5) in Section 7.3 and the identity $\cos(-x) = \cos x.$
$$\int \cos 6x\,\cos 7x\,dx = \frac{1}{2}\int (\cos x + \cos 13x)\,dx = \frac{1}{2}\sin x + \frac{1}{26}\sin 13x + C.$$

29.   If $\dfrac{x^2+2}{(x-3)(x+4)(x-5)} = \dfrac{A}{x-3} + \dfrac{B}{x+4} + \dfrac{C}{x-5}$, then $A = \dfrac{x^2+2}{(x+4)(x-5)}\Big|_{x=3} = \dfrac{9+2}{7(-2)} = \dfrac{-11}{14}$,

$B = \dfrac{x^2+2}{(x-3)(x-5)}\Big|_{x=-4} = \dfrac{16+2}{(-7)(-9)} = \dfrac{2}{7}$, and $C = \dfrac{x^2+2}{(x-3)(x+4)}\Big|_{x=5} = \dfrac{25+2}{(2)(9)} = \dfrac{3}{2}$. Therefore

$$\int \frac{x^2+2}{(x-3)(x+4)(x-5)}\,dx = \int \left(\frac{-11/14}{x-3} + \frac{2/7}{x+4} + \frac{3/2}{x-5}\right)dx = \frac{-11}{14}\ln|x-3| + \frac{2}{7}\ln|x+4| + \frac{3}{2}\ln|x-5| + K.$$

31.
$$\int x^3\sqrt{x^2-4}\,dx = \frac{1}{2}\int x^2\sqrt{x^2-4}\,(2x\,dx) \overset{x^2-4\equiv u}{=} \left(\frac{1}{2}\right)\int (u+4)\sqrt{u}\,du = \int\left(\frac{1}{2}u^{3/2}+2u^{1/2}\right)du$$

$$= \frac{1}{5}u^{5/2}+\frac{4}{3}u^{3/2}+C = \frac{1}{5}\left(x^2-4\right)^{5/2}+\frac{4}{3}\left(x^2-4\right)^{3/2}+C = \frac{\left(3x^2+8\right)\left(x^2-4\right)^{3/2}}{15}+C.$$

33.
$$\int \sec^4 x\,dx = \int (\tan^2 x+1)\sec^2 x\,dx = \int \tan^2 x\,(\sec^2 x\,dx) + \int \sec^2 x\,dx = \frac{1}{3}\tan^3 x + \tan x + C.$$

REMARK: Problem 7.2.54 yields answer $\frac{1}{3}\left(\sec^2 x+2\right)\tan x + C_*$.

35.
$$\int \frac{(x+1)^3}{x^3}\,dx = \int \frac{x^3+3x^2+3x+1}{x^3}\,dx = \int\left(1+\frac{3}{x}+\frac{3}{x^2}+\frac{1}{x^3}\right)dx = x+3\ln|x| - \frac{3}{x} - \frac{1}{2x^2}+C.$$

37. Solution 7.3.49 computes $\int_0^{\pi/2}\sin^6 x\,dx = \frac{5\pi}{32}$ in two ways. (Problem 7.3.51 is a generalization.)

39. See Problem 7.2.60(a) or 7.4.57.
$$\int_0^1 (x^2+1)^{-1/2}\,dx = \int_0^1 \frac{dx}{\sqrt{x^2+1}} = \ln\left(x+\sqrt{x^2+1}\right)\Big|_0^1 = \ln\left(1+\sqrt{2}\right) - \ln\left(0+\sqrt{1}\right) = \ln\left(1+\sqrt{2}\right).$$

41.
$$\int \frac{dx}{1+5\sin x} \overset{\tan(x/2)\equiv u}{=} \int \frac{2\,du}{(1+u^2)+5(2u)} = \int \frac{2\,du}{u^2+10u+1} = \int \frac{2\,du}{(u+5)^2-24}$$

$$= \frac{1}{2\sqrt{6}}\int\left(\frac{1}{u+5-2\sqrt{6}} - \frac{1}{u+5+2\sqrt{6}}\right)dx$$

$$= \frac{1}{2\sqrt{6}}\left(\ln\left|u+5-2\sqrt{6}\right| - \ln\left|u+5+2\sqrt{6}\right|\right)+C$$

$$= \frac{1}{2\sqrt{6}}\ln\left|\frac{\tan(x/2)+5-2\sqrt{6}}{\tan(x/2)+5+2\sqrt{6}}\right|+C.$$

43.
$$\int \frac{2x^3\,dx}{(1+x^4)^{4/3}} = \frac{1}{2}\int (1+x^4)^{-4/3}\,(4x^3\,dx) = \frac{1}{2}(-3)\,(1+x^4)^{-1/3}+C = \frac{-3}{2\sqrt[3]{1+x^4}}+C.$$

45. Let $u = 2+3x$, then $x = \frac{u-2}{3}$ and $x-2x^3 = \frac{1}{3}(u-2) - \frac{2}{27}(u-2)^3 = \frac{-2}{27}u^3 + \frac{4}{9}u^2 - \frac{5}{9}u - \frac{2}{27}$; thus
$$\int \frac{x-2x^3}{\sqrt{2+3x}}\,dx = \int\left(\frac{-2}{81}u^{5/2} + \frac{4}{27}u^{3/2} - \frac{5}{27}u^{1/2} - \frac{2}{81}u^{-1/2}\right)du$$

$$= \frac{-4}{567}u^{7/2} + \frac{8}{135}u^{5/2} - \frac{10}{81}u^{3/2} - \frac{4}{81}u^{1/2}+C$$

$$= \frac{-4}{567}(2+3x)^{7/2} + \frac{8}{135}(2+3x)^{5/2} - \frac{10}{81}(2+3x)^{3/2} - \frac{4}{81}(2+3x)^{1/2}+C$$

$$= \frac{2}{2835}\left(-27x^3+216x^2+123x-164\right)\sqrt{2+3x}+C.$$

47. Let $I = \int e^{-x}\cos(x/3)\,dx$. Two integrations by parts yields an equation which can be solved for I.
$$I = 3\int e^{-x}\,d\left(\sin\frac{x}{3}\right) = 3e^{-x}\sin\frac{x}{3} - 3\int \sin\frac{x}{3}\,d\left(e^{-x}\right) = 3e^{-x}\sin\frac{x}{3} - 9\int e^{-x}\,d\left(\cos\frac{x}{3}\right)$$

$$= 3e^{-x}\sin\frac{x}{3} - 9\left(e^{-x}\cos\frac{x}{3} - \int \cos\frac{x}{3}\,d\left(e^{-x}\right)\right) = 3e^{-x}\left(\sin\frac{x}{3} - 3\cos\frac{x}{3}\right) - 9I + C_*,$$

therefore $(1+9)I = 3e^{-x}\left(\sin\frac{x}{3} - 3\cos\frac{x}{3}\right) + C_*$ and $\int e^{-x}\cos\frac{x}{3}\,dx = I = \frac{3}{10}e^{-x}\left(\sin\frac{x}{3} - 3\cos\frac{x}{3}\right) + C.$

49. Let $I = \int \cosh x\,\cosh 3x\,dx$. Then
$$I = \frac{1}{3}\int \cosh x\,d(\sinh 3x) = \frac{1}{3}\cosh x\,\sinh 3x - \frac{1}{3}\int \sinh 3x\,d(\cosh x)$$

$$= \frac{1}{3}\cosh x\,\sinh 3x - \frac{1}{9}\int \sinh x\,d(\cosh 3x) = \frac{1}{3}\cosh x\,\sinh 3x - \frac{1}{9}\left(\sinh x\,\cosh 3x - I\right)+C.$$

Therefore $\left(1-\frac{1}{9}\right)I = \frac{1}{3}\cosh x\,\sinh 3x - \frac{1}{9}\sinh x\,\cosh 3x + C$ and
$$\int \cosh x\,\cosh 3x\,dx = \frac{3}{8}\cosh x\,\sinh 3x - \frac{1}{8}\sinh x\,\cosh 3x + C_1.$$

REMARK:

$$\int \cosh x \, \cosh 3x \, dx = \int \left(\frac{e^x + e^{-x}}{2} \right) \left(\frac{e^{3x} + e^{-3x}}{2} \right) dx = \frac{1}{4} \int \left(e^{4x} + e^{2x} + e^{-2x} + e^{-4x} \right) dx$$

$$= \frac{1}{4} \left(\frac{e^{4x}}{4} + \frac{e^{2x}}{2} - \frac{e^{-2x}}{2} - \frac{e^{-4x}}{4} \right) + C = \frac{1}{8} \left(\frac{e^{4x} - e^{-4x}}{2} \right) + \frac{1}{4} \left(\frac{e^{2x} - e^{-2x}}{2} \right) + C$$

$$= \frac{1}{8} \sinh 4x + \frac{1}{4} \sinh 2x + C.$$

51.

$$\int x^2 e^{-x} \, dx = - \int x^2 \, d\left(e^{-x}\right) = - \left(x^2 e^{-x} - \int e^{-x} \, d(x^2) \right) = -x^2 e^{-x} + 2 \int x e^{-x} \, dx$$

$$= -x^2 e^{-x} + 2(-1) \left(x e^{-x} - \int e^{-x} \, dx \right) = (-x^2 - 2x) e^{-x} + 2 \int e^{-x} \, dx$$

$$= (-x^2 - 2x - 2) e^{-x} + C.$$

53.

$$\int \frac{\sin \sqrt{x} \, \cos \sqrt{x}}{\sqrt{x}} \, dx \overset{x \equiv u^2}{=} \int \frac{\sin u \, \cos u}{u} \, (2u \, du) = 2 \int \sin u \, \cos u \, du = \sin^2 u + C = \sin^2 \sqrt{x} + C.$$

55. Let $u = (3/2)x^{-1}$, then $du = (-3/2)x^{-2} \, dx$, $9 - 4x^2 = 9 - (2x)^2 = 9 - (3/u)^2 = 9(u^2 - 1)/u^2$, and

$$\int \frac{dx}{x\sqrt{9 - 4x^2}} = \frac{-1}{3} \int \frac{du}{\sqrt{u^2 - 1}} = \frac{-1}{3} \cosh^{-1} u + C = \frac{-1}{3} \ln \left| u + \sqrt{u^2 - 1} \right| + C$$

$$= \frac{-1}{3} \cosh^{-1} \left(\frac{3}{2x} \right) + C = \frac{-1}{3} \ln \left| \frac{3 + \sqrt{9 - 4x^2}}{x} \right| + C.$$

REMARK:

$$\int \frac{dx}{x\sqrt{9 - 4x^2}} \overset{2x \equiv 3\cos\theta}{=} \left(\frac{-1}{3} \right) \int \sec \theta \, d\theta = \frac{-1}{3} \ln |\sec \theta + \tan \theta| + C = \frac{-1}{3} \ln \left| \frac{3}{2x} + \frac{\sqrt{1 - (2x/3)^2}}{2x/3} \right| + C.$$

57. Use the substitution $x = 4 \sec \theta$, and let $S = \text{sgn}(x) = x/|x| = \text{sgn}(\tan \theta)$.

$$\int \frac{\sqrt{x^2 - 16}}{x^2} \, dx = \int \frac{4|\tan \theta|}{16 \sec^2 \theta} (4 \sec \theta \tan \theta \, d\theta) = S \int \frac{\tan^2 \theta}{\sec \theta} \, d\theta = S \int \frac{\sin^2 \theta}{\cos \theta} \, d\theta = S \int \frac{1 - \cos^2 \theta}{\cos \theta} \, d\theta$$

$$= S \int (\sec \theta - \cos \theta) \, d\theta = S \ln |\sec \theta + \tan \theta| - S \sin \theta + C_*$$

$$= S \ln \left| \frac{x}{4} + \frac{S\sqrt{1 - (4/x)^2}}{4/x} \right| - S\sqrt{1 - (4/x)^2} + C_* = \ln \left| x + \sqrt{x^2 - 16} \right| - \frac{\sqrt{x^2 - 16}}{x} + C.$$

59. $u = 2x^2$ implies $du = 4x \, dx$ and $V = \sin u$ implies $dV = \cos u \, du$, therefore

$$\int x \sin^{3/4} \left(2x^2 \right) \cos \left(2x^2 \right) \, dx = \frac{1}{4} \int \sin^{3/4} \left(2x^2 \right) \left[\cos \left(2x^2 \right) (4x \, dx) \right]$$

$$= \frac{1}{4} \int \sin^{3/4} u \, [\cos u \, du] = \frac{1}{4} \int V^{3/4} \, dV = \frac{1}{7} V^{7/4} + C$$

$$= \frac{1}{7} (\sin u)^{7/4} + C = \frac{1}{7} \left(\sin \left(2x^2 \right) \right)^{7/4} + C.$$

61. Use the reduction formula of Example 7.2.7.

$$\int_0^{\pi/2} \sin^7 x \, dx = \frac{-1}{7} \sin^6 x \, \cos x \Big|_0^{\pi/2} + \frac{6}{7} \int_0^{\pi/2} \sin^5 x \, dx$$

$$= 0 + \frac{6}{7} \cdot \frac{-1}{5} \sin^4 x \, \cos x \Big|_0^{\pi/2} + \frac{6}{7} \cdot \frac{4}{5} \int_0^{\pi/2} \sin^3 x \, dx$$

$$= 0 + \frac{6}{7} \cdot \frac{4}{5} \cdot \frac{-1}{3} \sin^2 x \, \cos x \Big|_0^{\pi/2} + \frac{6}{7} \cdot \frac{4}{5} \cdot \frac{2}{3} \int_0^{\pi/2} \sin x \, dx$$

$$= 0 + \frac{6}{7} \cdot \frac{4}{5} \cdot \frac{2}{3} \cdot (-\cos x) \Big|_0^{\pi/2} = \frac{6}{7} \cdot \frac{4}{5} \cdot \frac{2}{3} = \frac{16}{35}.$$

REMARK: Problem 7.3.52 generalizes this result.

63.
$$\int \sinh^2(3x)\,dx = \int \left(\frac{e^{3x}-e^{-3x}}{2}\right)^2 dx = \int \left(\frac{1}{4}\left(e^{6x}+e^{-6x}\right)-\frac{1}{2}\right)dx$$
$$= \frac{1}{24}\left(e^{6x}-e^{-6x}\right) - \frac{x}{2} + C = \frac{1}{12}\sinh 6x - \frac{x}{2} + C.$$

REMARK:
$$\int \sinh^2 3x\,dx = I = \frac{1}{3}\int \sinh 3x\,d(\cosh 3x) = \frac{1}{3}\sinh 3x\,\cosh 3x - \frac{1}{3}\int \cosh 3x\,d(\sinh 3x)$$
$$= \frac{1}{3}\sinh 3x\,\cosh 3x - \int \cosh^2 3x\,dx = \frac{1}{3}\sinh 3x\,\cosh 3x - \int \left(1+\sinh^2 3x\right)dx$$
$$= \frac{1}{3}\sinh 3x\,\cosh 3x - \int 1\,dx - \int \sinh^2(3x)\,dx = \frac{1}{3}\sinh 3x\,\cosh 3x - x - I,$$

hence
$$\int \sinh^2 3x\,dx = I = \frac{1}{2}(2I) = \frac{1}{2}\left(\frac{1}{3}\sinh 3x\,\cosh 3x - x\right) + C = \frac{1}{6}\sinh 3x\,\cosh 3x - \frac{x}{2} + C.$$

65.   Let $u = x^3$, then $\int x^2 e^{-3x^3}\sin 2x^3\,dx = \frac{1}{3}\int e^{-3u}\sin 2u\,du$. Let $I = \int e^{-3u}\sin 2u\,du$. Two integrations by parts will produce either the equation
$$I = -e^{-3u}\left(\frac{3}{4}\sin 2u + \frac{1}{2}\cos 2u\right) - \frac{9}{4}I$$

or the equation
$$I = -e^{-3u}\left(\frac{1}{3}\sin 2u + \frac{2}{9}\cos 2u\right) - \frac{4}{9}I.$$

These equations have the same solution and
$$\int x^2 e^{-3x^3}\sin 2x^3\,dx = \frac{1}{3}I = -e^{-3x^3}\left(\frac{1}{13}\sin 2x^3 + \frac{2}{39}\cos 2x^3\right) + C.$$

67.   Let $u = 3x$ and use the answer for Problem 33.
$$\int \sec^4 3x\,dx = \frac{1}{3}\int \sec^4 u\,du = \frac{1}{3}\left(\frac{1}{3}\tan^3 u + \tan u\right) + C = \frac{1}{9}\tan^3 3x + \frac{1}{3}\tan 3x + C.$$

REMARK: The reduction formula of Problem 7.2.54 yields $\frac{1}{3}\left(\frac{1}{3}\tan 3x\,\sec^2 3x + \frac{2}{3}\tan 3x\right) + C.$

69.   Let $I = \int e^{2x}\sinh 3x\,dx$. Then two integrations by parts yields one of the following equations:
$$I = \left(\frac{1}{2}\sinh 3x - \frac{3}{4}\cosh 3x\right)e^{2x} + \frac{9}{4}I,$$
$$I = e^{2x}\left(\frac{1}{3}\cosh 3x - \frac{2}{9}\sinh 3x\right) + \frac{4}{9}I.$$

Each equation has solution $\int e^{2x}\sinh 3x\,dx = I = \frac{1}{5}e^{2x}\left(3\cosh 3x - 2\sinh 3x\right) + C.$

71. $\int \tanh^2 3x\,dx = \int \left(1 - \operatorname{sech}^2 3x\right)dx = \int dx - \frac{1}{3}\int \operatorname{sech}^2(3x)\,d(3x) = x - \frac{1}{3}\tanh(3x) + C.$

73.   If $u = \tan(2x/2) = \tan x$, then $dx = \dfrac{du}{1+u^2}$ and $\cos 2x = \dfrac{1-u^2}{1+u^2}$; therefore
$$\int \frac{dx}{4-3\cos 2x} = \int \frac{du}{4(1+u^2)-3(1-u^2)} = \int \frac{du}{1+7u^2} = \frac{1}{\sqrt{7}}\tan^{-1}\left(\sqrt{7}\,u\right) + C = \frac{1}{\sqrt{7}}\tan^{-1}\left(\sqrt{7}\tan x\right) + C.$$

75.   If $u = \tan(2x/2) = \tan x$, then $\sin 2x = \dfrac{2u}{1+u^2}$ and
$$\int \frac{dx}{1+\sin 2x} = \int \frac{du/(1+u^2)}{1+2u/(1+u^2)} = \int \frac{du}{1+u^2+2u} = \int (1+u)^{-2}\,du = -(1+u)^{-1} + C = \frac{-1}{1+\tan x} + C.$$

REMARK:

$$\int \frac{dx}{1 + \sin 2x} = \int \left(\frac{1 - \sin 2x}{1 - \sin 2x}\right)\left(\frac{1}{1 + \sin 2x}\right) dx$$

$$= \int \frac{1 - \sin 2x}{\cos^2 2x} dx = \frac{1}{2} \int (\sec^2 2x - \sec 2x \tan 2x) \, d(2x) = \frac{1}{2}(\tan 2x - \sec 2x) + C_*.$$

77.

$$\int x^3 \cos 4x \, dx = \frac{1}{4}\int x^3 \, d(\sin 4x) = \frac{1}{4}x^3 \sin 4x - \frac{3}{4}\int x^2 \sin 4x \, dx = \frac{1}{4}x^3 \sin 4x + \frac{3}{16}\int x^2 \, d(\cos 4x)$$

$$= \frac{1}{4}x^3 \sin 4x + \frac{3}{16}x^2 \cos 4x - \frac{3}{8}\int x \cos 4x \, dx$$

$$= \frac{1}{4}x^3 \sin 4x + \frac{3}{16}x^2 \cos 4x - \frac{3}{32}\int x \, d(\sin 4x)$$

$$= \left(\frac{1}{4}x^3 - \frac{3}{32}x\right)\sin 4x + \frac{3}{16}x^2 \cos 4x + \frac{3}{32}\int \sin 4x \, dx$$

$$= \left(\frac{x^3}{4} - \frac{3x}{32}\right)\sin 4x + \left(\frac{3x^2}{16} - \frac{3}{128}\right)\cos 4x + C.$$

REMARK: This took about the same amount of work as would be needed to obtain a general formula analogous to the one in Problem 7.2.49.

79.

$$\int \frac{x^5 \, dx}{9 + x^{12}} = \frac{1}{6}\int \frac{6x^5 \, dx}{3^2 + (x^6)^2} \overset{x^6 \equiv u}{=} \frac{1}{6}\int \frac{du}{3^2 + u^2} = \frac{1}{6}\cdot\frac{1}{3}\tan^{-1}\left(\frac{u}{3}\right) + C = \frac{1}{18}\tan^{-1}\left(\frac{x^6}{3}\right) + C.$$

81. If $u = x^6$, then $\int x^5 \sqrt{9 + x^{12}} \, dx = \frac{1}{6}\int \sqrt{9 + u^2} \, du$. Now let $u = 3\tan V$ and use Example 7.2.6,

$$\frac{1}{6}\int \sqrt{9 + u^2} \, du = \frac{1}{6}\int (3 \sec V)(3\sec^2 V) \, dV = \frac{3}{2}\int \sec^3 V \, dV = \frac{3}{4}(\sec V \tan V + \ln|\sec V + \tan V|) + C_*$$

$$= \frac{3}{4}\left(\sqrt{1 + \left(\frac{u}{3}\right)^2}\left(\frac{u}{3}\right) + \ln\left|\sqrt{1 + \left(\frac{u}{3}\right)^2} + \frac{u}{3}\right|\right) + C_*$$

$$= \frac{1}{12}u\sqrt{9 + u^2} + \frac{3}{4}\ln\left|\sqrt{9 + u^2} + u\right| + C = \frac{1}{12}x^6\sqrt{9 + x^{12}} + \frac{3}{4}\ln\left|\sqrt{9 + x^{12}} + x^6\right| + C.$$

83. Suppose $\dfrac{x^2 - 3}{x^3 - 1} = \dfrac{A}{x - 1} + \dfrac{Bx + C}{x^2 + x + 1}$. Then $A = \left.\dfrac{x^2 - 3}{x^2 + x + 1}\right|_{x=1} = \dfrac{-2}{3}$ and

$$Bx + C = \frac{(x^2 - 3) - (-2/3)(x^2 + x + 1)}{x - 1} = \frac{5}{3}x + \frac{7}{3}, \text{ hence } B = \frac{5}{3} \text{ and } C = \frac{7}{3}. \text{ Therefore}$$

$$\int \frac{x^2 - 3}{x^3 - 1} \, dx = \int \left(\frac{-2/3}{x - 1} + \frac{(5/3)x + 7/3}{x^2 + x + 1}\right) dx$$

$$= \frac{-2}{3}\ln|x - 1| + \int\left(\frac{5}{6}\cdot\frac{2x + 1}{x^2 + x + 1} + \frac{3}{2}\cdot\frac{1}{(x + 1/2)^2 + 3/4}\right) dx$$

$$= \frac{-2}{3}\ln|x - 1| + \frac{5}{6}\ln|x^2 + x + 1| + \frac{3}{2}\cdot\frac{1}{\sqrt{3/4}}\tan^{-1}\left(\frac{x + 1/2}{\sqrt{3/4}}\right) + K$$

$$= \frac{-2}{3}\ln|x - 1| + \frac{5}{6}\ln|x^2 + x + 1| + \sqrt{3}\tan^{-1}\left(\frac{2x + 1}{\sqrt{3}}\right) + K.$$

85.

$$\int \frac{2x - 3}{(x - 1)^2} \, dx = \int \left(\frac{2}{x - 1} - \frac{1}{(x - 1)^2}\right) dx = 2\ln|x - 1| + \frac{1}{x - 1} + C.$$

87. Let $u = x$ and $a = 4$ in table entry 20: $\displaystyle\int \frac{dx}{16 - x^2} = \frac{1}{8}\ln\left|\frac{4 + x}{4 - x}\right| + C.$

89. Let $u = 3x$ and $a = 5$ in entry 87: $\displaystyle\int \frac{x \, dx}{\sqrt{25 - 9x^2}} = \frac{1}{9}\int \frac{(3x) \, d(3x)}{\sqrt{5^2 - (3x)^2}} = \left(\frac{-1}{9}\right)\sqrt{25 - 9x^2} + C.$

91.   Let $u = 3x$ and $a = 4$ in entry 69:

$$\int \frac{9x^2 \, dx}{\sqrt{9x^2 + 16}} = \frac{1}{3} \int \frac{(3x)^2 \, d(3x)}{\sqrt{(3x)^2 + 4^2}} = \frac{1}{3} \left(\frac{1}{2}(3x)\sqrt{(3x)^2 + 4^2} - \frac{4^2}{2} \ln \left| 3x + \sqrt{(3x)^2 + 4^2} \right| \right) + C$$

$$= \frac{x}{2}\sqrt{9x^2 + 16} - \frac{8}{3} \ln \left| 3x + \sqrt{9x^2 + 16} \right| + C.$$

93.   Let $u = x$ and $a = \frac{1}{3}$ in entry 161: $\int x^3 \cos^{-1}(3x) \, dx = \frac{x^{3+1}}{3+1} \cos^{-1}(3x) + \frac{1}{3+1} \int \frac{x^{3+1}}{\sqrt{(1/3)^2 - x^2}} \, dx.$

Now use integration by parts:

$$\int x^4 \left(\frac{1}{9} - x^2 \right)^{-1/2} dx = - \int x^3 \, d\left(\sqrt{\frac{1}{9} - x^2} \right) = -x^3 \sqrt{\frac{1}{9} - x^2} + 3 \int x^2 \sqrt{\frac{1}{9} - x^2} \, dx.$$

Tackle the remaining integral with $u = x$, $a = 1/3$ in table entry 93 and use the fact that $\sin^{-1} W = -\cos^{-1} W + \pi/2$,

$$\int x^2 \sqrt{\frac{1}{9} - x^2} \, dx = \frac{-1}{4} x \left(\frac{1}{9} - x^2 \right)^{3/2} + \frac{1}{72} x \sqrt{\frac{1}{9} - x^2} - \frac{1}{648} \cos^{-1}(3x) + C.$$

Fit these pieces together to obtain

$$\int x^3 \cos^{-1} 3x \, dx = \frac{x^4}{4} \cos^{-1} 3x - \frac{x^3}{4}\sqrt{\frac{1}{9} - x^2} + \frac{3}{4} \left(\frac{-x}{4} \left(\frac{1}{9} - x^2 \right)^{3/2} + \frac{x}{72}\sqrt{\frac{1}{9} - x^2} - \frac{1}{648} \cos^{-1} 3x \right) + C$$

$$= \frac{1}{4} \left(x^4 - \frac{1}{216} \right) \cos^{-1} 3x - \frac{1}{96} \left(6x^3 + x \right) \sqrt{\frac{1}{9} - x^2} + C.$$

95.   Let $u = x^3$ and $a = 1$ in entry 158:

$$\int x^2 \tan^{-1} x^3 \, dx = \frac{1}{3} \int \tan^{-1}(x^3) \, d(x^3) = \frac{1}{3} \left(x^3 \tan^{-1} x^3 - \frac{1}{2} \ln(x^6 + 1) \right) + C.$$

97. $\displaystyle\int \frac{x}{(x-4)^3} \, dx = \int \frac{(x-4) + 4}{(x-4)^3} \, dx = \int \left(\frac{1}{(x-4)^2} + \frac{4}{(x-4)^3} \right) dx = \frac{-1}{x-4} + \frac{-2}{(x-4)^2} + C.$

99. $\displaystyle\int \frac{x^3 \, dx}{x^4 - x^2} = \int \frac{x \, dx}{x^2 - 1} = \frac{1}{2} \int \frac{d(x^2 - 1)}{x^2 - 1} = \frac{1}{2} \ln |x^2 - 1| + C.$

REMARK: The solution above seems easier to do than searching the integral table to discover entry 62.

101.   A moderate bit of partial fractions followed by two uses of table entry 19 yields

$$\int \frac{dx}{x^4 - 5x^2 + 4} = \int \frac{dx}{(x^2 - 4)(x^2 - 1)} = \frac{1}{3} \int \left(\frac{1}{x^2 - 4} - \frac{1}{x^2 - 1} \right) dx = \frac{1}{3} \left(\frac{1}{4} \ln \left| \frac{x-2}{x+2} \right| - \frac{1}{2} \ln \left| \frac{x-1}{x+1} \right| \right) + C.$$

103. $\displaystyle\int \frac{x^3 - 1}{x^2 + x + 1} \, dx = \int (x - 1) \, dx = \frac{x^2}{2} - x + C.$

105. $\displaystyle\int_{\pi/2}^{\pi} \ln \sin \theta \, d\theta \overset{\theta \equiv t + \frac{\pi}{2}}{=} \int_0^{\pi/2} \ln \sin \left(t + \frac{\pi}{2} \right) d\left(t + \frac{\pi}{2} \right) = \int_0^{\pi/2} \ln \cos t \, dt$

implies

$$\int_0^{\pi} \ln \sin \theta \, d\theta = I = \int_0^{\pi/2} (\ln \sin \theta + \ln \cos \theta) \, d\theta = \int_0^{\pi/2} \ln (\sin \theta \cdot \cos \theta) \, d\theta$$

$$= \int_0^{\pi/2} \ln \left(\frac{1}{2} \cdot \sin 2\theta \right) d\theta \overset{2\theta \equiv u}{=} \int_0^{\pi} \ln \left(\frac{1}{2} \cdot \sin u \right) \frac{du}{2}$$

$$= \frac{1}{2} \left(\int_0^{\pi} \ln \left(\frac{1}{2} \right) du + \int_0^{\pi} \ln \sin u \, du \right) = \frac{1}{2} (-\ln 2) \pi + \frac{1}{2} I,$$

thus

$$\int_0^{\pi} \ln \sin \theta \, d\theta = I = 2 \left(1 - \frac{1}{2} \right) I = 2 \left(\frac{1}{2} (-\ln 2) \pi \right) = -\pi \ln 2.$$

107.
$$\text{Area} = \int_0^1 \frac{x^3}{(x^2+1)^3}\,dx = \int_0^1 \frac{((x^2+1)-1)\,x}{(x^2+1)^3}\,dx = \int_0^1 \left(\frac{x}{(x^2+1)^2} - \frac{x}{(x^2+1)^3} \right) dx$$

$$= \left(\frac{-1}{2}(x^2+1)^{-1} + \frac{1}{4}(x^2+1)^{-2} \right)\Big|_0^1 = \frac{-1}{4} \cdot \frac{2x^2+1}{(x^2+1)^2}\Big|_0^1 = \frac{-1}{4}\left(\frac{3}{4} - \frac{1}{1} \right) = \left(\frac{-1}{4} \right)^2 = \frac{1}{16}.$$

109. Since $a < b$, the graph of $y = \sqrt{(x-a)(b-x)}$ is a semicircle centered at $\left(\frac{a+b}{2}, 0 \right)$ with radius $r = \frac{b-a}{2}$; use the substitution $u = x - \frac{a+b}{2}$ and then apply entry 91 of the integral table.

$$\int_a^b \sqrt{(x-a)(b-x)}\,dx = \int_a^b \sqrt{\left(\frac{b-a}{2} \right)^2 - \left(x - \frac{a+b}{2} \right)^2}\,dx = \int_{-r}^r \sqrt{r^2 - u^2}\,du$$

$$= \left(\frac{u\sqrt{r^2-u^2}}{2} + \frac{r^2}{2}\sin^{-1}\frac{u}{r} \right)\Big|_{-r}^r = \frac{r^2}{2}\left(\frac{\pi}{2} - \frac{-\pi}{2} \right) = \frac{1}{2}\pi r^2 = \frac{1}{8}\pi(b-a)^2.$$

111.
$$\int_0^1 e^{x^3}\,dx \approx \left(\frac{1}{2} \right)\left(\frac{1}{4} \right)\left(e^0 + 2e^{1/64} + 2e^{1/8} + 2e^{27/64} + e^1 \right) \approx 1.3832137466.$$

113.
$$\int_0^1 \frac{dx}{\sqrt{1+x^4}} \approx \frac{1}{2} \cdot \frac{1}{6}\left(1 + 2 \cdot \frac{36}{\sqrt{2}} + 2 \cdot \frac{36}{\sqrt{17}} + 2 \cdot \frac{36}{\sqrt{82}} + 2 \cdot \frac{36}{\sqrt{257}} + 2 \cdot \frac{36}{\sqrt{626}} + \frac{1}{\sqrt{2}} \right) \approx 0.9253959267.$$

115.
$$\int_0^{\pi/2} \cos\sqrt{x}\,dx \approx \frac{1}{3} \cdot \frac{\pi}{12}\left[\cos 0 + 4\cos\sqrt{\frac{\pi}{12}} + 2\cos\sqrt{\frac{\pi}{6}} + 4\cos\sqrt{\frac{\pi}{4}} + 2\cos\sqrt{\frac{\pi}{3}} + 4\cos\sqrt{\frac{5\pi}{12}} + \cos\sqrt{\frac{\pi}{2}} \right]$$
$$\approx 1.0057092571.$$

REMARK:
$$\int_0^{\pi/2} \cos\sqrt{x}\,dx \overset{x\equiv u^2}{=} \int_0^{\sqrt{\pi/2}} \cos u \cdot (2u\,du) = 2(\cos u + u\sin u)\big|_0^{\sqrt{\pi/2}}$$

$$= 2\left(\cos\sqrt{\frac{\pi}{2}} + \sqrt{\frac{\pi}{2}} \cdot \sin\sqrt{\frac{\pi}{2}} \right) - 2 \approx 1.005709234.$$

---

117. If $f(x) = e^{x^3}$, then $|f''(x)| = \left| (6x + 9x^4)\,e^{x^3} \right| \leq |f''(1)| = 15e$ on the interval $[0,1]$; therefore pick integer n so that $0.01 > \frac{15e(1-0)^3}{12 \cdot n^2}$, i.e., $n > \sqrt{\frac{15e}{12 \cdot 0.01}} \approx \sqrt{339.79} \approx 18.4$; choose $n = 19$ intervals to use with trapezoidal rule.

119. If $f(x) = x^{-2}$, then $\left| f^{(4)}(x) \right| = \left| (-2)(-3)(-4)(-5)x^{-6} \right| = 120\,x^{-6} \leq f^{(4)}(1) = 120$ on $[1,2]$. Choose integer n such that $0.0001 > 120(2-1)^5/180n^4$, i.e., $n > (6666.67)^{1/4} \approx 9.036$. Apply Simpson's rule with $n = 10$ subintervals, the resulting approximation is 0.5000124699. The exact value is $\int_1^2 x^{-2}\,dx = -x^{-1}\big|_1^2 = -0.5 + 1 = 0.5$ and $\epsilon_{10}^S \approx -0.0000125$, about 8 times smaller than the bound for $\left| \epsilon_{10}^S \right|$.

Solutions 7.C   *Computer Exercises*   (pages 537-8)

1. a. $\int_0^1 \frac{dx}{1+x^2} = \tan^{-1}x\big|_0^1 = \frac{\pi}{4}.$ ∎

b. If $f(x) = (1+x^2)^{-1}$, then $f^{(4)}(x) = 24(1+x^2)^{-3} - 288x^2(1+x^2)^{-4} + 384x^4(1+x^2)^{-5}$. Examination of $f^{(5)}(x) = 240x(x^2-3)(3x^2-1)(1+x^2)^{-6}$ shows 0 and $\sqrt{1/3}$ to be the only critical points of $f^{(4)}$ in $[0,1]$; therefore $f^{(4)}(0) = 24$ is the maximum of $\left| f^{(4)}(x) \right|$ on $[0,1]$.

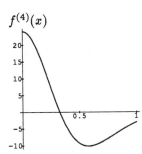

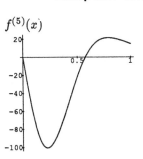

c.   Equation (7.9.12) implies $\left|\epsilon_n^S\right| \le (24)(1-0)^5/\left(180n^4\right) = (2/15)n^{-4}$, this error bound is less than 10^{-11} if and only if $n \ge \sqrt[4]{(2/15)10^{11}} = 200\sqrt{5}/\sqrt[4]{3} \approx 339.81$; the next largest even integer yields $n = 340$.

d.   Simpson's rule with $n = 340$ yields $\displaystyle\int_0^1 \frac{dx}{1+x^2} \approx .78539\ 81633\ 97448\ 30319\ 37249\ 60694$. Multiply by 4 to obtain the approximation $\pi \approx 3.14159\ 26535\ 89793\ 21277\ 48998\ 4278$ with error $2.57 \cdot 10^{-17}$.

3.   If $f(x) = x^{-1}$, then $f^{(4)}(x) = 24x^{-5}$. x^{-5} is a positive decreasing function on $(0,\infty)$. We choose n using the bound given by equation (7.9.12). For interval $[1,100]$ we have $\left|\epsilon_n^S\right| \le \left((24)1^{-5}\right)(100-1)^5/\left(180n^4\right) = (2/15)\left(99^5\right)n^{-4}$; for $[1/100,1]$ we have $\left|\epsilon_n^S\right| \le \left((24)(1/100)^{-5}\right)(1-1/100)^5/\left(180n^4\right) = (2/15)\left(99^5\right)n^{-4}$. In either case, the error is less than 10^{-11} if we choose $n \ge 106115.46$; the next largest even integer yields $n = 106116$. Simpson's rule with that choice of n yields $\ln 100 = \displaystyle\int_{1/100}^1 \frac{dx}{x} = \int_1^{100} \frac{dx}{x} \approx 4.60517\ 01859\ 88116\ 612$ with error $2.5 \cdot 10^{-14}$.

REMARK: A wise choice for this problem is to split the interval at $10 = \sqrt{100}$. We can bound the error on $[1,10]$ and on $[10,100]$ by $(1/2)10^{-11}$ if we use $n = 6300$; Simpson's rule yields $\displaystyle\int_1^{10} \frac{dx}{x} = \int_{10}^{100} \frac{dx}{x} \approx 2.30258\ 50929\ 941845$ and $\displaystyle\int_1^{100} \frac{dx}{x} = \int_1^{10} \frac{dx}{x} + \int_{10}^{100} \frac{dx}{x} \approx (2)(2.30258\ 50929\ 941845) \approx 4.60517\ 01859\ 88369$ with error $2.8 \cdot 10^{-13}$. This two-part computation takes only $1/9$ the time needed for the previous one.

5. $\dfrac{x+4}{(x-2)^2\left(x^2+x+1\right)(x+3)}$ has a partial fraction expansion of the form

$$\frac{a}{(x-2)^2} + \frac{b}{x-2} + \frac{c}{x+3} + \frac{d(2x+1)}{x^2+x+1} + \frac{e}{x^2+x+1}$$

which has antiderivative

$$\frac{-a}{x-2} + b\ln|x-2| + c\ln|x+3| + d\ln\left(x^2+x+1\right) + \frac{2e}{\sqrt{3}}\tan^{-1}\left(\frac{x+1/2}{2\sqrt{3}}\right) + C.$$

The actual antiderivative is

$$\frac{-6}{35(x-2)} - \frac{157}{1225}\ln|x-2| + \frac{1}{175}\ln|x+3| + \frac{3}{49}\ln\left(x^2+x+1\right) + \frac{16}{49\sqrt{3}}\tan^{-1}\left(\frac{x+1/2}{2\sqrt{3}}\right) + C.$$

8

Conic Sections and Polar Coordinates

Solutions 8.1   *The ellipse and Translation of axes*

1. $1 = \dfrac{x^2}{16} + \dfrac{y^2}{25} = \left(\dfrac{x}{4}\right)^2 + \left(\dfrac{y}{5}\right)^2$. The line segment from $(-4, 0)$ to $(4, 0)$ is the minor axis, the line segment from $(0, -5)$ to $(0, 5)$ is the major axis; the second one is the major axis because $5 > 4$. The vertices of the ellipse are $(0, -5)$ and $(0, 5)$, the endpoints of the major axis. The two axes intersect at $(0, 0)$, the center of the ellipse. $c^2 = b^2 - a^2 = 25 - 16 = 9$; the foci are $(0, -3)$ and $(0, 3)$.

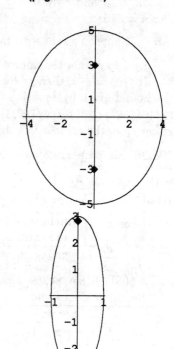

3. $1 = x^2 + \dfrac{y^2}{9} = \left(\dfrac{x}{1}\right)^2 + \left(\dfrac{y}{3}\right)^2$. The line segment from $(-1, 0)$ to $(1, 0)$ is the minor axis, the line segment from $(0, -3)$ to $(0, 3)$ is the major axis. The vertices of the ellipse are $(0, -3)$ and $(0, 3)$, the endpoints of the major axis. The two axes intersect at $(0, 0)$, the center of the ellipse. $c^2 = b^2 - a^2 = 9 - 1 = 8$; the foci are $\left(0, -\sqrt{8}\right)$ and $\left(0, \sqrt{8}\right)$.

5. $16 = x^2 + 4y^2 \iff 1 = \dfrac{x^2}{16} + \dfrac{y^2}{4} = \left(\dfrac{x}{4}\right)^2 + \left(\dfrac{y}{2}\right)^2$.
$(-4, 0)$ and $(4, 0)$ are the vertices, the line segment between them is the major axis; the line segment between $(0, -2)$ and $(0, 2)$ is the minor axis. The two axes intersect at the center $(0, 0)$. $c^2 = a^2 - b^2 = 16 - 4 = 12$; the foci are $\left(-\sqrt{12}, 0\right)$ and $\left(\sqrt{12}, 0\right)$.

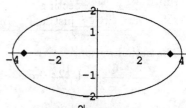

7. $1 = \dfrac{(x-1)^2}{16} + \dfrac{(y+3)^2}{25} = \left(\dfrac{x-1}{4}\right)^2 + \left(\dfrac{y-(-3)}{5}\right)^2$.
$(1, -3)$ is the center of the ellipse; $(1, -3-5) = (1, -8)$ and $(1, -3+5) = (1, 2)$ are the vertices and the major axis is the line segment between them; the minor axis is the line segment between $(1-4, -3) = (-3, -3)$ and $(1+4, -3) = (5, -3)$. $c^2 = b^2 - a^2 = 25 - 16 = 9$; because the foci are on the major axis, they are $(1, -3-3) = (1, -6)$ and $(1, -3+3) = (1, 0)$.

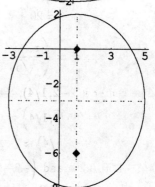

9. $2 = 2x^2 + 2y^2 \iff 1 = x^2 + y^2$. This is an equation for the unit circle, its center is $(0,0)$ and its radius is 1. (See Example 0.3.2.) All chords through the center have the same length, there are no special chords to single out and call major or minor axes; similarly there is no special pair of points to be called vertices. $c^2 = 1 - 1 = 0$; for this special ellipse the foci coincide and are equal to the center.

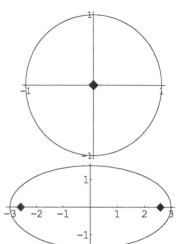

11. $9 = x^2 + 4y^2 \iff 1 = \left(\dfrac{x}{3}\right)^2 + \left(\dfrac{y}{3/2}\right)^2$. The center is $(0,0)$; the major axis is the segment between vertices $(\pm 3, 0)$; the minor axis is the segment between $(0, \pm 3/2)$. $c^2 = a^2 - b^2 = 9 - 9/4 = 27/4$, the foci are $(\pm 3\sqrt{3}/2, 0)$.

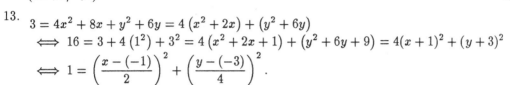

13. $3 = 4x^2 + 8x + y^2 + 6y = 4\left(x^2 + 2x\right) + \left(y^2 + 6y\right)$
$\iff 16 = 3 + 4\left(1^2\right) + 3^2 = 4\left(x^2 + 2x + 1\right) + \left(y^2 + 6y + 9\right) = 4(x+1)^2 + (y+3)^2$
$\iff 1 = \left(\dfrac{x - (-1)}{2}\right)^2 + \left(\dfrac{y - (-3)}{4}\right)^2.$

The center is $(-1, -3)$; the major axis is the segment between vertices $(-1, -3 - 4) = (-1, -7)$ and $(-1, -3 + 4) = (-1, 1)$. the minor axis is the segment between $(-1 - 2, -3) = (-3, -3)$ and $(-1 + 2, -3) = (1, -3)$. $c^2 = b^2 - a^2 = 16 - 4 = 12$, the foci are $\left(-1, -3 \pm 2\sqrt{3}\right)$.

REMARK: This ellipse can be obtained by rotating the graph for Problem 5 one-quarter turn and shifting the center to $(-1, -3)$.

15. $3 = 4x^2 + 8x + y^2 - 6y = 4\left(x^2 + 2x\right) + \left(y^2 - 6y\right)$
$\iff 16 = 3 + 4\left(1^2\right) + 3^2 = 4\left(x^2 + 2x + 1\right) + \left(y^2 - 6y + 9\right) = 4(x+1)^2 + (y-3)^2$
$\iff 1 = \left(\dfrac{x - (-1)}{2}\right)^2 + \left(\dfrac{y - 3}{4}\right)^2.$

The center is $(-1, 3)$; the major axis is the segment between vertices $(-1, 3 - 4) = (-1, -1)$ and $(-1, 3 + 4) = (-1, 7)$. the minor axis is the segment between $(-1 - 2, 3) = (-3, 3)$ and $(-1 + 2, 3) = (1, 3)$. $c^2 = b^2 - a^2 = 16 - 4 = 12$, the foci are $\left(-1, 3 \pm 2\sqrt{3}\right)$.

REMARK: Reflecting this graph in the x-axis yields the graph for Problem 13; alternatively, just lower this ellipse 6 units.

17. $20 = 3x^2 + 12x + 8y^2 - 4y = 3\left(x^2 + 4x\right) + 8\left(y^2 - \dfrac{1}{2}y\right)$
$\iff \dfrac{65}{2} = 20 + 3(2)^2 + 8\left(\dfrac{1}{4}\right)^2 = 3(x+2)^2 + 8\left(y - \dfrac{1}{4}\right)^2$
$\iff 1 = \left(\dfrac{x - (-2)}{\sqrt{65/6}}\right)^2 + \left(\dfrac{y - 1/4}{\sqrt{65}/4}\right)^2.$

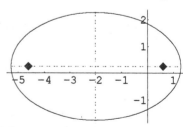

The center is $(-2, 1/4)$; the major axis is the segment between vertices $\left(-2 - \sqrt{65/6}, 1/4\right) \approx (-5.29, 0.25)$ and $\left(-2 + \sqrt{65/6}, 1/4\right) \approx (1.29, 0.25)$; the minor axis is the segment from $\left(-2, 1/4 - \sqrt{65}/4\right) \approx (-2, -1.77)$ to $\left(-2, 1/4 + \sqrt{65}/4\right) \approx (-2, 2.27)$. $c^2 = a^2 - b^2 = 65/6 - 65/16 = 325/48$, the foci are $\left(-2 - (5/4)\sqrt{13/3}, 1/4\right) \approx (-4.60, 0.25)$ and $\left(-2 + (5/4)\sqrt{13/3}, 1/4\right) \approx (0.60, 0.25)$.

19. The ellipse of Problem 1 has eccentricity equal to $c/b = 3/5 = 0.6$.

21. The ellipse of Problem 11 has eccentricity equal to $c/a = \left(3\sqrt{3}/2\right)/3 = \sqrt{3}/2 \approx 0.87$.

23. The ellipse of Problem 17 has eccentricity equal to $c/a = \sqrt{325/48}\Big/\sqrt{65/6} = \sqrt{5/8} \approx 0.79$.

25.
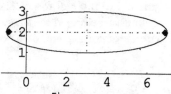

$(x-3)^2 + 16(y-2)^2 = 16 \iff \left(\dfrac{x-3}{4}\right)^2 + \left(\dfrac{y-2}{1}\right)^2 = 1$. The center of

this ellipse is $(3,2)$ and the vertices are $(3+4,2) = (7,2)$, $(3-4,2) = (-1,2)$, $(3,2+1) = (3,3)$, $(3,2-1) = (3,1)$. Furthermore, $c = \sqrt{4^2 - 1^2} = \sqrt{15}$ and the foci are $\left(3 \pm \sqrt{15}, 2\right)$.

27.
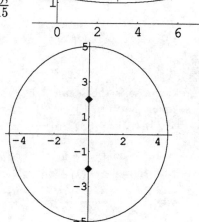

$25x^2 + 21y^2 = 525 \iff \left(\dfrac{x-0}{\sqrt{21}}\right)^2 + \left(\dfrac{y-0}{5}\right)^2 = 1$. The center

of this ellipse is $(0,0)$; the vertices are $\left(0 \pm \sqrt{21}, 0\right) = \left(\pm\sqrt{21}, 0\right)$

and $(0, 0 \pm 5) = (0, \pm 5)$. Furthermore, $c = \sqrt{5^2 - \sqrt{21}^2} = 2$ and the foci are $(0, 0 \pm 2) = (0, \pm 2)$.

29.
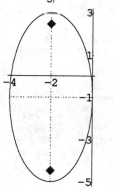

$16(x+2)^2 + 4(y+1)^2 = 64 \iff \left(\dfrac{x-(-2)}{2}\right)^2 + \left(\dfrac{y-(-1)}{4}\right)^2 = 1$.

The center of this ellipse is $(-2,-1)$; the vertices are $(-2+2,-1) = (0,-1)$, $(-2-2,-1) = (-4,-1)$, $(-2,-1+4) = (-2,3)$, $(-2,-1-4) = (-2,-5)$. Furthermore, $c = \sqrt{4^2 - 2^2} = 2\sqrt{3}$ and the foci are $\left(-2, -1 \pm 2\sqrt{3}\right)$.

31.
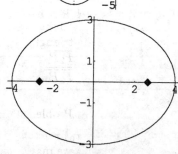

$9x^2 + 16y^2 = 144 \iff \left(\dfrac{x-0}{4}\right)^2 + \left(\dfrac{y-0}{3}\right)^2 = 1$. The center of this

ellipse is $(0,0)$; the vertices are $(0 \pm 4, 0) = (\pm 4, 0)$ and $(0, 0 \pm 3) = (0, \pm 3)$. Furthermore, $c = \sqrt{4^2 - 3^2} = \sqrt{7}$ and the foci are $\left(0 \pm \sqrt{7}, 0\right) = \left(\pm\sqrt{7}, 0\right)$.

33. The center of the ellipse is $(0,0)$, the midpoint of the line segment between the vertices $(0,-5)$ and $(0,5)$. $b = 5$ (half of the distance between the vertices) and $c = 4$ (half of the distance between the foci); therefore $a = \sqrt{b^2 - c^2} = \sqrt{25 - 16} = 3$. The ellipse satisfies the equation

$$1 = \left(\frac{x-0}{3}\right)^2 + \left(\frac{y-0}{5}\right)^2 = \frac{x^2}{9} + \frac{y^2}{25}.$$

REMARK: Also note that the center of the ellipse is the midpoint of the line segment between the foci.

---

35. Use implicit differentiation. If $2x^2 + 3y^2 = 14$, then $4x + 6y\dfrac{dy}{dx} = 0$ and $\dfrac{dy}{dx}\bigg|_{(1,2)} = \dfrac{-4(1)}{6(2)} = \dfrac{-1}{3}$. The

line tangent at $(1,2)$ satisfies the equation $y = \left(\dfrac{-1}{3}\right)(x-1) + 2 = \dfrac{-x+7}{3}$.

37. The ellipse $\dfrac{x^2}{a^2} + \dfrac{y^2}{b^2} = 1$ is symmetric with respect to the x-axis, thus its area is

$$\int_{-|a|}^{|a|} 2\sqrt{b^2\left(1 - \frac{x^2}{a^2}\right)}\, dx \overset{x \equiv |a|\sin\theta}{=} \int_{-\pi/2}^{\pi/2} 2|b|\sqrt{1 - \sin^2\theta}\ (|a|\cos\theta\, d\theta)$$

$$= |ab| \int_{-\pi/2}^{\pi/2} 2\cos^2\theta\, d\theta = |ab|\, (\theta + \sin\theta\ \cos\theta)\Big|_{-\pi/2}^{\pi/2} = |ab|\,\pi.$$

39. The distance between the foci is $2\sqrt{(36.2/2)^2 - (9.1/2)^2} = 2\sqrt{306.9075} \approx 35.04$ and the eccentricity is $2\sqrt{306.9075}/36.2 \approx 0.9679$.

41. $T^2 = a^3$ (Kepler's third law) implies $a = T^{2/3}$. If $T = 8.4$ years, then the length of the major axis is $2 \cdot a = 2 \cdot 8.4^{2/3} \approx 8.26$ astronomical units.

43. Place the origin of a coordinate system at highway level between lanes 3 and 4. The roof of the tunnel coincides with the upper half of the ellipse given by the equation $(x/56)^2 + (y/20)^2 = 1$; therefore $y = 20\sqrt{1 - (x/56)^2}$. This function is symmetric about 0 and it decreases on the interval $[0, 56]$. The minimum height on lane 3 and lane 4 is $y(14) = 20\sqrt{1 - (14/56)^2} = 5\sqrt{15} \approx 19.36$ feet. The minimum height on lane 2 and lane 5 is $y(28) = 20\sqrt{1 - (28/56)^2} = 10\sqrt{3} \approx 17.32$ feet. The minimum height on lane 1 and lane 6 is $y(42) = 20\sqrt{1 - (42/56)^2} = 5\sqrt{7} \approx 13.23$ feet.

45.
$$\left(x^2 + Ax\right) + 2\left(y^2 + \frac{B}{2}y\right) = x^2 + Ax + 2y^2 + By = C$$

$$\Longleftrightarrow \quad \left(x^2 + Ax + \frac{A^2}{4}\right) + 2\left(y^2 + \frac{B}{2}y + \frac{B^2}{16}\right) = C + \left(\frac{A}{2}\right)^2 + 2\left(\frac{B}{4}\right)^2$$

$$\Longleftrightarrow \quad \left(x^2 + \frac{A}{2}\right)^2 + 2\left(y + \frac{B}{4}\right)^2 = C + \frac{A^2}{4} + \frac{B^2}{8}.$$

Let D denote $C + (A^2/4) + (B^2/8)$. Because $(x + A/2)^2 + 2(y + B/4)^2 \geq 0$ for all (x, y), the equation

a. is an equation for an ellipse centered at $(-A/2, -B/4)$ if $D > 0$,

b. has the unique solution $(-A/2, -B/4)$ if $D = 0$,

c. has no solutions if $D < 0$.

47. Without loss of generality, we may assume the ellipse is centered at the origin with its foci on the x-axis; a symmetry argument also shows we may assume our point on the ellipse lies in the first quadrant. The figure to the right shows the notation we will use.

The following discussion shows angles α and β are of equal size because the right triangles $P_0 F_1 P_1$ and $P_0 F_2 P_2$ are similar triangles; we show $\dfrac{\overline{F_1 P_1}}{\overline{F_1 P_0}} = \dfrac{\overline{F_2 P_2}}{\overline{F_2 P_0}}.$

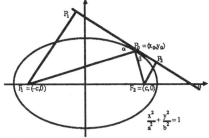

The result of Problem 33 implies the line tangent to the ellipse $\dfrac{x^2}{a^2} + \dfrac{y^2}{b^2} = 1$ at the point (x_0, y_0) satisfies the equation $(b^2 x_0)\, x + (a^2 y_0)\, y - a^2 b^2 = 0$. If $x_0 = 0$, then both focal radii are perpendicular to the vertical tangent line; thus we may assume $x_0 \neq 0$ for the subsequent discussion. The result of Problem 0.3.63 implies the distances from the foci $(\pm c, 0)$ to that tangent line are $\dfrac{\left|(b^2 x_0)(\pm c) - a^2 b^2\right|}{\sqrt{(b^2 x_0)^2 + (a^2 y_0)^2}} = \dfrac{b^2 \left|a^2 \mp cx_0\right|}{\sqrt{b^4 x_0^2 + a^4 y_0^2}}.$ Thus

$$\frac{\overline{F_1 P_1}}{\overline{F_1 P_0}} = \frac{\overline{F_2 P_2}}{\overline{F_2 P_0}} \Longleftrightarrow \frac{\left|a^2 + cx_0\right|}{\sqrt{(x_0 + c)^2 + y_0^2}} = \frac{\left|a^2 - cx_0\right|}{\sqrt{(x_0 - c)^2 + y_0^2}}$$

$$\Longleftrightarrow \left|a^2 + cx_0\right|^2 \left[(x_0 - c)^2 + y_0^2\right] = \left|a^2 - cx_0\right|^2 \left[(x_0 + c)^2 + y_0^2\right]$$

$$\Longleftrightarrow \left((a^4 + c^2 x_0^2) + 2a^2 cx_0\right)\left((x_0^2 + c^2 + y_0^2) - 2cx_0\right) = \left((a^4 + c^2 x_0^2) - 2a^2 cx_0\right)\left((x_0^2 + c^2 + y_0^2) + 2cx_0\right)$$

$$\Longleftrightarrow \left(a^2 (x_0^2 + c^2 + y_0^2) - (a^4 + c^2 x_0^2)\right) 2cx_0 = \left((a^4 + c^2 x_0^2) - a^2 (x_0^2 + c^2 + y_0^2)\right) 2cx_0$$

$$\Longleftrightarrow a^2 (x_0^2 + c^2 + y_0^2) - (a^4 + c^2 x_0^2) = 0 \Longleftrightarrow (a^2 - c^2) x_0^2 + a^2 y_0^2 = a^4 - a^2 c^2 = a^2 (a^2 - c^2)$$

$$\Longleftrightarrow b^2 x_0^2 + a^2 y_0^2 = a^2 b^2 \Longleftrightarrow \frac{x_0^2}{a^2} + \frac{y_0^2}{b^2} = 1.$$

8.1 — The ellipse and Translation of axes

This final equation is true because the point (x_0, y_0) lies on the ellipse, thus the chain of equivalences implies the first equation is also true. ∎

Solutions 8.2   *The parabola*   (pages 558-9)

1.   See Example 1 and use formula (1). If $4cy = x^2 = 16y$, then $4c = 16$ and $c = 4$. The focus is $(0, 4)$ and the directrix is the line $y = -4$. Because $(-x)^2 = x^2$, the parabola is symmetric about the y-axis which is also the axis of the parabola; that axis meets the parabola at $(0, 0)$, the vertex of the parabola.

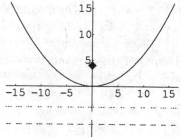

3.   Reflect the graph of $x^2 = 16y$ (Problem 1) in the x-axis to get the graph of $x^2 = -16y$. If $4c = -16$, then $c = -4$; the focus is $(0, -4)$ and the directrix is the line $y = -(-4) = 4$. The axis of the parabola is the y-axis and the vertex is $(0, 0)$.

5. $2x^2 = 3y \implies x^2 = \left(\dfrac{3}{2}\right)y \implies 4c = \dfrac{3}{2} \implies c = \dfrac{3}{8}$. The focus is $\left(0, \dfrac{3}{8}\right)$, the directrix has equation $y = \dfrac{-3}{8}$; the parabola's axis is the y-axis; the vertex is $(0, 0)$.

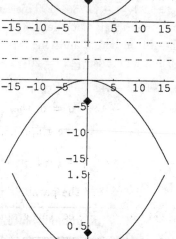

7. $4x^2 = -9y \implies x^2 = \left(\dfrac{-9}{4}\right)y \implies 4c = \dfrac{-9}{4} \implies c = \dfrac{-9}{16}$.

The focus is $\left(0, \dfrac{-9}{16}\right)$; the directrix has equation $y = -\left(\dfrac{-9}{16}\right) = \dfrac{9}{16}$; the parabola's axis is the y-axis; the vertex is $(0, 0)$.

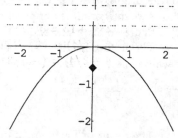

9.   The parabola with equation $(x - 1)^2 = -16(y + 3) = -16\left(y - (-3)\right)$ has vertex $(1, -3)$ and $4c = -16$. Thus $c = -4$ and the focus is $(1, (-3) + c) = \left(1, (-3) + (-4)\right) = (1, -7)$; the directrix has equation $y = (-3) - c = -3 + 4 = 1$; the axis of the parabola is the line $x = 1$ because $(x - 1)^2 = (1 - x)^2$.

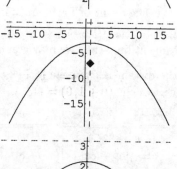

11.   If $x^2 + 4y = 9$, then $(x - 0)^2 = x^2 = -4y + 9 = -4(y - 9/4)$; the graph has vertex $(0, 9/4)$. Because $c = -4/4 = -1$, the focus is $(0, 9/4 + c) = \left(0, (9/4) + (-1)\right) = (0, 5/4)$; the directrix has equation $y = 9/4 - c = 9/4 + 1 = 13/4$; the axis of the parabola is the line $x = 0$, the y-axis.

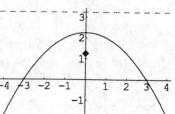

13.   If $x^2 + 2x + y + 1 = 0$, then $(x+1)^2 = x^2 + 2x + 1 = -y$; the graph has vertex $(-1, 0)$. Because $c = -1/4$, the focus is $(-1, 0 + c) = (-1, -1/4)$; the directrix has equation $y = 0 - c = 1/4$; the axis of the parabola is the line $x = -1$.

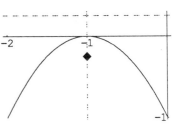

15.   If $x^2 + 4x + y = 0$, then $\left(x - (-2)\right)^2 = (x+2)^2 = x^2 + 4x + 4 = -y + 4 = -(y - 4)$; the graph has vertex $(-2, 4)$. Because $c = \dfrac{-1}{4}$, the focus is $\left(-2, 4 + \dfrac{-1}{4}\right) = \left(-2, \dfrac{15}{4}\right)$; the directrix has equation $y = 4 - \dfrac{-1}{4} = \dfrac{17}{4}$; the axis of the parabola is the line $x = -2$.

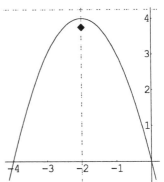

17.   If $x^2 + 4x - y = 0$, then $\left(x - (-2)\right)^2 = (x+2)^2 = x^2 + 4x + 4 = y + 4 = y - (-4)$; the graph has vertex $(-2, -4)$. Because $c = \dfrac{1}{4}$, the focus is $\left(-2, -4 + \dfrac{1}{4}\right) = \left(-2, \dfrac{-15}{4}\right)$; the directrix has equation $y = -4 - \dfrac{1}{4} = \dfrac{-17}{4}$; the axis of the parabola is the line $x = -2$.

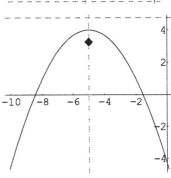

REMARK: Reflect this graph in the x-axis to get the graph for Problem 15.

19. $x^2 + 10x + 3y + 13 = 0 \iff (x+5)^2 = x^2 + 10x + 25 = (-3y - 13) + 25 = -3(y - 4) = 4(-3/4)(y - 4)$. The vertex of this parabola is $(-5, 4)$ and the focus is $(-5, 4 - 3/4) = (-5, 13/4)$.

19.

21. $4x - y^2 = 0 \iff y^2 = 4x$. The vertex of this parabola is $(0, 0)$ and the focus is $(0 + 1, 0) = (1, 0)$.

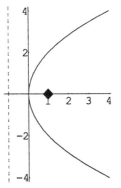

197

23. $2x + y^2 = 0 \iff y^2 = -2x = 4(-1/2)x$. The vertex of this parabola
is $(0,0)$ and the focus is $(0 - 1/2, 0) = (-1/2, 0)$.

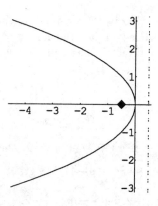

25. $y^2 - 2x - 4y + 6 = 0 \iff (y-2)^2 = y^2 - 4y + 4 = (2x - 6) + 4 = 2(x-1) = 4(1/2)(x-1)$. The vertex of this parabola is $(1,2)$ and the focus
is $(1 + 1/2, 2) = (3/2, 2)$.

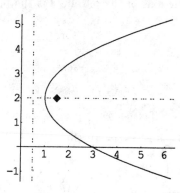

27.   The line $x = 0$, the y-axis, passes through the focus $(0,4)$ and is perpendicular to the directrix $y = -4$;
it is the axis of the parabola. The vertex lies on the axis and is midway between the focus and the directrix,
it is $(0,0)$. Therefore $c = 4 - 0 = 4$ and the parabola has equation $x^2 = (x-0)^2 = 4 \cdot 4 \cdot (y - 0) = 16y$.

29.   The parabola $(x^2 = 16y)$ of Problem 19 has $c = 4$; if its vertex is shifted from $(0,0)$ to $(-2,5)$, then it
will satisfy $(x+2)^2 = (x - (-2))^2 = 4c(y - 5) = 16(y - 5)$.

31. $x^2 = 9y \iff y = \dfrac{1}{9}x^2 \implies \dfrac{dy}{dx} = \dfrac{2}{9}x \implies \dfrac{dy}{dx}\Big|_{(3,1)} = \dfrac{2}{9} \cdot 3 = \dfrac{2}{3}$. Thus the line tangent to the parabola
at $(3,1)$ has equation $y = \dfrac{2}{3}(x - 3) + 1 = \dfrac{2}{3}x - 1$.

33.   Suppose the earth is located at the point $(0, c)$, the focus of the parabola $x^2 = 4cy$ with directrix $y = -c$;
for convenience, assume $c > 0$. The distance from asteroid at point (x, y) on the parabola to the earth is equal
to $y + c$, the distance from (x, y) to the directrix. At one moment the asteroid traveling along the parabola
passes through a point (x, y) such that $y + c = 150,000$ km and the angle from $\overline{(0, y)\,(0, c)} = \overline{(0, c)\,(x, y)}$ is
$30°$. Therefore $y - c = (y + c) - 2c = 150,000 - 2c$ and $\dfrac{\sqrt{3}}{2} = \cos 30° = \dfrac{|y - c|}{150,000} = \pm\left(1 - \dfrac{2c}{150,000}\right)$; thus
$c = \dfrac{150,000}{2}\left(1 \mp \dfrac{\sqrt{3}}{2}\right)$. The distance from the earth to the parabola's vertex is c; the solution to Problem
28 shows the vertex is the point on the parabola closest to the focus.

If the asteroid's position on its orbit looks like

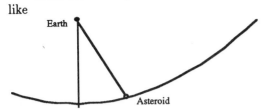

then its closest point to the Earth is

$$\frac{150,000}{2}\left(1+\frac{\sqrt{3}}{2}\right) \approx 139,952\,\text{km}.$$

On the other hand, if the problem statement refers to a position

the closest point is $\dfrac{150,000}{2}\left(1-\dfrac{\sqrt{3}}{2}\right) \approx 10,048\,\text{km}.$

35. Put the origin of a coordinate system at water level and midway between the two towers. We consider a parabola of the form $y = a + bx^2$ If this curve passes through the points $(-375/2, 100)$, $(0, 40)$, and $(375/2, 100)$, then $a = 40$ and $b = (100-40)(2/375)^2 = 16/9375$. Therefore, the height 60 feet (horizontally) from one of the towers is $y(375/2 - 60) = y(127.5) = 40 + (16/9375)(255/2)^2 = 67.744$ feet.

37. a. $4cy = x^2 \implies \dfrac{dy}{dx} = \dfrac{1}{4c}\cdot 2x \implies \dfrac{dy}{dx}\bigg|_{(x_0, y_0)} = \dfrac{x_0}{2c}$; thus the line tangent at $(x_0, y_0) = \left(x_0, \dfrac{x_0^2}{4c}\right)$

satisfies

$$y = \frac{x_0}{2c}(x - x_0) + y_0 = \frac{x_0}{2c}x - \frac{x_0^2}{2c} + y_0 = \frac{x_0}{2c}x - 2y_0 + y_0 = \frac{x_0}{2c}x - y_0 \iff y + y_0 = \frac{x_0}{2c}x$$
$$\iff 2c(y + y_0) = x_0 x. \ \blacksquare$$

b. The line $2c(y + y_0) = x_0 x$ meets the y-axis at the point $(0, -y_0)$; this point of intersection of the tangent at (x_0, y_0) and the parabola's axis is as far below the vertex as the point of tangency is above. (The remark following the solution to part (c) shows this relation holds for an arbitrary parabola.)

c. Suppose P is a point on a parabola. Let A be the point on the axis of the parabola such that the line on A and P is perpendicular to the axis; then let B be the point on the parabola's axis which is as far below the vertex as A is above. The line on B and P is tangent to the parabola at P.

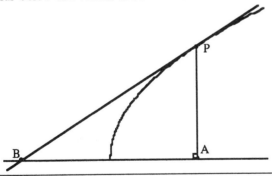
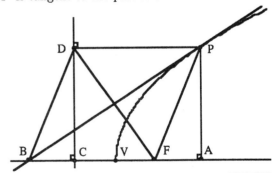

REMARK: Let D be the point on the parabola's directrix such that the line on D and P is perpendicular to the directrix. $\overline{DP} = \overline{CA} = \overline{CV} + \overline{VA} = \overline{VF} + \overline{BV} = \overline{BF}$ and those line segments are parallel; $\overline{BC} = \overline{BV} - \overline{CV} = \overline{VA} - \overline{VF} = \overline{FA}$, right triangles BCD and FAP are congruent, so $\overline{BD} = \overline{FP}$. The quadrilateral $PFBD$ has sides of equal length, thus it is a rhombus. The diagonals of a rhombus are perpendicular bisectors of each other, thus any point on the line BP is equidistant from D and F. P is the only point on that line such that its distance to the directrix is also its distance to D, thus P is the only point on that line equidistant from the focus and the directrix of the parabola; this proves the line BP meets the parabola only at the point P and it must be the desired tangent. $\blacksquare$

"Calculus" — a homily: The geometric construction discussed in the preceding remark works for a parabola but different special techniques are required for other conics (ellipses or hyperbolas). The analytic geometry of Descartes and the differential calculus of Leibniz and Newton provide us with a routine method to obtain a tangent to any polynomial curve. Our preference for one method over the other may be influenced by how the curve is presented to us, but using one routine computational procedure instead of a bag of miscellaneous tricks does seem to simplify our life.

39.   Consider a parabola with equation $x^2 = 4cy$. The line tangent at $P = (x_0, y_0)$ has equation $2c(y + y_0) = x_0 x$ (see Problem 27a); this line meets the parabola's directrix at the point $Q = \left(\dfrac{2c(y_0 - c)}{x_0}, -c \right)$.

The line between focus $F = (0, c)$ and point P has slope $\dfrac{y_0 - c}{x_0}$; the line between F and Q has slope $\dfrac{c - (-c)}{0 - 2c(y_0 - c)/x_0} = \dfrac{-x_0}{y_0 - c}$. The product of these two slopes is -1, thus the two lines are perpendicular. ∎

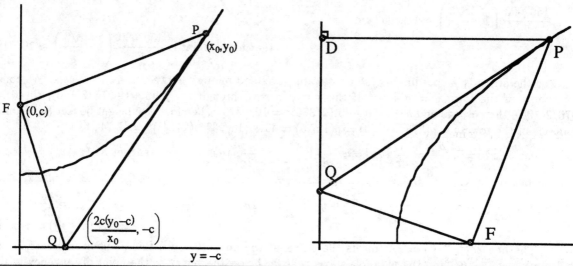

REMARK: Let D be the point on the directrix closest to P. The definition of a parabola implies $\overline{PD} = \overline{PF}$. The remark following Solution 37c shows line QP bisects the angle DPF. Therefore, "side-angle-side", triangles DPQ and FPQ are congruent; since angle QDP is a right angle, then so is angle QFP. ∎

41.   Suppose $P = (x_0, y_0)$ is a point on the parabola $4cy = x^2$; the focus of this parabola is $F = (0, c)$. Solution 27 shows the line tangent at P meets the parabola's axis at $Q = (0, -y_0)$. Triangle QFP is isosceles because

$$\overline{PF} = \sqrt{(x_0 - 0)^2 + (y_0 - c)^2} = \sqrt{x_0^2 + (y_0^2 - 2cy_0 + c^2)} = \sqrt{4cy_0 + (y_0^2 - 2cy_0 + c^2)} = \sqrt{(y_0 + c)^2}$$
$$= |y_0 + c| = \overline{QF}.$$

Therefore $\beta = \gamma$; on the other hand, because L_2 is parallel to the axis, $\gamma = \alpha$. Thus $\beta = \alpha$. ∎

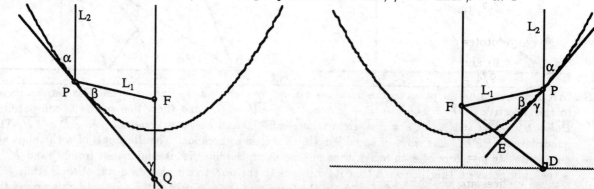

REMARK: Let $D = (x_0, -c)$. Triangle FPD is isosceles because $\overline{FP} \equiv \overline{FD}$. The tangent at P has slope $\dfrac{dy}{dx} = \dfrac{x_0}{2c}$ and is perpendicular to the line DF since that line has slope $\dfrac{c - (-c)}{0 - x_0} = \dfrac{-2c}{x_0}$. Therefore triangles EPF and EPD are congruent right triangles; thus $\beta = \gamma$; furthermore, $\gamma = \alpha$ since the angles are opposite each other at an intersection of two lines. ∎

Solutions 8.3   *The hyperbola*   (pages 568-9)

1.
The hyperbola $1 = \dfrac{x^2}{16} - \dfrac{y^2}{25} = \left(\dfrac{x}{4}\right)^2 - \left(\dfrac{y}{5}\right)^2$ is centered at $(0,0)$, its asymptotes are $y = \dfrac{5x}{4}$ and $y = \dfrac{-5x}{4}$, its vertices are $(-4,0)$ and $(4,0)$, those vertices are joined by the transverse axis. $c = \sqrt{16+25} = \sqrt{41} \approx 6.40$, the foci are $\left(-\sqrt{41},0\right)$ and $\left(\sqrt{41},0\right)$.

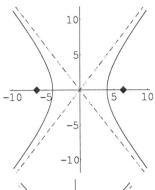

3.
The hyperbola $1 = \dfrac{y^2}{25} - \dfrac{x^2}{16} = \left(\dfrac{y}{5}\right)^2 - \left(\dfrac{x}{4}\right)^2$ is centered at $(0,0)$, its asymptotes are $y = \dfrac{5x}{4}$ and $y = \dfrac{-5x}{4}$, its vertices are $(0,-5)$ and $(0,5)$, those vertices are joined by the transverse axis. $c = \sqrt{25+16} = \sqrt{41} \approx 6.40$, the foci are $\left(0,-\sqrt{41}\right)$ and $\left(0,\sqrt{41}\right)$.

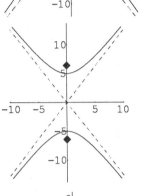

5.
The hyperbola $1 = y^2 - x^2$ is centered at $(0,0)$, its asymptotes are $y = -x$ and $y = x$, its vertices are $(0,-1)$ and $(0,1)$, those vertices are joined by the transverse axis. $c = \sqrt{1+1} = \sqrt{2} \approx 1.41$, the foci are $\left(0,-\sqrt{2}\right)$ and $\left(0,\sqrt{2}\right)$.

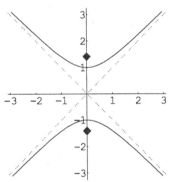

7.
If $9 = x^2 - 4y^2$, then $1 = \left(\dfrac{x}{3}\right)^2 - \left(\dfrac{y}{3/2}\right)^2$. The hyperbola's center is $(0,0)$, its asymptotes are $y = \dfrac{-(3/2)x}{3} = \dfrac{-x}{2}$ and $y = \dfrac{(3/2)x}{3} = \dfrac{x}{2}$, its vertices are $(-3,0)$ and $(3,0)$, those vertices are joined by the transverse axis. $c = \sqrt{9+9/4} = 3\sqrt{5}/2 \approx 3.35$, the foci are $\left(-3\sqrt{5}/2,0\right)$ and $\left(3\sqrt{5}/2,0\right)$.

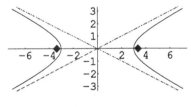

9.
If $9 = y^2 - 4x^2$, then $1 = \left(\dfrac{y}{3}\right)^2 - \left(\dfrac{x}{3/2}\right)^2$. The hyperbola's center is $(0,0)$, its asymptotes are $y = \dfrac{-3x}{3/2} = -2x$ and $y = \dfrac{3x}{3/2} = 2x$, its vertices are $(0,-3)$ and $(0,3)$, those vertices are joined by the transverse axis. $c = \sqrt{9+9/4} = 3\sqrt{5}/2 \approx 3.35$, the foci are $\left(0,-3\sqrt{5}/2\right)$ and $\left(0,3\sqrt{5}/2\right)$.

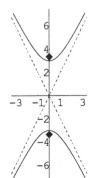

REMARK: Rotate this hyperbola one-quarter turn about the origin to obtain the graph for Problem 7.

11.

If $4 = 2x^2 - 3y^2$, then $1 = \left(\dfrac{x}{\sqrt{2}}\right)^2 - \left(\dfrac{y}{2/\sqrt{3}}\right)^2$. The hyperbola's center is $(0,0)$, its asymptotes are $y = \dfrac{-(2/\sqrt{3})\,x}{\sqrt{2}} = -\sqrt{\dfrac{2}{3}}\,x$ and $y = \dfrac{(2/\sqrt{3})\,x}{\sqrt{2}} = \sqrt{\dfrac{2}{3}}\,x$, its vertices are $(-\sqrt{2},0)$ and $(\sqrt{2},0)$, those vertices are joined by the transverse axis. $c = \sqrt{2 + 4/3} = \sqrt{10/3} \approx 1.83$, the foci are $\left(-\sqrt{10/3},0\right)$ and $\left(\sqrt{10/3},0\right)$.

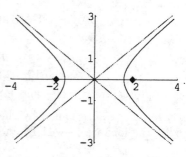

13.

If $4 = 2y^2 - 3x^2$, then $1 = \left(\dfrac{y}{\sqrt{2}}\right)^2 - \left(\dfrac{x}{2/\sqrt{3}}\right)^2$. The hyperbola's center is $(0,0)$, its asymptotes are $y = \dfrac{-\sqrt{2}\,x}{2/\sqrt{3}} = -\sqrt{\dfrac{3}{2}}\,x$ and $y = \dfrac{\sqrt{2}\,x}{2/\sqrt{3}} = \sqrt{\dfrac{3}{2}}\,x$, its vertices are $(0,-\sqrt{2})$ and $(0,\sqrt{2})$, those vertices are joined by the transverse axis. $c = \sqrt{2 + 4/3} = \sqrt{10/3} \approx 1.83$, the foci are $\left(0,-\sqrt{10/3}\right)$ and $\left(0,\sqrt{10/3}\right)$.

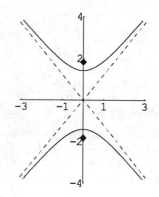

REMARK: Reflect the graph for Problem 11 in the line $y = x$ to obtain this hyperbola; alternatively, rotate it one-quarter turn about the origin.

15.

If $4 = (x-1)^2 - 4(y+2)^2$, then $1 = \left(\dfrac{x-1}{2}\right)^2 - \left(\dfrac{y-(-2)}{1}\right)^2$. The hyperbola's center is $(1,-2)$, its vertices are $(1-2,-2) = (-1,-2)$ and $(1+2,-2) = (3,-2)$, the principal axis is the line $y = -2$. $c = \sqrt{4+1} = \sqrt{5} \approx 2.24$, the foci are $(1-\sqrt{5},-2) \approx (-1.24,-2)$ and $(1+\sqrt{5},-2) \approx (3.24,-2)$. The asymptotes satisfy $(x-1)^2 - 4(y+2)^2 = 0$; they are the lines $y = \left(\dfrac{-1}{2}\right)(x-1) + (-2) = \left(\dfrac{-1}{2}\right)x - \dfrac{3}{2} = \dfrac{-(x+3)}{2}$ and $y = \left(\dfrac{1}{2}\right)(x-1) + (-2) = \left(\dfrac{1}{2}\right)x - \dfrac{5}{2} = \dfrac{x-5}{2}$.

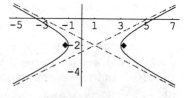

17.

$21 = 4x^2 + 8x - y^2 - 6y = 4\left(x^2 + 2x\right) - \left(y^2 + 6y\right)$

$\iff 16 = 21 + 4 \cdot 1^2 - 3^2 = 4\left(x^2 + 2x + 1\right) - \left(y^2 + 6y + 9\right) = 4(x+1)^2 - (y+3)^2$

$\iff 1 = \left(\dfrac{x-(-1)}{2}\right)^2 - \left(\dfrac{y-(-3)}{4}\right)^2$.

This hyperbola is centered at $(-1,-3)$, its vertices are $(-1-2,-3) = (-3,-3)$ and $(-1+2,-3) = (1,-3)$, the principal axis is the line $y = -3$. $c = \sqrt{4+16} = 2\sqrt{5} \approx 4.47$, the foci are $(-1-2\sqrt{5},-3) \approx (-5.47,-3)$ and $(-1+2\sqrt{5},-3) \approx (3.47,-3)$. The asymptotes satisfy $\left|\dfrac{x+1}{2}\right| = \left|\dfrac{y+3}{4}\right|$; they are the lines $y = -2(x+1) + (-3) = -2x - 5$ and $y = 2(x+1) + (-3) = 2x - 1$.

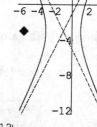

19.

$45 = 2x^2 - 16x - 3y^2 + 12y = 2\left(x^2 - 8x\right) - 3\left(y^2 - 4y\right)$

$\iff 65 = 45 + 2 \cdot (-4)^2 - 3 \cdot (-2)^2 = 2(x-4)^2 - 3(y-2)^2$

$\iff 1 = \left(\dfrac{x-4}{\sqrt{65/2}}\right)^2 - \left(\dfrac{y-2}{\sqrt{65/3}}\right)^2$.

This hyperbola is centered at $(4,2)$, its principal axis is the line $y = 2$, the vertices are $\left(4 - \sqrt{65/2},2\right) \approx (-1.70,2)$ and $\left(4 + \sqrt{65/2},2\right) \approx (9.70,2)$. $c = \sqrt{65/2 + 65/3} = \sqrt{325/6} \approx 7.36$, the foci are $\left(4 - \sqrt{325/6},2\right) \approx (-3.36,2)$ and $\left(4 + \sqrt{325/6},2\right) \approx (11.36,2)$. The asymptotes satisfy $\left|\dfrac{y-2}{\sqrt{65/3}}\right| = \left|\dfrac{x-4}{\sqrt{65/2}}\right|$, they are the lines $y = \pm\sqrt{\dfrac{2}{3}}\,(x-4) + 2$.

21.   The midpoint of the line segment between the foci $(\pm 4, 0)$ is $(0, 0)$, the center of the hyperbola. If $c = 4$ and $a = 3$, then $b = \sqrt{c^2 - a^2} = \sqrt{4^2 - 3^2} = \sqrt{7}$. The standard equation for the hyperbola is $1 = \left(\dfrac{x - 0}{3}\right)^2 - \left(\dfrac{y - 0}{\sqrt{7}}\right)^2 = \dfrac{x^2}{9} - \dfrac{y^2}{7}$.

23.   The midpoint of the line segment between the foci $(0, \pm 3)$ is $(0, 0)$, the center of this hyperbola. If $c = 3$ and $b = 2$, then $a = \sqrt{c^2 - b^2} = \sqrt{3^2 - 2^2} = \sqrt{5}$. The standard equation for this hyperbola is $1 = \left(\dfrac{y - 0}{2}\right)^2 - \left(\dfrac{x - 0}{\sqrt{5}}\right)^2 = \dfrac{y^2}{4} - \dfrac{x^2}{5}$.

25.   The midpoint of the line segment between the foci $(-1, 1)$ and $(5, 1)$ is $(2, 1)$, the center of this hyperbola. We see that $c = (5 - (-1))/2 = 5 - 2 = 3$ and $a = (4 - 0)/2 = 4 - 2 = 2$, therefore $b = \sqrt{c^2 - a^2} = \sqrt{3^2 - 2^2} = \sqrt{5}$. The standard equation for this hyperbola is $1 = \left(\dfrac{x - 2}{2}\right)^2 - \left(\dfrac{y - 1}{\sqrt{5}}\right)^2 = \dfrac{(x - 2)^2}{4} - \dfrac{(y - 1)^2}{5}$.

27.   The midpoint of the line segment between the vertices $(\pm a, 0) = (\pm 2, 0)$ is $(0, 0)$, the center of this hyperbola. If the asymptotes are $y = \pm x$, then $b = a = 2$. The standard equation for this hyperbola is $1 = \left(\dfrac{x - 0}{2}\right)^2 - \left(\dfrac{y - 0}{2}\right)^2 = \dfrac{x^2}{4} - \dfrac{y^2}{4}$.

29.   The midpoint of the line segment between the vertices $(1, 1)$ and $(5, 1)$ is $(3, 1)$, the center of this hyperbola. We also see that $a = (5 - 1)/2 = 5 - 3 = 2$. The asymptotes pass through the center and satisfy equations of the form $y - 1 = \pm(b/a)(x - 3)$; since the specified asymptotes have slopes ± 2, we see that $b = 2a = 4$. The standard equation for this hyperbola is $1 = \left(\dfrac{x - 3}{2}\right)^2 - \left(\dfrac{y - 1}{4}\right)^2 = \dfrac{(x - 3)^2}{4} - \dfrac{(y - 1)^2}{16}$.

31.   The hyperbola of Problem 1 has eccentricity equal to $\sqrt{1 + \left(\dfrac{b}{a}\right)^2} = \sqrt{1 + \dfrac{25}{16}} = \dfrac{\sqrt{41}}{4} \approx 1.60$.

33.   The hyperbola of Problem 7 has eccentricity equal to $\sqrt{1 + \left(\dfrac{b}{a}\right)^2} = \sqrt{1 + \dfrac{9/4}{9}} = \dfrac{\sqrt{5}}{2} \approx 1.12$.

35.   The hyperbola of Problem 3 has eccentricity equal to $\sqrt{1 + \left(\dfrac{a}{b}\right)^2} = \sqrt{1 + \dfrac{16}{25}} = \dfrac{\sqrt{41}}{5} \approx 1.28$.

37.   The hyperbola of Problem 15 has eccentricity equal to $\sqrt{1 + \left(\dfrac{b}{a}\right)^2} = \sqrt{1 + \dfrac{1}{4}} = \dfrac{\sqrt{5}}{2} \approx 1.12$.

39.   The hyperbola of Problem 19 has eccentricity $\sqrt{1 + \left(\dfrac{b}{a}\right)^2} = \sqrt{1 + \dfrac{65/3}{65/2}} = \sqrt{1 + \dfrac{2}{3}} = \sqrt{\dfrac{5}{3}} \approx 1.29$.

41.   The hyperbola with foci $(\pm 5, 0)$ and vertices $(\pm 4, 0)$ has center $(0, 0)$, $c = 5$, and $a = 4$. Therefore $b^2 = c^2 - a^2 = 25 - 16 = 9$ and the hyperbola satisfies $\dfrac{x^2}{16} - \dfrac{y^2}{9} = 1$.

43.   Any hyperbola centered at $(0, 0)$ with vertices $(\pm 2, 0)$ satisfies an equation of the form $\dfrac{x^2}{4} - \dfrac{y^2}{b^2} = 1$; the asymptotes of this curve are $y = \pm \dfrac{bx}{2}$. These asymptotes are the lines $y = \pm 3x$ if and only if $|b| = 6$; hence the hyperbola satisfies $\dfrac{x^2}{4} - \dfrac{y^2}{36} = 1$.

45.   Obtain the equation of the curve shifted 5 units to the right and 2 units down by replacing x with $x - 5$ and y with $y - (-2) = y + 2$. The shifted curve satisfies $21 = 4(x - 5)^2 + 8(x - 5) - (y + 2)^2 - 6(y + 2) = 4x^2 - 32x - y^2 - 10y + 44$.

REMARK: $1 = \left(\dfrac{(x - 5) + 1}{2}\right)^2 - \left(\dfrac{(y + 2) + 3}{4}\right)^2 = \left(\dfrac{x - 4}{2}\right)^2 - \left(\dfrac{y + 5}{4}\right)^2$; the center is $(4, -5)$.

47.   Let d_1 be the distance between (x, y) and $(1, -2)$; let d_2 be the distance between (x, y) and $(4, 3)$.

$$5 = |d_1 - d_2| = \left| \sqrt{(x-1)^2 + (y+2)^2} - \sqrt{(x-4)^2 + (y-3)^2} \right|$$

$$\iff 5 \pm \sqrt{(x-1)^2 + (y+2)^2} = \sqrt{(x-4)^2 + (y-3)^2}$$

$$\iff 25 \pm 10\sqrt{(x-1)^2 + (y+2)^2} + (x-1)^2 + (y+2)^2 = (x-4)^2 + (y-3)^2$$

$$\iff \pm 10\sqrt{(x-1)^2 + (y+2)^2} = -6x - 10y - 5$$

$$\iff 100\left((x-1)^2 + (y+2)^2\right) = (6x + 10y + 5)^2$$

$$\iff 64x^2 - 120xy - 260x + 300y + 475 = 0.$$

REMARK:
     The center of this hyperbola is $(5/2, 1/2)$, the midpoint of the line segment (transverse axis) between the foci $(1, -2)$ and $(4, 3)$. If we rewrite the preceding equation in terms of $x - 5/2$ and $y - 1/2$ we obtain

$$120\left(y - \frac{1}{2}\right) = \frac{225}{x - 5/2} + 64\left(x - \frac{5}{2}\right).$$

The asymptotes for this curve are the straight lines $x = 5/2$ and $120(y - 1/2) = 64(x - 5/2)$.

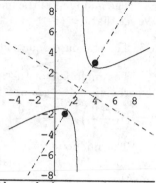

49.   Three points in space determine a plane — our plane will be the one passing through the two submarines and the point where the depth charge explodes; within that plane, orient a coordinate system so that the x-axis passes through the positions of the two submarines with sub A at point $(c, 0)$ and sub B at point $(-c, 0)$. We are told the submarines are 4 kilometers apart, therefore $c = 4/2 = 2$. The location of the depth charge lies on a hyperbola with equation $(x/a)^2 - (y/b)^2 = 1$ and foci at $(\pm c, 0)$. Let v be the speed of sound in sea water; let t_A and t_B be the times for the sound of the explosion to reach submarines A and B (we are told that $t_B = t_A + 2$). Since sound travels $t \cdot v$ in t seconds, we infer $2a = t_B \cdot v - t_A \cdot v = (t_B - t_A)v = 2v$ which then implies $a = v$. Therefore $b = \sqrt{c^2 - a^2} = \sqrt{c^2 - v^2} = \sqrt{2^2 - 1.533^2} = \sqrt{1.649911} \approx 1.2844$ kilometers. Since submarine A is closer to the explosion, we know the explosion point is on the right-hand branch of the hyperbola, i.e., $x = a\sqrt{1 + y^2/b^2} = 1.533\sqrt{1 + y^2/1.649911}$.

REMARK: The equation $2a = 2v$ involves correct magnitudes but different units; more precisely, if a and v were dimensionless, $2a$ km $= (2 \sec)v$ km/ sec.

51. $\left(\dfrac{x}{a}\right)^2 - \left(\dfrac{y}{b}\right)^2 = 1 \implies \dfrac{2x}{a^2} - \dfrac{2y}{b^2} \cdot \dfrac{dy}{dx} = 0 \implies \dfrac{dy}{dx} = \dfrac{2x}{a^2} \cdot \dfrac{b^2}{2y} = \dfrac{b^2 x}{a^2 y}$; the line tangent at the point

(x_0, y_0) satisfies $\dfrac{b^2 x_0}{a^2 y_0} = \dfrac{y - y_0}{x - x_0} \iff \dfrac{x_0}{a^2}(x - x_0) = \dfrac{y_0}{b^2}(y - y_0) \iff \dfrac{x_0 \cdot x}{a^2} - \dfrac{y_0 \cdot y}{b^2} = \dfrac{x_0^2}{a^2} - \dfrac{y_0^2}{b^2} = 1.$ ∎

53.   Fix two foci F_1 and F_2; suppose P is any point not collinear with F_1 and F_2. Let $d_1 = \overline{PF_1}$, $d_2 = \overline{PF_2}$, and $2c = \overline{F_1 F_2}$. Because the length of one side of a triangle is less than the sum of the lengths of the other two sides, $d_1 < d_2 + 2c$ and $d_2 < d_1 + 2c$. If $0 < a$, then the set

$$\{P : |d_2 - d_1| = 2a\} = \begin{cases} \emptyset & \text{if } c < a, \\ \text{two rays extending out from } F_1 \text{ and } F_2 & \text{if } c = a, \\ \text{"genuine" hyperbola} & \text{if } c > a. \end{cases}$$

The vertices of a hyperbola are on the line segment between the foci; thus $d_1 + d_2 = 2c$ for them. In order that $|d_2 - d_1| = 2a$ also, it is necessary that $d_1 = c \mp a$; these values for d_1 locate the two vertices and the distance between them is $(c + a) - (c - a) = 2a$. Therefore the eccentricity is $\dfrac{2c}{2a} = \dfrac{c}{a}$ which exceeds 1 because $a < c$ for any hyperbola. ∎

55.   a. $c = x^2 + 4x - 3y^2 + 6y = (x^2 + 4x) - 3(y^2 - 2y)$

$$\iff c + 1 = c + 2^2 - 3 \cdot 1^2 = (x^2 + 4x + 4) - 3(y^2 - 2y + 1) = (x + 2)^2 - 3(y - 1)^2.$$

If $c + 1 > 0$, this is the equation for an hyperbola with vertices at $(-2 \pm \sqrt{c+1}, 1)$; if $c + 1 < 0$, it is for one with vertices $\left(-2, 1 \pm \sqrt{\dfrac{|c+1|}{3}}\right)$. In either of these cases, the center is $(-2, 1)$.

b.   If $c + 1 = 0$, the equation becomes $(x + 2)^2 = 3(y - 1)^2$ which is equivalent to $|x + 2| = \sqrt{3}\,|y - 1|$; this is equivalent to the pair of linear equations $y = \dfrac{\pm 1}{\sqrt{3}}(x + 2) + 1$.

REMARK: The two straight lines for the case $c = -1$ are the asymptotes of the hyperbolæ in all of the other cases.

57.   In the left figure below, suppose F_- and F_+ are the foci of an ellipse and an hyperbola; at a point common to those curves suppose T_e is the line tangent to the ellipse and T_h is the line tangent to the hyperbola. The reflection property of an ellipse (Problem 8.1.47) implies $\alpha = \beta$; the reflection property of an hyperbola (Problem 52 above) implies $\gamma = \delta$. Therefore $\alpha + \gamma = \beta + \delta = \dfrac{\pi}{2}$ which means $T_e \perp T_h$. ∎

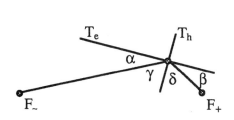

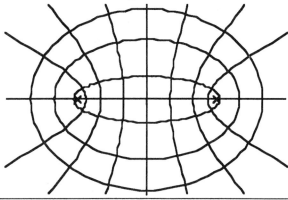

REMARK: Any pair of points serves as foci for a family of ellipses and an orthogonal family of hyperbolas; examine the right-hand figure above.

Solutions 8.4   *Second-degree equations and Rotation of axes*   (pages 575-6)

1.
See Example 1. If $\theta = \dfrac{\pi}{4}$, then
$\begin{aligned} x &= x' \cos\theta - y' \sin\theta = (x' - y')/\sqrt{2} \\ y &= x' \sin\theta + y' \cos\theta = (x' + y')/\sqrt{2}. \end{aligned}$

Thus

$$45 = 4x^2 - 2xy + 4y^2 = 4\left(\frac{x'-y'}{\sqrt{2}}\right)^2 - 2\left(\frac{x'-y'}{\sqrt{2}}\right)\left(\frac{x'+y'}{\sqrt{2}}\right) + 4\left(\frac{x'+y'}{\sqrt{2}}\right)^2$$
$$= 2\left((x')^2 - 2x'y' + (y')^2\right) - \left((x')^2 - (y')^2\right) + 2\left((x')^2 + 2x'y' + (y')^2\right)$$
$$= (2 - 1 + 2)(x')^2 + (-4 + 4)x'y' + (2 + 1 + 2)(y')^2 = 3(x')^2 + 5(y')^2.$$

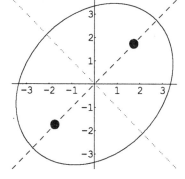

In the new coordinate system this is the equation for an ellipse centered at $(0,0)$ with $a = \sqrt{45/3} = \sqrt{15}$, $b = \sqrt{45/5} = \sqrt{9} = 3$, $c = \sqrt{15 - 9} = \sqrt{6}$, vertices $(\pm\sqrt{15}, 0)$, and foci $(\pm\sqrt{6}, 0)$.

3.   If $\theta = \dfrac{\pi}{6}$, then
$\begin{aligned} x &= x' \cos\theta - y' \sin\theta = \left(\sqrt{3}\,x' - y'\right)/2 \\ y &= x' \sin\theta + y' \cos\theta = \left(x' + \sqrt{3}\,y'\right)/2. \end{aligned}$ Thus

$$6 = 2x - 3y = 2\left(\frac{\sqrt{3}\,x' - y'}{2}\right) - 3\left(\frac{x' + \sqrt{3}\,y'}{2}\right) = \left(\sqrt{3} - \frac{3}{2}\right)x' - \left(1 + \frac{3\sqrt{3}}{2}\right)y'.$$

5.   The x-axis is the axis of the parabola $y^2 = -12x$. The line $y = \sqrt{3}\,x$ has slope $\sqrt{3}$ and its angle with the x-axis is $\tan^{-1}\sqrt{3} = \dfrac{\pi}{3}$. If we rotate the coordinate axes through an angle $\theta = \dfrac{\pi}{3}$, then $x = x' \cos\theta - y' \sin\theta = \left(x' - \sqrt{3}\,y'\right)/2$ and $y = x' \sin\theta + y' \cos\theta = \left(\sqrt{3}\,x' + y'\right)/2$. Therefore

$$y^2 = -12x \iff \left(\frac{\sqrt{3}\,x' + y'}{2}\right)^2 = -12\left(\frac{x' - \sqrt{3}\,y'}{2}\right) \iff \left(\sqrt{3}\,x' + y'\right)^2 = -24\left(x' - \sqrt{3}\,y'\right)$$

$$\iff 3(x')^2 + 2\sqrt{3}\,x'y' + (y')^2 = -24x' + 24\sqrt{3}\,y'.$$

8.4 — Second-degree equations and Rotation of axes

If we rotate these x'-y' axes through an additional angle π for a total rotation of $\dfrac{4\pi}{3}$, the parabola's axis will again coincide with the old line $y = \sqrt{3}\,x$. We have $x'' = -x'$ and $y'' = -y'$, the equation of the parabola becomes $3(x'')^2 + 2\sqrt{3}\,x''y'' + (y'')^2 = 24x'' - 24\sqrt{3}\,y''$.

7.   Use equation (9) to pick angle θ.

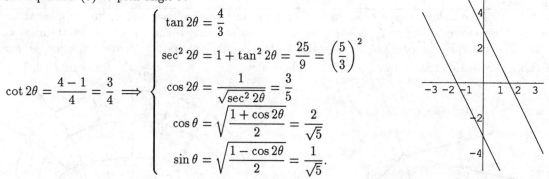

$$\cot 2\theta = \frac{4-1}{4} = \frac{3}{4} \implies \begin{cases} \tan 2\theta = \dfrac{4}{3} \\[2mm] \sec^2 2\theta = 1 + \tan^2 2\theta = \dfrac{25}{9} = \left(\dfrac{5}{3}\right)^2 \\[2mm] \cos 2\theta = \dfrac{1}{\sqrt{\sec^2 2\theta}} = \dfrac{3}{5} \\[2mm] \cos\theta = \sqrt{\dfrac{1+\cos 2\theta}{2}} = \dfrac{2}{\sqrt{5}} \\[2mm] \sin\theta = \sqrt{\dfrac{1-\cos 2\theta}{2}} = \dfrac{1}{\sqrt{5}}. \end{cases}$$

Rotating the axes through angle $\theta = \dfrac{1}{2}\tan^{-1}\left(\dfrac{4}{3}\right) \approx 26.57°$ yields
$$x = x'\cos\theta - y'\sin\theta = (2x' - y')/\sqrt{5},$$
$$y = x'\sin\theta + y'\cos\theta = (x' + 2y')/\sqrt{5},$$
and

$$9 = 4x^2 + 4xy + y^2 = 4\frac{(2x'-y')^2}{5} + 4\frac{(2x'-y')(x'+2y')}{5} + \frac{(x'+2y')^2}{5}$$
$$= \frac{(16+8+1)(x')^2 + (-16+12+4)x'y' + (4-8+4)(y')^2}{5} = 5(x')^2.$$

The graph consists of the two lines which have equations $x' = \pm\dfrac{3}{\sqrt{5}}$.

REMARK: In terms of the original coordinates, $9 = 4x^2 + 4xy + y^2 = (2x+y)^2 \iff 3 = |2x+y| = \pm(2x+y)$.

9.   Rotating the axes through angle θ yields
$$5 = 3x^2 - 2xy = 3\left(x'\cos\theta - y'\sin\theta\right)^2 - 2\left(x'\cos\theta - y'\sin\theta\right)\left(x'\sin\theta + y'\cos\theta\right)$$
$$= \left(3\cos^2\theta - 2\cos\theta\sin\theta\right)(x')^2 + \left(-6\cos\theta\sin\theta - 2\cos^2\theta + 2\sin^2\theta\right)x'y' + \left(3\sin^2\theta + 2\sin\theta\cos\theta\right)(y')^2.$$
Use equation (9) to pick angle θ.

$$\cot 2\theta = \frac{3-0}{-2} = \frac{-3}{2} \implies \begin{cases} \sin 2\theta = \dfrac{1}{\sqrt{1+\cot^2 2\theta}} = \dfrac{1}{\sqrt{13/4}} = \dfrac{2}{\sqrt{13}} \\[2mm] \cos 2\theta = \cot 2\theta \cdot \sin 2\theta = \dfrac{-3}{\sqrt{13}} \\[2mm] \cos\theta = \sqrt{\dfrac{1+\cos 2\theta}{2}} = \sqrt{\dfrac{\sqrt{13}-3}{2\sqrt{13}}} \\[2mm] \sin\theta = \sqrt{\dfrac{1-\cos 2\theta}{2}} = \sqrt{\dfrac{\sqrt{13}+3}{2\sqrt{13}}}. \end{cases}$$

Using this particular rotation we have $\theta = \dfrac{1}{2}\cot^{-1}\dfrac{-3}{2} = \dfrac{1}{2}\left(\dfrac{\pi}{2} - \tan^{-1}\dfrac{-3}{2}\right) = \dfrac{1}{2}\cos^{-1}\dfrac{-3}{\sqrt{13}} \approx 73.2°$ and

$$3\cos^2\theta - 2\cos\theta\sin\theta = 3\cdot\frac{\sqrt{13}-3}{2\sqrt{13}} - 2\cdot\sqrt{\frac{13-3^2}{2^2\cdot 13}} = \frac{3-\sqrt{13}}{2},$$

$$-6\cos\theta\sin\theta - 2\left(\cos^2\theta - \sin^2\theta\right) = -6\cdot\sqrt{\frac{13-3^2}{2^2\cdot 13}} - 2\cdot\left(\frac{\sqrt{13}-3}{2\sqrt{13}} - \frac{\sqrt{13}+3}{2\sqrt{13}}\right) = 0,$$

$$3\sin^2\theta + 2\sin\theta\cos\theta = 3\cdot\frac{\sqrt{13}+3}{2\sqrt{13}} + 2\cdot\sqrt{\frac{13-3^2}{2^2\cdot 13}} = \frac{3+\sqrt{13}}{2}.$$

Thus $5 = \left(\dfrac{3-\sqrt{13}}{2}\right)(x')^2 + \left(\dfrac{3+\sqrt{13}}{2}\right)(y')^2$ or $1 \approx \left(\dfrac{y'}{1.23}\right)^2 - \left(\dfrac{x'}{4.06}\right)^2$; this describes an hyperbola.

$$a^2 = 5 \cdot \frac{2}{\sqrt{13}-3} = \frac{5\left(\sqrt{13}+3\right)}{2} \approx 4.06^2$$

$$b^2 = 5 \cdot \frac{2}{\sqrt{13}+3} = \frac{5\left(\sqrt{13}-3\right)}{2} \approx 1.23^2$$

$$c^2 = a^2 + b^2 = 5\sqrt{13} \approx 4.25^2$$

$$\frac{b}{a} = \sqrt{\frac{\sqrt{13}-3}{\sqrt{13}+3}} = \sqrt{\frac{(\sqrt{13}-3)^2}{13-3^2}} = \frac{\sqrt{13}-3}{2} \approx 0.30$$

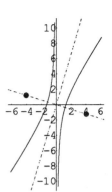

In the rotated coordinate system we have vertices $(0,\pm1.23)$ and foci $(0,\pm4.25)$, the asymptotes are $y' = \pm0.30x'$. (In the original coordinate system the foci are $(\pm4.06, \mp1.23)$ and the asymptotes are the lines $x = 0$ and $3x - 2y = 0$.)

If these axes are rotated through angle $-90°$ for a net rotation of $\frac{1}{2}\tan^{-1}\frac{-2}{3} \approx -16.8°$, we obtain the equation $10 = \left(3 - \sqrt{13}\right)(y'')^2 + \left(3 + \sqrt{13}\right)(x'')^2$.

REMARK: Some parts of this solution involved "nested radicals", but the coefficients of the transformed expression were less complicated; such simplification will always occur. If $B \neq 0$, rotate the axes through angle $\theta = \frac{1}{2}\cot^{-1}\left(\frac{A-C}{B}\right)$; let $W = \sqrt{B^2 + (A-C)^2}$ and $S = \begin{cases} -1 & \text{if } B < 0, \\ 1 & \text{if } B > 0. \end{cases}$ Since csc is positive on the range of $\cot^{-1}$, if $y = \cot^{-1}x$, then $\cot y = x$ and $\csc y = \sqrt{1 + \cot^2 y} = \sqrt{1 + x^2}$; hence $\sin\cot^{-1}x = \frac{1}{\csc\cot^{-1}x} = \frac{1}{\sqrt{1+x^2}}$ and $\cos\cot^{-1}x = \left(\sin\cot^{-1}x\right)\left(\cot\cot^{-1}x\right) = \frac{x}{\sqrt{1+x^2}}$. Thus

$$\sin 2\theta = \frac{1}{\sqrt{1 + ((A-C)/B)^2}} = \frac{|B|}{W} = \left(\frac{B}{W}\right) \cdot S,$$

$$\cos 2\theta = \frac{(A-C)/B}{\sqrt{1 + ((A-C)/B)^2}} = \left(\frac{A-C}{W}\right)\left(\frac{|B|}{B}\right) = \left(\frac{A-C}{W}\right) \cdot S.$$

For any rotation we have

$$Ax^2 + Bxy + Cy^2$$
$$= A\left(x'\cos\theta - y'\sin\theta\right)^2 + B\left(x'\cos\theta - y'\sin\theta\right)\left(x'\sin\theta + y'\cos\theta\right) + C\left(x'\sin\theta + y'\cos\theta\right)^2$$
$$= \left(A\cos^2\theta + B\cos\theta\sin\theta + C\sin^2\theta\right)(x')^2 + \left(B\left(\cos^2\theta - \sin^2\theta\right) + 2(C-A)\cos\theta\sin\theta\right)x'y'$$
$$\quad + \left(A\sin^2\theta - B\cos\theta\sin\theta + C\cos^2\theta\right)(y')^2$$

With our particular choice of θ the coefficient of $(x')^2$ is

$$A\cos^2\theta + B\cos\theta\sin\theta + C\sin^2\theta = A\left(\frac{1 + \cos 2\theta}{2}\right) + B\left(\frac{\sin 2\theta}{2}\right) + C\left(\frac{1 - \cos 2\theta}{2}\right)$$
$$= \frac{(A+C) + (A-C)\cos 2\theta + B\sin 2\theta}{2}$$
$$= \frac{(A+C) + (A-C)^2/W \cdot S + B^2/W \cdot S}{2}$$
$$= \frac{(A+C) + \left((A-C)^2 + B^2\right)/W \cdot S}{2} = \frac{A + C + W \cdot S}{2}.$$

Solution 21 will show the coefficient of $x'y'$ is zero. We now compute the coefficient of $(y')^2$ to be

$$A\sin^2\theta - B\cos\theta\sin\theta + C\cos^2\theta = A\left(\frac{1 - \cos 2\theta}{2}\right) - B\left(\frac{\sin 2\theta}{2}\right) + C\left(\frac{1 + \cos 2\theta}{2}\right)$$
$$= \frac{(A+C) + (-A+C)\cos 2\theta - B\sin 2\theta}{2}$$
$$= \frac{(A+C) - \left((A-C)^2 + B^2\right)/W \cdot S}{2} = \frac{A + C - W \cdot S}{2}.$$

Therefore

$$Ax^2 + Bxy + Cy^2 = \left(\frac{A + C + W \cdot S}{2}\right)(x')^2 + \left(\frac{A + C - W \cdot S}{2}\right)(y')^2.$$

8.4 — Second-degree equations and Rotation of axes

11. See Example 2. Rotate the axes through angle $\theta = \dfrac{\pi}{4}$; then

$$a = xy = \left(\frac{x' - y'}{\sqrt{2}}\right)\left(\frac{x' + y'}{\sqrt{2}}\right) = \frac{(x')^2 - (y')^2}{2} \iff 1 = \left(\frac{x'}{\sqrt{2a}}\right)^2 - \left(\frac{y'}{\sqrt{2a}}\right)^2.$$

The graph is an hyperbola with vertices $(\pm\sqrt{2a}, 0)$, foci $(\pm 2\sqrt{a}, 0)$, and asymptotes $y' = \pm x'$.

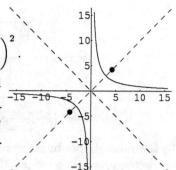

REMARK: In the original coordinate system, the vertices are $(\pm\sqrt{a}, \pm\sqrt{a})$, the foci are $(\pm 2\sqrt{a}, \pm 2\sqrt{a})$, and the asymptotes are the x- and y-axes.

13. Problem 7 has the same second-degree expression, $4x^2 + 4xy + y^2$. Hence the same rotation, $\theta = \dfrac{1}{2}\cot^{-1}\left(\dfrac{4-1}{4}\right) \approx 26.6°$, works for both problems transforming $4x^2 + 4xy + y^2$ into $5(x')^2$. Thus

$$0 = 4x^2 + 4xy + y^2 + 20x - 10y$$
$$= 5(x')^2 + 20\left(\frac{2x' - y'}{\sqrt{5}}\right) - 10\left(\frac{x' + 2y'}{\sqrt{5}}\right)$$
$$= 5(x')^2 + 6\sqrt{5}\,x' - 8\sqrt{5}\,y'$$
$$\iff \left(\frac{8}{\sqrt{5}}\right)y' = (x')^2 + \left(\frac{6}{\sqrt{5}}\right)x' = \left(x' + \frac{3}{\sqrt{5}}\right)^2 - \frac{9}{5}$$
$$\iff \frac{8}{\sqrt{5}}\left(y' + \frac{9}{8\sqrt{5}}\right) = \left(\frac{8}{\sqrt{5}}\right)y' + \frac{9}{5} = \left(x' + \frac{3}{\sqrt{5}}\right)^2.$$

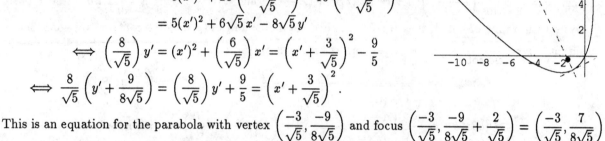

This is an equation for the parabola with vertex $\left(\dfrac{-3}{\sqrt{5}}, \dfrac{-9}{8\sqrt{5}}\right)$ and focus $\left(\dfrac{-3}{\sqrt{5}}, \dfrac{-9}{8\sqrt{5}} + \dfrac{2}{\sqrt{5}}\right) = \left(\dfrac{-3}{\sqrt{5}}, \dfrac{7}{8\sqrt{5}}\right)$.

15. Rotate through angle $\theta = \dfrac{1}{2}\cot^{-1}\dfrac{2-1}{1} = \dfrac{\pi}{8} = 22.5°$. The Remark following Solution 9 shows that with this choice of θ and letting $W = \sqrt{1^2 + (2-1)^2} = \sqrt{2}$ we have

$$4 = 2x^2 + xy + y^2 = \left(\frac{3 + \sqrt{2}}{2}\right)(x')^2 + \left(\frac{3 - \sqrt{2}}{2}\right)(y')^2$$
$$\iff 1 \approx \left(\frac{x'}{1.35}\right)^2 + \left(\frac{y'}{2.25}\right)^2.$$

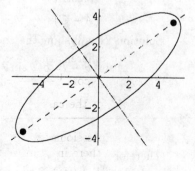

We obtain an equation describing an ellipse with $(a')^2 = \dfrac{8}{3 + \sqrt{2}} \approx 1.81 \approx 1.35^2$,

$(b')^2 = \dfrac{8}{3 - \sqrt{2}} \approx 5.04 \approx 2.25^2$, and $(c')^2 = (b')^2 - (a')^2 = \dfrac{16\sqrt{2}}{7} \approx 3.23 \approx 1.80^2$; the foci are $(0, \pm 1.80)$.

17. Use $\theta = \dfrac{1}{2}\cot^{-1}\dfrac{3-5}{-6} \approx 35.8°$. The Remark following Solution 9 shows that with this choice of θ and letting $W = \sqrt{(-6)^2 + (3-5)^2} = 2\sqrt{10}$ we have

$$36 = 3x^2 - 6xy + 5y^2$$
$$= \left(\frac{3 + 5 - 2\sqrt{10}}{2}\right)(x')^2 + \left(\frac{3 + 5 + 2\sqrt{10}}{2}\right)(y')^2$$
$$= \left(4 - \sqrt{10}\right)(x')^2 + \left(4 + \sqrt{10}\right)(y')^2$$
$$\iff 1 \approx \left(\frac{x'}{6.56}\right)^2 + \left(\frac{y'}{2.24}\right)^2.$$

We obtain an equation describing an ellipse with $(a')^2 = \dfrac{36}{4 - \sqrt{10}} \approx 42.97 \approx 6.56^2$, $(b')^2 = \dfrac{36}{4 + \sqrt{10}} \approx$

$5.03 \approx 2.24^2$, and $(c')^2 = (a')^2 - (b')^2 = 12\sqrt{10} \approx 37.95 \approx 6.16^2$; the foci are $(\pm 6.16, 0)$.

19.

Use $\theta = \frac{1}{2}\cot^{-1}\frac{0-3}{-4} \approx 26.6°$, the same rotation as used in Solution 7. The Remark following Solution 9 also shows some simplifications which occur with this choice of θ. Letting $W = \sqrt{(-4)^2 + (0-3)^2} = 5$ we have

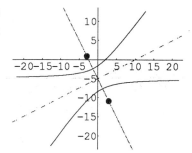

$$
\begin{aligned}
-40 &= 3y^2 - 4xy + 30y - 20x \\
&= \left(\frac{3-5}{2}\right)(x')^2 + \left(\frac{3+5}{2}\right)(y')^2 + 30\left(\frac{x'+2y'}{\sqrt{5}}\right) - 20\left(\frac{2x'-y'}{\sqrt{5}}\right) \\
&= -(x')^2 - 2\sqrt{5}\,x' + 4(y')^2 + 16\sqrt{5}\,y' \\
&= \left[-\left(x'+\sqrt{5}\right)^2 + 5\right] + \left[4\left(y'+2\sqrt{5}\right)^2 - 80\right].
\end{aligned}
$$

Therefore $1 = \left(\dfrac{y'+2\sqrt{5}}{\sqrt{35}/2}\right)^2 - \left(\dfrac{x'+\sqrt{5}}{\sqrt{35}}\right)^2$; this describes an hyperbola centered at $(-\sqrt{5}, -2\sqrt{5})$ with

$(a')^2 = 35$, $(b')^2 = \dfrac{35}{4}$, $(c')^2 = (a')^2 + (b')^2 = \dfrac{175}{4}$, and asymptotes $2\left(y'+2\sqrt{5}\right) = \pm\left(x'+\sqrt{5}\right)$.

REMARK: In terms of the original coordinates the center is $(0,-5)$, the asymptotes are $y = -5$ and $y = (4/3)x - 5$, the original equation can be rewritten in the form $(y+5)\cdot\big(3(y+5) - 4x\big) = 35$.

---

21. If $B = 0$, then no rotation is needed to "eliminate" the xy-term.

If the axes are rotated through an angle θ, then

$$
\begin{aligned}
0 &= Ax^2 + Bxy + Cy^2 + Dx + Ey + F \\
&= A\left(x'\cos\theta - y'\sin\theta\right)^2 + B\left(x'\cos\theta - y'\sin\theta\right)\left(x'\sin\theta + y'\cos\theta\right) + C\left(x'\sin\theta + y'\cos\theta\right)^2 \\
&\quad + D\left(x'\cos\theta - y'\sin\theta\right) + E\left(x'\sin\theta + y'\cos\theta\right) + F \\
&= \left(A\cos^2\theta + B\cos\theta\sin\theta + C\sin^2\theta\right)(x')^2 \\
&\quad + \left(-2A\cos\theta\sin\theta + B\left(\cos^2\theta - \sin^2\theta\right) + 2C\sin\theta\cos\theta\right)x'y' \\
&\quad + \left(A\sin^2\theta - B\cos\theta\sin\theta + C\cos^2\theta\right)(y')^2 \\
&\quad + \left(D\cos\theta + E\sin\theta\right)x' + \left(-D\sin\theta + E\cos\theta\right)y' + F.
\end{aligned}
$$

The $x'y'$-term is "absent" if and only if

$$
\begin{aligned}
0 &= -2A\cos\theta\sin\theta + B\left(\cos^2\theta - \sin^2\theta\right) + 2C\sin\theta\cos\theta \\
&= B\left(\cos^2\theta - \sin^2\theta\right) - (A-C)\cdot 2\sin\theta\cos\theta \\
&= B\cdot\cos 2\theta - (A-C)\cdot\sin 2\theta;
\end{aligned}
$$

if $B \neq 0$, this is equivalent to $\dfrac{A-C}{B} = \dfrac{\cos 2\theta}{\sin 2\theta} = \cot 2\theta$. ∎

REMARK: The restriction $A \neq C$ is not needed for this result using cot. On the other hand, if $A \neq C$, then $B\cdot\cos 2\theta - (A-C)\cdot\sin 2\theta = 0 \iff \dfrac{B}{A-C} = \dfrac{\sin 2\theta}{\cos 2\theta} = \tan 2\theta$ so we could choose $\theta = \dfrac{1}{2}\tan^{-1}\dfrac{B}{A-C}$. Because the range of $\tan^{-1}$ is $(-\pi/2, \pi/2)$ while the range of $\cot^{-1}$ is $(0, \pi)$, these relations may recommend two different choices for θ (both of which work). The result of Problem 6.8.83 offers us a third expression, $\dfrac{1}{2}\cot^{-1}\dfrac{A-C}{B} = \dfrac{\pi}{4} - \dfrac{1}{2}\tan^{-1}\dfrac{A-C}{B}$, which has the dual advantage of not requiring special treatment for the case $A = C$ while being computable using a "hand calculator" with a $\boxed{\tan^{-1}}$ key.

---

23. General expressions for A', B', C' appear in Solution 21; the Remark following Solution 9 includes re-expression of them in terms of $\cos 2\theta$ and $\sin 2\theta$.

a.
$$
\begin{aligned}
A' + C' &= \left(A\cos^2\theta + B\cos\theta\sin\theta + C\sin^2\theta\right) + \left(A\sin^2\theta - B\cos\theta\sin\theta + C\cos^2\theta\right) \\
&= (A+C)\left(\cos^2\theta + \sin^2\theta\right) = A+C. \quad ∎
\end{aligned}
$$

b.
$$\begin{aligned}
(B')^2 - 4A'C' &= \left(B\left(\cos^2\theta - \sin^2\theta\right) + 2(C-A)\sin\theta\cos\theta\right)^2 \\
&\quad - 4\left(A\cos^2\theta + B\cos\theta\sin\theta + C\sin^2\theta\right)\left(A\sin^2\theta - B\cos\theta\sin\theta + C\cos^2\theta\right) \\
&= \left(B\cos 2\theta - (A-C)\sin 2\theta\right)^2 \\
&\quad - \left((A+C)+(A-C)\cos 2\theta + B\sin 2\theta\right)\left((A+C)-(A-C)\cos 2\theta - B\sin 2\theta\right) \\
&= \left(B\cos 2\theta - (A-C)\sin 2\theta\right)^2 \\
&\quad - \left((A+C)^2 - [(A-C)\cos 2\theta + B\sin 2\theta]^2\right) \\
&= B^2\cos^2 2\theta - 2B(A-C)\cos 2\theta\sin 2\theta + (A-C)^2\sin^2 2\theta \\
&\quad - (A+C)^2 + (A-C)^2\cos^2 2\theta + 2(A-C)B\cos 2\theta\sin 2\theta + B^2\sin^2 2\theta \\
&= \left(B^2 + (A-C)^2\right)\left(\cos^2 2\theta + \sin^2 2\theta\right) - (A+C)^2 \\
&= B^2 + (A-C)^2 - (A+C)^2 = B^2 - 4AC. \quad\blacksquare
\end{aligned}$$

25.   Apply Theorem 1 to the second degree equation $0\cdot x^2 + 1\cdot xy + 0\cdot y^2 + 0\cdot x + 0\cdot y - k = 0$. The discriminant is $B^2 - 4AC = 1^2 - 4\cdot 0\cdot 0 = 1 > 0$; since the discriminant is positive, the graph of $xy - k = 0$ is an hyperbola or a pair of intersecting lines (a degenerate hyperbola). $\blacksquare$

REMARK: Example 2 provides the key for an alternative proof and also shows the case "a pair of intersecting lines" does not occur.

Solutions 8.5   *The polar coordinate system*   (pages 580-1)

REMARK: To help you distinguish which type of coordinates are being used, polar coordinates are written with subscript "p" while rectangular coordinates are written with subscript "r": e.g., $(r,\theta)_p$ or $(x,y)_r$.

1.   If $(r,\theta)_p = (3,0)_p$, then $x = r\cos\theta = 3\cos 0 = 3$, $y = r\sin\theta = 3\sin 0 = 0$, and $(x,y)_r = (3,0)_r$.

     1: $(3,0)_p = (3,0)_r$ 3: $(4,\pi/4)_p = \left(2\sqrt{2}, 2\sqrt{2}\right)_r$

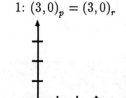 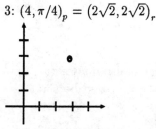

3. $\left(4, \dfrac{\pi}{4}\right)_p = \left(4\cos\dfrac{\pi}{4}, 4\sin\dfrac{\pi}{4}\right)_r = \left(2\sqrt{2}, 2\sqrt{2}\right)_r.$

5. $\left(6, \dfrac{7\pi}{6}\right)_p = \left(6\cos\dfrac{7\pi}{6}, 6\sin\dfrac{7\pi}{6}\right)_r = \left(6\cdot\dfrac{\sqrt{3}}{2}, 6\cdot\dfrac{-1}{2}\right)_r = \left(-3\sqrt{3}, -3\right)_r.$

     5: $(6, 7\pi/6)_p = \left(-3\sqrt{3}, -3\right)_r$ 7: $(5,\pi)_p = (-5,0)_r$

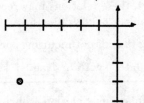 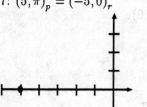

7. $(5,\pi)_p = (5\cos\pi, 5\sin\pi)_r = (-5,0)_r.$

9. $\left(5, \dfrac{-\pi}{3}\right)_p = \left(5\cos\dfrac{-\pi}{3}, 5\sin\dfrac{-\pi}{3}\right)_r = \left(5\cdot\dfrac{1}{2}, 5\cdot\dfrac{-\sqrt{3}}{2}\right)_r = \left(\dfrac{5}{2}, \dfrac{-5\sqrt{3}}{2}\right)_r.$

9: $(5, -\pi/3)_p = (5/2, -5\sqrt{3}/2)_r$

11: $(-2, 3\pi/4)_p = (\sqrt{2}, -\sqrt{2})_r$

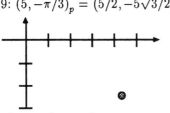

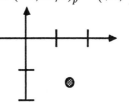

11. $\left(-2, \dfrac{3\pi}{4}\right)_p = \left(-2\cos\dfrac{3\pi}{4}, -2\sin\dfrac{3\pi}{4}\right)_r = \left(-2 \cdot \dfrac{-1}{\sqrt{2}}, -2 \cdot \dfrac{1}{\sqrt{2}}\right)_r = \left(\sqrt{2}, -\sqrt{2}\right)_r.$

13. $\left(2, \dfrac{7\pi}{4}\right)_p = \left(2\cos\dfrac{7\pi}{4}, 2\sin\dfrac{7\pi}{4}\right)_r = \left(2 \cdot \dfrac{1}{\sqrt{2}}, 2 \cdot \dfrac{-1}{\sqrt{2}}\right)_r = \left(\sqrt{2}, -\sqrt{2}\right)_r.$

13: $(2, 7\pi/4)_p = (\sqrt{2}, -\sqrt{2})_r$

15: $(1, 3\pi/2)_p = (0, -1)_r$

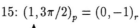

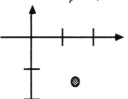

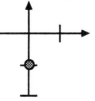

15. $\left(1, \dfrac{3\pi}{2}\right)_p = \left(\cos\dfrac{3\pi}{2}, \sin\dfrac{3\pi}{2}\right)_r = (0, -1)_r.$

17. $\left(-1, \dfrac{\pi}{2}\right)_p = \left(-1 \cdot \cos\dfrac{\pi}{2}, -1 \cdot \sin\dfrac{\pi}{2}\right)_r = (-1 \cdot 0, -1 \cdot 1)_r = (0, -1)_r.$

17: $(-1, \pi/2)_p = (0, -1)_r$

19: $(0, 0)_r = (0, 0)_p$

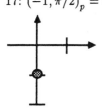

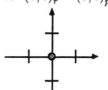

19.   The origin $(0, 0)_r$ is also the pole and can be represented as $(0, \theta)_p$ for any choice of θ; the simplest such choice is $(0, 0)_p$.

Note: The following generalization of the arctangent function is used to compute θ when converting rectangular coordinates to polar coordinates.

$$\mathrm{ArcTan}(x, y) = \begin{cases} 0 & \text{if } x = 0, \\ \tan^{-1}(y/x) & \text{if } x > 0 \text{ and } y > 0, \\ \pi + \tan^{-1}(y/x) & \text{if } x < 0, \\ 2\pi + \tan^{-1}(y/x) & \text{if } x > 0 \text{ and } y < 0. \end{cases}$$

---

21.   If $(x, y)_r = (3, 0)_r$, then the point is on the positive x-axis which is also the polar axis. $r = \sqrt{3^2 + 0^2} = 3$ and $\theta = \tan^{-1}\dfrac{0}{3} = 0$. Thus $(r, \theta)_p = (3, 0)_p$.

21: $(3, 0)_p = (3, 0)_r$

23: $(0, 1)_r = (1, \pi/2)_p$

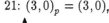

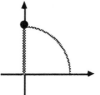

23.   If $(x, y)_r = (0, 1)_r$, then the point is on the positive y-axis which is also on the line $\theta = \dfrac{\pi}{2}$. $r = \sqrt{0^2 + 1^2} = 1$ and $\theta = \cot^{-1}\dfrac{0}{1} = \dfrac{\pi}{2}$. Thus $(r, \theta)_p = \left(1, \dfrac{\pi}{2}\right)_p.$

25. $(1,1)_r = \left(\sqrt{1^2 + 1^2}, \tan^{-1} \dfrac{1}{1} \right)_p = \left(\sqrt{2}, \dfrac{\pi}{4} \right)_p.$

25: $(1,1)_p = (\sqrt{2}, \pi/4)_r$

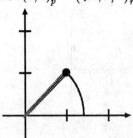

27: $(3,-3)_r = (3\sqrt{2}, 7\pi/4)_p$

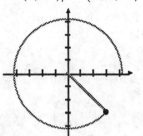

27. $(3,-3)_r = \left(\sqrt{3^2 + (-3)^2}, 2\pi + \tan^{-1} \dfrac{-3}{3} \right)_p = \left(3\sqrt{2}, \dfrac{7\pi}{4} \right)_p.$

29. $\left(2, 2\sqrt{3} \right)_r = \left(\sqrt{2^2 + (2\sqrt{3})^2}, \tan^{-1} \dfrac{2\sqrt{3}}{2} \right)_p = \left(4, \dfrac{\pi}{3} \right)_p.$

29: $(2, 2\sqrt{3})_p = (4, \pi/3)_r$

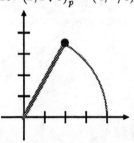

31: $(5, -5\sqrt{3})_r = (10, 5\pi/3)_p$

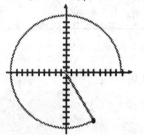

31. $\left(5, -5\sqrt{3} \right)_r = \left(\sqrt{5^2 + (-5\sqrt{3})^2}, 2\pi + \tan^{-1} \dfrac{-5\sqrt{3}}{5} \right)_p = \left(10, \dfrac{5\pi}{3} \right)_p.$

33. $\left(\sqrt{3}, 1 \right)_r = \left(\sqrt{(\sqrt{3})^2 + 1^2}, \tan^{-1} \dfrac{1}{\sqrt{3}} \right)_p = \left(2, \dfrac{\pi}{6} \right)_p.$

33: $(\sqrt{3}, 1)_p = (2, \pi/6)_r$

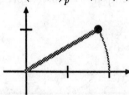

35: $(-4\sqrt{3}, 4)_r = (8, 5\pi/6)_p$

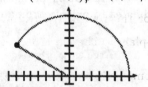

35. $\left(-4\sqrt{3}, 4 \right)_r = \left(\sqrt{(-4\sqrt{3})^2 + 4^2}, \pi + \tan^{-1} \dfrac{4}{-4\sqrt{3}} \right)_p = \left(8, \dfrac{5\pi}{6} \right)_p.$

37. $(-r, \theta + \pi)_p = ((-r) \cdot \cos(\theta + \pi), (-r) \cdot \sin(\theta + \pi))_r = ((-r) \cdot (-\cos\theta), (-r) \cdot (-\sin\theta))_r$
$= (r\cos\theta, r\sin\theta)_r = (r, \theta)_p.$ ∎

Solutions 8.6 *Graphing in polar coordinates* (pages 589-90)

1. See Example 1. Points $(r,\theta)_p = (5,\theta)_p$ are on the circle of radius 5 centered at the pole; the curve is symmetric about the pole and about each line $\theta = $ constant.

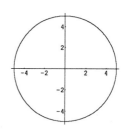

3. See Example 2. Points $(r,\theta)_p = \left(r, \dfrac{-\pi}{6}\right)_p$ are on the straight line through the pole making angle $\dfrac{\pi}{6}$ clockwise from the polar axis. The curve is symmetric about the pole and the orthogonal line $\theta = \dfrac{-\pi}{6} + \dfrac{\pi}{2} = \dfrac{\pi}{3}$.

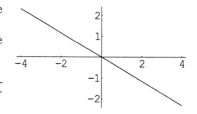

REMARK: In rectangular coordinates, the straight line satisfies the equation $y = \tan\left(\dfrac{-\pi}{6}\right) \cdot x = \dfrac{-x}{\sqrt{3}}$.

5. See Example 3 and Problem 90.
$$r = 5\sin\theta \iff r^2 = r\cdot(5\sin\theta) \iff x^2+y^2 = 5y \iff x^2+\left(y-\frac{5}{2}\right)^2 = \left(\frac{5}{2}\right)^2,$$
this is an equation for the circle of radius 5/2 that is centered in the xy-plane at the point $\left(0, \dfrac{5}{2}\right)_r$. The curve is symmetric about the vertical line $\theta = \dfrac{\pi}{2}$; the curve is completely traced out for $\theta \in \left[\dfrac{-\pi}{2}, \dfrac{\pi}{2}\right]$ or $\theta \in [0, \pi]$.

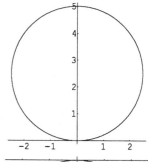

7. See Example 3 and Problem 90.
$$r = -5\sin\theta \iff r^2 = r\cdot(-5\sin\theta) \iff x^2+y^2 = -5y \iff x^2+\left(y+\frac{5}{2}\right)^2 = \left(\frac{5}{2}\right)^2$$
this is an equation for the circle of radius 5/2 that is centered at the point $\left(0, \dfrac{-5}{2}\right)_r$ in the xy-plane. The curve is symmetric about the vertical line $\theta = \dfrac{\pi}{2}$; the curve is completely traced out for $\theta \in \left[\dfrac{-\pi}{2}, \dfrac{\pi}{2}\right]$ or $\theta \in [0, \pi]$.

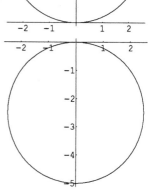

9. $$r = |\cos\theta| \iff r^2 = r\cdot|\cos\theta| \iff x^2+y^2 = \pm y \iff x^2+\left(y \mp \frac{1}{2}\right)^2 = \left(\frac{1}{2}\right)^2.$$

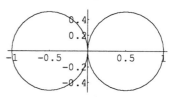

The graph consists of a pair of circles of radius $\dfrac{1}{2}$; their centers are $\left(0, \dfrac{\pm 1}{2}\right)_r$.

The curve is symmetric about the polar axis and the vertical line $\theta = \dfrac{\pi}{2}$; the upper half of the curve is traversed for $\theta \in [0, \pi]$ and the lower half for $\theta \in [\pi, 2\pi]$ (the right circle is traversed for $\theta \in [-\pi/2, \pi/2]$ and the left circle for $\theta \in [\pi/2, 3\pi/2]$).

REMARK: These circles are said to be *osculating* because they are tangent at their common point.

8.6 — Graphing in polar coordinates

11. $r = 5\cos\theta + 5\sin\theta \iff r^2 = 5r\cos\theta + 5r\sin\theta$

$$\iff x^2 + y^2 = 5x + 5y \iff \left(x - \frac{5}{2}\right)^2 + \left(y - \frac{5}{2}\right)^2 = 2\left(\frac{5}{2}\right)^2.$$

This is an equation for the circle of radius $\frac{5}{\sqrt{2}}$ that is centered at $\left(\frac{5}{2}, \frac{5}{2}\right)_r$. The curve is symmetric with respect to the line $\theta = \frac{\pi}{4}$; the curve is traversed for θ in any closed interval of length π.

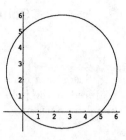

13.   The graph of $r = 2 \cdot (1 + \sin\theta)$ is a cardioid which is twice the size of the one discussed in Example 4; it too is symmetric about the vertical line $\theta = \frac{\pi}{2}$; the curve is traced out for $\theta \in [0, 2\pi]$.

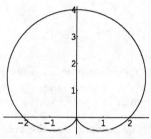

15.   Because $\sin(-\theta) = -\sin\theta$, if the graph in Solution 13 is reflected through the polar axis, the result is the graph of $r = 2 \cdot (1 - \sin\theta)$; both cardioids are symmetric about the vertical line $\theta = \frac{\pi}{2}$; both are traced out for $\theta \in [0, 2\pi]$.

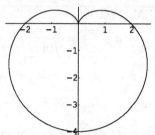

REMARK: Since $\sin(\theta \pm \pi) = -\sin\theta$, a half-rotation about the pole (origin) also produces one curve from the other.

17.   If $(a, b)_p$ is on $r = -2 - 2\sin\theta$, then $(-a, b - \pi)_p$ is on $r = 2 - 2\sin\theta$ (Problem 15) because $2 - 2\sin(b - \pi) = 2 - 2(-\sin b) = 2 + 2\sin b = -(-2 - 2\sin b) = -a$. Since $(a, b)_p$ and $(-a, b - \pi)_p$ are polar coordinates for the same point, the graph for this problem looks the same as the one shown in Solution 15 (but points are traced differently).

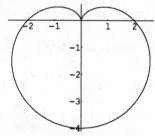

19.   Because $\sin(\pi - \theta) = \sin\theta$, the graph of $r = 1 + 3\sin\theta$ is symmetric about the line $\theta = \frac{\pi}{2}$. As discussed after Example 6, this curve is a limaçon with a loop; the graph is traversed for $\theta \in [0, 2\pi]$.

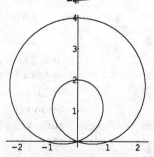

21.   The graph of $r = -2 + 4\cos\theta$ is symmetric about the polar axis because $\cos(-\theta) = \cos\theta$. As discussed after Example 6, this curve is a limaçon with a loop; the graph is traversed for $\theta \in [0, 2\pi]$.

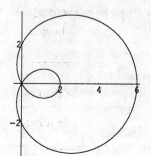

23. The graph of $r = -3 - 4\cos\theta$ is symmetric about the polar axis, this curve is a limaçon with a loop; the graph is traversed for $\theta \in [0, 2\pi]$.

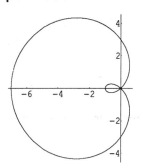

25. Because $\cos\left(\theta - \dfrac{\pi}{2}\right) = \sin\theta$, rotating the graph $r = -3 - 4\cos\theta$ (Problem 23) a quarter-turn counter-clockwise produces the graph of $r = -3 - 4\sin\theta$, a limaçon with a loop which is symmetric about the vertical line $\theta = \dfrac{\pi}{2}$; the graph is traversed for $\theta \in [0, 2\pi]$.

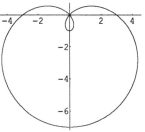

27. The graph of $r = 4 - 3\cos\theta$ is symmetric about the polar axis, it is a concave limaçon without a loop; the graph is traversed for $\theta \in [0, 2\pi]$. (See Example 5.)

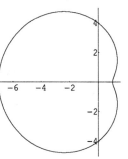

29. The graph of $r = 4 + 3\sin\theta$ is symmetric about the vertical line $\theta = \dfrac{\pi}{2}$, it is a concave limaçon without a loop; the graph is traversed for $\theta \in [0, 2\pi]$.

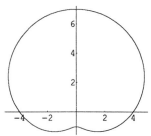

31. The graph of $r = 3\sin 2\theta$ is symmetric with respect to the pole because $3\sin 2(\theta + \pi) = 3\sin(2\theta + 2\pi) = 3\sin 2\theta$. If $a = 3\sin 2b$, then $3\sin 2(\pi - b) = 3\sin(-2b) = -3\sin 2b = -a$; because $(-a, \pi - b)_p$ and $(a, -b)_p$ are polar coordinates for the same point, the graph is symmetric about the polar axis since $(a, b)_p$ and $(a, -b)_p$ are on the graph. Similarly $3\sin 2(2\pi - b) = -a$, $(-a, 2\pi - b)_p$ and $(a, \pi - b)_p$ are polar coordinates for the same point, therefore the curve is symmetric about the line $\theta = \dfrac{\pi}{2}$ because $(a, b)_p$ and $(a, \pi - b)_p$ are on the graph. This four-leafed rose is traversed for $\theta \in [0, 2\pi]$.

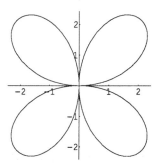

REMARK: The graph is also symmetric about the line $\theta = \dfrac{\pi}{4}$ because $\sin 2\left(\dfrac{\pi}{4} - t\right) = \cos 2t = \sin 2\left(\dfrac{\pi}{4} + t\right)$; it is symmetric about $\theta = \dfrac{3\pi}{4}$ since $\sin 2\left(\dfrac{3\pi}{4} - t\right) = -\cos 2t = \sin 2\left(\dfrac{3\pi}{4} + t\right)$.

8.6 — Graphing in polar coordinates

33.   Because $3\sin 2(-\theta) = -3\sin 2\theta$, reflecting the graph of $r = 3\sin 2\theta$ (Problem 31) through the polar axis will produce the graph of $r = -3\sin 2\theta$. Since the first graph is symmetric with respect to the polar axis (see Solution 31), it coincides with its reflection. Therefore the graph of $r = -3\sin 2\theta$ is the same four-leafed rose as the graph of $r = 3\sin 2\theta$ (but traversed differently).

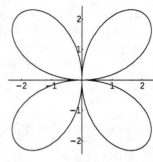

35.   The graph of $r = 5\sin 3\theta$ is symmetric about the vertical line $\theta = \dfrac{\pi}{2}$ because $5\sin 3(\pi - \theta) = 5\sin(3\pi - 3\theta) = -5\sin(-3\theta) = 5\sin 3\theta$.
        Note that this three-leafed rose is fully traced out as θ varies from 0 to π because $5\sin 3(\theta + \pi) = -5\sin 3\theta$ and the same point is represented by the polar coordinates $(a, b)_p$ and $(-a, b + \pi)_p$.

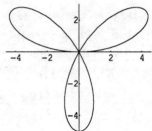

37.   Since $-5\sin 3\theta = 5\sin 3(-\theta)$, the graph of $r = -5\sin 3\theta$ is the reflection of the graph of $r = 5\sin 3\theta$ (Solution 35) in the polar axis; this reflection does not destroy the symmetry about the vertical line $\theta = \dfrac{\pi}{2}$.

REMARK: A half-turn rotation of Solution 35 about the pole also does the trick since $5\sin 3(\theta + \pi) = -5\sin 3\theta$.

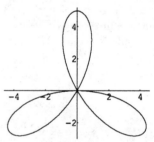

39.   The graph of $r = 2\cos 4\theta$ is symmetric about the polar axis, the vertical line $\theta = \dfrac{\pi}{2}$, and the pole because each of $2\cos 4(-\theta)$, $2\cos 4(\pi - \theta)$, and $2\cos 4(\theta + \pi)$ equals $2\cos 4\theta$. This eight-leafed rose is traversed for $\theta \in [0, 2\pi]$.

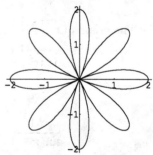

41.   Here are three views of the Archimedean spiral $r = -5\theta$, $\theta \geq 0$; the curve is not periodic, it also has no axis or point of symmetry.

$r = -5\theta, \ 0 \leq \theta \leq 2\pi + .3$

$r = -5\theta, \ 0 \leq \theta \leq 4\pi + .7$

$r = -5\theta, \ 0 \leq \theta \leq 7\pi + 1$

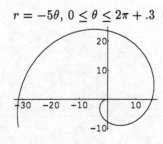
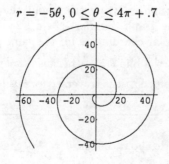
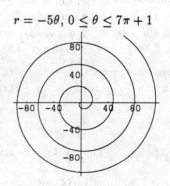

43.   Here are three views of the logarithmic spiral $r = e^{\theta}$; the curve is not periodic, it also has no axis or point of symmetry.

$r = e^\theta, \; -2\pi \le \theta \le 0$ $r = e^\theta, \; 0 \le \theta \le \pi$ $r = e^\theta, \; 0 \le \theta \le 6.55$

   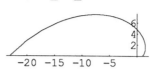   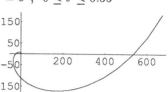

45. The graph of $r^2 = -4\sin 2\theta$ is a lemniscate symmetric about the pole, the line $\theta = \dfrac{\pi}{4}$, and the line $\theta = \dfrac{3\pi}{4}$. Using both $r = 2\sqrt{-\sin 2\theta}$ and $r = -2\sqrt{-\sin 2\theta}$, the whole graph is traced out for $\theta \in \left[\dfrac{\pi}{2}, \pi\right]$; alternatively, using just $r = 2\sqrt{-\sin 2\theta}$, the graph is traversed for $\theta \in \left[\dfrac{\pi}{2}, \pi\right] \cup \left[\dfrac{3\pi}{2}, 2\pi\right]$.

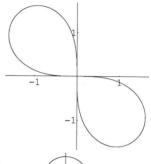

REMARK: Because $\sin 2\left(\theta - \dfrac{\pi}{2}\right) = -\sin 2\theta$, a quarter-turn counterclockwise rotation of the graph in Example 7 will produce this graph.

47. The graph of $r^2 = -25\cos 2\theta$ is a lemniscate symmetric about the pole, the polar axis, and the vertical line $\theta = \dfrac{\pi}{2}$ because each of $-25\cos 2(\theta + \pi)$, $-25\cos 2(-\theta)$, $-25\cos 2(\pi - \theta)$ equals $-25\cos 2\theta$. Using both $r = 5\sqrt{-\cos 2\theta}$ and $r = -5\sqrt{-\cos 2\theta}$, the whole graph is traced out for $\theta \in \left[\dfrac{\pi}{4}, \dfrac{3\pi}{4}\right]$; alternatively, using just $r = \sqrt{\cos 2\theta}$, the graph is traversed for $\theta \in \left[\dfrac{\pi}{4}, \dfrac{3\pi}{4}\right] \cup \left[\dfrac{5\pi}{4}, \dfrac{7\pi}{4}\right]$.

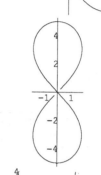

49. The graph of $r = \sin\theta\tan\theta$ is symmetric about the polar axis because $\sin(-\theta)\tan(-\theta)$

$= (-\sin\theta)(-\tan\theta) = \sin\theta\tan\theta$. The curve is traversed for $\theta \in \left(\dfrac{-\pi}{2}, \dfrac{\pi}{2}\right)$.

By rewriting the equation in the form $r\cos\theta = \sin^2\theta$, we can recognize this cissoid is asymptotic to the vertical line $x = r\cos\theta = 1$.

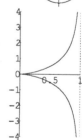

51. $\csc(\pi - \theta) = \dfrac{1}{\sin(\pi - \theta)} = \dfrac{1}{\sin\theta} = \csc\theta$ implies the graph of $r = 4 + 3\csc\theta$ is symmetric about the vertical line $\theta = \dfrac{\pi}{2}$. The graph is traversed for $\theta \in (0, \pi) \cup (\pi, 2\pi)$. By rewriting the equation in the form $r\sin\theta = 4\sin\theta + 3$, we can notice the graph is asymptotic to the horizontal line $y = r\sin\theta = 3$.

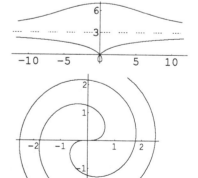

53. The parabolic spiral $r^2 = \theta$ is symmetric about the pole because $\left(\sqrt{\theta}, \theta\right)_p$ and $\left(\sqrt{\theta}, \theta + \pi\right)_p = \left(-\sqrt{\theta}, \theta\right)_p$ are both on the curve if $\theta \ge 0$; the curve is not periodic. The graph is traversed by $r = \sqrt{\theta}$ and $r = -\sqrt{\theta}$ for $\theta \in [0, \infty)$.

55. If $\left(-2, \dfrac{3}{2}\right)_r = \left(\dfrac{a}{2}, \dfrac{b}{2}\right)_r$, then $a = 2 \cdot (-2) = -4$ and $b = 2 \cdot \dfrac{3}{2} = 3$. With this choice of a and b we have $\dfrac{\sqrt{a^2 + b^2}}{2} = \dfrac{\sqrt{16 + 9}}{2} = \dfrac{5}{2}$. The conditions for Problem 91 are satisfied, hence $r = a\cos\theta + b\sin\theta = -4\cos\theta + 3\sin\theta$ is a polar equation for the specified circle.

REMARK: The circles discussed in Problems 89-91 have the special feature that they pass through the origin (pole); in general we could proceed as follows:

$$\left(x - (-2)\right)^2 + \left(y - \frac{3}{2}\right)^2 = \left(\frac{5}{2}\right)^2 \iff x^2 + 4x + y^2 - 3y = \frac{25}{4} - 4 - \frac{9}{4} = 0$$

$$\iff r^2 = x^2 + y^2 = -4x + 3y = -4r\cos\theta + 3r\sin\theta$$

$$\iff r = -4\cos\theta + 3\sin\theta \text{ or } r = 0.$$

57.   The graph of $r = 1$ is a circle of radius 1 centered at the pole, the graph of $r = \sin\theta$ is a circle of radius $\frac{1}{2}$ centered at $\left(\frac{1}{2}, \frac{\pi}{2}\right)_p$. These circles intersect at the point $\left(1, \frac{\pi}{2}\right)_p$.

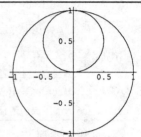

59.   The graphs of $r_1 = \sin\theta$ and $r_2 = \cos\theta$ are circles of radius $\frac{1}{2}$, the first centered at $\left(\frac{1}{2}, \frac{\pi}{2}\right)_p$ and the second at $\left(\frac{1}{2}, 0\right)_p$. These circles intersect at the pole: $(0,0)_p = \left(0, \frac{\pi}{2}\right)_p$. They also intersect in the first quadrant: $r_1 = r_2 \iff \sin\theta = \cos\theta \implies \theta = \tan^{-1} 1 = \frac{\pi}{4}$; this point of intersection is $\left(\frac{1}{\sqrt{2}}, \frac{\pi}{4}\right)_p$.

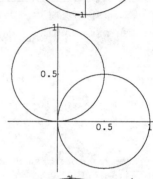

61.   The graph of $r = 3$ is a circle of radius 3 centered at the pole. The graph of $\theta = \frac{\pi}{4}$ is a straight line through the pole which meets the circle at two points: $\left(3, \frac{\pi}{4}\right)_p$ and $\left(-3, \frac{\pi}{4}\right)_p = \left(3, \frac{5\pi}{4}\right)_p$.

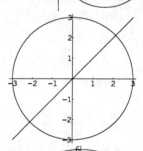

63.   The graph of $r_1 = -2 + 2\cos\theta$ is a cardioid and the graph of $r_2 = -4 + 2\sin\theta$ is a convex limaçon without a loop; each is traversed counterclockwise as θ varies from 0 to 2π. The equation $-2 + 2\cos\theta = r_1 = r_2 = -4 + 2\sin\theta$ is equivalent to $1 = \sin\theta - \cos\theta = \sqrt{2}\sin\left(\theta - \frac{\pi}{4}\right)$; its only solutions in $[0, 2\pi)$ are $\theta = \frac{\pi}{2}$ and $\theta = \pi$, the corresponding points of intersection are $\left(-2, \frac{\pi}{2}\right)_p$ and $(-4, \pi)_p$. The pole is not on the limaçon because $-4 + 2\sin\theta \leq -2$; hence it cannot be a point of intersection.

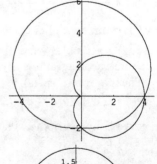

65.   The circles $r_1 = \sqrt{3}\sin\theta$ and $r_2 = \cos\theta$ are traversed counter-clockwise as θ varies from 0 to π. They meet at the pole: $(0,0)_p = \left(0, \frac{\pi}{2}\right)_p$. They also intersect in the first quadrant: $r_1 = r_2 \iff \sqrt{3}\sin\theta = \cos\theta \implies \theta = \tan^{-1}\frac{1}{\sqrt{3}} = \frac{\pi}{6}$; this point of intersection is $\left(\frac{\sqrt{3}}{2}, \frac{\pi}{6}\right)_p$.

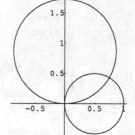

67. The graph of $r = \dfrac{3}{2}$ is a circle centered at the pole (and, hence, not passing through the pole) while the graph of $r = 3\cos 2\theta$ is a four-leafed rose; each graph is traversed as θ increases from 0 to 2π. If $\dfrac{3}{2} = 3\cos 2\theta$,

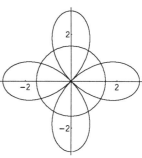

then $2\theta = 2k\pi \pm \cos^{-1}\left(\dfrac{1}{2}\right) = 2k\pi \pm \dfrac{\pi}{3}$. There are four such choices of θ satisfying $0 \le \theta < 2\pi$; the corresponding points of intersection are $\left(\dfrac{3}{2}, \dfrac{n\pi}{6}\right)_p$ for $n \in \{1,5,7,11\}$. The circle is also the graph of $r = \dfrac{-3}{2}$;

if $\dfrac{-3}{2} = 3\cos 2\theta$, then $2\theta = 2k\pi \pm \cos^{-1}\left(\dfrac{-1}{2}\right) = 2k\pi \pm \dfrac{2\pi}{3}$. This yields the four remaining points of

intersection: $\left(\dfrac{-3}{2}, \dfrac{m\pi}{6}\right)_p$ for $m \in \{2,4,8,10\}$.

69. The graph of $r^2 = \cos\theta$ is a fat "lazy-eight"; its two loops, $r = \sqrt{\cos\theta}$ and $r = -\sqrt{\cos\theta}$, are traced-out as θ varies from $\dfrac{-\pi}{2}$ to $\dfrac{\pi}{2}$. The graph

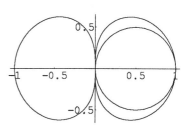

of $r = \cos\theta$ is a circle; it too is fully traversed as θ increases from $\dfrac{-\pi}{2}$ to $\dfrac{\pi}{2}$. This circle is inside the right-hand loop of $r^2 = \cos\theta$ because $c \le \sqrt{c}$ if $0 \le c \le 1$; since $c = \sqrt{c}$ if and only if $c = 0$ or $c = 1$, the only points where the curves meet are the pole $= \left(0, \dfrac{\pi}{2}\right)_p$ and $(1,0)_p$.

71. The spirals $r = \theta$ and $r^2 = \theta$ meet at $(0,0)_p$, the pole. The equation $r = r^2$ has solutions $r = 0$ and $r = 1$ so the graphs also meet at $(1,1)_p$.

If we restrict r and θ so that $0 \le r \le 2$ and $0 \le \theta$, there is no non-zero integer k such that $r^2 = \theta + 2k\pi$ and $r = \theta$ hold simultaneously because they imply $r^2 - r = 2k\pi$ but $r^2 - r = r(r-1)$ is bounded between -2 and 2 while $2k\pi$ is outside that interval. Hence the graphs meet only at the pole and $(1,1)_p$.

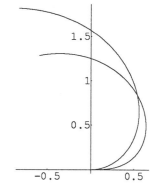

REMARK: If we allowed r to be larger, the curves next meet at the point

with $r = \theta - 2\pi = \dfrac{1 + 4\pi + \sqrt{1 + 8\pi}}{2} \approx 3.056$.

73. $r_1 = \sin 2\theta$ and $r_2 = \cos 2\theta$ are congruent four-leafed roses; each is traversed for $\theta \in [0, 2\pi]$ and each passes through the pole.

$$r_1 = r_2 \ne 0 \iff \tan 2\theta = 1 \iff 2\theta = k\pi + \tan^{-1} 1 = k\pi + \dfrac{\pi}{4};$$

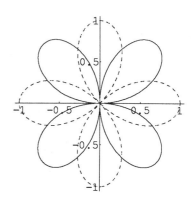

the corresponding points with $0 \le \theta < 2\pi$ are $\left(\dfrac{1}{\sqrt{2}}, \dfrac{\pi}{8}\right)_p$, $\left(\dfrac{-1}{\sqrt{2}}, \dfrac{5\pi}{8}\right)_p$,

$\left(\dfrac{1}{\sqrt{2}}, \dfrac{9\pi}{8}\right)_p$, and $\left(\dfrac{-1}{\sqrt{2}}, \dfrac{13\pi}{8}\right)_p$. If $\sin 2\theta = -\cos 2(\theta + \pi) = -\cos 2\theta$,

$2\theta = k\pi + \tan^{-1}(-1) = k\pi - \dfrac{\pi}{4}$; this yields four more points of in-

tersection: $\left(\dfrac{1}{\sqrt{2}}, \dfrac{3\pi}{8}\right)_p = \left(\dfrac{-1}{\sqrt{2}}, \dfrac{11\pi}{8}\right)_p$, $\left(\dfrac{-1}{\sqrt{2}}, \dfrac{7\pi}{8}\right)_p = \left(\dfrac{1}{\sqrt{2}}, \dfrac{15\pi}{8}\right)_p$,

$\left(\dfrac{1}{\sqrt{2}}, \dfrac{11\pi}{8}\right)_p = \left(\dfrac{-1}{\sqrt{2}}, \dfrac{3\pi}{8}\right)_p$, $\left(\dfrac{-1}{\sqrt{2}}, \dfrac{15\pi}{8}\right)_p = \left(\dfrac{1}{\sqrt{2}}, \dfrac{7\pi}{8}\right)_p$.

75.

The graph of $r = \sin\theta$ is a circle of radius $\frac{1}{2}$ centered at $\left(\frac{1}{2}, \frac{\pi}{2}\right)_p$, the graph of $r = \sec\theta$ is a vertical line through $(1,0)_p$; these graphs have no points in common.

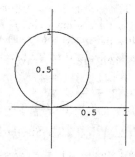

REMARK: $\sin\theta = \pm\sec\theta = \dfrac{\pm 1}{\cos\theta}$ is equivalent to $\sin\theta \cdot \cos\theta = \pm 1$. Because the sine and cosine functions are both bounded by -1 and 1 and because when one of them is ± 1, the other is 0, the equation $\sin\theta \cdot \cos\theta = \pm 1$ has no solutions.

77.

The curves $r_1 = 1+\sin\theta$ and $r_2 = 1+2\cos\theta$ are traversed for $\theta \in [0, 2\pi]$. The cardioid $r_1 = 1+\sin\theta$ passes through the pole as $\left(0, \dfrac{3\pi}{2}\right)_p$; the limaçon $r_2 = 1+2\cos\theta$ passes through the pole as $\left(0, \dfrac{2\pi}{3}\right)_p = \left(0, \dfrac{4\pi}{3}\right)_p$. $r_1 = r_2 \neq 0 \Longleftrightarrow \tan\theta = 2 \Longleftrightarrow \theta = k\pi + \tan^{-1} 2$; the choices for θ in $[0, 2\pi)$ are $\tan^{-1} 2$ and $\pi + \tan^{-1} 2$, the corresponding points of intersection are $\left(1 + \dfrac{2}{\sqrt{5}}, \tan^{-1} 2\right)_p$ and $\left(1 - \dfrac{2}{\sqrt{5}}, \pi + \tan^{-1} 2\right)_p$. Furthermore,

$$
\begin{aligned}
r_1(\theta) = -r_2(\theta + \pi) \neq 0 &\Longleftrightarrow 1 + \sin\theta = -\left(1 + 2\cos(\theta + \pi)\right) = -1 + 2\cos\theta \\
&\Longleftrightarrow 2 = 2\cos\theta - \sin\theta \\
&\Longleftrightarrow \frac{2}{\sqrt{5}} = \frac{2}{\sqrt{5}} \cdot \cos\theta - \frac{1}{\sqrt{5}} \cdot \sin\theta = \cos\left(\theta + \cos^{-1}\frac{2}{\sqrt{5}}\right) \\
&\Longleftrightarrow 2k\pi \pm \cos^{-1}\frac{2}{\sqrt{5}} = \theta + \cos^{-1}\frac{2}{\sqrt{5}}.
\end{aligned}
$$

The solutions for θ in $[0, 2\pi)$ are $\theta = 0$ and $\theta = 2\pi - 2\cos^{-1}\dfrac{2}{\sqrt{5}} = 2\pi - \cos^{-1}\dfrac{3}{5}$, the corresponding points where the curves intersect are $(1,0)_p = (-1, \pi)_p$ and $\left(\dfrac{1}{5}, 2\pi - \cos^{-1}\dfrac{3}{5}\right)_p = \left(\dfrac{-1}{5}, \pi - \cos^{-1}\dfrac{3}{5}\right)_p$.

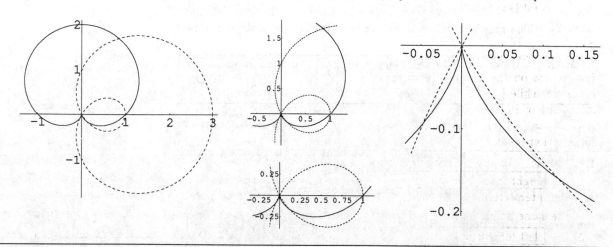

REMARK: Two points of intersection, shown in the right-hand detail figure, lie very close to the pole and are apt to be overlooked if you ignore the facts that $\theta = \tan^{-1} 2$ is not the only solution to $\tan\theta = 2$ and $\theta = 0$ is not the only solution to $\dfrac{2}{\sqrt{5}} = \cos\left(\theta + \cos^{-1}\dfrac{2}{\sqrt{5}}\right)$.

79. The graph of $r = 1$ does not go through the pole. In applying Problem 95, it is convenient to let $f(\theta)$ equal $\tan 2\theta$ because then $g(\theta) = 1 = g(\theta + n\pi)$ implies parts (b) and (c) collapse into one. Each graph is traced-out as θ increases from 0 to 2π (avoiding odd multiples of $\pi/4$ for $\tan 2\theta$).

$$\tan 2\theta = 1 \iff 2\theta = k\pi + \tan^{-1} 1 = k\pi + \frac{\pi}{4};$$ there are four such

choices of θ in $[0, 2\pi)$, the associated points of intersection are $\left(1, \dfrac{k\pi}{8}\right)_p$

for $k \in \{1, 5, 9, 13\}$. Similarly, there are four solutions in $[0, 2\pi)$ of $\tan 2\theta = -g(\theta + n\pi) = -1$; the corresponding intersection points are $\left(-1, \dfrac{k\pi}{8}\right)_p$ for

$k \in \{3, 7, 11, 15\}$.

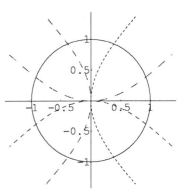

81. The parabola $r_1 = \dfrac{6}{1 - \sin\theta}$ is traversed as θ increases from $\dfrac{\pi}{2}$ to $\dfrac{5\pi}{2}$,
it does not go through the pole; the cardioid $r_2 = 8(1 + \sin\theta)$ is traced-out as θ varies on any interval of length 2π; therefore, only parts (a) and (c) of Problem 95 need to be investigated further.

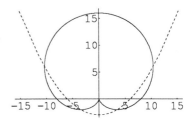

a.
$$r_1(\theta) = r_2(\theta) \iff \frac{6}{1 - \sin\theta} = 8(1 + \sin\theta)$$
$$\iff \frac{6}{8} = 1 - \sin^2\theta = \cos^2\theta$$
$$\iff \cos\theta = \pm\frac{\sqrt{3}}{2};$$

if $\dfrac{\pi}{2} < \theta < \dfrac{5\pi}{2}$, this implies $\theta = \dfrac{k\pi}{6}$ where $k \in \{5, 7, 11, 13\}$. The corresponding points of intersection

are $\left(12, \dfrac{5\pi}{6}\right)_p$, $\left(4, \dfrac{7\pi}{6}\right)_p$, $\left(4, \dfrac{11\pi}{6}\right)_p$, $\left(12, \dfrac{13\pi}{6}\right)_p = \left(12, \dfrac{\pi}{6}\right)_p$.

c. $r_1(\theta) = -r_2(\theta + \pi) \iff \dfrac{6}{1 - \sin\theta} = -8\left(1 + \sin(\theta + \pi)\right) = -8(1 - \sin\theta) \iff \dfrac{-6}{8} = (1 - \sin\theta)^2$;
this equation has no solutions.

83. The graph of $r_1^2 = -\cos\theta$ is a lazy-eight, it coincides with the graph of $r^2 = \cos\theta$ (see Solution 69) except that it is traversed as θ varies from $\dfrac{\pi}{2}$ to $\dfrac{3\pi}{2}$. The graph of $r_2 = \sec\theta$ is the vertical line $x = r\cos\theta = 1$, it too is traversed as θ varies from $\dfrac{\pi}{2}$ to $\dfrac{3\pi}{2}$. Because $\sec\theta$ is never zero, the pole is not on the graph of $r = \sec\theta$. In fact, $|r| \geq 1$ on the graph of $r = \sec\theta$ while $|r| \leq 1$ on the graph of $r^2 = -\cos\theta$. Therefore $|r| = 1$ at any point of intersection. The only such point of the graph of $r = \sec\theta$ is $(1, 0)_p = (-1, \pi)_p$, this point is also on $r^2 = -\cos\theta$ so it is the unique intersection point.

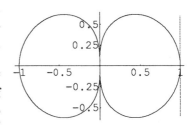

REMARK: To apply Problem 95 we split the analysis of $r^2 = -\cos\theta$ into cases where r is a function of θ.

1. $\boxed{\text{right loop: } r = -\sqrt{-\cos\theta}}$ If $-\sqrt{-\cos\theta} = \sec\theta$, then $\cos^3\theta = -1$ and $\cos\theta = -1$; we have found the intersection point $(-1, \pi)_p$. Because $\sec\theta$ has period 2π, solving $-\sqrt{-\cos\theta} = \sec(\theta + 2n\pi)$ locates the same point; because $\sec(\theta + \pi) = -\sec\theta$, so does solving $-\sqrt{-\cos\theta} = -\sec\left(\theta + (1 + 2n)\pi\right)$.

2. $\boxed{\text{left loop: } r = \sqrt{-\cos\theta}}$ $\sqrt{-\cos\theta}$ is defined if and only if $\cos\theta \leq 0$. If $\cos\theta = 0$, then $\sec\theta$ is undefined. If $\cos\theta < 0$, then $\sec\theta$ is also negative and can not equal the positive square root of anything. Hence there are no solutions to $\sqrt{-\cos\theta} = \sec\theta$. Similarly, neither $\sec(\theta + 2n\pi)$ nor $-\sec\left(\theta + (1 + 2n)\pi\right)$ ever equals $\sqrt{-\cos\theta}$.

85. The graph of $r = \dfrac{1}{2}\tan\theta$ is traced-out as θ varies from $\dfrac{-\pi}{2}$ to $\dfrac{3\pi}{2}$ (skipping $\pi/2$); the circle $r = \dfrac{1}{\sqrt{3}}\sin\theta$
is traversed as θ varies on any interval of length π. Both graphs pass through the pole when $\theta = 0$.

a. $\dfrac{1}{2}\tan\theta = \dfrac{1}{\sqrt{3}}\sin\theta \iff \sin\theta = 0$ or $\cos\theta = \dfrac{\sqrt{3}}{2}$. Each curve goes through the pole when $\sin\theta = 0$.

If $\cos\theta = \dfrac{\sqrt{3}}{2}$ and $\dfrac{-\pi}{2} < \theta < \dfrac{3\pi}{2}$, then $\theta = \pm\dfrac{\pi}{6}$; these solutions correspond to intersection points $\left(\dfrac{1}{2\sqrt{3}}, \dfrac{\pi}{6}\right)_p$ and $\left(\dfrac{-1}{2\sqrt{3}}, \dfrac{-\pi}{6}\right)_p$.

b. $\dfrac{1}{2}\tan\theta = \dfrac{1}{\sqrt{3}}\sin(\theta + 2n\pi) = \dfrac{1}{\sqrt{3}}\sin\theta$; solving this equation locates the points already found.

c. $\dfrac{1}{2}\tan\theta = -\dfrac{1}{\sqrt{3}}\sin(\theta + (1+2n)\pi) = \dfrac{1}{\sqrt{3}}\sin\theta$; we're back to part (a) again.

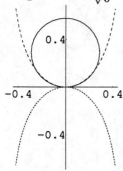

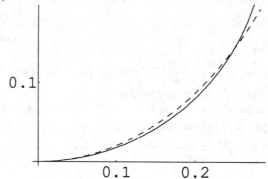

87. $\theta = 2\theta \iff \theta = 0$; the two spirals $r_1 = \theta$ and $r_2 = 2\theta$ do meet at the pole. $\theta = r_1(\theta) = r_2(\theta + 2n\pi) = 2(\theta + 2n\pi) \iff \theta = (-n)4\pi$. If we consider the spirals only for $\theta \geq 0$, we discover the additional points of intersection $(4k\pi, 4k\pi)_p$ where k is a positive integer. Since each of these is on the polar axis, we may summarize the preceding two parts by saying that $(4k\pi, 0)_p$ is a point on both spirals for $k = 0, 1, 2, 3, \ldots$. Because $\theta \geq 0$ implies $r_1 = \theta \geq 0$ and $r_2 = 2\theta \geq 0$, we do not need to apply part (c) of Problem 95.

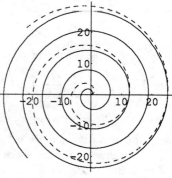

REMARK: If we interchange the two functions and solve $2\theta = \theta + 2n\pi$, we locate $(4n\pi, 2n\pi)_p$ for $n = 0, 1, 2, 3, \ldots$; these are the same points of intersection we've already found.

89. The pole is on the graph of $r = a\cos\theta$ as $\left(0, \dfrac{\pi}{2}\right)_p$, therefore multiplying this equation by r does not introduce a spurious solution:

$$r = a\cos\theta \iff r^2 - a(r\cos\theta) = 0 \iff (x^2 + y^2) - ax = 0 \iff \left(x - \dfrac{a}{2}\right)^2 + y^2 = \left(\dfrac{a}{2}\right)^2.$$

This is an equation for the circle centered at $\left(\dfrac{a}{2}, 0\right)_r$ with radius $\sqrt{\left(\dfrac{a}{2}\right)^2} = \dfrac{|a|}{2}$. ∎

91. The graph of $r = a\cos\theta + b\sin\theta$ passes through $(0, A(-a, b))_p$ where $A(y, x) = \begin{cases} \dfrac{\pi}{2} & \text{if } x = 0, \\ \tan^{-1}\dfrac{y}{x} & \text{if } x \neq 0. \end{cases}$

Therefore multiplying this equation by r does not introduce a spurious solution:

$$r = a\cos\theta + b\sin\theta \iff r^2 = r(a\cos\theta + b\sin\theta) = a(r\cos\theta) + b(r\sin\theta)$$

$$\iff (x^2 + y^2) - ax - by = 0 \iff \left(x - \dfrac{a}{2}\right)^2 + \left(y - \dfrac{b}{2}\right)^2 = \left(\dfrac{a}{2}\right)^2 + \left(\dfrac{b}{2}\right)^2.$$

This is an equation for the circle centered at $\left(\dfrac{a}{2}, \dfrac{b}{2}\right)_r$ with radius $\sqrt{\left(\dfrac{a}{2}\right)^2 + \left(\dfrac{b}{2}\right)^2} = \dfrac{\sqrt{a^2 + b^2}}{2}$. ∎

93. The translation of $r\sin\theta = a$ into rectangular coordinates is $y = a$ since $y = r\sin\theta$; the graph of $y = a$ is the horizontal line passing through $(0, a)_r$. ∎

95. Suppose $a = f(b)$; therefore the point $P = (a, b)_p$ is on the polar graph of $r = f(\theta)$. The full list of polar coordinates for this point P is

a. $(a, b)_p$,

b. $(a, b \pm 2\pi)_p$, $(a, b \pm 4\pi)_p$, $\ldots$, $(a, b \pm 2n\pi)_p$, $\ldots$

c. $(-a, b \pm \pi)_p$, $(-a, b \pm 3\pi)_p$, $\ldots$, $(-a, b \pm (1 + 2n)\pi)_p$, $\ldots$.

The polar graph of $r = g(\theta)$ passes through the point P if and only if there is some choice of θ such that $(g(\theta), \theta)_p$ matches one of those polar coordinates:

a. $\theta = b$ and $g(\theta) = a$,

b.   there is an integer n such that $\theta = b + 2n\pi$ and $a = g(\theta) = g(b + 2n\pi)$,

c.   there is an integer n such that $\theta = b + (1 + 2n)\pi$ and $-a = g(\theta) = g\left(b + (1 + 2n)\pi\right)$. ∎

97. a.   If we assume r is a function of θ, we may also consider $x = r\cos\theta$ and $y = r\sin\theta$ to be functions of θ. Therefore, if we avoid the pole, $r^2 = x^2 + y^2 \implies 2r\dfrac{dr}{d\theta} = 2x\dfrac{dx}{d\theta} + 2y\dfrac{dy}{d\theta} \implies \dfrac{dr}{d\theta} = \cos\theta\dfrac{dx}{d\theta} + \sin\theta\dfrac{dy}{d\theta}$.

In general, the lines satisfying $a_1 x + b_1 y = c_1$ and $a_2 x + b_2 y = c_2$ are perpendicular if and only if $a_1 a_2 + b_1 b_2 = 0$. We get the radial line by choosing $a_1 = \sin\theta$, $b_1 = -\cos\theta$, and $c_1 = 0$; we get a line parallel to the tangent if we choose $a_2 = \dfrac{dy}{d\theta}$ and $b_2 = -\dfrac{dx}{d\theta}$. The radial and tangential lines are perpendicular if and only if $0 = a_1 a_2 + b_1 b_2 = \sin\theta\dfrac{dy}{d\theta} + (-\cos\theta)\left(-\dfrac{dx}{d\theta}\right) = \cos\theta\dfrac{dx}{d\theta} + \sin\theta\dfrac{dy}{d\theta} = \dfrac{dr}{d\theta}$. ∎

b.   Consider a line tangent to the polar graph $r = f(\theta)$. Let α be the angle between the tangent and the polar axis, let β be the angle between the tangent and the radial line; therefore $\alpha = \theta + \beta$, $\tan\alpha = \dfrac{dy}{dx} = \dfrac{dy/d\theta}{dx/d\theta}$ and $\tan\theta = \dfrac{y}{x}$. Hence, if $f'(\theta) \neq 0$, then

$$\tan\beta = \tan(\alpha - \theta) = \frac{\tan\alpha - \tan\theta}{1 + \tan\alpha \cdot \tan\theta} = \frac{\dfrac{dy/d\theta}{dx/d\theta} - \dfrac{y}{x}}{1 + \dfrac{dy/d\theta}{dx/d\theta} \cdot \dfrac{y}{x}} = \frac{x\dfrac{dy}{d\theta} - y\dfrac{dx}{d\theta}}{x\dfrac{dx}{d\theta} + y\dfrac{dy}{d\theta}} = \frac{r^2}{r\dfrac{dr}{d\theta}} = \frac{r}{f'(\theta)}. \quad ∎$$

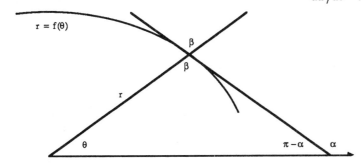

$$\tan\beta = \frac{r\, d\theta}{dr}$$

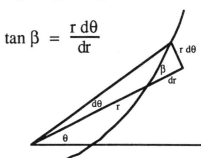

Solutions 8.7   *Areas in polar coordinates*   (pages 592-3)

1.   Area $= \dfrac{1}{2}\displaystyle\int_0^{\pi/2} \theta^2\, d\theta = \dfrac{\theta^3}{6}\Big|_0^{\pi/2} = \dfrac{\pi^3}{48}$.

3.   Area $= \dfrac{1}{2}\displaystyle\int_{-\pi/2}^{\pi/2}(2\cos\theta)^2\, d\theta = \int_{-\pi/2}^{\pi/2}(1 + \cos 2\theta)\, d\theta = \left(\theta + \dfrac{1}{2}\sin 2\theta\right)\Big|_{-\pi/2}^{\pi/2} = \dfrac{\pi}{2} - \dfrac{-\pi}{2} = \pi$.

5.   Area $= \dfrac{1}{2}\displaystyle\int_{-\pi/10}^{\pi/10}(a\cos 5\theta)^2\, d\theta = \dfrac{a^2}{4}\int_{-\pi/10}^{\pi/10}(1 + \cos 10\theta)\, d\theta = \dfrac{a^2}{4}\left(\theta + \dfrac{1}{10}\sin 10\theta\right)\Big|_{-\pi/10}^{\pi/10} = \dfrac{\pi a^2}{20}$.

7.   Area $= \dfrac{1}{2}\displaystyle\int_0^{\pi/2}\left(2\theta^2\right)^2 d\theta = 2\int_0^{\pi/2}\theta^4\, d\theta = \dfrac{2}{5}\theta^5\Big|_0^{\pi/2} = \dfrac{\pi^5}{80}$.

9.   Area $= \dfrac{1}{2}\displaystyle\int_0^{\pi}\left(5\theta^4\right)^2 d\theta = \dfrac{25}{2}\int_0^{\pi}\theta^8\, d\theta = \dfrac{25}{18}\theta^9\Big|_0^{\pi} = \dfrac{25\pi^9}{18}$.

11.   The circle $r = a\sin\theta$ is swept out as θ varies from 0 to π, its area is

$$\dfrac{1}{2}\int_0^{\pi}(a\sin\theta)^2\, d\theta = \dfrac{a^2}{4}\int_0^{\pi}(1 - \cos 2\theta)\, d\theta = \dfrac{a^2}{4}\left(\theta - \dfrac{1}{2}\sin 2\theta\right)\Big|_0^{\pi} = \dfrac{\pi a^2}{4}.$$

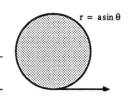

r = a sin θ

REMARK: The area of a circle with radius $\dfrac{a}{2}$ is $\pi\left(\dfrac{a}{2}\right)^2$.

13. The circle $r = a\left(\cos\theta + \sin\theta\right)$ is covered as θ varies from 0 to π, its area is

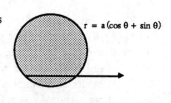

$$\frac{1}{2}\int_0^\pi \left(a\left(\cos\theta + \sin\theta\right)\right)^2 d\theta = \frac{a^2}{2}\int_0^\pi \left(1 + 2\sin\theta\cos\theta\right) d\theta$$
$$= \frac{a^2}{2}\left(\theta + \sin^2\theta\right)\Big|_0^\pi = \frac{\pi a^2}{2}.$$

15. The circle $r = a\left(\sin\theta - \cos\theta\right)$ is covered as θ varies from 0 to π, its area is

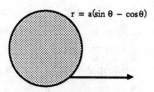

$$\frac{1}{2}\int_0^\pi \left(a\left(\sin\theta - \cos\theta\right)\right)^2 d\theta = \frac{a^2}{2}\int_0^\pi \left(1 - 2\sin\theta\cos\theta\right) d\theta$$
$$= \frac{a^2}{2}\left(\theta + \cos^2\theta\right)\Big|_0^\pi = \frac{\pi a^2}{2}.$$

17. The concave limaçon without a loop $r = 3 + 2\cos\theta$ is traversed as θ increases from 0 to 2π, its area is

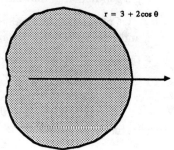

$$\frac{1}{2}\int_0^{2\pi} \left(3 + 2\cos\theta\right)^2 d\theta = \frac{1}{2}\int_0^{2\pi} \left(9 + 12\cos\theta + 4\cos^2\theta\right) d\theta$$
$$= \frac{1}{2}\int_0^{2\pi} \left(11 + 12\cos\theta + 2\cos 2\theta\right) d\theta$$
$$= \frac{1}{2}\left(11\theta + 12\sin\theta + \sin 2\theta\right)\Big|_0^{2\pi} = 11\pi.$$

19. The concave limaçon without a loop $r = 3 - 2\cos\theta$ is symmetric about the polar axis; the top half is traversed as θ increases from 0 to π; the total area is

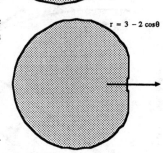

$$2\cdot\left(\frac{1}{2}\right)\int_0^\pi \left(3 - 2\cos\theta\right)^2 d\theta = \int_0^\pi \left(9 - 12\cos\theta + 4\cos^2\theta\right) d\theta$$
$$= \int_0^\pi \left(11 - 12\cos\theta + 2\cos 2\theta\right) d\theta$$
$$= \left(11\theta - 12\sin\theta + \sin 2\theta\right)\Big|_0^\pi = 11\pi.$$

REMARK: Problem 17 can be worked this way, and vice versa.

21. The four-leafed rose $r = a\sin 2\theta$ is traversed as θ increases from 0 to 2π; the area is

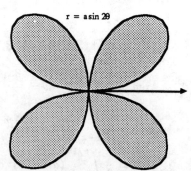

$$\frac{1}{2}\int_0^{2\pi} \left(a\sin 2\theta\right)^2 d\theta = \frac{a^2}{2}\int_0^{2\pi} \sin^2 2\theta\, d\theta$$
$$= \frac{a^2}{4}\int_0^{2\pi} \left(1 - \cos 4\theta\right) d\theta$$
$$= \frac{a^2}{4}\left(\theta - \frac{1}{4}\sin 4\theta\right)\Big|_0^{2\pi} = \frac{\pi a^2}{2}.$$

23. The five-leafed rose $r = a\cos 5\theta$ is traversed as θ increases from 0 to π; the area is

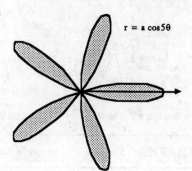

$$\frac{1}{2}\int_0^\pi \left(a\cos 5\theta\right)^2 d\theta = \frac{a^2}{2}\int_0^\pi \cos^2 5\theta\, d\theta$$
$$= \frac{a^2}{4}\int_0^\pi \left(1 + \cos 10\theta\right) d\theta$$
$$= \frac{a^2}{4}\left(\theta + \frac{1}{10}\sin 10\theta\right)\Big|_0^\pi = \frac{\pi a^2}{4}.$$

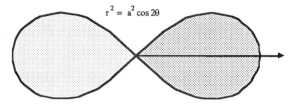

25. The lemniscate $r = a^2 \cos 2\theta$ has two branches, $r = a\sqrt{\cos 2\theta}$ and $r = -a\sqrt{\cos 2\theta}$, each of which is fully covered as θ increases from $-\pi/4$ to $\pi/4$; the area of each branch is $\dfrac{1}{2} \displaystyle\int_{-\pi/4}^{\pi/4} a^2 \cos 2\theta \, d\theta = \dfrac{a^2}{4} \sin 2\theta \Big|_{-\pi/4}^{\pi/4} = \dfrac{a^2}{2}$,

hence the total area is $2 \cdot \dfrac{a^2}{2} = a^2$.

27. The inner loop of the limaçon $r = 1 + 2\cos\theta$ is traversed as θ increases from $2\pi/3$ to $4\pi/3$; the area is

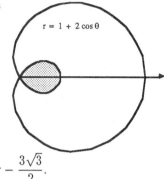

$$\dfrac{1}{2}\int_{2\pi/3}^{4\pi/3} (1 + 2\cos\theta)^2 \, d\theta = \dfrac{1}{2}\int_{2\pi/3}^{4\pi/3} \left(1 + 4\cos\theta + 4\cos^2\theta\right) d\theta$$

$$= \dfrac{1}{2}\int_{2\pi/3}^{4\pi/3} \left(3 + 4\cos\theta + 2\cos 2\theta\right) d\theta$$

$$= \dfrac{1}{2}\left(3\theta + 4\sin\theta + \sin 2\theta\right)\Big|_{2\pi/3}^{4\pi/3}$$

$$= \dfrac{1}{2}\left(4\pi - 2\sqrt{3} + \dfrac{\sqrt{3}}{2}\right) - \dfrac{1}{2}\left(2\pi + 2\sqrt{3} - \dfrac{\sqrt{3}}{2}\right) = \pi - \dfrac{3\sqrt{3}}{2}.$$

29. The bottom half of the inner loop of the limaçon $r = 2 + 3\cos\theta$ is swept out as θ increases from $\cos^{-1}(-2/3)$ to π. Let $\gamma = \cos^{-1}(-2/3)$, then $\sin\gamma = \sqrt{5}/3$ and the total area of the inner loop is

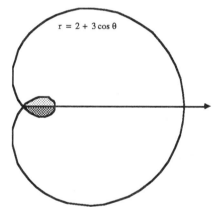

$$2 \cdot \dfrac{1}{2}\int_{\gamma}^{\pi} (2 + 3\cos\theta)^2 \, d\theta = \int_{\gamma}^{\pi} \left(4 + 12\cos\theta + 9\cos^2\theta\right) d\theta$$

$$= \left(4\theta + 12\sin\theta + \dfrac{9}{2}\left(\theta + \sin\theta\cos\theta\right)\right)\Big|_{\gamma}^{\pi}$$

$$= \left(\dfrac{17\theta}{2} + \dfrac{3}{2}(\sin\theta)(3\cos\theta + 8)\right)\Big|_{\gamma}^{\pi}$$

$$= \dfrac{17}{2}\left(\pi - \cos^{-1}\left(\dfrac{-2}{3}\right)\right) - 3\sqrt{5} \approx 0.44.$$

31. The upper half of the outer loop of the limaçon $r = 1 + 2\cos\theta$ is swept out as θ increases from 0 to $2\pi/3$. The area of the whole region, top half plus bottom half, within the outer loop is

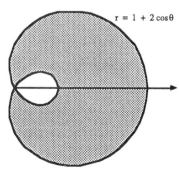

$$2\int_{0}^{2\pi/3} \dfrac{1}{2}(1 + 2\cos\theta)^2 \, d\theta = \int_{0}^{2\pi/3} \left(1 + 4\cos\theta + 4\cos^2\theta\right) d\theta$$

$$= \left(\theta + 4\sin\theta + 2\left(\theta + \sin\theta\cos\theta\right)\right)\Big|_{0}^{2\pi/3}$$

$$= \left(3\theta + 2\sin\theta(2 + \cos\theta)\right)\Big|_{0}^{2\pi/3}$$

$$= 2\pi + \dfrac{3\sqrt{3}}{2}.$$

We have just computed the area of a region which includes (!) the small inner loop; to obtain the area of the region between the two loops, we must subtract the answer to Problem 27 from the value just computed:

$$\text{Area between outer loop and inner loop} = \left(2\pi + \dfrac{3\sqrt{3}}{2}\right) - \left(\pi - \dfrac{3\sqrt{3}}{2}\right) = \pi + 3\sqrt{3}.$$

8.7 — Areas in polar coordinates

33. The upper half of the outer loop of the limaçon $r = 2 + 3\cos\theta$ is swept out as θ increases from 0 to $\cos^{-1}(-2/3)$. The area of the whole region, top half plus bottom half, within the outer loop is

$$2 \int_0^{\cos^{-1}(-2/3)} \frac{1}{2}(2 + 3\cos\theta)^2 \, d\theta$$

$$= \int_0^{\cos^{-1}(-2/3)} \left(4 + 12\cos\theta + 9\cos^2\theta\right) d\theta$$

$$= \left.\left(4\theta + 12\sin\theta + \frac{9}{2}\left(\theta + \sin\theta\cos\theta\right)\right)\right|_0^{\cos^{-1}(-2/3)}$$

$$= \left.\left(\frac{17}{2}\theta + 3\sin\theta\left(\frac{3}{2}\cos\theta + 4\right)\right)\right|_0^{\cos^{-1}(-2/3)}$$

$$= \frac{17}{2}\cos^{-1}\left(\frac{-2}{3}\right) + 3\sqrt{5}.$$

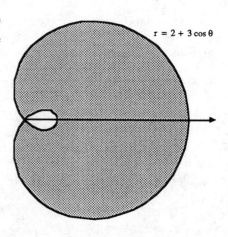

$r = 2 + 3\cos\theta$

We have just computed the area of a region which includes (!) the small inner loop; to obtain the area of the region between the two loops, we must subtract the answer to Problem 29 from the value just computed:

$$\text{Area between outer loop and inner loop} = \left(\frac{17}{2}\cos^{-1}\left(\frac{-2}{3}\right) + 3\sqrt{5}\right) - \left(\frac{17}{2}\left(\pi - \cos^{-1}\left(\frac{-2}{3}\right)\right) - 3\sqrt{5}\right)$$

$$= 17\left(\cos^{-1}\left(\frac{-2}{3}\right) - \frac{\pi}{2}\right) + 6\sqrt{5}.$$

35.

$$\text{Area} = \int_0^\pi \frac{1}{2}\left(a(1 + \sin\theta)\right)^2 d\theta - \int_0^\pi \frac{1}{2}a^2 \, d\theta$$

$$= a^2 \int_0^\pi \left(\sin\theta + \frac{1}{2}\sin^2\theta\right) d\theta$$

$$= \left.a^2\left(-\cos\theta + \frac{1}{4}(\theta - \sin\theta\cos\theta)\right)\right|_0^\pi$$

$$= a^2\left(1 + \frac{\pi}{4}\right) - a^2(-1)$$

$$= \left(2 + \frac{\pi}{4}\right)a^2.$$

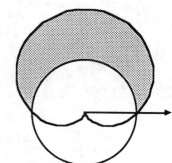

37.

$$\text{Area} = \int_0^{\tan^{-1}(4/3)} \frac{1}{2}(3\sin\theta)^2 \, d\theta + \int_{\tan^{-1}(4/3)}^{\pi/2} \frac{1}{2}(4\cos\theta)^2 \, d\theta$$

$$= \left.\frac{9}{4}(\theta - \sin\theta\cos\theta)\right|_0^{\tan^{-1}(4/3)} + \left.\frac{16}{4}(\theta + \sin\theta\cos\theta)\right|_{\tan^{-1}(4/3)}^{\pi/2}$$

$$= \frac{9}{4}\left[\tan^{-1}\frac{4}{3} - \frac{4}{5}\cdot\frac{3}{5}\right] + 4\left[\frac{\pi}{2} - \left(\tan^{-1}\frac{4}{3} + \frac{4}{5}\cdot\frac{3}{5}\right)\right]$$

$$= 2\pi - 3 - \frac{7}{4}\tan^{-1}\frac{4}{3} \approx 1.6604186757.$$

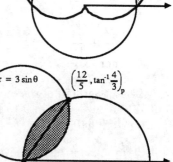

$r = 3\sin\theta$ $\left(\frac{12}{5}, \tan^{-1}\frac{4}{3}\right)_p$

$(0,0)_p = \left(0, \frac{\pi}{2}\right)_p$

$r = 4\cos\theta$

39. The pole lies on the circle $r_1 = \cos\theta$ and the four-leafed rose $r_2 = \cos 2\theta$; the curves meet at three other points: $\cos\theta = r_1 = r_2 = \cos 2\theta = 2\cos^2\theta - 1 \iff 0 = 2\cos^2\theta - \cos\theta - 1 = (2\cos\theta + 1)(\cos\theta - 1) \iff \cos\theta = -1/2$ or $\cos\theta = 1$. The nonpolar intersection points are $(1,0)_p$, $\left(\frac{-1}{2}, \frac{2\pi}{3}\right)_p$, and $\left(\frac{-1}{2}, \frac{4\pi}{3}\right)_p$. The right petal of the rose lies within the circle and is traversed for $\theta \in \left[\frac{-\pi}{4}, \frac{\pi}{4}\right]$; the lower common region is bounded by the circle for $\theta \in \left[\frac{\pi}{2}, \frac{2\pi}{3}\right]$ and by the rose for $\theta \in \left[\frac{2\pi}{3}, \frac{3\pi}{4}\right]$;

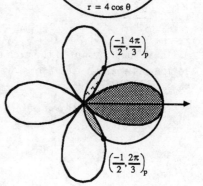

$\left(\frac{-1}{2}, \frac{4\pi}{3}\right)_p$

$\left(\frac{-1}{2}, \frac{2\pi}{3}\right)_p$

the upper common region is bounded by the rose for $\theta \in \left[\dfrac{5\pi}{4}, \dfrac{4\pi}{3}\right]$ and by the circle for $\theta \in \left[\dfrac{4\pi}{3}, \dfrac{3\pi}{2}\right]$. Since each curve is symmetric about the polar axis, so is their common region; the arithmetic is simplified by computing the area of the upper half of the right petal plus all of the lower sliver and then multiplying by 2.

$$\text{Area} = 2\left(\int_0^{\pi/4} \frac{1}{2}\cos^2 2\theta\, d\theta + \int_{\pi/2}^{2\pi/3} \frac{1}{2}\cos^2 \theta\, d\theta + \int_{2\pi/3}^{3\pi/4} \frac{1}{2}\cos^2 2\theta\, d\theta\right)$$

$$= \left(\frac{1}{2}\theta + \frac{1}{8}\sin 4\theta\right)\Big|_0^{\pi/4} + \left(\frac{1}{2}\theta + \frac{1}{4}\sin 2\theta\right)\Big|_{\pi/2}^{2\pi/3} + \left(\frac{1}{2}\theta + \frac{1}{8}\sin 4\theta\right)\Big|_{2\pi/3}^{3\pi/4}$$

$$= \frac{\pi}{8} + \left(\frac{\pi}{12} - \frac{\sqrt{3}}{8}\right) + \left(\frac{\pi}{24} - \frac{\sqrt{3}}{16}\right) = \frac{\pi}{4} - \frac{3\sqrt{3}}{16}.$$

41. The pole lies on the circles $r_1 = a\sin\theta$ and $r_2 = b\cos\theta$; the curves meet at one other point: $a\sin\theta = r_1 = r_2 = b\cos\theta \iff \tan\theta = \dfrac{b}{a}$ or $\cos\theta = 0$, the nonpolar point of intersection is $\left(\dfrac{ab}{\sqrt{a^2+b^2}}, \tan^{-1}\dfrac{b}{a}\right)_p$.

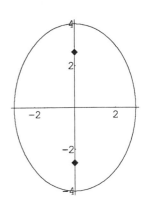

$$\text{Area} = \frac{1}{2}\int_0^{\tan^{-1}(b/a)} (a\sin\theta)^2\, d\theta + \frac{1}{2}\int_{\tan^{-1}(b/a)}^{\pi/2} (b\cos\theta)^2\, d\theta$$

$$= \frac{a^2}{4}(\theta - \sin\theta\cos\theta)\Big|_0^{\tan^{-1}(b/a)} + \frac{b^2}{4}(\theta + \sin\theta\cos\theta)\Big|_{\tan^{-1}(b/a)}^{\pi/2}$$

$$= \frac{a^2}{4}\left(\tan^{-1}\frac{b}{a} - \frac{ab}{a^2+b^2}\right) + \frac{b^2}{4}\left(\frac{\pi}{2} - \tan^{-1}\frac{b}{a} - \frac{ab}{a^2+b^2}\right) = \frac{1}{4}\left(a^2\tan^{-1}\frac{b}{a} + b^2\tan^{-1}\frac{a}{b} - ab\right).$$

REMARK: $\dfrac{\pi}{2} - \tan^{-1}\dfrac{b}{a} = \tan^{-1}\dfrac{a}{b}$ is implied by the result of Problem 6.8.72.

Solutions 8.R   *Review*   (pages 593-4)

1. $1 = \dfrac{x^2}{9} + \dfrac{y^2}{16} = \left(\dfrac{x}{3}\right)^2 + \left(\dfrac{y}{4}\right)^2$. The line segment from $(-3, 0)$ to $(3, 0)$ is the minor axis, the line segment from $(0, -4)$ to $(0, 4)$ is the major axis, the vertices of the ellipse are $(0, -4)$ and $(0, 4)$. The two axes intersect at $(0, 0)$, the center of the ellipse. $c^2 = b^2 - a^2 = 16 - 9 = 7$; the foci are $(0, -\sqrt{7})$ and $(0, \sqrt{7})$; the eccentricity is $\left|\dfrac{c}{b}\right| = \dfrac{\sqrt{7}}{4}$.

3. If $\dfrac{x^2}{9} - \dfrac{y}{16} = 0$, then $x^2 = \left(\dfrac{9}{16}\right)y = 4\left(\dfrac{9}{64}\right)y$. This is an equation for a parabola with vertex $(0, 0)$ and axis on the y-axis, its focus is $\left(0, \dfrac{9}{64}\right)$ and directrix is the line $y = \dfrac{-9}{64}$.

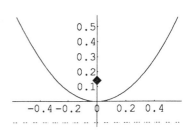

5.
The hyperbola $1 = \dfrac{y^2}{9} - \dfrac{x^2}{16} = \left(\dfrac{y}{3}\right)^2 - \left(\dfrac{x}{4}\right)^2$ is centered at $(0,0)$, its asymptotes are $y = \dfrac{3x}{4}$ and $y = \dfrac{-3x}{4}$, its vertices are $(0,-3)$ and $(0,3)$, those vertices are joined by the transverse axis. $c = \sqrt{9+16} = 5$, the foci are $(0,-5)$ and $(0,5)$; the eccentricity is $\left|\dfrac{c}{b}\right| = \dfrac{5}{3}$.

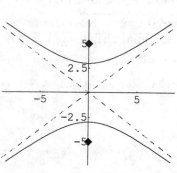

7.
$1 = \left(\dfrac{x-1}{2}\right)^2 + \left(\dfrac{y+1}{3}\right)^2$ is an equation for the ellipse centered at $(1,-1)$ with $a = 2$, $b = 3$, and $c = \sqrt{b^2 - a^2} = \sqrt{5}$. The minor axis is the line segment from $(1-2,-1) = (-1,-1)$ to $(1+2,-1) = (3,-1)$; the major axis is the line segment from $(1,-1-3) = (1,-4)$ to $(1,-1+3) = (1,2)$ and the vertices are its endpoints $(1,-4)$ and $(1,2)$; the foci are $(1,-1-\sqrt{5})$ and $(1,-1+\sqrt{5})$; the eccentricity is $\dfrac{c}{b} = \dfrac{\sqrt{5}}{3}$.

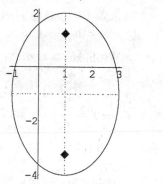

9.
The unique solution to $\left(\dfrac{x+2}{5}\right)^2 + \left(\dfrac{y-5}{5}\right)^2 = 0$ is $x = -2$ and $y = 5$; the graph consists of the single point $(-2,5)$, it can be considered a degenerate circle centered at $(-2,5)$ of radius 0.

11.
$$x^2 + 2x - y^2 + 2y = 0 \iff x^2 + 2x = y^2 - 2y$$
$$\iff x^2 + 2x + 1 = y^2 - 2y + 1$$
$$\iff (x+1)^2 = (y-1)^2$$
$$\iff |x+1| = |y-1|$$
$$\iff y = \begin{cases} -(x+1)+1 = -x \\ (x+1)+1 = x+2. \end{cases}$$

the graph consists of two straight lines, it is a degenerate hyperbola.

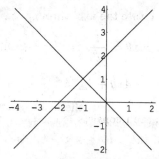

13.
$$\left(4x^2 + 4x\right) + 3\left(y^2 + 8y\right) = 5$$
$$\iff \left(4x^2 + 4x + 1\right) + 3\left(y^2 + 8y + 16\right) = 5 + 1^2 + 3 \cdot 4^2$$
$$\iff (2x+1)^2 + 3(y+4)^2 = 54$$
$$\iff \frac{(x+1/2)^2}{27/2} + \frac{(y+4)^2}{18} = 1.$$

This is an equation for the ellipse centered at $\left(\dfrac{-1}{2}, -4\right)$ with $a = \sqrt{\dfrac{27}{2}} = 3\sqrt{\dfrac{3}{2}}$, $b = \sqrt{18} = 3\sqrt{2}$, and $c = \sqrt{b^2 - a^2} = \dfrac{3}{\sqrt{2}}$. The vertices are $\left(\dfrac{-1}{2}, -4 \pm 3\sqrt{2}\right)$, the major axis is the line segment between them; minor

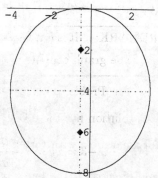

axis is the line segment between $\left(\dfrac{-1}{2} \pm 3\sqrt{\dfrac{3}{2}}, -4\right)$; foci are $\left(\dfrac{-1}{2}, -4 \pm \dfrac{3}{\sqrt{2}}\right)$; eccentricity is $\dfrac{c}{b} = \dfrac{1}{2}$.

15.
An ellipse with foci $(-3,0)$ and $(3,0)$ is centered at $(0,0)$, the midpoint of the line segment between the foci, and has $c^2 = 9 = a^2 - b^2$. If the eccentricity is 0.6, then $0.6 = \dfrac{c}{a} = \dfrac{3}{a}$; therefore $a = \dfrac{3}{0.6} = 5$ and

$b = \sqrt{a^2 - c^2} = \sqrt{25 - 9} = 4$. This ellipse satisfies the equation $1 = \left(\dfrac{x}{5}\right)^2 + \left(\dfrac{y}{4}\right)^2$.

REMARK: See Figure 3 of Section 8.1 for its graph.

17. An hyperbola with foci $(0, -3)$ and $(0, 3)$ is centered at $(0, 0)$ and has $c = 3$. If the vertices are $(0, \pm 2)$, then $b = 2$ and $a = \sqrt{c^2 - b^2} = \sqrt{5}$. We obtain the equation $1 = \dfrac{y^2}{4} - \dfrac{x^2}{5}$.

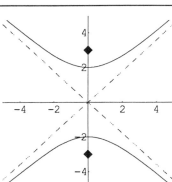

19. Use Problems 8.4.21-22. If the coordinate axes are rotated through angle $\dfrac{1}{2} \cot^{-1}\left(\dfrac{2-2}{1}\right) = \dfrac{\pi}{4}$, then $x = \dfrac{1}{\sqrt{2}}x' - \dfrac{1}{\sqrt{2}}y' = \dfrac{x' - y'}{\sqrt{2}}$, $y = \dfrac{1}{\sqrt{2}}x' + \dfrac{1}{\sqrt{2}}y' = \dfrac{x' + y'}{\sqrt{2}}$, and $10 = 2x^2 + xy + 2y^2 = \left(1 + \dfrac{1}{2} + 1\right)(x')^2 + \left(1 - \dfrac{1}{2} + 1\right)(y')^2 = \dfrac{5}{2}(x')^2 + \dfrac{3}{2}(y')^2$. Therefore, $1 = \dfrac{(x')^2}{4} + \dfrac{(y')^2}{20/3}$; this is an equation for an ellipse centered at $(0,0)$ with $a = 2$, $b = 2\sqrt{\dfrac{5}{3}}$, $c = \sqrt{b^2 - a^2} = 2\sqrt{\dfrac{2}{3}}$ and eccentricity $\dfrac{c}{b} = \sqrt{\dfrac{2}{5}}$.

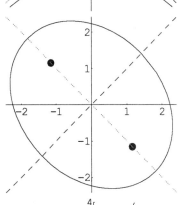

21. Rotate the axes through angle $\theta = \dfrac{1}{2}\cot^{-1}\left(\dfrac{1-4}{4}\right) = \dfrac{1}{2}\cos^{-1}\left(\dfrac{-3}{5}\right)$. Then $\cos\theta = \dfrac{1}{\sqrt{5}}$, $\sin\theta = \dfrac{2}{\sqrt{5}}$, and

$$9 = x^2 + 4xy + 4y^2 = \left(\dfrac{1}{5} + \dfrac{8}{5} + \dfrac{16}{5}\right)(x')^2 + \left(\dfrac{4}{5} - \dfrac{8}{5} + \dfrac{4}{5}\right)(y')^2 = 5(x')^2.$$

The graph consists of the two straight lines $x' = \dfrac{-3}{\sqrt{5}}$ and $x' = \dfrac{3}{\sqrt{5}}$.

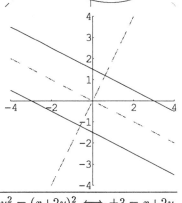

REMARK: It is not necessary to rotate the axes to observe $9 = x^2 + 4xy + 4y^2 = (x + 2y)^2 \iff \pm 3 = x + 2y$. The graph consists of the parallel lines $y = \dfrac{-1}{2}(x + 3)$ and $y = \dfrac{-1}{2}(x - 3)$.

23. Use Problem 8.4.22; since the text answer uses rotation through $\dfrac{\pi}{4}$, this solution has the details for the other choice: $\theta = \dfrac{-\pi}{4}$. This yields

$$-3 = xy = \left(\dfrac{x' + y'}{\sqrt{2}}\right)\left(\dfrac{-x' + y'}{\sqrt{2}}\right) = \dfrac{-(x')^2 + (y')^2}{2};$$

therefore $1 = \left(\dfrac{x'}{\sqrt{6}}\right)^2 - \left(\dfrac{y'}{\sqrt{6}}\right)^2$. This is an equation for the hyperbola with $a = \sqrt{6} = b$, $c = \sqrt{12} = 2\sqrt{3}$, eccentricity $\dfrac{c}{a} = \sqrt{2}$, and asymptotes $y' = \pm x'$.

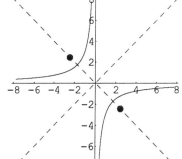

(x', y') coordinates (x, y) coordinates

center	$(0,0)'$	$(0,0)$
foci	$(\pm 2\sqrt{3},0)'$	$(-\sqrt{6},\sqrt{6}), (\sqrt{6},-\sqrt{6})$
vertices	$(\pm\sqrt{6},0)'$	$(-\sqrt{3},\sqrt{3}), (\sqrt{3},-\sqrt{3})$
asymptotes	$y'=\pm x'$	x-axis, y-axis

25.   The discriminant of $-2x^2+3xy+4y^2$ is $3^2-4\cdot(-2)\cdot 4=41>0$; Theorem 8.4.1 implies the graph is an hyperbola. Rotate the axes through angle $\theta=\dfrac{1}{2}\cot^{-1}\left(\dfrac{-2-4}{3}\right)\approx 76.7°$. Then, using the Remark following Solution 8.4.9, $W=\sqrt{3^2+(-2-4)^2}=3\sqrt{5}$, $S=\operatorname{sgn}(3)=1$, and

$$5=-2x^2+3xy+4y^2=\left(\frac{-2+4+3\sqrt{5}}{2}\right)(x')^2+\left(\frac{-2+4-3\sqrt{5}}{2}\right)(y')^2.$$

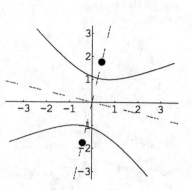

The graph is an hyperbola centered at $(0,0)'$ with vertices $(\pm a,0)'$ where $a^2=\dfrac{10}{2+3\sqrt{5}}\approx 1.07^2$, foci $(\pm c,0)'$ where $c^2=\dfrac{60\sqrt{5}}{41}\approx 1.81^2$, eccentricity $\dfrac{c}{a}=\sqrt{\dfrac{6\sqrt{5}}{3\sqrt{5}-2}}\approx 1.69$, and asymptotes $y'=\pm\sqrt{\dfrac{3\sqrt{5}-2}{3\sqrt{5}+2}}\,x'\approx\pm 1.36x'$.

27. $(2,0)_p=(2\cos 0, 2\sin 0)_r=(2,0)_r$.

29. $\left(-7,\dfrac{\pi}{2}\right)_p=\left(-7\cos\dfrac{\pi}{2}, -7\sin\dfrac{\pi}{2}\right)_r=(0,-7)_r$.

31. $\left(-1,\dfrac{-\pi}{2}\right)_p=\left(-\cos\dfrac{-\pi}{2}, -\sin\dfrac{-\pi}{2}\right)_r=(-0,-(-1))_r=(0,1)_r$.

33.   If $(x,y)_r=(2,0)_r$, then the point is on the positive x-axis which is also the polar axis. $r=\sqrt{2^2+0^2}=2$ and $\theta=\tan^{-1}\dfrac{0}{2}=0$. Thus $(r,\theta)_p=(2,0)_p$.

35. $\left(\sqrt{3},-1\right)_r=\left(\sqrt{(\sqrt{3})^2+(-1)^2}, 2\pi+\tan^{-1}\dfrac{-1}{\sqrt{3}}\right)_p=\left(2,\dfrac{11\pi}{6}\right)_p$.

37. $(-6,-6)_r=\left(\sqrt{(-6)^2+(-6)^2}, \pi+\tan^{-1}\dfrac{-6}{-6}\right)_p=\left(6\sqrt{2},\dfrac{5\pi}{4}\right)_p$.

39.   Each point with $r=8$ is on the circle of radius 8 centered at the pole. Any circle centered at the pole is symmetric about the pole and about any line through the pole.

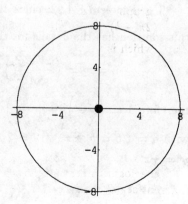

41. $r = 2\cos\theta \iff r^2 = r(2\cos\theta) = 2(r\cos\theta) \iff x^2 + y^2 = 2x \iff$ $(x-1)^2 + y^2 = 1$; this is an equation for the circle of radius 1 centered at $(1,0)_r = (1,0)_p$. This circle is symmetric about the polar axis.

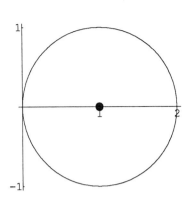

43. The graph of $r = 3 - 2\sin\theta$ is a limaçon without a loop; it is traversed for $\theta \in [0, 2\pi]$; it is symmetric about the vertical line $\theta = \dfrac{\pi}{2}$.

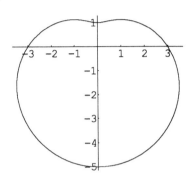

45. The graph of $r = 5\cos 2\theta$ is a four-leafed rose; it is traversed for $\theta \in [0, 2\pi]$; it is symmetric about the polar axis, the vertical line $\theta = \dfrac{\pi}{2}$, and the pole because each of $5\cos 2(-\theta)$, $5\cos 2(\pi - \theta)$, and $5\cos 2(\theta + \pi)$ equals $5\cos 2\theta$.

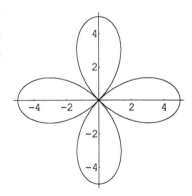

47. The eight-leafed rose $r = 3\sin 4\theta$ is symmetric about the pole because $3\sin 4(\theta + \pi) = 3\sin 4\theta$. If $a = 3\sin 4b$, then $3\sin 4(\pi - b) = -3\sin 4b = -a$. Because $(-a, \pi - b)_p = (a, -b)_p$, this means $(a, b)_p$ and $(a, -b)_p$ are both on the graph, hence the graph is symmetric about the polar axis. Similarly, $3\sin 4(2\pi - b) = -a$, the point $(-a, 2\pi - b)_p = (a, \pi - b)_p$ is also on the graph which is thus symmetric about the vertical line $\theta = \dfrac{\pi}{2}$.

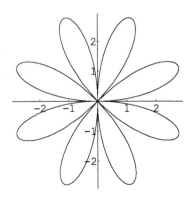

49. The logarithmic spiral $r = e^{2\theta}$ is not periodic and has no symmetries. The four views below have different scales, they correspond to θ in $[-\pi/4, \pi/4]$, $[-\pi/3, \pi/3]$, $[-\pi/2, \pi/2]$, and $[-2\pi/3, 2\pi/3]$.

$r = e^{2\theta},\ |\theta| \leq \pi/4$

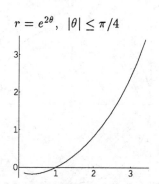

$r = e^{2\theta},\ |\theta| \leq \pi/3$

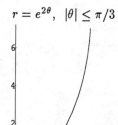

$r = e^{2\theta},\ |\theta| \leq \pi/3$

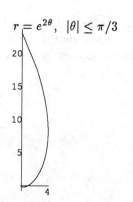

$r = e^{2\theta},\ |\theta| \leq 2\pi/3$

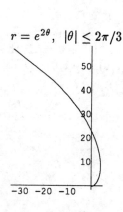

51.   The graph of $r \sin \theta = 4$ is the horizontal line $y = 4$; the curve is traced out for $\theta \in (0, \pi)$, it is symmetric about the vertical line $\theta = \dfrac{\pi}{2}$.

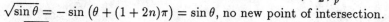

53. $r = 3\sin\theta + 4\cos\theta \iff r^2 = r(3\sin\theta + 4\cos\theta) = 3(r\sin\theta) + 4(r\cos\theta)$
$$\iff x^2 + y^2 = 3y + 4x$$
$$\iff (x-2)^2 + \left(y - \frac{3}{2}\right)^2 = (-2)^2 + \left(\frac{-3}{2}\right)^2 = \frac{25}{4} = \left(\frac{5}{2}\right)^2,$$
we have an equation for the circle of radius $\dfrac{5}{2}$ centered at the point $\left(2, \dfrac{3}{2}\right)_r$.

55.   The circle $r = \cos\theta$ is traced-out as θ increases from 0 to π; it meets the circle $r = 1$ only at the point $(1,0)_p = (-1, \pi)_p$ because $-1 \leq \cos\theta \leq 1$ for all θ.

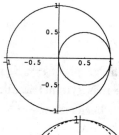

57.   There are two functions $r = \sqrt{\sin\theta}$ and $r = -\sqrt{\sin\theta}$ which generate the graph of $r^2 = \sin\theta$ as θ increases from 0 to π; the circle $r = \sin\theta$ is also traced-out for $\theta \in [0, \pi]$.

$\sqrt{\sin\theta} = \sin(\theta + 2n\pi) = \sin\theta \iff \sin\theta \geq 0$ and $\sin\theta = \sin^2\theta \iff \sin\theta = 0$ or $\sin\theta = 1$, the solution $\sin\theta = 0$ corresponds to the pole $(0,0)_p$ as a point of intersection and the solution $\sin\theta = 1$ corresponds to $\left(1, \dfrac{\pi}{2}\right)_p$ as another.

$\sqrt{\sin\theta} = -\sin\left(\theta + (1+2n)\pi\right) = \sin\theta$, no new point of intersection.
$-\sqrt{\sin\theta} = \sin(\theta + 2n\pi) = \sin\theta \iff \sin\theta = 0$, the pole is found again.
$-\sqrt{\sin\theta} = -\sin\left(\theta + (1+2n)\pi\right) = \sin\theta \iff \sin\theta = 0$ and we find the pole yet one last time.

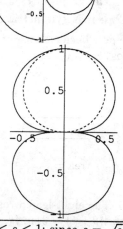

REMARK: The circle is inside the upper loop of $r^2 = \sin\theta$ because $s \leq \sqrt{s}$ if $0 \leq s \leq 1$; since $s = \sqrt{s}$ if and only if $s = 0$ or $s = 1$, the only points where the curves meet are the pole $(0,0)_p$ and $\left(1, \dfrac{\pi}{2}\right)_p$.

59.   The pole is on both of the spirals $r_1 = \theta$ and $r_2^2 = 9\theta$. Since the problem restricts $r \geq 0$, the second spiral is described by $r_2 = 3\sqrt{\theta}$ (which implicitly incorporates the other restriction, $\theta \geq 0$).

$\theta = r_1(\theta) = r_2(\theta + 2n\pi) = 3\sqrt{\theta + 2n\pi}$
$\iff \theta^2 = 9(\theta + 2n\pi)$ and $\theta \geq 0$
$\iff \left(\theta - \dfrac{9}{2}\right)^2 = \dfrac{81}{4} + 18n\pi = \left(\dfrac{9}{2}\right)^2\left(1 + \dfrac{8n\pi}{9}\right)$ and $\theta \geq 0$
$\iff \theta = \dfrac{9}{2}\left(1 + \sqrt{1 + \dfrac{8n\pi}{9}}\right)$.

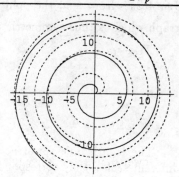

The point $\left(\dfrac{9}{2}\left(1+\sqrt{1+\dfrac{8n\pi}{9}}\right),\dfrac{9}{2}\left(1+\sqrt{1+\dfrac{8n\pi}{9}}\right)\right)_p$ lies on both spirals for each non-negative integer n. For $n=0$, we obtain the point $(9,9)_p$; for $n=1$, the point is approximately $(13.26,13.26)_p$.

61. The pole is not on the circle $r=1$.

$$\sqrt{3}\tan 2\theta = 1 \iff 2\theta = \tan^{-1}\frac{1}{\sqrt{3}}+k\pi \iff \theta = \frac{\pi}{12}+\frac{k\pi}{2};$$

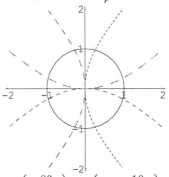

this locates four distinct points of intersection: $\left(1,\dfrac{\pi}{12}\right)_p$, $\left(1,\dfrac{7\pi}{12}\right)_p$,

$\left(1,\dfrac{13\pi}{12}\right)_p$, $\left(1,\dfrac{19\pi}{12}\right)_p$. $\sqrt{3}\tan 2\theta = -1 \iff 2\theta = \tan^{-1}\dfrac{-1}{\sqrt{3}}+k\pi \iff$

$\theta = \dfrac{-\pi}{12}+\dfrac{k\pi}{2}$; this too locates four distinct points of intersection:

$\left(-1,\dfrac{-\pi}{12}\right)_p = \left(1,\dfrac{11\pi}{12}\right)_p$, $\left(-1,\dfrac{-7\pi}{12}\right)_p = \left(1,\dfrac{5\pi}{12}\right)_p$, $\left(-1,\dfrac{-13\pi}{12}\right)_p = \left(1,\dfrac{23\pi}{12}\right)_p$, $\left(-1,\dfrac{-19\pi}{12}\right)_p = \left(1,\dfrac{17\pi}{12}\right)_p$.

63.
$$\text{Area} = \int_{\pi/6}^{\pi/2}\frac{1}{2}\theta^2\,d\theta = \frac{\theta^3}{6}\Big|_{\pi/6}^{\pi/2} = \frac{\pi^3}{6}\left(\frac{1}{8}-\frac{1}{216}\right) = \frac{13\pi^3}{648}.$$

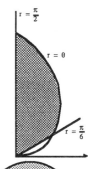

65.
$$\text{Area} = \int_0^\pi \frac{1}{2}\left(a(\sin\theta+\cos\theta)\right)^2 d\theta = \frac{a^2}{2}\int_0^\pi (1+2\sin\theta\cos\theta)\,d\theta$$
$$= \frac{a^2}{2}\left(\theta+\sin^2\theta\right)\Big|_0^\pi = \frac{a^2}{2}\pi.$$

REMARK: See Problem 8.6.91; we've computed the area of a circle with radius $a/\sqrt{2}$.

67.
$$\text{Area} = \int_0^{2\pi}\frac{1}{2}(-2-2\sin\theta)^2\,d\theta = \int_0^{2\pi}\left(2+4\sin\theta+2\sin^2\theta\right)d\theta$$
$$= \left(2\theta-4\cos\theta+(\theta-\sin\theta\cos\theta)\right)\Big|_0^{2\pi}$$
$$= (6\pi-4)-(-4) = 6\pi.$$

69. The upper half of the outer loop of the limaçon $r=3-4\cos\theta$ is traversed as θ increases from $\cos^{-1}\dfrac{3}{4}$ to π; the total area within the outer loop is

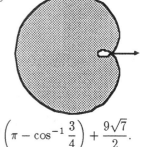

$$2\cdot\frac{1}{2}\int_{\cos^{-1}(3/4)}^{\pi}(3-4\cos\theta)^2\,d\theta = \int_{\cos^{-1}(3/4)}^{\pi}\left(9-24\cos\theta+16\cos^2\theta\right)d\theta$$
$$= \left(9\theta-24\sin\theta+8\left(\theta+\sin\theta\cos\theta\right)\right)\Big|_{\cos^{-1}(3/4)}^{\pi}$$
$$= \left(17\theta+8\sin\theta\left(\cos\theta-3\right)\right)\Big|_{\cos^{-1}(3/4)}^{\pi}$$
$$= 17\left(\pi-\cos^{-1}\frac{3}{4}\right)+8\left(0-\frac{\sqrt{7}}{4}\left(\frac{3}{4}-3\right)\right) = 17\left(\pi-\cos^{-1}\frac{3}{4}\right)+\frac{9\sqrt{7}}{2}.$$

The lower half of the inner loop is covered as θ increases from 0 to $\cos^{-1}\dfrac{3}{4}$; the total area of the inner loop is $2 \cdot \dfrac{1}{2} \displaystyle\int_0^{\cos^{-1}(3/4)} (3 - 4\cos\theta)^2 \, d\theta = \left(17\theta + 8\sin\theta \, (\cos\theta - 3)\right)\Big|_0^{\cos^{-1}(3/4)} = 17\cos^{-1}\dfrac{3}{4} - \dfrac{9\sqrt{7}}{2}$. The area of

the region between the two loops is the difference of these two areas, it equals $17\left(\pi - 2\cos^{-1}\dfrac{3}{4}\right) + 9\sqrt{7}$.

71.   Both of the circles $r = 2\sin\theta$ and $r = 2\cos\theta$ have radius 1; the first
is centered at $\left(1, \dfrac{\pi}{2}\right)_p$ and the second at $(1,0)_p$. They meet at the pole
and at $\left(\sqrt{2}, \dfrac{\pi}{4}\right)_p$. Their common region is bounded below by $r = 2\sin\theta$,
$0 \le \theta \le \pi/4$, and above by $r = 2\cos\theta$, $\pi/4 \le \theta \le \pi/2$; its area is

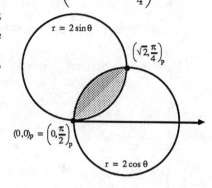

$$\dfrac{1}{2}\int_0^{\pi/4} (2\sin\theta)^2 \, d\theta + \dfrac{1}{2}\int_{\pi/4}^{\pi/2} (2\cos\theta)^2 \, d\theta$$

$$= (\theta - \sin\theta\cos\theta)\Big|_0^{\pi/4} + (\theta + \sin\theta\cos\theta)\Big|_{\pi/4}^{\pi/2}$$

$$= \left(\dfrac{\pi}{4} - \left(\dfrac{1}{\sqrt{2}}\right)^2\right) + \left(\dfrac{\pi}{4} - \left(\dfrac{1}{\sqrt{2}}\right)^2\right) = \dfrac{\pi}{2} - 1.$$

REMARK: $\sin\left(\dfrac{\pi}{2} - \theta\right) = \cos\theta$ and $\cos\left(\dfrac{\pi}{2} - \theta\right) = \sin\theta$, therefore reflection through the line $\theta = \dfrac{\pi}{4}$ carries
one circle into the other. This means their common region is symmetric about that line and its area is

$$2 \cdot \dfrac{1}{2}\int_0^{\pi/4} (2\sin\theta)^2 \, d\theta.$$

---

73.   The smallest positive solution to $\sin 10\theta = 0$ is $\theta = \dfrac{\pi}{10}$. Therefore, as
θ increases from 0 to $\dfrac{\pi}{10}$, one petal of the twenty-leafed rose $r = 4\sin 10\theta$ is
covered. The area of a single petal is

$$\dfrac{1}{2}\int_0^{\pi/10} (4\sin 10\theta)^2 \, d\theta = 8\int_0^{\pi/10} \sin^2 10\theta \, d\theta = 4\int_0^{\pi/10} (1 - \cos 20\theta) \, d\theta$$

$$= \left(4\theta - \dfrac{1}{5}\sin 20\theta\right)\Big|_0^{\pi/10} = \dfrac{2\pi}{5}.$$

REMARK: Compare this value to the answer for Problem 8.7.40 with $n = 10$.

Solutions 8.C   *Computer Exercises*   (page 594)

1.   If $\theta = (1/2)\cot^{-1}(2/3)$, then the rotation $x = x'\cos\theta - y'\sin\theta$ and $y = x'\sin\theta + y'\cos\theta$ transforms
$3x^2 + 3xy + y^2 = 1$ into $\left(2 + \sqrt{13}/2\right)x'^2 + \left(2 - \sqrt{13}/2\right)y'^2 = 1$. The curve is an ellipse because the discriminant,
$b^2 - 4ac = 3^2 - 4(3)(1) = -3$ is negative. The (x', y') parameterization $x' = \cos t \big/ \sqrt{2 + \sqrt{13}/2}$ and $y' = \sin t \big/ \sqrt{2 - \sqrt{13}/2}$ combines with our rotation to yield

$$x = \dfrac{\cos\theta}{\sqrt{2 + \sqrt{13}/2}}\cos t - \dfrac{\sin\theta}{\sqrt{2 - \sqrt{13}/2}}\sin t \approx 0.4521244333 \cos t - 1.0625049789 \sin t$$

$$y = \dfrac{\sin\theta}{\sqrt{2 + \sqrt{13}/2}}\cos t + \dfrac{\cos\theta}{\sqrt{2 - \sqrt{13}/2}}\sin t \approx 0.2419696535 \cos t + 1.9853087132 \sin t$$

which we use for plotting.

1: $3x^2 + 3xy + y^2 = 1$

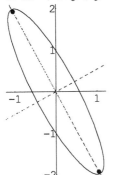

3: $r = 1 + \sec\theta$, $r = 3\cos\theta$

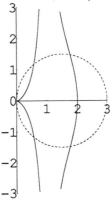

3.   The circle $r = 3\cos\theta$ is traversed for θ in an interval of length π; the conchoid $r = 1+\sec\theta$ meets the circle at points symmetric about the polar axis with $|\theta| \approx .5$. Newton's method applied to $f(\theta) = (1+\sec\theta)-3\cos\theta$ with $\theta_0 = .5$ converges to 0.695721. The region inside the circle and to the right of the conchoid has area

$$\int_{-0.695721}^{0.695721} \frac{(3\cos\theta)^2}{2}\, d\theta - \int_{-0.695721}^{0.695721} \frac{(1+\sec\theta)^2}{2}\, d\theta \approx 5.344652 - 3.050252 \approx 2.294399.$$

9
Indeterminate forms, Improper integrals, and Taylor's Theorem

Solutions 9.1 *The indeterminate form $\frac{0}{0}$ and L'Hôpital's rule* (pages 601-2)

1. $\dfrac{x + \sin 5x}{x - \sin 5x}$ has the indeterminate form $\dfrac{0}{0}$ at $x = 0$ and

$$\lim_{x \to 0} \frac{\frac{d}{dx}(x + \sin 5x)}{\frac{d}{dx}(x - \sin 5x)} = \lim_{x \to 0} \frac{1 + 5\cos 5x}{1 - 5\cos 5x} = \frac{1 + 5}{1 - 5} = \frac{-3}{2}.$$

Therefore $\lim\limits_{x \to 0} \dfrac{x + \sin 5x}{x - \sin 5x}$ exists and it equals $\dfrac{-3}{2}$.

REMARK: Alternatively, divide numerator and denominator by x, then use Theorem 2.5.1:

$$\frac{x + \sin 5x}{x - \sin 5x} = \frac{1 + \frac{\sin 5x}{x}}{1 - \frac{\sin 5x}{x}} = \frac{1 + 5\frac{\sin 5x}{5x}}{1 - 5\frac{\sin 5x}{5x}} \text{ and } \lim_{u \to 0} \frac{1 + 5\frac{\sin u}{u}}{1 - 5\frac{\sin u}{u}} = \frac{1 + 5\lim_{u \to 0}\frac{\sin u}{u}}{1 - 5\lim_{u \to 0}\frac{\sin u}{u}} = \frac{1 + 5}{1 - 5}.$$

3. $\dfrac{\cos 3x - 1}{\sin 5x}$ has the indeterminate form $\dfrac{0}{0}$ at $x = 0$ and $\lim\limits_{x \to 0} \dfrac{\frac{d}{dx}(\cos 3x - 1)}{\frac{d}{dx}(\sin 5x)} = \lim\limits_{x \to 0} \dfrac{-3\sin 3x}{5\cos 5x} = \dfrac{0}{5} = 0$.

Therefore $\lim\limits_{x \to 0} \dfrac{\cos 3x - 1}{\sin 5x}$ exists and it equals 0.

5. $\dfrac{\sin x}{\sqrt{x}}$ has indeterminate form $\dfrac{0}{0}$ at $x = 0^+$ and $\lim\limits_{x \to 0^+} \dfrac{\frac{d}{dx}(\sin x)}{\frac{d}{dx}(\sqrt{x})} = \lim\limits_{x \to 0^+} \dfrac{\cos x}{1/(2\sqrt{x})} = \lim\limits_{x \to 0^+} 2\sqrt{x}\cos x = 0$.

Therefore $\lim\limits_{x \to 0^+} \dfrac{\sin x}{\sqrt{x}}$ exists and it equals 0.

7. $\dfrac{x^4 - x^3 + x^2 - 1}{x^3 - x^2 + x - 1}$ has indeterminate form $\dfrac{0}{0}$ at $x = 1$ and

$$\lim_{x \to 1} \frac{\frac{d}{dx}(x^4 - x^3 + x^2 - 1)}{\frac{d}{dx}(x^3 - x^2 + x - 1)} = \lim_{x \to 1} \frac{4x^3 - 3x^2 + 2x}{3x^2 - 2x + 1} = \frac{4 - 3 + 2}{3 - 2 + 1} = \frac{3}{2}.$$

Therefore $\lim\limits_{x \to 1} \dfrac{x^4 - x^3 + x^2 - 1}{x^3 - x^2 + x - 1} = \dfrac{3}{2}$.

REMARK: $\lim\limits_{x \to 1} \dfrac{x^4 - x^3 + x^2 - 1}{x^3 - x^2 + x - 1} = \lim\limits_{x \to 1} \dfrac{(x - 1)(x^3 + x + 1)}{(x - 1)(x^2 + 1)} = \lim\limits_{x \to 1} \dfrac{x^3 + x + 1}{x^2 + 1} = \dfrac{1 + 1 + 1}{1 + 1}$.

9. $\sqrt{x_0}$ is undefined if $x_0 < 0$; if $x_0 = 0$, then every neighborhood of x_0 includes negative numbers at which $\sqrt{x}$ would be undefined; thus the limit can not exist if $x_0 \leq 0$, the following work presumes $x_0 > 0$. $\dfrac{\sqrt{x} - \sqrt{x_0}}{x - x_0}$ has indeterminate form $\dfrac{0}{0}$ at $x = x_0$ and $\lim\limits_{x \to x_0} \dfrac{\frac{d}{dx}(\sqrt{x} - \sqrt{x_0})}{\frac{d}{dx}(x - x_0)} = \lim\limits_{x \to x_0} \dfrac{\frac{1}{2}x^{-1/2}}{1} = \dfrac{1}{2\sqrt{x_0}}$.

Therefore $\lim\limits_{x \to x_0} \dfrac{\sqrt{x} - \sqrt{x_0}}{x - x_0} = \dfrac{1}{2\sqrt{x_0}}$.

REMARK: See the algebra in Example 1.5.5 for another approach to computing this limit.

11. $\dfrac{x^2}{\sin^{-1} x}$ has indeterminate form $\dfrac{0}{0}$ at $x = 0$ and $\displaystyle\lim_{x \to 0} \dfrac{\frac{d}{dx}\left(x^2\right)}{\frac{d}{dx}\left(\sin^{-1} x\right)} = \lim_{x \to 0} \dfrac{2x}{1/\sqrt{1-x^2}} = \lim_{x \to 0} 2x\sqrt{1-x^2} =$

0, hence $\displaystyle\lim_{x \to 0} \dfrac{x^2}{\sin^{-1} x} = 0$.

13. We use L'Hôpital's rule twice. $\dfrac{e^x - 1}{x^2} - \dfrac{1}{x} = \dfrac{e^x - 1 - x}{x^2}$ has indeterminate form $\dfrac{0}{0}$ as $x \to 0^+$ and so

does $\dfrac{\frac{d}{dx}\left(e^x - 1 - x\right)}{\frac{d}{dx}\left(x^2\right)} = \dfrac{e^x - 1}{2x}$. But, persisting, $\displaystyle\lim_{x \to 0^+} \dfrac{\frac{d}{dx}\left(e^x - 1\right)}{\frac{d}{dx}\left(2x\right)} = \lim_{x \to 0^+} \dfrac{e^x}{2} = \dfrac{1}{2}$, hence $\displaystyle\lim_{x \to 0^+} \dfrac{e^x - 1}{2x}$ exists

and equals $\dfrac{1}{2}$ and $\displaystyle\lim_{x \to 0^+} \left(\dfrac{e^x - 1}{x^2} - \dfrac{1}{x}\right) = \lim_{x \to 0^+} \dfrac{e^x - 1 - x}{x^2} = \lim_{x \to 0^+} \dfrac{e^x - 1}{2x} = \dfrac{1}{2}$.

REMARK: Sections 9.4-5 will give us more tools to examine a relation of the form $e^x = 1 + x + \dfrac{x^2}{2} + r(x)$

where $\displaystyle\lim_{x \to 0} \dfrac{r(x)}{x^2} = 0$; also see Example 10.10.1.

15. $\dfrac{e^x - 1}{x(3 + x)}$ has indeterminate form $\dfrac{0}{0}$ at $x = 0$ and $\displaystyle\lim_{x \to 0} \dfrac{\frac{d}{dx}\left(e^x - 1\right)}{\frac{d}{dx}\left(x(3+x)\right)} = \lim_{x \to 0} \dfrac{\frac{d}{dx}\left(e^x - 1\right)}{\frac{d}{dx}\left(3x + x^2\right)} = \lim_{x \to 0} \dfrac{e^x}{3 + 2x} =$

$\dfrac{1}{3}$, thus $\displaystyle\lim_{x \to 0} \dfrac{e^x - 1}{x(3 + x)} = \dfrac{1}{3}$.

17. $x \sin\left(\dfrac{1}{x}\right) = \dfrac{\sin(1/x)}{1/x}$ has indeterminate form $\dfrac{0}{0}$ as $x \to \infty$ and

$$\lim_{x \to \infty} \dfrac{\frac{d}{dx}\left(\sin(1/x)\right)}{\frac{d}{dx}\left(1/x\right)} = \lim_{x \to \infty} \dfrac{\cos(1/x) \cdot (-1/x^2)}{-1/x^2} = \lim_{x \to \infty} \cos\left(\dfrac{1}{x}\right) = \cos\left(\lim_{x \to \infty} \dfrac{1}{x}\right) = \cos 0 = 1,$$

thus $\displaystyle\lim_{x \to \infty} x \sin(1/x) = 1$.

REMARK: Let $u = \dfrac{1}{x}$, then $\displaystyle\lim_{x \to \infty} x \sin\left(\dfrac{1}{x}\right) = \lim_{u \to 0^+} \dfrac{\sin u}{u} = 1$.

19. $\dfrac{x^2 - 1}{\sqrt{1 - x}}$ has indeterminate form $\dfrac{0}{0}$ as $x \to 1^-$ and

$$\lim_{x \to 1^-} \dfrac{\frac{d}{dx}\left(x^2 - 1\right)}{\frac{d}{dx}\sqrt{1 - x}} = \lim_{x \to 1^-} \dfrac{2}{-1/\left(2\sqrt{1-x}\right)} = \lim_{x \to 1^-} -4x\sqrt{1 - x} = 0,$$

thus $\displaystyle\lim_{x \to 1^-} \dfrac{x^2 - 1}{\sqrt{1 - x}} = 0$.

21. $\dfrac{4^x - 3^x}{\sqrt{x}}$ has indeterminate form $\dfrac{0}{0}$ as $x \to 0^+$ and

$$\lim_{x \to 0^+} \dfrac{\frac{d}{dx}\left(4^x - 3^x\right)}{\frac{d}{dx}\sqrt{x}} = \lim_{x \to 0^+} \dfrac{4^x \ln 4 - 3^x \ln 3}{1/(2\sqrt{x})} = \lim_{x \to 0^+} 2\sqrt{x}\left(4^x \ln 4 - 3^x \ln 3\right) = 2(0)\left(\ln 4 - \ln 3\right) = 0,$$

thus $\displaystyle\lim_{x \to 0^+} \dfrac{4^x - 3^x}{\sqrt{x}} = 0$.

23. See Example 5. $Q(x) = \ln\left((1 + x)^{2/x}\right) = (2/x)\ln(1 + x) = \dfrac{2\ln(1 + x)}{x}$ has indeterminate form $\dfrac{0}{0}$ at

$x = 0$ and $\displaystyle\lim_{x \to 0} \dfrac{\frac{d}{dx}\left(2\ln(1 + x)\right)}{\frac{d}{dx} x} = \lim_{x \to 0} \dfrac{2/(1 + x)}{1} = \lim_{x \to 0} \dfrac{2}{1 + x} = 2$, thus $\displaystyle\lim_{x \to 0} Q(x) = 2$ and

$$\lim_{x \to 0}(1 + x)^{2/x} = \lim_{x \to 0} e^{Q(x)} = e^{\lim_{x \to 0} Q(x)} = e^2.$$

25. We apply L'Hôpital's rule twice.
$$\lim_{x \to -2} \dfrac{3x^3 + 16x^2 + 28x + 16}{x^5 + 4x^4 + 4x^3 + 3x^2 + 12x + 12} = \lim_{x \to -2} \dfrac{9x^2 + 32x + 28}{5x^4 + 16x^3 + 12x^2 + 6x + 12}$$
$$= \lim_{x \to -2} \dfrac{18x + 32}{20x^3 + 48x^2 + 24x + 6} = \dfrac{-4}{-10} = \dfrac{2}{5}.$$

9.1 — The indeterminate form $\frac{0}{0}$ and L'Hôpital's rule

Each of the first two quotients has the indeterminate form $\frac{0}{0}$ at $x = -2$, the third does not and its limit is $2/5$; therefore each of the preceding limits also exists and each equals $2/5$.

REMARK: $\lim\limits_{x \to -2} \dfrac{3x^3 + 16x^2 + 28x + 16}{x^5 + 4x^4 + 4x^3 + 3x^2 + 12x + 12} = \lim\limits_{x \to -2} \dfrac{(x+2)^2(3x+4)}{(x+2)^2(x^3+3)} = \lim\limits_{x \to -2} \dfrac{3x+4}{x^3+3} = \dfrac{-6+4}{-8+3}.$

27.   We apply L'Hôpital's rule twice.

$$\lim_{x \to 0} \frac{x - \sin x}{x^3} = \lim_{x \to 0} \frac{1 - \cos x}{3x^2} = \lim_{x \to 0} \frac{\sin x}{6x} = \frac{1}{6} \cdot \lim_{x \to 0} \frac{\sin x}{x} = \frac{1}{6} \cdot 1 = \frac{1}{6}.$$

Each of the first two quotients has the indeterminate form $\frac{0}{0}$ at $x = 0$, the third does not and its limit is $1/6$; therefore each of the preceding limits also exists and each equals $1/6$.

REMARK: Sections 9.4-5 will give us more tools to examine a relation of the form $\sin x = x - \dfrac{x^3}{6} + r(x)$ where $\lim\limits_{x \to 0} \dfrac{r(x)}{x^3} = 0.$

29. $\dfrac{(x - 4\pi)^2}{\ln \cos x}$ has the indeterminate form $\frac{0}{0}$ at $x = 4\pi$ and so does $\dfrac{\frac{d}{dx}\left((x-4\pi)^2\right)}{\frac{d}{dx}(\ln \cos x)} = \dfrac{2(x - 4\pi)}{-\tan x}$, but

$\lim\limits_{x \to 4\pi} \dfrac{\frac{d}{dx}\left(2(x-4\pi)\right)}{\frac{d}{dx}(-\tan x)} = \lim\limits_{x \to 4\pi} \dfrac{2}{-\sec^2 x} = \dfrac{2}{-1} = -2.$ Therefore, backing up one step at a time, $\lim\limits_{x \to 4\pi} \dfrac{2(x - 4\pi)}{-\tan x}$

exists and equals -2, hence $\lim\limits_{x \to 4\pi} \dfrac{(x - 4\pi)^2}{\ln \cos x}$ also exists and equals -2.

31. $\dfrac{\int_0^x \cos t^2 \, dt}{\int_0^x e^{t^2} \, dt}$ has the indeterminate form $\frac{0}{0}$ as $x \to 0$ because $\lim\limits_{x \to 0} \int_0^x f = \int_0^0 f = 0$ for every continuous

function f. Theorem 4.6.2 implies $\dfrac{\frac{d}{dx} \int_0^x \cos t^2 \, dt}{\frac{d}{dx} \int_0^x e^{t^2} \, dt} = \dfrac{\cos x^2}{e^{x^2}}$; we compute $\lim\limits_{x \to 0} \dfrac{\cos x^2}{e^{x^2}} = \dfrac{\cos 0}{e^0} = \dfrac{1}{1} = 1.$ Thus

$\lim\limits_{x \to 0} \dfrac{\int_0^x \cos t^2 \, dt}{\int_0^x e^{t^2} \, dt}$ exists and also equals 1.

33.   If f is a smooth function, then $\dfrac{f(x+h) - f(x-h)}{2h}$ has the indeterminate form $\frac{0}{0}$ as $h \to 0$ and

$$\lim_{h \to 0} \frac{\frac{d}{dh}\left(f(x+h) - f(x-h)\right)}{\frac{d}{dh}(2h)} = \lim_{h \to 0} \frac{f'(x+h) \cdot 1 - f'(x-h) \cdot (-1)}{2}$$

$$= \lim_{h \to 0} \frac{f'(x+h) + f'(x-h)}{2} = \frac{f'(x+0) + f'(x-0)}{2} = f'(x).$$

Therefore $\lim\limits_{h \to 0} \dfrac{f(x+h) - f(x-h)}{2h} = f'(x).$

REMARK: Problem 1.5.49 asked you to discuss whether such a limit would be a substitute definition of the derivative of f at x. In light of the computations shown here, a search for counterexamples should focus on functions not sufficiently smooth for L'Hôpital's rule to apply.

35.   If $b \neq 0$ and $x_0 > 0$, then $\dfrac{x^a - x_0^a}{x^b - x_0^b}$ has the indeterminate form $\frac{0}{0}$ at $x = x_0$ and $\lim\limits_{x \to x_0} \dfrac{\frac{d}{dx}\left(x^a - x_0^a\right)}{\frac{d}{dx}\left(x^b - x_0^b\right)} =$

$\lim\limits_{x \to x_0} \dfrac{ax^{a-1}}{bx^{b-1}} = \lim\limits_{x \to x_0} \dfrac{a}{b} \cdot x^{a-b} = \dfrac{a}{b} \cdot x_0^{a-b}.$ Therefore $\lim\limits_{x \to x_0} \dfrac{x^a - x_0^a}{x^b - x_0^b} = \dfrac{a}{b} \cdot x_0^{a-b}.$ ∎

37.   If $a \neq 0$, then $Q(x) = \ln\left((1+x)^{a/x}\right) = \dfrac{a\ln(1+x)}{x}$ is a quotient which has the indeterminate form $\frac{0}{0}$

at $x = 0$. We compute $\lim\limits_{x \to 0} \dfrac{\frac{d}{dx}\left(a\ln(1+x)\right)}{\frac{d}{dx} x} = \lim\limits_{x \to 0} \dfrac{a/(1+x)}{1} = \lim\limits_{x \to 0} \dfrac{a}{1+x} = a$; therefore $\lim\limits_{x \to 0} Q(x) = a$ and

$$\lim_{x \to 0}(1+x)^{a/x} = \lim_{x \to 0} e^{Q(x)} = e^{\lim\limits_{x \to 0} Q(x)} = e^a. \quad \blacksquare$$

REMARK: If $a = 0$, then $(1+x)^{a/x} = (1+x)^0 = 1 = e^0 = e^a.$

39. $$A(x) = \frac{1}{x-a} \cdot \int_a^x f = \frac{\int_a^x f}{x-a}$$ has the indeterminate form $\frac{0}{0}$ as $x \to a^+$ and $\displaystyle\lim_{x \to a^+} \frac{\frac{d}{dx} \int_a^x f}{\frac{d}{dx}(x-a)} = $

$\displaystyle\lim_{x \to a^+} \frac{f(x)}{1} = f(a)$; therefore $\displaystyle\lim_{x \to a^+} A(x) = \lim_{x \to a^+} \frac{\int_a^x f}{x-a} = f(a)$.

41. $$\frac{\int \left(\frac{1}{x-a-\epsilon} - \frac{1}{x-a} \right) dx}{\epsilon} = \frac{\ln|x-a-\epsilon| - \ln|x-a|}{\epsilon} + C = \frac{1}{\epsilon} \cdot \ln \left| \frac{x-a-\epsilon}{x-a} \right| + C = \frac{\ln \left| 1 - \frac{\epsilon}{x-a} \right|}{\epsilon} + C.$$

Apply L'Hôpital's rule to the non-constant term (remember to differentiate with respect to ϵ, not x or a.)

$\dfrac{\ln \left| 1 - \frac{\epsilon}{x-a} \right|}{\epsilon}$ has the indeterminate form $\dfrac{0}{0}$ at $\epsilon = 0$ and

$$\lim_{\epsilon \to 0} \frac{\frac{d}{d\epsilon} \ln \left| 1 - \frac{\epsilon}{x-a} \right|}{\frac{d}{d\epsilon} \epsilon} = \lim_{\epsilon \to 0} \left(\frac{1}{1 - \frac{\epsilon}{x-a}} \right) \left(\frac{-1}{x-a} \right) = \lim_{\epsilon \to 0} \frac{-1}{x-a-\epsilon} = \frac{-1}{x-a};$$

therefore $\displaystyle\lim_{\epsilon \to 0} \frac{\int \left(\frac{1}{x-a-\epsilon} - \frac{1}{x-a} \right) dx}{\epsilon} = \frac{-1}{x-a} + C.$

REMARK: $\displaystyle\lim_{\epsilon \to 0} \int \left(\frac{1}{\epsilon} \right) \left(\frac{1}{x-a-\epsilon} - \frac{1}{x-a} \right) dx = \lim_{\epsilon \to 0} \int \frac{dx}{(x-a)(x-a-\epsilon)} = \int \lim_{\epsilon \to 0} \left(\frac{1}{(x-a)(x-a-\epsilon)} \right) dx$

$= \displaystyle\int \frac{1}{(x-a)^2} dx = \frac{-1}{x-a} + C.$ This alternative calculation leads to the correct answer, but getting a general theorem to justify interchanging the limit and the integral has some technical difficulties which are best deferred to an advanced course and a discussion of uniform convergence.

43. If $p < 0$, then $\displaystyle\lim_{x \to 0} \frac{\cos x - 1}{x^p} = \lim_{x \to 0} x^{-p} \cdot (\cos x - 1) = 0 \cdot 0 = 0$; if $p = 0$, then $\displaystyle\lim_{x \to 0} \frac{\cos x - 1}{x^p} = \lim_{x \to 0} (\cos x - $

$1) = 0.$ If $0 < p$, then $\dfrac{\cos x - 1}{x^p}$ has the indeterminate form $\dfrac{0}{0}$ as $x \to 0$ and $\dfrac{\frac{d}{dx}(\cos x - 1)}{\frac{d}{dx}(x^p)} = \dfrac{-\sin x}{px^{p-1}}.$ If

$0 < p < 1$, then $\displaystyle\lim_{x \to 0} \frac{-\sin x}{px^{p-1}} = \lim_{x \to 0} \frac{x^{1-p}}{p} \cdot (-\sin x) = 0 \cdot 0 = 0$; if $p = 1$, we also get 0 as the limit.

If $1 < p$, then $\dfrac{-\sin x}{px^{p-1}}$ has indeterminate form $\dfrac{0}{0}$ at $x = 0$ and $\dfrac{\frac{d}{dx}(-\sin x)}{\frac{d}{dx}(px^{p-1})} = \dfrac{-\cos x}{p(p-1)x^{p-2}}.$ We now

observe that $\cos 0 \neq 0$ and we can find the limit of our current quotient directly. $\displaystyle\lim_{x \to 0} \frac{-\cos x}{p(p-1)x^{p-2}} = $

$\displaystyle\lim_{x \to 0} \frac{x^{2-p}}{p(p-1)} \cdot (-\cos x) = \begin{cases} 0 & \text{if } p < 2, \\ -1/2 & \text{if } p = 2, \\ \infty & \text{if } p > 2. \end{cases}$ L'Hôpital's rule now implies we have also found $\displaystyle\lim_{x \to 0} \frac{\cos x - 1}{x^p}$;

therefore 2 is the maximal p for which that limit is finite.

REMARK: Sections 9.4-5 will give us more tools to examine a relation of the form $\cos x = 1 - \dfrac{x^2}{2} + r(x)$

where $\displaystyle\lim_{x \to 0} \frac{r(x)}{x^3} = 0.$

Solutions 9.2   *Other indeterminate forms*   (pages 606-8)

1. $\dfrac{x^3 + 3x + 4}{2x^3 - 4x + 2}$ has the indeterminate form $\dfrac{\infty}{\infty}$ as $x \to \infty$ and so does $\dfrac{\frac{d}{dx}(x^3 + 3x + 4)}{\frac{d}{dx}(2x^3 - 4x + 2)} = \dfrac{3x^2 + 3}{6x^2 - 4}$,

but $\displaystyle\lim_{x \to \infty} \frac{\frac{d}{dx}(3x^2 + 3)}{\frac{d}{dx}(6x^2 - 4)} = \lim_{x \to \infty} \frac{6x}{12x} = \frac{1}{2}.$ Therefore $\displaystyle\lim_{x \to \infty} \frac{3x^2 + 3}{6x^2 - 4}$ exists and equals $\dfrac{1}{2}$; this, in turn, implies

$\displaystyle\lim_{x \to \infty} \frac{x^3 + 3x + 4}{2x^3 - 4x + 2}$ also exists and equals $\dfrac{1}{2}.$

REMARK: $\lim\limits_{x\to\infty} \dfrac{x^3 + 3x + 4}{2x^3 - 4x + 2} = \lim\limits_{x\to\infty} \dfrac{x^3 + 3x + 4}{2x^3 - 4x + 2} \cdot \left(\dfrac{1/x^{5/2}}{1/x^{5/2}} \right) = \lim\limits_{x\to\infty} \dfrac{1 + (3/x^2) + (4/x^3)}{2 - (4/x^2) + (2/x^3)} = \dfrac{1 + 0 + 0}{2 - 0 + 0}$.

3. $\dfrac{\ln x}{\sqrt{x}}$ has the indeterminate form $\dfrac{\infty}{\infty}$ as $x \to \infty$ and $\lim\limits_{x\to\infty} \dfrac{\frac{d}{dx}(\ln x)}{\frac{d}{dx}(\sqrt{x})} = \lim\limits_{x\to\infty} \dfrac{1/x}{1/(2\sqrt{x})} = \lim\limits_{x\to\infty} \dfrac{2}{\sqrt{x}} = 0$;

therefore $\lim\limits_{x\to\infty} \dfrac{\ln x}{\sqrt{x}} = 0$.

5. If $a > 0$, then $\dfrac{\ln x}{x^a}$ has the indeterminate form $\dfrac{\infty}{\infty}$ as $x \to \infty$ and $\lim\limits_{x\to\infty} \dfrac{\frac{d}{dx}(\ln x)}{\frac{d}{dx}(x^a)} = \lim\limits_{x\to\infty} \dfrac{x^{-1}}{a\,x^{a-1}} = $

$\lim\limits_{x\to\infty} \dfrac{1}{a\,x^a} = 0$; therefore $\lim\limits_{x\to\infty} \dfrac{\ln x}{x^a} = 0$.

REMARK: The ln function grows more slowly than any positive power of x; compare this result with the interpretation of Example 5 and Theorem 2.

7. $\dfrac{\ln x}{\cot x}$ has indeterminate form $\dfrac{-\infty}{\infty}$ as $x \to 0^+$ whereas $\dfrac{\frac{d}{dx}(\ln x)}{\frac{d}{dx}(\cot x)} = \dfrac{x^{-1}}{-\csc^2 x} = \dfrac{-\sin^2 x}{x}$ has indetermi-

nate form $\dfrac{0}{0}$, but $\lim\limits_{x\to 0^+} \dfrac{\frac{d}{dx}(-\sin^2 x)}{\frac{d}{dx} x} = \lim\limits_{x\to 0^+} \dfrac{-2\sin x \cos x}{1} = 0$; thus $0 = \lim\limits_{x\to 0^+} \dfrac{x^{-1}}{-\csc^2 x} = \lim\limits_{x\to 0^+} \dfrac{\ln x}{\cot x}$.

9. $x \tan \dfrac{1}{x} = \dfrac{\tan(1/x)}{1/x}$ has the indeterminate form $\dfrac{0}{0}$ as $x \to \infty$ and

$$\lim\limits_{x\to\infty} \dfrac{\frac{d}{dx}(\tan(1/x))}{\frac{d}{dx}(1/x)} = \lim\limits_{x\to\infty} \dfrac{(-1/x^2)\sec^2(1/x)}{-1/x^2} = \lim\limits_{x\to\infty} \sec^2 \dfrac{1}{x} = \sec^2 0 = 1;$$

hence $\lim\limits_{x\to\infty} x \tan \dfrac{1}{x} = 1$.

11.
$$\lim\limits_{x\to\infty} \dfrac{x}{e^{\sqrt{x}}} = \lim\limits_{x\to\infty} \dfrac{\frac{d}{dx}(x)}{\frac{d}{dx}(e^{\sqrt{x}})} = \lim\limits_{x\to\infty} \dfrac{1}{(1/(2\sqrt{x}))\,e^{\sqrt{x}}} = \lim\limits_{x\to\infty} \dfrac{2\sqrt{x}}{e^{\sqrt{x}}}$$

$$= \lim\limits_{x\to\infty} \dfrac{\frac{d}{dx}(2\sqrt{x})}{\frac{d}{dx}(e^{\sqrt{x}})} = \lim\limits_{x\to\infty} \dfrac{1/\sqrt{x}}{(1/(2\sqrt{x}))\,e^{\sqrt{x}}} = \lim\limits_{x\to\infty} \dfrac{2}{e^{\sqrt{x}}} = 0.$$

$\dfrac{x}{e^{\sqrt{x}}}$ and $\dfrac{2\sqrt{x}}{e^{\sqrt{x}}}$ have the indeterminate form $\dfrac{\infty}{\infty}$ as $x \to \infty$; the above work shows application of L'Hôpital's rule alternated with algebraic rearrangements.

13. Example 3 shows $\lim\limits_{x\to 0^+} x^x = 1$; Theorem 2.5.1 shows $\lim\limits_{x\to 0} \dfrac{\sin x}{x} = 1$ (an easy application of L'Hôpital's rule would remind you of that fact). Since those two limits exist, so does $\lim\limits_{x\to 0^+} x^{\sin x} = \lim\limits_{x\to 0^+} (x^x)^{(\sin x)/x}$;

that limit is $\left(\lim\limits_{x\to 0^+} x^x \right)^{\left(\lim\limits_{x\to 0^+} (\sin x)/x \right)} = 1^1 = 1$.

REMARK: $Q(x) = \ln\left(x^{\sin x}\right) = \sin x \ln x = \dfrac{\ln x}{\csc x}$ has the indeterminate of form $\dfrac{\infty}{\infty}$ as $x \to 0^+$. Using miscellaneous old results together with L'Hôpital's rule, we find

$$\lim\limits_{x\to 0^+} \dfrac{\frac{d}{dx}(\ln x)}{\frac{d}{dx}(\csc x)} = \lim\limits_{x\to 0^+} \dfrac{1/x}{-\csc x \cot x} = \lim\limits_{x\to 0^+} \dfrac{\sin x}{x} \cdot (-\tan x) = 1 \cdot 0 = 0;$$

therefore $\lim\limits_{x\to 0^+} Q(x) = 0$, and $\lim\limits_{x\to 0^+} x^{\sin x} = \lim\limits_{x\to 0^+} e^{Q(x)} = e^{\lim\limits_{x\to 0^+} Q(x)} = e^0 = 1$.

The computation of $\lim Q$ could also have been finished by applying L'Hôpital's rule to $\dfrac{-\sin^2 x}{x \cos x}$.

15. $x \cdot \ln \sin x = \dfrac{\ln \sin x}{x^{-1}}$ has indeterminate form $\dfrac{-\infty}{\infty}$ as $x \to 0^+$ and

$$\lim_{x \to 0^+} \frac{\frac{d}{dx}\left(\ln \sin x\right)}{\frac{d}{dx}\left(x^{-1}\right)} = \lim_{x \to 0^+} \frac{\cos x / \sin x}{-x^{-2}} = \frac{\lim_{x \to 0^+}(-x\cos x)}{\lim_{x \to 0^+} \frac{\sin x}{x}} = \frac{0}{1} = 0,$$

therefore $\lim_{x \to 0^+} x \cdot \ln \sin x = 0$.

17. $Q(x) = \ln\left(1 + \dfrac{1}{2x}\right)^{x^2} = x^2 \ln\left(1 + \dfrac{1}{2x}\right) = \dfrac{\ln(1 + 1/2x)}{x^{-2}}$ has indeterminate form $\dfrac{0}{0}$ as $x \to \infty$ and

$$\lim_{x \to \infty} \frac{\frac{d}{dx}\left(\ln(1 + 1/2x)\right)}{\frac{d}{dx}\left(x^{-2}\right)} = \lim_{x \to \infty} \frac{-1/(2x^2 + x)}{-2x^{-3}} = \lim_{x \to \infty} \frac{x}{4 + 2/x} = \frac{\infty}{4 + 0} = \infty,$$

therefore $\lim_{x \to \infty} Q(x) = \infty$ and $\lim_{x \to \infty}\left(1 + \dfrac{1}{2x}\right)^{x^2} = \lim_{x \to \infty} e^{Q(x)} = \infty$.

19. $Q(x) = \ln(1 + \sinh x)^{1/x} = \dfrac{\ln(1 + \sinh x)}{x}$ has indeterminate form $\dfrac{0}{0}$ as $x \to 0^+$ and L'Hôpital's rule

implies $\lim_{x \to 0^+} Q(x) = \lim_{x \to 0^+} \dfrac{\frac{d}{dx}\left(\ln(1 + \sinh x)\right)}{\frac{d}{dx}\left(x\right)} = \lim_{x \to 0^+} \dfrac{\cosh x}{1 + \sinh x} = 1$, therefore $\lim_{x \to 0^+}(1 + \sinh x)^{1/x} =$

$\lim_{x \to 0^+} e^{Q(x)} = e^{\lim_{x \to 0^+} Q(x)} = e^1 = e.$

21. $\csc 2x \sin 3x = \dfrac{\sin 3x}{\sin 2x}$ has indeterminate form $\dfrac{0}{0}$ as $x \to 0^+$ and L'Hôpital's rule implies

$$\lim_{x \to 0^+} \csc 2x \sin 3x = \lim_{x \to 0^+} \frac{\sin 3x}{\sin 2x} = \lim_{x \to 0^+} \frac{\frac{d}{dx}\left(\sin 3x\right)}{\frac{d}{dx}\left(\sin 2x\right)} = \lim_{x \to 0^+} \frac{3\cos 3x}{2\cos 2x} = \frac{3}{2}.$$

23. $(\sin x)^{\sin x - x} = \dfrac{(\sin x)^{\sin x}}{(\sin x)^x}$; let

$$A(x) = \ln\left((\sin x)^{\sin x}\right) = (\sin x) \cdot (\ln \sin x) = \frac{\ln \sin x}{\csc x}$$

and

$$B(x) = \ln\left((\sin x)^x\right) = x \cdot (\ln \sin x) = \frac{\ln \sin x}{x^{-1}}.$$

Both A and B have the indeterminate form $\dfrac{-\infty}{\infty}$ as $x \to 0^+$. L'Hôpital's rule implies

$$\lim_{x \to 0^+} A(x) = \lim_{x \to 0^+} \frac{\frac{d}{dx}\left(\ln \sin x\right)}{\frac{d}{dx}\left(\csc x\right)} = \lim_{x \to 0^+} \frac{\cot x}{-\csc x \cot x} = \lim_{x \to 0^+}(-\sin x) = 0,$$

and

$$\lim_{x \to 0^+} B(x) = \lim_{x \to 0^+} \frac{\frac{d}{dx}\left(\ln \sin x\right)}{\frac{d}{dx}\left(x^{-1}\right)} = \lim_{x \to 0^+} \frac{\cot x}{-x^{-2}} = \lim_{x \to 0^+} \frac{-x \cos x}{(\sin x)/x} = \frac{0}{1} = 0.$$

Therefore $\lim_{x \to 0^+} (\sin x)^{\sin x - x} = \lim_{x \to 0^+} \dfrac{(\sin x)^{\sin x}}{(\sin x)^x} = \lim_{x \to 0^+} \dfrac{e^{A(x)}}{e^{B(x)}} = \dfrac{e^0}{e^0} = 1.$

25. Let $Q(x) = \ln\left(x^{1/(2-x)}\right) = \dfrac{\ln x}{2 - x}$. Because $\ln 2 > 0$ and $\lim_{x \to 2^+} \dfrac{1}{2 - x} = -\infty$, we see that $\lim_{x \to 2^+} Q(x) =$

$-\infty$ and $\lim_{x \to 2^+} x^{1/(2-x)} = \lim_{x \to 2^+} e^{Q(x)} = 0$. On the other hand, $\lim_{x \to 2^-} \dfrac{1}{2 - x} = \infty$, $\lim_{x \to 2^-} Q(x) = \infty$, and

$\lim_{x \to 2^-} x^{1/(2-x)} = \lim_{x \to 2^-} e^{Q(x)} = \infty$. Therefore $\lim_{x \to 2} x^{1/(2-x)}$ does not exist.

REMARK: L'Hôpital's rule does not apply to this problem.

27. $Q(x) = \ln\left(\left(\cos \dfrac{1}{x}\right)^x\right) = x \ln\left(\cos \dfrac{1}{x}\right) = \dfrac{\ln \cos x^{-1}}{x^{-1}}$ has the indeterminate form $\dfrac{0}{0}$ as $x \to \infty$

and L'Hôpital's rule implies $\lim_{x \to \infty} Q(x) = \lim_{x \to \infty} \dfrac{\ln \cos x^{-1}}{x^{-1}} = \lim_{x \to \infty} \dfrac{\frac{d}{dx}\left(\ln \cos x^{-1}\right)}{\frac{d}{dx}\left(x^{-1}\right)} = \lim_{x \to \infty} \dfrac{x^{-2} \tan x^{-1}}{-x^{-2}} =$

$$\lim_{x \to \infty} -\tan \frac{1}{x} = -\tan 0 = 0; \text{ therefore } \lim_{x \to \infty} \left(\cos \frac{1}{x}\right)^x = \lim_{x \to \infty} e^{Q(x)} = e^0 = 1.$$

29. $\dfrac{1}{1-x} - \dfrac{1}{\ln x} = \dfrac{x - 1 + \ln x}{(1-x)\ln x}$ has the indeterminate form $\dfrac{0}{0}$ as $x \to 1^-$ and $\lim\limits_{x \to 1^-} \dfrac{\frac{d}{dx}\left(x - 1 + \ln x\right)}{\frac{d}{dx}\left((1-x)\ln x\right)} =$

$\lim\limits_{x \to 1^-} \dfrac{1 + 1/x}{-\ln x + (1-x)/x} = \infty$; thus $\lim\limits_{x \to 1^-} \left(\dfrac{1}{1-x} - \dfrac{1}{\ln x}\right) = \infty$.

31.
$$\sqrt{x^2 + x} - x = \left(\sqrt{x^2 + x} - x\right)\left(\frac{\sqrt{x^2 + x} + x}{\sqrt{x^2 + x} + x}\right) = \frac{x}{\sqrt{x^2 + x} + x}$$

$$= \left(\frac{x}{\sqrt{x^2 + x} + x}\right)\left(\frac{1/x}{1/x}\right) = \frac{1}{\sqrt{1 + 1/x} + 1},$$

thus $\lim\limits_{x \to \infty} \left(\sqrt{x^2 + x} - x\right) = \lim\limits_{x \to \infty} \dfrac{1}{\sqrt{1 + 1/x} + 1} = \dfrac{1}{\sqrt{1 + 0} + 1} = \dfrac{1}{2}$.

REMARK: If c is constant, then $\sqrt{x(x + c)} - x = \dfrac{cx}{\sqrt{x(x + c)} + x}$ has indeterminate form $\dfrac{\infty}{\infty}$ as $x \to \infty$

and $\lim\limits_{x \to \infty} \dfrac{\frac{d}{dx}(cx)}{\frac{d}{dx}\left(\sqrt{x(x + c)} + x\right)} = \lim\limits_{x \to \infty} \dfrac{c}{\frac{2x + c}{2\sqrt{x(x + c)}} + 1} = \lim\limits_{x \to \infty} \dfrac{c}{\frac{2 + c/x}{2\sqrt{1 + c/x}} + 1} = \lim\limits_{x \to \infty} \dfrac{c}{\frac{2}{2} + 1} = \dfrac{c}{2}$.

(See Problem 1.4.30.)

---

33. a. $A(x) = \ln\left(x^{-k/\ln x}\right) = \left(\dfrac{-k}{\ln x}\right) \cdot \ln x = -k$, thus $x^{-k/\ln x} = e^{A(x)} = e^{-k}$ for all $x > 0$ and

$\lim\limits_{x \to 0^+} x^{-k/\ln x} = \lim\limits_{x \to 0^+} e^{A(x)} = e^{-k}$.

b. $B(x) = \ln\left(x^{-k/\sqrt{-\ln x}}\right) = \left(\dfrac{-k}{\sqrt{-\ln x}}\right) \cdot \ln x = k\sqrt{-\ln x}$; thus $\lim\limits_{x \to 0^+} B(x) = \begin{cases} \infty & \text{if } k > 0, \\ -\infty & \text{if } k < 0, \end{cases}$ and

$\lim\limits_{x \to 0^+} x^{-k/\sqrt{-\ln x}} = \lim\limits_{x \to 0^+} e^{B(x)} = \begin{cases} \infty & \text{if } k > 0, \\ 0 & \text{if } k < 0. \end{cases}$

REMARK: An indeterminate expression of the form 0^0 does not necessarily have limit 1. See *Indeterminate forms of exponential type* by Baxley and Hayashi in *The American Mathematical Monthly*, volume 85 (1978), pages 484-486, for further discussion and references.

---

35. Solution 5 contains the proof for this problem.

37. a. $\lim\limits_{x \to \infty} \dfrac{x + \cos x}{x + \sin x} = \lim\limits_{x \to \infty} \dfrac{1 + \frac{\cos x}{x}}{1 + \frac{\sin x}{x}} = \dfrac{\lim\limits_{x \to \infty}\left(1 + \frac{\cos x}{x}\right)}{\lim\limits_{x \to \infty}\left(1 + \frac{\sin x}{x}\right)} = \dfrac{1 + 0}{1 + 0} = 1$.

b. $\cos x$ equals -1 at every odd multiple of π and hence $\dfrac{1 - \sin x}{1 + \cos x}$ is undefined at each of those points.

Because each set of the form $\{x : N < x\}$ contains at least one odd multiple of π, we infer $\lim\limits_{x \to \infty} \dfrac{1 - \sin x}{1 + \cos x}$ does not exist.

L'Hôpital's rule applies only to quotients $\dfrac{f}{g}$ where $\dfrac{f'}{g'}$ has a limit, it makes no statement when $\dfrac{f'}{g'}$ does not have a limit.

REMARK: See *A Calculus Counterexample* by N.W. Rickert, *The American Mathematical Monthly*, volume 75 (1968), page 166, for another example.

---

39. $\dfrac{e^{-1/x^2}}{x} = \dfrac{x^{-1}}{e^{1/x^2}}$ has indeterminate form $\dfrac{\infty}{\infty}$ as $x \to 0^+$ and $\lim\limits_{x \to 0^+} \dfrac{\frac{d}{dx}\left(x^{-1}\right)}{\frac{d}{dx}\left(e^{1/x^2}\right)} = \lim\limits_{x \to 0^+} \dfrac{-x^{-2}}{-2x^{-3}e^{1/x^2}} =$

$\dfrac{1}{2}\lim\limits_{x \to 0^+}\left(x\, e^{-1/x^2}\right) = 0$; therefore $\lim\limits_{x \to 0^+} \dfrac{e^{-1/x^2}}{x} = 0$. The same result holds for the other one-sided limit, thus

$\lim\limits_{x \to 0} \dfrac{e^{-1/x^2}}{x} = 0$. ∎

REMARK: Let $Q(x) = \ln \left| \dfrac{e^{-1/x^2}}{x} \right| = \ln e^{-1/x^2} - \ln|x| = \dfrac{-1}{x^2} - \ln|x| = \dfrac{-(1 + x^2 \ln|x|)}{x^2}$. We examine

$x^2 \ln|x|$ which can be written as $\dfrac{\ln|x|}{x^{-2}}$, an indeterminate of form $\dfrac{\infty}{\infty}$ as $x \to 0$, and we see that

$\lim\limits_{x \to 0} \dfrac{\frac{d}{dx} \ln|x|}{\frac{d}{dx} x^{-2}} = \lim\limits_{x \to 0} \dfrac{x^{-1}}{-2x^{-3}} = \lim\limits_{x \to 0} \dfrac{-x^2}{2} = 0$; therefore $\lim\limits_{x \to 0} (1 + x^2 \ln|x|) = 1 + 0 = 1$, $\lim\limits_{x \to 0} Q(x) =$

$\lim\limits_{x \to 0} \dfrac{-(1 + x^2 \ln|x|)}{x^2} = -\infty$, $\lim\limits_{x \to 0} \left| \dfrac{e^{-1/x^2}}{x} \right| = \lim\limits_{x \to 0} e^{Q(x)} = 0$, and $\dfrac{e^{-1/x^2}}{x}$ has limit 0 as $x \to 0$ because its

absolute value does. ∎

Alternatively, use the substitution $u = x^{-2}$ and apply Example 5: $\lim\limits_{x \to 0} \dfrac{e^{-1/x^2}}{x} = \lim\limits_{u \to \infty} \pm\sqrt{u}\, e^{-u} = 0$.

41.   We begin our analysis by focusing on intervals excluding 0, thus $f(x) = e^{-1/x^2}$. Let $\mathcal{P}(n)$ represent the following assertion: $f^{(n)}(x)$ is a linear combination of terms of the form $x^{-m} e^{-1/x^2}$ where $n + 2 \le m \le 3n$. $\mathcal{P}(1)$ is true because $f^{(1)}(x) = f'(x) = 2x^{-3} e^{-1/x^2}$ and $1 + 2 = 3 = 3 \cdot 1$. Suppose $\mathcal{P}(k)$ were true. Then $f^{(k+1)}(x) = \dfrac{d}{dx} f^{(k)}(x)$ is the sum of terms like $\dfrac{d}{dx}\left(x^{-m} e^{-1/x^2} \right) = \left(-m x^{-(m+1)} + 2x^{-(m+3)} \right) e^{-1/x^2}$ and $(k+1) + 2 = (k + 2) + 1 \le m + 1 < m + 3 \le 3k + 3 = 3(k + 1)$; hence $\mathcal{P}(k+1)$ would also be true. $\mathcal{P}(1)$ is true and $\mathcal{P}(k)$ implies $\mathcal{P}(k+1)$, therefore $\mathcal{P}(n)$ is true for each positive integer n.

It now remains only to tidy up our analysis of the behavior of the various derivatives at $x = 0$. By definition of f we have $f^{(0)}(0) = f(0) = 0$. Suppose $f^{(k)}(0) = 0$. Then, using the full result of Problem 40,

$f^{(k+1)}(0) = \lim\limits_{x \to 0} \dfrac{f^{(k)}(x) - f^{(k)}(0)}{x - 0} = \lim\limits_{x \to 0} \dfrac{f^{(k)}(x)}{x} = $ sum of terms like $\lim\limits_{x \to 0} x^{-m} e^{-1/x^2} = 0$. ∎

REMARK: This result is used in Example 10.10.3.

Solutions 9.3   *Improper integrals*   (pages 616-8)

1. $\displaystyle\int_0^\infty \dfrac{dx}{x^3} = \int_0^1 x^{-3}\, dx + \int_1^\infty x^{-3}\, dx = \lim\limits_{\epsilon \to 0+} \int_\epsilon^1 x^{-3}\, dx + \lim\limits_{N \to \infty} \int_1^N x^{-3}\, dx$

$= \lim\limits_{\epsilon \to 0+} \left(\dfrac{-1}{2} \right)(1 - \epsilon^{-2}) + \lim\limits_{N \to \infty} \left(\dfrac{-1}{2} \right)(N^{-2} - 1) = \left(\dfrac{-1}{2} \right)(-\infty) + \left(\dfrac{-1}{2} \right)(-1) = \infty$;

the (doubly) improper integral diverges.

3. $\displaystyle\int_0^1 \dfrac{dx}{x^{3/2}} = \lim\limits_{\epsilon \to 0+} \int_\epsilon^1 x^{-3/2}\, dx = \lim\limits_{\epsilon \to 0+} (-2)x^{-1/2}\Big|_\epsilon^1 = \lim\limits_{\epsilon \to 0+} (-2)\left(1 - \dfrac{1}{\sqrt{\epsilon}} \right) = \infty$; the improper integral diverges.

5. $\displaystyle\int_2^3 \dfrac{dx}{(x - 2)^{1/3}} = \lim\limits_{\epsilon \to 0+} \int_{2+\epsilon}^3 (x - 2)^{-1/3}\, dx = \lim\limits_{\epsilon \to 0+} \left(\dfrac{3}{2} \right)(x - 2)^{2/3}\Big|_{2+\epsilon}^3 = \lim\limits_{\epsilon \to 0+} \left(\dfrac{3}{2} \right)\left(1 - \epsilon^{2/3} \right) = \dfrac{3}{2}$.

7. $\displaystyle\int_0^\infty e^{-2x}\, dx = \lim\limits_{N \to \infty} \int_0^N e^{-2x}\, dx = \lim\limits_{N \to \infty} \left(\dfrac{-1}{2} \right)e^{-2x}\Big|_0^N = \lim\limits_{N \to \infty} \left(\dfrac{-1}{2} \right)\left(\epsilon^{-2N} - 1 \right) = \left(\dfrac{-1}{2} \right)(0 - 1) = \dfrac{1}{2}$.

9. $\displaystyle\int_{-\infty}^0 e^x\, dx = \lim\limits_{M \to \infty} \int_{-M}^0 e^x\, dx = \lim\limits_{M \to \infty} e^x\Big|_{-M}^0 = \lim\limits_{M \to \infty} \left(1 - \epsilon^{-M} \right) = 1$.

11. $\displaystyle\int_0^\infty x\, e^{-2x}\, dx = \lim\limits_{N \to \infty} \int_0^N x\, e^{-2x}\, dx = \lim\limits_{N \to \infty} \left(\dfrac{-1}{4} \right)(2x + 1)\, e^{-2x}\Big|_0^N$

$= \lim\limits_{N \to \infty} \left(\dfrac{-1}{4} \right)(2N + 1)\, e^{-2N} + \left(\dfrac{1}{4} \right) = \dfrac{1}{4}$.

13.
$$\int_{-\infty}^{\infty} x^3 e^{-x^4} dx = \lim_{M \to \infty} \int_{-M}^{0} x^3 e^{-x^4} dx + \lim_{N \to \infty} \int_{0}^{N} x^3 e^{-x^4} dx$$

$$= \lim_{M \to \infty} \left(\frac{-1}{4}\right)\left(1 - e^{-M^4}\right) + \lim_{N \to \infty} \left(\frac{-1}{4}\right)\left(e^{-N^4} - 1\right)$$

$$= \left(\frac{-1}{4}\right)(1 - 0) + \left(\frac{-1}{4}\right)(0 - 1) = 0.$$

15.
$$\int_{-\infty}^{1} \frac{dx}{\sqrt{4 - x}} = \lim_{M \to \infty} \int_{-M}^{1} (4 - x)^{-1/2} dx = \lim_{M \to \infty} (-2)\sqrt{4 - x}\,\Big|_{-M}^{1} = -2\sqrt{3} + \lim_{M \to \infty} 2\sqrt{4 + M} = \infty, \text{ this}$$
improper integral diverges.

17. We may assume $b > 0$ without loss of generality.

$$\int_{a}^{\infty} \frac{dx}{b^2 + x^2} = \lim_{N \to \infty} \frac{1}{b} \int_{a}^{N} \frac{1}{1 + (x/b)^2} \, d\left(\frac{x}{b}\right) = \lim_{N \to \infty} \frac{1}{b}\left(\tan^{-1} \frac{N}{b} - \tan^{-1}\frac{a}{b}\right) = \left(\frac{1}{b}\right)\left(\frac{\pi}{2} - \tan^{-1}\frac{a}{b}\right).$$

19.
$$\int_{-\infty}^{\infty} \frac{x^2}{x^2 + 1} dx = \lim_{M \to \infty} \int_{-M}^{0} \left(1 - \frac{1}{x^2 + 1}\right) dx + \lim_{N \to \infty} \int_{0}^{N} \left(1 - \frac{1}{x^2 + 1}\right) dx$$

$$= \lim_{M \to \infty} (x - \tan^{-1} x)\Big|_{-M}^{0} + \lim_{N \to \infty} (x - \tan^{-1} x)\Big|_{0}^{N}$$

$$= \lim_{M \to \infty} (M - \tan^{-1} M) + \lim_{N \to \infty} (N - \tan^{-1} N) = \infty;$$

this improper integral diverges.

21.
$$\int_{0}^{\pi/2} \tan x \, dx = \lim_{\epsilon \to 0+} \int_{0}^{\pi/2 - \epsilon} \tan x \, dx = \lim_{\epsilon \to 0+} -\ln |\cos x|\,\Big|_{0}^{\pi/2 - \epsilon} = \lim_{\epsilon \to 0+} -\ln \left|\cos\left(\frac{\pi}{2} - \epsilon\right)\right| = \infty.$$

23.
$$\int_{0}^{\pi/4} \csc x \, dx = \lim_{\epsilon \to 0+} \int_{\epsilon}^{\pi/4} \csc x \, dx = \lim_{\epsilon \to 0+} \ln \left|\tan \frac{x}{2}\right|\,\Big|_{\epsilon}^{\pi/4} = \ln \tan \frac{\pi}{8} - \lim_{\epsilon \to 0+} \ln \tan \frac{\epsilon}{2} = \infty, \text{ the integral}$$
diverges.

25.
$$\int_{0}^{2} \frac{dx}{x - 1} = \lim_{\epsilon_1 \to 0+} \int_{0}^{1 - \epsilon_1} \frac{dx}{x - 1} + \lim_{\epsilon_2 \to 0+} \int_{1 + \epsilon_2}^{2} \frac{dx}{x - 1}$$

$$= \lim_{\epsilon_1 \to 0+} \ln |x - 1|\,\Big|_{0}^{1 - \epsilon_1} + \lim_{\epsilon_2 \to 0+} \ln |x - 1|\,\Big|_{1 + \epsilon_2}^{2} = \lim_{\epsilon_1 \to 0+} \ln \epsilon_1 - \lim_{\epsilon_2 \to 0+} \ln \epsilon_2;$$

because both limits are infinite, the improper integral diverges.

REMARK: Example 12 is similar to this problem.

27.
$$\int_{10}^{\infty} \frac{dx}{x^2 - 1} = \lim_{N \to \infty} \frac{1}{2} \int_{10}^{N} \left(\frac{1}{x - 1} - \frac{1}{x + 1}\right) dx = \lim_{N \to \infty} \frac{1}{2} \ln \left|\frac{x - 1}{x + 1}\right|\,\Big|_{10}^{N}$$

$$= \lim_{N \to \infty} \frac{1}{2} \ln \frac{N - 1}{N + 1} - \frac{1}{2} \ln \frac{9}{11} = \frac{1}{2} \ln 1 - \frac{1}{2} \ln \frac{9}{11} = -\frac{1}{2} \ln \frac{9}{11} = \frac{1}{2} \ln 11 - \ln 3.$$

29.
$$\lim_{N \to \infty} \int_{0}^{N} \sin x \, dx = 1 - \lim_{N \to \infty} \cos N, \text{ but } \lim_{N \to \infty} \cos N \text{ does not exist; thus } \int_{0}^{\infty} \sin x \, dx \text{ diverges.}$$

31.
$$\int_{-\infty}^{0} e^x \cos x \, dx = \lim_{M \to \infty} \int_{-M}^{0} e^x \cos x \, dx = \lim_{M \to \infty} \frac{1}{2} e^x (\sin x + \cos x)\,\Big|_{-M}^{0}$$

$$= \frac{1}{2} - \frac{1}{2} \lim_{M \to \infty} e^{-M}(\cos M - \sin M) = \frac{1}{2} - 0 = \frac{1}{2}.$$

33.
$$\int_{0}^{5} \frac{dx}{\sqrt{25 - x^2}} = \lim_{\epsilon \to 0+} \int_{0}^{5 - \epsilon} \frac{dx}{\sqrt{25 - x^2}} = \lim_{\epsilon \to 0+} \sin^{-1}\left(\frac{x}{5}\right)\Big|_{0}^{5 - \epsilon} = \lim_{\epsilon \to 0+} \sin^{-1}\left(1 - \frac{\epsilon}{5}\right) - \sin^{-1} 0 = \frac{\pi}{2} - 0 = \frac{\pi}{2}.$$

35.
$$\int_{-2}^{0} \frac{dx}{(x + 2)^5} = \lim_{\epsilon \to 0+} \int_{-2 + \epsilon}^{0} (x + 2)^{-5} dx = \lim_{\epsilon \to 0+} \left(\frac{-1}{4}\right)(x + 2)^{-4}\Big|_{-2 + \epsilon}^{0} = \frac{-1}{64} + \left(\frac{1}{4}\right) \lim_{\epsilon \to 0+} \epsilon^{-4} = \infty, \text{ the}$$
improper integral diverges.

REMARK: The substitution $u = x + 2$ transforms this integral into one which has already been analyzed by Example 11: $\int_{-2}^{0} \frac{dx}{(x + 2)^5} = \int_{0}^{2} \frac{du}{u^5}$.

37.
$$\int_2^4 \frac{dx}{(x-3)^7} = \lim_{\epsilon_1 \to 0+} \int_2^{3-\epsilon_1} (x-3)^{-7} dx + \lim_{\epsilon_2 \to 0+} \int_{3+\epsilon_2}^4 (x-3)^{-7} dx$$
$$= \lim_{\epsilon_1 \to 0+} \left(\frac{-1}{6} \right) \left((-\epsilon_1)^{-6} - 1 \right) + \lim_{\epsilon_2 \to 0+} \left(\frac{-1}{6} \right) \left(1 - \epsilon_2^{-6} \right);$$
both limits are infinite, the improper integral diverges.

39.   Observe that $\dfrac{1}{1 - \sin x} = \dfrac{1 + \sin x}{1 - \sin^2 x} = \dfrac{1 + \sin x}{\cos^2 x} = \sec^2 x + \sec x \tan x$. Therefore
$$\int_0^{\pi/2} \frac{dx}{1 - \sin x} = \lim_{\epsilon \to 0+} \int_0^{\pi/2-\epsilon} \left(\sec^2 x + \sec x \tan x \right) dx$$
$$= \lim_{\epsilon \to 0+} \left(\tan x + \sec x \right) \Big|_0^{\pi/2-\epsilon} = \lim_{\epsilon \to 0+} \left(\tan \left(\frac{\pi}{2} - \epsilon \right) + \sec \left(\frac{\pi}{2} - \epsilon \right) - 1 \right) = \infty;$$
the improper integral diverges.

REMARK: Use the $u = \tan(x/2)$ substitution introduced in Example 7.5.2 to obtain $\displaystyle\int_0^{\pi/2} \frac{dx}{1 - \sin x} =$
$\displaystyle\lim_{\epsilon \to 0+} \left. \frac{2}{1 - \tan(x/2)} \right|_0^{\pi/2-\epsilon}$.   Integral 104 (in the text's table of integrals) leads to the same result.

41.
$$\int_5^\infty \frac{dx}{x^2 - 6x + 8} = \lim_{N \to \infty} \int_5^N \frac{1}{2} \left(\frac{1}{x-4} - \frac{1}{x-2} \right) dx = \lim_{N \to \infty} \frac{1}{2} \ln \left| \frac{x-4}{x-2} \right| \Big|_5^N$$
$$= \lim_{N \to \infty} \frac{1}{2} \ln \left| \frac{N-4}{N-2} \right| - \frac{1}{2} \ln \frac{1}{3} = \frac{1}{2} \ln 3.$$

43.
$$\int_1^\infty \frac{dx}{x \ln x} = \int_1^e \frac{dx}{x \ln x} + \int_e^\infty \frac{dx}{x \ln x} = \lim_{\epsilon \to 0+} \int_{1+\epsilon}^e \frac{dx}{x \ln x} + \lim_{N \to \infty} \int_e^N \frac{dx}{x \ln x}$$
$$= \lim_{\epsilon \to 0+} \ln \ln x \Big|_{1+\epsilon}^e + \lim_{N \to \infty} \ln \ln x \Big|_e^N = \lim_{\epsilon \to 0+} - \ln \ln (1+\epsilon) + \lim_{N \to \infty} \ln \ln N = \infty + \infty = \infty;$$
the doubly improper integral diverges.

45. $\displaystyle\int_0^1 \ln x \, dx = \lim_{\epsilon \to 0+} \int_\epsilon^1 \ln x \, dx = \lim_{\epsilon \to 0+} \left(-x + x \ln x \right) \Big|_\epsilon^1 = \lim_{\epsilon \to 0+} \left(-1 + \epsilon - \epsilon \ln \epsilon \right) = -1 - 0 + 0 = -1.$

REMARK: We have used the result of Example 9.2.2: $\lim_{\epsilon \to 0+} \epsilon \ln \epsilon = 0$.

---

47. $\boxed{\text{Disk method}}$ $V = \displaystyle\int_0^\infty \pi \left(e^{-x} \right)^2 dx = \lim_{N \to \infty} \int_0^N \pi e^{-2x} dx = \lim_{N \to \infty} \frac{-\pi}{2} \left(e^{-2N} - 1 \right) = \frac{-\pi}{2} (0 - 1) = \frac{\pi}{2}.$

$\boxed{\text{Cylindrical shell method}}$ $V = \displaystyle\int_0^1 2\pi y(-\ln y) \, dy = \lim_{\epsilon \to 0+} \int_\epsilon^1 -2\pi y \ln y \, dy = \lim_{\epsilon \to 0+} (-\pi) y^2 \left(\ln y - \frac{1}{2} \right) \Big|_\epsilon^1 =$

$\dfrac{\pi}{2} + \pi \displaystyle\lim_{\epsilon \to 0+} \left(\epsilon^2 \ln \epsilon - \frac{\epsilon^2}{2} \right)$, Example 9.2.2 implies $\lim_{\epsilon \to 0+} \epsilon^2 \ln \epsilon = 0$ so we obtain $\dfrac{\pi}{2}$ again.

49. $\displaystyle\int_4^\infty x^{-3/2} dx = \lim_{N \to \infty} (-2) x^{-1/2} \Big|_4^N = \lim_{N \to \infty} \frac{-2}{\sqrt{N}} + 1 = 1$; if $4 \le x$, then
$$-1 \le \sin x \implies 0 \le 1 + \sin x \implies 0 \le x^3 \le x^3 + 1 + \sin x \implies 0 \le x^{3/2} = \sqrt{x^3} \le \sqrt{x^3 + 1 + \sin x}$$
$$\implies 0 \le \frac{1}{\sqrt{x^3 + 1 + \sin x}} \le x^{-3/2}.$$

Therefore part (a) of Problem 66 implies $\displaystyle\int_4^\infty \frac{dx}{\sqrt{x^3 + 1 + \sin x}}$ converges. ∎

51.   Let $f(x) = x - \ln(1 + x)$. Since $f(0) = 0$ and $f'(x) = 1 - \dfrac{1}{1 + x} = \dfrac{x}{1 + x} > 0$ if $x > 0$, we infer f is
strictly increasing throughout $(0, \infty)$; since $x > 0$ implies $\ln(1 + x) > 0$, we see $x - \ln(1 + x) > 0$ implies
$\dfrac{1}{\ln(1 + x)} > \dfrac{1}{x} > 0$. Furthermore, $\displaystyle\int_1^\infty \frac{dx}{x} = \lim_{N \to \infty} \ln |x| \Big|_1^N = \lim_{N \to \infty} \ln N = \infty$; part (b) of Problem 66 now
implies $\displaystyle\int_1^\infty \frac{dx}{\ln(1 + x)}$ diverges because $\displaystyle\int_1^\infty \frac{dx}{x}$ does. ∎

53. $\sin^2 x \leq 1$ for all x; if $x > 0$, then $0 \leq \dfrac{\sin^2 x}{x^2} \leq \dfrac{1}{x^2}$. Example 7 (or direct computation) shows that $\displaystyle\int_1^\infty \dfrac{dx}{x^2}$ converges, Problem 66(a) then implies $\displaystyle\int_1^\infty \dfrac{\sin^2 x}{x^2}\,dx$ converges. $\blacksquare$

55.   Use the substitution $t = 1 - x$ and the result of Problem 9.1.36 to obtain

$$\int_0^1 \frac{1 - (1-x)^N}{x}\,dx = \int_1^0 \frac{1 - t^N}{1 - t}(-\,dt) = \int_0^1 \left(1 + t + t^2 + \cdots + t^{N-1}\right) dt = 1 + \frac{1}{2} + \frac{1}{3} + \cdots + \frac{1}{N}.$$

REMARK: Do you think the original integral is improper? Although $\dfrac{1 - (1-x)^N}{x}$ is undefined if $x = 0$, it is bounded between N and 1 while x is between 0 and 1.

57.   If $f(x) = 1$, then for $s > 0$ we have

$$\hat{f}(s) = \int_0^\infty e^{-st} f(t)\,dt = \lim_{N\to\infty} \int_0^N e^{-st}\,dt = \lim_{N\to\infty} \frac{-1}{s} e^{-st}\Big|_0^N = \frac{-1}{s}\left(\lim_{N\to\infty} e^{-sN} - 1\right) = \frac{1}{s}.$$

59.   If $F(x) = e^x$, then for $s > 1$ we have

$$\hat{F}(s) = \int_0^\infty e^{-st} e^t\,dt = \lim_{N\to\infty} \int_0^N e^{-st+t}\,dt = \lim_{N\to\infty} \frac{e^{(1-s)t}}{1-s}\Big|_0^N = \lim_{N\to\infty} \frac{e^{(1-s)N}}{1-s} - \frac{1}{1-s} = \frac{1}{s-1}.$$

61. $\displaystyle\int_0^8 x^{-1/3}\,dx = \lim_{\epsilon\to 0^+} \int_\epsilon^8 x^{-1/3}\,dx = \lim_{\epsilon\to 0^+} \frac{3}{2} x^{2/3}\Big|_\epsilon^8 = 6 - \frac{3}{2}\lim_{\epsilon\to 0^+} \epsilon^{2/3} = 6 - 0 = 6.$

If $x = u^3$, then $dx = 3u^2\,du$ and $\displaystyle\int_0^8 x^{-1/3}\,dx = \int_0^2 u^{-1}\left(3u^2\,du\right) = \int_0^2 3u\,du = \frac{3}{2}u^2\Big|_0^2 = \frac{3}{2}\cdot 2^2 = 6.$

Although algebra is simplified by focusing on the relation $x = u^3$, our substitution really is $u = x^{1/3}$. Because $\dfrac{d}{dx} u = \dfrac{1}{3} x^{-2/3}$ is unbounded as $x \to 0^+$, we actually only transform $\displaystyle\int_\epsilon^8 x^{-1/3}\,dx$ into $\displaystyle\int_{\sqrt[3]{\epsilon}}^2 u^{-1}\cdot 3u^2\,du$ and we still need to compute a limit as $\epsilon \to 0^+$. Since $u^{-1}\cdot u^2 = u$ and u has no singularities at the origin, it is easy to see that $\displaystyle\lim_{\epsilon\to 0^+} \int_{\sqrt[3]{\epsilon}}^2 u^{-1}\cdot 3u^2\,du = \lim_{\epsilon\to 0^+} \int_{\sqrt[3]{\epsilon}}^2 3u\,du = \int_0^2 3u\,du.$

63.   Solution 45 shows $\int_0^1 \ln x\,dx$ converges; hence, in parts (a) and (b) here, we are on solid logical ground when we behave as if that improper integral does represent a real number. (Its value will be recomputed in part (c).)

   a.   On the interval $(0, 1]$, $\ln x$ is negative and strictly increasing. Partition $(0, 1]$ into N subintervals of equal length; on each subinterval, the maximum value of $\ln x$ is taken at the right-hand endpoint; hence $\displaystyle\int_0^1 \ln x\,dx < \sum_{k=1}^N \left(\ln \frac{k}{N}\right)\cdot \frac{1}{N}$. Use essentially the same partition on $\left[\dfrac{1}{N}, 1\right]$; on each subinterval, the minimum value of $\ln x$ is taken at the left-hand endpoint; thus $\displaystyle\int_{1/N}^1 \ln x\,dx > \sum_{k=1}^{N-1} \left(\ln \frac{k}{N}\right)\cdot \frac{1}{N}$. Therefore, since $\ln \dfrac{N}{N} = \ln 1 = 0$, the sum is unchanged if we add $\left(\ln \dfrac{N}{N}\right)\cdot \dfrac{1}{N}$ and we conclude

$$\int_0^1 \ln x\,dx < \frac{1}{N} \sum_{k=1}^N \ln \frac{k}{N} < \int_{1/N}^1 \ln x\,dx. \quad \blacksquare$$

   b.   Part (a) implies $\displaystyle\int_0^1 \ln x\,dx \leq \lim_{N\to\infty} \frac{1}{N}\sum_{k=1}^N \ln \frac{k}{N} \leq \lim_{N\to\infty} \int_{1/N}^1 \ln x\,dx$. Since $\displaystyle\lim_{N\to\infty} \frac{1}{N} = 0$, we have

$$\lim_{N\to\infty} \int_{1/N}^1 = \int_0^1 \text{ and we infer } \lim_{N\to\infty} \frac{1}{N}\sum_{k=1}^N \ln \frac{k}{N} = \int_0^1 \ln x\,dx. \quad \blacksquare$$

   c.   Example 9.2.2 shows $\displaystyle\lim_{x\to 0^+} x \ln x = 0$, hence $\displaystyle\int_0^1 \ln t\,dt = -1 - \lim_{\epsilon\to 0^+} \epsilon(-1 + \ln \epsilon) = -1$. Therefore

$$\lim_{N\to\infty} \frac{\sqrt[N]{N!}}{N} = \lim_{N\to\infty} \sqrt[N]{\frac{N!}{N^N}} = \lim_{N\to\infty} e^{\frac{1}{N}\ln\left(\frac{N!}{N^N}\right)} = \lim_{N\to\infty} e^{\frac{1}{N}\sum_{k=1}^N \ln \frac{k}{N}} = e^{\int_0^1 \ln t\,dt} = e^{-1} = \frac{1}{e}. \quad \blacksquare$$

REMARK: Problem 6.4.60 outlines an alternative sequence of steps to show this result.

65.   Integration-by-parts shows that

$$\int t^\alpha e^{-t}\,dt = \int t^\alpha\,d\left(-e^{-t}\right) = t^\alpha\left(-e^{-t}\right) - \int \left(-e^{-t}\right)d\left(t^\alpha\right) = -t^\alpha e^{-t} + \alpha \int t^{\alpha-1}e^{-t}\,dt.$$

Example 9.2.5 shows $\lim\limits_{x\to\infty} x^a e^{-x} = 0$ for any real a. Therefore, if $\alpha > 0$, then

$$\Gamma(\alpha+1) = \int_0^\infty t^\alpha e^{-t}\,dt = \lim_{N\to\infty} -t^\alpha e^{-t}\Big|_0^N + \alpha\int_0^\infty t^{\alpha-1}e^{-t}\,dt = (0+0^\alpha) + \alpha\cdot\Gamma(\alpha) = \alpha\cdot\Gamma(\alpha).\ \blacksquare$$

67.   If $st \ge 0$, then $0 < e^{-st} \le 1$. Thus $-|f(t)| \le -e^{-st}|f(t)| \le e^{-st}f(t) \le e^{-st}|f(t)| \le |f(t)|$ and $0 \le e^{-st}f(t) + |f(t)| \le 2|f(t)|$. If $\int_0^\infty |f|$ converges, then so does $\int_0^\infty 2|f|$; now use Problem 66 to infer that $\int_0^\infty \left(e^{-st}f(t) + |f(t)|\right)dt$ also converges. Subtract $\int_0^\infty |f|$ from this to obtain the convergent improper integral $\int_0^\infty e^{-st}f(t)\,dt = \hat{f}(s)$ if $s \ge 0$. $\blacksquare$

69.   Suppose $a < b$ and $f(x) = \begin{cases} 0 & \text{if } x < a, \\ 1/(b-a) & \text{if } a \le x \le b, \\ 0 & \text{if } b < x. \end{cases}$ Since $a < b$, it is obvious that $0 < b - a$ and $f(x) \ge 0$ for all x. We also compute $\int_{-\infty}^\infty f = \int_{-\infty}^a 0 + \int_a^b \frac{1}{b-a} + \int_b^\infty 0 = 0 + \frac{1}{b-a}\cdot(b-a) + 0 = 1$. Therefore f is a probability density. $\blacksquare$

71. $e^{-t^2/2} > 0$ for all t; if $t \ge 1$, then $e^{-t^2/2} \le te^{-t^2/2}$; furthermore $\int_1^\infty te^{-t^2/2}dt = \lim\limits_{N\to\infty}\left(-e^{-N^2/2} + e^{-1/2}\right)$

$= e^{-1/2}$. Hence, using Problem 66(a), we infer $\int_1^\infty e^{-t^2/2}dt$ converges. Because $e^{-t^2/2}$ is continuous on the finite interval $[0,1]$, the proper integral $\int_0^1 e^{-t^2/2}dt$ exists; thus $\int_0^\infty e^{-t^2/2}dt = \int_0^1 e^{-t^2/2}dt + \int_1^\infty e^{-t^2/2}dt$

converges. Since the function $e^{-t^2/2}$ is symmetric with respect to $t = 0$, $\int_{-M}^0 e^{-t^2/2}dt = \int_0^M e^{-t^2/2}dt$ for

every choice of M. Therefore $\int_{-\infty}^\infty e^{-t^2/2}dt = \int_{-\infty}^0 e^{-t^2/2}dt + \int_0^\infty e^{-t^2/2}dt = 2\int_0^\infty e^{-t^2/2}dt$ converges.

Multiplication by a constant, $\dfrac{1}{\sqrt{2\pi}}$, does not affect this convergence. $\blacksquare$

REMARK: The assertion that $\dfrac{1}{\sqrt{2\pi}}\int_{-\infty}^\infty e^{-t^2/2}dt = 1$ is justified in Example 14.4.5.

73. $\mu = \int_{-\infty}^\infty t\,f(t)\,dt = \int_{-\infty}^0 t\cdot 0\,dt + \int_0^\infty t\cdot ae^{-at}\,dt = \left(\dfrac{-1}{a} - t\right)e^{-at}\Big|_0^\infty = 0 - \dfrac{-1}{a} = \dfrac{1}{a}.$

75.   Solution 72 shows $\mu = \dfrac{a+b}{2}$, thus

$$\sigma^2 = \int_{-\infty}^\infty (t-\mu)^2\,f(t)\,dt = \int_a^b \left(t - \frac{a+b}{2}\right)^2\left(\frac{1}{b-a}\right)dt$$

$$= \frac{1}{3(b-a)}\left(t - \frac{a+b}{2}\right)^3\Big|_a^b = \frac{1}{3(b-a)}\left(\left(\frac{b-a}{2}\right)^3 - \left(\frac{a-b}{2}\right)^3\right) = \frac{(b-a)^2}{12}.$$

Thus $\sigma = \dfrac{|b-a|}{2\sqrt{3}}$.

77.   Solution 74 computes $\mu = 0$, thus $(t-\mu)^2 = t^2$. One integration-by-parts shows that

$$\int t^2 e^{-t^2/2}\,dt = \int t\cdot\left(te^{-t^2/2}\,dt\right) = \int t\cdot d\left(-e^{-t^2/2}\right) = -te^{-t^2/2} + \int e^{-t^2/2}\,dt.$$

9.3 — Improper Integrals

The result of Example 9.2.5 implies $\lim\limits_{N \to \infty} -Ne^{-N^2/2} = 0$ and $\lim\limits_{M \to \infty} -(-M)e^{-(-M)^2/2} = 0$. Therefore

$$\int_{-\infty}^{\infty} t^2 e^{-t^2/2}\, dt = \int_{-\infty}^{\infty} e^{-t^2/2}\, dt,$$

$$\sigma^2 = \int_{-\infty}^{\infty} t^2 \frac{1}{\sqrt{2\pi}} e^{-t^2/2}\, dt = \int_{-\infty}^{\infty} \frac{1}{\sqrt{2\pi}} e^{-t^2/2}\, dt = 1,$$

and $\sigma = \sqrt{1} = 1$.

The convergence of $\int_{-\infty}^{\infty} \frac{1}{\sqrt{2\pi}} e^{-t^2/2}\, dt$ is shown in Solution 71; the work above shows that convergence is inherited by $\int_{-\infty}^{\infty} t^2 \frac{1}{\sqrt{2\pi}} e^{-t^2/2}\, dt$. The fact that $\int_{-\infty}^{\infty} \frac{1}{\sqrt{2\pi}} e^{-t^2/2}\, dt = 1$ is implied by the announcement in Problem 71 that the integrand is a probability density; Example 14.4.5 verifies that claim.

Solutions 9.4   *Taylor's theorem and Taylor polynomials*   (pages 625-6)

REMARK: A Taylor polynomial for the function f at the point a will be written only in terms of powers of $(x - a)$. Theorem 9.5.1 and the Examples in Section 9.5 show that values of a Taylor polynomial are close to those of f provided $|x - a|$ is small (and f is "well-behaved"). Therefore, reexpressing a Taylor polynomial using powers of x, i.e., of $(x - 0)$, is not appropriate and is not a simplification in this context.

1.

n	0	1	2	3	4	5	6
$f^{(n)}(x)$	$\cos x$	$-\sin x$	$-\cos x$	$\sin x$	$\cos x$	$-\sin x$	$-\cos x$
$f^{(n)}\left(\frac{\pi}{4}\right)$	$\frac{1}{\sqrt{2}}$	$\frac{-1}{\sqrt{2}}$	$\frac{-1}{\sqrt{2}}$	$\frac{1}{\sqrt{2}}$	$\frac{1}{\sqrt{2}}$	$\frac{-1}{\sqrt{2}}$	$\frac{-1}{\sqrt{2}}$

$$P_6(x) = \frac{1}{\sqrt{2}} + \left(\frac{-1}{\sqrt{2}}\right)\left(x - \frac{\pi}{4}\right) + \left(\frac{-1}{\sqrt{2}}\right)\frac{(x - \pi/4)^2}{2!} + \left(\frac{1}{\sqrt{2}}\right)\frac{(x - \pi/4)^3}{3!} + \left(\frac{1}{\sqrt{2}}\right)\frac{(x - \pi/4)^4}{4!}$$
$$+ \left(\frac{-1}{\sqrt{2}}\right)\frac{(x - \pi/4)^5}{5!} + \left(\frac{-1}{\sqrt{2}}\right)\frac{(x - \pi/4)^6}{6!}.$$

3.

n	0	1	2	3	4	5
$f^{(n)}(x)$	$\ln x$	x^{-1}	$-x^{-2}$	$2x^{-3}$	$-6x^{-4}$	$24x^{-5}$
$f^{(n)}(e)$	1	$\frac{1}{e}$	$\frac{-1}{e^2}$	$\frac{2}{e^3}$	$\frac{-6}{e^4}$	$\frac{24}{e^5}$

$$P_5(x) = 1 + \left(\frac{1}{e}\right)(x - e) + \left(\frac{-1}{e^2}\right)\frac{(x - e)^2}{2!} + \left(\frac{2}{e^3}\right)\frac{(x - e)^3}{3!} + \left(\frac{-6}{e^4}\right)\frac{(x - e)^4}{4!} + \left(\frac{24}{e^5}\right)\frac{(x - e)^5}{5!}$$
$$= 1 + \left(\frac{x}{e} - 1\right) - \frac{1}{2}\left(\frac{x}{e} - 1\right)^2 + \frac{1}{3}\left(\frac{x}{e} - 1\right)^3 - \frac{1}{4}\left(\frac{x}{e} - 1\right)^4 + \frac{1}{5}\left(\frac{x}{e} - 1\right)^5.$$

REMARK: Because $\ln x = 1 + \ln\left(\frac{x}{e}\right)$, the alternate expression in powers of $\left(\frac{x}{e} - 1\right)$ should not surprise.

5.

n	0	1	2	3	4
$f^{(n)}(x)$	x^{-1}	$-x^{-2}$	$2x^{-3}$	$-6x^{-4}$	$24x^{-5}$
$f^{(n)}(1)$	1	-1	2	-6	24

$$P_4(x) = 1 + (-1)(x - 1) + (2)\frac{(x - 1)^2}{2!} + (-6)\frac{(x - 1)^3}{3!} + (24)\frac{(x - 1)^4}{4!}$$
$$= 1 - (x - 1) + (x - 1)^2 - (x - 1)^3 + (x - 1)^4.$$

7.

n	0	1	2	3	4
$f^{(n)}(x)$	$\tan x$	$\sec^2 x$	$2\sec^2 x \tan x$	$4\sec^2 x \tan^2 x + 2\sec^4 x$	$8\sec^2 x \tan^3 x + 16\sec^4 x \tan x$
$f^{(n)}(0)$	0	1	0	2	0

$$P_4(x) = 0 + (1)\,x + (0)\frac{x^2}{2!} + (2)\frac{x^3}{3!} + (0)\frac{x^4}{4!} = x + \frac{x^3}{3}.$$

9. The function $f(x) = \tan x$ has period π, therefore so do each of its derivatives and we may apply computations done for Solution 7: $f^{(n)}(\pi) = f^{(n)}(0)$ and

$$P_4(x) = 0 + (x - \pi) + 0 + 2\frac{(x-\pi)^3}{3!} + 0 = (x - \pi) + \frac{(x-\pi)^3}{3}.$$

REMARK: If $f(x) = \tan x$, then $f'(x) = \sec^2 x = 1 + \tan^2 x = 1 + f^2(x)$. Continuing in this way, we see:

n	0	1	2	3	4
$f^{(n)}(x)$	$\tan x$	$1 + f^2(x)$	$2f(x)f'(x)$	$2\left(f'(x)\right)^2 + 2f(x)f''(x)$	$6f'(x)f''(x) + 2f(x)f'''(x)$
$f^{(n)}(\pi)$	0	1	0	2	0

$$P_4(x) = 0 + (1)(x - \pi) + (0)\frac{(x-\pi)^2}{2!} + (2)\frac{(x-\pi)^3}{3!} + (0)\frac{(x-\pi)^4}{4!} = (x - \pi) + \frac{(x-\pi)^3}{3}.$$

11.

n	0	1	2	3	4
$f^{(n)}(x)$	$\dfrac{1}{1+x^2}$	$\dfrac{-2x}{(1+x^2)^2}$	$\dfrac{2(3x^2-1)}{(1+x^2)^3}$	$\dfrac{-24x(x^2-1)}{(1+x^2)^4}$	$\dfrac{24(5x^4-10x^2+1)}{(1+x^2)^5}$
$f^{(n)}(0)$	1	0	-2	0	24

$$P_4(x) = 1 + (0)\,x + (-2)\frac{x^2}{2!} + (0)\frac{x^3}{3!} + (24)\frac{x^4}{4!} = 1 - x^2 + x^4.$$

13.

n	0	1	2	3	4
$f^{(n)}(x)$	$\sinh x$	$\cosh x$	$\sinh x$	$\cosh x$	$\sinh x$
$f^{(n)}(0)$	0	1	0	1	0

$$P_4(x) = 0 + (1)\,x + (0)\frac{x^2}{2!} + (1)\frac{x^3}{3!} + (0)\frac{x^4}{4!} = x + \frac{x^3}{6}.$$

15.

n	0	1	2	3
$f^{(n)}(x)$	$\ln \sin x$	$\cot x$	$-\csc^2 x$	$2\csc^2 x \cot x$
$f^{(n)}\left(\dfrac{\pi}{2}\right)$	0	0	-1	0

$$P_3(x) = 0 + (0)\left(x - \frac{\pi}{2}\right) + (-1)\frac{(x-\pi/2)^2}{2!} + (0)\frac{(x-\pi/2)^3}{3!} = \frac{-1}{2}\left(x - \frac{\pi}{2}\right)^2.$$

17.

n	0	1	2	3	4
$f^{(n)}(x)$	$(4-x)^{-1/2}$	$\dfrac{1}{2}(4-x)^{-3/2}$	$\dfrac{3}{4}(4-x)^{-5/2}$	$\dfrac{15}{8}(4-x)^{-7/2}$	$\dfrac{105}{16}(4-x)^{-9/2}$
$f^{(n)}(0)$	$\dfrac{1}{2}$	$\dfrac{1}{16}$	$\dfrac{3}{128}$	$\dfrac{15}{1024}$	$\dfrac{105}{8192}$

$$P_4(x) = \frac{1}{2} + \left(\frac{1}{16}\right)x + \left(\frac{3}{128}\right)\frac{x^2}{2!} + \left(\frac{15}{1024}\right)\frac{x^3}{3!} + \left(\frac{105}{8192}\right)\frac{x^4}{4!} = \frac{1}{2} + \frac{x}{16} + 3\left(\frac{x}{16}\right)^2 + 10\left(\frac{x}{16}\right)^3 + 35\left(\frac{x}{16}\right)^4.$$

19. If $f(x) = e^{\beta x}$, then $f^{(n)}(x) = \beta \cdot f^{(n-1)}(x) = \beta^n \cdot e^{\beta x}$ and $f^{(n)}(0) = \beta^n$. Therefore

$$P_6(x) = 1 + \beta x + \beta^2 \frac{x^2}{2!} + \beta^3 \frac{x^3}{3!} + \beta^4 \frac{x^4}{4!} + \beta^5 \frac{x^5}{5!} + \beta^6 \frac{x^6}{6!} = \sum_{k=0}^{6} \frac{(\beta x)^k}{k!}.$$

21.

n	0	1	2	3
$f^{(n)}(x)$	$\sin^{-1} x$	$(1-x^2)^{-1/2}$	$x\left(1-x^2\right)^{-3/2}$	$(1+2x^2)\left(1-x^2\right)^{-5/2}$
$f^{(n)}(0)$	0	1	0	1

$$P_3(x) = 0 + (1)\,x + (0)\frac{x^2}{2!} + (1)\frac{x^3}{3!} = x + \frac{x^3}{6}.$$

23.

n	0	1	2	3	4	5	6	7	8	9	10
$f^{(n)}(x)$	$a_0+a_1x+a_2x^2+a_3x^3$	$a_1+2a_2x+3a_3x^2$	$2a_2+6a_3x$	$6a_3$	0	0	0	0	0	0	0
$f^{(n)}(1)$	$a_0+a_1+a_2+a_3$	$a_1+2a_2+3a_3$	$2a_2+6a_3$	$6a_3$	0	0	0	0	0	0	0

$$P_{10}(x) = (a_0+a_1+a_2+a_3) + (a_1+2a_2+3a_3)(x-1) + (2a_2+6a_3)\frac{(x-1)^2}{2!} + (6a_3)\frac{(x-1)^3}{3!}$$
$$+ (0)\frac{(x-1)^4}{4!} + (0)\frac{(x-1)^5}{5!} + \cdots + (0)\frac{(x-1)^9}{9!} + (0)\frac{(x-1)^{10}}{10!}$$
$$= (a_0+a_1+a_2+a_3) + (a_1+2a_2+3a_3)(x-1) + (a_2+3a_3)(x-1)^2 + a_3(x-1)^3.$$

25.

n	0	1	2	3	4
$f^{(n)}(x)$	$\sin x^2$	$2x\cos x^2$	$2\cos x^2-4x^2\sin x^2$	$-8x^3\cos x^2-12x\sin x^2$	$-48x^2\cos x^2+4(4x^4-3)\sin x^2$
$f^{(n)}(0)$	0	0	2	0	0

$$P_4(x) = 0 + (0)\,x + (2)\frac{x^2}{2!} + (0)\frac{x^3}{3!} + (0)\frac{x^4}{4!} = x^2.$$

REMARK: If Q_2 is the second-degree Taylor polynomial for $\sin x$ at $x=0$, then $P_4(x) = Q_2(x^2)$.

27. **Lemma.** *If $f(x) = (1-x)^{-1}$, then $f^{(n)}(x) = n! \cdot (1-x)^{-(n+1)}$ for any positive integer n.*
Proof of Lemma: Let $\mathcal{P}(n)$ represent the conclusion of this Lemma. $\mathcal{P}(1)$ is true because

$$f^{(1)}(x) = f'(x) = (-1)(1-x)^{-1-1} \cdot \frac{d}{dx}(1-x) = (-1)(1-x)^{-2} \cdot (-1) = (1-x)^{-2} = 1 \cdot (1-x)^{-(1+1)}.$$

If $\mathcal{P}(k)$ were true, then

$$f^{(k+1)}(x) = \frac{d}{dx} f^{(k)}(x) = \frac{d}{dx}\left(k! \cdot (1-x)^{-(k+1)}\right)$$
$$= k! \cdot (-(k+1))(1-x)^{-(k+1)-1} \cdot (-1) = (k+1)! \cdot (1-x)^{-[(k+1)+1]},$$

so $\mathcal{P}(k+1)$ would also be true. The Principle of Mathematical Induction guarantees the Lemma is proven. ∎

$$f(0) = (1-0)^{-1} = 1 \text{ and } r^{(n)}(0) = n! \cdot (1-0)^{-(n+1)} = n!, \text{ thus } P_n(x) = \sum_{j=0}^{n} j! \frac{(x-0)^j}{j!} = \sum_{j=0}^{n} x^j. \quad ∎$$

29. Let $f(x) = \sin x$. Solution 2.7.31 proves $f^{(n)}(x) = \begin{cases} \sin x & \text{if } n = 4k, \\ \cos x & \text{if } n = 4k+1, \\ -\sin x & \text{if } n = 4k+2, \\ -\cos x & \text{if } n = 4k+3, \end{cases}$ where k is a nonnegative integer. Thus $f^{(2j)}(x) = (-1)^j \sin x$ and $f^{(2j)}(0) = 0$. Furthermore, $f^{(2j+1)}(x) = (-1)^j \cos x$ and $f^{(2j+1)}(0) = (-1)^j$. Therefore if $N = 2n+2 = 2(n+1)$, then

$$P_N(x) = \sum_{i=0}^{N} f^{(i)}(0) \cdot \frac{(x-0)^i}{i!} = \sum_{i=0}^{2(n+1)} f^{(i)}(0) \cdot \frac{x^i}{i!} = \sum_{\substack{0 \le i \le 2(n+1) \\ i \text{ is even}}} f^{(i)}(0) \cdot \frac{x^i}{i!} + \sum_{\substack{0 \le i \le 2(n+1) \\ i \text{ is odd}}} f^{(i)}(0) \cdot \frac{x^i}{i!}$$

$$= \sum_{j=0}^{n+1} f^{(2j)}(0) \cdot \frac{x^{2j}}{(2j)!} + \sum_{j=0}^{n} f^{(2j+1)}(0) \cdot \frac{x^{2j+1}}{(2j+1)!}$$

$$= \sum_{j=0}^{n+1} 0 \cdot \frac{x^{2j}}{(2j)!} + \sum_{j=0}^{n} (-1)^j \cdot \frac{x^{2j+1}}{(2j+1)!}$$

$$= \sum_{j=0}^{n} (-1)^j \cdot \frac{x^{2j+1}}{(2j+1)!}.$$

We finish by noting $f^{(N)}(0) = 0$ since N is even, thus $P_N(x) - P_{N-1}(x) = f^{(N)}(0) \cdot \frac{x^N}{N!} = 0 \cdot \frac{x^N}{N!} = 0$ and

$$P_{2n+1}(x) = P_{2n+2}(x) = \sum_{j=0}^{n} (-1)^j \cdot \frac{x^{2j+1}}{(2j+1)!}. \quad ∎$$

31.   Let $x \in [a, b]$ be fixed and let $h(t) = f(x) - \sum_{k=0}^{n} f^{(k)}(t) \frac{(x-t)^k}{k!} - \frac{R_n(x)}{(x-a)^{n+1}} (x-t)^{n+1}$. If we differentiate the right-hand side with respect to t and apply Problem 30, the result is a telescoping sum:

$$h'(t) = \frac{d}{dt} \left(f(x) - f(t) \right) - \frac{d}{dt} \left(\sum_{k=1}^{n} f^{(k)}(t) \frac{(x-t)^k}{k!} \right) - \frac{d}{dt} \left(\frac{R_n(x)}{(x-a)^{n+1}} (x-t)^{n+1} \right)$$

$$= (0 - f'(t)) - \sum_{k=1}^{n} \left(\frac{f^{(k+1)}(t)}{k!} \cdot (x-t)^k - \frac{f^{(k)}(t)}{(k-1)!} \cdot (x-t)^{k-1} \right) - \frac{R_n(x)}{(x-a)^{n+1}} \cdot (n+1)(x-t)^n(-1)$$

$$= - \left(f'(t) + \sum_{k=1}^{n} \frac{f^{(k+1)}(t)}{k!} \cdot (x-t)^k \right) + \sum_{k=1}^{n} \frac{f^{(k)}(t)}{(k-1)!} \cdot (x-t)^{k-1} + \frac{(n+1)R_n(x)}{(x-a)^{n+1}} \cdot (x-t)^n$$

$$= - \sum_{k=0}^{n} \frac{f^{(k+1)}(t)}{k!} \cdot (x-t)^k + \sum_{j=0}^{n-1} \frac{f^{(j+1)}(t)}{j!} \cdot (x-t)^j + \frac{(n+1)R_n(x)}{(x-a)^{n+1}} \cdot (x-t)^n$$

$$= - \frac{f^{(n+1)}(t)}{n!} \cdot (x-t)^n + \frac{(n+1)R_n(x)}{(x-a)^{n+1}} \cdot (x-t)^n. \ \blacksquare$$

33.   Suppose the function f and its first $n+1$ derivatives exist and are continuous on $[a, b]$. Let P_n be the nth-degree Taylor polynomial of f centered at a and let $R_n(x) = f(x) - P_n(x)$. Theorem 1, augmented with Problems 30-1, implies there is a number c such that $a \leq c \leq x$ and $R_n(x) = f^{(n+1)}(c) \cdot \frac{(x-a)^{n+1}}{(n+1)!}$. Since $f^{(n+1)}$ is continuous, $\lim_{x \to a^+} f^{(n+1)}(c) = f^{(n+1)}(a)$; therefore

$$\lim_{x \to a^+} \frac{R_n(x)}{(x-a)^n} = \lim_{x \to a^+} \frac{f^{(n+1)}(c)}{(n+1)!} \cdot \frac{(x-a)^{n+1}}{(x-a)^n} = \frac{1}{(n+1)!} \left(\lim_{x \to a^+} f^{(n+1)}(c) \right) \left(\lim_{x \to a^+} (x-a) \right)$$

$$= \frac{1}{(n+1)!} f^{(n+1)}(a) \cdot 0 = 0. \ \blacksquare$$

35.   Suppose that $f(x) = P_n(x) + R_n(x) = Q_n(x) + S_n(x)$ where $P_n(x)$ is the nth-degree Taylor polynomial, $Q_n(x)$ is another n^{th}-degree polynomial, and $S_n(x)$ satisfies $\lim_{x \to a^+} \frac{S_n(x)}{(x-a)^n} = 0$. Let $D(x) = P_n(x) - Q_n(x) = S_n(x) - R_n(x)$; since D is the difference of nth-degree polynomials, $D(x)$ is also an nth-degree polynomial and can be written in the form $D(x) = d_0 + d_1(x-a) + d_2(x-a)^2 + \cdots + d_n(x-a)^n$. For each integer k between 0 and n, we have $\lim_{x \to a^+} \frac{D(x)}{(x-a)^k} = \lim_{x \to a^+} \frac{R_n(x)}{(x-a)^k} - \lim_{x \to a^+} \frac{S_n(x)}{(x-a)^k} = 0 - 0 = 0$. Therefore

$$D(x) = \sum_{j=0}^{n} d_j(x-a)^j = d_0 + d_1(x-a) + d_2(x-a)^2 + \cdots + d_n(x-a)^n \implies d_0 = D(a) = \lim_{x \to a^+} D(x) = 0$$

$$\implies D(x) = \sum_{j=1}^{n} d_j(x-a)^j = d_1(x-a) + d_2(x-a)^2 + \cdots + d_n(x-a)^n \implies d_1 = \lim_{x \to a^+} \frac{D(x)}{x-a} = 0$$

$$\implies D(x) = \sum_{j=2}^{n} d_j(x-a)^j = d_2(x-a)^2 + \cdots + d_n(x-a)^n \implies d_2 = \lim_{x \to a^+} \frac{D(x)}{(x-a)^2} = 0$$

$$\vdots \qquad\qquad \vdots \qquad\qquad \vdots$$

$$\implies D(x) = \sum_{j=k}^{n} d_j(x-a)^j = d_k(x-a)^k + \cdots + d_n(x-a)^n \implies d_k = \lim_{x \to a^+} \frac{D(x)}{(x-a)^k} = 0.$$

The induction step is to suppose we have shown $d_0 = d_1 = \cdots = d_k = 0$ where $0 \leq k < n$. Then

$$d_{k+1} = \lim_{x \to a^+} \left(d_{k+1} + d_{k+2}(x-a) + \cdots + d_n(x-a)^{n-k-1} \right) = \lim_{x \to a^+} \frac{D(x)}{(x-a)^{k+1}} = 0.$$

Therefore each coefficient of D is zero, this implies $D(x) = 0$ for all x in $[a, b]$ which is equivalent to $P_n(x) = Q_n(x)$ for all x in $[a, b]$. $\blacksquare$

9.5 — Approximation using Taylor polynomials

Solutions 9.5 *Approximation using Taylor polynomials* (pages 632-4)

1. If $f(x) = \sin x$, then $f^{(7)}(x) = -\cos x$ and $\left|f^{(7)}(c)\right| = |\cos c| \le 1$. Therefore

$$\left|R_6(x)\right| = \left|f^{(7)}(c)\right| \cdot \frac{|x - \pi/4|^7}{7!} \le \max_c |-\cos c| \cdot \frac{|x - \pi/4|^7}{7!} \le 1 \cdot \frac{|x - \pi/4|^7}{7!} = \frac{|x - \pi/4|^7}{7!}.$$

REMARK: If $x \in \left[0, \frac{\pi}{2}\right]$, then $\left|x - \frac{\pi}{4}\right| \le \frac{\pi}{4}$ and $\left|R_6(x)\right| \le \frac{(\pi/4)^7}{7!} \approx 3.658 \cdot 10^{-5}$.

3. If $f(x) = \sqrt{x} = x^{1/2}$, then $f^{(5)}(x) = \frac{105}{32} \cdot x^{-9/2}$. Therefore $x \in \left[\frac{1}{4}, 4\right]$ implies

$$\left|R_4(x)\right| = \frac{105}{32} c^{-9/2} \frac{|x-1|^5}{5!} \le \begin{cases} \dfrac{105}{32}\left(\dfrac{1}{4}\right)^{-9/2} \dfrac{|x-1|^5}{5!} & \text{if } \dfrac{1}{4} \le x \le 1 \\ \dfrac{105}{32} \cdot 1^{-9/2} \cdot \dfrac{|x-1|^5}{5!} & \text{if } 1 \le x \le 4 \end{cases} = \begin{cases} 14 \cdot (1-x)^5 & \text{if } \dfrac{1}{4} \le x \le 1, \\ \dfrac{7}{256} \cdot (x-1)^5 & \text{if } 1 \le x \le 4. \end{cases}$$

REMARK: If $x \in \left[\frac{1}{4}, 4\right]$, then

$$\left|R_4(x)\right| \le \begin{cases} 14 \cdot |x-1|^5 \\ \dfrac{7}{256} \cdot |x-1|^5 \end{cases} \le \begin{cases} 14 \cdot \left|\dfrac{1}{4} - 1\right|^5 \\ \dfrac{7}{256} \cdot |4-1|^5 \end{cases} \le \begin{cases} \dfrac{1701}{512} & \text{if } \dfrac{1}{4} \le x \le 1 \\ \dfrac{1701}{256} & \text{if } 1 \le x \le 4 \end{cases} \le \dfrac{1701}{256} \approx 6.64.$$

If we are less careful, we obtain a very crude bound: $\left|R_4(x)\right| \le \dfrac{105}{32}\left(\dfrac{1}{4}\right)^{-9/2} \dfrac{|4-1|^5}{5!} = 3402.$

5. If $f(x) = e^{\beta x}$, then $f^{(5)}(x) = \beta^5 e^{\beta x}$. The exponential function is monotonic, but we do not know the sign of β, hence we describe the bound for $\left|f^{(5)}(x)\right|$ on the interval $[0,1]$ as $\max\left\{\left|f^{(5)}(0)\right|, \left|f^{(5)}(1)\right|\right\} = |\beta|^5 \cdot \max\left\{1, e^\beta\right\}$. Therefore $\left|R_4(x)\right| \le |\beta|^5 \cdot \max\left\{1, e^\beta\right\} \cdot \dfrac{|1-0|^5}{5!} = \dfrac{|\beta|^5 \cdot \max\left\{1, e^\beta\right\}}{120}.$

7. Let $f(x) = \tan x$. Solution 9.4.7 shows the first four derivatives of $\tan x$, hence

$$f^{(5)}(x) = \frac{d}{dx}\left(8\sec^2 x \tan^3 x + 16\sec^4 x \tan x\right)$$
$$= 8\sec^2 x \left(2\sec^4 x + 11\sec^2 x \tan^2 x + 2\tan^4 x\right) = 8\left(1 + \tan^2 x\right)\left(2 + 15\tan^2 x + 15\tan^4 x\right).$$

Since $\tan x$ increases from 0 to 1 on $\left[0, \frac{\pi}{4}\right]$, we see that $\left|f^{(5)}(x)\right| \le 8 \cdot (1+1) \cdot (2 + 15 + 15) = 512$; thus $\left|R_4(x)\right| \le 512 \cdot \dfrac{|x - 0|^5}{5!} = \dfrac{64}{15}|x|^5$ and we can get a constant bound: $\left|R_4(x)\right| \le \dfrac{64}{15}\left(\dfrac{\pi}{4}\right)^5 = \dfrac{\pi^5}{240} \approx 1.275.$

REMARK: In the spirit of the Remark following Solution 9.4.9, if we let $T = \tan x$ for brevity, then

$$f'(x) = T' = \sec^2 x = 1 + T^2,$$
$$f''(x) = 2T \cdot T' = 2T\left(1 + T^2\right) = 2\left(T + T^3\right),$$
$$f'''(x) = 2\left(1 + 3T^2\right) \cdot T' = 2\left(1 + 3T^2\right)\left(1 + T^2\right) = 2\left(1 + 4T^2 + 3T^4\right),$$
$$f^{(4)}(x) = 2\left(8T + 12T^3\right) \cdot T' = 2 \cdot 4\left(2T + 3T^3\right)\left(1 + T^2\right) = 8\left(2T + 5T^3 + 3T^5\right),$$
$$f^{(5)}(x) = 8\left(2 + 15T^2 + 15T^4\right) \cdot T' = 8\left(2 + 15T^2 + 15T^4\right)\left(1 + T^2\right).$$

9. If $f(x) = e^{x^2}$, then $f^{(5)}(x) = 8x\left(15 + 20x^2 + 4x^4\right)e^{x^2}$. Since $f^{(5)}(x)$ is the product of positive functions which increase when $x \ge 0$, on the interval $\left[0, \frac{1}{3}\right]$ we have $\left|R_4(x)\right| \le f^{(5)}\left(\dfrac{1}{3}\right) \cdot \dfrac{|x - 0|^5}{5!} = \dfrac{1399\, e^{1/9}}{3645} x^5$ and, if we want a constant bound, we get $\left|R_4(x)\right| = 1399 \cdot 3^{-11} \cdot 5^{-1} \cdot e^{1/9} \approx 0.001765.$

11.   Let $f(x) = \sin x$ and $a = \frac{\pi}{6} + 0.2$. For every n, $\left|f^{(n)}(x)\right| \le 1$, thus $\left|R_n\left(\frac{\pi}{6} + 0.2\right)\right| \le \frac{(0.2)^{n+1}}{(n+1)!}$. If $n = 1$,

then $\frac{(0.2)^{n+1}}{(n+1)!} = \frac{0.04}{2} = 0.02$; if $n = 2$, it equals $\frac{0.008}{6} \approx 0.0013$; if $n = 3$, $\frac{(0.2)^4}{4!} = \frac{0.0016}{24} \approx 0.0000667$.

Hence, with error bounded by $6.67 \cdot 10^{-5}$,

$$\sin\left(\frac{\pi}{6} + 0.2\right) \approx \sin\frac{\pi}{6} + \left(\cos\frac{\pi}{6}\right)(0.2) - \left(\sin\frac{\pi}{6}\right)\frac{(0.2)^2}{2!} - \left(\cos\frac{\pi}{6}\right)\frac{(0.2)^3}{3!} = \frac{1}{2} + \frac{\sqrt{3}}{10} - \frac{1}{100} - \frac{\sqrt{3}}{1500} \approx 0.66205.$$

13.   To approximate $e = e^1$ we will obtain an appropriate Taylor polynomial for $f(x) = e^x$ centered at $a = 0$ and then evaluate that polynomial at $x = 1$. Using the facts $f^{(n+1)}(x) = e^x$, e^x increases, and $e < 3$, we

see $\left|R_n(1)\right| \le e^1 \cdot \frac{|1 - 0|^{n+1}}{(n+1)!} < \frac{3}{(n+1)!}$. To approximate e with error bounded by 0.0001, we want n such

that $\frac{3}{(n+1)!} \le 0.0001$; that inequality is equivalent to $(n+1)! \ge \frac{3}{0.0001} = 30{,}000$; since $7! = 5{,}040$ and

$8! = 40{,}320$, we choose $n = 8 - 1 = 7$. Therefore, with error bounded by $\frac{3}{8!} \approx 7.44 \cdot 10^{-5}$, we approximate

$$e = e^1 \approx \sum_{k=0}^{7} \frac{1}{k!} = \frac{658}{252} \approx 2.71825.$$

REMARK: Instead of determining in advance how many terms to use, we might prefer to just accumulate terms while using some testing procedure to decide when we can safely stop. (Theorem 10.7.3 will give justification for one such procedure in the setting of some convergent alternating series.)

    We have error bound $B_{n-1} = \max\left\{\left|f^{(n)}(c)\right| : a \le c \le x\right\} \cdot \frac{|x - a|^n}{n!}$ for using the $(n-1)$-degree

Taylor polynomial centered at a to approximate $f(x)$; the term we would add in order to have the nth-

degree Taylor polynomial is $T_n = f^{(n)}(a)\frac{(x - a)^n}{n!}$. For many "nice" functions, we can find a positive

constant r such that $r_n = \frac{|T_n|}{B_{n-1}} \ge r$ for all n. Then $|T_n| < r\epsilon$ implies $B_{n-1} \le \frac{|T_n|}{r} < \epsilon$. We accumulate

the terms until we get one with magnitude less than $r\epsilon$; it is usually a good idea to also add that term because we will then have an even better approximation.

    For Problem 13 we have $r_n = \frac{f^{(n)}(0)}{f^{(n)}(1)} = \frac{1}{e} > \frac{1}{3}$; we continue until $|T_n| < \frac{0.0001}{3} \approx 3.33 \cdot 10^{-5}$.

n	4	5	6	7	8
$\frac{1}{n!}$	$\frac{1}{24} \approx .04$	$\frac{1}{120} \approx 8.3 \cdot 10^{-3}$	$\frac{1}{720} \approx 1.4 \cdot 10^{-3}$	$\frac{1}{5040} \approx 2.0 \cdot 10^{-4}$	$\frac{1}{40320} \approx 2.5 \cdot 10^{-5}$
$\sum_{k=0}^{n} \frac{1}{k!}$	$\frac{65}{24} \approx 2.70833$	$\frac{163}{60} \approx 2.71667$	$\frac{1957}{720} \approx 2.71806$	$\frac{685}{252} \approx 2.71825$	$\frac{109601}{40320} \approx 2.71828$

15. $33° = 30° + 3° = \frac{\pi}{6} + \frac{\pi}{60}$. Let $f(x) = \sin x$ and $a = \frac{\pi}{6}$. Since $\left|f^{(n)}(x)\right| \le 1$, we need n such that

$\left|R_n\left(\frac{\pi}{60}\right)\right| \le B_n = 1 \cdot \frac{(\pi/60)^{n+1}}{(n+1)!} \le 0.001$. We compute $B_1 \approx 0.00137$ and $B_2 \approx 0.0000239$. Thus we use P_2:

$$\sin 33° \approx P_2\left(\frac{\pi}{60}\right) = \sin\left(\frac{\pi}{6}\right) + \cos\left(\frac{\pi}{6}\right) \cdot \frac{\pi}{60} - \sin\left(\frac{\pi}{6}\right) \cdot \frac{(\pi/60)^2}{2} = \frac{1}{2} + \frac{\sqrt{3}}{2} \cdot \frac{\pi}{60} - \frac{1}{4}\left(\frac{\pi}{60}\right)^2 \approx 0.54466.$$

17.   Solution 9.4.7 shows $x + \frac{x^3}{3}$ is the 4th-degree Taylor polynomial for $\tan x$ centered at $a = 0$, thus

$$\int_0^{\pi/4} \tan x \, dx \approx \int_0^{\pi/4}\left(x + \frac{x^3}{3}\right) dx = \left(\frac{x^2}{2} + \frac{x^4}{12}\right)\Big|_0^{\pi/4} = \frac{(\pi/4)^2}{2} + \frac{(\pi/4)^4}{12} \approx 0.34013.$$

Solution 7 of this section shows that if $0 \le x \le \frac{\pi}{4}$, then the absolute error in approximating $\tan x$ by $x + \frac{x^3}{3}$

has an upper bound of $512\dfrac{(x-0)^5}{5!} = \dfrac{64x^5}{15}$. Thus we have a bound for the error of our integral estimate:

$$\left| \int_0^{\pi/4} \tan x \, dx - \int_0^{\pi/4} \left(x + \frac{x^3}{3} \right) dx \right| \le \int_0^{\pi/4} \left| \tan x - \left(x + \frac{x^3}{3} \right) \right| dx$$

$$= \int_0^{\pi/4} |R_4(x)| \, dx \le \int_0^{\pi/4} \frac{64x^5}{15} \, dx = \frac{32x^6}{45} \Big|_0^{\pi/4} = \frac{\pi^6}{5760} \approx 0.16691.$$

REMARK: If our bound for $|R_4(x)|$ was too complicated as a function of x to integrate, then we might simplify further by integrating a constant bound: $0 \le x \le \dfrac{\pi}{4} \implies |R_4(x)| \le \dfrac{64x^5}{15} \le \dfrac{64(\pi/4)^5}{15} = \dfrac{\pi^5}{240}$;

therefore $\displaystyle\int_0^{\pi/4} |R_4(x)| \, dx \le \int_0^{\pi/4} \frac{\pi^5}{240} \, dx = \frac{\pi^6}{960} \approx 1.0014$ (6 times larger than bound obtained above).

$\displaystyle\int_0^{\pi/4} \tan x \, dx = -\ln \cos x \Big|_0^{\pi/4} = \frac{1}{2}\ln 2 \approx 0.34657$ and the actual error is roughly 0.00644.

---

19. Solution 9.4.24 shows $1 + x^2 + \dfrac{x^4}{2}$ is the 4th-degree Taylor polynomial for e^{x^2} centered at $a = 0$, thus

$$\int_0^{1/3} e^{x^2} \, dx \approx \int_0^{1/3} \left(1 + x^2 + \frac{x^4}{2} \right) dx = \left(x + \frac{x^2}{3} + \frac{x^5}{10} \right) \Big|_0^{1/3} = \frac{841}{2430} \approx 0.34609.$$

Solution 9 of this section shows that if $0 \le x \le \dfrac{1}{3}$, then the absolute error in approximating e^{x^2} by $1 + x^2 + \dfrac{x^4}{2}$ has an upper bound of $\dfrac{1399\, e^{1/9}}{3645}\, x^5$. Thus we have a bound for the error of our integral estimate:

$$\left| \int_0^{1/3} e^{x^2} \, dx - \int_0^{1/3} \left(1 + x^2 + \frac{x^4}{2} \right) dx \right| \le \int_0^{1/3} \left| e^{x^2} - \left(1 + x^2 + \frac{x^4}{2} \right) \right| dx$$

$$= \int_0^{1/3} |R_4(x)| \, dx \le \int_0^{1/3} \frac{1399\, e^{1/9}}{3645}\, x^5 \, dx \approx 9.81 \cdot 10^{-5}.$$

REMARK: $\displaystyle\int_0^{1/3} e^{x^2} \, dx \approx 0.3461017$ and the actual error is roughly $-1.12 \cdot 10^{-5}$.

---

21. Let $f(x) = e^{x^3}$, each derivative of f has the form of a polynomial times f; the following table displays those polynomials for the first five derivatives of f:

n	1	2	3	4	5
$\dfrac{f^{(n)}(x)}{e^{x^3}}$	$3x^2$	$3x(2 + 3x^3)$	$3(2 + 18x^3 + 9x^6)$	$9x^2(20 + 36x^3 + 9x^6)$	$9x(40 + 240x^3 + 180x^6 + 27x^9)$

Since $f^{(4)}(0) = 0 = f^{(5)}(0)$, we see that $1 + x^3$ is the 3rd, 4th, and 5th-degree Taylor polynomial for f at $a = 0$. Each of the derivatives of f is a positive increasing function, hence if $0 \le x \le \dfrac{1}{5}$, then

$$|R_4(x)| \le f^{(5)}\left(\frac{1}{5} \right) \cdot \frac{|x - 0|^5}{5!} \approx 0.63402\, x^5 \text{ and}$$

$$\left| \int_0^{1/5} e^{x^3} \, dx - \int_0^{1/5} (1 + x^3) \, dx \right| \le \int_0^{1/5} \left| e^{x^3} - (1 + x^3) \right| dx = \int_0^{1/5} |R_4(x)| \, dx \le \int_0^{1/5} 0.63402\, x^5 \, dx \approx 6.76 \cdot 10^{-6}.$$

Therefore

$$\int_0^{1/5} e^{x^3} \, dx \approx \int_0^{1/5} (1 + x^3) \, dx = \left(x + \frac{x^4}{4} \right) \Big|_0^{1/5} = 0.2004.$$

REMARK: $\displaystyle\int_0^{1/5} e^{x^3} \, dx \approx 0.20040092$ so the actual error is roughly $9.2 \cdot 10^{-9}$.

---

23.   Let $f(x) = \cos x^2$. We will obtain a Taylor polynomial P_n for f centered at $a = 0$ and compute $\int_0^{\pi/6} P_n$ as an approximation of $\int_0^{\pi/6} \cos x^2 \, dx$. If M_n is an upper bound for $\left|f^{(n+1)}(x)\right|$ on the interval $\left[0, \frac{\pi}{6}\right]$, then $\left|f(x) - P_n(x)\right| = \left|R_n(x)\right| \le M_n \cdot \dfrac{x^{n+1}}{(n+1)!}$ and we obtain an upper bound for $\left|\int_0^{\pi/6} f - \int_0^{\pi/6} P_n\right|$ as

$$B_n = \int_0^{\pi/6} M_n \cdot \frac{x^{n+1}}{(n+1)!} \, dx = M_n \cdot \frac{(\pi/6)^{n+2}}{(n+2)!}.$$

In the following table we report bounds M_n which were found quickly by liberal use of the triangle inequality ($|a \pm b| \le |a| + |b|$), identification of monotonic subexpressions of $f^{(n+1)}$ and evaluation of them at an endpoint of the interval $[0, \pi/6]$.

n	$f^{(n)}(x)$	$f^{(n)}(0)$	M_n	B_n
0	$\cos x^2$	1	$2 \cdot \frac{\pi}{6} \cdot \sin\frac{\pi^2}{36} \approx 0.28$	0.03886
1	$-2x \sin x^2$	0	$2\left(\frac{2\pi^2}{36} + \sin\frac{\pi^2}{36}\right) \approx 1.64$	0.03919
2	$-2\left(2x^2 \cos x^2 + \sin x^2\right)$	0	$\frac{4\pi}{6}\left(3 + \frac{2\pi^2}{36}\sin\frac{\pi^2}{36}\right) \approx 6.59$	0.02065
3	$-4x\left(3\cos x^2 - 2x^2 \sin x^2\right)$	0	$4\left(3 + \frac{12\pi^2}{36}\sin\frac{\pi^2}{36}\right) \approx 15.56$	0.00510
4	$4\left((4x^4 - 3)\cos x^2 + 12x^2 \sin x^2\right)$	-12	$\frac{8\pi}{6}\left(\frac{20\pi^2}{36} + 15\sin\frac{\pi^2}{36}\right) \approx 39.98$	0.00114
5	$8x\left(20x^2 \cos x^2 + (15 - 4x^4)\sin x^2\right)$	0	$8\left(\frac{2\pi^2}{36}(45) + 15\sin\frac{\pi^2}{36}\right) \approx 801.61$	0.00172
6	$8\left(2x^2(45 - 4x^4)\cos x^2 + 15(1 - 4x^4)\sin x^2\right)$	0	$\frac{16\pi}{6}\left(105 + \frac{2\pi^2}{36}(105)\sin\frac{\pi^2}{36}\right) \approx 1010.23$	$1.42 \cdot 10^{-4}$
7	$16x\left((105 - 84x^4)\cos x^2 + 2x^2(4x^4 - 105)\sin x^2\right)$	0		

The computations summarized in the above table allow us to infer

$$\int_0^{\pi/6} \cos x^2 \, dx \approx \int_0^{\pi/6} P_6(x) \, dx = \int_0^{\pi/6}\left(1 + (-12)\frac{x^4}{4!}\right) dx = \int_0^{\pi/6}\left(1 - \frac{x^4}{2}\right) dx = \frac{\pi}{6} - \frac{\pi^5}{10 \cdot 6^5} \approx 0.51966$$

with $\left|\int_0^{\pi/6} \cos x^2 \, dx - \int_0^{\pi/6} P_6(x) \, dx\right| \le \int_0^{\pi/6} |R_6(x)| \, dx \le \int_0^{\pi/6} 1010.23 \, \frac{x^7}{7!} \, dx \approx 1.42 \cdot 10^{-4}$.

REMARK: $\int_0^{\pi/6} \cos x^2 \, dx \approx 0.51967701$ and the actual error is roughly $1.37 \cdot 10^{-5}$.

$f^{(5)}(x)$ is increasing on the interval $[0, \pi/6]$ so we can use $M_4 = f^{(5)}(\pi/6) \approx 38.77968$, but this only yields $B_4 \approx 0.0011098$. It is also true that $f^{(6)}(x)$ is increasing on the interval, using $M_5 = f^{(6)}(\pi/6) \approx 182.25$ yields $B_5 \approx 0.00039$ so we could have stopped one step earlier.

25. a. $1.03^3 = (1 + 0.03)^3 = 1 + 3 \cdot (0.03) + \frac{3 \cdot 2}{2!} \cdot (0.03)^2 + \frac{3!}{3!} \cdot (0.03)^3 = 1 + 0.09 + 0.0027 + 0.000027 = 1.092727$.

   b. $0.97^4 = \left(1 + (-0.03)\right)^4 = 1 + 4(-0.03) + \frac{4 \cdot 3}{2!}(-0.03)^2 + \frac{4 \cdot 3 \cdot 2}{3!}(-0.03)^3 + \frac{4!}{4!}(-0.03)^4$
$$= 1 - 0.12 + 0.0054 - 0.000108 + 0.00000081 = 0.88529281.$$

   c. $1.2^4 = \left(1 + 0.2\right)^4 = 1 + 4(0.2) + 6(0.2)^2 + 4(0.2)^3 + (0.2)^4$
$$= 1 + 0.8 + 0.24 + 0.032 + 0.0016 = 2.0736.$$

   d. $0.8^5 = \left(1 + (-0.2)\right)^5$
$$= 1 + 5(-0.2) + \frac{5 \cdot 4}{2!}(-0.2)^2 + \frac{5 \cdot 4 \cdot 3}{3!}(-0.2)^3 + \frac{5 \cdot 4 \cdot 3 \cdot 2}{4!}(-0.2)^4 + \frac{5!}{5!}(-0.2)^5$$
$$= 1 - 1 + 0.4 - 0.08 + 0.008 - 0.00032 = 0.32768.$$

27.   Problem 28 implies $\pi = 4 \cdot \dfrac{\pi}{4} = 4 \cdot \tan^{-1} 1 = 4 \cdot \left(4\tan^{-1}\dfrac{1}{5} - \tan^{-1}\dfrac{1}{239}\right) = 16\tan^{-1}\dfrac{1}{5} - 4\tan^{-1}\dfrac{1}{239}$

Problem 34(b) states an inequality which we can apply to accumulate terms of the Taylor polynomials for $\tan x$ stopping when the next term to be added is smaller (in absolute value) than our chosen error bound.

n	0	1	2	3	4
$T_n = 16\dfrac{(-1)^n(1/5)^{2n+1}}{2n+1}$	3.2	$-4.27\cdot 10^{-2}$	$1.02\cdot 10^{-3}$	$-2.93\cdot 10^{-5}$	$9.10\cdot 10^{-7}$
$\displaystyle\sum_{k=0}^{n} T_k$	3.2	3.1573333333	3.1583573333	3.1583280762	3.1583289864

n	0	1
$T_n = 4\dfrac{(-1)^n(1/239)^{2n+1}}{2n+1}$	$1.67\cdot 10^{-2}$	$-9.77\cdot 10^{-8}$
$\displaystyle\sum_{k=0}^{n} T_k$	0.0167364017	0.0167363040

Thus $\pi \approx 3.1583280762 - 0.0167364017 \approx 3.1415916745$ with error bounded by $9.1022\cdot 10^{-7} + 9.776\cdot 10^{-8} \approx 1.008\cdot 10^{-6}$. (We also have $\pi \approx 3.1583289864 - 0.0167363040 \approx 3.1445926824$ with unknown error bound.)

29. Let $f(x) = (1+x)^n$ where n is a positive integer. Then the induction discussion of Solution 2.7.29 shows that

$$f'(x) = n\cdot(1+x)^{n-1}$$
$$f''(x) = n\cdot(n-1)\cdot(1+x)^{n-2}$$
$$\vdots$$
$$f^{(k)} = n\cdot(n-1)\cdots(n-k+1)\cdot(1+x)^{n-k} \qquad \text{for any integer } k \text{ between 1 and } n$$
$$\vdots$$
$$f^{(n)}(x) = n\cdot(n-1)\cdots 3\cdot 2\cdot 1 = n!$$

a. $f^{(n)}(x)$ is constant, therefore $f^{(n+1)}(x) = \dfrac{d}{dx}f^{(n)}(x) = \dfrac{d}{dx}n! = 0$. ∎

b. $f(0) = (1+0)^n = 1^n = 1$, $f'(0) = n$, $f''(0) = n(n-1)$, $f'''(0) = n(n-1)(n-2)$, $\cdots$, $f^{(n-k)}(0) = n(n-1)(n-2)\cdots(n-k+1)$, $\cdots$, $f^{(n)}(0) = n!$, and $f^{(k)}(0) = 0$ for $k > n$. Therefore the nth-degree Taylor polynomial for $(1+x)^n$ at $a = 0$ is

$$1 + nx + \frac{n(n-1)}{2!}x^2 + \frac{n(n-1)(n-2)}{3!}x^3 + \cdots + \frac{n(n-1)(n-2)\cdots(n-k+1)}{k!}x^k + \cdots + x^n.$$

Since $f^{(n+1)}(x) = 0$ for all x, Theorem 1 implies $|R_n(x)| \le 0$ which then implies $R_n(x) = 0$ and the nth-degree Taylor polynomial equals $(1+x)^n$. ∎

REMARK: The coefficient of x^k in this Taylor polynomial is $\dfrac{n(n-1)(n-2)\cdots(n-k+1)}{k!} = \dfrac{n!}{k!\cdot(n-k)!}$.

31. $0 \le t \le 1 \implies 1 \le 1+t \implies 0 < \dfrac{1}{1+t} \le 1$, therefore $0 \le t \le x \le 1$ implies

$$|R_n(x)| = \left|(-1)^n \int_0^x \frac{t^n}{1+t}\,dt\right| = \int_0^x \frac{1}{1+t}\cdot t^n\,dt \le \int_0^x t^n\,dt = \frac{x^{n+1}}{n+1} \le \frac{1}{n+1}. \quad ∎$$

33. a. Solution 31 shows that $0 \le x \le 1$ implies $0 \le |R_n(x)| \le \dfrac{1}{n+1}$; Solution 32(c) shows that $-1 < x \le 0$ implies $0 \le |R_n(x)| \le \dfrac{(-x)^{n+1}}{(1+x)(n+1)} \le \dfrac{1}{(1+x)(n+1)}$. Since $\displaystyle\lim_{N\to\infty}\frac{1}{n+1} = 0$, the Squeezing Theorem (1.4.1) then implies $\displaystyle\lim_{N\to\infty}|R_n(x)| = 0$. ∎

b. Solutions 31 and 32(c) actually show $|R_n(x)| \le \begin{cases} \dfrac{x^{n+1}}{n+1} & \text{if } 0 \le x \le 1, \\[2ex] \dfrac{(-x)^{n+1}}{(1+x)(n+1)} & \text{if } -1 < x \le 0. \end{cases}$ Division by

$|x^n|$ yields the inequalities $\left|\dfrac{R_n(x)}{x^n}\right| \le \begin{cases} \dfrac{x}{n+1} & \text{if } 0 \le x \le 1, \\[2ex] \dfrac{|x|}{(1+x)(n+1)} & \text{if } -1 < x \le 0. \end{cases}$ Because $\displaystyle\lim_{x\to 0^+}\frac{x}{n+1} = 0$ and

$\lim\limits_{x\to 0^-}\dfrac{|x|}{(1+x)(n+1)}=0$, the Squeezing Theorem implies $\lim\limits_{x\to 0^+}\left|\dfrac{R_n(x)}{x^n}\right|$ and $\lim\limits_{x\to 0^-}\left|\dfrac{R_n(x)}{x^n}\right|$ both exist and equal zero; therefore the two-sided limit also exists and equals zero. ∎

35.   Let N be an integer such that $N>2\,|x|$; then $|x|<\dfrac{N}{2}$. If $k>0$, then

$$\frac{|x|}{N+k}<\frac{N/2}{N+k}=\frac{1}{2}\left(\frac{N}{N+k}\right)<\frac{1}{2}.$$

Therefore, if $n>N$, we see that

$$\left|\frac{x^n}{n!}\right|=\frac{|x|^n}{n!}=\overbrace{\frac{|x|\cdot|x|\cdots|x|}{1\cdot 2\cdots N}}^{N\text{ times}}\cdot\overbrace{\frac{|x|}{N+1}\cdot\frac{|x|}{N+2}\cdots\frac{|x|}{n}}^{n-N\text{ terms, each }<\frac{1}{2}}<\frac{|x|^N}{N!}\cdot\left(\frac{1}{2}\right)^{n-N}=\frac{|x|^N}{N!}\cdot 2^N\cdot\left(\frac{1}{2}\right)^n.$$

Since x and N are fixed, the preceding inequality implies one of the form $0<\left|\dfrac{x^n}{n!}\right|\le M\cdot\left(\dfrac{1}{2}\right)^n$ where M is a positive constant. Since $\lim\limits_{N\to\infty}\left(\dfrac{1}{2}\right)^n=0$, the Squeezing Theorem implies $\lim\limits_{N\to\infty}\left|\dfrac{x^n}{n!}\right|=0=\lim\limits_{N\to\infty}\dfrac{x^n}{n!}$. ∎

Solutions 9.R   *Review*   (pages 634-5)

1. $\dfrac{\sin 3x}{2x}$ has the indeterminate form $\dfrac{0}{0}$ at $x=0$ and $\lim\limits_{x\to 0}\dfrac{\frac{d}{dx}\left(\sin 3x\right)}{\frac{d}{dx}\left(2x\right)}=\lim\limits_{x\to 0}\dfrac{3\cos 3x}{2}=\dfrac{3}{2}$. Therefore $\lim\limits_{x\to 0}\dfrac{\sin 3x}{2x}$ exists and it equals $\dfrac{3}{2}$.

3.   If $0<x<1$, then $0<x-x^3=x\left(1-x^2\right)$; thus $\lim\limits_{x\to 0^+}\dfrac{2x^3+3x+4}{x-x^3}=+\infty$. On the other hand, if $-1<x<0$, then $x-x^3<0$; hence $\lim\limits_{x\to 0^-}\dfrac{2x^3+3x+4}{x-x^3}=-\infty$. Therefore the two-sided limit as $x\to 0$ does not exist. (Note that L'Hôpital's rule is not relevant here.)

5. $3\,x\,e^{-x}=\dfrac{3x}{e^x}$ has the indeterminate form $\dfrac{\infty}{\infty}$ as $x\to\infty$ and $\lim\limits_{x\to\infty}\dfrac{\frac{d}{dx}\left(3x\right)}{\frac{d}{dx}\left(e^x\right)}=\lim\limits_{x\to\infty}\dfrac{3}{e^x}=0$. Therefore $\lim\limits_{x\to\infty}\dfrac{3x}{e^x}$ exists and it equals 0.

7. $\dfrac{\sqrt{x}}{\sin x}$ has the indeterminate form $\dfrac{0}{0}$ as $x\to 0^+$ and $\lim\limits_{x\to 0^+}\dfrac{\frac{d}{dx}\left(\sqrt{x}\right)}{\frac{d}{dx}\left(\sin x\right)}=\lim\limits_{x\to 0^+}\dfrac{1/(2\sqrt{x})}{\cos x}=\infty$. Therefore $\lim\limits_{x\to 0^+}\dfrac{\sqrt{x}}{\sin x}$ exists and it equals ∞.

9. $Q(x)=\ln\left(\left(\dfrac{\pi}{4}-x\right)^{\cos 2x}\right)=\cos 2x\cdot\ln\left(\dfrac{\pi}{4}-x\right)=\dfrac{\ln(\pi/4-x)}{\sec 2x}$ has indeterminate form $\dfrac{0}{0}$ as $x\to\dfrac{\pi}{4}^-$ and so does

$$\frac{\frac{d}{dx}\left(\ln(\pi/4-x)\right)}{\frac{d}{dx}\left(\sec 2x\right)}=\frac{-(\pi/4-x)^{-1}}{2\sec 2x\,\tan 2x}=\frac{\cos^2 2x}{2(x-\pi/4)\sin 2x},$$

but

$$\lim\limits_{x\to \pi/4^-}\frac{\frac{d}{dx}\left(\cos^2 2x\right)}{\frac{d}{dx}\left(2(x-\pi/4)\sin 2x\right)}=\lim\limits_{x\to \pi/4^-}\frac{-2\cos 2x\,\sin 2x}{2\sin 2x+4(x-\pi/4)\cos 2x}=\frac{0}{2}=0.$$

Therefore $\lim\limits_{x\to \pi/4^-}\left(\dfrac{\pi}{4}-x\right)^{\cos 2x}=\lim\limits_{x\to \pi/4^-}e^{Q(x)}=e^{\lim\limits_{x\to \pi/4^-}Q(x)}=e^0=1$.

11. $\dfrac{\sin x}{x^3}$ has the indeterminate form $\dfrac{0}{0}$ at $x=0$ and $\lim\limits_{x\to 0}\dfrac{\frac{d}{dx}\left(\sin x\right)}{\frac{d}{dx}\left(x^3\right)}=\lim\limits_{x\to 0}\dfrac{\cos x}{3x^2}=\infty$. Therefore $\lim\limits_{x\to 0}\dfrac{\sin x}{x^3}$ exists and is ∞.

REMARK: We can write $\dfrac{\sin x}{x^3} = \left(\dfrac{1}{x^2}\right)\left(\dfrac{\sin x}{x}\right)$ and use familiar results: $\lim\limits_{x\to 0}\dfrac{1}{x^2} = \infty$ and $\lim\limits_{x\to 0}\dfrac{\sin x}{x} = 1$.

13. $\dfrac{\tan x}{\sec x} = \dfrac{\sin x/\cos x}{1/\cos x} = \sin x$, thus $\lim\limits_{x\to\pi/2-}\dfrac{\tan x}{\sec x} = \lim\limits_{x\to\pi/2-}\sin x = 1$.

REMARK: L'Hôpital's rule would not help here because $\dfrac{\frac{d}{dx}\tan x}{\frac{d}{dx}\sec x} = \dfrac{\sec^2 x}{\sec x\tan x} = \left(\dfrac{\tan x}{\sec x}\right)^{-1}$ only lets us infer that **if** the limit exists, then it satisfies the equation $L = L^{-1}$.

15. $x - \sqrt{x^2+x} = \left(x - \sqrt{x^2+x}\right)\left(\dfrac{x+\sqrt{x^2+x}}{x+\sqrt{x^2+x}}\right) = \dfrac{-x}{x+\sqrt{x^2+x}}$ has the indeterminate form $\dfrac{-\infty}{\infty}$ as $x\to\infty$ and

$$\lim_{x\to\infty}\frac{\frac{d}{dx}(-x)}{\frac{d}{dx}\left(x+\sqrt{x^2+x}\right)} = \lim_{x\to\infty}\frac{-1}{1+\dfrac{2x+1}{2\sqrt{x^2+x}}} = \frac{-1}{1+\lim\limits_{x\to\infty}\dfrac{1+1/2x}{\sqrt{1+1/x}}} = \frac{-1}{1+\dfrac{1+0}{\sqrt{1+0}}} = \frac{-1}{1+1} = \frac{-1}{2},$$

thus $\lim\limits_{x\to\infty}\left(x - \sqrt{x^2+x}\right) = \dfrac{-1}{2}$.

REMARK: Solutions 1.4.25, 26, 38 show that we do not need to use L'Hôpital's rule here.

17. $e > 2$, thus $\dfrac{e}{2} > 1$ and $\lim\limits_{x\to\infty}\dfrac{e^x}{2^x} = \lim\limits_{x\to\infty}\left(\dfrac{e}{2}\right)^x = \infty$. This limit statement can be justified using the inequality $\left(\dfrac{e}{2}\right)^x \geq 1 + \left(\dfrac{e}{2}-1\right)x$ for $x \geq 1$ which is implied by Bernoulli's inequality (Problem 6.4.54).

19. $\lim\limits_{x\to 0} x^{10}\,e^{-1/x^2} = \left(\lim\limits_{x\to 0}x^{10}\right)\left(\lim\limits_{x\to 0}e^{-1/x^2}\right) = (0)(0) = 0.$

21. $\displaystyle\int_0^\infty e^{-3x}\,dx = \lim_{N\to\infty}\int_0^N e^{-3x}\,dx = \lim_{N\to\infty}\left.\dfrac{e^{-3x}}{-3}\right|_0^N = \lim_{N\to\infty}\left(\dfrac{-1}{3}\right)\left(e^{-3N}-1\right) = \dfrac{1}{3}.$

23. Using integration by parts or integral 166 we find

$$\int_0^\infty x^3 e^{-7x}\,dx = \lim_{N\to\infty}\int_0^N x^3 e^{-7x}\,dx = \lim_{N\to\infty}\left.\left(\dfrac{e^{-7x}}{-7}\right)\left(x^3 + \dfrac{3x^2}{7} + \dfrac{6x}{49} + \dfrac{6}{343}\right)\right|_0^N.$$

Example 9.2.5 implies $\lim\limits_{N\to\infty} e^{-7N}\cdot N^k = 0$ for $k \in \{3,2,1,0\}$ so the limit is $-\left(\dfrac{-1}{7}\right)\left(\dfrac{6}{343}\right) = \dfrac{6}{2401}.$

25. $\displaystyle\int_{50}^\infty \dfrac{dx}{x^2-4} = \lim_{N\to\infty}\int_{50}^N \left(\dfrac{1}{4}\right)\left(\dfrac{1}{x-2} - \dfrac{1}{x+2}\right)dx = \lim_{N\to\infty}\left.\left(\dfrac{1}{4}\right)\ln\left|\dfrac{x-2}{x+2}\right|\right|_{50}^N$

$\qquad = \lim\limits_{N\to\infty}\left(\dfrac{1}{4}\right)\left(\ln\left|\dfrac{N-2}{N+2}\right| - \ln\dfrac{48}{52}\right) = \left(\dfrac{1}{4}\right)\left(\ln 1 - \ln\dfrac{12}{13}\right) = -\left(\dfrac{1}{4}\right)\ln\dfrac{12}{13} = \left(\dfrac{1}{4}\right)\ln\dfrac{13}{12}.$

27. $\displaystyle\int_0^1 \dfrac{dx}{x^4} = \lim_{\epsilon\to 0+}\int_\epsilon^1 x^{-4}\,dx = \lim_{\epsilon\to 0+}\left.\left(\dfrac{1}{3}\right)x^{-3}\right|_\epsilon^1 = \lim_{\epsilon\to 0+}\left(\dfrac{1}{3}\right)\left(1-\epsilon^{-3}\right) = \infty$, this improper integral diverges. (See Example 9.3.11.)

29. $\displaystyle\int_{-\infty}^0 \dfrac{dx}{(1-x)^{5/2}} = \lim_{M\to\infty}\int_{-M}^0 (1-x)^{-5/2}\,dx = \lim_{M\to\infty}\left.\left(\dfrac{2}{3}\right)(1-x)^{-3/2}\right|_{-M}^0$

$\qquad = \left(\dfrac{2}{3}\right)\left(1 - \lim\limits_{M\to\infty}\dfrac{1}{(1+M)^{3/2}}\right) = \left(\dfrac{2}{3}\right)(1-0) = \dfrac{2}{3}.$

31. $\displaystyle\int_0^{\pi/2}\csc x\,dx = \lim_{\epsilon\to 0+}\int_\epsilon^{\pi/2}\csc x\,dx = \lim_{\epsilon\to 0+}\left.\ln\left|\tan\dfrac{x}{2}\right|\right|_\epsilon^{\pi/2} = \ln\tan\dfrac{\pi}{4} - \lim_{\epsilon\to 0+}\ln\left|\tan\dfrac{\epsilon}{2}\right| = 0 - (-\infty) = \infty,$ this improper integral diverges.

33. $\displaystyle\int_2^4 \dfrac{dx}{\sqrt{x-2}} = \lim_{\epsilon\to 0+}\int_{2+\epsilon}^4 \dfrac{dx}{\sqrt{x-2}} = \lim_{\epsilon\to 0+}\left.2\sqrt{x-2}\right|_{2+\epsilon}^4 = 2\sqrt{4-2} - 2\lim_{\epsilon\to 0+}\sqrt{\epsilon} = 2\sqrt 2 - 0 = 2^{3/2}.$

35.   If $x \geq 1$, then $x^3 + 1 \geq 2$ and $\dfrac{x^3}{x^3 + 1} = 1 - \dfrac{1}{x^3 + 1} \geq 1 - \dfrac{1}{2} = \dfrac{1}{2}$. This implies

$$\int_0^N \frac{x^3}{x^3 + 1}\, dx \geq \int_1^N \frac{x^3}{x^3 + 1}\, dx \geq \int_1^N \frac{1}{2}\, dx = \frac{N-1}{2};$$

hence $\displaystyle\int_0^\infty \frac{x^3}{x^3+1}\, dx$ diverges.

Alternative for Masochists: Apply the techniques of Sections 7.5-7 to find

$$\int \frac{x^3}{x^3+1}\, dx = x - \frac{1}{3}\ln|x+1| + \frac{1}{6}\ln\left|x^2 - x + 1\right| - \frac{1}{\sqrt{3}}\tan^{-1}\left(\frac{2x-1}{\sqrt{3}}\right) + C$$

$$= x + \frac{1}{6}\ln\left|\frac{x^2 - x + 1}{x^2 + 2x + 1}\right| - \frac{1}{\sqrt{3}}\tan^{-1}\left(\frac{2x-1}{\sqrt{3}}\right) + C.$$

Then use the information that $\displaystyle\lim_{N\to\infty}\ln\left|\frac{N^2 - N + 1}{N^2 + 2N + 1}\right| = \ln 1 = 0$, $\displaystyle\lim_{N\to\infty}\tan^{-1}\left(\frac{2N-1}{\sqrt{3}}\right) = \frac{\pi}{2}$, and $\displaystyle\lim_{N\to\infty}(N + C) = \infty$ to infer the improper integral diverges.

---

37. $\displaystyle\int_2^\infty \frac{dx}{(x-1)^{1/100}} = \lim_{N\to\infty}\int_2^N (x-1)^{-1/100}\, dx = \left(\frac{100}{99}\right)\lim_{N\to\infty}\left((N-1)^{99/100} - 1\right) = \infty$, this improper integral diverges.

39.   The graph of $f(x) = \dfrac{1}{x^2 - 3x + 2} = \dfrac{1}{x-2} - \dfrac{1}{x-1}$ has vertical asymptotes at $x = 1$ and $x = 2$. The improper integral $\displaystyle\int_0^\infty f$ exists if and only if all five of the following limits exist and are finite: $\displaystyle\lim_{\alpha\to 0^+}\int_0^{1-\alpha} f$, $\displaystyle\lim_{\beta\to 0^+}\int_{1+\beta}^{3/2} f$, $\displaystyle\lim_{\gamma\to 0^+}\int_{3/2}^{2-\gamma} f$, $\displaystyle\lim_{\delta\to 0^+}\int_{2+\delta}^{3} f$, and $\displaystyle\lim_{N\to\infty}\int_3^N f$. Because $\displaystyle\int \frac{dx}{x^2 - 3x + 2} = \ln\left|\frac{x-2}{x-1}\right| + C$, $\displaystyle\lim_{w\to 0^+}\ln w = -\infty$, and $\displaystyle\lim_{w\to\infty}\ln w = \infty$, each of the first four limits diverges; therefore $\displaystyle\int_0^\infty \frac{dx}{x^2 - 3x + 2}$ also diverges.

41.   If $f(x) = e^x$, then $f^{(n)}(x) = e^x$ for every n. Therefore, if $a = 0$, then $f^{(n)}(a) = e^0 = 1$ and $P_3(x) = 1 + x + \dfrac{x^2}{2!} + \dfrac{x^3}{3!}$.

43.

n	0	1	2	3
$f^{(n)}(x)$	$\sin x$	$\cos x$	$-\sin x$	$-\cos x$
$f^{(n)}\left(\frac{\pi}{6}\right)$	$\dfrac{1}{2}$	$\dfrac{\sqrt{3}}{2}$	$\dfrac{-1}{2}$	$\dfrac{-\sqrt{3}}{2}$

$$P_3(x) = \frac{1}{2} + \left(\frac{\sqrt{3}}{2}\right)\left(x - \frac{\pi}{6}\right) + \left(\frac{-1}{2}\right)\frac{(x - \pi/6)^2}{2!} + \left(\frac{-\sqrt{3}}{2}\right)\frac{(x - \pi/6)^3}{3!}.$$

45.

n	0	1	2	3	4
$f^{(n)}(x)$	$\cot x$	$-\csc^2 x$	$2\csc^2 x \cot x$	$-4\csc^2 x \cot^2 x - 2\csc^4 x$	$8\csc^2 x \cot^3 x + 16\csc^4 x \cot x$
$f^{(n)}\left(\frac{\pi}{2}\right)$	0	-1	0	-2	0

$$P_4(x) = 0 + (-1)\left(x - \frac{\pi}{2}\right) + (0)\frac{(x - \pi/2)^2}{2!} + (-2)\frac{(x - \pi/2)^3}{3!} + (0)\frac{(x - \pi/2)^4}{4!} = -\left(x - \frac{\pi}{2}\right) - \left(\frac{1}{3}\right)\left(x - \frac{\pi}{2}\right)^3.$$

47.

n	0	1	2	3	4	5	6	7	8
$f^{(n)}(x)$	$x^3 - x^2 + 2x + 3$	$3x^2 - 2x + 2$	$6x - 2$	6	0	0	0	0	0
$f^{(n)}(0)$	3	2	-2	6	0	0	0	0	0

$$P_8(x) = 3 + (2)x + (-2)\frac{x^2}{2!} + (6)\frac{x^3}{3!} + (0)\frac{x^4}{4!} + (0)\frac{x^5}{5!} + (0)\frac{x^6}{6!} + (0)\frac{x^7}{7!} + (0)\frac{x^8}{8!} = 3 + 2x - x^2 + x^3.$$

49.   If $f(x) = \cos x$, then $\left|f^{(n)}(x)\right| \le 1$ for all n and x. Hence, for the 5th-degree Taylor polynomial for f centered at $a = \dfrac{\pi}{6}$, we have $\left|R_5(x)\right| = \left|-\cos c\right| \cdot \dfrac{|x - \pi/6|^6}{6!} \le \dfrac{|x - \pi/6|^6}{6!}$. If $x \in \left[0, \dfrac{\pi}{2}\right]$, then $\left|x - \dfrac{\pi}{6}\right| \le \max\left\{\left|0 - \dfrac{\pi}{6}\right|, \left|\dfrac{\pi}{2} - \dfrac{\pi}{6}\right|\right\} = \dfrac{\pi}{3}$; therefore $\left|R_5(x)\right| \le \dfrac{(\pi/3)^6}{6!} \approx 1.83 \cdot 10^{-3}$.

51.   If $f(x) = e^x$, then $f^{(n)}(x) = e^x$ and $f^{(n)}(0) = e^0 = 1$ for all n. For any x there is some c between 0 and x such that $\left|R_6(x)\right| = e^c \cdot \dfrac{|x - 0|^7}{7!}$. We know e^x is an increasing function; if $x \in [-\ln e, \ln e] = [-1, 1]$, then $\left|R_6(x)\right| \le e^1 \cdot \dfrac{|x|^7}{7!} \le e^1 \cdot \dfrac{1^7}{7!} = \dfrac{e}{7!} \approx 5.39 \cdot 10^{-4}$.

53. $\boxed{\text{Disk method}}$ $V = \displaystyle\int_0^\infty \pi\left((x+1)^{-2}\right)^2 dx = \lim_{N\to\infty} \dfrac{\pi}{-3}(x+1)^{-3}\Big|_0^N = \lim_{N\to\infty} \dfrac{-\pi}{3}\left((N+1)^{-3} - 1\right) = \dfrac{\pi}{3}$.

   $\boxed{\text{Cylindrical shell method}}$ $V = \displaystyle\lim_{\epsilon\to 0+} \int_\epsilon^1 2\pi y \left(\dfrac{1}{\sqrt{y}} - 1\right) dy = 2\pi \int_0^1 \left(\sqrt{y} - y\right) dy = 2\pi\left(\dfrac{2}{3} - \dfrac{1}{2}\right) = \dfrac{\pi}{3}$.

55. $\boxed{\text{Cylindrical shell method}}$

$$V = \int_0^\infty 2\pi x\, e^{-x}\, dx = \lim_{N\to\infty} -2\pi(1+x)e^{-x}\Big|_0^N = -2\pi \lim_{N\to\infty}(1+N)e^{-N} + 2\pi(1+0)e^0 = 2\pi.$$

   $\boxed{\text{Disk method}}$ Example 9.2.2 implies $\lim_{\epsilon\to 0+}\epsilon \ln \epsilon = 0 = \lim_{\epsilon\to 0+}\epsilon(\ln \epsilon)^2$. Thus

$$V = \lim_{\epsilon\to 0+}\int_\epsilon^1 \pi(-\ln y)^2 dy = \lim_{\epsilon\to 0+}\pi\left(2y - 2y\ln y + y(\ln y)^2\right)\Big|_\epsilon^1$$
$$= \pi(2 - 0 + 0) - \pi\lim_{\epsilon\to 0+}\left(2\epsilon - 2\epsilon\ln\epsilon + \epsilon(\ln\epsilon)^2\right) = 2\pi.$$

57.   We adapt the computations shown in Solution 9.5.23. The 4th-degree Taylor polynomial for $f(x) = \cos x^2$ centered at $a = 0$ is $P_4(x) = 1 - \dfrac{x^4}{2}$ and we compute

$$\int_0^{1/2} \cos x^2\, dx \approx \int_0^{1/2}\left(1 - \dfrac{x^4}{2}\right) dx = \left(x - \dfrac{x^5}{10}\right)\Big|_0^{1/2} = \dfrac{1}{2} - \dfrac{1}{320} = \dfrac{159}{320} = 0.496875.$$

   $f^{(5)}(x) = 8x\left(20x^2\cos x^2 + (15 - 4x^4)\sin x^2\right)$ and we obtain a quick bound over the interval $[0, 1/2]$:

$$\left|f^{(5)}(x)\right| \le 8 \cdot \dfrac{1}{2}\left(20\left(\dfrac{1}{2}\right)^2 \cos 0^2 + (15 - 4\cdot 0^4)\sin\left(\dfrac{1}{2}\right)^2\right) = 4\left(5 + 15\sin 0.25\right) \approx 34.844. \text{ Therefore}$$

$$\left|\int_0^{1/2}\left(\cos x^2 - P_4(x)\right) dx\right| \le \int_0^{1/2}\left|R_4(x)\right| dx \le \int_0^{1/2} 34.844\left(\dfrac{x^5}{5!}\right) dx \approx 7.56 \cdot 10^{-4}.$$

REMARK: $\displaystyle\int_0^{1/2}\cos x^2\, dx \approx 0.49688403$ so the actual error is roughly $9.03 \cdot 10^{-6}$.

59. $0 \le x < \dfrac{\pi}{2} \implies \cos x \ge 1 - \dfrac{2x}{\pi} > 0 \implies \sqrt{\cos x} \ge \sqrt{1 - \dfrac{2x}{\pi}} > 0 \implies \dfrac{1}{\sqrt{1 - \frac{2x}{\pi}}} \ge \dfrac{1}{\sqrt{\cos x}}$.

$$\int_0^{\pi/2}\dfrac{dx}{\sqrt{1 - \frac{2x}{\pi}}} = \lim_{\epsilon\to 0+}\int_0^{\pi/2-\epsilon}\dfrac{dx}{\sqrt{1 - \frac{2x}{\pi}}} = \lim_{\epsilon\to 0+} -\pi\sqrt{1 - \dfrac{2x}{\pi}}\,\Big|_0^{\pi/2-\epsilon} = -\lim_{\epsilon\to 0+}\sqrt{2\pi\epsilon} - (-\pi) = 0 + \pi = \pi.$$

Therefore $\displaystyle\int_0^{\pi/2-\epsilon}\dfrac{dx}{\sqrt{\cos x}}$ is an increasing but bounded function of ϵ, hence it converges to $\displaystyle\int_0^{\pi/2}\dfrac{dx}{\sqrt{\cos x}}$. ∎

61.   a.   By L'Hôpital's rule, $p > 0$ implies $\displaystyle\lim_{x\to 0+} x^p \cdot \ln x = \lim_{x\to 0+}\dfrac{\frac{d}{dx}\ln x}{\frac{d}{dx}x^{-p}} = \lim_{x\to 0+}\dfrac{x^{-1}}{-px^{-p-1}} = \lim_{x\to 0+}\dfrac{x^p}{-p} = 0$.

   Therefore $\displaystyle\lim_{x\to 0+}(x^p \cdot \ln x)^N = 0$. This implies there is some $a > 0$ such that if $0 < x < a < 1$, then

$|\ln x|^N < x^{-1/2}$ (since, choosing $p = 1/(2N)$, we know $\left|\sqrt{x}(\ln x)^N\right|$ is eventually less than 1). Hence

$$\int_0^1 |\ln t|^N \, dt = \lim_{\epsilon \to 0+} \int_\epsilon^a |\ln t|^N \, dt + \int_a^1 |\ln t|^N \, dt < \lim_{\epsilon \to 0+} \int_\epsilon^a t^{-1/2} dt + \int_a^1 |\ln a|^N \, dt = \frac{\sqrt{a}}{2} + |\ln a|^N (1-a).$$

b.   Using integration by parts,

$$I_N = \int_0^1 (\ln t)^N \, dt = \lim_{\epsilon \to 0+} t \cdot (\ln t)^N \Big|_\epsilon^1 - \int_0^1 t \cdot N(\ln T)^{N-1} \frac{dt}{t} = 0 - \int_0^1 N(\ln t)^{N-1} \, dt = -N \cdot I_{N-1}.$$

Since $I_0 = 1$, we conclude that $I_N = (-1)^N \cdot N!$.

REMARK: $\displaystyle\int_0^1 (\ln t)^N \, dt \overset{\ln t \equiv -u}{=} \int_\infty^0 (-u)^N \left(-e^{-u} \, du\right) = (-1)^N \int_0^\infty u^N e^{-u} \, du = (-1)^N \cdot N!.$

Solutions 9.C   *Computer Exercises*   (page 636)

1. a. $0 < x^3 < x^3 + x + 1$ implies $0 < \left(x^3 + x + 1\right)^{-1} < x^{-3}$ on $(0, \infty)$; because $\displaystyle\int_1^\infty x^{-3} \, dx$ converges (to

$1/2$), we can infer $\displaystyle\int_1^\infty \frac{dx}{x^3 + x + 1}$ converges. (Invoke the comparison theorem in Problem 9.3.66a.)

b. $\displaystyle\int_1^{10} \frac{dx}{x^3 + x + 1} \approx .28647\ 08548\ 84757\ 29554,\quad \int_1^{100} \frac{dx}{x^3 + x + 1} \approx .29139\ 40522\ 92480\ 00588,$

$\displaystyle\int_1^{1000} \frac{dx}{x^3 + x + 1} \approx .29144\ 35497\ 72899\ 72919,\quad$ and $\quad\displaystyle\int_1^\infty \frac{dx}{x^3 + x + 1} \approx .29144\ 40497\ 72649\ 52935.$

c. $0 < \displaystyle\int_1^\infty \frac{dx}{x^3 + x + 1} - \int_1^{1000} \frac{dx}{x^3 + x + 1} = \int_{1000}^\infty \frac{dx}{x^3 + x + 1} < \int_{1000}^\infty \frac{dx}{x^3} = 5.0 \cdot 10^{-7}.$

REMARK: $\left(x^3 + x + 1\right)^{-1}$ does have an "elementary" antiderivative which a good computer algebra system can find; calculations in part (b) are numerical approximations using such an exact expression. If a numerical integration procedure such as Simpson's rule were used, then additional sources of error enter as the result of computational strategy and fineness of partition mesh.

3. I:a.   If $f(x) = x - \sin x$ then $f'(x) = 1 - \cos x \geq 0$. Therefore f is an increasing function. Because $f(0) = 0 - \sin 0 = 0 - 0 = 0$, we can infer $f(x) \geq 0$ if $x \geq 0$; that is equivalent to $x \geq 0 \implies x \geq \sin x$. ∎

I:b. $0 \leq x - \sin x$ on $(0, \infty)$ implies $0 \leq \displaystyle\int_0^x (t - \sin t) \, dt = \frac{x^2}{2} + \cos x - 1$ which implies $1 - \frac{x^2}{2} \leq \cos x$.

(This is a stronger result than what we were asked to prove.) ∎

I:c.   If $x \in \left[0, \sqrt{2}\right)$, then $0 < 1 - \frac{x^2}{2} \leq \cos x$ which implies $0 < \frac{1}{\cos x} \leq \frac{1}{1 - x^2/2}$. This and the result of

part (I:a) implies $\tan x = (\sin x)\left(\frac{1}{\cos x}\right) \leq (x)\left(\frac{1}{1 - x^2/2}\right) = \frac{x}{1 - x^2/2}$ for $x \in \left[0, \sqrt{2}\right)$. ∎

II:a.   If $g(x) = \tan x$, then

$$g^{(12)}(x) = \sec^2 x\ \big(22368256 \tan x + 287468544 \tan^3 x + 1148590080 \tan^5 x$$
$$+ 2001162240 \tan^7 x + 1596672000 \tan^9 x + 479001600 \tan^{11} x\big).$$

Since $\sec^2 x = 1 + \tan^2 x$, we can infer $g^{(12)}(x)$ can be written as a polynomial in $\tan x$ with positive coefficients. This implies $g^{(12)}(x) \geq 0$ if $\tan x \geq 0$, therefore $g^{(11)}(x)$ increases on $(0, \pi/2)$.

II:b.   The maximum value of $g^{(11)}(x)$ on $[0, .1]$ is $g^{(11)}(.1)$ because $g^{(11)}$ increases on that interval.

$$g^{(11)}(x) = 353792 + 11184128 \tan^2 x + 71867136 \tan^4 x + 191431680 \tan^6 x$$
$$+ 250145280 \tan^8 x + 159667200 \tan^{10} x + 39916800 \tan^{12} x$$

combines with the result of part (I:c) to imply $g^{(11)}(.1) \leq 474292.3$ Thus $|R_{10}(.1)| \leq (474292.3)\frac{.1^{11}}{11!} \approx 1.1882 \cdot 10^{-13}$. ∎

II:c.   The tenth-order Maclaurin polynomial for $\tan x$ is $P_{10}(x) = x + \frac{x^3}{3} + \frac{2x^5}{15} + \frac{17x^7}{315} + \frac{62x^9}{2835}$ and

$P_{10}(.1) \approx 0.10033\ 46720\ 85361\ 55203$ (with actual error $8.9 \cdot 10^{-14}$).

Sequences and Series

Solutions 10.1 *Sequences of real numbers* (page 644)

1.

n	1	2	3	4	5
$\dfrac{1}{3^n}$	$\dfrac{1}{3}$	$\dfrac{1}{9}$	$\dfrac{1}{27}$	$\dfrac{1}{81}$	$\dfrac{1}{243}$

3.

n	1	2	3	4	5
$1-\dfrac{1}{4^n}$	$\dfrac{3}{4}$	$\dfrac{15}{16}$	$\dfrac{63}{64}$	$\dfrac{255}{256}$	$\dfrac{1023}{1024}$

5.

n	1	2	3	4	5
$\sin\dfrac{n\pi}{2}$	1	0	-1	0	1

7.

1	2	3	4	5	n
$\dfrac{1}{2}=\dfrac{1}{1+1}$	$\dfrac{2}{3}=\dfrac{2}{2+1}$	$\dfrac{3}{4}=\dfrac{3}{3+1}$	$\dfrac{4}{5}=\dfrac{4}{4+1}$	$\dfrac{5}{6}=\dfrac{5}{5+1}$	$a_n=\dfrac{n}{n+1}$

9.

1	2	3	4	5	n
$1=1\cdot5^{1-1}$	$2\cdot5=2\cdot5^{2-1}$	$3\cdot5^2=3\cdot5^{3-1}$	$4\cdot5^3=4\cdot5^{4-1}$	$5\cdot5^4=5\cdot5^{5-1}$	$a_n=n\cdot5^{n-1}$

11. For any positive ϵ let $N=\left\lceil\dfrac{17^2}{\epsilon^2}\right\rceil$ where $\lceil x\rceil$ is the least integer equal to or greater than x. Thus

$$n>N\geq\frac{17^2}{\epsilon^2}\implies\sqrt{n}>\frac{17}{\epsilon}>0\implies\epsilon>\frac{17}{\sqrt{n}}>0;$$

therefore $\displaystyle\lim_{n\to\infty}\frac{17}{\sqrt{n}}=0$.

13. The function $\sin x$ has period 2π; furthermore $\sin x=1$ if $x=\dfrac{\pi}{2}+k(2\pi)$ and $\sin x=-1$ if $x=\dfrac{3\pi}{2}+k(2\pi)$ where k is an integer. Since $2\pi<7$, within each set of 7 consecutive integers there is at least one, call it p, which is no more than 0.5 away from a number of the form $\dfrac{\pi}{2}+k(2\pi)$; the value of $\sin p$ must be at least $0.87\approx\sin\left(\dfrac{\pi}{2}-0.5\right)=\sin\left(\dfrac{\pi}{2}+0.5\right)$. Similarly, within the same set of 7 consecutive integers, there must be at least one, call it q, which is no more than 0.5 away from a number of the form $\dfrac{3\pi}{2}+k(2\pi)$; the value of $\sin q$ must be less than $-0.87\approx\sin\left(\dfrac{3\pi}{2}-0.5\right)=\sin\left(\dfrac{3\pi}{2}+0.5\right)$. Therefore, the values of $\sin n$ can not stay close to any fixed number L as n becomes large; the sequence $\{\sin n\}$ does not converge.

REMARK: It is fairly easy to show $\displaystyle\lim_{x\to\infty}\sin x$ does not exist, but Theorem 5 provides no guidance here.

15. For any positive ϵ let $N=\left\lceil\dfrac{3}{5\epsilon}\right\rceil$, then $n>N\geq\dfrac{3}{5\epsilon}>0\implies\epsilon>\dfrac{3}{5n}>0$; therefore $\displaystyle\lim_{n\to\infty}\frac{3}{5n}=0$.

17. $\displaystyle\lim_{n\to\infty}\frac{n^5+3n^2+1}{n^6+4n}=\lim_{n\to\infty}\left(\frac{n^5+3n^2+1}{n^6+4n}\right)\left(\frac{n^{-6}}{n^{-6}}\right)=\lim_{n\to\infty}\frac{n^{-1}+3n^{-4}+n^{-6}}{1+4n^{-5}}=\frac{0+0+0}{1+0}=0.$

REMARK: Apply Example 12 with $m=5$ and $r=6$.

19.   Example 10 reminds us $\lim\limits_{x \to \infty} \left(1 + \frac{1}{x}\right)^x = e$ (shown by Example 9.1.5); that result and Theorem 5 imply

$$\lim_{n \to \infty} \left(1 + \frac{4}{n}\right)^n = \lim_{n \to \infty} \left(\left(1 + \frac{4}{n}\right)^{n/4}\right)^4 = \lim_{n \to \infty} \left(\left(1 + \frac{1}{n/4}\right)^{n/4}\right)^4 = \left(\lim_{n \to \infty} \left(1 + \frac{1}{n/4}\right)^{n/4}\right)^4 = (e)^4 = e^4.$$

REMARK: Apply Problem 35.

21.   L'Hôpital's rule implies $\lim\limits_{x \to \infty} \frac{\sqrt{x}}{\ln x} = \lim\limits_{x \to \infty} \frac{\frac{d}{dx}\sqrt{x}}{\frac{d}{dx}\ln x} = \lim\limits_{x \to \infty} \frac{1/(2\sqrt{x})}{1/x} = \lim\limits_{x \to \infty} \frac{\sqrt{x}}{2} = \infty$; Theorem 5 then

implies $\lim\limits_{n \to \infty} \frac{\sqrt{n}}{\ln n} = \infty$.

23. $0 < \frac{2^n}{n!} = \frac{2 \cdot 2 \cdot 2 \cdots 2 \cdot 2}{1 \cdot 2 \cdot 3 \cdots n-1 \cdot n} = \left(\frac{2}{1}\right)\left(\frac{2}{2}\right)\left(\frac{2}{3}\right) \cdots \left(\frac{2}{n-1}\right)\left(\frac{2}{n}\right) \le (2)(1)(1) \cdots (1)\left(\frac{2}{n}\right) = \frac{4}{n}$ if

$n \ge 2$; we see that $\lim\limits_{n \to \infty} \frac{4}{n} = 0$, Theorem 4 then implies $\lim\limits_{n \to \infty} \frac{2^n}{n!} = 0$.

25. $\lim\limits_{n \to \infty} \frac{n+1}{n^{5/2}} = \lim\limits_{n \to \infty} \frac{n+1}{n} \cdot \frac{1}{n^{3/2}} = \left(\lim\limits_{n \to \infty} \frac{n+1}{n}\right) \cdot \left(\lim\limits_{n \to \infty} \frac{1}{n^{3/2}}\right) = 1 \cdot 0 = 0.$

27. $\cos n\pi = (-1)^n$ for any integer n, thus $\lim\limits_{n \to \infty} (-1)^n \cdot \cos n\pi = \lim\limits_{n \to \infty} (-1)^n \cdot (-1)^n = \lim\limits_{n \to \infty} (-1)^{2n} = \lim\limits_{n \to \infty} 1 = 1.$

29.   Suppose $a_{n+1} = a_n - a_n^2$ and $\lim\limits_{n \to \infty} a_n$ exists; let $L = \lim\limits_{n \to \infty} a_n$. Then $\lim\limits_{n \to \infty} a_{n+1}$ also exists and $L = $

$\lim\limits_{n \to \infty} a_{n+1} = \lim\limits_{n \to \infty} (a_n - a_n^2) = \left(\lim\limits_{n \to \infty} a_n\right) - \left(\lim\limits_{n \to \infty} a_n\right)^2 = L - L^2$. The only solution to $L = L - L^2$ is $L = 0$;
therefore the sequence $\{a_n\}$ converges to 0 **if** it converges.

    Suppose a_0 is chosen and $a_{n+1} = a_n - a_n^2$ for all $n \ge 0$. If a_0 is 0 or 1, then a_1 and all later terms of the sequence are 0. Choosing a_0 different from 0 or 1 produces a strictly decreasing sequence because $x - x^2 < x$ if $x \ne 0$ and $x - x^2 = 0$ only if x is 0 or 1. If $a_0 < 0$, then all later terms are less than a_0; this means the sequence can not converge to 0 and hence does not converge. Similarly, if $a_0 > 1$, then $a_1 < 0$, all later terms are less than a_1, and the sequence does not converge. If $0 < x < 1$, then $x^2 < x$ and $0 < x - x^2 < x$. Therefore, if $0 < a_0 < 1$, the sequence is strictly decreasing and is bounded below; such a sequence converges to its greatest lower bound (see Theorem 10.2.2). We already know 0 is the only possible limit value, so we conclude that $\lim\limits_{n \to \infty} a_n = 0$ if and only if $0 \le a_0 \le 1$.

REMARK: See *Geometric convergence* by M.S. Klamkin, *The American Mathematical Monthly*, volume 60 (1953), pages 256-9, for a graphical discussion of some sequences defined by a recursion relation of the form $a_{n+1} = f(a_n)$.

31. $\lim\limits_{x \to \infty} f(x) = L \iff$ for any $\epsilon > 0$ there is an M such that x is real and $x > M$ implies $|f(x) - L| < \epsilon$

           $\implies$ for any $\epsilon > 0$ there is an N such that n is integer and $n \ge N$ implies $|f(n) - L| < \epsilon$

           $\iff \lim\limits_{n \to \infty} f(n) = L.$ ∎

(For the middle implication, let $N = 1 + \lceil M \rceil$.)

33.   If $|b_n|$ converges to 0, then so does $-|b_n|$. If $|a_n| \le |b_n|$, then $-|b_n| \le a_n \le |b_n|$. Theorem 4 then

implies $\lim\limits_{n \to \infty} a_n = 0$, therefore $\lim\limits_{n \to \infty} |a_n| = \left|\lim\limits_{n \to \infty} a_n\right| = |0| = 0.$ ∎

REMARK: $|a_n| \le |b_n|$ implies $\{|a_n|\}$ is squeezed between $\{0\}$, the sequence which is constantly zero, and
      $\{b_n\}$. Since $\lim\limits_{n \to \infty} 0 = 0 = \lim\limits_{n \to \infty} |b_n|$, Theorem 4 implies $\lim\limits_{n \to \infty} |a_n| = 0.$ ∎

      We can go beyond the Problem's conclusion: $|a_n - 0| = \big||a_n| - 0\big|$ gets arbitrarily small as n
      increases, i.e., $\lim\limits_{n \to \infty} a_n = 0.$

35.   We extend Example 9.1.5 and apply Theorem 5. $\dfrac{\ln(1 + \beta/x)}{1/x}$ has indeterminate form $\dfrac{0}{0}$ as $x \to \infty$ and

$$\lim_{x \to \infty} \frac{\frac{d}{dx}\left(\ln(1 + \beta/x)\right)}{\frac{d}{dx}\left(1/x\right)} = \lim_{x \to \infty} \frac{(1 + \beta/x)^{-1} \cdot (-\beta)x^{-2}}{-x^{-2}} = \lim_{x \to \infty} \frac{\beta}{1 + \beta/x} = \beta. \text{ Therefore}$$

$$\lim_{n \to \infty} \left(1 + \frac{\beta}{n}\right)^n = \lim_{x \to \infty} \left(1 + \frac{\beta}{x}\right)^x = \lim_{x \to \infty} e^{x \cdot \ln(1 + \beta/x)} = e^{\lim\limits_{x \to \infty} x \cdot \ln(1 + \beta/x)} = e^\beta. \text{ ∎}$$

10.1 — Sequences of real numbers

37. Solution 9.2.5 shows that $\ln x$ "blows up more slowly" than any positive power of x as $x \to \infty$; more precisely, if $\epsilon > 0$, then $\lim\limits_{x \to \infty} \dfrac{x^\epsilon}{\ln x} = \infty$.

Let $a_n = \dfrac{1}{n \cdot \ln n}$ and $p = 1$. Then $\lim\limits_{n \to \infty} n^p \cdot a_n = \lim\limits_{n \to \infty} \dfrac{1}{\ln n} = 0$, but $\lim\limits_{n \to \infty} n^{p+\epsilon} \cdot a_n = \lim\limits_{n \to \infty} \dfrac{n^\epsilon}{\ln n} = \infty$. This counterexample disproves the statement given in this problem.

REMARK: This is problem Q569 in *Mathematics Magazine*, volume 46 (1973), page 167; a different counterexample is given there.

Solutions 10.2   *Bounded and monotonic sequences*   (pages 651-2)

1. $a_n = \dfrac{1}{n+1}$ is a strictly decreasing positive function of n, hence $|a_n| = a_n = \dfrac{1}{n+1} \le a_1 = \dfrac{1}{1+1} = \dfrac{1}{2}$ for all $n \ge 1$.

3. $a_n = \cos n\pi = \begin{cases} 1 & \text{if } n \text{ is even} \\ -1 & \text{if } n \text{ is odd} \end{cases} = (-1)^n$; therefore $|a_n| = |\cos n\pi| \le a_2 = 1$ for all $n \ge 1$.

5. $\left|\dfrac{2^n}{1+2^n}\right| = \left|1 - \dfrac{1}{1+2^n}\right| \le 1$; 1 is the least upper bound because $\lim\limits_{n \to \infty} \dfrac{1}{1+2^n} = 0$.

7. $a_n = \dfrac{1}{n!}$ is a strictly decreasing positive function of n, hence $\left|\dfrac{1}{n!}\right| \le \dfrac{1}{1!} = 1$.

9. $\left\{\dfrac{n^2}{n!}\right\} = \left\{1, 2, \dfrac{3}{2}, \dfrac{2}{3}, \dfrac{5}{24}, \dfrac{1}{20}, \dfrac{7}{720}, \ldots\right\}$; it appears the sequence is strictly decreasing after $n = 2$ and that $\left|\dfrac{n^2}{n!}\right| = \dfrac{n^2}{n!} \le \dfrac{2^2}{2!} = 2$ (so 2 is the least upper bound we're to find). This is true because

$$\frac{(n+1)^2}{(n+1)!} < \frac{n^2}{n!} \iff \frac{n+1}{n!} < \frac{n}{(n-1)!} \iff n+1 < n \cdot \frac{n!}{(n-1)!} = n^2 \iff \frac{5}{4} < \left(n - \frac{1}{2}\right)^2$$

and $n \ge 2$ easily implies $\left(n - \dfrac{1}{2}\right)^2 \ge \left(\dfrac{3}{2}\right)^2 = \dfrac{9}{4} > \dfrac{5}{4}$.

11. Solution 9.2.5 shows $\lim\limits_{x \to \infty} \dfrac{\ln x}{x} = 0$, therefore $\lim\limits_{n \to \infty} \dfrac{\ln n}{n} = 0$ and $\left\{\dfrac{\ln n}{n}\right\}$ is bounded. The function $f(x) = \dfrac{\ln x}{x}$ is strictly decreasing on the interval (e, ∞) because $f'(x) = \dfrac{1 - \ln x}{x^2}$ is negative if $n \ge e$. $f(1) = 0 < f(2) \approx 0.347 < f(3) \approx 0.366$ and $n > 3 \implies f(n) < f(3)$. Therefore $\left|\dfrac{\ln n}{n}\right| = \dfrac{\ln n}{n} \le \dfrac{\ln 3}{3}$.

13. $a_n = \sin n\pi = 0$, the sequence $\{\sin n\pi\}$ is constant. (Since $0 \le 0$ and $0 \ge 0$, the sequence is also monotone increasing and monotone decreasing.)

15. $a_{n+1} = \left(\dfrac{n+1}{25}\right)^{1/3} = \left(\dfrac{n+1}{n}\right)^{1/3}\left(\dfrac{n}{25}\right)^{1/3} = \left(1 + \dfrac{1}{n}\right)^{1/3} a_n$; since $1 + \dfrac{1}{n}$ and its cube-root are greater than 1 for any positive n, we see that $a_{n+1} > a_n$ and the sequence $\left\{\left(\dfrac{n}{25}\right)^{1/3}\right\}$ is strictly increasing.

17. The sequences $\left\{\dfrac{1}{n}\right\}$ and $\left\{\dfrac{1}{n^2}\right\}$ are both strictly decreasing, therefore so is $\left\{\dfrac{1}{n} + \dfrac{1}{n^2}\right\} = \left\{\dfrac{n+1}{n^2}\right\}$. The square-root function preserves ordering because it is strictly increasing, therefore $n \ge 1$ implies

$$a_{n+1} = \frac{\sqrt{(n+1)+1}}{n+1} = \sqrt{\frac{(n+1)+1}{(n+1)^2}} < \sqrt{\frac{n+1}{n^2}} = \frac{\sqrt{n+1}}{n} = a_n;$$

hence the sequence $\left\{\dfrac{\sqrt{n+1}}{n}\right\}$ is strictly decreasing.

19. $\dfrac{(n+1)^{n+1}}{(n+1)!} = \dfrac{(n+1)^n}{n!} > \dfrac{n^n}{n!}$, the sequence $\left\{\dfrac{n^n}{n!}\right\}$ is strictly increasing.

REMARK: If $a_n = \dfrac{n^n}{n!}$, then $\dfrac{a_{n+1}}{a_n} = \dfrac{(n+1)^{n+1}}{n^n} \cdot \dfrac{n!}{(n+1)!} = \left(\dfrac{n+1}{n}\right)^n \cdot \dfrac{(n+1)!}{(n+1)!} = \left(1 + \dfrac{1}{n}\right)^n$; this ratio is always greater than 1 (in fact, its limit exists and equals e).

21.   Let $f(x) = x + \cos x$, then $f'(x) = 1 - \sin x$. Since $f'(x) > 0$ except at the separated points $x = \dfrac{\pi}{2} + 2k\pi$, we can infer $f(x+1) > f(x)$ for all x; therefore the sequence $\{n + \cos n\}$ is strictly increasing.

23. $\left\{\dfrac{\sqrt{n}-1}{n}\right\} \approx \{0.0000,\ 0.2071,\ 0.2440,\ 0.2500,\ 0.2472,\ 0.2416,\ 0.2351,\ 0.2286,\ 0.2222,\ 0.2162,\ \ldots\}$ is not monotonic; the sequences increases to $\dfrac{1}{4}$ for $n = 4$ and then decreases.

REMARK: If $f(x) = \dfrac{\sqrt{x}-1}{x}$, then $f'(x) = \dfrac{2-\sqrt{x}}{2x^2}$ is negative if $x > 4$.

25. $\dfrac{3n}{n+1} = 3 - \dfrac{3}{n+1}$ is strictly increasing, the logarithm function preserves ordering because it is an increasing function, therefore $\left\{\ln\left(\dfrac{3n}{n+1}\right)\right\}$ is a strictly increasing sequence.

27. $\left\{\left(1 - \dfrac{3}{n}\right)^n\right\} = \left\{-2,\ \dfrac{1}{4},\ 0,\ \dfrac{1}{256},\ \dfrac{32}{3125},\ \dfrac{1}{64},\ \ldots\right\}$ is not monotonic (since $a_1 < a_2$ but $a_2 > a_3$); however it is strictly increasing for $n \geq 3$. (See Problem 10.1.35; this sequence converges to e^{-3}.)

REMARK: If $f(x) = \left(1 - \dfrac{3}{x}\right)^x$ for $x > 3$, then $f'(x) = f(x) \cdot g(x)$ where $g(x) = \ln\left(1 - \dfrac{3}{x}\right) + \dfrac{3}{x-3}$.

Since $g'(x) = \dfrac{-9}{x(x-3)^2} < 0$ and $\lim\limits_{x \to \infty} g(x) = 0$, we see that $g(x) > 0$ and $f'(x) > 0$ if $x > 3$.

29.   a. $y = e^{-x}$ and $y = x$ $\qquad\qquad\qquad\qquad$ $y = e^{-x} - x$

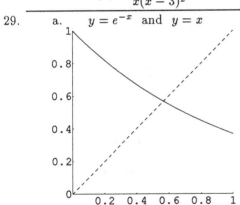
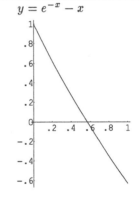

b.

n	x_n	x_{n+1}	$x_{n+1} - x_n$	n	x_n	x_{n+1}	$x_{n+1} - x_n$
1	1.00000000	0.36787944	-0.63212056	16	0.56706790	0.56718605	0.00011815
2	0.36787944	0.69220063	0.32432119	17	0.56718605	0.56711904	-0.00006701
3	0.69220063	0.50047350	-0.19172713	18	0.56711904	0.56715704	0.00003800
4	0.50047350	0.60624354	0.10577003	19	0.56715704	0.56713549	-0.00002155
5	0.60624354	0.54539579	-0.06084775	20	0.56713549	0.56714771	0.00001222
6	0.54539579	0.57961234	0.03421655	21	0.56714771	0.56714078	-0.00000693
7	0.57961234	0.56011546	-0.01949687	22	0.56714078	0.56714471	0.00000393
8	0.56011546	0.57114312	0.01102765	23	0.56714471	0.56714248	-0.00000223
9	0.57114312	0.56487935	-0.00626377	24	0.56714248	0.56714375	0.00000126
10	0.56487935	0.56842873	0.00354938	25	0.56714375	0.56714303	-0.00000072
11	0.56842873	0.56641473	-0.00201399	26	0.56714303	0.56714344	0.00000041
12	0.56641473	0.56755664	0.00114190	27	0.56714344	0.56714321	-0.00000023
13	0.56755664	0.56690891	-0.00064773	28	0.56714321	0.56714334	0.00000013
14	0.56690891	0.56727623	0.00036732	29	0.56714334	0.56714326	-0.00000007
15	0.56727623	0.56706790	-0.00020833	30	0.56714326	0.56714331	0.00000004

31.

n	x_n	x_{n+1}	$x_{n+1} - x_n$	n	x_n	x_{n+1}	$x_{n+1} - x_n$
1	6.00000000	8.95879735	2.95879735	1	12345.0000000	47.1050320	−12297.89496799
2	8.95879735	10.96317997	2.00438262	2	47.1050320	19.2618992	−27.84313285
3	10.96317997	11.97271191	1.00953195	3	19.2618992	14.7906450	−4.47125412
4	11.97271191	12.41315026	0.44043835	4	14.7906450	13.4699745	−1.32067059
5	12.41315026	12.59378208	0.18063181	5	13.4699745	13.0023155	−0.46765898
6	12.59378208	12.66601603	0.07223395	6	13.0023155	12.8256373	−0.17667819
7	12.66601603	12.69461252	0.02859649	7	12.8256373	12.7572304	−0.06840687
8	12.69461252	12.70588846	0.01127594	8	12.7572304	12.7304910	−0.02673939
9	12.70588846	12.71032772	0.00443926	9	12.7304910	12.7199999	−0.01049109
10	12.71032772	12.71207435	0.00174663	10	12.7199999	12.7158778	−0.00412216
11	12.71207435	12.71276139	0.00068704	11	12.7158778	12.7142571	−0.00162061
12	12.71276139	12.71303161	0.00027022	12	12.7142571	12.7136199	−0.00063728
13	12.71303161	12.71313789	0.00010628	13	12.7136199	12.7133692	−0.00025062
14	12.71313789	12.71317969	0.00004180	14	12.7133692	12.7132707	−0.00009857
15	12.71317969	12.71319613	0.00001644	15	12.7132707	12.7132319	−0.00003876
16	12.71319613	12.71320260	0.00000647	16	12.7132319	12.7132167	−0.00001525
17	12.71320260	12.71320514	0.00000254	17	12.7132167	12.7132107	−0.00000600
18	12.71320514	12.71320614	0.00000100	18	12.7132107	12.7132083	−0.00000236
19	12.71320614	12.71320653	0.00000039	19	12.7132083	12.7132074	−0.00000093
20	12.71320653	12.71320669	0.00000015	20	12.7132074	12.7132070	−0.00000036

REMARK: $x = 1.2958555090953692$ is also a fixed point for $f(x) = 5 \ln x$ but the above iteration procedure will not find it because $f'(x) = 5/x > 1$ if $0 < x < 5$.

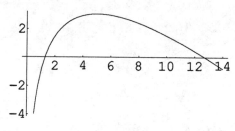

33. Let $F(x) = x^2 + x - 2$.

a. $F(x) = x \iff x^2 + x - 2 = x \iff x^2 - 2 = 0 \iff x = \pm\sqrt{2}$.

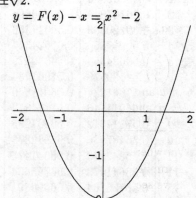

b. $F(0) = 0^2 + 0 - 2 = 0 + 0 - 2 = -2$ and $F(-2) = (-2)^2 + (-2) - 2 = 4 - 2 - 2 = 0$. Therefore, if $x_1 = 0$, then $F(x_{k+2}) = F(x_k) \neq F(x_{k+1})$ for each positive integer k. ∎

REMARK: There are choices of x_1 close to either fixed point such that $\{x_k\}$ diverges to ∞; other choices near $\pm\sqrt{2}$ produce sequences which cycle through neighborhoods of -2.247, 0.802, and -0.554.

---

35. a. Problem 9.5.35 implies $\displaystyle\lim_{n\to\infty}\frac{x^n}{n!}=0$ for any choice of x; Theorem 1 then implies the sequence $\left\{\dfrac{x^n}{n!}\right\}$

   is bounded. Therefore $\left\{\dfrac{5^n}{n!}\right\}$ is bounded.

   b. The maximum value of the sequence is $\dfrac{5^4}{4!}=\dfrac{625}{24}\approx 26.04$; choose $M=27$.

37. $(2^n+3^n)^{1/n}=2\left(1+\left(\dfrac{3}{2}\right)^n\right)^{1/n}$, this sequence is strictly decreasing because $1<a$ implies

   $$\left(1+a^{n+1}\right)^n<\left(a+a^{n+1}\right)^n=a^n\left(1+a^n\right)^n<\left(1+a^n\right)^{n+1}\iff\left(1+a^{n+1}\right)^{1/(n+1)}<\left(1+a^n\right)^{1/n}.$$

   The sequence is bounded above by $\left(2^1+3^1\right)^{1/1}=5$ and below by 0. Theorem 2 guarantees the sequence $\left\{(2^n+3^n)^{1/n}\right\}$ is convergent. ∎

REMARK: Let $Q(x)=\dfrac{\ln\left(2^x+3^x\right)}{x}$.

$$\lim_{x\to\infty}\frac{\frac{d}{dx}\ln\left(2^x+3^x\right)}{\frac{d}{dx}x}=\lim_{x\to\infty}\frac{2^x\ln 2+3^x\ln 3}{2^x+3^x}=\lim_{x\to\infty}\frac{(2/3)^x\ln 2+\ln 3}{(2/3)^x+1}=\frac{0\cdot\ln 2+\ln 3}{0+1}=\ln 3,$$

thus $\displaystyle\lim_{x\to\infty}\left(2^x+3^x\right)^{1/x}=\lim_{x\to\infty}e^{Q(x)}=e^{\ln 3}=3$. Theorem 9.1.5 now assures us that $\displaystyle\lim_{n\to\infty}\left(2^n+3^n\right)^{1/n}$ also exists and equals 3. ∎

---

39. $a_n=\dfrac{n!}{n^n}$ is always positive so the sequence is bounded below by 0;

   $$\frac{a_{n+1}}{a_n}=\frac{(n+1)!}{n!}\cdot\frac{n^n}{(n+1)^{n+1}}=\frac{(n+1)!}{(n+1)!}\cdot\frac{n^n}{(n+1)^n}=\left(\frac{n}{n+1}\right)^n=\left(1-\frac{1}{n+1}\right)^n<1,$$

   thus the sequence is strictly decreasing and is bounded above by $a_1=\dfrac{1!}{1^1}=1$. ∎

41. a. $\left\{\dfrac{\ln n}{n}:n\geq 1\right\}\approx\{0,\ 0.347,\ 0.366,\ 0.347,\ 0.322,\ 0.299,\ \ldots\}$ is not monotonic because $a_1<a_2<a_3$ but $a_3>a_4$. ∎

   b. Solution 11 shows $\dfrac{\ln x}{x}$ is strictly decreasing on (e,∞), thus $\left\{\dfrac{\ln n}{n}:n\geq 3\right\}=\left\{\dfrac{\ln n}{n}:n>2\right\}$ is a strictly decreasing sequence. ∎

43. $n\geq 2\implies\dfrac{n}{2n-1}=\dfrac{1}{2}+\dfrac{1}{2(2n-1)}\leq\dfrac{1}{2}+\dfrac{1}{2(4-1)}=\dfrac{2}{3}$, therefore

   $$a_n=\frac{2\cdot n!}{1\cdot 3\cdot 5\cdot 7\cdots(2n-1)}=2\left(\frac{1}{1}\right)\left(\frac{2}{3}\right)\left(\frac{3}{5}\right)\left(\frac{4}{7}\right)\cdots\left(\frac{n}{2n-1}\right)\leq 2\left(\frac{2}{3}\right)^{n-1}$$

   Since $\displaystyle\lim_{n\to\infty}2\left(\frac{2}{3}\right)^{n-1}=0$, Theorem 10.1.4 implies $\displaystyle\lim_{n\to\infty}a_n=0$.

45. Suppose u and v are distinct fixed points for F. If F is continuously differentiable in the interval with endpoints at u and v, then the mean value theorem implies there is some c between u and v such that $F'(c)=\dfrac{F(u)-F(v)}{u-v}=\dfrac{u-v}{u-v}=1$. Therefore v can not be in an interval centered at u where $|F'|<1$, i.e., in an interval where $|F'|<1$, there can be at most one fixed point. ∎

47. $M<1$ implies $|x_{n+1}-u|=|E_{n+1}|\leq M\cdot|E_n|<|E_n|=|x_n-u|$. Because $u-c<A<u+c\iff|A-u|<c$, we see that if $x_n\in(u-c,u+c)$, then $x_{n+1}\in(u-c,u+c)\subset\operatorname{dom}F$. ∎

49. If $0<M<1$, then $\lim_{n\to\infty}M^n=0$ and $\lim_{n\to\infty}M^{n-1}\cdot|E_1|=0$. Therefore, $|E_n|\leq M^{n-1}\cdot|E_1|$ and the Squeezing Theorem for limits implies $0=\lim_{n\to\infty}E_n$. Since $E_n=x_n-u$, we see that $\lim_{n\to\infty}(x_n-u)=0$ which then implies $\lim_{n\to\infty}x_n$ exists and equals u. ∎

51.   Suppose F is continuously differentiable on $(u-c, u+c)$. If $x_n \in (u-c, u+c)$ and $x_n \neq u$, then the mean value theorem implies there is some $\delta \in (u-c, u+c)$ such that $F'(\delta)(x_n - u) = F(x_n) - F(u) = x_{n+1} - u$. If $|F'| \geq 1$ on $(u-c, u+c)$, then $|x_{n+1} - u| \geq |x_n - u| \geq |x_1 - u| > 0$; thus $\{x_n\}$ does not converge to u. ∎

Solutions 10.3   *Geometric series*   (pages 656-7)

1. $$1 + 3 + 9 + 27 + 81 + 243 = 1 + 3 + 3^2 + 3^3 + 3^4 + 3^5 = \sum_{k=0}^{5} 3^k = \frac{1 - 3^6}{1 - 3} = \frac{-728}{-2} = 364.$$

3. $$1 - 5 + 25 - 125 + 625 - 3125 = 1 + (-5) + (-5)^2 + (-5)^3 + (-5)^4 + (-5)^5$$
$$= \sum_{k=0}^{5} (-5)^k = \frac{1 - (-5)^6}{1 - (-5)} = \frac{-15624}{6} = -2604.$$

5. $$0.3^2 - 0.3^3 + 0.3^4 - 0.3^5 + 0.3^6 - 0.3^7 + 0.3^8 = \sum_{k=2}^{8} (-0.3)^k = (-0.3)^2 \sum_{j=0}^{6} (-0.3)^j$$
$$= (-0.3)^2 \left(\frac{1 - (-0.3)^7}{1 - (-0.3)} \right) = 0.06924591.$$

7. $$1 - \frac{1}{b^2} + \frac{1}{b^4} - \frac{1}{b^6} + \frac{1}{b^8} - \frac{1}{b^{10}} + \frac{1}{b^{12}} - \frac{1}{b^{14}} = \sum_{k=0}^{7} \left(\frac{-1}{b^2} \right)^k = \frac{1 - \left(\frac{-1}{b^2} \right)^8}{1 - \frac{-1}{b^2}} = \frac{1 - \frac{1}{b^{16}}}{1 + \frac{1}{b^2}} = \frac{b^{16} - 1}{b^{16} + b^{14}}.$$

9. $$1 + \sqrt{2} + 2 + 2^{3/2} + 4 + 2^{5/2} + 8 + 2^{7/2} + 16 = \sum_{k=0}^{8} \left(\sqrt{2} \right)^k = \frac{1 - \left(\sqrt{2} \right)^9}{1 - \sqrt{2}} = \frac{1 - 16\sqrt{2}}{1 - \sqrt{2}} = 31 + 15\sqrt{2}.$$

11. $$-16 + 64 - 256 + 1024 - 4096 = (-16) \cdot (1 - 4 + 16 - 64 + 256)$$
$$= (-16) \cdot \sum_{k=0}^{4} (-4)^k = (-16) \cdot \frac{1 - (-4)^5}{1 - (-4)} = (-16) \cdot \frac{1025}{5} = -3280.$$

13. $$1 + \frac{1}{4} + \frac{1}{4^2} + \frac{1}{4^3} + \cdots = \sum_{k=0}^{\infty} \left(\frac{1}{4} \right)^k = \frac{1}{1 - \frac{1}{4}} = \frac{4}{3}.$$

15. $$1 - \frac{1}{3} + \frac{1}{9} - \frac{1}{27} + \frac{1}{81} - \cdots = \sum_{k=0}^{\infty} \left(\frac{-1}{3} \right)^k = \frac{1}{1 - \left(\frac{-1}{3} \right)} = \frac{3}{4}.$$

17. $$1 + 0.7 + 0.7^2 + 0.7^3 + \cdots = \sum_{k=0}^{\infty} (0.7)^k = \frac{1}{1 - 0.7} = \frac{10}{3}.$$

19. $$\frac{1}{4} + \frac{1}{16} + \frac{1}{64} + \cdots = \frac{1}{4} \left(1 + \frac{1}{4} + \frac{1}{16} + \cdots \right) = \frac{1}{4} \sum_{k=0}^{\infty} \left(\frac{1}{4} \right)^k = \frac{1}{4} \left(\frac{1}{1 - \frac{1}{4}} \right) = \frac{1}{3}.$$

REMARK: $$\frac{1}{4} + \frac{1}{16} + \frac{1}{64} + \cdots = \sum_{k=1}^{\infty} \left(\frac{1}{4} \right)^k = - \left(\frac{1}{4} \right)^0 + \sum_{k=0}^{\infty} \left(\frac{1}{4} \right)^k = -1 + \frac{1}{1 - \frac{1}{4}} = -1 + \frac{4}{3}.$$

21. $\left(\frac{1}{2} \right)^n < 0.01 = \frac{1}{100} \iff 100 < 2^n$; since $2^6 = 64 < 100 < 128 = 2^7$, we see $\left(\frac{1}{2} \right)^n < 0.01 \iff n \geq 7$.

23.   Because $\ln x$ is an increasing function of x, $0.99^n < 0.01 \iff n \cdot \ln 0.99 = \ln 0.99^n < \ln 0.01$; since $\ln 0.99 < 0$, this is equivalent to $n > \frac{\ln 0.01}{\ln 0.99} \approx \frac{-4.60517}{-0.01005} \approx 458.2$. Therefore $0.99^n < 0.01 \iff n \geq 459$.

REMARK: Formula 6.5 (7) shows that common logarithms (base ten) yield the same result.

25. $1 < x \iff 0 < \frac{1}{x} < 1$, Theorem 2 then implies convergence of the geometric series:
$$1 + \frac{1}{x} + \frac{1}{x^2} + \frac{1}{x^3} + \cdots = 1 + \frac{1}{x} + \left(\frac{1}{x} \right)^2 + \left(\frac{1}{x} \right)^3 + \cdots = \sum_{k=0}^{\infty} \left(\frac{1}{x} \right)^k = \frac{1}{1 - \frac{1}{x}} = \frac{x}{x - 1}. \quad ∎$$

Solutions 10.4   *Infinite series*   (pages 667-9)

1.
$$0.333\ldots = 0.3+0.03+0.003+\cdots = \frac{3}{10}+\frac{3}{100}+\frac{3}{1000}+\cdots = \frac{3}{10}\left(1+\frac{1}{10}+\frac{1}{100}+\cdots\right)$$
$$= \frac{3}{10}\sum_{k=0}^{\infty}\left(\frac{1}{10}\right)^k = \frac{3}{10}\cdot\frac{1}{1-\frac{1}{10}} = \frac{3}{10}\cdot\frac{10}{9} = \frac{1}{3}.$$

3.
$$0.353535\ldots = 0.35+0.0035+0.000035+\cdots = \frac{35}{100}+\frac{35}{10000}+\frac{35}{1000000}+\cdots$$
$$= \frac{35}{100}\left(1+\frac{1}{100}+\frac{1}{10000}+\cdots\right) = \frac{35}{100}\sum_{k=0}^{\infty}\left(\frac{1}{100}\right)^k = \frac{35}{100}\cdot\frac{1}{1-\frac{1}{100}} = \frac{35}{100}\cdot\frac{100}{99} = \frac{35}{99}.$$

5.
$$0.717171\ldots = 0.71+0.0071+0.000071+\cdots = 0.71\left(1+0.01+0.0001+\cdots\right)$$
$$= 0.71\sum_{k=0}^{\infty}(0.01)^k = 0.71\cdot\frac{1}{1-0.01} = \frac{71}{100}\cdot\frac{100}{99} = \frac{71}{99}.$$

7.
$$0.501501501\ldots = 0.501\sum_{k=0}^{\infty}(0.001)^k = 0.501\cdot\frac{1}{1-0.001} = \frac{501}{1000}\cdot\frac{1000}{999} = \frac{501}{999} = \frac{167}{333}.$$

9.
$$0.11362362362\ldots = 0.11+0.00362362362\ldots = 0.11+0.00362\left(1+0.001+0.000001+\cdots\right)$$
$$= 0.11+0.00362\sum_{k=0}^{\infty}(0.001)^k$$
$$= 0.11+0.00362\cdot\frac{1}{1-0.001} = \frac{11}{100}+\frac{1}{100}\cdot\frac{362}{1000}\cdot\frac{1000}{999} = \frac{11}{100}+\frac{362}{99900} = \frac{11351}{99900}.$$

11.
$$\sum_{k=0}^{\infty}\frac{9}{10^k} = 9\sum_{k=0}^{\infty}\left(\frac{1}{10}\right)^k = 9\cdot\frac{1}{1-\frac{1}{10}} = 9\cdot\frac{10}{9} = 10.$$

13.
$$\sum_{k=2}^{\infty}\frac{1}{2^k} = \left(\frac{1}{2}\right)^2\sum_{j=0}^{\infty}\left(\frac{1}{2}\right)^j = \frac{1}{4}\cdot\frac{1}{1-\frac{1}{2}} = \frac{1}{4}\cdot 2 = \frac{1}{2}.$$

REMARK: $\displaystyle\sum_{k=2}^{\infty}\frac{1}{2^k} = -\left(\left(\frac{1}{2}\right)^0+\left(\frac{1}{2}\right)^1\right)+\sum_{k=0}^{\infty}\left(\frac{1}{2}\right)^k = -\left(1+\frac{1}{2}\right)+\frac{1}{1-\frac{1}{2}} = \frac{-3}{2}+2.$

15.
$$\sum_{k=0}^{\infty}\left(\frac{-2}{3}\right)^k = \frac{1}{1-\left(\frac{-2}{3}\right)} = \frac{1}{1+\frac{2}{3}} = \frac{3}{5}.$$

17.
$$\sum_{k=0}^{\infty}\frac{100}{5^k} = 100\sum_{k=0}^{\infty}\left(\frac{1}{5}\right)^k = 100\cdot\frac{1}{1-\frac{1}{5}} = 100\cdot\frac{5}{4} = 125.$$

19.   See Example 2. $\displaystyle\sum_{k=2}^{\infty}\frac{1}{k(k+1)} = -\left(\frac{1}{1\cdot(1+1)}\right)+\sum_{k=1}^{\infty}\frac{1}{k(k+1)} = \frac{-1}{2}+1 = \frac{1}{2}.$

21.   Let $j=k+1$ and use the result of Example 2. $\displaystyle\sum_{k=0}^{\infty}\frac{1}{(k+1)(k+2)} = \sum_{j=1}^{\infty}\frac{1}{j(j+1)} = 1.$

23.
$$\sum_{k=2}^{\infty}\frac{2^{k+3}}{3^k} = \sum_{k=2}^{\infty}2^3\left(\frac{2}{3}\right)^k = 2^3\sum_{k=2}^{\infty}\left(\frac{2}{3}\right)^2\cdot\left(\frac{2}{3}\right)^{k-2} = 2^3\left(\frac{2}{3}\right)^2\sum_{j=0}^{\infty}\left(\frac{2}{3}\right)^j = 8\left(\frac{4}{9}\right)\cdot\frac{1}{1-\frac{2}{3}} = \frac{32}{9}\cdot 3 = \frac{32}{3}.$$

REMARK: $\displaystyle\sum_{k=2}^{\infty}\frac{2^{k+3}}{3^k} = 8\sum_{k=2}^{\infty}\left(\frac{2}{3}\right)^k = 8\left(\sum_{k=0}^{\infty}\left(\frac{2}{3}\right)^k-\left(1+\frac{2}{3}\right)\right) = 8\left(3-\frac{5}{3}\right) = 8\cdot\frac{4}{3}.$

25.
$$\sum_{k=4}^{\infty}\frac{5^{k-2}}{6^{k+1}} = \sum_{k=4}^{\infty}\frac{5^{-2}}{6^1}\cdot\frac{5^k}{6^k} = \frac{1}{5^2\cdot 6}\sum_{k=4}^{\infty}\left(\frac{5}{6}\right)^k = \frac{1}{5^2\cdot 6}\sum_{k=4}^{\infty}\left(\frac{5}{6}\right)^4\left(\frac{5}{6}\right)^{k-4}$$
$$= \frac{1}{5^2\cdot 6}\left(\frac{5}{6}\right)^4\sum_{j=0}^{\infty}\left(\frac{5}{6}\right)^j = \frac{5^2}{6^5}\cdot\frac{1}{1-\frac{5}{6}} = \frac{5^2}{6^5}\cdot 6 = \frac{5^2}{6^4} = \frac{25}{1296}.$$

27. $\displaystyle\sum_{k=0}^{\infty}\left(\left(\frac{1}{3}\right)^{k}+\left(\frac{2}{3}\right)^{k}\right)=\sum_{k=0}^{\infty}\left(\frac{1}{3}\right)^{k}+\sum_{k=0}^{\infty}\left(\frac{2}{3}\right)^{k}=\frac{1}{1-\frac{1}{3}}+\frac{1}{1-\frac{2}{3}}=\frac{3}{2}+3=\frac{9}{2}.$

29. $\displaystyle\sum_{k=0}^{\infty}\left(\frac{3}{5^{k}}-\frac{7}{4^{k}}\right)=\sum_{k=0}^{\infty}\frac{3}{5^{k}}-\sum_{k=0}^{\infty}\frac{7}{4^{k}}=3\sum_{k=0}^{\infty}\left(\frac{1}{5}\right)^{k}-7\sum_{k=0}^{\infty}\left(\frac{1}{4}\right)^{k}=3\cdot\frac{1}{1-\frac{1}{5}}-7\cdot\frac{1}{1-\frac{1}{4}}=\frac{15}{4}-\frac{28}{3}=\frac{-67}{12}.$

31. $\displaystyle\sum_{k=1}^{\infty}\left(\frac{8}{5^{k}}-\frac{7}{(k+3)(k+4)}\right)=8\sum_{k=1}^{\infty}\frac{1}{5^{k}}-7\sum_{k=1}^{\infty}\frac{1}{(k+3)(k+4)}=8\sum_{j=0}^{\infty}\left(\frac{1}{5}\right)^{j+1}-7\sum_{k=1}^{\infty}\left(\frac{1}{k+3}-\frac{1}{k+4}\right)$

$$=8\left(\frac{1}{5}\right)\sum_{j=0}^{\infty}\left(\frac{1}{5}\right)^{j}-7\lim_{n\to\infty}\sum_{k=1}^{n}\left(\frac{1}{k+3}-\frac{1}{k+4}\right)$$

$$=\frac{8}{5}\cdot\frac{1}{1-\frac{1}{5}}-7\lim_{n\to\infty}\left(\frac{1}{1+3}-\frac{1}{n+4}\right)=\frac{8}{5}\cdot\frac{5}{4}-7\cdot\frac{1}{4}=2-\frac{7}{4}=\frac{1}{4}.$$

33.   The hint for this problem uses the same idea as the solution to Problem 10.3.26. The minute hand has caught up with the hour hand at time

$$1+\frac{1}{12}+\frac{1}{12^{2}}+\frac{1}{12^{3}}+\cdots=\sum_{k=0}^{\infty}\left(\frac{1}{12}\right)^{k}=\frac{1}{1-\frac{1}{12}}=\frac{12}{11}\approx 1:05:27\text{ PM}.$$

REMARK: Let x be the number of minutes past 1:00 PM when the hands meet. The hour hand starts at 5 and moves $\frac{1}{12}$ as fast as the minute hand, they meet when $x=5+\frac{x}{12}$, i.e., $x=\frac{60}{11}\approx 5.4545$ minutes.

35.   The ball travels 8 m before it first hits the ground; thereafter it alternately bounds some distance into the air and falls back to the ground, the distance travelled between successive bounces is twice the height of the rebound. Since one height is $\frac{2}{3}$ times the previous height, the total distance travelled is

$$8+2\left(\frac{2}{3}\right)8+2\left(\frac{2}{3}\right)^{2}8+2\left(\frac{2}{3}\right)^{3}8+\cdots=8+2\left(\frac{2}{3}\right)8\left(1+\frac{2}{3}+\left(\frac{2}{3}\right)^{2}+\cdots\right)=8+\frac{32}{3}\left(\frac{1}{1-\frac{2}{3}}\right)=8+32=40\text{ m}.$$

REMARK:
$$8+2\left(\frac{2}{3}\right)8+2\left(\frac{2}{3}\right)^{2}8+2\left(\frac{2}{3}\right)^{3}8+\cdots=-8+2\cdot 8\left(1+\frac{2}{3}+\left(\frac{2}{3}\right)^{2}+\left(\frac{2}{3}\right)^{3}+\cdots\right)$$

$$=-8+16\left(\frac{1}{1-\frac{2}{3}}\right)=-8+48.$$

37.   Let $a_{k}=\frac{1}{k}$ for all $k\geq 1$, let $b_{k}=0$ for $1\leq k\leq 6$ and $b_{k}=\frac{1}{k}$ for $k\geq 7$; these two sequences differ only in their first six terms. Problem 42 implies $\displaystyle\sum_{k=1}^{\infty}b_{k}$ diverges because $\displaystyle\sum_{k=1}^{\infty}a_{k}=\sum_{k=1}^{\infty}\frac{1}{k}$ is already known to diverge. But $\displaystyle\sum_{k=1}^{\infty}b_{k}=0+0+0+0+0+0+\sum_{k=7}^{\infty}\frac{1}{k}=\sum_{j=1}^{\infty}\frac{1}{j+6}$, so $\displaystyle\sum_{j=1}^{\infty}\frac{1}{j+6}$ also diverges.

39. $\displaystyle\sum_{k=0}^{\infty}(-1)^{k}x^{k}=\sum_{k=0}^{\infty}(-x)^{k}$; if $|x|=|-x|<1$, this geometric series converges to $\displaystyle\frac{1}{1-(-x)}=\frac{1}{1+x}.$ ∎

41.   Let $c=\frac{b}{a}$. If $\displaystyle\sum_{k=1}^{\infty}\frac{a}{b\,k}$ converged, then Theorem 1 would imply $\displaystyle c\sum_{k=1}^{\infty}\frac{a}{b\,k}=\sum_{k=1}^{\infty}c\cdot\frac{a}{b\,k}=\sum_{k=1}^{\infty}\frac{1}{k}$ also converged. Example 3 shows that $\displaystyle\sum_{k=1}^{\infty}\frac{1}{k}$ does not converge, this contradiction means that $\displaystyle\sum_{k=1}^{\infty}\frac{a}{b\,k}$ also does not converge. ∎

43.   If $\displaystyle\sum_{k=1}^{\infty}a_{k}$ and $\displaystyle\sum_{k=1}^{\infty}(a_{k}+b_{k})$ converge, then $\displaystyle(-1)\sum_{k=1}^{\infty}a_{k}=\sum_{k=1}^{\infty}(-1)a_{k}$ and $\displaystyle\sum_{k=1}^{\infty}\left((a_{k}+b_{k})+(-1)a_{k}\right)=\sum_{k=1}^{\infty}b_{k}$ also converge. Thus if $\displaystyle\sum_{k=1}^{\infty}a_{k}$ converges and $\displaystyle\sum_{k=1}^{\infty}b_{k}$ diverges, then $\displaystyle\sum_{k=1}^{\infty}(a_{k}+b_{k})$ does not converge. ∎

45.　If $n(n-1)a_n = (n-1)(n-2)a_{n-1} - (n-3)a_{n-2}$ for $n \geq 2$, then $2a_2 = 0 \cdot a_1 - (-1)a_0 = a_0$ and $3 \cdot 2 \cdot a_3 = 2 \cdot 1 \cdot a_2 - 0 \cdot a_1 = 2a_2 = a_0$. Hence, for $n \geq 2$, a_n is determined by a_0 alone. The conjecture that $a_n = \dfrac{a_0}{n!}$ is proved by induction since

$$n(n-1)a_n = (n-1)(n-2) \cdot \frac{a_0}{(n-1)!} - (n-3) \cdot \frac{a_0}{(n-2)!} = \big((n-2)-(n-3)\big) \cdot \frac{a_0}{(n-2)!} = \frac{a_0}{(n-2)!}.$$

Therefore

$$\sum_{k=0}^{\infty} a_k = a_0 + a_1 + \sum_{k=2}^{\infty} \frac{a_0}{k!} = a_0 + a_1 + a_0 \sum_{k=0}^{\infty} \frac{1}{k!} - a_0 \left(\frac{1}{0!} + \frac{1}{1!}\right) = a_0 + a_1 + a_0 \cdot e - a_0 \cdot 2 = a_1 + a_0(e-1).$$

REMARK:　Example 10.5.3 shows $\sum_{k=0}^{\infty} \dfrac{1}{k!}$ converges (Examples 10.6.1 and 10.8.4 also show this fact), Example 10.9.5 extends Example 9.4.3 and implies $\sum_{k=0}^{\infty} \dfrac{1}{k!} = e$.

Solutions 10.5　　*Two comparison tests & the integral test*　　(pages 675-6)

1.　$\displaystyle\sum_{k=1}^{n}(-1)^k = \begin{cases} -1 & \text{if } n \text{ is odd,} \\ 0 & \text{if } n \text{ is even,} \end{cases}$, therefore $\displaystyle\sum_{k=1}^{\infty}(-1)^k = \lim_{n\to\infty} \sum_{k=1}^{n}(-1)^k$ does not exist.

REMARK:　Although the partial sums are bounded, Theorem 1 does **not** imply this series converges. Example 10.4.5 also discusses this series.

3.　$\displaystyle\sum_{k=4}^{\infty} \frac{1}{5k+50} = \frac{1}{5}\sum_{k=4}^{\infty}\frac{1}{k+10} = \frac{1}{5}\sum_{j=14}^{\infty}\frac{1}{j}$; Example 10.4.3 and Problem 10.4.42 imply $\displaystyle\sum_{j=14}^{\infty}\frac{1}{j}$ diverges, thus $\displaystyle\sum_{k=4}^{\infty}\frac{1}{5k+50}$ diverges.

REMARK:　The proof of Theorem 3 is easily adapted to show $\displaystyle\sum_{k=4}^{\infty}\frac{1}{5k+50}$ converges if and only if $\displaystyle\int_{4}^{\infty}\frac{dx}{5x+50}$ converges. But $\displaystyle\int_{4}^{\infty}\frac{dx}{5x+50} = \frac{1}{5}\lim_{N\to\infty}\ln\left(\frac{N+10}{14}\right) = \infty$, hence $\displaystyle\sum_{k=4}^{\infty}\frac{1}{5k+50}$ diverges.

5.　$1 \leq k \implies 1 < k^2 < k^2+1 \implies 0 < \dfrac{1}{k^2+1} < \dfrac{1}{k^2}$; Example 1 shows $\displaystyle\sum_{k=1}^{\infty}\frac{1}{k^2}$ converges, Theorem 2(i) then implies $\displaystyle\sum_{k=1}^{\infty}\frac{1}{k^2+1}$ also converges.

REMARK:　$\displaystyle\int_{1}^{\infty}\frac{dx}{x^2+1} = \lim_{N\to\infty}\tan^{-1}N - \tan^{-1}1 = \frac{\pi}{2} - \frac{\pi}{4} = \frac{\pi}{4}$; Theorem 3 implies $\displaystyle\sum_{k=1}^{\infty}\frac{1}{k^2+1}$ converges.

7.　Let $a_k = \dfrac{1}{\sqrt{k^2+2k}}$ and $b_k = \dfrac{1}{k}$. Then $\displaystyle\lim_{k\to\infty}\frac{a_k}{b_k} = \lim_{k\to\infty}\frac{k}{\sqrt{k^2+2k}} = \lim_{k\to\infty}\frac{1}{\sqrt{1+2/k}} = 1 > 0$. $\displaystyle\sum_{k=1}^{\infty}b_k = \sum_{k=1}^{\infty}\frac{1}{k}$ diverges, hence the Limit Comparison Test implies $\displaystyle\sum_{k=1}^{\infty}a_k = \sum_{k=1}^{\infty}\frac{1}{\sqrt{k^2+2k}}$ also diverges.

9.　$\displaystyle\int_{0}^{\infty}x\,e^{-x}\,dx = \lim_{N\to\infty} -(N+1)e^{-N} + 1 = 1$, the Integral Test implies $\displaystyle\sum_{k=0}^{\infty}k\,e^{-k} = \sum_{k=1}^{\infty}k\,e^{-k}$ converges.

11.　$\displaystyle\int_{2}^{\infty}\frac{4\,dx}{x\ln x} = \lim_{N\to\infty} 4\ln(\ln N) - 4\ln(\ln 2) = \infty$; the Integral Test implies $\displaystyle\sum_{k=2}^{\infty}\frac{4}{k\ln k}$ diverges.

13.　$\displaystyle\int_{5}^{\infty}\frac{dx}{x(\ln x)^3} = \frac{-1}{2}\left(\lim_{N\to\infty}\frac{1}{(\ln N)^2} - \frac{1}{(\ln 5)^2}\right) = \frac{1}{2(\ln 5)^2}$, therefore $\displaystyle\sum_{k=5}^{\infty}\frac{1}{k(\ln k)^3}$ converges.

15. $1 \leq k \implies 1 \leq k < 3k - 1 \implies 1 \leq k^{3/2} < (3k - 1)^{3/2} \implies 0 < \dfrac{1}{(3k-1)^{3/2}} < \dfrac{1}{k^{3/2}}$. Example 4 implies

$\displaystyle\sum_{k=1}^{\infty} \dfrac{1}{k^{3/2}}$ converges, the Comparison Test then guarantees $\displaystyle\sum_{k=1}^{\infty} \dfrac{1}{(3k-1)^{3/2}}$ also converges.

17. $\displaystyle\int_{2}^{\infty} \dfrac{dx}{x\sqrt{\ln x}} = \lim_{N \to \infty} 2\sqrt{\ln N} - 2\sqrt{\ln 2} = \infty$, the Integral Test implies $\displaystyle\sum_{k=2}^{\infty} \dfrac{1}{k\sqrt{\ln k}}$ diverges.

19. Let $Q(x) = \ln\left(\dfrac{x}{x+1}\right)^x = x\ln\left(\dfrac{x}{x+1}\right) = \dfrac{\ln\left(\dfrac{x}{x+1}\right)}{x^{-1}}$; L'Hôpital's rule implies

$$\lim_{x \to \infty} Q(x) = \lim_{x \to \infty} \dfrac{x^{-1} - (x+1)^{-1}}{-x^{-2}} = \lim_{x \to \infty} \dfrac{-x}{x+1} = \lim_{x \to \infty} \dfrac{-1}{1 + 1/x} = -1.$$

Thus $\displaystyle\lim_{k \to \infty} \left(\dfrac{k}{k+1}\right)^k = \lim_{x \to \infty} \left(\dfrac{x}{x+1}\right)^x = \lim_{x \to \infty} e^{Q(x)} = e^{-1}$. Since the individual terms do not tend to zero,

the Corollary to Theorem 10.4.2 implies $\displaystyle\sum_{k=3}^{\infty} \left(\dfrac{k}{k+1}\right)^k$ diverges.

REMARK: Problem 10.1.35 implies $\left(\dfrac{k}{k+1}\right)^k = \left(1 + \dfrac{-1}{k+1}\right)^k = \left(1 + \dfrac{-1}{k+1}\right)^{k+1} \cdot \left(\dfrac{k+1}{k}\right) \to e^{-1} \cdot 1.$

---

21. $e < 4 \leq k \implies 1 < \ln k < k \implies 0 < \ln(\ln k) < \ln k \implies 0 < \dfrac{1}{k \ln k} < \dfrac{1}{k \ln(\ln k)}$. Since

$\displaystyle\int_{4}^{\infty} \dfrac{dx}{x \ln x} = \lim_{N \to \infty} \ln(\ln N) - \ln(\ln 2) = \infty$, the Integral Test shows $\displaystyle\sum_{k=4}^{\infty} \dfrac{1}{k \ln k}$ diverges; the Comparison

Test then shows $\displaystyle\sum_{k=4}^{\infty} \dfrac{1}{k \ln(\ln k)}$ also diverges.

23. $\displaystyle\int_{1}^{\infty} \dfrac{dx}{(x+2)\sqrt{\ln(x+2)}} = \lim_{N \to \infty} 2\sqrt{\ln(N+2)} - 2\sqrt{\ln 3} = \infty$ implies $\displaystyle\sum_{k=1}^{\infty} \dfrac{1}{(k+2)\sqrt{\ln(k+2)}}$ diverges;

since $k \geq 1$ implies $\dfrac{1}{\sqrt{\ln(k+2)}} < \dfrac{1}{\sqrt{\ln(k+1)}}$, the Comparison Test shows $\displaystyle\sum_{k=1}^{\infty} \dfrac{1}{(k+2)\sqrt{\ln(k+1)}}$ diverges.

25. $\displaystyle\lim_{k \to \infty} \tan^{-1} k = \dfrac{\pi}{2}$; therefore, $\displaystyle\sum_{k=1}^{\infty} \tan^{-1} k$ diverges since the individual terms do not tend to zero

27. Since $\displaystyle\lim_{k \to \infty} \dfrac{\ln k}{k} = \lim_{x \to \infty} \dfrac{\ln x}{x} = 0$, there is an M such that $k \geq M \implies 0 < \dfrac{\ln k}{k} < 1 \implies 0 < \dfrac{\ln k}{k^3} < \dfrac{1}{k^2}$.

Therefore $\displaystyle\sum_{k=1}^{\infty} \dfrac{1}{k^2}$ converges $\implies \displaystyle\sum_{k=M}^{\infty} \dfrac{1}{k^2}$ converges $\implies \displaystyle\sum_{k=M}^{\infty} \dfrac{\ln k}{k^3}$ converges $\implies \displaystyle\sum_{k=1}^{\infty} \dfrac{\ln k}{k^3}$ converges.

29. $0 < \operatorname{sech} x = \dfrac{2}{e^x + e^{-x}} < \dfrac{2}{e^x} = 2e^{-x}$. Our series is bounded by a convergent geometric series:

$\displaystyle\sum_{k=1}^{\infty} 2e^{-k} = \dfrac{2}{e} \sum_{j=0}^{\infty} \left(\dfrac{1}{e}\right)^j = \dfrac{2}{e} \cdot \dfrac{1}{1 - \frac{1}{e}} = \dfrac{2}{e-1}$; the Comparison Theorem then implies $\displaystyle\sum_{k=1}^{\infty} \operatorname{sech} k$ converges.

31. Since $\displaystyle\lim_{x \to \infty} \dfrac{\ln x}{x^\epsilon} = 0$ if $\epsilon > 0$ (see Solution 9.2.5), we may infer $\dfrac{|\ln x|}{x^{1/20}} < 1$ and $|\ln x| < x^{1/20}$ provided x is large enough (90^{20} is large enough). Therefore there is some large M such that

$$x > M \implies \sqrt{x} \cdot (\ln x)^{10} < \sqrt{x} \cdot \left(x^{1/20}\right)^{10} = x^{1/2 + 10/20} = x \implies \dfrac{1}{x} < \dfrac{1}{\sqrt{x} \cdot (\ln x)^{10}}.$$

Since $\displaystyle\sum_{k=M}^{\infty} \dfrac{1}{k}$ diverges, so do $\displaystyle\sum_{k=M}^{\infty} \dfrac{1}{\sqrt{x} \cdot (\ln x)^{10}}$ and $\displaystyle\sum_{k=2}^{\infty} \dfrac{1}{\sqrt{x} \cdot (\ln x)^{10}}$.

33.　$1 \le k \implies 1 < (k+1)^{7/8} < k+1$ and $2 \le k^3 + k^2 < k^3 + k^2 + 3 \implies 0 < \dfrac{(k+1)^{7/8}}{k^3 + k^2 + 3} < \dfrac{k+1}{k^3 + k^2} = \dfrac{1}{k^2}$.

Since $\displaystyle\sum_{k=1}^{\infty} \dfrac{1}{k^2}$ converges, so does $\displaystyle\sum_{k=1}^{\infty} \dfrac{(k+1)^{7/8}}{k^3 + k^2 + 3}$.

35.　If $\displaystyle\sum_{k=0}^{\infty} a_k$ is a convergent series, then $\displaystyle\lim_{k \to \infty} a_k = 0$ (Theorem 10.4.2). We're also given that $a_k \ge 0$; hence
there is an M such that $k \ge M \implies 0 \le a_k < 1 \implies 0 \le a_k^2 < a_k < 1$. Therefore

$$\sum_{k=0}^{\infty} a_k \text{ converges} \implies \sum_{k=M}^{\infty} a_k \text{ converges} \implies \sum_{k=M}^{\infty} a_k^2 \text{ converges} \implies \sum_{k=0}^{\infty} a_k^2 \text{ converges. } \blacksquare$$

37.　Suppose $p(x)$ is a polynomial of degree n and $q(x)$ is a polynomial of degree $m \le n + 1$, both with
positive coefficients. Then $a_k = \dfrac{p(k)}{q(k)} > 0$ and we use $b_k = \dfrac{1}{k^{m-n}}$ for comparison. $\dfrac{a_k}{b_k} = \dfrac{p(k) \cdot k^m}{q(k) \cdot k^n}$ is the
quotient of two polynomials of the same degree; Problem 1.5.32(c) implies $\displaystyle\lim_{k \to \infty} \dfrac{a_k}{b_k}$ is the quotient of the
leading coefficients of p and q; in particular, this limit is positive so $\displaystyle\sum_{k=0}^{\infty} a_k$ and $\displaystyle\sum_{k=0}^{\infty} b_k$ converge or
diverge together. $m \le n + 1 \implies m - n \le 1 \implies \displaystyle\sum_{k=0}^{\infty} \dfrac{1}{k^{m-n}} = \sum_{k=0}^{\infty} b_k$ diverges (Example 4). The
Limit Comparison theorem thus implies $\displaystyle\sum_{k=0}^{\infty} a_k = \sum_{k=0}^{\infty} \dfrac{p(k)}{q(k)}$ diverges. $\blacksquare$

39.　Let $a_k = \left(\dfrac{1}{k}\right)^{1+1/k}$ and $b_k = \dfrac{1}{k}$, then we apply Example 9.2.7 to infer $\displaystyle\lim_{k \to \infty} \dfrac{a_k}{b_k} = \lim_{k \to \infty} \left(\dfrac{1}{k}\right)^{1/k} =$
$\dfrac{1}{\lim_{x \to \infty} x^{1/x}} = \dfrac{1}{1} = 1 > 0$. The harmonic series $\displaystyle\sum_{k=0}^{\infty} b_k = \sum_{k=0}^{\infty} \dfrac{1}{k}$ diverges, so does $\displaystyle\sum_{k=0}^{\infty} a_k = \sum_{k=0}^{\infty} \left(\dfrac{1}{k}\right)^{1+1/k}$.

41.　If $k > 0$, $\dfrac{1}{x}$ is a strictly decreasing function on the interval $[k, k+1]$, thus $\dfrac{1}{k} > \displaystyle\int_k^{k+1} \dfrac{dx}{x} > \dfrac{1}{k+1}$.

$$S_N = \sum_{k=1}^{N} \frac{1}{k} > \sum_{k=1}^{N} \int_k^{k+1} \frac{dx}{x} = \int_1^{n+1} \frac{dx}{x} = \ln(N+1)$$
$$> \sum_{k=1}^{N} \frac{1}{k+1} = \sum_{j=2}^{N+1} \frac{1}{j} = S_{N+1} - 1.$$

The relation $N \ge 1 \implies \ln(N+1) > S_{N+1} - 1$ can also be stated $n \ge 2 \implies \ln n > S_n - 1$; hence
$1 + \ln n > S_n > \ln(n+1)$ for all $n \ge 2$. $\blacksquare$

43.　We will exploit the relation $\gamma \approx S_n - \ln n$ expressed in the form $n \approx e^{S_n - \gamma}$ to approximate the number
of terms needed to obtain a sum of a chosen size. Dividing that approximated n by one million $(1.0 \cdot 10^6)$
will yield the time in seconds for our hypothesized computer to add that many terms; further division will
convert that time to the desired unit (hours, days, years).
　a.　$n_{25} \approx e^{25 - \gamma} \approx 4.04278 \cdot 10^{10}$;　time to add $\approx n_{25} / \left(1.0 \cdot 10^6\right) / 60^2 \approx 11.22995$ hours.
　b.　$n_{30} \approx e^{30 - \gamma} \approx 6.00002 \cdot 10^{12}$;　time to add $\approx n_{30} / \left(1.0 \cdot 10^6\right) / 60^2 / 24 \approx 69.44470$ days.
　c.　$n_{35} \approx e^{35 - \gamma} \approx 8.90482 \cdot 10^{14}$;　time to add $\approx n_{35} / \left(1.0 \cdot 10^6\right) / 60^2 / 24 / 365.25 \approx 28.21768$ years.

REMARK:　The preceding computations use the seven-figure value $\gamma \approx 0.5772157$; a 24 decimal place
value is given on page 3 of the marvelous *Handbook of Mathematical Functions with Formulas, Graphs,
and Mathematical Tables* edited by Milton Abramowitz and Irene A. Stegun. Nevertheless, the above
answers do not need to be revised even if we use the extra accuracy.

10.6 — The ratio and root tests

1. If $a_k = \dfrac{2^k}{k^2}$, then $\lim\limits_{k\to\infty} \dfrac{a_{k+1}}{a_k} = \lim\limits_{k\to\infty} 2\left(\dfrac{k}{k+1}\right)^2 = 2 > 1$; according to Theorem 1 (the Ratio Test), this means $\sum\limits_{k=1}^{\infty} \dfrac{2^k}{k^2}$ diverges.

3. If $a_k = \dfrac{r^k}{k^r}$ where $0 < r < 1$, then $\lim\limits_{k\to\infty} \dfrac{a_{k+1}}{a_k} = \lim\limits_{k\to\infty} r\left(\dfrac{k}{k+1}\right)^r = r \cdot 1^r = r < 1$; the Ratio Test implies $\sum\limits_{k=1}^{\infty} \dfrac{r^k}{k^r}$ converges.

5. $a_k = \dfrac{k!}{k^k} \implies \lim\limits_{k\to\infty} \dfrac{a_{k+1}}{a_k} = \lim\limits_{k\to\infty} \dfrac{(k+1)!}{k!} \cdot \dfrac{k^k}{(k+1)^{k+1}} = \lim\limits_{k\to\infty} \left(\dfrac{k}{k+1}\right)^k = \dfrac{1}{\lim_{k\to\infty}(1+1/k)^k} = \dfrac{1}{e} < 1$; the Ratio Test implies $\sum\limits_{k=2}^{\infty} \dfrac{k!}{k^k}$ converges.

REMARK: See Example 3 and Solutions 10.2.33-34, 10.5.19.

7. The Ratio Test shows $\sum\limits_{k=1}^{\infty} \dfrac{e^k}{k^5}$ diverges because $a_k = \dfrac{e^k}{k^5}$ implies $\lim\limits_{k\to\infty} \dfrac{a_{k+1}}{a_k} = \lim\limits_{k\to\infty} e\left(\dfrac{k}{k+1}\right)^5 = e \cdot 1^5 = e$.

9. If $a_k = \dfrac{k^{2/3}}{10^k}$, then $\lim\limits_{k\to\infty} \dfrac{a_{k+1}}{a_k} = \lim\limits_{k\to\infty} \dfrac{1}{10}\left(\dfrac{k+1}{k}\right)^{2/3} = \dfrac{1}{10} < 1$; therefore $\sum\limits_{k=1}^{\infty} \dfrac{k^{2/3}}{10^k}$ converges.

11. Example 9.2.7 implies $\lim\limits_{k\to\infty} k^{1/k} = 1$. Hence, if $a_k = \dfrac{k}{(\ln k)^k}$, then

$$\lim_{k\to\infty} a_k^{1/k} = \lim_{k\to\infty} \frac{k^{1/k}}{\ln k} = \left(\lim_{k\to\infty} k^{1/k}\right)\left(\lim_{k\to\infty} \frac{1}{\ln k}\right) = 1 \cdot 0 = 0 < 1;$$

Theorem 2 (the Root Test) implies $\sum\limits_{k=2}^{\infty} \dfrac{k}{(\ln k)^k}$ converges.

13. Since $\lim\limits_{k\to\infty}\left(1 + \dfrac{1}{k}\right)^k = e$, the individual terms do not converge to zero, hence $\sum\limits_{k=2}^{\infty}\left(1+\dfrac{1}{k}\right)^k$ diverges.

REMARK: $1 < \left(1 + \dfrac{1}{k}\right)^k$ for every $k \geq 1$ so a simple comparison indicates divergence; however, the Ratio and Root tests are both inconclusive.

15. If $A_k = \dfrac{3^{4k}}{k^k}$, then $\lim\limits_{k\to\infty} A_k^{1/k} = \lim\limits_{k\to\infty} \dfrac{3^4}{k} = 0$; the Root Test implies $\sum\limits_{k=1}^{\infty} \dfrac{3^{4k}}{k^k}$ converges, therefore so does $3^5 \sum\limits_{k=1}^{\infty} \dfrac{3^{4k}}{k^k} = \sum\limits_{k=1}^{\infty} \dfrac{3^{4k+5}}{k^k}$.

17. If $A_k = \dfrac{k^k}{a^{mk}}$ where $a > 1$, then $\lim\limits_{k\to\infty} A_k^{1/k} = \lim\limits_{k\to\infty} \dfrac{k}{a^m} = \infty$; the Root Test implies $\sum\limits_{k=1}^{\infty} \dfrac{k^k}{a^{mk}}$ diverges, therefore so does $a^{-b} \sum\limits_{k=1}^{\infty} \dfrac{k^k}{a^{mk}} = \sum\limits_{k=1}^{\infty} \dfrac{k^k}{a^{mk+b}}$.

19. If $a_k = \dfrac{k^2\, k!}{(2k)!}$, then the Ratio Test implies $\sum\limits_{k=1}^{\infty} a_k = \sum\limits_{k=1}^{\infty} \dfrac{k^2\, k!}{(2k)!}$ converges because

$$\lim_{k\to\infty} \frac{a_{k+1}}{a_k} = \lim_{k\to\infty} \left(\frac{k+1}{k}\right)^2 \cdot \frac{(k+1)! \cdot (2k)!}{k! \cdot (2k+2)!} = \lim_{k\to\infty} \left(\frac{k+1}{k}\right)^2 \cdot \frac{1}{(2k+1)2} = 1 \cdot 0 = 0 < 1.$$

21. Solution 5 shows $\sum\limits_{k=1}^{\infty} \dfrac{k!}{k^k}$ converges, Theorem 10.4.2 then implies $\lim\limits_{k\to\infty} \dfrac{k!}{k^k} = 0$. Therefore, if $a_k = \left(\dfrac{k!}{k^k}\right)^k$,

then $\lim\limits_{k\to\infty} a_k^{1/k} = \lim\limits_{k\to\infty} \dfrac{k!}{k^k} = 0$; the Root Test shows that $\sum\limits_{k=1}^{\infty} \left(\dfrac{k!}{k^k}\right)^k$ converges.

23. If $a_k = \dfrac{e^k}{(\ln k)^k}$, then $\lim\limits_{k\to\infty} a_k^{1/k} = \lim\limits_{k\to\infty} \dfrac{e}{\ln k} = 0$; the Root Test implies $\sum\limits_{k=2}^{\infty} \dfrac{e^k}{(\ln k)^k}$ converges.

25. If $a_k = \left(\dfrac{k}{3k+2}\right)^k$, then $\lim\limits_{k\to\infty} a_k^{1/k} = \lim\limits_{k\to\infty} \dfrac{k}{3k+2} = \lim\limits_{k\to\infty} \dfrac{1}{3+2/k} = \dfrac{1}{3} < 1$; the Root Test implies

$\sum\limits_{k=1}^{\infty} a_k = \sum\limits_{k=1}^{\infty} \left(\dfrac{k}{3k+2}\right)^k$ converges.

REMARK: Since $\dfrac{k}{3k+2} = \dfrac{1}{3+2/k} < \dfrac{1}{3}$, we reach the same conclusion using the geometric series

$\sum_{k=1}^{\infty} \left(\dfrac{1}{3}\right)^k$ for direct comparison.

27. Let $a_k = \begin{cases} 3 \cdot k^{-2} & \text{if } k \text{ is even,} \\ k^{-2} & \text{if } k \text{ is odd.} \end{cases}$ Since $0 \le a_k \le \dfrac{3}{k^2}$ for all k and since $\sum\limits_{k=1}^{\infty} \dfrac{3}{k^2} = 3\sum\limits_{k=1}^{\infty} \dfrac{1}{k^2}$ converges, the

regular Comparison Test implies that $\sum\limits_{k=1}^{\infty} a_k$ also converges. However $\dfrac{a_{k+1}}{a_k} = \begin{cases} \dfrac{1}{3}\left(\dfrac{k}{k+1}\right)^2 & \text{if } k \text{ is even,} \\ 3\left(\dfrac{k}{k+1}\right)^2 & \text{if } k \text{ is odd.} \end{cases}$

Since $\lim\limits_{k\to\infty} \left(\dfrac{k}{k+1}\right)^2 = 1$, we see $\dfrac{a_{k+1}}{a_k}$ oscillates between a neighborhood of 3 and a neighborhood of $\dfrac{1}{3}$,

hence $\lim\limits_{k\to\infty} \dfrac{a_{k+1}}{a_k}$ does not exist.

29. If $x = 0$, then $\dfrac{x^k}{k!} = \dfrac{0^k}{k!} = \dfrac{0}{k!} = 0$ and $\sum\limits_{k=0}^{\infty} \dfrac{x^k}{k!} = \sum\limits_{k=0}^{\infty} 0 = 0$. On the other hand, if $x \ne 0$, then

$\lim\limits_{k\to\infty} \dfrac{a_{k+1}}{a_k} = \lim\limits_{k\to\infty} \dfrac{x^{k+1}}{x^k} \cdot \dfrac{k!}{(k+1)!} = \lim\limits_{k\to\infty} \dfrac{x}{k+1} = 0$; the Ratio Test implies $\sum\limits_{k=0}^{\infty} \dfrac{x^k}{k!}$ converges. ∎

REMARK: Example 10.9.5 shows $\sum_{k=1}^{\infty} \dfrac{x^k}{k!} = e^x$.

31. Suppose $\{a_n\}$ is a sequence of nonnegative terms such that $\lim\limits_{n\to\infty} a_n^{1/n} = R < 1$. Let $\epsilon = \dfrac{1-R}{2}$; there

is an integer M such that $n \ge M$ implies $\left|a_n^{1/n} - R\right| < \epsilon$. Our choice of ϵ means that $n \ge M$ implies

$0 < a_n^{1/n} < R + \epsilon = 1 - \epsilon < 1$ and thus $0 < a_n < (R+\epsilon)^n = (1-\epsilon)^n$. The geometric series $\sum\limits_{n=M}^{\infty} (1-\epsilon)^n$

converges to $\dfrac{(1-\epsilon)^M}{\epsilon}$; the Comparison Test (Theorem 10.5.2) then assures us that $\sum\limits_{n=M}^{\infty} a_n$ must converge,

therefore $\sum\limits_{n=1}^{M-1} a_n + \sum\limits_{n=M}^{\infty} a_n = \sum\limits_{n=1}^{\infty} a_n$ also converges. ∎

33. Suppose $a_k = 1$; it is obvious both that $\sum\limits_{k=1}^{\infty} 1$ diverges and that $\lim\limits_{k\to\infty} a_k^{1/k} = \lim\limits_{k\to\infty} 1^{1/k} = \lim\limits_{k\to\infty} 1 = 1$. On

the other hand, if we let $b_k = \dfrac{1}{k^2}$, then $\sum\limits_{k=1}^{\infty} b_k = \sum\limits_{k=1}^{\infty} \dfrac{1}{k^2}$ converges (shown by Example 10.5.1) while Example

9.2.7 implies $\lim\limits_{k\to\infty} b_k^{1/k} = \dfrac{1}{\left(\lim_{k\to\infty} k^{1/k}\right)^2} = \dfrac{1}{1^2} = 1$.

These two examples are sufficient to show that $c_n \geq 0$ and $c_n^{1/n} \to 1$ does not provide information about the convergence or divergence of $\sum_{k=1}^{\infty} c_n$, i.e., the Root Test is inconclusive for this case. ∎

35.   Suppose $a_n > 0$ for all n and $\lim_{n \to \infty} \frac{a_{n+1}}{a_n} = \infty$. Then there is some M such that $n > M$ implies $\frac{a_{n+1}}{a_n} > 2$. Therefore $a_{n+1} > 2a_n \geq 2^{n-M} a_{M+1} > a_{M+1} > 0$. This, in turn, implies the sequence $\{a_n\}$ can not converge to zero; therefore, using the Corollary to Theorem 10.4.2, we infer $\sum_{k=1}^{\infty} a_k$ must diverge.

Solutions 10.7   *Absolute & conditional convergence: Alternating series*   (pages 689-91)

REMARK: The Alternating Series Test applies to series $-a_1 + a_2 - a_3 + \cdots$ as well as $a_1 - a_2 + a_3 - \cdots$, i.e., $\sum_{k=1}^{\infty} (-1)^k a_k$ converges if and only if $(-1) \sum_{k=1}^{\infty} (-1)^k a_k = \sum_{k=1}^{\infty} (-1)^{k+1} a_k$ converges. Hence, I will use an easy extension of Theorem 2 which says both series converge if $a_k \geq 0$, $a_k \geq a_{k+1}$, and $\lim_{k \to \infty} a_k = 0$.

---

1.   The sequence $\{(-1)^n\}$ does not converge to 0, hence $\sum_{k=1}^{\infty} (-1)^k$ diverges (use the Corollary to Theorem 10.4.2).

REMARK: The partial sums $S_n = \sum_{k=1}^{n} (-1)^k$ do not converge because they oscillate between -1 and 0.

---

3.   The sequence $\left\{ \cos \frac{k\pi}{2} : k \geq 0 \right\} = \{1, 0, -1, 0, 1, 0, \cdots\}$ does not converge to 0, thus $\sum_{k=0}^{\infty} \cos \frac{k\pi}{2}$ diverges.

5. $\frac{\sqrt{k}}{k+3} > 0$, the sequence $\left\{ \frac{\sqrt{k}}{k+3} \right\}$ decreases for $k \geq 3$ because

$$\frac{\sqrt{k}}{k+3} > \frac{\sqrt{k+1}}{k+4} \iff k(k+4)^2 > (k+1)(k+3)^2$$

$$\iff k(k+4)^2 - (k+1)(k+3)^2 = k^2 + k - 9 > 0 \iff \left(k + \frac{1}{2}\right)^2 > 9 + \frac{1}{4},$$

and $\lim_{k \to \infty} \frac{\sqrt{k}}{k+3} = \lim_{k \to \infty} \frac{1}{\sqrt{k} + 3/\sqrt{k}} = 0$; therefore $\sum_{k=1}^{\infty} \frac{(-1)^k \sqrt{k}}{k+3}$ converges. On the other hand, comparison with $\sum_{k=1}^{\infty} \frac{1}{\sqrt{k}}$ and use of Example 10.5.4 shows $\sum_{k=1}^{\infty} \frac{\sqrt{k}}{k+3}$ diverges; hence $\sum_{k=1}^{\infty} \frac{(-1)^k \sqrt{k}}{k+3}$ is conditionally convergent.

7. $k \ln k$ is positive and increasing for $k \geq 2$, thus $\frac{1}{k \ln k}$ is positive and decreasing; since $\lim_{k \to \infty} \frac{1}{k \ln k} = 0$, we conclude $\sum_{k=2}^{\infty} \frac{(-1)^k}{k \ln k}$ converges. This series is conditionally convergent since $\int_2^{\infty} \frac{dx}{x \ln x} = \lim_{N \to \infty} \ln \left| \frac{\ln N}{\ln 2} \right| = \infty$ implies $\sum_{k=2}^{\infty} \frac{1}{k \ln k}$ diverges.

9. $\{k\sqrt{\ln k}\}$ is strictly increasing, hence $\left\{ \frac{1}{k\sqrt{\ln k}} \right\}$ is strictly decreasing; $\lim_{k \to \infty} \frac{1}{k\sqrt{\ln k}} = 0$, thus $\sum_{k=2}^{\infty} \frac{(-1)^k}{k\sqrt{\ln k}}$ converges. Solution 10.5.17 shows $\sum_{k=2}^{\infty} \frac{1}{k\sqrt{\ln k}}$ is divergent, hence the alternating series is only conditionally convergent.

11. $\{5k - 4\}$ is increasing and positive for $k \geq 1$, thus $\left\{\dfrac{1}{5k-4}\right\}$ is positive and decreasing; $\lim\limits_{k \to \infty} \dfrac{1}{5k-4} = 0$,

hence $\sum\limits_{k=1}^{\infty} \dfrac{(-1)^{k+1}}{5k-4}$ converges. This series converges conditionally because $\dfrac{1}{5k} < \dfrac{1}{5k-4}$ and the divergence

of $\sum\limits_{k=1}^{\infty} \dfrac{1}{5k}$ imply $\sum\limits_{k=1}^{\infty} \dfrac{1}{5k-4}$ diverges.

13. $\left|\dfrac{k!}{(-3)^k}\right| = \dfrac{k!}{3^k} \geq \left(\dfrac{5!}{3^5}\right) \cdot 2^{k-5}$ if $k \geq 6$ and $\lim\limits_{k \to \infty} \dfrac{k!}{3^k} = \infty$. Therefore $\sum\limits_{k=1}^{\infty} \dfrac{k!}{(-3)^k}$ diverges because the

sequence $\left\{\dfrac{k!}{(-3)^k}\right\}$ does not converge to 0.

15. L'Hôpital's Rule implies $\lim\limits_{k \to \infty} \dfrac{3^k}{k^2} = \lim\limits_{x \to \infty} \dfrac{3^x}{x^2} = \lim\limits_{x \to \infty} \dfrac{3^x \ln 3}{2x} = \lim\limits_{x \to \infty} \dfrac{3^x (\ln 3)^2}{2} = \infty$, therefore $\left\{\dfrac{(-3)^k}{k^2}\right\}$

does not converge to 0 and $\sum\limits_{k=1}^{\infty} \dfrac{(-3)^k}{k^2}$ diverges.

17. $\sum\limits_{k=2}^{\infty} \dfrac{(-1)^k k^2}{k^3 + 1}$ cannot be absolutely convergent because $\dfrac{k^2}{k^3+1} > \dfrac{1}{2k}$ and $\sum\limits_{k=2}^{\infty} \dfrac{1}{2k}$ diverges. On the other

hand, $\lim\limits_{k \to \infty} \dfrac{k^2}{k^3+1} = \lim\limits_{k \to \infty} \dfrac{1}{k + 1/k^2} = 0$. The sequence $\left\{\dfrac{k^2}{k^3+1}\right\}$ is strictly decreasing since

$$\dfrac{k^2}{k^3+1} > \dfrac{(k+1)^2}{(k+1)^3+1} \iff k^2\left((k+1)^3+1\right) > (k+1)^2\left(k^3+1\right) \iff (k^2-1)\left((k+1)^2+1\right) > -1$$

and this last inequality is true for every $k \geq 1$; therefore $\sum\limits_{k=2}^{\infty} \dfrac{(-1)^k k^2}{k^3+1}$ is conditionally convergent.

19. Example 10.5.4 implies $\sum\limits_{k=3}^{\infty} \dfrac{1}{k^3}$ converges; since $|\sin x| \leq 1$ for all x, the Comparison Test shows

$\sum\limits_{k=3}^{\infty} \dfrac{\sin(k\pi/7)}{k^3}$ is absolutely convergent.

21. $\dfrac{k+2}{k(k+1)} > \dfrac{k+3}{(k+1)(k+2)} \iff k^2+4k+4 = (k+2)^2 > k(k+3) = k^2+3k \iff k+4 > 0$, hence the

sequence $\left\{\dfrac{k+2}{k(k+1)}\right\}$ is strictly decreasing, $\lim\limits_{k \to \infty} \dfrac{k+2}{k(k+1)} = \lim\limits_{k \to \infty} \left(\dfrac{1}{k}\right)\left(1+\dfrac{1}{k+1}\right) = 0 \cdot 1 = 0$, therefore

$\sum\limits_{k=1}^{\infty} \dfrac{(-1)^k (k+2)}{k(k+1)}$ converges. That convergence is only conditional since $\dfrac{k+2}{k(k+1)} > \dfrac{1}{k}$ and $\sum\limits_{k=1}^{\infty} \dfrac{1}{k}$ diverges.

23. If $a_k = \dfrac{k(k+1)}{(k+2)^4}$ and $b_k = \dfrac{1}{k^2}$, then $\lim\limits_{k \to \infty} \dfrac{a_k}{b_k} = \lim\limits_{k \to \infty} \dfrac{k^3(k+1)}{(k+2)^4} = \lim\limits_{k \to \infty} \left(\dfrac{k}{k+2}\right)^3 \left(\dfrac{k+1}{k+2}\right) = 1 > 0$;

since $\sum\limits_{k=2}^{\infty} b_k = \sum\limits_{k=2}^{\infty} \dfrac{1}{k^2}$ converges, we infer $\sum\limits_{k=2}^{\infty} \dfrac{(-1)^k k(k+1)}{(k+2)^4}$ is absolutely convergent.

25. L'Hôpital's Rule implies $\lim\limits_{k \to \infty} \dfrac{2^k}{k} = \lim\limits_{x \to \infty} \dfrac{2^x}{x} = \lim\limits_{x \to \infty} \dfrac{2^x \ln 2}{1} = \infty$, therefore $\left\{\dfrac{(-1)^k 2^k}{k}\right\}$ does not

converge to zero and $\sum\limits_{k=1}^{\infty} \dfrac{(-1)^k 2^k}{k}$ diverges.

27. $\dfrac{(-1)^k k^2}{k^2 + 4} = \dfrac{(-1)^k}{1 + (4/k^2)}$ oscillates between a neighborhood of -1 and a neighborhood of 1; therefore,

since $\left\{\dfrac{(-1)^k k^2}{k^2+4}\right\}$ does not converge to 0, we infer $\sum\limits_{k=1}^{\infty} \dfrac{(-1)^k k^2}{k^2+4}$ diverges.

29.   The sequence $\left\{\dfrac{k^2+3}{k^3+4}\right\}$ is strictly decreasing for all $k \geq 1$ because $\dfrac{d}{dx}\left(\dfrac{x^2+3}{x^3+4}\right) = \dfrac{(8-(9+x^2)x)\,x}{(x^3+4)^2}$

is negative on $[1,\infty)$. Thus, since $\lim\limits_{k\to\infty}\dfrac{k^2+3}{k^3+4} = \lim\limits_{k\to\infty}\dfrac{1+3/k^2}{k+4/k^2} = 0$, we infer $\sum\limits_{k=2}^{\infty}\dfrac{(-1)^k(k^2+3)}{k^3+4}$ converges.

This converge is only conditional because $\dfrac{k^2+3}{k^3+4} > \dfrac{1}{k}$ for all $k \geq 2$ and $\sum\limits_{k=2}^{\infty}\dfrac{1}{k}$ diverges.

31.   If $a_k = \dfrac{(-1)^{k+1}}{k!}$, the bound of Theorem 3 for $|S - S_n|$ is $\dfrac{1}{(n+1)!}$; that is less than 0.001 if $n \geq 6$. Thus

$S_6 = \dfrac{1}{1} - \dfrac{1}{2} + \dfrac{1}{6} - \dfrac{1}{24} + \dfrac{1}{120} - \dfrac{1}{720} = \dfrac{91}{144} \approx 0.631944$ and $|S - S_6| \leq \dfrac{1}{5040} \approx 1.98 \cdot 10^{-4}$.

The sequence $\{|a_n - a_{n+1}|\} = \left\{\dfrac{1}{n!} - \dfrac{1}{(n+1)!}\right\} = \left\{\dfrac{n}{(n+1)!}\right\} = \left\{\dfrac{1}{(n-1)!\cdot(n+1)}\right\}$ is monotone

decreasing for $n \geq 2$, hence we may also use Theorem 4. The bound of Theorem 4 for $|S - T_n|$ is $\dfrac{n}{2(n+1)!}$,

that is less than 0.001 if $n \geq 6$. We compute $T_6 = \dfrac{1}{1} - \dfrac{1}{2} + \dfrac{1}{6} - \dfrac{1}{24} + \dfrac{1}{120} - \dfrac{1}{1440} = \dfrac{911}{1440} \approx 0.632639$ and

$|S - T_6| \leq \dfrac{1}{1680} \approx 5.95 \cdot 10^{-4}$.

REMARK: $S = \sum\limits_{k=1}^{\infty}\dfrac{(-1)^{k+1}}{k!} = 1 - e^{-1}$, therefore $S - S_6 \approx 1.76 \cdot 10^{-4}$ and $S - T_6 \approx -5.18 \cdot 10^{-4}$.

As a procedural matter, if we verify the bounds asserted by Theorem 4 are a monotone decreasing sequence, then we have also verified that the additional hypothesis of Theorem 4 is satisfied. In the following work, I'll obtain that bound and just proceed if it is obviously monotone decreasing; if that is not obvious (or is false), I'll slow down to comment more fully.

33.   If $a_k = \dfrac{(-1)^{k+1}}{k^4}$, the bound of Theorem 3 for $|S - S_n|$ is $\dfrac{1}{(n+1)^4}$; that is less than 0.0001 if $n \geq 9$.

Thus $S_9 \approx 0.947093$ and $|S - S_9| \leq \dfrac{1}{10^4} = 0.0001$.

The bound of Theorem 4 for $|S - T_n|$ is $\dfrac{4n^3 + 6n^2 + 4n + 1}{2n^4(n+1)^4}$ which is less than 0.0001 if $n \geq 7$. We

compute $T_7 = \dfrac{1}{1^4} - \dfrac{1}{2^4} + \dfrac{1}{3^4} - \dfrac{1}{4^4} + \dfrac{1}{5^4} - \dfrac{1}{6^4} + \dfrac{1}{2\cdot 7^4} \approx 0.946976$ and $|S - T_7| \leq 8.62 \cdot 10^{-5}$.

REMARK: $S = \sum\limits_{k=1}^{\infty}\dfrac{(-1)^{k+1}}{k^4} = \dfrac{7\pi^4}{720}$, therefore $S - S_9 \approx -5.98 \cdot 10^{-5}$ and $S - T_7 \approx 5.68 \cdot 10^{-5}$.

35.   If $a_k = \dfrac{(-1)^{k+1}}{k^k}$, the bound of Theorem 3 for $|S - S_n|$ is $\dfrac{1}{(n+1)^{n+1}}$; it is less than 0.0001 if $n \geq 5$.

Thus we compute $S_5 = 1 - \dfrac{1}{2^2} + \dfrac{1}{3^3} - \dfrac{1}{4^4} + \dfrac{1}{5^5} = \dfrac{16922537}{21600000} \approx 0.783451$ and $|S - S_5| \leq 2.14 \cdot 10^{-5}$.

REMARK: $\sum\limits_{k=1}^{\infty}\dfrac{(-x)^{k-1}}{k^k} = \int_0^1 t^{xt}\,dt$ for any x; therefore $S \approx 0.78343051$ and $S - S_5 \approx -2.03 \cdot 10^{-5}$.

There are two reasons why the above work does not include a T_n approximation. Firstly, it is not obvious that $\{|a_n - a_{n+1}|\}$ is a monotone decreasing sequence. Secondly, $\{S_n\}$ has tighter error bounds than $\{T_n\}$. Since $\{a_n\}$ is decreasing and positive, we see $\{T_n\}$ is better than $\{S_n\}$ if and only if

$\dfrac{1}{2}|a_n - a_{n+1}| < |a_{n+1}| \iff \dfrac{1}{2}(a_n - a_{n+1}) < a_{n+1} \iff a_n < 3a_{n+1} \iff \dfrac{a_n}{3} < a_{n+1} < a_n$.

For our particular series, we have

$a_{n+1} = \dfrac{1}{(n+1)^{n+1}} = \dfrac{1}{(n+1)\cdot(n+1)^n}\cdot\dfrac{n^n}{n^n} = \dfrac{1}{n+1}\cdot\dfrac{n^n}{(n+1)^n}\cdot\dfrac{1}{n^n} = \left(\dfrac{1}{n+1}\right)\left(\dfrac{n}{n+1}\right)^n a_n.$

Since $\lim\limits_{n\to\infty}\left(\dfrac{n}{n+1}\right)^n = e^{-1} \approx \dfrac{1}{2.72}$ and $\dfrac{1}{n+1} < \dfrac{1}{3}$ if $n \geq 3$, we have $0 < a_{n+1} \ll \dfrac{a_n}{3}$ for this problem.

37. Let $a_k = \dfrac{(-1)^{k+1}}{k}$. The following table lists the first sixteen terms of a rearrangement $\{b_n\}$ such that $\sum_{k=1}^{\infty} b_k$ converges to 0. At each stage, we choose from the unused terms that one of largest magnitude which will displace the current partial sum towards zero.

n	1	2	3	4	5	6	7	8	9	10	11	12	13	14	15	16
b_n	1	$\frac{-1}{2}$	$\frac{-1}{4}$	$\frac{-1}{6}$	$\frac{-1}{8}$	$\frac{1}{3}$	$\frac{-1}{10}$	$\frac{-1}{12}$	$\frac{-1}{14}$	$\frac{-1}{16}$	$\frac{1}{5}$	$\frac{-1}{18}$	$\frac{-1}{20}$	$\frac{-1}{22}$	$\frac{-1}{24}$	$\frac{1}{7}$
$\sum_{k=1}^{n} b_k$	1	.5	.25	.083	$-.042$	.292	.192	.108	.037	$-.026$	.174	.119	.069	.023	$-.018$	.125

39. Example 10.5.4 implies that if we pick α such that $\dfrac{1}{2} < \alpha \le 1$ and let $a_k = \dfrac{1}{k^{\alpha}}$, then $\sum_{k=1}^{\infty} a_k$ diverges

but $\sum_{k=1}^{\infty} a_k^2 = \sum_{k=1}^{\infty} \dfrac{1}{k^{2\alpha}}$ converges since $2\alpha > 1$.

REMARK: Problems 10.5.11, 10.5.17, 10.5.21 involve series which we could also use for examples here.

41. If $\sum_{k=1}^{\infty} a_k$ converges, then $a_k \to 0$. Therefore there is a M such that $k \ge M$ implies $|a_k| < 1$ which then implies $|a_k^p| = |a_k|^p < |a_k|$ for any integer $p \ge 1$. If the original series is absolutely convergent, then $\sum_{k=1}^{\infty} |a_k|$ converges and we can use it for direct comparison to infer $\sum_{k=M}^{\infty} |a_k^p|$ and $\sum_{k=1}^{\infty} |a_k^p|$ converge. Finally, invoking Theorem 1, we conclude $\sum_{k=1}^{\infty} a_k^p$ converges. ∎

43. Suppose $a_k \ge a_{k+1} \ge 0$ for all $k \ge 1$ and suppose $\sum_{k=1}^{\infty}(-1)^{k+1}a_k$ converges to S. Let S_m be the partial sum of the first m terms: $S_m = \sum_{k=1}^{m}(-1)^{k+1}a_k$. Then
$$S_{2n+2} - S_{2n} = (-1)^{2n+2}a_{2n+1} + (-1)^{2n+3}a_{2n+2} = a_{2n+1} - a_{2n+2} \ge 0$$
(recall that $\{a_k\}$ decreases), hence $S_{2n+2} \ge S_{2n}$. Since these even-numbered partial sums converge to S, that limit must be the least upper bound of the increasing sequence $\{S_{2n}\}$; thus $S \ge S_{2n}$ for all n. Similarly,
$$S_{2n+1} - S_{2n-1} = (-1)^{2n+1}a_{2n} + (-1)^{2n+2}a_{2n+1} = -(a_{2n} - a_{2n+1}) \le 0,$$
$S_{2n+1} \le S_{2n-1} = S_{2(n-1)+1}$, S is the greatest lower bound of the decreasing sequence $\{S_{2n+1}\}$, and $S \le S_{2n+1}$. Therefore
$$S_{2n} \le S_{2n+2} \le S \le S_{2n+1} \le S_{2n-1}.$$
If $m = 2n$, then $S_{2n} \le S \le S_{2n+1}$ implies $0 \le S - S_m \le S_{m+1} - S_m = a_{m+1}$; if $m = 2n + 1$, then $S_{2n+2} \le S \le S_{2n+1}$ implies $-a_{m+1} = S_{m+1} - S_m \le S - S_m \le 0$. Therefore $-a_{m+1} \le S - S_m \le a_{m+1}$ for all m; this is equivalent to $|S - S_m| \le a_{m+1}$. ∎

REMARK: The size of an alternating series whose terms decrease in magnitude is bounded by the first term (a simplified version of the preceding argument shows this since $0 \le S \le S_1 = a_1$). The absolute value of the nth remainder is also an alternating series:
$$|R_n| = |S - S_n| = a_{n+1} - a_{n+2} + a_{n+3} - a_{n+4} + \cdots.$$
Therefore $|S - S_n| = |R_n| \le a_{n+1}$. ∎

45. $\sum_{k=1}^{\infty} \dfrac{(-1)^k}{k^2}$ is absolutely convergent (see Examples 1, 3, and 10.5.1). Therefore, using the text's Remark following equation (5), rearrangement will not alter the limit value. Since $\sum_{k=1}^{\infty} \dfrac{(-1)^k}{k^2} = -\dfrac{\pi^2}{12}$, no rearrangement will converge to -1.

47. Rearrange the alternating harmonic series so that two positive terms are followed by one negative term:
$$1 + \frac{1}{3} - \frac{1}{2} + \frac{1}{5} + \frac{1}{7} - \frac{1}{4} + \frac{1}{9} + \frac{1}{11} - \frac{1}{6} + \cdots$$

If we let p_n be the number of positive terms in the first n of this rearrangement, then $\lim_{n \to \infty} \dfrac{p_n}{n} = \dfrac{2}{3}$. The Theorem below implies this rearranged series converges to $\ln 2 + \dfrac{1}{2} \ln \left(\dfrac{2/3}{1/3} \right) = \ln 2 + \dfrac{1}{2} \ln 2 = \dfrac{3}{2} \ln 2$.

10.7 — Absolute & conditional convergence: Alternating series

The following Theorem and proof appear in *Rearranging the alternating harmonic series* by C.C. Cowen, K.R. Davidson, R.P. Kaufman in *The American Mathematical Monthly*, volume 87 (1980), pages 817-9. The term "simple rearrangement" refers to one with the subsequence of positive terms and the subsequence of negative terms both appearing in their original order. As above, p_n counts the number of positive terms among the first n terms of the rearranged series.

Theorem. *A simple rearrangement of the alternating harmonic series converges to* $\ln 2 + \dfrac{1}{2}\ln\left(\dfrac{\alpha}{1-\alpha}\right)$ *if and only if* $\alpha = \lim\limits_{n\to\infty} \dfrac{p_n}{n}$.

$\boxed{\text{Proof}}$ Suppose $\displaystyle\sum_{k=1}^{\infty} a_k$ is a simple rearrangement of the alternating harmonic series and let $q_n = n - p_n$

(the number of negative terms in the first n). Thus $\displaystyle\sum_{k=1}^{n} a_k = \sum_{j=1}^{p_n} \frac{1}{2j-1} - \sum_{i=1}^{q_n} \frac{1}{2i}$. Let $E_n = -\ln n + \displaystyle\sum_{k=1}^{n} \frac{1}{k}$.

Solution 10.5.42 shows the sequence $\{E_n\}$ is a positive decreasing sequence with limit γ. If we look at just the even terms, we have $\displaystyle\sum_{i=1}^{q_n} \frac{1}{2i} = \frac{1}{2}\sum_{i=1}^{q_n} \frac{1}{i} = \frac{1}{2}\left(\ln q_n + E_n\right)$. On the other hand, the odd terms by themselves

give us $\displaystyle\sum_{j=1}^{p_n} \frac{1}{2j-1} = \sum_{j=1}^{2p_n} \frac{1}{j} - \sum_{j=1}^{p_n} \frac{1}{2j} = \left(\ln(2p_n) + E_{2p_n}\right) - \frac{1}{2}\left(\ln p_n + E_{p_n}\right)$. Therefore

$$\lim_{n\to\infty} \sum_{k=1}^{n} a_k = \lim_{n\to\infty} \left(\ln(2p_n) - \frac{1}{2}\ln p_n - \frac{1}{2}\ln q_n + E_{2p_n} - \frac{1}{2}E_{p_n} - \frac{1}{2}E_{q_n} \right)$$

$$= \ln 2 + \frac{1}{2}\lim_{n\to\infty} \ln\left(\frac{p_n}{q_n}\right) + \gamma - \frac{1}{2}\gamma - \frac{1}{2}\gamma = \ln 2 + \frac{1}{2}\ln\left(\frac{\alpha}{1-\alpha}\right). \blacksquare$$

49. If $\lim\limits_{k\to\infty} 2^k a_k = 0$, then there is some M such that $k \geq M \implies \left|2^k a_k\right| < 1 \implies |a_k| < \dfrac{1}{2^k}$. Now we can

use $\displaystyle\sum_{k=M}^{\infty} \frac{1}{2^k} = 2^{1-M}$ for direct comparison to see $\displaystyle\sum_{k=M}^{\infty} |a_k|$ converges; therefore $\displaystyle\sum_{k=0}^{\infty} |a_k|$ converges and $\displaystyle\sum_{k=0}^{\infty} a_k$ is absolutely convergent.

If $\displaystyle\sum_{k=0}^{\infty} a_k$ is conditionally convergent, then $\displaystyle\sum_{k=0}^{\infty} 2^k a_k$ must diverge; otherwise, we would have $\lim\limits_{k\to\infty} 2^k a_k = 0$ which, as shown above, then implies absolute convergence.

51. $\left\{ \dfrac{(-1)^k}{k\left(2 + (-1)^k\right)} \right\} = \left\{ -1, \dfrac{1}{2\cdot 3}, \dfrac{-1}{3}, \dfrac{1}{4\cdot 3}, \dfrac{-1}{5}, \dfrac{1}{6\cdot 3}, \dfrac{-1}{7}, \ldots \right\} = \left\{ -1, \dfrac{1}{6}, \dfrac{-1}{3}, \dfrac{1}{12}, \dfrac{-1}{5}, \dfrac{1}{18}, \dfrac{-1}{7}, \ldots \right\}$.

The negative (odd) terms decay (approach zero) more slowly than the positive (even) terms; the series will diverge to $-\infty$. We can start to get a more precise notion of the speed of that divergence by considering the sum of two adjacent terms — suppose $k = 2j-1$:

$$\frac{(-1)^k}{k\left(2 + (-1)^k\right)} + \frac{(-1)^{k+1}}{(k+1)\left(2 + (-1)^{k+1}\right)} = \frac{-1}{k} + \frac{1}{(k+1)\cdot 3} = \frac{-2k-3}{3k(k+1)} < \frac{-2k-2}{3k(k+1)} = \frac{-2}{3k} = \frac{-1}{3(j-1/2)} < \frac{-1}{3j}.$$

If $N = 2n$, then

$$\sum_{k=1}^{N} \frac{(-1)^k}{k\left(2 + (-1)^k\right)} = \sum_{j=1}^{n} \left(\frac{(-1)^{2j-1}}{(2j-1)\left(2 + (-1)^{2j-1}\right)} + \frac{(-1)^{2j}}{(2j)\left(2 + (-1)^{2j}\right)} \right) < \sum_{j=1}^{n} \frac{-1}{3j}.$$

Therefore, a subsequence of partial sums is dominated by a series known to diverge to $-\infty$, so it must diverge to $-\infty$ also. Since the series would converge if and only every subsequence of partial sums converges to the same value, the existence of a divergent subsequence is sufficient to imply the series diverges. $\blacksquare$

Solutions 10.8   *Power series*   (page 696)

1.   If $a_k = \dfrac{1}{6^k}$, then $\dfrac{a_{k+1}}{a_k} = \dfrac{1}{6} = a_k^{1/k}$ and the radius of convergence (according to Theorem 2) is $\dfrac{1}{1/6} = 6$.

Because neither $\left\{\dfrac{6^k}{6^k}\right\}$ nor $\left\{\dfrac{(-6)^k}{6^k}\right\}$ converge to 0, the power series $\displaystyle\sum_{k=0}^{\infty} \dfrac{x^k}{6^k}$ does not converge if $x = \pm 6$, hence its interval of convergence is $(-6, 6)$.

REMARK: Alternatively, observe that $\displaystyle\sum_{k=0}^{\infty} \dfrac{x^k}{6^k} = \sum_{k=0}^{\infty} \dfrac{(x/2)^k}{3^k}$, let $u = \dfrac{x}{2}$, and apply Example 1.

---

3.   Let $u = x + 1$ and apply Example 1: $(-3, 3)$ is the interval of convergence for $\displaystyle\sum_{k=0}^{\infty} \dfrac{u^k}{3^k}$;

$-3 < x + 1 < 3 \iff -4 < x < 2$, therefore $(-4, 2)$ is the interval of convergence for $\displaystyle\sum_{k=0}^{\infty} \dfrac{(x+1)^k}{3^k}$.

5. $(3x)^k = 3^k x^k$, $\displaystyle\lim_{k\to\infty} \left|\dfrac{3^{k+1}}{3^k}\right| = 3 = \lim_{k\to\infty} |3^k|^{1/k}$ and $\dfrac{1}{3}$ is the radius of convergence for $\displaystyle\sum_{k=0}^{\infty}(3x)^k$. The power series diverges if $x = \dfrac{-1}{3}$ or $x = \dfrac{1}{3}$ because neither $\{(-1)^k\}$ nor $\{1^k\}$ converge to zero; the interval of convergence is $\left(\dfrac{-1}{3}, \dfrac{1}{3}\right)$.

7.   Let $u = x - 1$. The radius of convergence of $\displaystyle\sum_{k=0}^{\infty} \dfrac{(x-1)^k}{k^3 + 3} = \sum_{k=0}^{\infty} \dfrac{u^k}{k^3 + 3}$ is $\dfrac{1}{1} = 1$ because $\displaystyle\lim_{k\to\infty} \dfrac{a_{k+1}}{a_k} =$

$\displaystyle\lim_{k\to\infty} \dfrac{k^3 + 1}{(k+1)^3 + 1} = \lim_{k\to\infty} \dfrac{1 + 1/k^3}{(1 + 1/k)^3 + 1/k^3} = \dfrac{1 + 0}{(1 + 0)^3 + 0} = 1$. Since $\dfrac{1}{k^3 + 3} < \dfrac{1}{k^3}$ and $\displaystyle\sum_{k=1}^{\infty} \dfrac{1}{k^3}$ converges,

we infer $\displaystyle\sum_{k=0}^{\infty} \dfrac{u^k}{k^3 + 3}$ converges (absolutely) if $u = \pm 1$. Since $-1 \le u \le 1$ is equivalent to $0 \le x \le 2$, we see

$[0, 2]$ is the interval of convergence for $\displaystyle\sum_{k=0}^{\infty} \dfrac{(x-1)^k}{k^3 + 3}$.

9.   Let $u = x + 17$. Example 4 implies $\displaystyle\sum_{k=0}^{\infty} \dfrac{u^k}{k!}$ converges for any u in $(-\infty, \infty)$; this means $\displaystyle\sum_{k=0}^{\infty} \dfrac{(x+17)^k}{k!}$ converges for each x in $(-\infty, \infty)$ and the radius of convergence is infinite.

11.   Since $x^{2k} = \left(x^2\right)^k$, Theorem 10.3.2 implies $\displaystyle\sum_{k=0}^{\infty} x^{2k}$ converges to $\dfrac{1}{1 - x^2}$ if $x^2 < 1$ and diverges if $x \ge 1$. The interval of convergence is $(-1, 1)$ and the radius of convergence is 1. (Reread Example 6.)

13.   Theorem 2 does not apply to $\displaystyle\sum_{k=1}^{\infty} \dfrac{(-1)^k x^{2k}}{k^k}$ because each odd power of x has a zero coefficient; nevertheless, we may apply the Root Test directly: $\displaystyle\lim_{k\to\infty} \left|\dfrac{(-1)^k x^{2k}}{k^k}\right|^{1/k} = \lim_{k\to\infty} \dfrac{x^2}{k} = 0$ for all x, hence $\displaystyle\sum_{k=1}^{\infty} \dfrac{(-x^2)^k}{k^k}$ is absolutely convergent throughout $(-\infty, \infty)$ and the radius of convergence is ∞.

15.   If $a_k = \dfrac{(-1)^k k}{\sqrt{k+1}}$ then $\displaystyle\lim_{k\to\infty} \left|\dfrac{a_{k+1}}{a_k}\right| = \lim_{k\to\infty} \dfrac{k+1}{k} \cdot \sqrt{\dfrac{k+1}{k+2}} = 1 \cdot 1 = 1$; the radius of convergence is 1.

neither $\left\{\dfrac{k}{\sqrt{k+1}}\right\}$ nor $\left\{\dfrac{(-1)^k k}{\sqrt{k+1}}\right\}$ converges to zero, thus $\displaystyle\sum_{k=0}^{\infty} \dfrac{(-1)^k k x^k}{\sqrt{k+1}}$ diverges if $x = \pm 1$; therefore the interval of convergence is $(-1, 1)$.

17. $\displaystyle\lim_{k\to\infty} \left|\dfrac{1}{(\ln k)^k}\right|^{1/k} = \lim_{k\to\infty} \dfrac{1}{\ln k} = 0$, thus $\displaystyle\sum_{k=2}^{\infty} \dfrac{x^k}{(\ln k)^k}$ converges on $(-\infty, \infty)$ and $R = \infty$.

19. If $a_k = \dfrac{(-2)^k}{k^4}$, then $\lim\limits_{k\to\infty}\left|\dfrac{a_{k+1}}{a_k}\right| = \lim\limits_{k\to\infty} 2\cdot\left(\dfrac{k}{k+1}\right)^4 = 2$; thus $R = \dfrac{1}{2}$. If $x = \dfrac{-1}{2}$, then $\sum\limits_{k=1}^{\infty}\dfrac{(-2x)^k}{k^4} =$

$\sum\limits_{k=1}^{\infty}\dfrac{1}{k^4}$ which Example 10.5.4 shows to be convergent; that implies $x = \dfrac{1}{2}$ yields a convergent alternating

series. The interval of convergence is $\left[\dfrac{-1}{2},\dfrac{1}{2}\right]$.

21. $\sum\limits_{k=0}^{\infty}\dfrac{(2x+3)^k}{5^k} = \sum\limits_{k=0}^{\infty}\left(\dfrac{2x+3}{5}\right)^k$ is a geometric series converging to $\dfrac{1}{1-\frac{2x+3}{5}} = \dfrac{5}{2(1-x)}$ if $|\frac{2x+3}{5}| < 1$

and diverging otherwise. $|\frac{2x+3}{5}| < 1 \iff -5 < 2x+3 < 5 \iff -4 < x < 1$, so $(-4,1)$ is the interval of

convergence and $\dfrac{1-(-4)}{2} = \dfrac{5}{2}$ is the radius of convergence.

23. $a_k = \left(\dfrac{k}{15}\right)^k \implies \lim\limits_{k\to\infty}|a_k|^{1/k} = \lim\limits_{k\to\infty}\dfrac{k}{15} = \infty$, hence $R = 0$ and $\sum\limits_{k=0}^{\infty}\left(\dfrac{k}{15}\right)^k x^k$ converges only for $x = 0$.

25. $\sum\limits_{k=0}^{\infty}(-1)^k x^{2k+1} = x\sum\limits_{k=0}^{\infty}(-x^2)^k$ is a geometric series converging to $x\cdot\dfrac{1}{1-(-x^2)} = \dfrac{x}{1+x^2}$ if $|-x^2| < 1$

and diverging otherwise. The radius of convergence is 1 and the interval of convergence is $(-1,1)$.

REMARK: Theorem 2 does not apply immediately to this series because each even-power term has coefficient equal to 0.

27. $\lim\limits_{k\to\infty}\left|k^k\right|^{1/k} = \lim\limits_{k\to\infty} k = \infty$, hence $R = 0$ and $\sum\limits_{k=1}^{\infty} k^k(x+1)^k$ converges only for $x+1 = 0$, i.e., $x = -1$.

29. If $a_k = \dfrac{1}{(k+1)3^k}$, then $\lim\limits_{k\to\infty}\dfrac{a_{k+1}}{a_k} = \lim\limits_{k\to\infty}\left(\dfrac{k+1}{k+2}\right)\cdot\dfrac{1}{3} = \dfrac{1}{3}$; therefore $R = 3$. If $x+10 = 3$, then

$\sum\limits_{k=0}^{\infty}\dfrac{(x+10)^k}{(k+1)3^k} = \sum\limits_{k=0}^{\infty}\dfrac{1}{k+1}$ which diverges; if $x+10 = -3$, then $\sum\limits_{k=0}^{\infty}\dfrac{(x+10)^k}{(k+1)3^k} = \sum\limits_{k=0}^{\infty}\dfrac{(-1)^k}{k+1} = \sum\limits_{n=1}^{\infty}\dfrac{(-1)^{n-1}}{n}$

which converges. Since $-3 \le x+10 < 3 \iff -13 \le x < 7$, the interval of convergence is $[-13,7)$.

31. $1+(-1)^k = \begin{cases} 2 & \text{if } k \text{ is even,} \\ 0 & \text{if } k \text{ is odd.} \end{cases}$ Therefore $\sum\limits_{k=0}^{n}\left(1+(-1)^k\right)x^k = \sum\limits_{j=0}^{\lfloor n/2\rfloor} 2\, x^{2j}$ and $\sum\limits_{k=0}^{\infty}\left(1+(-1)^k\right)x^k =$

$2\sum\limits_{j=0}^{\infty}\left(x^2\right)^j = \dfrac{2}{1-x^2}$ if $x^2 < 1$; the series diverges otherwise. $(-1,1)$ is the interval of convergence and $R = 1$.

33. Because $\left|\dfrac{x^k}{k}\right| \le |x^k|$ for $k \ge 1$, Solution 31 and the Comparison Test imply $\sum\limits_{k=1}^{\infty}\dfrac{1+(-1)^k}{k}\cdot x^k$ converges

on $(-1,1)$. If $x = \pm 1$, the series becomes $\sum\limits_{j=1}^{\infty}\dfrac{2(\pm 1)^{2j}}{2j} = \sum\limits_{j=1}^{\infty}\dfrac{1}{j}$ which diverges; Theorem 1 then implies the

power series diverges for all x such that $|x| > 1$. Hence the interval of convergence is just $(-1,1)$ and $R = 1$.

35. $\sum\limits_{k=0}^{\infty}\dfrac{(ax-b)^k}{c^k} = \sum\limits_{k=0}^{\infty}\left(\dfrac{ax-b}{c}\right)^k$ is a geometric series which converges if $|\frac{ax-b}{c}| < 1$ and diverges

otherwise. If $a > 0$ and $c > 0$, then

$|\dfrac{ax-b}{c}| < 1 \iff -1 < \dfrac{ax-b}{c} < 1 \iff -c < ax-b < c \iff \dfrac{b-c}{a} < x < \dfrac{b+c}{a}$;

the interval of convergence is $\left(\dfrac{b-c}{a},\dfrac{b+c}{a}\right)$. ∎

37. The following proof applies Theorem 10.6.1 (the Ratio Test) and Theorem 10.7.1 (absolute convergence

implies convergence). We examine the series $\sum\limits_{k=0}^{\infty} A_k$ where $A_k = a_k x^k$. Suppose $x \neq 0$, then $\left|\dfrac{A_{k+1}}{A_k}\right| =$

$\left|\dfrac{a_{n+1}x^{n+1}}{a_n x^n}\right| = \left|\dfrac{a_{n+1}}{a_n}\right|\cdot|x| = r_n\cdot|x|$ where we let $r_n = \left|\dfrac{a_{n+1}}{a_n}\right|$.

$\boxed{L = \infty}$ If $\lim\limits_{n\to\infty} r_n = \infty$, then for any x, $\lim\limits_{n\to\infty} r_n|x| = \infty$ and $\sum_{k=0}^{\infty} a_k |x|^k$ diverges. If there were some $b < 0$ such that $\sum_{k=0}^{\infty} a_k b^k$ converged, Theorem 1 implies $\sum_{k=0}^{\infty} a_k |b/2|^k$ would also converge, but this is impossible; therefore $\sum_{k=0}^{\infty} a_k x^k$ diverges for all nonzero x. ∎

$\boxed{L = 0}$ If $\lim\limits_{n\to\infty} r_n = 0$, then for any x, $\lim\limits_{n\to\infty} r_n|x| = 0 < 1$ and $\sum_{k=0}^{\infty} a_k |x|^k$ converges; this further implies $\sum_{k=0}^{\infty} a_k x^k$ converges. ∎

$\boxed{0 < L < \infty}$ Suppose there is a finite positive L such that $\lim\limits_{n\to\infty} r_n = L$. If $|x| < \dfrac{1}{L}$, then $\lim\limits_{n\to\infty} r_n|x| = L \cdot |x| < 1$ so both $\sum_{k=0}^{\infty} a_k |x|^k$ and $\sum_{k=0}^{\infty} a_k x^k$ converge while $|x| > \dfrac{1}{L}$ implies $\lim\limits_{n\to\infty} r_n|x| = L \cdot |x| > 1$ and $\sum_{k=0}^{\infty} a_k x^k$ diverges. ∎

39. Suppose $\sum_{k=0}^{\infty} a_k x^k$ has radius of convergence R. Then, for any positive integer m,

$$\sum_{k=0}^{\infty} a_k \left(x^m\right)^k \text{ converges} \iff |x|^m = |x^m| < R \iff |x| < R^{1/m},$$

$$\sum_{k=0}^{\infty} a_k \left(x^m\right)^k \text{ diverges} \iff |x|^m = |x^m| > R \iff |x| > R^{1/m}.$$

Thus $R^{1/m}$ is the radius of convergence for $\sum_{k=0}^{\infty} a_k \left(x^m\right)^k = \sum_{k=0}^{\infty} a_k x^{mk}$. ∎

Solutions 10.9   *Differentiation and integration of power series*   (pages 704-5)

1. $\dfrac{1}{1+x^2} = 1 - x^2 + \left(x^2\right)^2 - \left(x^2\right)^3 + \left(x^2\right)^4 - \cdots = \sum_{k=0}^{\infty} \left(-x^2\right)^k = \sum_{k=0}^{\infty} (-1)^k x^{2k}$; this series converges to $\dfrac{1}{1+x^2}$ provided $|x^2| < 1$, i.e., if $|x| < 1$.

3. $\tan^{-1} x = \displaystyle\int_0^x \dfrac{dt}{1+t^2} = \int_0^x \left(\sum_{k=0}^{\infty}(-1)^k t^{2k}\right) dt = \sum_{k=0}^{\infty}(-1)^k \int_0^x t^{2k}\, dt = \sum_{k=0}^{\infty} \dfrac{(-1)^k x^{2k+1}}{2k+1} = x - \dfrac{x^3}{3} + \dfrac{x^5}{5} - \cdots$.

5. $\dfrac{1}{x} = \dfrac{1}{1+(x-1)} = 1 - (x-1) + (x-1)^2 - (x-1)^3 + (x-1)^4 - \cdots = \sum_{k=0}^{\infty}(-1)^k (x-1)^k$; this equation holds for $|x-1| < 1$, i.e., $0 < x < 2$.

REMARK: $\dfrac{1}{x} = \dfrac{1}{1-(1-x)} = \displaystyle\sum_{k=0}^{\infty}(1-x)^k$ if $|1-x| < 1$.

---

7. See the last example (on page 477) of section 9.3.

$$\dfrac{\pi}{4} = \tan^{-1} 1 = \int_0^1 \dfrac{dx}{1+x^2} = \int_0^1 \sum_{k=0}^{\infty}(-1)^k x^{2k}\, dx = \sum_{k=0}^{\infty}(-1)^k \int_0^1 x^{2k}\, dx = \sum_{k=0}^{\infty} \dfrac{(-1)^k}{2k+1}.$$

Theorem 10.7.3 implies $\left|\pi - 4\displaystyle\sum_{k=0}^{n} \dfrac{(-1)^k}{2k+1}\right| \le \dfrac{4}{2n+3}$, this bound is less than 0.005 if $n \ge 399$. We compute $S_{399} \approx 3.139092$ with $|\pi - S_{399}| \approx 4.99 \cdot 10^{-3}$ (after all that work, we might as well compute $T_{399} \approx 3.141596$ with error bounded by $6.25 \cdot 10^{-6}$). Theorem 10.7.4 implies $|\pi - T_n| \le \dfrac{4}{(2n+1)(2n+3)}$, this bound is less than 0.005 if $n \ge 14$; $T_{14} \approx 3.139220$ with $|\pi - T_{14}| \le 4.45 \cdot 10^{-3}$ (actual error is about $2.37 \cdot 10^{-3}$).

REMARK:

$$\dfrac{\pi}{6} = \tan^{-1}\left(\dfrac{1}{\sqrt{3}}\right) = \sum_{k=0}^{\infty} \dfrac{(-1)^k}{2k+1}\left(\dfrac{1}{\sqrt{3}}\right)^{2k+1} = \dfrac{1}{\sqrt{3}}\sum_{k=0}^{\infty} \dfrac{(-1/3)^k}{2k+1}$$

10.9 — Differentiation and integration of power series

$$\left| \pi - 2\sqrt{3} \sum_{k=0}^{n} \frac{(-1/3)^k}{2k+1} \right| \le \frac{2\sqrt{3}/3^{n+1}}{2n+3}; \text{ this bound is less than } 0.005 \text{ if } n \ge 3. \text{ We compute } S_3^* \approx 3.137853$$

with error bounded by $4.75 \cdot 10^{-3}$.

$$\frac{\pi}{12} = \tan^{-1}\left(2 - \sqrt{3}\right) = \sum_{k=0}^{\infty} \frac{(-1)^k}{2k+1}\left(2 - \sqrt{3}\right)^{2k+1} = \left(2 - \sqrt{3}\right) \sum_{k=0}^{\infty} \frac{\left(4\sqrt{3} - 7\right)^k}{2k+1}$$

$$\left| \pi - 12\left(2 - \sqrt{3}\right) \sum_{k=0}^{n} \frac{\left(4\sqrt{3} - 7\right)^k}{2k+1} \right| \le 12\left(2 - \sqrt{3}\right) \cdot \frac{\left|4\sqrt{3} - 7\right|^{n+1}}{2n+3}; \ S_1^{**} \approx 3.138439 \text{ with error bounded}$$

by $3.31 \cdot 10^{-3}$ (Note that $(-1) \cdot \left(2 - \sqrt{3}\right)^2 = 4\sqrt{3} - 7 \approx -0.071797$, this series converges quickly.).

---

9. See Example 8. The error bound $\dfrac{1}{(2n+3) \cdot (n+1)!}$ is less than 0.01 if $n \ge 3$. We compute $S_3 =$

$1 - \dfrac{1}{3} + \dfrac{1}{10} - \dfrac{1}{42} = \dfrac{26}{35} \approx 0.742857$ with error $\le \dfrac{1}{216} \approx 4.63 \cdot 10^{-3}$.

$T_3 = 1 - \dfrac{1}{3} + \dfrac{1}{10} - \dfrac{1}{84} = \dfrac{317}{420} \approx 0.754762$ with error $\le \dfrac{29}{3024} \approx 9.59 \cdot 10^{-3}$.

11. See Example 7 and use formula (14) to obtain

$$\int_0^{1/2} \cos t^2\, dt = \int_0^{1/2} \sum_{k=0}^{\infty} \frac{(-1)^k (t^2)^{2k}}{(2k)!}\, dt = \sum_{k=0}^{\infty} \frac{(-1)^k}{(2k)!} \int_0^{1/2} t^{4k}\, dt = \sum_{k=0}^{\infty} \frac{(-1)^k (1/2)^{4k+1}}{(2k)! \cdot (4k+1)}.$$

n	0	1	2
$\dfrac{(-1)^n (1/2)^{4n+1}}{(2n)! \cdot (4n+1)}$	$\dfrac{1}{2} = 0.5$	$\dfrac{-1}{320} = -3.125 \cdot 10^{-3}$	$\dfrac{1}{110592} \approx 9.04 \cdot 10^{-6}$
$\displaystyle\sum_{k=0}^{n} \dfrac{(-1)^k (1/2)^{4k+1}}{(2k)! \cdot (4k+1)}$	$\dfrac{1}{2} = 0.5$	$\dfrac{159}{320} = 0.496875$	$\dfrac{274757}{552960} \approx 0.496884$

13. See Example 7 and use formula (14) to obtain

$$\int_0^1 \cos \sqrt{t}\, dt = \int_0^1 \sum_{k=0}^{\infty} \frac{(-1)^k \left(t^{1/2}\right)^{2k}}{(2k)!}\, dt = \sum_{k=0}^{\infty} \frac{(-1)^k}{(2k)!} \int_0^1 t^k\, dt = \sum_{k=0}^{\infty} \frac{(-1)^k}{(2k)! \cdot (k+1)}.$$

n	0	1	2	3
$\dfrac{(-1)^n}{(2n)! \cdot (n+1)}$	1	$\dfrac{-1}{4} = -0.25$	$\dfrac{1}{72} \approx 1.39 \cdot 10^{-2}$	$\dfrac{-1}{2880} \approx -3.47 \cdot 10^{-4}$
$\displaystyle\sum_{k=0}^{n} \dfrac{(-1)^k}{(2k)! \cdot (k+1)}$	1	$\dfrac{3}{4} = 0.75$	$\dfrac{55}{72} \approx 0.763889$	$\dfrac{733}{960} \approx 0.763542$

REMARK: $\displaystyle\int_0^1 \cos \sqrt{t}\, dt \overset{t \equiv u^2}{=} \int_0^1 \cos u (2u\, du) = 2(u \sin u + \cos u)\Big|_0^1 = 2(\cos 1 + \cos 1 - 1) \approx 0.763547.$

---

15. $\displaystyle\int_0^1 t^2 \cdot e^{-t^2}\, dt = \int_0^1 t^2 \cdot \sum_{k=0}^{\infty} \frac{(-t^2)^k}{k!}\, dt = \sum_{k=0}^{\infty} \frac{(-1)^k}{k!} \int_0^1 t^{2k+2}\, dt = \sum_{k=0}^{\infty} \frac{(-1)^k}{k! \cdot (2k+3)}.$

n	0	1	2	3	4
$\dfrac{(-1)^n}{n! \cdot (2n+3)}$	$\dfrac{1}{3} \approx 0.333$	$\dfrac{-1}{5} = -0.2$	$\dfrac{1}{14} \approx 7.14 \cdot 10^{-2}$	$\dfrac{-1}{54} \approx -1.85 \cdot 10^{-2}$	$\dfrac{1}{264} \approx 3.79 \cdot 10^{-3}$
$\displaystyle\sum_{k=0}^{n} \dfrac{(-1)^k}{k! \cdot (2k+3)}$	$\dfrac{1}{3} \approx 0.333333$	$\dfrac{2}{15} \approx 0.133333$	$\dfrac{43}{210} \approx 0.204762$	$\dfrac{176}{945} \approx 0.186243$	$\dfrac{15803}{83160} \approx 0.190031$

17. Example 7 implies $\dfrac{1}{\sqrt{2\pi}} \displaystyle\int_0^{0.25} e^{-t^2/2}\, dt = \dfrac{1}{\sqrt{2\pi}} \sum_{k=0}^{\infty} \frac{(-1)^k \, 0.25^{2k+1}}{(2k+1) 2^k k!}$. The absolute value of the third term

($k = 2$) is approximately $9.739802 \cdot 10^{-6}$; hence the integral is approximately $0.09973557 - 0.001038912 \approx$
0.09869666.

19. Example 7 implies $\dfrac{1}{\sqrt{2\pi}}\displaystyle\int_0^2 e^{-t^2/2}\,dt = \dfrac{1}{\sqrt{2\pi}}\sum_{k=0}^{\infty}\dfrac{(-1)^k\,2^{2k+1}}{(2k+1)2^k\,k!}$. The absolute value of the twelth term

$(k = 11)$ is approximately $1.779869 \cdot 10^{-6}$; hence the integral is approximately $7.978846 \cdot 10^{-1} - 5.319230 \cdot 10^{-1} + 3.191538 \cdot 10^{-1} - 1.519780 \cdot 10^{-1} + 5.910256 \cdot 10^{-2} - 1.934266 \cdot 10^{-2} + 5.455621 \cdot 10^{-3} - 1.350916 \cdot 10^{-3} + 2.979961 \cdot 10^{-4} - 5.925069 \cdot 10^{-5} + 1.072157 \cdot 10^{-5} \approx 0.4772514$.

21.
$$\int_0^1 t\,e^t\,dt = \begin{cases} (t-1)\,e^t\big|_0^1 = 0\cdot e - (-1)\cdot 1 = 1, \\[4pt] \displaystyle\int_0^1 t\left(\sum_{k=0}^{\infty}\frac{t^k}{k!}\right)dt = \int_0^1\left(\sum_{k=0}^{\infty}\frac{t^{k+1}}{k!}\right)dt = \sum_{k=0}^{\infty}\frac{1}{k!}\int_0^1 t^{k+1}\,dt = \sum_{k=0}^{\infty}\frac{1}{k!\cdot(k+2)}; \end{cases}$$

therefore $1 = \displaystyle\int_0^1 te^t\,dt = \sum_{k=0}^{\infty}\dfrac{1}{k!\cdot(k+2)}$. ■

REMARK: $\displaystyle\sum_{k=0}^{n}\frac{1}{k!\cdot(k+2)} = \sum_{k=0}^{n}\frac{k+1}{(k+2)!} = \sum_{k=0}^{n}\left(\frac{1}{(k+1)!} - \frac{1}{(k+2)!}\right) = \sum_{j=1}^{n+1}\frac{1}{j!} - \sum_{i=2}^{n+2}\frac{1}{i!} = 1 - \frac{1}{(n+2)!}$.

23.
$$-\ln|1-x| = \int_0^x \frac{1}{1-t}\,dt = \int_0^x\sum_{k=0}^{\infty}t^k\,dt = \sum_{k=0}^{\infty}\int_0^x t^k\,dt = \sum_{k=0}^{\infty}\frac{x^{k+1}}{k+1},$$

$$\ln|1+x| = \int_0^x \frac{dt}{1+t} = \int_0^x\sum_{k=0}^{\infty}(-t)^k dt = \sum_{k=0}^{\infty}(-1)^k\int_0^x t^k\,dt = \sum_{k=0}^{\infty}\frac{(-1)^k x^{k+1}}{k+1},$$

therefore

$$\ln\left|\frac{1+x}{1-x}\right| = -\ln|1-x| + \ln|1+x| = \sum_{k=0}^{\infty}\frac{x^{k+1}}{k+1} + \sum_{k=0}^{\infty}\frac{(-1)^k x^{k+1}}{k+1} = \sum_{k=0}^{\infty}\frac{1+(-1)^k}{k+1}x^{k+1} = \sum_{j=0}^{\infty}\frac{2\,x^{2j+1}}{2j+1}.$$

The various power series appearing above do converge if $-1 < x < 1$. Since the power series for $\ln\left|\dfrac{1+x}{1-x}\right|$ contains only odd powers of x, the sequence of partial sums is a monotonic sequence. Therefore

$$\left|\ln\left|\frac{1+x}{1-x}\right| - \sum_{k=0}^{n}\frac{2\,x^{2k+1}}{2k+1}\right| = 2\sum_{k=n+1}^{\infty}\frac{|x|^{2k+1}}{2k+1} \le \frac{2}{2n+3}\sum_{k=n+1}^{\infty}|x|^{2k+1} = \frac{2}{2n+3}\cdot\frac{|x|^{2n+3}}{1-x^2}.$$

We can devise an effective procedure for accumulating sufficiently many terms that the truncation error is bounded by ϵ. (See the Remark following Solution 9.5.13.) Let $\epsilon^* = (1-x^2)\cdot\epsilon$. Keep adding terms $\dfrac{2\,x^{2k+1}}{2k+1}$ until encountering one whose absolute value is less than ϵ^*. The partial sum computed so far is "good enough"; however, since all terms have the same sign, we should add this last term before stopping.

The function $f(x) = \dfrac{1+x}{1-x}$ maps the interval $(-1,1)$ onto $(0,\infty)$ in a 1–1 fashion. For any positive t, we have $f^{-1}(t) = \dfrac{t-1}{t+1} = 1 - \dfrac{2}{t+1}$. Also note that $f^{-1}\left(\dfrac{1}{t}\right) = \dfrac{(1/t)-1}{(1/t)+1} = \dfrac{1-t}{1+t} = -\dfrac{t-1}{t+1} = -f^{-1}(t)$; this corresponds to a well-known property of the logarithm function: $\ln\left(\dfrac{1}{t}\right) = -\ln t$.

To approximate $\ln 1.5$ within $\epsilon = 5.0 \cdot 10^{-5}$, we let $x = f^{-1}(1.5) = \dfrac{1.5-1}{1.5+1} = \dfrac{1}{5} = 0.2$ and $\epsilon^* = (1-0.04)\cdot(5.0\cdot 10^{-5}) = 4.80\cdot 10^{-5}$. The following table summarizes work yielding $\ln 1.5 \approx 0.405461$ with truncation error $\le 3.66 \cdot 10^{-6}$; I prefer to use the improved approximation $\ln 1.5 \approx 0.405465$.

n	0	1	2	3
$\dfrac{2\,(0.2)^{2n+1}}{2n+1}$	$\dfrac{2}{5} = 0.4$	$\dfrac{2}{375} \approx 5.33\cdot 10^{-3}$	$\dfrac{2}{15625} \approx 1.28\cdot 10^{-4}$	$\dfrac{2}{546875} \approx 3.66\cdot 10^{-6}$
$\displaystyle\sum_{k=0}^{n}\dfrac{2\,(0.2)^{2k+1}}{2k+1}$	$\dfrac{2}{5} = 0.4$	$\dfrac{152}{375} \approx 0.405333$	$\dfrac{19006}{15635} \approx 0.405461$	$\dfrac{665216}{1640625} \approx 0.405465$

To approximate $\ln 0.5$, we let $x = f^{-1}(0.5) = \dfrac{0.5-1}{0.5+1} = \dfrac{-1}{3}$ and $\epsilon^* = \dfrac{8}{9}\cdot(5.0\cdot 10^{-5}) = 4.44\cdot 10^{-5}$. We compute $\ln 0.5 \approx -0.693135$ with truncation error $\le 1.13\cdot 10^{-5}$; I prefer to use the improved approximation

$\ln 0.5 \approx -0.693146$.

n	0	1	2	3	4
$\dfrac{2(-1/3)^{2n+1}}{2n+1}$	$\dfrac{-2}{3}$	$\dfrac{-2}{81}$	$\dfrac{-2}{1215}$	$\dfrac{-2}{15309} \approx -1.31 \cdot 10^{-4}$	$\dfrac{-2}{177147} \approx -1.13 \cdot 10^{-5}$
$\displaystyle\sum_{k=0}^{n} \dfrac{2(-1/3)^{2k+1}}{2k+1}$	-0.667	-0.691	-0.6930	-0.693135	-0.693146

To approximate $\ln 2$, we let $x = f^{-1}(2) = \dfrac{2-1}{2+1} = \dfrac{1}{3}$. Changing the sign of our just-obtained approximation for $\ln 0.5$, we now have $\ln 2 \approx 0.693146$ with error bounded by $-1.13 \cdot 10^{-5}$.

We have just computed some good approximations using a power series truncated at $n = 3$ or $n = 4$. For comparison, suppose we use equation (5) of Example 4: the Alternating Series Test (or Problem 9.5.31) implies we would need $n = 10$ to approximate $\ln 1.5$ and $n = 20,000$ for $\ln 2$ while the bound of Problem 9.5.32 would require $n = 11$ to approximate $\ln 0.5$. (Since $\ln 2 = -\ln 0.5$, the "need" for a twenty-thousand term summation is more apparent than real.) Furthermore, if we let $w = x^2$, then we have

$\ln\left|\dfrac{1+x}{1-x}\right| \approx \displaystyle\sum_{k=0}^{n} \dfrac{2x^{2k+1}}{2k+1} = (2x)\sum_{k=0}^{n} \dfrac{x^{2k}}{2k+1} = (2x)\sum_{k=0}^{n} \dfrac{w^k}{2k+1}$. Although use of this approximation requires some preparatory work (i.e., to compute $\ln t$, we first need to find $x = f^{-1}(t)$, compute adjusted error bound ϵ^*, and compute $w = x^2$), we then obtain a power series which converges much faster than the simpler one.

REMARK: The algebra shown at the beginning of this solution is not the only path to the power series we have been analyzing. If $x^2 < 1$, then (presuming we can swap integration and summation)

$$\ln\left|\dfrac{1+x}{1-x}\right| = \int_0^x \left(\dfrac{1}{1+t} + \dfrac{1}{1-t}\right) dt = \int_0^x \dfrac{2}{1-t^2}\, dt = \int_0^x 2\sum_{k=0}^{\infty} t^{2k}\, dt = 2\sum_{k=0}^{\infty} \int_0^x t^{2k}\, dt = 2\sum_{k=0}^{\infty} \dfrac{x^{2k+1}}{2k+1}.$$

Solutions 10.10   *Taylor and Maclaurin series*   (pages 714-5)

1. If $f(x) = e^2$, then $f^{(n)}(x) = e^x$ and $f^{(n)}(1) = e$ for all n. Therefore $\displaystyle\sum_{k=0}^{\infty} \dfrac{f^{(k)}(1)}{k!} \cdot (x-1)^k = \sum_{k=0}^{\infty} \dfrac{e \cdot (x-1)^k}{k!}$ is the Taylor series centered at 1 for e^x.

REMARK: The series does converge to e^x everywhere because Examples 1 and 4 imply $e^{x-1} = \displaystyle\sum_{k=0}^{\infty} \dfrac{(x-1)^k}{k!}$ for all x and $e^x = e \cdot e^{x-1} = e \cdot \displaystyle\sum_{k=0}^{\infty} \dfrac{(x-1)^k}{k!}$.

3. If $f(x) = \cos x$, then $f\left(\dfrac{\pi}{4}\right) = \dfrac{1}{\sqrt{2}}$, $f'\left(\dfrac{\pi}{4}\right) = -\sin\left(\dfrac{\pi}{4}\right) = \dfrac{-1}{\sqrt{2}}$, $f''\left(\dfrac{\pi}{4}\right) = -\cos\left(\dfrac{\pi}{4}\right) = \dfrac{-1}{\sqrt{2}}$, $f'''\left(\dfrac{\pi}{4}\right) = \sin\left(\dfrac{\pi}{4}\right) = \dfrac{1}{\sqrt{2}}$, and $f^{(n+4)}\left(\dfrac{\pi}{4}\right) = f^{(n)}\left(\dfrac{\pi}{4}\right)$ for all n. The Taylor series for $\cos x$ at $\dfrac{\pi}{4}$ begins as follows

$$\dfrac{1}{\sqrt{2}} - \dfrac{1}{\sqrt{2}}\left(x - \dfrac{\pi}{4}\right) - \dfrac{1/\sqrt{2}}{2!}\left(x - \dfrac{\pi}{4}\right)^2 + \dfrac{1/\sqrt{2}}{3!}\left(x - \dfrac{\pi}{4}\right)^3 + \dfrac{1/\sqrt{2}}{4!}\left(x - \dfrac{\pi}{4}\right)^4 - \cdots.$$

REMARK: The greatest integer function can be used to obtain a more compact expression for this series. The function $\left\lfloor \dfrac{n+1}{2} \right\rfloor$ transforms the sequence $\{0, 1, 2, 3, 4, 5, \ldots\}$ into $\{0, 1, 1, 2, 2, 3, \ldots\}$. Hence the series is $\displaystyle\sum_{k=0}^{\infty} (-1)^{\lfloor(n+1)/2\rfloor} \cdot \dfrac{(x - \pi/4)^k}{k! \cdot \sqrt{2}}$.

5. $f(x) = e^{\beta x} \implies f^{(n)}(x) = \beta^n \cdot e^{\beta x} \implies f^{(n)}(0) = \beta^n$. The Maclaurin series for $e^{\beta x}$ is $\displaystyle\sum_{k=0}^{\infty} \dfrac{\beta^k \cdot x^k}{k!}$.

REMARK: The result of Examples 1 and 4 implies $e^{\beta x} = \sum_{k=0}^{\infty} \frac{(\beta x)^k}{k!}$.

7. $f(x) = xe^x \implies f^{(n)}(x) = xe^x + ne^x \implies f^{(n)}(0) = n$, the Maclaurin series for xe^x is

$$\sum_{k=0}^{\infty} \frac{k \cdot x^k}{k!} = 0 + \sum_{k=1}^{\infty} \frac{k \cdot x^k}{k!} = \sum_{k=1}^{\infty} \frac{x^k}{(k-1)!} = \sum_{j=0}^{\infty} \frac{x^{j+1}}{j!}.$$

REMARK: $xe^x = x \sum_{k=0}^{\infty} \frac{x^k}{k!} = \sum_{k=0}^{\infty} x \cdot \frac{x^k}{k!} = \sum_{k=0}^{\infty} \frac{x^{k+1}}{k!}$.

9. Equation 10.9.(15) presents the Maclaurin series for $\sin x$, we use that to obtain

$$\frac{\sin x}{x} = x^{-1} \cdot \sin x = x^{-1} \cdot \sum_{k=0}^{\infty} (-1)^k \frac{x^{2k+1}}{(2k+1)!} = \sum_{k=0}^{\infty} (-1)^k \frac{x^{2k}}{(2k+1)!} = 1 - \frac{x^2}{3!} + \frac{x^4}{5!} - \frac{x^6}{7!} + \cdots.$$

REMARK: Note how easy it now is to compute $\lim_{x \to 0} \frac{\sin x}{x}$. (Compare with the proof of Theorem 2.5.1.)

11. Example 7 shows that $\ln x = \sum_{k=1}^{\infty} (-1)^{k+1} \frac{(x-1)^k}{k}$ on $(0, 2]$; therefore, on the same interval,

$$(x-1) \ln x = (x-1) \sum_{k=1}^{\infty} (-1)^{k+1} \frac{(x-1)^k}{k} = \sum_{k=1}^{\infty} (-1)^{k+1} \frac{(x-1)^{k+1}}{k} = \sum_{j=2}^{\infty} \frac{(1-x)^j}{j-1}.$$

REMARK: The steps leading to equation (10) show that if there is an $R > 0$ such that $\sum_{k=0}^{\infty} a_k (x-x_0)^k = \sum_{k=0}^{\infty} b_k (x-x_0)^k$ for all x in $(x_0 - R, x_0 + R)$, then $a_k = b_k$ for all k. Hence, if hard work or sleight of hand produces a power series in x which converges to $f(x)$ on $(x_0 - R, x_0 + R)$, that series must be the Taylor series centered at x_0 for f.

13.

n	0	1	2	3
$f^{(n)}(x)$	$\csc x$	$-\csc x \cot x$	$\csc x \cot^2 x + \csc^3 x$	$-\csc x \cos^3 x - 5 \cos^2 x \cot x$
$f^{(n)}\left(\frac{\pi}{2}\right)$	1	0	1	0

$1 + 0 + \frac{(x - \pi/2)^2}{2} + 0$ is the sum of the first four terms of the Taylor series for $\csc x$ at $\frac{\pi}{2}$.

$(0, \pi)$ is the largest interval containing $\pi/2$ on which $\csc x$ is continuous. $\csc x$ does have continuous derivatives of all orders on $(0, \pi)$; the job of showing $\csc x$ is analytic on $(0, \pi)$ is best left to an advanced calculus course (or to a course on complex variables).

15. If $f(x) = \sqrt{x} = x^{1/2}$, then $f'(x) = \frac{1}{2} x^{-1/2}$, $f''(x) = \left(\frac{1}{2}\right)\left(\frac{-1}{2}\right) x^{-3/2}$, $f'''(x) = \left(\frac{1}{2}\right)\left(\frac{-1}{2}\right)\left(\frac{-3}{2}\right) x^{-5/2}$.

In general,

$$f^{(n)}(x) = \left(\frac{1}{2}\right)\left(\frac{-1}{2}\right)\left(\frac{-3}{2}\right) \cdots \left(\frac{1}{2} - (n-1)\right) x^{1/2-n} = (-1)^{n-1} \cdot \frac{1 \cdot 3 \cdots (2n-3)}{2^n} x^{-(2n-1)/2},$$

$$f^{(n)}(4) = (-1)^{n-1} \cdot \frac{1 \cdot 3 \cdot 5 \cdots (2n-3)}{2^n} 2^{-(2n-1)} = (-1)^{n-1} \cdot \frac{1 \cdot 3 \cdot 5 \cdots (2n-3)}{2^{3n-1}}.$$

The Taylor series centered at 4 for $\sqrt{x}$ converges provided

$$\lim_{n \to \infty} \left| \frac{f^{(n+1)}(4) (x-4)^{n+1}/(n+1)!}{f^{(n)}(4) (x-4)^n/n!} \right| = \lim_{n \to \infty} \frac{1 \cdot 3 \cdot 5 \cdots (2n-3)(2n-1)}{1 \cdot 3 \cdot 5 \cdots (2n-3)} \cdot \frac{2^{3n-1}}{2^{3n+2}} \cdot \frac{n!}{(n+1)!} \cdot \frac{|x-4|^{n+1}}{|x-4|^n}$$

$$= \lim_{n \to \infty} \frac{(2n-1)|x-4|}{2^3(n+1)} = \frac{|x-4|}{4} < 1.$$

Therefore $\sqrt{x} = 2 + \sum_{k=1}^{\infty} \frac{(-1)^{k+1} (1 \cdot 3 \cdots (2k-3))}{2^{3k-1} \cdot k!} (x-4)^k$ if $0 < x < 8$; the radius of convergence is 4.

17. $\sin u = \sum_{k=0}^{\infty} \dfrac{(-1)^k}{(2k+1)!} \cdot u^{2k+1}$ for all real u, therefore

$$\sin\left(x^2\right) = \sum_{k=0}^{\infty} \dfrac{(-1)^k}{(2k+1)!} \cdot \left(x^2\right)^{2k+1} = \sum_{k=0}^{\infty} \dfrac{(-1)^k}{(2k+1)!} \cdot x^{4k+2}$$

for all real x; this series, in powers of $x - 0$, which always converges to $\sin x^2$ must be the Maclaurin series for that function.

19. $\tan^{-1} x = \displaystyle\int_0^x \dfrac{dt}{1+t^2} = \int_0^x \sum_{k=0}^{\infty} \left(-t^2\right)^k \, dt = \sum_{k=0}^{\infty} (-1)^k \int_0^x t^{2k} \, dt = \sum_{k=0}^{\infty} \dfrac{(-1)^k}{2k+1} \cdot x^{2k+1}$;

the limit of the absolute ratio of successive terms is x^2, the radius of convergence is $\sqrt{1} = 1$.

REMARK: See the final example of Section 9.5 on page 631 of the text.

21. $\sqrt[4]{1+x} = (1+x)^{1/4} = 1 + \displaystyle\sum_{k=1}^{\infty} \dfrac{(1/4)(-3/4)\cdots(1/4 - k + 1)}{k!} x^k = 1 + \sum_{k=1}^{\infty} (-1)^{k-1} \cdot \dfrac{1\cdot 3\cdot 7\cdots(4k-5)}{4^k \cdot k!} \cdot x^k$;

the limit of the absolute ratio of successive terms is $|x|$, the interval of convergence is $(0,2)$.

23. $\displaystyle\int_0^{1/2} \sqrt[4]{1+x^3} \, dx = \int_0^{1/2} \left(1 + \sum_{k=1}^{\infty} (-1)^{k-1} \cdot \dfrac{1\cdot 3\cdot 7\cdots(4k-5)}{4^k \cdot k!} \cdot x^{3k} \right) dx$

$\displaystyle = \dfrac{1}{2} + \sum_{k=1}^{\infty} (-1)^{k-1} \cdot \dfrac{1\cdot 3\cdot 7\cdots(4k-5)}{4^k \cdot k! \cdot (3k+1)} \left(\dfrac{1}{2}\right)^{3k+1}$

$\displaystyle = \dfrac{1}{2} + \sum_{k=1}^{\infty} (-1)^{k-1} \cdot \dfrac{1\cdot 3\cdot 7\cdots(4k-5)}{2^{5k+1} \cdot k! \cdot (3k+1)}.$

This is essentially an alternating series so we can accumulate partial sums, stopping when the next term is small enough.

n	0	1	2	3
a_n	$\dfrac{1}{2} = 0.5$	$\dfrac{1}{256} \approx 3.91 \cdot 10^{-3}$	$\dfrac{-3}{28672} \approx -1.04 \cdot 10^{-4}$	$\dfrac{7}{1310720} \approx 5.34 \cdot 10^{-6}$
$\displaystyle\sum_{k=0}^{n} a_k$	$\dfrac{1}{2} = 0.5$	$\dfrac{129}{256} \approx 0.503906$	$\dfrac{14445}{28672} \approx 0.503802$	$\dfrac{4622449}{9175040} \approx 0.503807$

25. a. $\mathrm{erf}(x) = \dfrac{2}{\sqrt{\pi}} \displaystyle\int_0^x e^{-t^2} \, dx = \dfrac{2}{\sqrt{\pi}} \int_0^x \sum_{k=0}^{\infty} \dfrac{\left(-t^2\right)^k}{k!} \, dx = \dfrac{2}{\sqrt{\pi}} \sum_{k=0}^{\infty} \dfrac{(-1)^k}{k!} \int_0^x x^{2k} \, dx = \dfrac{2}{\sqrt{\pi}} \sum_{k=0}^{\infty} \dfrac{(-1)^k x^{2k+1}}{k! \cdot (2k+1)}.$

b.   As a practical matter, to approximate a specified value using this alternating series, it is convenient to defer multiplication by the constant $2/\sqrt{\pi}$ until a sufficient number of terms have been accumulated. For this purpose, we use an adjusted error bound: $\epsilon^* = (\sqrt{\pi}/2) \cdot 0.0001 \approx 8.862 \cdot 10^{-5}$.

n	0	1	2	3	4	5	6	7
$\dfrac{(-1)^n}{n!\,(2n+1)}$	1	$\dfrac{-1}{3}$	$\dfrac{1}{10}$	$\dfrac{-1}{42}$	$\dfrac{1}{216}$	$\dfrac{-1}{1320}$	$\dfrac{1}{9360} \approx 1.07 \cdot 10^{-4}$	$\dfrac{-1}{75600} \approx -1.32 \cdot 10^{-5}$
$\displaystyle\sum_{k=0}^{n} \dfrac{(-1)^k}{k!\,(2k+1)}$	1	$\dfrac{2}{3}$	$\dfrac{23}{30}$	$\dfrac{26}{35}$	$\dfrac{5651}{7560}$	$\dfrac{31049}{41580}$	$\dfrac{1614779}{2162160} \approx 0.746836$	$\dfrac{1009219}{1351350} \approx 0.746823$

$\mathrm{erf}(1) \approx \dfrac{2}{\sqrt{\pi}} \cdot 0.746823 \approx 0.842699$ with error bounded by $\dfrac{2}{\sqrt{\pi}} \cdot \left|-1.32 \cdot 10^{-5}\right| \approx 1.49 \cdot 10^{-5}$.

n	0	1	2	3	4
$\dfrac{(-1)^n (0.5)^{2n+1}}{n!\,(2n+1)}$	$\dfrac{1}{2}$	$\dfrac{-1}{24}$	$\dfrac{1}{320}$	$\dfrac{-1}{5376} \approx -1.86 \cdot 10^{-4}$	$\dfrac{1}{110592} \approx 9.04 \cdot 10^{-6}$
$\displaystyle\sum_{k=0}^{n} \dfrac{(-1)^k (0.5)^{2k+1}}{k!\,(2k+1)}$	$\dfrac{1}{2}$	$\dfrac{11}{24}$	$\dfrac{443}{960} \approx 0.461458$	$\dfrac{4133}{8960} \approx 0.461272$	$\dfrac{1785491}{3870720} \approx 0.461281$

$\mathrm{erf}(0.5) \approx \dfrac{2}{\sqrt{\pi}} \cdot 0.461281 \approx 0.520500$ with error bounded by $\dfrac{2}{\sqrt{\pi}} \cdot \left|9.04 \cdot 10^{-6}\right| \approx 1.02 \cdot 10^{-5}$.

27. Examples 2 and 5 prove that $\cos x = \sum_{k=0}^{\infty} \frac{(-1)^k}{(2k)!} x^{2k}$ for all real x. Theorem 10.9.1 then implies

$$\sin x = \int_0^x \cos t \, dt = \int_0^x \sum_{k=0}^{\infty} \frac{(-1)^k}{(2k)!} t^{2k} \, dt = \sum_{k=0}^{\infty} \frac{(-1)^k}{(2k)!} \int_0^x t^{2k} \, dt = \sum_{k=0}^{\infty} \frac{(-1)^k}{(2k)!} \cdot \frac{x^{2k+1}}{2k+1} = \sum_{k=0}^{\infty} \frac{(-1)^k}{(2k+1)!} \cdot x^{2k+1}. \blacksquare$$

REMARK: Problem 2.7.31 implies $\left.\frac{d^n}{dx^n} \sin x\right|_{x=0} = \begin{cases} 0 & \text{if } n \text{ is even,} \\ (-1)^{(n-1)/2} & \text{if } n \text{ is odd.} \end{cases}$ Theorem 9.4.1 implies that for every real number x and every positive integer n, there is a number $c = c(x, n)$ between 0 and x such that (letting $f(x) = \sin x$)

$$\sin x = \sum_{k=0}^{n} \frac{f^{(k)}(0)}{k!} \cdot (x-0)^k + \frac{f^{(n+1)}(c)}{(n+1)!} \cdot (x-0)^{n+1} = \sum_{j=0}^{\lfloor (n-1)/2 \rfloor} \frac{(-1)^j}{(2j+1)!} \cdot x^{2j+1} + \frac{f^{(n+1)}(c)}{(n+1)!} \cdot x^{n+1}.$$

$\left|f^{(m)}(w)\right| \le 1$ for all m and w, thus $\left|\frac{f^{(n+1)}(c)}{(n+1)!} \cdot x^{n+1}\right| \le \frac{|x|^{n+1}}{(n+1)!}$ for all n and x. Solution 10.1.24 implies

$$\lim_{n\to\infty} \frac{|x|^{n+1}}{(n+1)!} = 0 \text{ for all } x. \text{ Therefore, } \sin x = \lim_{n\to\infty} \sum_{j=0}^{\lfloor (n-1)/2 \rfloor} \frac{(-1)^j}{(2j+1)!} \cdot x^{2j+1} = \sum_{j=0}^{\infty} \frac{(-1)^j}{(2j+1)!} \cdot x^{2j+1}. \blacksquare$$

---

29. $$\frac{d}{dx} \sin x = \frac{d}{dx} \sum_{k=0}^{\infty} \frac{(-1)^k x^{2k+1}}{(2k+1)!} = \sum_{k=0}^{\infty} \frac{d}{dx} \frac{(-1)^k x^{2k+1}}{(2k+1)!} = \sum_{k=0}^{\infty} (2k+1) \cdot \frac{(-1)^k x^{2k}}{(2k+1)!} = \sum_{k=0}^{\infty} \frac{(-1)^k x^{2k}}{(2k)!}$$
$$= \cos x. \blacksquare$$

REMARK: If you are worried about logical circularity because integration of the series for $\cos x$ produced the series for $\sin x$ which we have just differentiated, you may want to invoke the independent proof shown in the Remark following Solution 27.

---

31. If $f(x) = \frac{1}{1-x} = (1-x)^{-1}$, then $f^{(n)}(x) = n! \cdot (1-x)^{-n-1}$ and $f^{(n)}(0) = n!$. Thus $\sum_{k=0}^{\infty} \frac{f^{(k)}(0)}{k!} (x-0)^k = \sum_{k=0}^{\infty} \frac{k!}{k!} x^k = \sum_{k=0}^{\infty} x^k$ is the Maclaurin series, i.e., Taylor series centered at 0, for f. The absolute ratio of successive terms is $|x|$, thus the series converges provided $|x| < 1$, i.e., the interval of convergence is $(-1, 1)$. $\blacksquare$

33. Let $a_k = \frac{r(r-1)\cdots(r-k+1)}{k!}$.
If r is a positive integer, then $a_k = 0$ if $k > r$ and $(1+x)^r = 1 + \sum_{k=1}^{\infty} a_k x^k = 1 + \sum_{k=1}^{r} a_k x^k$ for all x.
If r is not a positive integer, then $a_k \neq 0$ for all k. We apply the Ratio Test to the Binomial Series:
$\lim_{k\to\infty} \left|\frac{a_{k+1} x^{k+1}}{a_k x^k}\right| = \lim_{k\to\infty} \left|\frac{r-k}{k+1}\right| \cdot |x| = |x|$, the Binomial Series converges if $|x| < 1$ and diverges if $|x| > 1$. $\blacksquare$

35. a. $f(x) = \begin{cases} \frac{\sin x}{x} & \text{if } x \neq 0 \\ 1 & \text{if } x = 0 \end{cases}$ is continuous everywhere. The Second Fundamental Theorem of Calculus (Theorem 4.6.2) implies $\text{Si}(x) = \int_0^x \frac{\sin t}{t} \, dt = \int_0^x f(t) \, dt$ is everywhere continuous and differentiable.

b. Let $a_k = \int_{(k-1)\pi}^{k\pi} \frac{\sin t}{t} \, dt$. The sequence $\{a_k\}$ is alternating because $a_k = (-1)^{k-1} \int_{(k-1)\pi}^{k\pi} \frac{|\sin t|}{t} \, dt$; the sequence satisfies $|a_{k+1}| < |a_k|$ because

$$|a_{k+1}| = \int_{k\pi}^{(k+1)\pi} \frac{|\sin t|}{t} \, dt \overset{t=u+\pi}{=} \int_{(k-1)\pi}^{k\pi} \frac{|\sin(u+\pi)|}{u+\pi} \, du = \int_{(k-1)\pi}^{k\pi} \frac{|-\sin u|}{u+\pi} \, du = \int_{(k-1)\pi}^{k\pi} \frac{|\sin u|}{u+\pi} \, du$$
$$< \int_{(k-1)\pi}^{k\pi} \frac{|\sin u|}{u} \, du = |a_k|;$$

and $\lim_{k\to\infty} a_k = 0$ because $|a_k| = \int_{(k-1)\pi}^{k\pi} \frac{|\sin t|}{t} \, dt < \int_{(k-1)\pi}^{k\pi} \frac{1}{t} \, dt = \ln \frac{k}{k-1} \to \ln 1 = 0$. Therefore $\sum_{k=0}^{\infty} a_k$ is a convergent alternating series. For any positive x, the value of $\text{Si}(x)$ lies between two consecutive partial sums for that convergent series; hence $\lim\limits_{x \to \infty} \text{Si}(x) = \lim\limits_{n \to \infty} \sum\limits_{k=1}^{n} a_k$ exists and is finite. ∎

REMARK: $\lim\limits_{x \to \infty} \text{Si}(x) = \dfrac{\pi}{2}$.

c.　We integrate the Maclaurin series for $(\sin t)/t$ obtained in Solution 9.

$$\text{Si}(x) = \int_0^x \frac{\sin t}{t}\,dt = \int_0^x \sum_{k=1}^{\infty} (-1)^k \frac{t^{2k}}{(2k+1)!}\,dt = \sum_{k=1}^{\infty} \frac{(-1)^k}{(2k+1)!} \int_0^x t^{2k}\,dt = \sum_{k=1}^{\infty} \frac{(-1)^k}{(2k+1)\cdot(2k+1)!} x^{2k+1}.$$

The interval of convergence for this series is $(-\infty, \infty)$ because

$$\lim_{k \to \infty} \left| \frac{(-1)^{k+1}/(2k+3)(2k+3)!}{(-1)^k/(2k+1)(2k+1)!} \right| = \lim_{k \to \infty} \frac{1}{(2k+2)\cdot(2k+3)^2} = 0.$$

d.

n	0	1	2	3
$\dfrac{(-1)^n\,(0.5)^{2n+1}}{(2n+1)\cdot(2n+1)!}$	$\dfrac{1}{2}$	$\dfrac{-1}{144} \approx -6.94\cdot10^{-3}$	$\dfrac{1}{19200} \approx 5.21\cdot10^{-5}$	$\dfrac{-1}{4515840} \approx -2.21\cdot10^{-7}$
$\displaystyle\sum_{k=0}^{n} \dfrac{(-1)^k\,(0.5)^{2k+1}}{(2k+1)\cdot(2k+1)!}$	$\dfrac{1}{2}$	$\dfrac{71}{144} \approx 0.493056$	$\dfrac{28403}{57600} \approx 0.493108$	

$\text{Si}(0.5) \approx 0.493108$ with absolute error bounded by $2.21\cdot10^{-7}$.

n	0	1	2	3
$\dfrac{(-1)^n}{(2n+1)\cdot(2n+1)!}$	1	$\dfrac{-1}{18} \approx -5.56\cdot10^{-2}$	$\dfrac{1}{600} \approx 1.67\cdot10^{-3}$	$\dfrac{-1}{35280} \approx -2.83\cdot10^{-5}$
$\displaystyle\sum_{k=0}^{n} \dfrac{(-1)^k}{(2k+1)\cdot(2k+1)!}$	1	$\dfrac{17}{18} \approx 0.944444$	$\dfrac{1703}{1800} \approx 0.946111$	$\dfrac{166889}{176400} \approx 0.946083$

$\text{Si}(1) \approx 0.946083$ with absolute error bounded by $2.83\cdot10^{-5}$.

37.　Taylor's theorem implies $f(b) = f(a) + f'(a)(b-a) + \dfrac{f''(c)}{2}(b-a)^2$ for some c between a and b. If f'' is continuous and $f''(a) \neq 0$, the intermediate value theorem implies f'' does not change sign in some open neighborhood of a (i.e., $f''(a)$ and $f''(c)$ are both positive or both negative). Suppose $f'(a) = 0$ and $b \neq a$.

$$f(b) = f(a) + f''(c)\cdot\frac{(b-a)^2}{2} = \begin{cases} f(a) - |f''(c)|\cdot\dfrac{|b-a|^2}{2} < f(a) & \text{if } f''(a) < 0, \\[2mm] f(a) + |f''(c)|\cdot\dfrac{|b-a|^2}{2} > f(a) & \text{if } f''(a) > 0. \end{cases}$$

Therefore, $f'(a) = 0$ and $f''(a) < 0$ implies $f(a)$ is a local maximum for f; $f'(a) = 0$ and $f''(a) > 0$ implies $f(a)$ is a local minimum. ∎

39.　Pick a positive ϵ. Suppose $|f''(x)| \leq 1$ and $|f(x)| \leq 1$ for $-1 \leq x \leq 1$. Let $w = x + 1$, then $-1 \leq x \leq 1$ implies $0 \leq w \leq 2$ and

$$0 \leq |f''(x)w^2 + f'(x)w + f(x)| \leq |f''(x)|\,w^2 + |f'(x)|\,w + |f(x)| \leq w^2 + |f'(x)|\,w + 1 < w^2 + |f'(x)|\,w + (1+\epsilon).$$

A necessary and sufficient condition for the quadratic $aw^2 + bw + c$ to have no real roots is that its discriminant $b^2 - 4ac$ be strictly negative. Therefore,

$$0 < w^2 + |f'(x)|\,w + (1+\epsilon) \implies |f'(x)|^2 - 4(1+\epsilon) < 0 \implies |f'(x)| \leq 2\sqrt{1+\epsilon}.$$

Since that holds for any choice of ϵ, we infer $|f'(x)| \leq 2$. ∎

Solutions 10.R　　*Review*　　(pages 716-7)

1.

n	1	2	3	4	5
$\dfrac{n-2}{n}$	-1	0	$\dfrac{1}{3}$	$\dfrac{1}{2}$	$\dfrac{3}{5}$

3.

1	2	3	4	n
$\dfrac{1}{8} = \dfrac{2\cdot1-1}{2^{1+2}}$	$\dfrac{3}{16} = \dfrac{2\cdot2-1}{2^{2+2}}$	$\dfrac{5}{32} = \dfrac{2\cdot3-1}{2^{3+2}}$	$\dfrac{7}{64} = \dfrac{2\cdot4-1}{2^{4+2}}$	$a_n = \dfrac{2\cdot n-1}{2^{n+2}}$

5. $\lim\limits_{n\to\infty}\dfrac{-7}{n}=0$; the sequence $\left\{\dfrac{-7}{n}\right\}$ converges to zero.

7. Solution 9.2.3 together with Theorem 10.1.5 imply $\lim\limits_{n\to\infty}\dfrac{\ln n}{\sqrt{n}}=0$; hence $\left\{\dfrac{\ln n}{\sqrt{n}}\right\}$ converges to zero.

9. Solution 10.1.35 implies $\left\{\left(1-\dfrac{2}{n}\right)^n\right\}=\left\{\left(1+\left(\dfrac{-2}{n}\right)\right)^n\right\}$ converges to $\lim\limits_{n\to\infty}\left(1+\left(\dfrac{-2}{n}\right)\right)^n=e^{-2}$.

11. $\{\sqrt{n}\}$ is unbounded. From each set of seven consecutive integers, it is possible to choose p and q such that $\cos q\le -0.75$ and $0.75\le\cos p$. (Adapt Solution 10.1.13 to prove this.) Therefore $\{\sqrt{n}\ \cos n\}$ is neither bounded nor monotonic, the sequence does not converge.

13. $\{1+2^n\}$ is strictly increasing, thus $\left\{\dfrac{1}{1+2^n}\right\}$ is strictly decreasing and $\left\{1-\dfrac{1}{1+2^n}\right\}=\left\{\dfrac{2^n}{1+2^n}\right\}$

is strictly increasing. The sequence is bounded below by $\dfrac{2^1}{1+2^1}=\dfrac{2}{3}$ and above by $\lim\limits_{n\to\infty}\dfrac{2^n}{1+2^n}=$

$\lim\limits_{n\to\infty}\dfrac{1}{2^{-n}+1}=\dfrac{1}{0+1}=1$.

15. $\dfrac{\sqrt{n}+1}{n}=\dfrac{1}{\sqrt{n}}+\dfrac{1}{n}$; each of $\left\{\dfrac{1}{\sqrt{n}}\right\}$ and $\left\{\dfrac{1}{n}\right\}$ is bounded and strictly decreasing, therefore so is

$\left\{\dfrac{\sqrt{n}+1}{n}\right\}$. Furthermore, $\lim\limits_{n\to\infty}\dfrac{\sqrt{n}+1}{n}=\lim\limits_{n\to\infty}\dfrac{1}{\sqrt{n}}+\lim\limits_{n\to\infty}\dfrac{1}{n}=0+0=0$.

17. $\left\{\dfrac{n-7}{n+4}\right\}=\left\{1-\dfrac{11}{n+4}\right\}$ is a strictly increasing sequence, bounded below by $\dfrac{-6}{5}$ and above by

$\lim\limits_{n\to\infty}\dfrac{n-7}{n+4}=1$.

19. $\displaystyle\sum_{k=2}^{10}4^k=4^2\sum_{j=0}^{8}4^j=4^2\cdot\dfrac{1-4^9}{1-4}=16\cdot\dfrac{-262143}{-3}=16\cdot 87381=1,398,096$.

REMARK: $\displaystyle\sum_{k=2}^{10}4^k=-(4^0+4^1)+\sum_{k=0}^{10}4^k=-(1+4)+\dfrac{1-4^{11}}{1-4}=-5+\dfrac{-4194303}{-3}=-5+1398101$.

21.
$$\sum_{k=3}^{\infty}\left(\dfrac{3}{4}\right)^k=-\left(\left(\dfrac{3}{4}\right)^0+\left(\dfrac{3}{4}\right)^1+\left(\dfrac{3}{4}\right)^2\right)+\sum_{k=0}^{\infty}\left(\dfrac{3}{4}\right)^k$$
$$=-\left(1+\dfrac{3}{4}+\dfrac{9}{16}\right)+\dfrac{1}{1-\frac{3}{4}}=\dfrac{-37}{16}+4=\dfrac{27}{16},$$
$$\sum_{k=3}^{\infty}\left(\dfrac{2}{5}\right)^k=-\left(1+\dfrac{2}{5}+\dfrac{4}{25}\right)+\sum_{k=0}^{\infty}\left(\dfrac{2}{5}\right)^k=\dfrac{-39}{25}+\dfrac{5}{3}=\dfrac{8}{75},$$
$$\sum_{k=3}^{\infty}\left(\left(\dfrac{3}{4}\right)^k-\left(\dfrac{2}{5}\right)^k\right)=\sum_{k=3}^{\infty}\left(\dfrac{3}{4}\right)^k-\sum_{k=3}^{\infty}\left(\dfrac{2}{5}\right)^k=\dfrac{27}{16}-\dfrac{8}{75}=\dfrac{1897}{1200}.$$

23.
$$0.797979\ldots=0.79+0.0079+0.000079+\cdots=\dfrac{79}{100}+\dfrac{79}{100^2}+\dfrac{79}{100^3}+\cdots$$
$$=\dfrac{79}{100}\sum_{k=0}^{\infty}\left(\dfrac{1}{100}\right)^k=\dfrac{79}{100}\cdot\dfrac{1}{1-\frac{1}{100}}=\dfrac{79}{100}\cdot\dfrac{100}{99}=\dfrac{79}{99}.$$

25. If $a_k=\dfrac{1}{k^3-5}$ and $b_k=\dfrac{1}{k^3}$, then $\lim\limits_{k\to\infty}\dfrac{a_k}{b_k}=\lim\limits_{k\to\infty}\dfrac{k^3}{k^3-5}=\lim\limits_{k\to\infty}\dfrac{1}{1-5/k^3}=1>0$; Example 10.5.4

implies $\displaystyle\sum_{k=1}^{\infty}\dfrac{1}{k^3}$ converges, the Limit Comparison Test now implies $\displaystyle\sum_{k=1}^{\infty}\dfrac{1}{k^3-5}$ also converges.

27. If $a_k=\dfrac{1}{\sqrt{k^3+4}}$ and $b_k=\dfrac{1}{k^{3/2}}$, then $\lim\limits_{k\to\infty}\dfrac{a_k}{b_k}=\lim\limits_{k\to\infty}\sqrt{\dfrac{k^3}{k^3+4}}=\lim\limits_{k\to\infty}\sqrt{\dfrac{1}{1+4/k^3}}=\sqrt{1}>0$; Example

10.5.4 implies $\displaystyle\sum_{k=1}^{\infty}\dfrac{1}{k^{3/2}}$ converges, the Limit Comparison Test now implies $\displaystyle\sum_{k=1}^{\infty}\dfrac{1}{\sqrt{k^3+4}}$ also converges.

29. If $a_k = \dfrac{1}{\sqrt[3]{k^3 + 50}}$ and $b_k = \dfrac{1}{k}$, then $\lim\limits_{k\to\infty} \dfrac{a_k}{b_k} = \lim\limits_{k\to\infty} \sqrt[3]{\dfrac{k^3}{k^3 + 50}} = \lim\limits_{k\to\infty} \sqrt[3]{\dfrac{1}{1 + 50/k^3}} = \sqrt[3]{1} > 0$;

Example 10.5.4 implies $\sum\limits_{k=4}^{\infty} \dfrac{1}{k}$ diverges, the Limit Comparison Test now implies $\sum\limits_{k=4}^{\infty} \dfrac{1}{\sqrt[3]{k^3 + 50}}$ also diverges.

31. If $a_k = \dfrac{10^k}{k^5}$, then $\lim\limits_{k\to\infty} \dfrac{a_{k+1}}{a_k} = \lim\limits_{k\to\infty} 10 \cdot \left(\dfrac{k}{k+1}\right)^5 = 10 \cdot 1^5 = 10$. Since this limit is greater than 1, the

Ratio Test implies $\sum\limits_{k=2}^{\infty} \dfrac{10^k}{k^5}$ diverges.

33. $\dfrac{\sqrt{k}\,\ln(k+3)}{k^2 + 2} < \dfrac{\sqrt{k}\,\ln(k+3)}{k^2} = \dfrac{\ln(k+3)}{k^{3/2}} \leq \dfrac{\ln(2k)}{k^{3/2}}$ if $k \geq 3$. The Integral Test says $\sum\limits_{k=1}^{\infty} \dfrac{\ln(2k)}{k^{3/2}}$

converges because $\displaystyle\int_1^{\infty} \dfrac{\ln(2x)}{x^{3/2}}\,dx = \lim\limits_{N\to\infty} \dfrac{-2}{\sqrt{N}}\left(2 + \ln(2N)\right) + 2(2 + \ln 2) = 0 + 2(2 + \ln 2)$ is finite. Now use
the Comparison Test to infer the original series also converges.

REMARK: The Ratio and Root tests are inconclusive for this series (and the following one).

35. $\dfrac{e^{1/k}}{k^{3/2}} < \dfrac{3}{k^{3/2}}$ for all $k \geq 1$. Example 10.5.4 implies $\sum\limits_{k=2}^{\infty} \dfrac{3}{k^{3/2}} = 3\sum\limits_{k=2}^{\infty} \dfrac{1}{k^{3/2}}$ converges, therefore $\sum\limits_{k=2}^{\infty} \dfrac{e^{1/k}}{k^{3/2}}$
also converges.

37. $\sum\limits_{k=1}^{\infty} \dfrac{(-1)^{k+1}}{50k}$ converges because $\left\{\dfrac{1}{50k}\right\}$ is strictly decreasing and $\lim\limits_{k\to\infty} \dfrac{1}{50k} = 0$, the series is conditionally

convergent because $\sum\limits_{k=1}^{\infty} \dfrac{1}{50k}$ diverges.

39. $\left\{\dfrac{1}{\sqrt{k(k-1)}}\right\}$ is strictly decreasing and $\lim\limits_{k\to\infty} \dfrac{1}{\sqrt{k(k-1)}} = 0$, thus $\sum\limits_{k=2}^{\infty} \dfrac{(-1)^{k+1}}{\sqrt{k(k-1)}}$ converges. It is only

conditionally convergent because $\dfrac{1}{\sqrt{k(k-1)}} > \dfrac{1}{k-1}$ and $\sum\limits_{k=2}^{\infty} \dfrac{1}{k-1} = \sum\limits_{j=1}^{\infty} \dfrac{1}{j}$ diverges.

41. $\dfrac{k^2}{k^4 + 1} < \dfrac{k^2}{k^4} = \dfrac{1}{k^2}$ and $\sum\limits_{k=2}^{\infty} \dfrac{1}{k^2}$ converges, hence $\sum\limits_{k=2}^{\infty} \dfrac{(-1)^{k+1}k^2}{k^4 + 1}$ is absolutely convergent.

43. $\left\{\dfrac{(k+2)(k+3)}{(k+1)^3}\right\} = \left\{\dfrac{1}{k+1} + \dfrac{3}{(k+1)^2} + \dfrac{2}{(k+1)^3}\right\}$ is a strictly decreasing sequence with limit equal

to zero, therefore $\sum\limits_{k=3}^{\infty} \dfrac{(-1)^{k+1}(k+2)(k+3)}{(k+1)^3}$ converges. The series is not absolutely convergent because

$\dfrac{(k+2)(k+3)}{(k+1)^3} > \dfrac{1}{k+1}$ and $\sum\limits_{k=3}^{\infty} \dfrac{1}{k+1}$ diverges, hence the original series is conditionally convergent.

45. $n^n \geq n!$ for all $n \geq 1$, thus $\dfrac{n^n}{n!} \geq 1$ and $\left\{\dfrac{(-1)^k k^k}{k!}\right\}$ does not converge to zero, hence $\sum\limits_{k=1}^{\infty} \dfrac{(-1)^k k^k}{k!}$
diverges.

47. $\lim\limits_{k\to\infty} \left(1 + \dfrac{1}{k}\right)^k = e \neq 0$, therefore $\sum\limits_{k=1}^{\infty} (-1)^k \left(1 + \dfrac{1}{k}\right)^k$ diverges.

49. $\dfrac{1}{(n+1)^3} < 0.001 \iff 10^3 = \dfrac{1}{0.001} < (n+1)^3 \iff 10 < n+1 \iff 10 \leq n$, hence we approximate

$\sum\limits_{k=1}^{\infty} \dfrac{(-1)^{k+1}}{k^3}$ with $S_{10} = \sum\limits_{k=1}^{10} \dfrac{(-1)^{k+1}}{k^3} \approx 0.901116$ and error bound $11^{-3} \approx 7.51 \cdot 10^{-4}$.

$T_6 \approx 0.902097$ with error bounded by $8.57 \cdot 10^{-4}$.

51. See Problem 10.4.33 and its solution. The minute hand catches the hour hand at time
$$9 + \frac{9}{12} + \frac{9}{12^2} + \cdots = 9 \sum_{k=0}^{\infty} \left(\frac{1}{12}\right)^k = 9 \cdot \left(\frac{12}{11}\right) \approx 9 : 49 : 05 \text{ P.M.}$$

53. $\sum_{k=0}^{\infty} \left(\frac{x}{3}\right)^k = \frac{1}{1 - \frac{x}{3}} = \frac{3}{3 - x}$ provided $\left|\frac{x}{3}\right| < 1$ and this geometric series diverges otherwise, the interval of convergence is $(-3, 3)$ and the radius of convergence is 3.

55. If $a_k = \frac{1}{k^2 + 2}$, then $\lim_{k \to \infty} \left|\frac{a_{k+1}}{a_k}\right| = \lim_{k \to \infty} \frac{(k+1)^2 + 2}{k^2 + 2} = 1$; therefore the radius of convergence is $\frac{1}{1} = 1$.

For the cases with $|x| = 1$, we observe that $\frac{1}{k^2 + 2} < \frac{1}{k^2}$ and $\sum_{k=0}^{\infty} \frac{1}{k^2}$ converges, thus the power series is absolutely convergent when $|x| = 1$ and the interval of convergence is $[-1, 1]$.

57. If $a_k = \frac{1}{(2 \ln k)^k}$, then $\lim_{k \to \infty} |a_k|^{1/k} = \lim_{k \to \infty} \frac{1}{2 \ln k} = 0$; therefore $\sum_{k=2}^{\infty} \frac{x^k}{(2 \ln k)^k}$ converges everywhere and the radius of convergence is ∞.

59. The geometric series $\sum_{k=0}^{\infty} \frac{(3x - 5)^k}{3^k} = \sum_{k=0}^{\infty} \left(\frac{3x - 5}{3}\right)^k = \sum_{k=0}^{\infty} \left(x - \frac{5}{3}\right)^k$ converges if and only if $\left|x - \frac{5}{3}\right| < 1$.

This implies the interval of convergence is $\left(\frac{5}{3} - 1, \frac{5}{3} + 1\right) = \left(\frac{2}{3}, \frac{8}{3}\right)$ and the radius of convergence is 1.

61. $\sum_{k=0}^{\infty} (-1)^k x^{3k} = \sum_{k=0}^{\infty} (-x^3)^k$ is a geometric series which converges provided $\left|-x^3\right| < 1$ and diverges otherwise. The interval of convergence is $(-1, 1)$ and its radius is 1.

63. $\int_0^{1/2} e^{-t^2} \, dt = \int_0^{1/2} \sum_{k=0}^{\infty} \frac{(-1)^k t^{2k}}{k!} \, dt = \sum_{k=0}^{\infty} \frac{(-1)^k}{k! \cdot (2k + 1)} \left(\frac{1}{2}\right)^{2k+1} = \sum_{k=0}^{\infty} \frac{(-1)^k}{2^{2k+1} \cdot k! \cdot (2k + 1)}.$

n	0	1	2	3	4
a_n	$\frac{1}{2}$	$\frac{-1}{24}$	$\frac{1}{320} \approx 3.13 \cdot 10^{-3}$	$\frac{-1}{5376} \approx -1.86 \cdot 10^{-4}$	$\frac{1}{110592} \approx 9.04 \cdot 10^{-6}$
$\sum_{k=0}^{n} a_k$	$\frac{1}{2}$	$\frac{11}{24}$	$\frac{443}{960} \approx 0.461458$	$\frac{4133}{8960} \approx 0.461272$	$\frac{1785491}{3870720} \approx 0.461281$

If we truncate the alternating series at $n = 3$, then our approximation is 0.461272 with an absolute error bounded by $9.04 \cdot 10^{-6}$.

65. $\int_0^{1/2} t^3 e^{-t^3} \, dt = \int_0^{1/2} t^3 \cdot \sum_{k=0}^{\infty} \frac{(-t^3)^k}{k!} \, dt = \sum_{k=0}^{\infty} \frac{(-1)^k}{k!} \int_0^{1/2} t^{3k+3} \, dt = \sum_{k=0}^{\infty} \frac{(-1)^k}{k! \cdot (3k + 4)} \left(\frac{1}{2}\right)^{3k+4}.$

Our approximation for this alternating series is $\frac{1}{1! \cdot 4 \cdot 2^4} - \frac{1}{2! \cdot 7 \cdot 2^7} = \frac{1}{64} - \frac{1}{896} \approx 0.014509$ with absolute error bounded by $\frac{1}{3! \cdot 10 \cdot 2^{10}} = \frac{1}{20480} \approx 4.88 \cdot 10^{-5}$.

67. $x^2 e^x = x^2 \sum_{k=0}^{\infty} \frac{x^k}{k!} = \sum_{k=0}^{\infty} \frac{x^{k+2}}{k!}.$

69. $\cos^2 x = \frac{1}{2}(1 + \cos 2x) = \frac{1}{2}\left(1 + \sum_{k=0}^{\infty} (-1)^k \frac{(2x)^{2k}}{(2k)!}\right) = \frac{1}{2}\left(1 + 1 + \sum_{k=1}^{\infty} (-1)^k \frac{2^{2k} x^{2k}}{(2k)!}\right)$
$$= 1 + \sum_{k=1}^{\infty} (-1)^k \frac{2^{2k-1} x^{2k}}{(2k)!} = 1 - x^2 + \frac{x^4}{3} - \frac{2x^6}{45} + \cdots$$

1. a.   Example 10.5.1 shows $\sum_{n=1}^{\infty} \frac{1}{n^2}$ converges with the tactic of bounding partial sums by those of a convergent geometric series; Example 10.5.4 uses the integral test to show convergence. Alternatively, observe that

$$1 - \frac{1}{k+1} = \int_1^{k+1} \frac{dx}{x^2} < \sum_{n=1}^{k} \frac{1}{n^2} = 1 + \sum_{n=2}^{k} \frac{1}{n^2} < 1 + \int_1^{k} \frac{dx}{x^2} = 2 - \frac{1}{k}.$$

The sequence of partial sums is monotonically increasing and is bounded, hence it converges. ∎

b. $\sum_{n=1}^{10} \frac{1}{n^2} = \frac{1968329}{1270080} \approx 1.54976\,77312;$ $\sum_{n=1}^{100} \frac{1}{n^2} \approx 1.63498\,39002;$ $\sum_{n=1}^{500} \frac{1}{n^2} \approx 1.64293\,60655.$

c. $\sum_{n=501}^{\infty} \frac{1}{n^2} < \int_{500}^{\infty} \frac{dx}{x^2} < \sum_{n=500}^{\infty} \frac{1}{n^2}.$

d. $\left| \sum_{n=1}^{\infty} \frac{1}{n^2} - \sum_{n=1}^{500} \frac{1}{n^2} \right| = \sum_{n=501}^{\infty} \frac{1}{n^2} < \int_{500}^{\infty} \frac{dx}{x^2} = \frac{1}{500} = 0.002.$

3. a.   On the interval $[k, k+1]$, the region shaded in Figure 2 has area

$$\int_k^{k+1} \left(\frac{1}{k} - \frac{1}{x} \right) dx = \frac{1}{k} - \ln(k+1) + \ln k = \frac{1}{k} + \ln \left(\frac{k}{k+1} \right).$$

The sum of these terms for the first n intervals is

$$\sum_{k=1}^{n} \left(\frac{1}{k} - \ln(k+1) + \ln k \right) = \sum_{k=1}^{n} \frac{1}{k} - \sum_{k=1}^{n} \left(\ln(k+1) - \ln k \right) = \sum_{k=1}^{n} \frac{1}{k} - \int_1^{n+1} \frac{dx}{x} = \sum_{k=1}^{n} \frac{1}{k} - \ln(n+1) = \gamma_n.$$

b.   Part (a) shows γ_n is the sum of positive terms, hence the sequence $\{\gamma_n\}$ is strictly increasing. This sequence is bounded above by 1 because the shaded regions can be shifted horizontally to the left so they fit in the square $[1, 2] \times [0, 1]$ without overlap.

c. $\gamma_{100} \approx 0.57225\,70008;$ $\gamma_{500} \approx 0.57621\,73290;$ $\gamma_{1000} \approx 0.57671\,60812.$

d.   On each interval $[k, k+1]$, the shaded region lies between the graphs of $y = 1/(x-1)$ and $y = 1/x$. The approximation error of γ_n is bounded by $\int_{n+1}^{\infty} \left(\frac{1}{x-1} - \frac{1}{x} \right) dx = \ln \frac{n+1}{n}.$ The error in using γ_{1000} is bounded by $\ln 1.001 \approx 0.0009995.$

e.   Our approximation bound, $\ln(1 + 1/n)$, is less than 10^{-10} if and only if $n > 1/ \left(e^{10^{-10}} - 1 \right) \approx 10^{10}.$

5. a.   Let $f(x) = \tan x$. The first three distinct partial sums of its Maclaurin series are $g_1(x) = x$, $g_3(x) = x + x^3/3$, and $g_5(x) = x + x^3/3 + 2x^5/15$.

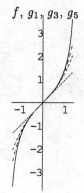
f, g_1, g_3, g_5

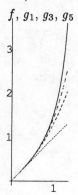
f, g_1, g_3, g_5

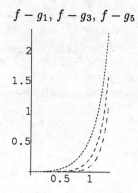
$f - g_1, f - g_3, f - g_5$

b.

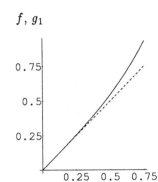 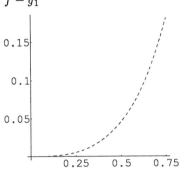 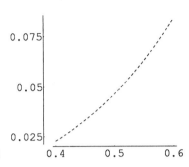

$g_1(x) = x$ is a "good" approximation to $\tan x$ on $[0, .5)$.

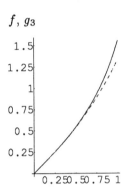 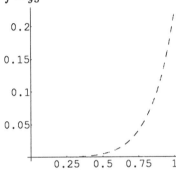 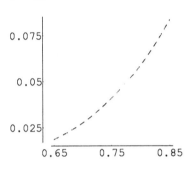

$g_3(x) = x + x^3/3$ is a "good" approximation to $\tan x$ on $[0, .75)$.

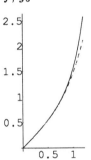 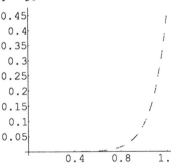 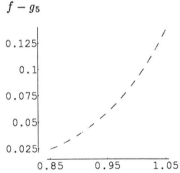

$g_5(x) = x + x^3/3 + 2x^5/15$ is a "good" approximation to $\tan x$ on $[0, .95)$.

c. $f - g_1$ is positive and increasing on $[0, 1]$; Newton's method started at .5 shows $f - g_1$ reaches 0.05 when $x \approx 0.51204$. Hence we choose $a_1 = 0.51204$.

   $f - g_3$ is positive and increasing on $[0, 1]$; Newton's method started at .75 shows $f - g_3$ reaches 0.05 when $x \approx 0.77706$. Hence we choose $a_3 = 0.77706$.

   $f - g_5$ is positive and increasing on $[0, 1]$; Newton's method started at .95 shows $f - g_5$ reaches 0.05 when $x \approx 0.93001$. Hence we choose $a_5 = 0.93001$.

11
Vectors in the Plane and in Space

Solutions 11.1 *Vectors and vector operations* (pages 728-9)

1. If $\mathbf{v} = (2,5) = \overrightarrow{PQ} = \mathbf{q} - \mathbf{p}$ and $P = (1,-2)$, then $\mathbf{q} = \mathbf{v} + \mathbf{p} = (2,5) + (1,-2) = (3,3)$.

1: $Q = (3,3)$ 3: $Q = (4,4)$ 5: $Q = (-2,-5)$

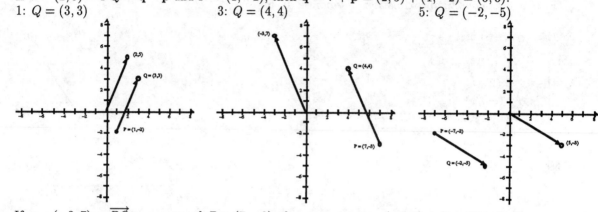

3. If $\mathbf{v} = (-3,7) = \overrightarrow{PQ} = \mathbf{q} - \mathbf{p}$ and $P = (7,-3)$, then $\mathbf{q} = \mathbf{v} + \mathbf{p} = (-3,7) + (7,-3) = (4,4)$.

5. If $\mathbf{v} = (5,-3) = \overrightarrow{PQ} = \mathbf{q} - \mathbf{p}$ and $P = (-7,-2)$, then $\mathbf{q} = \mathbf{v} + \mathbf{p} = (5,-3) + (-7,-2) = (-2,-5)$.

7. If $\mathbf{v} = (4,4)$, then $|\mathbf{v}| = \sqrt{4^2 + 4^2} = 4\sqrt{2}$ and $\theta = \tan^{-1}(4/4) = \tan^{-1} 1 = \pi/4$.

9. If $\mathbf{v} = (4,-4)$, then $|\mathbf{v}| = \sqrt{4^2 + (-4)^2} = 4\sqrt{2}$ and $\theta = \tan^{-1}((-4)/4) + 2\pi = \tan^{-1}(-1) + 2\pi = 7\pi/4$.

11. If $\mathbf{v} = (\sqrt{3},1)$, then $|\mathbf{v}| = \sqrt{(\sqrt{3})^2 + 1^2} = \sqrt{3+1} = 2$ and $\theta = \tan^{-1}(1/\sqrt{3}) = \pi/6$.

13. If $\mathbf{v} = (-1,\sqrt{3})$, then $|\mathbf{v}| = \sqrt{(-1)^2 + (\sqrt{3})^2} = \sqrt{1+3} = 2$ and $\theta = \tan^{-1}(\sqrt{3}/(-1)) + \pi = 2\pi/3$.

15. If $\mathbf{v} = (-1,-\sqrt{3})$, then $|\mathbf{v}| = \sqrt{(-1)^2 + (-\sqrt{3})^2} = \sqrt{1+3} = 2$ and $\theta = \tan^{-1}((-\sqrt{3})/(-1)) + \pi = 4\pi/3$.

17. If $\mathbf{v} = (-5,8)$, then $|\mathbf{v}| = \sqrt{(-5)^2 + 8^2} = \sqrt{89} \approx 9.43$ and $\theta = \tan^{-1}(8/(-5)) + \pi = \tan^{-1}(-1.6) + \pi \approx 2.129$.

19. If $P = (1,2)$ and $Q = (1,3)$, then $\mathbf{v} = \overrightarrow{PQ} = \mathbf{q} - \mathbf{p} = (1,3) - (1,2) = (0,1) = \mathbf{j}$.

19: $\mathbf{v} = \mathbf{j}$ 21: $\mathbf{v} = -6\mathbf{i} + \mathbf{j}$

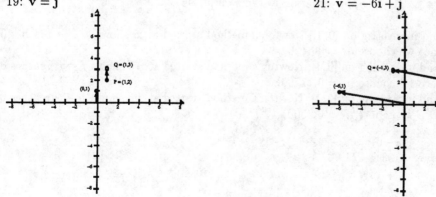

21. If $P = (5,2)$ and $Q = (-1,3)$, then $\mathbf{v} = \overrightarrow{PQ} = \mathbf{q} - \mathbf{p} = (-1,3) - (5,2) = (-6,1) = -6\mathbf{i} + \mathbf{j}$.

23. If $P = (7, -1)$ and $Q = (-2, 4)$, then $\mathbf{v} = \overrightarrow{PQ} = \mathbf{q} - \mathbf{p} = (-2, 4) - (7, -1) = (-9, 5) = -9\mathbf{i} + 5\mathbf{j}$.

23: $\mathbf{v} = -9\mathbf{i} + 5\mathbf{j}$ 25: $\mathbf{v} = -5\mathbf{i} + 5\mathbf{j}$

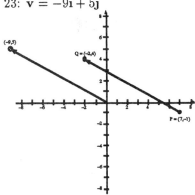 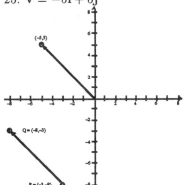

25. If $P = (-3, -8)$ and $Q = (-8, -3)$, then $\mathbf{v} = \overrightarrow{PQ} = \mathbf{q} - \mathbf{p} = (-8, -3) - (-3, -8) = (-5, 5) = -5\mathbf{i} + 5\mathbf{j}$.

27. If $\mathbf{u} = (2, 3)$ and $\mathbf{v} = (-5, 4)$, then

   a. $3\mathbf{u} = (3 \cdot 2, 3 \cdot 3) = (6, 9)$,

   b. $\mathbf{u} + \mathbf{v} = (2, 3) + (-5, 4) = (2 - 5, 3 + 4) = (-3, 7)$,

   c. $\mathbf{v} - \mathbf{u} = (-5, 4) - (2, 3) = (-5 - 2, 4 - 3) = (-7, 1)$,

   d. $2\mathbf{u} - 7\mathbf{v} = (4, 6) - (-35, 28) = (4 + 35, 6 - 28) = (39, -22)$.

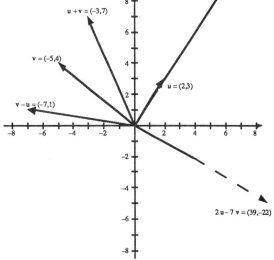

29. If $\mathbf{v} = 2\mathbf{i} + 3\mathbf{j}$, then $|\mathbf{v}|^2 = 2^2 + 3^2 = 13$ and $\mathbf{u} = \mathbf{v}/|\mathbf{v}| = (2/\sqrt{13})\mathbf{i} + (3/\sqrt{13})\mathbf{j}$.

31. If $\mathbf{v} = 3\mathbf{i} + 4\mathbf{j}$, then $|\mathbf{v}|^2 = 3^2 + 4^2 = 25 = 5^2$ and $\mathbf{u} = \mathbf{v}/|\mathbf{v}| = (3/5)\mathbf{i} + (4/5)\mathbf{j}$.

33. If $\mathbf{v} = -3\mathbf{i} + 4\mathbf{j}$, then $|\mathbf{v}|^2 = (-3)^2 + 4^2 = 25 = 5^2$ and $\mathbf{u} = \mathbf{v}/|\mathbf{v}| = (-3/5)\mathbf{i} + (4/5)\mathbf{j}$.

35. If $\mathbf{v} = 2\mathbf{i} - 3\mathbf{j}$, then $|\mathbf{v}|^2 = 2^2 + (-3)^2 = 13$; therefore $\cos\theta = 2/\sqrt{13}$ and $\sin\theta = -3/\sqrt{13}$.

37. $\mathbf{v} = \mathbf{i} + \mathbf{j}$ implies $|\mathbf{v}|^2 = 1^2 + 1^2 = 2$; the unit vector with direction opposite to $\mathbf{v}$ is $-\mathbf{v}/|\mathbf{v}| = -(1/\sqrt{2})\mathbf{i} - (1/\sqrt{2})\mathbf{j}$.

39. $\mathbf{v} = -3\mathbf{i} + 4\mathbf{j}$ implies $|\mathbf{v}|^2 = (-3)^2 + 4^2 = 25$; the unit vector with direction opposite to $\mathbf{v}$ is $-\mathbf{v}/|\mathbf{v}| = (3/5)\mathbf{i} - (4/5)\mathbf{j}$.

41. $\mathbf{v} = -3\mathbf{i} - 4\mathbf{j}$ implies $|\mathbf{v}|^2 = (-3)^2 + (-4)^2 = 25$; the unit vector with direction opposite to $\mathbf{v}$ is $-\mathbf{v}/|\mathbf{v}| = (3/5)\mathbf{i} + (4/5)\mathbf{j}$.

43. If $\mathbf{u} = 2\mathbf{i} - 3\mathbf{j}$ and $\mathbf{v} = -\mathbf{i} + 2\mathbf{j}$, then

   a. $\mathbf{u} + \mathbf{v} = (2 - 1)\mathbf{i} + (-3 + 2)\mathbf{j} = \mathbf{i} - \mathbf{j}$ and $|\mathbf{u} + \mathbf{v}|^2 = 1^2 + (-1)^2 = 2$, therefore $(\mathbf{u} + \mathbf{v})/|\mathbf{u} + \mathbf{v}| = (1/\sqrt{2})\mathbf{i} - (1/\sqrt{2})\mathbf{j}$.

   b. $\mathbf{u} - \mathbf{v} = (2 + 1)\mathbf{i} + (-3 - 2)\mathbf{j} = 3\mathbf{i} - 5\mathbf{j}$ and $|\mathbf{u} - \mathbf{v}|^2 = 3^2 + (-5)^2 = 34$, therefore $(\mathbf{u} - \mathbf{v})/|\mathbf{u} - \mathbf{v}| = (3/\sqrt{34})\mathbf{i} - (5/\sqrt{34})\mathbf{j}$.

   c. $2\mathbf{u} - 3\mathbf{v} = (4 + 3)\mathbf{i} + (-6 - 6)\mathbf{j} = 7\mathbf{i} - 12\mathbf{j}$ and $|2\mathbf{u} - 3\mathbf{v}|^2 = 7^2 + (-12)^2 = 193$, therefore $(2\mathbf{u} - 3\mathbf{v})/|2\mathbf{u} - 3\mathbf{v}| = (7/\sqrt{193})\mathbf{i} - (12/\sqrt{193})\mathbf{j}$.

   d. $3\mathbf{u} + 8\mathbf{v} = (6 - 8)\mathbf{i} + (-9 + 16)\mathbf{j} = -2\mathbf{i} + 7\mathbf{j}$ and $|3\mathbf{u} + 8\mathbf{v}|^2 = (-2)^2 + 7^2 = 53$, therefore $(3\mathbf{u} + 8\mathbf{v})/|3\mathbf{u} + 8\mathbf{v}| = (-2/\sqrt{53})\mathbf{i} + (7/\sqrt{53})\mathbf{j}$.

45. If $|\mathbf{v}| = 3$ and $\theta = \pi/6$, then $\mathbf{v} = 3\left(\cos(\pi/6)\mathbf{i} + \sin(\pi/6)\mathbf{j}\right) = 3\left((\sqrt{3}/2)\mathbf{i} + (1/2)\mathbf{j}\right) = (3\sqrt{3}/2)\mathbf{i} + (3/2)\mathbf{j}$.

47.   If $|\mathbf{v}| = 7$ and $\theta = \pi$, then $\mathbf{v} = 7(\cos\pi\,\mathbf{i} + \sin\pi\,\mathbf{j}) = 7\left((-1)\mathbf{i} + (0)\mathbf{j}\right) = -7\mathbf{i}$.

49.   If $|\mathbf{v}| = 1$ and $\theta = \pi/4$, then $\mathbf{v} = \cos(\pi/4)\mathbf{i} + \sin(\pi/4)\mathbf{j} = (1/\sqrt{2})\mathbf{i} + (1/\sqrt{2})\mathbf{j}$.

51.   If $|\mathbf{v}| = 8$ and $\theta = 3\pi/2$, then $\mathbf{v} = 8\left(\cos(3\pi/2)\mathbf{i} + \sin(3\pi/2)\mathbf{j}\right) = 8\left((0)\mathbf{i} + (-1)\mathbf{j}\right) = -8\mathbf{j}$.

53.   A unit vector has length 1. $|\mathbf{i}| = |(1,0)| = \sqrt{1^2 + 0^2} = \sqrt{1} = 1$, $|\mathbf{j}| = |(0,1)| = \sqrt{0^2 + 1^2} = \sqrt{1} = 1$. ∎

55.   If $\mathbf{v} = a\mathbf{i} + b\mathbf{j} \neq \mathbf{0}$, then $a \neq 0$ or $b \neq 0$; therefore $|\mathbf{v}| = \sqrt{a^2 + b^2} \geq \max\{|a|, |b|\} > 0$ and, invoking the remark preceding Example 5 with $\alpha = 1/|\mathbf{v}|$, we see that $\mathbf{u} = \mathbf{v}/|\mathbf{v}|$ has the same direction as $\mathbf{v}$. Furthermore

$$|\mathbf{u}| = \left|\frac{a}{\sqrt{a^2+b^2}}\mathbf{i} + \frac{b}{\sqrt{a^2+b^2}}\mathbf{j}\right| = \sqrt{\left(\frac{a}{\sqrt{a^2+b^2}}\right)^2 + \left(\frac{b}{\sqrt{a^2+b^2}}\right)^2} = \sqrt{\frac{a^2}{a^2+b^2} + \frac{b^2}{a^2+b^2}} = \sqrt{\frac{a^2+b^2}{a^2+b^2}} = 1. \blacksquare$$

57.   If $P = (c, d)$ and $Q = (c + a, d + b)$, then $\overrightarrow{PQ} = \mathbf{q} - \mathbf{p} = (c + a, d + b) - (c, d) = (a, b)$; hence $\overrightarrow{PQ}$ and (a, b) are two expressions for the same vector — having the same length and direction is a truism. ∎

59. $0 \leq (u_1 v_2 - u_2 v_1)^2 = (u_1 v_2)^2 - 2(u_1 v_2)(u_2 v_1) + (u_2 v_1)^2$

$\implies 2(u_1 v_1)(u_2 v_2) \leq u_1^2 v_2^2 + u_2^2 v_1^2$

$\implies 2(u_1 v_1)(u_2 v_2) + \left((u_1 v_1)^2 + (u_2 v_2)^2\right) \leq \left(u_1^2 v_2^2 + u_2^2 v_1^2\right) + \left((u_1 v_1)^2 + (u_2 v_2)^2\right)$

$\implies (u_1 v_1 + u_2 v_2)^2 \leq \left(u_1^2 + u_2^2\right)\left(v_1^2 + v_2^2\right) = |\mathbf{u}|^2 |\mathbf{v}|^2$

$\implies u_1 v_1 + u_2 v_2 \leq |u_1 v_1 + u_2 v_2| \leq |\mathbf{u}|\,|\mathbf{v}|$

$\implies 2(u_1 v_1 + u_2 v_2) + \left(|\mathbf{u}|^2 + |\mathbf{v}|^2\right) \leq 2|\mathbf{u}|\,|\mathbf{v}| + \left(|\mathbf{u}|^2 + |\mathbf{v}|^2\right) = \left(|\mathbf{u}| + |\mathbf{v}|\right)^2$

$\implies \left(u_1^2 + 2u_1 v_1 + v_1^2\right) + \left(u_2^2 + 2u_2 v_2 + v_2^2\right) \leq \left(|\mathbf{u}| + |\mathbf{v}|\right)^2$

$\implies |\mathbf{u} + \mathbf{v}|^2 = (u_1 + v_1)^2 + (u_2 + v_2)^2 \leq \left(|\mathbf{u}| + |\mathbf{v}|\right)^2$

$\implies |\mathbf{u} + \mathbf{v}| \leq |\mathbf{u}| + |\mathbf{v}|$. ∎

Solutions 11.2   *The dot product*   (pages 736-8)

1.   If $\mathbf{u} = \mathbf{i} + \mathbf{j}$ and $\mathbf{v} = \mathbf{i} - \mathbf{j}$, then $\mathbf{u} \cdot \mathbf{v} = (1)(1) + (1)(-1) = 1 - 1 = 0$; therefore $\cos\theta = 0/(|\mathbf{u}|\,|\mathbf{v}|) = 0$.

3.   If $\mathbf{u} = -5\mathbf{i}$ and $\mathbf{v} = 18\mathbf{j}$, then $\mathbf{u} \cdot \mathbf{v} = (-5)(0) + (0)(18) = 0 + 0 = 0$; therefore $\cos\theta = 0/(|\mathbf{u}|\,|\mathbf{v}|) = 0$.

5.   If $\mathbf{u} = 2\mathbf{i} + 5\mathbf{j}$ and $\mathbf{v} = 5\mathbf{i} + 2\mathbf{j}$, then $\mathbf{u} \cdot \mathbf{v} = (2)(5) + (5)(2) = 10 + 10 = 20$, $|\mathbf{u}| = \sqrt{2^2 + 5^2} = \sqrt{29}$, and $|\mathbf{v}| = \sqrt{5^2 + 2^2} = \sqrt{29}$; therefore $\cos\theta = 20/(\sqrt{29}\sqrt{29}) = 20/29$.

7.   If $\mathbf{u} = -3\mathbf{i} + 4\mathbf{j}$ and $\mathbf{v} = -2\mathbf{i} - 7\mathbf{j}$, then $\mathbf{u} \cdot \mathbf{v} = (-3)(-2) + (4)(-7) = 6 - 28 = -22$, $|\mathbf{u}| = \sqrt{(-3)^2 + 4^2} = 5$, and $|\mathbf{v}| = \sqrt{(-2)^2 + (-7)^2} = \sqrt{53}$; therefore $\cos\theta = -22/(5\sqrt{53})$.

9.   If $\mathbf{u} = 11\mathbf{i} - 8\mathbf{j}$ and $\mathbf{v} = 4\mathbf{i} - 7\mathbf{j}$, then $\mathbf{u} \cdot \mathbf{v} = (11)(4) + (-8)(-7) = 44 + 56 = 100$, $|\mathbf{u}| = \sqrt{11^2 + (-8)^2} = \sqrt{185}$, and $|\mathbf{v}| = \sqrt{4^2 + (-7)^2} = \sqrt{65}$; therefore $\cos\theta = 100/(\sqrt{185}\sqrt{65}) = 100/\sqrt{(5 \cdot 37)(5 \cdot 13)} = 20/\sqrt{481}$.

11.   If $\mathbf{u} = 3\mathbf{i} + 5\mathbf{j}$ and $\mathbf{v} = -6\mathbf{i} - 10\mathbf{j}$, then $\mathbf{v} = (-2)\mathbf{u}$ so the two vectors are parallel.

13.   Let $\mathbf{u} = 2\mathbf{i} + 3\mathbf{j}$ and $\mathbf{v} = 6\mathbf{i} + 4\mathbf{j}$. These vectors are not orthogonal because $\mathbf{u} \cdot \mathbf{v} = (2)(6) + (3)(4) = 12 + 12 = 24 \neq 0$; they are not parallel because $\cos\theta = 24/\left(\sqrt{2^2 + 3^2}\sqrt{6^2 + 4^2}\right) = 24/26 \neq \pm1$.

11: parallel   13: neither   15: orthogonal   17: parallel

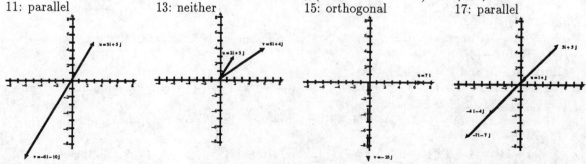

15.   If $\mathbf{u} = 7\mathbf{i}$ and $\mathbf{v} = -23\mathbf{j}$, then $\mathbf{u} \cdot \mathbf{v} = (7)(0) + (0)(-23) = 0 + 0 = 0$ so the two vectors are orthogonal.

17.   If $\mathbf{u} = \mathbf{i} + \mathbf{j}$ and $\mathbf{v} = \alpha\mathbf{i} + \alpha\mathbf{j} = \alpha(\mathbf{i} + \mathbf{j}) = \alpha\mathbf{u}$, then $\mathbf{u}$ and $\mathbf{v}$ are parallel for any nonzero choice of α.

19.   If $\mathbf{u} = 3\mathbf{i}$ and $\mathbf{v} = \mathbf{i} + \mathbf{j}$, then $(\mathbf{u} \cdot \mathbf{v})/|\mathbf{v}|^2 = \left((3)(1) + (0)(1)\right)/\left(1^2 + 1^2\right) = 3/2$ and $\text{Proj}_{\mathbf{v}}\mathbf{u} = (3/2)\mathbf{v} = (3/2)\mathbf{i} + (3/2)\mathbf{j}$.

21. If $\mathbf{u} = 2\mathbf{i} + \mathbf{j}$ and $\mathbf{v} = \mathbf{i} - 2\mathbf{j}$, then $(\mathbf{u} \cdot \mathbf{v})/|\mathbf{v}|^2 = ((2)(1) + (1)(-2)) / \left(1^2 + (-2)^2\right) = 0$ and $\mathrm{Proj}_{\mathbf{v}}\mathbf{u} = (0)\mathbf{v} = \mathbf{0}$.

23. If $\mathbf{u} = \mathbf{i} + \mathbf{j}$ and $\mathbf{v} = 2\mathbf{i} - 3\mathbf{j}$, then $(\mathbf{u} \cdot \mathbf{v})/|\mathbf{v}|^2 = ((1)(2) + (1)(-3)) / \left(2^2 + (-3)^2\right) = -1/13$ and $\mathrm{Proj}_{\mathbf{v}}\mathbf{u} = (-1/13)\mathbf{v} = (-2/13)\mathbf{i} + (3/13)\mathbf{j}$.

25. If $\mathbf{u} = 4\mathbf{i} + 5\mathbf{j}$ and $\mathbf{v} = 2\mathbf{i} + 4\mathbf{j}$, then $(\mathbf{u} \cdot \mathbf{v})/|\mathbf{v}|^2 = ((4)(2) + (5)(4)) / \left(2^2 + 4^2\right) = 28/20 = 7/5$ and $\mathrm{Proj}_{\mathbf{v}}\mathbf{u} = (7/5)\mathbf{v} = (14/5)\mathbf{i} + (28/5)\mathbf{j}$.

27. If $\mathbf{u} = -4\mathbf{i} + 5\mathbf{j}$ and $\mathbf{v} = 2\mathbf{i} - 4\mathbf{j}$, then $(\mathbf{u} \cdot \mathbf{v})/|\mathbf{v}|^2 = ((-4)(2) + (5)(-4)) / \left(2^2 + (-4)^2\right) = -28/20 = -7/5$ and $\mathrm{Proj}_{\mathbf{v}}\mathbf{u} = (-7/5)\mathbf{v} = (-14/5)\mathbf{i} + (28/5)\mathbf{j}$.

29. If $\mathbf{u} = \alpha\mathbf{i} + \beta\mathbf{j}$ and $\mathbf{v} = \mathbf{i} + \mathbf{j}$, then $(\mathbf{u} \cdot \mathbf{v})/|\mathbf{v}|^2 = ((\alpha)(1) + (\beta)(1)) / \left(1^2 + 1^2\right) = (\alpha + \beta)/2$ and $\mathrm{Proj}_{\mathbf{v}}\mathbf{u} = ((\alpha + \beta)/2)\,\mathbf{v} = ((\alpha + \beta)/2)\,\mathbf{i} + ((\alpha + \beta)/2)\,\mathbf{j}$.

31. If $\mathbf{u} = \alpha\mathbf{i} - \beta\mathbf{j}$ and $\mathbf{v} = \mathbf{i} + \mathbf{j}$, then $(\mathbf{u} \cdot \mathbf{v})/|\mathbf{v}|^2 = ((\alpha)(1) + (-\beta)(1)) / \left(1^2 + 1^2\right) = (\alpha - \beta)/2$ and $\mathrm{Proj}_{\mathbf{v}}\mathbf{u} = ((\alpha - \beta)/2)\,\mathbf{v} = ((\alpha - \beta)/2)\,\mathbf{i} + ((\alpha - \beta)/2)\,\mathbf{j}$.

33. Let $\mathbf{u} = 3\mathbf{i} + 4\mathbf{j}$ and $\mathbf{v} = \mathbf{i} + \alpha\mathbf{j}$; then $\mathbf{u} \cdot \mathbf{v} = 3 + 4\alpha$, $|\mathbf{u}| = \sqrt{3^2 + 4^2} = 5$, and $|\mathbf{v}| = \sqrt{1 + \alpha^2}$. The parts of this problem can be worked by solving equations of the form $\cos\theta = t$, hence we'll begin by obtaining a general expression.

$$t = \cos\theta = \frac{\mathbf{u} \cdot \mathbf{v}}{|\mathbf{u}|\,|\mathbf{v}|} = \frac{3 + 4\alpha}{5\sqrt{1 + \alpha^2}}$$
$$\implies 25t^2\left(1 + \alpha^2\right) = (3 + 4\alpha)^2 = 9 + 24\alpha + 16\alpha^2$$
$$\iff \left(25t^2 - 16\right)\alpha^2 - 24\alpha + \left(25t^2 - 9\right) = 0$$
$$\iff \alpha = \frac{24 \pm \sqrt{24^2 - 4(25t^2 - 16)(25t^2 - 9)}}{2(25t^2 - 16)} = \frac{24 \pm \sqrt{2500t^2(1 - t^2)}}{2(25t^2 - 16)} = \frac{12 \pm 25|t|\sqrt{1 - t^2}}{25t^2 - 16}.$$

Note that this chain of implications involves a non-reversable step (squaring), so the formula produces solution-candidates which must be checked by explicit substitution.

a. $\mathbf{u} \perp \mathbf{v} \iff 0 = \mathbf{u} \cdot \mathbf{v} = 3 + 4\alpha \iff \alpha = -3/4$. ($t = 0 \implies \alpha = 12/(-16)$.)

b. $\mathbf{u} \parallel \mathbf{v} \iff \pm 1 = \cos\theta$. Using our formula with $|t| = 1$ we compute $\alpha = (12 \pm 25(1)(0))/(25 - 16) = 12/9 = 4/3$ and $\mathbf{v} = \mathbf{i} + \alpha\mathbf{j} = \mathbf{i} + (4/3)\mathbf{j} = (1/3)(3\mathbf{i} + 4\mathbf{j}) = (1/3)\mathbf{u}$.

c & d. $\cos(2\pi/3) = -1/2$ and $\cos(\pi/3) = 1/2$; in either case, $|t| = 1/2$ implies $\alpha = \left(-48 \mp 25\sqrt{3}\right)/39$. Choosing $\alpha = \left(-48 - 25\sqrt{3}\right)/39 \approx -2.341$ yields $\cos\theta = -1/2 = \cos(2\pi/3)$; on the other hand, $\alpha = \left(-48 + 25\sqrt{3}\right)/39 \approx -0.120$ yields $\cos\theta = 1/2 = \cos(\pi/3)$.

35. Let $A = (1, 3)$, $B = (4, -2)$, and $C = (-3, 6)$. Then

$$\overrightarrow{AB} = \mathbf{b} - \mathbf{a} = 3\mathbf{i} - 5\mathbf{j} = -\overrightarrow{BA} \qquad \text{and} \qquad |\overrightarrow{AB}| = \sqrt{3^2 + 5^2} = \sqrt{34},$$
$$\overrightarrow{BC} = \mathbf{c} - \mathbf{b} = -7\mathbf{i} + 8\mathbf{j} = -\overrightarrow{CB} \qquad \text{and} \qquad |\overrightarrow{BC}| = \sqrt{7^2 + 8^2} = \sqrt{113},$$
$$\overrightarrow{CA} = \mathbf{a} - \mathbf{c} = 4\mathbf{i} - 3\mathbf{j} = -\overrightarrow{AC} \qquad \text{and} \qquad |\overrightarrow{CA}| = \sqrt{4^2 + 3^2} = 5.$$

Therefore

$$\cos A = \overrightarrow{AB} \cdot \overrightarrow{AC}/(|\overrightarrow{AB}|\,|\overrightarrow{AC}|) = (3\mathbf{i} - 5\mathbf{j}) \cdot (-4\mathbf{i} + 3\mathbf{j})/(\sqrt{34} \cdot 5) = -27/5\sqrt{34} \approx -0.9261;$$
$$\cos B = \overrightarrow{BA} \cdot \overrightarrow{BC}/(|\overrightarrow{BA}|\,|\overrightarrow{BC}|) = (-3\mathbf{i} + 5\mathbf{j}) \cdot (-7\mathbf{i} + 8\mathbf{j})/(\sqrt{34}\,\sqrt{113}) = 61/\sqrt{3842} \approx 0.9841;$$
$$\cos C = \overrightarrow{CA} \cdot \overrightarrow{CB}/(|\overrightarrow{CA}|\,|\overrightarrow{CB}|) = (4\mathbf{i} - 3\mathbf{j}) \cdot (7\mathbf{i} - 8\mathbf{j})/(5\sqrt{113}) = 52/5\sqrt{113} \approx 0.9783.$$

37. If $P = (2, 3)$ and $Q = (5, 5)$, then $\overrightarrow{PQ} = 3\mathbf{i} + 2\mathbf{j}$; if $R = (2, -3)$ and $S = (1, 2)$, then $\overrightarrow{RS} = -\mathbf{i} + 5\mathbf{j}$. Hence $\overrightarrow{PQ} \cdot \overrightarrow{PQ} = 3^2 + 2^2 = 13$, $\overrightarrow{PQ} \cdot \overrightarrow{RS} = (3)(-1) + (2)(5) = 7 = \overrightarrow{RS} \cdot \overrightarrow{PQ}$, and $\overrightarrow{RS} \cdot \overrightarrow{RS} = (-1)^2 + 5^2 = 26$. Therefore $\mathrm{Proj}_{\overrightarrow{PQ}}\overrightarrow{RS} = \dfrac{7}{13}(3\mathbf{i} + 2\mathbf{j})$ and $\mathrm{Proj}_{\overrightarrow{RS}}\overrightarrow{PQ} = \dfrac{7}{26}(-\mathbf{i} + 5\mathbf{j})$.

39.   We will apply Remark 4. The distance from point $P = (2,3)$ to the line on the points $Q = (-1,7)$ and $R = (3,5)$ is the length of $\overrightarrow{PQ} - \text{Proj}_{\overrightarrow{QR}}\overrightarrow{PQ}$. Since $\overrightarrow{PQ} = -3\mathbf{i} + 4\mathbf{j}$ and

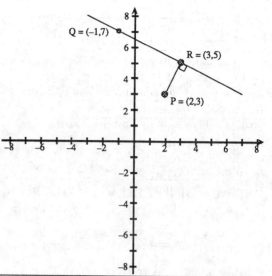

$\overrightarrow{QR} = 4\mathbf{i} - 2\mathbf{j}$, thus $\overrightarrow{PQ} \cdot \overrightarrow{QR} = (-3)(4) + (4)(-2) = -20$, $\overrightarrow{QR} \cdot \overrightarrow{QR} = 4^2 + (-2)^2 = 20$, and $\text{Proj}_{\overrightarrow{QR}}\overrightarrow{PQ} = \dfrac{-20}{20}\overrightarrow{QR} = -\overrightarrow{QR}$.

Therefore $\overrightarrow{PQ} - \text{Proj}_{\overrightarrow{QR}}\overrightarrow{PQ} = \overrightarrow{PQ} + \overrightarrow{QR} = \overrightarrow{PR} = \mathbf{i} + 2\mathbf{j}$ and the distance between point and line is $\sqrt{1^2 + 2^2} = \sqrt{5}$.

REMARK: An alternative approach would be to compute the length of $\overrightarrow{PR} - \text{Proj}_{\overrightarrow{QR}}\overrightarrow{PR} = \overrightarrow{PR} - 0 = \overrightarrow{PR}$.

41. $-\mathbf{i}$ is the unit vector from the right and $-\mathbf{j}$ is the unit vector from above. Therefore a force of 2 N from the right and 5 N from above has resultant $\mathbf{r} = 2(-\mathbf{i}) + 5(-\mathbf{j}) = (-2\mathbf{i} - 5\mathbf{j})$ N. To keep the object at rest, apply force $-\mathbf{r} = (2\mathbf{i} + 5\mathbf{j})$ N.

43.   The resultant force is $\mathbf{r} = 3\mathbf{i} + 5(-\mathbf{i}) + 3(-\mathbf{j}) = (-2\mathbf{i} - 3\mathbf{j})$ N. Apply force $-\mathbf{r} = (2\mathbf{i} + 3\mathbf{j})$ N to keep the object at rest.

45.   The resultant is $\mathbf{r} = 5(-\mathbf{j}) + 4(-1)(\cos\pi/6\,\mathbf{i} + \sin\pi/6\,\mathbf{j}) = -5\mathbf{j} - 4\left(\sqrt{3}/2\,\mathbf{i} + 1/2\,\mathbf{j}\right) = \left(-2\sqrt{3}\mathbf{i} - 7\mathbf{j}\right)$ N. Apply force $-\mathbf{r} = \left(2\sqrt{3}\mathbf{i} + 7\mathbf{j}\right)$ N to keep the object at rest.

47.   The resultant is $\mathbf{r} = 2(-\mathbf{j}) + 3(-1)\left(\cos(3\pi/4)\mathbf{i} + \sin(3\pi/4)\mathbf{j}\right) = \left((3/\sqrt{2})\mathbf{i} - (2 + 3/\sqrt{2})\mathbf{j}\right)$ N. Apply force $-\mathbf{r} = \left(-(3/\sqrt{2})\mathbf{i} + (2 + 3/\sqrt{2})\mathbf{j}\right)$ N to keep the object at rest.

49. $7\cos(\pi/6) + 7\cos(\pi/3) + 14\cos(5\pi/4) = (7/2)\left(\sqrt{3} + 1 - 2\sqrt{2}\right) = 7\sin(\pi/6) + 7\sin(\pi/3) + 14\sin(5\pi/4)$. The resultant force is $\mathbf{r} = -(7/2)\left(\sqrt{3} + 1 - 2\sqrt{2}\right)(\mathbf{i} + \mathbf{j})$ N and the force $-\mathbf{r} = (7/2)\left(\sqrt{3} + 1 - 2\sqrt{2}\right)(\mathbf{i} + \mathbf{j})$ N keeps things at rest.

51. $\mathbf{F} = 2\left[\cos(\pi/2)\mathbf{i} + \sin(\pi/2)\mathbf{j}\right] = 2\mathbf{j}$ and $\overrightarrow{PQ} = (1-5)\mathbf{i} + (1-7)\mathbf{j} = -4\mathbf{i} - 6\mathbf{j}$. The work is $W = \mathbf{F} \cdot \overrightarrow{PQ} = (2\mathbf{j}) \cdot (-4\mathbf{i} - 6\mathbf{j}) = (0)(-4) + (2)(-6) = -12$ joules.

53. $\mathbf{F} = 4\left[\cos(\pi/6)\mathbf{i} + \sin(\pi/6)\mathbf{j}\right] = 2\sqrt{3}\mathbf{i} + 2\mathbf{j}$ and $\overrightarrow{PQ} = (3+1)\mathbf{i} + (4-2)\mathbf{j} = 4\mathbf{i} + 2\mathbf{j}$. The work is $W = \mathbf{F} \cdot \overrightarrow{PQ} = (2\sqrt{3})(4) + (2)(2) = 8\sqrt{3} + 4$ joules.

55. $\mathbf{F} = 3\left[\cos(3\pi/4)\mathbf{i} + \sin(3\pi/4)\mathbf{j}\right] = (3/\sqrt{2})(-\mathbf{i} + \mathbf{j})$ and $\overrightarrow{PQ} = (1-2)\mathbf{i} + (2-1)\mathbf{j} = -\mathbf{i} + \mathbf{j}$. The work is $W = \mathbf{F} \cdot \overrightarrow{PQ} = (3/\sqrt{2})\left((-1)^2 + 1^2\right) = 3\sqrt{2}$ joules.

57. $\mathbf{F} = 4(2\mathbf{i} + 3\mathbf{j})/\sqrt{13}$ and $\overrightarrow{PQ} = (-1-2)\mathbf{i} + (3-0)\mathbf{j} = -3\mathbf{i} + 3\mathbf{j}$. The work is $W = \mathbf{F} \cdot \overrightarrow{PQ} = (4/\sqrt{13})(2\mathbf{i} + 3\mathbf{j}) \cdot (-3\mathbf{i} + 3\mathbf{j}) = (4/\sqrt{13})(-6 + 9) = 12/\sqrt{13}$ joules.

59.   The force applied by tugboat #1 is $\mathbf{F}_1 = 500\left(\cos 20°\,\mathbf{i} + \sin 20°\,\mathbf{j}\right)$ and the force applied by tugboat #2 is $\mathbf{F}_2 = x\left(\cos(-30°)\mathbf{i} + \sin(-30°)\mathbf{j}\right)$. The resultant $\mathbf{F}_1 + \mathbf{F}_2$ is purely horizontal if and only if its $\mathbf{j}$ component is zero, i.e., $500\sin 20° + x\sin(-30°) = 0$. This is equivalent to $x = -500\sin 20° / \sin(-30°) = 500\sin 20° / \sin 30° \approx 342.02$ N.

61.   The work done by tugboat #1 is $W_1 = \mathbf{F}_1 \cdot 750\mathbf{i} = (500\cos 20°)(750) \approx 352384.7$ joules; the work done by tugboat #2 is $W_2 = \mathbf{F}_2 \cdot 750\mathbf{i} = [500(\sin 20° / \sin 30°)\cos 30°](750) = (500)(750)(\sin 20°)(\cot 30°) \approx 222148.6$ joules.

63. $(\alpha\mathbf{i} + \beta\mathbf{j}) \cdot (\beta\mathbf{i} - \alpha\mathbf{j}) = (\alpha)(\beta) + (\beta)(-\alpha) = 0$; Theorem 4 implies the vectors $\mathbf{u} = \alpha\mathbf{i} + \beta\mathbf{j}$ and $\mathbf{v} = \beta\mathbf{i} - \alpha\mathbf{j}$ are orthogonal. ∎

65.   There is no choice of α for which $\mathbf{i} + \alpha\mathbf{j}$ has a negative first component (i.e., negative coefficient of $\mathbf{i}$), therefore $\mathbf{i} + \alpha\mathbf{j}$ and $3\mathbf{i} + 4\mathbf{j}$ can not have opposite directions.

REMARK: The formula in Solution 33 with $t = \pm 1$ yields $\alpha = 4/3$ which then implies $t = 1$, thus $t = \cos \theta = -1$ is impossible.

67. $\mathbf{v}$ and $\text{Proj}_{\mathbf{v}}\mathbf{u} = c\mathbf{v}$ have the same direction if and only if $c = (\mathbf{u} \cdot \mathbf{v})/|\mathbf{v}|^2 > 0$. If $\mathbf{v} \neq \mathbf{0}$, then this condition reduces to $\mathbf{u} \cdot \mathbf{v} > 0$. Therefore, if $\mathbf{u} = a_1\mathbf{i} + b_1\mathbf{j}$ and $\mathbf{v} = a_2\mathbf{i} + b_2\mathbf{j}$, our full conditions are (1) at least one of a_2, b_2 is non-zero and (2) $a_1a_2 + b_1b_2 > 0$.

REMARK: The zero vector does not have a direction (see text Remark 11.1.1).

69. The first half of Solution 11.1.59 (reproduced below) shows $|\mathbf{u} \cdot \mathbf{v}| \leq |\mathbf{u}|\,|\mathbf{v}|$ and that sequence of algebraic steps is easily reversed to show equality holds if and only if one vector is a scalar multiple of the other. Therefore, if $\mathbf{u}$ and $\mathbf{v}$ are both nonzero, $\theta = k\pi \iff |\cos \theta| = 1 \iff |\mathbf{u} \cdot \mathbf{v}| = |\mathbf{u}|\,|\mathbf{v}| \iff \mathbf{u} = \alpha\mathbf{v}$ for some nonzero α. ∎

REMARK: Suppose all four of u_1, u_2, v_1, v_2 are nonzero (just to simplify the expression for α at the end).

$$0 \leq (u_1v_2 - u_2v_1)^2 = (u_1v_2)^2 - 2(u_1v_2)(u_2v_1) + (u_2v_1)^2$$
$$\implies 2(u_1v_1)(u_2v_2) \leq u_1^2v_2^2 + u_2^2v_1^2$$
$$\implies 2(u_1v_1)(u_2v_2) + \left((u_1v_1)^2 + (u_2v_2)^2\right) \leq \left(u_1^2v_2^2 + u_2^2v_1^2\right) + \left((u_1v_1)^2 + (u_2v_2)^2\right)$$
$$\implies (u_1v_1 + u_2v_2)^2 \leq \left(u_1^2 + u_2^2\right)\left(v_1^2 + v_2^2\right) = |\mathbf{u}|^2\,|\mathbf{v}|^2$$
$$\implies |\mathbf{u} \cdot \mathbf{v}| = |u_1v_1 + u_2v_2| \leq |\mathbf{u}|\,|\mathbf{v}|.$$

If $|\mathbf{u} \cdot \mathbf{v}| = |\mathbf{u}|\,|\mathbf{v}|$, then these steps are easily reversed to show $u_1v_2 - u_2v_1 = 0$; we let $\alpha = u_1/v_1 = u_2/v_2$.

71. A vector is orthogonal to a line if and only if it is orthogonal to every vector lying in the line. Suppose $\mathbf{v} = a\mathbf{i} + b\mathbf{j}$ and also suppose $P = (p_1, p_2)$ and $Q = (q_1, q_2)$ are two distinct points of the line satisfying $ax + by + c = 0$. Then $\mathbf{v} \cdot \overrightarrow{PQ} = (a\mathbf{i} + b\mathbf{j}) \cdot ((q_1 - p_1)\mathbf{i} + (q_2 - p_2)\mathbf{j}) = a(q_1 - p_1) + b(q_2 - p_2) = (aq_1 + bq_2) - (ap_1 + bp_2) = (-c) - (-c) = 0$. Now invoke Theorem 4 to infer $\mathbf{v} \perp \overrightarrow{PQ}$ for any choice of P and Q on the line, therefore $\mathbf{v}$ is orthogonal to the line. ∎

73. $|\mathbf{a} + \mathbf{b}|^2 + |\mathbf{a} - \mathbf{b}|^2 = (\mathbf{a} + \mathbf{b}) \cdot (\mathbf{a} + \mathbf{b}) + (\mathbf{a} - \mathbf{b}) \cdot (\mathbf{a} - \mathbf{b})$
$$= (\mathbf{a} \cdot \mathbf{a} + 2\mathbf{a} \cdot \mathbf{b} + \mathbf{b} \cdot \mathbf{b}) + (\mathbf{a} \cdot \mathbf{a} - 2\mathbf{a} \cdot \mathbf{b} + \mathbf{b} \cdot \mathbf{b}) = 2\mathbf{a} \cdot \mathbf{a} + 2\mathbf{b} \cdot \mathbf{b} = 2|\mathbf{a}|^2 + 2|\mathbf{b}|^2.$$ ∎

75. Suppose $\mathbf{a} = (a_1, a_2)$ and $\mathbf{b} = (b_1, b_2)$ are nonzero.

$$1 \geq |\cos \theta| = \frac{|\mathbf{a} \cdot \mathbf{b}|}{|\mathbf{a}|\,|\mathbf{b}|} = \frac{|a_1b_1 + a_2b_2|}{\sqrt{a_1^2 + a_2^2}\,\sqrt{b_1^2 + b_2^2}} \iff \sqrt{a_1^2 + a_2^2}\,\sqrt{b_1^2 + b_2^2} \geq |a_1b_1 + a_2b_2|$$

$$\iff \left(\sum_{i=1}^{2} a_i^2\right)^{1/2}\left(\sum_{i=1}^{2} b_i^2\right)^{1/2} \geq \left|\sum_{i=1}^{2} a_ib_i\right|.$$ ∎

Equality holds if one of the vectors is zero or if $\cos \theta = \pm 1$, i.e., if the vectors have the same or opposite directions.

REMARK: The Remark for Solution 69 has an alternative proof; here is a third proof.

If $\mathbf{a}$ or $\mathbf{b}$ is the zero vector, then $\mathbf{a} \cdot \mathbf{b}$ and $|\mathbf{a}|\,|\mathbf{b}|$ both equal zero. Therefore, for the remainder of this discussion, we can assume both $\mathbf{a}$ and $\mathbf{b}$ are nonzero. For any real numbers s and t, we have $|s\mathbf{a} - t\mathbf{b}| \geq 0$. Hence $0 \leq |s\mathbf{a} - t\mathbf{b}|^2 = (s\mathbf{a} - t\mathbf{b}) \cdot (s\mathbf{a} - t\mathbf{b}) = s^2|\mathbf{a}|^2 - (2st)(\mathbf{a} \cdot \mathbf{b}) + t^2|\mathbf{b}|^2$ which then implies $(2st)(\mathbf{a} \cdot \mathbf{b}) \leq s^2|\mathbf{a}|^2 + t^2|\mathbf{b}|^2$. For the particular choice $s = |\mathbf{b}|$ and $t = |\mathbf{a}|$ this result specializes to the statement that $(2|\mathbf{a}|\,|\mathbf{b}|)(\mathbf{a} \cdot \mathbf{b}) \leq 2|\mathbf{a}|^2|\mathbf{b}|^2$. Since $\mathbf{a}$ and $\mathbf{b}$ are nonzero, we know $2|\mathbf{a}|\,|\mathbf{b}|$ is positive so we can divide the inequality by it to see that $\mathbf{a} \cdot \mathbf{b} \leq |\mathbf{a}|\,|\mathbf{b}|$. Replacing $\mathbf{a}$ with $-\mathbf{a}$ yields $-(\mathbf{a} \cdot \mathbf{b}) = (-\mathbf{a}) \cdot \mathbf{b} \leq |-\mathbf{a}|\,|\mathbf{b}| = |\mathbf{a}|\,|\mathbf{b}|$. Thus $|\mathbf{a} \cdot \mathbf{b}| \leq |\mathbf{a}|\,|\mathbf{b}|$. Furthermore, this is a strict inequality unless $0 = s\mathbf{a} - t\mathbf{b}$, i.e., unless $\mathbf{b} = (s/t)\mathbf{a}$. ∎

77. Let P be a point, F be its foot on the line, and A any other point on the line. Then the Pythagorean Theorem implies $\overline{PF}^2 = \overline{PA}^2 - \overline{AF}^2 < \overline{PA}^2$. ∎

REMARK: If $\mathbf{v}$ is a vector along the line and if $\mathbf{u} = \overrightarrow{PA}$, then $\mathbf{u} = (\text{Proj}_{\mathbf{v}}\mathbf{u}) + (\mathbf{u} - \text{Proj}_{\mathbf{v}}\mathbf{u})$. The triangle inequality implies $|\mathbf{u}| \geq |\mathbf{u} - \text{Proj}_{\mathbf{v}}\mathbf{u}|$ and text Remark 4(ii) implies $\mathbf{u} - \text{Proj}_{\mathbf{v}}\mathbf{u}$ is orthogonal to the line.

Solutions 11.3   *The rectangular coordinate system in space*   **(pages 742-3)**

1: $(1, 4, 2)$ 3: $(-1, 5, 7)$ 5: $(-2, 1, -2)$ 7: $(3, 2, -5)$

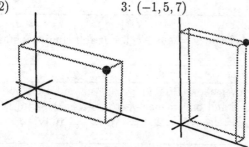 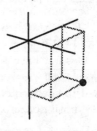 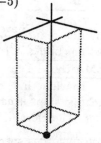

9: $(2, 0, 4)$ 11: $(0, 4, 7)$ 13: $(3, 0, 0)$ 15: $(0, 0, -7)$

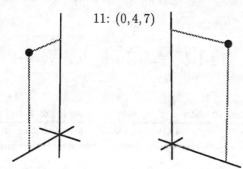 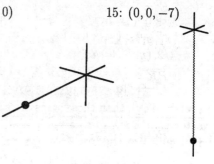

17.   If $P = (8, 1, 6)$ and $Q = (8, 1, 4)$, then $\overline{PQ} = \sqrt{(8-8)^2 + (1-1)^2 + (6-4)^2} = \sqrt{0+0+4} = 2$.

19. $P = (3, -4, 7)$, $Q = (3, -4, 9)$ implies $\overline{PQ} = \sqrt{(3-3)^2 + (-4-(-4))^2 + (7-9)^2} = \sqrt{0+0+4} = 2$.

21. $P = (2, -7, 5)$, $Q = (8, -7, -1)$, $\overline{PQ} = \sqrt{(2-8)^2 + (-7-(-7))^2 + (5-(-1))^2} = \sqrt{36+0+36} = 6\sqrt{2}$.

23.   If $P = (3, 1, 2)$ and $Q = (1, 2, 3)$, then $\overline{PQ} = \sqrt{(3-1)^2 + (1-2)^2 + (2-3)^2} = \sqrt{4+1+1} = \sqrt{6}$.

25. $P = (-1, -7, -2)$, $Q = (-4, 3, -5)$ implies $\overline{PQ} = \sqrt{(-1-(-4))^2 + (-7-3)^2 + (-2-(-5))^2} = \sqrt{9+100+9} = \sqrt{118}$.

27.   The sphere centered at $(2, -1, 4)$ with radius 2 satisfies the equation
$$4 = 2^2 = (x-2)^2 + (y-(-1))^2 + (z-4)^2 = (x-2)^2 + (y+1)^2 + (z-4)^2.$$

29.   Let $P = (3, 0, 1)$, $Q = (0, -4, 0)$, $R = (6, 4, 2)$. Then
$$\overline{PQ} = \sqrt{(3-0)^2 + (0+4)^2 + (1-0)^2} = \sqrt{9+16+1} = \sqrt{26} = \overline{QP},$$
$$\overline{QR} = \sqrt{(0-6)^2 + (-4-4)^2 + (0-2)^2} = \sqrt{36+64+4} = \sqrt{104} = 2\sqrt{26} = \overline{RQ},$$
$$\overline{PR} = \sqrt{(3-6)^2 + (0-4)^2 + (1-2)^2} = \sqrt{9+16+1} = \sqrt{26} = \overline{RP}.$$

The point P lies on the straight line segment between Q and R because $\overline{QR} = \overline{QP} + \overline{PR}$. ∎

REMARK: P is the midpoint of the line segment between Q and R; the result of Problem 39 provides easier computations than the ones shown above.

31.
$$x^2 + y^2 + z^2 - 4x - 4y + 8z + 8 = 0$$
$$\Longleftrightarrow (x^2 - 4x) + (y^2 - 4y) + (z^2 + 8z) = -8$$
$$\Longleftrightarrow (x^2 - 4x + 4) + (y^2 - 4y + 4) + (z^2 + 8z + 16) = -8 + (-4/2)^2 + (-4/2)^2 + (8/2)^2 = 16$$
$$\Longleftrightarrow (x-2)^2 + (y-2)^2 + (z+4)^2 = 4^2.$$

Points satisfying one of these equations lie 4 units from the point $(2, 2, -4)$; the surface is the sphere of radius 4 centered at $(2, 2, -4)$.

33.
$$x^2 - 2x + y^2 - 4y + z^2 + z - 2 = 0$$
$$\Longleftrightarrow (x^2 - 2x + 1) + (y^2 - 4y + 4) + (z^2 + z + 1/4) = 2 + (-2/2)^2 + (-4/2)^2 + (1/2)^2 = 29/4$$
$$\Longleftrightarrow (x-1)^2 + (y-2)^2 + (z+1/2)^2 = (\sqrt{29}/2)^2.$$

This sphere is centered at $(1, 2, -1/2)$; the inscribed sphere of radius 1 satisfies the equation
$$(x-1)^2 + (y-2)^2 + (z+1/2)^2 = 1^2.$$

35.
$$x^2 - 2x + y^2 + 8y + z^2 - 5z + \alpha = 0$$
$$\iff \left(x^2 - 2x + 1\right) + \left(y^2 + 8y + 16\right) + \left(z^2 - 5z + 25/4\right) = -\alpha + (-2/2)^2 + (8/2)^2 + (-5/2)^2$$
$$\iff (x-1)^2 + (y+4)^2 + (z-5/2)^2 = 93/4 - \alpha.$$

These equations have a unique solution ($x = 1$, $y = -4$, $z = 5/2$) if and only if $\alpha = 93/4$.

37.   Let $E = d - (a/2)^2 - (b/2)^2 - (c/2)^2$.
$$x^2 + ax + y^2 + by + z^2 + cz + d = 0$$
$$\iff \left(x^2 + ax\right) + \left(y^2 + by\right) + \left(z^2 + cz\right) = -d$$
$$\iff \left(x^2 + ax + a^2/4\right) + \left(y^2 + by + b^2/4\right) + \left(z^2 + cz + c^2/4\right) = -d + (a/2)^2 + (b/2)^2 + (c/2)^2$$
$$\iff (x + a/2)^2 + (y + b/2)^2 + (z + c/2)^2 = -E.$$

These equations specify a sphere centered at $(-a/2, -b/2, -c/2)$ with radius $\sqrt{-E}$ if and only if $-E \geq 0$, i.e., $E \leq 0$; if $E = 0$, then such a sphere of radius 0 is just the point $(-a/2, -b/2, -c/2)$. There are no solutions if $-E < 0$, i.e., if $E > 0$. ∎

39.   Let $P = (x_1, y_1, z_1)$ and $Q = (x_2, y_2, z_2)$. Then $\overline{PQ} = \sqrt{(x_1 - x_2)^2 + (y_1 - y_2)^2 + (z_1 - z_2)^2}$. If we let $R = ((x_1 + x_2)/2, (y_1 + y_2)/2, (z_1 + z_2)/2)$, then
$$\overline{PR} = \sqrt{(x_1 - ((x_1 + x_2)/2))^2 + (y_1 - ((y_1 + y_2)/2))^2 + (z_1 - ((z_1 + z_2)/2))^2}$$
$$= \sqrt{((x_1 - x_2)/2)^2 + ((y_1 - y_2)/2)^2 + ((z_1 - z_2)/2)^2}$$
$$= (1/2)\sqrt{(x_1 - x_2)^2 + (y_1 - y_2)^2 + (z_1 - z_2)^2} = (1/2)\overline{PQ},$$
$$\overline{RQ} = \sqrt{(((x_1 + x_2)/2) - x_2)^2 + (((y_1 + y_2)/2) - y_2)^2 + (((z_1 + z_2)/2) - z_2)^2}$$
$$= \sqrt{((x_1 - x_2)/2)^2 + ((y_1 - y_2)/2)^2 + ((z_1 - z_2)/2)^2}$$
$$= (1/2)\sqrt{(x_1 - x_2)^2 + (y_1 - y_2)^2 + (z_1 - z_2)^2} = (1/2)\overline{PQ}.$$

The points are collinear because $\overline{PQ} = \overline{PR} + \overline{RQ}$; R is the midpoint of the line segment between P and Q because $\overline{PR} = \overline{RQ}$. ∎

Solutions 11.4   *Vectors in* $\mathbb{R}^3$ (pages 749-50)

NOTE: The direction cosines of the nonzero vector $\mathbf{v}$ are the components of the unit vector $\mathbf{v}/|\mathbf{v}|$.

1.   If $\mathbf{v} = 3\mathbf{j} = 0\mathbf{i} + 3\mathbf{j} + 0\mathbf{k}$, then $|\mathbf{v}| = \sqrt{0^2 + 3^2 + 0^2} = 3$ and $\mathbf{v}/|\mathbf{v}| = \mathbf{j}$.

3.   If $\mathbf{v} = 14\mathbf{k} = 0\mathbf{i} + 0\mathbf{j} + 14\mathbf{k}$, then $|\mathbf{v}| = \sqrt{0^2 + 0^2 + 14^2} = 14$ and $\mathbf{v}/|\mathbf{v}| = \mathbf{k}$.

5.   If $\mathbf{v} = 4\mathbf{i} - \mathbf{j}$, then $|\mathbf{v}| = \sqrt{4^2 + (-1)^2 + 0^2} = \sqrt{17}$ and $\mathbf{v}/|\mathbf{v}| = (4/\sqrt{17})\mathbf{i} + (-1/\sqrt{17})\mathbf{j}$.

7.   If $\mathbf{v} = -2\mathbf{i} + 3\mathbf{j}$, then $|\mathbf{v}| = \sqrt{(-2)^2 + 3^2 + 0^2} = \sqrt{13}$ and $\mathbf{v}/|\mathbf{v}| = (-2/\sqrt{13})\mathbf{i} + (3/\sqrt{13})\mathbf{j}$.

9.   If $\mathbf{v} = \mathbf{i} - \mathbf{j} + \mathbf{k}$, then $|\mathbf{v}| = \sqrt{1^2 + (-1)^2 + 1^2} = \sqrt{3}$ and $\mathbf{v}/|\mathbf{v}| = (1/\sqrt{3})\mathbf{i} + (-1/\sqrt{3})\mathbf{j} + (1/\sqrt{3})\mathbf{k}$.

11.   If $\mathbf{v} = -\mathbf{i} + \mathbf{j} + \mathbf{k}$, then $|\mathbf{v}| = \sqrt{(-1)^2 + 1^2 + 1^2} = \sqrt{3}$ and $\mathbf{v}/|\mathbf{v}| = (-1/\sqrt{3})\mathbf{i} + (1/\sqrt{3})\mathbf{j} + (1/\sqrt{3})\mathbf{k}$.

13.   If $\mathbf{v} = -\mathbf{i} + \mathbf{j} - \mathbf{k}$, then $|\mathbf{v}| = \sqrt{(-1)^2 + 1^2 + (-1)^2} = \sqrt{3}$ and $\mathbf{v}/|\mathbf{v}| = (-1/\sqrt{3})\mathbf{i} + (1/\sqrt{3})\mathbf{j} + (-1/\sqrt{3})\mathbf{k}$.

15.   If $\mathbf{v} = -\mathbf{i} - \mathbf{j} - \mathbf{k}$, then $|\mathbf{v}| = \sqrt{(-1)^2 + (-1)^2 + (-1)^2} = \sqrt{3}$ and $\mathbf{v}/|\mathbf{v}| = (-1/\sqrt{3})\mathbf{i} + (-1/\sqrt{3})\mathbf{j} + (-1/\sqrt{3})\mathbf{k}$.

17.   If $\mathbf{v} = -7\mathbf{i} + 2\mathbf{j} - 13\mathbf{k}$, then $|\mathbf{v}| = \sqrt{(-7)^2 + 2^2 + (-13)^2} = \sqrt{222}$ and $\mathbf{v}/|\mathbf{v}| = (-7/\sqrt{222})\mathbf{i} + (2/\sqrt{222})\mathbf{j} + (-13/\sqrt{222})\mathbf{k}$.

19.   If $\mathbf{v} = -3\mathbf{i} - 3\mathbf{j} + 8\mathbf{k}$, then $|\mathbf{v}| = \sqrt{(-3)^2 + (-3)^2 + 8^2} = \sqrt{82}$ and $\mathbf{v}/|\mathbf{v}| = (-3/\sqrt{82})\mathbf{i} + (-3/\sqrt{82})\mathbf{j} + (8/\sqrt{82})\mathbf{k}$.

NOTE: In Problems 21-40, we have $\mathbf{u} = 2\mathbf{i} - 3\mathbf{j} + 4\mathbf{k}$, $\mathbf{v} = -2\mathbf{i} - 3\mathbf{j} + 5\mathbf{k}$, $\mathbf{w} = \mathbf{i} - 7\mathbf{j} + 3\mathbf{k}$, and $\mathbf{t} = 3\mathbf{i} + 4\mathbf{j} + 5\mathbf{k}$.

21. $\mathbf{u} + \mathbf{v} = (2\mathbf{i} - 3\mathbf{j} + 4\mathbf{k}) + (-2\mathbf{i} - 3\mathbf{j} + 5\mathbf{k}) = (2 - 2)\mathbf{i} + (-3 - 3)\mathbf{j} + (4 + 5)\mathbf{k} = -6\mathbf{j} + 9\mathbf{k}$.

23. $-18\mathbf{u} = -18(2\mathbf{i} - 3\mathbf{j} + 4\mathbf{k}) = -36\mathbf{i} + 54\mathbf{j} - 72\mathbf{k}$.

25. $\mathbf{t} + 3\mathbf{w} - \mathbf{v} = (3\mathbf{i} + 4\mathbf{j} + 5\mathbf{k}) + 3(\mathbf{i} - 7\mathbf{j} + 3\mathbf{k}) - (-2\mathbf{i} - 3\mathbf{j} + 5\mathbf{k}) = (3 + 3 + 2)\mathbf{i} + (4 - 21 + 3)\mathbf{j} + (5 + 9 - 5)\mathbf{k} = 8\mathbf{i} - 14\mathbf{j} + 9\mathbf{k}$.

27. $2\mathbf{v} + 7\mathbf{t} - \mathbf{w} = 2(-2\mathbf{i} - 3\mathbf{j} + 5\mathbf{k}) + 7(3\mathbf{i} + 4\mathbf{j} + 5\mathbf{k}) - (\mathbf{i} - 7\mathbf{j} + 3\mathbf{k}) = (-4 + 21 - 1)\mathbf{i} + (-6 + 28 + 7)\mathbf{j} + (10 + 35 - 3)\mathbf{k} = 16\mathbf{i} + 29\mathbf{j} + 42\mathbf{k}$.

29. $|\mathbf{w}| = |\mathbf{i} - 7\mathbf{j} + 3\mathbf{k}| = \sqrt{1^2 + (-7)^2 + 3^2} = \sqrt{1 + 49 + 9} = \sqrt{59}$.

31. $\quad \cos\theta = (\mathbf{u}\cdot\mathbf{w})/(|\mathbf{u}||\mathbf{w}|) = ((2)(1) + (-3)(-7) + (4)(3)) \,/\, (\sqrt{2^2 + 3^2 + 4^2}\sqrt{1^2 + 7^2 + 3^2}) = 35/\,(\sqrt{29}\sqrt{59})$
≈ 0.8461; therefore $\theta \approx \cos^{-1} 0.8461 \approx 0.5621$.

33. $\quad \cos\theta = (\mathbf{v}\cdot\mathbf{t})/(|\mathbf{v}||\mathbf{t}|) = ((-2)(3) + (-3)(4) + (5)(5)) \,/\, (\sqrt{2^2 + 3^2 + 5^2}\sqrt{3^2 + 4^2 + 5^2}) = 7/\,(\sqrt{38}\sqrt{50}) =$
$7/(10\sqrt{19}) \approx 0.1606$; therefore $\theta = \cos^{-1}\left(7/(10\sqrt{19})\right) \approx 1.4095$.

35. $\quad \dfrac{\mathbf{u}\cdot\mathbf{v}}{|\mathbf{v}|^2} = \dfrac{(2\mathbf{i} - 3\mathbf{j} + 4\mathbf{k})\cdot(-2\mathbf{i} - 3\mathbf{j} + 5\mathbf{k})}{|-2\mathbf{i} - 3\mathbf{j} + 5\mathbf{k}|^2} = \dfrac{(2)(-2) + (-3)(-3) + (4)(5)}{(-2)^2 + (-3)^2 + 5^2} = \dfrac{-4 + 9 + 20}{4 + 9 + 25} = \dfrac{25}{38};$

therefore $\text{Proj}_\mathbf{v}\mathbf{u} = \dfrac{25}{38}\mathbf{v} = \dfrac{25}{38}(-2\mathbf{i} - 3\mathbf{j} + 5\mathbf{k}) = \dfrac{-25}{19}\mathbf{i} + \dfrac{-75}{38}\mathbf{j} + \dfrac{125}{38}\mathbf{k}.$

37. $\quad \dfrac{\mathbf{w}\cdot\mathbf{t}}{|\mathbf{t}|^2} = \dfrac{(\mathbf{i} - 7\mathbf{j} + 3\mathbf{k})\cdot(3\mathbf{i} + 4\mathbf{j} + 5\mathbf{k})}{|3\mathbf{i} + 4\mathbf{j} + 5\mathbf{k}|^2} = \dfrac{(1)(3) + (-7)(4) + (3)(5)}{3^2 + 4^2 + 5^2} = \dfrac{3 - 28 + 15}{9 + 16 + 25} = \dfrac{-10}{50} = \dfrac{-1}{5};$

therefore $\text{Proj}_\mathbf{t}\mathbf{w} = \dfrac{-1}{5}\mathbf{t} = \dfrac{-1}{5}(3\mathbf{i} + 4\mathbf{j} + 5\mathbf{k}) = -\dfrac{3}{5}\mathbf{i} - \dfrac{4}{5}\mathbf{j} - \mathbf{k}.$

39. $\quad \dfrac{\mathbf{u}\cdot\mathbf{w}}{|\mathbf{w}|^2} = \dfrac{(2\mathbf{i} - 3\mathbf{j} + 4\mathbf{k})\cdot(\mathbf{i} - 7\mathbf{j} + 3\mathbf{k})}{|\mathbf{i} - 7\mathbf{j} + 3\mathbf{k}|^2} = \dfrac{(2)(1) + (-3)(-7) + (4)(3)}{1^2 + (-7)^2 + 3^2} = \dfrac{2 + 21 + 12}{1 + 49 + 9} = \dfrac{35}{59};$

therefore $\text{Proj}_\mathbf{w}\mathbf{u} = \dfrac{35}{59}\mathbf{w} = \dfrac{35}{59}(\mathbf{i} - 7\mathbf{j} + 3\mathbf{k}) = \dfrac{35}{59}\mathbf{i} - \dfrac{245}{59}\mathbf{j} + \dfrac{105}{59}\mathbf{k}.$

41. $\quad \mathbf{v} = a\mathbf{i} + a\mathbf{j} + a\mathbf{k}$ and $|\mathbf{v}| = 1$ implies $1 = |\mathbf{v}|^2 = a^2 + a^2 + a^2 = 3a^2$ which then implies $a^2 = 1/3$. If the angle(s) are between 0 and $\pi/2$, then $a = 1/\sqrt{3}$ and $\mathbf{v} = (1/\sqrt{3})\,(\mathbf{i} + \mathbf{j} + \mathbf{k})$.

43. $\quad$ If $P = (2, 1, 4)$ and $Q = (3, -2, 8)$, then $\overrightarrow{PQ} = (3 - 2)\mathbf{i} + (-2 - 1)\mathbf{j} + (8 - 4)\mathbf{k} = \mathbf{i} - 3\mathbf{j} + 4\mathbf{k}$ and $\overline{PQ} = \sqrt{1^2 + (-3)^2 + 4^2} = \sqrt{1 + 9 + 16} = \sqrt{26}$. Therefore, the unit vector in the direction of $\overrightarrow{PQ}$ is $\mathbf{u} = (1/\overline{PQ})\,\overrightarrow{PQ} = (1/\sqrt{26})\,\mathbf{i} - (3/\sqrt{26})\,\mathbf{j} + (4/\sqrt{26})\,\mathbf{k}.$

45. $\quad \cos^2\left(\dfrac{\pi}{6}\right) + \cos^2\left(\dfrac{\pi}{3}\right) + \cos^2\left(\dfrac{\pi}{4}\right) = \left(\dfrac{\sqrt{3}}{2}\right)^2 + \left(\dfrac{1}{2}\right)^2 + \left(\dfrac{1}{\sqrt{2}}\right)^2 = \dfrac{3}{4} + \dfrac{1}{4} + \dfrac{1}{2} = \dfrac{3}{2} \neq 1;$

this is incompatible with equation (3), hence there is no vector with direction angles $\pi/6, \pi/3, \pi/4$. ∎

47. $\quad$ Let $P = (-3, 1, 7)$, $Q = (8, 1, 7)$, and $R = (x, y, z)$.

$\overrightarrow{PR} \perp \overrightarrow{PQ} \iff 0 = \overrightarrow{PR}\cdot\overrightarrow{PQ}$
$\iff 0 = ((x - (-3))\mathbf{i} + (y - 1)\mathbf{j} + (z - 7)\mathbf{k})\cdot((8 - (-3))\mathbf{i} + (1 - 1)\mathbf{j} + (7 - 7)\mathbf{k}) = (x + 3)\,11$
$\iff -3 = x.$

49. $\quad$ Suppose $P = (2, 1, 3)$, $Q = (-1, 1, 2)$, $R = (6, 0, 1)$. Let $\mathbf{u} = \overrightarrow{PQ} = -3\mathbf{i} - \mathbf{k}$ and $\mathbf{v} = \overrightarrow{QR} = 7\mathbf{i} - \mathbf{j} - \mathbf{k}$. The projection of $\mathbf{u}$ on $\mathbf{v}$ is

$$\text{Proj}_\mathbf{v}\mathbf{u} = \left(\dfrac{\mathbf{u}\cdot\mathbf{v}}{|\mathbf{v}|^2}\right)\mathbf{v} = \left(\dfrac{-21 + 1}{49 + 1 + 1}\right)(7\mathbf{i} - \mathbf{j} - \mathbf{k}) = \left(\dfrac{-20}{51}\right)(7\mathbf{i} - \mathbf{j} - \mathbf{k}).$$

The distance from point P to the line on Q and R is the length of

$$\mathbf{w} = \mathbf{u} - \text{Proj}_\mathbf{v}\mathbf{u} = (-3\mathbf{i} - \mathbf{k}) - \left(\dfrac{-20}{51}\right)(7\mathbf{i} - \mathbf{j} - \mathbf{k}) = \left(\dfrac{1}{51}\right)(-13\mathbf{i} - 20\mathbf{j} - 71\mathbf{k}).$$

We finish by computing $|\mathbf{w}| = (1/51)\sqrt{13^2 + 20^2 + 71^2} = (1/51)\sqrt{5610} = \sqrt{110/51}.$

REMARK: $\quad |\mathbf{w}|^2 = |\mathbf{u}|^2 - 2\left(\dfrac{\mathbf{u}\cdot\mathbf{v}}{|\mathbf{v}|^2}\right)(\mathbf{u}\cdot\mathbf{v}) + \left(\dfrac{\mathbf{u}\cdot\mathbf{v}}{|\mathbf{v}|^2}\right)^2|\mathbf{v}|^2 = |\mathbf{u}|^2 - \dfrac{(\mathbf{u}\cdot\mathbf{v})^2}{|\mathbf{v}|^2} = |\mathbf{u}|^2 - |\text{Proj}_\mathbf{v}\mathbf{u}|^2 = 10 - \dfrac{20^2}{51}.$
(This equation also follows directly from the Pythagorean Theorem: $\mathbf{w} \perp \mathbf{v}$ implies $\mathbf{w} \perp \text{Proj}_\mathbf{v}\mathbf{u}$, thus $|\mathbf{u}|^2 = |\mathbf{w} + \text{Proj}_\mathbf{v}\mathbf{u}|^2 = |\mathbf{w}|^2 + |\text{Proj}_\mathbf{v}\mathbf{u}|^2.$)

51. $\quad$ Suppose $P = (3, 5, 6)$, $Q = (1, 2, 7)$, $R = (6, 1, 0)$; then let $\overrightarrow{PQ} = -2\mathbf{i} - 3\mathbf{j} + \mathbf{k}$, $\overrightarrow{QR} = 5\mathbf{i} - \mathbf{j} - 7\mathbf{k}$, $\overrightarrow{RP} = -3\mathbf{i} + 4\mathbf{j} + 6\mathbf{k}$. One way to verify the triangle with vertices at P, Q, R is a right-triangle is to examine the various dot-products, we look for a zero value as an indication of an orthogonal pair of sides.

$\overrightarrow{PQ}\cdot\overrightarrow{QR} = (-2\mathbf{i} - 3\mathbf{j} + \mathbf{k})\cdot(5\mathbf{i} - \mathbf{j} - 7\mathbf{k}) = (-2)(5) + (-3)(-1) + (1)(-7) = -10 + 3 - 7 = -14,$

$\overrightarrow{QR}\cdot\overrightarrow{RP} = (5\mathbf{i} - \mathbf{j} - 7\mathbf{k})\cdot(-3\mathbf{i} + 4\mathbf{j} + 6\mathbf{k}) = (5)(-3) + (-1)(4) + (-7)(6) = -15 - 4 - 42 = -61,$

$\overrightarrow{RP}\cdot\overrightarrow{PQ} = (-3\mathbf{i} + 4\mathbf{j} + 6\mathbf{k})\cdot(-2\mathbf{i} - 3\mathbf{j} + \mathbf{k}) = (-3)(-2) + (4)(-3) + (6)(1) = 6 - 12 + 6 = 0.$

Therefore $\overrightarrow{RP} \perp \overrightarrow{PQ}$ and the triangle has a right angle at vertex P. ∎

REMARK: An alternate tactic is to compute the squared-lengths of the vectors and look for a Pythagorean arrangement.

$$\overline{PQ}^2 = (-2)^2 + (-3)^2 + 1^2 = 4 + 9 + 1 = 14,$$
$$\overline{QR}^2 = 5^2 + (-1)^2 + (-7)^2 = 25 + 1 + 49 = 75,$$
$$\overline{RP}^2 = (-3)^2 + 4^2 + 6^2 = 9 + 16 + 36 = 61.$$

We finish by noting that $\overline{QR}^2 = 75 = 61 + 14 = \overline{RP}^2 + \overline{PQ}^2$.

53. Let $\mathbf{u} = \mathbf{i} + 2\mathbf{j} + 3\mathbf{k}$, $\mathbf{v} = -4\mathbf{i} + \mathbf{j} + 5\mathbf{k}$, and $\mathbf{w} = x\mathbf{i} + y\mathbf{j} + z\mathbf{k}$. The requirement that $\mathbf{w} \perp \mathbf{u}$ and $\mathbf{w} \perp \mathbf{v}$ leads to a system of two equations in three unknowns — algebraic manipulations produce a system with two of those unknowns expressed in terms of the third, selecting a value for that third variable then fixes the other two, division by $|\mathbf{w}|$ produces one unit vector and multiplication of that vector by -1 yields another unit vector.

$$\begin{array}{l}\mathbf{w} \perp \mathbf{u} \\ \mathbf{w} \perp \mathbf{v}\end{array} \iff \begin{array}{l} 0 = \mathbf{u} \cdot \mathbf{w} = x + 2y + 3z \\ 0 = \mathbf{v} \cdot \mathbf{w} = -4x + y + 5z \end{array} \iff \begin{array}{l} x = -2y - 3z \\ y = 4x - 5z \end{array} \iff \begin{array}{l} x = -2(4x - 5z) - 3z = -8x + 7z \\ y = 4(-2y - 3z) - 5z = -8y - 17z \end{array}$$
$$\iff \begin{array}{l} 9x = 7z \\ 9y = -17z \end{array} \iff \begin{array}{l} x = (7/9)z \\ y = (-17/9)z \end{array}$$

Choose $z = 9$ to get $\mathbf{w} = 7\mathbf{i} - 17\mathbf{j} + 9\mathbf{k}$ which is orthogonal to both $\mathbf{u} = \mathbf{i} + 2\mathbf{j} + 3\mathbf{k}$ and $\mathbf{v} = -4\mathbf{i} + \mathbf{j} + 5\mathbf{k}$. The length of $\mathbf{w}$ is $\sqrt{7^2 + (-17)^2 + 9^2} = \sqrt{419}$; the unit vectors $\pm(1/\sqrt{419})(7\mathbf{i} - 17\mathbf{j} + 9\mathbf{k})$ are orthogonal to $\mathbf{u}$ and $\mathbf{v}$.

REMARK: Linear combinations, $a\mathbf{u} + b\mathbf{v}$, of $\mathbf{u}$ and $\mathbf{v}$ lie on a plane through the origin; there is a unique line which passes through the origin and is orthogonal to that plane. We found the two unit vectors lying in that line.

55. a. Let $\mathbf{u} = u_1\mathbf{i} + u_2\mathbf{j} + u_3\mathbf{k}$ and $\mathbf{v} = v_1\mathbf{i} + v_2\mathbf{j} + v_3\mathbf{k}$. If either of these vectors is $\mathbf{0}$, then $\sum_{i=1}^{3} u_i v_i = \mathbf{u} \cdot \mathbf{v} = 0$ and $\left(\sum_{i=1}^{3} u_i^3\right)\left(\sum_{i=1}^{3} v_i^3\right) = |\mathbf{u}|^2|\mathbf{v}|^2 = 0$. If both vectors are nonzero, then

$$1 \geq \cos^2\theta = \left(\frac{\mathbf{u} \cdot \mathbf{v}}{|\mathbf{u}|\,|\mathbf{v}|}\right)^2 = \frac{(\mathbf{u} \cdot \mathbf{v})^2}{|\mathbf{u}|^2\,|\mathbf{v}|^2} = \frac{\left(\sum_{i=1}^{3} u_i v_i\right)^2}{\left(\sum_{i=1}^{3} u_i^3\right)\left(\sum_{i=1}^{3} v_i^3\right)} \iff \left(\sum_{i=1}^{3} u_i^3\right)\left(\sum_{i=1}^{3} v_i^3\right) \geq \left(\sum_{i=1}^{3} u_i v_i\right)^2. \text{∎}$$

b. Equality holds if and only if $\cos\theta = 1$, i.e., the vectors have the same or opposite directions. ∎

REMARK: See the Remark following Solution 11.2.75 for an elegant proof that works here (and in $\mathbb{R}^n$).

57. Let $P = (a_1, b_1, c_1)$ and $Q = (a_2, b_2, c_2)$. For any choice of points R and S, we have $\overrightarrow{PQ} = \overrightarrow{PR} + \overrightarrow{RS} + \overrightarrow{SQ}$; our result will follow by making nice choices which give simple forms for the three vectors on the right.

Let $R = (a_2, b_1, c_1)$. If $a_2 \geq a_1$, then $\overrightarrow{PR}$ has length $a_2 - a_1$ and has the same direction as $\mathbf{i}$; that implies $\overrightarrow{PR} = (a_2 - a_1)\mathbf{i}$. If $a_2 < a_1$, then $\overrightarrow{PR}$ has length $a_1 - a_2$ and has direction opposite to $\mathbf{i}$; that implies $\overrightarrow{PR} = -(a_1 - a_2)\mathbf{i} = (a_2 - a_1)\mathbf{i}$. In either case we see that $\overrightarrow{PR} = (a_2 - a_1)\mathbf{i}$.

Let $S = (a_2, b_2, c_1)$. Reasoning as above we see that $\overrightarrow{RS} = (b_2 - b_1)\mathbf{j}$ and $\overrightarrow{SQ} = (c_2 - c_1)\mathbf{k}$.

Therefore $\overrightarrow{PQ} = \overrightarrow{PR} + \overrightarrow{RS} + \overrightarrow{SQ} = (a_2 - a_1)\mathbf{i} + (b_2 - b_1)\mathbf{j} + (c_2 - c_1)\mathbf{k}$. ∎

REMARK: There are 5 easy variants of this proof; these 6 paths from P to Q go along edges of a rectangular solid with faces parallel to the coordinate planes.

59. Three noncollinear points in space determine a unique plane which passes through them; therefore metric results (involving measurements of distance or angle) from plane geometry will apply to a triangle or the vectors lying along the sides of a triangle.

The law of cosines (Problem A1.4.11) implies $|\mathbf{v} - \mathbf{u}|^2 = |\mathbf{u}|^2 + |\mathbf{v}|^2 - 2|\mathbf{u}|\,|\mathbf{v}|\cos\theta$. On the other hand, $|\mathbf{v} - \mathbf{u}|^2 = (\mathbf{v} - \mathbf{u}) \cdot (\mathbf{v} - \mathbf{u}) = (\mathbf{u} \cdot \mathbf{u}) - 2(\mathbf{u} \cdot \mathbf{v}) + (\mathbf{v} \cdot \mathbf{v}) = |\mathbf{u}|^2 - 2(\mathbf{u} \cdot \mathbf{v}) + |\mathbf{v}|^2$. Therefore $|\mathbf{u}|\,|\mathbf{v}|\cos\theta = \mathbf{u} \cdot \mathbf{v}$ which then implies $\cos\theta = (\mathbf{u} \cdot \mathbf{v})/(|\mathbf{u}||\mathbf{v}|)$. ∎

Solutions 11.5   *Lines in* $\mathbb{R}^3$ (pages 755-6)

1.   If $P = (2, 1, 3)$ and $Q = (1, 2, -1)$, then $\mathbf{v} = \vec{PQ} = \mathbf{q} - \mathbf{p} = (\mathbf{i} + 2\mathbf{j} - \mathbf{k}) - (2\mathbf{i} + \mathbf{j} + 3\mathbf{k}) = -\mathbf{i} + \mathbf{j} - 4\mathbf{k}$. The line passing through P and Q has

        vector equation : $x\mathbf{i} + y\mathbf{j} + z\mathbf{k} = \mathbf{p} + t\mathbf{v} = (2\mathbf{i} + \mathbf{j} + 3\mathbf{k}) + t(-\mathbf{i} + \mathbf{j} - 4\mathbf{k})$;

        parametric equations : $x = 2 - t, \quad y = 1 + t, \quad z = 3 - 4t$;

        symmetric equations : $\dfrac{x - 2}{-1} = \dfrac{y - 1}{1} = \dfrac{z - 3}{-4}$.

3.   If $P = (1, 3, 2)$ and $Q = (2, 4, -2)$, then $\mathbf{v} = \vec{PQ} = \mathbf{q} - \mathbf{p} = (2\mathbf{i} + 4\mathbf{j} - 2\mathbf{k}) - (\mathbf{i} + 3\mathbf{j} + 2\mathbf{k}) = \mathbf{i} + \mathbf{j} - 4\mathbf{k}$. The line passing through P and Q has

        vector equation : $x\mathbf{i} + y\mathbf{j} + z\mathbf{k} = \mathbf{p} + t\mathbf{v} = (\mathbf{i} + 3\mathbf{j} + 2\mathbf{k}) + t(\mathbf{i} + \mathbf{j} - 4\mathbf{k})$;

        parametric equations : $x = 1 + t, \quad y = 3 + t, \quad z = 2 - 4t$;

        symmetric equations : $\dfrac{x - 1}{1} = \dfrac{y - 3}{1} = \dfrac{z - 2}{-4}$.

5.   If $P = (-4, 1, 3)$ and $Q = (-4, 0, 1)$, then $\mathbf{v} = \vec{PQ} = \mathbf{q} - \mathbf{p} = (-4\mathbf{i} + \mathbf{k}) - (-4\mathbf{i} + \mathbf{j} + 3\mathbf{k}) = -\mathbf{j} - 2\mathbf{k}$. The line passing through P and Q has

        vector equation : $x\mathbf{i} + y\mathbf{j} + z\mathbf{k} = \mathbf{p} + t\mathbf{v} = (-4\mathbf{i} + \mathbf{j} + 3\mathbf{k}) + t(-\mathbf{j} - 2\mathbf{k})$;

        parametric equations : $x = -4 + 0t = -4, \quad y = 1 - t, \quad z = 3 - 2t$;

        symmetric equations : $x = -4, \quad \dfrac{y - 1}{-1} = \dfrac{z - 3}{-2}$.

7.   If $P = (1, 2, 3)$ and $Q = (3, 2, 1)$, then $\mathbf{v} = \vec{PQ} = \mathbf{q} - \mathbf{p} = (3\mathbf{i} + 2\mathbf{j} + \mathbf{k}) - (\mathbf{i} + 2\mathbf{j} + 3\mathbf{k}) = 2\mathbf{i} - 2\mathbf{k}$. The line passing through P and Q has

        vector equation : $x\mathbf{i} + y\mathbf{j} + z\mathbf{k} = \mathbf{p} + t\mathbf{v} = (\mathbf{i} + 2\mathbf{j} + 3\mathbf{k}) + t(2\mathbf{i} - 2\mathbf{k})$;

        parametric equations : $x = 1 + 2t, \quad y = 2 + 0t = 2, \quad z = 3 - 2t$;

        symmetric equations : $y = 2, \quad \dfrac{x - 1}{2} = \dfrac{z - 3}{-2}$.

9.   If $P = (1, 2, 4)$ and $Q = (1, 2, 7)$, then $\mathbf{v} = \vec{PQ} = \mathbf{q} - \mathbf{p} = (\mathbf{i} + 2\mathbf{j} + 7\mathbf{k}) - (\mathbf{i} + 2\mathbf{j} + 4\mathbf{k}) = 3\mathbf{k}$. The line passing through P and Q has

        vector equation : $x\mathbf{i} + y\mathbf{j} + z\mathbf{k} = \mathbf{p} + t\mathbf{v} = (\mathbf{i} + 2\mathbf{j} + 4\mathbf{k}) + t(3\mathbf{k})$;

        parametric equations : $x = 1 + 0t = 1, \quad y = 2 + 0t = 2, \quad z = 4 + 3t$;

        symmetric equations : $x = 1, \quad y = 2$.

11.   The line passing through $P = (2, 2, 1)$ and parallel to $\mathbf{v} = 2\mathbf{i} - \mathbf{j} - \mathbf{k}$ has

        vector equation : $x\mathbf{i} + y\mathbf{j} + z\mathbf{k} = \mathbf{p} + t\mathbf{v} = (2\mathbf{i} + 2\mathbf{j} + \mathbf{k}) + t(2\mathbf{i} - \mathbf{j} - \mathbf{k})$;

        parametric equations : $x = 2 + 2t, \quad y = 2 - t, \quad z = 1 - t$;

        symmetric equations : $\dfrac{x - 2}{2} = \dfrac{y - 2}{-1} = \dfrac{z - 1}{-1}$.

13.   The line passing through $P = (1, 0, 3)$ and parallel to $\mathbf{v} = \mathbf{i} - \mathbf{j}$ has

        vector equation : $x\mathbf{i} + y\mathbf{j} + z\mathbf{k} = \mathbf{p} + t\mathbf{v} = (\mathbf{i} + 3\mathbf{k}) + t(\mathbf{i} - \mathbf{j})$;

        parametric equations : $x = 1 + t, \quad y = 0 - t = -t, \quad z = 3 + 0t = 3$;

        symmetric equations : $\dfrac{x - 1}{1} = \dfrac{y - 0}{-1}, \quad z = 3$.

15.   The line passing through $P = (-1, -2, 5)$ and parallel to $\mathbf{v} = -3\mathbf{j} + 4\mathbf{k}$ has

        vector equation : $x\mathbf{i} + y\mathbf{j} + z\mathbf{k} = \mathbf{p} + t\mathbf{v} = (-\mathbf{i} - 2\mathbf{j} + 5\mathbf{k}) + t(-3\mathbf{j} + 4\mathbf{k})$;

        parametric equations : $x = -1 + 0t = -1, \quad y = -2 - 3t, \quad z = 5 + 4t$;

        symmetric equations : $x = -1, \quad \dfrac{y - (-2)}{-3} = \dfrac{z - 5}{4}$.

17. The line passing through $P = (-1, -3, 1)$ and parallel to $\mathbf{v} = -7\mathbf{j}$ has

vector equation : $\quad x\mathbf{i} + y\mathbf{j} + z\mathbf{k} = \mathbf{p} + t\mathbf{v} = (-\mathbf{i} - 3\mathbf{j} + \mathbf{k}) + t(-7\mathbf{j})$;

parametric equations : $\quad x = -1 + 0t = -1, \quad y = -3 - 7t, \quad z = 1 + 0t = 1$;

symmetric equations : $\quad x = -1, \quad z = 1$.

19. The line passing through $P = (a, b, c)$ and parallel to $\mathbf{v} = d\mathbf{i} + e\mathbf{j}$ has

vector equation : $\quad x\mathbf{i} + y\mathbf{j} + z\mathbf{k} = \mathbf{p} + t\mathbf{v} = (a\mathbf{i} + b\mathbf{j} + c\mathbf{k}) + t(d\mathbf{i} + e\mathbf{j})$;

parametric equations : $\quad x = a + dt, \quad y = b + et, \quad z = c + 0t$;

symmetric equations : $\quad \dfrac{x - a}{d} = \dfrac{y - b}{e}, \quad z = c.$

21. $\mathbf{v} = 3\mathbf{i} + 6\mathbf{j} + 2\mathbf{k}$ is parallel to the line $\dfrac{x - 2}{3} = \dfrac{y + 1}{6} = \dfrac{z - 5}{2}$ and to any line parallel to this specified line ($\mathbf{v} \parallel L_0$ and $L_0 \parallel L_1 \implies \mathbf{v} \parallel L_1$). The line passing through $P = (4, 1, -6)$ and parallel to $\mathbf{v}$ has

vector equation : $\quad x\mathbf{i} + y\mathbf{j} + z\mathbf{k} = \mathbf{p} + t\mathbf{v} = (4\mathbf{i} + \mathbf{j} - 6\mathbf{k}) + t(3\mathbf{i} + 6\mathbf{j} + 2\mathbf{k})$;

parametric equations : $\quad x = 4 + 3t, \quad y = 1 + 6t, \quad z = -6 + 2t$;

symmetric equations : $\quad \dfrac{x - 4}{3} = \dfrac{y - 1}{6} = \dfrac{z - (-6)}{2}.$

23. Suppose (a, b, c) is a point on lines

$$L_1(t): \quad x = 1 + t, \quad y = -3 + 2t, \quad z = -2 - t;$$
$$L_2(s): \quad x = 17 + 3s, \quad y = 4 + s, \quad z = -8 - s.$$

$$
\begin{aligned}
a &= \quad 1 + t = 17 + 3s \\
b &= -3 + 2t = \quad 4 + s \\
c &= \quad -2 - t = -8 - s
\end{aligned}
\iff
\begin{aligned}
-3s + t &= 16 \\
-s + 2t &= 7 \\
s - t &= -6
\end{aligned}
\iff
\begin{aligned}
-3s + t &= 16 \\
t &= 1 \\
s - t &= -6
\end{aligned}
\iff
\begin{aligned}
-3s + t &= 16 \\
t &= 1 \\
s &= -5
\end{aligned}
\iff
\begin{aligned}
t &= \quad 1 \\
s &= -5
\end{aligned}
$$

The lines meet at the point $(2, -1, -3) = (1 + 1, -3 + 2, -2 - 1) = (17 - 15, 4 - 5, -8 + 5)$.

25. Suppose (a, b, c) is a point on lines

$$L_1(t): \quad x = 2 + t, \quad y = -1 + 2t, \quad z = 3 + 4t;$$
$$L_2(s): \quad x = 9 + s, \quad y = -2 - s, \quad z = 1 - 2s.$$

$$
\begin{aligned}
a &= \quad 2 + t = \quad 9 + s \\
b &= -1 + 2t = -2 - s \\
c &= \quad 3 + 4t = 1 - 2s
\end{aligned}
\iff
\begin{aligned}
-s + t &= 7 \\
s + 2t &= -1 \\
2s + 4t &= -2
\end{aligned}
\iff
\begin{aligned}
-s + t &= 7 \\
s + 2t &= -1
\end{aligned}
\iff
\begin{aligned}
3t &= 6 \\
s + 2t &= -1
\end{aligned}
\iff
\begin{aligned}
t &= 2 \\
s + 2t &= -1
\end{aligned}
\iff
\begin{aligned}
t &= \quad 2 \\
s &= -5
\end{aligned}
$$

The lines meet at the point $(4, 3, 11) = (2 + 2, -1 + 4, 3 + 8) = (9 - 5, -2 + 5, 1 + 10)$.

27. Line $L_1 : \dfrac{x - 4}{-3} = \dfrac{y - 1}{7} = \dfrac{z + 2}{-8}$ has parametric equations $x = 4 - 3t, \; y = 1 + 7t, \; z = -2 - 8t$. Line $L_2 : \dfrac{x - 5}{1} = \dfrac{y - 3}{-1} = \dfrac{z - 1}{2}$ has parametric equations $x = 5 + s, \; y = 3 - s, \; z = 1 + 2s$. Suppose point (a, b, c) is on both lines.

$$
\begin{aligned}
a &= \quad 4 - 3t = 5 + s \\
b &= \quad 1 + 7t = 3 - s \\
c &= -2 - 8t = 1 + 2s
\end{aligned}
\iff
\begin{aligned}
-s - 3t &= 1 \\
s + 7t &= 2 \\
-2s - 8t &= 3
\end{aligned}
\iff
\begin{aligned}
4t &= 3 \\
s + 7t &= 2 \\
6t &= 7
\end{aligned}
\implies 3/4 = t = 7/6, \quad \text{impossible.}
$$

Assuming a common point leads to a contradiction; therefore the lines do not meet.

REMARK: The equations for a and b are solved by $t = 3/4$ and $s = -13/4$; the lines do not meet because $z_1(t = 3/4) = -2 - 6 = -8 \neq -11/2 = 1 - 26/4 = z_2(s = -13/4)$.

The equations for a and c are solved by $t = -1/2$ and $s = 1/2$; the lines do not meet because $y_1(t = -1/2) = 1 - 7/2 = -5/2 \neq 5/2 = 3 - 1/2 = y_2(s = 1/2)$.

The equations for b and c are solved by $t = 7/6$ and $s = -37/6$; the lines do not meet because $x_1(t = 7/6) = 4 - 21/6 = 1/2 \neq -7/6 = 5 - 37/6 = x_2(s = -37/6)$.

29. Suppose (a, b, c) is a point on lines $\quad \begin{aligned} L_1(t): \quad & x = 4 - t, \quad y = 7 + 5t, \quad z = 2 - 3t\ ; \\ L_2(s): \quad & x = 1 + 2s, \quad y = 6 - 2s, \quad z = 10 + 3s. \end{aligned}$

$$
\begin{aligned}
a &= 4 - t = 1 + 2s \\
b &= 7 + 5t = 6 - 2s
\end{aligned}
\iff
\begin{aligned}
-2s - t &= -3 \\
2s + 5t &= -1
\end{aligned}
\iff
\begin{aligned}
-2s - t &= -3 \\
4t &= -4
\end{aligned}
\iff
\begin{aligned}
-2s &= -4 \\
t &= -1
\end{aligned}
\iff
\begin{aligned}
s &= \quad 2 \\
t &= -1
\end{aligned}
$$

However, $z_1(t = -1) = 2 + 3 = 5 \neq 16 = 10 + 6 = z_2(s = 2)$; the lines do not meet.

REMARK: The equations for a and c are solved by $t = -25/3$ and $s = 17/3$; the lines do not meet because $y_1(t = -25/3) = 7 - 125/3 = -104/3 \neq -16/3 = 6 - 34/3 = y_2(s = 17/3)$.

The equations for b and c are solved by $t = 13/9$ and $s = -37/9$; the lines do not meet because $x_1(t = 13/9) = 4 - 13/9 = 23/9 \neq -64/9 = 1 - 74/9 = x_2(s = -37/9)$.

31. The geometric fact that two distinct points determine a unique line implies we can show $L_1: \dfrac{x-1}{1} = \dfrac{y+3}{2} = \dfrac{z+3}{3}$ and $L_2: \dfrac{x-3}{3} = \dfrac{y-1}{6} = \dfrac{z-3}{9}$ are equations for the same line by exhibiting a pair of points satisfying both sets of equations.

$(1, -3, -3)$ is obviously on L_1; it is also on L_2 because $\dfrac{1-3}{3} = \dfrac{-3-1}{6} = \dfrac{-3-3}{9} = \dfrac{-2}{3}$.

$(3, 1, 3)$ is obviously on L_2; it is also on L_1 because $\dfrac{3-1}{1} = \dfrac{1+3}{2} = \dfrac{3+3}{3} = 2$. ∎

33. Let $\mathbf{r} = \mathbf{r}(t) = \mathbf{p} + t\mathbf{v}$ where $\mathbf{v}$ is a nonzero vector.
$$\mathbf{r} \perp \mathbf{v} \iff 0 = \mathbf{r} \cdot \mathbf{v} = (\mathbf{p} + t\mathbf{v}) \cdot \mathbf{v} = (\mathbf{p} \cdot \mathbf{v}) + t(\mathbf{v} \cdot \mathbf{v}) \iff t = -(\mathbf{p} \cdot \mathbf{v})/|\mathbf{v}|^2.$$

REMARK: If $t = -(\mathbf{p} \cdot \mathbf{v})/|\mathbf{v}|^2$, then $\mathbf{r} = \mathbf{p} - \text{Proj}_{\mathbf{v}}\mathbf{p}$. (See Theorem 11.4.5 and equation 11.4.9.)

35. The point $P = (2, -3, 1)$ is on the line $L: \dfrac{x+2}{4} = \dfrac{y-1}{-4} = \dfrac{z+2}{3}$. Line L has direction $\mathbf{v} = 4\mathbf{i} - 4\mathbf{j} + 3\mathbf{k}$; any line of the form $(2\mathbf{i} - 3\mathbf{j} + \mathbf{k}) + t\mathbf{w}$ where $\mathbf{v} \cdot \mathbf{w} = 0$ will pass through P and be perpendicular to L. Two easy choices of vectors orthogonal to v are $\mathbf{w}_1 = \mathbf{i} + \mathbf{j}$ and $\mathbf{w}_2 = 3\mathbf{j} + 4\mathbf{k}$; any linear combination $a\mathbf{w}_1 + b\mathbf{w}_2 = a\mathbf{i} + (a + 3b)\mathbf{j} + 4b\mathbf{k}$ also works.

REMARK: Solutions for this problem lie on a plane. If the point P did not lie on line L, then there would be only one line through P and perpendicular to L.

37. Line $L_1: \dfrac{x-3}{2} = \dfrac{y+1}{4} = \dfrac{z-2}{-1}$ has direction $\dfrac{\mathbf{u}}{|\mathbf{u}|}$ where $\mathbf{u} = 2\mathbf{i} + 4\mathbf{j} - \mathbf{k}$. Line $L_2: \dfrac{x-3}{5} = \dfrac{y+1}{-2} = \dfrac{z-3}{2}$ has direction $\dfrac{\mathbf{v}}{|\mathbf{v}|}$ where $\mathbf{v} = 5\mathbf{i} - 2\mathbf{j} + 2\mathbf{k}$. We can easily confirm that $\mathbf{u}$ and $\mathbf{v}$ are orthogonal because $\mathbf{u} \cdot \mathbf{v} = (2\mathbf{i} + 4\mathbf{j} - \mathbf{k}) \cdot (5\mathbf{i} - 2\mathbf{j} + 2\mathbf{k}) = 10 - 8 - 2 = 0$; this implies $a\mathbf{u} \perp b\mathbf{v}$ for any choices of a and b. ∎

REMARK: The lines do not meet, hence it is not correct to say the lines are orthogonal.

39. a. The line through $(x_1, y_1, 0)$ and $(x_2, y_2, 0)$ has vector equation
$$x\mathbf{i} + y\mathbf{j} + z\mathbf{k} = (x_1\mathbf{i} + y_1\mathbf{j}) + t\left((x_2 - x_1)\mathbf{i} + (y_2 - y_1)\mathbf{j}\right)$$
and symmetric equations $\dfrac{x - x_1}{x_2 - x_1} = \dfrac{y - y_1}{y_2 - y_1}$, $z = 0$.

b. $\dfrac{x - x_1}{x_2 - x_1} = \dfrac{y - y_1}{y_2 - y_1} \iff y - y_1 = (y_2 - y_1)\dfrac{x - x_1}{x_2 - x_1} \iff y = \left(\dfrac{y_2 - y_1}{x_2 - x_1}\right)x + \left(\dfrac{x_2 y_1 - x_1 y_2}{x_2 - x_1}\right)$. ∎

Solutions 11.6   *The cross product of two vectors*   (pages 765-6)

1. $(\mathbf{i} - 2\mathbf{j}) \times 3\mathbf{k} = \begin{vmatrix} \mathbf{i} & \mathbf{j} & \mathbf{k} \\ 1 & -2 & 0 \\ 0 & 0 & 3 \end{vmatrix} = \begin{vmatrix} -2 & 0 \\ 0 & 3 \end{vmatrix}\mathbf{i} - \begin{vmatrix} 1 & 0 \\ 0 & 3 \end{vmatrix}\mathbf{j} + \begin{vmatrix} 1 & -2 \\ 0 & 0 \end{vmatrix}\mathbf{k} = -6\mathbf{i} - 3\mathbf{j}.$

3. $(\mathbf{i} - \mathbf{j}) \times (\mathbf{j} + \mathbf{k}) = \begin{vmatrix} \mathbf{i} & \mathbf{j} & \mathbf{k} \\ 1 & -1 & 0 \\ 0 & 1 & 1 \end{vmatrix} = \begin{vmatrix} -1 & 0 \\ 1 & 1 \end{vmatrix}\mathbf{i} - \begin{vmatrix} 1 & 0 \\ 0 & 1 \end{vmatrix}\mathbf{j} + \begin{vmatrix} 1 & -1 \\ 0 & 1 \end{vmatrix}\mathbf{k} = -\mathbf{i} - \mathbf{j} + \mathbf{k}.$

5. $(-2\mathbf{i} + 3\mathbf{j}) \times (7\mathbf{i} + 4\mathbf{k}) = \begin{vmatrix} \mathbf{i} & \mathbf{j} & \mathbf{k} \\ -2 & 3 & 0 \\ 7 & 0 & 4 \end{vmatrix} = \begin{vmatrix} 3 & 0 \\ 0 & 4 \end{vmatrix}\mathbf{i} - \begin{vmatrix} -2 & 0 \\ 7 & 4 \end{vmatrix}\mathbf{j} + \begin{vmatrix} -2 & 3 \\ 7 & 0 \end{vmatrix}\mathbf{k} = 12\mathbf{i} + 8\mathbf{j} - 21\mathbf{k}.$

7. $(a\mathbf{i} + b\mathbf{k}) \times (c\mathbf{i} + d\mathbf{k}) = \begin{vmatrix} \mathbf{i} & \mathbf{j} & \mathbf{k} \\ a & 0 & b \\ c & 0 & d \end{vmatrix} = \begin{vmatrix} 0 & b \\ 0 & d \end{vmatrix}\mathbf{i} - \begin{vmatrix} a & b \\ c & d \end{vmatrix}\mathbf{j} + \begin{vmatrix} a & 0 \\ c & 0 \end{vmatrix}\mathbf{k} = (bc - ad)\mathbf{j}.$

9. $(2\mathbf{i} - 3\mathbf{j} + \mathbf{k}) \times (\mathbf{i} + 2\mathbf{j} + \mathbf{k}) = \begin{vmatrix} \mathbf{i} & \mathbf{j} & \mathbf{k} \\ 2 & -3 & 1 \\ 1 & 2 & 1 \end{vmatrix} = \begin{vmatrix} -3 & 1 \\ 2 & 1 \end{vmatrix}\mathbf{i} - \begin{vmatrix} 2 & 1 \\ 1 & 1 \end{vmatrix}\mathbf{j} + \begin{vmatrix} 2 & -3 \\ 1 & 2 \end{vmatrix}\mathbf{k} = -5\mathbf{i} - \mathbf{j} + 7\mathbf{k}.$

11. $(-3\mathbf{i} - 2\mathbf{j} + \mathbf{k}) \times (6\mathbf{i} + 4\mathbf{j} - 2\mathbf{k}) = \begin{vmatrix} \mathbf{i} & \mathbf{j} & \mathbf{k} \\ -3 & -2 & 1 \\ 6 & 4 & -2 \end{vmatrix} = \begin{vmatrix} -2 & 1 \\ 4 & -2 \end{vmatrix}\mathbf{i} - \begin{vmatrix} -3 & 1 \\ 6 & -2 \end{vmatrix}\mathbf{j} + \begin{vmatrix} -3 & -2 \\ 6 & 4 \end{vmatrix}\mathbf{k} = 0.$

REMARK: $(6\mathbf{i} + 4\mathbf{j} - 2\mathbf{k}) = -2(-3\mathbf{i} - 2\mathbf{j} + \mathbf{k})$, use part (vii) of Theorem 2.

13. $(\mathbf{i} - 7\mathbf{j} - 3\mathbf{k}) \times (-\mathbf{i} + 7\mathbf{j} - 3\mathbf{k}) = \begin{vmatrix} \mathbf{i} & \mathbf{j} & \mathbf{k} \\ 1 & -7 & -3 \\ -1 & 7 & -3 \end{vmatrix} = \begin{vmatrix} -7 & -3 \\ 7 & -3 \end{vmatrix}\mathbf{i} - \begin{vmatrix} 1 & -3 \\ -1 & -3 \end{vmatrix}\mathbf{j} + \begin{vmatrix} 1 & -7 \\ -1 & 7 \end{vmatrix}\mathbf{k} = 42\mathbf{i} + 6\mathbf{j}.$

REMARK: Let $\mathbf{w} = \mathbf{i} - 7\mathbf{j}$, we computed $(\mathbf{w} - 3\mathbf{k}) \times (-\mathbf{w} - 3\mathbf{k}) = 3\mathbf{k} \times \mathbf{w} - 3\mathbf{w} \times \mathbf{k} = 6\mathbf{k} \times \mathbf{w} = -6\mathbf{w} \times \mathbf{k}$.

15. $(10\mathbf{i} + 7\mathbf{j} - 3\mathbf{k}) \times (-3\mathbf{i} + 4\mathbf{j} - 3\mathbf{k}) = \begin{vmatrix} \mathbf{i} & \mathbf{j} & \mathbf{k} \\ 10 & 7 & -3 \\ -3 & 4 & -3 \end{vmatrix} = \begin{vmatrix} 7 & -3 \\ 4 & -3 \end{vmatrix}\mathbf{i} - \begin{vmatrix} 10 & -3 \\ -3 & -3 \end{vmatrix}\mathbf{j} + \begin{vmatrix} 10 & 7 \\ -3 & 4 \end{vmatrix}\mathbf{k} = -9\mathbf{i} +$

$39\mathbf{j} + 61\mathbf{k}.$

17. $(2\mathbf{i} - \mathbf{j} + \mathbf{k}) \times (4\mathbf{i} + 2\mathbf{j} + 2\mathbf{k}) = \begin{vmatrix} \mathbf{i} & \mathbf{j} & \mathbf{k} \\ 2 & -1 & 1 \\ 4 & 2 & 2 \end{vmatrix} = \begin{vmatrix} -1 & 1 \\ 2 & 2 \end{vmatrix}\mathbf{i} - \begin{vmatrix} 2 & 1 \\ 4 & 2 \end{vmatrix}\mathbf{j} + \begin{vmatrix} 2 & -1 \\ 4 & 2 \end{vmatrix}\mathbf{k} = -4\mathbf{i} + 8\mathbf{k}.$

19. $(a\mathbf{i} + a\mathbf{j} + a\mathbf{k}) \times (b\mathbf{i} + b\mathbf{j} + b\mathbf{k}) = \begin{vmatrix} \mathbf{i} & \mathbf{j} & \mathbf{k} \\ a & a & a \\ b & b & b \end{vmatrix} = \begin{vmatrix} a & a \\ b & b \end{vmatrix}\mathbf{i} - \begin{vmatrix} a & a \\ b & b \end{vmatrix}\mathbf{j} + \begin{vmatrix} a & a \\ b & b \end{vmatrix}\mathbf{k} = 0.$

REMARK: If $a \neq 0$ and $b \neq 0$, then $a\mathbf{i} + a\mathbf{j} + a\mathbf{k}$ and $b\mathbf{i} + b\mathbf{j} + b\mathbf{k}$ are parallel.

21. If $\mathbf{u} = 2\mathbf{i} - 3\mathbf{j}$ and $\mathbf{v} = 4\mathbf{j} + 3\mathbf{k}$, then let

$$\mathbf{w} = \mathbf{u} \times \mathbf{v} = \begin{vmatrix} \mathbf{i} & \mathbf{j} & \mathbf{k} \\ 2 & -3 & 0 \\ 0 & 4 & 3 \end{vmatrix} = \begin{vmatrix} -3 & 0 \\ 4 & 3 \end{vmatrix}\mathbf{i} - \begin{vmatrix} 2 & 0 \\ 0 & 3 \end{vmatrix}\mathbf{j} + \begin{vmatrix} 2 & -3 \\ 0 & 4 \end{vmatrix}\mathbf{k} = -9\mathbf{i} - 6\mathbf{j} + 8\mathbf{k}.$$

$\mathbf{w}$ is orthogonal to both $\mathbf{u}$ and $\mathbf{v}$; the unit vectors are $\pm\mathbf{w}/|\mathbf{w}| = \pm(-9\mathbf{i} - 6\mathbf{j} + 8\mathbf{k})/\sqrt{9^2 + 6^2 + 8^2} = \pm(-9\mathbf{i} - 6\mathbf{j} + 8\mathbf{k})/\sqrt{181}$.

23. If $\mathbf{u} = 2\mathbf{i} + \mathbf{j} - \mathbf{k}$ and $\mathbf{v} = -3\mathbf{i} - 2\mathbf{j} + 4\mathbf{k}$, then

$$\mathbf{u} \times \mathbf{v} = \begin{vmatrix} \mathbf{i} & \mathbf{j} & \mathbf{k} \\ 2 & 1 & -1 \\ -3 & -2 & 4 \end{vmatrix} = \begin{vmatrix} 1 & -1 \\ -2 & 4 \end{vmatrix}\mathbf{i} - \begin{vmatrix} 2 & -1 \\ -3 & 4 \end{vmatrix}\mathbf{j} + \begin{vmatrix} 2 & 1 \\ -3 & -2 \end{vmatrix}\mathbf{k} = 2\mathbf{i} - 5\mathbf{j} - \mathbf{k}.$$

Therefore, invoking Theorem 3, $\sin\phi = \dfrac{|\mathbf{u} \times \mathbf{v}|}{|\mathbf{u}|\,|\mathbf{v}|} = \dfrac{\sqrt{2^2 + 5^2 + 1^2}}{\sqrt{2^2 + 1^2 + 1^2}\,\sqrt{3^2 + 2^2 + 4^2}} = \dfrac{\sqrt{30}}{\sqrt{6}\,\sqrt{29}} = \sqrt{\dfrac{5}{29}}.$

25. Line $L_1 : \dfrac{x + 2}{-3} = \dfrac{y - 1}{4} = \dfrac{z}{-5}$ is parallel to $\mathbf{u} = -3\mathbf{i} + 4\mathbf{j} - 5\mathbf{k}$. Line $L_2 : \dfrac{x - 3}{7} = \dfrac{y + 2}{-2} = \dfrac{z - 8}{3}$ is parallel to $\mathbf{v} = 7\mathbf{i} - 2\mathbf{j} + 3\mathbf{k}$.

$$\mathbf{u} \times \mathbf{v} = \begin{vmatrix} \mathbf{i} & \mathbf{j} & \mathbf{k} \\ -3 & 4 & -5 \\ 7 & -2 & 3 \end{vmatrix} = \begin{vmatrix} 4 & -5 \\ -2 & 3 \end{vmatrix}\mathbf{i} - \begin{vmatrix} -3 & -5 \\ 7 & 3 \end{vmatrix}\mathbf{j} + \begin{vmatrix} -3 & 4 \\ 7 & -2 \end{vmatrix}\mathbf{k} = 2\mathbf{i} - 26\mathbf{j} - 22\mathbf{k}$$

The line $\dfrac{x - 1}{2} = \dfrac{y - (-3)}{-26} = \dfrac{z - 2}{-22}$ passes through the point $(1, -3, 2)$ and its direction is orthogonal to the directions of lines L_1 and L_2.

27. Line $L_1 : (3 - 2t, 4 + 3t, -7 + 5t)$ is parallel to $\mathbf{u} = -2\mathbf{i} + 3\mathbf{j} + 5\mathbf{k}$. Line $L_2 : (-2 + 4s, 3 - 2s, 3 + s)$ is parallel to $\mathbf{v} = 4\mathbf{i} - 2\mathbf{j} + \mathbf{k}$.

$$\mathbf{u} \times \mathbf{v} = \begin{vmatrix} \mathbf{i} & \mathbf{j} & \mathbf{k} \\ -2 & 3 & 5 \\ 4 & -2 & 1 \end{vmatrix} = \begin{vmatrix} 3 & 5 \\ -2 & 1 \end{vmatrix}\mathbf{i} - \begin{vmatrix} -2 & 5 \\ 4 & 1 \end{vmatrix}\mathbf{j} + \begin{vmatrix} -2 & 3 \\ 4 & -2 \end{vmatrix}\mathbf{k} = 13\mathbf{i} + 22\mathbf{j} - 8\mathbf{k}.$$

The line $(-2\mathbf{i} + 3\mathbf{j} + 4\mathbf{k}) + t(13\mathbf{i} + 22\mathbf{j} - 8\mathbf{k})$ passes through the point $(-2, 3, 4)$ and its direction is orthogonal to the directions of lines L_1 and L_2.

11.6 — The cross product of two vectors

NOTE for 29-34:   If P, Q, R are distinct and non-collinear points, there are three parallelograms with these points as vertices; the areas of these parallelograms are equal. (The facts $\mathbf{q} \times \mathbf{q} = \mathbf{r} \times \mathbf{r} = \mathbf{p} \times \mathbf{p} = 0$ have been used to simplify the results displayed below.)

$$\overrightarrow{PQ} \times \overrightarrow{QR} = (\mathbf{q} - \mathbf{p}) \times (\mathbf{r} - \mathbf{q}) = \mathbf{q} \times \mathbf{r} - \mathbf{p} \times \mathbf{r} + \mathbf{p} \times \mathbf{q}$$

$$\overrightarrow{QR} \times \overrightarrow{RP} = (\mathbf{r} - \mathbf{q}) \times (\mathbf{p} - \mathbf{r}) = \mathbf{r} \times \mathbf{p} - \mathbf{q} \times \mathbf{p} + \mathbf{q} \times \mathbf{r}$$

$$\overrightarrow{RP} \times \overrightarrow{PQ} = (\mathbf{p} - \mathbf{r}) \times (\mathbf{q} - \mathbf{p}) = \mathbf{p} \times \mathbf{q} - \mathbf{r} \times \mathbf{q} + \mathbf{r} \times \mathbf{p}$$

Each cross products of vectors along consecutive sides of a parallelogram equals $\mathbf{p} \times \mathbf{q} + \mathbf{q} \times \mathbf{r} + \mathbf{r} \times \mathbf{p}$. Therefore, we work Problems 29-34 (and 35-38) with arbitrary assignment of labels P, Q, R to the specified points.

Calculations such as

$$\overrightarrow{PQ} \times \overrightarrow{PR} = \overrightarrow{PQ} \times \left(\overrightarrow{PQ} + \overrightarrow{QR} \right) = \overrightarrow{PQ} \times \overrightarrow{PQ} + \overrightarrow{PQ} \times \overrightarrow{QR} = 0 + \overrightarrow{PQ} \times \overrightarrow{QR} = \overrightarrow{PQ} \times \overrightarrow{QR}$$

provide an alternate path to the same conclusion.

---

29.   Let $P = (1, -2, 3)$, $Q = (2, 0, 1)$, $R = (0, 4, 0)$. Then $\overrightarrow{PQ} = \mathbf{i} + 2\mathbf{j} - 2\mathbf{k}$, $\overrightarrow{QR} = -2\mathbf{i} + 4\mathbf{j} - \mathbf{k}$, and

$$\overrightarrow{PQ} \times \overrightarrow{QR} = \begin{vmatrix} \mathbf{i} & \mathbf{j} & \mathbf{k} \\ 1 & 2 & -2 \\ -2 & 4 & -1 \end{vmatrix} = \begin{vmatrix} 2 & -2 \\ 4 & -1 \end{vmatrix} \mathbf{i} - \begin{vmatrix} 1 & -2 \\ -2 & -1 \end{vmatrix} \mathbf{j} + \begin{vmatrix} 1 & 2 \\ -2 & 4 \end{vmatrix} \mathbf{k} = 6\mathbf{i} + 5\mathbf{j} + 8\mathbf{k}.$$

The area of the parallelogram is $|6\mathbf{i} + 5\mathbf{j} + 8\mathbf{k}| = \sqrt{6^2 + 5^2 + 8^2} = \sqrt{125} = 5\sqrt{5}$.

31.   Let $P = (-2, 1, 0)$, $Q = (1, 4, 2)$, $R = (-3, 1, 5)$. Then $\overrightarrow{PQ} = 3\mathbf{i} + 3\mathbf{j} + 2\mathbf{k}$, $\overrightarrow{QR} = -4\mathbf{i} - 3\mathbf{j} + 3\mathbf{k}$, and

$$\overrightarrow{PQ} \times \overrightarrow{QR} = \begin{vmatrix} \mathbf{i} & \mathbf{j} & \mathbf{k} \\ 3 & 3 & 2 \\ -4 & -3 & 3 \end{vmatrix} = \begin{vmatrix} 3 & 2 \\ -3 & 3 \end{vmatrix} \mathbf{i} - \begin{vmatrix} 3 & 2 \\ -4 & 3 \end{vmatrix} \mathbf{j} + \begin{vmatrix} 3 & 3 \\ -4 & -3 \end{vmatrix} \mathbf{k} = 15\mathbf{i} - 17\mathbf{j} + 3\mathbf{k}.$$

The area of the parallelogram is $|15\mathbf{i} - 17\mathbf{j} + 3\mathbf{k}| = \sqrt{15^2 + 17^2 + 3^2} = \sqrt{523}$.

33.   Let $P = (a, 0, 0)$, $Q = (0, b, 0)$, $R = (0, 0, c)$. Then $\overrightarrow{PQ} = -a\mathbf{i} + b\mathbf{j}$, $\overrightarrow{QR} = -b\mathbf{j} + c\mathbf{k}$, and

$$\overrightarrow{PQ} \times \overrightarrow{QR} = \begin{vmatrix} \mathbf{i} & \mathbf{j} & \mathbf{k} \\ -a & b & 0 \\ 0 & -b & c \end{vmatrix} = \begin{vmatrix} b & 0 \\ -b & c \end{vmatrix} \mathbf{i} - \begin{vmatrix} -a & 0 \\ 0 & c \end{vmatrix} \mathbf{j} + \begin{vmatrix} -a & b \\ 0 & -b \end{vmatrix} \mathbf{k} = bc\mathbf{i} + ac\mathbf{j} + ab\mathbf{k}.$$

The area of the parallelogram is $|bc\mathbf{i} + ac\mathbf{j} + ab\mathbf{k}| = \sqrt{(bc)^2 + (ac)^2 + (ab)^2}$.

35.   Let $P = (2, 1, -4)$, $Q = (1, 7, 2)$, $R = (3, -2, 3)$. The triangle with vertices P, Q, R has area equal to one-half of the area of a parallelogram with the same vertices. $\overrightarrow{PQ} = -\mathbf{i} + 6\mathbf{j} + 6\mathbf{k}$, $\overrightarrow{QR} = 2\mathbf{i} - 9\mathbf{j} + \mathbf{k}$,

$$\overrightarrow{PQ} \times \overrightarrow{QR} = \begin{vmatrix} \mathbf{i} & \mathbf{j} & \mathbf{k} \\ -1 & 6 & 6 \\ 2 & -9 & 1 \end{vmatrix} = \begin{vmatrix} 6 & 6 \\ -9 & 1 \end{vmatrix} \mathbf{i} - \begin{vmatrix} -1 & 6 \\ 2 & 1 \end{vmatrix} \mathbf{j} + \begin{vmatrix} -1 & 6 \\ 2 & -9 \end{vmatrix} \mathbf{k} = 60\mathbf{i} + 13\mathbf{j} - 3\mathbf{k}$$

and the area of the triangle is $|60\mathbf{i} + 13\mathbf{j} - 3\mathbf{k}|/2 = \sqrt{60^2 + 13^2 + 3^2}/2 = \sqrt{3778}/2$.

37.   Let $P = (1, 0, 0)$, $Q = (0, 1, 0)$, $R = (0, 0, 1)$. The triangle with vertices P, Q, R has area equal to one-half of the area of a parallelogram with the same vertices. $\overrightarrow{PQ} = -\mathbf{i} + \mathbf{j}$, $\overrightarrow{QR} = -\mathbf{j} + \mathbf{k}$,

$$\overrightarrow{PQ} \times \overrightarrow{QR} = \begin{vmatrix} \mathbf{i} & \mathbf{j} & \mathbf{k} \\ -1 & 1 & 0 \\ 0 & -1 & 1 \end{vmatrix} = \begin{vmatrix} 1 & 0 \\ -1 & 1 \end{vmatrix} \mathbf{i} - \begin{vmatrix} -1 & 0 \\ 0 & 1 \end{vmatrix} \mathbf{j} + \begin{vmatrix} -1 & 1 \\ 0 & -1 \end{vmatrix} \mathbf{k} = \mathbf{i} + \mathbf{j} + \mathbf{k}$$

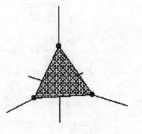

and the area of the triangle is $|\mathbf{i} + \mathbf{j} + \mathbf{k}|/2 = \sqrt{1^2 + 1^2 + 1^2}/2 = \sqrt{3}/2$.

39.   If $\mathbf{u} = 2\mathbf{i} - \mathbf{j} + \mathbf{k}$ and $\mathbf{v} = 3\mathbf{i} + 2\mathbf{j} - 2\mathbf{k}$, then

$$\mathbf{u} \times \mathbf{v} = \begin{vmatrix} \mathbf{i} & \mathbf{j} & \mathbf{k} \\ 2 & -1 & 1 \\ 3 & 2 & -2 \end{vmatrix} = \begin{vmatrix} -1 & 1 \\ 2 & -2 \end{vmatrix} \mathbf{i} - \begin{vmatrix} 2 & 1 \\ 3 & -2 \end{vmatrix} \mathbf{j} + \begin{vmatrix} 2 & -1 \\ 3 & 2 \end{vmatrix} \mathbf{k} = 7\mathbf{j} + 7\mathbf{k};$$

if $\mathbf{w} = 3\mathbf{i} + 2\mathbf{j}$, then $(\mathbf{u} \times \mathbf{v}) \cdot \mathbf{w} = (7\mathbf{j} + 7\mathbf{k}) \cdot (3\mathbf{i} + 2\mathbf{j}) = 0 + 14 + 0 = 14$ and the volume of the parallelepiped determined by $\mathbf{u}$, $\mathbf{v}$, $\mathbf{w}$ is $|(\mathbf{u} \times \mathbf{v}) \cdot \mathbf{w}| = |14| = 14$.

REMARK: Problems 51 and 47 imply the volume is the absolute value of

$$\begin{vmatrix} 2 & -1 & 1 \\ 3 & 2 & -2 \\ 3 & 2 & 0 \end{vmatrix} = (2)\begin{vmatrix} 2 & -2 \\ 2 & 0 \end{vmatrix} - (-1)\begin{vmatrix} 3 & -2 \\ 3 & 0 \end{vmatrix} + (1)\begin{vmatrix} 3 & 2 \\ 3 & 2 \end{vmatrix} = (2)(4) - (-1)(6) + (1)(0) = 8 + 6 + 0 = 14.$$

41.　If $\mathbf{u} = \mathbf{i} - \mathbf{j}$ and $\mathbf{v} = 3\mathbf{i} + 2\mathbf{k}$, then

$$\mathbf{u} \times \mathbf{v} = \begin{vmatrix} \mathbf{i} & \mathbf{j} & \mathbf{k} \\ 1 & -1 & 0 \\ 3 & 0 & 2 \end{vmatrix} = \begin{vmatrix} -1 & 0 \\ 0 & 2 \end{vmatrix}\mathbf{i} - \begin{vmatrix} 1 & 0 \\ 3 & 2 \end{vmatrix}\mathbf{j} + \begin{vmatrix} 1 & -1 \\ 3 & 0 \end{vmatrix}\mathbf{k} = -2\mathbf{i} - 2\mathbf{j} + 3\mathbf{k};$$

if $\mathbf{w} = -7\mathbf{j} + 3\mathbf{k}$, then $(\mathbf{u} \times \mathbf{v}) \cdot \mathbf{w} = (-2\mathbf{i} - 2\mathbf{j} + 3\mathbf{k}) \cdot (-7\mathbf{j} + 3\mathbf{k}) = 0 + 14 + 9 = 23$ and the volume of the parallelepiped determined by $\mathbf{u}$, $\mathbf{v}$, $\mathbf{w}$ is $|(\mathbf{u} \times \mathbf{v}) \cdot \mathbf{w}| = |23| = 23$.

REMARK: Problems 51 and 47 imply the volume is the absolute value of

$$\begin{vmatrix} 1 & -1 & 0 \\ 3 & 0 & 2 \\ 0 & -7 & 3 \end{vmatrix} = (1)\begin{vmatrix} 0 & 2 \\ -7 & 3 \end{vmatrix} - (-1)\begin{vmatrix} 3 & 2 \\ 0 & 3 \end{vmatrix} + (0)\begin{vmatrix} 3 & 0 \\ 0 & -7 \end{vmatrix} = (1)(14) - (-1)(9) + (0)(-21) = 14 + 9 + 0 = 23.$$

43.　Line $L_1 : \dfrac{x-2}{3} = \dfrac{y-5}{2} = \dfrac{z-1}{-1}$ is parallel to $3\mathbf{i} + 2\mathbf{j} - \mathbf{k}$ and passes through $P = (2, 5, 1)$; line $L_2 : \dfrac{x-4}{-4} = \dfrac{y-5}{4} = \dfrac{z+2}{1}$ is parallel to $-4\mathbf{i} + 4\mathbf{j} + \mathbf{k}$ and passes through $Q = (4, 5, -2)$. Then $\overrightarrow{PQ} = 2\mathbf{i} - 3\mathbf{k}$ and the distance d between the lines is the length of this vector projected onto a vector perpendicular to both lines:

$$\mathbf{v} = (3\mathbf{i} + 2\mathbf{j} - \mathbf{k}) \times (-4\mathbf{i} + 4\mathbf{j} + \mathbf{k}) = \begin{vmatrix} \mathbf{i} & \mathbf{j} & \mathbf{k} \\ 3 & 2 & -1 \\ -4 & 4 & 1 \end{vmatrix} = \begin{vmatrix} 2 & -1 \\ 4 & 1 \end{vmatrix}\mathbf{i} - \begin{vmatrix} 3 & -1 \\ -4 & 1 \end{vmatrix}\mathbf{j} + \begin{vmatrix} 3 & 2 \\ -4 & 4 \end{vmatrix}\mathbf{k} = 6\mathbf{i} + \mathbf{j} + 20\mathbf{k}.$$

$$d = \left|\text{Proj}_{\mathbf{v}}\overrightarrow{PQ}\right| = \frac{|(2\mathbf{i} - 3\mathbf{k}) \cdot (6\mathbf{i} + \mathbf{j} + 20\mathbf{k})|}{|6\mathbf{i} + \mathbf{j} + 20\mathbf{k}|} = \frac{|12 + 0 - 60|}{\sqrt{6^2 + 1^2 + 20^2}} = \frac{48}{\sqrt{437}}.$$

45.　Line $L_1 : (2 - 3t, 1 + 2t, -2 - t)$ is parallel to $-3\mathbf{i} + 2\mathbf{j} - \mathbf{k}$ and passes through $P = (2, 1, -2)$; line $L_2 : (1 + 4s, -2 - s, 3 + s)$ is parallel to $4\mathbf{i} - \mathbf{j} + \mathbf{k}$ and passes through $Q = (1, -2, 3)$. Then $\overrightarrow{PQ} = -\mathbf{i} - 3\mathbf{j} + 5\mathbf{k}$ and the distance d between the lines is the length of this vector projected onto a vector perpendicular to both lines:

$$\mathbf{v} = (-3\mathbf{i} + 2\mathbf{j} - \mathbf{k}) \times (4\mathbf{i} - \mathbf{j} + \mathbf{k}) = \begin{vmatrix} \mathbf{i} & \mathbf{j} & \mathbf{k} \\ -3 & 2 & -1 \\ 4 & -1 & 1 \end{vmatrix} = \begin{vmatrix} 2 & -1 \\ -1 & 1 \end{vmatrix}\mathbf{i} - \begin{vmatrix} -3 & -1 \\ 4 & 1 \end{vmatrix}\mathbf{j} + \begin{vmatrix} -3 & 2 \\ 4 & -1 \end{vmatrix}\mathbf{k} = \mathbf{i} - \mathbf{j} - 5\mathbf{k}.$$

$$d = \left|\text{Proj}_{\mathbf{v}}\overrightarrow{PQ}\right| = \frac{|(-\mathbf{i} - 3\mathbf{j} + 5\mathbf{k}) \cdot (\mathbf{i} - \mathbf{j} - 5\mathbf{k})|}{|\mathbf{i} - \mathbf{j} - 5\mathbf{k}|} = \frac{|-1 + 3 - 25|}{\sqrt{1^2 + 1^2 + 5^2}} = \frac{23}{\sqrt{27}} = \frac{23}{3\sqrt{3}}.$$

47.　Let $\mathbf{a} = a_1\mathbf{i} + a_2\mathbf{j} + a_3\mathbf{k}$, $\mathbf{b} = b_1\mathbf{i} + b_2\mathbf{j} + b_3\mathbf{k}$, $\mathbf{c} = c_1\mathbf{i} + c_2\mathbf{j} + c_3\mathbf{k}$.

$$\mathbf{a} \cdot (\mathbf{b} \times \mathbf{c}) = \mathbf{a} \cdot \begin{vmatrix} \mathbf{i} & \mathbf{j} & \mathbf{k} \\ b_1 & b_2 & b_3 \\ c_1 & c_2 & c_3 \end{vmatrix} = (a_1\mathbf{i} + a_2\mathbf{j} + a_3\mathbf{k}) \cdot \left(\begin{vmatrix} b_2 & b_3 \\ c_2 & c_3 \end{vmatrix}\mathbf{i} - \begin{vmatrix} b_1 & b_3 \\ c_1 & c_3 \end{vmatrix}\mathbf{j} + \begin{vmatrix} b_1 & b_2 \\ c_1 & c_2 \end{vmatrix}\mathbf{k}\right)$$

$$= a_1\begin{vmatrix} b_2 & b_3 \\ c_2 & c_3 \end{vmatrix} - a_2\begin{vmatrix} b_1 & b_3 \\ c_1 & c_3 \end{vmatrix} + a_3\begin{vmatrix} b_1 & b_2 \\ c_1 & c_2 \end{vmatrix} = \begin{vmatrix} a_1 & a_2 & a_3 \\ b_1 & b_2 & b_3 \\ c_1 & c_2 & c_3 \end{vmatrix}. \ \blacksquare$$

49.　If $\mathbf{a}$ and $\mathbf{b}$ lie on two sides of a triangle, then $\mathbf{c} = \mathbf{a} - \mathbf{b}$ lies on the third side.

$$\mathbf{0} = \mathbf{c} \times \mathbf{c} = (\mathbf{a} - \mathbf{b}) \times \mathbf{c} = \mathbf{a} \times \mathbf{c} - \mathbf{b} \times \mathbf{c} \implies \mathbf{b} \times \mathbf{c} = \mathbf{a} \times \mathbf{c} \implies |\mathbf{b} \times \mathbf{c}| = |\mathbf{a} \times \mathbf{c}|$$
$$\implies |\mathbf{b}||\mathbf{c}|\sin A = |\mathbf{a}||\mathbf{c}|\sin B \implies |\mathbf{b}|/\sin B = |\mathbf{a}|/\sin A;$$
$$\mathbf{0} = \mathbf{a} \times \mathbf{a} = (\mathbf{b} + \mathbf{c}) \times \mathbf{a} = \mathbf{b} \times \mathbf{a} + \mathbf{c} \times \mathbf{a} \implies -(\mathbf{b} \times \mathbf{a}) = \mathbf{c} \times \mathbf{a} \implies |\mathbf{b} \times \mathbf{a}| = |-(\mathbf{b} \times \mathbf{a})| = |\mathbf{c} \times \mathbf{a}|$$
$$\implies |\mathbf{b}||\mathbf{a}|\sin C = |\mathbf{c}||\mathbf{a}|\sin B \implies |\mathbf{b}|/\sin B = |\mathbf{c}|/\sin C.$$

Therefore $|\mathbf{a}|/\sin A = |\mathbf{b}|/\sin B = |\mathbf{c}|/\sin C$. $\blacksquare$

REMARK: If $\mathbf{a}$, $\mathbf{b}$, $\mathbf{c}$ are vectors lying on the sides of a triangle such that $\mathbf{a} + \mathbf{b} + \mathbf{c} = \mathbf{0}$, then $\mathbf{a} \times \mathbf{b} = \mathbf{b} \times \mathbf{c} = \mathbf{c} \times \mathbf{a}$ and the Law of Sines follows "as above".

51. Use a nonrectangular coordinate system with origin at the center of the parallelepiped P and axes parallel to the edges of P so that the 8 vertices of P are $(\pm 1, \pm 1, \pm 1)$ and the volume of P in each octant is 1. The 6 center points, vertices of an octahedron Q, have coordinates $(\pm 1, 0, 0)$ or $(0, \pm 1, 0)$ or $(0, 0, \pm 1)$; the volume of this solid in each octant is $1/6$ (see Problem 4.6.13), so the total volume of Q is $1/6$ the total volume of P.

Solutions 11.7   *Planes*   (pages 772-3)

1. The plane through the point $P = (0,0,0)$ and normal to the vector $\mathbf{N} = \mathbf{i}$ satisfies the equation
$0 = \mathbf{N} \cdot \overrightarrow{PQ} = \mathbf{i} \cdot ((x - 0)\mathbf{i} + (y - 0)\mathbf{j} + (z - 0)\mathbf{k}) = \mathbf{i} \cdot (x\mathbf{i} + y\mathbf{j} + z\mathbf{k}) = x$ (the yz-plane).

1: $P = (0,0,0)$, $\mathbf{N} = \mathbf{i}$ 3: $P = (0,0,0)$, $\mathbf{N} = \mathbf{k}$

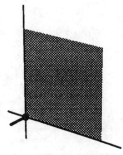
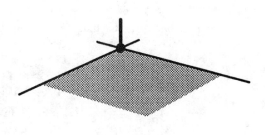

3. The plane through the point $P = (0,0,0)$ and normal to the vector $\mathbf{N} = \mathbf{k}$ satisfies the equation
$0 = \mathbf{N} \cdot \overrightarrow{PQ} = \mathbf{k} \cdot ((x - 0)\mathbf{i} + (y - 0)\mathbf{j} + (z - 0)\mathbf{k}) = \mathbf{k} \cdot (x\mathbf{i} + y\mathbf{j} + z\mathbf{k}) = z$ (the xy-plane).

5. The plane through the point $P = (1,2,3)$ and normal to the vector $\mathbf{N} = \mathbf{i} + \mathbf{k}$ satisfies the equations
$0 = \mathbf{N} \cdot \overrightarrow{PQ} = (\mathbf{i} + \mathbf{k}) \cdot ((x - 1)\mathbf{i} + (y - 2)\mathbf{j} + (z - 3)\mathbf{k}) = (x - 1) + (z - 3) \iff 4 = x + z$.

5: $P = (1,2,3)$, $\mathbf{N} = \mathbf{i} + \mathbf{k}$ 7: $P = (2,-1,6)$, $\mathbf{N} = 3\mathbf{i} - \mathbf{j} + 2\mathbf{k}$

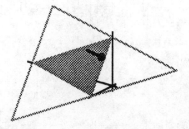

7. The plane through the point $P = (2,-1,6)$ and normal to the vector $\mathbf{N} = 3\mathbf{i} - \mathbf{j} + 2\mathbf{k}$ satisfies the equations $0 = \mathbf{N} \cdot \overrightarrow{PQ} = (3\mathbf{i} - \mathbf{j} + 2\mathbf{k}) \cdot ((x - 2)\mathbf{i} + (y + 1)\mathbf{j} + (z - 6)\mathbf{k}) = 3(x - 2) - (y + 1) + 2(z - 6) \iff 19 = 3x - y + 2z$.

9. The plane through the point $P = (-3, 11, 2)$ and normal to the vector $\mathbf{N} = 4\mathbf{i} + \mathbf{j} - 7\mathbf{k}$ satisfies the equations $0 = \mathbf{N} \cdot \overrightarrow{PQ} = (4\mathbf{i} + \mathbf{j} - 7\mathbf{k}) \cdot ((x + 3)\mathbf{i} + (y - 11)\mathbf{j} + (z - 2)\mathbf{k}) = 4(x + 3) + (y - 11) - 7(z - 2) \iff -15 = 4x + y - 7z$.

9: $P = (-3, 11, 2)$, $\mathbf{N} = 4\mathbf{i} + \mathbf{j} - 7\mathbf{k}$

11: $P = (4, -7, -3)$, $\mathbf{N} = -\mathbf{i} - \mathbf{j} - \mathbf{k}$

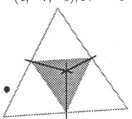

11.   The plane through the point $P = (4, -7, -3)$ and normal to the vector $\mathbf{N} = -\mathbf{i} - \mathbf{j} - \mathbf{k}$ satisfies the equations $0 = \mathbf{N} \cdot \overrightarrow{PQ} = (-\mathbf{i} - \mathbf{j} - \mathbf{k}) \cdot ((x-4)\mathbf{i} + (y+7)\mathbf{j} + (z+3)\mathbf{k}) = -(x-4) - (y+7) - (z+3) \iff 6 = -x - y - z \iff -6 = x + y + z$.

13.   If $P = (1, 2, -4)$, $Q = (2, 3, 7)$, $R = (4, -1, 3)$, then $\overrightarrow{PQ} = \mathbf{i} + \mathbf{j} + 11\mathbf{k}$ and $\overrightarrow{QR} = 2\mathbf{i} - 4\mathbf{j} - 4\mathbf{k}$. Therefore

$$\mathbf{N} = \overrightarrow{PQ} \times \overrightarrow{QR} = \begin{vmatrix} \mathbf{i} & \mathbf{j} & \mathbf{k} \\ 1 & 1 & 11 \\ 2 & -4 & -4 \end{vmatrix} = \begin{vmatrix} 1 & 11 \\ -4 & -4 \end{vmatrix} \mathbf{i} - \begin{vmatrix} 1 & 11 \\ 2 & -4 \end{vmatrix} \mathbf{j} + \begin{vmatrix} 1 & 1 \\ 2 & -4 \end{vmatrix} \mathbf{k} = 40\mathbf{i} + 26\mathbf{j} - 6\mathbf{k}$$

and the plane through P, Q, R satisfies the equations

$$0 = \mathbf{N} \cdot \overrightarrow{Q\,(x, y, z)} = \mathbf{N} \cdot ((x-2)\mathbf{i} + (y-3)\mathbf{j} + (z-7)\mathbf{k}) = 40(x-2) + 26(y-3) - 6(z-7)$$
$$\iff 116 = 40x + 26y - 6z \iff 58 = 20x + 13y - 3z.$$

13: $(1, 2, -4)$, $(2, 3, 7)$, $(4, -1, 3)$ 15: $(1, 0, 0)$, $(0, 1, 0)$, $(0, 0, 1)$

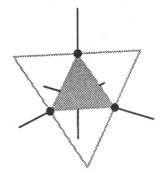

15.   If $P = (1, 0, 0)$, $Q = (0, 1, 0)$, $R = (0, 0, 1)$, then $\overrightarrow{PQ} = -\mathbf{i} + \mathbf{j}$ and $\overrightarrow{QR} = -\mathbf{j} + \mathbf{k}$. Therefore

$$\mathbf{N} = \overrightarrow{PQ} \times \overrightarrow{QR} = \begin{vmatrix} \mathbf{i} & \mathbf{j} & \mathbf{k} \\ -1 & 1 & 0 \\ 0 & -1 & 1 \end{vmatrix} = \begin{vmatrix} 1 & 0 \\ -1 & 1 \end{vmatrix} \mathbf{i} - \begin{vmatrix} -1 & 0 \\ 0 & 1 \end{vmatrix} \mathbf{j} + \begin{vmatrix} -1 & 1 \\ 0 & -1 \end{vmatrix} \mathbf{k} = \mathbf{i} + \mathbf{j} + \mathbf{k}$$

and the plane through P, Q, R satisfies the equations

$$0 = \mathbf{N} \cdot \overrightarrow{Q\,(x, y, z)} = \mathbf{N} \cdot ((x-0)\mathbf{i} + (y-1)\mathbf{j} + (z-0)\mathbf{k}) = x + (y-1) + z$$
$$\iff 1 = x + y + z.$$

17.   If $\Pi_1 : x + y + z = 2$ and $\Pi_2 : 2x + 2y + 2z = 4$, then the equation for Π_2 is 2 times the equation for Π_1 so the planes coincide.

19.   If $\Pi_1 : 2x - y + z = 3$ and $\Pi_2 : x + y - z = 7$, then $\mathbf{N}_1 \cdot \mathbf{N}_2 = (2\mathbf{i} - \mathbf{j} + \mathbf{k}) \cdot (\mathbf{i} + \mathbf{j} - \mathbf{k}) = 2 - 1 - 1 = 0$; the planes are orthogonal.

21.   If $\Pi_1 : 3x - 2y + 7z = 4$ and $\Pi_2 : -2x + 4y + 2z = 16$, then $\mathbf{N}_1 \cdot \mathbf{N}_2 = (3\mathbf{i} - 2\mathbf{j} + 7\mathbf{k}) \cdot (-2\mathbf{i} + 4\mathbf{j} + 2\mathbf{k}) = -6 - 8 + 14 = 0$ so the planes are orthogonal.

23.   If $\Pi_1 : -4x + 4y - 6z = 6$ and $\Pi_2 : 2x - 2y + 3z = -3$, then $\mathbf{N}_1 \cdot \mathbf{N}_2 = (-4\mathbf{i} + 4\mathbf{j} - 6\mathbf{k}) \cdot (2\mathbf{i} - 2\mathbf{j} + 3\mathbf{k}) = -34 \neq 0$ implies the planes are not orthogonal; $\mathbf{N}_1 \times \mathbf{N}_2 = \mathbf{0}$ so the planes are parallel. ($\mathbf{N}_1 = (-2)\mathbf{N}_2$ and $6 = (-2)(-3)$ so the planes are coincident.)

25. Suppose (x, y, z) is on both $\Pi_1 : x + y + z = 1$ and $\Pi_2 : x - y - z = -3$.

$$\begin{array}{l} x + y + z = 1 \\ x - y - z = -3 \end{array} \iff \begin{array}{l} 2x = -2 \\ x - y - z = -3 \end{array} \iff \begin{array}{l} x = -1 \\ -y - z = -2 \end{array} \iff \begin{array}{l} x = -1 \\ y + z = 2 \end{array}$$

Planes Π_1 and Π_2 meet in a line which has parametric equation $(x, y, z) = (-1, 2 - t, t)$.

REMARK: $(-1, 2, 0)$ is on both planes and $\mathbf{N}_1 \times \mathbf{N}_2 = (\mathbf{i} + \mathbf{j} + \mathbf{k}) \times (\mathbf{i} - \mathbf{j} - \mathbf{k}) = 2\mathbf{j} - 2\mathbf{k}$, hence the common line is given by $x\mathbf{i} + y\mathbf{j} + z\mathbf{k} = (-\mathbf{i} + 2\mathbf{j}) + s(2\mathbf{j} - 2\mathbf{k})$.

27. Suppose (x, y, z) is on both $\Pi_1 : 3x - y + 4z = 3$ and $\Pi_2 : -4x - 2y + 7z = 8$.

$$\begin{array}{l} 3x - y + 4z = 3 \\ -4x - 2y + 7z = 8 \end{array} \iff \begin{array}{l} 3x - y + 4z = 3 \\ -10x - z = 2 \end{array} \iff \begin{array}{l} -37x - y = 11 \\ -10x - z = 2 \end{array} \iff \begin{array}{l} -11 - 37x = y \\ -2 - 10x = z \end{array}$$

Planes Π_1 and Π_2 meet in a line which has parametric equation $(x, y, z) = (t, -11 - 37t, -2 - 10t)$.

REMARK: $(0, -11, -2)$ is on both planes and $\mathbf{N}_1 \times \mathbf{N}_2 = (3\mathbf{i} - \mathbf{j} + 4\mathbf{k}) \times (-4\mathbf{i} - 2\mathbf{j} + 7\mathbf{k}) = \mathbf{i} - 37\mathbf{j} - 10\mathbf{k}$, hence the common line is given by $x\mathbf{i} + y\mathbf{j} + z\mathbf{k} = (-11\mathbf{j} - 2\mathbf{k}) + s(\mathbf{i} - 37\mathbf{j} - 10\mathbf{k})$.

29. The acute angle between vectors $\mathbf{a}$ and $\mathbf{b}$ is the smaller of $\cos^{-1}\left((\pm\mathbf{a} \cdot \mathbf{b})/(|\pm\mathbf{a}|\,|\mathbf{b}|)\right)$ and can be computed directly as $\cos^{-1}\left(|\mathbf{a} \cdot \mathbf{b}|/(|\mathbf{a}|\,|\mathbf{b}|)\right)$.

If $\mathbf{N}_1 = \mathbf{i} + \mathbf{j} + \mathbf{k}$ and $\mathbf{N}_2 = \mathbf{i} - \mathbf{j} - \mathbf{k}$, then $\cos\theta = |\mathbf{N}_1 \cdot \mathbf{N}_2|/(|\mathbf{N}_1|\,|\mathbf{N}_2|) = |-1|/(\sqrt{3}\sqrt{3}) = 1/3$ and $\theta = \cos^{-1}(1/3) \approx 1.231$ radians.

31. If $\mathbf{N}_1 = 3\mathbf{i} - \mathbf{j} + 4\mathbf{k}$ and $\mathbf{N}_2 = -4\mathbf{i} - 2\mathbf{j} + 7\mathbf{k}$, then $\cos\theta = |\mathbf{N}_1 \cdot \mathbf{N}_2|/(|\mathbf{N}_1|\,|\mathbf{N}_2|) = |18|/(\sqrt{26}\sqrt{69}) = \sqrt{54/299}$ and $\theta = \cos^{-1}\sqrt{54/299} \approx 1.1319$ radians.

33. Suppose $\mathbf{u} = 2\mathbf{i} - 3\mathbf{j} + 4\mathbf{k}$, $\mathbf{v} = 7\mathbf{i} - 2\mathbf{j} + 3\mathbf{k}$, and $\mathbf{w} = 9\mathbf{i} - 5\mathbf{j} + 7\mathbf{k}$. Then $\mathbf{u} \cdot (\mathbf{v} \times \mathbf{w}) = (2\mathbf{i} - 3\mathbf{j} + 4\mathbf{k}) \cdot (\mathbf{i} - 22\mathbf{j} - 17\mathbf{k}) = 2 + 66 - 68 = 0$ and Problem 52 implies the vectors have coplanar endpoints.

$\mathbf{N} = (\mathbf{u} - \mathbf{v}) \times (\mathbf{w} - \mathbf{v}) = (-5\mathbf{i} - \mathbf{j} + \mathbf{k}) \times (2\mathbf{i} - 3\mathbf{j} + 4\mathbf{k}) = -\mathbf{i} + 22\mathbf{j} + 17\mathbf{k}$ is normal to the plane; an equation for the plane is $-x + 22y + 17z = \mathbf{N} \cdot \mathbf{v} = 0$.

35. Suppose $\mathbf{u} = 2\mathbf{i} + \mathbf{j} - 2\mathbf{k}$, $\mathbf{v} = 2\mathbf{i} - \mathbf{j} - 2\mathbf{k}$, and $\mathbf{w} = 2\mathbf{i} - \mathbf{j} + 2\mathbf{k}$. Then $\mathbf{u} \cdot (\mathbf{v} \times \mathbf{w}) = (2\mathbf{i} + \mathbf{j} - 2\mathbf{k}) \cdot (-4\mathbf{i} - 8\mathbf{j}) = -8 - 8 + 0 = -16 \neq 0$ and Problem 52 implies the vectors do not have coplanar endpoints.

37. The distance between point $P = (2, -1, 4)$ and plane $\Pi : 3x - y + 7z - 2 = 0$ is

$$D = \frac{|3(2) - (-1) + 7(4) - 2|}{\sqrt{3^2 + (-1)^2 + 7^2}} = \frac{|6 + 1 + 28 - 2|}{\sqrt{9 + 1 + 49}} = \frac{33}{\sqrt{59}}.$$

39. The distance between point $P = (-7, -2, -1)$ and plane $\Pi : -2x + 8z + 5 = 0$ is

$$D = \frac{|-2(-7) + 8(-1) + 5|}{\sqrt{(-2)^2 + 0^2 + 8^2}} = \frac{|14 - 8 + 5|}{\sqrt{4 + 0 + 64}} = \frac{11}{\sqrt{68}}.$$

41: $x = 1$ 43: $z = -2$ 45: $x + y + z = 1$ 47: $x - y - z = 1$ 49: $-2x + 3y + 5z = 10$

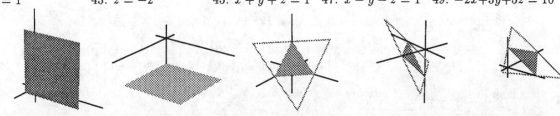

51. $\mathbf{N} = \mathbf{v} \times \mathbf{w}$ is normal (orthogonal) to the plane Π through the origin containing $\mathbf{v}$ and $\mathbf{w}$. Therefore $\mathbf{u}$ lies in plane Π if and only if it is orthogonal to normal $\mathbf{N}$, i.e., $\mathbf{u} \perp \mathbf{N} \iff 0 = \mathbf{u} \cdot \mathbf{N} = \mathbf{u} \cdot (\mathbf{v} \times \mathbf{w})$. ∎

53. Let $A = (x, y, z)$, $P = (x_1, y_1, z_1)$, $Q = (x_2, y_2, z_2)$, $R = (x_3, y_3, z_3)$. Subtract the first row of the determinant from each of the others and then expand by the last column:

$$\begin{vmatrix} x & y & z & 1 \\ x_1 & y_1 & z_1 & 1 \\ x_2 & y_2 & z_2 & 1 \\ x_3 & y_3 & z_3 & 1 \end{vmatrix} = \begin{vmatrix} x & y & z & 1 \\ x_1 - x & y_1 - y & z_1 - z & 0 \\ x_2 - x & y_2 - y & z_2 - z & 0 \\ x_3 - x & y_3 - y & z_3 - z & 0 \end{vmatrix} = (-1)^{1+4}(1) \begin{vmatrix} x_1 - x & y_1 - y & z_1 - z \\ x_2 - x & y_2 - y & z_2 - z \\ x_3 - x & y_3 - y & z_3 - z \end{vmatrix} = -\overrightarrow{AP} \cdot \left(\overrightarrow{AQ} \times \overrightarrow{AR}\right).$$

Problem 51 implies the triple product is zero if and only if vectors $\overrightarrow{AP}$, $\overrightarrow{AQ}$, $\overrightarrow{AR}$ are coplanar. ∎

55.   Although the geometry of the text figure seems fairly clear, a couple items deserve clarification. Suppose Π is a plane, $\mathbf{N}$ is a vector normal to Π, P and R are points on plane Π, Q is an arbitrary point. Therefore, $\overrightarrow{PR}$ lies in the plane and is orthogonal to $\mathbf{N}$; thus

$$\overrightarrow{PQ} \cdot \mathbf{N} = \left(\overrightarrow{PR} + \overrightarrow{RQ} \right) \cdot \mathbf{N} = \left(\overrightarrow{PR} \cdot \mathbf{N} \right) + \left(\overrightarrow{RQ} \cdot \mathbf{N} \right) = 0 + \left(\overrightarrow{RQ} \cdot \mathbf{N} \right) = \overrightarrow{RQ} \cdot \mathbf{N}$$

and the value of $\text{Proj}_{\mathbf{N}} \overrightarrow{PQ}$ is invariant to choice of point P on Π. If S is chosen on Π such that $\overrightarrow{SQ} = \text{Proj}_{\mathbf{N}} \overrightarrow{PQ}$, then $\overrightarrow{PS} \perp \overrightarrow{SQ}$ and $|\overrightarrow{PQ}|^2 = |\overrightarrow{PS}|^2 + |\overrightarrow{SQ}|^2 \geq |\overrightarrow{SQ}|^2$ which implies the minimal distance from Q to a point on plane Π is

$$\left| \overrightarrow{SQ} \right| = \left| \text{Proj}_{\mathbf{N}} \overrightarrow{PQ} \right| = \frac{|\overrightarrow{PQ} \cdot \mathbf{N}|}{|\mathbf{N}|}. \ \blacksquare$$

Solutions 11.8   *Quadric surfaces*   (pages 780-1)

Note: The term "x-section" will be used to mean "cross section with constant x value; "y-section" and "z-section" will be used analogously.

1: $y = \sin x$, generatrix is z-axis

3: $y = \cos x$, generatrix is z-axis

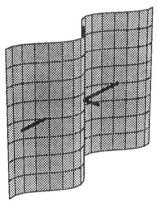

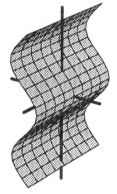

5: $z = x^3$, generatrix is y-axis

7: $|y| + |z - 5| = 1$, generatrix is x-axis

9. $\sqrt{x^2 + y^2}$ is the distance from (x, y, z) to the z-axis. The graph of $x^2 + y^2 = 4$ is the right circular cylinder of radius 2 centered on the z-axis.

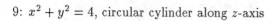

9: $x^2 + y^2 = 4$, circular cylinder along z-axis   11: $x^2 + z^2 = 4$, circular cylinder along y-axis

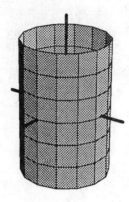   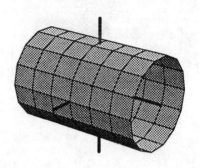

11. $\sqrt{x^2 + z^2}$ is the distance from (x, y, z) to the y-axis. The graph of $x^2 + z^2 = 4$ is the right circular cylinder of radius 2 centered on the y-axis.

13. $1 = x^2 + 4z^2 = x^2 + (z/0.5)^2$; y-sections are ellipses centered at $(0, y, 0)$. The surface is an elliptic cylinder with the y-axis as generatrix.

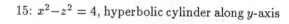

13: $x^2 + 4z^2 = 1$, ellipsoid along y-axis   15: $x^2 - z^2 = 4$, hyperbolic cylinder along y-axis

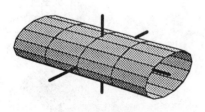   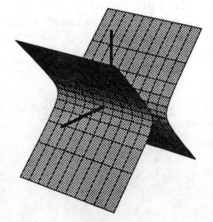

15. y-sections of $x^2 - z^2 = 1$ are hyperbolas. The surface is an hyperbolic cylinder with the y-axis as generatrix; the planes $z = \pm x$ are asymptotes.

17. $\sqrt{x^2 + y^2 + z^2}$ is the distance from (x, y, z) to the origin The graph of $x^2 + y^2 + z^2 = 1$ is the sphere of radius 1 centered at the origin.

17: $x^2+y^2+z^2=1$, sphere centered at origin

19: $x^2+(y+1)^2=2+z^2$, hyperboloid of one sheet

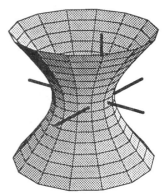

19. $y^2+2y-z^2+x^2=1 \iff x^2+y^2+2y=1+z^2 \iff x^2+(y+1)^2=2+z^2$. Hence z-sections are circles centered at $(0,-1,z)$, all x- and y-sections are hyperbolas; the surface is an hyperboloid of revolution with one sheet.

REMARK: A surface is said to be a **ruled surface** if each point is on a straight line which lies entirely on the surface. Planes, cylinders,and cones are simple examples of ruled surfaces; a hyperboloid with one sheet is also a ruled surface.

Each point of $x^2+(y+1)^2-z^2=2$ is on two straight lines which lie entirely on the surface. One family of lines (rulings) is the intersection of planes $y+1-z=\alpha(\sqrt{2}-x)$ and $\alpha(y+1+z)=\sqrt{2}+x$; the other family of rulings satisfy $y+1-z=\beta(\sqrt{2}+x)$ and $\beta(y+1+z)=\sqrt{2}-x$.

We can construct a model of a hyperboloid of revolution with one sheet in a way which exhibits these rulings. Cut out two circular disks of equal radius from cardboard, punch out 12 small holes at equally spaced intervals around the edges of both disks, thread 12 equal lengths of string through corresponding holes and attach them to the disks (e.g., by knotting or with tape). Hold one disk horizontally with the other disk suspended below on the strings (the disks will be parallel, their centers will be on a vertical line, and the strings will lie on a right circular cylinder). Twist the lower disk in one direction — the strings will now show one family of rulings of the hyperboloid; twist the disk in the other direction to see the other family of rulings.

21. $x+2y^2+3z^2=4 \iff 2y^2+3z^2=4-x$. There are no points on the surface for which $4-x<0$, i.e., $x>4$. If $x\leq 4$, then x-sections are ellipses centered at $(x,0,0)$; since the equation can also be written as $x=4-2y^2-3z^2$, we see that y- and z-sections are parabolas. The surface is an elliptic paraboloid.

21: $2y^2+3z^2=4-x$, elliptic paraboloid

23: $(x+2)^2+(y+3)^2=(z+4)^2-1$, hyperboloid of 2 sheets

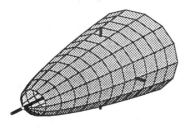

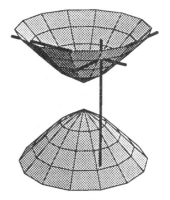

23. $x^2+4x+y^2+6y-z^2-8z=2 \iff x^2+4x+y^2+6y=z^2+8z+2 \iff (x+2)^2+(y+3)^2=(z+4)^2-1$. There are no points on the surface if $|z+4|<1$; if $|z+4|\geq 1$, then z-sections are circles centered at $(-2,-3,z)$. All x- and y-sections are hyperbolas. The surface is an hyperboloid of two sheets.

25. $5x^2+7y^2+8z^2=8z \iff 5x^2+7y^2+8(z-1/2)^2=8(1/2)^2=2 \iff \dfrac{x^2}{2/5}+\dfrac{y^2}{2/7}+\dfrac{(z-1/2)^2}{1/4}=1.$

11.8 — Quadric surfaces

The surface is an ellipsoid centered at $(0, 0, 1/2)$.

25: $5x^2 + 7y^2 + 8(z - 1/2)^2 = 2$, ellipsoid

27: $2y^2 + 3z^2 = x^2 - 4$, hyperboloid of two sheets

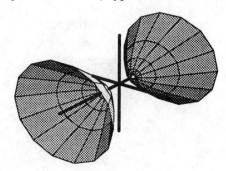

27. $\quad x^2 - 2y^2 - 3z^2 = 4 \iff 2y^2 + 3z^2 = x^2 - 4$. There are no points on the surface if $|x| < 2$; if $|x| \geq 2$, then x-sections are ellipses centered at $(x, 0, 0)$. All y- and z-sections are hyperbolas. The surface is an hyperboloid of two sheets.

29. $\quad y^2 - 3x^2 - 3z^2 = 27 \iff 3x^2 + 3z^2 = y^2 - 27 \iff x^2 + z^2 = y^2/3 - 9$. There are no points on the surface if $|y| < \sqrt{27}$; if $|y| \geq \sqrt{27}$, then y-sections are circles centered at $(0, y, 0)$. All x- and z-sections are hyperbolas. The surface is an hyperboloid of two sheets.

29: $3x^2 + 3z^2 = y^2 - 27$, hyperboloid of two sheets

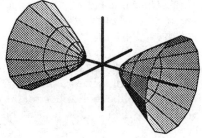

31: $x^2 + y^2 + 4z^2 = 4$, ellipsoid

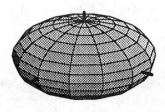

31. $\quad 4x^2 + 4y^2 + 16z^2 = 16 \iff x^2 + y^2 + 4z^2 = 4 \iff (x/2)^2 + (y/2)^2 + z^2 = 1$. Hence z-sections are circles if $|z| \leq 1$ while x- and y-sections are non-circular ellipses if $|x| \leq 2$ and $|y| \leq 2$, The surface is an ellipsoid centered at the origin.

33. $\quad z + x^2 - y^2 = 0 \iff z = y^2 - x^2$. Clearly x- and y-sections are parabolas while all z-sections are hyperbolas; the surface is an hyperbolic paraboloid.

REMARK: There are two families of lines (rulings) which lie on this surface. One family is the intersection of planes $y + x = \alpha$ and $z = \alpha(y - x)$; the other family is $y - x = \beta$ and $z = \beta(y + x)$.

33: $z = y^2 - x^2$, hyperbolic paraboloid

35: $y = (x - .5)^2 + (z - .5)^2 - .5$, circular paraboloid

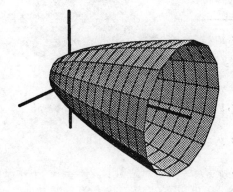

35. $\quad x + y + z = x^2 + z^2 \iff y = x^2 - x + z^2 - z = (x - 1/2)^2 + (z - 1/2)^2 - 1/2$. Hence y-sections are circles centered at $(1/2, y, 1/2)$ if $y \geq -1/2$ while x- and z-sections are parabolas. The surface is a circular paraboloid.

37. Revolving a curve in the xy-plane about the y-axis generates a surface — the algebraic step can be stated colloquially as "replace x^2 by $x^2 + z^2$". Therefore revolving the ellipse $1 = (x/a)^2 + (y/b)^2 = x^2/a^2 + y^2/b^2$ about the y-axis yields the ellipsoid $1 = (x^2 + z^2)/a^2 + y^2/b^2$.

37a: $x^2 + (y/2)^2 = 1$ rotated about y-axis $\qquad\qquad$ 37b: $(x/2)^2 + y^2 = 1$ rotated about y-axis

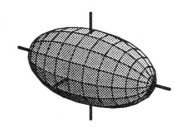 

Solutions 11.9 $\qquad$ *Cylindrical and spherical Coordinates* $\qquad$ (page 786)

Note: To help you distinguish which type of coordinates are being used in this section, rectangular coordinates will be written with subscript "r", cylindrical coordinates will be written with subscript "c", spherical coordinates will be written with subscript "s": e.g., $(x, y, z)_r$ or $(r, \theta, z)_c$ or $(\rho, \theta, \phi)_s$.

1. If $(r, \theta, z)_c = \left(2, \dfrac{\pi}{3}, 5\right)_c$, then $(x, y, z)_r = (r\cos\theta, r\sin\theta, z)_r = \left(2\cos\dfrac{\pi}{3}, 2\sin\dfrac{\pi}{3}, 5\right)_r = (1, \sqrt{3}, 5)_r$.

3. If $(r, \theta, z)_c = \left(8, \dfrac{2\pi}{3}, 1\right)_c$, then $(x, y, z)_r = (r\cos\theta, r\sin\theta, z)_r = \left(8\cos\dfrac{2\pi}{3}, 8\sin\dfrac{2\pi}{3}, 1\right)_r = (-4, 4\sqrt{3}, 1)_r$.

5. If $(r, \theta, z)_c = \left(3, \dfrac{3\pi}{4}, 2\right)_c$, then $(x, y, z)_r = (r\cos\theta, r\sin\theta, z)_r = \left(3\cos\dfrac{3\pi}{4}, 3\sin\dfrac{3\pi}{4}, 2\right)_r = \left(\dfrac{-3}{\sqrt{2}}, \dfrac{3}{\sqrt{2}}, 2\right)_r$.

7. If $(r, \theta, z)_c = (10, \pi, -3)_c$, then $(x, y, z)_r = (r\cos\theta, r\sin\theta, z)_r = (10\cos\pi, 10\sin\pi, -3)_r = (-10, 0, -3)_r$.

9. If $(r, \theta, z)_c = \left(7, \dfrac{5\pi}{4}, 2\right)_c$, then $(x, y, z)_r = (r\cos\theta, r\sin\theta, z)_r = \left(7\cos\dfrac{5\pi}{4}, 7\sin\dfrac{5\pi}{4}, 2\right)_r = \left(\dfrac{-7}{\sqrt{2}}, \dfrac{-7}{\sqrt{2}}, 2\right)_r$.

Note: The following generalization of the arctangent function is used to compute θ when converting rectangular coordinates to cylindrical or spherical coordinates.

$$\text{ArcTan}(x, y) = \begin{cases} 0 & \text{if } x = 0, \\ \tan^{-1}(y/x) & \text{if } x > 0 \text{ and } y > 0, \\ \pi + \tan^{-1}(y/x) & \text{if } x < 0, \\ 2\pi + \tan^{-1}(y/x) & \text{if } x > 0 \text{ and } y < 0. \end{cases}$$

11. If $(x, y, z)_r = (1, 0, 0)_r$, then $(r, \theta, z)_c = \left(\sqrt{1^2 + 0^2}, \tan^{-1}\dfrac{0}{1}, 0\right)_c = (1, 0, 0)_c$.

13. If $x = y = 0$, then θ is arbitrary. Hence, if $(x, y, z)_r = (0, 0, 1)_r$, then $(r, \theta, z)_c = \left(\sqrt{0^2 + 0^2}, \theta, 1\right)_c = (0, \theta, 1)_c$.

15. If $(x, y, z)_r = (-1, 1, 4)_r$, then $(r, \theta, z)_c = \left(\sqrt{(-1)^2 + 1^2}, \pi + \tan^{-1}\dfrac{1}{-1}, 4\right)_c = \left(\sqrt{2}, \dfrac{3\pi}{4}, 4\right)_c$.

17. If $(x, y, z)_r = (2\sqrt{3}, -2, 8)_r$, then $(r, \theta, z)_c = \left(\sqrt{(2\sqrt{3})^2 + (-2)^2}, 2\pi + \tan^{-1}\dfrac{-2}{2\sqrt{3}}, 8\right)_c = \left(4, \dfrac{11\pi}{6}, 8\right)_c$.

19. If $(x, y, z)_r = (-2\sqrt{3}, -2, 1)_r$, then $(r, \theta, z)_c = \left(\sqrt{(-2\sqrt{3})^2 + (-2)^2}, \pi + \tan^{-1}\dfrac{-2}{-2\sqrt{3}}, 1\right)_c = \left(4, \dfrac{7\pi}{6}, 1\right)_c$.

21. If $(\rho, \theta, \phi)_s = \left(2, 0, \dfrac{\pi}{3}\right)_s$, then $(x, y, z)_r = \left(2\cos 0 \sin\dfrac{\pi}{3}, 2\sin 0 \sin\dfrac{\pi}{3}, 2\cos\dfrac{\pi}{3}\right)_r = (\sqrt{3}, 0, 1)_r$.

23. If $(\rho, \theta, \phi)_s = \left(6, \dfrac{\pi}{2}, \dfrac{\pi}{3}\right)_s$, then $(x, y, z)_r = \left(6\cos\dfrac{\pi}{2}\sin\dfrac{\pi}{3}, 6\sin\dfrac{\pi}{2}\sin\dfrac{\pi}{3}, 6\cos\dfrac{\pi}{3}\right)_r = (0, 3\sqrt{3}, 3)_r$.

25. If $(\rho, \theta, \phi)_s = \left(7, \dfrac{7\pi}{4}, \dfrac{3\pi}{4}\right)_s$, then

$$(x, y, z)_r = \left(7\cos\dfrac{7\pi}{4}\sin\dfrac{3\pi}{4}, 7\sin\dfrac{7\pi}{4}\sin\dfrac{3\pi}{4}, 7\cos\dfrac{3\pi}{4}\right)_r = \left(\dfrac{7}{2}, \dfrac{-7}{2}, \dfrac{-7}{\sqrt{2}}\right)_r.$$

27. If $(\rho, \theta, \phi)_s = \left(4, \dfrac{\pi}{3}, \dfrac{2\pi}{3}\right)_s$, then $(x, y, z)_r = \left(4\cos\dfrac{\pi}{3}\sin\dfrac{2\pi}{3}, 4\sin\dfrac{\pi}{3}\sin\dfrac{2\pi}{3}, 4\cos\dfrac{2\pi}{3}\right)_r = (\sqrt{3}, 3, -2)_r.$

29. If $(\rho, \theta, \phi)_s = \left(5, \dfrac{11\pi}{6}, \dfrac{5\pi}{6}\right)_s$, then

$$(x, y, z)_r = \left(5\cos\dfrac{11\pi}{6}\sin\dfrac{5\pi}{6}, 5\sin\dfrac{11\pi}{6}\sin\dfrac{5\pi}{6}, 5\cos\dfrac{5\pi}{6}\right)_r = \left(\dfrac{5\sqrt{3}}{4}, \dfrac{-5}{4}, \dfrac{-5\sqrt{3}}{2}\right)_r.$$

31. If $(x, y, z)_r = (1, 1, 0)_r$, then $(\rho, \theta, \phi)_s = \left(\sqrt{1^2 + 1^2 + 0^2}, \tan^{-1}\dfrac{1}{1}, \cos^{-1}\dfrac{0}{\sqrt{1^2 + 1^2 + 0^2}}\right)_s = \left(\sqrt{2}, \dfrac{\pi}{4}, \dfrac{\pi}{2}\right)_s.$

33. If $(x, y, z)_r = (1, -1, \sqrt{2})_r$, then

$$(\rho, \theta, \phi)_s = \left(\sqrt{1^2 + (-1)^2 + (\sqrt{2})^2}, \; 2\pi + \tan^{-1}\dfrac{-1}{1}, \; \cos^{-1}\dfrac{\sqrt{2}}{\sqrt{1^2 + (-1)^2 + (\sqrt{2})^2}}\right)_s = \left(2, \dfrac{7\pi}{4}, \dfrac{\pi}{4}\right)_s.$$

35. If $(x, y, z)_r = (1, -\sqrt{3}, 2)_r$, then

$$(\rho, \theta, \phi)_s = \left(\sqrt{1^2 + (-\sqrt{3})^2 + 2^2}, \; 2\pi + \tan^{-1}\dfrac{-\sqrt{3}}{1}, \; \cos^{-1}\dfrac{2}{\sqrt{1^2 + (-\sqrt{3})^2 + 2^2}}\right)_s = \left(2\sqrt{2}, \dfrac{5\pi}{3}, \dfrac{\pi}{4}\right)_s.$$

37. If $(x, y, z)_r = (2, \sqrt{3}, 4)_r$, then

$$(\rho, \theta, \phi)_s = \left(\sqrt{2^2 + (\sqrt{3})^2 + 4^2}, \; \tan^{-1}\dfrac{\sqrt{3}}{2}, \; \cos^{-1}\dfrac{4}{\sqrt{2^2 + (\sqrt{3})^2 + 4^2}}\right)_s = \left(\sqrt{23}, \tan^{-1}\dfrac{\sqrt{3}}{2}, \cos^{-1}\dfrac{4}{\sqrt{23}}\right)_s$$

$$\approx (4.796, \; 0.7137, \; 0.5844)_s.$$

39. If $(x, y, z)_r = (-\sqrt{3}, -2, -4)_r$, then

$$(\rho, \theta, \phi)_s = \left(\sqrt{(-\sqrt{3})^2 + (-2)^2 + (-4)^2}, \; \pi + \tan^{-1}\dfrac{-2}{-\sqrt{3}}, \; \cos^{-1}\dfrac{-4}{\sqrt{(-\sqrt{3})^2 + (-2)^2 + (-4)^2}}\right)_s$$

$$= \left(\sqrt{23}, \pi + \tan^{-1}\dfrac{2}{\sqrt{3}}, \cos^{-1}\dfrac{-4}{\sqrt{23}}\right)_s \approx (4.796, \; 3.999, \; 2.557)_s.$$

41. $25 = x^2 + y^2 + z^2 = (x^2 + y^2) + z^2 = r^2 + z^2$ so the equation in cylindrical coordinates is $r^2 + z^2 = 25$. The equation in spherical coordinates is $\rho^2 = 25$ which can be simplified to $\rho = 5$.

43. $r = 9\sin\theta \implies r^2 = 9r\sin\theta \iff x^2 + y^2 = 9y.$

45. The hyperboloid $x^2 + y^2 - z^2 = 1$ has cylindrical coordinate equation $r^2 - z^2 = 1$. The spherical coordinate equation is a bit messier: $x^2 + y^2 - z^2 = 1 \iff x^2 + y^2 + z^2 = 1 + 2z^2 \iff \rho^2 = 1 + 2(\rho\cos\phi)^2 = 1 + 2\rho^2\cos^2\phi.$

47. $z = r^2 \iff z = x^2 + y^2.$

49. $1 = \rho^2 \sin\phi\cos\phi = (\rho\cos\phi)(\rho\sin\phi) = zr = z\sqrt{x^2 + y^2}.$

51. $z = r^2 \sin 2\theta = r^2(2\cos\theta\,\sin\theta) = 2(r\cos\theta)(r\sin\theta) = 2xy.$

53. $9 = x^2 + (y - 3)^2 + z^2 = x^2 + (y^2 - 6y + 9) + z^2 \iff 6y = x^2 + y^2 + z^2 \iff 6\rho\sin\theta\sin\phi = \rho^2 \implies 6\sin\theta\,\sin\phi = \rho.$

55. ϕ is the angle between the unit vector in the direction of the positive z-axis and the unit radial vector in the direction from the origin to the point. Therefore

$$\cos\phi = \mathbf{k} \cdot \dfrac{\mathbf{r}}{|\mathbf{r}|} = \mathbf{k} \cdot \left(\dfrac{x\mathbf{i} + y\mathbf{j} + z\mathbf{k}}{\rho}\right) = \dfrac{z}{\rho}. \quad \blacksquare$$

Solutions 11.R   *Review*   (pages 786-9)

1.   Let $\mathbf{u} = (2, 1)$ and $\mathbf{v} = (-3, 4)$.
   a. $5\mathbf{u} = 5(2, 1) = (10, 5)$.
   b. $\mathbf{u} - \mathbf{v} = (2, 1) - (-3, 4) = (2 + 3, 1 - 4) = (5, -3)$.
   c. $-8\mathbf{u} + 5\mathbf{v} = -8(2, 1) + 5(-3, 4) = (-16, -8) + (-15, 20) = (-16 - 15, -8 + 20) = (-31, 12)$.
   d. $3\mathbf{u} + 2\mathbf{v} = 3(2, 1) + 2(-3, 4) = (6, 3) + (-6, 8) = (6 - 6, 3 + 8) = (0, 11)$.

3.   Let $\mathbf{u} = (1, -3, 5)$ and $\mathbf{v} = (5, 0, 2)$.
   a. $3\mathbf{v} = 3(5, 0, 2) = (15, 0, 6)$.
   b. $\mathbf{v} - \mathbf{u} = (5, 0, 2) - (1, -3, 5) = (5 - 1, 0 - (-3), 2 - 5) = (4, 3, -3)$.
   c. $2\mathbf{u} - 2\mathbf{v} = 2(1, -3, 5) - 2(5, 0, 2) = (2, -6, 10) - (10, 0, 4) = (2 - 10, -6 - 0, 10 - 4) = (-8, -6, 6)$.
   d. $5\mathbf{u} - \mathbf{v} = 5(1, -3, 5) - (5, 0, 2) = (5, -15, 25) - (5, 0, 2) = (5 - 5, -15 - 0, 25 - 2) = (0, -15, 23)$.

5.   If $\mathbf{u} = (3, 3)$, then $|\mathbf{u}| = |(3, 3)| = \sqrt{3^2 + 3^2} = 3\sqrt{2}$, $\mathbf{u}/|\mathbf{u}| = (3, 3)/3\sqrt{2} = (1/\sqrt{2}, 1/\sqrt{2})$, and $\theta = \tan^{-1}(3/3) = \pi/4$.

7.   If $\mathbf{u} = -12\mathbf{i} - 12\mathbf{j}$, then $|\mathbf{u}| = |-12\mathbf{i} - 12\mathbf{j}| = \sqrt{(-12)^2 + (-12)^2} = 12\sqrt{2}$, $\mathbf{u}/|\mathbf{u}| = (-12\mathbf{i} - 12\mathbf{j})/12\sqrt{2} = (1/\sqrt{2})(-\mathbf{i} - \mathbf{j})$, and $\theta = \pi + \tan^{-1}((-12)/(-12)) = 5\pi/4$.

9.   If $\mathbf{w} = (\sqrt{3}, 1, 0)$, then $|\mathbf{w}| = |(\sqrt{3}, 1, 0)| = \sqrt{(\sqrt{3})^2 + 1^2 + 0^2} = 2$, and $\mathbf{w}/|\mathbf{w}| = (\sqrt{3}, 1, 0)/2 = (\sqrt{3}/2, 1/2, 0)$.

11.   If $\mathbf{w} = 7\mathbf{i} - 3\mathbf{j}$, then $|\mathbf{w}| = |7\mathbf{i} - 3\mathbf{j}| = \sqrt{7^2 + (-3)^2 + 0^2} = \sqrt{58}$, and $\mathbf{w}/|\mathbf{w}| = (7\mathbf{i} - 3\mathbf{j})/\sqrt{58} = (7/\sqrt{58})\mathbf{i} + (-3/\sqrt{58})\mathbf{j}$.

13.   If $\mathbf{w} = -3\mathbf{i} + 4\mathbf{j} + 5\mathbf{k}$, then $|\mathbf{w}| = |-3\mathbf{i} + 4\mathbf{j} + 5\mathbf{k}| = \sqrt{(-3)^2 + 4^2 + 5^2} = 5\sqrt{2}$, and $\mathbf{w}/|\mathbf{w}| = (-3\mathbf{i} + 4\mathbf{j} + 5\mathbf{k})/5\sqrt{2} = (-3/5\sqrt{2})\mathbf{i} + (4/5\sqrt{2})\mathbf{j} + (1/\sqrt{2})\mathbf{k}$.

15.   If $P = (3, 1, 2)$ and $Q = (-1, -3, -4)$, then
$$\left|\overrightarrow{PQ}\right| = \sqrt{(-1 - 3)^2 + (-3 - 1)^2 + (-4 - 2)^2} = \sqrt{16 + 16 + 36} = \sqrt{68} = 2\sqrt{17}.$$

15: $(3, 1, 2)$ and $(-1, -3, -4)$ 17: $(-2, 4, -8)$ and $(0, 0, 6)$

17.   If $P = (-2, 4, -8)$ and $Q = (0, 0, 6)$, then
$$\left|\overrightarrow{PQ}\right| = \sqrt{(0 + 2)^2 + (0 - 4)^2 + (6 + 8)^2} = \sqrt{4 + 16 + 196} = \sqrt{216} = 6\sqrt{6}.$$

19.   If $P = (2, 3)$ and $Q = (4, 5)$, then $\mathbf{u} = \overrightarrow{PQ} = \mathbf{q} - \mathbf{p} = (4\mathbf{i} + 5\mathbf{j}) - (2\mathbf{i} + 3\mathbf{j}) = (4 - 2)\mathbf{i} + (5 - 3)\mathbf{j} = 2\mathbf{i} + 2\mathbf{j}$.

21.   If $P = (-1, -6)$ and $Q = (3, -4)$, then $\overrightarrow{PQ} = \mathbf{q} - \mathbf{p} = (3\mathbf{i} - 4\mathbf{j}) - (-\mathbf{i} - 6\mathbf{j}) = (3 - (-1))\mathbf{i} + (-4 - (-6))\mathbf{j} = 4\mathbf{i} + 2\mathbf{j}$.

23.   If $P = (2, -3, 0)$ and $Q = (-2, 3, 0)$, then $\overrightarrow{PQ} = \mathbf{q} - \mathbf{p} = (-2\mathbf{i} + 3\mathbf{j}) - (2\mathbf{i} - 3\mathbf{j}) = (-2 - 2)\mathbf{i} + (3 - (-3))\mathbf{j} = -4\mathbf{i} + 6\mathbf{j} = 2\mathbf{q}$.

25.   If $P = (-3, 5, 12)$ and $Q = (-1, 1, 8)$, then $\overrightarrow{PQ} = \mathbf{q} - \mathbf{p} = (-\mathbf{i} + \mathbf{j} + 8\mathbf{k}) - (-3\mathbf{i} + 5\mathbf{j} + 12\mathbf{k}) = (-1 - (-3))\mathbf{i} + (1 - 5)\mathbf{j} + (8 - 12)\mathbf{k} = 2\mathbf{i} - 4\mathbf{j} - 4\mathbf{k}$.

27.   If $\mathbf{v} = \mathbf{i} + \mathbf{j}$, then $|\mathbf{v}| = \sqrt{1^2 + 1^2} = \sqrt{2}$ and $\mathbf{u} = -\mathbf{v}/|\mathbf{v}| = -(\mathbf{i}+\mathbf{j})/\sqrt{2} = \left(-1/\sqrt{2}\right)\mathbf{i} + \left(-1/\sqrt{2}\right)\mathbf{j}$.

29.   If $\mathbf{v} = 10\mathbf{i} - 7\mathbf{j}$, then $|\mathbf{v}| = \sqrt{10^2 + (-7)^2} = \sqrt{149}$ and $\mathbf{u} = -\mathbf{v}/|\mathbf{v}| = -(10\mathbf{i}-7\mathbf{j})/\sqrt{149} = \left(-10/\sqrt{149}\right)\mathbf{i} + \left(7/\sqrt{149}\right)\mathbf{j}$.

31.   If $\mathbf{v} = 3\mathbf{j} + 11\mathbf{k}$, then $|\mathbf{v}| = \sqrt{0^2 + 3^2 + 11^2} = \sqrt{130}$ and $\mathbf{u} = -\mathbf{v}/|\mathbf{v}| = -(3\mathbf{j} + 11\mathbf{k})/\sqrt{130} = \left(-3/\sqrt{130}\right)\mathbf{j} + \left(-11/\sqrt{130}\right)\mathbf{k}$.

33.   If $\mathbf{v} = \mathbf{i} - 2\mathbf{j} - 3\mathbf{k}$, then $|\mathbf{v}| = \sqrt{1^2 + (-2)^2 + (-3)^2} = \sqrt{14}$ and $\mathbf{u} = -\mathbf{v}/|\mathbf{v}| = -(\mathbf{i} - 2\mathbf{j} - 3\mathbf{k})/\sqrt{14} = \left(-1/\sqrt{14}\right)\mathbf{i} + \left(2/\sqrt{14}\right)\mathbf{j} + \left(3/\sqrt{14}\right)\mathbf{k}$.

35.   If $P = (3, -1)$ and $Q = (-4, 1)$, then $\mathbf{v} = \overrightarrow{PQ} = (-4\mathbf{i}+\mathbf{j})-(3\mathbf{i}-\mathbf{j}) = -7\mathbf{i}+2\mathbf{j}$, $|\mathbf{v}| = \sqrt{(-7)^2 + 2^2} = \sqrt{53}$ and $\mathbf{u} = -\mathbf{v}/|\mathbf{v}| = -(-7\mathbf{i} + 2\mathbf{j})/\sqrt{53} = \left(7/\sqrt{53}\right)\mathbf{i} + \left(-2/\sqrt{53}\right)\mathbf{j}$.

37.   If $|\mathbf{v}| = 2$ and $\theta = \pi/3$, then $\mathbf{v} = (2\cos\pi/3, \, 2\sin\pi/3) = (1, \sqrt{3})$.

39.   If $|\mathbf{v}| = 4$ and $\theta = \pi$, then $\mathbf{v} = (4\cos\pi, \, 4\sin\pi) = (-4, 0)$.

41.   If $|\mathbf{v}| = 5$ and the direction is $(0.6, 0.8, 0)$, then $\mathbf{v} = 5\,(0.6, 0.8, 0) = (3, 4, 0)$.

43.   If $|\mathbf{v}| = \sqrt{2}$ and the direction is $\left(\dfrac{1}{2}, \dfrac{-1}{\sqrt{2}}, \dfrac{-1}{2}\right)$, then $\mathbf{v} = \sqrt{2}\,\left(\dfrac{1}{2}, \dfrac{-1}{\sqrt{2}}, \dfrac{-1}{2}\right) = \left(\dfrac{1}{\sqrt{2}}, -1, \dfrac{-1}{\sqrt{2}}\right)$.

45. $\mathbf{u} \cdot \mathbf{v} = (\mathbf{i}-\mathbf{j}) \cdot (\mathbf{i}+2\mathbf{j}) = (1)(1) + (-1)(2) = 1 - 2 = -1$; $|\mathbf{u}| = \sqrt{1^2 + (-1)^2} = \sqrt{2}$, $|\mathbf{v}| = \sqrt{1^2 + 2^2} = \sqrt{5}$, and $\cos\theta = (\mathbf{u} \cdot \mathbf{v})/(|\mathbf{u}|\,|\mathbf{v}|) = -1/\left(\sqrt{2}\,\sqrt{5}\right) = -1/\sqrt{10}$.

47. $\mathbf{u} \cdot \mathbf{v} = (4\mathbf{i} - 7\mathbf{j}) \cdot (5\mathbf{i} + 6\mathbf{j}) = (4)(5) + (-7)(6) = 20 - 42 = -22$; $|\mathbf{u}| = \sqrt{4^2 + (-7)^2} = \sqrt{65}$, $|\mathbf{v}| = \sqrt{5^2 + 6^2} = \sqrt{61}$, and $\cos\theta = (\mathbf{u} \cdot \mathbf{v})/(|\mathbf{u}|\,|\mathbf{v}|) = -22/\left(\sqrt{65}\,\sqrt{61}\right)$.

49. $\mathbf{u} \cdot \mathbf{v} = (\mathbf{i} + 2\mathbf{k}) \cdot (2\mathbf{j} - \mathbf{k}) = (1)(0) + (0)(2) + (2)(-1) = 0 + 0 - 2 = -2$; $|\mathbf{u}| = \sqrt{1^2 + 0^2 + 2^2} = \sqrt{5}$, $|\mathbf{v}| = \sqrt{0^2 + 2^2 + (-1)^2} = \sqrt{5}$, and $\cos\theta = (\mathbf{u} \cdot \mathbf{v})/(|\mathbf{u}|\,|\mathbf{v}|) = -2/\left(\sqrt{5}\,\sqrt{5}\right) = -2/5$.

51. $\mathbf{u} \cdot \mathbf{v} = (\mathbf{i}+\mathbf{j}+\mathbf{k}) \cdot (2\mathbf{i}-\mathbf{j}-\mathbf{k}) = (1)(2) + (1)(-1) + (1)(-1) = 2 - 1 - 1 = 0$; thus $\cos\theta = (\mathbf{u} \cdot \mathbf{v})/(|\mathbf{u}|\,|\mathbf{v}|) = 0/(|\mathbf{u}|\,|\mathbf{v}|) = 0$.

53.   If $\mathbf{u} = 2\mathbf{i} - 6\mathbf{j}$ and $\mathbf{v} = -\mathbf{i} + 3\mathbf{j}$, then $\mathbf{u} = (-2)\mathbf{v}$ so the vectors are parallel.

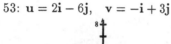

53: $\mathbf{u} = 2\mathbf{i} - 6\mathbf{j}$, $\mathbf{v} = -\mathbf{i} + 3\mathbf{j}$

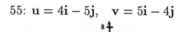

55: $\mathbf{u} = 4\mathbf{i} - 5\mathbf{j}$, $\mathbf{v} = 5\mathbf{i} - 4\mathbf{j}$

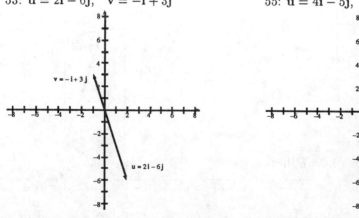

55.   Let $\mathbf{u} = 4\mathbf{i} - 5\mathbf{j}$ and $\mathbf{v} = 5\mathbf{i} - 4\mathbf{j}$. Then $\mathbf{u} \cdot \mathbf{v} = (4)(5) + (-5)(-4) = 20 + 20 = 40 \neq 0$ so the vectors are not orthogonal; $\cos\theta = (\mathbf{u} \cdot \mathbf{v})/(|\mathbf{u}|\,|\mathbf{v}|) = 40/\left(\sqrt{4^2 + (-5)^2}\,\sqrt{5^2 + (-4)^2}\right) = 40/41 \neq \pm 1$ so the vectors are not parallel.

57.   Let $\mathbf{u} = 3\mathbf{i} - 4\mathbf{j} + 5\mathbf{k}$ and $\mathbf{v} = -5\mathbf{i} + 4\mathbf{j} + 3\mathbf{k}$. Then $\mathbf{u} \cdot \mathbf{v} = (3)(-5) + (-4)(4) + (5)(3) = -15 - 16 + 15 = -16 \neq 0$ so the vectors are not orthogonal; $\cos\theta = (\mathbf{u} \cdot \mathbf{v})/(|\mathbf{u}|\,|\mathbf{v}|) = -16/\left(\sqrt{3^2 + (-4)^2 + 5^2}\,\sqrt{(-5)^2 + 4^2 + 3^2}\right) = -16/50 \neq \pm 1$ so the vectors are not parallel.

57: $\mathbf{u} = 3\mathbf{i} - 4\mathbf{j} + 5\mathbf{k}, \quad \mathbf{v} = -5\mathbf{i} + 4\mathbf{j} + 3\mathbf{k}$ 59: $\mathbf{u} = 2\mathbf{i}+\mathbf{j}-3\mathbf{k}, \quad \mathbf{v} = 6\mathbf{i}+3\mathbf{j}+9\mathbf{k}$

59.   Let $\mathbf{u} = 2\mathbf{i}+\mathbf{j}- 3\mathbf{k}$ and $\mathbf{v} = 6\mathbf{i}+3\mathbf{j}+9\mathbf{k}$. Then $\mathbf{u}\cdot\mathbf{v} = (2)(6)+(1)(3)+(-3)(9) = 12+3-27 = -12 \neq 0$ so the vectors are not orthogonal; $\cos\theta = (\mathbf{u}\cdot\mathbf{v})/(|\mathbf{u}|\,|\mathbf{v}|) = -12/\left(\sqrt{2^2 + 1^2 + (-3)^2}\,\sqrt{6^2 + 3^2 + 9^2}\right) = -12/42 \neq \pm 1$ so the vectors are not parallel.

61.   If $\mathbf{u} = 14\mathbf{i}$ and $\mathbf{v} = \mathbf{i} + \mathbf{j}$, then
$$\text{Proj}_{\mathbf{v}}\mathbf{u} = \frac{\mathbf{u}\cdot\mathbf{v}}{|\mathbf{v}|^2}\,\mathbf{v} = \frac{(14\mathbf{i})\cdot(\mathbf{i}+\mathbf{j})}{|\mathbf{i}+\mathbf{j}|^2}(\mathbf{i}+\mathbf{j}) = \frac{14+0}{1+1}(\mathbf{i}+\mathbf{j}) = 7(\mathbf{i}+\mathbf{j}) = 7\mathbf{i} + 7\mathbf{j}.$$

63.   If $\mathbf{u} = 3\mathbf{i} - 2\mathbf{j}$ and $\mathbf{v} = 3\mathbf{i} + 2\mathbf{j}$, then
$$\text{Proj}_{\mathbf{v}}\mathbf{u} = \frac{\mathbf{u}\cdot\mathbf{v}}{|\mathbf{v}|^2}\,\mathbf{v} = \frac{(3\mathbf{i}-2\mathbf{j})\cdot(3\mathbf{i}+2\mathbf{j})}{|3\mathbf{i}+2\mathbf{j}|^2}(3\mathbf{i}+2\mathbf{j}) = \frac{9-4}{9+4}(3\mathbf{i}+2\mathbf{j}) = \left(\frac{5}{13}\right)(3\mathbf{i}+2\mathbf{j}) = \left(\frac{15}{13}\right)\mathbf{i} + \left(\frac{10}{13}\right)\mathbf{j}.$$

65.   If $\mathbf{u} = 2\mathbf{i} - 5\mathbf{j} + \mathbf{k}$ and $\mathbf{v} = -3\mathbf{i} - 7\mathbf{j} + \mathbf{k}$, then
$$\text{Proj}_{\mathbf{v}}\mathbf{u} = \frac{\mathbf{u}\cdot\mathbf{v}}{|\mathbf{v}|^2}\,\mathbf{v} = \frac{(2\mathbf{i}-5\mathbf{j}+\mathbf{k})\cdot(-3\mathbf{i}-7\mathbf{j}+\mathbf{k})}{|-3\mathbf{i}-7\mathbf{j}+\mathbf{k}|^2}(-3\mathbf{i}-7\mathbf{j}+\mathbf{k}) = \frac{-6+35+1}{9+49+1}(-3\mathbf{i}-7\mathbf{j}+\mathbf{k})$$
$$= \left(\frac{30}{59}\right)(-3\mathbf{i}-7\mathbf{j}+\mathbf{k}) = \left(\frac{-90}{59}\right)\mathbf{i} + \left(\frac{-210}{59}\right)\mathbf{j} + \left(\frac{30}{59}\right)\mathbf{k}.$$

67.   If $\mathbf{u} = \mathbf{i} - 2\mathbf{j} + 3\mathbf{k}$ and $\mathbf{v} = 2\mathbf{i} - 4\mathbf{j} + \mathbf{k}$, then
$$\text{Proj}_{\mathbf{v}}\mathbf{u} = \frac{\mathbf{u}\cdot\mathbf{v}}{|\mathbf{v}|^2}\,\mathbf{v} = \frac{(\mathbf{i}-2\mathbf{j}+3\mathbf{k})\cdot(2\mathbf{i}-4\mathbf{j}+\mathbf{k})}{|2\mathbf{i}-4\mathbf{j}+\mathbf{k}|^2}(2\mathbf{i}-4\mathbf{j}+\mathbf{k}) = \frac{2+8+3}{4+16+1}(2\mathbf{i}-4\mathbf{j}+\mathbf{k})$$
$$= \left(\frac{13}{21}\right)(2\mathbf{i}-4\mathbf{j}+\mathbf{k}) = \left(\frac{26}{21}\right)\mathbf{i} + \left(\frac{-52}{21}\right)\mathbf{j} + \left(\frac{13}{21}\right)\mathbf{k}.$$

69. $(7\mathbf{j}) \times (\mathbf{i} - \mathbf{k}) = \begin{vmatrix} \mathbf{i} & \mathbf{j} & \mathbf{k} \\ 0 & 7 & 0 \\ 1 & 0 & -1 \end{vmatrix} = \begin{vmatrix} 7 & 0 \\ 0 & -1 \end{vmatrix}\mathbf{i} - \begin{vmatrix} 0 & 0 \\ 1 & -1 \end{vmatrix}\mathbf{j} + \begin{vmatrix} 0 & 7 \\ 1 & 0 \end{vmatrix}\mathbf{k} = -7\mathbf{i} - 7\mathbf{k}.$

71. $(4\mathbf{i} - \mathbf{j} + 7\mathbf{k}) \times (-7\mathbf{i} + \mathbf{j} - 2\mathbf{k}) = \begin{vmatrix} \mathbf{i} & \mathbf{j} & \mathbf{k} \\ 4 & -1 & 7 \\ -7 & 1 & -2 \end{vmatrix} = \begin{vmatrix} -1 & 7 \\ 1 & -2 \end{vmatrix}\mathbf{i} - \begin{vmatrix} 4 & 7 \\ -7 & -2 \end{vmatrix}\mathbf{j} + \begin{vmatrix} 4 & -1 \\ -7 & 1 \end{vmatrix}\mathbf{k} =$

$-5\mathbf{i} - 41\mathbf{j} - 3\mathbf{k}.$

73.   If $P = (3, -1, 4)$ and $Q = (-1, 6, 2)$, then $\mathbf{v} = \overrightarrow{PQ} = \mathbf{q}-\mathbf{p} = (-\mathbf{i}+6\mathbf{j}+2\mathbf{k})-(3\mathbf{i}-\mathbf{j}+4\mathbf{k}) = -4\mathbf{i}+7\mathbf{j}-2\mathbf{k}$. The line passing through P and Q has

vector equation :	$x\mathbf{i} + y\mathbf{j} + z\mathbf{k} = \mathbf{p} + t\mathbf{v} = (3\mathbf{i} - \mathbf{j} + 4\mathbf{k}) + t(-4\mathbf{i} + 7\mathbf{j} - 2\mathbf{k});$
parametric equations :	$x = 3 - 4t, \quad y = -1 + 7t, \quad z = 4 - 2t;$
symmetric equations :	$\dfrac{x - 3}{-4} = \dfrac{y - (-1)}{7} = \dfrac{z - 4}{-2}.$

75.   The line passing through $P = (3, 1, 2)$ and parallel to $\mathbf{v} = 3\mathbf{i} - \mathbf{j} - \mathbf{k}$ has

vector equation :	$x\mathbf{i} + y\mathbf{j} + z\mathbf{k} = \mathbf{p} + t\mathbf{v} = (3\mathbf{i} + \mathbf{j} + 2\mathbf{k}) + t(3\mathbf{i} - \mathbf{j} - \mathbf{k});$
parametric equations :	$x = 3 + 3t, \quad y = 1 - t, \quad z = 2 - t;$
symmetric equations :	$\dfrac{x - 3}{3} = \dfrac{y - 1}{-1} = \dfrac{z - 2}{-1}.$

77.   The plane through the point $P = (-1, 1, 1)$ and normal to the vector $\mathbf{N} = \mathbf{j} - \mathbf{k}$ satisfies the equation
$0 = \mathbf{N} \cdot \overrightarrow{PQ} = (\mathbf{j} - \mathbf{k}) \cdot ((x + 1)\mathbf{i} + (y - 1)\mathbf{j} + (z - 1)\mathbf{k}) = (y - 1) - (z - 1) = y - z.$

79.   The plane through the point $P = (1, -4, 6)$ and normal to the vector $\mathbf{N} = 2\mathbf{j} - 3\mathbf{k}$ satisfies the equation
$0 = \mathbf{N} \cdot \overrightarrow{PQ} = (2\mathbf{j} - 3\mathbf{k}) \cdot ((x-1)\mathbf{i} + (y+4)\mathbf{j} + (z-6)\mathbf{k}) = 2(y+4) - 3(z-6) = 2y - 3z + 26.$

Note: Rectangular coordinates will be written with subscript "r", cylindrical coordinates with subscript "c", and spherical coordinates with subscript "s": e.g., $(x, y, z)_r$ or $(r, \theta, z)_c$ or $(\rho, \theta, \phi)_s$.

81.   If $(r, \theta, z)_c = \left(3, \dfrac{\pi}{6}, -1\right)_c$, then $(x, y, z)_r = (r\cos\theta, r\sin\theta, z)_r = \left(3\cos\dfrac{\pi}{6}, 3\sin\dfrac{\pi}{6}, -1\right)_r = \left(\dfrac{3\sqrt{3}}{2}, \dfrac{3}{2}, -1\right)_r$.

83.   If $(x, y, z)_r = (2, 2, -4)_r$, then $(r, \theta, z)_c = \left(\sqrt{2^2 + 2^2},\ \tan^{-1}\dfrac{2}{2},\ -4\right)_c = \left(2\sqrt{2},\ \dfrac{\pi}{4},\ -4\right)_c$.

85.   If $(\rho, \theta, \phi)_s = \left(3, \dfrac{\pi}{3}, \dfrac{\pi}{4}\right)_s$, then $(x, y, z)_r = \left(3\cos\dfrac{\pi}{3}\sin\dfrac{\pi}{4}, 3\sin\dfrac{\pi}{3}\sin\dfrac{\pi}{4}, 3\cos\dfrac{\pi}{4}\right)_r = \left(\dfrac{3}{2\sqrt{2}}, \dfrac{3\sqrt{3}}{2\sqrt{2}}, \dfrac{3}{\sqrt{2}}\right)_r$.

87.   If $(x, y, z)_r = (-1, 1, -\sqrt{2})_r$, then

$$(\rho, \theta, \phi)_s = \left(\sqrt{(-1)^2 + 1^2 + (-\sqrt{2})^2},\ \pi + \tan^{-1}\dfrac{1}{-1},\ \cos^{-1}\dfrac{-\sqrt{2}}{\sqrt{(-1)^2 + 1^2 + (-\sqrt{2})^2}}\right)_s = \left(2, \dfrac{3\pi}{4}, \dfrac{3\pi}{4}\right)_s.$$

89. $25 = x^2 + y^2 + z^2 = r^2 + z^2 = \rho^2.$

91.   cylindrical: $1 = x^2 - y^2 + z^2 = (r\cos\theta)^2 - (r\sin\theta)^2 + z^2 = r^2\left(\cos^2\theta - \sin^2\theta\right) + z^2.$
spherical: $1 = x^2 - y^2 + z^2 = (x^2 + y^2 + z^2) - 2y^2 = \rho^2 - 2(\rho\sin\theta\sin\phi)^2 = \rho^2\left(1 - 2\sin^2\theta\sin^2\phi\right).$

93: $y = 3 - 5x$ 95: $y = z^2$

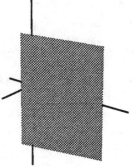

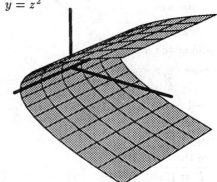

97.   If $\mathbf{u} = \mathbf{i} - \mathbf{j}$ and $\mathbf{w} = a\mathbf{i} + b\mathbf{j}$, then $\mathbf{u} \perp \mathbf{w} \iff 0 = \mathbf{u} \cdot \mathbf{w} = a - b$. Hence, $(\mathbf{i} + \mathbf{j})/\sqrt{2}$ and $(-\mathbf{i} - \mathbf{j})/\sqrt{2}$ are the unit vectors in $\mathbb{R}^2$ orthogonal to $\mathbf{i} - \mathbf{j}$.

99.   Let $\mathbf{u} = 2\mathbf{i} + 3\mathbf{j}$ and $\mathbf{v} = 4\mathbf{i} + \alpha\mathbf{j}$.

a. $\mathbf{u} \parallel \mathbf{v}$ if and only if there is a real k such that $\mathbf{v} = k\mathbf{u}$. We see that $\mathbf{v} = k\mathbf{u} \iff 4 = 2k$ and $\alpha = 3k \iff k = 2$ and $\alpha = 6$; hence $\mathbf{u}$ and $\mathbf{v}$ are parallel if $\alpha = 6$.

b. $\mathbf{u} \perp \mathbf{v} \iff 0 = \mathbf{u} \cdot \mathbf{v} = (2\mathbf{i} + 3\mathbf{j}) \cdot (4\mathbf{i} + \alpha\mathbf{j}) = 8 + 3\alpha \iff \alpha = -8/3.$

Prologue to parts c & d: Suppose $\mathbf{u} = a\mathbf{i} + b\mathbf{j}$ and $\mathbf{v} = p\mathbf{i} + q\mathbf{j}$. Also suppose a, b, p are constants and our task is to find q so that $\mathbf{u}$ and $\mathbf{v}$ have a specified angle between them.

$$t = \cos\theta = \frac{\mathbf{u} \cdot \mathbf{v}}{|\mathbf{u}|\,|\mathbf{v}|} = \frac{ap + bq}{\sqrt{a^2 + b^2}\,\sqrt{p^2 + q^2}}$$

$$\implies (a^2 + b^2)\,(p^2 + q^2)\,t^2 = (ap + bq)^2 = a^2p^2 + 2abpq + b^2q^2$$

$$\iff \left[(a^2 + b^2)t^2 - b^2\right]q^2 - 2abpq + \left[(a^2 + b^2)t^2 - a^2\right]p^2 = 0$$

$$\iff q = \frac{2abp \pm \sqrt{(-2abp)^2 - 4\left[(a^2 + b^2)t^2 - b^2\right]\left[(a^2 + b^2)t^2 - a^2\right]p^2}}{2\left[(a^2 + b^2)t^2 - b^2\right]}$$

$$= \frac{2abp \pm \sqrt{4(a^2 + b^2)^2 p^2 t^2(1 - t^2)}}{2\left[(a^2 + b^2)t^2 - b^2\right]}$$

$$= p\,\frac{ab \pm (a^2 + b^2)|t|\sqrt{1 - t^2}}{(a^2 + b^2)t^2 - b^2} = \frac{p}{2}\,\frac{2ab \pm (a^2 + b^2)|\sin 2\theta|}{a^2\cos^2\theta - b^2\sin^2\theta}.$$

This chain of implications involves a non-reversable step (squaring), so the formula produces solution-candidates which must be checked by explicit substitution. As an extreme instance, note that $t = -1$ and $t = 1$ both imply $q = pb/a$; that value of q can produce at most one of ± 1 as the value of $\cos\theta$.

For this problem, $a = 2$, $b = 3$, and $p = 4$; our formula to compute α is $\alpha = 4\dfrac{6 \pm 13|t|\sqrt{1 - t^2}}{13t^2 - 9}$.

c. $\cos(\pi/4) = 1/\sqrt{2} = t$ implies $\alpha = 4\dfrac{6 \pm 13/2}{13/2 - 9} = \begin{cases} -20 & (+) \\ 4/5 & (-) \end{cases}$. If $\alpha = -20$, then $\cos\theta = -1/\sqrt{2}$;

however, if $\alpha = 4/5$, then $\cos\theta = 1/\sqrt{2}$ Therefore, $\alpha = 4/5$ is the unique solution.

d. $\cos(\pi/6) = \sqrt{3}/2 = t$ implies $\alpha = 4\dfrac{6 \pm 13\sqrt{3}/4}{39/4 - 9} = \dfrac{4}{3}\left(24 \pm 13\sqrt{3}\right) \approx \begin{cases} 62.022 & (+) \\ 1.978 & (-) \end{cases}$. Numeric

computations show that both of these values for α work.

REMARK: A quick sketch and computation will show why part (c) has only one value for α. Vectors $\mathbf{v} = 4\mathbf{i} + \alpha\mathbf{j}$ go from the origin to the vertical line $x = 4$. Vector $\mathbf{u} = 2\mathbf{i} + 3\mathbf{j}$ has angle $\tan^{-1}(2/3) \approx 0.59$ with the vertical and this is less than $\pi/4$; hence a counterclockwise rotation of $\pi/4$ from $\mathbf{u}$ ends in the second quadrant where the line $x = 4$ does not go.

101. If $P = (-2, 4, 1)$, $Q = (3, -7, 5)$, $R = (-1, -2, -1)$, then $\overrightarrow{PQ} = 5\mathbf{i} - 11\mathbf{j} + 4\mathbf{k}$ and $\overrightarrow{QR} = -4\mathbf{i} + 5\mathbf{j} - 6\mathbf{k}$. Therefore

$$\mathbf{N} = \overrightarrow{PQ} \times \overrightarrow{QR} = \begin{vmatrix} \mathbf{i} & \mathbf{j} & \mathbf{k} \\ 5 & -11 & 4 \\ -4 & 5 & -6 \end{vmatrix} = \begin{vmatrix} -11 & 4 \\ 5 & -6 \end{vmatrix}\mathbf{i} - \begin{vmatrix} 5 & 4 \\ -4 & -6 \end{vmatrix}\mathbf{j} + \begin{vmatrix} 5 & -11 \\ -4 & 5 \end{vmatrix}\mathbf{k} = 46\mathbf{i} + 14\mathbf{j} - 19\mathbf{k}$$

and the plane through P, Q, R satisfies the equations

$$0 = \mathbf{N} \cdot \overrightarrow{Q(x,y,z)} = \mathbf{N} \cdot ((x - 3)\mathbf{i} + (y - (-7))\mathbf{j} + (z - 5)\mathbf{k}) = 46(x - 3) + 14(y + 7) - 19(z - 5)$$
$$\Longleftrightarrow -55 = 46x + 14y - 19z.$$

103. Suppose (x, y, z) is on both $\Pi_1 : -x + y + z = 3$ and $\Pi_2 : -4x + 2y - 7z = 5$.

$$\begin{array}{l} -x + y + z = 3 \\ -4x + 2y - 7z = 5 \end{array} \Longleftrightarrow \begin{array}{l} x - y - z = -3 \\ -2y - 11z = -7 \end{array} \Longleftrightarrow \begin{array}{l} x + 9/2\,z = 1/2 \\ y + 11/2\,z = 7/2 \end{array} \Longleftrightarrow \begin{array}{l} x = 1/2 - 9/2\,z \\ y = 7/2 - 11/2\,z \end{array}$$

Planes Π_1 and Π_2 meet in a line which has parametric equation $(x, y, z) = (1/2, 7/2, 0) + (-9/2, -11/2, 1)t$.

REMARK: $(5, 9, -1)$ is on both planes and $\mathbf{N}_1 \times \mathbf{N}_2 = (-\mathbf{i} + \mathbf{j} + \mathbf{k}) \times (-4\mathbf{i} + 2\mathbf{j} - 7\mathbf{k}) = -9\mathbf{i} - 11\mathbf{j} + 2\mathbf{k}$, hence the common line is also given by $x\mathbf{i} + y\mathbf{j} + z\mathbf{k} = (5\mathbf{i} + 9\mathbf{j} - \mathbf{k}) + s(-9\mathbf{i} - 11\mathbf{j} + 2\mathbf{k})$.

105. Let $P = (2, 3)$, $Q = (7, -3)$, and $\mathbf{v} = 2\mathbf{i} - \mathbf{j}$. $\mathbf{v}$ is orthogonal to $\mathbf{i} + 2\mathbf{j}$ and to the line with equation $(x, y) = Q + t(1, 2)$; hence, the distance from P to that line is the length of the projection of $\overrightarrow{PQ}$ on $\mathbf{v}$:

$$\left|\operatorname{Proj}_{\mathbf{v}} \overrightarrow{PQ}\right| = \frac{|(5\mathbf{i} - 6\mathbf{j}) \cdot (2\mathbf{i} - \mathbf{j})|}{|2\mathbf{i} - \mathbf{j}|} = \frac{|10 + 6|}{\sqrt{4 + 1}} = \frac{16}{\sqrt{5}}.$$

107. The distance from $(1, -2, 3)$ to the plane $2x - y - z - 6 = 0$ is (invoking problem 11.7.45 or 11.7.46)

$$\frac{|(2)(1) + (-1)(-2) + (-1)(3) + (-6)|}{\sqrt{2^2 + (-1)^2 + (-1)^2}} = \frac{|2 + 2 - 3 - 6|}{\sqrt{4 + 1 + 1}} = \frac{5}{\sqrt{6}}.$$

109. $\mathbf{n}_1 = 4\mathbf{i} + 3\mathbf{j} - 2\mathbf{k}$ is parallel to line $L_1 : (x - 1)/4 = (y + 6)/3 = z/(-2)$ and $\mathbf{n}_2 = 5\mathbf{i} + \mathbf{j} + 4\mathbf{k}$ is parallel to line $L_2 : (x + 3)/5 = y - 1 = (z + 3)/4$; therefore $(4\mathbf{i} + 3\mathbf{j} - 2\mathbf{k}) \times (5\mathbf{i} + \mathbf{j} + 4\mathbf{k}) = \begin{vmatrix} \mathbf{i} & \mathbf{j} & \mathbf{k} \\ 4 & 3 & -2 \\ 5 & 1 & 4 \end{vmatrix} =$

$\begin{vmatrix} 3 & -2 \\ 1 & 4 \end{vmatrix}\mathbf{i} - \begin{vmatrix} 4 & -2 \\ 5 & 4 \end{vmatrix}\mathbf{j} + \begin{vmatrix} 4 & 3 \\ 5 & 1 \end{vmatrix}\mathbf{k} = 14\mathbf{i} - 26\mathbf{j} - 11\mathbf{k}$ and line $L_3 : (x+1)/14 = (y-2)/(-26) = (z-4)/(-11)$

passes through $(-1, 2, 4)$ and is orthogonal to $\mathbf{n}_1$ and $\mathbf{n}_2$.

REMARK: Line L_3 does not meet line L_1 or line L_2.

111. If $P = (1, 3, 0)$, $Q = (3, -1, -2)$, $R = (-1, 7, 2)$, then $\overrightarrow{PQ} = 2\mathbf{i} - 4\mathbf{j} - 2\mathbf{k}$ and $\overrightarrow{QR} = -4\mathbf{i} + 8\mathbf{j} + 4\mathbf{k}$. Since $\overrightarrow{QR} = (-2)\overrightarrow{PQ}$, the points are collinear.

REMARK: P is the midpoint of $\overline{QR}$.

113.   If $P = (1, 4, -2)$, $Q = (-3, 1, 6)$, $R = (1, -2, 3)$, then $\overrightarrow{PQ} = -4\mathbf{i} - 3\mathbf{j} + 8\mathbf{k}$, $\overrightarrow{QR} = 4\mathbf{i} - 3\mathbf{j} - 3\mathbf{k}$, $\overrightarrow{PQ} \times \overrightarrow{QR} = 33\mathbf{i} + 20\mathbf{j} + 24\mathbf{k}$, and a parallelogram with adjacent vertices at P, Q, R (in some order) has area $|\overrightarrow{PQ} \times \overrightarrow{QR}| = \sqrt{33^2 + 20^2 + 24^2} = \sqrt{2065}$.

REMARK: $\overrightarrow{QR} \times \overrightarrow{RP} = \overrightarrow{QR} \times (\overrightarrow{RQ} + \overrightarrow{QP}) = (\overrightarrow{QR} \times \overrightarrow{RQ}) + (\overrightarrow{QR} \times \overrightarrow{QP}) = \mathbf{0} + (\overrightarrow{QR} \times \overrightarrow{QP}) = (\overrightarrow{QR} \times (-\overrightarrow{PQ})) = -((-\overrightarrow{PQ}) \times \overrightarrow{QR}) = \overrightarrow{PQ} \times \overrightarrow{QR}$; similarly, $\overrightarrow{RP} \times \overrightarrow{PQ} = \overrightarrow{PQ} \times \overrightarrow{QR}$.

115: $x^2 + y^2 = 3^2$ 117: $x^2 + \frac{z^2}{9} = 1 + \frac{y^2}{4}$ 119: $9(x^2 + z^2) = 16y^2 - 25$

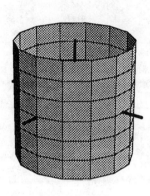
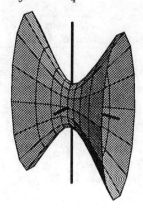
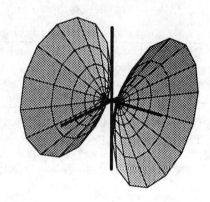

121.   Suppose point (a, b, c) is on both lines.

$$3 - 2t = a = -3 + s \qquad -s - 2t = -6 \qquad s + 2t = 6$$
$$4 + t = b = 2 - 4s \iff 4s + t = -2 \iff -7t = -26 \implies 26/7 = t = 39/19, \quad \text{impossible.}$$
$$-2 + 7t = c = 1 + 6s \qquad -6s + 7t = 3 \qquad 19t = 39$$

Assuming a common point leads to a contradiction; therefore the lines do not meet.

12

Vector Functions, Vector Differentiation, and Parametric Equations

Solutions 12.1   *Vector functions and Parametric equations*   (pages 793-5)

1. $\operatorname{dom}\dfrac{1}{t} = \mathbb{R}-\{0\}$ and $\operatorname{dom}\dfrac{1}{t-1} = \mathbb{R}-\{1\}$; therefore, if $\mathbf{f}(t) = \dfrac{1}{t}\mathbf{i}+\dfrac{1}{t-1}\mathbf{j}$, then $\operatorname{dom}\mathbf{f} = \big(\mathbb{R}-\{0\}\big)\cap$
$\big(\mathbb{R}-\{1\}\big) = \mathbb{R}-\{0,1\} = (-\infty,0)\cup(0,1)\cup(1,\infty)$.

3. $\operatorname{dom}\dfrac{1}{s^2-1} = \mathbb{R}-\{-1,1\}$ and $\operatorname{dom}(s^2-1) = \mathbb{R}$; therefore, if $\mathbf{f}(s) = \dfrac{1}{s^2-1}\mathbf{i}+(s^2-1)\mathbf{j}$, then
$\operatorname{dom}\mathbf{f} = \big(\mathbb{R}-\{-1,1\}\big)\cap\mathbb{R} = \mathbb{R}-\{-1,1\} = (-\infty,-1)\cup(-1,1)\cup(1,\infty)$.

5. $\operatorname{dom}(\ln r) = (0,\infty)$, $\operatorname{dom}\big(\ln(1-r)\big) = (-\infty,1)$, and $\operatorname{dom}\big(\ln(1+r)\big) = (-1,\infty)$; therefore, if $\mathbf{f}(r) =$
$(\ln r)\mathbf{i}+\big(\ln(1-r)\big)\mathbf{j}+\big(\ln(1+r)\big)\mathbf{k}$, then $\operatorname{dom}\mathbf{f} = (0,\infty)\cap(-\infty,1)\cap(-1,\infty) = (0,1)$.

7. $\operatorname{dom}(\sec t) = \operatorname{dom}(1/\cos t) = \mathbb{R}-\{(i+1/2)\pi\}$, $\operatorname{dom}(\csc t) = \operatorname{dom}(1/\sin t) = \mathbb{R}-\{j\pi\}$, $\operatorname{dom}(\cos 2t) =$
$\mathbb{R}$; therefore, if $\mathbf{f}(t) = (\sec t)\mathbf{i}+(\csc t)\mathbf{j}+(\cos 2t)\mathbf{k}$, then $\operatorname{dom}\mathbf{f} = \big(\mathbb{R}-\{(i+1/2)\pi\}\big)\cap\big(\mathbb{R}-\{j\pi\}\big)\cap\mathbb{R} =$
$\mathbb{R}-\{(k/2)\pi\}$.

9.   Let $\mathbf{f}(t) = x\mathbf{i}+y\mathbf{j} = t^2\mathbf{i}+2t\mathbf{j}$. Then $y = 2t \implies y/2 = t$ and $x = t^2 = (y/2)^2 = y^2/4$.

9: $x = y^2/4$

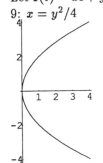

11: $x = y^{2/3}$

11.   Let $\mathbf{f}(t) = x\mathbf{i}+y\mathbf{j} = t^2\mathbf{i}+t^3\mathbf{j}$. Then $y = t^3 \implies y^{1/3} = t$ and $x = t^2 = [y^{1/3}]^2 = y^{2/3}$.

REMARK: $x = t^2 \implies (x \geq 0 \text{ and } \pm\sqrt{x} = t)$ and $y = t^3 = [\pm\sqrt{x}]^3 = \pm x^{3/2}$.

13.   If $\mathbf{f}(t) = x\mathbf{i}+y\mathbf{j} = (2t-1)\mathbf{i}+(4t+3)\mathbf{j}$, then $x = 2t-1 \implies (x+1)/2 = t$ and $y = 4t+3 = 4[(x+1)/2]+3 =$
$2(x+1)+3 = 2x+5$.

13: $y = 2x+5$

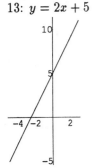

15: $x = y^2+y+1, y \geq 0$

15.   If $\mathbf{f}(t) = x\mathbf{i}+y\mathbf{j} = (t^4+t^2+1)\mathbf{i}+t^2\mathbf{j}$, then $y \geq 0$ and $x = t^4+t^2+1 = (t^2)^2+t^2+1 = y^2+y+1$.

17.   If $\mathbf{f}(t) = x\mathbf{i} + y\mathbf{j} = t^3\mathbf{i} + (t^9 - 1)\mathbf{j}$, then $y = t^9 - 1 = (t^3)^3 - 1 = x^3 - 1$.

17: $y = x^3 - 1$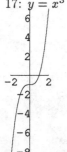

19: $y = (\ln x)^2$

19.   If $\mathbf{f}(t) = x\mathbf{i} + y\mathbf{j} = e^t\mathbf{i} + t^2\mathbf{j}$, then ($x > 0$ and $\ln x = t$) and $y = t^2 = (\ln x)^2$.

21.   Let $\mathbf{f}(t) = x\mathbf{i} + y\mathbf{j} = e^t \sin t\,\mathbf{i} + e^t \cos t\,\mathbf{j}$. Thus $x^2 + y^2 = (e^t \sin t)^2 + (e^t \cos t)^2 = e^{2t}\left(\sin^2 t + \cos^2 t\right) = e^{2t}$ which implies $t = (1/2)\ln\left(x^2 + y^2\right)$ and $x/y = (e^t \sin t)/(e^t \cos t) = \tan t = \tan\left[(1/2)\ln\left(x^2 + y^2\right)\right]$.

REMARK: $(1/2)\ln\left(x^2 + y^2\right) = \ln\sqrt{x^2 + y^2} = \ln r$; this curve, like Solution 8.6.43, is a logarithmic spiral.

21: $x/y = \tan\left[(1/2)\ln\left(x^2 + y^2\right)\right]$

23: $y = x^2$, $x \geq 0$

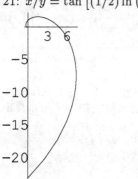

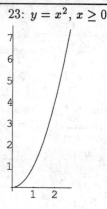

23.   If $\mathbf{f}(t) = x\mathbf{i} + y\mathbf{j} = e^t\mathbf{i} + e^{2t}\mathbf{j}$, then $x = e^t > 0$ and $y = e^{2t} = (e^t)^2 = x^2$.

25.   If $\mathbf{f}(t) = x\mathbf{i} + y\mathbf{j} + z\mathbf{k} = 2t\mathbf{i} - 3t\mathbf{j} + t\mathbf{k}$, then $t = z = x/2 = y/(-3)$; the curve is a straight line through the origin.

25: $z = x/2 = y/(-3)$

27: $y = \cos x$, $z = \sin x$

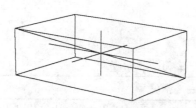

27.   If $\mathbf{f}(t) = x\mathbf{i} + y\mathbf{j} + z\mathbf{k} = t\mathbf{i} + \cos t\,\mathbf{j} + \sin t\,\mathbf{k}$, then $y = \cos x$ and $z = \sin x$.

REMARK: $y^2 + z^2 = \cos^2 t + \sin^2 t = 1$, the curve is a circular helix.

29.   The line through $(2, 4)$ and $(1, 6)$ has parametric equations $x = 2 + t(1 - 2) = 2 - t$ and $y = 4 + t(6 - 4) = 4 + 2t$.

31.   The line through $(3, 5)$ and $(-1, -7)$ has parametric equations $x = 3 + t(-1 - 3) = 3 - 4t$ and $y = 5 + t(-7 - 5) = 5 - 12t$.

33.   The line through $(-2, 3)$ and $(4, 7)$ has parametric equations $x = -2 + t(4 - (-2)) = -2 + 6t$ and $y = 3 + t(7 - 3) = 3 + 4t$.

35.   The line through $(1, 3, 5)$ and $(2, 4, 6)$ has parametric equations $x = 1 + t(2 - 1) = 1 + t$, $y = 3 + t(4 - 3) = 3 + t$, and $z = 5 + t(6 - 5) = 5 + t$.

37.   The line through $(-3, 0, 7)$ and $(-5, 3, 0)$ has parametric equations $x = -3 + t(-5 - (-3)) = -3 - 2t$, $y = 0 + t(3 - 0) = 3t$, and $z = 7 + t(0 - 7) = 7 - 7t$.

39.   If $x = a\cos t$ and $y = b\sin t$, then the range of $x(t)$ is $[-|a|,|a|]$, the range of $y(t)$ is $[-|b|,|b|]$, and $x^2/a^2 + y^2/b^2 = \cos^2 t + \sin^2 t = 1$. ∎

REMARK: The ellipse is "traced out" once for $t \in [0, 2\pi)$.

41.   a. $\overrightarrow{OR} + \overrightarrow{RC} + \overrightarrow{CP}$ corresponds to a directed path which goes from O to P (by way of intermediate points R and C), hence the vector sum equals $\overrightarrow{OP}$. ∎

   b.   An arc of size α radians on a circle with radius r has length αr. If the wheel rolls without slipping, then $\overrightarrow{OR}$ has length equal to that arc length; the wheel rolls along the positive x-axis, hence $\overrightarrow{OR} = \alpha r \mathbf{i}$. ∎

   c. P is distance s from the circle's center, that does not change as the wheel rolls; θ is the angle from horizontal to P, hence $\overrightarrow{CP} = s\cos\theta\mathbf{i} + s\sin\theta\mathbf{j} = s(\cos\theta\mathbf{i} + \sin\theta\mathbf{j})$. ∎

   d. $\overrightarrow{OP} = \overrightarrow{OR} + \overrightarrow{RC} + \overrightarrow{CP} = (\alpha r\mathbf{i}) + (r\mathbf{j}) + s(\cos\theta\mathbf{i} + \sin\theta\mathbf{j}) = (\alpha r + s\cos\theta)\mathbf{i} + (r + s\sin\theta)\mathbf{j}$. ∎

   e.   The wheel rolls counter-clockwise so θ decreases as α increases. Those two angles change by equal amounts so $\theta + \alpha$ is constant. In the initial position we have $\alpha = 0$ and $\theta = 3\pi/2$, therefore $\theta + \alpha = 3\pi/2$ and $\theta = 3\pi/2 - \alpha$ for all α. ∎

   f. $\cos\theta = \cos(3\pi/2 - \alpha) = \cos(3\pi/2)\cos\alpha + \sin(3\pi/2)\sin\alpha = (0)\cos\alpha + (-1)\sin\alpha = -\sin\alpha$; $\sin\theta = \sin(3\pi/2 - \alpha) = \sin(3\pi/2)\cos\alpha - \cos(3\pi/2)\sin\alpha = (-1)\cos\alpha - (0)\sin\alpha = -\cos\alpha$.

   g. $x\mathbf{i} + y\mathbf{j} = \overrightarrow{OP} = (\alpha r + s\cos\theta)\mathbf{i} + (r + s\sin\theta)\mathbf{j} = (r\alpha - s\sin\alpha)\mathbf{i} + (r - s\cos\theta)\mathbf{j}$; hence $x = r\alpha - s\sin\alpha$ and $y = r - s\cos\alpha$. ∎

REMARK: Note that $y = \frac{d}{d\alpha}x$.

43.   a.   Let $s = r$ in equation (8): $x = r\alpha - r\sin\alpha = r(\alpha - \sin\alpha)$ and $y = r - r\cos\alpha = r(1 - \cos\alpha)$.

   b. $y = r(1 - \cos\alpha) \implies y \in r[1 - 1, 1 - (-1)] = [0, 2r]$ and $y/r = 1 - \cos\alpha$; if $\alpha \in [0, \pi]$, then $\alpha = \cos^{-1}(1 - y/r)$. Solution 6.8.69 shows $\sin\cos^{-1}w = \sqrt{1 - w^2}$ for all $w \in [-1, 1]$; hence $r\sin\alpha = r\sin\cos^{-1}(1 - y/r) = r\sqrt{1 - (1 - y/r)^2} = \sqrt{2ry - y^2}$. Therefore $x = r\alpha - r\sin\alpha = r\cos^{-1}(1 - y/r) - \sqrt{2ry - y^2}$.

45.   If the small circle rolls without slipping, then arc QR on the large circle and arc PR on the small circle have the same length; thus $a\theta = b\alpha$. The angle from horizontal to radius CP is $\theta - \alpha = \theta - (a/b)\theta = (1 - a/b)\theta$. Point P on the hypocycloid has path

$$x\mathbf{i} + y\mathbf{j} = \overrightarrow{OP} = \overrightarrow{OC} + \overrightarrow{CP} = [(a - b)\cos\theta\mathbf{i} + (a - b)\sin\theta\mathbf{j}] + [b\cos(\theta - \alpha)\mathbf{i} + b\sin(\theta - \alpha)\mathbf{j}]$$
$$= [(a - b)\cos\theta + b\cos(\theta - \alpha)]\mathbf{i} + [(a - b)\sin\theta + b\sin(\theta - \alpha)]\mathbf{j}$$
$$= \left[(a - b)\cos\theta + b\cos\left(1 - \frac{a}{b}\right)\theta\right]\mathbf{i} + \left[(a - b)\sin\theta + b\sin\left(1 - \frac{a}{b}\right)\theta\right]\mathbf{j}$$
$$= \left[(a - b)\cos\theta + b\cos\left(\frac{a}{b} - 1\right)\theta\right]\mathbf{i} + \left[(a - b)\sin\theta - b\sin\left(\frac{a}{b} - 1\right)\theta\right]\mathbf{j}. \ ∎$$

47.   Solution 46 implies that $a = 4b$ yields the hypocycloid of four cusps satisfying

$$x = b(3\cos\theta + \cos 3\theta) = b\left(3\cos\theta + (\cos^3\theta - 3\sin^2\theta\cos\theta)\right) = b\left(4\cos^3\theta\right) = a\cos^3\theta,$$
$$y = b(3\sin\theta - \sin 3\theta) = b\left(3\sin\theta - (3\sin\theta\cos^2\theta - \sin^3\theta)\right) = b\left(4\sin^3\theta\right) = a\sin^3\theta,$$
$$x^{2/3} + y^{2/3} = a^{2/3}\cos^2\theta + a^{2/3}\sin^2\theta = a^{2/3}\left(\cos^2\theta + \sin^2\theta\right) = a^{2/3}. \ ∎$$

49.   Suppose P, Q, R are three distinct points which lie on a straight line.

   a. $\overrightarrow{PR}$ and $\overrightarrow{PQ}$ have the same or opposite direction, hence one vector is a scalar multiple of the other; since the points are distinct, there is a nonzero t such that $\overrightarrow{PR} = t\overrightarrow{PQ}$. ∎

   b.   For any choice of points P and R we have $\overrightarrow{OR} = \overrightarrow{OP} + \overrightarrow{PR}$. If P, Q, R are collinear, then part (a) implies there is some t such that $\overrightarrow{PR} = t\overrightarrow{PQ}$. Therefore $\overrightarrow{OR} = \overrightarrow{OP} + \overrightarrow{PR} = \overrightarrow{OP} + t\overrightarrow{PQ}$. ∎

   c.   Let $P = (x_1, y_1)$, $Q = (x_2, y_2)$, $R = (x, y)$. Then $\overrightarrow{OP} = x_1\mathbf{i} + y_1\mathbf{j}$, $\overrightarrow{PQ} = (x_2 - x_1)\mathbf{i} + (y_2 - y_1)\mathbf{j}$, and $x\mathbf{i} + y\mathbf{j} = \overrightarrow{OR} = \overrightarrow{OP} + \overrightarrow{PR} = \overrightarrow{OP} + t\overrightarrow{PQ} = (x_1\mathbf{i} + y_1\mathbf{j}) + t\left((x_2 - x_1)\mathbf{i} + (y_2 - y_1)\mathbf{j}\right) = (x_1 + t(x_2 - x_1))\mathbf{i} + (y_1 + t(y_2 - y_1))\mathbf{j}$. Hence $x = x_1 + t(x_2 - x_1)$ and $y = y_1 + t(y_2 - y_1)$. ∎

Solutions 12.2 *Equation of tangent line to a plane curve & Smoothness* (pages 798-9)

NOTE: Since equation (2) is stated in terms of a limit for a quotient, $m = \lim\limits_{t \to t_0} \dfrac{y'(t)}{x'(t)}$, I will algebraically simplify such a quotient by cancelling common factors before computing a numerical value.

1. $x = t^3 \implies \dfrac{d}{dt} x = 3t^2$, $y = t^4 - 5 \implies \dfrac{d}{dt} y = 4t^3$; thus $\dfrac{dy}{dx} = \dfrac{y'}{x'} = \dfrac{4t^3}{3t^2} = \dfrac{4t}{3}$ and $m(t = -1) = \dfrac{-4}{3}$.

3. If $x = \cos\theta$ and $y = \sin\theta$, then $\dfrac{dy}{dx} = \dfrac{dy/d\theta}{dx/d\theta} = \dfrac{\cos\theta}{-\sin\theta} = -\cot\theta$ and $m(\theta = \pi/4) = -1$.

5. If $x = \sec\theta$ and $y = \tan\theta$, then $\dfrac{dy}{dx} = \dfrac{dy/d\theta}{dx/d\theta} = \dfrac{\sec^2\theta}{\sec\theta\,\tan\theta} = \dfrac{\sec\theta}{\tan\theta} = \dfrac{1}{\sin\theta}$ and $m(\theta = \pi/4) = \sqrt{2}$.

7. If $x = 8\cos\theta$ and $y = -3\sin\theta$, then $\dfrac{dy}{dx} = \dfrac{dy/d\theta}{dx/d\theta} = \dfrac{-3\cos\theta}{-8\sin\theta} = \dfrac{3}{8}\cot\theta$ and $m(\theta = 2\pi/3) = \dfrac{-\sqrt{3}}{8}$.

9. If $x = \theta$ and $y = 1/\theta = \theta^{-1}$, then $\dfrac{dy}{dx} = \dfrac{dy/d\theta}{dx/d\theta} = \dfrac{-\theta^{-2}}{1} = \dfrac{-1}{\theta^2}$ and $m(\theta = \pi/4) = \dfrac{-16}{\pi^2}$.

11. If $x = t^2 - 2$ and $y = 4t$, then $\dfrac{dy}{dx} = \dfrac{4}{2t} = \dfrac{2}{t}$. Thus $x(3) = 3^2 - 2 = 7$, $y(3) = 4(3) = 12$, $m(3) = \dfrac{2}{3}$, and the tangent line has equation $y = 12 + \dfrac{2}{3}(x - 7) = \dfrac{2}{3}x + \dfrac{22}{3}$.

13. If $x = e^{2t}$ and $y = e^{-2t}$, then $\dfrac{dy}{dx} = \dfrac{-2e^{-2t}}{2e^{2t}} = -e^{-4t}$. Thus $x(1) = e^2$, $y(1) = e^{-2}$, $m(1) = -e^{-4}$, and the tangent line has equation $y = e^{-2} - e^{-4}(x - e^2) = -e^{-4}x + 2e^{-2}$.

15. If $x = \sqrt{1 - \sin\theta}$ and $y = \sqrt{1 + \cos\theta}$, then $\dfrac{dy}{dx} = \dfrac{-\sin\theta/2\sqrt{1 + \cos\theta}}{-\cos\theta/2\sqrt{1 - \sin\theta}} = \tan\theta\,\sqrt{\dfrac{1 - \sin\theta}{1 + \cos\theta}}$. Thus $x(0) = 1$, $y(0) = \sqrt{2}$, $m(0) = 0$, and the tangent line has equation $y = \sqrt{2} + (0)(x - 1) = \sqrt{2}$.

17. If $x = t^2 - 1$ and $y = 4 - t^2$, then $\dfrac{dy}{dx} = \dfrac{-2t}{2t} = -1$; this (-1) is always defined and is never zero, so there are no vertical and no horizontal tangents.

REMARK: $y = 4 - t^2 = 4 - (x + 1) = 3 - x$ and $x \geq -1$.

19. If $x = \sinh t$ and $y = \cosh t$, then $\dfrac{dy}{dx} = \dfrac{\sinh t}{\cosh t} = \tanh t$. This is zero if and only if $t = 0$, so the curve has its sole horizontal tangent at $(\sinh 0, \cosh 0) = (0, 1)$; there are no vertical tangents.

REMARK: The curve is one branch of a hyperbola: $y^2 - x^2 = 1$ and $y \geq 1$.

21. If $x = 2\cos\theta$ and $y = 3\sin\theta$, then $\dfrac{dy}{dx} = \dfrac{3\cos\theta}{-2\sin\theta}$. The curve is traversed for $\theta \in [-\pi, \pi)$. Tangents are horizontal if $\theta = \pm\pi/2$, the points are $(0, \pm 3)$; tangents are vertical if $\theta = -\pi$ or $\theta = 0$, the points are $(\mp 2, 0)$.

REMARK: The curve is the ellipse satisfying $x^2/4 + y^2/9 = 1$.

23. If $x = \ln(1 + t^2)$ and $y = \ln(1 + t^3)$, then $\dfrac{dy}{dx} = \dfrac{3t^2/(1 + t^3)}{2t/(1 + t^2)} = \dfrac{3t(1 + t^2)}{2(1 + t^3)}$. $1 + t^2 \geq 1$ so the only horizontal tangent corresponds to $t = 0$ and is at the point $(\ln 1, \ln 1) = (0, 0)$. The domain of y excludes t for which $1 + t^3 \leq 0$, hence there are no vertical tangents.

23: $\left(\ln(1 + t^2), \ln(1 + t^3)\right)$

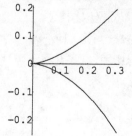

25: $(\sin 3\theta, \cos 5\theta)$

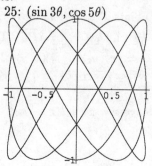

25. If $x = \sin 3\theta$ and $y = \cos 5\theta$, then $\dfrac{dy}{dx} = \dfrac{-5\sin 5\theta}{3\cos 3\theta}$. Tangents are horizontal if $\theta = k\pi/5$, the points on the curve are $\left(\sin k(3/5)\pi, (-1)^k\right)$. Tangents are vertical if $\theta = (2n+1)\pi/6$, the points on the curve are $\left((-1)^n, \cos(5/6)(2n+1)\pi\right)$.

REMARK: The curve is traversed for $\theta \in [0, 2\pi)$; there are 10 points with horizontal tangents and 6 points with vertical tangents. (Some mathematicians would call this a *Lissajou* curve, others would refer to a *Bowditch* curve.)

27. If $r = 5\sin\theta$, then $x = r\cos\theta = 5\sin\theta\cos\theta$, $y = r\sin\theta = 5\sin^2\theta$, and $\dfrac{dy}{dx} = \dfrac{10\sin\theta\cos\theta}{5\cos^2\theta - 5\sin^2\theta} = \dfrac{\sin 2\theta}{\cos 2\theta}$.

Therefore $m(\pi/6) = \tan(\pi/3) = \sqrt{3}$.

29. If $r = -4 + 2\cos\theta$, then $x = r\cos\theta = (-4 + 2\cos\theta)\cos\theta = -4\cos\theta + 2\cos^2\theta$, $y = r\sin\theta = (-4 + 2\cos\theta)\sin\theta = -4\sin\theta + 2\sin\theta\cos\theta$, and $\dfrac{dy}{dx} = \dfrac{-4\cos\theta + 2(\cos^2\theta - \sin^2\theta)}{4\sin\theta - 4\sin\theta\cos\theta} = \dfrac{-4\cos\theta + 2\cos 2\theta}{4\sin\theta - 2\sin 2\theta}$. Therefore $m(\pi/3) = \dfrac{-2 - 1}{2\sqrt{3} - \sqrt{3}} = -\sqrt{3}$.

31. If $r = 3\sin 2\theta$, then $x = r\cos\theta = 3\sin 2\theta\cos\theta$, $y = r\sin\theta = 3\sin 2\theta\sin\theta$, and
$$\frac{dy}{dx} = \frac{6\cos 2\theta\sin\theta + 3\sin 2\theta\cos\theta}{6\cos 2\theta\cos\theta - 3\sin 2\theta\sin\theta} = \frac{2\cos 2\theta\sin\theta + \sin 2\theta\cos\theta}{2\cos 2\theta\cos\theta - \sin 2\theta\sin\theta}.$$
Therefore $m(\pi/6) = \dfrac{1/2 + 3/4}{\sqrt{3}/2 - \sqrt{3}/4} = \dfrac{5}{\sqrt{3}}$.

33. If $r = e^{\theta/2}$, then $x = r\cos\theta = e^{\theta/2}\cos\theta$, $y = r\sin\theta = e^{\theta/2}\sin\theta$, and $\dfrac{dy}{dx} = \dfrac{(1/2)e^{\theta/2}\sin\theta + e^{\theta/2}\cos\theta}{(1/2)e^{\theta/2}\cos\theta - e^{\theta/2}\sin\theta} = \dfrac{\sin\theta + 2\cos\theta}{\cos\theta - 2\sin\theta}$. Therefore $m(0) = \dfrac{0 + 2}{1 - 0} = 2$.

35. $11 = x = t^3 - 2t^2 - 3t + 11 \iff 0 = t^3 - 2t^2 - 3t = t(t^2 - 2t - 3) = t(t + 1)(t - 3) \iff t \in \{0, -1, 3\}$; $-2 = y = t^2 - 2t - 5 \iff 0 = t^2 - 2t - 3 = (t + 1)(t - 3) \iff t = -1$ or $t = 3$. The point $(11, -2)$ is on the curve for $t = -1$ and $t = 3$. $\dfrac{dy}{dx} = \dfrac{y'}{x'} = \dfrac{2t - 2}{3t^2 - 4t - 3}$. Thus $m(-1) = \dfrac{-2 - 2}{3 + 4 - 3} = -1$ and the tangent line is $y = -2 + (-1)(x - 11) = -x + 9$; $m(3) = \dfrac{6 - 2}{27 - 12 - 3} = \dfrac{1}{3}$ and this tangent line is $y = -2 + (1/3)(x - 11) = (x - 17)/3$.

37. $x = r(\alpha - \sin\alpha)$ and $y = r(1 - \cos\alpha)$, therefore $\dfrac{dy}{dx} = \dfrac{dy/d\alpha}{dx/d\alpha} = \dfrac{r\sin\alpha}{r(1 - \cos\alpha)} = \dfrac{\sin\alpha}{1 - \cos\alpha}$. The denominator, $1 - \cos\alpha$, is zero if and only if α is a multiple of 2π. However, the numerator, $\sin\alpha$, is also zero for those α. To prepare for a possible application of L'Hopital's Rule, consider $\dfrac{y''}{x''} = \dfrac{\cos\alpha}{\sin\alpha} = \cot\alpha$. Since $\lim_{\alpha\to k2\pi^-}\cot\alpha = -\infty$ and $\lim_{\alpha\to k2\pi^+}\cot\alpha = \infty$, L'Hopital's Rule implies both one-sided limits also exist for $\dfrac{dy}{dx}$. Thus the vertical line $x = rk2\pi$ is tangent to the curve at the point $\left(x(2k\pi), y(2k\pi)\right) = (rk2\pi, 0)$. ∎

REMARK: If $\alpha \neq k\pi$, then $\dfrac{\sin\alpha}{1 - \cos\alpha} = \dfrac{\sin\alpha(1 + \cos\alpha)}{1 - \cos^2\alpha} = \dfrac{1 + \cos\alpha}{\sin\alpha}$. Our information about one-sided limits of $\dfrac{dy}{dx}$ can be obtained from this expression without need for L'Hopital's Rule. (Although use of L'Hopital's Rule requires care about logic, the computations are fairly mechanical; the manipulation of trigonometric identities uses one flavor of ingenuity.) Solution 12.3.42 shows a third approach.

39. If $r = f(\theta)$, then $x = r\cos\theta = f(\theta)\cos\theta$ and $y = r\sin\theta = f(\theta)\sin\theta$. ∎

12.3 — Differentiation & Integration of a vector function

Solutions 12.3   *Differentiation & integration of a vector function*   (page 805)

1. If $\mathbf{f}(t) = t\mathbf{i} - t^5\mathbf{j}$, then $\mathbf{f}'(t) = \mathbf{i} - 5t^4\mathbf{j}$ and $\mathbf{f}''(t) = -20t^3\mathbf{j}$. All have domain $\mathbb{R}$.

3. If $\mathbf{f}(t) = \sin 2t\mathbf{i} + \cos 3t\mathbf{j}$, then $\mathbf{f}'(t) = 2\cos 2t\mathbf{i} - 3\sin 3t\mathbf{j}$ and $\mathbf{f}''(t) = -4\sin 2t\mathbf{i} - 9\cos 3t\mathbf{j}$; all have domain $\mathbb{R}$.

5. If $\mathbf{f}(t) = \ln t\mathbf{i} + e^{3t}\mathbf{j}$, then $\mathbf{f}'(t) = t^{-1}\mathbf{i} + 3e^{3t}\mathbf{j}$ and $\mathbf{f}''(t) = -t^{-2}\mathbf{i} + 9e^{3t}\mathbf{j}$; all have domain $(0,\infty)$.

REMARK: It is always the case that $\operatorname{dom}\mathbf{f}'' \subseteq \operatorname{dom}\mathbf{f}' \subseteq \operatorname{dom}\mathbf{f}$. Since $\operatorname{dom}\mathbf{f} = (0,\infty)$, it is incorrect to say $\operatorname{dom}\mathbf{f}'' = \operatorname{dom}\mathbf{f}' = \mathbb{R} - \{0\}$.

On the other hand, if $\mathbf{g}(t) = \ln|t|\mathbf{i} + e^{3t}\mathbf{j}$, then $\mathbf{g}'(t) = t^{-1}\mathbf{i} + 3e^{3t}\mathbf{j}$ and $\mathbb{R} - \{0\} = \operatorname{dom}\mathbf{g} = \operatorname{dom}\mathbf{g}'$.

7. If $\mathbf{f}(t) = \tan t\mathbf{i} + \sec t\mathbf{j}$, then $\mathbf{f}'(t) = \sec^2 t\mathbf{i} + \sec t\tan t\mathbf{j}$ and $\mathbf{f}''(t) = 2\sec^2 t\tan t\mathbf{i} + (\sec t\tan^2 t + \sec^3 t)\mathbf{j}$; all have domain $\mathbb{R} - \{(k+1/2)\pi\}$.

9. If $\mathbf{f}(t) = \ln\cos t\mathbf{i} + \ln\sin t\mathbf{j}$, then $\mathbf{f}'(t) = \dfrac{-\sin t}{\cos t}\mathbf{i} + \dfrac{\cos t}{\sin t}\mathbf{j} = -\tan t\mathbf{i} + \cot t\mathbf{j}$ and $\mathbf{f}''(t) = -\sec^2 t\mathbf{i} - \csc^2 t\mathbf{j}$; $\operatorname{dom}\mathbf{f} = \{t : \cos t > 0, \sin t > 0\} = \{t : 0 < t - k2\pi < \pi/2\} = \operatorname{dom}\mathbf{f}' = \operatorname{dom}\mathbf{f}''$.

11. If $\mathbf{f}(t) = t^2\mathbf{i} + t^3\mathbf{j}$, then $\mathbf{f}'(t) = 2t\mathbf{i} + 3t^2\mathbf{j}$; therefore $\mathbf{f}'(1) = 2\mathbf{i} + 3\mathbf{j}$, $|\mathbf{f}'(1)| = \sqrt{2^2 + 3^2} = \sqrt{13}$, and $\mathbf{T}(1) = \dfrac{\mathbf{f}'(1)}{|\mathbf{f}'(1)|} = \dfrac{2}{\sqrt{13}}\mathbf{i} + \dfrac{3}{\sqrt{13}}\mathbf{j}$.

13. If $\mathbf{f}(t) = \cos t\mathbf{i} + \sin t\mathbf{j}$, then $\mathbf{f}'(t) = -\sin t\mathbf{i} + \cos t\mathbf{j}$, $|\mathbf{f}'(t)| = \sqrt{\sin^2 t + \cos^2 t} = 1$ and $\mathbf{T}(t) = \mathbf{f}'(t)$; therefore $\mathbf{T}(0) = \mathbf{f}'(0) = \mathbf{j}$.

15. If $\mathbf{f}(t) = \cos t\mathbf{i} + \sin t\mathbf{j}$, then $\mathbf{f}'(t) = -\sin t\mathbf{i} + \cos t\mathbf{j}$, $|\mathbf{f}'(t)| = \sqrt{\sin^2 t + \cos^2 t} = 1$ and $\mathbf{T}(t) = \mathbf{f}'(t)$; therefore $\mathbf{T}(\pi/4) = \mathbf{f}'(\pi/4) = (-1/\sqrt{2})\mathbf{i} + (1/\sqrt{2})\mathbf{j}$.

17. If $\mathbf{f}(t) = \tan t\mathbf{i} + \sec t\mathbf{j}$, then $\mathbf{f}'(t) = \sec^2 t\mathbf{i} + \sec t\tan t\mathbf{j}$; therefore $\mathbf{f}'(0) = \mathbf{i}$, $|\mathbf{f}'(0)| = \sqrt{1^2 + 0^2} = 1$, and $\mathbf{T}(0) = \dfrac{\mathbf{f}'(0)}{|\mathbf{f}'(0)|} = \mathbf{f}'(0) = \mathbf{i}$.

19. If $\mathbf{f}(t) = \dfrac{t}{t+1}\mathbf{i} + \dfrac{t+1}{t}\mathbf{j} = (1 - (t+1)^{-1})\mathbf{i} + (1 + t^{-1})\mathbf{j}$, then $\mathbf{f}'(t) = (t+1)^{-2}\mathbf{i} - t^{-2}\mathbf{j}$; therefore $\mathbf{f}'(2) = \dfrac{1}{9}\mathbf{i} - \dfrac{1}{4}\mathbf{j}$, $|\mathbf{f}'(2)| = \sqrt{\dfrac{1}{81} + \dfrac{1}{16}} = \dfrac{\sqrt{97}}{36}$, and $\mathbf{T}(2) = \dfrac{\mathbf{f}'(2)}{|\mathbf{f}'(2)|} = \dfrac{1/9}{\sqrt{97}/36}\mathbf{i} - \dfrac{1/4}{\sqrt{97}/36}\mathbf{j} = \dfrac{4}{\sqrt{97}}\mathbf{i} - \dfrac{9}{\sqrt{97}}\mathbf{j}$.

21. If $\mathbf{g}(t) = t\mathbf{i} + t^2\mathbf{j} + t^3\mathbf{k}$, then $\mathbf{g}'(t) = \mathbf{i} + 2t\mathbf{j} + 3t^2\mathbf{k}$; therefore $\mathbf{g}'(1) = \mathbf{i} + 2\mathbf{j} + 3\mathbf{k}$, $|\mathbf{g}'(1)| = \sqrt{1 + 4 + 9} = \sqrt{14}$, and $\mathbf{T}(1) = \dfrac{\mathbf{g}'(1)}{|\mathbf{g}'(1)|} = \dfrac{1}{\sqrt{14}}\mathbf{i} + \dfrac{2}{\sqrt{14}}\mathbf{j} + \dfrac{3}{\sqrt{14}}\mathbf{k}$.

23. If $\mathbf{g}(t) = t\mathbf{i} + e^t\mathbf{j} + e^{-t}\mathbf{k}$, then $\mathbf{g}'(t) = \mathbf{i} + e^t\mathbf{j} - e^{-t}\mathbf{k}$; therefore $\mathbf{g}'(0) = \mathbf{i} + \mathbf{j} - \mathbf{k}$, $|\mathbf{g}'(0)| = \sqrt{1 + 1 + 1} = \sqrt{3}$, and $\mathbf{T}(0) = \dfrac{\mathbf{g}'(0)}{|\mathbf{g}'(0)|} = \dfrac{1}{\sqrt{3}}\mathbf{i} + \dfrac{1}{\sqrt{3}}\mathbf{j} - \dfrac{1}{\sqrt{3}}\mathbf{k}$.

25. If $\mathbf{g}(t) = 4\cos 2t\mathbf{i} + 9\sin 2t\mathbf{j} + t\mathbf{k}$, then $\mathbf{g}'(t) = -8\sin 2t\mathbf{i} + 18\cos 2t\mathbf{j} + \mathbf{k}$; therefore $\mathbf{g}'(\pi/4) = -8\mathbf{i} + \mathbf{k}$, $|\mathbf{g}'(\pi/4)| = \sqrt{64 + 0 + 1} = \sqrt{65}$, and $\mathbf{T}(\pi/4) = \dfrac{\mathbf{g}'(\pi/4)}{|\mathbf{g}'(\pi/4)|} = \dfrac{-8}{\sqrt{65}}\mathbf{i} + \dfrac{1}{\sqrt{65}}\mathbf{k}$.

27. $\int (\sin 2t\mathbf{i} + e^t\mathbf{j})\,dt = (-1/2)\cos 2t\mathbf{i} + e^t\mathbf{j} + \mathbf{C} = ((-1/2)\cos 2t + c_x)\mathbf{i} + (e^t + c_y)\mathbf{j}$.

29. $\displaystyle\int_0^{\pi/4} (\cos 2t\mathbf{i} - \sin 2t\mathbf{j})\,dt = \left(\frac{1}{2}\sin 2t\mathbf{i} + \frac{1}{2}\cos 2t\mathbf{j}\right)\Big|_0^{\pi/4} = \frac{1}{2}(1 - 0)\mathbf{i} + \frac{1}{2}(0 - 1)\mathbf{j} = \frac{1}{2}\mathbf{i} - \frac{1}{2}\mathbf{j}$.

REMARK: Alternatively, we compute $\left(\dfrac{1}{2}\sin 2t\mathbf{i} + \dfrac{1}{2}\cos 2t\mathbf{j}\right)\Big|_0^{\pi/4} = \left(\dfrac{1}{2}\mathbf{i} + 0\mathbf{j}\right) - \left(0\mathbf{i} + \dfrac{1}{2}\mathbf{j}\right) = \dots$

31. $\int (\ln t\mathbf{i} + te^t\mathbf{j})\,dt = t(\ln t - 1)\mathbf{i} + e^t(t-1)\mathbf{j} + \mathbf{C} = (t(\ln t - 1) + c_x)\mathbf{i} + (e^t(t-1) + c_y)\mathbf{j}$.

33. $\displaystyle\int_0^2 (t^2\mathbf{i} - t^3\mathbf{j} + t^4\mathbf{k})\,dt = \left(\frac{t^3}{3}\mathbf{i} - \frac{t^4}{4}\mathbf{j} + \frac{t^5}{5}\mathbf{k}\right)\Big|_0^2 = \frac{8}{3}\mathbf{i} - 4\mathbf{j} + \frac{32}{5}\mathbf{k}$.

35. $\int 2t\left(\sin t^2\mathbf{i} + \cos t^2\mathbf{j} + e^{t^2}\mathbf{k}\right)dt = \int \left(\sin t^2\mathbf{i} + \cos t^2\mathbf{j} + e^{t^2}\mathbf{k}\right)d(t^2) = -\cos t^2\mathbf{i} + \sin t^2\mathbf{j} + e^{t^2}\mathbf{k} + \mathbf{C}$.

37. $\mathbf{f}(t) = \mathbf{f}(0) + \int_0^t \mathbf{f}' = (2\mathbf{i} + 5\mathbf{j}) + \int_0^t (s^3\mathbf{i} - s^5\mathbf{j})\,ds = (2\mathbf{i} + 5\mathbf{j}) + ((1/4)t^4\mathbf{i} - (1/6)t^6\mathbf{j}) = (2 + t^4/4)\mathbf{i} + (5 - t^6/6)\mathbf{j}$.

39. $\mathbf{f}(t) = \mathbf{f}(\pi/2) + \int_{\pi/2}^t \mathbf{f}' = \mathbf{i} + \int_{\pi/2}^t (\cos s\mathbf{i} + \sin s\mathbf{k})\,ds = \mathbf{i} + ((\sin t - 1)\mathbf{i} + (-\cos t)\mathbf{k}) = \sin t\mathbf{i} - \cos t\mathbf{k}$.

41. $\mathbf{f}(\theta) = a\cos\theta\mathbf{i} + b\sin\theta\mathbf{j}$ implies $\mathbf{f}'(\theta) = -a\sin\theta\mathbf{i} + b\cos\theta\mathbf{j}$ and $\mathbf{f}'(\pi/4) = (-a/\sqrt{2})\mathbf{i} + (b/\sqrt{2})\mathbf{j} = (-a\mathbf{i} + b\mathbf{j})/\sqrt{2}$. Therefore $|\mathbf{f}'(\pi/4)| = \sqrt{a^2/2 + b^2/2} = \sqrt{a^2 + b^2}/\sqrt{2}$ and $\mathbf{T}(\pi/4) = (-a\mathbf{i} + b\mathbf{j})/\sqrt{a^2 + b^2}$.

43. If $a = 5$ and $b = 2$, then the hypocycloid of Problem 12.1.49 is the path of $\mathbf{f}(\theta) = \big(3\cos\theta + 2\cos(3\theta/2)\big)\mathbf{i} + \big(3\sin\theta - 2\sin(3\theta/2)\big)\mathbf{j}$. Hence $\mathbf{f}'(\theta) = \big(-3\sin\theta - 3\sin(3\theta/2)\big)\mathbf{i} + \big(3\cos\theta - 3\cos(3\theta/2)\big)\mathbf{j}$,

$$\mathbf{f}'\left(\frac{\pi}{6}\right) = -3\left(\sin\frac{\pi}{6} + \sin\frac{\pi}{4}\right)\mathbf{i} + 3\left(\cos\frac{\pi}{6} - \cos\frac{\pi}{4}\right)\mathbf{j} = -3\left(\frac{1 + \sqrt{2}}{2}\right)\mathbf{i} + 3\left(\frac{\sqrt{3} - \sqrt{2}}{2}\right)\mathbf{j},$$

and $|\mathbf{f}'(\pi/6)| = (3/2)\sqrt{\left(1 + \sqrt{2}\right)^2 + \left(\sqrt{3} - \sqrt{2}\right)^2} = (3/2)\sqrt{8 + 2\sqrt{2} - 2\sqrt{6}}$. Therefore

$$\mathbf{T}\left(\frac{\pi}{6}\right) = \frac{-1 - \sqrt{2}}{\sqrt{8 + 2\sqrt{2} - 2\sqrt{6}}}\mathbf{i} + \frac{\sqrt{3} - \sqrt{2}}{\sqrt{8 + 2\sqrt{2} - 2\sqrt{6}}}\mathbf{j}.$$

REMARK: Here is an alternative computation of the tangent's length.

$$|\mathbf{f}'(\theta)| = 3\sqrt{\big(-\sin\theta - \sin(3\theta/2)\big)^2 + \big(\cos\theta - \cos(3\theta/2)\big)^2}$$
$$= 3\sqrt{\big(\sin^2\theta + \cos^2\theta\big) + \big(\sin^2(3\theta/2) + \cos^2(3\theta/2)\big) - 2\big(\cos\theta\cos(3\theta/2) - \sin\theta\sin(3\theta/2)\big)}$$
$$= 3\sqrt{1 + 1 - 2\cos(\theta + 3\theta/2)} = 3\sqrt{2}\sqrt{1 - \cos(5\theta/2)}.$$

A half-angle formula shows $\cos 5\pi/12 = (1/2)\sqrt{2 - \sqrt{3}}$, thus $|\mathbf{f}'(\pi/6)| = 3\sqrt{2 - \sqrt{2 - \sqrt{3}}}$.

45. $\mathbf{s}(0) = \mathbf{0}$ if the cannon is at the origin. $\mathbf{v}_0 = 1300\big(\cos(\pi/4)\mathbf{i} + \sin(\pi/4)\mathbf{j}\big) = (1300/\sqrt{2})(\mathbf{i} + \mathbf{j})$ ft/sec. The acceleration due to gravity is $\mathbf{a} = -32\mathbf{j}$ ft/sec^2, hence $\mathbf{v}(t) = (1300/\sqrt{2})\mathbf{i} + \big((1300/\sqrt{2}) - 32t\big)\mathbf{j}$ ft/sec. Therefore $\mathbf{s}(t) = (1300/\sqrt{2})t\mathbf{i} + \big((1300/\sqrt{2})t - 16t^2\big)\mathbf{j}$ feet. The cannonball hits the ground when $t = (1300/\sqrt{2})/16 = 325/\big(4\sqrt{2}\big)$; the horizontal component of $\mathbf{s}$ is then $325^2/2 = 52812.5$ feet.

REMARK: The text is inconsistent in its value(s) for g: 32 ft/sec^2 in example 4.2.6 on page 272, but the more precise 32.2 is used for the answer on page A-129 to this problem. Multiply $325^2/2$ by $32/32.2 = 16/16.1$ to transform the result of this solution into the text's answer.

47. If the initial path is 30° below the horizontal, then the initial angle is $-\pi/6$ and $\mathbf{v}(t) = 100\big(\cos(-\pi/6)\mathbf{i} + \sin(-\pi/6)\mathbf{j}\big) + (-9.81t\mathbf{j}) = 50\sqrt{3}\mathbf{i} + (-50 - 9.81t)\mathbf{j}$ m/sec. If the origin is at the base of the building, then $\mathbf{s}(0) = 150\mathbf{j}$ and $\mathbf{s}(t) = 50\sqrt{3}t\mathbf{i} + \big(150 - 50t - 9.81t^2/2\big)\mathbf{j}$ feet. The positive root of $150 - 50t - 9.81t^2/2 = 0$ is $t = \big(-50 + \sqrt{5443}\big)/9.81 \approx 2.4237$; this implies the object hits the ground $50\sqrt{3}\big(-50 + \sqrt{5443}\big)/9.81 \approx 209.9$ meters from the base of the building.

Solutions 12.4   *Some differentiation formulas*   (pages 810-1)

1. $\dfrac{d}{dt}\left[(2t\mathbf{i} + \cos t\mathbf{j}) + (\tan t\mathbf{i} - \sec t\mathbf{j})\right] = (2\mathbf{i} - \sin t\mathbf{j}) + \big(\sec^2 t\mathbf{i} - \sec t\tan t\mathbf{j}\big)$
$$= \big(2 + \sec^2 t\big)\mathbf{i} + (-\sin t - \sec t\tan t)\mathbf{j}.$$

3. $\dfrac{d}{dt}\left[(t^3\mathbf{i} - t^5\mathbf{j}) + (-t^3\mathbf{i} + t^5\mathbf{j})\right] = (3t^2\mathbf{i} - 5t^4\mathbf{j}) + (-3t^2\mathbf{i} + 5t^4\mathbf{j})$
$$= \big(3t^2 - 3t^2\big)\mathbf{i} + \big(-5t^4 + 5t^4\big)\mathbf{j} = 0\mathbf{i} + 0\mathbf{j} = \mathbf{0}.$$

REMARK: The work shown above is drill for part (i) of Theorem 1; the problem has a simpler solution: $(t^3\mathbf{i} - t^5\mathbf{j}) + (-t^3\mathbf{i} + t^5\mathbf{j}) = (t^3 - t^3)\mathbf{i} + (-t^5 + t^5)\mathbf{j} = \mathbf{0}$ and the derivative of this constant function is $\mathbf{0}$.

5. $\dfrac{d}{dt}\left[(\sin^{-1} t\mathbf{i} + \cos^{-1} t\mathbf{j}) + (\cos t\mathbf{i} + \sin t\mathbf{j})\right] = \left(\big(1 - t^2\big)^{-1/2}\mathbf{i} - \big(1 - t^2\big)^{-1/2}\mathbf{j}\right) + (-\sin t\mathbf{i} + \cos t\mathbf{j})$
$$= \left(\big(1 - t^2\big)^{-1/2} - \sin t\right)\mathbf{i} + \left(-\big(1 - t^2\big)^{-1/2} + \cos t\right)\mathbf{j}.$$

12.4 — Some differentiation formulas

7.
$$\frac{d}{dt}\left[(2\mathbf{i}+t\mathbf{j}-t^2\mathbf{k})\times(3t\mathbf{i}+5\mathbf{k})\right]$$
$$=\left[\frac{d}{dt}\left(2\mathbf{i}+t\mathbf{j}-t^2\mathbf{k}\right)\right]\times(3t\mathbf{i}+5\mathbf{k})+(2\mathbf{i}+t\mathbf{j}-t^2\mathbf{k})\times\left[\frac{d}{dt}\left(3t\mathbf{i}+5\mathbf{k}\right)\right]$$
$$=(\mathbf{j}-2t\mathbf{k})\times(3t\mathbf{i}+5\mathbf{k})+(2\mathbf{i}+t\mathbf{j}-t^2\mathbf{k})\times(3\mathbf{i})$$
$$=\begin{vmatrix}\mathbf{i}&\mathbf{j}&\mathbf{k}\\0&1&-2t\\3t&0&5\end{vmatrix}+\begin{vmatrix}\mathbf{i}&\mathbf{j}&\mathbf{k}\\2&t&-t^2\\3&0&0\end{vmatrix}=(5\mathbf{i}-6t^2\mathbf{j}-3t\mathbf{k})+(-3t^2\mathbf{j}-3t\mathbf{k})=5\mathbf{i}-9t^2\mathbf{j}-6t\mathbf{k}.$$

REMARK: $(2\mathbf{i}+t\mathbf{j}-t^2\mathbf{k})\times(3t\mathbf{i}+5\mathbf{k})=\begin{vmatrix}\mathbf{i}&\mathbf{j}&\mathbf{k}\\2&t&-t^2\\3t&0&5\end{vmatrix}=5t\mathbf{i}-\left(10+3t^3\right)\mathbf{j}-3t^2\mathbf{k}.$

---

9. $\mathbf{g}'(t)=\dfrac{d}{dt}\left(\cos t\,\mathbf{i}+\sin t\,\mathbf{j}+t\mathbf{k}\right)=-\sin t\,\mathbf{i}+\cos t\,\mathbf{j}+\mathbf{k}$ and $|\mathbf{g}'(t)|=\sqrt{\sin^2 t+\cos^2 t+1}=\sqrt{2}$, thus $\mathbf{T}(t)=(-\sin t\,\mathbf{i}+\cos t\,\mathbf{j}+\mathbf{k})/\sqrt{2}$ and $\mathbf{T}(\pi/4)=\left(-(1/\sqrt{2})\mathbf{i}+(1/\sqrt{2})\mathbf{j}+\mathbf{k}\right)/\sqrt{2}=(-1/2)\mathbf{i}+(1/2)\mathbf{j}+(1/\sqrt{2})\mathbf{k}.$

11. $\mathbf{g}'(t)=\dfrac{d}{dt}\left(3\cos t\,\mathbf{i}+4\sin t\,\mathbf{j}+t\mathbf{k}\right)=-3\sin t\,\mathbf{i}+4\cos t\,\mathbf{j}+\mathbf{k}$ and $|\mathbf{g}'(t)|=\sqrt{9\sin^2 t+16\cos^2 t+1}=\sqrt{10+7\cos^2 t}$, thus $\mathbf{T}(t)=(-3\sin t\,\mathbf{i}+4\cos t\,\mathbf{j}+\mathbf{k})/\sqrt{10+7\cos^2 t}$ and $\mathbf{T}(0)=(4\mathbf{j}+\mathbf{k})/\sqrt{17}.$

13. $\mathbf{g}'(t)=\dfrac{d}{dt}\left(\mathbf{i}+t\mathbf{j}+t^2\mathbf{k}\right)=\mathbf{j}+2t\mathbf{k}$ and $|\mathbf{g}'(t)|=\sqrt{0+1+4t^2}=\sqrt{1+4t^2}$, thus $\mathbf{T}(t)=(\mathbf{j}+2t\mathbf{k})/\sqrt{1+4t^2}$ and $\mathbf{T}(1)=(\mathbf{j}+2\mathbf{k})/\sqrt{5}.$

15. $\mathbf{g}'(t)=\dfrac{d}{dt}\left(e^t\cos 2t\,\mathbf{i}+e^t\sin 2t\,\mathbf{j}+e^t\mathbf{k}\right)=e^t(\cos 2t-2\sin 2t)\mathbf{i}+e^t(\sin 2t+2\cos 2t)\mathbf{j}+e^t\mathbf{k}.$ Therefore
$$|\mathbf{g}'(t)|=\sqrt{e^{2t}(\cos 2t-2\sin 2t)^2+e^{2t}(\sin 2t+2\cos 2t)^2+e^{2t}}$$
$$=e^t\sqrt{(\cos^2 2t-4\cos 2t\sin 2t+4\sin^2 2t)+(\sin^2 2t+4\sin 2t\cos 2t+4\cos^2 2t)+1}=e^t\sqrt{6},$$
$$\mathbf{T}(t)=(e^t(\cos 2t-2\sin 2t)\mathbf{i}+e^t(\sin 2t+2\cos 2t)\mathbf{j}+e^t\mathbf{k})/e^t\sqrt{6}$$
$$=((\cos 2t-2\sin 2t)\mathbf{i}+(\sin 2t+2\cos 2t)\mathbf{j}+\mathbf{k})/\sqrt{6},$$
and $\mathbf{T}(0)=(\mathbf{i}+2\mathbf{j}+\mathbf{k})/\sqrt{6}.$

17. $\mathbf{f}'(t)=\dfrac{d}{dt}\left(\cos 3t\,\mathbf{i}+\sin 3t\,\mathbf{j}\right)=-3\sin 3t\,\mathbf{i}+3\cos 3t\,\mathbf{j}$ and $|\mathbf{f}'(t)|=\sqrt{9\sin^2 3t+9\cos^2 3t}=3$ imply $\mathbf{T}(t)=\mathbf{f}'(t)/|\mathbf{f}'(t)|=-\sin 3t\,\mathbf{i}+\cos 3t\,\mathbf{j}.$ Hence $\mathbf{T}'(t)=-3\cos 3t\,\mathbf{i}-3\sin 3t\,\mathbf{j}$ and $|\mathbf{T}'(t)|=\sqrt{9\cos^2 3t+9\sin^2 3t}=3$ which imply $\mathbf{n}(t)=\mathbf{T}'(t)/|\mathbf{T}'(t)|=-\cos 3t\,\mathbf{i}-\sin 3t\,\mathbf{j}=-\mathbf{f}(t).$ Thus $\mathbf{T}(0)=\mathbf{j}$ and $\mathbf{n}(0)=-\mathbf{i}.$

17: $\cos 3t\,\mathbf{i}+\sin 3t\,\mathbf{j}$

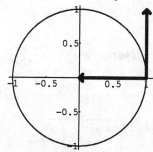

19: $2\cos 4t\,\mathbf{i}+2\sin 4t\,\mathbf{j}$

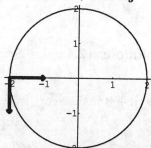

19. $\mathbf{f}'(t)=\dfrac{d}{dt}\left(2\cos 4t\,\mathbf{i}+2\sin 4t\,\mathbf{j}\right)=-8\sin 4t\,\mathbf{i}+8\cos 4t\,\mathbf{j}$ and $|\mathbf{f}'(t)|=\sqrt{64\sin^2 4t+64\cos^2 4t}=8$ imply $\mathbf{T}(t)=\mathbf{f}'(t)/|\mathbf{f}'(t)|=-\sin 4t\,\mathbf{i}+\cos 4t\,\mathbf{j}.$ Thus $\mathbf{T}'(t)=-4\cos 4t\,\mathbf{i}-4\sin 4t\,\mathbf{j}$ and $|\mathbf{T}'(t)|=\sqrt{16\cos^2 4t+16\sin^2 4t}=4$ which imply $\mathbf{n}(t)=\mathbf{T}'(t)/|\mathbf{T}'(t)|=-\cos 4t\,\mathbf{i}-\sin 4t\,\mathbf{j}.$ Hence $\mathbf{T}(\pi/4)=-\mathbf{j}$ and $\mathbf{n}(\pi/4)=\mathbf{i}.$

21. $\mathbf{f}'(t)=\dfrac{d}{dt}\left(8\cos t\,\mathbf{i}+8\sin t\,\mathbf{j}\right)=-8\sin t\,\mathbf{i}+8\cos t\,\mathbf{j}$ and $|\mathbf{f}'(t)|=\sqrt{64\sin^2 t+64\cos^2 t}=8$ imply $\mathbf{T}(t)=\mathbf{f}'(t)/|\mathbf{f}'(t)|=-\sin t\,\mathbf{i}+\cos t\,\mathbf{j}.$ Thus $\mathbf{T}'(t)=-\cos t\,\mathbf{i}-\sin t\,\mathbf{j}$ and $|\mathbf{T}'(t)|=\sqrt{\cos^2 t+\sin^2 t}=1$ which imply $\mathbf{n}(t)=\mathbf{T}'(t)/|\mathbf{T}'(t)|=-\cos t\,\mathbf{i}-\sin t\,\mathbf{j}.$ Hence $\mathbf{T}(\pi/4)=(-\mathbf{i}+\mathbf{j})/\sqrt{2}$ and $\mathbf{n}(\pi/4)=(-\mathbf{i}-\mathbf{j})/\sqrt{2}.$

21: $8\cos t\,\mathbf{i} + 8\sin t\,\mathbf{j}$

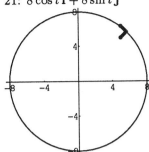

23: $(2+3t)\mathbf{i} + (8-5t)\mathbf{j}$

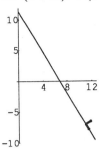

23. $\mathbf{f}'(t) = \dfrac{d}{dt}\left((2+3t)\mathbf{i} + (8-5t)\mathbf{j}\right) = 3\mathbf{i} - 5\mathbf{j}$ and $|\mathbf{f}'(t)| = \sqrt{9+25} = \sqrt{34}$ imply $\mathbf{T}(t) = \mathbf{f}'(t)/|\mathbf{f}'(t)| = (3\mathbf{i}-5\mathbf{j})/\sqrt{34} = \mathbf{T}(3)$. Since $\mathbf{T}(t)$ is constant, $\mathbf{T}'(t) = \mathbf{0}$ so equation (8) can not be invoked to find a unit normal. Problem 11.2.41 states a fact which can serve as partial support for an alternative procedure: $(a\mathbf{i}+b\mathbf{j}) \perp (b\mathbf{i}-a\mathbf{j})$; furthermore, if $\mathbf{u}$ and $\mathbf{v}$ are in $\mathbb{R}^2$ and $a\mathbf{i}+b\mathbf{j} = \mathbf{u} \perp \mathbf{v}$, then there is a constant c such that $\mathbf{v} = c(b\mathbf{i}-a\mathbf{j})$. Therefore $\mathbf{n}(t) = \pm(5\mathbf{i}+3\mathbf{j})/\sqrt{34} = \mathbf{n}(3)$.

25. $\mathbf{f}'(t) = \dfrac{d}{dt}\left((a+bt)\mathbf{i} + (c+dt)\mathbf{j}\right) = b\mathbf{i} + d\mathbf{j}$ and $|\mathbf{f}'(t)| = \sqrt{b^2+d^2}$ imply $\mathbf{T}(t) = \mathbf{f}'(t)/|\mathbf{f}'(t)| = (b\mathbf{i}+d\mathbf{j})/\sqrt{b^2+d^2} = \mathbf{T}(t_0)$. Since $\mathbf{T}(t)$ is constant, $\mathbf{T}'(t) = \mathbf{0}$; let $\mathbf{n}(t) = \pm(d\mathbf{i}-b\mathbf{j})/\sqrt{b^2+d^2} = \mathbf{n}(t_0)$.

27. $\mathbf{f}'(t) = \dfrac{d}{dt}\left((t-\cos t)\mathbf{i} + (1-\sin t)\mathbf{j}\right) = (1+\sin t)\mathbf{i} - \cos t\,\mathbf{j}$ and $|\mathbf{f}'(t)| = \sqrt{(1+\sin t)^2 + \cos^2 t} = \sqrt{2(1+\sin t)}$ imply $\mathbf{T}(t) = \dfrac{1}{\sqrt{2}}\left(\sqrt{1+\sin t}\,\mathbf{i} - \dfrac{\cos t}{\sqrt{1+\sin t}}\mathbf{j}\right)$. Therefore

$$\mathbf{T}'(t) = \frac{1}{\sqrt{2}}\left(\frac{\cos t}{2\sqrt{1+\sin t}}\mathbf{i} + \frac{2\sin t\,(1+\sin t) + \cos^2 t}{2(1+\sin t)^{3/2}}\mathbf{j}\right) = \frac{1}{2\sqrt{2}}\left(\frac{\cos t}{\sqrt{1+\sin t}}\mathbf{i} + \sqrt{1+\sin t}\,\mathbf{j}\right),$$

$|\mathbf{T}'(t)| = 1/2$, and $\mathbf{n}(t) = 2\mathbf{T}'(t) = \left(\cos t\,\mathbf{i} + (1+\sin t)\mathbf{j}\right)/\sqrt{2(1+\sin t)}$. Thus $\mathbf{T}(\pi/2) = \mathbf{i}$ and $\mathbf{n}(\pi/2) = \mathbf{j}$.

27: $(t-\cos t)\mathbf{i} + (1-\sin t)\mathbf{j}, t = \pi/2$

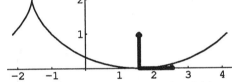

29: $(t-\cos t)\mathbf{i} + (1-\sin t)\mathbf{j}, t = \pi/4$

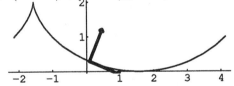

29. Solution 27 derives expressions for $\mathbf{T}(t)$ and $\mathbf{n}(t)$. Thus

$$\mathbf{T}\left(\frac{\pi}{4}\right) = \frac{1}{\sqrt{2}}\left(\sqrt{1+\frac{1}{\sqrt{2}}}\,\mathbf{i} - \frac{1/\sqrt{2}}{\sqrt{1+1/\sqrt{2}}}\mathbf{j}\right) \approx .9239\,\mathbf{i} - .3827\,\mathbf{j}$$

$$\mathbf{n}\left(\frac{\pi}{4}\right) = \frac{1}{\sqrt{2}}\left(\frac{1/\sqrt{2}}{\sqrt{1+1/\sqrt{2}}}\,\mathbf{i} + \sqrt{1+\frac{1}{\sqrt{2}}}\,\mathbf{j}\right) \approx .3827\,\mathbf{i} + .9239\,\mathbf{j}.$$

Prologue for 31-36: Let $\mathbf{f} = f_1\mathbf{i} + f_2\mathbf{j} + f_3\mathbf{k}$, $\mathbf{g} = g_1\mathbf{i} + g_2\mathbf{j} + g_3\mathbf{k}$, $\mathbf{v} = v_1\mathbf{i} + v_2\mathbf{j} + v_3\mathbf{k}$; apply Theorem 12.3.1 (and its obvious generalization to $\mathbb{R}^3$) frequently.

31. $[\mathbf{f} + \mathbf{g}]' = [(f_1+g_1)\mathbf{i} + (f_2+g_2)\mathbf{j} + (f_3+g_3)\mathbf{k}]' = (f_1+g_1)'\mathbf{i} + (f_2+g_2)'\mathbf{j} + (f_3+g_3)'\mathbf{k} = (f_1'+g_1')\mathbf{i} + (f_2'+g_2')\mathbf{j} + (f_3'+g_3')\mathbf{k} = (f_1'\mathbf{i} + f_2'\mathbf{j} + f_3'\mathbf{k}) + (g_1'\mathbf{i} + g_2'\mathbf{j} + g_3'\mathbf{k}) = \mathbf{f}' + \mathbf{g}'$. ∎

33. $[\mathbf{v} \cdot \mathbf{f}]' = [v_1 f_1 + v_2 f_2 + v_3 f_3]' = (v_1 f_1)' + (v_2 f_2)' + (v_3 f_3)' = v_1 f_1' + v_2 f_2' + v_3 f_3' = \mathbf{v} \cdot \mathbf{f}'$. ∎

35. $[\mathbf{f} \cdot \mathbf{g}]' = [f_1 g_1 + f_2 g_2 + f_3 g_3]' = (f_1 g_1)' + (f_2 g_2)' + (f_3 g_3)' = (f_1' g_1 + f_1 g_1') + (f_2' g_2 + f_2 g_2') + (f_3' g_3 + f_3 g_3') = (f_1' g_1 + f_2' g_2 + f_3' g_3) + (f_1 g_1' + f_2 g_2' + f_3 g_3') = \mathbf{f}' \cdot \mathbf{g} + \mathbf{f} \cdot \mathbf{g}'$. ∎

37. $\mathbf{f}'(t) = \dfrac{d}{dt}\left(a\cos t\,\mathbf{i} + a\sin t\,\mathbf{j} + t\mathbf{k}\right) = -a\sin t\,\mathbf{i} + a\cos t\,\mathbf{j} + \mathbf{k}$ and $|\mathbf{f}'(t)| = \sqrt{a^2\sin^2 t + a^2\sin^2 t + 1^2} = \sqrt{a^2+1}$ imply $\mathbf{T}(t) = (-a\sin t\,\mathbf{i} + a\cos t\,\mathbf{j} + \mathbf{k})/\sqrt{a^2+1}$. The angle between this unit tangent and the z-axis is the ArcCosine of $\mathbf{T} \cdot \mathbf{k} = (0+0+1)/\sqrt{a^2+1} = 1/\sqrt{a^2+1}$, a constant. ∎

12.5 — Arc length revisited

1. Arc length of the curve $(x, y) = (t^3, t^2)$ on the interval $[-1, 1]$ is

$$s = \int_{-1}^{1} \sqrt{(3t^2)^2 + (2t)^2}\, dt = \int_{-1}^{1} |t|\sqrt{9t^2 + 4}\, dt = 2\int_{0}^{1} t\sqrt{9t^2 + 4}\, dt = \frac{1}{9}\int_{0}^{1} \sqrt{9t^2 + 4}\, d(9t^2 + 4)$$

$$= \frac{2}{27}\left(9t^2 + 4\right)^{3/2}\Big|_{0}^{1} = \frac{2}{27}\left(13^{3/2} - 8\right).$$

3. Arc length of the curve $(x, y) = (t^3 + 1, 3t^2 + 2)$ on the interval $[0, 2]$ is

$$s = \int_{0}^{2} \sqrt{(3t^2)^2 + (6t)^2}\, dt = \int_{0}^{2} 3t\sqrt{t^2 + 4}\, dt = \frac{3}{2}\int_{0}^{2}\sqrt{t^2+4}\, d(t^2 + 4) = (t^2 + 4)^{3/2}\Big|_{0}^{2} = 8\left(2^{3/2} - 1\right).$$

5. Arc length of the curve $(x, y) = \left(a(\theta - \sin\theta), a(1 - \cos\theta)\right)$ on the interval $[0, 2\pi]$ is

$$s = \int_{0}^{2\pi} |a|\sqrt{(1 - \cos\theta)^2 + \sin^2\theta}\, d\theta = |a|\int_{0}^{2\pi}\sqrt{2(1 - \cos\theta)}\, d\theta = |a|\int_{0}^{2\pi}\sqrt{4\sin^2(\theta/2)}\, d\theta$$

$$= |a|\int_{0}^{2\pi} 2|\sin(\theta/2)|\, d\theta = -4|a|\cos(\theta/2)\Big|_{0}^{2\pi} = 8|a|.$$

7. Arc length of the curve $(x, y) = \left(1/\sqrt{1+t}, t/2(1+t)\right) = \left((1+t)^{-1/2}, (1/2)(1 - (1+t)^{-1})\right)$ on the interval $[0, 4]$ is

$$s = \int_{0}^{4}\sqrt{\left(\frac{-1}{2(1+t)^{3/2}}\right)^2 + \left(\frac{1}{2(1+t)^2}\right)^2}\, dt = \int_{0}^{4}\frac{1}{2}\sqrt{(1+t)^{-3} + (1+t)^{-4}}\, dt = \frac{1}{2}\int_{0}^{4}\frac{\sqrt{2+t}}{(1+t)^2}\, dt$$

$$\overset{t+2\equiv u^2}{=} \frac{1}{2}\int_{\sqrt 2}^{\sqrt 6}\frac{u}{(u^2 - 1)^2}(2u\, du) = \frac{1}{4}\int_{\sqrt 2}^{\sqrt 6}\left(\frac{1}{u-1} + \frac{1}{(u-1)^2} + \frac{-1}{u+1} + \frac{1}{(u+1)^2}\right) du$$

$$= \frac{1}{4}\left(\ln|u-1| - \frac{1}{u+1} - \ln|u+1| - \frac{1}{u-1}\right)\Big|_{\sqrt 2}^{\sqrt 6} = \left(\frac{1}{4}\ln\left|\frac{u-1}{u+1}\right| - \frac{u}{2(u^2-1)}\right)\Big|_{\sqrt 2}^{\sqrt 6}$$

$$= \left(\frac{1}{4}\ln\frac{\sqrt 6 - 1}{\sqrt 6 + 1} - \frac{\sqrt 6}{10}\right) - \left(\frac{1}{4}\ln\frac{\sqrt 2 - 1}{\sqrt 2 + 1} - \frac{\sqrt 2}{2}\right).$$

9. Arc length of the curve $(x, y) = (e^t \cos t, e^t \sin t)$ on the interval $[0, \pi/2]$ is

$$s = \int_{0}^{\pi/2}\sqrt{e^{2t}(\cos t - \sin t)^2 + e^{2t}(\sin t + \cos t)^2}\, dt = \sqrt 2\int_{0}^{\pi/2} e^t\, dt = \sqrt 2\, e^t\Big|_{0}^{\pi/2} = \sqrt 2\left(e^{\pi/2} - 1\right).$$

11. Arc length of the curve $(x, y) = \left(\sin^2 t, \cos^2 t\right)$ on the interval $[0, \pi/2]$ is

$$s = \int_{0}^{\pi/2}\sqrt{(2\sin t\cos t)^2 + (-2\cos t\sin t)^2}\, dt = \int_{0}^{\pi/2} 2\sqrt 2\, |\sin t\cos t|\, dt = \sqrt 2\int_{0}^{\pi/2} 2\sin t\cos t\, dt$$

$$= \sqrt 2\,\sin^2 t\Big|_{0}^{\pi/2} = \sqrt 2.$$

13. Arc length of the polar-coordinate curve $r = a\sin\theta$ on the interval $[0, \pi/2]$ is

$$s = \int_{0}^{\pi/2}\sqrt{(a\sin\theta)^2 + (a\cos\theta)^2}\, d\theta = \int_{0}^{\pi/2}|a|\, d\theta = \frac{\pi\,|a|}{2}.$$

REMARK: The curve $r = a\sin\theta$ for θ in any interval of length π is a circle with radius $|a|/2$; we have computed the arc length for a semicircle.

---

15. Arc length of the polar-coordinate curve $r = \theta^2$ on the interval $[0, \pi]$ is

$$s = \int_{0}^{\pi}\sqrt{(\theta^2)^2 + (2\theta)^2}\, d\theta = \int_{0}^{\pi}|\theta|\sqrt{\theta^2 + 4}\, d\theta = \int_{0}^{\pi}\theta\sqrt{\theta^2 + 4}\, d\theta = \frac{1}{3}\left(\theta^2 + 4\right)^{3/2}\Big|_{0}^{\pi} = \frac{1}{3}\left((\pi^2 + 4)^{3/2} - 8\right).$$

17.   The cardioid $r = a(1 + \sin\theta)$ is traversed for θ in any interval of length 2π; since this cardioid is symmetric about the vertical line thru the origin, the interval $[-\pi/2, 3\pi/2]$ is convenient. The arc length is

$$s = \int_{-\pi/2}^{3\pi/2} \sqrt{\left(a(1+\sin\theta)\right)^2 + (a\cos\theta)^2}\, d\theta = \sqrt{2}\,|a| \int_{-\pi/2}^{3\pi/2} \sqrt{1+\sin\theta}\, d\theta = \sqrt{2}\,|a| \int_{-\pi/2}^{3\pi/2} \sqrt{1+\sin\theta}\, \frac{\sqrt{1-\sin\theta}}{\sqrt{1-\sin\theta}}\, d\theta$$

$$= \sqrt{2}\,|a| \int_{-\pi/2}^{3\pi/2} \frac{|\cos\theta|}{\sqrt{1-\sin\theta}}\, d\theta = \sqrt{2}\,|a| \int_{-\pi/2}^{\pi/2} \frac{\cos\theta}{\sqrt{1-\sin\theta}}\, d\theta + \sqrt{2}\,|a| \int_{\pi/2}^{3\pi/2} \frac{-\cos\theta}{\sqrt{1-\sin\theta}}\, d\theta$$

$$= -2\sqrt{2}\,|a|\sqrt{1-\sin\theta}\,\Big|_{-\pi/2}^{\pi/2} + 2\sqrt{2}\,|a|\sqrt{1-\sin\theta}\,\Big|_{\pi/2}^{3\pi/2} = -2\sqrt{2}\,|a|(0-\sqrt{2}) + 2\sqrt{2}\,|a|(\sqrt{2}-0) = 8|a|.$$

REMARK: $\sqrt{2(1+\sin\theta)} = \sqrt{2\left(1+\cos(\theta-\pi/2)\right)} = \sqrt{4\cos^2(\theta/2 - \pi/4)} = 2\,|\cos(\theta/2 - \pi/4)|$; note that $\cos(\theta/2 - \pi/4) \geq 0$ for $-\pi/2 \leq \theta \leq 3\pi/2$, therefore

$$\int_{-\pi/2}^{3\pi/2} \sqrt{2(1+\sin\theta)}\, d\theta = \int_{-\pi/2}^{3\pi/2} 2\cos\left(\frac{\theta}{2} - \frac{\pi}{4}\right) d\theta = 4\sin\left(\frac{\theta}{2} - \frac{\pi}{4}\right)\Big|_{-\pi/2}^{3\pi/2} = 8.$$

19.   Arc length of the polar-coordinate curve $r = e^\theta$ on the interval $[0,3]$ is

$$s = \int_0^3 \sqrt{\left(e^\theta\right)^2 + \left(e^\theta\right)^2}\, d\theta = \int_0^3 \sqrt{2}\, e^\theta\, d\theta = \sqrt{2}\, e^\theta\,\Big|_0^3 = \sqrt{2}\left(e^3 - 1\right).$$

21.   Arc length of the curve $(x,y) = (3u^2, 2u^3)$ on the interval $[0,t]$ is, if $t \geq 0$,

$$s = \int_0^t \sqrt{(6u)^2 + (6u^2)^2}\, du = \int_0^t 6|u|\sqrt{1+u^2}\, du = \int_0^t 6u(1+u^2)^{1/2} du = 2(1+u^2)^{3/2}\Big|_0^t = 2\left((1+t^2)^{3/2} - 1\right).$$

Thus s is a monotonically increasing function of t so its inverse exists; we solve for t as a function of s:

$$s = 2\left((1+t^2)^{3/2} - 1\right) \iff 1 + s/2 = (1+t^2)^{3/2} \iff (1+s/2)^{2/3} = 1 + t^2 \iff t = \sqrt{(1+s/2)^{2/3} - 1}$$

(positive square root because we initially assumed $t \geq 0$). Therefore $\mathbf{f} = 3t^2\mathbf{i} + 2t^3\mathbf{j} = 3\left((1+s/2)^{2/3} - 1\right)\mathbf{i} + 2\left((1+s/2)^{2/3} - 1\right)^{3/2}\mathbf{j}$.

    We verify equation (12) holds:

$$\left(\frac{dx}{ds}\right)^2 + \left(\frac{dy}{ds}\right)^2 = \left((1+s/2)^{-1/3}\right)^2 + \left(\left((1+s/2)^{2/3} - 1\right)^{1/2}(1+s/2)^{-1/3}\right)^2$$

$$= \frac{1}{(1+s/2)^{2/3}} + \frac{-1 + (1+s/2)^{2/3}}{(1+s/2)^{2/3}} = 1. \quad\blacksquare$$

REMARK: The work shown above can be extended to cover the case for $t < 0$. Note that $ds \geq 0$, but integrating ds in the negative direction yields a negative value. The negative sign for the resulting $s(t)$ corresponds to the fact that t decreases from zero; the fact that $s(t)$ is a monotonically increasing function of t will be preserved. s and t will have the same sign. The signum function is defined as

$$\mathrm{sgn}(w) = \begin{cases} 1 & \text{if } w > 0, \\ 0 & \text{if } w = 0, \\ -1 & \text{if } w < 0. \end{cases}$$ Let $\sigma = \mathrm{sgn}(t) = \mathrm{sgn}(s)$. Then $\sigma t = |t|$, $\sigma s = |s|$, $\sigma^2 = 1$ if $t \neq 0$,

$s = \int_0^t 6\sigma\, u(1+u^2)^{1/2} du = 2\sigma\left((1+t^2)^{3/2} - 1\right)$, and $t = \sigma\sqrt{(1+|s|/2)^{2/3} - 1}$.

    The problem statement does not restrict t to be positive so, strictly construed, the full solution requires the paragraph above (plus adjustment of the formula for $\mathbf{f}(s)$ and subsequent verification of equation (12)). On the other hand, the work preceding this remark is a bit complicated. In some of the following solutions, I'll make the same compromise as here: main solution for $t \geq 0$, remark contains extension to whole curve. (The definition here for σ will be used throughout.)

23.   Arc length of the curve $(x,y) = (u^3 + 1, u^2 - 1)$ on the interval $[0,t]$ is, if $t \geq 0$,

$$s = \int_0^t \sqrt{(3u^2)^2 + (2u)^2}\, du = \int_0^t |u|\sqrt{9u^2 + 4}\, du = \int_0^t u(9u^2+4)^{1/2} du = \frac{1}{27}(9u^2+4)^{3/2}\Big|_0^t = \frac{(9t^2+4)^{3/2} - 8}{27}.$$

We solve for t as a function of s:

$$s = \frac{1}{27}\left((9t^2+4)^{3/2} - 8\right) \iff 27s + 8 = (9t^2+4)^{3/2} \iff (27s+8)^{2/3} = 9t^2 + 4 \iff t = \frac{1}{3}\sqrt{(27s+8)^{2/3} - 4}.$$

Thus $\mathbf{f} = (t^3+1)\mathbf{i} + (t^2-1)\mathbf{j} = \left((1/27)\left((27s+8)^{2/3}-4\right)^{3/2}+1\right)\mathbf{i} + \left((1/9)\left((27s+8)^{2/3}-4\right)-1\right)\mathbf{j}.$

We verify equation (12) holds:

$$\left(\frac{dx}{ds}\right)^2 + \left(\frac{dy}{ds}\right)^2 = \left(\left((27s+8)^{2/3}-4\right)^{1/2}(27s+8)^{-1/3}\right)^2 + \left(2(27s+8)^{-1/3}\right)^2$$

$$= \frac{-4+(27s+8)^{2/3}}{(27s+8)^{2/3}} + \frac{4}{(27s+8)^{2/3}} = 1. \ \blacksquare$$

REMARK: $\quad s = \int_0^t \sigma u(9u^2+4)^{1/2}\,du = (1/27)\sigma\left((9t^2+4)^{3/2}-8\right)$ and $t = (1/3)\sigma\sqrt{(27|s|+8)^{2/3}-4}.$

---

25. Arc length of the curve $(x,y) = (3\cos u, 3\sin u)$ is

$$s = \int_0^t \sqrt{(-3\sin u)^2 + (3\cos u)^2}\,du = \int_0^t 3\,du = 3u\big|_0^t = 3t.$$

We solve easily for t as a function of s: $s = 3t \iff t = s/3$. Therefore $\mathbf{f} = 3\cos t\,\mathbf{i} + 3\sin t\,\mathbf{j} = 3\cos(s/3)\mathbf{i} + 3\sin(s/3)\mathbf{j}$. We verify equation (12) holds: $\left(\frac{dx}{ds}\right)^2 + \left(\frac{dy}{ds}\right)^2 = (-\sin(s/3))^2 + (\cos(s/3))^2 = 1. \ \blacksquare$

REMARK: The curve is a circle of radius 3 and $\mathbf{f} = 3(\cos t\,\mathbf{i} + \sin t\,\mathbf{j})$, so $s = 3t$ is immediate.

---

27. Arc length of the curve $(x,y) = (a\cos u, -a\sin u)$ is, assuming $a > 0$,

$$s = \int_0^t \sqrt{(-a\sin u)^2 + (-a\cos u)^2}\,du = \int_0^t |a|\,du = |a|\,u\big|_0^t = |a|\,t = at.$$

Therefore $t = s/a$ and $\mathbf{f} = a\cos t\,\mathbf{i} - a\sin t\,\mathbf{j} = a\cos(s/a)\mathbf{i} - a\sin(s/a)\mathbf{j}$. We verify equation (12) holds:
$\left(\frac{dx}{ds}\right)^2 + \left(\frac{dy}{ds}\right)^2 = (-\sin(s/a))^2 + (-\cos(s/a))^2 = 1. \ \blacksquare$

29. Arc length of the curve $(x,y) = (a+b\cos u, c+b\sin u)$ is, assuming $b > 0$,

$$s = \int_0^t \sqrt{(-b\sin u)^2 + (b\cos u)^2}\,du = \int_0^t |b|\,du = |b|\,u\big|_0^t = |b|\,t = bt.$$

Therefore $t = s/b$ and $\mathbf{f} = (a+b\cos(s/b))\mathbf{i} + (c+b\sin(s/b))\mathbf{j}$. We verify equation (12): $\left(\frac{dx}{ds}\right)^2 + \left(\frac{dy}{ds}\right)^2 = (-\sin(s/b))^2 + (\cos(s/b))^2 = 1. \ \blacksquare$

31. Arc length of the curve $(x,y) = (a\cos^3 u, a\sin^3 u)$ on the interval $[0, \pi/2]$ is, assuming $a > 0$,

$$s = \int_0^t \sqrt{(-3a\cos^2 u\sin u)^2 + (3a\sin^2 u\cos u)^2}\,du = \int_0^t 3\,|a\cos u\sin u|\,du = \int_0^t 3a\cos u\sin u\,du$$

$$= \frac{3a}{2}\sin^2 u\,\bigg|_0^t = \frac{3a}{2}\sin^2 t.$$

Therefore $\sin^2 t = \dfrac{2s}{3a}$, $t = \sin^{-1}\sqrt{\dfrac{2s}{3a}}$, $\cos^2 t = 1 - \sin^2 t = 1 - \dfrac{2s}{3a}$, and $\mathbf{f} = a\cos^3 t\,\mathbf{i} + a\sin^3 t\,\mathbf{j} = a\left(1-\dfrac{2s}{3a}\right)^{3/2}\mathbf{i} + a\left(\dfrac{2s}{3a}\right)^{3/2}\mathbf{j}$. We verify equation (12):

$$\left(\frac{dx}{ds}\right)^2 + \left(\frac{dy}{ds}\right)^2 = \left(-\left(1-\frac{2s}{3a}\right)^{1/2}\right)^2 + \left(\left(\frac{2s}{3a}\right)^{1/2}\right)^2 = \left(1-\frac{2s}{3a}\right) + \frac{2s}{3a} = 1. \ \blacksquare$$

33. The ellipse $(x/a)^2 + (y/b)^2 = 1$ is traversed by the curve $(a\cos\theta, b\sin\theta)$ for $0 \le \theta \le 2\pi$. The ellipse is symmetric about both axes, hence its arc length is four times the length of the arc in the first quadrant:

$$s = 4\int_0^{\pi/2} \sqrt{(-a\sin\theta)^2 + (b\cos\theta)^2}\,d\theta = 4\int_0^{\pi/2}\sqrt{a^2\sin^2\theta + b^2\cos^2\theta}\,d\theta.$$

35. A general result on integration by substitution can be stated as $\int_A^B h(p)\,dp = \int_{g^{-1}(A)}^{g^{-1}(B)} h(g(r))\,g'(r)\,dr$

where g is smooth and invertible; we will use a mild rephrasing: $\int_{g(\alpha)}^{g(\beta)} h(p)\,dp = \int_\alpha^\beta h(g(r))\,g'(r)\,dr$

Suppose $t_0 < t_1$. If $x = x(t)$ is differentiable and monotonic with $x(t_0) = a < b = x(t_1)$, then $\dfrac{dx}{dt} > 0$.

If y can be expressed as a smooth function of x, the chain rule implies $\dfrac{dy}{dx}\dfrac{dx}{dt} = \dfrac{dy}{dt}$. Therefore

$$\int_a^b \sqrt{1 + \left(\frac{dy}{dx}\right)^2}\,dx = \int_{x(t_0)}^{x(t_1)} \sqrt{1 + \left(\frac{dy}{dx}\right)^2}\,dx$$

$$= \int_{t_0}^{t_1} \sqrt{1 + \left(\frac{dy}{dx}\right)^2}\,\frac{dx}{dt}\,dt = \int_{t_0}^{t_1} \sqrt{\left(\frac{dx}{dt}\right)^2 + \left(\frac{dy}{dx}\right)^2\left(\frac{dx}{dt}\right)^2}\,dt$$

$$= \int_{t_0}^{t_1} \sqrt{\left(\frac{dx}{dt}\right)^2 + \left(\frac{dy}{dx}\frac{dx}{dt}\right)^2}\,dt = \int_{t_0}^{t_1} \sqrt{\left(\frac{dx}{dt}\right)^2 + \left(\frac{dy}{dt}\right)^2}\,dt. \ \blacksquare$$

REMARK: Let $t = x$ in $\sqrt{(dx/dt)^2 + (dy/dt)^2}\,dt$ to obtain $\sqrt{1 + (dy/dx)^2}\,dx$; if we know arc length is a well-defined notion which is invariant to parameterization of the curve, then we might consider our work done. The computations shown above can be viewed as support for our subsequent knowledge.

37. Example 6 shows $x = 4\sqrt{1+s} - 4 - s$ and $y = (8/3)\left(\sqrt{1+s} - 1\right)^{3/2}$, therefore

$$\left(\frac{dx}{ds}\right)^2 + \left(\frac{dy}{ds}\right)^2 = \left(\frac{2}{\sqrt{1+s}} - 1\right)^2 + \left(\frac{2}{\sqrt{1+s}}\left(\sqrt{1+s} - 1\right)^{1/2}\right)^2$$

$$= \left(\frac{4}{1+s} - \frac{4}{\sqrt{1+s}} + 1\right) + \frac{4\left(\sqrt{1+s} - 1\right)}{1+s} = 1. \ \blacksquare$$

Solutions 12.6   *Curvature and the acceleration vector*   (pages 828-9)

Prologue: To locate the center C of (the circle of) curvature we travel a distance ρ along the unit normal from the point on the curve: $\mathbf{C} = \mathbf{f} + \rho\mathbf{n}$. The following computations obtain a general expression for $\mathbf{C}$ which will be applied in Solutions 1-22.

Suppose $\mathbf{f}(t) = x(t)\mathbf{i} + y(t)\mathbf{j}$ and let $'$ denote differentiation with respect to t. Then $s' = \left|\frac{d}{dt}\mathbf{f}\right| = |\mathbf{f}'| = \sqrt{(x')^2 + (y')^2}$ and $\mathbf{T} = \dfrac{\mathbf{f}'}{|\mathbf{f}'|} = \dfrac{x'\mathbf{i} + y'\mathbf{j}}{\sqrt{(x')^2 + (y')^2}} = \dfrac{x'\mathbf{i} + y'\mathbf{j}}{s'}$. Let $D = x'y'' - y'x''$; Theorem 2

states curvature $\kappa(t) = \left|\dfrac{d\mathbf{f}}{ds}\right| = \dfrac{|D|}{(s')^3}$ and implies radius of curvature $\rho(t) = \dfrac{1}{\kappa(t)} = \dfrac{(s')^3}{|D|}$. Therefore

$$\mathbf{T}' = \frac{x''\mathbf{i} + y''\mathbf{j}}{\sqrt{(x')^2 + (y')^2}} + \frac{(x'\mathbf{i} + y'\mathbf{j})(-1/2)(2x'x'' + 2y'y'')}{\left((x')^2 + (y')^2\right)^{3/2}} = \frac{\left(x''(y')^2 - x'y'y''\right)\mathbf{i} + \left((x')^2 y'' - x'x''y'\right)\mathbf{j}}{\left((x')^2 + (y')^2\right)^{3/2}}$$

$$= \frac{y'(x''y' - x'y'')\mathbf{i} + x'(x'y'' - x''y')\mathbf{j}}{(s')^3} = \left(\frac{x'y'' - y'x''}{(s')^3}\right)\left(-y'\mathbf{i} + x'\mathbf{j}\right)$$

$$= \left(\frac{D}{(s')^3}\right)\left(-y'\mathbf{i} + x'\mathbf{j}\right) = \mathrm{sgn}(D)\,\kappa(t)\left(-y'\mathbf{i} + x'\mathbf{j}\right),$$

$$\mathbf{n} = \frac{\mathbf{T}'}{|\mathbf{T}'|} = \frac{\mathrm{sgn}(D)}{s'}\left(-y'\mathbf{i} + x'\mathbf{j}\right),$$

$$\rho\mathbf{n} = \frac{(s')^3}{|D|}\frac{\mathrm{sgn}(D)}{s'}\left(-y'\mathbf{i} + x'\mathbf{j}\right) = \frac{(s')^2}{D}\left(-y'\mathbf{i} + x'\mathbf{j}\right),$$

$$\mathbf{C} = \mathbf{f} + \rho\mathbf{n} = \left(x - y'\frac{(s')^2}{D}\right)\mathbf{i} + \left(y + x'\frac{(s')^2}{D}\right)\mathbf{j}.$$

12.6 — Curvature and the acceleration vector

1. Suppose $\mathbf{f} = x\mathbf{i} + y\mathbf{j} = t\mathbf{i} + t^2\mathbf{j}$, Then $x' = 1$, $y' = 2t$, $s' = \sqrt{1^2 + (2t)^2} = \sqrt{1 + 4t^2}$, and $\mathbf{T} = (\mathbf{i} + 2t\mathbf{j})/\sqrt{1 + 4t^2}$. Furthermore $x'' = 0$, $y'' = 2$, $D = (1)(2) - (2t)(0) = 2$, $\kappa = 2/(1 + 4t^2)^{3/2}$, $\rho = (1 + 4t^2)^{3/2}/2$, and $\mathbf{C} = \left(t - (2t)(1 + 4t^2)/2\right)\mathbf{i} + \left(t^2 + (1)(1 + 4t^2)/2\right)\mathbf{j} = -4t^3\mathbf{i} + (3t^2 + 1/2)\mathbf{j}$.

$\mathbf{f}(1) = \mathbf{i} + \mathbf{j}$, $\mathbf{T}(1) = (\mathbf{i} + 2\mathbf{j})/\sqrt{5}$, $\kappa(1) = 2/5^{3/2}$, $\rho(1) = 5^{3/2}/2$, and $\mathbf{C}(1) = -4\mathbf{i} + (7/2)\mathbf{j}$.

1: $t\mathbf{i} + t^2\mathbf{j}$

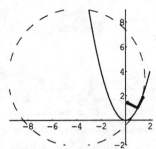

3: $2\cos t\,\mathbf{i} + 2\sin t\,\mathbf{j}$, $t = \pi/4$

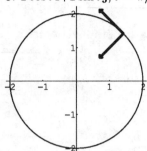

3. Suppose $\mathbf{f} = x\mathbf{i} + y\mathbf{j} = 2\cos t\,\mathbf{i} + 2\sin t\,\mathbf{j}$, Then $x' = -2\sin t$, $y' = 2\cos t$, $s' = \sqrt{(-2\sin t)^2 + (2\cos t)^2} = 2$, and $\mathbf{T} = -\sin t\,\mathbf{i} + \cos t\,\mathbf{j}$. Also $x'' = -2\cos t$, $y'' = -2\sin t$, $D = (-2\sin t)(-2\sin t) - (2\cos t)(-2\cos t) = 4$, $\kappa = 4/(4)^{3/2} = 1/2$, $\rho = 2$, and $\mathbf{C} = (2\cos t - (2\cos t)(4)/(4))\mathbf{i} + (2\sin t + (-2\sin t)(4)/(4))\mathbf{j} = 0\mathbf{i} + 0\mathbf{j} = 0$. (A circle is its own circle of curvature at each of its points.)

$\mathbf{f}(\pi/4) = \sqrt{2}\mathbf{i} + \sqrt{2}\mathbf{j}$, $\mathbf{T}(\pi/4) = (-\mathbf{i} + \mathbf{j})/\sqrt{2}$, $\kappa(\pi/4) = 1/2$, $\rho(\pi/4) = 2$, and $\mathbf{C}(\pi/4) = 0$.

5. Suppose $\mathbf{f} = x\mathbf{i} + y\mathbf{j} = 3\sin t\,\mathbf{i} + 4\cos t\,\mathbf{j}$, Then $x' = 3\cos t$, $y' = -4\sin t$, $s' = \sqrt{(3\cos t)^2 + (-4\sin t)^2} = \sqrt{9 + 7\sin^2 t}$, and $\mathbf{T} = (3\cos t\,\mathbf{i} - 4\sin t\,\mathbf{j})/\sqrt{9 + 7\sin^2 t}$. Furthermore $x'' = -3\sin t$, $y'' = -4\cos t$, $D = (3\cos t)(-4\cos t) - (-4\sin t)(-3\sin t) = -12$, $\kappa = 12/(9 + 7\sin^2 t)^{3/2}$, $\rho = (9 + 7\sin^2 t)^{3/2}/12$, and

$$\mathbf{C} = \left(3\sin t - (-4\sin t)(9 + 7\sin^2 t)/(-12)\right)\mathbf{i} + \left(4\cos t + (3\cos t)(9 + 7\sin^2 t)/(-12)\right)\mathbf{j}$$
$$= (-7/3)\sin^3 t\,\mathbf{i} + (7/4)\cos^3 t\,\mathbf{j}.$$

$\mathbf{f}(0) = 4\mathbf{j}$, $\mathbf{T}(0) = \mathbf{i}$, $\kappa(0) = 4/9$, $\rho(0) = 9/4$, and $\mathbf{C}(0) = (7/4)\mathbf{j}$.

5: $3\sin t\,\mathbf{i} + 4\cos t\,\mathbf{j}$, $t = 0$

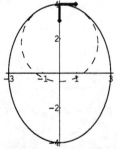

7: $3\cos t\,\mathbf{i} + 4\sin t\,\mathbf{j}$, $t = \pi/4$

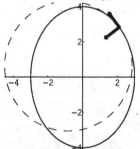

7. Suppose $\mathbf{f} = x\mathbf{i} + y\mathbf{j} = 3\cos t\,\mathbf{i} + 4\sin t\,\mathbf{j}$, Then $x' = -3\sin t$, $y' = 4\cos t$, $s' = \sqrt{(-3\sin t)^2 + (4\cos t)^2} = \sqrt{9 + 7\cos^2 t}$, and $\mathbf{T} = (-3\sin t\,\mathbf{i} + 4\cos t\,\mathbf{j})/\sqrt{9 + 7\cos^2 t}$. Furthermore $x'' = -3\cos t$, $y'' = -4\sin t$, $D = (-3\sin t)(-4\sin t) - (4\cos t)(-3\cos t) = 12$, $\kappa = 12/(9 + 7\cos^2 t)^{3/2}$, $\rho = (9 + 7\cos^2 t)^{3/2}/12$, and

$$\mathbf{C} = \left(3\cos t - (4\cos t)(9 + 7\cos^2 t)/(12)\right)\mathbf{i} + \left(4\sin t + (-3\sin t)(9 + 7\cos^2 t)/(12)\right)\mathbf{j}$$
$$= (-7/3)\cos^3 t\,\mathbf{i} + (7/4)\sin^3 t\,\mathbf{j}.$$

$\mathbf{f}(\pi/4) = (3\mathbf{i} + 4\mathbf{j})/\sqrt{2}$, $\mathbf{T}(\pi/4) = (-3\mathbf{i} + 4\mathbf{j})/5$, $\kappa(\pi/4) = 24\sqrt{2}/125$, $\rho(\pi/4) = 125/24\sqrt{2}$, and $\mathbf{C}(\pi/4) = (-7/6\sqrt{2})\mathbf{i} + (7/8\sqrt{2})\mathbf{j}$.

9. Suppose $\mathbf{f} = x\mathbf{i} + y\mathbf{j} = (t - \sin t)\mathbf{i} + (1 - \cos t)\mathbf{j}$, Then $x' = 1 - \cos t$, $y' = \sin t$, $s' = \sqrt{(1 - \cos t)^2 + (\sin t)^2} = \sqrt{2(1 - \cos t)}$, and $\mathbf{T} = ((1 - \cos t)\mathbf{i} + \sin t\,\mathbf{j})/\sqrt{2(1 - \cos t)}$. Furthermore $x'' = \sin t$, $y'' = \cos t$, $D = (1 - \cos t)(\cos t) - (\sin t)(\sin t) = \cos t - 1$, $\kappa = |\cos t - 1|/(2(1 - \cos t))^{3/2} = 1/2\sqrt{2(1 - \cos t)}$, $\rho = 2\sqrt{2(1 - \cos t)} = 4|\sin(t/2)|$. Note that $(s')^2 = -2D$, thus $\mathbf{C} = (x + 2y')\mathbf{i} + (y - 2x')\mathbf{j} = (t + \sin t)\mathbf{i} - (1 - \cos t)\mathbf{j}$.

$\mathbf{f}(\pi/3) = (\pi/3 - \sqrt{3}/2)\mathbf{i} + (1/2)\mathbf{j}$, $\mathbf{T}(\pi/3) = (\mathbf{i} + \sqrt{3}\mathbf{j})/2$, $\kappa(\pi/3) = 1/2$, $\rho(\pi/3) = 2$, and $\mathbf{C}(\pi/3) = (\pi/3 + \sqrt{3}/2)\mathbf{i} - (1/2)\mathbf{j}$.

340

9: $(t-\sin t)\mathbf{i}+(1-\cos t)\mathbf{j}, t = \pi/3$

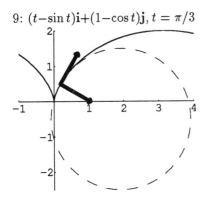

11: $xy = 1$

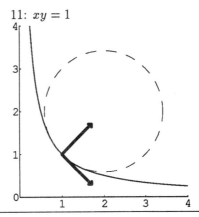

Prologue for 11-22: If the curve is specified in the form $y = g(x)$, then we can apply the general parametric results to the special case $x = t$ and $y = g(t)$. If we now use $'$ to denote differentiation with respect to x, then $x' = 1$, $s' = \sqrt{1+(y')^2}$, $x'' = 0$, $D = x'y'' - x''y' = y''$, $\kappa = |y''|/(1+(y')^2)^{3/2}$, $\rho = (1+(y')^2)^{3/2}/|y''|$, and $C = \left(x - y'(1+(y')^2)/y'', y + (1+(y')^2)/y''\right)$.

11.   If $xy = 1$, then $y = 1/x$ and $x \neq 0$. Then $y' = -x^{-2}$, $s' = \sqrt{1+(-x^{-2})^2} = \sqrt{1+x^{-4}} = x^{-2}\sqrt{x^4+1}$, and $\mathbf{T} = \left(\mathbf{i} - x^{-2}\mathbf{j}\right)/\sqrt{1+x^{-4}} = \left(x^2\mathbf{i} - \mathbf{j}\right)/\sqrt{x^4+1}$. Furthermore $y'' = 2x^{-3}$, $\kappa = 2|x|^{-3}/(1+x^{-4})^{3/2} = 2\left(x^2 + x^{-2}\right)^{-3/2}$, $\rho = (1/2)|x|^3(1+x^{-4})^{3/2} = (1/2)\left(x^2 + x^{-2}\right)^{3/2}$, $(s')^2/y'' = (x^3 + x^{-1})/2$, and $C = \left(x - y'(x^3 + x^{-1})/2, y + (x^3 + x^{-1})/2\right) = \left((3x + x^{-3})/2, (3x^{-1} + x^3)/2\right)$.

$y(1) = 1$, $\mathbf{T}(1) = (\mathbf{i}-\mathbf{j})/\sqrt{2}$, $\kappa(1) = 1/\sqrt{2}$, $\rho(1) = \sqrt{2}$, and $C(1) = (2,2)$.

REMARK: $y' = -y^2$, $s' = \sqrt{1+y^4}$, $y'' = 2y^3$, $\rho = (1/2)\left(x^2 + y^2\right)^{3/2}$, and $C = \left((3x + y^3)/2, (x^3 + 3y)/2\right)$.

13.   Suppose $y = e^x$. Then $y' = e^x$, $s' = \sqrt{1+(e^x)^2} = \sqrt{1+e^{2x}}$, and $\mathbf{T} = (\mathbf{i}+e^x\mathbf{j})/\sqrt{1+e^{2x}}$. Furthermore $y'' = e^x$, $\kappa = e^x/(1+e^{2x})^{3/2}$, $\rho = (1+e^{2x})^{3/2}/e^x$, $(s')^2/y'' = (1+e^{2x})/e^x = e^{-x} + e^x$, and $C = \left(x - y'(e^{-x} + e^x), y + (e^{-x} + e^x)\right) = \left(x - 1 - e^{2x}, e^{-x} + 2e^x\right)$.

$y(0) = 1$, $\mathbf{T}(0) = (\mathbf{i}+\mathbf{j})/\sqrt{2}$, $\kappa(0) = 2^{-3/2}$, $\rho(0) = 2^{3/2}$, and $C(0) = (-2, 3)$.

13: $y = e^x$

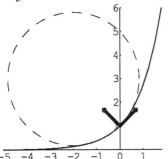

15: $y = \ln x$

15.   Suppose $y = \ln x$. Then $y' = x^{-1}$, $s' = \sqrt{1+(x^{-1})^2} = \sqrt{1+x^{-2}}$, and $\mathbf{T} = \left(\mathbf{i}+x^{-1}\mathbf{j}\right)/\sqrt{1+x^{-2}} = (x\mathbf{i}+\mathbf{j})/\sqrt{x^2+1}$. Furthermore $y'' = -x^{-2}$, $\kappa = x^{-2}/(1+x^{-2})^{3/2}$, $\rho = (1+x^{-2})^{3/2}/x^{-2} = x^2(1+x^{-2})^{3/2}$, $(s')^2/y'' = (1+x^{-2})/(-x^{-2}) = -x^2-1$, and $C = \left(x - y'(-x^2 - 1), y + (-x^2 - 1)\right) = \left(2x + 1/x, \ln x - x^2 - 1\right)$.

$y(e) = 1$, $\mathbf{T}(e) = (e\mathbf{i}+\mathbf{j})/\sqrt{e^2+1}$, $\kappa(e) = e^{-2}\left(1+e^{-2}\right)^{-3/2}$, $\rho(e) = e^2\left(1+e^{-2}\right)^{3/2}$, and $C(e) = (2e + e^{-1}, -e^2)$.

17.   Let $y = \cos x$; then $y' = -\sin x$, $s' = \sqrt{1+(-\sin x)^2} = \sqrt{1+\sin^2 x}$, and $\mathbf{T} = (\mathbf{i} - \sin x\mathbf{j})/\sqrt{1+\sin^2 x}$. Furthermore $y'' = -\cos x$, $\kappa = |\cos x|/(1+\sin^2 x)^{3/2}$, $\rho = (1+\sin^2 x)^{3/2}/|\cos x|$, $(s')^2/y'' = \cos x - 2\sec x$, and $C = \left(x - y'(\cos x - 2\sec x), y + (\cos x - 2\sec x)\right) = \left(x - \sin x \cos x - 2\tan x, 2\cos x - 2\sec x\right)$.

$y(\tfrac{\pi}{3}) = 1/2$, $\mathbf{T}(\tfrac{\pi}{3}) = (2\mathbf{i} - \sqrt{3}\mathbf{j})/\sqrt{7}$, $\kappa(\tfrac{\pi}{3}) = 4/7\sqrt{7}$, $\rho(\tfrac{\pi}{3}) = 7\sqrt{7}/4$, and $C(\tfrac{\pi}{3}) = \left(\pi/3 - 7\sqrt{3}/4, -3\right)$.

17: $y = \cos x$

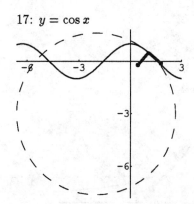

21: $y = \sin^{-1} x$

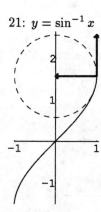

19.   Suppose $y = ax^2 + bx + c$ and $a \neq 0$; then $y' = 2ax+b$, $s' = \sqrt{1 + (2ax+b)^2}$, and $\mathbf{T} = (\mathbf{i} + (2ax + b)\mathbf{j})/s'$. Furthermore $y'' = 2a$, $\kappa = 2|a|/\left(1 + (2ax + b)^2\right)^{3/2}$, and $\rho = \left(1 + (2ax + b)^2\right)^{3/2}/2|a|$.

  $y(0) = c$, $\mathbf{T}(0) = (\mathbf{i} + b\mathbf{j})/\sqrt{1 + b^2}$, $\kappa(0) = 2|a|\left(1 + b^2\right)^{-3/2}$, $\rho(0) = \left(1 + b^2\right)^{3/2}/2|a|$, $\left((s')^2/y''\right)(0) = (1 + b^2)/2a$, and $C(0) = (-b(1 + b^2)/2a, c + (1 + b^2)/2a)$.

21.   Suppose $y = \sin^{-1} x$ and $-1 < x < 1$; then $y' = (1 - x^2)^{-1/2}$, $s' = \sqrt{(2 - x^2)/(1 - x^2)}$, and $\mathbf{T} = (\mathbf{i} + y'\mathbf{j})/s' = (\sqrt{1 - x^2}\,\mathbf{i} + \mathbf{j})/\sqrt{2 - x^2}$. Furthermore $y'' = x(1-x^2)^{-3/2}$, $\kappa = |y''|/(s')^3 = (2-x^2)^{3/2}/x$, $\rho = x(2-x^2)^{-3/2}$, $\left((s')^2/y''\right)(x) = (2-x^2)\sqrt{1 - x^2}/x$, and $C(x) = (-2(1 - x^2)/x, \sin^{-1} x + (2 - x^2)\sqrt{1 - x^2}/x)$.

  Since 1 is the right endpoint of the domain of $\sin^{-1} x$, we consider limits from the left: $y(1) = \pi/2$, $\lim_{x \to 1^-} \mathbf{T}(x) = \mathbf{j}$, $\lim_{x \to 1^-} \kappa(x) = 1$, $\lim_{x \to 1^-} \rho(x) = 1$, and $\lim_{x \to 1^-} C(x) = (0, \pi/2)$.

23.   Suppose $r = 2a \cos \theta$; then $r' = dr/d\theta = -2a \sin \theta$ and $r'' = d^2r/d\theta^2 = -2a \cos \theta$. These imply $r^2 + (r')^2 = (2a \cos \theta)^2 + (-2a \sin \theta)^2 = 4a^2(\cos^2 \theta + \sin^2 \theta) = 4a^2$ and $(r')^2 - rr'' = (-2a \sin \theta)^2 - (2a \cos \theta)(-2a \cos \theta) = 4a^2$. Therefore

$$\kappa(\theta) = \frac{r^2 + 2(r')^2 - rr''}{(r^2 + (r')^2)^{3/2}} = \frac{(r^2 + (r')^2) + ((r')^2 - rr'')}{(r^2 + (r')^2)^{3/2}} = \frac{4a^2 + 4a^2}{(4a^2)^{3/2}} = \frac{8a^2}{8|a|^3} = \frac{1}{|a|}$$

and $\rho(\theta) = |a|$.

REMARK: Points on the curve satisfy the Cartesian equation $(x - a)^2 + y^2 = a^2$; a circle has constant curvature κ and ρ is everywhere equal to the circle's radius.

---

25.   Suppose $r = a(1 + \sin \theta)$; then $r' = dr/d\theta = a \cos \theta$ and $r'' = d^2r/d\theta^2 = -a \sin \theta$. These imply $r^2 + (r')^2 = \left(a(1 + \sin \theta)\right)^2 + (a \cos \theta)^2 = 2a^2(1 + \sin \theta)$ and $(r')^2 - rr'' = (a \cos \theta)^2 - \left(a(1 + \sin \theta)\right)(-a \sin \theta) = a^2(1 + \sin \theta)$. Therefore $\kappa(\theta) = \dfrac{r^2 + 2(r')^2 - rr''}{(r^2 + (r')^2)^{3/2}} = \dfrac{3a^2(1 + \sin \theta)}{(2a^2(1 + \sin \theta))^{3/2}} = \dfrac{3}{2|a|\sqrt{2(1 + \sin \theta)}}$; hence $\kappa(\pi/2) = 3/(4|a|)$ and $\rho(\pi/2) = 4|a|/3$.

27.   Suppose $r = a\theta$; then $r' = dr/d\theta = a$ and $r'' = d^2r/d\theta^2 = 0$. These imply $r^2 + (r')^2 = a^2(\theta^2 + 1)$ and $(r')^2 - rr'' = a^2$. Therefore $\kappa(\theta) = \dfrac{r^2 + 2(r')^2 - rr''}{(r^2 + (r')^2)^{3/2}} = \dfrac{a^2(\theta^2 + 2)}{(a^2(\theta^2 + 1))^{3/2}} = \dfrac{\theta^2 + 2}{|a|(\theta^2 + 1)^{3/2}}$; hence $\kappa(1) = 3/\left(|a|2^{3/2}\right)$ and $\rho(1) = \left(|a|2^{3/2}\right)/3$.

Prelude to Solutions 29-36:   Parameterizing a curve in terms of arc-length s does provide an intrinsic parameter and nice properties; however, computations are usually done more easily (and directly) by differentiating with respect to t and then invoking the chain rule. The following facts suggest a convenient sequence for our calculations (and yield an alternative formula for κ).

$$s' = \frac{d}{dt}s(t) = \left|\frac{d}{dt}\mathbf{g}(t)\right| = |\mathbf{g}'|$$

$$\mathbf{T} = \frac{d}{ds}\mathbf{g}(s) = \frac{d\mathbf{g}/dt}{ds/dt} = \frac{\mathbf{g}'}{|\mathbf{g}'|}$$

$$\kappa\mathbf{n} = \frac{d}{ds}\mathbf{T} = \frac{d\mathbf{T}/dt}{ds/dt} = \frac{\mathbf{T}'}{|\mathbf{g}'|} = \frac{|\mathbf{T}'|}{|\mathbf{g}'|}\frac{\mathbf{T}'}{|\mathbf{T}'|}$$

$$\kappa = \frac{|\mathbf{T}'|}{|\mathbf{g}'|} \quad \text{and} \quad \mathbf{n} = \frac{\mathbf{T}'}{|\mathbf{T}'|}$$

$$\mathbf{B} = \mathbf{T} \times \mathbf{n}.$$

29. If $\mathbf{g} = \mathbf{i} + t\mathbf{j} + t^2\mathbf{k}$, then $\mathbf{g}' = \mathbf{j} + 2t\mathbf{k}$, $s' = |\mathbf{g}'| = \sqrt{1+4t^2}$, $\mathbf{T} = \mathbf{g}'/|\mathbf{g}'| = (\mathbf{j}+2t\mathbf{k})/\sqrt{1+4t^2}$, $\mathbf{T}' = (-4t\mathbf{j}+2\mathbf{k})/(1+4t^2)^{3/2}$, $|\mathbf{T}'| = 2/(1+4t^2)$, $\kappa = |\mathbf{T}'|/|\mathbf{g}'| = 2/(1+4t^2)^{3/2}$, $\mathbf{n} = \mathbf{T}'/|\mathbf{T}'| = (-2t\mathbf{j}+\mathbf{k})/\sqrt{1+4t^2}$, and $\mathbf{B} = \mathbf{T} \times \mathbf{n} = \mathbf{i}$.
$\mathbf{T}(1) = (\mathbf{j}+2\mathbf{k})/\sqrt{5}$, $\mathbf{n}(1) = (-2\mathbf{j}+\mathbf{k})/\sqrt{5}$, $\mathbf{B}(1) = \mathbf{i}$, and $\kappa(1) = 2/5^{3/2}$.

31. If $\mathbf{g}(t) = a\sin t\,\mathbf{i} + a\cos t\,\mathbf{j} + t\mathbf{k}$ and $a > 0$, then $\mathbf{g}' = a\cos t\,\mathbf{i} - a\sin t\,\mathbf{j} + \mathbf{k}$, $s' = |\mathbf{g}'| = \sqrt{1+a^2}$, $\mathbf{T} = \mathbf{g}'/|\mathbf{g}'| = (a\cos t\,\mathbf{i} - a\sin t\,\mathbf{j} + \mathbf{k})/\sqrt{1+a^2}$, $\mathbf{T}' = (-a\sin t\,\mathbf{i} - a\cos t\,\mathbf{j})/\sqrt{1+a^2}$, $|\mathbf{T}'| = a/\sqrt{1+a^2}$, $\kappa = |\mathbf{T}'|/|\mathbf{g}'| = a/(1+a^2)$, $\mathbf{n} = \mathbf{T}'/|\mathbf{T}'| = -\sin t\,\mathbf{i} - \cos t\,\mathbf{j}$, and $\mathbf{B} = \mathbf{T} \times \mathbf{n} = (\cos t\,\mathbf{i} - \sin t\,\mathbf{j} - a\mathbf{k})/\sqrt{1+a^2}$.
$\mathbf{T}(\pi/4) = ((a/\sqrt{2})\mathbf{i} - (a/\sqrt{2})\mathbf{j} + \mathbf{k})/\sqrt{1+a^2}$, $\kappa(\pi/4) = a/(1+a^2)$, $\mathbf{n}(\pi/4) = (-\mathbf{i}-\mathbf{j})/\sqrt{2}$, and $\mathbf{B}(\pi/4) = ((1/\sqrt{2})\mathbf{i} - (1/\sqrt{2})\mathbf{j} - a\mathbf{k})/\sqrt{1+a^2}$.

33. If $\mathbf{g}(t) = a\cos t\,\mathbf{i} + b\sin t\,\mathbf{j} + t\mathbf{k}$, $a > 0$ and $b > 0$, then

$$\mathbf{g}' = -a\sin t\,\mathbf{i} + b\cos t\,\mathbf{j} + \mathbf{k}$$

$$s' = |\mathbf{g}'| = \sqrt{1 + a^2\sin^2 t + b^2\cos^2 t}$$

$$\mathbf{T} = \mathbf{g}'/|\mathbf{g}'| = (-a\sin t\,\mathbf{i} + b\cos t\,\mathbf{j} + \mathbf{k})/\sqrt{1 + a^2\sin^2 t + b^2\cos^2 t}$$

$$\mathbf{T}' = \left(-a(1+b^2)\cos t\,\mathbf{i} - b(1+a^2)\sin t\,\mathbf{j} + (b^2-a^2)\sin t\,\cos t\,\mathbf{k}\right)/\left(1 + a^2\sin^2 t + b^2\cos^2 t\right)^{3/2}$$

$$|\mathbf{T}'| = \left(a^2b^2 + a^2\cos^2 t + b^2\sin^2 t\right)^{1/2}\left(1 + a^2\sin^2 t + b^2\cos^2 t\right)^{-1}$$

$$\kappa = |\mathbf{T}'|/|\mathbf{g}'| = \left(a^2b^2 + a^2\cos^2 t + b^2\sin^2 t\right)^{1/2}\left(1 + a^2\sin^2 t + b^2\cos^2 t\right)^{-3/2}$$

$$\mathbf{n} = \frac{\mathbf{T}'}{|\mathbf{T}'|} = \frac{-a(1+b^2)\cos t\,\mathbf{i} - b(1+a^2)\sin t\,\mathbf{j} + (b^2-a^2)\sin t\,\cos t\,\mathbf{k}}{\sqrt{\left(1 + a^2\sin^2 t + b^2\cos^2 t\right)\left(a^2b^2 + a^2\cos^2 t + b^2\sin^2 t\right)}}$$

$$\mathbf{B} = \mathbf{T} \times \mathbf{n} = (b\sin t\,\mathbf{i} - a\cos t\,\mathbf{j} + ab\mathbf{k})/\sqrt{a^2b^2 + a^2\cos^2 t + b^2\sin^2 t}.$$

$\mathbf{T}(0) = (b\mathbf{j}+\mathbf{k})/\sqrt{1+b^2}$, $\mathbf{n}(0) = -\mathbf{i}$, $\mathbf{B}(0) = (-\mathbf{j}+b\mathbf{k})/\sqrt{1+b^2}$, and $\kappa(0) = a/(1+b^2)$.

35. If $\mathbf{g}(t) = e^t\cos 2t\,\mathbf{i} + e^t\sin 2t\,\mathbf{j} + e^t\mathbf{k}$, then

$$\mathbf{g}' = e^t(\cos 2t - 2\sin 2t)\mathbf{i} + e^t(\sin 2t + 2\cos 2t)\mathbf{j} + e^t\mathbf{k}$$

$$s' = |\mathbf{g}'| = \sqrt{6}e^t$$

$$\mathbf{T} = \mathbf{g}'/|\mathbf{g}'| = ((\cos 2t - 2\sin 2t)\mathbf{i} + (\sin 2t + 2\cos 2t)\mathbf{j} + \mathbf{k})/\sqrt{6}$$

$$\mathbf{T}' = ((-2\sin 2t - 4\cos 2t)\mathbf{i} + (2\cos 2t - 4\sin 2t)\mathbf{j})/\sqrt{6}$$

$$|\mathbf{T}'| = \sqrt{10/3}$$

$$\kappa = |\mathbf{T}'|/|\mathbf{g}'| = \left(\sqrt{5}/3\right)e^{-t}$$

$$\mathbf{n} = \mathbf{T}'/|\mathbf{T}'| = ((-\sin 2t - 2\cos 2t)\mathbf{i} + (\cos 2t - 2\sin 2t)\mathbf{j})/\sqrt{5}$$

$$\mathbf{B} = \mathbf{T} \times \mathbf{n} = ((-\cos 2t + 2\sin 2t)\mathbf{i} - (\sin 2t + 2\cos 2t)\mathbf{j} + 5\mathbf{k})/\sqrt{30}.$$

$\mathbf{T}(0) = (\mathbf{i}+2\mathbf{j}+\mathbf{k})/\sqrt{6}$, $\mathbf{n}(0) = (-2\mathbf{i}+\mathbf{j})/\sqrt{5}$, $\mathbf{B}(0) = (-\mathbf{i}-2\mathbf{j}+5\mathbf{k})/\sqrt{30}$, and $\kappa(0) = \sqrt{5}/3$.

Prelude to Solutions 37-44: The results of Theorems 4 and 5 can be rearranged for computational purposes and then simplified for $\mathbf{f} = x\mathbf{i} + y\mathbf{j}$:

$$a_T = \frac{d^2s}{dt^2} = \frac{d}{dt}s' = \frac{d}{dt}\sqrt{\mathbf{f}'\cdot\mathbf{f}'} = (\mathbf{f}'\cdot\mathbf{f}'')/\sqrt{\mathbf{f}'\cdot\mathbf{f}'} = (x'x'' + y'y'')/\sqrt{(x')^2 + (y')^2},$$

$$a_n = (s')^2\kappa = |\mathbf{f}' \times \mathbf{f}''|/s' = |x'y'' - x''y'|/\sqrt{(x')^2 + (y')^2}.$$

12.6 — Curvature and the acceleration vector

37. If $\mathbf{f} = \cos 2t\,\mathbf{i} + \sin 2t\,\mathbf{j}$, then $\mathbf{f}' = -2\sin 2t\,\mathbf{i} + 2\cos 2t\,\mathbf{j}$, $s' = |\mathbf{f}'| = \sqrt{4\sin^2 2t + 4\cos^2 2t} = 2$, and $\mathbf{f}'' = -4\cos 2t\,\mathbf{i} - 4\sin 2t\,\mathbf{j}$. Thus $a_T = (\mathbf{f}\cdot\mathbf{f}'')/s' = 0/2 = 0$ and $a_n = |\mathbf{f}'\times\mathbf{f}''|/s' = |8\sin^2 2t + 8\cos^2 2t|/2 = 8/2 = 4$.

39. If $\mathbf{f} = t\mathbf{i} + t^2\mathbf{j}$, then $\mathbf{f}' = \mathbf{i} + 2t\mathbf{j}$, $s' = |\mathbf{f}'| = \sqrt{1+4t^2}$, and $\mathbf{f}'' = 2\mathbf{j}$. Thus $a_T = (\mathbf{f}'\cdot\mathbf{f}'')/s' = ((1)(0) + (2t)(2))/s' = 4t/\sqrt{1+4t^2}$ and $a_n = |\mathbf{f}'\times\mathbf{f}''|/s' = |(1)(2) - (2t)(0)|/s' = 2/\sqrt{1+4t^2}$.

41. If $\mathbf{f} = t\mathbf{i} + \cos t\,\mathbf{j}$, then $\mathbf{f}' = \mathbf{i} - \sin t\,\mathbf{j}$, $s' = |\mathbf{f}'| = \sqrt{1+\sin^2 t}$, and $\mathbf{f}'' = -\cos t\,\mathbf{j}$. Thus $a_T = (\mathbf{f}'\cdot\mathbf{f}'')/s' = ((1)(0) + (-\sin t)(-\cos t))/s' = \sin t\cos t/\sqrt{1+\sin^2 t}$ and $a_n = |\mathbf{f}'\times\mathbf{f}''|/s' = |(1)(-\cos t) - (-\sin t)(0)|/s' = |\cos t|/\sqrt{1+\sin^2 t}$.

43. If $\mathbf{f} = e^{-t}\mathbf{i} + e^t\mathbf{j}$, then $\mathbf{f}' = -e^{-t}\mathbf{i} + e^t\mathbf{j}$, $s' = |\mathbf{f}'| = \sqrt{e^{-2t}+e^{2t}}$, and $\mathbf{f}'' = \mathbf{f}$. Thus $a_T = (\mathbf{f}\cdot\mathbf{f}'')/s' = (-e^{-2t}+e^{2t})/\sqrt{e^{-2t}+e^{2t}}$ and $a_n = |\mathbf{f}'\times\mathbf{f}''|/s' = |-e^0 - e^0|/s' = 2/\sqrt{e^{-2t}+e^{2t}}$.

45. If $y = \ln x$, then $x > 0$, $y' = x^{-1}$, $y'' = -x^{-2}$, and

$$\kappa = \frac{|y''|}{(1+(y')^2)^{3/2}} = \frac{|-x^{-2}|}{(1+(x^{-1})^2)^{3/2}} = \frac{x^{-2}}{(1+x^{-2})^{3/2}} = \frac{|x|}{(x^2+1)^{3/2}} = \frac{x}{(x^2+1)^{3/2}}.$$

Therefore $\kappa' = (1 - 2x^2)(x^2+1)^{-5/2}$ which is zero at $x = \sqrt{1/2}$. Since κ changes sign from positive to negative there, $\kappa\left(\sqrt{1/2}\right) = 2\cdot 3^{-3/2}$ is the maximum curvature; the corresponding point on the curve is $\left(\sqrt{1/2}, -(1/2)\ln 2\right)$.

47. $F(v) = mv^2\kappa$ implies $F(v/M) = m(v/M)^2\kappa = mv^2\kappa/M^2 = F(v)/M^2$. Reducing the velocity by factor M reduces the required frictional force by factor M^2.

49. If $y = x^2 - x$, then $y' = 2x - 1$ and $y'' = 2$. Therefore $\kappa = |y''|/\left(1+(y')^2\right)^{3/2} = 2\left(1+(2x-1)^2\right)^{-3/2} = 2^{-1/2}\left(2x^2 - 2x + 1\right)^{-3/2}$ and $\kappa(0) = 2^{-1/2}$.

   a.
$$F(0) = mv^2\kappa(0) = (10000\,\text{kg})\left(\frac{80000\,\text{m/hr}}{3600\,\text{sec/hr}}\right)^2\left(\frac{1}{\sqrt{2}}\right) = \frac{4\cdot 10^8}{81\sqrt{2}}\,\text{N}.$$

   b.   The truck will remain on the track provided $mv^2\kappa \le \mu mg$; i.e., $v \le \sqrt{\mu g/\kappa} = \sqrt{(2.5)(9.81)\sqrt{2}} \approx 5.89\,\text{m/sec} \approx 21.2\,\text{km/hr}$.

51. We can not apply Problem 62 here because its formula would involve division by zero. The text definitions omit information we need to compute torsion for the straight line — the following is only suggestive.
If $\mathbf{f}(t) = t\mathbf{i} + (1-t)\mathbf{j} + (2+3t)\mathbf{k}$, then $\mathbf{f}'(t) = \mathbf{i} - \mathbf{j} + 3\mathbf{k}$, $\mathbf{f}''(t) = 0 = \mathbf{f}'''(t)$, and $\mathbf{T}(t) = (\mathbf{i} - \mathbf{j} + 3\mathbf{k})/\sqrt{11}$. Theorem 5 implies $\kappa = |\mathbf{f}'\times\mathbf{f}''|/|\mathbf{f}'|^3 = 0/11^{3/2} = 0$. Because $\kappa = 0$, we can not use equation (11) to define $\mathbf{n}$. Since $\mathbf{T}$ is constant, $\mathbf{n}$ would also need to be constant; then $\mathbf{B} = \mathbf{T}\times\mathbf{n}$ would be constant, $d\mathbf{B}/ds = \mathbf{0}$; hence Problem 60 would imply $\tau = 0$.

53. If $\mathbf{f}(t) = \cos t\,\mathbf{i} + \sin t\,\mathbf{j} + t\mathbf{k}$, then

$$\mathbf{f}' = -\sin t\,\mathbf{i} + \cos t\,\mathbf{j} + \mathbf{k},$$
$$\mathbf{f}'' = -\cos t\,\mathbf{i} - \sin t\,\mathbf{j},$$
$$\mathbf{f}'\times\mathbf{f}'' = \sin t\,\mathbf{i} - \cos t\,\mathbf{j} + \mathbf{k}, \qquad |\mathbf{f}'\times\mathbf{f}''| = \sqrt{2},$$
$$\mathbf{f}''' = \sin t\,\mathbf{i} - \cos t\,\mathbf{j}, \qquad \mathbf{f}'''\cdot(\mathbf{f}'\times\mathbf{f}'') = 1,$$
$$\tau = \frac{\mathbf{f}'''\cdot(\mathbf{f}'\times\mathbf{f}'')}{|\mathbf{f}'\times\mathbf{f}''|^2} = \frac{1}{2}.$$

55. Suppose $y = f(x)$ is a smooth curve. Let $t = x$; then $\dot{x} = \frac{dx}{dt} = 1$, $\ddot{x} = \frac{d^2x}{dt^2} = 0$, $\dot{y} = \frac{dy}{dt} = \frac{dy}{dx} = y'$, $\ddot{y} = \frac{d^2y}{dt^2} = \frac{d^2y}{dx^2} = y''$, and equation (4) implies

$$\kappa = \frac{|\dot{x}\ddot{y} - \dot{y}\ddot{x}|}{\left((\dot{x})^2 + (\dot{y})^2\right)^{3/2}} = \frac{|(1)(y'') - (y')(0)|}{\left((1)^2 + (y')^2\right)^{3/2}} = \frac{|y''|}{\left(1+(y')^2\right)^{3/2}}.\ \blacksquare$$

57. The discussion of equation (18) points out that the tangential component of acceleration a_T equals $\frac{dv}{dt}$.

If a particle moves with constant speed v, then $a_T = \frac{dv}{dt} = 0$. $\blacksquare$

59.
$$0 = \mathbf{B} \cdot \mathbf{T} \implies 0 = \frac{d}{ds}(\mathbf{B} \cdot \mathbf{T}) = \frac{d\mathbf{B}}{ds} \cdot \mathbf{T} + \mathbf{B} \cdot \frac{d\mathbf{T}}{ds} = \frac{d\mathbf{B}}{ds} \cdot \mathbf{T} + \mathbf{B} \cdot (\kappa \mathbf{n}) = \frac{d\mathbf{B}}{ds} \cdot \mathbf{T} + 0 = \frac{d\mathbf{B}}{ds} \cdot \mathbf{T} \implies \frac{d\mathbf{B}}{ds} \perp \mathbf{T};$$
$$1 = \mathbf{B} \cdot \mathbf{B} \implies 0 = \frac{d}{ds}(\mathbf{B} \cdot \mathbf{B}) = 2\,\mathbf{B} \cdot \frac{d\mathbf{B}}{ds} \implies \mathbf{B} \perp \frac{d\mathbf{B}}{ds}. \ \blacksquare$$

61. We begin with the expression for $\mathbf{f}''$ established by Theorem 4, then apply the results of equation (11) and Problem 60.

$$\mathbf{f}''' = \frac{d}{ds}\mathbf{f}'' = \frac{d}{ds}\left(\frac{d^2 s}{dt^2}\mathbf{T} + \left(\frac{ds}{dt}\right)^2 \kappa\,\mathbf{n}\right)$$

$$= \frac{d^3 s}{dt^3}\mathbf{T} + \frac{d^2 s}{dt^2}\left(\frac{d\mathbf{T}}{ds}\frac{ds}{dt}\right) + 2\left(\frac{ds}{dt}\right)\frac{d^2 s}{dt^2}\kappa\,\mathbf{n} + \left(\frac{ds}{dt}\right)^2\frac{d\kappa}{dt}\mathbf{n} + \left(\frac{ds}{dt}\right)^2\kappa\left(\frac{d\mathbf{n}}{ds}\frac{ds}{dt}\right)$$

$$= \frac{d^3 s}{dt^3}\mathbf{T} + \frac{d^2 s}{dt^2}\left((\kappa\mathbf{n})\frac{ds}{dt}\right) + 2\left(\frac{ds}{dt}\right)\frac{d^2 s}{dt^2}\kappa\,\mathbf{n} + \left(\frac{ds}{dt}\right)^2\frac{d\kappa}{dt}\mathbf{n} + \left(\frac{ds}{dt}\right)^2\kappa\left((-\kappa\mathbf{T} + \tau\mathbf{B})\frac{ds}{dt}\right)$$

$$= \left(\frac{d^3 s}{dt^3} - \kappa^2\left(\frac{ds}{dt}\right)^3\right)\mathbf{T} + \left(3\frac{ds}{dt}\frac{d^2 s}{dt^2}\kappa + \left(\frac{ds}{dt}\right)^2\frac{d\kappa}{dt}\right)\mathbf{n} + \tau\kappa\left(\frac{ds}{dt}\right)^3\mathbf{B}. \ \blacksquare$$

63. If $\mathbf{f}$ lies in a plane, then so do all of its derivatives. Therefore the scalar triple product $\mathbf{f}''' \cdot (\mathbf{f}' \times \mathbf{f}'')$ must equal zero; the result of Problem 62 then implies $\tau = 0$. $\blacksquare$

Solutions 12.R   *Review*   (pages 830-1)

1. If $\mathbf{f}(t) = x\mathbf{i} + y\mathbf{j} = t\mathbf{i} + 2t\mathbf{j}$, then $x = t \implies y = 2t = 2x$.

1: $t\,\mathbf{i} + 2t\,\mathbf{j}$
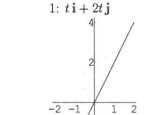

3: $t^2\mathbf{i} + (2t - 6)\mathbf{j}$
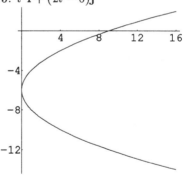

3. If $\mathbf{f}(t) = x\mathbf{i} + y\mathbf{j} = t^2\mathbf{i} + (2t - 6)\mathbf{j}$, then $y = 2t - 6 \implies t = (y + 6)/2$ and $x = t^2 = ((y + 6)/2)^2$.

5. If $\mathbf{f}(t) = x\mathbf{i} + y\mathbf{j} = \cos 4t\,\mathbf{i} + \sin 4t\,\mathbf{j}$, then $x^2 + y^2 = \cos^2 4t + \sin^2 4t = 1$. (The curve is the unit circle centered at the origin.)

5: $\cos 4t\,\mathbf{i} + \sin 4t\,\mathbf{j}$
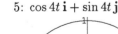
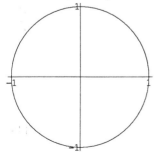

7: $t^6\mathbf{i} + t^2\mathbf{j}$
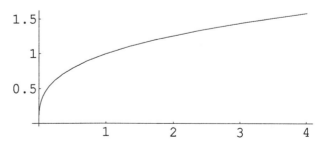

7. If $\mathbf{f}(t) = x\mathbf{i} + y\mathbf{j} = t^6\mathbf{i} + t^2\mathbf{j}$, then $y = t^2 \geq 0$ and $x = t^6 = (t^2)^3 = y^3$.

9. $x = t^2 \implies x' = 2t$ and $y = 6t \implies y' = 6$, therefore $m = dy/dx = y'/x' = 6/(2t) = 3/t$ and $m(1) = 3/1 = 3$. The curve has a vertical tangent at the point $(x(0), y(0)) = (0,0)$ because $\lim_{t \to 0^-} 3/t = -\infty$ and $\lim_{t \to 0^+} 3/t = \infty$; there are no other vertical tangents. The curve has no horizontal tangents because $m(t) = 3/t \neq 0$ for all t.

9: $(t^2, 6t)$

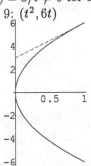

11: $(\sin 5\theta, \cos 5\theta)$

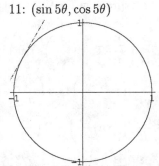

11. $x = \sin 5\theta \implies x' = 5\cos t\theta$ and $y = \cos t\theta \implies y' = -5\sin 5\theta$, therefore $m = dy/dx = y'/x' = (-5\sin 5\theta)/(5\cos t\theta) = -\tan 5\theta$ and $m(\pi/3) = -\tan(5\pi/3) = \sqrt{3}$. If $x' = 5\cos t\theta = 0$, then $x = \sin 5\theta = \pm 1$ and $y = \cos t\theta = 0$ (and $y' = \mp 1$); the curve has vertical tangents at the points $(\pm 1, 0)$. If $y' = -5\sin 5\theta = 0$, then $y = \cos t\theta = \pm 1 = x'$ and $x = \sin 5\theta = 0$; the curve has horizontal tangents at the points $(0, \pm 1)$.

13. $x = \cosh t \implies x' = \sinh t$ and $y = \sinh t \implies y' = \cosh t$, therefore $m = dy/dx = y'/x' = \cosh t/\sinh t = \coth t$ and $m(0)$ is undefined because $x'(0) = \sinh t(0) = 0$. The curve has no horizontal tangents because $\cosh t \geq 1$. The curve has its only vertical tangent at $(x(0), y(0)) = (1, 0)$.

REMARK: The curve is the right half of the hyperbola $x^2 - y^2 = 1$.

13: $(\cosh t, \sinh t)$

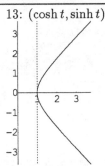

15: $(3\cos\theta, 4\sin\theta)$

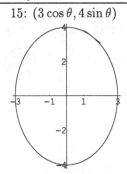

15. $x = 3\cos\theta \implies x' = -3\sin\theta$ and $y = 4\sin\theta \implies y' = 4\cos\theta$, therefore $m = dy/dx = y'/x' = (4\cos\theta)/(-3\sin\theta) = (-4/3)\cot\theta$ and $m(\pi/3) = -4/3^{3/2}$. The curve has vertical tangents at $(x(k\pi), y(k\pi)) = (3(-1)^k, 0) = (\pm 3, 0)$ and horizontal tangents at $(x((k+1/2)\pi), y((k+1/2)\pi)) = (0, 4(-1)^k) = (0, \pm 4)$.

REMARK: The curve is on the ellipse $(x/3)^2 + (y/4)^2 = 1$.

17.   If $\mathbf{f}(t) = 2t\mathbf{i} - t^2\mathbf{j}$, then $\mathbf{f}'(t) = 2\mathbf{i} - 2t\mathbf{j}$ and $\mathbf{f}''(t) = -2\mathbf{j}$.

19.   If $\mathbf{f}(t) = \cos 5t\,\mathbf{i} + 2\sin t\,\mathbf{j}$, then $\mathbf{f}'(t) = -5\sin 5t\,\mathbf{i} + 2\cos t\,\mathbf{j}$ and $\mathbf{f}''(t) = -25\cos 5t\,\mathbf{i} - 2\sin t\,\mathbf{j}$.

21.   If $\mathbf{g}(t) = t^3\mathbf{i} - t^2\mathbf{j} + t\mathbf{k}$, then $\mathbf{g}'(t) = 3t^2\mathbf{i} - 2t\mathbf{j} + \mathbf{k}$ and $\mathbf{g}''(t) = 6t\mathbf{i} - 2\mathbf{j}$.

23. $$\int_0^3 (t^2\mathbf{i} + t^5\mathbf{j})\,dt = \left(\frac{t^3}{3}\mathbf{i} + \frac{t^6}{6}\mathbf{j}\right)\Big|_0^3 = 9\mathbf{i} + \frac{243}{2}\mathbf{j}.$$

25. $$\int_0^{\pi/2} (\cos t\,\mathbf{i} + \sin t\,\mathbf{j} + t\mathbf{k})\,dt = \left(\sin t\,\mathbf{i} - \cos t\,\mathbf{j} + \frac{t^2}{2}\mathbf{k}\right)\Big|_0^{\pi/2} = \mathbf{i} + \mathbf{j} + \frac{\pi^2}{8}\mathbf{k}.$$

27. $$\frac{d}{dt}\left((2t\mathbf{i} + \sqrt{t}\mathbf{j})\cdot(t\mathbf{i} - 3\mathbf{j})\right) = \frac{d}{dt}\left(2t^2 - 3\sqrt{t}\right) = 4t - \frac{3}{2\sqrt{t}}.$$

REMARK: If $\mathbf{a} = 2t\mathbf{i} + \sqrt{t}\mathbf{j}$ and $\mathbf{b} = t\mathbf{i} - 3\mathbf{j}$, then
$$\frac{d}{dt}(\mathbf{a}\cdot\mathbf{b}) = \frac{d\mathbf{a}}{dt}\cdot\mathbf{b} + \mathbf{a}\cdot\frac{d\mathbf{b}}{dt} = \left(2\mathbf{i} + \frac{1}{2\sqrt{t}}\mathbf{j}\right)\cdot(t\mathbf{i} - 3\mathbf{j}) + (2t\mathbf{i} + \sqrt{t}\mathbf{j})\cdot(\mathbf{i}) = \left(2t + \frac{-3}{2\sqrt{t}}\right) + (2t).$$

29. $\dfrac{d}{dt}\left((3\mathbf{i}+4\mathbf{j})\times(\cos t\,\mathbf{i}+\sin t\,\mathbf{j}+t\mathbf{k})\right)=\dfrac{d}{dt}\left(4t\mathbf{i}-3t\mathbf{j}+(3\sin t-4\cos t)\mathbf{k}\right)=4\mathbf{i}-3\mathbf{j}+(3\cos t+4\sin t)\mathbf{k}.$

REMARK: If $\mathbf{a}=3\mathbf{i}+4\mathbf{j}$ and $\mathbf{b}=\cos t\,\mathbf{i}+\sin t\,\mathbf{j}+t\mathbf{k}$, then $d\mathbf{a}/dt=0$ and

$$\dfrac{d}{dt}\left(\mathbf{a}\times\mathbf{b}\right)=\left(\dfrac{d\mathbf{a}}{dt}\times\mathbf{b}\right)+\left(\mathbf{a}\times\dfrac{d\mathbf{b}}{dt}\right)=\mathbf{a}\times\dfrac{d\mathbf{b}}{dt}=(3\mathbf{i}+4\mathbf{j})\times(-\sin t\,\mathbf{i}+\cos t\,\mathbf{j}+\mathbf{k}).$$

31. If $\mathbf{f}(t)=t^4\mathbf{i}+t^5\mathbf{j}$, then $\mathbf{f}'=4t^3\mathbf{i}+5t^4\mathbf{j}$, $|\mathbf{f}'|=|t|^3\sqrt{16+25t^2}$, $\mathbf{T}=\dfrac{\mathbf{f}'}{|\mathbf{f}'|}=\dfrac{4\operatorname{sgn}(t)\mathbf{i}+5|t|\mathbf{j}}{\sqrt{16+25t^2}}$, and $\mathbf{T}(1)=$

$\dfrac{4\mathbf{i}+5\mathbf{j}}{\sqrt{41}}$. Furthermore $\mathbf{T}'=\dfrac{-100|t|\mathbf{i}+80\operatorname{sgn}(t)\mathbf{j}}{(16+25t^2)^{3/2}}$, $\mathbf{n}=\dfrac{\mathbf{T}'}{|\mathbf{T}'|}=\dfrac{-5|t|\mathbf{i}+4\operatorname{sgn}(t)\mathbf{j}}{\sqrt{25t^2+16}}$, and $\mathbf{n}(1)=\dfrac{-5\mathbf{i}+4\mathbf{j}}{\sqrt{41}}$.

33. If $\mathbf{f}(t)=\ln t\,\mathbf{i}+\sqrt{t}\,\mathbf{j}$ and $t>0$, then $\mathbf{f}'=\dfrac{1}{t}\mathbf{i}+\dfrac{1}{2\sqrt{t}}\mathbf{j}$, $|\mathbf{f}'|=\dfrac{\sqrt{4+t}}{2t}$, $\mathbf{T}=\dfrac{2\mathbf{i}+\sqrt{t}\,\mathbf{j}}{\sqrt{4+t}}$, and $\mathbf{T}(1)=\dfrac{2\mathbf{i}+\mathbf{j}}{\sqrt{5}}$.

Furthermore $\mathbf{T}'=-(4+t)^{-3/2}\mathbf{i}+2t^{-1/2}(4+t)^{-3/2}\mathbf{j}$, $\mathbf{n}=\dfrac{-\sqrt{t}\,\mathbf{i}+2\mathbf{j}}{\sqrt{4+t}}$, and $\mathbf{n}(1)=\dfrac{-\mathbf{i}+2\mathbf{j}}{\sqrt{5}}$.

35. Arc length of the curve $(x,y)=(\cos4\theta,\sin4\theta)$ on the interval $[0,\pi/12]$ is

$$s=\int_0^{\pi/12}\sqrt{(-4\sin4\theta)^2+(4\cos4\theta)^2}\,d\theta=\int_0^{\pi/12}4\sqrt{\sin^2 4\theta+\cos^2 4\theta}\,d\theta=4\int_0^{\pi/12}d\theta=4\left(\dfrac{\pi}{12}-0\right)=\dfrac{\pi}{3}.$$

REMARK: $0\le\theta\le\pi/12\iff0\le4\theta\le\pi/3=(2\pi)/6$; s is one-sixth the circumference of the unit circle.

37. The cardioid $r=2(1+\cos\theta)$ is traversed for θ in the interval $[-\pi,\pi]$; the arc length is

$$s=\int_{-\pi}^{\pi}\sqrt{r^2+(r')^2}\,d\theta=\int_{-\pi}^{\pi}\sqrt{4(1+\cos\theta)^2+(-2\sin\theta)^2}\,d\theta=\int_{-\pi}^{\pi}2\sqrt{2+2\cos\theta}\,d\theta=\int_{-\pi}^{\pi}2\sqrt{4\cos^2(\theta/2)}\,d\theta$$

$$=\int_{-\pi}^{\pi}4\,|\cos(\theta/2)|\;d\theta=8\sin(\theta/2)\Big|_{-\pi}^{\pi}=16.$$

39. Arc length of the curve $(x,y,z)=(-2t,\cos2t,\sin2t)$ on the interval $[0,\pi/6]$ is

$$s=\int_0^{\pi/6}\sqrt{(-2)^2+(-2\sin2t)^2+(2\cos2t)^2}\,dt=\int_0^{\pi/6}\sqrt{4+4}\,dt=2\sqrt{2}\,(\pi/6-0)=\sqrt{2}\,\pi/3.$$

41. If $\mathbf{f}(t)=\cos2t\,\mathbf{i}+\sin2t\,\mathbf{j}$, then $\mathbf{v}(t)=\mathbf{f}'(t)=-2\sin2t\,\mathbf{i}+2\cos2t\,\mathbf{j}$ and $\mathbf{a}(t)=\mathbf{f}''(t)=-4\cos2t\,\mathbf{i}-4\sin2t\,\mathbf{j}$. The speed is $|\mathbf{f}'(t)|=2$ and the acceleration scalar is $|\mathbf{f}''(t)|=4$. For the specified value of $t=\pi/6$ we have $\mathbf{f}(\pi/6)=(1/2)\mathbf{i}+(\sqrt{3}/2)\mathbf{j}$, $\mathbf{v}(\pi/6)=-\sqrt{3}\mathbf{i}+\mathbf{j}$, and $\mathbf{a}(\pi/6)=-2\mathbf{i}-2\sqrt{3}\mathbf{j}$.

43. If $\mathbf{f}(t)=(2^t+e^{-t})\mathbf{i}+2t\mathbf{j}$, then $\mathbf{v}(t)=\mathbf{f}'(t)=\left((\ln2)2^t-e^{-t}\right)\mathbf{i}+2\mathbf{j}$ and $\mathbf{a}(t)=\mathbf{f}''(t)=\left((\ln2)^2 2^t+e^{-t}\right)\mathbf{i}$. For the specified value of $t=0$ we have $\mathbf{f}(0)=2\mathbf{i}$, $\mathbf{v}(0)=(\ln2-1)\mathbf{i}+2\mathbf{j}$, and $\mathbf{a}(0)=\left((\ln2)^2+1\right)\mathbf{i}$; the speed is $|\mathbf{v}(0)|=\sqrt{(\ln2-1)^2+4}$ and the acceleration scalar is $|\mathbf{a}(0)|=(\ln2)^2+1$.

41: $\cos2t\,\mathbf{i}+\sin2t\,\mathbf{j}$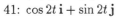

43: $(2^t+e^{-t})\mathbf{i}+2t\,\mathbf{j}$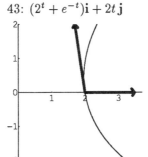

45: $2\sinh t\,\mathbf{i}+4t\,\mathbf{j}$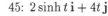

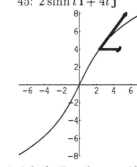

45. If $\mathbf{f}(t)=2\sinh t\,\mathbf{i}+4t\mathbf{j}$, then $\mathbf{v}(t)=\mathbf{f}'(t)=2\cosh t\,\mathbf{i}+4\mathbf{j}$ and $\mathbf{a}(t)=\mathbf{f}''(t)=2\sinh t\,\mathbf{i}$. For the specified value of $t=1$ we have $\mathbf{f}(1)=2\sinh1\,\mathbf{i}+4\mathbf{j}$, $\mathbf{v}(1)=2\cosh1\,\mathbf{i}+4\mathbf{j}$, and $\mathbf{a}(1)=2\sinh1\,\mathbf{i}$; the speed is $|\mathbf{v}(1)|=2\sqrt{\cosh^2 1+4}$ and the acceleration scalar is $|\mathbf{a}(1)|=2\sinh1$.

47. If $\mathbf{f}(t)=2\sin t\,\mathbf{i}+2\cos t\,\mathbf{j}$, then $\mathbf{f}'(t)=2\cos t\,\mathbf{i}-2\sin t\,\mathbf{j}$ and $s'=|\mathbf{f}'(t)|=2$. Thus $a_T=\dfrac{d}{dt}s'=\dfrac{d}{dt}2=0$, therefore $a_n=|\mathbf{a}|=|-2\sin t\,\mathbf{i}-2\cos t\,\mathbf{j}|=2$.

49.   If $\mathbf{f}(t) = x\mathbf{i} + y\mathbf{j} = 3t^2\mathbf{i} + 2t^3\mathbf{j}$, then $\mathbf{v} = \mathbf{f}'(t) = 6t\mathbf{i} + 6t^2\mathbf{j}$, $|\mathbf{v}| = \sqrt{36t^2 + 36t^4} = 6|t|\sqrt{1+t^2}$, and $\mathbf{a} = \mathbf{f}''(t) = 6\mathbf{i} + 12t\mathbf{j}$. Thus $a_T = (\mathbf{v} \cdot \mathbf{a})/|\mathbf{v}| = (36t + 72t^3)/6|t|\sqrt{1+t^2} = 6\left(1 + 2t^2\right)\operatorname{sgn}(t)/\sqrt{1+t^2}$ and $a_n = |x'y'' - x''y'|/|\mathbf{v}| = |72t^2 - 36t^2|/|\mathbf{v}| = 6|t|/\sqrt{1+t^2}$.

51.   Suppose $\mathbf{f} = x\mathbf{i} + y\mathbf{j} = \cos 2t\mathbf{i} + \sin 2t\mathbf{j}$, Then $x' = -2\sin 2t$, $y' = 2\cos 2t$, $s' = \sqrt{(-2\sin 2t)^2 + (2\cos 2t)^2} = 2$, and $\mathbf{T} = (-2\sin 2t\mathbf{i} + 2\cos 2t\mathbf{j})/2 = -\sin 2t\mathbf{i} + \cos 2t\mathbf{j}$. Furthermore $x'' = -4\cos 2t$, $y'' = -4\sin 2t$, $D = x'y'' - y'x'' = (-2\sin 2t)(-4\sin 2t) - (2\cos 2t)(-4\cos 2t) = 8$, $\kappa = 8/(2^2)^{3/2} = 1$, $\rho = 1/\kappa = 1$, and $\mathbf{C} = \left(\cos 2t - (2\cos 2t)(2^2/8)\right)\mathbf{i} + \left(\sin 2t + (-2\sin 2t)(2^2/8)\right)\mathbf{j} = \mathbf{0}$.
     $\mathbf{f}(\pi/3) = (-\mathbf{i} + \sqrt{3}\mathbf{j})/2$, $\mathbf{T}(\pi/3) = (-\sqrt{3}\mathbf{i} - \mathbf{j})/2$.

REMARK: A circle is its own circle-of-curvature at each point, the center-of-curvature is the circle's center.

53.   Suppose $\mathbf{f} = x\mathbf{i} + y\mathbf{j} = 4\cos t\mathbf{i} + 9\sin t\mathbf{j}$, Then $x' = -4\sin t$, $y' = 9\cos t$, $s' = \sqrt{(-4\sin t)^2 + (9\cos t)^2} = \sqrt{16 + 65\cos^2 t}$, and $\mathbf{T} = (-4\sin t\mathbf{i} + 9\cos t\mathbf{j})/\sqrt{16 + 65\cos^2 t}$. Furthermore $x'' = -4\cos t$, $y'' = -9\sin t$, $D = x'y'' - y'x'' = (-4\sin t)(-9\sin t) - (9\cos t)(-4\cos t) = 36$, $\rho = (16 + 65\cos^2 t)^{3/2}/36$, $\kappa = 36(16 + 65\cos^2 t)^{-3/2}$, and $\mathbf{C} = \left(4\cos t - (9\cos t)(16 + 65\cos^2 t)/(36)\right)\mathbf{i} + \left(9\sin t + (-4\sin t)(16 + 65\cos^2 t)/(36)\right)\mathbf{j} = (-65/4)\cos^3 t\mathbf{i} + (65/9)\sin^3 t\mathbf{j}$.
     $\mathbf{f}(\pi/4) = 2\sqrt{2}\mathbf{i} + (9/\sqrt{2})\mathbf{j}$, $\mathbf{T}(\pi/4) = (-4\mathbf{i} + 9\mathbf{j})/\sqrt{97}$. $\rho(\pi/4) = 97^{3/2}/72\sqrt{2}$, $\kappa(\pi/4) = 72\sqrt{2}\,97^{-3/2}$, and $\mathbf{C}(\pi/4) = (-65\sqrt{2}/16)\mathbf{i} + (65\sqrt{2}/36)\mathbf{j}$.

51: $\cos 2t\,\mathbf{i} + \sin 2t\,\mathbf{j}$ 53: $4\cos t\,\mathbf{i} + 9\sin t\,\mathbf{j}$ 55: $y = 2x^2$

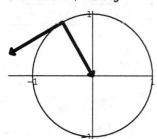 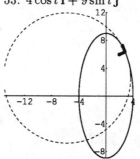 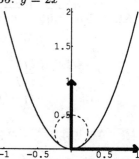

55.   Suppose $y = 2x^2$. Then $y' = 4x$, $s' = \sqrt{1 + (4x)^2} = \sqrt{1 + 16x^2}$, and $\mathbf{T} = (\mathbf{i} + 4x\mathbf{j})/\sqrt{1 + 16x^2}$. Furthermore $y'' = 4$, $\kappa = 4\left(1 + 16x^2\right)^{-3/2}$, $\rho = (1/4)\left(1 + 16x^2\right)^{3/2}$, and $\mathbf{C} = -16x^3\mathbf{i} + (6x^2 + 1/4)\mathbf{j}$.
     $y(0) = 0$, $\mathbf{T}(0) = \mathbf{i}$, $\kappa(0) = 4$, $\rho(0) = 1/4$, and $\mathbf{C}(0) = (1/4)\mathbf{j}$.

57.   Suppose $y = e^{-x}$. Then $y' = -e^{-x}$, $s' = \sqrt{1 + (-e^{-x})^2} = \sqrt{1 + e^{-2x}}$, and $\mathbf{T} = (\mathbf{i} - e^{-x}\mathbf{j})/\sqrt{1 + e^{-2x}}$. $y'' = e^{-x}$, $\kappa = e^{-x}\left(1 + e^{-2x}\right)^{-3/2}$, $\rho = e^x\left(1 + e^{-2x}\right)^{3/2}$, and $\mathbf{C} = (1 + x + e^{-2x})\mathbf{i} + (e^x + e^{-2x})\mathbf{j}$.
     $y(1) = 1/e$, $\mathbf{T}(1) = (\mathbf{i} - (1/e)\mathbf{j})/\sqrt{1 + 1/e^2} = (e\mathbf{i} - \mathbf{j})/\sqrt{e^2 + 1}$, $\kappa(1) = (1/e)(1 + 1/e^2)^{-3/2}$, $\rho(1) = e(1 + 1/e^2)^{3/2}$, and $\mathbf{C}(1) = (2 + e^{-2})\mathbf{i} + (e + e^{-2})\mathbf{j}$.

57: $y = e^{-x}$ 59: $r = 1 + \sin\theta$

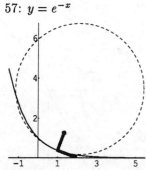 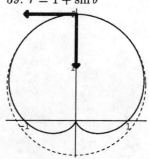

59.   Suppose $r = 1 + \sin\theta$; then $r' = dr/d\theta = \cos\theta$ and $r'' = d^2r/d\theta^2 = -\sin\theta$. These imply $r^2 + (r')^2 = (1 + 2\sin\theta + \sin^2\theta) + \cos^2\theta = 2(1 + \sin\theta)$ and $(r')^2 - rr'' = \cos^2\theta - (1 + \sin\theta)(-\sin\theta) = 1 + \sin\theta$. Hence

$$\kappa(\theta) = \frac{\left|r^2 + 2(r')^2 - rr''\right|}{\left(r^2 + (r')^2\right)^{3/2}} = \frac{3(1 + \sin\theta)}{\left(2(1 + \sin\theta)\right)^{3/2}} = \frac{3}{2\sqrt{2(1 + \sin\theta)}}.$$

Therefore $\kappa(\pi/2) = 3/4$ and $\rho(\pi/2) = 4/3$.

We can find the circle of curvature by switching to rectangular coordinates. If $\mathbf{f} = x\mathbf{i} + y\mathbf{j} = r\cos\theta\mathbf{i} + r\sin\theta\mathbf{j} = (1+\sin\theta)\cos\theta\mathbf{i} + (1+\sin\theta)\sin\theta\mathbf{j}$, then $\mathbf{f}' = (1+\sin\theta)(1-2\sin\theta)\mathbf{i} + \cos\theta(1+2\sin\theta)\mathbf{j}$ and $x'y'' - x''y' = 3(1+\sin\theta)$. Thus $\mathbf{C} = (1/3)\cos\theta(1-\sin\theta)\mathbf{i} + (1/3)(1+\sin\theta)(2-\sin\theta)\mathbf{j}$ and $\mathbf{C}(\pi/2) = (2/3)\mathbf{j}$.

REMARK: $\mathbf{C} = (2/3)\mathbf{j} + (1/3)(1-\sin\theta)(\cos\theta\mathbf{i} + \sin\theta\mathbf{j})$; the evolute of the original cardioid is also a cardioid — centered at $(0, 2/3)$, one-third the original size, inverted.

61. The function $\mathbf{f}(t) = 3t\mathbf{i} + 4t^{3/2}\mathbf{j}$ is defined only for $t \geq 0$; the arc length on the interval $[0, t]$ is

$$s = \int_0^t \sqrt{3^2 + (6\sqrt{u})^2}\, du = \int_0^t 3\sqrt{1+4u}\, du = \frac{1}{2}(1+4u)^{3/2}\Big|_0^t = \frac{1}{2}(1+4t)^{3/2} - \frac{1}{2}.$$

Solving for t as a function of s:

$$s = \frac{(1+4t)^{3/2} - 1}{2} \iff 2s+1 = (1+4t)^{3/2} \iff (2s+1)^{2/3} = 1+4t \iff t = \frac{(2s+1)^{2/3} - 1}{4}.$$

Therefore $\mathbf{f}(s) = x\mathbf{i} + y\mathbf{j} = 3\left(\frac{(2s+1)^{2/3} - 1}{4}\right)\mathbf{i} + 4\left(\frac{(2s+1)^{2/3} - 1}{4}\right)^{3/2}\mathbf{j}$ and

$$\left(\frac{dx}{ds}\right)^2 + \left(\frac{dy}{ds}\right)^2 = \left(\frac{1}{(2s+1)^{1/3}}\right)^2 + \left(\frac{\sqrt{(2s+1)^{2/3} - 1}}{(2s+1)^{1/3}}\right)^2 = \frac{1}{(2s+1)^{2/3}} + \frac{(2s+1)^{2/3} - 1}{(2s+1)^{2/3}} = 1. \blacksquare$$

63. The arc length of $\mathbf{f}(t) = 2\cos 3t\mathbf{i} + 2\sin 3t\mathbf{j}$ on the interval $[0, t]$ is $s = \int_0^t \sqrt{(-6\sin 3u)^2 + (6\cos 3u)^2}\, du =$

$\int_0^t 6\, du = 6u\Big|_0^t = 6t$. Since $s = 6t \iff t = s/6$, it is immediate that $\mathbf{f}(s) = x\mathbf{i} + y\mathbf{j} = 2\cos(s/2)\mathbf{i} + 2\sin(s/2)\mathbf{j}$

and $\left(\frac{dx}{ds}\right)^2 + \left(\frac{dy}{ds}\right)^2 = \sin^2(s/2) + \cos^2(s/2) = 1$.

65. If $\mathbf{g}(t) = \mathbf{i} + t\mathbf{j} + (1/2)t^2\mathbf{k}$, then

$$\mathbf{g}' = \mathbf{j} + t\mathbf{k}$$
$$s' = |\mathbf{g}'| = \sqrt{1+t^2}$$
$$\mathbf{T} = \mathbf{g}'/|\mathbf{g}'| = (\mathbf{j} + t\mathbf{k})/\sqrt{1+t^2}$$
$$\mathbf{T}' = (-t\mathbf{j} + \mathbf{k})/(1+t^2)^{3/2}$$
$$|\mathbf{T}'| = (1+t^2)^{-1}$$
$$\kappa = |\mathbf{T}'|/|\mathbf{g}'| = (1+t^2)^{-3/2}$$
$$\mathbf{n} = \mathbf{T}'/|\mathbf{T}'| = (-t\mathbf{j} + \mathbf{k})/\sqrt{1+t^2}$$
$$\mathbf{B} = \mathbf{T} \times \mathbf{n} = \mathbf{i}$$

$\mathbf{T}(2) = (\mathbf{j} + 2\mathbf{k})/\sqrt{5}$, $\mathbf{n}(2) = (-2\mathbf{j} + \mathbf{k})/\sqrt{5}$, $\mathbf{B}(2) = \mathbf{i}$, and $\kappa(2) = 5^{-3/2}$.

67. If $\mathbf{g}(t) = 2\cos t\mathbf{i} + 2\sin t\mathbf{j} + t\mathbf{k}$, then

$$\mathbf{g}' = -2\sin t\mathbf{i} + 2\cos t\mathbf{j} + \mathbf{k}$$
$$s' = |\mathbf{g}'| = \sqrt{5}$$
$$\mathbf{T} = \mathbf{g}'/|\mathbf{g}'| = (-2\sin t\mathbf{i} + 2\cos t\mathbf{j} + \mathbf{k})/\sqrt{5}$$
$$\mathbf{T}' = (-2\cos t\mathbf{i} - 2\sin t\mathbf{j})/\sqrt{5}$$
$$|\mathbf{T}'| = 2/\sqrt{5}$$
$$\kappa = |\mathbf{T}'|/|\mathbf{g}'| = 2/5$$
$$\mathbf{n} = \mathbf{T}'/|\mathbf{T}'| = (-\cos t\mathbf{i} - \sin t\mathbf{j})$$
$$\mathbf{B} = \mathbf{T} \times \mathbf{n} = (\sin t\mathbf{i} - \cos t\mathbf{j} + 2\mathbf{k})/\sqrt{5}$$

$\mathbf{T}(\pi/6) = (-\mathbf{i} + \sqrt{3}\mathbf{j} + \mathbf{k})/\sqrt{5}$, $\mathbf{n}(\pi/6) = (-\sqrt{3}\mathbf{i} - \mathbf{j})/2$, $\mathbf{B}(\pi/6) = (\mathbf{i} - \sqrt{3}\mathbf{j} + 4\mathbf{k})/2\sqrt{5}$, $\kappa(\pi/6) = 2/5$.

12.C — Computer Exercises

1. On the brachistochrome $(x, y) = (\pi t - \sin(\pi t), -1 + \cos(\pi t))$, we compute $\sqrt{\dfrac{(x')^2 + (y')^2}{-2y}} = \pi$ so the

finish time is $\dfrac{1}{\sqrt{g}} \displaystyle\int_0^1 \pi \, dt = \dfrac{\pi}{\sqrt{g}}$. On the straight line $(x, y) = (\pi t, -2t)$, we compute $\sqrt{\dfrac{(x')^2 + (y')^2}{-2y}} =$

$\sqrt{\dfrac{4 + \pi^2}{2t}}$ so the finish time is $\sqrt{\dfrac{4 + \pi^2}{2g}} \displaystyle\int_0^1 \dfrac{1}{\sqrt{t}} \, dt = \sqrt{\dfrac{4 + \pi^2}{g}} \sqrt{t} \, \Big|_0^1 = \sqrt{\dfrac{4 + \pi^2}{g}}$. On the parabolic arch

$(x, y) = (\pi t, 2(t - 1)^2 - 2) = (\pi t, -4t + 2t^2)$, we compute $\sqrt{\dfrac{(x')^2 + (y')^2}{-2y}} = \dfrac{1}{2} \sqrt{\dfrac{\pi^2 + 16(t - 2)^2}{t(2 - t)}}$. This

function is undefined at 0; many numerical integration routines will not be able to decide whether the resulting improper integral actually converges. Therefore we will integrate over $[a, 1]$ where $0 < a \approx 0$. Choosing a to be smaller than 10^{-15} yields time-to-finish $\approx 3.27633/\sqrt{g}$.

The path of quickest descent is the brachistochrome (this can be proven using the *calculus of variations*).

3. If $\mathbf{f} = (3 \cos t, \sin t, t)$, then $\mathbf{f}' = (-3 \sin t, \cos t, 1)$ and $\mathbf{f}'' = (-3 \cos t, -\sin t, 0)$. Hence $\mathbf{f}' \times \mathbf{f}'' = (\sin t, -3 \cos t, 3)$, $|\mathbf{f}' \times \mathbf{f}''| = \sqrt{10 + 8 \cos^2 t}$, $|\mathbf{f}'| = \sqrt{2 + 8 \sin^2 t}$, and

$$\kappa = \frac{|\mathbf{f}' \times \mathbf{f}''|}{|\mathbf{f}'|^3} = \frac{\sqrt{5 + 4 \cos^2 t}}{2(1 + 4 \sin^2 t)^{3/2}} = \frac{\sqrt{7 + 2 \cos 2t}}{2(3 - 2 \cos 2t)^{3/2}}.$$

Therefore κ has period π and

$$\kappa' = \frac{-8 \sin t \cos t \, (5 + 2 \cos^2 t)}{\sqrt{5 + 4 \cos^2 t} \, (1 + 4 \sin^2 t)^{5/2}} = \frac{-4 \sin 2t \, (6 + \cos 2t)}{\sqrt{7 + 2 \cos 2t} \, (3 - 2 \cos 2t)^{5/2}}.$$

The extrema of κ occur when t is a multiple of $\pi/2$; the maximum value is $\kappa(0) = 3/2$ and the minimum value is $\kappa(\pi/2) = 1/10$. The helix has maximal curvature at points $((-1)^k 3, 0, k\pi)$ and minimal curvature at $(0, (-1)^k, (2k + 1)\pi/2)$.

13
Differentiation of Functions of Two & Three Variables

Solutions 13.1   *Functions of two and three variables*   (pages 840-42)

1. The maximal domain of $f(x,y) = \sin(x+y)$ is $\mathbb{R}^2$ and the range is $[-1,1]$.

3. The maximal domain of $f(x,y) = \sqrt{x^2 + y^2}$ is $\mathbb{R}^2$ and the range is $[0,\infty)$.

5. The maximal domain of $f(x,y) = x/y$ is $\{(x,y) : y \neq 0\}$ and the range is $\mathbb{R}$.

7. The maximal domain of $f(x,y) = \sqrt{1 - x^2 - 4y^2}$ is $\{(x,y) : x^2 + 4y^2 \leq 1\}$ (an ellipse and its interior) and the range is $[0,1]$.

9. The maximal domain of $f(x,y) = \ln\left(1 + x^2 - y^2\right)$ is $\{(x,y) : y^2 < 1 + x^2\}$ and the range is $\mathbb{R}$.

11. $(x+y)/(x-y)$ is defined if and only if $x \neq y$; since $x \neq y \iff x - y \neq 0 \iff (x-y)^2 > 0$, we see that $0 \leq (x+y)/(x-y) \implies 0 \leq (x-y)^2 \cdot (x+y)/(x-y) = x^2 - y^2 \iff |y| \leq |x|$. The maximal domain of $f(x,y) = \sqrt{(x+y)/(x-y)}$ is $\{(x,y) : |y| \leq |x| \text{ and } y \neq x\}$ and the range is $[0,\infty)$.

13. The maximal domain of $f(x,y) = \sin^{-1}(x+y)$ is $\{(x,y) : -1 \leq x + y \leq 1\}$ and the range is $[-\pi/2, \pi/2]$.

15. The maximal domain of $f(x,y) = (x^2 - y^2)/(x+y)$ is $\{(x,y) : y \neq -x\}$ and the range is $\mathbb{R}$.

17. The maximal domain of $g(x,y,z) = x + y + z$ is $\mathbb{R}^3$ and the range is $\mathbb{R}$.

19. The maximal domain of $g(x,y,z) = e^{xy+z}$ is $\mathbb{R}^3$ and the range is $(0,\infty)$.

21. The maximal domain of $g(x,y,z) = xyz$ is $\mathbb{R}^3$ and the range is $\mathbb{R}$.

23. The maximal domain of $g(x,y,z) = xy/z$ is $\{(x,y,z) : z \neq 0\} = \mathbb{R}^3 - \{z = 0\}$ and the range is $\mathbb{R}$.

25. The maximal domain of $g(x,y,z) = \tan^{-1}\left((x+z)/y\right)$ is $\{(x,y,z) : y \neq 0\}$ and the range is $(-\pi/2, \pi/2)$.

27. The maximal domain of $g(x,y,z) = \sin(x + y - z)$ is $\mathbb{R}^3$ and the range is $[-1,1]$.

29. The maximal domain of $g(x,y,z) = \ln(x + y - z)$ is $\{(x,y,z) : 0 < x + y - z\}$ and the range is $\mathbb{R}$.

31. The maximal domain of $g(x,y,z) = 1/\sqrt{x^2 + y^2 + z^2}$ is $\mathbb{R}^3 - \{(0,0,0)\}$ and the range is $(0,\infty)$.

33. The maximal domain of $g(x,y,z) = 1/\sqrt{x^2 - y^2 - z^2}$ is $\{(x,y,z) : y^2 + z^2 < x^2\}$ and the range is $(0,\infty)$.

35: $z = x - y^2$, parabolic cylinder

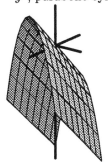

37: $y = x^2 + 4z^2$, elliptic paraboloid

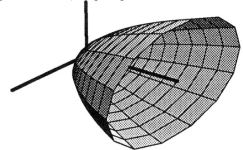

13.1 — Functions of two and three variables

39: $z = x^2 - 4y^2$, hyperbolic paraboloid

41: $y = \sqrt{x^2 - 4z^2 + 4}$, hyperboloid of one sheet (right 1/2)

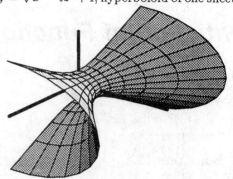

43.   If $z = \sqrt{1 + x + y}$ is constant, then $y = (z^2 - 1) - x$; the level curves are parallel lines with slope -1.

43: $z = \sqrt{1 + x + y}$; $z = 0, 1, 5, 10$

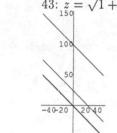

45: $z = \sqrt{1 - x^2 - 4y^2}$; $z = 0, 1/4, 1/2$

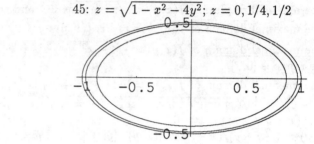

45.   If $z = \sqrt{1 - x^2 - 4y^2}$ is constant, then $x^2 + 4y^2 = 1 - z^2$; the level curves are concentric ellipses. The curve for $z = 1$ is the single point $(0, 0)$.

47.   If $z = \cos^{-1}(x - y)$ is constant, then $y = x - \cos z$; the level curves are parallel lines with slope 1 and intercepts in the interval $[-1, 1]$.

47: $z = \cos^{-1}(x - y)$; $z = 0, \pi/6, \pi/3, \pi/2$

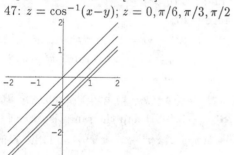

49: $z = \tan(x + y)$; $z = 0, \pm 1, \sqrt{3}$

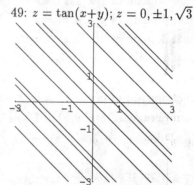

49.   If $z = \tan(x + y)$ is constant, then $y = -x + \left(\tan^{-1} z + k\pi\right)$; the level curve for a single z is a family of lines with slope -1.

51: $x^2 + 4y^2 = T - 20$; $T = 50, 60, 70$

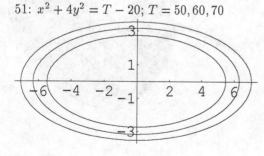

53: $2x^2 + 3y^2 = P - 100$, $x, y \geq 0$; $P = 100, 200, 300$

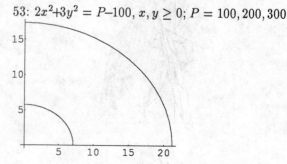

55.   Let $F(L, K) = 500\, L^{1/3} K^{2/3}$.

a. $F(250, 150) = 500 \cdot 250^{1/3} \cdot 150^{2/3} = 500 \cdot 50^{1/3 + 2/3} \cdot 5^{1/3} \cdot 3^{2/3} = 25000 \cdot 5^{1/3} \cdot 3^{2/3} \approx 88922.3$.

b. $F(2L, K/2) = 500 \cdot (2L)^{1/3} \cdot (K/2)^{2/3} = 500 \cdot 2^{1/3-2/3} \cdot L^{1/3} \cdot K^{2/3} = 2^{-1/3} \cdot 500 \cdot L^{1/3} \cdot K^{2/3} = 2^{-1/3} \cdot F(L, K) \approx 0.79 \cdot F(L, K)$. (I.e., output decreases 21%.)

c. $F(L/2, 2K) = 500 \cdot (L/2)^{1/3} \cdot (2K)^{2/3} = 500 \cdot 2^{-1/3+2/3} \cdot L^{1/3} \cdot K^{2/3} = 2^{1/3} \cdot 500 \cdot L^{1/3} \cdot K^{2/3} = 2^{1/3} \cdot F(L, K) \approx 1.26 \cdot F(L, K)$. (I.e., output increases 26%.)

57. If p_1 is the selling price for ordinary watches and p_2 is the selling price for digital watches, then $q_1 = 80 - 2.5p_1 + 0.8p_2$ ordinary watches will be sold and $q_2 = 120 + p_1 - 1.8p_2$ digital watches will be sold. If an ordinary watch costs the jeweler \$8 and quantity q_1 sells at unit price p_1, then the profit on ordinary watches is $(p_1 - 8)q_1$. Similarly, if a digital watch costs \$25 and q_2 sell at unit price p_2, then the profit on digital watches is $(p_2 - 25)q_2$. Therefore the total profit is $(p_1 - 8)q_1 + (p_2 - 25)q_2 = (p_1 - 8)(80 - 2.5p_1 + 0.8p_2) + (p_2 - 25)(120 + p_1 - 1.8p_2)$.

If $p_1 = \$30$ and $p_2 = \$40$, the total profit is $\$22 \cdot 37 + \$15 \cdot 78 = \$814 + \$1170 = \$1984$.

Prologue to Solutions 58-69: Alternative plots of the surfaces are shown together with contour plots. Note that the surface plots are oriented as in the text (positive x-axis to SouthWest and positive y-axis to SouthEast), while the contour plots have the standard orientation for xy-plots; also note that the same domain for (x, y) was used in both plots for a problem.

---

59. The y-sections of $z = \ln x + e^{-y}$ go to $-\infty$ as $x \to 0^+$, text figure (f) is the only one with that property.

59 (f) : $z = \ln x + e^{-y}$

 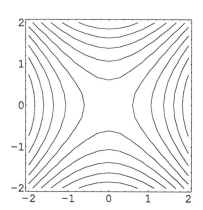

61. The y-sections of $z = y^2 - x^2$ are upward-opening parabolas and the x-sections are downward-opening parabolas, that implies figure (a). Level curves are hyperbolas, the surface is a hyperbolic paraboloid.

61 (a) : $z = y^2 - x^2$

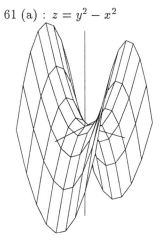

63. The y-sections of $z = (y - 1.5x^2)(y - 0.5x^2)$ are positive and increasing for $y < 0$ but have a hump and change sign for $y > 0$, that implies figure (b).

63 (b) : $z = (y - 1.5x^2)(y - 0.5x^2)$

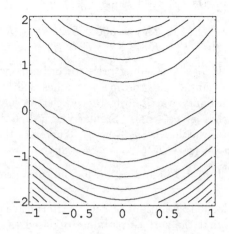

65.   The x-sections of $z = \sin x \tan y$ are tangent curves which tend to $-\infty$ as y approaches one end of $(-\pi/2, \pi/2)$ and tends to ∞ as y approaches the other end, that implies figure (g).

65 (g) : $z = \sin x \tan y$

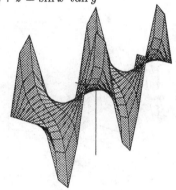

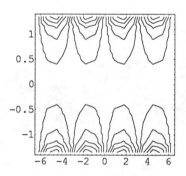

67.   The graph of $z = x - 2y + 4$ is a plane, hence figure (l) is the answer.

67 (l) : $z = x - 2y + 4$

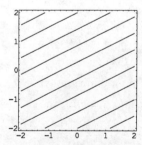

69. $(x+0.15)^2 + y^2 + 0.2$ has its minimum at $(-0.15, 0)$ and $(x-0.15)^2 + y^2 + 0.2$ has its minimum at $(0.15, 0)$; the surface $z = \left((x+0.15)^2 + y^2 + 0.2\right)^{-1} - \left((x-0.15)^2 + y^2 + 0.2\right)^{-1}$ has a positive maximum at the first point and a negative minimum at the second point, it also flattens out to zero as $r \to 0$; figure (i) shows this surface.

69 (i) : $z = \left((x+0.15)^2 + y^2 + 0.2\right)^{-1} - \left((x-0.15)^2 + y^2 + 0.2\right)^{-1}$

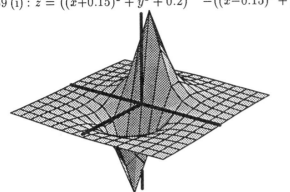

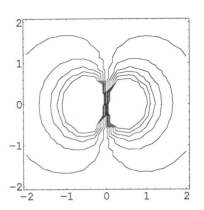

Solutions 13.2   *Limits and continuity*   (pages 851-2)

1: center $(3, 2)$, radius 3; interior of circle

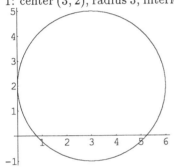

3: center $(1, 0, 0)$, radius 1; interior of ball

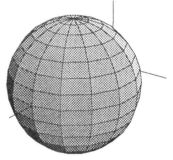

5. $|3x + y - 5| = |3(x - 1) + (y - 2)| \leq \sqrt{3^2 + 1^2} \, \sqrt{(x - 1)^2 + (y - 2)^2} = \sqrt{10} \, \sqrt{(x - 1)^2 + (y - 2)^2}$. Let $\delta = \epsilon/\sqrt{10}$; then $\sqrt{(x - 1)^2 + (y - 2)^2} < \delta$ implies $|3x + y - 5| < \epsilon$. Therefore $\lim_{(x,y)\to(1,2)}(3x + y) = 5$. ∎

7. $|(ax + by) - (5a - 2b)| = |a(x - 5) + b(y - (-2))| \leq \sqrt{a^2 + b^2} \, \sqrt{(x - 5)^2 + (y + 2)^2}$. If $\delta = \epsilon/\sqrt{a^2 + b^2}$, then $\sqrt{(x - 5)^2 + (y + 2)^2} < \delta$ implies $|(ax + by) - (5a - 2b)| < \epsilon$. Thus $\lim_{(x,y)\to(5,-2)}(ax + by) = 5a - 2b$. ∎

9. $x^2 \leq x^2 + y^2$ is true for all x and y. If $(x, y) \neq (0, 0)$, then $x^2 + y^2 > 0$ and
$$\left|\frac{2x^2 y}{x^2 + y^2} - 0\right| = \left|\frac{x^2}{x^2 + y^2}\right| \, |2y| \leq |2y| \leq 2\sqrt{x^2 + y^2}.$$
Choose $\delta = \epsilon/2$, then $0 < \sqrt{x^2 + y^2} < \delta$ implies $\left|\dfrac{2x^2 y}{x^2 + y^2} - 0\right| < \epsilon$. Therefore $\displaystyle\lim_{(x,y)\to(0,0)} \frac{2x^2 y}{x^2 + y^2} = 0$. ∎

11.   The line $y = x$ passes through every neighborhood of the origin and $\dfrac{x + y}{x - y}$ is undefined on that line. Hence the first sentence of the definition can not be satisfied.

   Furthermore, if $m \neq 1$ and $y = mx$, then $\displaystyle\lim_{(x,y)\to(0,0)} \frac{x + y}{x - y} = \lim_{x\to 0} \frac{x + mx}{x - mx} = \frac{1 + m}{1 - m}$ depends on m.

13.   If $y = mx$, then $\displaystyle\lim_{(x,y)\to(0,0)} \frac{xy}{x^2 + y^2} = \lim_{x\to 0} \frac{x \cdot mx}{x^2 + (mx)^2} = \frac{m}{1 + m^2}$ which is not independent of m.

15. $x^3 + y^3 = (x + y)(x^2 - xy + y^2)$, hence $\dfrac{xy}{x^3 + y^3}$ is undefined on the line $y = -x$ which passes through every neighborhood of the origin. The first sentence of the definition can not be satisfied.

   Furthermore, if $m \notin \{-1, 0\}$ and $y = mx$, then $\displaystyle\lim_{(x,y)\to(0,0)} \frac{xy}{x^3 + y^3} = \lim_{x\to 0} \frac{x \cdot mx}{x^3 + (mx)^3} = \frac{m}{1 + m^3} \lim_{x\to 0} \frac{1}{x}$ does not exist.

17.   If $y = mx$, then $\lim\limits_{(x,y)\to(0,0)} \dfrac{x^2 - 2y}{y^2 + 2x} = \lim\limits_{x\to 0} \dfrac{x^2 - 2mx}{m^2x^2 + 2x} = \lim\limits_{x\to 0} \dfrac{x - 2m}{m^2x + 2} = -m$ which depends on m.

Furthermore, the parabola $y^2 + 2x = 0$ passes through every neighborhood of the origin and $\dfrac{x^2 - 2y}{y^2 + 2x}$ is undefined on that curve; hence the first sentence of the definition is never satisfied.

19.   If $y = px$ and $z = qx$, then $\lim\limits_{(x,y,z)\to(0,0,0)} \dfrac{xy + 2xz + 3yz}{x^2 + y^2 + z^2} = \lim\limits_{x\to 0} \dfrac{px^2 + 2qx^2 + 3pqx^2}{x^2 + (px)^2 + (qx)^2} = \dfrac{p + 2q + 3pq}{1 + p^2 + q^2}$ which depends on p and q.

21. $f(x,y) = xy + 4y^2x^3$ is a polynomial function, Theorem 1 (i) implies it is continuous throughout $\mathbb{R}^2$. Therefore $\lim\limits_{(x,y)\to(-1,2)} f(x,y) = f(-1,2) = (-1)(2) + 4(2)^2(-1)^3 = -2 - 16 = -18$.

23. $\dfrac{1 + xy}{1 - xy}$ is a rational function which Theorem 1 (ii) says is continuous except where its denominator is zero. Therefore $\lim\limits_{(x,y)\to(-4,3)} \dfrac{1 + xy}{1 - xy} = \dfrac{1 + (-4)(3)}{1 - (-4)(3)} = \dfrac{-11}{13}$.

25. $x + y$ is a polynomial function which is continuous throughout $\mathbb{R}^2$, hence $\ln(1 + e^{x+y})$ is continuous since it is composed of continuous functions. Thus $\lim\limits_{(x,y)\to(1,2)} \ln(1 + e^{x+y}) = \ln(1 + e^{1+2}) = \ln(1 + e^3)$.

27. $x - y$ is continuous everywhere, therefore so are $\sin(x - y)$ and $\cos(x - y)$; furthermore, $\dfrac{\sin(x - y)}{\cos(x - y)} = \tan(x - y)$ is continous except where $\cos(x - y) = 0$. Thus $\lim\limits_{(x,y)\to(1,1)} \dfrac{\sin(x - y)}{\cos(x - y)} = \dfrac{\sin(1 - 1)}{\cos(1 - 1)} = \dfrac{0}{1} = 0$.

29. $\dfrac{xy^2 - 4xz^2 + 5yz}{3z^2 - 8z^3y^7x^4 + 7x - y + 2}$ is a rational function which is continuous everywhere its denominator is nonzero. Thus

$$\lim\limits_{(x,y,z)\to(1,1,3)} \dfrac{xy^2 - 4xz^2 + 5yz}{3z^2 - 8z^3y^7x^4 + 7x - y + 2} = \dfrac{(1)(1)^2 - 4(1)(3)^2 + 5(1)(3)}{3(3)^2 - 8(3)^3(1)^7(1)^4 + 7(1) - (1) + 2} = \dfrac{-20}{-181} = \dfrac{20}{181}.$$

31. $f(x,y) = \sqrt{x - y}$ is continuous on $\{(x,y) : x > y\}$ (the interior of its maximal domain).

33. $f(x,y) = \dfrac{x^3 + 4xy^6 - 7x^4}{x^3 - y^3} = \dfrac{N}{D}$ where N and D are everywhere continuous. $D = x^3 - y^3 = (x - y)(x^2 + xy + y^2)$ and $x^2 + xy + y^2 = (x + y/2)^2 + 3y^2/4 = 3x^2/4 + (x/2 + y)^2$; $D = 0$ if and only if $y = x$. Therefore f is continuous on $\{(x,y) : y \neq x\}$.

35. $f(x,y) = \ln(3x + 2y + 6)$ is continuous on $\{(x,y) : 3x + 2y + 6 > 0\}$.

37. $f(x,y) = e^{xy+2}$ is continuous on $\mathbb{R}^2$ (since $xy + 2$ and the exponential function are everywhere continuous).

39. $f(x,y) = \dfrac{x}{\sqrt{1 - (x/2)^2 - y^2}}$ is continuous on $\{(x,y) : (x/2)^2 + y^2 < 1\}$.

41. $f(x,y,z) = \dfrac{xyz^2 + yzx^2 - 3x^3yz^5}{x - y + 2z + 4}$ is continuous on $\{(x,y,z) : x - y + 2z + 4 \neq 0\}$.

43. $f(x,y,z) = z + \cos^{-1}(x^2 - y)$ is continuous on $\{(x,y,z) : -1 < x^2 - y < 1\}$ (the interior of its maximal domain).

45.   If $x \neq y$, then $\dfrac{x^2 - y^2}{x - y} = \dfrac{(x - y)(x + y)}{x - y} = x + y$; furthermore, if $x = y$, then $x + y = 2x$. Therefore,

$$f(x,y) = \begin{cases} \dfrac{x^2 - y^2}{x - y} & \text{if } x \neq y \\ 2x & \text{if } x = y \end{cases}$$ is everywhere continuous.

47. $0 \leq (x \pm y)^2 = x^2 \pm 2xy + y^2 \implies |xy| \leq (x^2 + y^2)/2$; similarly, $(|x| + |y|)^2 = x^2 + 2|xy| + y^2 \geq x^2 + y^2 \implies |x| + |y| \geq \sqrt{x^2 + y^2} \implies 1/(|x| + |y|) \leq 1/\sqrt{x^2 + y^2}$. Therefore $\left| \dfrac{xy}{|x| + |y|} \right| \leq \dfrac{(x^2 + y^2)/2}{\sqrt{x^2 + y^2}} = \dfrac{\sqrt{x^2 + y^2}}{2}$ and, invoking a "Squeezing Theorem", we see that $\lim\limits_{(x,y)\to(0,0)} \dfrac{xy}{|x| + |y|} = 0$. Let c be this limit, i.e., let $c = 0$.

49. $0 \le (x-y)^2 = x^2 - 2xy + y^2 \implies xy \le (x^2+y^2)/2 \implies |xy| \le (x^2+y^2)/2.$ Therefore $\left| \dfrac{3xy}{\sqrt{x^2+y^2}} \right| \le$

$\dfrac{3(x^2+y^2)/2}{\sqrt{x^2+y^2}} = \dfrac{3\sqrt{x^2+y^2}}{2}$ and, invoking a "Squeezing Theorem", we see that $\displaystyle\lim_{(x,y)\to(0,0)} \dfrac{3xy}{\sqrt{x^2+y^2}} = 0.$

51. $x^2 \le x^2+y^2$ and $y^2 \le x^2+y^2$; therefore $x^3+y^3 = (x^2)^{3/2} + (y^2)^{3/2} \le 2(x^2+y^2)^{3/2}$ and $\left| \dfrac{x^3+y^3}{x^2+y^2} \right| \le$

$\left| \dfrac{2(x^2+y^2)^{3/2}}{x^2+y^2} \right| = 2\sqrt{x^2+y^2}.$ Since $\displaystyle\lim_{(x,y)\to(0,0)} 2\sqrt{x^2+y^2} = 0,$ we can now infer $\displaystyle\lim_{(x,y)\to(0,0)} \dfrac{x^3+y^3}{x^2+y^2} = 0.$

REMARK: $\left| \dfrac{x^3+y^3}{x^2+y^2} \right| = \left| \dfrac{x^3}{x^2+y^2} + \dfrac{y^3}{x^2+y^2} \right| \le \left| \dfrac{x^3}{x^2+y^2} \right| + \left| \dfrac{y^3}{x^2+y^2} \right| \le \left| \dfrac{x^3}{x^2} \right| + \left| \dfrac{y^3}{y^2} \right| = |x| + |y|.$

53. $x^2 - y^2 = (x-y)(x+y)$ so $f(x,y) = \dfrac{x-y}{x^2-y^2}$ is undefined on the lines $y = \pm x.$ The line $y = x$ passes

through every neighborhood of $(1,1)$ so there is no deleted neighborhood of $(1,1)$ in which f is defined.

Solutions 13.3 *Partial derivatives* (pages 857-9)

1. If $z = x^2 y$, then $\dfrac{\partial z}{\partial x} = 2xy$ and $\dfrac{\partial z}{\partial y} = x^2.$

3. If $z = 3e^{xy^3}$, then $\dfrac{\partial z}{\partial x} = 3e^{xy^3} \cdot \dfrac{\partial}{\partial x}(xy^3) = 3y^3 e^{xy^3}$ and $\dfrac{\partial z}{\partial y} = 3e^{xy^3} \cdot \dfrac{\partial}{\partial y}(xy^3) = 9xy^2 e^{xy^3}.$

5. If $z = \dfrac{4x}{y^5}$, then $\dfrac{\partial z}{\partial x} = \dfrac{4}{y^5}$ and $\dfrac{\partial z}{\partial y} = \dfrac{-20x}{y^6}.$

7. If $z = \ln(x^3 y^5 - 2)$, then $\dfrac{\partial z}{\partial x} = \dfrac{3x^2 y^5}{x^3 y^5 - 2}$ and $\dfrac{\partial z}{\partial y} = \dfrac{5x^3 y^4}{x^3 y^5 - 2}.$

9. If $z = (x + 5y\sin x)^{4/3}$, then $\dfrac{\partial z}{\partial x} = \dfrac{4}{3}(x + 5y\sin x)^{1/3}(1 + 5y\cos x)$ and $\dfrac{\partial z}{\partial y} = \dfrac{20}{3}(x + 5y\sin x)^{1/3}\sin x.$

11. If $z = \sin^{-1}(x-y)$, then $\dfrac{\partial z}{\partial x} = \dfrac{1}{\sqrt{1-(x-y)^2}}$ and $\dfrac{\partial z}{\partial y} = \dfrac{-1}{\sqrt{1-(x-y)^2}}.$

13. Suppose $z = xz + yz^3.$ $\dfrac{\partial z}{\partial x} = z + x\dfrac{\partial z}{\partial x} + 3yz^2 \dfrac{\partial z}{\partial x}$ implies $\dfrac{\partial z}{\partial x} = \dfrac{z}{1-x-3yz^2}$ and $\dfrac{\partial z}{\partial y} = x\dfrac{\partial z}{\partial y} + z^3 + 3yz^2\dfrac{\partial z}{\partial y}$

implies $\dfrac{\partial z}{\partial y} = \dfrac{z^3}{1-x-3yz^2}.$

REMARK: If $z = 0$, then $\dfrac{\partial z}{\partial x} = 0 = \dfrac{\partial z}{\partial y}$; if $z \ne 0$, then $z = xz + yz^3 \implies 1 = x + yz^2.$ Therefore

$0 = 1 + 2yz\dfrac{\partial z}{\partial x} \implies \dfrac{\partial z}{\partial x} = \dfrac{-1}{2yz}$ and $0 = 0 + z^2 + 2yz\dfrac{\partial z}{\partial y} \implies \dfrac{\partial z}{\partial y} = \dfrac{-z^2}{2yz} = \dfrac{-z}{2y}.$

15. Suppose $\sin(z-x) = yz^{-1}.$ $\cos(z-x)\left(\dfrac{\partial z}{\partial x} - 1\right) = -yz^{-2}\dfrac{\partial z}{\partial x}$ implies $\dfrac{\partial z}{\partial x} = \dfrac{\cos(z-x)}{yz^{-2} + \cos(z-x)}$ and

$\cos(z-x)\dfrac{\partial z}{\partial y} = z^{-1} - yz^{-2}\dfrac{\partial z}{\partial x}$ implies $\dfrac{\partial z}{\partial y} = \dfrac{z^{-1}}{yz^{-2} + \cos(z-x)}.$

17. If $f(x,y) = x^3 - y^4$, then $f_x(x,y) = 3x^2$ and $f_x(1,-1) = 3 \cdot 1^2 = 3.$

19. If $f(x,y) = \sin(x+y)$, then $f_x(x,y) = \cos(x+y)$ and $f_x(\pi/6, \pi/3) = \cos(\pi/6 + \pi/3) = \cos(\pi/2) = 0.$

21. If $f(x,y) = \sinh(x-y)$, then $f_x(x,y) = \cosh(x-y)$ and $f_x(3,3) = \cosh(0) = 1.$

23. If $f(x,y) = \dfrac{x^2-y^2}{x^2+y^2} = \dfrac{2x^2}{x^2+y^2} - 1$, then $f_y(x,y) = \dfrac{-4x^2 y}{(x^2+y^2)^2}$ and $f_y(2,-3) = \dfrac{48}{169}.$

25. If $w = \sqrt{x+y+z}$, then $\dfrac{\partial w}{\partial x} = \dfrac{1}{2\sqrt{x+y+z}} = \dfrac{\partial w}{\partial y} = \dfrac{\partial w}{\partial z}.$

27. If $w = e^{x+2y+3z}$, then $\dfrac{\partial w}{\partial x} = e^{x+2y+3z} = w$, $\dfrac{\partial w}{\partial y} = 2e^{x+2y+3z} = 2w$, and $\dfrac{\partial w}{\partial z} = 3e^{x+2y+3z} = 3w.$

29.   If $w = \dfrac{x+y}{z}$, then $\dfrac{\partial w}{\partial x} = \dfrac{1}{z} = \dfrac{\partial w}{\partial y}$ and $\dfrac{\partial w}{\partial z} = \dfrac{-(x+y)}{z^2}$.

31.   If $w = \ln(x^3 + y^2 + z)$, then $\dfrac{\partial w}{\partial x} = \dfrac{3x^2}{x^3 + y^2 + z}$, $\dfrac{\partial w}{\partial y} = \dfrac{2y}{x^3 + y^2 + z}$, and $\dfrac{\partial w}{\partial z} = \dfrac{1}{x^3 + y^2 + z}$.

33.   If $f(x,y,z) = xyz$, then $f_x(x,y,z) = yz$ and $f_x(2,3,4) = 3 \cdot 4 = 12$.

35.   If $f(x,y,z) = \dfrac{x-y}{z}$, then $f_z(x,y,z) = \dfrac{-(x-y)}{z^2} = \dfrac{y-x}{z^2}$ and $f_z(-3,-1,2) = \dfrac{-1+3}{4} = \dfrac{1}{2}$.

37.   If $f(x,y,z) = \ln(x + 2y + 3z)$, then $f_z(x,y,z) = \dfrac{3}{x + 2y + 3z}$ and $f_z(2,2,5) = \dfrac{3}{2 + 4 + 15} = \dfrac{1}{7}$.

39.   If $f(x,y,z) = \dfrac{y^3 - z^5}{x^2 y + z}$, then $f_y(x,y,z) = \dfrac{3y^2}{x^2 y + z} - \dfrac{(y^3 - z^5)x^2}{(x^2 y + z)^2}$ and $f_y(4,0,1) = \dfrac{0}{1} - \dfrac{(-1)(16)}{1} = 16$.

41.   If $f(x,y,z) = e^{xy}(\cosh z - \sinh z)$, then $f_z(x,y,z) = e^{xy}(\sinh z - \cosh z)$ and $f_z(2,3,0) = e^6(\sinh 0 - \cosh 0) = -e^6$.

43.   a. $\dfrac{\partial}{\partial y}(x^3 - 4y^3) = -12y^2$; the line in the plane $x = 1$ which is tangent to $z = x^3 - 4y^3$ at $(1,-1,5)$ has equations $x = 1$ and $z - 5 = -12(-1)^2\left(y - (-1)\right) = -12(y+1)$.

     b. $\dfrac{\partial}{\partial x}(x^3 - 4y^3) = 3x^2$; the line in the plane $y = -1$ which is tangent to $z = x^3 - 4y^3$ at $(1,-1,5)$ has equations $y = -1$ and $z - 5 = 3(1)^2(x-1) = 3(x-1)$.

45.   The intersection of the surface $x^2 + 4y^2 + 4z^2 = 9$ and the plane $y = 1$ is a curve such that $0 = \dfrac{\partial}{\partial x}9 = \dfrac{\partial}{\partial x}(x^2 + 4y^2 + 4z^2) = 2x + 0 + 8z\dfrac{\partial z}{\partial x}$ which then implies $\dfrac{\partial z}{\partial x} = \dfrac{-2x}{8z} = \dfrac{-x}{4z}$. The line tangent to that curve at the point $(1,1,1)$ has equations $y = 1$ and $z - 1 = (-1/4)(x-1)$.

47.   If x units of A and y units of B cost $C(x,y) = \dfrac{50}{2+x} + \dfrac{125}{(3+y)^2}$, then the marginal cost for product A is $\dfrac{\partial}{\partial x}C(x,y) = \dfrac{-50}{(2+x)^2}$ and the marginal cost for product B is $\dfrac{\partial}{\partial y}C(x,y) = \dfrac{-250}{(3+y)^3}$.

49.   Suppose $z = e^{(x+\sqrt{3}y)/4} - 4x - 2y - 4 - 2\sqrt{3}$; let $w = e^{(x+\sqrt{3}y)/4}$. Then $z = w - 4x - 2y - 4 - 2\sqrt{3}$, $\dfrac{\partial z}{\partial x} = \dfrac{w}{4} - 4$, $\dfrac{\partial z}{\partial y} = \dfrac{\sqrt{3}w}{4} - 2$, $\dfrac{\partial z}{\partial x} + \sqrt{3}\dfrac{\partial z}{\partial y} = w - 4 - 2\sqrt{3}$, and $\dfrac{\partial z}{\partial x} + \sqrt{3}\dfrac{\partial z}{\partial y} - z = -(-4x - 2y) = 4x + 2y$. ∎

51.   If $P = \dfrac{nRT}{V}$, then $\dfrac{\partial P}{\partial V} = \dfrac{-nRT}{V^2}$ and $\dfrac{\partial P}{\partial V}(n = 10, T = 293, V = 2) = \dfrac{-10 \cdot R \cdot 293}{2^2} = -732.5\,R$ where $R \approx 8315$ joules per kg–mole–degree.

53.   If $C = (5/2)(T - t)w^{-2/3}$, then $C_t = -(5/2)w^{-2/3}$. Thus $C_t(w = 10, T = 23, t = 5) = -2.5 \cdot 10^{-2/3}$.

55.   Suppose g and h are differentiable functions of a single variable, let $f(x,y) = g(x)\,h(y)$.

$$\dfrac{\partial f}{\partial x} = \lim_{\Delta x \to 0}\dfrac{f(x+\Delta x, y) - f(x,y)}{\Delta x} = \lim_{\Delta x \to 0}\dfrac{g(x+\Delta x)h(y) - g(x)h(y)}{\Delta x} = \lim_{\Delta x \to 0}\dfrac{g(x+\Delta x) - g(x)}{\Delta x}h(y) = g'(x)h(y),$$

$$\dfrac{\partial f}{\partial y} = \lim_{\Delta y \to 0}\dfrac{f(x, y+\Delta y) - f(x,y)}{\Delta y} = \lim_{\Delta y \to 0}\dfrac{g(x)h(y+\Delta y) - g(x)h(y)}{\Delta y} = \lim_{\Delta y \to 0}g(x)\dfrac{h(y+\Delta y) - h(y)}{\Delta y} = g(x)h'(y).$$ ∎

57.   Let $h = f + g$.

$$\begin{aligned}
\dfrac{\partial h}{\partial x}(x,y) &= \lim_{\Delta x \to 0}\dfrac{h(x+\Delta x, y) - h(x,y)}{\Delta x} \\
&= \lim_{\Delta x \to 0}\dfrac{f(x+\Delta x, y) + g(x+\Delta x, y) - f(x,y) - g(x,y)}{\Delta x} \\
&= \lim_{\Delta x \to 0}\left(\dfrac{f(x+\Delta x, y) - f(x,y)}{\Delta x} + \dfrac{g(x+\Delta x, y) - g(x,y)}{\Delta x}\right) \\
&= \lim_{\Delta x \to 0}\dfrac{f(x+\Delta x, y) - f(x,y)}{\Delta x} + \lim_{\Delta x \to 0}\dfrac{g(x+\Delta x, y) - g(x,y)}{\Delta x} = \dfrac{\partial f}{\partial x}(x,y) + \dfrac{\partial g}{\partial x}(x,y).
\end{aligned}$$

Therefore $\dfrac{\partial(f+g)}{\partial x} = \dfrac{\partial f}{\partial x} + \dfrac{\partial g}{\partial x}$. ∎

59. Let $h = f \cdot g$.
$$\frac{\partial h}{\partial x}(x, y) = \lim_{\Delta x \to 0} \frac{h(x+\Delta x, y) - h(x, y)}{\Delta x} = \lim_{\Delta x \to 0} \frac{f(x+\Delta x, y) \cdot g(x+\Delta x, y) - f(x, y) \cdot g(x, y)}{\Delta x}$$
$$= \lim_{\Delta x \to 0} \frac{f(x+\Delta x, y) \cdot g(x+\Delta x, y) - f(x, y) \cdot g(x+\Delta x, y) + f(x, y) \cdot g(x+\Delta x, y) - f(x, y) \cdot g(x, y)}{\Delta x}$$
$$= \lim_{\Delta x \to 0} \left(\frac{f(x+\Delta x, y) - f(x, y)}{\Delta x} \cdot g(x+\Delta x, y) + f(x, y) \cdot \frac{g(x+\Delta x, y) - g(x, y)}{\Delta x} \right)$$
$$= \frac{\partial f}{\partial x}(x, y) \cdot g(x, y) + f(x, y) \cdot \frac{\partial g}{\partial x}(x, y).$$

Therefore $\dfrac{\partial (f \cdot g)}{\partial x} = \dfrac{\partial f}{\partial x} \cdot g + f \cdot \dfrac{\partial g}{\partial x}.$ ∎

61. a. $\dfrac{x+y}{x-y}$ is undefined on the line $y = x$ which passes through every neighborhood of $(0,0)$, therefore

$\lim\limits_{(x,y)\to(0,0)} \dfrac{x+y}{x-y}$ does not exist. Were we unconvinced by that, then considering limits along lines $y = mx$

we obtain $\lim\limits_{(x,y)\to(0,0)} \dfrac{x+y}{x-y} = \lim\limits_{x\to 0} \dfrac{x+mx}{x-mx} = \dfrac{1+m}{1-m}$ which is not independent of m. Since the limit at

$(0,0)$ does not exist, no choice of value for $f(0,0)$ yields a continuous function.)

  b. $f(0,0) = 0$ but $f(x,0) = x/x = 1$ for $x \neq 0$, thus $f(x,0)$ is not a continuous function of x at
$x = 0$; that implies $f_x(0,0)$ does not exist. Similarly $f(0,y) = y/(-y) = -1$ if $y \neq 0$ so $f(0,y)$ is not a
continuous function of y at $y = 0$ and thus $f_y(0,0)$ does not exist.

Solutions 13.4 *Higher-order partial derivatives* (pages 864-5)

1. If $f(x,y) = x^2 y$, then $f_x = 2xy$ and $f_y = x^2$. Therefore $f_{xx} = \frac{\partial}{\partial x}(2xy) = 2y$, $f_{xy} = \frac{\partial}{\partial y}(2xy) = 2x = \frac{\partial}{\partial x}(x^2) = f_{yx}$, and $f_{yy} = \frac{\partial}{\partial y}(x^2) = 0$.

3. If $f(x,y) = e^{xy^3}$, then $f_x = 3y^3 e^{xy^3}$ and $f_y = 9xy^2 e^{xy^3}$. Therefore $f_{xx} = \frac{\partial}{\partial x}\left(3y^3 e^{xy^3}\right) = 3y^6 e^{xy^3}$, $f_{xy} = \frac{\partial}{\partial y}\left(3y^3 e^{xy^3}\right) = (9y^2 + 9xy^5)e^{xy^3} = \frac{\partial}{\partial x}\left(9xy^2 e^{xy^3}\right) = f_{yx}$, and $f_{yy} = \frac{\partial}{\partial y}\left(9xy^2 e^{xy^3}\right) = (18xy + 27x^2 y^4)e^{xy^3}$.

5. If $f(x,y) = 4xy^{-5}$, then $f_x = 4y^{-5}$ and $f_y = -20xy^{-6}$. Therefore $f_{xx} = \frac{\partial}{\partial x}\left(4y^{-5}\right) = 0$, $f_{xy} = \frac{\partial}{\partial y}\left(4y^{-5}\right) = -20y^{-6} = \frac{\partial}{\partial x}\left(-20xy^{-6}\right) = f_{yx}$, and $f_{yy} = \frac{\partial}{\partial y}\left(-20xy^{-6}\right) = 120xy^{-7}$.

7. If $f(x,y) = \ln(x^3 y^5 - 2)$, then $f_x = 3x^2 y^5 (x^3 y^5 - 2)^{-1}$ and $f_y = 5x^3 y^4 (x^3 y^5 - 2)^{-1}$. Therefore
$f_{xx} = \frac{\partial}{\partial x}\left(3x^2 y^5 (x^3 y^5 - 2)^{-1}\right) = 6xy^5 (x^3 y^5 - 2)^{-1} - 9x^4 y^{10}(x^3 y^5 - 2)^{-2} = -3xy^5 (x^3 y^5 + 4)(x^3 y^5 - 2)^{-2}$,
$f_{xy} = \frac{\partial}{\partial y}\left(3x^2 y^5 (x^3 y^5 - 2)^{-1}\right) = 15x^2 y^4 (x^3 y^5 - 2)^{-1} - 15x^5 y^9 (x^3 y^5 - 2)^{-2} = -30x^2 y^4 (x^3 y^5 - 2)^{-2} = \frac{\partial}{\partial x}\left(5x^3 y^4 (x^3 y^5 - 2)^{-1}\right) = f_{yx}$, and $f_{yy} = \frac{\partial}{\partial y}\left(5x^3 y^4 (x^3 y^5 - 2)^{-1}\right) = 20x^3 y^3 (x^3 y^5 - 2)^{-1} - 25x^6 y^8 (x^3 y^5 - 2)^{-2} = -5x^3 y^3 (x^3 y^5 + 8)(x^3 y^5 - 2)^{-2}$.

9. If $f(x,y) = (x + 5y\sin x)^{4/3}$, then $f_x = \frac{4}{3}(x + 5y\sin x)^{1/3}(1 + 5y\cos x)$ and $f_y = \frac{20}{3}(x + 5y\sin x)^{1/3}\sin x$.
Therefore
$$f_{xx} = \frac{\partial}{\partial x}\left(\frac{4}{3}(x + 5y\sin x)^{1/3}(1 + 5y\cos x)\right) = \frac{4(1 + 5y\cos x)^2}{9(x + 5y\sin x)^{2/3}} - \frac{20(x + 5y\sin x)^{1/3} y\sin x}{9},$$
$$f_{xy} = \frac{\partial}{\partial y}\left(\frac{4}{3}(x + 5y\sin x)^{1/3}(1 + 5y\cos x)\right) = \frac{20\sin x(1 + 5y\cos x)}{9(x + 5y\sin x)^{2/3}} + \frac{20(x + 5y\sin x)^{1/3}\cos x}{3}$$
$$= \frac{\partial}{\partial x}\left(\frac{20}{3}(x + 5y\sin x)^{1/3}\sin x\right) = f_{yx},$$
$$f_{yy} = \frac{\partial}{\partial x}\left(\frac{20}{3}(x + 5y\sin x)^{1/3}\sin x\right) = \frac{100\sin^2 x}{9(x + 5y\sin x)^{2/3}}.$$

11. If $f(x,y) = \sin^{-1}(x - y)$, then $f_x = (1 - (x-y)^2)^{-1/2}$ and $f_y = -(1 - (x-y)^2)^{-1/2}$. Therefore
$f_{xx} = \frac{\partial}{\partial x}\left((1 - (x-y)^2)^{-1/2}\right) = (x - y)(1 - (x-y)^2)^{-3/2}$, $f_{xy} = \frac{\partial}{\partial y}\left((1 - (x-y)^2)^{-1/2}\right) = -(x - y)(1 - (x-y)^2)^{-3/2} = \frac{\partial}{\partial x}\left(-(1 - (x-y)^2)^{-1/2}\right) = f_{yx}$, and $f_{yy} = \frac{\partial}{\partial y}\left(-(1 - (x-y)^2)^{-1/2}\right) = (x - y)(1 - (x-y)^2)^{-3/2}$.

13.   If $f(x, y, z) = xyz$, then $f_x = yz$, $f_y = xz$, $f_z = xy$. Therefore $f_{xx} = \frac{\partial}{\partial x}(yz) = 0$, $f_{yy} = \frac{\partial}{\partial y}(xz) = 0$, $f_{zz} = \frac{\partial}{\partial z}(xy) = 0$, $f_{xy} = \frac{\partial}{\partial y}(yz) = z = \frac{\partial}{\partial x}(xz) = f_{yx}$, $f_{yz} = \frac{\partial}{\partial z}(xz) = x = \frac{\partial}{\partial y}(xy) = f_{zy}$, $f_{zx} = \frac{\partial}{\partial x}(xy) = y = \frac{\partial}{\partial z}(yz) = f_{xz}$.

15.   If $f(x, y, z) = x^2 y^3 z^4$, then $f_x = 2xy^3 z^4$, $f_y = 3x^2 y^2 z^4$, $f_z = 4x^2 y^3 z^3$. Therefore $f_{xx} = \frac{\partial}{\partial x}(2xy^3 z^4) = 2y^3 z^4$, $f_{yy} = \frac{\partial}{\partial y}(3x^2 y^2 z^4) = 6x^2 y z^4$, $f_{zz} = \frac{\partial}{\partial z}(4x^2 y^3 z^3) = 12x^2 y^3 z^2$, and

$$f_{xy} = \frac{\partial}{\partial y}(2xy^3 z^4) = 6xy^2 z^4 = \frac{\partial}{\partial x}(3x^2 y^2 z^4) = f_{yx},$$

$$f_{yz} = \frac{\partial}{\partial z}(3x^2 y^2 z^4) = 12x^2 y^2 z^3 = \frac{\partial}{\partial y}(4x^2 y^3 z^3) = f_{zy},$$

$$f_{zx} = \frac{\partial}{\partial x}(4x^2 y^3 z^3) = 8xy^3 z^3 = \frac{\partial}{\partial z}(2xy^3 z^4) = f_{xz}.$$

17.   If $f(x, y, z) = (x + y)z^{-1}$, then $f_x = z^{-1}$, $f_y = z^{-1}$, $f_z = -(x + y)z^{-2}$. Therefore $f_{xx} = \frac{\partial}{\partial x}(z^{-1}) = 0$, $f_{yy} = \frac{\partial}{\partial y}(z^{-1}) = 0$, $f_{zz} = \frac{\partial}{\partial z}(-(x + y)z^{-2}) = 2(x + y)z^{-3}$, $f_{xy} = \frac{\partial}{\partial y}(z^{-1}) = 0 = \frac{\partial}{\partial x}(z^{-1}) = f_{yx}$, $f_{yz} = \frac{\partial}{\partial z}(z^{-1}) = -z^{-2} = \frac{\partial}{\partial y}(-(x + y)z^{-2}) = f_{zy}$, $f_{zx} = \frac{\partial}{\partial x}(-(x + y)z^{-2}) = -z^{-2} = \frac{\partial}{\partial z}(z^{-1}) = f_{xz}$.

19.   If $f(x, y, z) = e^{3xy} \cos z$, then $f_x = 3ye^{3xy} \cos z$, $f_y = 3xe^{3xy} \cos z$, $f_z = -e^{3xy} \sin z$. Thus $f_{xx} = \frac{\partial}{\partial x}(3ye^{3xy} \cos z) = 9y^2 e^{3xy} \cos z$, $f_{yy} = \frac{\partial}{\partial y}(3xe^{3xy} \cos z) = 9x^2 e^{3xy} \cos z$, $f_{zz} = \frac{\partial}{\partial z}(-e^{3xy} \sin z) = -e^{3xy} \cos z$,

$$f_{xy} = \frac{\partial}{\partial y}(3ye^{3xy} \cos z) = (3 + 9xy)e^{3xy} \cos z = \frac{\partial}{\partial x}(3xe^{3xy} \cos z) = f_{yx},$$

$$f_{yz} = \frac{\partial}{\partial z}(3xe^{3xy} \cos z) = -3xe^{3xy} \sin z = \frac{\partial}{\partial y}(-e^{3xy} \sin z) = f_{zy},$$

$$f_{zx} = \frac{\partial}{\partial x}(-e^{3xy} \sin z) = -3ye^{3xy} \sin z = \frac{\partial}{\partial z}(3ye^{3xy} \cos z) = f_{xz}.$$

21.   If $f(x, y) = x^2 y^3 + 2y$, then $f_x = 2xy^3$, $f_{xy} = 6xy^2$, and $f_{xyx} = 6y^2$.

23.   If $f(x, y) = \ln(3x - 2y)$, then $f_y = -2(3x - 2y)^{-1}$, $f_{yx} = 6(3x - 2y)^{-2}$, and $f_{yxy} = 24(3x - 2y)^{-3}$.

25.   If $f(x, y, z) = x^2 y + y^2 z - 3\sqrt{xz}$, then $f_x = 2xy - (3/2)\sqrt{z/x}$, $f_{xy} = 2x$, and $f_{xyz} = 0$.

27.   If $y(x, t) = ((x - ct)^2 + (x + ct)^2)/2$, then $y(x, t) = x^2 + c^2 t^2$. Therefore $y_x = 2x$, $y_{xx} = 2$, $y_t = 2c^2 t$, and $y_{tt} = 2c^2 = c^2 2 = c^2 y_{xx}$. ∎

29.   If $T(x, t) = \frac{1}{\sqrt{t}} \cdot e^{-x^2/(4\delta t)}$, then

$$T_x = \frac{1}{\sqrt{t}} \cdot \frac{-2x}{4\delta t} e^{-x^2/(4\delta t)} = \frac{-x}{2\delta t^{3/2}} e^{-x^2/(4\delta t)},$$

$$T_{xx} = \frac{-1}{2\delta t^{3/2}} e^{-x^2/(4\delta t)} + \frac{-x}{2\delta t^{3/2}} \cdot \frac{-2x}{4\delta t} e^{-x^2/(4\delta t)} = \left(\frac{-1}{2\delta t^{3/2}} + \frac{x^2}{4\delta^2 t^{5/2}} \right) e^{-x^2/(4\delta t)}$$

$$T_t = \frac{-1}{2t^{3/2}} \cdot e^{-x^2/(4\delta t)} + \frac{1}{\sqrt{t}} \cdot \frac{x^2}{4\delta t^2} e^{-x^2/(4\delta t)} = \left(\frac{-1}{2t^{3/2}} + \frac{x^2}{4\delta t^{5/2}} \right) e^{-x^2/(4\delta t)} = \delta\, T_{xx}. ∎$$

31.   If $f(x, y) = x^2 - y^2$, then $f_x = 2x$, $f_{xx} = 2$, $f_y = -2y$, $f_{yy} = -2$, and $f_{xx} + f_{yy} = 2 + (-2) = 0$. ∎

33.   If $f(x, y) = \ln(x^2 + y^2)$, then $f_x = \frac{2x}{x^2 + y^2}$, $f_{xx} = \frac{2}{x^2 + y^2} - \frac{4x^2}{(x^2 + y^2)^2} = \frac{2(y^2 - x^2)}{(x^2 + y^2)^2}$, $f_y = \frac{2y}{x^2 + y^2}$,

$f_{yy} = \frac{2}{x^2 + y^2} - \frac{4y^2}{(x^2 + y^2)^2} = \frac{2(x^2 - y^2)}{(x^2 + y^2)^2}$, and $f_{xx} + f_{yy} = \frac{2(y^2 - x^2) + 2(x^2 - y^2)}{(x^2 + y^2)^2} = 0$. ∎

35.   If $f(x, y, z) = x^2 + y^2 - 2z^2$, then $f_x = 2x$, $f_{xx} = 2$, $f_y = 2y$, $f_{yy} = 2$, $f_z = -4z$, $f_{zz} = -4$, and $f_{xx} + f_{yy} + f_{zz} = 2 + 2 + (-4) = 0$. ∎

Solutions 13.5   *Differentiability and the Gradient*   (pages 874-5)

1. $\nabla(x + y)^2 = 2(x + y)\mathbf{i} + 2(x + y)\mathbf{j}$.

3. $\nabla e^{\sqrt{xy}} = e^{\sqrt{xy}}(1/2)\sqrt{y/x}\,\mathbf{i} + e^{\sqrt{xy}}(1/2)\sqrt{x/y}\,\mathbf{j}$; at $(1, 1)$ we compute $\nabla e^{\sqrt{xy}} = (e/2)\mathbf{i} + (e/2)\mathbf{j}$.

5. $\nabla\sqrt{x^2 + y^2} = x(x^2 + y^2)^{-1/2}\mathbf{i} + y(x^2 + y^2)^{-1/2}\mathbf{j}$.

7. $\nabla y \tan(y - x) = -y \sec^2(y - x)\mathbf{i} + (\tan(y - x) + y \sec^2(y - x))\mathbf{j}$.

9. $\nabla \sec(x+3y) = \sec(x+3y)\tan(x+3y)\mathbf{i} + 3\sec(x+3y)\tan(x+3y)\mathbf{j}$; at $(0,1)$ we compute $\nabla \sec(x+3y) = \sec 3 \tan 3\mathbf{i} + 3\sec 3 \tan 3\mathbf{j}$.

11. $\nabla \left(\dfrac{x^2 - y^2}{x^2 + y^2} \right) = \dfrac{4xy^2}{(x^2+y^2)^2}\mathbf{i} - \dfrac{4x^2 y}{(x^2+y^2)^2}\mathbf{j}$.

13. $\nabla \sin x \cos y \tan z = \cos x \cos y \tan z\,\mathbf{i} - \sin x \sin y \tan z\,\mathbf{j} + \sin x \cos y \sec^2 z\,\mathbf{k}$; at $(\pi/6, \pi/4, \pi/3)$ we compute $\nabla \sin x \cos y \tan z = \left(\dfrac{\sqrt 3}{2} \cdot \dfrac{1}{\sqrt 2} \cdot \sqrt 3 \right)\mathbf{i} - \left(\dfrac{1}{2} \cdot \dfrac{1}{\sqrt 2} \cdot \sqrt 3 \right)\mathbf{j} + \left(\dfrac{1}{2} \cdot \dfrac{1}{\sqrt 2} \cdot 4 \right)\mathbf{k} = \dfrac{3}{2\sqrt 2}\mathbf{i} - \dfrac{\sqrt 3}{2\sqrt 2}\mathbf{j} + \sqrt 2\,\mathbf{k}$.

15. $\nabla \left(\dfrac{x^2 - y^2 + z^2}{3xy} \right) = \dfrac{x^2 + y^2 - z^2}{3x^2 y}\mathbf{i} - \dfrac{x^2 + y^2 + z^2}{3xy^2}\mathbf{j} + \dfrac{2z}{xy}\mathbf{k}$; at $(1,2,0)$ we find $\nabla \left(\dfrac{x^2 - y^2 + z^2}{3xy} \right) = (5/6)\mathbf{i} - (5/12)\mathbf{j} + 0\mathbf{k}$.

17. $\nabla y^2 (x + z^3) = y^2\mathbf{i} + 2y(x + z^3)\,\mathbf{j} + 3y^2 z^2\,\mathbf{k}$; at $(2,3,-1)$ we compute $\nabla y^2 (x + z^3) = 9\mathbf{i} + 6\mathbf{j} + 27\mathbf{k}$.

19. $\nabla \dfrac{x - z}{\sqrt{1 - y^2 + x^2}} = \dfrac{1 - y^2 + xz}{(1 - y^2 + x^2)^{3/2}}\mathbf{i} + \dfrac{(x - z)y}{(1 - y^2 + x^2)^{3/2}}\mathbf{j} + \dfrac{-1}{\sqrt{1 - y^2 + x^2}}\mathbf{k}$; at the point $(0,0,1)$ we compute $\nabla \dfrac{x - z}{\sqrt{1 - y^2 + x^2}} = \mathbf{i} - \mathbf{k}$.

21.   If $f(x,y) = x^2 + y^2$, then $\nabla f(x,y) = 2x\mathbf{i} + 2y\mathbf{j}$ and
$$g(\Delta x, \Delta y) = f(x + \Delta x, y + \Delta y) - f(x,y) - \nabla f(x,y) \cdot (\Delta x\mathbf{i} + \Delta y\mathbf{j})$$
$$= (x + \Delta x)^2 + (y + \Delta y)^2 - (x^2 + y^2) - (2x\mathbf{i} + 2y\mathbf{j}) \cdot (\Delta x\mathbf{i} + \Delta y\mathbf{j})$$
$$= (\Delta x)^2 + (\Delta y)^2.$$

Since $g(\Delta x, \Delta y) = |\Delta x\mathbf{i} + \Delta y\mathbf{j}|^2$, it is clear that $\displaystyle\lim_{(\Delta x, \Delta y) \to (0,0)} \dfrac{g(\Delta x, \Delta y)}{|\Delta x\mathbf{i} + \Delta y\mathbf{j}|} = \lim_{(\Delta x, \Delta y) \to (0,0)} |\Delta x\mathbf{i} + \Delta y\mathbf{j}| = 0$; therefore f is differentiable everywhere in $\mathbb{R}^2$.

23.   Suppose $f = f(x,y)$ is a polynomial function, then f_x and f_y are defined everywhere and they also are polynomial functions. Theorem 13.2.1 (i) says any polynomial function is continuous throughout $\mathbb{R}^2$. Theorem 1 of this section then implies f is differentiable everywhere because f, f_x, f_y are defined and continuous everywhere. ∎

25.   Suppose f and g are differentiable at $\mathbf{x}$ and let $h = f + g$. Then $h_x = \frac{\partial}{\partial x}(f + g) = \frac{\partial}{\partial x}f + \frac{\partial}{\partial x}g = f_x + g_x$, $h_y = \frac{\partial}{\partial y}(f + g) = \frac{\partial}{\partial y}f + \frac{\partial}{\partial y}g = f_y + g_y$, and $h_z = \frac{\partial}{\partial z}(f + g) = \frac{\partial}{\partial z}f + \frac{\partial}{\partial z}g = f_z + g_z$ imply $\nabla(f + g) = \nabla h = (\nabla f) + (\nabla g)$. Furthermore, since f and g are differentiable at $\mathbf{x}$, so is $h = f + g$:
$$\lim_{\Delta\mathbf{x} \to 0} \dfrac{h(\mathbf{x} + \Delta\mathbf{x}) - h(\mathbf{x}) - \nabla h(\mathbf{x}) \cdot \Delta\mathbf{x}}{|\Delta\mathbf{x}|}$$
$$= \lim_{\Delta\mathbf{x} \to 0} \dfrac{(f + g)(\mathbf{x} + \Delta\mathbf{x}) - (f + g)(\mathbf{x}) - \nabla(f + g)(\mathbf{x}) \cdot \Delta\mathbf{x}}{|\Delta\mathbf{x}|}$$
$$= \lim_{\Delta\mathbf{x} \to 0} \dfrac{(f(\mathbf{x} + \Delta\mathbf{x}) - f(\mathbf{x}) - \nabla f(\mathbf{x}) \cdot \Delta\mathbf{x}) + (g(\mathbf{x} + \Delta\mathbf{x}) - g(\mathbf{x}) - \nabla g(\mathbf{x}) \cdot \Delta\mathbf{x})}{|\Delta\mathbf{x}|}$$
$$= \lim_{\Delta\mathbf{x} \to 0} \dfrac{f(\mathbf{x} + \Delta\mathbf{x}) - f(\mathbf{x}) - \nabla f(\mathbf{x}) \cdot \Delta\mathbf{x}}{|\Delta\mathbf{x}|} + \lim_{\Delta\mathbf{x} \to 0} \dfrac{g(\mathbf{x} + \Delta\mathbf{x}) - g(\mathbf{x}) - \nabla g(\mathbf{x}) \cdot \Delta\mathbf{x}}{|\Delta\mathbf{x}|} = 0 + 0 = 0. ∎$$

REMARK: If equations (10) and (11) had used a notation such as $R[f]$ instead of g, the above work would finish $\displaystyle\lim_{\Delta\mathbf{x} \to 0} \dfrac{R[f](\Delta\mathbf{x})}{|\Delta\mathbf{x}|} + \lim_{\Delta\mathbf{x} \to 0} \dfrac{R[g](\Delta\mathbf{x})}{|\Delta\mathbf{x}|} = 0 + 0 = 0$. Such notation avoids a glut of things named "g".

27.   If $f(x,y) = c$, then $f_x = 0 = f_y$ and $\nabla f(x,y) = 0\mathbf{i} + 0\mathbf{j} = \mathbf{0}$. Conversely, suppose $\nabla f = \mathbf{0}$ in a neighborhood of (x,y); that implies $f_x = 0 = f_y$ throughout the neighborhood. The mean-value result summarized in equation (18) implies there is a x_* between x and $x + \Delta x$ and a y_* between y and $y + \Delta y$ such that $f(x + \Delta x, y + \Delta y) - f(x,y) = f_x(x_*, y)\,\Delta x + f_y(x + \Delta x, y_*)\,\Delta y = 0\,\Delta x + 0\,\Delta y = 0 + 0 = 0$. Therefore $f(x + \Delta x, y + \Delta y) = f(x,y)$ which means f is constant throughout the neighborhood. ∎

29.   a. $f_x(0,0) = \displaystyle\lim_{\Delta x \to 0} \dfrac{f(\Delta x, 0) - f(0,0)}{\Delta x} = \lim_{\Delta x \to 0} \dfrac{(\Delta x)^2 \sin(1/|\Delta x|)}{\Delta x} = \lim_{\Delta x \to 0} \Delta x \sin\left(\dfrac{1}{|\Delta x|} \right) = 0$; similarly, $f_y(0,0) = 0$.

      b.   If $(x,y) \neq (0,0)$, then $f_x(x,y) = 2x \sin\left((x^2 + y^2)^{1/2} \right) - (x^2 + y^2)^{-1/2} \cos\left((x^2 + y^2)^{1/2} \right)$. Since $2x \sin\left((x^2 + y^2)^{1/2} \right)$ is bounded but $(x^2 + y^2)^{-1/2} \cos\left((x^2 + y^2)^{1/2} \right)$ is unbounded near $(0,0)$, we see that $\lim_{(x,y)\to(0,0)} f_x(x,y)$ does not exist. Similarly, $f_y(x,y)$ is unbounded near $(0,0)$ so it too can not be continuous there.

c. The gradient of f at $(0,0)$ exists and is $\mathbf{0}$ because $|\sin\theta| \le 1$ implies

$$\lim_{\Delta\mathbf{x}\to 0} \frac{f(\Delta\mathbf{x}) - f(0) - 0\cdot\Delta\mathbf{x}}{|\Delta\mathbf{x}|} = \lim_{\Delta\mathbf{x}\to 0} \frac{|\Delta\mathbf{x}|^2 \sin(1/|\Delta\mathbf{x}|) - 0 - 0}{|\Delta\mathbf{x}|} = \lim_{\Delta\mathbf{x}\to 0} |\Delta\mathbf{x}| \sin\left(\frac{1}{|\Delta\mathbf{x}|}\right) = 0.$$

29: $-.24 \le x \le .26, -.26 \le y \le .24$ $-.099 \le x \le .101, -.101 \le y \le .99$

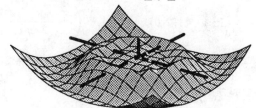 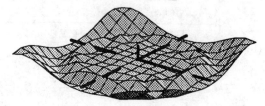

31. The function $g(x,y) = \left(x^2 + y^2\right)/2$ has the property that $g_x(x,y) = x$, $g_y(x,y) = y$, and $\nabla g(x,y) = x\mathbf{i} + y\mathbf{j}$; i.e., $\nabla g(\mathbf{x}) = \mathbf{x}$. Now invoke the result of Problem 28 to infer that $\nabla f(\mathbf{x}) = \mathbf{x}$ implies there is a constant c such that $f(x,y) = g(x,y) + c = \left(x^2 + y^2\right)/2 + c$.

Solutions 13.6   *The chain rule*   (pages 879-81)

1. $z = xy$, $x = e^t$, $y = e^{2t}$ imply

$$\frac{dz}{dt} = \frac{\partial z}{\partial x}\frac{dx}{dt} + \frac{\partial z}{\partial y}\frac{dy}{dt} = (y)(e^t) + (x)(2e^{2t}) = (e^{2t})(e^t) + (e^t)(2e^{2t}) = e^{3t} + 2e^{3t} = 3e^{3t}.$$

On the other hand, $z = xy = (e^t)(e^{2t}) = e^{3t}$ also implies $dz/dt = 3e^{3t}$.

3. $z = y/x$, $x = t^2$, $y = t^3$ imply

$$\frac{dz}{dt} = \frac{\partial z}{\partial x}\frac{dx}{dt} + \frac{\partial z}{\partial y}\frac{dy}{dt} = \left(\frac{-y}{x^2}\right)(2t) + \left(\frac{1}{x}\right)(3t^2) = \left(\frac{-t^3}{(t^2)^2}\right)(2t) + \left(\frac{1}{t^2}\right)(3t^2) = -2 + 3 = 1.$$

$z = y/x = t^3/t^2 = t$ also implies $dz/dt = 1$.

5. $z = \tan^{-1}(y/x)$, $x = \cos 3t$, $y = \sin 5t$ imply

$$\frac{dz}{dt} = \frac{\partial z}{\partial x}\frac{dx}{dt} + \frac{\partial z}{\partial y}\frac{dy}{dt} = \left(\frac{-y}{x^2 + y^2}\right)(-3\sin 3t) + \left(\frac{x}{x^2 + y^2}\right)(5\cos 5t)$$

$$= \left(\frac{-\sin 5t}{\cos^2 3t + \sin^2 5t}\right)(-3\sin 3t) + \left(\frac{\cos 3t}{\cos^2 3t + \sin^2 5t}\right)(5\cos 5t).$$

$z = \tan^{-1}\left(\frac{y}{x}\right) = \tan^{-1}\left(\frac{\sin 5t}{\cos 3t}\right)$ also implies $\dfrac{dz}{dt} = \left(\dfrac{1}{1 + (\sin 5t/\cos 3t)^2}\right)\left(\dfrac{5\cos 5t}{\cos 3t} - \dfrac{\sin 5t(-3\sin 3t)}{\cos^2 3t}\right).$

7. $w = x^2 + y^2 + z^2$, $x = \cos t$, $y = \sin t$, $z = t$ imply

$$\frac{dw}{dt} = \frac{\partial w}{\partial x}\frac{dx}{dt} + \frac{\partial w}{\partial y}\frac{dy}{dt} + \frac{\partial w}{\partial z}\frac{dz}{dt} = (2x)(-\sin t) + (2y)(\cos t) + (2z)(1)$$

$$= (2\cos t)(-\sin t) + (2\sin t)(\cos t) + (2t)(1) = (-2+2)\cos t\sin t + 2t = 2t.$$

$w = x^2 + y^2 + z^2 = \cos^2 t + \sin^2 t + t^2 = 1 + t^2$ also implies $dw/dt = 2t.$

9. $w = xy - yz + zx$, $x = e^t$, $y = e^{2t}$, $z = e^{3t}$ imply

$$\frac{dw}{dt} = \frac{\partial w}{\partial x}\frac{dx}{dt} + \frac{\partial w}{\partial y}\frac{dy}{dt} + \frac{\partial w}{\partial z}\frac{dz}{dt} = (y+z)(e^t) + (x-z)(2e^{2t}) + (-y+x)(3e^{3t})$$

$$= (e^{2t} + e^{3t})(e^t) + (e^t - e^{3t})(2e^{2t}) + (-e^{2t} + e^t)(3e^{3t})$$

$$= (e^{3t} + e^{4t}) + (2e^{3t} - 2e^{5t}) + (-3e^{5t} + 3e^{4t}) = 3e^{3t} + 4e^{4t} - 5e^{5t}.$$

$w = xy - yz + zx = (e^t)(e^{2t}) - (e^{2t})(e^{3t}) + (e^{3t})(e^t) = e^{3t} - e^{5t} + e^{4t}$ also implies $dw/dt = 3e^{3t} - 5e^{5t} + 4e^{4t}.$

11.　$z = xy$, $x = r + s$, $y = r - s$ imply

$$\frac{\partial z}{\partial r} = \frac{\partial z}{\partial x}\frac{\partial x}{\partial r} + \frac{\partial z}{\partial y}\frac{\partial y}{\partial r} = (y)(1) + (x)(1) = (r - s) + (r + s) = 2r,$$

$$\frac{\partial z}{\partial s} = \frac{\partial z}{\partial x}\frac{\partial x}{\partial s} + \frac{\partial z}{\partial y}\frac{\partial y}{\partial s} = (y)(1) + (x)(-1) = (r - s) - (r + s) = -2s.$$

13.　$z = y/x$, $x = e^r$, $y = e^s$ imply

$$\frac{\partial z}{\partial r} = \frac{\partial z}{\partial x}\frac{\partial x}{\partial r} + \frac{\partial z}{\partial y}\frac{\partial y}{\partial r} = \left(\frac{-y}{x^2}\right)(e^r) + \left(\frac{1}{x}\right)(0) = \left(\frac{-e^s}{e^{2r}}\right)e^r = -e^{s-r},$$

$$\frac{\partial z}{\partial s} = \frac{\partial z}{\partial x}\frac{\partial x}{\partial s} + \frac{\partial z}{\partial y}\frac{\partial y}{\partial s} = \left(\frac{-y}{x^2}\right)(0) + \left(\frac{1}{x}\right)(e^s) = \left(\frac{1}{e^r}\right)e^s = e^{s-r}.$$

15.　$z = \dfrac{e^{x+y}}{e^{x-y}} = e^{(x+y)-(x-y)} = e^{2y}$, $x = \ln(rs) = \ln r + \ln s$, $y = \ln(r/s) = \ln r - \ln s$ imply

$$\frac{\partial z}{\partial r} = \frac{\partial z}{\partial x}\frac{\partial x}{\partial r} + \frac{\partial z}{\partial y}\frac{\partial y}{\partial r} = (0)\left(\frac{1}{r}\right) + (2e^{2y})\left(\frac{1}{r}\right) = 0 + \left(2e^{2\ln(r/s)}\right)\left(\frac{1}{r}\right) = 2\left(\frac{r}{s}\right)^2\left(\frac{1}{r}\right) = \frac{2r}{s^2},$$

$$\frac{\partial z}{\partial s} = \frac{\partial z}{\partial x}\frac{\partial x}{\partial s} + \frac{\partial z}{\partial y}\frac{\partial y}{\partial s} = (0)\left(\frac{1}{s}\right) + (2e^{2y})\left(\frac{-1}{s}\right) = 0 + \left(2e^{2\ln(r/s)}\right)\left(\frac{-1}{s}\right) = 2\left(\frac{r}{s}\right)^2\left(\frac{-1}{s}\right) = \frac{-2r^2}{s^3}.$$

17.　$w = x + y + z$, $x = rs$, $y = r + s$, $z = r - s$ imply

$$\frac{\partial w}{\partial r} = \frac{\partial w}{\partial x}\frac{\partial x}{\partial r} + \frac{\partial w}{\partial y}\frac{\partial y}{\partial r} + \frac{\partial w}{\partial z}\frac{\partial z}{\partial r} = (1)(s) + (1)(1) + (1)(1) = s + 2,$$

$$\frac{\partial w}{\partial s} = \frac{\partial w}{\partial x}\frac{\partial x}{\partial s} + \frac{\partial w}{\partial y}\frac{\partial y}{\partial s} + \frac{\partial w}{\partial z}\frac{\partial z}{\partial s} = (1)(r) + (1)(1) + (1)(-1) = r.$$

19.　$w = xy/z$, $x = r+s$, $y = t-r$, $z = s+2t$ imply

$$\frac{\partial w}{\partial r} = \frac{\partial w}{\partial x}\frac{\partial x}{\partial r} + \frac{\partial w}{\partial y}\frac{\partial y}{\partial r} + \frac{\partial w}{\partial z}\frac{\partial z}{\partial r} = \left(\frac{y}{z}\right)(1) + \left(\frac{x}{z}\right)(-1) + \left(\frac{-xy}{z^2}\right)(0) = \frac{y - x}{z} = \frac{(t-r) - (r+s)}{s+2t},$$

$$\frac{\partial w}{\partial s} = \frac{\partial w}{\partial x}\frac{\partial x}{\partial s} + \frac{\partial w}{\partial y}\frac{\partial y}{\partial s} + \frac{\partial w}{\partial z}\frac{\partial z}{\partial s} = \left(\frac{y}{z}\right)(1) + \left(\frac{x}{z}\right)(0) + \left(\frac{-xy}{z^2}\right)(1) = \frac{y}{z} - \frac{xy}{z^2} = \frac{t-r}{s+2t} - \frac{(r+s)(t-r)}{(s+2t)^{-2}},$$

$$\frac{\partial w}{\partial t} = \frac{\partial w}{\partial x}\frac{\partial x}{\partial t} + \frac{\partial w}{\partial y}\frac{\partial y}{\partial t} + \frac{\partial w}{\partial z}\frac{\partial z}{\partial t} = \left(\frac{y}{z}\right)(0) + \left(\frac{x}{z}\right)(1) + \left(\frac{-xy}{z^2}\right)(2) = \frac{x}{z} - \frac{2xy}{z^2} = \frac{r+s}{s+2t} - \frac{2(r+s)(t-r)}{(s+2t)^2}.$$

21.　$w = \sinh(x + 2y + 3z)$, $x = \sqrt{r+s}$, $y = \sqrt[3]{s-t}$, $z = 1/(r+t)$ imply

$$w_s = w_x x_s + w_y y_s + w_z z_s$$

$$= \cosh(x+2y+3z)\left(\frac{1}{2\sqrt{r+s}}\right) + 2\cosh(x+2y+3z)\left(\frac{1}{3(s-t)^{2/3}}\right) + 3\cosh(x+2y+3z)(0)$$

$$= \left(\frac{1}{2\sqrt{r+s}} + \frac{2}{3(s-t)^{2/3}}\right)\cosh\left(\sqrt{r+s} + 2\sqrt[3]{s-t} + \frac{3}{r+t}\right).$$

23.　$w = \ln(x + 2y + 3z)$, $x = r^2 + t^2$, $y = s^2 - t^2$, $z = r^2 + s^2$ imply

$$w_r = w_x x_r + w_y y_r + w_z z_r = \left(\frac{1}{x+2y+3z}\right)(2r) + \left(\frac{2}{x+2y+3z}\right)(0) + \left(\frac{3}{x+2y+3z}\right)(2r)$$

$$= \left(\frac{1+3}{x+2y+3z}\right)(2r) = \frac{(1+3)(2r)}{(r^2+t^2) + 2(s^2-t^2) + 3(r^2+s^2)} = \frac{8r}{4r^2 + 5s^2 - t^2},$$

$$w_s = w_x x_s + w_y y_s + w_z z_s = \left(\frac{1}{x+2y+3z}\right)(0) + \left(\frac{2}{x+2y+3z}\right)(2s) + \left(\frac{3}{x+2y+3z}\right)(2s)$$

$$= \left(\frac{2+3}{x+2y+3z}\right)(2s) = \frac{(2+3)(2s)}{(r^2+t^2) + 2(s^2-t^2) + 3(r^2+s^2)} = \frac{10s}{4r^2 + 5s^2 - t^2},$$

$$w_t = w_x x_t + w_y y_t + w_z z_t = \left(\frac{1}{x+2y+3z}\right)(2t) + \left(\frac{2}{x+2y+3z}\right)(-2t) + \left(\frac{3}{x+2y+3z}\right)(0)$$

$$= \left(\frac{1-2}{x+2y+3z}\right)(2t) = \frac{(1-2)(2t)}{(r^2+t^2) + 2(s^2-t^2) + 3(r^2+s^2)} = \frac{-2t}{4r^2 + 5s^2 - t^2},$$

25. $w = r - \cos\theta + t$, $r = \sqrt{x^2 + y^2}$, $\theta = \tan^{-1}(y/x)$, $t = z$ imply $\sin\theta = y/\sqrt{x^2 + y^2}$ and

$$w_x = w_r r_x + w_\theta \theta_x + w_t t_x = (1)\left(\frac{x}{\sqrt{x^2+y^2}}\right) + (\sin\theta)\left(\frac{-y/x^2}{1+(y/x)^2}\right) + (1)(0)$$

$$= \frac{x}{\sqrt{x^2+y^2}} + \left(\frac{y}{\sqrt{x^2+y^2}}\right)\left(\frac{-y}{x^2+y^2}\right) = \frac{x}{\sqrt{x^2+y^2}} - \frac{y^2}{(x^2+y^2)^{3/2}},$$

$$w_y = w_r r_y + w_\theta \theta_y + w_t t_y = (1)\left(\frac{y}{\sqrt{x^2+y^2}}\right) + (\sin\theta)\left(\frac{1/x}{1+(y/x)^2}\right) + (1)(0)$$

$$= \frac{y}{\sqrt{x^2+y^2}} + \left(\frac{y}{\sqrt{x^2+y^2}}\right)\left(\frac{x}{x^2+y^2}\right) = \frac{y}{\sqrt{x^2+y^2}} + \frac{xy}{(x^2+y^2)^{3/2}},$$

$$w_z = w_r r_z + w_\theta \theta_z + w_t t_z = (1)(0) + (\sin\theta)(0) + (1)(1) = 0 + 0 + 1 = 1.$$

27. $V = \dfrac{\pi r^2 h}{3}$ implies $\dfrac{dV}{dt} = \dfrac{\partial V}{\partial r}\dfrac{dr}{dt} + \dfrac{\partial V}{\partial h}\dfrac{dh}{dt} = \dfrac{2\pi r h}{3}\dfrac{dr}{dt} + \dfrac{\pi r^2}{3}\dfrac{dh}{dt}$. If $\dfrac{dr}{dt} = 7$ inch/min and $\dfrac{dh}{dt} = -20$
inch/min, then $dV/dt = (\pi/3)\left(14rh - 20r^2\right) = (\pi/3)(14h - 20r)r$. Therefore $dV/dt(r = 45, h = 100) =$
$(\pi/3) \cdot (1400 - 900) \cdot 45 = 7500\pi$ cubic-inches per minute; since $V'(45, 100) > 0$, the volume is increasing.

29. $PV = nRT$ implies $\dfrac{dP}{dt}V + P\dfrac{dV}{dt} = \dfrac{d(PV)}{dt} = \dfrac{d(nRT)}{dt} = nR\dfrac{dT}{dt}$ and $\dfrac{dP}{dt} = \dfrac{1}{V}\left(nR\dfrac{dT}{dt} - P\dfrac{dV}{dt}\right)$.
If $n = 10$, $dV/dt = -25$ cm^3/min and $dT/dt = 1$ °C/min, then $dP/dt = (10R + 25P)/V$. Therefore
$dP/dt(V = 1000, P = 3) = (10R + 75)/1000$ N per square-cm per second; $R > 0$ implies $P'(1000, 3) > 0$, the
pressure is increasing.

REMARK: Centigrade and Kelvin temperature scales differ in their zero-point but use the same unit of
measure, hence $dT/dt = 1$ °C/min $= 1$ °K/min.

31.   Let $\overline{BC} = a$, $\overline{AC} = b$, $\overline{AB} = c$; the law of cosines implies $a^2 = b^2 + c^2 - 2bc\cos A$. Therefore $2a\dfrac{da}{dt} = (2b -$
$2c\cos A)\dfrac{db}{dt} + (2c - 2b\cos A)\dfrac{dc}{dt} + 2bc\sin A\dfrac{dA}{dt}$ and $\dfrac{da}{dt} = \dfrac{1}{a}\left((b - c\cos A)\dfrac{db}{dt} + (c - b\cos A)\dfrac{dc}{dt} + bc\sin A\dfrac{dA}{dt}\right)$.
If $dA/dt = 3(\pi/180) = \pi/60$ radian/sec and $dc/dt = 1$ cm/sec, and $db/dt = -2$ cm/sec, then $da/dt =$
$(1/a)\left(-2(b - c\cos A) + (c - b\cos A) + (\pi/60)bc\sin A\right)$. If $A = 30(\pi/180) = \pi/6$, then $\cos A = \sqrt{3}/2$ and
$\sin A = 1/2$. If $c = 10$ and $b = 24$, then $a = \sqrt{10^2 + 24^2 - 2(10)(24)(\sqrt{3}/2)} = \sqrt{676 - 240\sqrt{3}} \approx 16.134$ and
$da/dt = (1/a)\left(-2(24 - 5\sqrt{3}) + (10 - 12\sqrt{3}) + 2\pi\right) = \left(2\pi - 38 - 2\sqrt{3}\right)/\sqrt{676 - 240\sqrt{3}} \approx -2.18$ cm/sec;
the length of side BC is decreasing.

33. $w = g(x, y, z)$, $x = r\cos\theta$, $y = r\sin\theta$, $z = t$ imply

$w_r = w_x x_r + w_y y_r + w_z z_r = w_x(\cos\theta) + w_y(\sin\theta) + w_z(0) = g_x(x, y, z)\cos\theta + g_y(x, y, z)\sin\theta$,

$w_\theta = w_x x_\theta + w_y y_\theta + w_z z_\theta = w_x(-r\sin\theta) + w_y(r\cos\theta) + w_z(0) = r\left(-g_x(x, y, z)\sin\theta + g_y(x, y, z)\cos\theta\right)$,

$w_t = w_x x_t + w_y y_t + w_z z_t = w_x(0) + w_y(0) + w_z(1) = g_z(x, y, z)$.

35.   If $y(x, t) = \left(f(x+ct) + f(x-ct)\right)/2$, then $y_x = \left(f'(x+ct) + f'(x-ct)\right)/2$,
$y_{xx} = \left(f''(x+ct) + f''(x-ct)\right)/2$, $y_t = \left(cf'(x+ct) - cf'(x-ct)\right)/2$, and

$$y_{tt} = \left(c^2 f''(x+ct) + c^2 f''(x-ct)\right)/2 = c^2\left(f''(x+ct) + f''(x-ct)\right)/2 = c^2 y_{xx}.\blacksquare$$

37.   If $z = F(x, y)$, $x = X(r, s)$, and $y = Y(r, s)$, then

$$\frac{\partial z}{\partial r} = \frac{\partial z}{\partial x}\frac{\partial x}{\partial r} + \frac{\partial z}{\partial y}\frac{\partial y}{\partial r} = z_r,$$

$$\frac{\partial^2 z}{\partial r^2} = \frac{\partial}{\partial r}\frac{\partial z}{\partial r} = \frac{\partial z_r}{\partial x}\frac{\partial x}{\partial r} + \frac{\partial z_r}{\partial y}\frac{\partial y}{\partial r} = \frac{\partial}{\partial x}\left(\frac{\partial z}{\partial x}\frac{\partial x}{\partial r} + \frac{\partial z}{\partial y}\frac{\partial y}{\partial r}\right)\frac{\partial x}{\partial r} + \frac{\partial}{\partial y}\left(\frac{\partial z}{\partial x}\frac{\partial x}{\partial r} + \frac{\partial z}{\partial y}\frac{\partial y}{\partial r}\right)\frac{\partial y}{\partial r}.\blacksquare$$

39.   Let $u = y + ax$ and $v = y - ax$.

$$0 = \frac{\partial z}{\partial y} - \frac{1}{a}\frac{\partial z}{\partial x} = \left(\frac{\partial z}{\partial u}\frac{\partial u}{\partial y} + \frac{\partial z}{\partial v}\frac{\partial v}{\partial y}\right) - \frac{1}{a}\left(\frac{\partial z}{\partial u}\frac{\partial u}{\partial x} + \frac{\partial z}{\partial v}\frac{\partial v}{\partial x}\right) = \left(\frac{\partial z}{\partial u} + \frac{\partial z}{\partial v}\right) - \frac{1}{a}\left(\frac{\partial z}{\partial u}(a) + \frac{\partial z}{\partial v}(-a)\right) = 2\frac{\partial z}{\partial v}.$$

Therefore z is independent of v and is a function of u alone, i.e., $z = g(y + ax)$.

Solutions 13.7   *Tangent planes, Normal lines, and Gradients*   (pages 884-5)

1.   If $F(x,y,z) = x^2 + y^2 + z^2 - 1$, then $\nabla F(x,y,z) = 2x\mathbf{i} + 2y\mathbf{j} + 2z\mathbf{k}$ and $\nabla F(0,1,0) = 2\mathbf{j}$. The plane tangent to the surface at $(0,1,0)$ has equation $0 = \nabla F(0,1,0) \cdot ((x-0)\mathbf{i} + (y-1)\mathbf{j} + (z-0)\mathbf{k}) = 2\mathbf{j} \cdot (x\mathbf{i} + (y-1)\mathbf{j} + z\mathbf{k}) = 2(y-1)$ and the normal line has equation $x\mathbf{i} + y\mathbf{j} + z\mathbf{k} = (\mathbf{j}) + (2\mathbf{j})t = (1 + 2t)\mathbf{j}$.

3.   If $F(x,y,z) = (x/a)^2 + (y/b)^2 + (z/c)^2 - 3$, then $\nabla F(x,y,z) = (2x/a^2)\mathbf{i} + (2y/b^2)\mathbf{j} + (2z/c^2)\mathbf{k}$ and $\nabla F(a,b,c) = (2/a)\mathbf{i} + (2/b)\mathbf{j} + (2/c)\mathbf{k}$. The plane tangent to the surface at (a,b,c) has equation $0 = \nabla F(a,b,c) \cdot ((x-a)\mathbf{i} + (y-b)\mathbf{j} + (z-c)\mathbf{k}) = 2(x/a + y/b + z/c - 3)$ and the normal line has equation $x\mathbf{i} + y\mathbf{j} + z\mathbf{k} = (a\mathbf{i} + b\mathbf{j} + c\mathbf{k}) + ((2/a)\mathbf{i} + (2/b)\mathbf{j} + (2/c)\mathbf{k}) t = (a + 2t/a)\mathbf{i} + (b + 2t/b)\mathbf{j} + (c + 2t/c)\mathbf{k}$.

5.   If $F(x,y,z) = \sqrt{x} + \sqrt{y} + \sqrt{z} - 6$, then $\nabla F(x,y,z) = (1/2)\left(x^{-1/2}\mathbf{i} + y^{-1/2}\mathbf{j} + z^{-1/2}\mathbf{k}\right)$ and $\nabla F(4,1,9) = (1/4)\mathbf{i} + (1/2)\mathbf{j} + (1/6)\mathbf{k}$. The plane tangent to the surface at $(4,1,9)$ has equation $0 = \nabla F(4,1,9) \cdot ((x-4)\mathbf{i} + (y-1)\mathbf{j} + (z-9)\mathbf{k}) = (1/4)(x-4) + (1/2)(y-1) + (1/6)(z-9)$ and the normal line has equation $x\mathbf{i} + y\mathbf{j} + z\mathbf{k} = (4\mathbf{i} + \mathbf{j} + 9\mathbf{k}) + ((1/4)\mathbf{i} + (1/2)\mathbf{j} + (1/6)\mathbf{k}) t = (4 + t/4)\mathbf{i} + (1 + t/2)\mathbf{j} + (9 + t/6)\mathbf{k}$.

7.   If $F(x,y,z) = xyz - 4$, then $\nabla F(x,y,z) = yz\mathbf{i} + xy\mathbf{j} + xy\mathbf{k}$ and $\nabla F(1,2,2) = 4\mathbf{i} + 2\mathbf{j} + 2\mathbf{k}$. The plane tangent to the surface at $(1,2,2)$ has equation $0 = (4\mathbf{i} + 2\mathbf{j} + 2\mathbf{k}) \cdot ((x-1)\mathbf{i} + (y-2)\mathbf{j} + (z-2)\mathbf{k}) = 2(2x + y + z - 6)$ and the normal line has equation $x\mathbf{i} + y\mathbf{j} + z\mathbf{k} = (\mathbf{i} + 2\mathbf{j} + 2\mathbf{k}) + (4\mathbf{i} + 2\mathbf{j} + 2\mathbf{k}) t = (1 + 4t)\mathbf{i} + (2 + 2t)\mathbf{j} + (2 + 2t)\mathbf{k}$.

9.   If $F(x,y,z) = 4x^2 - y^2 - 5z^2 - 15$, then $\nabla F(x,y,z) = 8x\mathbf{i} - 2y\mathbf{j} - 10z\mathbf{k}$ and $\nabla F(3,1,-2) = 24\mathbf{i} - 2\mathbf{j} + 20\mathbf{k}$. The plane tangent to the surface at $(3,1,-2)$ has equation $0 = (24\mathbf{i} - 2\mathbf{j} + 20\mathbf{k}) \cdot ((x-3)\mathbf{i} + (y-1)\mathbf{j} + (z-(-2))\mathbf{k}) = 2(12x - y + 10z - 15)$ and the normal line has equation $x\mathbf{i} + y\mathbf{j} + z\mathbf{k} = (3\mathbf{i} + \mathbf{j} - 2\mathbf{k}) + (24\mathbf{i} - 2\mathbf{j} + 20\mathbf{k}) t = (3 + 24t)\mathbf{i} + (1 - 2t)\mathbf{j} + (-2 + 20t)\mathbf{k}$.

11.   If $F(x,y,z) = \sin xy - 2\cos yz$, then $\nabla F(x,y,z) = y\cos xy\mathbf{i} + (x\cos xy + 2z\sin yz)\mathbf{j} + 2y\sin yz\mathbf{k}$ and $\nabla F(\pi/2,1,\pi/3) = (\pi/\sqrt{3})\mathbf{j} + \sqrt{3}\mathbf{k}$. The plane tangent to the surface at $(\pi/2,1,\pi/3)$ has equation

$$0 = \left((\pi/\sqrt{3})\mathbf{j} + \sqrt{3}\mathbf{k}\right) \cdot ((x - \pi/2)\mathbf{i} + (y-1)\mathbf{j} + (z - \pi/3)\mathbf{k}) = (\pi/\sqrt{3})(y-1) + \sqrt{3}(z - \pi/3)$$

and the normal line has equation

$$x\mathbf{i} + y\mathbf{j} + z\mathbf{k} = ((\pi/2)\mathbf{i} + \mathbf{j} + (\pi/3)\mathbf{k}) + \left((\pi/\sqrt{3})\mathbf{j} + \sqrt{3}\mathbf{k}\right) t = (\pi/2)\mathbf{i} + \left(1 + (\pi/\sqrt{3})t\right)\mathbf{j} + \left((\pi/3) + \sqrt{3}\,t\right)\mathbf{k}.$$

13.   If $F(x,y,z) = e^{eyz} - 5$, then $\nabla F(x,y,z) = e^{eyz}(yz\mathbf{i} + xz\mathbf{j} + xy\mathbf{k})$ and $\nabla F(1,1,\ln 5) = 5\ln 5\mathbf{i} + 5\ln 5\mathbf{j} + 5\mathbf{k}$. The plane tangent to the surface at $(1,1,\ln 5)$ has equation $0 = (5\ln 5\mathbf{i} + 5\ln 5\mathbf{j} + 5\mathbf{k}) \cdot ((x-1)\mathbf{i} + (y-1)\mathbf{j} + (z - \ln 5)\mathbf{k}) = 5(\ln 5\,x + \ln 5\,y + z - 3\ln 5)$ and the normal line has equation $x\mathbf{i} + y\mathbf{j} + z\mathbf{k} = (\mathbf{i} + \mathbf{j} + \ln 5\mathbf{k}) + (5\ln 5\mathbf{i} + 5\ln 5\mathbf{j} + 5\mathbf{k}) t = (1 + (5\ln 5)t)\mathbf{i} + (1 + (5\ln 5)t)\mathbf{j} + (\ln 5 + 5t)\mathbf{k}$.

15. $z(1,1) = 1 \cdot 1^2 = 1$, $\nabla xy^2 = y^2\mathbf{i} + 2xy\mathbf{j}$ and $\nabla z(1,1) = \mathbf{i} + 2\mathbf{j}$. The plane tangent to the surface at $(1,1,1)$ has equation $z = 1 + (\mathbf{i} + 2\mathbf{j}) \cdot ((x-1)\mathbf{i} + (y-1)\mathbf{j}) = -2 + x + 2y$ and the normal line has equation $x\mathbf{i} + y\mathbf{j} + z\mathbf{k} = (\mathbf{i} + \mathbf{j} + \mathbf{k}) + (\mathbf{i} + 2\mathbf{j} - \mathbf{k}) t = (1 + t)\mathbf{i} + (1 + 2t)\mathbf{j} + (1 - t)\mathbf{k}$.

17. $z(\pi/8,\pi/20) = \sin(\pi/2) = 1$, $\nabla\sin(2x+5y) = \cos(2x+5y)(2\mathbf{i} + 5\mathbf{j})$ and $\nabla z(\pi/8,\pi/20) = (0)(2\mathbf{i} + 5\mathbf{j}) = 0$. The plane tangent to the surface at $(\pi/8,\pi/20,1)$ has equation $z = 1 + 0 \cdot ((x - \pi/8)\mathbf{i} + (y - \pi/20)\mathbf{j}) = 1$ and the normal line has equation $x\mathbf{i} + y\mathbf{j} + z\mathbf{k} = ((\pi/8)\mathbf{i} + (\pi/20)\mathbf{j} + \mathbf{k}) + (0 - \mathbf{k})t = (\pi/8)\mathbf{i} + (\pi/20)\mathbf{j} - t\mathbf{k}$.

19. $z(-2,2) = \tan^{-1}(2/(-2)) = -\pi/4$, $\nabla\tan^{-1}(y/x) = (x^2+y^2)^{-1}(-y\mathbf{i} + x\mathbf{j})$ and $\nabla z(-2,2) = (-1/4)(\mathbf{i}+\mathbf{j})$. The plane tangent to the surface at $(-2,2,-\pi/4)$ has equation

$$z = -\pi/4 + ((-1/4)(\mathbf{i}+\mathbf{j})) \cdot ((x - (-2))\mathbf{i} + (y-2)\mathbf{j}) = -(1/4)(\pi + x + y)$$

and the normal line has equation $x\mathbf{i} + y\mathbf{j} + z\mathbf{k} = (-2\mathbf{i} + 2\mathbf{j} - \pi/4\mathbf{k}) + ((-1/4)(\mathbf{i}+\mathbf{j}) - \mathbf{k}) t = (-2 - t/4)\mathbf{i} + (2 + t/4)\mathbf{j} + (-\pi/4 - t)\mathbf{k}$.

21. $z(\pi/2,\pi/6) = \sec(\pi/2 - \pi/6) = 2$, $\nabla\sec(x-y) = \sec(x-y)\tan(x-y)(\mathbf{i}-\mathbf{j})$ and $\nabla z(\pi/2,\pi/6) = 2\sqrt{3}(\mathbf{i}-\mathbf{j})$. The plane tangent to the surface at $(\pi/2,\pi/6,2)$ has equation

$$z = 2 + \left(2\sqrt{3}(\mathbf{i}-\mathbf{j})\right) \cdot ((x - \pi/2)\mathbf{i} + (y - \pi/6)\mathbf{j}) = 2 + 2\sqrt{3}(x - \pi/2) - 2\sqrt{3}(y - \pi/6)$$

and the normal line has equation $x\mathbf{i} + y\mathbf{j} + z\mathbf{k} = ((\pi/2)\mathbf{i} + (\pi/6)\mathbf{j} + 2\mathbf{k}) + (2\sqrt{3}(\mathbf{i} - \mathbf{j}) - \mathbf{k}) t = (\pi/2 + 2\sqrt{3}t)\mathbf{i} + (\pi/6 - 2\sqrt{3}t)\mathbf{j} + (2 - t)\mathbf{k}$.

23. $\nabla(xy - 5) = y\mathbf{i} + x\mathbf{j}$; therefore $\nabla(xy - 5)(1,5) = 5\mathbf{i} + \mathbf{j}$ is normal to the graph of $xy = 5$ at $(1,5)$ and $0 = (5\mathbf{i}+\mathbf{j}) \cdot ((x-1)\mathbf{i} + (y-5)\mathbf{j}) = 5(x-1) + (y-5) = 5x + y - 10$ is an equation for the line tangent there.

25.   If $f(x,y) = (x+y)/(x-y) - 7$, then $\nabla f(x,y) = (x-y)^{-2}(-2y\mathbf{i} + 2x\mathbf{j})$; therefore $\nabla f(4,3) = -6\mathbf{i} + 8\mathbf{j}$ is normal to the graph of $(x+y)/(x-y) = 7$ at $(4,3)$ and $0 = (-6\mathbf{i} + 8\mathbf{j}) \cdot ((x-4)\mathbf{i} + (y-3)\mathbf{j}) = -6(x-4) + 8(y-3) = -6x + 8y$ is an equation for the line tangent there.

27.   If $f(x,y) = (x/2)^2 + (y/4)^2 - 1$, then $\nabla f(x,y) = (x/2)\mathbf{i} + (y/8)\mathbf{j}$; therefore $\nabla f(\sqrt{2}, 2\sqrt{2}) = (1/\sqrt{2})\mathbf{i} + (1/2\sqrt{2})\mathbf{j}$ is normal to the graph of $(x/2)^2 + (y/4)^2 = 1$ at $(\sqrt{2}, 2\sqrt{2})$ and $0 = ((1/\sqrt{2})\mathbf{i} + (1/2\sqrt{2})\mathbf{j}) \cdot ((x-\sqrt{2})\mathbf{i} + (y - 2\sqrt{2})\mathbf{j}) = x/\sqrt{2} + y/2\sqrt{2} - 2$ is an equation for the line tangent there.

29.   The line $(x-3)/1 = (y+1)/(-1) = (z+2)/(-2)$ can be parameterized $(x,y,z) = (3,-1,-2) + (1,-1,-2)t = (3+t, -1-t, -2-2t)$; it is parallel to the vector $\mathbf{v} = \mathbf{i} - \mathbf{j} - 2\mathbf{k}$. A point on this line is also on the surface $z = x^2 + y^2$ if and only if $-2 - 2t = (3+t)^2 + (-1-t)^2$; this equation is equivalent to $0 = 2(t^2 + 5t + 6) = 2(t+3)(t+2)$. If $t = -3$, the intersection point is $(0,2,4)$; if $t = -2$, the intersection point is $(1,1,2)$.

$\mathbf{N}(x,y,z) = \nabla(x^2 + y^2 - z) = 2x\mathbf{i} + 2y\mathbf{j} - \mathbf{k}$ is normal to the surface $x^2 + y^2 - z = 0$. At $(0,2,4)$, we have $\mathbf{N}(0,2,4) = 4\mathbf{j} - \mathbf{k}$; the cosine of the angle between it and the line is $\dfrac{(4\mathbf{j} - \mathbf{k}) \cdot (\mathbf{i} - \mathbf{j} - 2\mathbf{k})}{|4\mathbf{j} - \mathbf{k}|\,|\mathbf{i} - \mathbf{j} - 2\mathbf{k}|} = \dfrac{0 - 4 + 2}{\sqrt{0 + 16 + 1}\,\sqrt{1 + 1 + 4}} = \dfrac{-2}{\sqrt{17}\,\sqrt{6}}$. At $(1,1,2)$, we have $\mathbf{N}(1,1,2) = 2\mathbf{i} + 2\mathbf{j} - \mathbf{k}$; the cosine of the angle between it and the line is $\dfrac{(2\mathbf{i} + 2\mathbf{j} - \mathbf{k}) \cdot (\mathbf{i} - \mathbf{j} - 2\mathbf{k})}{|2\mathbf{i} + 2\mathbf{j} - \mathbf{k}|\,|\mathbf{i} - \mathbf{j} - 2\mathbf{k}|} = \dfrac{2 - 2 + 2}{\sqrt{4 + 4 + 1}\,\sqrt{1 + 1 + 4}} = \dfrac{2}{3\sqrt{6}}$.

31.   Let $F(x,y,z) = ax^2 + by^2 + cz^2 - d$, then $\nabla F(x,y,z) = 2ax\mathbf{i} + 2by\mathbf{j} + 2cz\mathbf{k}$. The plane tangent to the surface at (x_0, y_0, z_0) satisfies the equation

$$
\begin{aligned}
0 = \nabla F(x_0, y_0, z_0) \cdot ((x - x_0)\mathbf{i} + (y - y_0)\mathbf{j} + (z - z_0)\mathbf{k}) &= 2ax_0(x - x_0) + 2by_0(y - y_0) + 2cz_0(z - z_0) \\
&= 2(ax_0 x + by_0 y + cz_0 z) - 2(ax_0^2 + by_0^2 + cz_0^2) \\
&= 2(ax_0 x + by_0 y + cz_0 z) - 2(d)
\end{aligned}
$$

which is equivalent to $d = ax_0 x + by_0 y + cz_0 z$. ∎

33.   Let $F(x,y,z) = a(x^2 + y^2) - z^2$; $\nabla F(x,y,z) = 2(ax\mathbf{i} + ay\mathbf{j} - z\mathbf{k})$. The normal line at (x_0, y_0, z_0) satisfies the equation $x\mathbf{i} + y\mathbf{j} + z\mathbf{k} = (x_0\mathbf{i} + y_0\mathbf{j} + z_0\mathbf{k}) + \nabla F(x_0, y_0, z_0)t = x_0(1 + 2at)\mathbf{i} + y_0(1 + 2at)\mathbf{j} + z_0(1 - 2t)\mathbf{k}$. Choosing $t = -1/(2a)$ yields the point $\big(0, 0, z_0(1 + 1/a)\big)$ which is on the normal line and the z-axis. ∎

35.   Theorem 11.4.4 (ii) states $\mathbf{a} \perp \mathbf{b} \iff \mathbf{a} \cdot \mathbf{b} = 0$ (assuming both $\mathbf{a}$ and $\mathbf{b}$ are nonzero). The normal to $F(x,y,z) = 0$ at (x_0, y_0, z_0) is parallel to $\nabla F(x_0, y_0, z_0)$ and the normal to $G(x,y,z) = 0$ at (x_0, y_0, z_0) is parallel to $\nabla G(x_0, y_0, z_0)$. Those vectors are orthogonal if and only if $\nabla F(x_0, y_0, z_0) \cdot \nabla G(x_0, y_0, z_0) = 0$; the angle between those vectors is defined to be the angle between the surfaces at (x_0, y_0, z_0), hence the surface meet orthogonally at (x_0, y_0, z_0) (i.e., at right angles) if and only if $\nabla F(x_0, y_0, z_0) \cdot \nabla G(x_0, y_0, z_0) = 0$. ∎

37.   Suppose F is a differentiable functions such that $F(x,y) = 0$ defines a smooth curve parameterized as $\mathbf{v}(t) = x(t)\mathbf{i} + y(t)\mathbf{j}$. The chain rule implies

$$
0 = \frac{\partial}{\partial t} F(x,y) = \frac{\partial F}{\partial x}\frac{dx}{dt} + \frac{\partial F}{\partial y}\frac{dy}{dt} = F_x x' + F_y y' = (F_x\mathbf{i} + F_y\mathbf{j}) \cdot (x'\mathbf{i} + y'\mathbf{j}) = \nabla F \cdot \mathbf{v}'
$$

which then implies $\nabla F \perp \mathbf{v}'$. Since $\mathbf{v}'$ is tangent to the curve, this shows ∇F is orthogonal to the curve. ∎

39.   Let $F(x,y) = x/y - a$ and $G(x,y) = x^2 + y^2 - r^2$. Then $\nabla F \cdot \nabla G = ((1/y)\mathbf{i} - (x/y^2)\mathbf{j}) \cdot (2x\mathbf{i} + 2y\mathbf{j}) = 2x/y - 2xy/y^2 = 0$ which is equivalent to $\nabla F \perp \nabla G$. Since the normals to the curves are orthogonal to their tangents, this then implies the tangents are also orthogonal — that is equivalent to saying the curves are orthogonal. ∎

Solutions 13.8   *Directional derivatives and the Gradient*   (pages 890-1)

1.   If $f(x,y) = xy$, then $\nabla f(x,y) = y\mathbf{i} + x\mathbf{j}$ and $\nabla f(2,3) = 3\mathbf{i} + 2\mathbf{j}$. Let $\mathbf{v} = \mathbf{i} + 3\mathbf{j}$ and $\mathbf{u} = \mathbf{v}/|\mathbf{v}|$. Therefore $f'_{\mathbf{u}}(2,3) = \nabla f(2,3) \cdot \mathbf{v}/|\mathbf{v}| = (3\mathbf{i} + 2\mathbf{j}) \cdot (\mathbf{i} + 3\mathbf{j})/\sqrt{1^2 + 3^2} = (3 + 6)/\sqrt{10} = 9/\sqrt{10}$.

3.   If $f(x,y) = \ln(x + 3y)$, then $\nabla f(x,y) = (x + 3y)^{-1}(\mathbf{i} + 3\mathbf{j})$ and $\nabla f(2,4) = (1/14)(\mathbf{i} + 3\mathbf{j})$. Let $\mathbf{v} = \mathbf{i} + \mathbf{j}$ and $\mathbf{u} = \mathbf{v}/|\mathbf{v}|$. Therefore $f'_{\mathbf{u}}(2,4) = \nabla f(2,4) \cdot \mathbf{v}/|\mathbf{v}| = (1/14)(\mathbf{i} + 3\mathbf{j}) \cdot (\mathbf{i} + \mathbf{j})/\sqrt{1^2 + 1^2} = (1/14)(1 + 3)/\sqrt{2} = \sqrt{2}/7$.

5.   If $f(x,y) = \tan^{-1}(y/x)$, then $\nabla f(x,y) = (x^2 + y^2)^{-1}(-y\mathbf{i} + x\mathbf{j})$ and $\nabla f(2,2) = (1/4)(-\mathbf{i} + \mathbf{j})$. Let $\mathbf{v} = 3\mathbf{i} - 2\mathbf{j}$ and $\mathbf{u} = \mathbf{v}/|\mathbf{v}|$. Therefore $f'_{\mathbf{u}}(2,2) = \nabla f(2,2) \cdot \mathbf{v}/|\mathbf{v}| = (1/4)(-\mathbf{i} + \mathbf{j}) \cdot (3\mathbf{i} - 2\mathbf{j})/\sqrt{3^2 + (-2)^2} = (1/4)(-3 - 2)/\sqrt{13} = -5/(4\sqrt{13})$.

7.   If $f(x,y) = xe^y + ye^x$, then $\nabla f(x,y) = (e^y + ye^x)\mathbf{i} + (xe^y + e^x)\mathbf{j}$ and $\nabla f(1,2) = (e^2 + 2e)\mathbf{i} + (e^2 + e)\mathbf{j}$. Let $\mathbf{v} = \mathbf{i} + \mathbf{j}$ and $\mathbf{u} = \mathbf{v}/|\mathbf{v}|$. Therefore $f'_\mathbf{u}(1,2) = \nabla f(1,2) \cdot \mathbf{v}/|\mathbf{v}| = \left((e^2 + 2e)\mathbf{i} + (e^2 + e)\mathbf{j}\right) \cdot (\mathbf{i} + \mathbf{j})/\sqrt{1^2 + 1^2} = \left((e^2 + 2e) + (e^2 + e)\right)/\sqrt{2} = \left(2e^2 + 3e\right)/\sqrt{2}$.

9.   If $f(x,y,z) = xy + yz + xz$, then $\nabla f(x,y,z) = (y+z)\mathbf{i} + (x+z)\mathbf{j} + (y+z)\mathbf{k}$ and $\nabla f(1,1,1) = 2\mathbf{i} + 2\mathbf{j} + 2\mathbf{k}$. Let $\mathbf{v} = \mathbf{i} + \mathbf{j} + \mathbf{k}$ and $\mathbf{u} = \mathbf{v}/|\mathbf{v}|$. Thus $f'_\mathbf{u}(1,1,1) = \nabla f(1,1,1) \cdot \mathbf{v}/|\mathbf{v}| = (2\mathbf{i} + 2\mathbf{j} + 2\mathbf{k}) \cdot (\mathbf{i} + \mathbf{j} + \mathbf{k})/\sqrt{1^2 + 1^2 + 1^2} = (2 + 2 + 2)/\sqrt{3} = 6/\sqrt{3} = 2\sqrt{3}$.

11.   If $f(x,y,z) = xe^{yz}$, then $\nabla f(x,y,z) = e^{yz}(\mathbf{i} + xz\mathbf{j} + xy\mathbf{k})$ and $\nabla f(2,0,-4) = \mathbf{i} - 8\mathbf{j}$. Let $\mathbf{v} = -\mathbf{i} + 2\mathbf{j} + 5\mathbf{k}$ and $\mathbf{u} = \mathbf{v}/|\mathbf{v}|$. Thus $f'_\mathbf{u}(2,0,-4) = \nabla f(2,0,-4) \cdot \mathbf{v}/|\mathbf{v}| = (\mathbf{i} - 8\mathbf{j}) \cdot (-\mathbf{i} + 2\mathbf{j} + 5\mathbf{k})/\sqrt{(-1)^2 + 2^2 + 5^2} = (-1 - 16 + 0)/\sqrt{30} = -17/\sqrt{30}$.

13.   If $f(x,y,z) = e^{-(x^2+y^2+z^2)}$, then $\nabla f(x,y,z) = -2e^{-(x^2+y^2+z^2)}(x\mathbf{i} + y\mathbf{j} + z\mathbf{k})$ and $\nabla f(1,1,1) = -2e^{-3}(\mathbf{i} + \mathbf{j} + \mathbf{k})$. Let $\mathbf{v} = \mathbf{i} + 3\mathbf{j} - 5\mathbf{k}$ and $\mathbf{u} = \mathbf{v}/|\mathbf{v}|$. Therefore $f'_\mathbf{u}(1,1,1) = \nabla f(1,1,1) \cdot \mathbf{v}/|\mathbf{v}| = \left(-2e^{-3}(\mathbf{i} + \mathbf{j} + \mathbf{k})\right) \cdot (\mathbf{i} + 3\mathbf{j} - 5\mathbf{k})/\sqrt{1^2 + 3^2 + (-5)^2} = -2e^{-3}(1 + 3 - 5)/\sqrt{35} = 2e^{-3}/\sqrt{35}$.

15.   If $f(x,y) = 2x^2 y$, then $\nabla f(x,y) = 4xy\mathbf{i} + 2x^2\mathbf{j}$ and $\nabla f(-1,4) = -16\mathbf{i} + 2\mathbf{j}$. Let $\mathbf{v} = (2,-5) - (-1,4) = 3\mathbf{i} - 9\mathbf{j}$ and $\mathbf{u} = \mathbf{v}/|\mathbf{v}|$. Therefore $f'_\mathbf{u}(-1,4) = \nabla f(-1,4) \cdot \mathbf{v}/|\mathbf{v}| = (-16\mathbf{i} + 2\mathbf{j}) \cdot (3\mathbf{i} - 9\mathbf{j})/\sqrt{3^2 + (-9)^2} = (-48 - 18)/(3\sqrt{10}) = -22/\sqrt{10}$.

17.   If $f(x,y,z) = xy^2 z + x^3 yz^2$, then $\nabla f(x,y,z) = (y^2 z + 3x^2 yz^2)\mathbf{i} + (2xyz + x^3 z^2)\mathbf{j} + (xy^2 + 2x^3 yz)\mathbf{k}$ and $\nabla f(1,2,-1) = 2\mathbf{i} - 3\mathbf{j}$. Let $\mathbf{v} = (0,-2,4) - (1,2,-1) = -\mathbf{i} - 4\mathbf{j} + 5\mathbf{k}$ and $\mathbf{u} = \mathbf{v}/|\mathbf{v}|$. Therefore $f'_\mathbf{u}(1,2,-1) = \nabla f(1,2,-1) \cdot \mathbf{v}/|\mathbf{v}| = (2\mathbf{i} - 3\mathbf{j}) \cdot (-\mathbf{i} - 4\mathbf{j} + 5\mathbf{k})/\sqrt{(-1)^2 + (-4)^2 + 5^2} = (-2 + 12 + 0)/\sqrt{42} = 10/\sqrt{42}$.

19.   If $f(x,y) = x + 2y$, then $\nabla f(x,y) = \mathbf{i} + 2\mathbf{j}$ and $|\nabla f| = |\mathbf{i} + 2\mathbf{j}| = \sqrt{1^2 + 2^2} = \sqrt{5}$ at every point.

21.   If $f(x,y) = \tan^{-1}(y/x)$, then $\nabla f(x,y) = (x^2 + y^2)^{-1}(-y\mathbf{i} + x\mathbf{j})$ and $|\nabla f(x,y)| = (x^2 + y^2)^{-1}|-y\mathbf{i} + x\mathbf{j}| = 1/\sqrt{x^2 + y^2}$; therefore $|\nabla f(-1,-1)| = 1/\sqrt{2}$.

23.   If $v(x,y,z) = \left(0.02 + \sqrt{x^2 + y^2 + z^2}\right)^{-1}$, then

$$\nabla v(x,y,z) = \frac{1}{\left(0.02 + \sqrt{x^2 + y^2 + z^2}\right)^2 \sqrt{x^2 + y^2 + z^2}}(-x\mathbf{i} - y\mathbf{j} - z\mathbf{k}).$$

$\nabla v(1,-1,2)$ is a positive multiple of $-\mathbf{i} + \mathbf{j} - 2\mathbf{k}$ and voltage increases most rapidly at $(1,-1,2)$ in that direction.

REMARK: The equipotential surfaces are spheres centered at the orign and ∇v points to that center.

25.   a. $T(x,y,z) = 100\left(1 + x^2 + y^2 + z^2\right)^{-1} = 100\left(1 + r^2\right)^{-1}$ is a decreasing function of $|r|$; its largest value occurs where $r = 0$, i.e., at the origin.

   b. $\nabla T(x,y,z) = 200\left(1 + x^2 + y^2 + z^2\right)^{-2}(-x\mathbf{i} - y\mathbf{j} - z\mathbf{k})$; $\nabla T(3,-1,2) = (8/9)(-3\mathbf{i} + \mathbf{i} - 2\mathbf{j})$, the direction of greatest decrease is along $-\nabla T(3,-1,2) = (8/9)(3\mathbf{i} - \mathbf{i} + 2\mathbf{j})$.

   c. $\nabla T(x,y,z) = 200\left(1 + r^2\right)^{-2}(-\mathbf{r})$, a positive multiple of $-\mathbf{r}$; the direction of greatest increase points toward the origin.

27.   Suppose the particle moves on curve $\mathbf{v}(t) = (x(t), y(t))$. If the particle always moves in the direction of the gradient of $T(x,y) = 1 - (x/a)^2 - (y/b)^2$, then $x'\mathbf{i} + y'\mathbf{j} = \mathbf{v}' = k\nabla T(x,y) = k\left(-2(x/a^2)\mathbf{i} - 2(y/b^2)\mathbf{j}\right)$ for some positive proportionality factor k (which may depend on t). This leads to

$$\frac{x'}{-2(x/a^2)} = k = \frac{y'}{-2(y/b^2)} \iff a^2\frac{x'}{x} = b^2\frac{y'}{y} \iff a^2 \ln|x| = b^2 \ln|y| + c.$$

Since the curve passes thru $(-a,b)$, the equation reduces to $a^2 \ln|x/a| = b^2 \ln|y/b|$, i.e., $|x/a|^{a^2} = |y/b|^{b^2}$.

REMARK: If the factor k is constant and $\mathbf{v}(0) = (-a,b)$, then $x = -a\,e^{-2kt/a^2}$ and $y = b\,e^{-2kt/b^2}$.

29.   a.   The unit vector in the direction of $\mathbf{f}' = x'\mathbf{i} + y'\mathbf{j}$ is

$$\frac{\mathbf{f}'}{|\mathbf{f}'|} = \frac{x'\mathbf{i} + y'\mathbf{j}}{|x'\mathbf{i} + y'\mathbf{j}|} = \frac{x'\mathbf{i} + y'\mathbf{j}}{\sqrt{x'^2 + y'^2}} = \frac{x'}{\sqrt{x'^2 + y'^2}}\mathbf{i} + \frac{y'}{\sqrt{x'^2 + y'^2}}\mathbf{j}. \; ∎$$

   b.   The unit vector in the direction of $\nabla V = -2x\mathbf{i} - 8y\mathbf{j}$ is

$$\frac{\nabla V}{|\nabla V|} = \frac{-2x\mathbf{i} - 8y\mathbf{j}}{|-2x\mathbf{i} - 8y\mathbf{j}|} = \frac{-2x\mathbf{i} - 8y\mathbf{j}}{\sqrt{(-2x)^2 + (-8y)^2}} = \frac{-2x\mathbf{i} - 8y\mathbf{j}}{\sqrt{4x^2 + 64y^2}} = \frac{-2x}{\sqrt{4x^2 + 64y^2}}\mathbf{i} + \frac{-8y}{\sqrt{4x^2 + 64y^2}}\mathbf{j}. \; ∎$$

13.8 — Directional derivatives and the Gradient

c.
$$\frac{\mathbf{f}'}{|\mathbf{f}'|} = \frac{\nabla V}{|\nabla V|} \iff \frac{x'}{\sqrt{x'^2 + y'^2}} = \frac{-2x}{\sqrt{4x^2 + 64y^2}} \text{ and } \frac{y'}{\sqrt{x'^2 + y'^2}} = \frac{-8y}{\sqrt{4x^2 + 64y^2}}$$

$$\implies \frac{y'}{y} = \frac{-8\sqrt{x'^2 + y'^2}}{\sqrt{4x^2 + 64y^2}} = 4\frac{-2\sqrt{x'^2 + y'^2}}{\sqrt{4x^2 + 64y^2}} = 4\frac{x'}{x}. \blacksquare$$

d. $\dfrac{y'}{y} = 4\dfrac{x'}{x} \implies \ln|y| = 4\ln|x| + c = \ln|x|^4 + c. \blacksquare$

e. $\ln|y| = \ln|x|^4 + c \implies |y| = e^{\ln|x|^4 + c} = e^c|x|^4 = e^c x^4$. Since y'/y is undefined if $y = 0$ and since y is continuous, we conclude y does not change sign. Therefore that fixed sign can be absorbed into a constant $k = \text{sgn}(y)e^c$ so we can write $y = kx^4$. $\blacksquare$

f. If $y = kx^4$ passes through $(1, -2)$ as in Example 5, then $k = (-2)/1^4 = -2$. $\blacksquare$

REMARK: Solution 27 demonstrates a slightly simpler, but equivalent, approach to steps a–d.

31. $f(x, y, z) = f(\mathbf{x})$ is differentiable at $\mathbf{a}$ if and only if f is defined in a neighborhood of $\mathbf{a}$ and there exists a function ∇f from $\mathbb{R}^3$ to $\mathbb{R}^3$ such that $\lim\limits_{\Delta \mathbf{x} \to 0} \dfrac{f(\mathbf{a} + \Delta\mathbf{x}) - f(\mathbf{a}) - \nabla f(\mathbf{a}) \cdot \Delta\mathbf{x}}{|\Delta\mathbf{x}|} = 0$.

Let $R[f](\mathbf{a}, \Delta\mathbf{x}) = f(\mathbf{a} + \Delta\mathbf{x}) - f(\mathbf{a}) - \nabla f(\mathbf{a}) \cdot \Delta\mathbf{x}$. If $\Delta\mathbf{x} = h\mathbf{u}$, then $\nabla f(\mathbf{a}) \cdot \Delta\mathbf{x} = h\,\nabla f(\mathbf{a}) \cdot \mathbf{u}$ and $\dfrac{f(\mathbf{a} + h\mathbf{u}) - f(\mathbf{a})}{h} = \nabla f(\mathbf{a}) \cdot \mathbf{u} + \dfrac{R[f](\mathbf{a}, h\mathbf{u})}{h}$. If $\mathbf{u}$ is a unit vector, then $|h\mathbf{u}| = |h|$ and the differentiability of f implies $\lim\limits_{h \to 0} \dfrac{R[f](\mathbf{a}, h\mathbf{u})}{|h|} = 0$ and $\lim\limits_{h \to 0} \dfrac{R[f](\mathbf{a}, h\mathbf{u})}{h} = 0$; thus $\lim\limits_{h \to 0} \dfrac{f(\mathbf{a} + h\mathbf{u}) - f(\mathbf{a})}{h} = \nabla f(\mathbf{a}) \cdot \mathbf{u}$. $\blacksquare$

Solutions 13.9   *The total differential and Approximation*   (pages 893-4)

1. If $f(x, y) = xy^3$, then $df = \nabla f \cdot \Delta\mathbf{x} = (y^3\mathbf{i} + 3xy^2\mathbf{j}) \cdot (\Delta x\mathbf{i} + \Delta y\mathbf{j}) = y^3\Delta x + 3xy^2\Delta y$.

3. If $f(x, y) = \sqrt{(x - y)/(x + y)} = (x - y)^{1/2}(x + y)^{-1/2}$, then
$df = \nabla f \cdot \Delta\mathbf{x} = (x - y)^{-1/2}(x + y)^{-3/2}(y\mathbf{i} - x\mathbf{j}) \cdot (\Delta x\mathbf{i} + \Delta y\mathbf{j}) = (x - y)^{-1/2}(x + y)^{-3/2}(y\,\Delta x - x\,\Delta y)$.

5. If $f(x, y) = \ln(2x + 3y)$, then $df = \nabla f \cdot \Delta\mathbf{x} = (2x + 3y)^{-1}(2\mathbf{i} + 3\mathbf{j}) \cdot (\Delta x\mathbf{i} + \Delta y\mathbf{j}) = (2x + 3y)^{-1}(2\,\Delta x + 3\,\Delta y)$.

7. If $f(x, y, z) = xy^2z^5$, then $df = \nabla f \cdot \Delta\mathbf{x} = (y^2z^5\mathbf{i} + 2xyz^5\mathbf{j} + 5xy^2z^4\mathbf{k}) \cdot (\Delta x\mathbf{i} + \Delta y\mathbf{j} + \Delta z\mathbf{k}) = y^2z^5\,\Delta x + 2xyz^5\,\Delta y + 5xy^2z^4\,\Delta z$.

9. If $f(x, y, z) = \cosh(xy - z)$, then $df = \nabla f \cdot \Delta\mathbf{x} = \sinh(xy - z)(y\mathbf{i} + x\mathbf{j} - \mathbf{k}) \cdot (\Delta x\mathbf{i} + \Delta y\mathbf{j} + \Delta z\mathbf{k}) = \sinh(xy - z)(y\Delta x + x\Delta y - \Delta z)$.

11. Let $f(x, y) = xy^2$.

a.
$$df = \nabla f \cdot \Delta\mathbf{x} = (y^2\mathbf{i} + xy\mathbf{j}) \cdot (\Delta x\mathbf{i} + \Delta y\mathbf{j}) = y^2\Delta x + xy\Delta y,$$
$$\Delta f = f(x + \Delta x, y + \Delta y) - f(x, y) = (x + \Delta x)(y + \Delta y)^2 - xy^2$$
$$= 2xy\Delta y + x\Delta y^2 + y^2\Delta x + 2y\Delta x\Delta y + \Delta x\Delta y^2,$$
$$\Delta f - df = x\Delta y^2 + 2y\Delta x\Delta y + \Delta x\Delta y^2.$$

b. $f(1, 2) = 4$, $f(1 - .01, 2 + .03) = 4.079691$, $\Delta f = 4.079691 - 4 = .079691$, $df = 4(-.01) + 4(.03) = .08$,
$\Delta f - df = 1(.03)^2 + 2(2)(-.01)(.03) + (-.01)(.03)^2 = .0009 - .0012 - .000009 = -.000309$.

13. a. $V(r, h) = \pi r^2 h$, thus $V(10, 20) = \pi \cdot 10^2 \cdot 20 = 2000\pi$ cm^3.

b. $dV = 2\pi rh\,\Delta r + \pi r^2\Delta h$. If $|\Delta r| \leq .03$ and $|\Delta h| \leq .07$, then $|dV| \leq 2\pi rh\,|\Delta r| + \pi r^2|\Delta h| \leq 2\pi(10)(20)(.03) + \pi(10^2)(.07) = (12 + 7)\pi = 19\pi$ cm^3.

15. a. $R = \left(R_1^{-1} + R_2^{-1} + R_3^{-1}\right)^{-1} = (1/6 + 1/8 + 1/12)^{-1} = (3/8)^{-1} = 8/3\ \Omega$.

b. $dR = R^2\left(R_1^{-2}\Delta R_1 + R_2^{-2}\Delta R_2 + R_3^{-2}\Delta R_3\right)$. If $|\Delta R_1| \leq .1$, $|\Delta R_2| \leq .03$, and $|\Delta R_3| \leq .15$, then $|dR| \leq R^2\left(R_1^{-2}|\Delta R_1| + R_2^{-2}|\Delta R_2| + R_3^{-2}|\Delta R_3|\right) \leq (8/3)^2\left((.1)/6^2 + (.03)/8^2 + (.15)/12^2\right) \approx .03049\ \Omega$.

Solutions 13.10 *Maxima and minima for a function of two variables* (pages 902-3)

1. If $f(x,y) = 7x^2 - 8xy + 3y^2 + 1$, then $f_x = 14x - 8y$ and $f_y = -8x + 6y$. The unique solution of the system $f_x = 0$, $f_y = 0$ is $x = 0$ and $y = 0$.
$f_{xx} = 14$, $f_{xy} = -8 = f_{yx}$, $f_{yy} = 6$, and $D = f_{xx}f_{yy} - f_{xy}f_{yx} = (14)(6) - (-8)^2 = 20$. Since $D > 0$ and $f_{xx} > 0$, we infer f has a local minimum at $(0,0)$.

REMARK: $7x^2 - 8xy + 3y^2 + 1 = 7\left(x^2 - (8/7)xy + (4/7)^2 y^2\right) + (3 - 16/7)\,y^2 + 1 = 7(x - 4y/7)^2 + (5/7)y^2 + 1$ has a global minimum of 1 where $x - 4y/7 = 0$ and $y = 0$.

3. If $f(x,y) = x^2 + 3y^2 + 4x - 6y + 3$, then $f_x = 2x + 4$ and $f_y = 6y - 6$. The unique solution of the system $f_x = 0$, $f_y = 0$ is $x = -2$ and $y = 1$. $f_{xx} = 2$, $f_{xy} = 0 = f_{yx}$, $f_{yy} = 6$, and $D = f_{xx}f_{yy} - f_{xy}f_{yx} = (2)(6) - (0)^2 = 12$. Since $D > 0$ and $f_{xx} > 0$, we infer f has a local minimum at $(-2,1)$.

REMARK: $x^2 + 3y^2 + 4x - 6y + 3 = (x+2)^2 + 3(y-1)^2 - 4$ has a global minimum of -4 at $(-2,1)$.

5. If $f(x,y) = x^2 + y^2 + 4x - 2y + 3$, then $f_x = 2x + 4$ and $f_y = 2y - 2$. The unique solution of the system $f_x = 0$, $f_y = 0$ is $x = -2$ and $y = 1$. $f_{xx} = 2$, $f_{xy} = 0 = f_{yx}$, $f_{yy} = 2$, and $D = f_{xx}f_{yy} - f_{xy}f_{yx} = (2)(2) - (0)^2 = 4$. Since $D > 0$ and $f_{xx} > 0$, we infer f has a local minimum at $(-2,1)$.

REMARK: $x^2 + y^2 + 4x - 2y + 3 = (x+2)^2 + (y-1)^2 - 2$ has a global minimum of -2 at $(-2,1)$.

7. If $f(x,y) = x^3 + 3xy^2 + 3y^2 - 15x + 2$, then $f_x = 3x^2 + 3y^2 - 15$ and $f_y = 6xy + 6y$. $0 = f_y = 6y(x+1)$ implies $y = 0$ or $x = -1$; $0 = f_x \iff x^2 + y^2 = 5$. The system $f_x = 0$, $f_y = 0$ has four solutions: $(\pm\sqrt{5}, 0)$ and $(-1, \pm 2)$. $f_{xx} = 6x$, $f_{xy} = 6y = f_{yx}$, $f_{yy} = 6x + 6$, and $D = f_{xx}f_{yy} - f_{xy}f_{yx} = (6x)(6x+6) - (6y)^2 = 36(x^2 + x - y^2)$. f has saddle points at $(-1, \pm 2)$ because $D = -144 < 0$ at each. $D(\pm\sqrt{5}, 0) = 180 \pm \sqrt{5} > 0$ and $f_{xx}(\pm\sqrt{5}, 0) = \pm\sqrt{5}$; f has a local minimum at $(\sqrt{5}, 0)$ and a local maximum at $(-\sqrt{5}, 0)$.

9. If $f(x,y) = y^{-1} - x^{-1} - 4x + y$, then $f_x = x^{-2} - 4$ and $f_y = -y^{-2} + 1$. $f_x = 0$ implies $x = \pm 1/2$ and $f_y = 0$ implies $y = \pm 1$; the system $f_x = 0$, $f_y = 0$ has four solutions: $(\pm 1/2, \pm 1)$.
$f_{xx} = -2x^{-3}$, $f_{xy} = 0 = f_{yx}$, $f_{yy} = 2y^{-3}$, and $D = f_{xx}f_{yy} - f_{xy}f_{yx} = (-2x^{-3})(2y^{-3}) - (0)^2 = -4x^{-3}y^{-3}$. f has saddle points at $(1/2, 1)$ and $(-1/2, -1)$ because $D = -32 < 0$ there; $D = 32 > 0$ at the other two points, f has a local maximum at $(1/2, -1)$ because $f_{xx}(1/2, -1) = -16 < 0$ while f has a local minimum at $(-1/2, 1)$ because $f_{xx}(-1/2, 1) = 16 > 0$.

11. If $f(x,y) = x^2 - xy + y^2 + 2x + 2y$, then $f_x = 2x - y + 2$ and $f_y = -x + 2y + 2$. The unique solution of the system $f_x = 0$, $f_y = 0$ is $x = -2$ and $y = -2$. $f_{xx} = 2$, $f_{xy} = -1 = f_{yx}$, $f_{yy} = 2$, and $D = f_{xx}f_{yy} - f_{xy}f_{yx} = (2)(2) - (-1)^2 = 3$. f has a local minimum at $(-2, -2)$ since $D > 0$ and $f_{xx} > 0$.

REMARK: $x^2 - xy + y^2 + 2x + 2y = (x - y/2 + 1)^2 + (3/4)(y+2)^2 - 4$.

13. If $f(x,y) = (4 - x - y)xy = 4xy - x^2 y - xy^2$, then $f_x = 4y - 2xy - y^2 = y(4 - 2x - y)$ and $f_y = 4x - x^2 - 2xy = x(4 - x - 2y)$. The system $f_x = 0$, $f_y = 0$ has 4 solutions: $(0,0)$, $(0,4)$, $(4,0)$, and $(4/3, 4/3)$.
$f_{xx} = -2y$, $f_{xy} = 4 - 2x - 2y = f_{yx}$, $f_{yy} = -2x$, and $D = f_{xx}f_{yy} - f_{xy}f_{yx} = (-2y)(-2x) - (4 - 2x - 2y)^2 = 4\left(xy - (2 - x - y)^2\right)$. f has saddle points at $(0,0)$, $(0,4)$, $(4,0)$ because $D = -16 < 0$ at each; f has a local maximum at $(4/3, 4/3)$ because $D(4/3, 4/3) = 16/3 > 0$ and $f_{xx}(4/3, 4/3) = -8/3 < 0$.

REMARK: You may find it only mildly harder to analyze the slightly more general function $(a - x - y)xy$.

15. If $f(x,y) = 2x^2 + y^2 + 2x^{-2}y^{-1}$, then $f_x = 4x - 4x^{-3}y^{-1}$ and $f_y = 2y - 2x^{-2}y^{-2}$. $f_x = 0 \iff y = x^{-4}$, then $0 = f_x = 2(y - x^{-2}y^{-2}) = 2\left(x^{-4} - x^6\right) = 2x^{-4}(1 + x)(1 - x)\left(1 + x + x^2 + x^3 + x^4\right)\left(1 - x + x^2 - x^3 + x^4\right)$ implies $x = \pm 1$ and $y = 1$. $f_{xx} = 4 + 12x^{-4}y^{-1}$, $f_{xy} = 4x^{-3}y^{-2} = f_{yx}$, $f_{yy} = 2 + 4x^{-2}y^{-3}$, and $D = f_{xx}f_{yy} - f_{xy}f_{yx} = (4 + 12x^{-4}y^{-1})(2 + 4x^{-2}y^{-3}) - (4x^{-3}y^{-2})^2$. f has local minimums at $(\pm 1, 1)$ since $D(\pm 1, 1) = 80 > 0$ and $f_{xx}(\pm 1, 1) = 16 > 0$.

17. If $f(x,y) = x^{25} - y^{25}$, then $f_x = 25x^{24}$ and $f_y = -25y^{24}$. The unique critical point is $(0,0)$. $f_{xx} = 600x^{23}$, $f_{xy} = 0 = f_{yx}$, $f_{yy} = -600y^{23}$, and $D = f_{xx}f_{yy} - f_{xy}f_{yx} = (600x^{23})(-600y^{23}) - (0)^2 = -600^2(xy)^{23}$. Theorem 3 is uninformative about the nature of the critical points of f because $D(0,0) = 0$. However, once we recognize that $f(x,0) = x^{25}$ changes sign on the line $y = 0$ which passes through the critical point, we can classify $(0,0)$ as a saddle point.

13.10 — Maxima and minima for a function of two variables

19.   Suppose $x+y+z = 50$ and let $P = xyz = xy(50-x-y) = 50xy-x^2y-xy^2$. Then $P_x = 50y-2xy-y^2 = y(50 - 2x - y)$ and $P_y = 50x - x^2 - 2xy = x(50 - x - 2y)$; the critical points are $(0,0)$, $(50,0)$, $(0,50)$, and $(50/3,50/3)$. $D = f_{xx}f_{yy} - f_{xy}f_{yx} = (-2y)(-2x) - (50 - 2x - 2y)^2 = 4\left(xy - (25 - x - y)^2\right)$. The points $(0,0)$, $(50,0)$, $(0,50)$ are saddle points because $D = -2500 < 0$ there; the point $(50/3,50/3)$ is a local maximum because $D = 2500/3 > 0$ and $f_{xx} = -100/3 < 0$ there. Since we have the additional restriction that x, y, z are positive, the domain of P is the triangle with vertices $(0,0)$, $(50,0)$, $(0,50)$ and $P = 0$ on the triangle's boundary; therefore $P(50/3,50/3) = (50/3)^3$ is the global maximum for P.

REMARK: A non-calculus analysis can use a geometric-arithmetic mean inequality.

$$0 \le (x - y)^2 \iff 4xy \le x^2 + 2xy + y^2 \iff xy \le \left(\frac{x+y}{2}\right)^2$$

with equality holding if and only if $x = y$. Therefore

$$(wx)(yz) \le \left(\left(\frac{w+x}{2}\right)\left(\frac{y+z}{2}\right)\right)^2 \le \left(\left(\frac{(w+x)/2 + (y+z)/2}{2}\right)^2\right)^2 = \left(\frac{w+x+y+z}{4}\right)^4$$

with equality holding iff $w = x = y = z$. For the special case of $w = (x+y+z)/3$, we have $w+x+y+z = (4/3)(x + y + z)$ and

$$\left(\frac{x+y+z}{3}\right)xyz \le \left(\frac{x+y+z}{3}\right)^4 \iff xyz \le \left(\frac{x+y+z}{3}\right)^3.$$

For this problem we have $x + y + z = 50$, so $w = 50/3$ and the maximum product is $(50/3)^3$.
   Solution 13.11.1 reanalyzes this problem.

21.   Suppose x, y, z are positive such that $x + y + z = 50$ and let $P = xy^2z^3 = (50 - y - z)y^2z^3$. Then $P_y = 100yz^3 - 3y^2z^3 - 2yz^4 = yz^3(100 - 3y - 2z)$ and $P_z = 150y^2z^2 - 3y^3z^2 - 4y^2z^3 = y^2z^2(150 - 3y - 4z)$. The system $P_y = 0 = P_z$ has solutions $y = 0$ and $z = 0$ which we can discard because we assumed y and z were positive, the remaining critical point is $(y,z) = (50/3,50/2)$. $D = P_{yy}P_{zz} - P_{yz}P_{zy} = \left(100z^3 - 6yz^3 - 2z^4\right)\left(300y^2z - 6y^3z - 12y^2z^2\right) - \left(300yz^2 - 9y^2z^2 - 8yz^3\right)^2$. P has a local maximum at $(50/3,50/2)$ because $D(50/3,50/2) > 10^{11} > 0$ and $P_{yy}(50/3,50/2) = -781250 < 0$. Therefore $x = 50 - 50/3 - 50/2 = 50/6$ and $(x,y,z) = (50/6,50/3,50/2)$.

REMARK: $(50/3,50/2)$ is the only critical point of P, $P > 0$, and $P \to 0$ as y or z approach the boundary of the domain, thus we conclude $P(50/3,50/2)$ is global maximum (without appealing to Theorem 3).
   Solution 13.11.3 reanalyzes this problem.

23.   The square of the distance between (x,y,x) and $(1,-1,2)$ is $F(x,y,z) = (x - 1)^2 + (y + 1)^2 + (z - 2)^2$; if the point (x,y,x) is on the plane $x + 2y - z = 4$, then we can analyze $f(x,y) = F(x,y,x + 2y - 4) = (x - 1)^2 + (y + 1)^2 + (x + 2y - 6)^2$. The solution to $f_x = 4x + 4y - 14 = 2(2x + 2y - 7) = 0$ and $f_y = 4x + 10y - 22 = 2(2x + 5y - 11) = 0$ is $(13/6,4/3)$. This unique critical point must correspond to the global minimum of f since f is large when any one of x, y, z are. The minimum value is $f(13/6,4/3) = 49/6$. (Our conclusion is supported by the facts $D = f_{xx}f_{yy} - f_{xy}f_{yx} = 40 - 16 = 24 > 0$ and $f_{xx} = 4 > 0$.)

25.   An open-top rectangular box of width $x > 0$, length $y > 0$, height $z > 0$ has fixed surface area $xy + 2xz + 2yz = \beta$ and volume $V = xyz = xy\dfrac{(\beta - xy)}{2(x + y)}$. Thus $V_x = \dfrac{\beta y - 2xy^2}{2(x + y)} - \dfrac{xy(\beta - xy)}{2(x + y)^2}$ and $V_y = \dfrac{\beta x - 2x^2y}{2(x + y)} - \dfrac{xy(\beta - xy)}{2(x + y)^2}$. Therefore $V_x = 0 = V_y$ implies $0 = V_x - V_y = \dfrac{\beta(y - x) - 2(xy^2 - x^2y)}{2(x + y)} = \dfrac{(\beta - 2xy)(y - x)}{2(x + y)}$ which then implies $xy = \beta/2$ or $x = y$. If $xy = \beta/2$, then $V_x = (-1/8)\beta^2(x+y)^{-2}$ which is never zero; thus $xy = \beta/2$ yields no critical points. If $x = y$, then $V_x = \left(\beta - 3x^2\right)/8 = V_y$ so $x = \sqrt{\beta/3} = y$ is the unique critical point for V; furthermore, $z = (\beta - xy)/2(x + y) = (\beta - \beta/3)/4\sqrt{\beta/3} = \sqrt{\beta/12}$. Since $V \to 0$ as either x or y approach 0, $V = (\sqrt{\beta/3})^2\sqrt{\beta/12} = \beta^{3/2}/6\sqrt{3}$ is the global maximum for V.

REMARK:
$$D = V_{xx}V_{yy} - V_{xy}V_{yx} = \left(\frac{-y^2(y^2 + \beta)}{(x + y)^3}\right)\left(\frac{-x^2(x^2 + \beta)}{(x + y)^3}\right) - \left(\frac{-xy(x^2 + 3xy + y^2 - \beta)}{(x + y)^3}\right)^2$$
$$= \frac{-x^2y^2(x^2 + 4xy + y^2 - 3\beta)}{(x + y)^4}.$$

At the critical point, $D = \beta/16 > 0$ and $V_{xx} = -\sqrt{\beta/12} < 0$.

27.   The profit is $P(x,y) = 40U(x,y) - 10x - 4y = 40\left(8xy + 32x + 40y - 4x^2 - 6y^2\right) - 10x - 4y = 320xy + 1270x + 1596y - 160x^2 - 240y^2$. Thus $P_x = 320y + 1270 - 320x$ and $P_y = 320x + 1596 - 480y$. The solution of $P_x = 0 = P_y$ is $x = 3501/160$ and $y = 1433/80$. Since $D = (-320)(-480) - 320^2 = 51200 > 0$ and $P_{xx} = -320 < 0$, that critical point yields a local maximum for P; that maximal profit is $4510203/160 \approx 28188.77$.

REMARK: If we constrain x and y to be integers, it is difficult to show $(18, 20)$ yields the maximum profit.

29. a.   If $N(a,d)$ sell at unit profit $150 = 250 - 100$ as a result of spending a on advertising and d on development, then the profit is $P(a,d) = 150N(a,d) - a - d = 150\left(300a/(a+3) + 160d/(d+5)\right) - a - d$.

    b. $P_a = 135000(a+3)^{-2} - 1$ and $P_d = 120000(d+5)^{-2} - 1$. The unique critical point is $a = -3 + \sqrt{135000}$ and $d = -5 + \sqrt{120000}$. Since $D = 64800000000(a+3)^{-3}(d+5)^{-3} > 0$ and $P_{aa} = -270000(a+3)^{-3} < 0$, this critical point corresponds to a local maximum for P.

31.   For the points $(-1,3)$, $(1,2)$, $(2,0)$, $(4,-2)$ we have the following: $n = 4$, $\sum x = -1 + 1 + 2 + 4 = 6$, $\sum x^2 = 1 + 1 + 4 + 16 = 22$, $\sum y = 3 + 2 + 0 - 2 = 3$, $\sum xy = -3 + 2 + 0 - 8 = -9$. Therefore

$$n\sum x^2 - \left(\sum x\right)^2 = 4 \cdot 22 - 6^2 = 88 - 36 = 52,$$

$$m = \frac{n(\sum xy) - (\sum x)(\sum y)}{n\sum x^2 - (\sum x)^2} = \frac{4 \cdot (-9) - 6 \cdot 3}{52} = \frac{-36 - 18}{52} = \frac{-54}{52} = \frac{-27}{26},$$

$$b = \frac{(\sum x^2)(\sum y) - (\sum x)(\sum xy)}{n\sum x^2 - (\sum x)^2} = \frac{22 \cdot 3 - 6 \cdot (-9)}{52} = \frac{66 + 54}{52} = \frac{120}{52} = \frac{30}{13}.$$

The regression line has equation $y = (30/13) - (27/26)x$.

31: $\hat{y} = \frac{30}{13} - \frac{27}{26}x$

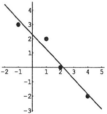

33. $f(x,y) = -(4x^2 - 4xy + y^2) + 5 = 5 - (2x - y)^2 \le 5$. f attains its global maximum value of 5 at every point on the line $y = 2x$. $D = (-8)(-2) - 4^2 = 0$ at every point.

35.   Since all spheres are similar, there is no loss of generality in assuming the sphere has equation $x^2 + y^2 + z^2 = 1$. We may also assume the box is rotated so its faces are parallel to the coordinate planes, i.e., the vertices are at points $(\pm x, \pm y, \pm z)$. The volume is maximized if and only if a positive multiple of its square is maximized, so we will analyze $F = (V/8)^2 = \left((2x)(2y)(2z)/8\right)^2 = (xyz)^2 = x^2y^2\left(1 - x^2 - y^2\right) = x^2y^2 - x^4y^2 - x^2y^4$. $F_x = 2xy^2 - 4x^3y^2 - 2xy^4 = 2xy^2(1 - 2x^2 - y^2)$ and $F_y = 2x^2y - 2x^4y - 4x^2y^3 = 2x^2y(1 - x^2 - 2y^2)$. In searching for maxima, we can ignore critical points with $x = 0$ or $y = 0$ since those correspond to a box with zero volume. Therefore $F_x = 0 = F_y \implies 1 - 2x^2 - y^2 = 1 - x^2 - 2y^2 \iff y^2 = x^2$. Thus $0 = F_x = 1 - x^2 - 2y^2 = 1 - 3x^2$ implies $|x| = \sqrt{1/3} = |y|$ and $|z| = \sqrt{1 - 1/3 - 1/3} = \sqrt{1/3}$. This critical point for F corresponds to a cubical box. Furthermore, at this critical point, $F_{xx} = F_{yy} = -8/9 < 0$, $F_{xy} = -4/9$, and $D = (-8/9)^2 - (-4/9)^2 = 48/81 > 0$ imply F has a local maximum. Since $F \to 0$ as either x or y approach 0, this turns out to be the global maximum for F (and V). ∎

Solutions 13.11   *Constrained maxima & minima — Lagrange multipliers*   (pages 910-2)

1.   Let $f = xyz$ and $g = x + y + z - 50$, then $\nabla f = yz\mathbf{i} + xz\mathbf{j} + xy\mathbf{k}$ and $\nabla g = \mathbf{i} + \mathbf{j} + \mathbf{k}$. Thus $\nabla f = \lambda \nabla g$ if and only if $\lambda = yz = xz = xy$; this implies $\lambda = (yz)(xz)/(xy) = z^2$ and $z = |z| = \sqrt{\lambda}$. Similarly, $x = y = \sqrt{\lambda} = z$. The constraint $g = 0$ implies $x = y = z = 50/3$ so the maximum value of f is $(50/3)^3$.

3.   Let $f = xy^2z^3$ and $g = x + y + z - 50$, then $\nabla f = y^2z^3\mathbf{i} + 2xyz^3\mathbf{j} + 3xy^2z^2\mathbf{k}$ and $\nabla g = \mathbf{i} + \mathbf{j} + \mathbf{k}$. Therefore $\nabla f = \lambda \nabla g$ if and only if $\lambda = y^2z^3 = 2xyz^3 = 3xy^2z^2$; this implies

$$1 = \frac{\lambda}{\lambda} = \frac{f_x}{f_y} = \frac{y^2z^3}{2xyz^3} = \frac{y}{2x} \qquad \text{and} \qquad 1 = \frac{\lambda}{\lambda} = \frac{f_x}{f_z} = \frac{y^2z^3}{3xy^2z^2} = \frac{z}{3x},$$

hence $x = \dfrac{y}{2} = \dfrac{z}{3}$. The constraint $g = 0$ implies $50 = x + (2x) + (3x) = 6x$ which then implies $x = 50/6$, $y = 50/3$, $z = 50/2$ so the maximum value of f is $50^6/(6 \cdot 9 \cdot 8)$.

REMARK: Let $f = x^a y^b z^c$; f is maximized if and only if $F = \ln f = a \ln x + b \ln y + c \ln z$ is maximized. $\nabla F = (a/x)\mathbf{i} + (b/y)\mathbf{j} + (c/z)\mathbf{k} = \lambda \nabla g$ if and only if $\lambda = a/x = b/y = c/z$. The constraint $g = 0$ implies $50 = a/\lambda + b/\lambda + c/\lambda = (a + b + c)/\lambda$, $\lambda = (a + b + c)/50$, and $x = a/\lambda = 50a/(a + b + c)$, $y = b/\lambda = 50b/(a + b + c)$, $z = c/\lambda = 50c/(a + b + c)$.

5. Let $f = (x-1)^2 + (y-2)^2$ and $g = 2x + 3y - 5$, then $\nabla f = 2(x-1)\mathbf{i} + 2(y-2)\mathbf{j}$ and $\nabla g = 2\mathbf{i} + 3\mathbf{j}$. Thus $\nabla f = \lambda \nabla g$ iff $x = 1 + \lambda$ and $y = 2 + 3\lambda/2$; the constraint $0 = g = 2(1 + \lambda) + 3(2 + 3\lambda/2) - 5 = 3 + 13\lambda/2$ implies $\lambda = -6/13$ which then yields $x = 1 - 6/13 = 7/13$ and $y = 2 + (3/2)(-6/13) = 17/13$ with minimal distance $\sqrt{f} = \sqrt{(-6/13)^2 + (-9/13)^2} = 3/\sqrt{13}$ between point $(1,2)$ and line $2x + 3y - 5 = 0$.

7. Let $f = (x - 1)^2 + (y + 1)^2 + (z - 2)^2$ and $g = x + y - z - 3$, then $\nabla f = 2(x - 1)\mathbf{i} + 2(y + 1)\mathbf{j} + 2(z - 2)\mathbf{k}$ and $\nabla g = \mathbf{i} + \mathbf{j} - \mathbf{k}$. Thus $\nabla f = \lambda \nabla g$ iff $x = 1 + \lambda/2$, $y = -1 + \lambda/2$, and $z = 2 - \lambda/2$; the constraint $0 = g = (1 + \lambda/2) + (-1 + \lambda/2) - (2 - \lambda/2) - 3 = -5 + (3/2)\lambda$ implies $\lambda = 10/3$ which then yields $x = 1 + (10/3)/2 = 8/3$, $y = -1 + (10/3)/2 = 2/3$, and $z = 2 - (10/3)/2 = 1/3$ with minimal distance $\sqrt{f} = \sqrt{(5/3)^2 + (5/3)^2 + (-5/3)^2} = 5/\sqrt{3}$ between point $(1, -1, 2)$ and plane $x + y - z - 3 = 0$.

9. Let $f = x^2 + y^2$ and $g = ax + by - d$, then $\nabla f = 2x\mathbf{i} + 2y\mathbf{j}$ and $\nabla g = a\mathbf{i} + b\mathbf{j}$. Thus $\nabla f = \lambda \nabla g$ iff $x = a\lambda/2$ and $y = b\lambda/2$; the constraint $0 = g = a(a\lambda/2) + b(b\lambda/2) - d = (a^2 + b^2)\lambda/2 - d$ implies $\lambda = 2d/(a^2 + b^2)$ which then yields $x = ad/(a^2 + b^2)$ and $y = bd/(a^2 + b^2)$ with minimal distance $\sqrt{f} = |d|/\sqrt{a^2 + b^2}$ between point $(0,0)$ and line $ax + by - d = 0$.

11. Let $f = (x - x_0)^2 + (y - y_0)^2$ and $g = ax + by - d$, then $\nabla f = 2(x - x_0)\mathbf{i} + 2(y - y_0)\mathbf{j}$ and $\nabla g = a\mathbf{i} + b\mathbf{j}$. Thus $\nabla f = \lambda \nabla g$ iff $x = x_0 + a\lambda/2$ and $y = y_0 + b\lambda/2$; the constraint $0 = g = a(x_0 + a\lambda/2) + b(y_0 + b\lambda/2) - d = (ax_0 + by_0 - d) + (a^2 + b^2)\lambda/2$ implies $\lambda = -2(ax_0 + by_0 - d)/(a^2 + b^2)$ which then yields minimal distance $\sqrt{f} = \sqrt{(a\lambda/2)^2 + (b\lambda/2)^2} = \sqrt{a^2 + b^2}\,|\lambda/2| = |ax_0 + by_0 - d|/\sqrt{a^2 + b^2}$ between point (x_0, y_0) and line $ax + by - d = 0$.

13. Let $f = x^2 + y^2$ and $g = x^3 + y^3 - 6xy$, then $\nabla f = 2x\mathbf{i} + 2y\mathbf{j}$ and $\nabla g = (3x^2 - 6y)\mathbf{i} + (3y^2 - 6x)\mathbf{j}$. Thus

$$\nabla f = \lambda \nabla g \implies \frac{x}{y} = \frac{f_x}{f_y} = \frac{g_x}{g_y} = \frac{3x^2 - 6y}{3y^2 - 6x} \implies 0 = x(3y^2 - 6x) - y(3x^2 - 6y) = 3(y - x)(xy + 2(y + x)).$$

If $y = x$, then $0 = g = 2x^3 - 6x^2 = 2x^2(x - 3)$. If $0 = x = y$, then $f = 0$, clearly the global minimum; if $3 = x = y$, then $f = 18$, a local maximum. If $xy = -2(y + x)$, then $0 = g = x^3 + y^3 + 12(y + x) = (y + x)(x^2 - xy + y^2 + 12) = (y + x)(x^2 + 2(y + x) + y^2 + 12) = (y + x)((x + 1)^2 + (y + 1)^2 + 10)$; if $y = -x$, then $0 = g = 6x^2$ implies $0 = x = y$ which has already been considered.

15. Let $f = x^2 + y^2 + z^2$ and $g = x^2 - z^2 - 1$, then $\nabla f = 2x\mathbf{i} + 2y\mathbf{j} + 2z\mathbf{k}$ and $\nabla g = 2x\mathbf{i} - 2z\mathbf{k}$. We will examine $\nabla f = \lambda \nabla g$ cautiously. First consider the $\mathbf{j}$ component: $2y = \lambda \cdot 0 = 0$, hence $y = 0$. Next consider the $\mathbf{i}$ component: $2x = \lambda \cdot 2x$ implies $2x(1 - \lambda) = 0$ which implies $x = 0$ or $\lambda = 1$; since $x^2 - 1 = z^2 \geq 0$, the alternative $x = 0$ is impossible, hence $\lambda = 1$. Now considering the $\mathbf{k}$ component, we see that $2z = \lambda \cdot (-2z) = -2z$ (remember that $\lambda = 1$) implies $z = 0$. Therefore $x^2 = z^2 + 1 = 0^2 + 1 = 1$, so $x = \pm 1$. $f(\pm 1, 0, 0) = 1$ is the minimum of f on $g = 0$; there is no maximum because f is unbounded on the surface $g = 0$ (where $f = 2x^2 + y^2 - 1$ and y is unconstrained).

17. Let $f = x + y + z$ and $g = x^2 + y^2 + z^2 - 1$, then $\nabla f = \mathbf{i} + \mathbf{j} + \mathbf{k}$ and $\nabla g = 2x\mathbf{i} + 2y\mathbf{j} + 2z\mathbf{k}$. Thus $\nabla f = \lambda \nabla g$ implies $1/(2\lambda) = x = y = z$. The constraint $0 = g = 3/(4\lambda^2) - 1$ then implies $\lambda = \pm\sqrt{3}/2$. The maximum value of f is $f\left(1/\sqrt{3}, 1/\sqrt{3}, 1/\sqrt{3}\right) = 3/\sqrt{3} = \sqrt{3}$; the minimum value is $f\left(-1/\sqrt{3}, -1/\sqrt{3}, -1/\sqrt{3}\right) = -3/\sqrt{3} = -\sqrt{3}$.

REMARK: The Cauchy-Schwarz inequality (Problem 11.4.55) implies

$$|x + y + z| = |1 \cdot x + 1 \cdot y + 1 \cdot z| \leq \sqrt{1^2 + 1^2 + 1^2}\,\sqrt{x^2 + y^2 + z^2} = \sqrt{3}\,\sqrt{x^2 + y^2 + z^2}$$

with equality holding iff $x = y = z$. The constraint for this problem implies $\sqrt{x^2 + y^2 + z^2} = 1$ so we have $|f| \leq \sqrt{3}$ with equality iff $x = y = z = \pm 1/\sqrt{3}$.

19. Let $f = xyz$ and $g = (x/a)^2 + (y/b)^2 + (z/c)^2 - 1$, also suppose $a > 0$, $b > 0$, $c > 0$. Then $\nabla f = yz\mathbf{i} + xz\mathbf{j} + xy\mathbf{k}$ and $\nabla g = (2x/a^2)\mathbf{i} + (2y/b^2)\mathbf{j} + (2z/c^2)\mathbf{k}$. Thus $\nabla f = \lambda \nabla g$ implies $xyz/(2\lambda) = (x/a)^2 = (y/b)^2 = (z/c)^2 = t$. The constraint $0 = g = 3t - 1$ then implies $t = 1/3$ and $x = \pm a/\sqrt{3}$, $y = \pm b/\sqrt{3}$, $z = \pm c/\sqrt{3} = \pm\sqrt{3}$. The maximum value of f is $abc/3\sqrt{3}$ at $(+, +, +)$, $(-, -, +)$, $(-, +, -)$, $(+, -, -)$; the minimum value is $-abc/3\sqrt{3}$ at $(-, -, -)$, $(+, +, -)$, $(+, -, +)$, $(-, +, +)$.

REMARK: The arithmetic-geometric mean inequality (see Remark for Solution 13.10.19) implies
$$f^2/(abc)^2 = (x/a)^2(y/b)^2(z/c)^2 \le \left(((x/a)^2 + (y/b)^2 + (z/c)^2)/3\right)^3 = (1/3)^3 = 1/27$$
which implies $|f| \le \sqrt{(abc)^2/27} = abc/3\sqrt{3}$.

21.   If $f = x^3+y^3+z^3$, $g = x+y+z-2$, and $h = x+y-z-3$, then $\nabla f = 3x^2\mathbf{i}+3y^2\mathbf{j}+3z^2\mathbf{k}$, $\nabla g = \mathbf{i}+\mathbf{j}+\mathbf{k}$, and $\nabla h = \mathbf{i}+\mathbf{j}-\mathbf{k}$. Thus $\nabla f = \lambda\nabla g + \mu\nabla h$ implies $x^2 = (\lambda+\mu)/2 = y^2$. If $x = -y$, then $0 = g = z-2$ and $0 = h = -z-3$ which implies $z = 2$ and $z = -3$; impossible! If $x = y$, then $0 = g = 2x+z-2$ and $0 = h = 2x-z-3$ which have solution $z = -1/2$ and $x = 5/4$. The critical point is $(5/4, 5/4, -1/2)$ and the minimum value of f is $2(5/4)^3 + (-1/2)^3 = 121/32$.

REMARK: $0 = g - h = 2z + 1$ implies $z = -1/2$, thus $0 = g = x + y - 5/2$ implies $y = 5/2 - x$; $x^3 + (5/2 - x)^3 + (-1/2)^3 = (15/2)x^2 - (75/4)x + 31/2 = (15/2)(x - 5/4)^2 + 121/32$.

23.   Let $f = x^2 + y^2 + z^2$ and $g = (x/a)^2 + (y/b)^2 + (z/c)^2 - 1$. Then $\nabla f = 2x\mathbf{i} + 2y\mathbf{j} + 2z\mathbf{k}$, and $\nabla g = 2(x/a^2)\mathbf{i}+2(y/b^2)\mathbf{j}+2(z/c^2)\mathbf{k}$. Hence $\nabla f = \lambda\nabla g$ implies $0 = 2x(1-\lambda/a^2) = 2y(1-\lambda/b^2) = 2z(1-\lambda/c^2)$. We can exclude $(0,0,0)$ since it is not on $g = 0$. If $0 = y = z$, then $0 = g = (x/a)^2$ implies $|x| = |a|$; if $0 = x = z$, then $0 = g = (y/b)^2$ implies $|y| = |b|$; if $0 = x = y$, then $0 = g = (z/c)^2$ implies $|z| = |c|$. There are other critical points iff at least two of $|a|$, $|b|$, $|c|$ are equal, e.g., if $x \ne 0$, $y \ne 0$, $z = 0$, then $0 = 1 - \lambda/a^2 = 1 - \lambda/b^2$ implies $a^2 = \lambda = b^2$. In such a case, the ellipsoid has a circular cross-section all of whose points are equidistant from the origin. If $0 < |a| \le |b| \le |c|$, then $|a|$ is the minimum distance and $|c|$ is the maximum distance.

25.   Let $f = x_1 + x_2 + \cdots + x_n$ and $g = x_1^2 + x_2^2 + \cdots + x_n^2 - 1$. The conditions $1 = \dfrac{\partial f}{\partial x_i} = \lambda\dfrac{\partial g}{\partial x_i} = 2\lambda x_i$ imply all x_i equal $1/(2\lambda)$, the constraint $0 = g = nx_i^2 - 1$ implies $x_i = \pm 1/\sqrt{n}$. The maximum is $n(1/\sqrt{n}) = \sqrt{n}$ and the minimum is $n(-1/\sqrt{n}) = -\sqrt{n}$.

REMARK: By the Cauchy-Schwarz inequality, if $g = 0$, then
$$|f|^2 = |x_1 + x_2 + \cdots + x_n|^2 \le (1 + 1 + \cdots + 1)\left(x_1^2 + x_2^2 + \cdots + x_4^2\right) = (n)(1) = n$$
with equality if and only if all x_i are equal.

27.   If the faces of the parallelepiped are parallel to the coordinate axes, then the vertices are $(\pm x, \pm y, \pm z)$ and we may assume x, y, z are positive; the volume is $V = (2x)(2y)(2z) = 8xyz$. The constraint is $0 = g = x^2 + 4y^2 + 9z^2 - 9$. $\nabla V = 8(yz\mathbf{i} + xz\mathbf{j} + xy\mathbf{k})$ and $\nabla g = 2x\mathbf{i} + 8y\mathbf{j} + 18z\mathbf{k}$. Thus $\nabla V = \lambda\nabla g$ implies $8xyz/(2\lambda) = x^2 = 4y^2 = 9z^2 = t$. The constraint $0 = g = 3t - 9$ implies $t = 3$, $x = \sqrt{3}$, $y = \sqrt{3}/2$, $z = 1/\sqrt{3}$; the maximal volume is $8(\sqrt{3})(\sqrt{3}/2)(1/\sqrt{3}) = 4\sqrt{3}$.

29.   A rectangular box of length x, width y, and height z has volume $V = xyz$. The cost constraint is $0 = g = (3)(xy) + (1)(2xz) + (1)(2yz) - 36$. Let $h = f - \lambda g$, then $\nabla f = \lambda\nabla g \iff \nabla h = \nabla f - \lambda\nabla g = 0$.
$$0 = h_x - h_y = (yz - \lambda(3y + 2z)) - (xz - \lambda(3x + 2z)) = (y - x)(z - 3\lambda),$$
$$0 = 2h_x - 3h_z = 2(yz - \lambda(3y + 2z)) - 3(xy - \lambda(2x + 2y)) = (2z - 3x)(y - 2\lambda),$$
$$0 = 2h_y - 3h_z = 2(xz - \lambda(3x + 2z)) - 3(xy - \lambda(2x + 2y)) = (2z - 3y)(x - 2\lambda).$$

Therefore $\nabla f = \lambda\nabla g$ implies $x = 2\lambda = y$ and $z = 3\lambda$. Hence $0 = g = 36(\lambda^2 - 1)$ which implies $\lambda = 1$ (since physical dimensions x, y, z are positive), $x = 2 = y$, $z = 3$, and the maximal volume is $2^2 \cdot 3 = 12 \text{ m}^3$.

REMARK: The three displayed equations are also solved if $y = x$ and $z = (3/2)x$ which could be stated as $y = 2\mu = x$ and $z = 3\mu$. The key point of this minor addendum is that paramter μ need not equal λ.

31.   Maximize $F = \alpha xy$ subject to the constraint $0 = g = 5000x + 2500y - 250000$. Hence $\nabla F = \alpha y\mathbf{i} + \alpha x\mathbf{j} = \lambda\nabla g = \lambda(5000\mathbf{i} + 2500\mathbf{j}) \implies x/y = (\alpha x)/(\alpha y) = (2500\lambda)/(5000\lambda) = 25/50 = 1/2$.

33. a.   Let $F = 250L^{.7}K^{.3}$ and $g = 200L + 350K - 25000$. Then $\nabla F = 175L^{-.3}K^{.3}\mathbf{i} + 75L^{.7}K^{-.7}\mathbf{j} = \lambda\nabla g = \lambda(200\mathbf{i}+350\mathbf{j}) \implies (175/200)L^{-.3}K^{.3} = \lambda = (75/350)75L^{.7}K^{-.7} \implies (175/200)K = (75/350)L \implies K = (200/175)(75/350)L = (12/49)L$. Therefore $0 = g = 200L + 350(12/49)L - 25000 = (2000/7)L - 25000$ implies $L = 175/2$ and $K = (12/49)(175/2) = 150/7$.

   b.   The maximal output is $250(175/2)^{.7}(150/7)^{.3}$.

   c. $\dfrac{\text{marginal productivity of labor}}{\text{marginal productivity of capital}} = \dfrac{\partial F/\partial L}{\partial F/\partial K} = \dfrac{175(175/2)^{-.3}(150/7)^{.3}}{75(175/2)^{.7}(150/7)^{-.7}} = \dfrac{175(150/7)}{75(175/2)} = \dfrac{4}{7} = \dfrac{200}{350}$.

13.11 — Constrained maxima & minima — Lagrange multipliers

REMARK: F is maximized iff $f = \ln F$ is maximized. $\nabla f = (7/10)L^{-1}\mathbf{i} + (3/10)K^{-1}\mathbf{j} = \lambda\nabla g$ implies $L = (7/2000)(1/\lambda)$ and $K = (3/3500)(1/\lambda)$. The constraint $0 = g = 1/\lambda - 25000$ implies $1/\lambda = 25000$, $L = (7/10)(1/200)25000 = 175/2$, and $K = (3/10)(1/350)25000 = 150/7$.

35. a.   Minimize $C = 30l + 16h$ subject to $0 = g = N - 80 = 4l^2 + 2.5h^2 - 80$. Then $30\mathbf{i} + 16\mathbf{j} = \nabla C = \lambda\nabla g = \lambda(8l\mathbf{i} + 5h\mathbf{j})$ implies $l = (15/4)(1/\lambda)$ and $h = (16/5)(1/\lambda)$. The constraint $0 = g = (1637/20)(1/\lambda)^2 - 80$ implies $1/\lambda = 40/\sqrt{1637}$, $l = 150/\sqrt{1637}$, $h = 128/\sqrt{1637}$. This computation has only located a critical point in the interior of the domain of f; endpoints are examined in part (b) and it turns out that the minimal cost can is produced using only horsemeat.

     b.   The cost at the critical point found in part (a) is $6548/\sqrt{1637} \approx 161.84$; this turns out to be the maximum of C. If $h = 0$, then $l = \sqrt{80/4} = \sqrt{20}$ and the cost is $30\sqrt{20} \approx 134.16$. If $l = 0$, then $h = \sqrt{80/2.5} = \sqrt{32}$ and the cost is $16\sqrt{32} \approx 90.51$. The least of these costs is for a can using only horsemeat.

37.   Maximize $S = 500A - 20A^2 + 300B - 10B^2$ subject to the constraint $0 = g = A + B - 20$. $\nabla S = (500 - 40A)\mathbf{i} + (300 - 20B)\mathbf{j} = \lambda\nabla g = \lambda(\mathbf{i} + \mathbf{j})$ implies $A = 25/2 - \lambda/40$ and $B = 15 - \lambda/20$. The constraint $0 = g = 15/2 - 3\lambda/40$ implies $\lambda = 100$, $A = 10$, and $B = 10$; $S(10, 10) = 5000$ is maximal.

REMARK: $S(A, 20 - A) = 500A - 20A^2 + 300(20 - A) - 10(20 - A)^2 = 5000 - 30(A - 10)^2$.

39.   Let x be the width of the rectangle and y be the length; maximize $A = xy$ subject to the constraint $0 = g = 2(x + y) - P$ where P is a constant. Then $\nabla A = y\mathbf{i} + x\mathbf{j} = \lambda\nabla g = \lambda(2\mathbf{i} + 2\mathbf{j})$ implies $y = 2\lambda$ and $x = 2\lambda$; since $y = x$, the rectangle is a square. x and y are in the interval $[0, P/2]$, $A \to 0$ as either x or y approach 0, hence this critical point corresponds to the maximal A. ∎

41.   Suppose $p + q + r = 1$ and restrict x, y, z to be positive. We will show the result holds on each level surface of $px + qy + rz$; that will show it holds in general.

     Let $f = \ln(x^p y^q z^r) = p\ln x + q\ln y + r\ln z$ and $g = px + qy + rz - C$. Then $\nabla f = px^{-1}\mathbf{i} + qy^{-1}\mathbf{j} + rz^{-1}\mathbf{k} = \lambda\nabla g = \lambda(p\mathbf{i} + q\mathbf{j} + r\mathbf{k})$ implies $x = y = z = 1/\lambda$; the level surface constraint $0 = g = (p + q + r)/\lambda - C$ implies $1/\lambda = C/(p + q + r)$. The condition $p + q + r = 1$ now implies $x = y = z = 1/\lambda = C$. The domain $(\{(x, y, z) : g(x, y, z) = 0, 0 \le x, 0 \le y, 0 \le z\})$ is bounded and $x^p y^q z^r = 0$ on the boundary but positive in the interior, therefore this critical point corresponds to maximal $x^p y^q z^r$. Therefore $x^p y^q z^r \le C^{p+q+r} = C^1 = C = px + qy + rz$. ∎

Solutions 13.12   *Newton's method for functions of two variables*   (pages 916-7)

1.   Let $f = x^2 - y$ and $g = y^2 - x$.

     a. $f(0, 0) = 0^2 - 0 = 0$, $g(0, 0) = 0^2 - 0 = 0$; $f(1, 1) = 1^2 - 1 = 0$, $g(1, 1) = 1^2 - 1 = 0$. These are the only solutions because $y = x^2 \implies x = y^2 = (x^2)^2 = x^4 \implies 0 = x^4 - x = x(x - 1)(x^2 + x + 1)$.

     b. $f_x = 2x$, $f_y = -1$, $g_x = -1$, $g_y = 2y$, $D = (2x)(2y) - (-1)(-1) = 4xy - 1$, $fg_y - f_y g = (x^2 - y)(2y) - (-1)(y^2 - x) = 2x^2y - y^2 - x$, $-fg_x + f_x g = -(x^2 - y)(-1) + (2x)(y^2 - x) = 2xy^2 - x^2 - y$. Therefore (x_{n+1}, y_{n+1}) is computed from (x_n, y_n) by the functions $x_* = x - (2x^2y - y^2 - x)/(4xy - 1)$ and $y_* = y - (2xy^2 - x^2 - y)/(4xy - 1)$.

n	x_n	y_n	$x_n - x_{n-1}$	$y_n - y_{n-1}$	$f(x_n)$	$g(x_n)$
0	.25	−.25			.3125	−.1875
1	−.025	−.075	−.275	.175	.075625	.030625
2	−.005573048	−.000346348	.019426952	.074653652	.00037741	.00557317

If we write results in fractional form, then we have $(x_0, y_0) = (1/4, -1/4)$, $(x_1, y_1) = (-1/40, -3/40)$, $(x_2, y_2) = (-177/31760, -11/31760)$.

     c.

n	x_n	y_n	$x_n - x_{n-1}$	$y_n - y_{n-1}$	$f(x_n)$	$g(x_n)$
0	.9	1.25			−.44	.6625
1	1.025	1.035	.125	−.215	.015625	.046225
2	1.0007765	1.0009669	−.0242235	−.0340331	.0005867	.0011582

1: $y = x^2$, $x = y^2$

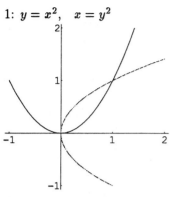

3: $2xy = 3$, $y = x^2 - 2$

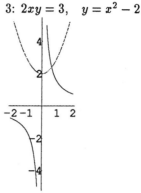

3. Let $f = 2xy - 3$ and $g = x^2 - y - 2$. $f_x = 2y$, $f_y = 2x$, $g_x = 2x$, $g_y = -1$, $D = (2y)(-1) - (2x)(2x) = -2y - 4x^2$, $fg_y - f_y g = (2xy - 3)(-1) - (2x)(x^2 - y - 2) = 3 + 4x - 2x^3$, $-fg_x + f_x g = -(2xy - 3)(2x) + (2y)(x^2 - y - 2) = 6x - 4y - 2y^2 - 2x^2 y$. Thus (x_{n+1}, y_{n+1}) is computed from (x_n, y_n) by the functions $x_* = x - (3 + 4x - 2x^3)/(-2y - 4x^2)$ and $y_* = y - (6x - 4y - 2y^2 - 2x^2 y)/(-2y - 4x^2)$.

n	x_n	y_n	$x_n - x_{n-1}$	$y_n - y_{n-1}$	$f(x_n)$	$g(x_n)$
0	1.5	.9			−.3	−.65
1	1.708333333	.875	.208333333	−.025	−.010416667	.0434027770
2	1.698062166	.8833096224	−.010271167	.0083096224	−.000170699	.0001054976

Continuing, we find approximate solution $(1.6980480624, .8833672222)$.

5. Let $f = 3x^2 - 2y^2 - 1$ and $g = x^2 - 2x + y^2 + 2y - 8$. $f_x = 6x$, $f_y = -4y$, $g_x = 2x - 2$, $g_y = 2y + 2$, $D = (6x)(2y + 2) - (-4y)(2x - 2) = 12x - 8y + 20xy$, $fg_y - f_y g = (3x^2 - 2y^2 - 1)(2y + 2) - (-4y)(x^2 - 2x + y^2 + 2y - 8) = -2 - 34y + 6x^2 + 4y^2 - 8xy + 10x^2 y$, $-fg_x + f_x g = -(3x^2 - 2y^2 - 1)(2x - 2) + (6x)(x^2 - 2x + y^2 + 2y - 8) = -2 - 46x - 6x^2 - 4y^2 + 12xy + 10xy^2$. Thus (x_{n+1}, y_{n+1}) is computed from (x_n, y_n) by the functions $x_* = x - (-2 - 34y + 6x^2 + 4y^2 - 8xy + 10x^2 y)/(12x - 8y + 20xy)$ and $y_* = y - (-2 - 46x - 6x^2 - 4y^2 + 12xy + 10xy^2)/(12x - 8y + 20xy)$.

i.

n	x_n	y_n	$x_n - x_{n-1}$	$y_n - y_{n-1}$	$f(x_n)$	$g(x_n)$
0	−1	1			0	−2
1	−1.2	1.3	−.2	.3	−.06	.13
2	−1.192857143	1.278571429	.007142857	−.021428571	−.000765306	.000510207

Continuing, we find approximate solution $(-1.1928730994, 1.2784441117)$.

ii.

n	x_n	y_n	$x_n - x_{n-1}$	$y_n - y_{n-1}$	$f(x_n)$	$g(x_n)$
0	3	−3.4			2.88	−.24
1	2.925	−3.5125	−.075	−.1125	−.00843750	.018281250
2	2.923492752	−3.510016747	−.001507248	.002483253	−.00000551	.000008433

Continuing, we find approximate solution $(2.9234921131, -3.5100155559)$.

The other solutions are approximately $(-1.9121604714, -2.2326075567)$ and $(1.7815414577, 2.0641790010)$.

5: $3x^2 - 2y^2 = 1$, $(x-1)^2 + (y+1)^2 = 10$

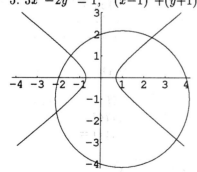

7: $y = 2x^3 - 12x - 1$, $x = 3y^2 - 6y - 3$

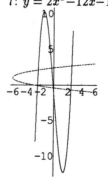

7. Let $f = 2x^3 - 12x - y - 1$ and $g = 3y^2 - 6y - x - 3$. $f_x = 6x^2 - 12$, $f_y = -1$, $g_x = -1$, $g_y = 6y - 6$, $D = (6x^2 - 12)(6y - 6) - (-1)(-1) = 71 - 72y - 36x^2 + 36x^2 y$, $P = fg_y - f_y g = (2x^3 - 12x - y - 1)(6y - 6) - (-1)(3y^2 - 6y - x - 3) = 3 + 71x - 6y - 3y^2 - 72xy - 12x^3 + 12x^3 y$, $Q = -fg_x + f_x g =$

13.12 — Newton's method for functions of two variables

$-(2x^3 - 12x - y - 1)(-1) + (6x^2 - 12)(3y^2 - 6y - x - 3) = 35 + 71y - 18x^2 - 36y^2 - 4x^3 - 36x^2y + 18x^2y^2$.
Thus (x_{n+1}, y_{n+1}) is computed from (x_n, y_n) by the functions $x_* = x - P/D$ and $y_* = y - Q/D$.

i.

n	x_n	y_n	$x_n - x_{n-1}$	$y_n - y_{n-1}$	$f(x_n)$	$g(x_n)$
0	2.5	2.5			-2.25	-1.75
1	2.596280088	2.705142232	.096280088	.205142232	.140832848	.126250012
2	2.590876452	2.692273937	$-.005403636$	$-.012868295$	.000454543	.000496788

Continuing, we find approximate solution $(2.5908585839, 2.6922232500)$.

ii.

n	x_n	y_n	$x_n - x_{n-1}$	$y_n - y_{n-1}$	$f(x_n)$	$g(x_n)$
0	2.5	-1			1.25	3.5
1	2.462540717	$-.7052117264$	$-.037459283$	.2947882736	.0209428664	.260700378
2	2.462726070	$-.6797490992$	.000185353	.0254626272	.0000004992	.001945039

Continuing, we find approximate solution $(2.4627339306, -.6795568791)$.

iii.

n	x_n	y_n	$x_n - x_{n-1}$	$y_n - y_{n-1}$	$f(x_n)$	$g(x_n)$
0	0	0			-1	-3
1	$-.0422535211$	$-.492957746$	$-.0422535211$	$-.492957746$	$-.000150875$	.72902202
2	$-.0491181274$	$-.410806881$	$-.0068646063$	.082150865	$-.000012594$	.02024629

Continuing, we find approximate solution $(-.0493209090, -.4083890432)$.

iv.

n	x_n	y_n	$x_n - x_{n-1}$	$y_n - y_{n-1}$	$f(x_n)$	$g(x_n)$
0	-2.5	2.5			-4.75	3.25
1	-2.327133479	2.158096280	.172866521	$-.341903720$	$-.437911030$	.350694459
2	-2.308094033	2.110366365	.019039446	$-.047729915$	$-.005047725$	.006834433

Continuing, we find approximate solution $(-2.3078910423, 2.1093705359)$.

v.

n	x_n	y_n	$x_n - x_{n-1}$	$y_n - y_{n-1}$	$f(x_n)$	$g(x_n)$
0	-2.5	0			-2.25	$-.5$
1	-2.415584416	$-.097402597$	.084415584	$-.097402597$	$-.105686793$	.028461798
2	-2.410834885	$-.093801326$	.004749531	.003601271	$-.000326714$	.000038907

Continuing, we find approximate solution $(-2.4108204382, -.0937975989)$.

vi.

n	x_n	y_n	$x_n - x_{n-1}$	$y_n - y_{n-1}$	$f(x_n)$	$g(x_n)$
0	0	2.5			-3.5	.75
1	$-.2821100917$	2.385321101	$-.2821100917$	$-.114678899$	$-.044904087$	.039453752
2	$-.2855592166$	2.380159497	$-.0034491249$	$-.005161604$	$-.000020217$	.000079927

Continuing, we find approximate solution $(-.2855601249, 2.3801497352)$.

9. Let $f = 7x^3 - 10x - y - 1$ and $g = 8y^3 - 11y + x - 1$. $f_x = 21x^2 - 10$, $f_y = -1$, $g_x = 1$, $g_y = 24y^2 - 11$, $D = (21x^2 - 10)(24y^2 - 11) - (-1)(1) = 111 - 231x^2 - 240y^2 + 504x^2y^2$, $P = fg_y - f_yg = (7x^3 - 10x - y - 1)(24y^2 - 11) - (-1)(8y^3 - 11y + x - 1) = 10 + 111x - 24y^2 - 77x^3 - 16y^3 - 240xy^2 + 168x^3y^2$, $Q = -fg_x + f_xg = -(7x^3 - 10x - y - 1)(1) + (21x^2 - 10)(8y^3 - 11y + x - 1) = 11 + 111y - 21x^2 + 14x^3 - 80y^3 - 231x^2y + 168x^2y^3$.
Thus (x_{n+1}, y_{n+1}) is computed from (x_n, y_n) by the functions $x_* = x - P/D$ and $y_* = y - Q/D$.

i.

n	x_n	y_n	$x_n - x_{n-1}$	$y_n - y_{n-1}$	$f(x_n)$	$g(x_n)$
0	0	0			-1	-1
1	$-.090090090$	$-.099099099$	$-.090090090$	$-.099099099$	$-.005118340$	$-.007785726$
2	$-.090533030$	$-.099863539$	$-.000442940$	$-.000764440$	$-.000000372$	$-.000001394$

Continuing, we find approximate solution $(-.0905330540, -.0998636709)$.

ii.

n	x_n	y_n	$x_n - x_{n-1}$	$y_n - y_{n-1}$	$f(x_n)$	$g(x_n)$
0	1	0			-4	0
1	1.366666667	.033333333	.366666667	.033333333	3.168407417	.0002962967
2	1.257907972	.023449154	$-.108758695$	$-.009884180$	.330471486	.0000704323

Continuing, we find approximate solution $(1.2433857533, .0221338638)$.

iii.

n	x_n	y_n	$x_n - x_{n-1}$	$y_n - y_{n-1}$	$f(x_n)$	$g(x_n)$
0	0	1			-2	-4
1	$-.232558140$	1.32558140	$-.232558140$	.32558140	$-.088042562$	2.820179340
2	$-.232283118$	1.23510097	.000275022	$-.09048042$	$-.000000369$	.254525612

Continuing, we find approximate solution $(-.2311447876, 1.2250007910)$.

iv.

n	x_n	y_n	$x_n - x_{n-1}$	$y_n - y_{n-1}$	$f(x_n)$	$g(x_n)$
0	-1	0			2	-2
1	-1.2	$-.2$	$-.2$	$-.2$	$-.896$	$-.064$
2	-1.155828803	$-.201974980$	$.044171197$	$-.001974980$	$-.048564310$	$-.000018784$

Continuing, we find approximate solution $(-1.1531141693, -.2017059939)$.

v.

n	x_n	y_n	$x_n - x_{n-1}$	$y_n - y_{n-1}$	$f(x_n)$	$g(x_n)$
0	0	-1			0	2
1	$.015503876$	-1.155038760	$.015503876$	$-.155038760$	$.000026087$	$-.606701774$
2	$.012604749$	-1.126036040	$-.002899127$	$.029002720$	$.000002566$	$-.023122511$

Continuing, we find approximate solution $(.0124851530, -1.1248379068)$.

vi.

n	x_n	y_n	$x_n - x_{n-1}$	$y_n - y_{n-1}$	$f(x_n)$	$g(x_n)$
0	1	1			-5	-3
1	1.472222222	1.194444444	$.472222222$	$.194444444$	5.419988846	$.966220842$
2	1.318631634	1.159478680	$-.153590588$	$-.034965764$	$.703963800$	$.034706004$
3	1.292067128	1.159095828	$-.026564506$	$-.000382852$	$.019409772$	$.000004078$

Continuing, we find approximate solution $(1.2912933376, 1.1591320580)$.

vii.

n	x_n	y_n	$x_n - x_{n-1}$	$y_n - y_{n-1}$	$f(x_n)$	$g(x_n)$
0	-1	1			1	-5
1	-1.055555556	1.388888889	$-.055555556$	$.388888889$	$-.066015094$	4.100137174
2	-1.059290585	1.272831325	$-.003735029$	$-.116057564$	$-.000309596$	$.436472815$
3	-1.060418864	1.257217723	$-.001128279$	$-.015613602$	$-.000028338$	$.007416676$

Continuing, we find approximate solution $(-1.0604369649, 1.2569429491)$.

viii.

n	x_n	y_n	$x_n - x_{n-1}$	$y_n - y_{n-1}$	$f(x_n)$	$g(x_n)$
0	1	-1			-3	3
1	1.25	-1.25	$.25$	$-.25$	1.421875	-1.625
2	1.190457759	-1.186432368	$-.059542241$	$.063567632$	$.091585898$	$-.119170371$
3	1.186097459	-1.181010294	$-.004360300$	$.005422074$	$.000474724$	$-.000835831$

Continuing, we find approximate solution $(1.1860751213, -1.1809721090)$.

ix.

n	x_n	y_n	$x_n - x_{n-1}$	$y_n - y_{n-1}$	$f(x_n)$	$g(x_n)$
0	-1	-1			3	1
1	-1.277777778	-1.055555556	$-.277777778$	$-.055555556$	-1.770404664	$-.075445818$
2	-1.204876078	-1.055393930	$.072901700$	$.000161626$	$-.139897860$	$-.000000672$
3	-1.198068333	-1.055826604	$.006807745$	$-.000432674$	$-.001170446$	$-.000004741$

Continuing, we find approximate solution $(-1.1980103894, -1.0558299813)$.

9: $y = 7x^3 - 10x - 1, \quad x = 1 + 11y - 8y^3$

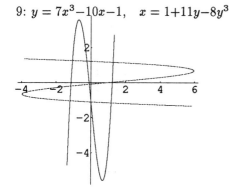

11. For any real a, b, c, d, we have the following computational facts:

$$\begin{pmatrix} a & b \\ c & d \end{pmatrix} \begin{pmatrix} d & -b \\ -c & a \end{pmatrix} = \begin{pmatrix} ad - bc & -ab + ba \\ cd - dc & -cb + da \end{pmatrix} = \begin{pmatrix} ad - bc & 0 \\ 0 & ad - bc \end{pmatrix} = (ad - bc) \begin{pmatrix} 1 & 0 \\ 0 & 1 \end{pmatrix},$$

$$\begin{pmatrix} d & -b \\ -c & a \end{pmatrix} \begin{pmatrix} a & b \\ c & d \end{pmatrix} = \begin{pmatrix} da - bc & db - bd \\ -ca + ac & -cb + ad \end{pmatrix} = \begin{pmatrix} ad - bc & 0 \\ 0 & ad - bc \end{pmatrix} = (ad - bc) \begin{pmatrix} 1 & 0 \\ 0 & 1 \end{pmatrix}.$$

13.12 — Newton's method for functions of two variables

If $ad - bc \neq 0$, then $\begin{pmatrix} a & b \\ c & d \end{pmatrix}^{-1}$ exists and equals $\dfrac{1}{ad-bc}\begin{pmatrix} d & -b \\ -c & a \end{pmatrix}$.

Therefore, if $J = \begin{pmatrix} f_x & f_y \\ g_x & g_y \end{pmatrix}$ and $D = f_x g_y - f_y g_x \neq 0$, then $J^{-1} = \dfrac{1}{D}\begin{pmatrix} g_y & -f_y \\ -g_x & f_x \end{pmatrix}$. ∎

13. If $f = x^2 - x + y^2 + z^2 - 5$, $g = x^2 + y^2 - y + z^2 - 4$, $h = x^2 + y^2 + z^2 + z - 6$, and $\mathbf{F} = (f, g, h)^t$, then $J =$
$\begin{pmatrix} 2x - 1 & 2y & 2z \\ 2x & 2y - 1 & 2z \\ 2x & 2y & 2z + 1 \end{pmatrix}$ and $J^{-1}\mathbf{F} = \dfrac{1}{2x + 2y - 2z - 1}\begin{pmatrix} -5 - x + 2y + 2z + x^2 - y^2 - z^2 + 2xy - 2xz \\ -4 - 2x - y + 4z - x^2 + y^2 - z^2 + 2xy - 2yz \\ 6 - 2x - 4y - z + x^2 + y^2 - z^2 + 2xz + 2yz \end{pmatrix}$.

(It's cheaper to compute the single vector $J^{-1}\mathbf{F}$ than to compute the full inverse J^{-1} and multiply $\mathbf{F}$ by it.)

a.

n	$\mathbf{X}_n$			$\mathbf{F}(\mathbf{X}_n)$			$(-J^{-1}\mathbf{F})(\mathbf{X}_n)$		
0	$-.8$	$.2$	1.8	$-.28$	$-.28$	$-.28$	$-.048276$	$-.048276$	$.048276$
1	$-.848276$	$.151724$	1.848276	$.006992$	$.006992$	$.006992$	$.001148$	$.001148$	$-.001148$
2	$-.847128$	$.152872$	1.847128	$.000004$	$.000004$	$.000004$	$.000001$	$.000001$	$-.000001$
3	$-.847127$	$.152873$	1.847127						

Continuing, we find approximate solution $(-.8471270884, .1528729116, 1.8471270884)$.

b.

n	$\mathbf{X}_n$			$\mathbf{F}(\mathbf{X}_n)$			$(-J^{-1}\mathbf{F})(\mathbf{X}_n)$		
0	1.2	2.2	$-.2$	$.12$	$.12$	$.12$	$-.019355$	$-.019355$	$.019355$
1	1.180645	2.180645	$-.180645$	$.001124$	$.001124$	$.001124$	$-.000185$	$-.000185$	$.000185$
2	1.180460	2.180460	$-.180460$	$.000000$	$.000000$	$.000000$			

Continuing, we find approximate solution $(1.1804604217, 2.1804604217, -.1804604217)$.

REMARK: These numerical tables show an alternative way to summarize the first several iterations of Newton's method while still presenting the same information as in the tables used for Solutions 1–10.

Both solutions are of the form $(r, 1+r, 1-r)$ where $3r^2 - r - 3 = 0$. Level surfaces of f, g, h are spheres (with different centers and radii), each pair of spheres intersects in a circle, the three circles for this problem $(f \cap g, g \cap h, h \cap f)$ intersect in two points; therefore we have the full solution set.

Solutions 13.R   Review   (pages 917-9)

1. If $f(x, y) = \sqrt{x^2 - y^2}$, the maximal domain of f is $\{(x, y) : |x| \geq |y|\}$ and the range is $[0, \infty)$.

3. If $f(x, y) = \cos(x + 3y)$, the maximal domain of f is $\mathbb{R}^2$ and the range is $[-1, 1]$.

5. If $f(x, y, z) = 1/\sqrt{x^2 + y^2 + z^2 - 1}$, the maximal domain of f is $\{(x, y, z) : x^2 + y^2 + z^2 > 1\}$ and the range is $(0, \infty)$.

7. $z = \sqrt{1 - x - y} \implies x + y = 1 - z^2$, the level curves of $\sqrt{1 - x - y}$ are parallel lines with slope -1.

9. $z = \ln(x - 3y) \implies y = (1/3)(x - e^z)$, the level curves of $\ln(x - 3y)$ are parallel lines with slope $1/3$.

7: $z = \sqrt{1 - x - y}$; $z = 0, 1, 3, 8$ 9: $z = \ln(x - 3y)$; $z = 0, 1, 2, 3$ 11: center $(-1, 2)$; radius 4

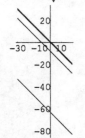
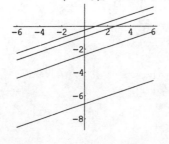
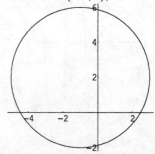

13. $y^2 - x^2 = (y - x)(y + x)$ is zero on the lines $y = \pm x$ which pass through $(0, 0)$; hence there is no deleted neighborhood of $(0, 0)$ in the domain of $xy/(y^2 - x^2)$.

Moreover — on the line $y = mx$, $\displaystyle \lim_{(x,y)\to(0,0)} \frac{xy}{y^2 - x^2} = \lim_{x\to 0} \frac{x(mx)}{(mx)^2 - x^2} = \lim_{x\to 0} \frac{m}{m^2 - 1} = \frac{m}{m^2 - 1}$; since the limit along these paths depends on m, $\lim_{(x,y)\to(0,0)}(xy)/(y^2 - x^2)$ does not exist. ∎

15. $0 \le (x \pm y^2)^2 = x^2 \pm 2xy^2 + y^4 \implies \left|2xy^2\right| \le x^2 + y^4$. Thus $\left|\dfrac{4xy^3}{x^2 + y^4}\right| = 2|y|\,\left|\dfrac{2xy^2}{x^2 + y^4}\right| \le 2|y|$. Hence $\lim_{(x,y)\to(0,0)} 2|y| = 0$ implies (via a Squeezing theorem) $\lim_{(x,y)\to(0,0)} \left(4xy^3/(x^2 + y^4)\right) = 0$. ∎

17. $(1 + x^2 y)/(2 - y)$ is a rational function whose denominator is not zero in a neighborhood of $(1, -2)$, hence it is continuous at that point and $\lim\limits_{(x,y)\to(1,-2)} \dfrac{1 + x^2 y}{2 - y} = \dfrac{1 + 1^2(-2)}{2 - (-2)} = \dfrac{1 - 2}{2 + 2} = \dfrac{-1}{4}$.

19. $\ln(1 - 2x + 3y)$ is continuous throughout its maximal domain, $\{(x, y) : 1 - 2x + 3y > 0\}$.

21. $1/\sqrt{1 - x^2 + y^2 - z^2}$ is continuous throughout its maximal domain, $\{(x, y, z) : 1 - x^2 + y^2 - z^2 > 0\}$.

23. If $f(x, y) = y/x$, then $f_x(x, y) = -y/x^2$ and $f_y(x, y) = 1/x$.

25. If $f(x, y) = \left(x^2 - y^2\right)^{-1/2}$, then $f_x(x, y) = -x\left(x^2 - y^2\right)^{-3/2}$ and $f_y(x, y) = y\left(x^2 - y^2\right)^{-3/2}$.

27. If $f(x, y, z) = \ln(x - y + 4z)$, then $f_x(x, y, z) = (x - y + 4z)^{-1}$, $f_y(x, y, z) = -(x - y + 4z)^{-1}$, and $f_z(x, y, z) = 4(x - y + 4z)^{-1}$.

29. If $f(x, y, z) = \cosh(x^{-2}y)$, then $f_x(x, y, z) = -2x^{-3}y\sinh(x^{-2}y)$, $f_y(x, y, z) = x^{-2}\sinh(x^{-2}y)$, and $f_z(x, y, z) = 0$.

31. Let $M = x^2 y - y^3 z^5 + x\sqrt{z}$, then $f(x, y, z) = M^{2/3}$ implies $f_x(x, y, z) = (2/3)M^{-1/3}\left(2xy + \sqrt{z}\right)$, $f_y(x, y, z) = (2/3)M^{-1/3}\left(x^2 - 3y^2 z^5\right)$, and $f_z(x, y, z) = (2/3)M^{-1/3}\left(-5y^3 z^4 + x/(2\sqrt{z})\right)$.

33. If $f(x, y, z, w) = \dfrac{x - z + w}{y + 2w - x}$, then $f_x(x, y, z, w) = \dfrac{y - z + 3w}{(y + 2w - x)^2}$, $f_y(x, y, z, w) = \dfrac{-(x - z + w)}{(y + 2w - x)^2}$, $f_z(x, y, z, w) = \dfrac{-1}{y + 2w - x}$, and $f_w(x, y, z, w) = \dfrac{-3x + y + 2z}{(y + 2w - x)^2}$.

35. If $f(x, y) = xy^3$, then $f_x = y^3$ and $f_y = 3xy^2$; therefore $f_{xx} = 0$, $f_{xy} = 3y^2 = f_{yx}$, and $f_{yy} = 6xy$.

37. If $f(x, y) = \left(x^2 - y^2\right)^{1/2}$, then $f_x = x\left(x^2 - y^2\right)^{-1/2}$ and $f_y = -y\left(x^2 - y^2\right)^{-1/2}$; therefore $f_{xy} = xy\left(x^2 - y^2\right)^{-3/2} = f_{yx}$, $f_{xx} = \left(x^2 - y^2\right)^{-1/2} - x^2\left(x^2 - y^2\right)^{-3/2} = -y^2\left(x^2 - y^2\right)^{-3/2}$, and $f_{yy} = -\left(x^2 - y^2\right)^{-1/2} - y^2\left(x^2 - y^2\right)^{-3/2} = -x^2\left(x^2 - y^2\right)^{-3/2}$.

39. If $f(x, y, z) = \ln(2 - 3x + 4y - 7z)$, then $f_x = -3(2 - 3x + 4y - 7z)^{-1}$, $f_y = 4(2 - 3x + 4y - 7z)^{-1}$, $f_z = -7(2 - 3x + 4y - 7z)^{-1}$; thus $f_{xx} = -9(2 - 3x + 4y - 7z)^{-2}$, $f_{yy} = -16(2 - 3x + 4y - 7z)^{-2}$, $f_{zz} = -49(2 - 3x + 4y - 7z)^{-2}$, $f_{xy} = 12(2 - 3x + 4y - 7z)^{-2} = f_{yx}$, $f_{yz} = 28(2 - 3x + 4y - 7z)^{-2} = f_{zy}$, $f_{zx} = -21(2 - 3x + 4y - 7z)^{-2} = f_{xz}$.

41. If $f(x, y, z) = x^2 y^3 - zx^5$, then $f_y = 3x^2 y^2$, $f_{yz} = 0$, and $f_{yzx} = 0$.

43. If $f(x, y) = x^2 - y^3$, then $\nabla f(x, y) = 2x\mathbf{i} - 3y^2\mathbf{j}$ and $\nabla f(1, 2) = 2\mathbf{i} - 12\mathbf{j}$.

45. If $f(x, y) = (x - y)(x + y)^{-1}$, then $\nabla f(x, y) = 2(x + y)^{-2}(y\mathbf{i} - x\mathbf{j})$ and $\nabla f(3, 2) = (2/25)(2\mathbf{i} - 3\mathbf{j})$.

47. If $f(x, y, z) = xy + yz^3$, then $\nabla f(x, y, z) = y\mathbf{i} + (x + z^3)\mathbf{j} + 3yz^2\mathbf{k}$ and $\nabla f(1, 2, -1) = 2\mathbf{i} + 6\mathbf{k}$.

49. If $f(x, y, z) = \left(x^2 + y^2 + z^2\right)^{-1/2}$, then $\nabla f(x, y, z) = -\left(x^2 + y^2 + z^2\right)^{-3/2}(x\mathbf{i} + y\mathbf{j} + z\mathbf{k})$ and $\nabla f(a, b, c) = -\left(a^2 + b^2 + c^2\right)^{-3/2}(a\mathbf{i} + b\mathbf{j} + c\mathbf{k})$.

51. If $z = 2xy$, $x = \cos t$, $y = \sin t$, then $z_t = z_x x_t + z_y y_t = (2y)(-\sin t) + (2x)(\cos t) = (2\sin t)(-\sin t) + (2\cos t)(\cos t) = 2\left(\cos^2 t - \sin^2 t\right)$.

53. If $z = xy^3$, $x = r/s$, $y = s^2/r$, then $z_r = z_x x_r + z_y y_r = (y^3)(1/s) + (3xy^2)(-s^2/r^2) = \left(s^6 r^{-3}\right)s^{-1} - 3\left(s^3 r^{-1}\right)s^2 r^{-2} = -2s^5 r^{-3}$.

55. If $w = xyz$, $x = rs$, $y = r/s$, $z = s^2 r^3$, then
$w_r = w_x x_r + w_y y_r + w_z z_r = (yz)(s) + (xz)(1/s) + (xy)(3s^2 r^2) = (sr^4)s + (s^3 r^4)s^{-1} + r^2(3s^2 r^2) = 5s^2 r^4$,
$w_s = w_x x_s + w_y y_s + w_z z_s = (yz)(r) + (xz)(-r/s^2) + (xy)(2sr^3) = sr^{4+1} - s^{3-2}r^{4+1} + 2sr^{2+3} = 2sr^5$.

57. If $f(x, y, z) = x^2 + y^2 + z^2 - 3$, then $\nabla f(x, y, z) = 2x\mathbf{i} + 2y\mathbf{j} + 2z\mathbf{k}$ and $\nabla f(1, 1, 1) = 2\mathbf{i} + 2\mathbf{j} + 2\mathbf{k}$. The plane tangent to the surface $f = 0$ at $(1, 1, 1)$ has equation $0 = \nabla f(1, 1, 1) \cdot ((x - 1)\mathbf{i} + (y - 1)\mathbf{j} + (z - 1)\mathbf{k}) = 2(x - 1) + 2(y - 1) + 2(z - 1)$ and the normal line has symmetric equations $(x - 1)/2 = (y - 1)/2 = (z - 1)/2$.

REMARK: The plane has equation $x + y + z = 3$ and the line has equation $x = y = z$.

59. If $f(x, y, z) = 3x - y + 5z - 15$, then $\nabla f(x, y, z) = 3\mathbf{i} - \mathbf{j} + 5\mathbf{k} = \nabla f(-1, 2, 4)$. The plane tangent to the surface $f = 0$ at $(-1, 2, 4)$ has equation $0 = \nabla f(-1, 2, 4) \cdot ((x - (-1))\mathbf{i} + (y - 2)\mathbf{j} + (z - 4)\mathbf{k}) = 3(x + 1) - (y - 2) + 5(z - 4)$ and the normal line has symmetric equations $(x - (-1))/3 = (y - 2)/(-1) = (z - 4)/5$.

61.   If $f(x, y, z) = xyz - 6$, then $\nabla f(x, y, z) = yz\mathbf{i} + xz\mathbf{j} + xy\mathbf{k}$ and $\nabla f(-2, 1, -3) = -3\mathbf{i} + 6\mathbf{j} - 2\mathbf{k}$. The plane tangent to the surface $f = 0$ at $(-2, 1, -3)$ has equation $0 = \nabla f(-2, 1, -3) \cdot ((x + 2)\mathbf{i} + (y - 1)\mathbf{j} + (z + 3)\mathbf{k}) = -3(x + 2) + 6(y - 1) - 2(z + 3) = -3x + 6y - 2z - 18$ and the normal line has symmetric equations $(x + 2)/(-3) = (y - 1)/6 = (z + 3)/(-2)$.

63.   If $f(x, y) = yx^{-1}$, then $\nabla f(x, y) = -yx^{-2}\mathbf{i} + x^{-1}\mathbf{j}$ and $\nabla f(1, 2) = -2\mathbf{i} + \mathbf{j}$. Let $\mathbf{v} = \mathbf{i} - \mathbf{j}$ and $\mathbf{u} = \mathbf{v}/|\mathbf{v}|$. Therefore $f_\mathbf{u}'(1, 2) = \nabla f(1, 2) \cdot \mathbf{v}/|\mathbf{v}| = (-2\mathbf{i} + \mathbf{j}) \cdot (\mathbf{i} - \mathbf{j})/\sqrt{1^2 + (-1)^2} = (-2 - 1)/\sqrt{2} = -3/\sqrt{2}$.

65.   If $f(x, y) = \tan^{-1}(y/x)$, then $\nabla f(x, y) = (x^2 + y^2)^{-1}(-y\mathbf{i} + x\mathbf{j})$ and $\nabla f(1, -1) = (1/2)(\mathbf{i} + \mathbf{j})$. Let $\mathbf{v} = -3\mathbf{i} + 2\mathbf{j}$ and $\mathbf{u} = \mathbf{v}/|\mathbf{v}|$. Thus $f_\mathbf{u}'(1, -1) = \nabla f(1, -1) \cdot \mathbf{v}/|\mathbf{v}| = (1/2)(\mathbf{i} + \mathbf{j}) \cdot (-3\mathbf{i} + 2\mathbf{j})/\sqrt{(-3)^2 + 2^2} = (1/2)(-3 + 2)/\sqrt{13} = -1/2\sqrt{13}$.

67.   If $f(x, y, z) = (x^2 + y^2 + z^2)^{-1/2}$, then $\nabla f(x, y, z) = -(x^2 + y^2 + z^2)^{-3/2}(x\mathbf{i} + y\mathbf{j} + z\mathbf{k})$ and $\nabla f(1, -1, 2) = 6^{-3/2}(-\mathbf{i} + \mathbf{j} - 2\mathbf{k})$. Let $\mathbf{v} = -2\mathbf{i} + \mathbf{j} - 3\mathbf{k}$ and $\mathbf{u} = \mathbf{v}/|\mathbf{v}|$. Therefore $f_\mathbf{u}'(1, -1, 2) = \nabla f(1, -1, 2) \cdot \mathbf{v}/|\mathbf{v}| = 6^{-3/2}(-\mathbf{i} + \mathbf{j} - 2\mathbf{k}) \cdot (-2\mathbf{i} + \mathbf{j} - 3\mathbf{k})/\sqrt{(-2)^2 + 1^2 + (-3)^2} = 6^{-2/3}(2 + 1 + 6)/\sqrt{14} = 3/4\sqrt{21} = \sqrt{21}/28$.

69.   If $f(x, y) = x^3 y^2$, then $df = \nabla f \cdot \Delta\mathbf{x} = (3x^2 y^2 \mathbf{i} + 2x^3 y \mathbf{j}) \cdot (\Delta x \mathbf{i} + \Delta y \mathbf{j}) = 3x^2 y^2 \Delta x + 2x^3 y \Delta y$.

71.   If $f(x, y) = (x + 1)^{1/2}(y - 1)^{-1/2}$, then $df = \nabla f \cdot \Delta\mathbf{x} = \dfrac{1}{2(x + 1)^{1/2}(y - 1)^{1/2}} \Delta x - \dfrac{(x + 1)^{1/2}}{2(y - 1)^{-3/2}} \Delta y$.

REMARK: Logarithmic differentiation yields $df/f = (1/2) d (\ln |x + 1| - \ln |y - 1|)$ and

$$df = \frac{1}{2}\sqrt{\frac{x + 1}{y - 1}} \left(\frac{\Delta x}{x + 1} - \frac{\Delta y}{y - 1} \right).$$

73.   If $f(x, y, z) = \ln(x - y + 4z)$, then $df = \nabla f \cdot \Delta\mathbf{x} = (x - y + 4z)^{-1}(\Delta x - \Delta y + 4\Delta z)$.

75.   If $f(x, y) = 6x^2 + 14y^2 - 16xy + 2$, then $f_x = 12x - 16y$ and $f_y = 28y - 16x$. The unique solution of the system $f_x = 0 = f_y$ is $x = 0$ and $y = 0$. $f_{xx} = 12$, $f_{xy} = -16 = f_{yx}$, $f_{yy} = 28$, and $D = f_{xx}f_{yy} - f_{xy}f_{yx} = (12)(28) - (-16)^2 = 80$. Since $D > 0$ and $f_{xx} > 0$, we infer f has a local minimum at $(0, 0)$.

REMARK: $6x^2 + 14y^2 - 16xy + 2 = 6(x - 4y/3)^2 + (10/3)y^2 + 2$ has a global minimum of 2 where $x - 4y/3 = 0$ and $y = 0$.

77.   If $f(x, y) = y^{-1} + 2x^{-1} + 2y + x + 4$, then $f_x = -2x^{-2} + 1$ and $f_y = -y^{-2} + 2$. The system $f_x = 0 = f_y$ has solutions $x = \pm\sqrt{2}$ and $y = \pm 1/\sqrt{2}$. $f_{xx} = 4x^{-3}$, $f_{xy} = 0 = f_{yx}$, $f_{yy} = 2y^{-3}$, and $D = f_{xx}f_{yy} - f_{xy}f_{yx} = (4x^{-3})(2y^{-3}) - (0)^2 = 8(xy)^{-3}$. Critical points $(\sqrt{2}, -1/\sqrt{2})$ and $(-\sqrt{2}, 1/\sqrt{2})$ are saddle points because $D = -8 < 0$ at each. $D = 8 > 0$ at the other two critical points; $f_{xx} = \sqrt{2} > 0$ at $(\sqrt{2}, 1/\sqrt{2})$ so that point is a local minimum, $f_{xx} = -\sqrt{2} < 0$ at $(-\sqrt{2}, -1/\sqrt{2})$ so that point is a local maximum.

79.   If $f(x, y) = x^2 + y^2 + 2x^{-1}y^{-2}$, then $f_x = 2x - 2x^{-2}y^{-2}$ and $f_y = 2y - 4x^{-1}y^{-3}$. Therefore $0 = f_y \implies x = 2y^{-4}$, then $0 = f_x \implies y^2 = x^{-3} = y^{12}/8 \implies y^{10} = 8 \implies y = \pm 8^{1/10} = \pm 2^{3/10}$ and $x = 2y^{-4} = 2 \cdot 2^{-12/10} = 2^{-1/5}$. $D = (2 + 4x^{-3}y^{-2})(2 + 12x^{-1}y^{-4}) - (4x^{-2}y^{-3})^2 = 4 + 32/(x^4y^6) + 8/(x^3y^2) + 24/(xy^4)$; thus $f_{xx}(2^{-1/5}, \pm 2^{3/10}) = 6 > 0$ and $D(2^{-1/5}, \pm 2^{3/10}) = (6)(8) - (\pm\sqrt{8})^2 = 40 > 0$ so each point yields a local minimum.

81.   Let $f = (x - 2)^2 + (y + 1)^2 + (z - 4)^2$ and $g = x - y + 3z - 7$, then $\nabla f = 2(x - 2)\mathbf{i} + 2(y + 1)\mathbf{j} + 2(z - 4)\mathbf{k}$ and $\nabla g = \mathbf{i} - \mathbf{j} + 3\mathbf{k}$. Therefore $\nabla f = \lambda \nabla g$ if and only if $x = 2 + \lambda/2$, $y = -1 - \lambda/2$, and $z = 4 + 3\lambda/2$. The constraint $0 = g = 8 + 11\lambda/2$ implies $\lambda = -16/11$ and $x = 14/11$, $y = -3/11$, $z = 20/11$; so the minimum value of f is $64/11$ and the minimum distance between point $(2, -1, 4)$ and line $x - y + 3z = 7$ is $\sqrt{64/11} = 8/\sqrt{11}$.

REMARK: If the numerical result is more important than reviewing techniques of this chapter, then the result of Problem 11.7.46 can be applied.

83.   Let x be length, y be width, and z be height of an open-top rectangular box. The volume is $V = xyz$ and our surface area constraint is $0 = g = xy + 2xz + 2yz - 10$. Thus $\nabla V = yz\mathbf{i} + xz\mathbf{j} + xy\mathbf{k}$ and $\nabla g = (y + 2z)\mathbf{i} + (x + 2z)\mathbf{j} + 2(x + y)\mathbf{k}$ Hence

$$\nabla V = \lambda \nabla g \implies \frac{y}{x} = \frac{yz}{xz} = \frac{V_x}{V_y} = \frac{\lambda g_x}{\lambda g_y} = \frac{g_x}{g_y} = \frac{y + 2z}{x + 2z} \implies 0 = y(x + 2z) - x(y + 2z) = 2(y - x)z.$$

Since $z > 0$, this implies $y = x$. Therefore $\nabla V = \lambda \nabla g$ also implies

$$\frac{z}{x} = \frac{yz}{xy} = \frac{V_x}{V_z} = \frac{\lambda g_x}{\lambda g_z} = \frac{g_x}{g_z} = \frac{y + 2z}{2(x + y)} = \frac{x + 2z}{4x} \implies 0 = (z)(4x) - (x)(x + 2z) = 2xz - x^2 = x(2z - x)$$

which implies $z = x/2$. The constraint $0 = g = xy + x(2z) + y(2z) - 10 = x^2 + x^2 + x^2 - 10$ implies $x = \sqrt{10/3}$ and the maximal volume is $V = xyz = x^3/2 = (1/2)(10/3)^{3/2}$.

REMARK: $xyz = xV_x = yV_y = zV_z$, hence

$$0 = xyz/\lambda - xyz/\lambda = xV_x/\lambda - yV_y/\lambda = xg_x - yg_y = x(y + 2z) - y(x + 2z) = 2(y - x)z$$

is an alternative path to the key inference $y = x$.

85.   Let $f = x^2 + y^2 + z^2$, $g = 2x + y + z - 2$, and $h = x - y - 3z - 4$. Then $\nabla f = \lambda \nabla g + \mu \nabla h$ implies $x = \lambda + \mu/2$, $y = \lambda/2 - \mu/2$, and $z = \lambda/2 - 3\mu/2$. Hence $0 = g = 3\lambda - \mu - 2$ and $0 = h = -\lambda + 11\mu/2 - 4$ implies $\lambda = 30/31$ and $\mu = 28/31$. Therefore $f(44/31, 1/31, -27/31) = 86/31$ is the maximal value of f subject to the constraints $g = 0 = h$.

87.   If $y = x$, then $0 = g = x - y + 2z - 2 = 2(z - 1)$ implies $z = 1$ and $f(x, -x, z) = x^4 \to \infty$ as $x \to \pm\infty$; if $y = -x$, then $0 = g = x - y + 2z - 2 = 2(x + z - 1)$ implies $z = 1 - x$ and $f(x, -x, z) = -x^4(1 - x)^2 \to -\infty$ as $x \to \pm\infty$. Hence f is neither bounded above nor below on the plane $x - y + 2z = 2$. ∎

Solutions 13.C   *Computer Exercises*   (page 919)

1.   Let $f(x, y) = \begin{cases} \dfrac{xy(x^2 - y^2)}{x^2 + y^2} & \text{if } (x, y) \neq (0, 0), \\ 0 & \text{if } (x, y) = (0, 0). \end{cases}$

a.   If $(x, y) \neq (0, 0)$, then $f_x(x, y) = \dfrac{y(x^4 + 4x^2y^2 - y^4)}{(x^2 + y^2)^2}$ and $f_y(x, y) = \dfrac{x(x^4 - 4x^2y^2 - y^4)}{(x^2 + y^2)^2}$. However

$$f_x(0, y) = \begin{cases} -y & \text{if } y \neq 0, \\ \lim_{h \to 0} \dfrac{f(h, 0) - f(0, 0)}{h} = \lim_{h \to 0} \dfrac{0 - 0}{h} = 0 & \text{if } y = 0, \end{cases}$$

$$f_y(x, 0) = \begin{cases} x & \text{if } x \neq 0, \\ \lim_{k \to 0} \dfrac{f(0, k) - f(0, 0)}{k} = \lim_{k \to 0} \dfrac{0 - 0}{k} = 0 & \text{if } x = 0. \end{cases}$$

b.   If $(x, y) \neq (0, 0)$, then $f_{xy}(x, y) = \dfrac{x^6 + 9x^4y^2 - 9x^2y^4 - y^6}{(x^2 + y^2)^3} = f_{yx}(x, y)$. However

$$f_{xy}(0, 0) = \lim_{k \to 0} \frac{f_x(0, k) - f_x(0, 0)}{k} = \lim_{k \to 0} \frac{-k - 0}{k} = -1,$$

$$f_{yx}(0, 0) = \lim_{h \to 0} \frac{f_y(h, 0) - f_y(0, 0)}{h} = \lim_{h \to 0} \frac{h - 0}{h} = 1.$$

c.   Theorem 13.4.1 requires both mixed partial derivatives to be continuous at the point. f_{xy} and f_{yx} are constant on lines through the origin, but the function value varies with the slope of the line. E.g., if $x \neq 0$, then $f_{xy}(x, x) = 0$ but $f_{xy}(x, 2x) = -171/125$. Hence $\lim_{(x,y)\to(0,0)} f_{xy}(x, y)$ does not exist.

1d.

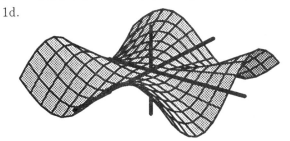

3c.

13.C — Computer Exercises

3.   Let $f(x,y) = 2x^3/3 - 2x^2y + x^2/2 + 2xy - x + y^2 + 2y$. Then $f_x(x,y) = 2x^2 - 4xy + x + 2y - 1 = (2x-1)(x-2y+1)$ and $f_y(x,y) = -2x^2 + 2x + 2y + 2 = 2(y - x^2 + x + 1)$.

   a.   The equations $f_x = 0 = f_y$ have three solutions. The critical points are $P_1 = (1/2, -5/4)$, $P_2 = \left(\dfrac{3 + \sqrt{33}}{4}, \dfrac{14 + 2\sqrt{33}}{16} \right) \approx (2.186, 1.593)$, and $P_3 = \left(\dfrac{3 - \sqrt{33}}{4}, \dfrac{14 - 2\sqrt{33}}{16} \right) \approx (-.686, .157)$.

   b.   The discriminant of f is $D(x,y) = (1 + 4x - 4y)(2) - (2 - 4x)^2 = 2(-1 + 12x - 8x^2 - 4y)$. f has a local minimum at P_1 because $D(P_1) = 16 > 0$ and $f_{xx}(P_1) = 8 > 0$; f has saddle points at P_2 and P_3 because $D(P_2) = -33 - \sqrt{33} < 0$ and $D(P_3) = -33 + \sqrt{33} < 0$.

5.   Let z be the height of the rectangular box, y be the width of the front and back panels, and x be the width of the left and right panels. The cost of constructing the box is $C(x,y,z) = xy + 4yz + 6xz + (2x + 2y + 4z)$; the volume constraint is equivalent to $xyz = 1$.

Using Lagrange multipliers leads to expressions involving roots of complicated sixth-degree polynomials; substituting $z = 1/(xy)$ into C and locating its critical points also involves zeros of messy sixth-degree polynomials. On the other hand, starting Newton's method for $C(x,y,1/(xy))$ with $x = y = z = 1$ produces an approximate minimum cost of \$16.621 when $x \approx 1.328177$, $y \approx 1.645504$, $z \approx .457557$.

14
Multiple Integration

Solutions 14.1 *Volume under a surface and the Double integral* **(pages 927-8)**

Prologue for solutions 1–14: Remember that $\sum_{k=1}^{n} 1 = n$, $\sum_{k=1}^{n} k = \dfrac{n(n+1)}{2}$, $\sum_{k=1}^{n} k^2 = \dfrac{n(n+1)(2n+1)}{6}$;

they imply $\sum_{k=1}^{n} \dfrac{1}{n} = 1$, $\sum_{k=1}^{n} \dfrac{k}{n} = \dfrac{n+1}{2}$, and $\sum_{k=1}^{n} \dfrac{k^2}{n^2} = \dfrac{(n+1)(2n+1)}{6n} = \dfrac{(n+1)\left(1+\frac{1}{2n}\right)}{3}$.

For notational convenience, we will abbreviate $\sum_{i=1}^{n}$ to just $\sum_i$, similarly, $\sum_j$ will denote $\sum_{j=1}^{n}$.

Prologue for solutions 1–8: As in Example 1, we simplify computations by working with convenient partitions and choices of x_i^*, y_j^*. Partition intervals $[0,3]$ and $[1,2]$ into n subintervals: $\Delta x = (3-0)/n = 3/n$ and $\Delta y = (2-1)/n = 1/n$. In each subinterval, choose the evaluation point to be the one with largest coordinates: $x_i^* = x_i = 0 + i\Delta x = 3i/n$ and $y_j^* = y_j = 1 + j\Delta y = 1 + j/n$.

1. $\displaystyle\sum_i\sum_j (2x_i^* + 3y_j^*)\Delta x\, \Delta y = \sum_i\sum_j \left(2\left(\frac{3i}{n}\right) + 3\left(1+\frac{j}{n}\right)\right)\left(\frac{3}{n}\right)\left(\frac{1}{n}\right) = \frac{3}{n^2}\sum_i\sum_j\left(3 + \frac{6i}{n} + \frac{3j}{n}\right).$

The first double sum is simple and the other two are not much harder:

$$\sum_i\sum_j 3 = 3\sum_i\sum_j 1 = 3\sum_i n = 3n\sum_i 1 = (3n)(n) = 3n^2,$$

$$\sum_i\sum_j \frac{6i}{n} = 6\sum_i i\sum_j \frac{1}{n} = 6\sum_i i\cdot 1 = 6\sum_i i = 6\cdot\frac{n(n+1)}{2} = 3n(n+1),$$

$$\sum_i\sum_j \frac{3j}{n} = 3\sum_i\sum_j \frac{j}{n} = 3\sum_i \frac{n+1}{2} = \frac{3(n+1)}{2}\sum_i 1 = \frac{3(n+1)}{2}\cdot n = \frac{3n(n+1)}{2}.$$

Hence $\displaystyle\sum_i\sum_j (2x_i^* + 3y_j^*)\Delta x\, \Delta y = \frac{3}{n^2}\left(3n^2 + 3n(n+1) + \frac{3n(n+1)}{2}\right) = \frac{3}{n^2}\left(\frac{15n^2 + 9n}{2}\right) = \frac{45}{2} + \frac{27}{2n}$ which

implies $\displaystyle\iint_\Omega (2x + 3y)\,dA = \lim_{n\to\infty}\left(\frac{45}{2} + \frac{27}{2n}\right) = \frac{45}{2}.$

3. $\displaystyle\sum_i\sum_j (y_j^* - x_i^*)\Delta x\, \Delta y = \sum_i\sum_j \left(\left(1 + \frac{j}{n}\right) - \left(\frac{3i}{n}\right)\right)\left(\frac{3}{n}\right)\left(\frac{1}{n}\right) = \frac{3}{n^2}\sum_i\sum_j\left(1 - \frac{3i}{n} + \frac{j}{n}\right).$ The

sums of the three pieces are $\sum_i\sum_j 1 = \sum_i n = n\sum_i 1 = n^2$,

$$\sum_i\sum_j \frac{-3i}{n} = -3\sum_i i\sum_j \frac{1}{n} = -3\sum_i i\cdot 1 = -3\sum_i i = -3\cdot\frac{n(n+1)}{2},$$

$$\sum_i\sum_j \frac{j}{n} = \sum_i \frac{n+1}{2} = \frac{n+1}{2}\sum_i 1 = \frac{n+1}{2}\cdot n = \frac{n(n+1)}{2}.$$

Hence $\displaystyle\sum_i\sum_j (y_j^* - x_i^*)\Delta x\, \Delta y = \frac{3}{n^2}\left(n^2 - 3\frac{n(n+1)}{2} + \frac{n(n+1)}{2}\right) = \frac{3}{n^2}(-n) = \frac{-3}{n},$

this implies $\displaystyle\iint_\Omega (y - x)\,dA = \lim_{n\to\infty}\frac{-3}{n} = 0.$

5.

$$\sum_i\sum_j\left(x_i^{*2}+y_j^{*2}\right)=\sum_i\sum_j\left(\left(\frac{3i}{n}\right)^2+\left(1+\frac{j}{n}\right)^2\right)$$

$$=\sum_i\sum_j\left(\frac{9i^2}{n^2}+1+\frac{2j}{n}+\frac{j^2}{n^2}\right)$$

$$=9\sum_i\sum_j\frac{i^2}{n^2}+\sum_i\sum_j 1+2\sum_i\sum_j\frac{j}{n}+\sum_i\sum_j\frac{j^2}{n^2}$$

$$=9\sum_i\frac{i^2}{n^2}\cdot n+\sum_i n+2\sum_i\frac{n+1}{2}+\sum_i\frac{(n+1)(2n+1)}{6n}$$

$$=9n\frac{(n+1)(2n+1)}{6n}+n^2+n(n+1)+\frac{(n+1)(2n+1)}{6n}n=\frac{16}{3}n^2+6n+\frac{5}{3}.$$

Hence $\sum_i\sum_j(x_i^{*2}+y_j^{*2})\Delta x\,\Delta y=\left(\frac{16}{3}n^2+6n+\frac{5}{3}\right)\left(\frac{3}{n}\right)\left(\frac{1}{n}\right)=16+\frac{18}{n}+\frac{5}{n^2}$, this implies

$$\iint_\Omega(x^2+y^2)\,dA=\lim_{n\to\infty}\left(16+\frac{18}{n}+\frac{5}{n^2}\right)=16+0+0=16.$$

7.

$$\sum_i\sum_j\left(2x_i^{*2}+3y_j^{*2}\right)=\sum_i\sum_j\left(2\left(\frac{3i}{n}\right)^2+3\left(1+\frac{j}{n}\right)^2\right)=\sum_i\sum_j\left(\frac{18i^2}{n^2}+3+\frac{6j}{n}+\frac{3j^2}{n^2}\right)$$

$$=18\sum_i\sum_j\frac{i^2}{n^2}+3\sum_i\sum_j 1+6\sum_i\sum_j\frac{j}{n}+3\sum_i\sum_j\frac{j^2}{n^2}$$

$$=18\sum_i\frac{i^2}{n^2}\cdot n+3\sum_i n+6\sum_i\frac{n+1}{2}+3\sum_i\frac{(n+1)(2n+1)}{6n}$$

$$=3(n+1)(2n+1)+3n^2+3n(n+1)+\frac{(n+1)(2n+1)}{2}=13n^2+\frac{27}{2}n+\frac{7}{2}.$$

Hence $\sum_i\sum_j(2x_i^{*2}+3y_j^{*2})\Delta x\,\Delta y=\left(13n^2+\frac{27}{2}n+\frac{7}{2}\right)\left(\frac{3}{n}\right)\left(\frac{1}{n}\right)=39+\frac{81}{2n}+\frac{21}{2n^2}$, this implies

$$\iint_\Omega(2x^2+3y^2)\,dA=\lim_{n\to\infty}\left(39+\frac{81}{2n}+\frac{21}{2n^2}\right)=39+0+0=39.$$

Prologue for solutions 9–14:   Partition intervals $[-1,0]$ and $[-2,3]$ into n subintervals: $\Delta x=(0-(-1))/n=1/n$ and $\Delta y=(3-(-2))/n=5/n$. In each subinterval, choose the evaluation point to be the one with largest coordinates: $x_i^*=x_i=-1+i\Delta x=-1+i/n$ and $y_j^*=y_j=-2+j\Delta y=-2+5j/n$.

9.

$$\sum_i\sum_j(x_i^*+y_j^*)=\sum_i\sum_j\left(\left(-1+\frac{i}{n}\right)+\left(-2+\frac{5j}{n}\right)\right)=\sum_i\sum_j\left(-3+\frac{i}{n}+5\frac{j}{n}\right)$$

$$=\sum_i\sum_j(-3)+\sum_i\sum_j\frac{i}{n}+5\sum_i\sum_j\frac{j}{n}=\sum_i(-3)n+\sum_i i+5\sum_i\frac{n+1}{2}$$

$$=-3n^2+\frac{n(n+1)}{2}+5\left(\frac{n+1}{2}\right)n=-3n^2+3n(n+1)=3n.$$

Hence $\sum_i\sum_j(x_i^*+y_j^*)\Delta x\,\Delta y=(3n)\left(\frac{1}{n}\right)\left(\frac{5}{n}\right)=\frac{15}{n}$, this implies $\iint_\Omega(x+y)\,dA=\lim_{n\to\infty}\left(\frac{15}{n}\right)=0.$

11.

$$\sum_i\sum_j(y_j^*-2x_i^*)=\sum_i\sum_j\left(\left(-2+\frac{5j}{n}\right)-2\left(-1+\frac{i}{n}\right)\right)=\sum_i\sum_j\left(-2\frac{i}{n}+5\frac{j}{n}\right)$$

$$=-2\sum_i\sum_j\frac{i}{n}+5\sum_i\sum_j\frac{j}{n}=-2\sum_i i+5\sum_i\frac{n+1}{2}$$

$$=-2\frac{n(n+1)}{2}+5\left(\frac{n+1}{2}\right)n=\frac{3}{2}n(n+1).$$

Hence $\sum_i\sum_j(y_j^*-2x_i^*)\Delta x\,\Delta y=\frac{3}{2}n(n+1)\left(\frac{1}{n}\right)\left(\frac{5}{n}\right)=\frac{15}{2}\left(\frac{n+1}{n}\right)$, this implies $\iint_\Omega(y-2x)\,dA=$

$$\lim_{n \to \infty} \frac{15}{2} \left(\frac{n+1}{n} \right) = \frac{15}{2}.$$

13.
$$\sum_i \sum_j \left(y_j^{*2} - x_i^{*2} \right) = \sum_i \sum_j \left(\left(-2 + \frac{5j}{n} \right)^2 - \left(-1 + \frac{i}{n} \right)^2 \right)$$

$$= \sum_i \sum_j \left(3 + 2\frac{i}{n} - \frac{i^2}{n^2} - 20\frac{j}{n} + 25\frac{j^2}{n^2} \right)$$

$$= 3 \sum_i \sum_j 1 + 2 \sum_i \sum_j \frac{i}{n} - \sum_i \sum_j \frac{i^2}{n^2} - 20 \sum_i \sum_j \frac{j}{n} + 25 \sum_i \sum_j \frac{j^2}{n^2}$$

$$= 3 \sum_i n + 2 \sum_i i - \sum_i \frac{i^2}{n} - 20 \sum_i \frac{n+1}{2} + 25 \sum_i \frac{(n+1)(2n+1)}{6n}$$

$$= 3n^2 + 2\frac{n(n+1)}{2} - \frac{(n+1)(2n+1)}{6} - 20\frac{n(n+1)}{2} + 25\frac{(n+1)(2n+1)}{6}$$

$$= 2n^2 + 3n + 4.$$

Hence $\sum_i \sum_j (y_j^{*2} - x_i^{*2}) \Delta x \, \Delta y = (2n^2 + 3n + 4) \frac{5}{n^2} = 10 + \frac{15}{2n} + \frac{20}{n^2}$, this implies $\iint_\Omega (y^2 - x^2) \, dA =$

$$\lim_{n \to \infty} \left(10 + \frac{15}{2n} + \frac{20}{n^2} \right) = 10 + 0 + 0 = 10.$$

REMARK: $\displaystyle \lim_{n \to \infty} \left(3n^2 + 2\frac{n(n+1)}{2} - \frac{(n+1)(2n+1)}{6} - 20\frac{n(n+1)}{2} + 25\frac{(n+1)(2n+1)}{6} \right) \frac{5}{n^2}$

$$= \left(3 + 1 - \frac{1}{3} - 10 + \frac{25}{3} \right) \cdot 5 = (2) \cdot 5 = 10.$$

---

15. $\iint_\Omega (x+y) \, dA = \iint_\Omega x \, dA + \iint_\Omega y \, dA = 2 + 7 = 9.$

17. $\iint_\Omega (3x + 5y) \, dA = \iint_\Omega (3x) \, dA + \iint_\Omega (5y) \, dA = 3 \iint_\Omega x \, dA + 5 \iint_\Omega y \, dA = 3(2) + 5(7) = 6 + 35 = 41.$

19. If $\Omega_1 \cup \Omega_2 = \Omega$ and $\Omega_1 \cap \Omega_2 = \emptyset$, then $\iint_\Omega f(x,y) \, dA = \iint_{\Omega_1} f(x,y) \, dA + \iint_{\Omega_2} f(x,y) \, dA = 3 + 8 = 11.$

21. The rectangle $\Omega = \{(x,y) : 0 \le x \le 1, 1 \le y \le 2\}$ has area $(1-0) \cdot (2-1) = 1 \cdot 1 = 1$. The function $f(x,y) = x^5 y^2 + xy$ has minimum value $f(0,1) = 0$ and maximum value $f(1,2) = 6$ on Ω. Therefore $0 = 0 \cdot 1 \le \iint_\Omega (x^5 y^2 + xy) \, dA \le 6 \cdot 1 = 6.$

23. The disk $\Omega = \{(x,y) : x^2 + y^2 \le 1\}$ has radius $\sqrt{1} = 1$ and area $\pi 1^2 = \pi$. We will get sloppy bounds for function $f(x,y) = (x-y)/(4-x^2-y^2)$ on Ω by considering numerator and denominator separately. The min value of $x-y$ is $0-1 = -1$ and the max value is $1-0 = 1$; the max value of $4-x^2-y^2 = 4-(x^2+y^2)$ is $4-0 = 4$ and the min value is $4 - 1 = 3$. Therefore $-1/4 \le f \le 1/3$ and $-\pi/4 \le \iint_\Omega (x-y)/(4 - x^2 - y^2) dA \le \pi/3.$

REMARK: Ω is symmetric, thus $\displaystyle \iint_\Omega \frac{x}{4 - x^2 - y^2} dA = \iint_\Omega \frac{y}{4 - x^2 - y^2} dA$ and $\displaystyle \iint_\Omega \frac{x-y}{4 - x^2 - y^2} dA = 0.$

---

25. Ω is the triangle with vertices $(0,0)$, $(1,0)$, $(1/2, 1/2)$; its area is $(1/2)bh = (1/2)(1)(1/2) = 1/4$. On this triangle we have $0 \le x + y \le 1$ which implies $0 = \ln(1+0) \le \ln(1 + x + y) \le \ln(1+1) = \ln 2$. Thus $0 \le \iint_\Omega \ln(1 + x + y) dA \le (\ln 2)(1/4).$

Solutions 14.2   *The calculation of double integrals*   (pages 938-40)

1. $\displaystyle \int_0^1 \int_0^2 xy^2 \, dx \, dy = \int_0^1 \left(\frac{x^2 y^2}{2} \Big|_{x=0}^2 \right) dy = \int_0^1 2y^2 \, dy = \frac{2y^3}{3} \Big|_0^1 = \frac{2}{3}.$

3. $\displaystyle \int_2^5 \int_0^4 e^{x-y} \, dx \, dy = \int_2^5 \left(e^{x-y} \Big|_{x=0}^4 \right) dy = \int_2^5 \left(e^{4-y} - e^{-y} \right) dy = \left(-e^{4-y} + e^{-y} \right) \Big|_2^5 = -e^{-1} + e^{-5} + e^2 - e^{-2}.$

REMARK: $e^{x-y} = e^x e^{-y}$ and $\int_2^5 \int_0^4 e^x e^{-y} \, dx \, dy = (e^4 - e^0)(-e^{-5} + e^{-2}) = \left(\int_0^4 e^x \, dx \right) \left(\int_2^5 e^{-y} \, dy \right)$.

5.
$$\int_2^4 \int_{1+y}^{2+3y} (x - y^2) \, dx \, dy = \int_2^4 \left(\left(\frac{1}{2}x^2 - xy^2 \right) \Big|_{x=1+y}^{2+3y} \right) dy = \int_2^4 \left(\frac{3}{2} + 5y + 3y^2 - 2y^3 \right) dy$$
$$= \left(\frac{3}{2}y + \frac{5}{2}y^2 + y^3 - \frac{1}{2}y^4 \right) \Big|_2^4 = -31.$$

7.
$$\int_0^3 \int_{-\sqrt{9-y^2}}^{\sqrt{9-y^2}} (x^2 y) \, dx \, dy = \int_0^3 \left(\frac{x^3 y}{3} \Big|_{x=-\sqrt{9-y^2}}^{\sqrt{9-y^2}} \right) dy = \int_0^3 \frac{2}{3} y \, (9 - y^2)^{3/2} \, dy = \frac{-2}{15} (9 - y^2)^{5/2} \Big|_0^3 = \frac{162}{5}.$$

9.
$$\int_{-1}^1 \int_1^2 (x^2 + y^2) \, dx \, dy = \int_{-1}^1 \left(\left(\frac{x^3}{3} + xy^2 \right) \Big|_{x=1}^2 \right) dy = \int_{-1}^1 \left(\frac{7}{3} + y^2 \right) dy = \left(\frac{7y + y^3}{3} \right) \Big|_{-1}^1 = \frac{16}{3}.$$

REMARK: $\int_1^2 \int_{-1}^1 (x^2 + y^2) \, dy \, dx = \int_1^2 \left(\left(x^2 y + \frac{y^3}{3} \right) \Big|_{y=-1}^1 \right) dx = \int_1^2 \left(2x^2 + \frac{2}{3} \right) dx = \frac{2}{3}(x^3 + x) \Big|_1^2 = \frac{16}{3}.$

11.
$$\int_0^1 \int_{-2}^2 (x - y)^2 \, dx \, dy = \int_0^1 \left(\frac{1}{3}(x - y)^3 \Big|_{x=-2}^2 \right) dy = \int_0^1 \frac{1}{3} \left((y + 2)^3 - (y - 2)^3 \right) dy$$
$$= \frac{1}{12} \left((y + 2)^4 - (y - 2)^4 \right) \Big|_0^1 = \frac{80}{12} = \frac{20}{3}.$$

REMARK: $\int_{-2}^2 \int_0^1 (x - y)^2 \, dy \, dx = \int_{-2}^2 \left(\frac{-1}{3}(x - y)^3 \Big|_{y=0}^1 \right) dx = \int_{-2}^2 \left(\frac{1}{3} - x + x^2 \right) dx = \left(\frac{x}{3} - \frac{x^2}{2} + \frac{x^3}{3} \right) \Big|_{-2}^2 = \frac{20}{3}.$

13.
$$\int_1^3 \int_0^4 xe^{x^2 + y} \, dx \, dy = \int_1^3 \int_0^4 xe^{x^2} e^y \, dx \, dy = \int_1^3 e^y \left(\frac{1}{2}e^{x^2} \Big|_{x=0}^4 \right) dy = \frac{1}{2} (e^{16} - 1) \int_1^3 e^y \, dy$$
$$= \left(\frac{1}{2} e^{x^2} \Big|_{x=0}^4 \right) \left(e^y \Big|_{y=1}^3 \right) = \frac{1}{2} (e^{16} - 1)(e^3 - e).$$

15.
$$\int_0^1 \int_0^{\sqrt{1-y^2}} (x^2 + y) \, dx \, dy = \int_0^1 \left(\left(\frac{1}{3}x^3 + xy \right) \Big|_{x=0}^{\sqrt{1-y^2}} \right) dy$$
$$= \int_0^1 \left(\frac{1}{3} (1 - y^2)^{3/2} + y\sqrt{1 - y^2} \right) dy$$
$$= \left(\frac{y(1 - y^2)^{3/2}}{12} + \frac{y(1 - y^2)^{1/2}}{8} + \frac{1}{8} \sin^{-1} y - \frac{1}{3} (1 - y^2)^{3/2} \right) \Big|_0^1 = \frac{\pi}{16} + \frac{1}{3}.$$

REMARK: Items 91 and 93 of the text's integral table yield the formula for $\int (1 - y^2)^{3/2} dy$ used above; item 93 yields the formula for $\int x^2 \sqrt{1 - x^2} dx$ used below.

$$\int_0^1 \int_0^{\sqrt{1-x^2}} (x^2 + y) \, dy \, dx = \int_0^1 \left(\left(x^2 y + \frac{1}{2}y^2 \right) \Big|_{y=0}^{\sqrt{1-x^2}} \right) dx$$
$$= \int_0^1 \left(x^2 \sqrt{1 - x^2} + \frac{1}{2} (1 - x^2) \right) dx$$
$$= \left(\frac{-x(1 - x^2)^{3/2}}{4} + \frac{x(1 - x^2)^{1/2}}{8} + \frac{1}{8} \sin^{-1} x + \frac{x}{2} - \frac{x^3}{6} \right) \Big|_0^1 = \frac{\pi}{16} + \frac{1}{3}.$$

17. The triangular region $\Omega = \{(x, y) : 0 \le x \le y \le 1 - x\}$ has vertices $(0, 0)$, $(0, 1)$, and $(1/2, 1/2)$.
$$\iint_\Omega (x + 2y) \, dA = \int_0^{1/2} \int_x^{1-x} (x + 2y) \, dy \, dx = \int_0^{1/2} \left((xy + y^2) \Big|_{y=x}^{1-x} \right) dx$$
$$= \int_0^{1/2} (1 - x - 2x^2) \, dx = \left(x - \frac{x}{2} - \frac{2x^3}{3} \right) \Big|_0^{1/2} = \frac{7}{24}.$$

19. The parabolas $y = x^2$ and $y = 1 - x^2$ meet in the first quadrant at the point $(1/\sqrt{2}, 1/2)$; the region Ω is $\{(x, y) : 0 \le x \le 1/\sqrt{2}, x^2 \le y \le 1 - x^2\}$.

$$\iint_\Omega (x^2 + y)\, dA = \int_0^{1/\sqrt{2}} \int_{x^2}^{1-x^2} (x^2 + y)\, dy\, dx = \int_0^{1/\sqrt{2}} \left(\left(x^2 y + \frac{1}{2} y^2 \right) \Big|_{y=x^2}^{1-x^2} \right) dx$$

$$= \int_0^{1/\sqrt{2}} \left(\frac{1}{2} - 2x^4 \right) dx = \left(\frac{1}{2} x - \frac{2}{5} x^5 \right) \Big|_0^{1/\sqrt{2}} = \frac{\sqrt{2}}{5}.$$

21. $1 \le x \le y$ and $1 \le y \le 2 \iff 1 \le x \le y \le 2 \iff 1 \le x \le 2$ and $x \le y \le 2$, therefore (with the aid of item 71 of the text's integral table)

$$\iint_\Omega \frac{y}{\sqrt{x^2 + y^2}}\, dA = \int_1^2 \int_x^2 \frac{y}{\sqrt{x^2 + y^2}}\, dy\, dx = \int_1^2 \left(\sqrt{x^2 + y^2} \Big|_{y=x}^2 \right) dx$$

$$= \int_1^2 2 \left(\sqrt{x^2 + 4} - \sqrt{2}\, x \right) dx$$

$$= \left(\frac{x\sqrt{x^2 + 4}}{2} + 2\ln\left(x + \sqrt{x^2 + 4} \right) - \sqrt{2}\, \frac{x^2}{2} \right) \Big|_1^2$$

$$= 2\ln\left(\frac{2 + 2\sqrt{2}}{1 + \sqrt{5}} \right) - \frac{\sqrt{5} - \sqrt{2}}{2}.$$

23. $\displaystyle \int_0^{\pi/2} \int_0^\pi \sin x \, \cos y \, dy \, dx = \int_0^{\pi/2} (\sin x) \left(\sin y \big|_{y=0}^\pi \right) dx = \int_0^{\pi/2} (\sin x) (0 - 0)\, dx = \int_0^{\pi/2} 0 \; dx = 0.$

25. $\displaystyle \int_{-1}^1 \int_{-2}^2 x^2 y \, dy \, dx = \int_{-1}^1 x^2 \left(\frac{y^2}{2} \Big|_{y=-2}^2 \right) dx = \int_{-1}^1 x^2 (2 - 2)\, dx = \int_{-1}^1 0 \; dx = 0.$

27. $\displaystyle \int_{-1}^1 \int_0^{\sqrt{1-x^2}} x^2 y \, dy \, dx = \int_{-1}^1 x^2 \left(\frac{y^2}{2} \Big|_{y=0}^{\sqrt{1-x^2}} \right) dx = \int_{-1}^1 \frac{1}{2} (x^2 - x^4)\, dx = \frac{1}{2} \left(\frac{x^3}{3} - \frac{x^5}{5} \right) \Big|_{-1}^1 = \frac{2}{15}.$

29. $\displaystyle \int_{-4}^4 \int_1^3 \frac{y}{x^2} \, dx \, dy = \int_{-4}^4 y \left(\frac{-1}{x} \Big|_{x=1}^3 \right) dy = \int_{-4}^4 \frac{2}{3} y \, dy = \frac{1}{3} y^2 \Big|_{-4}^4 = 0.$

31. $\displaystyle \int_0^2 \int_1^5 \frac{y}{1+y^2}\, \frac{y}{x^2} \, dx \, dy = \int_0^2 y \left(\frac{-1}{x} \Big|_{x=1+y^2}^5 \right) dy = \int_0^2 \left(\frac{-y}{5} + \frac{y}{1+y^2} \right) dy = \left(\frac{-y^2}{10} + \frac{1}{2} \ln(1 + y^2) \right) \Big|_0^2 =$

$\dfrac{-2}{5} + \dfrac{1}{5} \ln 5.$

33. $\displaystyle \int_{y=0}^2 \int_{x=-1}^3 dx \, dy = \int_{x=-1}^3 \int_{y=0}^2 dy \, dx = \int_{x=-1}^3 \left(y \big|_{y=0}^2 \right) dx = \int_{x=-1}^3 2 \, dx = 2\, x \big|_{x=-1}^3 = 2 \cdot 4 = 8.$

33: $-1 \le x \le 3,\; 0 \le y \le 2$

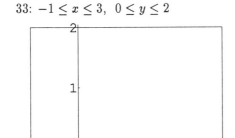

35: $1 \le x \le y,\; 2 \le y \le 4$

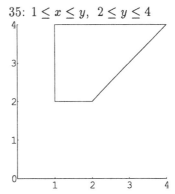

35.

$$\int_{y=2}^{4}\int_{x=1}^{y}\frac{y^3}{x^3}\,dx\,dy = \int_{y=2}^{4}\int_{x=1}^{2}\frac{y^3}{x^3}\,dx\,dy + \int_{y=2}^{4}\int_{x=2}^{y}\frac{y^3}{x^3}\,dx\,dy$$

$$= \int_{x=1}^{2}\int_{y=2}^{4}\frac{y^3}{x^3}\,dy\,dx + \int_{x=2}^{4}\int_{y=x}^{4}\frac{y^3}{x^3}\,dy\,dx$$

$$= \int_{x=1}^{2}\frac{60}{x^3}\,dx + \int_{x=2}^{4}\left(\frac{64}{x^3}-\frac{x}{4}\right)dx = \frac{-30}{x^2}\Big|_{1}^{2} + \left(\frac{-32}{x^2}-\frac{x^2}{8}\right)\Big|_{2}^{4} = \frac{45}{2}+\frac{9}{2}=27.$$

37.

$$\int_{x=0}^{1}\int_{y=x}^{1}dy\,dx = \int_{y=0}^{1}\int_{x=0}^{y}dx\,dy = \int_{y=0}^{1}y\,dy = \frac{1}{2}y^2\Big|_{0}^{1}=\frac{1}{2}.$$

39.

$$\int_{y=0}^{2}\int_{x=0}^{\sqrt{4-y^2}}\left(4-x^2\right)^{3/2}dx\,dy = \int_{x=0}^{2}\int_{y=0}^{\sqrt{4-x^2}}\left(4-x^2\right)^{3/2}dy\,dx$$

$$= \int_{x=0}^{2}\left(4-x^2\right)^2dx = \left(16x-\frac{8x^3}{3}+\frac{x^5}{5}\right)\Big|_{0}^{2}=\frac{256}{15}.$$

37: $0 \le x \le y \le 1$

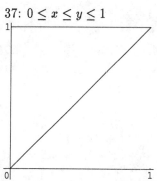

39: $0 \le x \le \sqrt{4-y^2},\ 0 \le y \le 2$

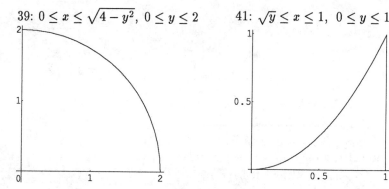

41: $\sqrt{y} \le x \le 1,\ 0 \le y \le 1$

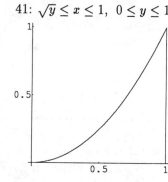

41.

$$\int_{y=0}^{1}\int_{x=\sqrt{y}}^{1}\sqrt{3-x^3}\,dx\,dy = \int_{x=0}^{1}\int_{y=0}^{x^2}\sqrt{3-x^3}\,dy\,dx$$

$$= \int_{x=0}^{1}x^2\sqrt{3-x^3}\,dx = \frac{-2}{9}\left(3-x^3\right)^{3/2}\Big|_{0}^{1}=\frac{2}{9}\left(3^{3/2}-2^{3/2}\right).$$

43.   See Example 8. We integrate $z = 3-x-y$ over the triangular region bounded by the axes and the line $x+y=3$ to compute the volume.

$$\int_{x=0}^{3}\int_{y=0}^{3-x}(3-x-y)\,dy\,dx = \int_{x=0}^{3}\left((3-x)y-\frac{y^2}{2}\right)\Big|_{y=0}^{3-x}dx = \int_{x=0}^{3}\frac{1}{2}(3-x)^2dx = \frac{-1}{6}(3-x)^3\Big|_{0}^{3}=\frac{9}{2}.$$

45.   On the cylinder $y^2 + z^2 = 4$ the height is $\sqrt{4-y^2}-\left(-\sqrt{4-y^2}\right)=2\sqrt{4-y^2}$; on the other cylinder $x^2+y^2=4$, we see x varies from $-\sqrt{4-y^2}$ to $\sqrt{4-y^2}$. The volume of the region enclosed by the two cylinders is $\displaystyle\int_{y=-2}^{2}\int_{x=-\sqrt{4-y^2}}^{\sqrt{4-y^2}}2\sqrt{4-y^2}\,dx\,dy = \int_{y=-2}^{2}4(4-y^2)\,dy = 4\left(4y-\frac{y^3}{3}\right)\Big|_{-2}^{2}=\frac{128}{3}.$

REMARK: The region is symmetric about the coordinate planes, so we can compute the volume in the first octant: $\displaystyle\int_{y=0}^{2}\int_{x=0}^{\sqrt{4-y^2}}\sqrt{4-y^2}\,dx\,dy$ and multiply by 8. The plane $z = x$ bisects the region in the first octant, so the total volume can be computed by $16\displaystyle\int_{y=0}^{2}\int_{x=0}^{\sqrt{4-y^2}}x\,dx\,dy = 8\int_{y=0}^{2}\left(4-y^2\right)dy = 8\cdot\frac{16}{3}.$

---

47. $x^2 + 4y^2 + 9z^2 = 36 \implies z = \pm(1/3)\sqrt{36 - x^2 - 4y^2}$; Ω is the ellipse $x^2 + 4y^2 = 36$ in the xy-plane.

$$\int_{y=-3}^{3} \int_{x=-2\sqrt{9-y^2}}^{2\sqrt{9-y^2}} 2\frac{\sqrt{36 - x^2 - 4y^2}}{3}\, dx\, dy \overset{x \equiv 2u}{=} \frac{2}{3} \int_{y=-3}^{3} \int_{u=-\sqrt{9-y^2}}^{\sqrt{9-y^2}} \sqrt{36 - 4u^2 - 4y^2}\,(2du)\, dy$$

$$= \frac{4}{3} \int_{y=-3}^{3} \int_{u=-\sqrt{9-y^2}}^{\sqrt{9-y^2}} 2\sqrt{9 - u^2 - y^2}\, du\, dy$$

$$= \frac{4}{3} \cdot \text{(volume of sphere with radius 3)} = \frac{4}{3} \cdot \frac{4\pi 3^3}{3} = 48\pi.$$

REMARK: A z-section is an ellipse: $x^2 + 4y^2 = 36 - 9z^2$; Problem 8.1.37 implies its area is $A(z) = \pi(36 - 9z^2)/2$. The volume of the ellipsoid is $\int_{-2}^{2} A(z)dz = \frac{\pi}{2} \int_{-2}^{2} (36 - 9z^2)dz = \frac{\pi}{2} (36z - 3z^3)\big|_{-2}^{2} = 48\pi.$

49. $\displaystyle\int_{0}^{\infty} \int_{0}^{\infty} e^{-(x+y)}dy\, dx = \int_{0}^{\infty} e^{-x} \left(-e^{-y}\big|_{y=0}^{\infty}\right) dx = \int_{0}^{\infty} e^{-x}dx = -e^{-x}\big|_{x=0}^{\infty} = 1.$

51. $\displaystyle\int_{y=-1}^{1} \int_{x=y^2}^{1} 2\sqrt{x - y^2}\, dx\, dy = \int_{y=-1}^{1} \frac{4}{3}(x - y^2)^{3/2}\Big|_{x=y^2}^{1} dy = \int_{y=-1}^{1} \frac{4}{3}(1 - y^2)^{3/2}dy$

$$= \left(\frac{1}{3}y(1 - y^2)^{3/2} + \frac{1}{2}y(1 - y^2)^{1/2} + \frac{1}{2}\sin^{-1} y\right)\Big|_{-1}^{1} = \frac{\pi}{2}.$$

REMARK: An x-section is a circle of radius $\sqrt{x}$ and area πx; the volume is $\int_{0}^{1} \pi x\, dx = \pi\frac{x^2}{2}\Big|_{0}^{1}.$

53. Let $f(x) = x^3 + 1$ and $g(x) = 3 - x^2$. Then $f = 0 \iff x = -1$ and $g = 0 \iff x = \pm\sqrt{3}$, the graphs meet if and only if $0 = f - g = (x^3 + 1) - (3 - x^2) = x^3 + x^2 - 2 = (x - 1)(x^2 + 2x + 2) \iff x = 1$.

The area of the left-hand region is

$$\int_{x=-\sqrt{3}}^{-1} \int_{y=0}^{3-x^2} dy\, dx + \int_{x=-1}^{1} \int_{y=x^3+1}^{3-x^2} dy\, dx = \int_{x=-\sqrt{3}}^{-1} (3 - x^2)\, dx + \int_{x=-1}^{1} (2 - x^2 - x^3)\, dx$$

$$= \left(3x - \frac{x^3}{3}\right)\Big|_{-\sqrt{3}}^{-1} + \left(2x - \frac{x^3}{3} - \frac{x^4}{4}\right)\Big|_{-1}^{1}$$

$$= \left(2\sqrt{3} - \frac{8}{3}\right) + \frac{10}{3} = 2\sqrt{3} + \frac{2}{3}.$$

The area of the right-hand region is

$$\int_{x=-1}^{1} \int_{y=0}^{x^3+1} dy\, dx + \int_{x=1}^{\sqrt{3}} \int_{y=0}^{3-x^2} dy\, dx = \int_{x=-1}^{1} (x^3 + 1)\, dx + \int_{x=1}^{\sqrt{3}} (3 - x^2)\, dx$$

$$= \left(\frac{x^4}{4} + x\right)\Big|_{-1}^{1} + \left(3x - \frac{x^3}{3}\right)\Big|_{1}^{\sqrt{3}} = 2 + \left(2\sqrt{3} - \frac{8}{3}\right) = 2\sqrt{3} - \frac{2}{3}.$$

55. The solid is a rectangular prism. Its base is the rectangle in the xy-plane with vertices at $(0, 1, 0), (0, 3, 0), (2, 3, 0), (2, 1, 0)$; its top is the planar region with vertices at $(0, 1, 3), (0, 3, 9), (2, 3, 11), (2, 1, 5)$.

57. $\displaystyle\int_{x=0}^{\infty} \int_{y=0}^{x} f\, dy\, dx = \lim_{M \to \infty} \int_{x=0}^{M} \int_{y=0}^{x} f\, dy\, dx = \lim_{M \to \infty} \int_{y=0}^{M} \int_{y=x}^{M} f\, dx\, dy = \int_{y=0}^{\infty} \int_{y=x}^{\infty} f\, dx\, dy.$ ∎

14.2 — The calculation of double integrals

59.   Suppose $p > 0$. Thus, building on the result of Example 9,

$$\frac{\pi}{2} = \int_{x=0}^{\infty} \frac{\sin x}{x}\, dx \overset{x=pt}{=} \int_{t=0}^{\infty} \frac{\sin pt}{pt}\,(p\,dt) = \int_{t=0}^{\infty} \frac{\sin pt}{t}\, dt.$$

Now integrate with respect to p over the interval $[0, q]$:

$$\frac{\pi}{2} q = \int_{p=0}^{q} \frac{\pi}{2}\, dp = \int_{p=0}^{q} \int_{t=0}^{\infty} \frac{\sin pt}{t}\, dt\, dp = \int_{t=0}^{\infty} \int_{p=0}^{q} \frac{\sin pt}{t}\, dp\, dt = \int_{t=0}^{\infty} \frac{-\cos pt}{t^2} \Big|_{p=0}^{q} dt = \int_{t=0}^{\infty} \frac{1 - \cos qt}{t^2}\, dt.$$

If $q = 2$, then $\frac{\pi}{2} q = \pi$ and $1 - \cos qt = 1 - \cos 2t = 2\sin^2 t$ imply $\pi = \int_{t=0}^{\infty} \frac{2\sin^2 t}{t^2}\, dt$ and $\int_{t=0}^{\infty} \left(\frac{\sin t}{t}\right)^2 dt = \frac{\pi}{2}$.

Solutions 14.3   *Density, Mass, and Center of mass*   **(page 944)**

1.
$$\mu = \int_{x=1}^{2} \int_{y=-1}^{1} (x^2 + y^2)\, dy\, dx = \int_{x=1}^{2} \left(x^2 y + \frac{y^3}{3}\right)\Big|_{y=-1}^{1} dx = \int_{x=1}^{2} \left(2x^2 + \frac{2}{3}\right) dx = \left(\frac{2x^2}{3} + \frac{2x}{3}\right)\Big|_{1}^{2} = \frac{16}{3},$$

$$M_y = \int_{x=1}^{2} \int_{y=-1}^{1} x\,(x^2 + y^2)\, dy\, dx = \int_{x=1}^{2} x\left(x^2 y + \frac{y^3}{3}\right)\Big|_{y=-1}^{1} dx = \int_{x=1}^{2} \left(2x^3 + \frac{2x}{3}\right) dx = \left(\frac{x^4}{4} + \frac{x^2}{3}\right)\Big|_{1}^{2} = \frac{17}{2},$$

$$M_x = \int_{x=1}^{2} \int_{y=-1}^{1} y\,(x^2 + y^2)\, dy\, dx = \int_{x=1}^{2} \left(\frac{x^2 y^2}{2} + \frac{y^4}{4}\right)\Big|_{y=-1}^{1} dx = \int_{x=1}^{2} 0\, dx = 0.$$

Therefore $(\bar{x}, \bar{y}) = \left(\dfrac{M_y}{\mu}, \dfrac{M_x}{\mu}\right) = \left(\dfrac{17/2}{16/3}, \dfrac{0}{16/3}\right) = \left(\dfrac{51}{32}, 0\right)$.

3.
$$\mu = \int_{x=0}^{\pi/6} \int_{y=0}^{\pi/18} \sin(2x + 3y)\, dy\, dx = \int_{x=0}^{\pi/6} \frac{-1}{3}\cos(2x + 3y)\Big|_{y=0}^{\pi/18} dx$$

$$= \frac{1}{3} \int_{x=0}^{\pi/6} (\cos 2x - \cos(2x + \pi/6))\, dx = \frac{1}{6}\left(\sin 2x - \sin(2x + \pi/6)\right)\Big|_{0}^{\pi/6} = \frac{\sqrt{3} - 1}{12},$$

$$M_y = \int_{x=0}^{\pi/6} \int_{y=0}^{\pi/18} x\sin(2x + 3y)\, dy\, dx = \int_{x=0}^{\pi/6} \frac{-1}{3}x\cos(2x + 3y)\Big|_{y=0}^{\pi/18} dx$$

$$= \frac{1}{3} \int_{x=0}^{\pi/6} x\,(\cos 2x - \cos(2x + \pi/6))\, dx$$

$$= \frac{1}{6}x\,(\sin 2x - \sin(2x + \pi/6))\Big|_{0}^{\pi/6} - \frac{1}{6}\int_{x=0}^{\pi/6} (\sin 2x - \sin(2x + \pi/6))\, dx$$

$$= \frac{\pi}{36}\left(\frac{\sqrt{3}}{2} - 1\right) + \frac{1}{12}\left(\cos 2x - \cos(2x + \pi/6)\right)\Big|_{0}^{\pi/6} = \frac{\pi}{36}\left(\frac{\sqrt{3}}{2} - 1\right) + \frac{\sqrt{3} - 1}{24},$$

$$M_x = \int_{x=0}^{\pi/6} \int_{y=0}^{\pi/18} y\sin(2x + 3y)\, dy\, dx = \int_{x=0}^{\pi/6} \left(\frac{-1}{3}y\cos(2x + 3y) + \frac{1}{9}\sin(2x + 3y)\right)\Big|_{y=0}^{\pi/18} dx$$

$$= \int_{x=0}^{\pi/6} \left(\frac{-\pi}{54}\cos(2x + \pi/6) + \frac{1}{9}\sin(2x + \pi/6) - \frac{1}{9}\sin 2x\right) dx$$

$$= \left(\frac{-\pi}{108}\sin(2x + \pi/6) - \frac{1}{18}\cos(2x + \pi/6) + \frac{1}{18}\cos 2x\right)\Big|_{0}^{\pi/6} = \frac{-\pi}{216} + \frac{\sqrt{3} - 1}{36}.$$

Therefore $(\bar{x}, \bar{y}) = \left(\dfrac{M_y}{\mu}, \dfrac{M_x}{\mu}\right) = \left(\dfrac{\pi(\sqrt{3} - 2)}{6(\sqrt{3} - 1)} + \dfrac{1}{2}, \dfrac{-\pi}{18(\sqrt{3} - 1)} + \dfrac{1}{3}\right)$.

5.
$$\mu = \int_{x=0}^{4} \int_{y=1}^{3} xe^{x-y} \, dy \, dx = \int_{x=0}^{4} -xe^{x-y} \Big|_{y=1}^{3} \, dx$$

$$= \int_{x=0}^{4} x \left(e^{x-1} - e^{x-3}\right) dx = (x-1)\left(e^{x-1} - e^{x-3}\right)\Big|_{0}^{4} = 3(e^3 - e) + (e^{-1} - e^{-3}),$$

$$M_y = \int_{x=0}^{4} \int_{y=1}^{3} x \left(xe^{x-y}\right) dy \, dx = \int_{x=0}^{4} -x^2 e^{x-y} \Big|_{y=1}^{3} \, dx$$

$$= \int_{x=0}^{4} x^2 \left(e^{x-1} - e^{x-3}\right) dx = (x^2 - 2x + 2)\left(e^{x-1} - e^{x-3}\right)\Big|_{0}^{4} = 10(e^3 - e) - 2(e^{-1} - e^{-3}),$$

$$M_x = \int_{x=0}^{4} \int_{y=1}^{3} y \left(xe^{x-y}\right) dy \, dx = \int_{x=0}^{4} -x(y+1)e^{x-y} \Big|_{y=1}^{3} \, dx$$

$$= \int_{x=0}^{4} 2x \left(e^{x-1} - e^{x-3}\right) dx = 2(x-1)\left(e^{x-1} - e^{x-3}\right)\Big|_{0}^{4} = 6e^3 - 12e + 2e^{-1} - 4e^{-3}.$$

Therefore $(\bar{x}, \bar{y}) = \left(\dfrac{M_y}{\mu}, \dfrac{M_x}{\mu}\right) = \left(\dfrac{10(e^3 - e) - 2(e^{-1} - e^{-3})}{3(e^3 - e) + (e^{-1} - e^{-3})}, \dfrac{6e^3 - 12e + 2e^{-1} - 4e^{-3}}{3(e^3 - e) + (e^{-1} - e^{-3})}\right).$

REMARK: The result of Problem 14.2.58 implies

$$\mu = \left(\int_{x=0}^{4} xe^x \, dx\right)\left(\int_{y=1}^{3} e^{-y} \, dy\right) = (3e^4 + 1)\left(e^{-1} - e^{-3}\right),$$

$$M_y = \left(\int_{x=0}^{4} x^2 e^x \, dx\right)\left(\int_{y=1}^{3} e^{-y} \, dy\right) = (10e^4 - 2)\left(e^{-1} - e^{-3}\right),$$

$$M_x = \left(\int_{x=0}^{4} xe^x \, dx\right)\left(\int_{y=1}^{3} ye^{-y} \, dy\right) = (3e^4 + 1)\left(2e^{-1} - 4e^{-3}\right).$$

Thus $(\bar{x}, \bar{y}) = \left(\dfrac{10e^4 - 2}{3e^4 + 1}, \dfrac{2e^{-1} - 4e^{-3}}{e^{-1} - e^{-3}}\right) = \left(3 + \dfrac{e^4 - 5}{3e^4 + 1}, 2 - \dfrac{2}{e^2 - 1}\right).$

7.
$$\mu = \int_{x=0}^{1} \int_{y=0}^{\sqrt{1-x^2}} (x^2 + y) \, dy \, dx = \int_{x=0}^{1} \left(x^2 y + \dfrac{y^2}{2}\right)\Big|_{y=0}^{\sqrt{1-x^2}} \, dx = \int_{x=0}^{1} \left(x^2\sqrt{1-x^2} + \dfrac{1}{2}(1-x^2)\right) dx$$

$$= \left(\dfrac{-x}{4}(1-x^2)^{3/2} + \dfrac{x}{8}\sqrt{1-x^2} + \dfrac{1}{8}\sin^{-1} t + \dfrac{x}{2} - \dfrac{x^3}{6}\right)\Big|_{0}^{1} = \dfrac{\pi}{16} + \dfrac{1}{3},$$

$$M_y = \int_{x=0}^{1} \int_{y=0}^{\sqrt{1-x^2}} x \left(x^2 + y\right) dy \, dx = \int_{x=0}^{1} x \left(x^2 y + \dfrac{y^2}{2}\right)\Big|_{y=0}^{\sqrt{1-x^2}} \, dx$$

$$= \int_{x=0}^{1} x \left(x^2\sqrt{1-x^2} + \dfrac{1}{2}(1-x^2)\right) dx \overset{x=\sin\theta}{=} \int_{\theta=0}^{\pi/2} \left(\sin^3\theta \cos^2\theta + \dfrac{1}{2}\sin\theta\cos^3\theta\right) d\theta$$

$$= \left(\dfrac{-1}{5}\sin^2\theta \cos^3\theta - \dfrac{2}{15}\cos^3\theta - \dfrac{1}{8}\cos^4\theta\right)\Big|_{0}^{\pi/2} = \dfrac{31}{120},$$

$$M_x = \int_{x=0}^{1} \int_{y=0}^{\sqrt{1-x^2}} y \left(x^2 + y\right) dy \, dx = \int_{x=0}^{1} \left(\dfrac{x^2 y^2}{2} + \dfrac{y^3}{3}\right)\Big|_{y=0}^{\sqrt{1-x^2}} \, dx$$

$$= \int_{x=0}^{1} \left(\dfrac{1}{2}x^2(1-x^2) + \dfrac{1}{3}(1-x^2)^{3/2}\right) dx = \dfrac{1}{2}\left(\dfrac{x^3}{3} - \dfrac{x^5}{5}\right)\Big|_{0}^{1} + \dfrac{1}{3}\int_{x=0}^{1}(1-x^2)^{3/2} \, dx$$

$$\overset{x=\sin\theta}{=} \dfrac{1}{15} + \dfrac{1}{3}\int_{\theta=0}^{\pi/2}\cos^4\theta \, d\theta = \dfrac{1}{15} + \dfrac{1}{3}\left(\dfrac{1}{4}\sin\theta\cos^3\theta + \dfrac{3}{8}\sin\theta\cos\theta + \dfrac{3}{8}\theta\right)\Big|_{0}^{\pi/2} = \dfrac{1}{15} + \dfrac{\pi}{16}.$$

Therefore $(\bar{x}, \bar{y}) = \left(\dfrac{M_y}{\mu}, \dfrac{M_x}{\mu}\right) = \left(\dfrac{62}{15\pi + 80}, \dfrac{15\pi + 16}{15\pi + 80}\right).$

9.
$$\mu = \int_{y=0}^{1/2}\int_{x=y}^{1-y}(x+2y)\,dx\,dy = \int_{y=0}^{1/2}\left(\frac{x^2}{2}+2xy\right)\Big|_{x=y}^{1-y}dy = \int_{y=0}^{1/2}\left(\frac{1}{2}+y-4y^2\right)dy$$
$$= \left(\frac{y}{2}+\frac{y^2}{2}-\frac{4y^3}{3}\right)\Big|_0^{1/2} = \frac{5}{24},$$
$$M_y = \int_{y=0}^{1/2}\int_{x=y}^{1-y}x(x+2y)\,dx\,dy = \int_{y=0}^{1/2}\left(\frac{x^3}{3}+x^2y\right)\Big|_{x=y}^{1-y}dy = \int_{y=0}^{1/2}\left(\frac{1}{3}-y^2-\frac{2y^3}{3}\right)dy$$
$$= \left(\frac{y}{3}-\frac{y^3}{3}-\frac{y^4}{6}\right)\Big|_0^{1/2} = \frac{11}{96},$$
$$M_x = \int_{y=0}^{1/2}\int_{x=y}^{1-y}y(x+2y)\,dx\,dy = \int_{y=0}^{1/2}y\left(\frac{x^2}{2}+2xy\right)\Big|_{x=y}^{1-y}dy = \int_{y=0}^{1/2}\left(\frac{y}{2}+y^2-4y^3\right)dy$$
$$= \left(\frac{y^2}{4}+\frac{y^3}{3}-y^4\right)\Big|_0^{1/2} = \frac{1}{24}.$$
Therefore $(\bar{x},\bar{y}) = \left(\dfrac{M_y}{\mu},\dfrac{M_x}{\mu}\right) = \left(\dfrac{11}{20},\dfrac{1}{5}\right).$

11.
$$\mu = \int_{y=0}^{\infty}\int_{x=0}^{\infty}e^{-y}(1+x)^{-3}dx\,dy = \int_{y=0}^{\infty}e^{-y}\frac{-1}{2(1+x)^2}\Big|_{x=0}^{\infty}dy = \int_{y=0}^{\infty}\frac{1}{2}e^{-y}dy = \frac{-1}{2}e^{-y}\Big|_0^{\infty} = \frac{1}{2},$$
$$M_y = \int_{y=0}^{\infty}\int_{x=0}^{\infty}e^{-y}\frac{x}{(1+x)^3}\,dx\,dy = \int_{y=0}^{\infty}e^{-y}\frac{-(1+2x)}{2(1+x)^2}\Big|_{x=0}^{\infty}dy = \int_{y=0}^{\infty}\frac{1}{2}e^{-y}dy = \frac{-1}{2}e^{-y}\Big|_0^{\infty} = \frac{1}{2},$$
$$M_x = \int_{y=0}^{\infty}\int_{x=0}^{\infty}\frac{ye^{-y}}{(1+x)^3}\,dx\,dy = \int_{y=0}^{\infty}ye^{-y}\frac{-1}{2(1+x)^2}\Big|_{x=0}^{\infty}dy = \int_{y=0}^{\infty}\frac{ye^{-y}}{2}dy = \frac{-(y+1)e^{-y}}{2}\Big|_0^{\infty} = \frac{1}{2}.$$
Therefore $(\bar{x},\bar{y}) = \left(\dfrac{M_y}{\mu},\dfrac{M_x}{\mu}\right) = (1,1).$

REMARK: A slight generalization of Problem 14.2.58 can be applied here.

13. The unit circle has area $A = \pi\cdot 1^2 = \pi$; the distance between the circle's center, the origin, and the line $x+y-4=0$ is $d = |0+0-4|/\sqrt{1^2+1^2} = 2\sqrt{2}$ (using the result of Problem 0.3.63). Since d is larger than the radius of the unit circle, the line does not intersect the circle and we may apply the First Theorem of Pappus. The volume of the solid of rotation is $V = 2\pi dA = 2\pi(2\sqrt{2})(\pi) = 4\sqrt{2}\,\pi^2.$

Solutions 14.4   *Double integrals in polar coordinates*   (page 949)

1. $$V = \int_{\theta=0}^{2\pi}\int_{r=0}^{a}r^n\,(r\,dr\,d\theta) = \int_{\theta=0}^{2\pi}\frac{r^{n+2}}{n+2}\Big|_{r=0}^{a}d\theta = \int_{\theta=0}^{2\pi}\frac{a^{n+2}}{n+2}d\theta = \frac{a^{n+2}}{n+2}\int_{\theta=0}^{2\pi}d\theta = \frac{a^{n+2}}{n+2}(2\pi).$$

3. $$V = \int_{\theta=0}^{2\pi}\int_{r=0}^{4(1-\cos\theta)}r^2\,(r\,dr\,d\theta) = \int_{\theta=0}^{2\pi}\frac{r^4}{4}\Big|_{r=0}^{4(1-\cos\theta)}d\theta = \int_{\theta=0}^{2\pi}64(1-\cos\theta)^4\,d\theta$$
$$= \left(16\cos^3\theta\sin\theta - \frac{256}{3}\cos^2\theta\sin\theta + 216\cos\theta\sin\theta - \frac{1280}{3}\sin\theta + 280\theta\right)\Big|_0^{2\pi} = 560\pi.$$

REMARK: In light of the fact that $2\displaystyle\int_{\theta=a}^{a+\pi}\int_{r=0}^{4(1-\cos\theta)}r^2\,(r\,dr\,d\theta) = 560\pi + \frac{1024}{3}\left(5+\cos^2 a\right)\sin a$, it is a non-trivial matter to explain why $2\displaystyle\int_{\theta=0}^{\pi}\int_{r=0}^{4(1-\cos\theta)}r^2\,(r\,dr\,d\theta)$ is more than coincidentally correct. (Hint: the function $z = r^2$ is symmetric about the pole and the cardioid is symmetric about at least one line through the pole.) Note, however, that $\displaystyle\int_{\theta=a}^{a+2\pi}\int_{r=0}^{4(1-\cos\theta)}r^2\,(r\,dr\,d\theta) = 560\pi$ for any choice of a.

5. The cardioid $r = 1-\cos\theta$ has area
$$\int_{\theta=0}^{2\pi}\int_{r=0}^{1-\cos\theta}r\,dr\,d\theta = \int_{\theta=0}^{2\pi}\frac{r^2}{2}\Big|_{r=0}^{1-\cos\theta}d\theta = \int_{\theta=0}^{2\pi}\frac{(1-\cos\theta)^2}{2}d\theta = \left(\frac{3\theta}{4}-\sin\theta+\frac{\cos\theta\sin\theta}{4}\right)\Big|_0^{2\pi} = \frac{3\pi}{2}.$$

7.   The limaçon with a loop $r = 1 + 2\cos\theta$ is plotted in text figure 8.6.8; the area of its outer loop is

$$\int_{\theta=-2\pi/3}^{2\pi/3}\int_{r=0}^{1+2\cos\theta} r\,dr\,d\theta = \int_{\theta=-2\pi/3}^{2\pi/3}\frac{(1+2\cos\theta)^2}{2}\,d\theta = \left(\frac{3\theta}{2} + 2\sin\theta + \cos\theta\sin\theta\right)\Bigg|_{-2\pi/3}^{2\pi/3} = 2\pi + \frac{3\sqrt{3}}{2}.$$

9.   The lemniscate $r^2 = \cos 2\theta$ is symmetric about the pole, the polar axis, and the vertical line $\theta = \pi/2$; the area of its right-hand loop is

$$\int_{\theta=-\pi/4}^{\pi/4}\int_{r=0}^{\sqrt{\cos 2\theta}} r\,dr\,d\theta = \int_{\theta=-\pi/4}^{\pi/4}\frac{\cos 2\theta}{2}\,d\theta = \frac{\sin 2\theta}{4}\Bigg|_{-\pi/4}^{\pi/4} = \frac{1}{2}.$$

Hence the area within both loops of the lemniscate is 1.

11.   If $a > b > 0$, the graph of $r = a + b\sin\theta$ is a limaçon without a loop; its area is

$$\int_{\theta=0}^{2\pi}\int_{r=0}^{a+b\sin\theta} r\,dr\,d\theta = \int_{\theta=0}^{2\pi}\frac{(a+b\sin\theta)^2}{2}\,d\theta = \left(\frac{a^2\theta}{2} - ab\cos\theta + \frac{b^2}{4}(\theta - \sin\theta\cos\theta)\right)\Bigg|_0^{2\pi} = \left(a^2 + \frac{b^2}{2}\right)\pi.$$

13. $4(1 + \sin\theta) \geq 6 \iff \sin\theta \geq 1/2$; if also $\theta \in [0, 2\pi)$, then $\theta \in [\pi/6, 5\pi/6]$. The enclosed region has area

$$\int_{\theta=\pi/6}^{5\pi/6}\int_{r=6}^{4(1+\sin\theta)} r\,dr\,d\theta = \int_{\theta=\pi/6}^{5\pi/6}(8(1+\sin\theta)^2 - 18)\,d\theta = (-6\theta - 16\cos\theta - 4\cos\theta\sin\theta)\big|_{\pi/6}^{5\pi/6} = 18\sqrt{3} - 4\pi.$$

15. $$\int_{\theta=0}^{2\pi}\int_{r=0}^{a}\sqrt{4a^2 - r^2}\,r\,dr\,d\theta = \int_{\theta=0}^{2\pi}\frac{-(4a^2-r^2)^{3/2}}{3}\Bigg|_{r=0}^{a} d\theta = \int_{\theta=0}^{2\pi}\frac{a^3}{3}\left(8 - 3\sqrt{3}\right)\,d\theta = \frac{a^3}{3}\left(8 - 3\sqrt{3}\right)(2\pi).$$

17.   The cone satisfies $z^2 = x^2 + y^2 = r^2$; the vertical distance between upper and lower surfaces is $2|z| = 2|r|$. The cylinder has equation $x^2 + y^2 = 4y$ which is equivalent to $r = 4\sin\theta$; the planar circle $r = 4\sin\theta$ is swept out (with positive r) for $\theta \in [0, \pi)$.

$$\int_{\theta=0}^{\pi}\int_{r=0}^{4\sin\theta}(2r)(r\,dr\,d\theta) = \int_{\theta=0}^{\pi}\frac{2r^3}{3}\Bigg|_{r=0}^{4\sin\theta} d\theta = \frac{128}{3}\int_{\theta=0}^{\pi}\sin^3\theta\,d\theta = \frac{128}{3}\left(\frac{-1}{3}\sin^2\theta\cos\theta - \frac{2}{3}\cos\theta\right)\Bigg|_0^{\pi} = \frac{512}{9}.$$

19.   A hyperboloid of one sheet similar to $z^2 = x^2 + y^2 - 1 = r^2 - 1$ is shown in table 11.8.1 of the text; the volume outside the hyperboloid but within the cylinder $3^2 = 9 = x^2 + y^2 = r^2$ is

$$\int_{\theta=0}^{2\pi}\int_{r=1}^{3} 2\sqrt{r^2-1}\,(r\,dr\,d\theta) = \int_{\theta=0}^{2\pi}\frac{2}{3}(r^2-1)\Bigg|_{r=1}^{3} d\theta = \int_{\theta=0}^{2\pi}\frac{32\sqrt{2}}{3}\,d\theta = \frac{64\sqrt{2}\pi}{3}.$$

21.   The circle $r = \cos\theta + 2\sin\theta$ is swept out for $\theta \in [0, \pi)$.

$$A = \int_{\theta=0}^{\pi}\int_{r=0}^{\cos\theta+2\sin\theta}(r\,dr\,d\theta) = \int_{\theta=0}^{\pi}\frac{1}{2}(\cos\theta + 2\sin\theta)^2\,d\theta = \left(\frac{-3}{4}\cos\theta\sin\theta + \frac{5}{4}\theta - \cos^2\theta\right)\Bigg|_0^{\pi} = \frac{5\pi}{4},$$

$$M_y = \int_{\theta=0}^{\pi}\int_{r=0}^{\cos\theta+2\sin\theta} r\cos\theta\,(r\,dr\,d\theta) = \int_{\theta=0}^{\pi}\frac{1}{3}(\cos\theta + 2\sin\theta)^3\cos\theta\,d\theta$$

$$= \left(\frac{-11}{12}\cos^3\theta\sin\theta - \frac{5}{8}\cos\theta\sin\theta + \frac{5}{8}\theta - \frac{1}{2}\cos^4\theta + \frac{2}{3}\sin^4\theta\right)\Bigg|_0^{\pi} = \frac{5\pi}{8},$$

$$M_y = \int_{\theta=0}^{\pi}\int_{r=0}^{\cos\theta+2\sin\theta} r\sin\theta\,(r\,dr\,d\theta) = \int_{\theta=0}^{\pi}\frac{1}{3}(\cos\theta + 2\sin\theta)^3\sin\theta\,d\theta$$

$$= \left(\frac{-1}{12}\cos^4\theta - \frac{1}{2}\cos^3\theta\sin\theta - \frac{3}{4}\cos\theta\sin\theta + \frac{5}{4}\theta + \sin^4\theta - \frac{2}{3}\cos\theta\sin^3\theta\right)\Bigg|_0^{\pi} = \frac{5\pi}{4}.$$

The centroid is $(\bar{x}, \bar{y}) = \left(\dfrac{M_y}{A}, \dfrac{M_x}{A}\right) = \left(\dfrac{5\pi/8}{5\pi/4}, \dfrac{5\pi/4}{5\pi/4}\right) = \left(\dfrac{1}{2}, 1\right).$

REMARK: $r = \cos\theta + 2\sin\theta \iff x^2 + y^2 = r^2 = r(\cos\theta + 2\sin\theta) = r\cos\theta + 2r\sin\theta = x + 2y \iff$
$(x - 1/2)^2 + (y - 1)^2 = 5/4.$

14.4 — Double integrals in polar coordinates

23.   If $a > b > 0$, then $r = a + b\cos\theta$ describes a limaçon without a loop which is swept out for $\theta \in [0, 2\pi)$.

$$A = \int_{\theta=0}^{2\pi} \int_{r=0}^{a+b\cos\theta} (r\,dr\,d\theta) = \int_{\theta=0}^{2\pi} \frac{1}{2}(a+b\cos\theta)^2 d\theta$$

$$= \left(\frac{a^2}{2}\theta + ab\sin\theta + \frac{b^2}{4}(\sin\theta\cos\theta + \theta)\right)\Big|_0^{2\pi} = \left(a^2 + \frac{b^2}{2}\right)\pi,$$

$$M_y = \int_{\theta=0}^{2\pi} \int_{r=0}^{a+b\cos\theta} r\cos\theta\,(r\,dr\,d\theta) = \int_{\theta=0}^{2\pi} \frac{1}{3}(a+b\cos\theta)^3\cos\theta\,d\theta$$

$$= \left(\frac{a^3}{3}\sin\theta + \frac{a^2 b}{2}(\sin\theta\cos\theta + \theta) + \frac{ab^2}{3}(\sin\theta\cos^2\theta + 2\sin\theta) + \frac{b^3}{24}(2\sin\theta\cos^3\theta + 3\sin\theta\cos\theta + 3\theta)\right)\Big|_0^{2\pi}$$

$$= \left(a^2 b + \frac{b^3}{4}\right)\pi,$$

$$M_x = \int_{\theta=0}^{2\pi} \int_{r=0}^{a+b\cos\theta} r\sin\theta\,(r\,dr\,d\theta) = \int_{\theta=0}^{2\pi} \frac{1}{3}(a+b\cos\theta)^3\sin\theta\,d\theta = \frac{-1}{12b}(a+b\cos\theta)^4\Big|_0^{2\pi} = 0.$$

The centroid is $(\bar{x}, \bar{y}) = \left(\dfrac{M_y}{A}, \dfrac{M_x}{A}\right) = \left(\dfrac{(a^2 b + b^3/4)\pi}{(a^2 + b^2/2)\pi}, \dfrac{0}{(a^2 + b^2/2)\pi}\right) = \left(\dfrac{b(b^2 + 4a^2)}{2(b^2 + 2a^2)}, 0\right).$

Solutions 14.5   *The triple integral*   (pages 955-6)

1.
$$\int_{y=0}^{1}\int_{x=0}^{y}\int_{z=0}^{x} y\,dz\,dx\,dy = \int_{y=0}^{1}\int_{x=0}^{y}\left(yz\big|_{z=0}^{x}\right)dx\,dy = \int_{y=0}^{1}\int_{x=0}^{y} yx\,dx\,dy$$

$$= \int_{y=0}^{1}\frac{yx^2}{2}\Big|_{x=0}^{y}dy = \int_{y=0}^{1}\frac{y^3}{2}dy = \frac{y^4}{8}\Big|_{y=0}^{1} = \frac{1}{8}.$$

3.
$$\int_{z=0}^{2}\int_{y=-z}^{z}\int_{x=y-z}^{y+z} 2xz\,dx\,dy\,dz = \int_{z=0}^{2}\int_{y=-z}^{z}\left(x^2 z\big|_{x=y-z}^{y+z}\right)dy\,dz = \int_{z=0}^{2}\int_{y=-z}^{z} 4yz^2\,dy\,dz$$

$$= \int_{z=0}^{2} 2y^2 z^2\big|_{y=-z}^{z}\,dz = \int_{z=0}^{2} 0\,dz = 0.$$

5.
$$\int_{y=0}^{\pi/2}\int_{z=0}^{\pi/2}\int_{x=0}^{z}\sin\left(\frac{x}{z}\right)dx\,dz\,dy = \int_{y=0}^{\pi/2}\int_{z=0}^{\pi/2}\left(-z\cos\left(\frac{x}{z}\right)\big|_{x=0}^{z}\right)dz\,dy = \int_{y=0}^{\pi/2}\int_{z=0}^{\pi/2}(1-\cos 1)z\,dz\,dy$$

$$= \int_{y=0}^{\pi/2}(1-\cos 1)\frac{z^2}{2}\Big|_{z=0}^{\pi/2}dy = \int_{y=0}^{\pi/2}(1-\cos 1)\frac{\pi^2}{8}dy = (1-\cos 1)\frac{\pi^3}{16}.$$

7.
$$\int_{x=0}^{1}\int_{y=0}^{\sqrt{1-x^2}}\int_{z=0}^{x} yz\,dz\,dy\,dx = \int_{x=0}^{1}\int_{y=0}^{\sqrt{1-x^2}}\frac{yz^2}{2}\Big|_{z=0}^{x}dy\,dx = \int_{x=0}^{1}\int_{y=0}^{\sqrt{1-x^2}}\frac{yx^2}{2}dy\,dx$$

$$= \int_{x=0}^{1}\frac{y^2 x^2}{4}\Big|_{y=0}^{\sqrt{1-x^2}}dx = \int_{x=0}^{1}\frac{(1-x^2)x^2}{4}dx = \left(\frac{x^3}{12} - \frac{x^5}{120}\right)\Big|_{x=0}^{1} = \frac{1}{30}.$$

9.   The region of integration for $I = \displaystyle\int_{y=0}^{1}\int_{x=0}^{y}\int_{z=0}^{x}$ satisfies inequalities $0 \le y \le 1, 0 \le x \le y, 0 \le z \le x$

which can be displayed as $\left\{\begin{array}{ccccc} 0 & & \le & y & \le & 1 \\ 0 & & \le & x & \le & y \\ 0 & \le & z & \le & x \end{array}\right\}$ and summarized as $0 \le z \le x \le y \le 1$.

a. $0 \le z \le x \le y \le 1 \iff \left\{\begin{array}{ccccc} 0 & \le & z & & \le & 1 \\ & & z & \le & y & \le 1 \\ & & z & \le & x & \le y \end{array}\right\}$; therefore $I = \displaystyle\int_{z=0}^{1}\int_{y=z}^{1}\int_{x=z}^{y}.$

b. $0 \le z \le x \le y \le 1 \iff \left\{\begin{array}{ccccc} 0 & & \le & x & \le & 1 \\ 0 & \le & z & \le & x \\ & & x & \le & y & \le 1 \end{array}\right\}$; therefore $I = \displaystyle\int_{x=0}^{1}\int_{z=0}^{x}\int_{y=x}^{1}.$

REMARK: The region of integration is the tetrahedron with vertices $(0,0,0), (0,1,0), (1,1,0)$, and $(1,1,1)$.

11.   The region of integration for $I = \int_{x=0}^{1} \int_{y=0}^{\sqrt{1-x^2}} \int_{z=0}^{x}$ satisfies inequalities $0 \le x \le 1$, $0 \le y \le \sqrt{1-x^2}$, $0 \le z \le x$ which can be rephrased as $0 \le x$, $0 \le y$, $x^2 + y^2 \le 1$, $0 \le z \le x \le 1$. The region lies in the first octant, within the cylinder $x^2 + y^2 = 1$, and below the plane $z = x$.

a.   The limits for the two innermost integrals involve only constants and functions of x, hence they can be interchanged: $I = \int_{x=0}^{1} \int_{z=0}^{x} \int_{y=0}^{\sqrt{1-x^2}}$

b. $0 \le z \le x \le 1$ and $x^2 + y^2 \le 1 \implies z \le x \le \sqrt{1-y^2}$; $0 \le z \le x \le 1 \implies \sqrt{1-x^2} \le \sqrt{1-z^2}$,

therefore $0 \le y$ and $x^2 + y^2 \le 1 \implies 0 \le y \le \sqrt{1-x^2} \le \sqrt{1-z^2}$. Hence $I = \int_{z=0}^{1} \int_{y=0}^{\sqrt{1-z^2}} \int_{x=z}^{\sqrt{1-y^2}}$.

13.
$$\iiint_S z \, dV = \int_{x=0}^{1} \int_{y=0}^{1-x} \int_{z=0}^{1-x-y} z \, dz \, dy \, dx = \int_{x=0}^{1} \int_{y=0}^{1-x} \frac{z^2}{2} \Big|_{z=0}^{1-x-y} dy \, dx$$
$$= \int_{x=0}^{1} \int_{y=0}^{1-x} \frac{(1-x-y)^2}{2} dy \, dx = \int_{x=0}^{1} \frac{-(1-x-y)^3}{6} \Big|_{y=0}^{1-x} dx$$
$$= \int_{x=0}^{1} \frac{(1-x)^3}{6} dx = \frac{-(1-x)^4}{24} \Big|_{x=0}^{1} = \frac{1}{24}.$$

15.
$$\iiint_S (x+y+z) \, dV = \int_{x=0}^{1} \int_{y=0}^{1-x} \int_{z=0}^{1-x-y} (x+y+z) \, dz \, dy \, dx = \int_{x=0}^{1} \int_{y=0}^{1-x} \left((x+y)z + \frac{z^2}{2} \right) \Big|_{z=0}^{1-x-y} dy \, dx$$
$$= \int_{x=0}^{1} \int_{y=0}^{1-x} \frac{1}{2} \left(1 - (x+y)^2 \right) dy \, dx = \int_{x=0}^{1} \left(\frac{y}{2} - \frac{(x+y)^3}{6} \right) \Big|_{y=0}^{1-x} dx$$
$$= \int_{x=0}^{1} \left(\frac{1}{3} - \frac{x}{2} + \frac{x^3}{6} \right) dx = \left(\frac{x}{3} - \frac{x^2}{4} + \frac{x^4}{24} \right) \Big|_{x=0}^{1} = \frac{1}{8}.$$

17.
$$\iiint_S z^2 \, dV = \int_{z=0}^{2} \int_{y=0}^{(6-3z)/2} \int_{x=0}^{6-3z-2y} z^2 \, dx \, dy \, dz = \int_{z=0}^{2} \int_{y=0}^{(6-3z)/2} z^2 x \Big|_{x=0}^{6-3z-2y} dy \, dz$$
$$= \int_{z=0}^{2} \int_{y=0}^{(6-3z)/2} z^2 (6 - 3z - 2y) \, dy \, dz = \int_{z=0}^{2} z^2 \left((6-3z)y - y^2 \right) \Big|_{y=0}^{(6-3z)/2} dz$$
$$= \int_{z=0}^{2} \frac{9}{4} z^2 (2-z)^2 dz = \left(3z^3 - \frac{9z^4}{4} + \frac{9z^5}{20} \right) \Big|_{z=0}^{2} = \frac{12}{5}.$$

19.   The plane $x+y+z = 1$ and the coordinate planes bound the tetrahedron with vertices $(0,0,0)$, $(1,0,0)$, $(0,1,0)$, $(0,0,1)$; the volume is
$$\int_{z=0}^{1} \int_{y=0}^{1-z} \int_{x=0}^{1-z-y} dx \, dy \, dz = \int_{z=0}^{1} \int_{y=0}^{1-z} (1 - z - y) \, dy \, dz = \int_{z=0}^{1} \frac{-(1-z-y)^2}{2} \Big|_{y=0}^{1-z} dz$$
$$= \int_{z=0}^{1} \frac{(1-z)^2}{2} dz = \frac{-(1-z)^3}{6} \Big|_{z=0}^{1} = \frac{1}{6}.$$

21.
$$V = \int_{x=0}^{3} \int_{y=0}^{4-x} \int_{z=0}^{\sqrt{9-x^2}} dz \, dy \, dx = \int_{x=0}^{3} \int_{y=0}^{4-x} \sqrt{9-x^2} \, dy \, dx = \int_{x=0}^{3} (4-x)\sqrt{9-x^2} \, dx$$
$$= \left(2x\sqrt{9-x^2} + 18\sin^{-1}\left(\frac{x}{3}\right) + \left(\frac{1}{3}\right)(9-x^2)^{3/2} \right) \Big|_{x=0}^{3} = 9(\pi - 1).$$

REMARK: We can use polar coordinates in the xz-plane: $x = r\cos\theta$ and $z = r\sin\theta$. Thus
$$V = \int_{x=0}^{3} \int_{z=0}^{\sqrt{9-x^2}} \int_{y=0}^{4-x} dy \, dz \, dx = \int_{x=0}^{3} \int_{z=0}^{\sqrt{9-x^2}} (4-x) \, dz \, dx = \int_{\theta=0}^{\pi/2} \int_{r=0}^{3} (4 - r\cos\theta) \, r \, dr \, d\theta$$
$$= \int_{\theta=0}^{\pi/2} \left(2r^2 - \frac{r^3}{3}\cos\theta \right) \Big|_{r=0}^{3} d\theta = \int_{\theta=0}^{\pi/2} (18 - 9\cos\theta) \, d\theta = (18\theta - 9\sin\theta)\big|_{\theta=0}^{\pi/2} = 9\pi - 9.$$

23.

$$V = \int_{z=2}^{4} \int_{y=-\sqrt{16-z^2}}^{\sqrt{16-z^2}} \int_{x=-\sqrt{16-z^2-y^2}}^{\sqrt{16-z^2-y^2}} dx\, dy\, dz = \int_{z=2}^{4} \int_{y=-\sqrt{16-z^2}}^{\sqrt{16-z^2}} 2\sqrt{16-z^2-y^2}\, dy\, dz$$

$$= \int_{z=2}^{4} \left(2y\sqrt{16-z^2-y^2} + \left(16-z^2\right)\sin^{-1}\frac{y}{\sqrt{16-z^2}} \right)\Bigg|_{y=-\sqrt{16-z^2}}^{\sqrt{16-z^2}} dz = \int_{z=2}^{4} \left(16-z^2\right)\pi\, dz = \frac{40\pi}{3}.$$

25. Suppose a, b, c are positive. The ellipsoid $(x/a)^2 + (y/b)^2 + (z/c)^2 \le 1$ has volume

$$V = \int_{z=-c}^{c} \int_{y=-b\sqrt{1-(z/c)^2}}^{b\sqrt{1-(z/c)^2}} \int_{x=-a\sqrt{1-(z/c)^2-(y/b)^2}}^{a\sqrt{1-(z/c)^2-(y/b)^2}} dx\, dy\, dz$$

$$= \int_{z=-c}^{c} \int_{y=-b\sqrt{1-(z/c)^2}}^{b\sqrt{1-(z/c)^2}} 2a\sqrt{1-(z/c)^2-(y/b)^2}\, dy\, dz$$

$$= \int_{z=-c}^{c} \left(ya\sqrt{1-(z/c)^2-(y/b)^2} + ab\left(1-\left(\frac{z}{c}\right)^2\right)\sin^{-1}\frac{y}{b\sqrt{1-(z/c)^2}} \right)\Bigg|_{y=-b\sqrt{1-(z/c)^2}}^{b\sqrt{1-(z/c)^2}} dz$$

$$= \int_{z=-c}^{c} ab\pi\left(1-\left(\frac{z}{c}\right)^2\right) dz = \frac{4}{3}\pi abc.$$

27.

$$\mu = \iiint_S \rho\, dV = \int_{x=0}^{1} \int_{y=0}^{1-x} \int_{z=0}^{1-x-y} x\, dz\, dy\, dx = \int_{x=0}^{1} \int_{y=0}^{1-x} x(1-x-y)\, dy\, dx$$

$$= \int_{x=0}^{1} x\left((1-x)y - \frac{y^2}{2}\right)\Bigg|_{y=0}^{1-x} dx = \int_{x=0}^{1} \frac{x(1-x)^2}{2}\, dx = \left(\frac{x^2}{4} - \frac{x^3}{3} + \frac{x^4}{8}\right)\Bigg|_{x=0}^{1} = \frac{1}{24}.$$

29.

$$\mu = \iiint_S \rho\, dV = \int_{x=0}^{3} \int_{y=0}^{4-x} \int_{z=0}^{\sqrt{9-x^2}} z\, dz\, dy\, dx = \int_{x=0}^{3} \int_{y=0}^{4-x} \frac{1}{2}\left(9-x^2\right) dy\, dx$$

$$= \int_{x=0}^{3} \frac{1}{2}\left(9-x^2\right)(4-x)\, dx = \left(18x - \frac{9x^2}{4} - \frac{2x^3}{3} + \frac{x^4}{8}\right)\Bigg|_{x=0}^{3} = \frac{207}{8}.$$

31. The plane $\dfrac{x}{a} + \dfrac{y}{b} + \dfrac{z}{c} = 1$ and the coordinate planes bound the tetrahedron with vertices $(0,0,0)$, $(a,0,0)$, $(0,b,0)$, $(0,0,c)$. The following computations use $\rho = 1$ but the centroid is the same for any other constant density.

$$\mu = \iiint_S dV = \int_{z=0}^{c} \int_{y=0}^{b(1-z/c)} \int_{x=0}^{a(1-z/c-y/b)} dx\, dy\, dz = \int_{z=0}^{c} \int_{y=0}^{b(1-z/c)} a\left(1-\frac{z}{c}-\frac{y}{b}\right) dy\, dz$$

$$= a\int_{z=0}^{c} \left(\left(1-\frac{z}{c}\right)y - \frac{y^2}{2b}\right)\Bigg|_{y=0}^{b(1-z/c)} dz = \frac{ab}{2}\int_{z=0}^{c}\left(1-\frac{z}{c}\right)^2 dz = \frac{-abc}{6}\left(1-\frac{z}{c}\right)^3\Bigg|_{z=0}^{c} = \frac{abc}{6},$$

$$M_{xy} = \iiint_S z\, dV = \int_{z=0}^{c} z\int_{y=0}^{b(1-z/c)} \int_{x=0}^{a(1-z/c-y/b)} dx\, dy\, dz = \frac{ab}{2}\int_{z=0}^{c} z\left(1-\frac{z}{c}\right)^2 dz$$

$$= ab\left(\frac{z^2}{4} - \frac{z^3}{3c} + \frac{z^4}{8c^2}\right)\Bigg|_{z=0}^{c} = \frac{abc^2}{24},$$

$$M_{xz} = \iiint_S y\, dV = \int_{y=0}^{b} y\int_{x=0}^{a(1-y/b)} \int_{z=0}^{c(1-y/b-x/a)} dz\, dx\, dy = \frac{ca}{2}\int_{y=0}^{b} y\left(1-\frac{y}{b}\right)^2 dy = \frac{ab^2c}{24},$$

$$M_{yz} = \iiint_S x\, dV = \int_{x=0}^{a} x\int_{z=0}^{c(1-x/a)} \int_{y=0}^{b(1-x/a-z/c)} dy\, dz\, dx = \frac{bc}{2}\int_{x=0}^{a} x\left(1-\frac{x}{a}\right)^2 dx = \frac{a^2bc}{24}.$$

Therefore $(\bar{x}, \bar{y}, \bar{z}) = \left(\dfrac{M_{yz}}{\mu}, \dfrac{M_{xz}}{\mu}, \dfrac{M_{xy}}{\mu}\right) = \left(\dfrac{a}{4}, \dfrac{b}{4}, \dfrac{c}{4}\right).$

33.
$$M_{yz} = \iiint_S x\,\rho\,dV = \int_{x=0}^1 x^2 \int_{y=0}^{1-x} \int_{z=0}^{1-x-y} dz\,dy\,dx = \int_{x=0}^1 x^2 \int_{y=0}^{1-x} z\Big|_{z=0}^{1-x-y} dy\,dx$$

$$= \int_{x=0}^1 x^2 \int_{y=0}^{1-x} (1-x-y)\,dy\,dx = \int_{x=0}^1 x^2 \left((1-x)y - \frac{y^2}{2} \right)\Big|_{y=0}^{1-x} dx$$

$$= \int_{x=0}^1 x^2 \frac{(1-x)^2}{2}\,dx = \left(\frac{x^3}{6} - \frac{x^4}{4} + \frac{x^5}{10} \right)\Big|_{x=0}^1 = \frac{1}{60},$$

$$M_{xz} = \iiint_S y\,\rho\,dV = \int_{x=0}^1 x \int_{y=0}^{1-x} y \int_{z=0}^{1-x-y} dz\,dy\,dx = \int_{x=0}^1 x \int_{y=0}^{1-x} y\,z\Big|_{z=0}^{1-x-y} dy\,dx$$

$$= \int_{x=0}^1 x \int_{y=0}^{1-x} y(1-x-y)\,dy\,dx = \int_{x=0}^1 x \left((1-x)\frac{y^2}{2} - \frac{y^3}{3} \right)\Big|_{y=0}^{1-x} dx$$

$$= \int_{x=0}^1 x \frac{(1-x)^3}{6}\,dx = \left(\frac{x^2}{12} - \frac{x^3}{6} + \frac{x^4}{8} - \frac{x^5}{30} \right)\Big|_{x=0}^1 = \frac{1}{120},$$

$$M_{xy} = \iiint_S z\,\rho\,dV = \int_{x=0}^1 x \int_{y=0}^{1-x} \int_{z=0}^{1-x-y} z\,dz\,dy\,dx = \int_{x=0}^1 x \int_{y=0}^{1-x} \frac{z^2}{2}\Big|_{z=0}^{1-x-y} dy\,dx$$

$$= \int_{x=0}^1 x \int_{y=0}^{1-x} \frac{(1-x-y)^2}{2}\,dy\,dx = \int_{x=0}^1 x \left(\frac{-1}{6}(1-x-y)^3 \right)\Big|_{y=0}^{1-x} dx$$

$$= \frac{1}{6} \int_{x=0}^1 x(1-x)^3\,dx = \frac{1}{6} \left(\frac{-(1-x)^4}{4} + \frac{(1-x)^5}{5} \right)\Big|_{x=0}^1 = \frac{1}{120} = M_{xz}.$$

Therefore $(\bar{x}, \bar{y}, \bar{z}) = \left(\dfrac{M_{yz}}{\mu}, \dfrac{M_{xz}}{\mu}, \dfrac{M_{xy}}{\mu} \right) = \left(\dfrac{1/60}{1/24}, \dfrac{1/120}{1/24}, \dfrac{1/120}{1/24} \right) = \left(\dfrac{24}{60}, \dfrac{24}{120}, \dfrac{24}{120} \right) = \left(\dfrac{2}{5}, \dfrac{1}{5}, \dfrac{1}{5} \right).$

35.
$$M_{yz} = \iiint_S x\rho\,dV = \int_{x=0}^3 x \int_{y=0}^{4-x} \int_{z=0}^{\sqrt{9-x^2}} z\,dz\,dy\,dx = \int_{x=0}^3 x \int_{y=0}^{4-x} \frac{1}{2}(9-x^2)\,dy\,dx$$

$$= \frac{1}{2} \int_{x=0}^3 x(9-x^2)(4-x)\,dx = \frac{1}{2} \left(18x^2 - 3x^3 - x^4 + \frac{x^5}{5} \right)\Big|_{x=0}^3 = \frac{243}{10},$$

$$M_{xz} = \iiint_S y\rho\,dV = \int_{x=0}^3 \int_{y=0}^{4-x} \int_{z=0}^{\sqrt{9-x^2}} z\,dz\,dy\,dx = \frac{1}{2} \int_{x=0}^3 \int_{y=0}^{4-x} y(9-x^2)\,dy\,dx$$

$$= \frac{1}{4} \int_{x=0}^3 (9-x^2)(4-x)^2\,dx = \frac{1}{4} \left(144x - 36x^2 - \frac{7x^3}{3} - 2x^4 + \frac{x^5}{5} \right)\Big|_{x=0}^3 = \frac{198}{5},$$

$$M_{xy} = \iiint_S z\rho\,dV = \int_{x=0}^3 \int_{y=0}^{4-x} \int_{z=0}^{\sqrt{9-x^2}} z^2\,dz\,dy\,dx = \int_{x=0}^3 \int_{y=0}^{4-x} \frac{1}{3}(9-x^2)^{3/2}\,dy\,dx$$

$$= \frac{1}{3} \int_{x=0}^3 (9-x^2)^{3/2}(4-x)\,dx$$

$$= \left(\frac{x(9-x^2)^{3/2}}{3} + \frac{9x(9-x^2)^{1/2}}{2} + \frac{81}{2}\sin^{-1}\frac{x}{3} + \frac{(9-x^2)^{5/2}}{15} \right)\Big|_{x=0}^3 = \frac{81\pi}{4} - \frac{81}{5}.$$

Therefore $(\bar{x}, \bar{y}, \bar{z}) = \left(\dfrac{M_{yz}}{\mu}, \dfrac{M_{xz}}{\mu}, \dfrac{M_{xy}}{\mu} \right) = \left(\dfrac{243/10}{207/8}, \dfrac{198/5}{207/8}, \dfrac{81(\pi/4 - 1/5)}{207/8} \right) = \left(\dfrac{108}{115}, \dfrac{176}{115}, \dfrac{18\pi}{23} - \dfrac{72}{115} \right).$

37. a. The solid in the first octant which is bounded by surfaces $y = 1$, $y = x$, $z = 0$, $z = xy$ consists of points satisfying the inequalities $0 \le x \le y \le 1$ and $0 \le z \le xy$.

$$\mu = \iiint_S \rho\,dV = \int_{y=0}^1 \int_{x=0}^y \int_{z=0}^{xy} (1+2z)\,dz\,dx\,dy = \int_{y=0}^1 \int_{x=0}^y (z+z^2)\Big|_{z=0}^{xy}\,dx\,dy$$

$$= \int_{y=0}^1 \int_{x=0}^y (xy + (xy)^2)\,dx\,dy = \int_{y=0}^1 \left(\frac{x^2 y}{2} + \frac{x^3 y^2}{3} \right)\Big|_{x=0}^y dy$$

$$= \int_{y=0}^1 \left(\frac{y^3}{2} + \frac{y^5}{3} \right) dy = \left(\frac{y^4}{8} + \frac{y^6}{18} \right)\Big|_{y=0}^1 = \frac{13}{72},$$

14.5 — The triple integral

$$M_{yz} = \iiint_S x\rho\, dV = \int_{y=0}^1 \int_{x=0}^y \int_{z=0}^{xy} x(1+2z)\, dz\, dx\, dy = \int_{y=0}^1 \int_{x=0}^y x\left(z+z^2\right)\Big|_{z=0}^{xy}\, dx\, dy$$

$$= \int_{y=0}^1 \int_{x=0}^y \left(x^2 y + x^3 y^2\right) dx\, dy = \int_{y=0}^1 \left(\frac{x^3 y}{3} + \frac{x^4 y^2}{4}\right)\Big|_{x=0}^y dy$$

$$= \int_{y=0}^1 \left(\frac{y^4}{3} + \frac{y^6}{4}\right) dy = \left(\frac{y^5}{15} + \frac{y^7}{28}\right)\Big|_{y=0}^1 = \frac{43}{420},$$

$$M_{xz} = \iiint_S y\rho\, dV = \int_{y=0}^1 \int_{x=0}^y \int_{z=0}^{xy} y(1+2z)\, dz\, dx\, dy = \int_{y=0}^1 \int_{x=0}^y y\left(z+z^2\right)\Big|_{z=0}^{xy}\, dx\, dy$$

$$= \int_{y=0}^1 \int_{x=0}^y \left(xy^2 + x^2 y^3\right) dx\, dy = \int_{y=0}^1 \left(\frac{x^2 y^2}{2} + \frac{x^3 y^3}{3}\right)\Big|_{x=0}^y dy$$

$$= \int_{y=0}^1 \left(\frac{y^4}{2} + \frac{y^6}{3}\right) dy = \left(\frac{y^5}{10} + \frac{y^7}{21}\right)\Big|_{y=0}^1 = \frac{31}{210},$$

$$M_{xy} = \iiint_S z\rho\, dV = \int_{y=0}^1 \int_{x=0}^y \int_{z=0}^{xy} z(1+2z)\, dz\, dx\, dy = \int_{y=0}^1 \int_{x=0}^y \left(\frac{z^2}{2} + \frac{2z^3}{3}\right)\Big|_{z=0}^{xy}\, dx\, dy$$

$$= \int_{y=0}^1 \int_{x=0}^y \left(\frac{x^2 y^2}{2} + \frac{2x^3 y^3}{3}\right) dx\, dy = \int_{y=0}^1 \left(\frac{x^3 y^2}{6} + \frac{x^4 y^3}{6}\right)\Big|_{x=0}^y dy$$

$$= \frac{1}{6}\int_{y=0}^1 \left(y^5 + y^7\right) dy = \frac{1}{6}\left(\frac{y^6}{6} + \frac{y^8}{8}\right)\Big|_{y=0}^1 = \frac{7}{144}.$$

Therefore $(\bar{x}, \bar{y}, \bar{z}) = \left(\dfrac{M_{yz}}{\mu}, \dfrac{M_{xz}}{\mu}, \dfrac{M_{xy}}{\mu}\right) = \left(\dfrac{43/420}{13/72}, \dfrac{31/210}{13/72}, \dfrac{7/144}{13/72}\right) = \left(\dfrac{258}{455}, \dfrac{372}{455}, \dfrac{7}{26}\right).$

b. $0 \le \dfrac{258}{455} \le \dfrac{372}{455} \le 1$ and $0 \le \dfrac{7}{26} \approx .269 < .464 \approx \dfrac{258}{455} \cdot \dfrac{372}{455}$; therefore $0 \le \bar{x} \le \bar{y} \le 1$ and $0 \le \bar{z} \le \bar{x}\bar{y}$ so $(\bar{x}, \bar{y}, \bar{z})$ lies inside the solid. ∎

Solutions 14.6   *Triple integral in cylindrical and spherical coordinates*   (pages 960-1)

1.   The cylinder has equation $1 = (x-1)^2 + y^2 = x^2 + y^2 - 2x + 1 = r^2 - 2x + 1 = r^2 - 2r\cos\theta + 1$ which is equivalent to $r = 2\cos\theta$ if $r \ne 0$; this surface is swept out for $\theta \in [0, \pi)$. The sphere has equation $2^2 = 4 = x^2 + y^2 + z^2 = r^2 + z^2$. Note that the cylinder intersects the xy-plane in a circle of radius 1 which lies within intersection of the sphere and the xy-plane. The region inside cylinder and sphere has volume

$$V = \int_{\theta=0}^\pi \int_{r=0}^{2\cos\theta} \int_{z=-\sqrt{4-r^2}}^{\sqrt{4-r^2}} r\, dz\, dr\, d\theta = \int_{\theta=0}^\pi \int_{r=0}^{2\cos\theta} 2r\sqrt{4-r^2}\, dr\, d\theta = \int_{\theta=0}^\pi \frac{-2}{3}\left(4-r^2\right)^{3/2}\Big|_{r=0}^{2\cos\theta} d\theta$$

$$= \int_{\theta=0}^\pi \frac{16}{3}\left(1 - (1-\cos^2\theta)^{3/2}\right) d\theta = \int_{\theta=0}^\pi \frac{16}{3}\left(1 - \sin^3\theta\right) d\theta = \frac{16}{3}\left(\theta + \cos\theta - \frac{1}{3}\cos^3\theta\right)\Big|_{\theta=0}^\pi$$

$$= \frac{16}{3}\left(\pi - \frac{4}{3}\right).$$

REMARK: The circle $r = 2\cos\theta$ is also swept out for $\theta \in [-\pi/2, \pi/2)$. If you work over this interval, remember that $\left(1 - \cos^2\theta\right)^{3/2} = |\sin\theta|^3$ and $\sin\theta$ changes sign at 0.

---

3.   The intersection of plane $z = y = r\sin\theta$ and paraboloid $z = x^2 + y^2 = r^2$ projects vertically to the circle $\sin\theta = r$ in the xy-plane; that circle is swept out for $\theta \in [0, \pi)$.

$$V = \int_{\theta=0}^\pi \int_{r=0}^{\sin\theta} \int_{z=r^2}^{r\sin\theta} r\, dz\, dr\, d\theta = \int_{\theta=0}^\pi \int_{r=0}^{\sin\theta} r\left(r^2\sin\theta - r^3\right) dr\, d\theta = \int_{\theta=0}^\pi \left(\frac{r^3}{3}\sin\theta - \frac{r^4}{4}\right)\Big|_{r=0}^{\sin\theta} d\theta$$

$$= \int_{\theta=0}^\pi \frac{1}{12}\sin^4\theta\, d\theta = \left(\frac{-1}{48}\sin^3\theta\cos\theta - \frac{1}{32}\sin\theta\cos\theta + \frac{\theta}{32}\right)\Big|_{\theta=0}^\pi = \frac{\pi}{32}.$$

5.

$$V = \int_{\theta=0}^{2\pi} \int_{z=0}^{1} \int_{r=0}^{z} r \, dr \, dz \, d\theta + \int_{\theta=0}^{2\pi} \int_{z=1}^{a} \int_{r=\sqrt{z^2-1}}^{z} r \, dr \, dz \, d\theta$$

$$= \int_{\theta=0}^{2\pi} \int_{z=0}^{1} \frac{r^2}{2}\Big|_{r=0}^{z} dz \, d\theta + \int_{\theta=0}^{2\pi} \int_{z=1}^{a} \frac{r^2}{2}\Big|_{r=\sqrt{z^2-1}}^{z} dz \, d\theta$$

$$= \int_{\theta=0}^{2\pi} \int_{z=0}^{1} \frac{z^2}{2} dz \, d\theta + \int_{\theta=0}^{2\pi} \int_{z=1}^{a} \frac{1}{2} dz \, d\theta = \int_{\theta=0}^{2\pi} \frac{z^3}{6}\Big|_{z=0}^{1} d\theta + \int_{\theta=0}^{2\pi} \frac{z}{2}\Big|_{z=1}^{a} d\theta$$

$$= \int_{\theta=0}^{2\pi} \frac{1}{6} d\theta + \int_{\theta=0}^{2\pi} \frac{a-1}{2} d\theta = \frac{\pi}{3} + (a-1)\pi = \left(a - \frac{2}{3}\right)\pi.$$

7.

$$\int_{x=0}^{1} \int_{y=0}^{\sqrt{1-x^2}} \int_{z=0}^{\sqrt{1-x^2-y^2}} \frac{1}{\sqrt{x^2+y^2+z^2}} dz \, dy \, dx = \int_{\theta=0}^{\pi/2} \int_{\phi=0}^{\pi/2} \int_{\rho=0}^{1} \frac{1}{\rho} \left(\rho^2 \sin\phi \, d\rho \, d\phi \, d\theta\right)$$

$$= \int_{\theta=0}^{\pi/2} \int_{\phi=0}^{\pi/2} \int_{\rho=0}^{1} \rho \sin\phi \, d\rho \, d\phi \, d\theta = \int_{\theta=0}^{\pi/2} \int_{\phi=0}^{\pi/2} \frac{\rho^2}{2}\Big|_{\rho=0}^{1} \sin\phi \, d\phi \, d\theta = \frac{1}{2} \int_{\theta=0}^{\pi/2} \int_{\phi=0}^{\pi/2} \sin\phi \, d\phi \, d\theta$$

$$= \frac{1}{2} \int_{\theta=0}^{\pi/2} (-\cos\phi)\Big|_{\phi=0}^{\pi/2} d\theta = \frac{1}{2} \int_{\theta=0}^{\pi/2} d\theta = \frac{\pi}{4}.$$

REMARK:
$$\int_{\theta=0}^{\pi/2} \int_{\phi=0}^{\pi/2} \int_{\rho=0}^{1} \rho \sin\phi \, d\rho \, d\phi \, d\theta = \left(\int_{\theta=0}^{\pi/2} d\theta\right)\left(\int_{\phi=0}^{\pi/2} \sin\phi \, d\phi\right)\left(\int_{\rho=0}^{1} \rho \, d\rho\right) = \left(\frac{\pi}{2}\right)(1)\left(\frac{1}{2}\right).$$

9. The region inside the sphere $\rho^2 = x^2 + y^2 + z^2 = 4 = 2^2$ satisfies $0 \le \rho \le 2$. If we consider "outside" the cone $z^2 = x^2 + y^2$ to be the region satisfying $z^2 \ge x^2 + y^2$, then that condition translates to $\rho^2 \cos^2\phi \ge \rho^2 \sin^2\phi$ which is equivalent to $1 \ge \tan^2\phi$, i.e., $\phi \in [0, \pi/4] \cup [3\pi/4, \pi]$. Since the sphere and cone are symmetric about the horizontal plane $\phi = \pi/2$, we'll integrate over the first interval and double the result.

$$V = 2\int_{\theta=0}^{2\pi} \int_{\phi=0}^{\pi/4} \int_{\rho=0}^{2} \rho^2 \sin\phi \, d\rho \, d\phi \, d\theta = 2\left(\int_{\theta=0}^{2\pi} d\theta\right)\left(\int_{\phi=0}^{\pi/4} \sin\phi \, d\phi\right)\left(\int_{\rho=0}^{2} \rho^2 \, d\rho\right)$$

$$= 2\left(\theta\big|_{\theta=0}^{2\pi}\right)\left(-\cos\phi\big|_{\phi=0}^{\pi/4}\right)\left(\frac{\rho^3}{3}\Big|_{\rho=0}^{2}\right) = 2(2\pi)\left(1 - \frac{1}{\sqrt{2}}\right)\left(\frac{8}{3}\right) = \frac{32}{3}\left(1 - \frac{1}{\sqrt{2}}\right)\pi.$$

REMARK: With the other interpretation of "outside", we compute the complementary volume:
$$\int_{\theta=0}^{2\pi} \int_{\phi=\pi/4}^{3\pi/4} \int_{\rho=0}^{2} \rho^2 \sin\phi \, d\rho \, d\phi \, d\theta = (2\pi)\left(\sqrt{2}\right)\left(\frac{8}{3}\right).$$

11.

$$V = \int_{\theta=a}^{a+\pi/6} \int_{\phi=0}^{\pi} \int_{\rho=0}^{1} \rho^2 \sin\phi \, d\rho \, d\phi \, d\theta = \left(\int_{\theta=a}^{a+\pi/6} d\theta\right)\left(\int_{\phi=0}^{\pi} \sin\phi \, d\phi\right)\left(\int_{\rho=0}^{1} \rho^2 \, d\rho\right)$$

$$= \left(\theta\big|_{\theta=a}^{a+\pi/6}\right)\left(-\cos\phi\big|_{\phi=0}^{\pi}\right)\left(\frac{\rho^3}{3}\Big|_{\rho=0}^{1}\right) = \left(\frac{\pi}{6}\right)(2)\left(\frac{1}{3}\right) = \frac{\pi}{9}.$$

REMARK:
$$\int_{\theta=0}^{2\pi} \int_{\phi=b}^{b+\pi/6} \int_{\rho=0}^{1} \rho^2 \sin\phi \, d\rho \, d\phi \, d\theta = \frac{\pi}{3}\left(\sin b + (2 - \sqrt{3})\cos b\right)$$
computes the volume within the unit sphere between two cones.

13. The region of integration is the sphere $x^2 + y^2 + z^2 \le 3^2$.

$$\int_{x=-3}^{3} \int_{y=-\sqrt{9-x^2}}^{\sqrt{9-x^2}} \int_{z=-\sqrt{9-x^2-y^2}}^{\sqrt{9-x^2-y^2}} (x^2 + y^2 + z^2)^{3/2} dz \, dy \, dx = \int_{\theta=0}^{2\pi} \int_{\phi=0}^{\pi} \int_{\rho=0}^{3} \rho^3 \left(\rho^2 \sin\phi \, d\rho \, d\phi \, d\theta\right)$$

$$= \left(\int_{\theta=0}^{2\pi} d\theta\right)\left(\int_{\phi=0}^{\pi} \sin\phi \, d\phi\right)\left(\int_{\rho=0}^{3} \rho^5 \, d\rho\right) = \left(\theta\big|_{\theta=0}^{2\pi}\right)\left(-\cos\phi\big|_{\phi=0}^{\pi}\right)\left(\frac{\rho^6}{6}\Big|_{\rho=0}^{3}\right) = (2\pi)(2)\left(\frac{3^6}{6}\right) = 486\pi.$$

15. Note that $\sin^{-1}(\cos\theta) = \pi/2 - \theta$ for $\theta \in [0, \pi]$.

$$M_{yz} = \int_{\theta=0}^{\pi} \int_{r=0}^{2\cos\theta} \int_{z=-\sqrt{4-r^2}}^{\sqrt{4-r^2}} (r\cos\theta)(r \, dz \, dr \, d\theta) = \int_{\theta=0}^{\pi} \cos\theta \int_{r=0}^{2\cos\theta} 2r^2\sqrt{4-r^2} \, dr \, d\theta$$

$$= \int_{\theta=0}^{\pi} \cos\theta \left(\frac{-1}{2} r \left(4 - r^2\right)^{3/2} + r \left(4 - r^2\right)^{1/2} + 4 \sin^{-1} \frac{r}{2} \right) \Big|_{r=0}^{2\cos\theta} d\theta$$

$$= \int_{\theta=0}^{\pi} \cos\theta \left(-\cos\theta \left(4 - 4\cos^2\theta\right)^{3/2} + 2\cos\theta \left(4 - 4\cos^2\theta\right)^{1/2} + 4\sin^{-1}(\cos\theta) \right) d\theta$$

$$= \int_{\theta=0}^{\pi} \left(-8\cos^2\theta \sin^3\theta + 4\cos^2\theta \sin\theta + 4 \left(\frac{\pi}{2} - \theta\right) \cos\theta \right) d\theta$$

$$= \left(-8 \left(\frac{\cos^5\theta}{5} - \frac{\cos^3\theta}{3} \right) - 4 \frac{\cos^3\theta}{3} + 2\pi \sin\theta - 4(\cos\theta + \theta \sin\theta) \right) \Big|_{\theta=0}^{\pi} = \frac{128}{15},$$

$$M_{xz} = \int_{\theta=0}^{\pi} \int_{r=0}^{2\cos\theta} \int_{z=-\sqrt{4-r^2}}^{\sqrt{4-r^2}} (r\sin\theta)(r\,dz\,dr\,d\theta)$$

$$= \int_{\theta=0}^{\pi} \sin\theta \left(-8\cos\theta \sin^3\theta + 4\cos\theta \sin\theta + 4\sin^{-1}(\cos\theta) \right) d\theta$$

$$= \left(\frac{-8\sin^5\theta}{5} + \frac{4\sin^3\theta}{3} - 4\cos\theta \sin^{-1}(\cos\theta) - 4\sqrt{1 - \cos^2\theta} \right) \Big|_{\theta=0}^{\pi} = 0,$$

$$M_{xy} = \int_{\theta=0}^{\pi} \int_{r=0}^{2\cos\theta} \int_{z=-\sqrt{4-r^2}}^{\sqrt{4-r^2}} (z)(r\,dz\,dr\,d\theta) = \int_{\theta=0}^{\pi} \int_{r=0}^{2\cos\theta} \frac{z^2}{2} \Big|_{z=-\sqrt{4-r^2}}^{\sqrt{4-r^2}} r\,dr\,d\theta$$

$$= \int_{\theta=0}^{\pi} \int_{r=0}^{2\cos\theta} 0 \cdot r\,dr\,d\theta = 0.$$

Thus $(\hat{x}, \hat{y}, \hat{z}) = \left(\dfrac{M_{yz}}{\mu}, \dfrac{M_{xz}}{\mu}, \dfrac{M_{xy}}{\mu} \right) = \left(\dfrac{128/15}{16(3\pi - 4)/9}, \dfrac{0}{16(3\pi - 4)/9}, \dfrac{0}{16(3\pi - 4)/9} \right) = \left(\dfrac{24}{5(3\pi - 4)}, 0, 0 \right).$

REMARK: Note that $\sin(\cos^{-1} w) = \sqrt{1 - w^2}$ for $w \in [-1, 1]$.

$$M_{yz} = \int_{r=0}^{2} \int_{\theta=-\cos^{-1}(r/2)}^{\cos^{-1}(r/2)} \int_{z=-\sqrt{4-r^2}}^{\sqrt{4-r^2}} (r\cos\theta)(r\,dz\,d\theta\,dr) = \int_{r=0}^{2} \sin\theta \Big|_{\theta=-\cos^{-1}(r/2)}^{\cos^{-1}(r/2)} 2r^2 \sqrt{4 - r^2}\,dr$$

$$= \int_{r=0}^{2} 2\sqrt{1 - (r/2)^2} \cdot 2r^2 \sqrt{4 - r^2}\,dr = \int_{r=0}^{2} 2r^2 \left(4 - r^2\right) dr = 2 \left(\frac{4r^3}{3} - \frac{r^5}{5} \right) \Big|_{r=0}^{2} = \frac{128}{15}.$$

---

17.   The density function has the form $c|z|$; without loss of generality, we may assume $c = 1$. Since the solid lies above the xy-plane, we will use z as density.

$$\mu = \int_{z=0}^{4} \int_{r=0}^{\sqrt{4-z}} \int_{\theta=0}^{2\pi} z\,(r\,d\theta\,dr\,dz) = \int_{z=0}^{4} \int_{r=0}^{\sqrt{4-z}} 2\pi z r\,dr\,dz$$

$$= \int_{z=0}^{4} \pi z r^2 \Big|_{r=0}^{\sqrt{4-z}} dz = \int_{z=0}^{4} \pi z(4 - z)\,dz = \pi \left(2z^2 - \frac{z^3}{3} \right) \Big|_{z=0}^{4} = \frac{32\pi}{3},$$

$$M_{xy} = \int_{z=0}^{4} \int_{r=0}^{\sqrt{4-z}} \int_{\theta=0}^{2\pi} z \cdot z\,(r\,d\theta\,dr\,dz) = \int_{z=0}^{4} z \cdot \pi z(4 - z)\,dz = \pi \left(\frac{4z^3}{3} - \frac{z^4}{4} \right) \Big|_{z=0}^{4} = \frac{64\pi}{3},$$

$$M_{xz} = \int_{z=0}^{4} \int_{r=0}^{\sqrt{4-z}} \int_{\theta=0}^{2\pi} r\sin\theta \cdot z\,(r\,d\theta\,dr\,dz) = \int_{z=0}^{4} \int_{r=0}^{\sqrt{4-z}} -\cos\theta \Big|_{\theta=0}^{2\pi} z r^2\,dr\,dz = 0,$$

$$M_{yz} = \int_{z=0}^{4} \int_{r=0}^{\sqrt{4-z}} \int_{\theta=0}^{2\pi} r\cos\theta \cdot z\,(r\,d\theta\,dr\,dz) = \int_{z=0}^{4} \int_{r=0}^{\sqrt{4-z}} \sin\theta \Big|_{\theta=0}^{2\pi} z r^2\,dr\,dz = 0.$$

Thus $(\bar{x}, \bar{y}, \bar{z}) = \left(\dfrac{M_{yz}}{\mu}, \dfrac{M_{xz}}{\mu}, \dfrac{M_{xy}}{\mu} \right) = \left(\dfrac{0}{32\pi/3}, \dfrac{0}{32\pi/3}, \dfrac{64\pi/3}{32\pi/3} \right) = (0, 0, 2).$

19.   The upper hemisphere of $0 = x^2 + y^2 + z^2 - z = r^2 + (z - 1/2)^2 - 1/4$ satisfies $z = 1/2 + \sqrt{1/4 - r^2}$; it

meets the cone $z^2 = r^2$ at the cylinder $r = 1/2$.

$$V = \int_{r=0}^{1/2} \int_{z=r}^{1/2+\sqrt{1/4-r^2}} \int_{\theta=0}^{2\pi} r \, d\theta \, dz \, dr = \int_{r=0}^{1/2} \int_{z=r}^{1/2+\sqrt{1/4-r^2}} 2\pi r \, dz \, dr$$

$$= \int_{r=0}^{1/2} 2\pi r \left(\frac{1}{2} + \sqrt{\frac{1}{4} - r^2} - r \right) dr = \pi \left(\frac{r^2}{2} - \frac{2}{3} \left(\frac{1}{4} - r^2 \right)^{3/2} - \frac{2r^3}{3} \right) \Bigg|_{r=0}^{1/2} = \frac{\pi}{8},$$

$$M_{yz} = \int_{r=0}^{1/2} \int_{z=r}^{1/2+\sqrt{1/4-r^2}} \int_{\theta=0}^{2\pi} r\cos\theta \cdot (r \, d\theta \, dz \, dr) = \int_{r=0}^{1/2} \int_{z=r}^{1/2+\sqrt{1/4-r^2}} \sin\theta \Big|_{\theta=0}^{2\pi} r \, dz \, dr = 0,$$

$$M_{xz} = \int_{r=0}^{1/2} \int_{z=r}^{1/2+\sqrt{1/4-r^2}} \int_{\theta=0}^{2\pi} r\sin\theta \cdot (r \, d\theta \, dz \, dr) = \int_{r=0}^{1/2} \int_{z=r}^{1/2+\sqrt{1/4-r^2}} -\cos\theta \Big|_{\theta=0}^{2\pi} r \, dz \, dr = 0,$$

$$M_{xy} = \int_{r=0}^{1/2} \int_{z=r}^{1/2+\sqrt{1/4-r^2}} \int_{\theta=0}^{2\pi} z \cdot (r \, d\theta \, dz \, dr) = \int_{r=0}^{1/2} \int_{z=r}^{1/2+\sqrt{1/4-r^2}} 2\pi z r \, dz \, dr$$

$$= \int_{r=0}^{1/2} \pi r z^2 \Big|_{z=r}^{1/2+\sqrt{1/4-r^2}} dr = \pi \int_{r=0}^{1/2} \left(\frac{r}{2} + r\sqrt{\frac{1}{4} - r^2} - 2r^3 \right) dr$$

$$= \pi \left(\frac{r^2}{4} - \frac{1}{3} \left(\frac{1}{4} - r^2 \right)^{3/2} - \frac{r^4}{4} \right) \Bigg|_{r=0}^{1/2} = \frac{7\pi}{96}.$$

Thus $(\bar{x}, \bar{y}, \bar{z}) = \left(\dfrac{M_{yz}}{\mu}, \dfrac{M_{xz}}{\mu}, \dfrac{M_{xy}}{\mu} \right) = \left(\dfrac{0}{\pi/8}, \dfrac{0}{\pi/8}, \dfrac{7\pi/96}{\pi/8} \right) = \left(0, 0, \dfrac{7}{12} \right).$

REMARK: The hemisphere can also be described as $\rho^2 = x^2 + y^2 + z^2 = z = \rho\cos\phi$, i.e., $\rho = \cos\phi$.

$$V = \int_{\phi=0}^{\pi/4} \int_{\rho=0}^{\cos\phi} \int_{\theta=0}^{2\pi} \rho^2 \sin\phi \, d\theta \, d\rho \, d\phi = \int_{\phi=0}^{\pi/4} \int_{\rho=0}^{\cos\phi} 2\pi\rho^2 \sin\phi \, d\rho \, d\phi = \int_{\phi=0}^{\pi/4} \frac{2\pi}{3} \cos^3\phi \sin\phi \, d\phi$$

$$= \frac{-\pi}{6} \cos^4\phi \Bigg|_{\phi=0}^{\pi/4} = \frac{\pi}{8},$$

$$M_{xy} = \int_{\phi=0}^{\pi/4} \int_{\rho=0}^{\cos\phi} \int_{\theta=0}^{2\pi} (\rho\cos\phi)(\rho^2 \sin\phi \, d\theta \, d\rho \, d\phi) = \int_{\phi=0}^{\pi/4} \int_{\rho=0}^{\cos\phi} 2\pi\rho^3 \cos\phi \sin\phi \, d\rho \, d\phi$$

$$= \int_{\phi=0}^{\pi/4} (2\pi) \frac{\cos^4\phi}{4} \cos\phi \sin\phi \, d\phi = \frac{-\pi}{12} \cos^6\phi \Bigg|_{\phi=0}^{\pi/4} = \frac{7\pi}{96}.$$

21. The solid region and the density function are symmetric about the origin, hence the origin is the center of mass.

23. Suppose $0 \le a < b$ and suppose $z = f(x)$ decreases on the interval $[a, b]$. We will compute the volume of the solid region generated by rotating $\{(x, z) : a \le x \le b, f(b) \le z \le f(a)\}$ about the z-axis.

$\boxed{\text{Cylindrical shell}}$ $\displaystyle\int_{r=a}^{b} \int_{z=f(b)}^{f(r)} \int_{\theta=0}^{2\pi} r \, d\theta \, dz \, dr = \int_{r=a}^{b} \int_{z=f(b)}^{f(r)} 2\pi r \, dz \, dr = \int_{r=a}^{b} 2\pi r \left(f(r) - f(b) \right) dr.$

$\boxed{\text{Disk}}$ $\displaystyle\int_{z=f(b)}^{f(a)} \int_{r=a}^{f^{-1}(z)} \int_{\theta=0}^{2\pi} r \, d\theta \, dr \, dz = \int_{z=f(b)}^{f(a)} \int_{r=a}^{f^{-1}(z)} 2\pi r \, dr \, dz = \int_{z=f(b)}^{f(a)} \pi \left((f^{-1}(z))^2 - a^2 \right) dz.$

Solutions 14.R   *Review*   (pages 962-3)

1. $\displaystyle\int_{y=0}^{1} \int_{x=0}^{2} x^2 y \, dx \, dy = \int_{y=0}^{1} \frac{x^3 y}{3} \Big|_{x=0}^{2} dy = \int_{y=0}^{1} \frac{8y}{3} dy = \frac{4y^2}{3} \Big|_{y=0}^{1} = \frac{4}{3}.$

3. $\displaystyle\int_{y=2}^{4} \int_{x=1+y}^{2+5y} (x - y^2) \, dx \, dy = \int_{y=2}^{4} \left(\frac{x^2}{2} - xy^2 \right) \Big|_{x=1+y}^{2+5y} dy = \int_{y=2}^{4} \left(\frac{3}{2} + 9y + 11y^2 - 4y^3 \right) dy$

$$= \left(\frac{3y}{2} + \frac{9y^2}{2} + \frac{11y^3}{3} - y^4 \right) \Bigg|_{y=2}^{4} = \frac{67}{3}.$$

5.
$$\int_{y=0}^{1}\int_{x=0}^{y}\int_{z=0}^{x} x^2\,dz\,dx\,dy = \int_{y=0}^{1}\int_{x=0}^{y} x^2 z\Big|_{z=0}^{x}\,dx\,dy = \int_{y=0}^{1}\int_{x=0}^{y} x^3\,dx\,dy$$
$$= \int_{y=0}^{1}\frac{x^4}{4}\Big|_{x=0}^{y}\,dy = \int_{y=0}^{1}\frac{y^4}{4}\,dy = \frac{y^5}{20}\Big|_{y=0}^{1} = \frac{1}{20}.$$

7.   The xy-domain of integration is the disk $x^2 + y^2 \le 4 = 2^2$. The functions giving the z-limits of integration cross above the circle $x^2 + y^2 = 2$ which lies within that disk — the multiple integral does not compute the volume of a solid.

$$\int_{x=-2}^{2}\int_{y=-\sqrt{4-x^2}}^{\sqrt{4-x^2}}\int_{z=\sqrt{x^2+y^2}}^{\sqrt{4-x^2-y^2}} z^2\,dz\,dy\,dx = \int_{\theta=0}^{2\pi}\int_{r=0}^{2}\int_{z=r}^{\sqrt{4-r^2}} z^2\,dz\,r\,dr\,d\theta = \int_{\theta=0}^{2\pi}\int_{r=0}^{2}\frac{z^3}{3}\Big|_{z=r}^{\sqrt{4-r^2}} r\,dr\,d\theta$$
$$= \frac{1}{3}\int_{\theta=0}^{2\pi}\int_{r=0}^{2}\left((4-r^2)^{3/2} - r^3\right) r\,dr\,d\theta$$
$$= \int_{\theta=0}^{2\pi}\frac{-1}{15}\left((4-r^2)^{5/2} + r^5\right)\Big|_{r=0}^{2}\,d\theta = \int_{\theta=0}^{2\pi} 0\,d\theta = 0.$$

9.
$$\iint_{\Omega} (y - x^2)\,dA = \int_{x=-3}^{3}\int_{y=0}^{2} (y - x^2)\,dy\,dx = \int_{x=-3}^{3}\left(\frac{y^2}{2} - x^2 y\right)\Big|_{y=0}^{2}\,dx$$
$$= \int_{x=-3}^{3} (2 - 2x^2)\,dx = 2\left(x - \frac{x^3}{3}\right)\Big|_{x=-3}^{3} = -24.$$

11.
$$\iint_{\Omega} (x + y^2)\,dA = \int_{\theta=0}^{\pi/2}\int_{r=0}^{1} (r\cos\theta + r^2\sin^2\theta)\,r\,dr\,d\theta = \int_{\theta=0}^{\pi/2}\left(\frac{r^3}{3}\cos\theta + \frac{r^4}{4}\sin^2\theta\right)\Big|_{r=0}^{1}\,d\theta$$
$$= \int_{\theta=0}^{\pi/2}\left(\frac{1}{3}\cos\theta + \frac{1}{4}\sin^2\theta\right)\,d\theta = \left(\frac{1}{3}\sin\theta + \frac{1}{8}(-\cos\theta\sin\theta + \theta)\right)\Big|_{\theta=0}^{\pi/2} = \frac{1}{3} + \frac{\pi}{16}.$$

13.
$$\int_{x=2}^{5}\int_{y=1}^{x} 3x^2 y\,dy\,dx = \int_{y=1}^{2}\int_{x=2}^{5} 3x^2 y\,dx\,dy + \int_{y=2}^{5}\int_{x=y}^{5} 3x^2 y\,dx\,dy$$
$$= \int_{y=1}^{2} x^3 y\Big|_{x=2}^{5}\,dy + \int_{y=2}^{5} x^3 y\Big|_{x=y}^{5}\,dy = \int_{y=1}^{2} 117y\,dy + \int_{y=2}^{5} (125y - y^4)\,dy$$
$$= \frac{117y^2}{2}\Big|_{y=1}^{2} + \left(\frac{125y^2}{2} - \frac{y^5}{5}\right)\Big|_{y=2}^{5} = \frac{351}{2} + \frac{6939}{10} = \frac{4347}{5}.$$

REMARK: $$\int_{x=2}^{5}\int_{y=1}^{x} 3x^2 y\,dy\,dx = \int_{x=2}^{5}\frac{3}{2}x^2 y^2\Big|_{y=1}^{x}\,dx = \int_{x=2}^{5}\frac{3}{2}x^2(x^2 - 1)\,dx = \frac{3}{2}\left(\frac{x^5}{5} - \frac{x^3}{3}\right)\Big|_{x=2}^{5}.$$

15.
$$\int_{x=0}^{\infty}\int_{y=x}^{\infty} f(x, y)\,dy\,dx = \int_{y=0}^{\infty}\int_{x=0}^{y} f(x, y)\,dx\,dy;\text{ the region of integration is } \{(x, y) : 0 \le x \le y\}.$$

17.
$$\iint_{\Omega} e^{-(x^2+y^2)}\,dA = \int_{\theta=0}^{2\pi}\int_{\rho=0}^{3} e^{-r^2} r\,dr\,d\theta = \left(\int_{\theta=0}^{2\pi} d\theta\right)\left(\int_{\rho=0}^{3} re^{-r^2}\,dr\right)$$
$$= (2\pi)\left(\frac{-1}{2}e^{-r^2}\Big|_{\rho=0}^{3}\right) = \pi(1 - e^{-9}).$$

19.
$$V = \int_{z=0}^{2}\int_{y=0}^{(6-3z)/2}\int_{x=0}^{6-3z-2y} dx\,dy\,dz = \int_{z=0}^{2}\int_{y=0}^{(6-3z)/2} (6 - 3z - 2y)\,dy\,dz$$
$$= \int_{z=0}^{2}\left((6 - 3z)y - y^2\right)\Big|_{y=0}^{(6-3z)/2}\,dz = \int_{z=0}^{2}\frac{9}{4}(2 - z)^2\,dz = \frac{-3}{4}(2 - z)^3\Big|_{z=0}^{2} = 6.$$

21.
$$V = \int_{x=0}^{3}\int_{y=-2}^{2}\int_{z=-\sqrt{4-y^2}}^{\sqrt{4-y^2}} dz\,dy\,dx = \int_{x=0}^{3}\int_{\theta=0}^{2\pi}\int_{r=0}^{2} r\,dr\,d\theta\,dx$$
$$= \left(\int_{x=0}^{3} dx\right)\left(\int_{\theta=0}^{2\pi} d\theta\right)\left(\int_{r=0}^{2} r\,dr\right) = (3)(2\pi)\left(\frac{2^2}{2}\right) = 12\pi.$$

23. In the xy-plane, the curves $x = y^2 + 0^2 = y^2$ and $y = x - 2$ intersect when $0 = (y+2) - y^2 = 2 + y - y^2 = (1+y)(2-y) = 9/4 - (y - 1/2)^2$, i.e., when $y = -1$ and $y = 2$. If $y \in [-1, 2]$, then $y + 2 \geq y^2$. The volume of the solid is

$$\int_{y=-1}^{2} \int_{x=y^2}^{y+2} \int_{z=-\sqrt{x-y^2}}^{\sqrt{x-y^2}} dz\, dx\, dy = \int_{y=-1}^{2} \int_{x=y^2}^{y+2} 2\sqrt{x - y^2}\, dx\, dy = \int_{y=-1}^{2} \frac{4}{3}\left(x - y^2\right)^{3/2} \Big|_{x=y^2}^{y+2} dy$$

$$= \int_{y=-1}^{2} \frac{4}{3}\left(2 + y - y^2\right)^{3/2} dy \overset{y - 1/2 \equiv (3/2)\sin t}{=} \int_{t=-\pi/2}^{\pi/2} \frac{27}{4}\cos^4 t\, dt$$

$$= \frac{27}{4}\left(\frac{1}{4}\sin t \cos^3 t + \frac{3}{8}\sin t \cos t + \frac{3}{8}t\right)\Big|_{t=-\pi/2}^{\pi/2} = \frac{81\pi}{32}.$$

REMARK: The volume between $x = y^2 + z^2$ and $x = y + 2$ equals that behind $x = (y + 2) - (y^2 + z^2) = 9/4 - (y - 1/2)^2 - z^2$. An x-section is a circle of radius $\sqrt{9/4 - x}$ and area $\pi(9/4 - x)$. The volume is

$$\int_{x=0}^{9/4} \pi\left(\frac{9}{4} - x\right) dx = \frac{-\pi}{2}\left(\frac{9}{4} - x\right)^2 \Big|_{x=0}^{9/4} = \frac{81\pi}{32}.$$

25. $$A = \int_{\theta=0}^{2\pi} \int_{r=0}^{2(1+\sin\theta)} r\, dr\, d\theta = \int_{\theta=0}^{2\pi} \frac{r^2}{2} \Big|_{r=0}^{2(1+\sin\theta)} d\theta = \int_{\theta=0}^{2\pi} 2(1 + \sin\theta)^2 d\theta = (3\theta - 4\cos\theta - \cos\theta\sin\theta)\Big|_{\theta=0}^{2\pi}$$
$$= 6\pi.$$

27. The cone $z^2 = x^2 + y^2 = r^2$ and the paraboloid $4z = x^2 + y^2 = r^2$ meet when $z = 0$ and $z = 4$; the first corresponds to the origin and the second to the circle $r = 4$. The volume above the paraboloid and below the cone is

$$\int_{\theta=0}^{2\pi} \int_{r=0}^{4} \int_{z=r^2/4}^{r} r\, dz\, dr\, d\theta = \int_{\theta=0}^{2\pi} \int_{r=0}^{4} r\left(r - \frac{r^2}{4}\right) dr\, d\theta = \left(d\Big|_{\theta=0}^{2\pi}\right)\left(\left(\frac{r^3}{3} - \frac{r^4}{16}\right)\Big|_{r=0}^{4}\right) = (2\pi)\left(\frac{16}{3}\right).$$

REMARK: Revolve the region $\{(x, z) : x^2/4 \leq z \leq x,\ 0 \leq x \leq 4\}$ about the z-axis; compute the volume by the cylindrical shell method of section 5.2: $\int_0^4 2\pi x\left(x - \frac{x^2}{4}\right) dx$.

29. $$V = \int_{y=-2}^{2} \int_{z=0}^{4-y^2} \int_{x=z/2}^{z} dx\, dz\, dy = \int_{y=-2}^{2} \int_{z=0}^{4-y^2} \frac{z}{2}\, dz\, dy = \int_{y=-2}^{2} \frac{z^2}{4}\Big|_{z=0}^{4-y^2} dy$$
$$= \int_{y=-2}^{2} \frac{1}{4}\left(4 - y^2\right) dy = \left(4y - \frac{2y^3}{3} - \frac{y^5}{20}\right)\Big|_{y=-2}^{2} = \frac{128}{15}.$$

31. The plane $z = x = r\cos\theta$ and paraboloid $z = x^2 + y^2 = r^2$ intersect above the circle $\cos\theta = r$ in the xy-plane. The volume below the plane and above the paraboloid is

$$\int_{\theta=-\pi/2}^{\pi/2} \int_{r=0}^{\cos\theta} \int_{z=r^2}^{r\cos\theta} r\, dz\, dr\, d\theta = \int_{\theta=-\pi/2}^{\pi/2} \int_{r=0}^{\cos\theta} r\left(r\cos\theta - r^2\right) dr\, d\theta = \int_{\theta=-\pi/2}^{\pi/2} \left(\frac{r^3}{3}\cos\theta - \frac{r^4}{4}\right)\Big|_{r=0}^{\cos\theta} d\theta$$

$$= \int_{\theta=-\pi/2}^{\pi/2} \frac{1}{12}\cos^4\theta\, d\theta = \left(\frac{\sin\theta\cos^3\theta}{48} + \frac{\sin\theta\cos\theta + \theta}{32}\right)\Big|_{\theta=-\pi/2}^{\pi/2} = \frac{\pi}{32}.$$

33. $$V = \int_{\theta=a}^{a+\pi/3} \int_{\phi=0}^{\pi} \int_{\rho=0}^{2} \rho^2 \sin\phi\, d\rho\, d\phi\, d\theta = \left(\int_{\theta=a}^{a+\pi/3} d\theta\right)\left(\int_{\phi=0}^{\pi} \sin\phi\, d\phi\right)\left(\int_{\rho=0}^{2} \rho^2\, d\rho\right)$$
$$= \left(\theta\Big|_{\theta=a}^{a+\pi/3}\right)\left(-\cos\phi\Big|_{\phi=0}^{\pi}\right)\left(\frac{\rho^3}{3}\Big|_{\rho=0}^{2}\right) = \left(\frac{\pi}{3}\right)(2)\left(\frac{8}{3}\right) = \frac{16\pi}{9}.$$

REMARK: $\dfrac{\text{volume of wedge}}{\text{volume of sphere}} = \dfrac{\pi/3}{2\pi} = \dfrac{1}{6}.$

35.

$$A = \int_{\theta=-\pi/4}^{\pi/4} \int_{r=0}^{2\sqrt{\cos 2\theta}} r\, dr\, d\theta = \int_{\theta=-\pi/4}^{\pi/4} \frac{r^2}{2}\Big|_{r=0}^{2\sqrt{\cos 2\theta}} d\theta = \int_{\theta=-\pi/4}^{\pi/4} 2\cos 2\theta\, d\theta = \sin 2\theta\Big|_{\theta=-\pi/4}^{\pi/4} = 2,$$

$$M_y = \int_{\theta=-\pi/4}^{\pi/4} \int_{r=0}^{2\sqrt{\cos 2\theta}} (r\cos\theta)r\, dr\, d\theta = \int_{\theta=-\pi/4}^{\pi/4} \frac{r^3}{3}\Big|_{r=0}^{2\sqrt{\cos 2\theta}} \cos\theta\, d\theta = \int_{\theta=-\pi/4}^{\pi/4} \frac{8}{3}(\cos 2\theta)^{3/2}\cos\theta\, d\theta$$

$$= \frac{8}{3}\int_{\theta=-\pi/4}^{\pi/4} \left(1 - 2\sin^2\theta\right)^{3/2}\cos\theta\, d\theta \stackrel{\sqrt{2}\sin\theta \equiv \sin t}{=} \frac{8}{3}\int_{t=-\pi/2}^{\pi/2} \left(1 - \sin^2 t\right)^{3/2} \frac{\cos t\, dt}{\sqrt{2}}$$

$$= \frac{8}{3\sqrt{2}}\int_{t=-\pi/2}^{\pi/2} \cos^4 t\, dt = \frac{8}{3\sqrt{2}}\left(\frac{1}{4}\sin t\cos^3 t + \frac{3}{8}(\sin t\cos t + t)\right)\Big|_{t=-\pi/2}^{\pi/2} = \frac{8}{3\sqrt{2}}\left(\frac{3\pi}{8}\right) = \frac{\pi}{\sqrt{2}},$$

$$M_x = \int_{\theta=-\pi/4}^{\pi/4} \int_{r=0}^{2\sqrt{\cos 2\theta}} (r\sin\theta)r\, dr\, d\theta = \int_{\theta=-\pi/4}^{\pi/4} \frac{8}{3}(\cos 2\theta)^{3/2}\sin\theta\, d\theta$$

$$= \frac{8}{3}\int_{\theta=-\pi/4}^{\pi/4} \left(2\cos^2\theta - 1\right)^{3/2}\sin\theta\, d\theta$$

$$= \left(\frac{-2\cos\theta}{3}\left(2\cos^2\theta - 1\right)^{3/2} + \cos\theta\sqrt{2\cos^2\theta - 1} - \frac{1}{\sqrt{2}}\ln\left(\sqrt{2}\cos\theta + \sqrt{2\cos^2\theta - 1}\right)\right)\Big|_{\theta=-\pi/4}^{\pi/4} = 0.$$

Therefore $(\bar{x}, \bar{y}) = \left(\dfrac{M_y}{A}, \dfrac{M_x}{A}\right) = \left(\dfrac{\pi/\sqrt{2}}{2}, \dfrac{0}{2}\right) = \left(\dfrac{\pi}{2\sqrt{2}}, 0\right).$

REMARK: The region is symmetric about the y-axis, hence $\hat{y} = 0$. Alternatively, note that computing M_x involves the integral of an odd function over an interval symmetric about 0.

37.

$$V = \int_{\theta=0}^{2\pi} \int_{\phi=0}^{\pi/2} \int_{\rho=0}^{4} \rho^2\sin\phi\, d\rho\, d\phi\, d\theta = \left(\int_{\theta=0}^{2\pi} d\theta\right)\left(\int_{\phi=0}^{\pi/2}\sin\phi\, d\phi\right)\left(\int_{\rho=0}^{4}\rho^2\, d\rho\right)$$

$$= \left(\theta\Big|_{\theta=0}^{2\pi}\right)\left(-\cos\phi\Big|_{\phi=0}^{\pi/2}\right)\left(\frac{\rho^3}{3}\Big|_{\rho=0}^{4}\right) = (2\pi)(1)\left(\frac{64}{3}\right) = \frac{128\pi}{3},$$

$$M_{yz} = \int_{\theta=0}^{2\pi} \int_{\phi=0}^{\pi/2} \int_{\rho=0}^{4} \rho\sin\phi\cos\theta\,(\rho^2\sin\phi\, d\rho\, d\phi\, d\theta) = \left(\int_{\theta=0}^{2\pi}\cos\theta\, d\theta\right)\left(\int_{\phi=0}^{\pi/2}\sin^2\phi\, d\phi\right)\left(\int_{\rho=0}^{4}\rho^3\, d\rho\right)$$

$$= \left(\sin\theta\Big|_{\theta=0}^{2\pi}\right)\left(\frac{\phi - \sin\phi\cos\phi}{2}\Big|_{\phi=0}^{\pi/2}\right)\left(\frac{\rho^4}{4}\Big|_{\rho=0}^{4}\right) = (0)\left(\frac{\pi}{4}\right)(64) = 0,$$

$$M_{xz} = \int_{\theta=0}^{2\pi} \int_{\phi=0}^{\pi/2} \int_{\rho=0}^{4} \rho\sin\phi\sin\theta\,(\rho^2\sin\phi\, d\rho\, d\phi\, d\theta) = \left(\int_{\theta=0}^{2\pi}\sin\theta\, d\theta\right)\left(\int_{\phi=0}^{\pi/2}\sin^2\phi\, d\phi\right)\left(\int_{\rho=0}^{4}\rho^3\, d\rho\right)$$

$$= \left(-\cos\theta\Big|_{\theta=0}^{2\pi}\right)\left(\frac{\phi - \sin\phi\cos\phi}{2}\Big|_{\phi=0}^{\pi/2}\right)\left(\frac{\rho^4}{4}\Big|_{\rho=0}^{4}\right) = (0)\left(\frac{\pi}{4}\right)(64) = 0,$$

$$M_{yz} = \int_{\theta=0}^{2\pi} \int_{\phi=0}^{\pi/2} \int_{\rho=0}^{4} \rho\cos\phi\,(\rho^2\sin\phi\, d\rho\, d\phi\, d\theta) = \left(\int_{\theta=0}^{2\pi} d\theta\right)\left(\int_{\phi=0}^{\pi/2}\cos\phi\sin\phi\, d\phi\right)\left(\int_{\rho=0}^{4}\rho^3\, d\rho\right)$$

$$= \left(\theta\Big|_{\theta=0}^{2\pi}\right)\left(\frac{\sin^2\phi}{2}\Big|_{\phi=0}^{\pi/2}\right)\left(\frac{\rho^4}{4}\Big|_{\rho=0}^{4}\right) = (2\pi)\left(\frac{1}{2}\right)(64) = 64\pi.$$

Therefore $(\bar{x}, \bar{y}, \bar{z}) = \left(\dfrac{M_{yz}}{V}, \dfrac{M_{xz}}{V}, \dfrac{M_{xy}}{V}\right) = \left(\dfrac{0}{128\pi/3}, \dfrac{0}{128\pi/3}, \dfrac{64\pi}{128\pi/3}\right) = \left(0, 0, \dfrac{3}{2}\right).$

39.

$$\mu = \int_{x=0}^{3}\int_{y=1}^{5}(3x^2 y)\,dy\,dx = \left(\int_{x=0}^{3}3x^2\,dx\right)\left(\int_{y=1}^{5}y\,dy\right) = \left(x^3\big|_{x=0}^{3}\right)\left(\frac{y^2}{2}\Big|_{y=1}^{5}\right) = (27)(12) = 324,$$

$$M_y = \int_{x=0}^{3}\int_{y=1}^{5}x\,(3x^2 y)\,dy\,dx = \left(\int_{x=0}^{3}3x^3\,dx\right)\left(\int_{y=1}^{5}y\,dy\right) = \left(\frac{3x^4}{4}\Big|_{x=0}^{3}\right)\left(\frac{y^2}{2}\Big|_{y=1}^{5}\right)$$

$$= \left(\frac{243}{4}\right)(12) = 729,$$

$$M_y = \int_{x=0}^{3}\int_{y=1}^{5}y\,(3x^2 y)\,dy\,dx = \left(\int_{x=0}^{3}3x^2\,dx\right)\left(\int_{y=1}^{5}y^2\,dy\right) = \left(x^3\big|_{x=0}^{3}\right)\left(\frac{y^3}{3}\Big|_{y=1}^{5}\right)$$

$$= (27)\left(\frac{124}{3}\right) = 1116.$$

Therefore $(\bar{x},\bar{y}) = \left(\dfrac{M_y}{\mu},\dfrac{M_x}{\mu}\right) = \left(\dfrac{729}{324},\dfrac{1116}{324}\right) = \left(\dfrac{243/4}{27},\dfrac{124/3}{12}\right) = \left(\dfrac{9}{4},\dfrac{31}{9}\right).$

41.

$$\int_{x=-(1/2)\sqrt{100-25z^2-y^2}}^{(1/2)\sqrt{100-25z^2-y^2}} x\cdot z^2\,dx = \frac{z^2}{2}x^2\bigg|_{x=-(1/2)\sqrt{100-25z^2-y^2}}^{(1/2)\sqrt{100-25z^2-y^2}} = 0;\ \text{therefore}$$

$$M_{yz} = \int_{z=-2}^{2}\int_{y=-\sqrt{100-25z^2}}^{\sqrt{100-25z^2}}\int_{x=-(1/2)\sqrt{100-25z^2-y^2}}^{(1/2)\sqrt{100-25z^2-y^2}} xz^2\,dx\,dy\,dz = 0.$$

Similarly, $\displaystyle\int_{y=-\sqrt{100-25z^2-4x^2}}^{\sqrt{100-25z^2-4x^2}} y\cdot z^2\,dy = \frac{z^2}{2}y^2\bigg|_{y=-\sqrt{100-25z^2-4x^2}}^{\sqrt{100-25z^2-4x^2}} = 0$ implies

$$M_{xz} = \int_{z=-2}^{2}\int_{x=-(1/2)\sqrt{100-25z^2}}^{(1/2)\sqrt{100-25z^2}}\int_{y=-\sqrt{100-25z^2-4x^2}}^{\sqrt{100-25z^2-4x^2}} yz^2\,dy\,dx\,dz = 0$$

and $\displaystyle\int_{z=-(1/5)\sqrt{100-4x^2-y^2}}^{(1/5)\sqrt{100-4x^2-y^2}} z\cdot z^2\,dz = \frac{1}{4}z^4\bigg|_{z=-(1/5)\sqrt{100-4x^2-y^2}}^{(1/5)\sqrt{100-4x^2-y^2}} = 0$ implies

$$M_{xy} = \int_{x=-5}^{5}\int_{y=-\sqrt{100-4x^2}}^{\sqrt{100-4x^2}}\int_{z=-(1/5)\sqrt{100-4x^2-y^2}}^{(1/5)\sqrt{100-4x^2-y^2}} z\cdot z^2\,dz\,dy\,dx = 0.$$

Therefore $(\bar{x},\bar{y},\bar{z}) = \left(\dfrac{M_{yz}}{\mu},\dfrac{M_{xz}}{\mu},\dfrac{M_{xy}}{\mu}\right) = \left(\dfrac{0}{\mu},\dfrac{0}{\mu},\dfrac{0}{\mu}\right) = (0,0,0).$

REMARK: The region and density are symmetric about the origin, hence the origin is the center of mass.

43.   A mild generalization will let us hide some irrelevant detail. Suppose the sphere has radius R and suppose the wedge is cut from the sphere by planes $\theta = a\pm b$; the density has form $c\rho$ and we lose no generality by assuming $c=1$.

$$\mu = \int_{\theta=a-b}^{a+b}\int_{\phi=0}^{\pi}\int_{\rho=0}^{2}\rho\,(\rho^2\sin\phi\,d\rho\,d\phi\,d\theta) = \left(\int_{\theta=a-b}^{a+b}d\theta\right)\left(\int_{\phi=0}^{\pi}\sin\phi\,d\phi\right)\left(\int_{\rho=0}^{2}\rho^3\,d\rho\right)$$

$$= \left(\theta\big|_{\theta=a-b}^{a+b}\right)\left(-\cos\phi\big|_{\phi=0}^{\pi}\right)\left(\frac{\rho^4}{4}\Big|_{\rho=0}^{2}\right) = (2b)(2)\left(\frac{R^4}{4}\right) = R^4 b,$$

$$M_{yz} = \int_{\theta=a-b}^{a+b}\int_{\phi=0}^{\pi}\int_{\rho=0}^{2}(\rho\sin\phi\cos\theta)(\rho)\,(\rho^2\sin\phi\,d\rho\,d\phi\,d\theta)$$

$$= \left(\int_{\theta=a-b}^{a+b}\cos\theta\,d\theta\right)\left(\int_{\phi=0}^{\pi}\sin^2\phi\,d\phi\right)\left(\int_{\rho=0}^{2}\rho^4\,d\rho\right)$$

$$= \left(\sin\theta\big|_{\theta=a-b}^{a+b}\right)\left(\frac{\phi-\sin\phi\cos\phi}{2}\Big|_{\phi=0}^{\pi}\right)\left(\frac{\rho^5}{5}\Big|_{\rho=0}^{2}\right) = (2\cos a\,\sin b)\left(\frac{\pi}{2}\right)\left(\frac{R^5}{5}\right) = \frac{\pi R^5}{5}\cos a\,\sin b,$$

$$M_{xz} = \int_{\theta=a-b}^{a+b}\int_{\phi=0}^{\pi}\int_{\rho=0}^{2}(\rho\sin\phi\sin\theta)(\rho)\,(\rho^2\sin\phi\,d\rho\,d\phi\,d\theta)$$

$$= \left(\int_{\theta=a-b}^{a+b} \sin\theta \, d\theta \right) \left(\int_{\phi=0}^{\pi} \sin^2\phi \, d\phi \right) \left(\int_{\rho=0}^{2} \rho^4 \, d\rho \right)$$

$$= \left(-\cos\theta \Big|_{\theta=a-b}^{a+b} \right) \left(\frac{\phi - \sin\phi\cos\phi}{2} \Big|_{\phi=0}^{\pi} \right) \left(\frac{\rho^4}{4} \Big|_{\rho=0}^{2} \right) = (2\sin a \sin b) \left(\frac{\pi}{2} \right) \left(\frac{R^5}{5} \right) = \frac{\pi R^5}{5} \sin a \, \sin b,$$

$$M_{yz} = \int_{\theta=a-b}^{a+b} \int_{\phi=0}^{\pi} \int_{\rho=0}^{2} (\rho\cos\phi)(\rho) \, (\rho^2 \sin\phi \, d\rho \, d\phi \, d\theta) = \left(\int_{\theta=a-b}^{a+b} d\theta \right) \left(\int_{\phi=0}^{\pi} \cos\phi \sin\phi \, d\phi \right) \left(\int_{\rho=0}^{2} \rho^4 \, d\rho \right)$$

$$= \left(\theta \Big|_{\theta=a-b}^{a+b} \right) \left(\frac{\sin^2\phi}{2} \Big|_{\phi=0}^{\pi} \right) \left(\frac{\rho^4}{4} \Big|_{\rho=0}^{2} \right) = (2b)(0) \left(\frac{R^5}{5} \right) = 0.$$

Therefore

$$(\bar{x}, \bar{y}, \bar{z}) = \left(\frac{M_{yz}}{\mu}, \frac{M_{xz}}{\mu}, \frac{M_{xy}}{\mu} \right) = \left(\frac{(1/5)\pi R^5 \cos a \sin b}{R^4 b}, \frac{(1/5)\pi R^5 \sin a \sin b}{R^4 b}, \frac{0}{R^4 b} \right)$$

$$= \left(\frac{\pi R \cos a \sin b}{5b}, \frac{\pi R \sin a \sin b}{5b}, 0 \right) = \frac{\pi \sin b}{5b} (R\cos a, R\sin a, 0).$$

If $b = \pi/3$ and $R = 2$, then $(\bar{x}, \bar{y}, \bar{z}) = \dfrac{3\sqrt{3}}{5} (\cos a, \sin a, 0).$

Solutions 14.C   *Computer Exercises*   (page 963)

1.   For any choice of a in $[-1, 1]$, we have

$$M_x = \int_a^1 \int_{-\sqrt{1-x^2}}^{\sqrt{1-x^2}} y \, dy \, dx = \int_a^1 0 \, dx = 0,$$

$$M_y = \int_a^1 \int_{-\sqrt{1-x^2}}^{\sqrt{1-x^2}} x \, dy \, dx = \int_a^1 2x\sqrt{1-x^2} \, dx = \left(\frac{2}{3} \right) (1-a^2)^{3/2},$$

$$\mu = \int_a^1 \int_{-\sqrt{1-x^2}}^{\sqrt{1-x^2}} dy \, dx = \int_a^1 2\sqrt{1-x^2} dx = \left(\frac{\pi}{2} \right) - \sin^{-1} a - a\sqrt{1-a^2}.$$

Our calculations show that $\bar{y} = 0$ (referring to symmetry about the x-axis would do as well). A numerical approximation to the root of $\bar{x}(a) = .5$ is $a \approx 0.13817$; the graph shows how $\bar{x}$ varies with a. ($\bar{x}(0) = 4/(3\pi) \approx 0.42$.)

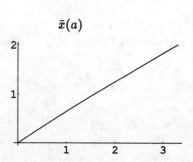

$\bar{x}(a)$

3.   For any positive choice of a, we have

$$M_x = \int_0^a \int_{-\tan^{-1}x}^{\tan^{-1}x} y \, dy \, dx = \int_0^a 0 \, dx = 0,$$

$$M_y = \int_0^a \int_{-\tan^{-1}x}^{\tan^{-1}x} x \, dy \, dx = \int_0^a 2x\tan^{-1}x \, dx = (1+a^2)\tan^{-1}a - a,$$

$$\mu = \int_0^a \int_{-\tan^{-1}x}^{\tan^{-1}x} dy \, dx = \int_0^a 2\tan^{-1}x \, dx = 2a\tan^{-1}a - \ln(1+a^2).$$

Clearly $\bar{y} = 0$. $a \approx 1.57107$ is an approximate root of $\bar{x}(a) = 1$ (this a is about 0.00028 away from $\pi/2$); the graph shows how $\bar{x}$ varies with a.

$\bar{x}(a)$

15
Introduction to Vector Analysis

Solutions 15.1   *Vector fields*   (page 968)

1. $\nabla \left(x^2 + y^2\right)^{-1/2} = (-1/2)\left(x^2 + y^2\right)^{-3/2} \nabla \left(x^2 + y^2\right) = -\left(x^2 + y^2\right)^{-3/2}(x\mathbf{i} + y\mathbf{j})$.

3. $\nabla(x + y)^2 = 2(x + y)\nabla(x + y) = 2(x + y)(\mathbf{i} + \mathbf{j})$.

5. $\nabla \cos(x - y) = -\sin(x - y)\nabla(x - y) = -\sin(x - y)(\mathbf{i} - \mathbf{j})$.

7. $\nabla\left(y\tan(y - x)\right) = -y\sec^2(y - x)\mathbf{i} + \left(\tan(y - x) + y\sec^2(y - x)\right)\mathbf{j}$.

9. $\nabla\sec(x + 3y) = \sec(x + 3y)\tan(x + 3y)\nabla(x + 3y) = \sec(x + 3y)\tan(x + 3y)(\mathbf{i} + 3\mathbf{j})$.

11. $\dfrac{x^2 - y^2}{x^2 + y^2} = 1 - \dfrac{2y^2}{x^2 + y^2} = \dfrac{2x^2}{x^2 + y^2} - 1$, thus $\nabla\left(\dfrac{x^2 - y^2}{x^2 + y^2}\right) = \dfrac{4xy^2}{\left(x^2 + y^2\right)^2}\mathbf{i} - \dfrac{4x^2 y}{\left(x^2 + y^2\right)^2}\mathbf{j} = \dfrac{4xy}{\left(x^2 + y^2\right)^2}(y\mathbf{i} - x\mathbf{j})$.

13. $\nabla\sqrt{x^2 + y^2 + z^2} = \dfrac{1}{2\sqrt{x^2 + y^2 + z^2}}\nabla\left(x^2 + y^2 + z^2\right) = \dfrac{1}{\sqrt{x^2 + y^2 + z^2}}(x\mathbf{i} + y\mathbf{j} + z\mathbf{k})$.

15. $\nabla(\sin x \cos y \tan z) = \cos x \cos y \tan z\,\mathbf{i} - \sin x \sin y \tan z\,\mathbf{j} + \sin x \cos y \sec^2 z\,\mathbf{k}$.

17. $\nabla(x \ln y - z \ln x) = \left(\ln y - \dfrac{z}{x}\right)\mathbf{i} + \left(\dfrac{x}{y}\right)\mathbf{j} - \ln x\,\mathbf{k}$.

19. $\nabla\left((y - z)e^{x+2y+3z}\right) = (y - z)e^{x+2y+3z}\mathbf{i} + \left(1 + 2(y - z)\right)e^{x+2y+3z}\mathbf{j} + \left(-1 + 3(y - z)\right)e^{x+2y+3z}\mathbf{k}$.

21.
If $F(x, y) = \ln\left(\dfrac{\sqrt{(x-1)^2 + y^2}}{\sqrt{(x+1)^2 + y^2}}\right) = \dfrac{\ln\left((x-1)^2 + y^2\right) - \ln\left((x+1)^2 + y^2\right)}{2}$,

then

$$\nabla F = \left(\dfrac{x - 1}{(x-1)^2 + y^2} - \dfrac{x + 1}{(x+1)^2 + y^2}\right)\mathbf{i} + \left(\dfrac{y}{(x-1)^2 + y^2} - \dfrac{y}{(x+1)^2 + y^2}\right)\mathbf{j}$$

$$= \dfrac{2}{\left((x-1)^2 + y^2\right)\left((x+1)^2 + y^2\right)}\left((x^2 - y^2 - 1)\mathbf{i} + 2xy\mathbf{j}\right).$$

23. $-xy$ is a potential function for $y\mathbf{i} + x\mathbf{j}$ because $-\nabla(-xy) = -(-y\mathbf{i} - x\mathbf{j}) = y\mathbf{i} + x\mathbf{j}$.

25.   If $G = g(x, y, z)$ and $H = h(t)$ are smooth functions, the chain rule implies $\nabla\left(H \circ G\right) = \left(H' \circ G\right)\nabla G$.
Let $\mathbf{w} = x\mathbf{i} + y\mathbf{j} + z\mathbf{k}$. Then $2|\mathbf{w}|\,\nabla|\mathbf{w}| = \nabla|\mathbf{w}|^2 = \nabla\left(x^2 + y^2 + z^2\right) = 2x\mathbf{i} + 2y\mathbf{j} + 2z\mathbf{k} = 2\mathbf{w}$ implies

$$\nabla|\mathbf{w}| = \dfrac{\mathbf{w}}{|\mathbf{w}|},$$

$$\nabla \ln|\mathbf{w}| = \dfrac{1}{|\mathbf{w}|}\nabla|\mathbf{w}| = \dfrac{1}{|\mathbf{w}|}\dfrac{\mathbf{w}}{|\mathbf{w}|} = \dfrac{\mathbf{w}}{|\mathbf{w}|^2},$$

$$\nabla\dfrac{1}{|\mathbf{w}|^n} = \dfrac{-n}{|\mathbf{w}|^{n+1}}\nabla|\mathbf{w}| = \dfrac{-n}{|\mathbf{w}|^{n+1}}\dfrac{\mathbf{w}}{|\mathbf{w}|} = -n\dfrac{\mathbf{w}}{|\mathbf{w}|^{n+2}}.$$

Hence $\alpha\ln|\mathbf{w}|$ is a potential function for $\dfrac{-\alpha\mathbf{w}}{|\mathbf{w}|^2}$ and $\dfrac{-\alpha}{(k - 2)|\mathbf{w}|^{k-2}}$ is a potential function for $\dfrac{-\alpha\mathbf{w}}{|\mathbf{w}|^k}$ if $k \neq 2$.

27.   Suppose $\mathbf{F} = y\mathbf{i} - x\mathbf{j}$ were conservative. Then there would be an f such that $y\mathbf{i} - x\mathbf{j} = \mathbf{F} = -\nabla f = -f_x\mathbf{i} - f_y\mathbf{j}$. Hence $y = -f_x$ which implies $f(x, y) = -xy + g(x)$ and $f_y = -x$, but that contradicts $-x = -f_y$. Therefore $y\mathbf{i} - x\mathbf{j}$ is not conservative. ∎

REMARK: If $\mathbf{F} = a\mathbf{i} + b\mathbf{j}$ is conservative with potential function f, then $a_y = -f_{xy}$ and $b_x = -f_{yx}$. If f is smooth enough for Theorem 13.4.1 (Equality of Mixed Partials) to apply, then $a_y = -f_{xy} = -f_{yx} = b_x$. In this problem we have $a = y$, $b = -x$, and $a_y = 1 \neq -1 = b_x$.

15.2 — Work and Line Integrals

1. $\mathbf{F} = |\mathbf{F}| \left((\cos\theta)\mathbf{i} + (\sin\theta)\mathbf{j} \right) = 3\left((\cos 0)\mathbf{i} + (\sin 0)\mathbf{j} \right) = 3\mathbf{i}$ and $\mathbf{d} = \mathbf{q} - \mathbf{p} = (\mathbf{i} + 7\mathbf{j}) - (2\mathbf{i} + 3\mathbf{j}) = -\mathbf{i} + 4\mathbf{j}$;
therefore $W = \mathbf{F} \cdot \mathbf{d} = (3\mathbf{i}) \cdot (-\mathbf{i} + 4\mathbf{j}) = -3$ joules.

3. $\mathbf{F} = |\mathbf{F}| \left(\cos\theta\,\mathbf{i} + \sin\theta\,\mathbf{j} \right) = 6 \left(\cos(\pi/4)\mathbf{i} + \sin(\pi/4)\mathbf{j} \right) = 3\sqrt{2}(\mathbf{i}+\mathbf{j})$ and $\mathbf{d} = \mathbf{q} - \mathbf{p} = (-\mathbf{i} + 4\mathbf{j}) - (2\mathbf{i} + 3\mathbf{j}) =$
$-3\mathbf{i} + \mathbf{j}$; therefore $W = \mathbf{F} \cdot \mathbf{d} = 3\sqrt{2}\,(\mathbf{i} + \mathbf{j}) \cdot (-3\mathbf{i} + \mathbf{j}) = -6\sqrt{2}$ joules.

5. $\mathbf{F} = |\mathbf{F}|\mathbf{u} = 4(2\mathbf{i} + 3\mathbf{j})/\sqrt{2^2 + 3^2} = (4/\sqrt{13})\,(2\mathbf{i} + 3\mathbf{j})$ and $\mathbf{d} = \mathbf{q} - \mathbf{p} = (-\mathbf{i} + 3\mathbf{j}) - (2\mathbf{i}) = -3\mathbf{i} + 3\mathbf{j}$;
therefore $W = \mathbf{F} \cdot \mathbf{d} = (4/\sqrt{13})\,(2\mathbf{i} + 3\mathbf{j}) \cdot (-3\mathbf{i} + 3\mathbf{j}) = 12/\sqrt{13}$ joules.

7.   If $\mathbf{F}(x,y) = xy\mathbf{i} + ye^x\mathbf{j}$ and C is the curve $\mathbf{x}(t) = (2 - t)\mathbf{i} + \mathbf{j}$ for $t \in [0, 2]$, then

$$\int_C \mathbf{F}(\mathbf{x}) \cdot d\mathbf{x} = \int_{t=0}^{2} \left((2 - t)\mathbf{i} + e^{2-t}\mathbf{j} \right) \cdot (-\mathbf{i})\, dt = \int_{t=0}^{2} -(2 - t)\, dt = \left. \frac{(2 - t)^2}{2} \right|_{t=0}^{2} = -2.$$

9.   Let $\mathbf{F}(x,y) = x^2\mathbf{i} + y^2\mathbf{j}$. If C is the straight line segment from $(0,0)$ to $(2,4)$, it can be parameterized
as the curve $\mathbf{x}(t) = t\,(2\mathbf{i} + 4\mathbf{j})$ for $t \in [0, 1]$, then

$$\int_C \mathbf{F}(\mathbf{x}) \cdot d\mathbf{x} = \int_{t=0}^{1} \left((2t)^2\mathbf{i} + (4t)^2\mathbf{j} \right) \cdot (2\mathbf{i} + 4\mathbf{j})\, dt = \int_{t=0}^{1} (8t^2 + 64t^2)\, dt = \int_{t=0}^{1} 72t^2\, dt = 24t^3\big|_{t=0}^{1} = 24.$$

11.   If $\mathbf{F}(x,y) = xy\,\mathbf{i} + (y - x)\mathbf{j}$ and C is the curve $\mathbf{x}(t) = t\mathbf{i} + (2t - 4)\mathbf{j}$ for $t \in [1, 2]$, then

$$\int_C \mathbf{F}(\mathbf{x}) \cdot d\mathbf{x} = \int_{t=1}^{2} \left((2t^2 - 4t)\mathbf{i} + (t - 4)\mathbf{j} \right) \cdot (\mathbf{i} + 2\mathbf{j})\, dt = \int_{t=1}^{2} (2t^2 - 2t - 8)\, dt = \left(\frac{2t^3}{3} - t^2 - 8t \right)\bigg|_{t=1}^{2} = \frac{-19}{3}.$$

13.   If $\mathbf{F}(x,y) = x\,y\,\mathbf{i} + (y - x)\mathbf{j}$ and C is the curve $\mathbf{x}(t) = \cos t\,\mathbf{i} + \sin t\,\mathbf{j}$ for $t \in [0, 2\pi]$, then

$$\int_C \mathbf{F}(\mathbf{x}) \cdot d\mathbf{x} = \int_{t=0}^{2\pi} \left(\cos t \sin t\,\mathbf{i} + (\sin t - \cos t)\mathbf{j} \right) \cdot (-\sin t\,\mathbf{i} + \cos t\,\mathbf{j})\, dt$$

$$= \int_{t=0}^{2\pi} (-\cos t\, \sin^2 t + \sin t\, \cos t - \cos^2 t)\, dt = \left(\frac{-\sin^3 t}{3} + \frac{\sin^2 t}{2} - \frac{\cos t \sin t + t}{2} \right)\bigg|_{t=0}^{2\pi} = -\pi.$$

15.   Let $\mathbf{F}(x,y) = x\,y\mathbf{i} + (y - x)\,\mathbf{j}$. Decompose the triangle C joining $(0,0)$, $(1,0)$, $(1,1)$ counterclockwise into
three straight line segments. $L_1 : \mathbf{x}(t) = t\,\mathbf{i}$, $L_2 : \mathbf{x}(t) = \mathbf{i} + t\,\mathbf{j}$, $L_3 : \mathbf{x}(t) = (1 - t)(\mathbf{i} + \mathbf{j})$; all parameterized
for $t \in [0, 1]$. Therefore

$$\int_{L_1} \mathbf{F}(\mathbf{x}) \cdot d\mathbf{x} = \int_{t=0}^{1} (-t\,\mathbf{j}) \cdot \mathbf{i}\, dt = \int_{t=0}^{1} 0\, dt = 0,$$

$$\int_{L_2} \mathbf{F}(\mathbf{x}) \cdot d\mathbf{x} = \int_{t=0}^{1} (t\,\mathbf{i} + (t - 1)\mathbf{j}) \cdot \mathbf{j}\, dt = \int_{t=0}^{1} (t - 1)\, dt = \left. \frac{(t - 1)^2}{2} \right|_{t=0}^{1} = \frac{-1}{2},$$

$$\int_{L_3} \mathbf{F}(\mathbf{x}) \cdot d\mathbf{x} = \int_{t=0}^{1} \left((1 - t)^2\mathbf{i} \right) \cdot (-\mathbf{i} - \mathbf{j})\, dt = \int_{t=0}^{1} -(1 - t)^2\, dt = \left. \frac{(1 - t)^3}{3} \right|_{t=0}^{1} = \frac{-1}{3};$$

$$\int_C \mathbf{F}(\mathbf{x}) \cdot d\mathbf{x} = \int_{L_1} \mathbf{F}(\mathbf{x}) \cdot d\mathbf{x} + \int_{L_2} \mathbf{F}(\mathbf{x}) \cdot d\mathbf{x} + \int_{L_3} \mathbf{F}(\mathbf{x}) \cdot d\mathbf{x} = 0 + \frac{-1}{2} + \frac{-1}{3} = \frac{-5}{6}.$$

17.   Let $\mathbf{F}(x,y) = (x^2 + 2y)\,\mathbf{i} - y^2\mathbf{j}$. The curve $\mathbf{x}(t) = -3\cos t\,\mathbf{i} + \sin t\,\mathbf{j}$ for $t \in [-\pi/2, \pi/2]$ lies on the left
half of the ellipse $x^2 + 9y^2 = 9$, this curve goes from $(0, -1)$ to $(0, 1)$ in the clockwise direction.

$$\int_C \mathbf{F}(\mathbf{x}) \cdot d\mathbf{x} = \int_{t=-\pi/2}^{\pi/2} \left((9\cos^2 t + 2\sin t)\mathbf{i} - \sin^2 t\,\mathbf{j} \right) \cdot (3\sin t\,\mathbf{i} + \cos t\,\mathbf{j})\, dt$$

$$= \int_{t=-\pi/2}^{\pi/2} (27\sin t\, \cos^2 t + 6\sin^2 t - \cos t\, \sin^2 t)\, dt$$

$$= \left(-9\cos^3 t - 3\cos t\, \sin t + 3t - \frac{\sin^3 t}{3} \right)\bigg|_{t=-\pi/2}^{\pi/2} = 3\pi - \frac{2}{3}.$$

19.   Let $\mathbf{F}(x,y) = e^{x+y}\mathbf{i} + e^{x-y}\mathbf{j}$. Decompose the triangle C joining $(0,0)$, $(1,0)$, $(0,1)$ counterclockwise into three straight line segments. $L_1 : \mathbf{x}(t) = t\,\mathbf{i}$, $L_2 : \mathbf{x}(t) = (1-t)\mathbf{i} + t\,\mathbf{j}$, $L_3 : \mathbf{x}(t) = (1-t)\mathbf{j}$; all parameterized for $t \in [0,1]$. Therefore

$$\int_{L_1} \mathbf{F}(\mathbf{x}) \cdot d\mathbf{x} = \int_{t=0}^{1} (e^t\mathbf{i} + e^t\mathbf{j}) \cdot \mathbf{i}\, dt = \int_{t=0}^{1} e^t\, dt = e^t\big|_{t=0}^{1} = e - 1,$$

$$\int_{L_2} \mathbf{F}(\mathbf{x}) \cdot d\mathbf{x} = \int_{t=0}^{1} (e\,\mathbf{i} + e^{1-2t}\mathbf{j}) \cdot (-\mathbf{i} + \mathbf{j})\, dt = \int_{t=0}^{1} (-e + e^{1-2t})\, dt = \left(-et - \frac{e^{1-2t}}{2} \right)\Big|_{t=0}^{1} = -e - \frac{e^{-1} - e}{2},$$

$$\int_{L_3} \mathbf{F}(\mathbf{x}) \cdot d\mathbf{x} = \int_{t=0}^{1} (e^{1-t}\mathbf{i} + e^{-1+t}\mathbf{j}) \cdot (-\mathbf{j})\, dt = \int_{t=0}^{1} -e^{-1+t}\, dt = -e^{-1+t}\big|_{t=0}^{1} = -1 + e^{-1};$$

$$\int_C \mathbf{F}(\mathbf{x}) \cdot d\mathbf{x} = \int_{L_1} \mathbf{F}(\mathbf{x}) \cdot d\mathbf{x} + \int_{L_2} \mathbf{F}(\mathbf{x}) \cdot d\mathbf{x} + \int_{L_3} \mathbf{F}(\mathbf{x}) \cdot d\mathbf{x}$$

$$= (e-1) + \left(-e - \frac{e^{-1} - e}{2} \right) + (-1 + e^{-1}) = -2 + \frac{1}{2}e^{-1} + \frac{1}{2}e = -2 + \cosh 1.$$

21.   Let $\mathbf{F}(x,y) = x^{-2}y\mathbf{i} + xy^{-2}\mathbf{j}$ and parameterize the straight line segment from $(2,1)$ to $(4,6)$ in the form $\mathbf{x}(t) = (2+2t)\mathbf{i} + (1+5t)\mathbf{j}$ for $t \in [0,1]$. Therefore

$$\int_C \mathbf{F}(\mathbf{x}) \cdot d\mathbf{x} = \int_{t=0}^{1} \left(\frac{1+5t}{(2+2t)^2}\mathbf{i} + \frac{2+2t}{(1+5t)^2}\mathbf{j} \right) \cdot (2\mathbf{i} + 5\mathbf{j})\, dt = \int_{t=0}^{1} \left(2\frac{1+5t}{(2+2t)^2} + 5\frac{2+2t}{(1+5t)^2} \right) dt$$

$$= \int_{t=0}^{1} \left(\frac{-2}{(1+t)^2} + \frac{5/2}{1+t} + \frac{8}{(1+5t)^2} + \frac{2}{1+5t} \right) dt$$

$$= \left(\frac{2}{1+t} + \frac{5}{2}\ln|1+t| - \frac{8/5}{1+5t} + \frac{2}{5}\ln|1+5t| \right)\Big|_{t=0}^{1} = \frac{1}{3} + \frac{5}{2}\ln 2 + \frac{2}{5}\ln 6.$$

23.   Let $\mathbf{F}(x,y,z) = x\mathbf{i} + y\mathbf{j} + z\mathbf{k}$ and C be the curve $\mathbf{x}(t) = t\mathbf{i} + t^2\mathbf{j} + t^3\mathbf{k}$ for $t \in [0,1]$. Therefore

$$\int_C \mathbf{F}(\mathbf{x}) \cdot d\mathbf{x} = \int_{t=0}^{1} (t\mathbf{i} + t^2\mathbf{j} + t^3\mathbf{k}) \cdot (\mathbf{i} + 2t\mathbf{j} + 3t^2\mathbf{k})\, dt = \int_{t=0}^{1} (t + 2t^3 + 3t^5)\, dt = \frac{t^2 + t^4 + t^6}{2}\Big|_{t=0}^{1} = \frac{3}{2}.$$

REMARK: If $\mathbf{F}(x,y) = f(x)\mathbf{i} + g(y)\mathbf{j} + h(z)\mathbf{k}$ and C is the smooth curve $\mathbf{w}(t) = x(t)\mathbf{i} + y(t)\mathbf{j} + z(t)\mathbf{k}$ from (x_0, y_0, z_0) to (x_1, y_1, z_1), then $\int_C \mathbf{F}(\mathbf{w}) \cdot d\mathbf{w} = \int_{x=x_0}^{x_1} f(x)dx + \int_{y=y_0}^{y_1} g(y)dy + \int_{z=z_0}^{z_1} h(z)dz$. (The substitutions $x = x(t)$, $y = y(t)$, $z = z(t)$ do not change the value of the integrals on the right-hand-side.) Therefore, we could compute

$$\int_C \mathbf{F}(\mathbf{w}) \cdot d\mathbf{w} = \int_{x=0}^{1} x\, dx + \int_{y=0}^{1} y\, dy + \int_{z=0}^{1} z\, dz = \frac{x^2}{2}\Big|_{x=0}^{1} + \frac{y^2}{2}\Big|_{y=0}^{1} + \frac{z^2}{2}\Big|_{z=0}^{1} = \frac{1}{2} + \frac{1}{2} + \frac{1}{2}.$$

25.   Let $\mathbf{F}(x,y,z) = x^2\mathbf{i} + y^2\mathbf{j} + z^2\mathbf{k}$ and C be the curve $\mathbf{x}(t) = \cos t\,\mathbf{i} + \sin t\,\mathbf{j} + t\mathbf{k}$ for $t \in [0, \pi/2]$. Therefore

$$\int_C \mathbf{F}(\mathbf{x}) \cdot d\mathbf{x} = \int_{t=0}^{\pi/2} (\cos^2 t\,\mathbf{i} + \sin^2 t\,\mathbf{j} + t^2\mathbf{k}) \cdot (-\sin t\,\mathbf{i} + \cos t\,\mathbf{j} + \mathbf{k})\, dt$$

$$= \int_{t=0}^{\pi/2} (-\sin t\,\cos^2 t + \cos t\,\sin^2 t + t^2)\, dt = \frac{\cos^3 t + \sin^3 t + t^3}{3}\Big|_{t=0}^{\pi/2} = \frac{\pi^3}{24}.$$

REMARK: As discussed in the Remark following Solution 23, we can compute

$$\int_C \mathbf{F}(\mathbf{x}) \cdot d\mathbf{x} = \int_{x=1}^{0} x^2 dx + \int_{y=0}^{1} y^2 dy + \int_{z=0}^{\pi/2} z^2 dz = \frac{x^3}{3}\Big|_{x=1}^{0} + \frac{y^3}{3}\Big|_{y=0}^{1} + \frac{z^3}{3}\Big|_{z=0}^{\pi/2} = \frac{-1}{3} + \frac{1}{3} + \frac{\pi^3/8}{3}.$$

27.   If $\mathbf{F}(x,y) = x^3\mathbf{i} + xy\mathbf{j}$ and C is the curve $\mathbf{x}(t) = \sin t\,\mathbf{i} + e^t\mathbf{j}$ for $t \in [0, \pi/2]$, then

$$W = \int_C \mathbf{F}(\mathbf{x}) \cdot d\mathbf{x} = \int_{t=0}^{\pi/2} (\sin^3 t\,\mathbf{i} + e^t \sin t\,\mathbf{j}) \cdot (\cos t\,\mathbf{i} + e^t\mathbf{j})\, dt$$

$$= \int_{t=0}^{\pi/2} (\cos t\,\sin^3 t + e^{2t} \sin t)\, dt = \left(\frac{\sin^4 t}{4} + \frac{e^{2t}}{5}(2\sin t - \cos t) \right)\Big|_{t=0}^{\pi/2} = \frac{9}{20} + \frac{2e^\pi}{5} \text{ joules}.$$

15.2 — Work and Line Integrals

29.   If $\mathbf{F}(x,y) = xy\,\mathbf{i} + (2x^3 - y)\mathbf{j}$ and C is the curve $\mathbf{x}(t) = \cos t\,\mathbf{i} + \sin t\,\mathbf{j}$ for $t \in [0, 2\pi]$, then

$$W = \int_C \mathbf{F}(\mathbf{x}) \cdot d\mathbf{x} = \int_{t=0}^{2\pi} \left(\cos t\,\sin t\,\mathbf{i} + (2\cos^3 t - \sin t)\mathbf{j}\right) \cdot (-\sin t\,\mathbf{i} + \cos t\,\mathbf{j})\,dt$$

$$= \int_{t=0}^{2\pi} \left(-\cos t\,\sin^2 t + 2\cos^4 t - \sin t\,\cos t\right)dt$$

$$= \left(\frac{-\sin^3 t}{3} + \frac{\sin t\,\cos^3 t}{2} + \frac{3\cos t\,\sin t + 3t}{4} + \frac{\cos^2 t}{2}\right)\Big|_{t=0}^{2\pi} = \frac{3\pi}{2}\text{ joules.}$$

31.   If $\mathbf{F}(x,y) = -xy^2\,\mathbf{i} + 2x\,\mathbf{j}$ and C is the curve $\mathbf{x}(t) = a\cos t\,\mathbf{i} + b\sin t\,\mathbf{j}$ for $t \in [0, 2\pi]$, then

$$W = \int_C \mathbf{F}(\mathbf{x}) \cdot d\mathbf{x} = \int_{t=0}^{2\pi} \left(-ab^2\cos t\,\sin^2 t\,\mathbf{i} + 2a\cos t\,\mathbf{j}\right) \cdot (-a\sin t\,\mathbf{i} + b\cos t\,\mathbf{j})\,dt$$

$$= \int_{t=0}^{2\pi} \left(a^2 b^2\cos t\,\sin^3 t + 2ab\cos^2 t\right)dt$$

$$= \left(\frac{a^2 b^2}{4}\sin^4 t + ab(\cos t\,\sin t - t)\right)\Big|_{t=0}^{2\pi} = 2ab\pi\text{ joules.}$$

33.   If $\mathbf{F}(x,y) = \alpha\left(x^2 + y^2\right)^{-3/2}(x\,\mathbf{i} + y\,\mathbf{j})$ and C is the curve $\mathbf{x}(t) = (1+t)\mathbf{i} - t\,\mathbf{j}$ for $t \in [0, 2]$, then

$$W = \int_C \mathbf{F}(\mathbf{x}) \cdot d\mathbf{x} = \int_{t=0}^{2} \left(\alpha\left((1+t)^2 + (-t)^2\right)^{-3/2}((1+t)\mathbf{i} - t\,\mathbf{j})\right) \cdot (\mathbf{i} - \mathbf{j})\,dt$$

$$= \int_{t=0}^{2} \alpha\left(1 + 2t + 2t^2\right)^{-3/2}(1 + 2t)\,dt$$

$$= -\alpha\left(1 + 2t + 2t^2\right)^{-1/2}\Big|_{t=0}^{2} = \alpha\left(1 - \frac{1}{\sqrt{13}}\right)\text{ joules.}$$

35.   If $\mathbf{F}(x,y,z) = x^2\,\mathbf{i} + y^2\,\mathbf{j} + 2xyz\,\mathbf{k}$ and C is the curve $\mathbf{x}(t) = \cos t\,\mathbf{i} + 4\sin t\,\mathbf{j} + t\,\mathbf{k}$ for $t \in [0, \pi/2]$, then

$$W = \int_C \mathbf{F}(\mathbf{x}) \cdot d\mathbf{x} = \int_{t=0}^{\pi/2} \left(\cos^2 t\,\mathbf{i} + 16\sin^2 t\,\mathbf{j} + 8t\,\sin t\,\cos t\,\mathbf{k}\right) \cdot (-\sin t\,\mathbf{i} + 4\cos t\,\mathbf{j} + \mathbf{k})\,dt$$

$$= \int_{t=0}^{\pi/2} \left(-\sin t\,\cos^2 t + 64\cos t\,\sin^2 t + 8t\,\sin t\,\cos t\right)dt$$

$$= \left(\frac{\cos^3 t}{3} + \frac{64\sin^3 t}{3} - 4t\,\cos^2 t + 2\cos t\,\sin t + 2t\right)\Big|_{t=0}^{\pi/2} = (21 + \pi)\text{ joules.}$$

Solutions 15.3   *Exact vector fields and Independence of path*   (pages 982-3)

1.   Let $\mathbf{F}(x,y) = 2xy\,\mathbf{i} + \left(x^2 + 1\right)\mathbf{j} = P(x,y)\mathbf{i} + Q(x,y)\mathbf{j}$. Therefore $P_y = 2x = Q_x$ implies $\mathbf{F}$ is exact. If $\nabla f = \mathbf{F}$, then

$$f = \int P\,dx = \int 2xy\,dx = x^2 y + g(y) \implies x^2 + g'(y) = f_y = Q = x^2 + 1$$

$$\implies g'(y) = 1 \implies g(y) = y + C \implies f(x,y) = x^2 y + y + C.$$

3.   Let $\mathbf{F}(x,y) = \left(4x^2 - 4y^2\right)\mathbf{i} + (8xy - \ln y)\mathbf{j} = P(x,y)\mathbf{i} + Q(x,y)\mathbf{j}$. Therefore $P_y = -8y \neq 8y = Q_x$ implies $\mathbf{F}$ is not exact.

5.   Let $\mathbf{F}(x,y) = 2x\cos y\,\mathbf{i} + x^2\sin y\,\mathbf{j} = P(x,y)\mathbf{i} + Q(x,y)\mathbf{j}$. Therefore $P_y = -2x\sin y \neq 2x\sin y = Q_x$ implies $\mathbf{F}$ is not exact.

7.   Let $\mathbf{F}(x,y) = (x - y\cos x)\mathbf{i} - \sin x\,\mathbf{j} = P(x,y)\mathbf{i} + Q(x,y)\mathbf{j}$. Therefore $P_y = -\cos x = Q_x$ implies $\mathbf{F}$ is exact. If $\nabla f = \mathbf{F}$, then

$$f = \int P\,dx = \int (x - y\cos x)\,dx = \frac{x^2}{2} - y\sin x + g(y)$$

$$\implies -\sin x + g'(y) = f_y = Q = -\sin x \implies g'(y) = 0 \implies g(y) = C \implies f(x,y) = \frac{x^2}{2} - y\sin x + C.$$

9.   Let $\mathbf{F}(x,y) = (3x\ln x + x^5 - y)\,\mathbf{i} - x\mathbf{j} = P\mathbf{i} + Q\mathbf{j}$. Therefore $P_y = -1 = Q_x$ implies $\mathbf{F}$ is exact. If $\nabla f = \mathbf{F}$, then

$$f = \int P\,dx = \int (3x\ln x + x^5 - y)\,dx = \frac{3x^2}{4}(2\ln x - 1) + \frac{x^6}{6} - xy + g(y)$$
$$\Longrightarrow -x + g'(y) = f_y = Q = -x \Longrightarrow g'(y) = 0 \Longrightarrow g(y) = C$$
$$\Longrightarrow f(x,y) = \frac{3x^2}{4}(2\ln x - 1) + \frac{x^6}{6} - xy + C.$$

11.   Let $\mathbf{F}(x,y) = (x^2 + y^2 + 1)\,\mathbf{i} - (xy + y)\mathbf{j} = P\mathbf{i} + Q\mathbf{j}$. Then $P_y = 2y \neq -y = Q_x$ implies $\mathbf{F}$ is not exact.

13.   Let $\mathbf{F}(x,y,z) = \mathbf{i} + \mathbf{j} + \mathbf{k} = P\mathbf{i} + Q\mathbf{j} + R\mathbf{k}$. Since $P = Q = R = 1$, all first partial derivatives are zero; that implies $\mathbf{F}$ is exact. If $\nabla f = \mathbf{F}$, then

$$f = \int f_x\,dx = \int P\,dx = \int dx = x + g(y,z) \Longrightarrow g_y(y,z) = 0 + g_y(y,z) = f_y = Q = 1$$
$$\Longrightarrow g = \int g_y\,dy = \int dy = y + h(z)$$
$$\Longrightarrow f = x + y + h(z)$$
$$\Longrightarrow h'(z) = 0 + 0 + h'(z) = f_z = R = 1$$
$$\Longrightarrow h = \int h'\,dz = \int dz = z + C$$
$$\Longrightarrow f(x,y,z) = x + y + z + C.$$

15.   Let $\mathbf{F}(x,y,z) = (e^{yz} + y)\,\mathbf{i} + (xze^{yz} - x)\mathbf{j} + (xye^{yz} + 2z)\,\mathbf{k} = P\mathbf{i} + Q\mathbf{j} + R\mathbf{k}$. Since $P_y = ze^{yz} + 1 \neq ze^{yz} - 1 = Q_x$, we infer $\mathbf{F}$ is not exact. (Note, incidentally, that $P_z = R_x$ and $Q_z = R_y$.)

17.   Solution 1 shows $\mathbf{F}(x,y) = 2xy\mathbf{i} + (x^2 + 1)\mathbf{j}$ is exact and $\nabla f = \mathbf{F}$ implies $f(x,y) = (x^2 + 1)\,y + C$. Therefore, if C is a smooth curve from $(0,1)$ to $(2,3)$, then $\displaystyle\int_C \mathbf{F}(\mathbf{x}) \cdot d\mathbf{x} = (x^2 + 1)\,y\,\big|_{(0,1)}^{(2,3)} = 15 - 1 = 14$.

19.   Let $\mathbf{F}(x,y) = (x\cos(x + y) + \sin(x + y))\,\mathbf{i} + x\cos(x + y)\,\mathbf{j}$. Thus $P_y = -x\sin(x + y) + \cos(x + y) = Q_x$ implies $\mathbf{F}$ is exact. Observing that $\mathbf{F} = \nabla x \sin(x + y)$, we conclude that if C is a smooth curve from $(0,0)$ to $(\pi/6, \pi/3)$, then $\displaystyle\int_C \mathbf{F}(\mathbf{x}) \cdot d\mathbf{x} = x\sin(x + y)\,\big|_{(0,0)}^{(\pi/6,\pi/3)} = \frac{\pi}{6}\,\sin\frac{\pi}{2} = \frac{\pi}{6}$.

21.   Let $\mathbf{F}(x,y) = 2x\cos y\,\mathbf{i} - x^2\sin y\,\mathbf{j}$. Thus $P_y = -2x\sin y = Q_x$ implies $\mathbf{F}$ is exact. Since $\mathbf{F} = \nabla x^2\cos y$, we conclude that if C is a smooth curve from $(0, \pi/2)$ to $(\pi/2, 0)$, then $\displaystyle\int_C \mathbf{F}(\mathbf{x}) \cdot d\mathbf{x} = x^2\cos y\,\big|_{(0,\pi/2)}^{(\pi/2,0)} = \frac{\pi^2}{4}$.

23.   Let $\mathbf{F}(x,y) = e^y\mathbf{i} + xe^y\mathbf{j}$. Thus $P_y = e^y = Q_x$ implies $\mathbf{F}$ is exact. Since $\mathbf{F} = \nabla xe^y$, we conclude that if C is a smooth curve from $(0,0)$ to $(5,7)$, then $\displaystyle\int_C \mathbf{F}(\mathbf{x}) \cdot d\mathbf{x} = xe^y\,\big|_{(0,0)}^{(5,7)} = 5e^7$.

25.   Let $\mathbf{F}(x,y,z) = y^2z^4\mathbf{i} + 2xyz^4\mathbf{j} + 4xy^2z^3\mathbf{k}$. Therefore $P_y = 2yz^4 = Q_x$, $Q_z = 8xyz^3 = R_y$, and $R_x = 4y^2z^3 = P_z$; that implies $\mathbf{F}$ is exact. Since $\mathbf{F} = \nabla xy^2z^4$, we conclude that if C is a smooth curve from $(0,0,0)$ to $(3,2,1)$, then $\displaystyle\int_C \mathbf{F}(\mathbf{x}) \cdot d\mathbf{x} = xy^2z^4\,\big|_{(0,0,0)}^{(3,2,1)} = 12$.

27.   If $P\mathbf{i} + Q\mathbf{j} = \nabla f = f_x\mathbf{i} + f_y\mathbf{j}$, and if f, $P = f_x$, $Q = f_y$, $P_y = f_{xy}$, $Q_x = f_{yx}$ are continuous, then Theorem 13.4.1 implies $\dfrac{\partial P}{\partial y} = f_{xy} = f_{yx} = \dfrac{\partial Q}{\partial x}$. ∎

29.   If $\mathbf{F} = -\nabla g$ and $\mathbf{x}_0 = \mathbf{x}_1$, then equation (5) implies $\displaystyle\int_C \mathbf{F}(\mathbf{x}) \cdot d\mathbf{x} = -g(\mathbf{x}_0) + g(\mathbf{x}_0) = 0$. ∎

Solutions 15.4   *Green's theorem in the plane*   **(pages 988-9)**

1.   If $\Omega = \{(x,y) : 0 \le x \le 1,\ 0 \le y \le 1\}$, then

$$\oint_{\partial\Omega} 3y\,dx + 5x\,dy = \iint_\Omega (5 - 3)\,dx\,dy = 2\iint_\Omega dx\,dy = 2 \cdot \text{area}(\Omega) = 2 \cdot (1 - 0) \cdot (1 - 0) = 2.$$

15.4 — Green's theorem in the plane

3. If Ω is the triangular region with vertices $(0,0)$, $(1,0)$, $(0,1)$, then

$$\oint_{\partial\Omega} e^x \cos y \, dx + e^x \sin y \, dy = \iint_\Omega \left(e^x \sin y - (-e^x \sin y)\right) dy \, dx = \int_{x=0}^1 \int_{y=0}^{1-x} 2e^x \sin y \, dy \, dx$$

$$= \int_{x=0}^1 -2e^x \cos y \Big|_{y=0}^{1-x} dx = \int_{x=0}^1 \left(2e^x - 2e^x \cos(1-x)\right) dx$$

$$= \left(2e^x + e^x \sin(1-x) - e^x \cos(1-x)\right)\Big|_{x=0}^1 = e - 2 - \sin 1 + \cos 1.$$

5. If Ω is the rectangular region with vertices $(0,0)$, $(2,0)$, $(2,1)$, $(0,1)$, then

$$\oint_{\partial\Omega} e^x \cos y \, dx + e^x \sin y \, dy = \iint_\Omega \left(e^x \sin y - (-e^x \sin y)\right) dy \, dx$$

$$= \int_{x=0}^2 \int_{y=0}^1 2e^x \sin y \, dy \, dx = 2\left(\int_{x=0}^2 e^x \, dx\right)\left(\int_{y=0}^1 \sin y \, dy\right)$$

$$= 2\left(e^x \Big|_{x=0}^2\right)\left(-\cos y \Big|_{y=0}^1\right) = 2\left(e^2 - 1\right)\left(1 - \cos 1\right).$$

7. If Ω is the unit disk, then $\oint_{\partial\Omega} (x^2 + y^2) \, dx - 2xy \, dy = \iint_\Omega (-2y - 2y) \, dy \, dx = \iint_\Omega -4y \, dy \, dx$. Since $-4y$ is an odd function in y and since Ω is symmetric about the x-axis, our integral must equal 0.

9. If $\Omega = \{(x,y) : 0 \le x \le \pi/4, \ 0 \le y \le \pi/3\}$, then

$$\oint_{\partial\Omega} \cos y \, dx + \cos x \, dy = \int_{x=0}^{\pi/4} \int_{y=0}^{\pi/3} (-\sin x + \sin y) \, dy \, dx = \int_{x=0}^{\pi/4} (-y \sin x - \cos y)\Big|_{y=0}^{\pi/3} dx$$

$$= \int_{x=0}^{\pi/4} \left(\frac{1}{2} - \frac{\pi}{3} \sin x\right) dx = \left(\frac{x}{2} + \frac{\pi}{3} \cos x\right)\Big|_{x=0}^{\pi/4} = \frac{\pi}{24}\left(4\sqrt{2} - 5\right).$$

11. If $\Omega = \left\{(x,y) : 1 \le y \le 3, \ e^y \le x \le e^{y^3}\right\}$, then

$$\oint_{\partial\Omega} y \ln x \, dy = \iint_\Omega \left(\frac{y}{x} - 0\right) dx \, dy = \int_{y=1}^3 \int_{x=e^y}^{e^{y^3}} \frac{y}{x} \, dx \, dy$$

$$= \int_{y=1}^3 y \ln x \Big|_{x=e^y}^{e^{y^3}} dy = \int_{y=1}^3 (y^4 - y^2) \, dy = \left(\frac{y^5}{5} - \frac{y^3}{3}\right)\Big|_{x=1}^3 = \frac{596}{15}.$$

13. If Ω satisfies the hypotheses of Green's theorem, then

$$\oint_{\partial\Omega} ay \, dx + bx \, dy = \iint_\Omega (b - a) \, dx \, dy = (b - a) \iint_\Omega dx \, dy = (b - a) \cdot (\text{area of } \Omega).$$

15. If Ω satisfies the hypotheses of Green's theorem, then

$$\oint_{\partial\Omega} \frac{-4x}{\sqrt{1+y^2}} \, dx + \frac{2x^2 y}{(1+y^2)^{3/2}} \, dy = \iint_\Omega \left(\frac{4xy}{(1+y^2)^{3/2}} - \frac{4xy}{(1+y^2)^{3/2}}\right) dx \, dy = \iint_\Omega 0 \, dx \, dy = 0.$$

REMARK: The integrand is the differential of $-2x^2 \left(1+y^2\right)^{-1/2}$.

17. Let Ω be the triangle with vertices $(0,0)$, $(5,2)$, $(-3,8)$. Then $\partial\Omega$ can be parameterized as

$$x\mathbf{i} + y\mathbf{j} = \begin{cases} 5t\mathbf{i} + 2t\mathbf{j} & \text{if } 0 \le t \le 1, \\ (5 - 8(t-1))\mathbf{i} + (2 + 6(t-1))\mathbf{j} = (13 - 8t)\mathbf{i} + (-4 + 6t)\mathbf{j} & \text{if } 1 \le t \le 2, \\ (-3 + 3(t-2))\mathbf{i} + (8 - 8(t-2))\mathbf{j} = (-9 + 3t)\mathbf{i} + (24 - 8t)\mathbf{j} & \text{if } 2 \le t \le 3. \end{cases}$$

Hence the area of the triangle is

$$\iint_\Omega dA = \oint_{\partial\Omega} x \, dy = \int_{t=0}^1 (5t)(2 \, dt) + \int_{t=1}^2 (13 - 8t)(6 \, dt) + \int_{t=2}^3 (-9 + 3t)(-8 \, dt)$$

$$= 5t^2 \Big|_{t=0}^1 + (78t - 24t^2)\Big|_{t=1}^2 + (72t - 12t^2)\Big|_{t=2}^3 = 5 + 6 + 12 = 23.$$

REMARK: Alternatively, we compute

$$\iint_\Omega dA = \oint_{\partial\Omega} -y \, dx = \int_{t=0}^1 -(2t)(5 \, dt) + \int_{t=1}^2 -(-4 + 6t)(-8 \, dt) + \int_{t=2}^3 -(24 - 8t)(3 \, dt)$$

$$= -5t^2 \Big|_{t=0}^1 + (-32t + 24t^2)\Big|_{t=1}^2 + (-72t + 12t^2)\Big|_{t=2}^3 = -5 + 40 - 12 = 23.$$

19.   Let Ω be the quadrilateral with vertices $(0,0)$, $(2,1)$, $(4,4)$, $(-1,3)$. Then $\partial\Omega$ can be parameterized as

$$x\,\mathbf{i} + y\,\mathbf{j} = \begin{cases} 2t\,\mathbf{i} + t\,\mathbf{j} & \text{if } 0 \le t \le 1, \\ \big(2 + 2(t-1)\big)\,\mathbf{i} + \big(1 + 3(t-1)\big)\,\mathbf{j} = 2t\,\mathbf{i} + (-2 + 3t)\mathbf{j} & \text{if } 1 \le t \le 2, \\ \big(4 - 5(t-2)\big)\,\mathbf{i} + \big(4 - (t-2)\big)\,\mathbf{j} = (14 - 5t)\mathbf{i} + (6 - t)\mathbf{j} & \text{if } 2 \le t \le 3, \\ \big(-1 + (t-3)\big)\,\mathbf{i} + \big(3 - 3(t-3)\big)\,\mathbf{j} = (-4 + t)\mathbf{i} + (12 - 3t)\mathbf{j} & \text{if } 3 \le t \le 4. \end{cases}$$

Hence the area of the triangle is

$$\iint_\Omega dA = \oint_{\partial\Omega} x\,dy = \int_{t=0}^{1} (2t)(dt) + \int_{t=1}^{2} (2t)(3\,dt) + \int_{t=2}^{3} (14 - 5t)(-\,dt) + \int_{t=3}^{4} (-4 + t)(-3\,dt)$$

$$= t^2\Big|_{t=0}^{1} + 3t^2\Big|_{t=1}^{2} + \left(-14t + \frac{5}{2}t^2\right)\Big|_{t=2}^{3} + \left(12t - \frac{3}{2}t^2\right)\Big|_{t=3}^{4} = 1 + 9 - \frac{3}{2} + \frac{3}{2} = 10.$$

REMARK:

$$-y\,dx + x\,dy = \begin{cases} -(t)(2\,dt) + (2t)(dt) = 0 & \text{if } 0 \le t \le 1, \\ -(-2 + 3t)(2\,dt) + (2t)(3\,dt) = 4\,dt & \text{if } 1 \le t \le 2, \\ -(6 - t)(-5\,dt) + (14 - 5t)(-\,dt) = 16\,dt & \text{if } 2 \le t \le 3, \\ -(12 - 3t)(dt) + (-4 + t)(-3\,dt) = 0 & \text{if } 3 \le t \le 4. \end{cases}$$

Thus $\displaystyle \iint_\Omega dA = \frac{1}{2}\oint_{\partial\Omega}(-y\,dx + x\,dy) = \frac{1}{2}\left(\int_{t=0}^{1} 0\,dt + \int_{t=1}^{2} 4\,dt + \int_{t=2}^{3} 16\,dt + \int_{t=3}^{4} 0\,dt\right) = \frac{20}{2} = 10.$

21.   If $\partial\Omega$ is the ellipse satisfying $(x/a)^2 + (y/b)^2 = 1$ and if $\mathbf{x}(s)$ is a representation of that curve parameterized by arc length s, then the unit tangent vector $\mathbf{T}(s)$ is given by $\mathbf{T}(s) = \dfrac{d\mathbf{x}}{ds}$ (see Theorem 12.5.3).

Therefore $\displaystyle \oint_{\partial\Omega} (-x\mathbf{i} - y\mathbf{j}) \cdot \mathbf{T}\,ds = \oint_{\partial\Omega} (-x\mathbf{i} - y\mathbf{j}) \cdot d\mathbf{x} = \iint_\Omega \left(\frac{\partial(-y)}{\partial x} - \frac{\partial(-x)}{\partial y}\right) dA = \iint_\Omega 0\,dA = 0.$ ∎

23.   If disk $\Omega = \{(x, y) : x^2 + y^2 \le a^2\}$, then $x^2 + y^2 = a^2$ on the circle $\partial\Omega$; if g is any function whose domain includes a^2, then

$$\oint_{\partial\Omega} \big(\alpha g\,(x^2 + y^2)\,dx + \beta g\,(x^2 + y^2)\,dy\big) = \oint_{\partial\Omega} \big(\alpha g\,(a^2)\,dx + \beta g\,(a^2)\,dy\big) = g\,(a^2)\oint_{\partial\Omega}(\alpha\,dx + \beta\,dy)$$

$$= g\,(a^2)\iint_\Omega \left(\frac{\partial\beta}{\partial x} - \frac{\partial\alpha}{\partial y}\right) dA = g\,(a^2)\iint_\Omega 0\,dA = 0. \quad ∎$$

REMARK: If g is continuously differentiable and if $P\,dx + Q\,dy = \alpha g\,(x^2 + y^2)\,dx + \beta g\,(x^2 + y^2)\,dy$, then $\dfrac{\partial Q}{\partial x} - \dfrac{\partial P}{\partial x} = 2(\beta x - \alpha y)\,g'(x^2 + y^2)$ is an odd function. Since Ω is symmetric about the origin, we infer

$$\oint_{\partial\Omega} \big(\alpha g\,(x^2 + y^2)\,dx + \beta g\,(x^2 + y^2)\,dy\big) = \iint_\Omega 2(\beta x - \alpha y)\,g'(x^2 + y^2)\,dA = 0. \quad ∎$$

Solutions 15.5   *Parametric representation of a surface, Surface area*   (pages 999-1000)

1.   Let Π be the plane $2x + 3y - z = 0$. Each vector on Π has the form

$$x\mathbf{i} + y\mathbf{j} + z\mathbf{k} = x\mathbf{i} + y\mathbf{j} + (2x + 3y)\mathbf{k} = x(\mathbf{i} + 2\mathbf{k}) + y(\mathbf{j} + 3\mathbf{k}) = x\mathbf{u} + y\mathbf{v}$$

where $\mathbf{u} = \mathbf{i} + 2\mathbf{k}$, $\mathbf{v} = \mathbf{j} + 3\mathbf{k}$, and $(x, y) \in \mathbb{R}^2$.

3.   Example 4 can be adapted to provide a parameterization of the hemisphere $x^2 + y^2 + z^2 = 2^2$, $z \ge 0$:

$$\mathbf{r}(\theta, \phi) = 2\cos\theta\,\sin\phi\,\mathbf{i} + 2\sin\theta\,\sin\phi\,\mathbf{j} + 2\cos\phi\,\mathbf{k}, \qquad 0 \le \theta \le 2\pi, \qquad 0 \le \phi \le \frac{\pi}{2}.$$

5. $\mathbf{r}(x, y) = x\mathbf{i} + y\mathbf{j} + \left(\dfrac{x^2 - y^2}{4}\right)\mathbf{k}, \quad 0 \le x \le 1, \quad 2 \le y \le 3.$

7. $\mathbf{p}(r, \theta) = r\cos\theta\,\mathbf{i} + r\sin\theta\,\mathbf{j} + r(4\cos\theta - \sin\theta)\mathbf{k}, \qquad 0 \le \theta \le 2\pi, \qquad 0 \le r \le 1$
lies on the plane $z = 4x - y$ and inside the cylinder $x^2 + y^2 = 1$.

9. $\mathbf{p}(\theta, \phi) = \cos\theta\,\sin\phi\,\mathbf{i} + \sin\theta\,\sin\phi\,\mathbf{j} + \cos\phi\,\mathbf{k}, \qquad 0 \le \theta \le 2\pi, \qquad 0 \le \phi \le \pi/4$
lies on the unit sphere $x^2 + y^2 + z^2 = 1$ and above the cone $z = \sqrt{x^2 + y^2}$.

15.5 — Parametric representation of a surface, Surface area

11. The plane $z = x + 2y$ can be parameterized $r(u, v) = u\mathbf{i} + v\mathbf{j} + (u + 2v)\mathbf{k}$. Therefore $\mathbf{r}_u = \mathbf{i} + \mathbf{k}$, $\mathbf{r}_v = \mathbf{j} + 2\mathbf{k}$, $\mathbf{n} = \mathbf{r}_u \times \mathbf{r}_v = -\mathbf{i} - 2\mathbf{j} + \mathbf{k}$, and $|\mathbf{n}| = \sqrt{6}$. The part of the plane corresponding to the region $\Omega = \{(u, v) : 0 \le u \le v \le 2\}$ has area

$$\iint_\Omega \sqrt{6} \, dA = \int_{v=0}^{2} \int_{u=0}^{v} \sqrt{6} \, du \, dv = \sqrt{6} \int_{v=0}^{2} v \, dv = \sqrt{6} \left. \frac{v^2}{2} \right|_{v=0}^{2} = 2\sqrt{6}.$$

13. The plane $z = ax + by$ can be parameterized $r(u, v) = u\mathbf{i} + v\mathbf{j} + (au + bv)\mathbf{k}$. Therefore $\mathbf{r}_u = \mathbf{i} + a\mathbf{k}$, $\mathbf{r}_v = \mathbf{j} + b\mathbf{k}$, $\mathbf{n} = \mathbf{r}_u \times \mathbf{r}_v = -a\mathbf{i} - b\mathbf{j} + \mathbf{k}$, and $|\mathbf{n}| = \sqrt{a^2 + b^2 + 1}$. If region Ω is any half of a circle with radius 1, then the corresponding part of the plane has area

$$\iint_\Omega \sqrt{a^2 + b^2 + 1} \, dA = \sqrt{a^2 + b^2 + 1} \iint_\Omega dA = \sqrt{a^2 + b^2 + 1} \, (\text{area of } \Omega) = \sqrt{a^2 + b^2 + 1} \left(\frac{\pi}{2} \right).$$

15. The surface $z = 3 + x^{2/3}$ can be parameterized $r(u, v) = u\mathbf{i} + v\mathbf{j} + \left(3 + u^{2/3}\right)\mathbf{k}$. Therefore $\mathbf{r}_u = \mathbf{i} + (2/3)u^{-1/3}\mathbf{k}$, $\mathbf{r}_v = \mathbf{j}$, $\mathbf{n} = \mathbf{r}_u \times \mathbf{r}_v = (-2/3)u^{-1/3}\mathbf{i} + \mathbf{k}$, and $|\mathbf{n}| = \sqrt{(4/9)u^{-2/3} + 1} = |u|^{-1/3}\sqrt{4/9 + u^{2/3}}$. The integrand $|\mathbf{n}|$ and the region $\Omega = \{(u, v) : -1 \le u \le 1, \ 0 \le v \le 2\}$ are symmetric about the line $u = 0$; although $|\mathbf{n}|$ is undefined along the line $u = 0$, our expression to compute surface area involves a convergent improper integral.

$$\iint_\Omega |\mathbf{n}| \, dA = \int_{v=0}^{2} \int_{u=-1}^{1} |u|^{-1/3}\sqrt{4/9 + u^{2/3}} \, du \, dv = \int_{v=0}^{2} 2 \int_{u=0}^{1} |u|^{-1/3}\sqrt{4/9 + u^{2/3}} \, du \, dv$$

$$= \left(\int_{v=0}^{2} 2 \, dv \right) \left(\lim_{\epsilon \to 0+} \int_{u=\epsilon}^{1} |u|^{-1/3}\sqrt{4/9 + u^{2/3}} \, du \right)$$

$$= 4 \lim_{\epsilon \to 0+} \left. \left(4/9 + u^{2/3}\right)^{3/2} \right|_{u=\epsilon}^{1} = 4 \lim_{\epsilon \to 0+} \left((13/9)^{3/2} - \left(4/9 + \epsilon^{2/3}\right)^{3/2} \right) = 4 \left(13^{3/2} - 8 \right) / 27.$$

17. The surface $z = (1/3)\left(y^2 + 2\right)^{3/2}$ can be parameterized $r(u, v) = u\mathbf{i} + v\mathbf{j} + (1/3)\left(v^2 + 2\right)^{3/2}\mathbf{k}$. Therefore $\mathbf{r}_u = \mathbf{i}$, $\mathbf{r}_v = \mathbf{j} + v\left(v^2 + 2\right)^{1/2}\mathbf{k}$, $\mathbf{n} = \mathbf{r}_u \times \mathbf{r}_v = -v\left(v^2 + 2\right)^{1/2}\mathbf{j} + \mathbf{k}$, and $|\mathbf{n}| = v^2 + 1$. The part of the surface corresponding to the region $\Omega = \{(u, v) : -4 \le u \le 7, \ 0 \le v \le 3\}$ has area

$$\iint_\Omega |\mathbf{n}| \, dA = \int_{v=0}^{3} \int_{u=-4}^{7} \left(v^2 + 1\right) du \, dv = \left(\int_{v=0}^{3} \left(v^2 + 1\right) dv \right) \left(\int_{u=-4}^{7} du \right)$$

$$= \left(\left. \left(\frac{v^3}{3} + v \right) \right|_{v=0}^{3} \right) \left(u \big|_{u=-4}^{7} \right) = (12)(11) = 132.$$

19. The surface $(z+1)^2 = 4x^3$ has two sheets $z = -1 \pm 2x^{3/2}$ for $x \ge 0$. They can be parameterized $r(u, v) = u\mathbf{i} + v\mathbf{j} + \left(-1 \pm 2u^{3/2}\right)\mathbf{k}$. Therefore $\mathbf{r}_u = \mathbf{i} \pm 3u^{1/2}\mathbf{k}$, $\mathbf{r}_v = \mathbf{j}$, $\mathbf{n} = \mathbf{r}_u \times \mathbf{r}_v = \mp 3u^{1/2}\mathbf{i} + \mathbf{k}$, and $|\mathbf{n}| = (9u+1)^{1/2}$. The part of each sheet of the surface corresponding to the region $\Omega = \{(u, v) : 0 \le u \le 1, \ 0 \le v \le 2\}$ has area

$$\iint_\Omega |\mathbf{n}| \, dA = \int_{v=0}^{2} \int_{u=0}^{1} (9u + 1)^{1/2} du \, dv = \left(\int_{v=0}^{2} dv \right) \left(\int_{u=0}^{1} (9u + 1)^{1/2} du \right)$$

$$= (2 - 0) \left. \frac{2}{27}(9u + 1)^{3/2} \right|_{u=0}^{1} = \frac{4}{27} \left(10^{3/2} - 1 \right);$$

double that to obtain the total area of the two sheets.

21. If $r(u, v) = u \cos v \, \mathbf{i} + u \sin v \, \mathbf{j} + (4u \cos v - u \sin v)\mathbf{k}$, then $\mathbf{r}_u = \cos v \, \mathbf{i} + \sin v \, \mathbf{j} + (4 \cos v - \sin v)\mathbf{k}$, $\mathbf{r}_v = -u \sin v \, \mathbf{j} + u \cos v \, \mathbf{j} + (-4u \sin v - u \cos v)\mathbf{k}$, $\mathbf{n} = \mathbf{r}_u \times \mathbf{r}_v = -4u\mathbf{i} + u\mathbf{j} + u\mathbf{k}$, and $|\mathbf{n}| = 3\sqrt{2}\,|u|$. The part of the plane corresponding to the region $\Omega = \{(u, v) : 0 \le u \le 1, \ 0 \le v \le 2\pi\}$ has area

$$\iint_\Omega |\mathbf{n}| \, dA = \int_{u=0}^{1} \int_{v=0}^{2\pi} 3\sqrt{2}\,|u| \, dv \, du = \int_{u=0}^{1} 6\sqrt{2}\,\pi u \, du = 3\sqrt{2}\,\pi u^2 \big|_{u=0}^{1} = 3\sqrt{2}\,\pi.$$

23. If $r(u, v) = \cos u \sin v \, \mathbf{i} + \sin u \sin v \, \mathbf{j} + \cos v \, \mathbf{k}$, then $\mathbf{r}_u = -\sin u \sin v \, \mathbf{i} + \cos u \sin v \, \mathbf{j}$, $\mathbf{r}_v = \cos u \cos v \, \mathbf{i} + \sin u \cos v \, \mathbf{j} - \sin v \, \mathbf{j}$, $\mathbf{n} = \mathbf{r}_u \times \mathbf{r}_v = -\cos u \sin^2 v \, \mathbf{i} - \sin u \sin^2 v \, \mathbf{j} - \sin v \cos v \, \mathbf{k}$, and $|\mathbf{n}| = |\sin v|$. The part of the sphere corresponding to the region $\Omega = \{(u, v) : 0 \le u \le 2\pi, \ 0 \le v \le \pi/4\}$ has area

$$\iint_\Omega |\mathbf{n}| \, dA = \int_{u=0}^{2\pi} \int_{v=0}^{\pi/4} |\sin v| \, dv \, du = \left(\int_{u=0}^{2\pi} du \right) \left(\int_{v=0}^{\pi/4} \sin v \, dv \right) = (2\pi - 0) \left(-\cos v \big|_{v=0}^{\pi/4} \right) = \left(2 - \sqrt{2} \right)\pi.$$

25. $y = \cos^3 v$ and $z = \sin^3 v$ implies $y^{2/3} + z^{2/3} = \cos^2 v + \sin^2 v = 1$. If $\mathbf{r}(u, v) = u\,\mathbf{i} + \cos^3 v\,\mathbf{j} + \sin^3 v\,\mathbf{k}$, then $\mathbf{r}_u = \mathbf{i}$, $\mathbf{r}_v = -3\cos^2 v \sin v\,\mathbf{j} + 3\sin^2 v \cos v\,\mathbf{k}$, $\mathbf{n} = \mathbf{r}_u \times \mathbf{r}_v = -3\sin^2 v \cos v\,\mathbf{j} - 3\cos^2 v \sin v\,\mathbf{k}$, and $|\mathbf{n}| = 3\,|\cos v \sin v|$. The part of the cylinder corresponding to $\Omega = \{(u, v) : 0 \le u \le 2,\ 0 \le v \le 2\pi\}$ has area

$$\iint_\Omega |\mathbf{n}|\,dA = \int_{u=0}^{2} \int_{v=0}^{2\pi} 3\,|\cos v \sin v|\,dv\,du = \left(\int_{u=0}^{2} 3\,du\right)\left(4\int_{v=0}^{\pi/2} \cos v \sin v\,dv\right)$$

$$= (6)\left(\frac{4}{2}\sin^2 v\,\Big|_{v=0}^{\pi/2}\right) = 12.$$

27.   Suppose $a > 0$. Using polar coordinates in the xy-plane, we see that the interior of the circle $x^2 + y^2 = ay$ is $\Omega = \{(x,y)_r : x^2 + y^2 \le ay\} = \{(r,\theta)_p : 0 \le r \le a \sin\theta,\ 0 \le \theta \le \pi\}$. The part of the sphere $x^2 + y^2 + z^2 = a^2$ which is inside the cylinder $x^2 + y^2 = ay$ has two patches, they can be parameterized $\mathbf{R}(r, \theta) = r\cos\theta\,\mathbf{i} + r\sin\theta\,\mathbf{j} \pm \sqrt{a^2 - r^2}\,\mathbf{k}$; then $\mathbf{R}_r = \cos\theta\,\mathbf{i} + \sin\theta\,\mathbf{j} \mp \dfrac{r}{\sqrt{a^2 - r^2}}\,\mathbf{k}$, $\mathbf{R}_\theta = -r\sin\theta\,\mathbf{i} + r\cos\theta\,\mathbf{j}$, $\mathbf{n} = \mathbf{R}_\theta \times \mathbf{R}_\phi = \dfrac{\pm r^2\cos\theta}{\sqrt{a^2 - r^2}}\,\mathbf{i} + \dfrac{\pm r^2\sin\theta}{\sqrt{a^2 - r^2}}\,\mathbf{j} + r\,\mathbf{k}$, and $|\mathbf{n}| = \dfrac{ar}{\sqrt{a^2 - r^2}}$. The two patches on the sphere have total area

$$2\iint_\Omega |\mathbf{n}|\,dA = 2\int_{\theta=0}^{\pi} \int_{r=0}^{a\sin\theta} \frac{ar}{\sqrt{a^2 - r^2}}\,dr\,d\theta = 2\int_{\theta=0}^{\pi} -a\sqrt{a^2 - r^2}\,\Big|_{r=0}^{a\sin\theta}\,d\theta$$

$$= 2\int_{\theta=0}^{\pi} a^2\,(1 - |\cos\theta|)\,d\theta = 2\int_{\theta=0}^{\pi/2} a^2\,(1 - \cos\theta)\,d\theta + 2\int_{\theta=\pi/2}^{\pi} a^2\,(1 + \cos\theta)\,d\theta$$

$$= 2\,a^2(\theta - \sin\theta)\big|_{\theta=0}^{\pi/2} + 2\,a^2(\theta + \sin\theta)\big|_{\theta=\pi/2}^{\pi} = a^2(\pi - 2) + a^2(\pi - 2) = 2a^2(\pi - 2).$$

29.   The sphere $16z = x^2 + y^2 + z^2 = r^2 + z^2$ meets the circular paraboloid $z = x^2 + y^2 = r^2$ at the origin and on the circle $r^2 = 15 = z$. The part of the sphere which is above the paraboloid can be parameterized $\mathbf{R}(r, \theta) = r\cos\theta\,\mathbf{i} + r\sin\theta\,\mathbf{j} + (8 + \sqrt{8^2 - r^2})\,\mathbf{k}$ on the region $\Omega = \{(r,\theta) : 0 \le r \le \sqrt{15},\ 0 \le \theta \le 2\pi\}$. Hence $\mathbf{R}_r = \cos\theta\,\mathbf{i} + \sin\theta\,\mathbf{j} - \dfrac{r}{\sqrt{8^2 - r^2}}\,\mathbf{k}$, $\mathbf{R}_\theta = -r\sin\theta\,\mathbf{i} + r\cos\theta\,\mathbf{j}$, $\mathbf{n} = \mathbf{R}_r \times \mathbf{R}_\theta = \dfrac{r^2\cos\theta}{\sqrt{8^2 - r^2}}\,\mathbf{i} + \dfrac{r^2\sin\theta}{\sqrt{8^2 - r^2}}\,\mathbf{j} + r\,\mathbf{k}$, and $|\mathbf{n}| = \dfrac{8r}{\sqrt{8^2 - r^2}}$. The area of the surface is

$$\iint_\Omega |\mathbf{n}|\,dA = \int_{\theta=0}^{2\pi} \int_{r=0}^{\sqrt{15}} \frac{8r}{\sqrt{8^2 - r^2}}\,dr\,d\theta = \int_{\theta=0}^{2\pi} -8\sqrt{8^2 - r^2}\,\Big|_{r=0}^{\sqrt{15}}\,d\theta = \int_{\theta=0}^{2\pi} 8\,d\theta = 16\,\pi.$$

31.   If $z = x^3 + y^3$ and Ω is the unit circle, then we can use equation (12). $z_x = 3x^2$, $z_y = 3y^2$, $\sqrt{1 + (z_x)^2 + (z_y)^2} = \sqrt{1 + 9x^4 + 9y^4}$; the surface area is $\displaystyle\int_{x=-1}^{1} \int_{y=-\sqrt{1-x^2}}^{\sqrt{1-x^2}} \sqrt{1 + 9x^4 + 9y^4}\,dy\,dx$.

33. $z = \sqrt{1 + x + y}$ and $\Omega = \{(x, y) : 0 \le x \le 2,\ x \le y \le 4 - x\}$; we will use equation (12). We see that $z_x = (1/2)(1 + x + y)^{-1/2} = z_y$ and $\sqrt{1 + (z_x)^2 + (z_y)^2} = \sqrt{1 + (1/2)(1 + x + y)^{-1}}$; the surface area is $\displaystyle\int_{x=0}^{2} \int_{y=x}^{4-x} \sqrt{1 + (1/2)(1 + x + y)^{-1}}\,dy\,dx$.

35.   Implicit differentiation of $(x/a)^2 + (y/b)^2 + (z/c)^2 = 1$ shows that $z_x = -(x/a^2)/(z/c^2)$ and $z_y = -(y/b^2)/(z/c^2)$. Therefore

$$1 + (z_x)^2 + (z_y)^2 = 1 + \frac{(x/a^2)^2}{(z/c^2)^2} + \frac{(y/b^2)^2}{(z/c^2)^2} = 1 + \frac{(c/a)^2(x/a)^2}{(z/c)^2} + \frac{(c/b)^2(y/b)^2}{(z/c)^2}$$

$$= \frac{(z/c)^2 + (c/a)^2(x/a)^2 + (c/b)^2(y/b)^2}{(z/c)^2} = \frac{1 + ((c/a)^2 - 1)\,(x/a)^2 + ((c/b)^2 - 1)\,(y/b)^2}{1 - (x/a)^2 - (y/b)^2}.$$

The ellipsoid $(x/a)^2 + (y/b)^2 + (z/c)^2 = 1$ is symmetric about each coordinate plane; its surface area is 8 times the area of the part lying in the first octant. Suppose a, b, c are positive; the surface area is

$$8\int_{x=0}^{a} \int_{y=0}^{b\sqrt{1-(x/a)^2}} \sqrt{\frac{1 + ((c/a)^2 - 1)\,(x/a)^2 + ((c/b)^2 - 1)\,(y/b)^2}{1 - (x/a)^2 - (y/b)^2}}\,dy\,dx.$$

37.   The plane on the points $(a,0,0)$, $(0,b,0)$, $(0,0,c)$ has equation $1 = x/a + y/b + z/c$ (if $a,b,c \neq 0$). The triangle with those points as vertices can be parameterized $\mathbf{r}(u,v) = au\,\mathbf{i} + bv\,\mathbf{j} + c(1-u-v)\,\mathbf{k}$ over the domain $\Omega = \{(u,v): 0 \leq u \leq 1,\ 0 \leq v \leq 1-u\}$. $\mathbf{r}_u = a\mathbf{i} - c\mathbf{k}$, $\mathbf{r}_v = b\mathbf{j} - c\mathbf{k}$, $\mathbf{n} = \mathbf{r}_u \times \mathbf{r}_v = bc\,\mathbf{i} + ca\,\mathbf{j} + ab\,\mathbf{k}$, and $|\mathbf{n}| = \sqrt{b^2c^2 + c^2a^2 + a^2b^2}$. The area of the triangle is

$$\iint_\Omega |\mathbf{n}|\,dA = \sqrt{b^2c^2 + c^2a^2 + a^2b^2} \int_{u=0}^{1} \int_{v=0}^{1-u} dv\,du = \sqrt{b^2c^2 + c^2a^2 + a^2b^2} \int_{u=0}^{1} (1-u)\,du$$

$$= \sqrt{b^2c^2 + c^2a^2 + a^2b^2}\ \frac{-(1-u)^2}{2}\bigg|_{u=0}^{1} = \frac{1}{2}\sqrt{b^2c^2 + c^2a^2 + a^2b^2}.$$

39.   Moving the arc of $z = g(x)$ for $x \in [a,b]$ parallel to the y-axis for $y \in [0,c]$ generates a surface which can be parameterized $\mathbf{r}(u,v) = x\,\mathbf{i} + y\,\mathbf{j} + g(x)\,\mathbf{k}$ over the domain $\Omega = \{(x,y): a \leq x \leq b,\ 0 \leq y \leq c\}$. $\mathbf{r}_x = \mathbf{i} + g'(x)\,\mathbf{k}$, $\mathbf{r}_y = \mathbf{j}$, $\mathbf{n} = \mathbf{r}_x \times \mathbf{r}_y = -g'(x)\mathbf{i} + \mathbf{k}$, and $|\mathbf{n}| = \sqrt{1 + g'(x)^2}$. The area of the surface is

$$\iint_\Omega |\mathbf{n}|\,dA = \int_{x=a}^{b} \int_{y=0}^{c} \sqrt{1 + g'(x)^2}\,dy\,dx = \int_{x=a}^{b} c\sqrt{1 + g'(x)^2}\,dx = c\left(\int_{x=a}^{b} \sqrt{1 + g'(x)^2}\,dx\right)$$

$$= (\text{distance moved})\,(\text{arc length}). \ \blacksquare$$

REMARK: The surface has the form $z = f(x,y) = g(x)$ and equation (12) integrates $\sqrt{1 + (z_x)^2 + (z_y)^2} = \sqrt{1 + g'(x)^2 + 0^2}$.

41.   Let $\mathbf{X} = c\mathbf{i} - a\mathbf{k}$ and $\mathbf{Y} = c\mathbf{j} - b\mathbf{k}$. Then $u\mathbf{X} + v\mathbf{Y} = uc\mathbf{i} + vc\mathbf{j} + (-ua - vb)\mathbf{k}$ is on the plane with equation $ax + by + cz = 0$ because $a(uc) + b(vc) + c(-ua - vb) = acu + bcv - acu - bcv = 0$. $\blacksquare$

43.   If $f(x) > 0$ for $x \in [a,b]$, then rotating that part of the graph of $y = f(x)$ about the x-axis generates a surface such that $\sqrt{y^2 + z^2} = f(x)$. Using angular parameter t, we can parameterize this surface in the form $\mathbf{r}(x,t) = x\mathbf{i} + f(x)\cos t\,\mathbf{j} + f(x)\sin t\,\mathbf{k}$ over the domain $\Omega = \{(x,t): a \leq x \leq b,\ 0 \leq t \leq 2\pi\}$.

45.   Suppose $h > 0$ and $a > 0$. Rotating the line segment from the origin to the point (h,a) about the x-axis generates a right-circular cone with height h and radius a at its base. That line segment is the graph of $y = f(x) = (a/h)x$ on the interval $[0,h]$. The lateral area of the cone is

$$2\pi \int_{x=0}^{h} f(x)\sqrt{f'(x)^2 + 1}\,dx = 2\pi \int_{x=0}^{h} \left(\frac{a}{h}\right) x\sqrt{\left(\frac{a}{h}\right)^2 + 1}\,dx = (2\pi)\left(\frac{a}{h}\right)\frac{\sqrt{a^2 + h^2}}{h}\ \frac{x^2}{2}\bigg|_{x=0}^{h} = \pi a\sqrt{a^2 + h^2}.$$

Solutions 15.6   *Surface integrals*   (pages 1007-8)

1. $z = x^2 \implies 1 + (z_x)^2 + (z_y)^2 = 1 + (2x)^2 + 0^2 = 1 + 4x^2$, therefore

$$\iint_S x\,d\sigma = \iint_\Omega x\sqrt{1 + 4x^2}\,dA = \int_{x=0}^{1} \int_{y=0}^{2} x\sqrt{1 + 4x^2}\,dy\,dx$$

$$= \int_{x=0}^{1} 2x\sqrt{1 + 4x^2}\,dx = \frac{1}{6}(1 + 4x^2)^{3/2}\bigg|_{x=0}^{1} = \frac{1}{6}\left(5^{3/2} - 1\right).$$

3. $z = x^2 \implies 1 + (z_x)^2 + (z_y)^2 = 1 + 4x^2$, therefore (using items 71 and 73 of the text's integral table)

$$\iint_S (x^2 - 2y^2)\,d\sigma = \iint_\Omega (x^2 - 2y^2)\sqrt{1 + 4x^2}\,dA = \int_{x=0}^{1} \sqrt{1 + 4x^2} \int_{y=0}^{2} (x^2 - 2y^2)\,dy\,dx$$

$$= \int_{x=0}^{1} \sqrt{1 + 4x^2}\left(x^2 y - \frac{2}{3}y^3\right)\bigg|_{y=0}^{2}\,dx = \int_{x=0}^{1}\left(2x^2 - \frac{16}{3}\right)\sqrt{1 + 4x^2}\,dx$$

$$= \left(\frac{1}{8}x(1 + 4x^2)^{3/2} - \frac{131}{48}x\sqrt{1 + 4x^2} - \frac{131}{96}\ln\left(2x + \sqrt{1 + 4x^2}\right)\right)\bigg|_{x=0}^{1}$$

$$= \frac{-101}{48}\sqrt{5} - \frac{131}{96}\ln\left(2 + \sqrt{5}\right).$$

5. $z = \sqrt{4 - x^2 - y^2} \implies 1 + (z_x)^2 + (z_y)^2 = 1 + (-x/z)^2 + (-y/z)^2 = (z^2 + x^2 + y^2)/z^2 = 4/z^2$, therefore

$$\iint_S x \, d\sigma = \iint_\Omega x \frac{2}{\sqrt{4 - x^2 - y^2}} \, dA = \int_{y=-2}^{2} \int_{x=-\sqrt{4-y^2}}^{\sqrt{4-y^2}} \frac{2x}{\sqrt{4 - x^2 - y^2}} \, dx \, dy$$

$$= \int_{y=-2}^{2} -2\sqrt{4 - x^2 - y^2} \Big|_{x=-\sqrt{4-y^2}}^{\sqrt{4-y^2}} \, dy = \int_{y=-2}^{2} 0 \, dy = 0.$$

7. $z = (x + 2y - 4)/3 \implies 1 + (z_x)^2 + (z_y)^2 = 1 + (1/3)^2 + (2/3)^2 = 14/9$, therefore

$$\iint_S (x + y) \, d\sigma = \iint_\Omega (x + y) \frac{\sqrt{14}}{3} \, dA = \frac{\sqrt{14}}{3} \int_{y=1}^{2} \int_{x=0}^{1} (x + y) \, dx \, dy = \frac{\sqrt{14}}{3} \int_{y=1}^{2} \left(\frac{x^2}{2} + xy\right)\Big|_{x=0}^{1} \, dy$$

$$= \frac{\sqrt{14}}{6} \int_{y=1}^{2} (1 + 2y) \, dy = \frac{\sqrt{14}}{6} \left(y + y^2\right)\Big|_{y=1}^{2} = \frac{2\sqrt{14}}{3}.$$

9. $z = (x + 2y - 4)/3 \implies 1 + (z_x)^2 + (z_y)^2 = 1 + (1/3)^2 + (2/3)^2 = 14/9$, therefore

$$\iint_S z^2 \, d\sigma = \iint_\Omega \left(\frac{x + 2y - 4}{3}\right)^2 \frac{\sqrt{14}}{3} \, dA = \frac{\sqrt{14}}{27} \int_{y=1}^{2} \int_{x=0}^{1} (x + 2y - 4)^2 \, dx \, dy$$

$$= \frac{\sqrt{14}}{81} \int_{y=1}^{2} (x + 2y - 4)^3 \big|_{x=0}^{1} \, dy = \frac{\sqrt{14}}{81} \int_{y=1}^{2} (12y^2 - 42y + 37) \, dy$$

$$= \frac{\sqrt{14}}{81} \left(4y^3 - 21y^2 + 37y\right)\Big|_{y=1}^{2} = \frac{2\sqrt{14}}{81}.$$

11. $z = 1 - 2x - 3y \implies 1 + (z_x)^2 + (z_y)^2 = 1 + (-2)^2 + (-3)^2 = 14$, therefore

$$\iint_S \cos z \, d\sigma = \iint_\Omega \cos(1 - 2x - 3y) \sqrt{14} \, dA = \sqrt{14} \int_{y=-1}^{2} \int_{x=0}^{1} \cos(1 - 2x - 3y) \, dx \, dy$$

$$= \frac{\sqrt{14}}{2} \int_{y=-1}^{2} -\sin(1 - 2x - 3y) \big|_{x=0}^{1} \, dy = \frac{\sqrt{14}}{2} \int_{y=-1}^{2} \left(\sin(1 + 3y) + \sin(1 - 3y)\right) dy$$

$$= \frac{\sqrt{14}}{6} \left(-\cos(1 + 3y) + \cos(1 - 3y)\right)\big|_{y=-1}^{2} = \frac{\sqrt{14}}{6} \left(-\cos 7 + \cos 5 + \cos 2 - \cos 4\right).$$

13. We adapt Theorem 1 and Solution 5 to a surface of the form $x = f(y, z)$. $x = \sqrt{4 - y^2 - z^2} \implies$
$1 + (x_y)^2 + (x_z)^2 = 1 + (-y/x)^2 + (-z/x)^2 = (x^2 + y^2 + z^2)/x^2 = 4/x^2$, therefore

$$\iint_S |x| \, d\sigma = \iint_\Omega |x| \frac{2}{|x|} \, dA = 2 \iint_\Omega dA = 2 \cdot (\text{area of } \Omega) = 2 \cdot \pi \, 2^2 = 8\pi.$$

15. $y = \sqrt{9 - x^2 - z^2} \implies 1 + (y_x)^2 + (y_z)^2 = 1 + (-x/y)^2 + (-z/y)^2 = (y^2 + x^2 + z^2)/y^2 = 9/y^2$. We use
polar coordinates in the xz-plane: $x = r \cos \theta$ and $z = r \sin \theta$. Therefore

$$\iint_S z^2 \, d\sigma = \iint_\Omega z^2 \frac{3}{\sqrt{9 - x^2 - z^2}} \, dA = \int_{\theta=0}^{2\pi} \int_{r=0}^{3} (r \sin \theta)^2 \frac{3}{\sqrt{9 - r^2}} \, r \, dr \, d\theta$$

$$= \left(\int_{r=0}^{3} \frac{3r^3}{\sqrt{9 - r^2}} \, dr\right) \left(\int_{\theta=0}^{2\pi} \sin^2 \theta \, d\theta\right)$$

$$= \left(-(18 + r^2)\sqrt{9 - r^2} \Big|_{r=0}^{3}\right) \left(\frac{1}{2}(-\cos \theta \sin \theta + \theta)\Big|_{\theta=0}^{2\pi}\right) = 54\pi.$$

REMARK: $\displaystyle \int \frac{3r^3}{\sqrt{9 - r^2}} \, dr = \int \left(\frac{27r}{\sqrt{9 - r^2}} - 3r\sqrt{9 - r^2}\right) dr = -27\sqrt{9 - r^2} + (9 - r^2)^{3/2} + C.$

---

17. $z = \sqrt{4 - x^2 - y^2} \implies d\sigma/dA = \sqrt{1 + (z_x)^2 + (z_y)^2} = \sqrt{1 + (-x/z)^2 + (-y/z)^2} = 2/z$. Using polar

417

coordinates we find $z = \sqrt{4 - r^2}$ and density $\rho = 25 - x^2 - y^2 = 25 - r^2$. Therefore

$$\mu = \iint_S \rho \, d\sigma = \iint_\Omega \rho \frac{2}{\sqrt{4 - r^2}} \, dA = \int_{r=0}^2 \int_{\theta=0}^{2\pi} (25 - r^2) \frac{2}{\sqrt{4 - r^2}} \, d\theta \, r \, dr$$

$$= 2\pi \int_{r=0}^2 \left(\frac{42r}{\sqrt{4 - r^2}} + 2r\sqrt{4 - r^2} \right) dr$$

$$= 2\pi \left(-42\sqrt{4 - r^2} - \frac{2}{3}(4 - r^2)^{3/2} \right) \Big|_{r=0}^2 = 2\pi \left(\frac{268}{3} \right) = \frac{536\pi}{3}.$$

19.   The plane $x + y + z = 1$ passes through $(1, 0, 0)$, $(0, 1, 0)$, $(0, 0, 1)$. Thus $d\sigma/dA = \sqrt{1 + (-1)^2 + (-1)^2} = \sqrt{3}$ and the total mass is

$$\mu = \iint_S \rho \, d\sigma = \iint_\Omega \alpha \sqrt{3} \, dA = \alpha \sqrt{3} \int_{x=0}^1 \int_{y=0}^{1-x} dy \, dx$$

$$= \alpha \sqrt{3} \int_{x=0}^1 (1 - x) \, dx = \alpha \sqrt{3} \left(\frac{-1}{2} \right) (1 - x)^2 \Big|_{x=0}^1 = \frac{\alpha \sqrt{3}}{2}.$$

21. $z = xy$ implies $z_x = y$ and $z_y = x$. If $\mathbf{F} = x^2 y \mathbf{i} - z\mathbf{j}$, then (invoking Problem 34)

$$\iint_S \mathbf{F} \cdot \mathbf{n} \, d\sigma = \iint_\Omega \left(-(x^2 y)(y) - (-xy)(x) + 0 \right) dy \, dx - \int_{x=0}^1 \int_{y=0}^2 (-x^2 y^2 + x^2 y) \, dy \, dx$$

$$= \int_{x=0}^1 x^2 \left(\frac{-y^3}{3} + \frac{y^2}{2} \right) \Big|_{y=0}^2 dx = \int_{x=0}^1 \frac{-2}{3} x^2 \, dx = \frac{-2}{9} x^3 \Big|_{x=0}^1 = \frac{-2}{9}.$$

23.   A radial vector is normal to the surface of a sphere; if $\mathbf{x}$ is on the unit sphere, then $\mathbf{n} = \mathbf{x}$ is the unit normal at $\mathbf{x}$. Hence, if $\mathbf{F} = x\mathbf{i} + y\mathbf{j} + z\mathbf{k} = \mathbf{x}$, then the flux is

$$\iint_S \mathbf{F} \cdot \mathbf{n} \, d\sigma = \iint_S (x^2 + y^2 + z^2) \, d\sigma = \iint_S 1 \, d\sigma = \text{area of unit sphere} = 4\pi \, 1^2 = 4\pi.$$

25.   A radial vector is normal to the surface of a sphere; if $\mathbf{x}$ is on the unit sphere, then $\mathbf{n} = \mathbf{x}$ is the unit normal at $\mathbf{x}$. Hence, if $\mathbf{F} = x\mathbf{i} + y\mathbf{j} + z\mathbf{k} = \mathbf{x}$, then the flux is

$$\iint_S \mathbf{F} \cdot \mathbf{n} \, d\sigma = \iint_S (x^2 + y^2 + z^2) \, d\sigma = \iint_S 1 \, d\sigma = \text{area of unit hemisphere} = 2\pi \, 1^2 = 2\pi.$$

27. $z = \sqrt{x^2 + y^2}$ implies $z_x = x/z$ and $z_y = y/z$. If $\mathbf{F} = x\mathbf{i} - y\mathbf{j} + xy\mathbf{k}$, then

$$\iint_S \mathbf{F} \cdot \mathbf{n} \, d\sigma = \iint_\Omega \left(-(x) \left(\frac{x}{\sqrt{x^2 + y^2}} \right) - (-y) \left(\frac{y}{\sqrt{x^2 + y^2}} \right) + xy \right) dA = \iint_\Omega \left(\frac{y^2 - x^2}{\sqrt{x^2 + y^2}} + xy \right) dA$$

$$= \int_{r=0}^1 \int_{\theta=0}^{2\pi} \left(\frac{r^2 (\sin^2 \theta - \cos^2 \theta)}{r} + r^2 \cos \theta \sin \theta \right) d\theta \, r \, dr$$

$$= \int_{r=0}^1 \left(r^2 (-\cos \theta \sin \theta) + r^3 \frac{\sin^2 \theta}{2} \right) \Big|_{\theta=0}^{2\pi} dr = \int_{r=0}^1 0 \, dr = 0.$$

29. $z = \sqrt{x^2 + y^2}$ implies $z_x = x/z$ and $z_y = y/z$. If $\mathbf{F} = x^2 \mathbf{i} + y^2 \mathbf{j} + z\mathbf{k}$, then

$$\iint_S \mathbf{F} \cdot \mathbf{n} \, d\sigma = \iint_\Omega \left(-(x^2) \left(\frac{x}{z} \right) - (y^2) \left(\frac{y}{z} \right) + z \right) dA = \iint_\Omega \left(\frac{-(x^3 + y^3)}{\sqrt{x^2 + y^2}} + \sqrt{x^2 + y^2} \right) dA$$

$$= \int_{\theta=0}^{\pi/2} \int_{r=0}^1 \left(\frac{-r^3 (\cos^3 \theta + \sin^3 \theta)}{r} + r \right) r \, dr \, d\theta = \int_{\theta=0}^{\pi/2} \left(\frac{-r^4}{4} (\cos^3 \theta + \sin^3 \theta) + \frac{r^3}{3} \right) \Big|_{r=0}^1 dr$$

$$= \int_{\theta=0}^{\pi/2} \left(\frac{-1}{4} (\cos^3 \theta + \sin^2 \theta) + \frac{1}{3} \right) d\theta$$

$$= \left(\frac{1}{12} (\sin^2 \theta \cos \theta + 2 \cos \theta - \cos^2 \theta \sin \theta - 2 \sin \theta) + \frac{\theta}{3} \right) \Big|_{\theta=0}^{\pi/2} = \frac{\pi}{6} - \frac{1}{3}.$$

31. The rectangular solid $S = \{(x, y, z) : 0 \le x \le 1,\ 0 \le y \le 2,\ 0 \le z \le 3\}$ has six faces.

$(x = 0)$ $\displaystyle\iint_{S_{x=0}} \mathbf{F} \cdot \mathbf{n}\, d\sigma = \iint_{S_{x=0}} (x^2 y\,\mathbf{i} - 2yz\,\mathbf{j} + x^3 y^2\,\mathbf{k}) \cdot (-\mathbf{i})\, d\sigma = \int_{y=0}^{2}\int_{z=0}^{3} -(0^2)\, y\, dz\, dy = 0,$

$(x = 1)$ $\displaystyle\iint_{S_{x=1}} \mathbf{F} \cdot \mathbf{n}\, d\sigma = \iint_{S_{x=1}} (x^2 y\,\mathbf{i} - 2yz\,\mathbf{j} + x^3 y^2\,\mathbf{k}) \cdot \mathbf{i}\, d\sigma = \int_{y=0}^{2}\int_{z=0}^{3} (1^2)\, y\, dz\, dy = 6,$

$(y = 0)$ $\displaystyle\iint_{S_{y=0}} \mathbf{F} \cdot \mathbf{n}\, d\sigma = \iint_{S_{y=0}} (x^2 y\,\mathbf{i} - 2yz\,\mathbf{j} + x^3 y^2\,\mathbf{k}) \cdot (-\mathbf{j})\, d\sigma = \int_{x=0}^{1}\int_{z=0}^{3} 2(0)\, z\, dz\, dx = 0,$

$(y = 2)$ $\displaystyle\iint_{S_{y=2}} \mathbf{F} \cdot \mathbf{n}\, d\sigma = \iint_{S_{y=2}} (x^2 y\,\mathbf{i} - 2yz\,\mathbf{j} + x^3 y^2\,\mathbf{k}) \cdot \mathbf{j}\, d\sigma = \int_{x=0}^{1}\int_{z=0}^{3} -2(2)\, z\, dz\, dx = -18,$

$(z = 0)$ $\displaystyle\iint_{S_{z=0}} \mathbf{F} \cdot \mathbf{n}\, d\sigma = \iint_{S_{z=0}} (x^2 y\,\mathbf{i} - 2yz\,\mathbf{j} + x^3 y^2\,\mathbf{k}) \cdot (-\mathbf{k})\, d\sigma = \int_{x=0}^{1}\int_{y=0}^{2} -x^3 y^2\, dy\, dx = \frac{-2}{3},$

$(z = 3)$ $\displaystyle\iint_{S_{z=3}} \mathbf{F} \cdot \mathbf{n}\, d\sigma = \iint_{S_{z=3}} (x^2 y\,\mathbf{i} - 2yz\,\mathbf{j} + x^3 y^2\,\mathbf{k}) \cdot \mathbf{k}\, d\sigma = \int_{x=0}^{1}\int_{y=0}^{2} x^3 y^2\, dy\, dx = \frac{2}{3}.$

Hence $\displaystyle\iint_S \mathbf{F} \cdot \mathbf{n}\, d\sigma = 0 + 6 + 0 - 18 - \frac{2}{3} + \frac{2}{3} = -12.$

33. If we suppose the surface to be of uniform thickness, then it is plausible for the density to be proportional to area. Hence, for a small patch where density is almost constant, the product of density at a point in the patch times the area of the patch is approximately proportional to the mass of the patch. Summing such products and passing to the limit as the patch mesh is refined yields the integral $\displaystyle\iint_S \rho\, d\sigma.$

35. Radial vectors are normal to the surface of a sphere, hence $\mathbf{n} = (x\mathbf{i} + y\mathbf{j} + z\mathbf{k})/r$ is the outward unit normal to the sphere $S = \{(x, y, z) : x^2 + y^2 + z^2 = r^2\}$ if $r > 0$. Hence $\mathbf{F}\cdot\mathbf{n} = (a\mathbf{i}+b\mathbf{j}+c\mathbf{k})\cdot(x\mathbf{i}+y\mathbf{j}+z\mathbf{k})/r = (ax + by + cz)/r$ and the flux of $\mathbf{F}$ over the sphere is

$$\iint_S \mathbf{F} \cdot \mathbf{n}\, d\sigma = \iint_S \frac{ax + by + cz}{r}\, d\sigma = \int_{\theta=0}^{2\pi}\int_{\phi=0}^{\pi} \frac{ar\cos\theta\sin\phi + br\sin\theta\sin\phi + cr\cos\phi}{r}\, r^2 \sin\phi\, d\phi\, d\theta$$

$$= \int_{\theta=0}^{2\pi} r^2 \left((a\cos\theta + b\sin\theta)\frac{\phi - \cos\phi\sin\phi}{2} + c\frac{\sin^2\phi}{2} \right)\Bigg|_{\phi=0}^{\pi} d\theta$$

$$= \int_{\theta=0}^{2\pi} r^2 \frac{\pi}{2}(a\cos\theta + b\sin\theta)\, d\theta = r^2 \frac{\pi}{2}(a\sin\theta - b\cos\theta)\Bigg|_{\theta=0}^{2\pi} = 0. \ \blacksquare$$

Solutions 15.7   *Divergence and Curl*   (pages 1013-4)

1. If $\mathbf{F} = x^2\mathbf{i} + y^2\mathbf{j} + z^2\mathbf{k}$, then div $\mathbf{F} = 2x + 2y + 2z$ and curl $\mathbf{F} = (0 - 0)\mathbf{i} + (0 - 0)\mathbf{j} + (0 - 0)\mathbf{k} = \mathbf{0}.$

3. If $\mathbf{F} = a\mathbf{i} + b\mathbf{j} + c\mathbf{k}$, then div $\mathbf{F} = 0 + 0 + 0 = 0$ and curl $\mathbf{F} = (0 - 0)\mathbf{i} + (0 - 0)\mathbf{j} + (0 - 0)\mathbf{k} = \mathbf{0}.$

5. If $\mathbf{F} = xy\mathbf{i} + yz\mathbf{j} + xz\,\mathbf{k}$, then div $\mathbf{F} = y + z + x$ and curl $\mathbf{F} = (0 - y)\mathbf{i} + (0 - z)\mathbf{j} + (0 - x)\mathbf{k} = -y\mathbf{i} - z\mathbf{j} - x\mathbf{k}.$

7. If $\mathbf{F} = e^{yz}\mathbf{i} + e^{xz}\mathbf{j} + e^{xy}\mathbf{k}$, then div $\mathbf{F} = 0 + 0 + 0 = 0$ and curl $\mathbf{F} = x\left(e^{xy} - e^{xz}\right)\mathbf{i} + y\left(e^{yz} - e^{xy}\right)\mathbf{j} + z\left(e^{xz} - e^{yz}\right)\mathbf{k}.$

9. If $\mathbf{F} = xy^{-1}\mathbf{i} + yz^{-1}\mathbf{j} + zx^{-1}\mathbf{k}$, then div $\mathbf{F} = y^{-1} + z^{-1} + x^{-1}$ and curl $\mathbf{F} = \left(0 - (-1)yz^{-2}\right)\mathbf{i} + \left(0 - (-1)zx^{-2}\right)\mathbf{j} + \left(0 - (-1)xy^{-2}\right)\mathbf{k} = yz^{-2}\mathbf{i} + zx^{-2}\mathbf{j} + xy^{-2}\mathbf{k}.$

11. Suppose $\mathbf{F} = ax\,\mathbf{i} + by\,\mathbf{j}$ and $\Omega = \{(x, y) : 0 \le x \le 1,\ 0 \le y \le 1\}.$

    a. curl $\mathbf{F} = (0 - 0)\mathbf{i} + (0 - 0)\mathbf{j} + (0 - 0)\mathbf{k} = \mathbf{0}.$

    b. $\displaystyle\oint_{\partial\Omega} \mathbf{F} \cdot \mathbf{T}\, ds = \iint_{\Omega} (\text{curl } \mathbf{F}) \cdot \mathbf{k}\, dx\, dy = \iint_{\Omega} 0\, dx\, dy = 0.$

    c. div $\mathbf{F} = a + b + 0 = a + b.$

    d. $\displaystyle\oint_{\partial\Omega} \mathbf{F} \cdot \mathbf{n}\, ds = \iint_{\Omega} \text{div } \mathbf{F}\, dx\, dy = \iint_{\Omega} (a + b)\, dx\, dy = (a + b)(\text{area of } \Omega) = a + b.$

13.   Suppose $\mathbf{F} = x^2\mathbf{i} + y^2\mathbf{j}$ and $\Omega = \{(x, y) : 0 \le x \le 1,\ 0 \le y \le 1\}$.

   a. $\text{curl}\,\mathbf{F} = (0 - 0)\mathbf{i} + (0 - 0)\mathbf{j} + (0 - 0)\mathbf{k} = \mathbf{0}$.

   b.

$$\oint_{\partial\Omega} \mathbf{F} \cdot \mathbf{T}\,ds = \iint_\Omega (\text{curl}\,\mathbf{F}) \cdot \mathbf{k}\,dx\,dy = \iint_\Omega 0\,dx\,dy = 0.$$

   c. $\text{div}\,\mathbf{F} = 2x + 2y + 0 = 2(x + y)$.

   d.

$$\oint_{\partial\Omega} \mathbf{F} \cdot \mathbf{n}\,ds = \iint_\Omega \text{div}\,\mathbf{F}\,dx\,dy = \int_{y=0}^1 \int_{x=0}^1 2(x + y)\,dx\,dy = \int_{y=0}^1 (1 + 2y)\,dy = 2.$$

15.   Suppose $\mathbf{F} = x\,\mathbf{i} + y\,\mathbf{j}$ and Ω is the unit disk.

   a. $\text{curl}\,\mathbf{F} = (0 - 0)\mathbf{i} + (0 - 0)\mathbf{j} + (0 - 0)\mathbf{k} = \mathbf{0}$.

   b.

$$\oint_{\partial\Omega} \mathbf{F} \cdot \mathbf{T}\,ds = \iint_\Omega (\text{curl}\,\mathbf{F}) \cdot \mathbf{k}\,dx\,dy = \iint_\Omega 0\,dx\,dy = 0.$$

   c. $\text{div}\,\mathbf{F} = 1 + 1 + 0 = 2$.

   d.

$$\oint_{\partial\Omega} \mathbf{F} \cdot \mathbf{n}\,ds = \iint_\Omega \text{div}\,\mathbf{F}\,dx\,dy = \iint_\Omega 2\,dx\,dy = 2\,(\text{area of } \Omega) = 2\pi.$$

17.   Suppose $\mathbf{F} = y^3\mathbf{i} + x^3\mathbf{j}$ and Ω is the unit disk.

   a. $\text{curl}\,\mathbf{F} = (0 - 0)\mathbf{i} + (0 - 0)\mathbf{j} + (3x^2 - 3y^2)\,\mathbf{k} = 3\left(x^2 - y^2\right)\mathbf{k}$.

   b.

$$\oint_{\partial\Omega} \mathbf{F} \cdot \mathbf{T}\,ds = \iint_\Omega (\text{curl}\,\mathbf{F}) \cdot \mathbf{k}\,dA = \iint_\Omega 3\left(x^2 - y^2\right)\,dA$$

$$= \int_{\theta=0}^{2\pi} \int_{r=0}^1 3\left(r^2 \cos^2\theta - r^2 \sin^2\theta\right) r\,dr\,d\theta$$

$$= \int_{\theta=0}^{2\pi} \frac{3}{4}\left(\cos^2\theta - \sin^2\theta\right) d\theta = \left.\frac{3}{4}\sin\theta\,\cos\theta\right|_{\theta=0}^{2\pi} = 0.$$

   c. $\text{div}\,\mathbf{F} = 0 + 0 + 0 = 0$.

   d.

$$\oint_{\partial\Omega} \mathbf{F} \cdot \mathbf{n}\,ds = \iint_\Omega \text{div}\,\mathbf{F}\,dx\,dy = \iint_\Omega 0\,dx\,dy = 0.$$

19. $\nabla^2(xyz) = \dfrac{\partial^2}{\partial x^2}(xyz) + \dfrac{\partial^2}{\partial y^2}(xyz) + \dfrac{\partial^2}{\partial z^2}(xyz) = \dfrac{\partial}{\partial x}(yz) + \dfrac{\partial}{\partial y}(xz) + \dfrac{\partial}{\partial z}(xy) = 0 + 0 + 0 = 0$; hence $f(x, y, z) = xyz$ defines a harmonic function.

21. $\nabla^2\left(2x^2 + 5y^2 + 3z^2\right) = \dfrac{\partial^2}{\partial x^2}\left(2x^2 + 5y^2 + 3z^2\right) + \dfrac{\partial^2}{\partial y^2}\left(2x^2 + 5y^2 + 3z^2\right) + \dfrac{\partial^2}{\partial z^2}\left(2x^2 + 5y^2 + 3z^2\right) = \dfrac{\partial}{\partial x}(4x) + \dfrac{\partial}{\partial y}(10y) + \dfrac{\partial}{\partial z}(6z) = 4 + 10 + 6 = 20 \ne 0$; hence $f(x, y, z) = 2x^2 + 5y^2 + 3z^2$ does not define a harmonic function.

REMARK: $\nabla^2\left(2x^2 + 5y^2 + 3z^2\right) = \text{div} \cdot \text{grad}\left(2x^2 + 5y^2 + 3z^2\right) = \text{div}(4x\mathbf{i} + 10y\mathbf{j} + 6z\mathbf{k}) = 4 + 10 + 6 = 20$.

23.   Let $\mathbf{F} = 2x\,\mathbf{i} + y\,\mathbf{j} - 3z\,\mathbf{k}$.

   a. $\text{div}\,\mathbf{F} = 2 + 1 - 3 = 0$.

   b.   Consider the vector field $\mathbf{G} = ayz\,\mathbf{i} + bxz\,\mathbf{j} + cxy\,\mathbf{k}$. The equation $\mathbf{F} = \text{curl}\,\mathbf{G} = (c - b)x\,\mathbf{i} + (a - c)y\,\mathbf{j} + (b - a)z\,\mathbf{k}$ has general solution $a = c + 1$ and $b = c - 2$. Hence, letting $c = 0$, we obtain $\mathbf{G} = yz\,\mathbf{i} - 2xz\,\mathbf{j}$. (Theorem 2 implies any other solution differs from this one by the gradient of a scalar function.)

REMARK: Suppose $\mathbf{F} = P\mathbf{i} + Q\mathbf{j} + R\mathbf{k}$ with $0 = \text{div}\,\mathbf{F} = P_x + Q_y + R_z$. We shall find $\mathbf{G} = A\mathbf{i} + B\mathbf{j}$ such that $P\mathbf{i} + Q\mathbf{j} + R\mathbf{k} = \mathbf{F} = \text{curl}\,\mathbf{G} = -B_z\mathbf{i} + A_z\mathbf{j} + (B_x - A_y)\mathbf{k}$. The $\mathbf{i}$ and $\mathbf{j}$ components will agree if we let $B = \displaystyle\int_{t=t_0}^z -P(x, y, t)\,dt$ and $A = c(x, y) + \displaystyle\int_{t=t_0}^z Q(x, y, t)\,dt$ where $c(x, y)$ is a function to be determined subsequently. The $\mathbf{k}$ components agree iff

$$R(x, y, z) = (B_x - A_y)(x, y, z) = \int_{t=t_0}^z -P_x(x, y, t)\,dt - \left(c_y(x, y, z) + \int_{t=t_0}^z Q_y(x, y, t)\,dt\right)$$

$$= -c_y(x, y, z) + \int_{t=t_0}^z -(P_x + Q_y)(x, y, t)\,dt$$

$$= -c_y(x, y, z) + \int_{t=t_0}^z R_z(x, y, t)\,dt = -c_y(x, y, z) + R(x, y, z) - R(x, y, t_0)$$

which is satisfied if we let $c(x, y, z) = \int_{u=u_0}^{y} -R(x, u, t_0)\, du.$ (The above solution uses $t_0 = 0$.)

25. If $\mathbf{F} = x^2 + y^2 + z^2$, then curl $\mathbf{F} = (0-0)\mathbf{i} + (0-0)\mathbf{j} + (0-0)\mathbf{k} = \mathbf{0}$ and div(curl $\mathbf{F}$) $= 0 + 0 + 0 = 0$. ∎

27. $\operatorname{curl}(x\,\mathbf{i} + y\,\mathbf{j}) = -\dfrac{\partial y}{\partial z}\mathbf{i} + \dfrac{\partial x}{\partial z}\mathbf{j} + \left(\dfrac{\partial y}{\partial x} - \dfrac{\partial x}{\partial y}\right)\mathbf{k} = 0\,\mathbf{i} + 0\,\mathbf{j} + 0\,\mathbf{k} = \mathbf{0}$, thus $\mathbf{F}(x, y) = x\,\mathbf{i} + y\,\mathbf{j}$ is irrotational. ∎

REMARK: Problem 42 generalizes this result.

29. If $\mathbf{F}(x, y) = y\sqrt{x^2 + y^2}\,\mathbf{i} - x\sqrt{x^2 + y^2}\,\mathbf{j}$, then div $\mathbf{F} = y\dfrac{x}{\sqrt{x^2+y^2}} - x\dfrac{y}{\sqrt{x^2+y^2}} = 0$; therefore $\mathbf{F}$ is incompressible. ∎

31.
$$\operatorname{div}\frac{\mathbf{x}}{|\mathbf{x}|^3} = \operatorname{div}\left(\frac{x}{(x^2+y^2+z^2)^{3/2}}\mathbf{i} + \frac{y}{(x^2+y^2+z^2)^{3/2}}\mathbf{j} + \frac{z}{(x^2+y^2+z^2)^{3/2}}\mathbf{k}\right)$$

$$= \left(\frac{1}{(x^2+y^2+z^2)^{3/2}} - \frac{3x^2}{(x^2+y^2+z^2)^{5/2}}\right) + \left(\frac{1}{(x^2+y^2+z^2)^{3/2}} - \frac{3y^2}{(x^2+y^2+z^2)^{5/2}}\right)$$

$$+ \left(\frac{1}{(x^2+y^2+z^2)^{3/2}} - \frac{3z^2}{(x^2+y^2+z^2)^{5/2}}\right)$$

$$= \frac{3}{(x^2+y^2+z^2)^{3/2}} - \frac{3(x^2+y^2+z^2)}{(x^2+y^2+z^2)^{5/2}} = 0.$$

If $\mathbf{E}(\mathbf{x}) = \dfrac{q}{4\pi\epsilon_0|\mathbf{x}|^3}\,\mathbf{x}$, then div $\mathbf{E} = 0$ for all $\mathbf{x} \neq \mathbf{0}$; thus $\lim_{\mathbf{x}\to 0}$ div $\mathbf{E} = 0$.

33. Suppose $\mathbf{F} = p\mathbf{i} + q\mathbf{j} + r\mathbf{k}$ and $\mathbf{G} = P\mathbf{i} + Q\mathbf{j} + R\mathbf{k}$. Then

$$\operatorname{div}(\mathbf{F}+\mathbf{G}) = \frac{\partial}{\partial x}(p+P) + \frac{\partial}{\partial y}(q+Q) + \frac{\partial}{\partial z}(r+R) = \left(\frac{\partial p}{\partial x} + \frac{\partial q}{\partial y} + \frac{\partial r}{\partial z}\right) + \left(\frac{\partial P}{\partial x} + \frac{\partial Q}{\partial y} + \frac{\partial R}{\partial z}\right) = \operatorname{div}\mathbf{F} + \operatorname{div}\mathbf{G}. ∎$$

REMARK: Using the notation of Theorem 1 together with several parts of Theorem 11.2.1, we have
$\operatorname{div}(\mathbf{F} + \mathbf{G}) = \nabla \cdot (\mathbf{F} + \mathbf{G}) = \nabla \cdot \mathbf{F} + \nabla \cdot \mathbf{G} = \operatorname{div}\mathbf{F} + \operatorname{div}\mathbf{G}$.

35. Suppose $f = f(x, y, z)$ is a scalar function and $\mathbf{G} = P\mathbf{i} + Q\mathbf{j} + R\mathbf{k}$. Then

$$\operatorname{div}(f\mathbf{G}) = \operatorname{div}(fP\mathbf{i} + fQ\mathbf{j} + fR\mathbf{k}) = \frac{\partial(fP)}{\partial x} + \frac{\partial(fQ)}{\partial y} + \frac{\partial(fR)}{\partial z}$$

$$= (f_x P + f P_x) + (f_y Q + f Q_y) + (f_z R + f R_z) = (f_x P + f_y Q + f_z R) + (f P_x + f Q_y + f R_z)$$

$$= (f_x\mathbf{i} + f_y\mathbf{j} + f_z\mathbf{k}) \cdot (P\mathbf{i} + Q\mathbf{j} + R\mathbf{k}) + (f)(P_x + Q_y + R_z) = (\nabla f) \cdot \mathbf{G} + (f)(\operatorname{div}\mathbf{G}). ∎$$

37. $\operatorname{curl}(\nabla f) = \operatorname{curl}(f_x\mathbf{i} + f_y\mathbf{j} + f_z\mathbf{k}) = (f_{zy} - f_{yz})\mathbf{i} + (f_{xz} - f_{zx})\mathbf{j} + (f_{yx} - f_{xy})\mathbf{k}$. If f and its partial derivatives are continuous, then the equality of mixed partials (Theorem 13.4.1) implies $\operatorname{curl}(\nabla f) = \mathbf{0}$. ∎

39. Suppose $\mathbf{F} = p\mathbf{i} + q\mathbf{j} + r\mathbf{k}$ and $\mathbf{G} = P\mathbf{i} + Q\mathbf{j} + R\mathbf{k}$.

$$\operatorname{div}(\mathbf{F} \times \mathbf{G}) = \operatorname{div}\left((qR - Qr)\mathbf{i} - (pR - Pr)\mathbf{j} + (pQ - Pq)\mathbf{k}\right)$$

$$= \frac{\partial}{\partial x}(qR - Qr) - \frac{\partial}{\partial y}(pR - Pr) + \frac{\partial}{\partial z}(pQ - Pq)$$

$$= (q_x R + q R_x - Q_x r - Q r_x) - (p_y R + p R_y - P_y r - P r_y) + (p_z Q + p Q_z - P_z q - P q_z)$$

$$= ((r_y - q_z)P + (p_z - r_x)Q + (q_x - p_y)R) - ((R_y - Q_z)p + (P_z - R_x)q + (Q_x - P_y)r)$$

$$= (\operatorname{curl}\mathbf{F}) \cdot \mathbf{G} - (\operatorname{curl}\mathbf{G}) \cdot \mathbf{F} = (\operatorname{curl}\mathbf{F}) \cdot \mathbf{G} - \mathbf{F} \cdot (\operatorname{curl}\mathbf{G}). ∎$$

41. Suppose $\mathbf{F} = P\mathbf{i} + Q\mathbf{j} + R\mathbf{k}$; also suppose P, Q, R and their partial derivatives (through second-order) are continuous.

$$\operatorname{curl}(\operatorname{curl}\mathbf{F}) = \operatorname{curl}\left((R_y - Q_z)\mathbf{i} + (P_z - R_x)\mathbf{j} + (Q_x - P_y)\mathbf{k}\right)$$

$$= \left((Q_x - P_y)_y - (P_z - R_x)_z\right)\mathbf{i} + \left((R_y - Q_z)_z - (Q_x - P_y)_x\right)\mathbf{j} + \left((P_z - R_x)_x - (R_y - Q_z)_y\right)\mathbf{k}$$

$$= (Q_{xy} - P_{yy} - P_{zz} + R_{xz})\mathbf{i} + (R_{yz} - Q_{zz} - Q_{xx} + P_{yx})\mathbf{j} + (P_{zx} - R_{xx} - R_{yy} + Q_{zy})\mathbf{k};$$

$$\nabla\operatorname{div}\mathbf{F} - \nabla^2\mathbf{F} = \nabla(P_x + Q_y + R_z) - (\nabla^2 P\mathbf{i} + \nabla^2 Q\mathbf{j} + \nabla^2 R\mathbf{k})$$

$$= \left((P_{xx} + Q_{yx} + R_{zx})\mathbf{i} + (P_{xy} + Q_{yy} + R_{zy})\mathbf{j} + (P_{xz} + Q_{yz} + R_{zz})\mathbf{k}\right)$$

$$- \left((P_{xx} + P_{yy} + P_{zz})\mathbf{i} + (Q_{xx} + Q_{yy} + Q_{zz})\mathbf{j} + (R_{xx} + R_{yy} + R_{zz})\mathbf{k}\right)$$

$$= (Q_{yx} + R_{zx} - P_{yy} - P_{zz})\mathbf{i} + (P_{xy} + R_{zy} - Q_{xx} - Q_{zz})\mathbf{j} + (P_{xz} + Q_{yz} - R_{xx} - R_{yy})\mathbf{k}.$$

Comparing the expressions and invoking Theorem 13.4.1 ("Equality of mixed partials"), we see that $\operatorname{curl}(\operatorname{curl}\mathbf{F}) = \nabla \operatorname{div}\mathbf{F} - \nabla^2\mathbf{F}$. ∎

43.　　　Suppose g is a smooth scalar function. Then $-y\,g(x^2+y^2)\mathbf{i} + x\,g(x^2+y^2)\mathbf{j}$ is incompressible because $\operatorname{div}\left(-y\,g(x^2+y^2)\mathbf{i} + x\,g(x^2+y^2)\mathbf{j}\right) = -y\,g'(x^2+y^2)(2x) + x\,g'(x^2+y^2)(2y) = (-2+2)\,x\,y\,g'(x^2+y^2) = 0$. ∎

45.　　　Suppose $\mathbf{F} = P\mathbf{i} + Q\mathbf{j}$ is smooth on Ω. If $\mathbf{T}(s)$ is the unit tangent to $\partial\Omega$, then $\mathbf{T} = \dfrac{d}{ds}\mathbf{x}$; therefore

$$\oint_{\partial\Omega} \mathbf{F}\cdot\mathbf{T}\,ds = \oint_{\partial\Omega} (P\,dx + Q\,dy) = \iint_{\Omega} (Q_x - P_y)\,dA = \iint_{\Omega} (\operatorname{curl}\mathbf{F})\cdot\mathbf{k}\,dA. \quad ∎$$

47.　　　Let $\mathbf{F} = \dfrac{y}{x^2+y^2}\mathbf{i} - \dfrac{x}{x^2+y^2}\mathbf{j}$.

a.　　$\mathbf{F} = \nabla\tan^{-1}(x/y) = \nabla\left(-\tan^{-1}(y/x)\right)$ (if $y\neq 0$ or $x\neq 0$, respectively). Theorem 1 then implies $\operatorname{curl}\mathbf{F} = 0$. ∎

b.　　If C is the unit circle oriented counterclockwise, then

$$\oint_C \mathbf{F}\cdot\mathbf{T}\,ds = \int_{\theta=0}^{2\pi} (\sin\theta\,\mathbf{i} - \cos\theta\,\mathbf{j})\cdot(-\sin\theta\,\mathbf{i} + \cos\theta\,\mathbf{j})\,d\theta = \int_{\theta=0}^{2\pi} (-\sin^2\theta - \cos^2\theta)\,d\theta = -2\pi.$$

c.　　$\mathbf{F}$ is undefined at the origin, therefore it is not smooth throughout the unit disk and does not satisfy the hypotheses of Problem 45.

Solutions 15.8　　*Stokes's theorem*　　(page 1019)

1.　　　If $\mathbf{F} = (x+y)\mathbf{i} + (z-2x+y)\mathbf{j} + (y-z)\mathbf{k}$, then $\operatorname{curl}\mathbf{F} = (1-1)\mathbf{i} + (0-0)\mathbf{j} + (-2-1)\mathbf{k} = -3\mathbf{k}$. If C is the unit circle in the plane $z=5$, then we let S be the unit disk in that plane and compute

$$\oint_C \mathbf{F}\cdot d\mathbf{x} = \iint_S (\operatorname{curl}\mathbf{F})\cdot\mathbf{n}\,d\sigma = \iint_S (-3\mathbf{k})\cdot\mathbf{k}\,d\sigma = \iint_S -3\,d\sigma = (-3)(\text{area of unit disk}) = -3\pi.$$

3.　　　If $\mathbf{F} = (z-2y)\mathbf{i} + (3x-4y)\mathbf{j} + (z+3y)\mathbf{k}$, then $\operatorname{curl}\mathbf{F} = (3-0)\mathbf{i} + (1-0)\mathbf{j} + (3+2)\mathbf{k} = 3\mathbf{i} + \mathbf{j} + 5\mathbf{k}$. If C is the boundary of the triangle with vertices $(2,0,0)$, $(0,2,0)$, $(0,0,2)$, then it lies in the plane $x+y+z=2$ which has unit normal $\mathbf{n} = (\mathbf{i}+\mathbf{j}+\mathbf{k})/\sqrt{3}$.

$$\oint_C \mathbf{F}\cdot d\mathbf{x} = \iint_S (\operatorname{curl}\mathbf{F})\cdot\mathbf{n}\,d\sigma = \iint_S (3\mathbf{i}+\mathbf{j}+5\mathbf{k})\cdot\left((\mathbf{i}+\mathbf{j}+\mathbf{k})/\sqrt{3}\right)d\sigma = \iint_S 3\sqrt{3}\,d\sigma$$

$$= 3\sqrt{3}\left(\text{area of equilateral triangle, side }2\sqrt{2}\right) = 3\sqrt{3}\left(\frac{(2\sqrt{2})^2\sqrt{3}}{4}\right) = 18.$$

REMARK:　　See Example 2; this answer is 2^2 times the value computed there.

5.　　　The curve $\mathbf{x}(t) = 2\cos t\,\mathbf{i} + 3\mathbf{j} + 2\sin t\,\mathbf{k}$, $t\in[0,2\pi]$, is a circle of radius 2 lying in the plane $y=3$ which has unit normal $\mathbf{n} = \mathbf{j}$. If $\mathbf{F} = x^3y^2\mathbf{i} + 2xyz^3\mathbf{j} + 3xy^2z^2\mathbf{k}$, then $\operatorname{curl}\mathbf{F} = (6xyz^2 - 6xyz^2)\mathbf{i} + (0-3y^2z^2)\mathbf{j} + (2yz^3 - 2x^3y)\mathbf{k} = -3y^2z^2\mathbf{j} + 2y(z^3-x^3)\mathbf{k}$; in the plane $y=3$, we have $\operatorname{curl}\mathbf{F} = -27z^2\mathbf{j} + 6(z^3-x^3)\mathbf{k}$.

$$\oint_C \mathbf{F}\cdot d\mathbf{x} = \iint_S (\operatorname{curl}\mathbf{F})\cdot\mathbf{n}\,d\sigma = \iint_S (-27z^2\mathbf{j} + 6(z^3-x^3)\mathbf{k})\cdot\mathbf{j}\,d\sigma = \iint_S -27z^2\,d\sigma$$

$$= -27\int_{\theta=0}^{2\pi}\int_{r=0}^{2} (r\sin\theta)^2\,r\,dr\,d\theta = -27\left(\frac{\theta - \cos\theta\sin\theta}{2}\Big|_{\theta=0}^{2\pi}\right)\left(\frac{r^4}{4}\Big|_{r=0}^{2}\right)$$

$$= -27(\pi)(4) = -108\pi.$$

7.　　　Let S be the part of the plane $x+y+z=3$ lying in the first octant, then $\mathbf{n} = (\mathbf{i}+\mathbf{j}+\mathbf{k})/\sqrt{3}$ is the unit normal to S and C is the boundary of S. Furthermore S is the surface $z=3-x-y$ over the domain $\Omega = \{(x,y): 0\le x\le 3,\ 0\le y\le 3-x\}$; on that surface we have $d\sigma = \sqrt{(-1)^2 + (-1)^2 + 1}\,dA = \sqrt{3}\,dA$. If

$\mathbf{F} = e^x\mathbf{i} + x\sin y\mathbf{j} + (y^2 - x^2)\mathbf{k}$, then $\operatorname{curl}\mathbf{F} = (2y-0)\mathbf{i} + (0-(-2x))\mathbf{j} + (\sin y - 0)\mathbf{k} = 2y\mathbf{i} + 2x\mathbf{j} + \sin y\mathbf{k}$.

$$\oint_C \mathbf{F}\cdot d\mathbf{x} = \iint_S (\operatorname{curl}\mathbf{F})\cdot\mathbf{n}\,d\sigma = \iint_S (2y\mathbf{i} + 2x\mathbf{j} + \sin y\mathbf{k})\cdot(\mathbf{i} + \mathbf{j} + \mathbf{k})/\sqrt{3}\,d\sigma$$

$$= \iint_S (2y + 2x + \sin y)/\sqrt{3}\,d\sigma = \iint_\Omega (2y + 2x + \sin y)\,dA$$

$$= \int_{x=0}^3 \int_{y=0}^{3-x} (2y + 2x + \sin y)\,dy\,dx = \int_{x=0}^3 (y^2 + 2xy - \cos y)\Big|_{y=0}^{3-x}\,dx$$

$$= \int_{x=0}^3 (10 - x^2 - \cos(3-x))\,dx = \left(10x - \frac{x^3}{3} + \sin(3-x)\right)\Big|_{x=0}^3 = 21 - \sin 3.$$

9.   Let $\mathbf{F} = z^2\mathbf{i} + x^2\mathbf{j} + y^2\mathbf{k}$. Let S be the part of the plane $x + y + z = 1$ in the first octant; it is convenient to split its triangular boundary curve C into three line segments:

line segment		interval	x	y	z	z^2	x^2	y^2	$\mathbf{F}\cdot d\mathbf{x}$
C_1	from $(1,0,0)$ to $(0,1,0)$	$0 \le t \le 1$	$1-t$	t	0	0	$(1-t)^2$	t^2	$(1-t)^2$
C_2	from $(0,1,0)$ to $(0,0,1)$	$1 \le t \le 2$	0	$2-t$	$t-1$	$(t-1)^2$	0	$(2-t)^2$	$(2-t)^2$
C_3	from $(0,0,1)$ to $(1,0,0)$	$2 \le t \le 3$	$t-2$	0	$3-t$	$(3-t)^2$	$(t-2)^2$	0	$(3-t)^2$

Therefore

$$\oint_C \mathbf{F}\cdot d\mathbf{x} = \int_{C_1}\mathbf{F}\cdot d\mathbf{x} + \int_{C_2}\mathbf{F}\cdot d\mathbf{x} + \int_{C_3}\mathbf{F}\cdot d\mathbf{x} = \int_0^1 (1-t)^2\,dt + \int_1^2 (2-t)^2\,dt + \int_2^3 (3-t)^2\,dt$$

$$= \frac{-(1-t)^3}{3}\Big|_0^1 + \frac{-(2-t)^3}{3}\Big|_1^2 + \frac{-(3-t)^3}{3}\Big|_2^3 = \frac{1}{3} + \frac{1}{3} + \frac{1}{3} = 1.$$

On the other hand, $\mathbf{n} = (\mathbf{i}+\mathbf{j}+\mathbf{k})/\sqrt{3}$ is the unit normal to S and $\operatorname{curl}\mathbf{F} = (2y-0)\mathbf{i}+(2z-0)\mathbf{j}+(2x-0)\mathbf{k} = 2(y\mathbf{i} + z\mathbf{j} + x\mathbf{k})$. Thus

$$\iint_S (\operatorname{curl}\mathbf{F})\cdot\mathbf{n}\,d\sigma = \iint_S 2(y\mathbf{i} + z\mathbf{j} + x\mathbf{k})\cdot(\mathbf{i} + \mathbf{j} + \mathbf{k})/\sqrt{3}\,d\sigma = \iint_S \frac{2(y+z+x)}{\sqrt{3}}\,d\sigma = \iint_S \frac{2}{\sqrt{3}}\,d\sigma$$

$$= \frac{2}{\sqrt{3}}\iint_S d\sigma = \frac{2}{\sqrt{3}}\left(\text{area of triangle, side }\sqrt{2}\right) = \frac{2}{\sqrt{3}}\left(\frac{\sqrt{3}}{2}\right) = 1.$$

11.   Let $\mathbf{F} = 2y\mathbf{i} + x^2\mathbf{j} + 3x\mathbf{k}$. The hemisphere $S = \{(x,y,z) : x^2 + y^2 + z^2 = 4^2,\ z \ge 0\}$ has circle $C = \{(x,y,0) : x^2 + y^2 = 4^2\}$ as its boundary curve. Hence

$$\oint_C \mathbf{F}\cdot d\mathbf{x} = \int_0^{2\pi} (2(4\sin\theta)\mathbf{i} + (4\cos\theta)^2\mathbf{j} + 3(4\cos\theta)\mathbf{k})\cdot(-4\sin\theta\,\mathbf{i} + 4\cos\theta\,\mathbf{j})d\theta$$

$$= \int_0^{2\pi} (-32\sin^2\theta + 64\cos^3\theta)\,d\theta = \left(-16(\theta - \cos\theta\sin\theta) + \frac{64}{3}(\cos^2\theta\sin\theta + 2\sin\theta)\right)\Big|_0^{2\pi} = -32\pi.$$

On the other hand, $\mathbf{n} = (x\mathbf{i}+y\mathbf{j}+z\mathbf{k})/4$ is the unit outward normal to S and $\operatorname{curl}\mathbf{F} = (0-0)\mathbf{i}+(0-3)\mathbf{j}+(2x-2)\mathbf{k} = -3\mathbf{j}+2(x-1)\mathbf{k}$. S is the surface $z = \sqrt{16 - x^2 - y^2}$ over the region $\Omega = \{(x,y) : x^2 + y^2 \le 16\}$ and $d\sigma = \sqrt{(-x/z)^2 + (-y/z)^2 + 1}\,dA = (4/z)dA$. Thus

$$\iint_S (\operatorname{curl}\mathbf{F})\cdot\mathbf{n}\,d\sigma = \iint_S (-3\mathbf{j}+2(x-1)\mathbf{k})\cdot(x\mathbf{i}+y\mathbf{j}+z\mathbf{k})/4\,d\sigma = \iint_S \frac{-3y + 2(x-1)z}{4}\,d\sigma$$

$$= \iint_\Omega \left(\frac{-3y + 2(x-1)z}{4}\right)\left(\frac{4}{z}\right)dA = \iint_\Omega \left(\frac{-3y}{\sqrt{16 - x^2 - y^2}} + 2x - 2\right)dA$$

$$= \iint_\Omega (\text{odd} + \text{odd} - 2)\,dA = 0 + 0 + (-2)(\text{area of disk, radius }4) = (-2)\pi\,4^2 = -32\pi.$$

13.

$$\text{voltage drop around } C = \oint_C \mathbf{E} \cdot d\mathbf{x}$$

$$= \iint_S (\nabla \times \mathbf{E}) \cdot \mathbf{n} \, d\sigma \qquad \text{(Stokes)}$$

$$= \iint_S \left(-\frac{\partial \mathbf{B}}{\partial t} \right) \cdot \mathbf{n} \, d\sigma \qquad \text{(Maxwell)}$$

$$= -\frac{\partial}{\partial t} \iint_S \mathbf{B} \cdot \mathbf{n} \, d\sigma \qquad \text{(Leibniz)}$$

$$= \text{time rate of decrease of flux of } \mathbf{B} \text{ over } S. \ \blacksquare$$

15. An equatorial circle C divides sphere S into hemispheres H_1 and H_2. Any direction on C corresponds to opposite directions for outbound normals to H_1 and H_2. Therefore

$$\iint_S \text{curl } \mathbf{F} \cdot \mathbf{n} \, d\sigma = \iint_{H_1} \text{curl } \mathbf{F} \cdot \mathbf{n}_1 \, d\sigma + \iint_{H_2} \text{curl } \mathbf{F} \cdot \mathbf{n}_2 \, d\sigma = \oint_C \mathbf{F} \cdot d\mathbf{x} - \oint_C \mathbf{F} \cdot d\mathbf{x} = 0. \ \blacksquare$$

Solutions 15.9 *The divergence theorem* (pages 1022-3)

1. If S is the unit sphere, then W is the unit ball; if $\mathbf{F} = x\mathbf{i} + y\mathbf{j} + z\mathbf{k}$, then $\text{div } \mathbf{F} = 3$. Therefore

$$\iint_S \mathbf{F} \cdot \mathbf{n} \, d\sigma = \iiint_W \text{div } \mathbf{F} \, dv = \iiint_W 3 \, dv = 3 \iiint_W dv = 3(\text{volume of unit ball}) = 3 \left(\frac{4\pi}{3} \right) = 4\pi.$$

REMARK: $\mathbf{n} = x\mathbf{i} + y\mathbf{j} + z\mathbf{k}$, $\mathbf{F} \cdot \mathbf{n} = x^2 + y^2 + z^2 = 1$, and $\iint_S \mathbf{F} \cdot \mathbf{n} \, d\sigma = \iint_S d\sigma = \text{area of sphere} = 4\pi.$

3. If $\mathbf{F} = x\mathbf{i} + y\mathbf{j} + z\mathbf{k}$, then $\text{div } \mathbf{F} = 3$. Therefore

$$\iint_S \mathbf{F} \cdot \mathbf{n} \, d\sigma = \iiint_W \text{div } \mathbf{F} \, dv = \iiint_W 3 \, dv = 3 \iiint_W dv = 3(\text{volume of cylinder}) = 3 \cdot \pi \cdot 2^2 \cdot 3 = 36\pi.$$

5. If $\mathbf{F} = (y^2 + z^2)^{3/2} \mathbf{i} + \sin\left((x^2 - z^5)^{4/3}\right) \mathbf{j} + e^{x^2 - y^2} \mathbf{k}$, then $\text{div } \mathbf{F} = 0$. Therefore $\iint_S \mathbf{F} \cdot \mathbf{n} \, d\sigma =$

$$\iiint_W \text{div } \mathbf{F} \, dv = \iiint_W 0 \, dv = 0.$$

7. If $\mathbf{F} = x^2\mathbf{i} + y^2\mathbf{j} - xy\,\mathbf{k}$, then $\text{div } \mathbf{F} = 2x + 2y - 0 = 2(x + y)$. Therefore

$$\iint_S \mathbf{F} \cdot \mathbf{n} \, d\sigma = \iiint_W \text{div } \mathbf{F} \, dv = \iiint_W 2(x + y) \, dv = \int_{x=0}^1 \int_{y=0}^1 \int_{z=0}^1 2(x + y) \, dz \, dy \, dx$$

$$= \int_{x=0}^1 \int_{y=0}^1 2(x + y) \, dy \, dx = \int_{x=0}^1 (2x + 1) \, dx = 2.$$

9. If $\mathbf{F} = 2x\mathbf{i} + 3y\mathbf{j} + z\mathbf{k}$, then $\text{div } \mathbf{F} = 2 + 3 + 1 = 6$. Therefore

$$\iint_S \mathbf{F} \cdot \mathbf{n} \, d\sigma = \iiint_W \text{div } \mathbf{F} \, dv = \iiint_W 6 \, dv = 6 \, (\text{volume of hemisphere, radius 3}) = 6 \cdot \frac{1}{2} \cdot \frac{4}{3}\pi 3^3 = 108\pi.$$

11. If $\mathbf{F} = x^2\mathbf{i} + y^2\mathbf{j} + z^2\mathbf{k}$, then $\text{div } \mathbf{F} = y + 2y + y = 4y$. Therefore

$$\iint_S \mathbf{F} \cdot \mathbf{n} \, d\sigma = \iiint_W \text{div } \mathbf{F} \, dv = \iiint_W 4y \, dv = \int_{x=0}^1 \int_{y=0}^{1-x} \int_{z=0}^{1-x-y} 4y \, dz \, dy \, dx$$

$$= \int_{x=0}^1 \int_{y=0}^{1-x} 4y(1 - x - y) \, dy \, dx = \int_{x=0}^1 \left(\frac{2}{3} - 2x + 2x^2 - \frac{2}{3}x^3 \right) dx = \frac{1}{6}.$$

13. If $\mathbf{F} = x(1 - \sin y)\mathbf{i} + (y - \cos y)\mathbf{j} + z\,\mathbf{k}$, then $\text{div } \mathbf{F} = (1 - \sin y) + (1 + \sin y) + 1 = 3$. Hence

$$\iint_S \mathbf{F} \cdot \mathbf{n} \, d\sigma = \iiint_W \text{div } \mathbf{F} \, dv = \iiint_W 3 \, dv = \int_{x=0}^1 \int_{y=0}^{1-x} \int_{z=0}^{1-x-y} 3 \, dz \, dy \, dx$$

$$= \int_{x=0}^1 \int_{y=0}^{1-x} 3(1 - x - y) \, dy \, dx = \int_{x=0}^1 \frac{3}{2}(1 - x)^2 dx = \frac{1}{2}.$$

15. If $\mathbf{F} = \left(x^2 + e^{y \cos z}\right)\mathbf{i} + \left(xy - \tan z^{1/3}\right)\mathbf{j} + \left(x - y^{3/5}\right)^{2/9}\mathbf{k}$, then div $\mathbf{F} = 2x + x + 0 = 3x$. Hence

$$\iint_S \mathbf{F} \cdot \mathbf{n}\, d\sigma = \iiint_W \operatorname{div} \mathbf{F}\, dv = \iiint_W 3x\, dv = \int_{y=-1}^1 \int_{z=0}^{1-y^2} \int_{x=0}^{2-z} 3x\, dx\, dz\, dy$$

$$= \int_{y=-1}^1 \int_{z=0}^{1-y^2} \frac{3}{2}(2-z)^2\, dz\, dy = \int_{y=-1}^1 \left(4 - \frac{1}{2}\left(1+y^2\right)^3\right) dx$$

$$= \left(\frac{7y}{2} - \frac{y^3}{2} - \frac{3y^5}{10} - \frac{y^7}{14}\right)\Big|_{y=-1}^1 = 7 - 1 - \frac{3}{5} - \frac{1}{7} = \frac{184}{35}.$$

17. If $\mathbf{F}$ is smooth (so Theorem 13.4.1 applies), then $\operatorname{div}(\operatorname{curl} \mathbf{F}) = 0$ (see Problem 15.7.38). Therefore, if S is a smooth closed surface, then

$$\iint_S \operatorname{curl} \mathbf{F} \cdot \mathbf{n}\, d\sigma = \iiint_W \operatorname{div}(\operatorname{curl} \mathbf{F})\, dv = \iiint_W 0\, dv = 0. \; \blacksquare$$

19. If $\mathbf{F} = x\mathbf{i} + y\mathbf{j} + z\mathbf{k}$, then div $\mathbf{F} = 1 + 1 + 1 = 3$. Hence, if S is a smooth closed surface, then

$$\frac{1}{3}\iint_S \mathbf{F} \cdot \mathbf{n}\, d\sigma = \frac{1}{3}\iiint_W \operatorname{div} \mathbf{F}\, dv = \frac{1}{3}\iiint_W 3\, dv = \iiint_W dv = \text{volume of } W. \; \blacksquare$$

Solutions 15.10   *Changing variables in multiple integrals, the Jacobian*   (pages 1029-30)

1. If $x = u + v$ and $y = u - v$, then $\dfrac{\partial(x,y)}{\partial(u,v)} = \begin{vmatrix} 1 & 1 \\ 1 & -1 \end{vmatrix} = -2.$

3. If $x = u^2 - v^2$ and $y = 2uv$, then $\dfrac{\partial(x,y)}{\partial(u,v)} = \begin{vmatrix} 2u & -2v \\ 2v & 2u \end{vmatrix} = 4u^2 + 4v^2.$

5. If $x = u + 3v - 1$ and $y = 2u + 4v + 6$, then $\dfrac{\partial(x,y)}{\partial(u,v)} = \begin{vmatrix} 1 & 3 \\ 2 & 4 \end{vmatrix} = 4 - 6 = -2.$

7. If $x = au + bv$ and $y = bu - av$, then $\dfrac{\partial(x,y)}{\partial(u,v)} = \begin{vmatrix} a & b \\ b & -a \end{vmatrix} = -a^2 - b^2.$

9. If $x = ue^v$ and $y = ve^u$, then $\dfrac{\partial(x,y)}{\partial(u,v)} = \begin{vmatrix} e^v & ue^v \\ ve^u & e^u \end{vmatrix} = e^v e^u - uve^u e^v = (1 - uv)e^{u+v}.$

11. If $x = \ln(u + v)$ and $y = \ln(uv)$, then $\dfrac{\partial(x,y)}{\partial(u,v)} = \begin{vmatrix} (u+v)^{-1} & (u+v)^{-1} \\ u^{-1} & v^{-1} \end{vmatrix} = \dfrac{u-v}{uv(u+v)}.$

13. If $x = u \sec v$ and $y = v \csc u$, then

$$\frac{\partial(x,y)}{\partial(u,v)} = \begin{vmatrix} \sec v & u \sec v \tan v \\ -v \csc u \cot u & \csc u \end{vmatrix} = \csc u \sec v + uv \csc u \cot u \sec v \tan v$$

$$= \csc u \sec v\,(1 + uv \cot u \tan v).$$

15. $x = u + v + w$, $y = u - v - w$, $z = -u + v + w$ imply $\dfrac{\partial(x,y,z)}{\partial(u,v,w)} = \begin{vmatrix} 1 & 1 & 1 \\ 1 & -1 & -1 \\ -1 & 1 & 1 \end{vmatrix} = 0.$

17. $x = u^2 + v^2 + w^2$, $y = u + v + w$, $z = uvw$ imply

$$\frac{\partial(x,y,z)}{\partial(u,v,w)} = \begin{vmatrix} 2u & 2v & 2w \\ 1 & 1 & 1 \\ vw & uw & uv \end{vmatrix} = 2u^2(v - w) - v^2(u - w) + w^2(u - v) = 2(u - v)(v - w)(u - w).$$

19. $x = e^u$, $y = e^v$, $z = e^w$ imply $\dfrac{\partial(x,y,z)}{\partial(u,v,w)} = \begin{vmatrix} e^u & 0 & 0 \\ 0 & e^v & 0 \\ 0 & 0 & e^w \end{vmatrix} = e^u e^v e^w = e^{u+v+w}.$

21. $x = u - v$ and $y = u + v$ imply $\dfrac{\partial(x,y)}{\partial(u,v)} = \begin{vmatrix} 1 & -1 \\ 1 & 1 \end{vmatrix} = 2$. The xy-region $\Omega = \{(x,y) : 0 \le y \le x \le 1\}$ is the triangle with vertices $(0,0)$, $(1,0)$, $(1,1)$. The inverse mapping is $u = (x+y)/2$ and $v = (y-x)/2$; it maps Ω into Σ which is the triangle with vertices $(0,0)$, $(1/2,-1/2)$, $(1,0)$.

$$\int_{y=0}^{1} \int_{x=y}^{1} x\, y\, dx\, dy = \iint_{\Omega} x\, y\, dx\, dy = \iint_{\Sigma} (u-v)(u+v)\, 2\, du\, dv = \int_{v=-1/2}^{0} \int_{u=-v}^{1+v} 2\left(u^2 - v^2\right) du\, dv = \frac{1}{8}.$$

23. $x = \sqrt{3}\, u + v/2$ and $y = 1 - u + (\sqrt{3}/2)v$ imply $\dfrac{\partial(x,y)}{\partial(u,v)} = \begin{vmatrix} \sqrt{3} & 1/2 \\ -1 & \sqrt{3}/2 \end{vmatrix} = 2$. The region of integration satisfies

$$16 \ge 7x^2 + 6\sqrt{3}\, x(y-1) + 13(y-1)^2$$
$$= 7\left(\sqrt{3}\, u + v/2\right)^2 + 6\sqrt{3}\left(\sqrt{3}\, u + v/2\right)\left(-u + (\sqrt{3}/2)v\right) + 13\left(-u + (\sqrt{3}/2)v\right)^2$$
$$= 16u^2 + 16v^2 = 16\left(u^2 + v^2\right).$$

Hence Σ is the unit circle in the uv-plane. We will evaluate the integral over Σ' using polar coordinates in the uv-plane.

$$\iint_{\Omega} y\, dx\, dy = \iint_{\Sigma}\left(1 - u + (\sqrt{3}/2)v\right)(2)\, du\, dv = \int_{\theta=0}^{2\pi}\int_{r=0}^{1}\left(2 - 2r\cos\theta + \sqrt{3}\, r\sin\theta\right) r\, dr\, d\theta$$
$$= \int_{\theta=0}^{2\pi}\left(1 - \frac{2}{3}\cos\theta + \frac{\sqrt{3}}{3}\sin\theta\right) d\theta = 2\pi.$$

REMARK: The composition of the uv and polar coordinate mappings yields $x = \sqrt{3}\, r\cos\theta + (1/2)r\sin\theta$ and $y = 1 - r\cos\theta + (\sqrt{3}/2)r\sin\theta$ with

$$\frac{\partial(x,y)}{\partial(r,\theta)} = \begin{vmatrix} \sqrt{3}\cos\theta + (1/2)\sin\theta & -\sqrt{3}r\sin\theta + (1/2)r\cos\theta \\ -\cos\theta + (\sqrt{3}/2)\sin\theta & r\sin\theta + (\sqrt{3}/2)r\cos\theta \end{vmatrix} = 2r$$

and region of integration Σ' satisfying $16 \ge 7x^2 + 6\sqrt{3}\, x(y-1) + 13(y-1)^2 = 16r^2$.

---

25. $x = v - u^2$ and $y = u + v$ imply $\dfrac{\partial(x,y)}{\partial(u,v)} = \begin{vmatrix} -2u & 1 \\ 1 & 1 \end{vmatrix} = -2u - 1$. The boundaries are given by

$x = -y^2$ $0 = x + y^2 = v(2u + v + 1)$
$y = x$ $0 = y - x = u(1 + u)$
$y = 2$ $0 = y - 2 = u + v - 2$

and we will use $v = 0$, $u = 0$, $u + v = 2$ as boundaries of Σ. The vertices of Σ are $(0,0)$, $(2,0)$, $(0,2)$; they are mapped to $(0,0)$, $(-4,2)$, $(2,2)$; the mapping reverses the orientation of the boundary curve so we use the minus sign of the $\pm$ in Theorem 1.

$$\iint_{\Omega}(y - x)\, dy\, dx = -\iint_{\Sigma}\left((u + v) - (v - u^2)\right)(-2u - 1)\, dv\, du$$
$$= \int_{u=0}^{2}\int_{v=0}^{2-u}\left(u + u^2\right)(2u + 1)\, dv\, du = \int_{u=0}^{2}\left(u + u^2\right)(2u + 1)(2 - u)\, du = \frac{128}{15}.$$

27.   The transformation $x = au$, $y = bv$, $z = cw$ maps the unit ball Σ in uvw-space onto the solid W enclosed by the ellipsoid $x^2/a^2 + y^2/b^2 + z^2/c^2 = 1$. Therefore $\dfrac{\partial(x,y,z)}{\partial(u,v,w)} = \begin{vmatrix} a & 0 & 0 \\ 0 & b & 0 \\ 0 & 0 & c \end{vmatrix} = abc$ and we compute

$$\text{volume of } W = \iiint_{W} dx\, dy\, dz = \iiint_{\Sigma} abc\, du\, dv\, dw = (abc)(\text{volume of unit ball}) = (abc)\left(\frac{4\pi}{3}\right).$$

Solutions 15.R   *Review*   (pages 1030-2)

1. $\nabla(x+y)^3 = 3(x+y)^2\,\nabla(x+y) = 3(x+y)^2\,(\mathbf{i}+\mathbf{j}).$

3. If $f(x,y) = \dfrac{x+y}{x-y} = 1 + \dfrac{2y}{x-y} = \dfrac{2x}{x-y} - 1$, then $\nabla f = \dfrac{-2y}{(x-y)^2}\mathbf{i} + \dfrac{2x}{(x-y)^2}\mathbf{j} = \dfrac{-2}{(x-y)^2}(y\,\mathbf{i} - x\,\mathbf{j}).$

5. $\nabla\left(x^2+y^2+z^2\right) = 2x\,\mathbf{i} + 2y\,\mathbf{j} + 2z\,\mathbf{k}.$

7. $\nabla(xy) = y\,\mathbf{i} + x\,\mathbf{j}.$

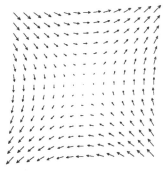

9. The curve $y = x^{3/2}$ from $(0,0)$ to $(1,1)$ can be parameterized $\mathbf{x}(t) = t^2\mathbf{i} + t^3\mathbf{j}$ for $t \in [0,1]$. If $\mathbf{F}(x,y) = x^2\mathbf{i} + y^2\mathbf{j}$, then

$$\int_C \mathbf{F}\cdot d\mathbf{x} = \int_{t=0}^1 \left(\left(t^2\right)^2\mathbf{i} + \left(t^3\right)^2\mathbf{j}\right)\cdot(2t\mathbf{i}+3t^2\mathbf{j})\,dt = \int_{t=0}^1 \left(2t^5+3t^8\right)dt = \left(\frac{t^6+t^9}{3}\right)\Bigg|_{t=0}^1 = \frac{2}{3}.$$

REMARK: $\mathbf{F} = \nabla\left(\dfrac{x^3+y^3}{3}\right)$, therefore $\displaystyle\int_C \mathbf{F}\cdot d\mathbf{x} = \dfrac{x^3+y^3}{3}\Bigg|_{(x,y)=(0,0)}^{(1,1)} = \dfrac{1+1}{3} - \dfrac{0+0}{3} = \dfrac{2}{3}.$

11. Let $O = (0,0)$, $A = (1,1)$, and $B = (0,1)$. If $\mathbf{F} = 3xy\,\mathbf{i} - y\,\mathbf{j}$, then

segment	$\mathbf{x}(t)$	t interval	$\int_{C_i}\mathbf{F}\cdot d\mathbf{x}$
$\overrightarrow{OA}$	$t\mathbf{i}+t\mathbf{j}$	$0\le t\le 1$	$\int_{\overrightarrow{OA}}\mathbf{F}\cdot d\mathbf{x} = \int_0^1 (3t^2\mathbf{i}-t\mathbf{j})\cdot(\mathbf{i}+\mathbf{j})\,dt = \int_0^1 (3t^2-t)\,dt = \frac{1}{2}$
$\overrightarrow{AB}$	$(2-t)\mathbf{i}+\mathbf{j}$	$1\le t\le 2$	$\int_{\overrightarrow{AB}}\mathbf{F}\cdot d\mathbf{x} = \int_1^2 (3(2-t)\mathbf{i}-\mathbf{j})\cdot(-\mathbf{i})\,dt = \int_1^2 -3(2-t)\,dt = \frac{-3}{2}$
$\overrightarrow{BO}$	$(3-t)\mathbf{j}$	$2\le t\le 3$	$\int_{\overrightarrow{BO}}\mathbf{F}\cdot d\mathbf{x} = \int_2^3 -(3-t)\mathbf{j}\cdot(-\mathbf{j})\,dt = \int_2^3 (3-t)\,dt = \frac{1}{2}$

Therefore $\displaystyle\int_C \mathbf{F}\cdot d\mathbf{x} = \frac{1}{2} + \frac{-3}{2} + \frac{1}{2} = \frac{-1}{2}.$

13. If $\mathbf{F} = x\,\mathbf{i} + y\,\mathbf{j} + z\,\mathbf{k} = \mathbf{x}$ and C is the curve $\mathbf{x}(t) = t^3\mathbf{i}+t^2\mathbf{j}+t\mathbf{k}$ for $t \in [0,1]$, then

$$\int_C \mathbf{F}\cdot d\mathbf{x} = \int_{t=0}^1 (t^3\mathbf{i}+t^2\mathbf{j}+t\mathbf{k})\cdot(3t^2\mathbf{i}+2t\mathbf{j}+\mathbf{k})\,dt = \int_{t=0}^1 (3t^5+2t^3+t)\,dt = \frac{t^6+t^4+t^2}{2}\Bigg|_{t=0}^1 = \frac{3}{2}.$$

REMARK: $\mathbf{F} = \nabla\left(\dfrac{x^2+y^2+z^2}{2}\right)$, therefore $\displaystyle\int_C \mathbf{F}\cdot d\mathbf{x} = \dfrac{x^2+y^2+z^2}{2}\Bigg|_{(x,y,z)=(0,0,0)}^{(1,1,1)} = \dfrac{3}{2}.$

15. If $\mathbf{F} = x^2y\,\mathbf{i} + \left(y^3+x^3\right)\mathbf{j}$ and C is the unit circle, then

$$\text{work} = \int_C \mathbf{F}\cdot d\mathbf{x} = \int_{\theta=0}^{2\pi} \left(\cos^2\theta\sin\theta\,\mathbf{i} + (\sin^3\theta+\cos^3\theta)\,\mathbf{j}\right)\cdot(-\sin\theta\,\mathbf{i}+\cos\theta\,\mathbf{j})\,d\theta$$

$$= \int_{\theta=0}^{2\pi} \left(-\cos^2\theta\sin^2\theta + \cos\theta\sin^3\theta + \cos^4\theta\right)d\theta$$

$$= \frac{1}{4}\left(2\sin\theta\cos^3\theta + \sin\theta\cos\theta + \theta + \sin^4\theta\right)\Bigg|_{\theta=0}^{2\pi} = \frac{\pi}{2}.$$

17. $\mathbf{F} = 3x^2y^2\mathbf{i} + 2x^3y\,\mathbf{j} = \nabla\left(x^3y^2\right)$, therefore $\displaystyle\int_C \mathbf{F}\cdot d\mathbf{x} = x^3y^2\Big|_{(x,y)=(1,2)}^{(3,-1)} = 3^3(-1)^2 - 1^3 2^2 = 27 - 4 = 23.$

19. $\mathbf{F} = \dfrac{-y}{z}\mathbf{i} - \dfrac{x}{z}\mathbf{j} + \dfrac{xy}{z^2}\mathbf{k} = \nabla\left(\dfrac{-xy}{z}\right)$, therefore $\displaystyle\int_C \mathbf{F}\cdot d\mathbf{x} = \dfrac{-xy}{z}\bigg|_{(x,y,z)=(1,1,1)}^{(2,-1,3)} = \dfrac{2}{3} - (-1) = \dfrac{5}{3}$.

21. $$\oint_{\partial\Omega}(x^2 y\, dx + xy^2\, dy) = \iint_\Omega \left(\frac{\partial(xy^2)}{\partial x} - \frac{\partial(x^2 y)}{\partial y}\right) dA$$
$$= \int_{x=0}^{1}\int_{y=x}^{1}(y^2 - x^2)\, dy\, dx = \int_{x=0}^{1}\left(\frac{1}{3} - x^2 + \frac{2}{3}x^3\right) dx = \frac{1}{6}.$$

23. $$\oint_{\partial\Omega}\sqrt{1+x^2}\, dy = \iint_\Omega \frac{\partial(\sqrt{1+x^2})}{\partial x}\, dA = \int_{x=-1}^{1}\int_{y=x^2}^{1} x\left(1+x^2\right)^{-1/2} dy\, dx$$
$$= \int_{x=-1}^{1}\left(x\left(1+x^2\right)^{-1/2} - x^3\left(1+x^2\right)^{-1/2}\right) dx$$
$$= \left(\frac{5}{3}\sqrt{1+x^2} - \frac{1}{3}x^2\sqrt{1+x^2}\right)\bigg|_{x=-1}^{1} = 0.$$

25. Let $\mathbf{F} = xy^2\mathbf{i} + x^2 y\mathbf{j}$. If Ω is a disk, then Stokes's (and Green's) Theorem can be applied.
   a. $\operatorname{curl}\mathbf{F} = (0-0)\mathbf{i} + (0-0)\mathbf{j} + (2xy - 2xy)\mathbf{k} = \mathbf{0}$.
   b. $\displaystyle\oint_{\partial\Omega}\mathbf{F}\cdot\mathbf{T}\, ds = \iint_\Omega \operatorname{curl}\mathbf{F}\cdot\mathbf{n}\, dA = \iint_\Omega \mathbf{0}\cdot\mathbf{k}\, dA = \iint_\Omega 0\, dA = 0.$
   c. $\operatorname{div}\mathbf{F} = (1)y^2 + x^2(1) = y^2 + x^2.$
   d. The result of Problem 15.7.46 implies
   $$\oint_{\partial\Omega}\mathbf{F}\cdot\mathbf{n}\, ds = \iint_\Omega \operatorname{div}\mathbf{F}\, dA = \iint_\Omega (y^2 + x^2)\, dA = \int_{\theta=0}^{2\pi}\int_{r=0}^{2} r^2\,(r\, dr\, d\theta) = (2\pi)\left(\frac{2^4}{4}\right) = 8\pi.$$

REMARK: $\mathbf{n}\, ds = (x\mathbf{i} + y\mathbf{j})\, d\theta$ on $\partial\Omega$, therefore
$$\oint_{\partial\Omega}\mathbf{F}\cdot\mathbf{n}\, ds = \oint_{\partial\Omega} 2x^2 y^2\, d\theta = \int_{\theta=0}^{2\pi} 32\cos^2\theta\,\sin^2\theta\, d\theta = (-8\sin\theta\,\cos^3\theta + 4\sin\theta\,\cos\theta + 4\theta)\big|_{\theta=0}^{2\pi} = 8\pi.$$

27. If $\mathbf{F} = \cos(x^2)\mathbf{i} + e^y\mathbf{j}$, then $\operatorname{curl}\mathbf{F} = (0-0)\mathbf{i} + (0-0)\mathbf{j} + (0-0)\mathbf{k} = \mathbf{0}$; hence $\mathbf{F}$ is irrotational. ∎

29. $2x - 3y + z = 0 \iff z = -2x + 3y$; hence every vector on the plane satisfying $2x - 3y + z = 0$ has the form $x\mathbf{i} + y\mathbf{j} + z\mathbf{k} = x\mathbf{i} + y\mathbf{j} + (-2x + 3y)\mathbf{k} = x(\mathbf{i} - 2\mathbf{k}) + y(\mathbf{j} + 3\mathbf{k})$.

31. $\mathbf{r}(x,y) = x\mathbf{i} + y\mathbf{j} + (x^2 - y^2)\mathbf{k}$, $\quad 0 \le x \le 2, \quad 2 \le y \le 4$.

33. $x = y^2 - x^2$ implies $\sigma = \displaystyle\iint_\Omega \sqrt{1 + (-2x)^2 + (2y)^2}\, dA = \iint_\Omega \sqrt{1 + 4x^2 + 4y^2}\, dA$.

35. An easy adaptation of Example 6 yields
$$\sigma = \iint_\Omega 4^2\sin\phi\, dA = \int_{\theta=0}^{2\pi}\int_{\phi=\pi/2}^{\pi} 4^2\sin\phi\, d\phi\, d\theta = 16(2\pi)\,(-\cos\phi)\big|_{\phi=\pi/2}^{\pi} = 32\pi.$$

37. $d\sigma = \sqrt{1 + 0^2 + (2y)^2}\, dA = \sqrt{1 + 4y^2}\, dA$, therefore
$$\iint_S y\, d\sigma = \int_{x=0}^{2}\int_{y=0}^{1} y\sqrt{1 + 4y^2}\, dy\, dx = \int_{x=0}^{2}\frac{1}{12}\left(1 + 4y^2\right)^{3/2}\bigg|_{y=0}^{1}\, dx = \int_{x=0}^{2}\frac{5^{3/2} - 1}{12}\, dx = \frac{5^{3/2} - 1}{6}.$$

39. $z = \sqrt{9 - x^2 - y^2} \implies 1 + (z_x)^2 + (z_y)^2 = 1 + (-x/z)^2 + (-y/z)^2 = (z^2 + x^2 + y^2)/z^2 = 9/z^2$. Therefore
$$\iint_S y^2\, d\sigma = \iint_\Omega y^2\frac{3}{\sqrt{9 - x^2 - y^2}}\, dA = \int_{\theta=0}^{2\pi}\int_{r=0}^{3}(r\sin\theta)^2\frac{3}{\sqrt{9 - r^2}}\, r\, dr\, d\theta$$
$$= \left(\int_{r=0}^{3}\frac{3r^3}{\sqrt{9 - r^2}}\, dr\right)\left(\int_{\theta=0}^{2\pi}\sin^2\theta\, d\theta\right)$$
$$= \left(-(18 + r^2)\sqrt{9 - r^2}\,\bigg|_{r=0}^{3}\right)\left(\frac{1}{2}(-\cos\theta\,\sin\theta + \theta)\bigg|_{\theta=0}^{2\pi}\right) = 54\pi.$$

REMARK: Compare with solution 15.6.15.

41. Let $O = (0,0,0)$, $A = (1,0,0)$, $B = (0,1,0)$, $C = (0,0,1)$.

$$\iint_{OAC} y\, d\sigma = \int_{x=0}^{1}\int_{z=0}^{1-x} 0\, dz\, dx = 0,$$

$$\iint_{OBC} y\, d\sigma = \int_{y=0}^{1}\int_{z=0}^{1-y} y\, dz\, dy = \int_{y=0}^{1} y(1-y)\, dy = \frac{1}{6},$$

$$\iint_{OAB} y\, d\sigma = \int_{x=0}^{1}\int_{y=0}^{1-x} y\, dy\, dx = \int_{x=0}^{1} \frac{1}{2}(1-x)^2 dx = \frac{1}{6},$$

$$\iint_{ABC} y\, d\sigma = \int_{x=0}^{1}\int_{y=0}^{1-x} y\sqrt{3}\, dy\, dx = \int_{x=0}^{1} \frac{\sqrt{3}}{2}(1-x)^2 dx = \frac{\sqrt{3}}{6};$$

therefore $\displaystyle\iint_{S} y\, d\sigma = \iint_{OAC} y\, d\sigma + \iint_{OBC} y\, d\sigma + \iint_{OAB} y\, d\sigma + \iint_{ABC} y\, d\sigma = 0 + \frac{1}{6} + \frac{1}{6} + \frac{\sqrt{3}}{6} = \frac{2+\sqrt{3}}{6}.$

43. The plane $z = 1 - x - y$ passes through $(1,0,0)$, $(0,1,0)$, $(0,0,1)$. If $\rho(x,y,z) = cy^2$, then

$$\mu = \iint_{S} cy^2\, d\sigma = \int_{x=0}^{1}\int_{y=0}^{1-x} cy^2\sqrt{1 + (-1)^2 + (-1)^2}\, dy\, dx = \frac{c\sqrt{3}}{3}\int_{x=0}^{1}(1-x)^3 dx = \frac{c\sqrt{3}}{12} = \frac{c}{4\sqrt{3}}.$$

45. If $S = \{(x,y,z) : z = 2xy,\ 0 \le x \le 1,\ 0 \le y \le 4\}$ and $\mathbf{F} = xy^2\mathbf{i} - 2z\mathbf{j}$, then the result of Problem 15.6.34 implies

$$\iint_{S} \mathbf{F}\cdot\mathbf{n}\, d\sigma = \iint_{\Omega}(-(xy^2)(2y) - (-2)(2xy)(2x) + 0)\, dA = \int_{x=0}^{1}\int_{y=0}^{4}(-2xy^3 + 8x^2 y)\, dy\, dx$$

$$= \int_{x=0}^{1}(-128x + 64x^2)\, dx = \frac{-128}{3}.$$

47. $S = \{(x,y,z) : z = \sqrt{x^2 + z^2},\ x^2 + z^2 \le 2^2\}$ and $\mathbf{F} = x\mathbf{i} - xz\mathbf{j} + 3z\mathbf{k}$ imply

$$\iint_{S}\mathbf{F}\cdot\mathbf{n}\, d\sigma = \iint_{\Omega}\left(-(x)\left(\frac{x}{y}\right) + (-xz) - (3z)\left(\frac{z}{y}\right)\right) dA = -\iint_{\Omega}\left(\frac{x^2 + 3z^2}{\sqrt{x^2 + z^2}} + xz\right) dA$$

$$= -\int_{\theta=0}^{2\pi}\int_{r=0}^{2}\left(\frac{r^2\cos^2\theta + 3r^2\sin^2\theta}{r} + r^2\cos\theta\sin\theta\right) r\, dr\, d\theta$$

$$= -\int_{\theta=0}^{2\pi}\left(\frac{8}{3}\cos^2\theta + 8\sin^2\theta + 4\cos\theta\sin\theta\right) d\theta$$

$$= -\left(\frac{-8}{3}\cos\theta\sin\theta + \frac{16}{3}\theta + 2\sin^2\theta\right)\Big|_{\theta=0}^{2\pi} = \frac{-32}{3}\pi.$$

49. If $\mathbf{F} = x\mathbf{i} + y\mathbf{j} + z\mathbf{k}$, then $\operatorname{div}\mathbf{F} = 1 + 1 + 1 = 3$ and $\operatorname{curl}\mathbf{F} = (0-0)\mathbf{i} + (0-0)\mathbf{j} + (0-0)\mathbf{k} = 0$.
51. If $\mathbf{F} = yz\mathbf{i} + xz\mathbf{j} + xy\mathbf{k}$, then $\operatorname{div}\mathbf{F} = 0+0+0 = 0$ and $\operatorname{curl}\mathbf{F} = (x-x)\mathbf{i} + (y-y)\mathbf{j} + (z-z)\mathbf{k} = 0$.
53. If $\mathbf{F} = e^{yz}\mathbf{i} + e^{xz}\mathbf{j} + e^{xy}\mathbf{k}$, then $\operatorname{div}\mathbf{F} = 0+0+0 = 0$ and $\operatorname{curl}\mathbf{F} = x\left(e^{xy} - e^{xz}\right)\mathbf{i} + y\left(e^{yz} - e^{xy}\right)\mathbf{j} + z\left(e^{xz} - e^{yz}\right)\mathbf{k}$.
55. If $\mathbf{F} = (x + 2y)\mathbf{i} + (y - 3z)\mathbf{j} + (z - x)\mathbf{k}$, then $\operatorname{curl}\mathbf{F} = 3\mathbf{i} + \mathbf{j} - 2\mathbf{k}$. If C is the unit circle in the $z = 2$ plane, then S is the unit disk in that plane and $\mathbf{n} = \mathbf{k}$. Therefore

$$\oint_C \mathbf{F}\cdot d\mathbf{x} = \iint_S \operatorname{curl}\mathbf{F}\cdot\mathbf{n}\, d\sigma = \iint_S (3\mathbf{i} + \mathbf{j} - 2\mathbf{k})\cdot\mathbf{k}\, d\sigma = \iint_S -2\, d\sigma = (-2)(\text{area of disk}) = -2\pi.$$

57. If $\mathbf{F} = y^2\mathbf{i} + x^2\mathbf{j} + z^2\mathbf{k}$, then $\operatorname{curl}\mathbf{F} = 2(x - y)\mathbf{k}$. The triangle S with vertices $(1,0,0)$, $(0,1,0)$, $(0,0,1)$ is on the plane $x + y + z = 1$ which has unit normal $\mathbf{n} = (\mathbf{i} + \mathbf{j} + \mathbf{k})/\sqrt{3}$; on that plane we have $d\sigma = \sqrt{1 + (-1)^2 + (-1)^2}\, dA = \sqrt{3}\, dA$. Therefore

$$\oint_C \mathbf{F}\cdot d\mathbf{x} = \iint_S \operatorname{curl}\mathbf{F}\cdot\mathbf{n}\, d\sigma = \iint_S 2(x-y)\mathbf{k}\cdot\frac{\mathbf{i}+\mathbf{j}+\mathbf{k}}{\sqrt{3}}\, d\sigma = \iint_S \frac{2(x-y)}{\sqrt{3}}\, d\sigma$$

$$= \int_{x=0}^{1}\int_{y=0}^{1-x} 2(x-y)\, dy\, dx = \int_{x=0}^{1}(-1 + 4x - 3x^2)\, dx = 0.$$

59. If $\mathbf{F} = x\mathbf{i} + 2y\mathbf{j} + 3z\mathbf{k}$, then $\operatorname{div}\mathbf{F} = 1 + 2 + 3 = 6$. Therefore

$$\iint_S \mathbf{F}\cdot\mathbf{n}\, d\sigma = \iiint_W \operatorname{div}\mathbf{F}\, dv = \iiint_W 6\, dv = (6)(\text{volume of unit ball}) = (6)\left(\frac{4\pi}{3}\right) = 8\pi.$$

61.   If $\mathbf{F} = x\mathbf{i} + 2y\mathbf{j} + 3z\mathbf{k}$, then div $\mathbf{F} = 1 + 2 + 3 = 6$. Therefore

$$\iint_S \mathbf{F} \cdot \mathbf{n} \, d\sigma = \iiint_W \text{div } \mathbf{F} \, dv = \iiint_W 6 \, dv = (6)(\text{volume of cylinder, radius 3, height 6})$$

$$= (6)\left(\pi \cdot 3^2 \cdot 6\right) = 324\pi.$$

63.   If $\mathbf{F} = x\mathbf{i} + 2y\mathbf{j} + 3z\mathbf{k}$, then div $\mathbf{F} = 1 + 2 + 3 = 6$. Therefore

$$\iint_S \mathbf{F} \cdot \mathbf{n} \, d\sigma = \iiint_W \text{div } \mathbf{F} \, dv = \iiint_W 6 \, dv = (6)(\text{volume of unit cube}) = 6 \cdot 1^3 = 6.$$

65.   If $x = u + 2v$ and $y = 2u - v$, then $\dfrac{\partial(x,y)}{\partial(u,v)} = \begin{vmatrix} 1 & 2 \\ 2 & -1 \end{vmatrix} = -5$.

67.   If $x = u \ln v$ and $y = v \ln u$, then $\dfrac{\partial(x,y)}{\partial(u,v)} = \begin{vmatrix} \ln v & u/v \\ v/u & \ln u \end{vmatrix} = (\ln u)(\ln v) - 1$.

69.   If $x = u/v$ and $y = v/u$, then $\dfrac{\partial(x,y)}{\partial(u,v)} = \begin{vmatrix} 1/v & -u/v^2 \\ -v/u^2 & 1/u \end{vmatrix} = 0$.

71. $x = u + v + w$, $y = u - 2v + 3w$, $z = -2u + v - 5w$ implies $\dfrac{\partial(x,y,z)}{\partial(u,v,w)} = \begin{vmatrix} 1 & 1 & 1 \\ 1 & -2 & 3 \\ -2 & 1 & -5 \end{vmatrix} = 3$.

73. $x = u + v$ and $y = u - v$ imply $\dfrac{\partial(x,y)}{\partial(u,v)} = \begin{vmatrix} 1 & 1 \\ 1 & -1 \end{vmatrix} = -2$. The xy-region $\Omega = \{(x,y) : 0 \le x \le y \le 1\}$
is the triangle with vertices $(0,0)$, $(1,1)$, $(0,1)$. The inverse mapping is $u = (x+y)/2$ and $v = (x-y)/2$;
it maps Ω into S which is the triangle with vertices $(0,0)$, $(1,0)$, $(1/2, -1/2)$; the mapping reverses the
orientation of the boundary curve so we use the minus sign of the $\pm$ in Theorem 15.10.1.

$$\int_{x=0}^1 \int_{y=x}^1 x \, y \, dy \, dx = \iint_\Omega x \, y \, dA = -\iint_S (u+v)(u-v)(-2) \, d\sigma$$

$$= \int_{v=-1/2}^0 \int_{u=-v}^{1+v} 2\left(u^2 - v^2\right) du \, dv = \int_{v=-1/2}^0 \left(\frac{2}{3} + 2v - \frac{8}{3}v^3\right) dv = \frac{1}{8}.$$

16
Ordinary Differential Equations

Prologue: The unique solution to the initial value problem $h(y)\,\dfrac{dy}{dx} = g(x)$, $y(a) = b$ can be written explicitly in the form $\displaystyle\int_b^y h = \int_a^x g$. In some situations we may write that integral equation as $\displaystyle\int_b^y h(y)\,dy = \int_a^x g(x)\,dx$ rather than the more obviously correct $\displaystyle\int_b^y h(u)\,du = \int_a^x g(v)\,dv$.

1. $\dfrac{dy}{dx} = -7x \implies \displaystyle\int dy = \int (-7x)\,dx \implies y = \dfrac{-7}{2}x^2 + C.$

3. $\dfrac{dx}{dt} = \sin x\,\cos t$, $x(\pi/2) = 3 \implies \displaystyle\int_3^x \csc x\,dx = \int_{\pi/2}^t \cos t\,dt \implies \ln|\tan(x/2)| - \ln|\tan(3/2)| =$
$\sin t - 1 \implies |\tan(x/2)| = \tan(3/2)e^{-1+\sin t} \implies x = 2\tan^{-1}\left(\tan(3/2)e^{-1+\sin t}\right).$

REMARK: The first part of item 99 in the text's Table of Integrals yields $\ln|\csc x - \cot x| - \ln|\csc 3 - \cot 3| = \sin t - 1$. The text answer results from the identity $\ln|\csc x - \cot x| = -\ln|\csc x + \cot x|$.

5. $x^3\left(y^2 - 1\right)\dfrac{dy}{dx} = (x+3)y^5 \implies \displaystyle\int\left(\dfrac{1}{y^3} - \dfrac{1}{y^5}\right)dy = \int\left(\dfrac{1}{x^2} + \dfrac{3}{x^3}\right)dx \implies \dfrac{-1}{2y^2} + \dfrac{1}{4y^4} = \dfrac{-1}{x} - \dfrac{3}{2x^2} + C.$

REMARK: Observe that $\dfrac{-1}{2y^2} + \dfrac{1}{4y^4} = \dfrac{1}{4}\left(\left(\dfrac{1}{y^2} - 1\right)^2 - 1\right)$, therefore $\dfrac{1}{y^2} = 1 \pm \sqrt{\dfrac{-4}{x} - \dfrac{6}{x^2} + C^*}$ and
$$y = \dfrac{\pm 1}{\sqrt{1 \pm \sqrt{\frac{-4}{x} - \frac{6}{x^2} + C^*}}}.$$

7. $\dfrac{dx}{dt} = e^x \sin t$, $x(0) = 1 \implies \displaystyle\int_1^x e^{-x}\,dx = \int_0^t \sin t\,dt \implies -e^{-x} + e^{-1} = -\cos t + 1 \implies x =$
$-\ln\left(-1 + e^{-1} + \cos t\right).$

9. $\dfrac{dx}{dt} = x\cdot(1-\cos 2t)$, $x(0) = 1 \implies \displaystyle\int_1^x \dfrac{dx}{x} = \int_0^t (1-\cos 2t)\,dt \implies \ln x = t - \dfrac{1}{2}\sin 2t \implies x = e^{t-(1/2)\sin 2t}.$

11. $\dfrac{dy}{dx} = x^2\left(1+y^2\right) \implies \displaystyle\int\dfrac{dy}{1+y^2} = \int x^2\,dx \implies \tan^{-1} y = \dfrac{x^3}{3} + C \implies y = \tan\left(\dfrac{x^3}{3} + C\right).$

13.   Let t be the time elapsed from 1968 and let $P(t)$ be the population at time t. If the growth rate is proportional to the population, then
$$\dfrac{dP}{dt} = kP \implies \int\dfrac{dP}{P} = \int k\,dt \implies \ln|P| = kt + C \implies P = C^* e^{kt} = P(0)e^{kt}.$$
The population in 1968 was $P(0) = 12,100,000$, the population in 1973 was $13,268,000 = P(1973 - 1968) = P(5) = P(0)e^{5k}$; this implies $e^k = \left(P(5)/P(0)\right)^{1/5} = (13268/12100)^{1/5} = (3317/3025)^{1/5}$ and $P(t) = (12100000)(3317/3025)^{t/5}$. This growth model leads to the following predictions:

population in 1978 = $P(1978 - 1968) = P(10) = P(0)(3317/3025)^2 \approx 14{,}548{,}746 \approx 14{,}549{,}000$
population in 1983 = $P(1983 - 1968) = P(15) = P(0)(3317/3025)^3 \approx 15{,}953{,}121 \approx 15{,}953{,}000$
population in 1988 = $P(1988 - 1968) = P(20) = P(0)(3317/3025)^4 \approx 17{,}493{,}058 \approx 17{,}493{,}000$

16.2 — First-order differential equation: Separation of variables

15.   a. $P_e = \lim_{t \to \infty} \dfrac{\beta}{\delta + ((\beta/P_0) - \delta) \, e^{-\beta t}} = \dfrac{\beta}{\delta + 0} = \dfrac{\beta}{\delta}.$

     b.   If $Q = \dfrac{P}{P_e} = \dfrac{\delta P}{\beta}$, then $\dfrac{dQ}{dt} = \dfrac{1}{P_e}\dfrac{dP}{dt} = \dfrac{1}{P_e}P(\beta - \delta P) = \beta\dfrac{P}{P_e}\left(1 - \dfrac{P}{P_e}\right) = \beta Q(1 - Q).$

17.   Let $D(t)$ be the temperature difference at time t. Newton's law of cooling implies $D' = k\,D$ which has solution $D = D_0\, e^{kt}$. If the air temperature remains at $30\,°\mathrm{C}$ and the object has initial temperature $10\,°\mathrm{C}$, then $D_0 = 30 - 10 = 20$. If the object warms to $14\,°\mathrm{C}$ in 1 hour, then $30 - 14 = 16 = D(1) = 20e^k$ implies $e^k = 16/20 = 0.8$ and $D(t) = 20 \cdot 0.8^t$.

     a. $D(2) = 20 \cdot 0.8^2 = 12.8$, hence the object's temperature will be $30 - 12.8 = 17.2\,°\mathrm{C}$.

     b.   If the object is $25\,°\mathrm{C}$, then $30 - 25 = 5 = D(t) = 20 \cdot 0.8^t$ which implies $t = \ln(5/20)/\ln 0.8 \approx$ 6.213 hours $\approx$ 6 hr 12 min 45 sec.

19.   a. $ma = mv' = mg - \alpha v^2 \iff v' = g - \dfrac{\alpha}{m}v^2 = g\left(1 - b^2 v^2\right)$ where $b = \sqrt{\dfrac{\alpha}{mg}}$. Since the downward direction is positive and since the object falls, we can use item 61 from the text's integral table:

$$\frac{dv}{dt} = g\left(1 - b^2 v^2\right), \; v(0) = 0 \implies \int_0^v \frac{dv}{1 - b^2 v^2} = \int_0^t g\,dt \implies \frac{1}{b}\tanh^{-1} bv = gt$$

$$\implies v = \frac{1}{b}\tanh bgt = \sqrt{\frac{mg}{\alpha}}\tanh\sqrt{\frac{\alpha g}{m}}\,t.$$

     b. $\tanh$ is an increasing function and $\lim_{x \to \infty}\tanh x = 1$; thus $\lim_{t \to \infty} v = \lim_{t \to \infty}\sqrt{\dfrac{mg}{\alpha}}\tanh\sqrt{\dfrac{\alpha g}{m}}\,t = \sqrt{\dfrac{mg}{\alpha}}.$ ∎

21.   a.   Let $a = \sqrt{\mu/\lambda}$, then (using item 61 of the text's integral table)

$$\frac{dc}{dt} = \mu - \lambda c^2 = \lambda\left(a^2 - c^2\right) \implies \int_{c_0}^c \frac{dc}{a^2 - c^2} = \int_0^t \lambda\,dt \implies \frac{1}{a}\left(\tanh^{-1}\frac{c}{a} - \tanh^{-1}\frac{c_0}{a}\right) = \lambda t$$

$$\implies c = a\tanh\left(a\lambda t + \tanh^{-1}\frac{c_0}{a}\right) = \sqrt{\frac{\mu}{\lambda}}\tanh\left(\sqrt{\mu\lambda}\,t + \tanh^{-1}\sqrt{\frac{\lambda}{\mu}}\,c_0\right).$$

     b.   Since $\lim_{x \to \infty}\tanh x = 1$ we infer $\lim_{t \to \infty} c(t) = \sqrt{\dfrac{\mu}{\lambda}}.$

23.   Suppose snow started to fall b hours before 11 a.m., let t be the time elapsed from 11 a.m., and let $s(t)$ be the distance traveled by the snowplow. If the snow falls at a constant rate, then the height of the accumulated snow at time t is proportional to $b + t$; if the snowplow's velocity is inversely proportional to that height, then $\dfrac{ds}{dt} = \dfrac{k}{b + t}$. Therefore, since $s(0) = 0$, we see

$$\int_0^s ds = \int_0^t \frac{k\,dt}{b + t} \implies s = k\ln\left(\frac{b + t}{b}\right) = k\ln\left(1 + \frac{t}{b}\right).$$

4 miles cleared by 2 p.m. implies $s(3) = 4$ and 2 additional miles cleared by 5 p.m. implies $s(6) = 6$. Thus

$$3\ln\left(1 + \frac{3}{b}\right) = \frac{12}{k} = 2\ln\left(1 + \frac{6}{b}\right) \iff \left(1 + \frac{3}{b}\right)^3 = \left(1 + \frac{6}{b}\right)^2 \iff b^2 + 3b - 9 = 0.$$

The positive solution to $b^2 + 3b - 9 = 0$ is $b = (3/2)\left(-1 + \sqrt{5}\right) \approx 1.854$ which implies the snow started to fall at 9:08:45 a.m.

25.   Let y be the depth of water and assume $\dfrac{dy}{dt} = k$; let V be the volume of water and let x be the radius of a horizontal circular cross-section. The velocity of water falling distance y is $\sqrt{2gy}$, Torricelli's law then implies $\dfrac{dV}{dt} = cy^{1/2}$ for some constant c. We can relate the area of the water surface to volume and flow in the following way: area of water surface $= \pi x^2 = \dfrac{dV}{dy} = \dfrac{dV}{dt}\dfrac{dt}{dy} = cy^{1/2}\dfrac{1}{k} = c^* y^{1/2}$ which implies $y = Cx^4$.

Solutions 16.3   *First-order linear differential equations*   **(pages 1049-50)**

Prologue: Equation $y' + ay = f$ has integrating factor $A = e^{\int a}$; the antiderivative of $A \cdot (y' + ay)$ is $A \cdot y$.

1. $\dfrac{dy}{dx} = 4x \implies y = \int dy = \int 4x \, dx = 2x^2 + C.$

3. $A = e^{\int -1 \, dt} = e^{-t}$ is an integrating factor for $x' - x = 1$; solutions to the differential equation satisfy $x \cdot e^{-t} = \int e^{-t} \, dt = -e^{-t} + C$, therefore $x = -1 + Ce^{t}$. The initial condition $1 = x(0) = -1 + C$ implies $C = 2$ and $x = -1 + 2e^{t}$.

5. $A = e^{\int dt} = e^{t}$ is an integrating factor for $x' + x = \sin t$; solutions to the differential equation satisfy $x \cdot e^{t} = \int e^{t} \sin t \, dt = (1/2)(-\cos t + \sin t)e^{t} + C$, therefore $x = (1/2)(-\cos t + \sin t) + Ce^{-t}$. The initial condition $1 = x(0) = -1/2 + C$ implies $C = 3/2$ and $x = (1/2)(-\cos t + \sin t)e^{t} + (3/2)e^{-t}$.

7. $A = e^{\int -\ln x \, dx} = e^{-x \ln x + x} = x^{-x} e^{x}$ is an integrating factor for $y' - y \ln x = x^{x}$; solutions to the differential equation satisfy $y \cdot x^{-x} e^{x} = \int (x^{-x} e^{x}) x^{x} \, dx = \int e^{x} \, dx = e^{x} + C$, therefore $y = x^{x}(1 + Ce^{-x})$.

9. $A = e^{\int -a \, dt} = e^{-at}$ is an integrating factor for $x' - ax = be^{at}$; solutions to the differential equation satisfy $x \cdot e^{-at} = \int e^{-at} be^{at} \, dt = \int b \, dt = bt + C$. Thus $x = (bt + C)e^{at}$.

11. $A = e^{\int -2 \tan 2x \, dx} = e^{\ln \cos 2x} = \cos 2x$ is an integrating factor for $y' - (2 \tan 2x)y = x$; solutions to the differential equation satisfy $y \cdot \cos 2x = \int x \cos 2x \, dt = \dfrac{1}{4} \cos 2x + \dfrac{x}{2} \sin 2x + C.$ Therefore $y = \dfrac{1}{4} + \dfrac{x}{2} \tan 2x + C \sec 2x.$

13. $\dfrac{d}{dx}\left((x^2 + 1)\,y\right) = (x^2 + 1)\dfrac{dy}{dx} + 2xy \implies (x^2 + 1)\,y = \dfrac{x^2}{2} + C \implies y = \dfrac{1}{2} + \dfrac{K}{2(x^2 + 1)}.$ The initial condition $1 = y(0) = 1/2 + K/2$ implies $K = 2(1 - 1/2) = 1$ and $y = \dfrac{1}{2} + \dfrac{1}{2(x^2 + 1)}.$

15. $A = e^{\int 4t(t^2 + 1)^{-1} \, dt} = e^{2\ln(t^2 + 1)} = (t^2 + 1)^{2}$ is an integrating factor for $x' + \dfrac{4t}{t^2 + 1}\,x = 3t$; solutions to the differential equation satisfy $x \cdot (t^2 + 1)^{2} = \int 3t(t^2 + 1)^{2} \, dt = \dfrac{1}{2}(t^2 + 1)^{3} + C.$ The initial condition $x(0) = 4$ implies $4 \cdot 1 = (1/2) \cdot 1^3 + C = 1/2 + C$, $C = 4 - 1/2 = 7/2$, and $x = \dfrac{t^2 + 1}{2} + \dfrac{7}{2(t^2 + 1)^2}.$

REMARK: $x \cdot (t^2 + 1)^{2} = \int 3t(t^2 + 1)^{2} \, dt = \dfrac{1}{2}(t^6 + 3t^4 + 3t^2) + D$, $4 \cdot 1 = (1/2)(0) + D = D$, and $x = \dfrac{t^6 + 3t^4 + 3t^2 + 8}{2(t^2 + 1)^{2}}.$

---

17.   a. $\dfrac{dG}{dt} = \text{in} - \text{out} = k - rG$; this can be rewritten in the form $\dfrac{dG}{dt} + rG = k.$

   b. $G' + rG = k$ has integrating factor $A = e^{\int r \, dt} = e^{rt}$; solutions to the differential equation satisfy $G \cdot e^{rt} = \int ke^{rt} dt = (k/r)e^{rt} + C$. If no glucose is present initially, then $G(0) = 0$ which implies $0 = 0 \cdot e^{0} = (k/r)e^{0} + C = k/r + C$, $C = -k/r$, and $G = (k/r)(1 - e^{-rt})$.

   c.   Because $r > 0$, the equilibrium value is $\lim\limits_{t \to \infty} G(t) = \lim\limits_{t \to \infty} \dfrac{k}{r}(1 - e^{-rt}) = \dfrac{k}{r}.$

19.   Equation (19) can be stated in the form $\dfrac{dI}{dt} + \dfrac{R}{L}I = \dfrac{E}{L}.$ If R, L, and E are constant, the differential equation has integrating factor $A = e^{\int (R/L) \, dt} = e^{Rt/L}$ and solutions satisfy $I \cdot e^{Rt/L} = \int (E/L)e^{Rt/L} dt = (E/R)e^{Rt/L} + C$. Thus $I(t) = E/R + Ce^{-Rt/L} = E/R + \left(I(0) - E/R\right)e^{-Rt/L}$. Therefore $R = 8$, $L = 1$, $E = 6$, $I(0) = 1$ implies $I(t) = 3/4 + (1/4)e^{-8t}$ and $I(.1) = 3/4 + (1/4)e^{-.8} \approx .862$ amperes.

16.3 — First-order linear differential equations

21. If $R = 10$, $C = 0.001$, and $E = 10\cos 60t$, then equation (24) implies $\dfrac{dQ}{dt} + 100\,Q = \cos 60t$; it has integrating factor $\mathcal{A} = e^{100t}$ and its solutions satisfy $Q \cdot e^{100t} = \int e^{100t}\cos 60t\,dt = (1/680)e^{100t}(5\cos 60t + 3\sin 60t) + K$. If $Q(0) = 0$, then $0 = 5/680 + K$, $K = -1/136$, and $Q(t) = (1/680)(5\cos 60t + 3\sin 60t) - (1/136)e^{-100t}$.

23. If $L = 2$, $R = 10$, and $E = 100\sin 60t$, then equation (19) implies $\dfrac{dI}{dt} + 5\,I = 50\sin 60t$; it has integrating factor $\mathcal{A} = e^{5t}$ and its solutions satisfy $I \cdot e^{5t} = \int e^{5t}50\sin 60t\,dt = (2/29)e^{5t}(-12\cos 60t + \sin 60t) + K$. If $I(0) = 0$, then $0 = -24/29 + K$, $I(t) = (2/29)(-12\cos 60t + \sin 60t) + (24/29)e^{-5t}$, and $I(.1) \approx -.31194$ A.

25. If $L = 1$, $R = 2$, and $E = 6e^{-.001\,t}$, then equation (19) implies $\dfrac{dI}{dt} + 2\,I = 6e^{-.001\,t}$; it has integrating factor $\mathcal{A} = e^{2t}$ and its solutions satisfy $I \cdot e^{2t} = \int 6e^{-.001\,t}e^{2t}\,dt = 6\int e^{1.999\,t}\,dt = \dfrac{6}{1.999}e^{1.999\,t} + K$. If $I(0) = 0$, then $0 = 6/1.999 + K$ and $I(t) = \dfrac{6}{1.999}\left(e^{-.001\,t} - e^{-2t}\right)$.

   Observe that $I(t) \approx 3e^{-.001\,t}$ for large t. Hence we can approximate a solution to $I(t) = .5$ by solving $3e^{-.001\,t} = .5$. This simpler equation yields $t = 10^3\ln 6 \approx 1791.759$ seconds $\approx .5$ hours. More careful analysis of $I(t) = .5$ (e.g., Newton's method) yields $t \approx 1792.25959$ seconds.

27. Suppose $R \neq 0$, $C \neq 0$, and E are constant. Equation (24) says $Q' + Q/(RC) = E/R$; since $I = Q'$, this implies $I' + I/(RC) = Q'' + Q'/(RC) = 0$. Hence $I' = -I/(RC)$ which implies $I(t) = I(0)e^{-t/(RC)}$. Therefore $I(t) = (1/2)I(0) \iff -t/(RC) = \ln\left(I(t)/I(0)\right) = \ln(1/2) \iff t = -RC\ln(1/2) = RC\ln 2$.

29. If the conditions for Theorem 6.1.4 are satisfied, then

$$y - x\frac{dy}{dx} = \frac{dy}{dx} \cdot y^2 e^y \iff y\frac{dx}{dy} - x = y^2 e^y \iff \frac{dx}{dy} - \frac{1}{y}x = ye^y.$$

The last differential equation has integrating factor $\mathcal{A} = e^{1.999\,t}-y^{-1}\,dy = e^{-\ln y} = \dfrac{1}{y}$. Thus $x \cdot \dfrac{1}{y} = \int e^y\,dy = e^y + C$ which implies $x(y) = y \cdot (e^y + C)$.

REMARK: If $y \neq 0$, then

$$y - x\frac{dy}{dx} = \frac{dy}{dx} \cdot y^2 e^y \iff 0 = \frac{1}{y} - \frac{x}{y^2}\frac{dy}{dx} - e^y\frac{dy}{dx} = \frac{d}{dx}\left(\frac{x}{y} - e^y\right) \iff C = \frac{x}{y} - e^y.$$

31. If y_1 and y_2 are solutions of $y' + ay = f$, then $(y_1 - y_2)' + a(y_1 - y_2) = (y_1' + ay_1) - (y_2' + ay_2) = f - f = 0$. ∎

33. The current in Problem 28 can be written $I = I_1 + I_2$ where $I_1 = \dfrac{E_0 C\omega}{1 + (RC\omega)^2}(-\sin\omega t + RC\omega\cos\omega t)$ is the steady-state term and the transient term is $I_2 = \dfrac{E_0/R}{1 + (RC\omega)^2}e^{-t/(RC)}$. The second-order Maclaurin expansion for the transient term is $\dfrac{E_0/R}{1 + (RC\omega)^2}\left(1 - \dfrac{t}{RC} + \left(\dfrac{t}{RC}\right)^2\right)$ which is larger than $\dfrac{E_0/R}{1 + 1} = \dfrac{E_0}{2R}$ if R and t are small.

REMARK: The second-order Maclaurin expansion for current I is $\dfrac{E_0}{R}\left(1 - \dfrac{t}{RC} + \dfrac{1 - (RC\omega)^2}{2}\left(\dfrac{t}{RC}\right)^2\right)$.

Solutions 16.4 2nd-order linear, homogeneous DE with constant coefficients   (page 1057)

Prologue: Case 3 can be restated without using complex numbers. If the auxiliary equation is $0 = (\lambda - \alpha)^2 + \beta^2$, then the general solution to the differential equation is $y = e^{\alpha x}(c_1\cos\beta x + c_2\sin\beta x)$.

1. $0 = y'' - 4y$ has auxiliary equation $0 = \lambda^2 - 4 = (\lambda - 2)(\lambda + 2)$ and general solution $y = c_1 e^{2x} + c_2 e^{-2x}$.

3. $0 = y'' + 2y' + 2y$ has auxiliary equation $0 = \lambda^2 + 2\lambda + 2 = (\lambda + 1)^2 + 1^2 = \left(\lambda - (-1)\right)^2 + 1^2$ and general solution $y = e^{-x}(c_1\cos x + c_2\sin x)$.

5. $0 = 8y'' + 4y' + y = 8\left(y'' + (1/2)y' + (1/8)y\right)$ has auxiliary equation $0 = \lambda^2 + (1/2)\lambda + 1/8 = (\lambda + 1/4)^2 + (1/4)^2$ and general solution $y = e^{-x/4}\left(c_1 \cos x/4 + c_2 \sin x/4\right)$. The initial conditions $0 = y(0) = c_1$ and $1 = y'(0) = (1/4)\left(-c_1 + c_2\right) = (1/4)c_2$ imply $y = 4e^{-x/4} \sin x/4$.

7. $0 = y'' + 5y' + 6y$ has auxiliary equation $0 = \lambda^2 + 5\lambda + 6 = (\lambda + 2)(\lambda + 3)$ and general solution $y = c_1 e^{-2x} + c_2 e^{-3x}$. Initial conditions $1 = y(0) = c_1 + c_2$ and $2 = y'(0) = -2c_1 - 3c_2$ imply $y = 5e^{-2x} - 4e^{-3x}$.

9. $0 = 4y'' + 20y' + 25y = 4\left(y'' + 5y' + (25/4)y\right)$ has auxiliary equation $0 = \lambda^2 + 5\lambda + 25/4 = (\lambda + 5/2)^2$ and general solution $y = (c_1 + c_2 x)\, e^{-5x/2}$. The initial conditions $1 = y(0) = c_1$ and $2 = y'(0) = (-5/2)c_1 + c_2 = -5/2 + c_2$ imply $y = (1 + 9x/2)e^{-5x/2}$.

11. $0 = y'' - y' + -6y$ has auxiliary equation $0 = \lambda^2 - \lambda + -6 = (\lambda + 2)(\lambda - 3)$ and general solution $y = c_1 e^{-2x} + c_2 e^{3x}$. The initial conditions $-1 = y(0) = c_1 + c_2$ and $1 = y'(0) = -2c_1 + 3c_2$ imply $y = (1/5)\left(-4e^{-2x} - e^{3x}\right)$.

13. $0 = y'' + 4y$ has auxiliary equation $0 = \lambda^2 + 4 = \lambda^2 + 2^2$ and general solution $y = c_1 \cos 2x + c_2 \sin 2x$. Initial conditions $1 = y(\pi/4) = c_2$ and $3 = y'(\pi/4) = -2c_1$ imply $y = (-3/2) \cos 2x + \sin 2x$.

15. $0 = y'' + 17y'$ has auxiliary equation $0 = \lambda^2 + 17\lambda = \lambda(\lambda + 17)$ and general solution $y = c_1 + c_2 e^{-17x}$. Initial conditions $1 = y(0) = c_1 + c_2$ and $0 = y'(0) = -17c_2$ imply $y = 1$.

17. $0 = y'' - 13y' + 42y$ has auxiliary equation $0 = \lambda^2 - 13\lambda + 42 = (\lambda - 6)(\lambda - 7)$ and general solution $y = c_1 e^{6x} + c_2 e^{7x}$.

19. $y'' = y$ has auxiliary equation $0 = \lambda^2 - 1 = (\lambda + 1)(\lambda - 1)$ and general solution $y = c_1 e^{-x} + c_2 e^{x}$. Initial conditions $2 = y(0) = c_1 + c_2$ and $-3 = y'(0) = -c_1 + c_2$ imply $y = (5/2)e^{-x} - (1/2)e^{x}$.

21.   If y_1 and y_2 are solutions to $y'' + ay' + by = 0$, then

$(c_1 y_1 + c_2 y_2)'' + a(c_1 y_1 + c_2 y_2)' + b(c_1 y_1 + c_2 y_2) = c_1\left(y_1'' + ay_1' + by_1\right) + c_2\left(y_2'' + ay_2' + by_2\right) = c_1 \cdot 0 + c_2 \cdot 0 = 0;$

therefore $c_1 y_1 + c_2 y_2$ is also a solution to $y'' + ay' + by = 0$. ∎

Solutions 16.5 2nd-order nonhomogeneous DE: undetermined coefficients   (pages 1063-4)

1.   The homogeneous equation $0 = y'' + y'$ has auxiliary equation $0 = \lambda^2 + \lambda = \lambda(\lambda + 1)$ and general solution $c_1 + c_2 e^{-x}$. Since $4x^2$ is quadratic and our differential equation has coefficient 0 for y, we search for a particular solution in the form of a cubic polynomial $P = a_3 x^3 + a_2 x^2 + a_1 x + a_0$.

$$4x^2 = P'' + P' = (6a_3 x + 2a_2) + \left(3a_3 x^2 + 2a_2 x + a_1\right) = 3a_3 x^2 + (6a_3 + 2a_2)\, x + (2a_2 + a_1)$$

$$\begin{aligned} 4 &= 3a_3 & a_3 &= 4/3 \\ \Longleftrightarrow \quad 0 &= 6a_3 + 2a_2 \quad \Longleftrightarrow \quad & a_2 &= -3a_3 = -3(4/3) = -4 \\ 0 &= 2a_2 + a_1 & a_1 &= -2a_2 = -2(-4) = 8 \end{aligned}$$

Hence the general solution to $y'' + y' = 4x^2$ is $y = (4/3)x^3 - 4x^2 + 8x + c_1 + c_2 e^{-x}$. Initial conditions $3 = y(0) = c_1 + c_2$ and $-1 = y'(0) = 8 - c_2$ imply $c_2 = 8 - (-1) = 9$, $c_1 = 3 - c_2 = 3 - 9 = -6$, and $y = (4/3)x^3 - 4x^2 + 8x - 6 + 9e^{-x}$.

REMARK: $\dfrac{4}{3}x^3 = \displaystyle\int_0^x 4x^2\, dx = \int_0^x (y'' + y')\, dx = (y' + y) - \left(y'(0) + y(0)\right) = (y' + y) - (-1 + 3)$ implies $y' + y = (4/3)x^2 + 2$. This first-order DE has integrating factor $\mathcal{A} = e^x$ and we compute

$$e^x y - 3 = e^x y - e^0 y(0) = \int_0^x e^x (y' + y)\, dx = \int_0^x e^x \left(\frac{4}{3}x^3 + 2\right) dx = e^x \left(\frac{4}{3}x^3 - 4x^2 + 8x - 6\right) + 6.$$

---

3.   The homogeneous equation $0 = y'' + 3y' + 2y$ has auxiliary equation $0 = \lambda^2 + 3\lambda + 2 = (\lambda + 2)(\lambda + 1)$ and general solution $c_1 e^{-2x} + c_2 e^{-x}$. We search for a particular solution in the form $P = ae^{4x}$:

$$3e^{4x} = P'' + 3P' + 2P = (16 + 12 + 2)ae^{4x} = 30ae^{4x} \iff a = \frac{3}{30} = \frac{1}{10}.$$

Hence the general solution to $y'' + 3y' + 2y = 3e^{4x}$ is $y = (1/10)e^{4x} + c_1 e^{-2x} + c_2 e^{-x}$.

5.   Homogeneous equation $0 = y'' + 6y$ has general solution $c_1 \cos \sqrt{6}\, x + c_2 \sin \sqrt{6}\, x$. We search for a particular solution in the form $P = a \cos x$: $4 \cos x = P'' + 6P = (-1 + 6)a \cos x = 5a \cos x \iff a = 4/5$. Hence the general solution to $y'' + 6y = 4 \cos x$ is $y = (4/5) \cos x + c_1 \cos \sqrt{6}\, x + c_2 \sin \sqrt{6}\, x$.

16.5 — 2nd-order nonhomogeneous DE: undetermined coefficients

7.   Homogeneous equation $0 = y'' + y$ has general solution $c_1 \cos x + c_2 \sin x$. We search for a particular solution in the form $P = x(a_1 \cos x + a_2 \sin x)$:

$$\cos x = P'' + P = -2a_1 \sin x + 2a_2 \cos x \iff \begin{matrix} 0 = -2a_1 \\ 1 = 2a_2 \end{matrix} \iff \begin{matrix} a_1 = 0 \\ a_2 = 1/2 \end{matrix}$$

Hence the general solution to $y'' + y = \cos x$ is $y = c_1 \cos x + (c_2 + x/2) \sin x$.

9.   Homogeneous equation $0 = y'' - 4y' + 5y$ has auxiliary equation $0 = \lambda^2 - 4\lambda + 5 = (\lambda - 1)^2 + 1^2$ and general solution $e^{2x}(c_1 \cos x + c_2 \sin x)$. We search for a particular solution in the form $P = a_1 \cos x + a_2 \sin x$:

$$\sin x = P'' - 4P' + 5P = 4(a_1 - a_2)\cos x + 4(a_1 + a_2)\sin x \iff \begin{matrix} 0 = 4(a_1 - a_2) \\ 1 = 4(a_1 + a_2) \end{matrix} \iff a_1 = 1/8 = a_2.$$

Hence the general solution to $y'' - 4y' + 5y = \sin x$ is $y = e^{2x}(c_1 \cos x + c_2 \sin x) + (1/8)(\cos x + \sin x)$.

11.   Homogeneous equation $0 = y'' - 2y' + y$ has auxiliary equation $0 = \lambda^2 - 2\lambda + 1 = (\lambda - 1)^2$ and general solution $(c_0 + c_1 x)e^x$. We search for a particular solution in the form $P = ae^{3x}$: $-2e^{3x} = P'' - 2P' + P = (9 - 6 + 1)P = 4ae^{3x} \iff a = -1/2$. Hence the general solution to $y'' - 2y' + y = -2e^{3x}$ is $y = -(1/2)e^{3x} + (c_0 + c_1 x)e^x$.

13.   Homogeneous equation $0 = y'' - 2y' + y$ has auxiliary equation $0 = \lambda^2 - 2\lambda + 1 = (\lambda - 1)^2$ and general solution $(c_0 + c_1 x)e^x$. We search for a particular solution in the form $P = x^2 ae^x$. Let $Q = P' - P = 2axe^x$, then $2e^x = P'' - 2P' + P = Q' - Q = 2ae^x \iff a = 1$. Hence the general solution to $y'' - 2y' + y = 2e^x$ is $y = (c_0 + c_1 x + x^2)e^x$.

REMARK: If $P(x) = f(x)e^{kx}$, then $P' - kP = (f'(x)e^{kx} + f(x)ke^{kx}) - kf(x)e^{kx} = f'(x)e^{kx}$.

---

15.   The homogeneous equation $0 = y'' + y' + y$ has auxiliary equation $0 = \lambda^2 + \lambda + 1 = (\lambda + 1/2)^2 + 3/4$ and general solution $e^{-x/2}\left(c_1 \cos\left(\frac{\sqrt{3}}{2}x\right) + c_2 \sin\left(\frac{\sqrt{3}}{2}x\right)\right)$. We search for a particular solution in the form

$$P = xe^{-x/2}\left(a_1 \cos\left(\frac{\sqrt{3}}{2}x\right) + a_2 \sin\left(\frac{\sqrt{3}}{2}x\right)\right):$$

$$e^{-x/2}\cos\left(\frac{\sqrt{3}}{2}x\right) = P'' + P' + P = \sqrt{3}e^{-x/2}\left(a_2 \cos\left(\frac{\sqrt{3}}{2}x\right) - a_1 \sin\left(\frac{\sqrt{3}}{2}x\right)\right) \iff \begin{matrix} a_1 = 0 \\ a_2 = 1/\sqrt{3} \end{matrix}$$

Hence the general solution to $y'' + y' + y = e^{-x/2}\cos\left(\frac{\sqrt{3}}{2}x\right)$ is

$$y = e^{-x/2}\left(c_1 \cos\left(\frac{\sqrt{3}}{2}x\right) + c_2 \sin\left(\frac{\sqrt{3}}{2}x\right)\right) + \frac{1}{\sqrt{3}}xe^{-x/2}\sin\left(\frac{\sqrt{3}}{2}x\right).$$

17.   The homogeneous equation $0 = y'' + y$ has general solution $c_1 \cos x + c_2 \sin x$. We decompose our search for a particular solution — let $P = a$ and $Q = x(b_1 \cos x + b_2 \sin x)$:

$$3 = P'' + P = 0 + a \iff a = 3,$$
$$-4\cos x = Q'' + Q = 2b_2 \cos x - 2b_1 \sin x \iff b_2 = -4/2 = -2 \text{ and } b_1 = 0.$$

Hence the general solution to $y'' + y = 3 - 4\cos x$ is $y = c_1 \cos x + c_2 \sin x + 3 - 2x \sin x$.

19.   The homogeneous equation $0 = y'' + 9y$ has general solution $c_1 \cos 3x + c_2 \sin 3x$. We decompose our search for a particular solution — let $P = x(a_1 \cos 3x + a_2 \sin 3x)$ and $Q = b_2 x^2 + b_1 x + b_0$:

$$2\sin 3x = P'' + 9P = 6a_2 \cos x - 6a_1 \sin x \iff a_1 = 2/(-6) = -1/3 \text{ and } a_2 = 0,$$
$$x^2 = Q'' + 9Q = 9b_2 x^2 + 9b_1 x + (2b_2 + 9b_0) \iff b_2 = 1/9 \text{ and } b_1 = 0 \text{ and } b_0 = -2b_2/9 = -2/81.$$

Hence the general solution to $y'' + 9y = 2\sin 3x + x^2$ is $y = c_1 \cos 3x + c_2 \sin 3x - \frac{x \cos x}{3} + \frac{x^2}{9} - \frac{2}{81}$.

21.   Equation (11) is equivalent to $Q'' + \frac{R}{L}Q' + \frac{1}{LC}Q = \frac{E}{L}$. If $L = 10$, $R = 250$, $C = 10^{-3}$, $E = 900$, then $Q'' + 25Q' + 100Q = 90$ which has general solution $Q = c_1 e^{-5t} + c_2 e^{-20t} + .9$. Initial conditions $0 = Q(0) = c_1 + c_2 + .9$ and $0 = I(0) = Q'(0) = -5c_1 - 20c_2$ imply $Q = -1.2e^{-5t} + .3e^{-20t} + .9$ and $I = Q' = 6e^{-5t} - 6e^{-20t}$.

436

23.   If $L = 10$, $R = 40$, $C = .025$, $E = 100\cos 5t$, then equation (11) is equivalent to $Q'' + 4Q' + 4Q = 10\cos 5t$ which has general solution $Q = (c_0 + c_1 t)\,e^{-2t} + \left(\dfrac{-210\cos 5t}{841} + \dfrac{200\sin 5t}{841}\right)$. Initial conditions $0 = Q(0) = c_0 - 210/841$ and $0 = I(0) = Q'(0) = -2c_0 + c_1 + 1000/841$ imply $Q = \left(\dfrac{210}{841} - \dfrac{20}{29}t\right)e^{-2t} + \dfrac{10}{841}(-21\cos 5t + 20\sin 5t)$ and $I = \dfrac{40}{29}\left(\dfrac{-25}{29} + t\right)e^{-2t} + \dfrac{50}{841}(21\sin 5t + 20\cos 5t)$. The transient component of I is $\dfrac{40}{29}\left(\dfrac{-25}{29} + t\right)e^{-2t}$ and the steady-state component is $\dfrac{50}{841}(21\sin 5t + 20\cos 5t)$.

25.   Let $L(y) = y'' + ay' + by$. If $L(y_1) = f_1$ and $L(y_2) = f_2$, then
$$L(y_1 + y_2) = (y_1 + y_2)'' + a(y_1 + y_2)' + b(y_1 + y_2) = (y_1'' + ay_1' + by_1) + (y_2'' + ay_2' + by_2) = f_1 + f_2$$
so $y_1 + y_2$ is a solution to $L(y) = f_1 + f_2$. ∎

Solutions 16.6   *Vibratory motion*   (page 1070)

1.   A 2-kg mass stretches the rubber band 32 cm to reach equilibrium position, hence Hooke's law implies $\sqrt{k/m} = \sqrt{g/32}$. If $x_0 = 16$ and $v_0 = 0$, then equation (5) yields $x(t) = 16\cos\sqrt{g/32}\,t$. Therefore $v(t) = x'(t) = -2\sqrt{2g}\sin\sqrt{g/32}\,t$. Since sin is 1 at the first positive place where cos is 0, the velocity when the mass first passes the equilibrium position is $-2\sqrt{2g} \approx -88.59$ cm/sec.

3.   If the bottom surface of the cylindrical block is x feet below the water surface, then the volume of displaced water is $\pi \cdot 1^2 \cdot x = \pi x$ and the upward force on the block is $62.4\pi x$ pounds. Since the block weighs 12.48 pounds, the equilibrium position of the block is at $x = 12.48/(62.4\pi) = 1/(5\pi)$ feet. The total force on the block is the sum of its weight (downward) and the water (upward); the block has a mass of 12.48 slugs, hence Newton's second law implies $12.48x'' = 12.48 - 62.4\pi x = 62.4\pi(1/(5\pi) - x)$. Therefore $(x - 1/(5\pi))'' = x'' = 5\pi(1/(5\pi) - x) = -5\pi(x - 1/(5\pi))$ which has solution $x - 1/(5\pi) = c_1\cos\sqrt{5\pi}\,t + c_2\sin\sqrt{5\pi}\,t$. Initial conditions $x(0) = 1$ and $x'(0) = 0$ imply $x = 1/(5\pi) + (1 - 1/(5\pi))\cos\sqrt{5\pi}\,t$. The period of $\cos\sqrt{5\pi}\,t$ is $2\pi/\sqrt{5\pi} = 2\sqrt{\pi/5} \approx 1.585$.

REMARK: Force at the equilibrium position x_e is zero and, in general, the water's force is proportional to the displacement from equilibrium, but we do not need to base our coordinate system there:
$$m(x - x_e)'' = mx'' = ma = F_{\text{net}} = w - kx = k\left(\frac{w}{k} - x\right) = k(x_e - x) = -k(x - x_e).$$

5.   In equations (9) and (14), we have $m = 10$ and $\mu = 1/2$. Therefore $\dfrac{\sqrt{4mk - \mu^2}}{2m} = \dfrac{\sqrt{40k - 1/4}}{20}$. The period of $c_1\cos\dfrac{\sqrt{40k - 1/4}}{20}t + c_2\sin\dfrac{\sqrt{40k - 1/4}}{20}t$ is $\dfrac{2\pi}{\frac{\sqrt{40k-1/4}}{20}} = \dfrac{40\pi}{\sqrt{40k - 1/4}}$. This period equals 8 seconds if and only if $k = \dfrac{(5\pi)^2 + 1/4}{40} = \dfrac{5\pi^2}{8} + \dfrac{1}{160} \approx 6.17$.

7.   Equation (15) with $m = 50$, $k = 50g/10 = 5g$, $\mu = 100\sqrt{17}$, $\alpha = 10$, $\omega = 2$ yields $x'' + 2\sqrt{17}x' + (g/10)x = (1/5)\sin 2t$. Hence $\dfrac{\sqrt{4mk - \mu^2}}{2m} = \sqrt{\dfrac{g}{10} - 17} \approx \sqrt{98.1 - 17} \approx 9$ and equation (14) yields homogeneous solution $e^{-\sqrt{17}t}\left(c_1\cos\sqrt{\dfrac{g}{10} - 17}\,t + c_2\sin\sqrt{\dfrac{g}{10} - 17}\,t\right)$. Equations (16) and (18) yield particular solution $\dfrac{-80\sqrt{17}\cos 2t + 2(g - 40)\sin 2t}{(g - 40)^2 + 27200}$. Therefore the equation of motion of the mass is
$$x(t) = e^{-\sqrt{17}t}\left(c_1\cos\sqrt{\frac{g}{10} - 17}\,t + c_2\sin\sqrt{\frac{g}{10} - 17}\,t\right) + \frac{-80\sqrt{17}\cos 2t + 2(g - 40)\sin 2t}{(g - 40)^2 + 27200}.$$

9.   Example 1, and thus Example 2, has $m = 5$ and $k = 5g/50 = g/10$. The motion in Example 2 is not damped harmonic motion if and only if $\mu^2 \geq 4mk = 2g \iff |\mu| \geq \sqrt{2g}$.

Solutions 16.7   *Numerical solution of DE: Euler's method*   (pages 1075-6)

Numerical tables will use y_E to label approximate values computed by Euler's method while y will denote the exact solution. Intermediate computations are done to high accuracy; results are rounded to low precision only for display. Problem 15-24 do not have solutions in terms of simple functions we already know about; y_{RK} denotes an approximate solution by means of a very accurate numerical procedure (a Fehlberg fourth-fifth order Runge-Kutta method).

1.   The initial-value problem $y' = -x$ and $y(0) = 1$ has exact solution $y = 1 - x^2/2$. Euler's method on the interval $[0, 1]$ with $h = .2$ yields

x	0	.2	.4	.6	.8	1.0
y_E	1	1.00	.96	.88	.76	.60
y	1	.98	.92	.82	.68	.50

3.   The initial-value problem $y' = -y$ and $y(0) = 1$ has exact solution $y = e^{-x}$. Euler's method on the interval $[0, 1]$ with $h = .2$ yields

x	0	.2	.4	.6	.8	1.
y_E	1	.8	.64	.512	.4096	.32768
y	1	.82	.67	.549	.4493	.36788

5.   The initial-value problem $y' = x + y$ and $y(0) = 1$ has exact solution $y = -1 - x + 2e^x$. Euler's method on the interval $[0, 1]$ with $h = .2$ yields

x	0	.2	.4	.6	.8	1.
y_E	1	1.2	1.48	1.856	2.3472	2.97664
y	1	1.24	1.58	2.044	2.6511	3.43656

7.   Initial-value problem $y' = (x - y)/(x + y)$ and $y(2) = 1$ has exact solution $y = -x + \sqrt{1 + 2x^2}$. Euler's method on the interval $[1, 2]$ with $h = -.2$ yields

x	2	1.8	1.6	1.4	1.2	1.
y_E	1	.93333	.86992	.81080	.75750	.71229
y	1	.93496	.87386	.81811	.76977	.73205

9.   Initial-value problem $y' = x\sqrt{1 + y^2}$ and $y(1) = 0$ has exact solution $y = \sinh\left((x^2 - 1)/2\right)$. Euler's method on the interval $[1, 3]$ with $h = .4$ yields

x	1	1.4	1.8	2.2	2.6	3.
y_E	0	.400	1.003	2.023	4.009	8.306
y	0	.499	1.369	3.337	8.879	27.290

11.   Initial-value problem $y' = y/x - (5/2)x^2y^3$ and $y(1) = 1/\sqrt{2}$ has exact solution $y = x/\sqrt{1 + x^5}$. Euler's method on the interval $[1, 2]$ with $h = .125$ yields

x	1.	1.125	1.25	1.375	1.5	1.625	1.75	1.875	2.
y_E	.70711	.68501	.63399	.57296	.51392	.46131	.41579	.37669	.34308
y	.70711	.67207	.62100	.56537	.51168	.46276	.41937	.38135	.34816

13.   Initial-value problem $y' = ye^x$ and $y(0) = 2$ has exact solution $y = 2e^{e^x - 1}$. Euler's method on the interval $[0, 2]$ with $h = .2$ yields

x	0	.2	.4	.6	.8	1.	1.2	1.4	1.6	1.8	2
y_E	2	2.4000	2.9863	3.8773	5.2903	7.645	11.802	19.639	35.57	70.8	156
y	2	2.4956	3.2704	4.5504	6.8116	11.150	20.354	42.450	104.18	311.9	1191

15. $y' = xy^2 + y^3$ and $y(0) = 1$.

x	0	.02	.04	.06	.08	.10
y_E	1	1.02	1.04164	1.06511	1.09064	1.11849
y_{RK}	1	1.02083	1.04345	1.06810	1.09503	1.12458

17. $y' = x + \sin(\pi y)$ and $y(1) = 0$.

x	1	1.2	1.4	1.6	1.8	2.
y_E	0	.2	.55756	1.03430	1.33279	1.51976
y_{RK}	0	.29727	.740774	1.08440	1.30809	1.50003

19. $y' = \sqrt{x^2 + y^2}$ and $y(1) = 5$.

x	1	.8	.6	.4	.2	0
y_E	5	3.98020	3.16824	2.52333	2.01236	1.60790
y_{RK}	5	4.07764	3.32663	2.71619	2.22053	1.81747

21. $y' = \sqrt{x + y^2}$ and $y(0) = 1$.

x	0	.2	.4	.6	.8	1.
y_E	1	1.2	1.45612	1.77363	2.16071	2.62842
y_{RK}	1	1.23055	1.52594	1.89479	2.34953	2.90658

23. $y' = \cos(xy)$ and $y(0) = 0$.

x	0	.78540	1.57080	2.35619	3.14159
y_E	0	.78540	1.42605	.93880	.46901
y_{RK}	0	.75718	1.05423	.82669	.57884

25. The linearization of g at a is the linear function $L(t) = g(a) + g'(a)(t - a)$; it is an approximation to g in a neighborhood of $(a, g(a))$. The differential equation $y' = f(x, y)$ can have a family $\mathcal{F}$ of solutions, an initial condition specifies a particular solution. Euler's method involves extrapolation by a succession of linearizations to members of $\mathcal{F}$.

Solutions 16.R Review (pages 1076-7)

1. $y' = 3x \implies y = \int 3x \, dx = (3/2)x^2 + C$.

3. $x' = e^x \cos t$ and $x(0) = 3 \implies -e^{-x} + e^{-3} = -e^{-x} \big|_3^x = \int_3^x e^{-x} \, dx = \int_0^t \cos t \, dt = \sin t - \sin 0 \implies$
$x = -\ln\left(e^{-3} - \sin t\right)$.

5. $y' + 3y = \cos x$ and $y(0) = 1$ imply
$$ye^{3x} - 1 = ye^{3x} \big|_0^x = \int_0^x (y' + 3y) e^{3x} \, dx = \int_0^x (\cos x) e^{3x} \, dx = \frac{3\cos x + \sin x}{10} e^{3x} - \frac{3}{10}$$
which then implies $y = (3\cos x + \sin x)/10 + (7/10)e^{-3x}$.

7. $x' = 3x + t^3 e^{3t}$ and $x(1) = 2 \implies xe^{-3t} - 2e^{-3} = xe^{-3t} \big|_1^t = \int_1^t (x' - 3x) e^{-3t} \, dt = \int_1^t t^3 \, dt = \frac{t^4 - 1}{4} \implies$
$x = \left(\frac{t^4 - 1}{4} + 2e^{-3}\right) e^{3t}$.

9. $y'' - 5y' + 4y = 0$ has auxiliary polynomial $\lambda^2 - 5\lambda + 4 = (\lambda - 1)(\lambda - 4)$ and general solution $y = c_1 e^x + c_2 e^{4x}$.

11. $y'' - 9y = 0$ has auxiliary polynomial $\lambda^2 - 9 = (\lambda + 3)(\lambda - 3)$ and general solution $y = c_1 e^{-3x} + c_2 e^{3x}$.

13. $y'' + 6y' + 9y = 0$ has auxiliary polynomial $\lambda^2 + 6\lambda + 9 = (\lambda + 3)^2$ and general solution $y = (c_0 + c_1 x) e^{-3x}$.

15. $y'' - 2y' + 2y = 0$ has auxiliary polynomial $\lambda^2 - 2\lambda + 2 = (\lambda - 1)^2 + 1^2$ and general solution $y = e^x (c_1 \cos x + c_2 \sin x)$. Initial conditions $0 = y(0) = c_1$ and $1 = y'(0) = c_1 + c_2 = c_2$ imply $y = e^x \sin x$.

17. Homogeneous equation $0 = y'' + 4y$ has auxiliary polynomial $\lambda^2 + 4 = \lambda^2 + 2^2$ and general solution $c_1 \cos 2x + c_2 \sin 2x$. We search for a particular solution in the form $P = a_1 \cos x + a_2 \sin x$:
$$2\sin x = P'' + 4P = -P + 4P = 3P = 3(a_1 \cos x + a_2 \sin x) \iff a_1 = 0 \text{ and } a_2 = \frac{2}{3}.$$
Hence the general solution to $y'' + 4y = 2\sin x$ is $y = (2/3)\sin x + c_1 \cos 2x + c_2 \sin 2x$.

19. Homogeneous equation $0 = y'' + y' + y$ has auxiliary polynomial $\lambda^2 + \lambda + 1 = (\lambda + 1/2)^2 + 3/4$ and general solution $e^{-x/2}\left(c_1 \cos \sqrt{3}\, x/2 + c_2 \sin \sqrt{3}\, x/2\right)$. We search for a particular solution in the form $P = xe^{-x/2}\left(a_1 \cos \sqrt{3}\, x/2 + a_2 \sin \sqrt{3}\, x/2\right)$:
$$e^{-x/2} \sin \frac{\sqrt{3}}{2} x = P'' + P' + P = \sqrt{3}e^{-x/2}\left(a_2 \cos \frac{\sqrt{3}}{2} x - a_1 \sin \frac{\sqrt{3}}{2} x\right) \iff a_2 = 0 \text{ and } a_1 = \frac{-1}{\sqrt{3}}.$$
The general solution to $y'' + y' + y = e^{-x/2} \sin \sqrt{3}\, x/2$ is $y = e^{-x/2}\left(\left(c_1 - \frac{x}{\sqrt{3}}\right) \cos \frac{\sqrt{3}}{2} x + c_2 \sin \frac{\sqrt{3}}{2} x\right)$.

REMARK: Observe that $(\lambda - a)^2 + b^2 = \lambda^2 - 2a\lambda + (a^2 + b^2)$. Let $L(y) = y'' - 2ay' + (a^2 + b^2)\, y$. If $P = e^{ax}\cos(bx)$, and $Q = e^{ax}\sin(bx)$, then $L(x \cdot P) = -2bQ$ and $L(x \cdot Q) = 2bP$.

21.   Homogeneous equation $y'' + y = 0$ has general solution $c_1 \cos x + c_2 \sin x$. Initial conditions $0 = y(0) = c_1$ and $0 = y'(0) = c_2$ imply $y(x) = 0$ for all x.

23.   Homogeneous equation $0 = y'' - 2y' + y$ has auxiliary polynomial $\lambda^2 - 2\lambda + 1 = (\lambda - 1)^2$ and general solution $(c_0 + c_1 x)\, e^x$. We search for a particular solution of $y'' - 2y' + y = e^{-x}$ in the form $P = ae^{-x}$:

$$e^{-x} = P'' - 2P' + P = \left((-1)^2 - 2(-1) + 1\right) P = 4ae^{-x} \iff a = 1/4.$$

The general solution to $y'' - 2y' + y = e^{-x}$ is $y = (c_0 + c_1 x)\, e^x + (1/4)e^{-x}$.

25.   Homogeneous equation $0 = y'' - y' - 6y$ has auxiliary polynomial $\lambda^2 - \lambda - 6 = (\lambda + 2)(\lambda - 3)$ and general solution $c_1 e^{-2x} + c_2 e^{3x}$. We search for a particular solution of $y'' - y' - 6y = e^x \cos x$ in the form $P = (a_1 \cos x + a_2 \sin x)\, e^x$:

$$e^x \cos x = P'' - P' - 6P = \left((-7a_1 + a_2)\cos x + (-a_1 - 7a_2)\sin x\right) e^x \iff \begin{array}{l} 1 = -7a_1 + a_2 \\ 0 = -a_1 - 7a_2 \end{array} \iff \begin{array}{l} a_1 = -7/50 \\ a_2 = 1/50 \end{array}$$

The general solution to $y'' - y' - 6y = e^x \cos x$ is $y = c_1 e^{-2x} + c_2 e^{3x} + (1/50)(-7\cos x + \sin x)e^x$.

27.   Homogeneous equation $0 = y'' + y$ has general solution $c_1 \cos x + c_2 \sin x$. We search for a particular solution of $y'' + y = x + e^x + \sin x$ in the form $P = a_0 + a_1 x + be^x + x \cdot (d_1 \cos x + d_2 \sin x)$:

$$x + e^x + \sin x = P'' + P = a_0 + a_1 x + 2be^x + 2(-d_1 \sin x + d_2 \cos x) \iff a_0 = 0,\ a_1 = 1,\ b = \frac{1}{2},\ d_1 = \frac{-1}{2},\ d_2 = 0.$$

The general solution to $y'' + y = x + e^x + \sin x$ is $y = x + (1/2)\,e^x + (c_1 - x/2)\cos x + c_2 \sin x$.

29.   The initial-value problem $y' = e^x/y$ and $y(0) = 2$ has exact solution $y = \sqrt{2(e^x + 1)}$. Euler's method on the interval $[0, 3]$ with $h = .5$ yields

x	0	.5	1.	1.5	2.	2.5	3.
y_E	2	2.25	2.61638	3.13586	3.85045	4.80996	6.07635
y	2	2.30162	2.72701	3.31110	4.09611	5.13470	6.49390

31.   The initial-value problem $y' = y/\sqrt{1 + x^2}$ and $y(0) = 1$ has exact solution $y = x + \sqrt{1 + x^2}$. Euler's method on the interval $[0, 3]$ with $h = .5$ yields

x	0	.5	1.	1.5	2.	2.5	3.
y_E	1	1.5	2.1708	2.9383	3.7533	4.5925	5.4453
y	1	1.6180	2.4142	3.3028	4.2361	5.1926	6.1623

33.   The initial-value problem $y' = y - xy^3$ and $y(0) = 1$ has exact solution $y = \sqrt{\dfrac{2}{3e^{-2x} + 2x - 1}}$. Euler's method on the interval $[0, 3]$ with $h = .5$ yields

x	0	.5	1.	1.5	2.	2.5	3.
y_E	1	1.5	1.40625	.71892	.79970	.68813	.62489
y	1	1.34618	1.19267	.96463	.80912	.70533	.63199

35. a.   Euler's method on $[0, 1]$ with $h = .2$ for the initial-value problem $y' = y^3$ and $y(0) = 1$ yields

x	0	.2	.4	.6	.8	1.
y_E	1	1.2	1.54560	2.28405	4.66718	24.9998

b. $y' = y^3$ and $y(0) = 1 \iff \dfrac{1 - y^{-2}}{2} = \displaystyle\int_1^y y^{-3}\, dy = \int_0^x dx = x \iff y^{-2} = 1 - 2x \implies y = (1 - 2x)^{-1/2}$. ∎

c. $(1 - 2x)^{-1/2}$ is defined if and only if $x < 1/2$; hence the solution to the initial value problem is undefined if $x \geq 1/2$.

37.   Equation (16.6.5) yields $x(t) = -10 \cos\sqrt{\dfrac{g}{60}}\, t + 5\sqrt{\dfrac{60}{g}} \sin\sqrt{\dfrac{g}{60}}\, t$ because the mass is $m = 10$, the spring constant is $k = 10g/60 = g/6$, initial position is $x_0 = -10$ and initial velocity is $v_0 = 5$. The period is $2\pi/\sqrt{g/60} = 4\pi\sqrt{15/g}$, the frequency is $\sqrt{g/60}/(2\pi) = \sqrt{15/g}/(4\pi)$, and the amplitude is

$$\sqrt{(-10)^2 + 5^2 \cdot \frac{60}{g}} = 10\sqrt{1 + \frac{15}{g}} \approx 10\sqrt{1 + \frac{15}{981}} \approx 10.0762 \text{ cm}.$$

39.   The mass would not oscillate if $\mu^2 \geq 4mk = 4 \cdot 10 \cdot g/6 = 20g/3 \approx 20 \cdot 981/3 = 6540$, i.e., if $\mu \geq \sqrt{20g/3} \approx 80.8702$.

A 1
Review of Trigonometry

Solutions A1.1 *Angles and radian measure* (page A3)

1. Use formula (3): $150° = \left(\dfrac{\pi \text{ radians}}{180°}\right) \cdot 150° = \left(\dfrac{150}{180}\right) \pi \text{ radians} = \dfrac{5\pi}{6} \text{ radians}.$

3. $300° = \left(\dfrac{\pi \text{ radians}}{180°}\right) \cdot 300° = \left(\dfrac{300}{180}\right) \pi \text{ radians} = \dfrac{5\pi}{3} \text{ radians}.$

5. $144° = \left(\dfrac{\pi \text{ radians}}{180°}\right) \cdot 144° = \left(\dfrac{144}{180}\right) \pi \text{ radians} = \dfrac{4\pi}{5} \text{ radians}.$

7. Use formula (2): $\dfrac{\pi}{12} \text{ radians} = \left(\dfrac{180°}{\pi \text{ radians}}\right) \cdot (\dfrac{\pi}{12}) \text{ radians} = \left(\dfrac{180}{12}\right)^{\circ} = 15°.$

9. $\dfrac{\pi}{8} \text{ radians} = \left(\dfrac{180°}{\pi \text{ radians}}\right) \cdot (\dfrac{\pi}{8}) \text{ radians} = \left(\dfrac{180}{8}\right)^{\circ} = 22.5°.$

11. $-\dfrac{\pi}{3} \text{ radians} = \left(\dfrac{180°}{\pi \text{ radians}}\right)\left(\dfrac{-\pi}{3}\right) \text{ radians} = -\left(\dfrac{180}{3}\right)^{\circ} = -60°.$

13. A circle of radius r has circumference $2\pi r$. An arc of length π is one-quarter of the circumference of a circle of radius 2; it corresponds to an angle which is one-quarter of that for a full circle, i.e., to $\dfrac{360°}{4} = 90°.$

Solutions A1.2 *The trigonometric functions and basic identities* (pages A6-7)

1. $\sin\theta$ has period 2π (see formula (4)), thus
$$\sin 6\pi = \sin(4\pi + 2\pi) = \sin 4\pi = \sin(2\pi + 2\pi) = \sin 2\pi = \sin(0 + 2\pi) = \sin 0 = 0.$$
Similarly, $\cos\theta$ has period 2π (formula (3)) and $\cos 6\pi = \cos 4\pi = \cos 2\pi = \cos 0 = 1.$

REMARK: The identity stated in formula (3) follows from two uses of item (iv) in Table 2; similarly the identity stated in formula (4) follows from two uses of item (v).

3. Use formula (8) and Table 1 to compute $\sin(\dfrac{7\pi}{6}) = \sin\left(\dfrac{\pi}{6} + \pi\right) = -\sin\dfrac{\pi}{6} = -\dfrac{1}{2}.$ Similarly, using formula (7) and Table 1, we find $\cos(\dfrac{7\pi}{6}) = \cos\left(\dfrac{\pi}{6} + \pi\right) = -\cos\dfrac{\pi}{6} = -\dfrac{\sqrt{3}}{2}.$

5. $75° = \dfrac{75}{180}\pi = \dfrac{5\pi}{12} = \left(\dfrac{\pi}{6} + \dfrac{\pi}{4}\right)$ radians. We use Theorem 1 and Table 1 to compute
$$\cos\dfrac{5\pi}{12} = \cos\left(\dfrac{\pi}{6} + \dfrac{\pi}{4}\right) = \left(\cos\dfrac{\pi}{6}\right)\left(\cos\dfrac{\pi}{4}\right) - \left(\sin\dfrac{\pi}{6}\right)\left(\sin\dfrac{\pi}{4}\right) = \left(\dfrac{\sqrt{3}}{2}\right)\left(\dfrac{1}{\sqrt{2}}\right) - \left(\dfrac{1}{2}\right)\left(\dfrac{1}{\sqrt{2}}\right) = \dfrac{\sqrt{3}-1}{2\sqrt{2}}.$$
Similar use of item (xiii) in Table 2 produces
$$\sin\dfrac{5\pi}{12} = \sin\left(\dfrac{\pi}{6} + \dfrac{\pi}{4}\right) = \left(\sin\dfrac{\pi}{6}\right)\left(\cos\dfrac{\pi}{4}\right) + \left(\cos\dfrac{\pi}{6}\right)\left(\sin\dfrac{\pi}{4}\right) = \left(\dfrac{1}{2}\right)\left(\dfrac{1}{\sqrt{2}}\right) + \left(\dfrac{\sqrt{3}}{2}\right)\left(\dfrac{1}{\sqrt{2}}\right) = \dfrac{1+\sqrt{3}}{2\sqrt{2}}.$$

7. $\dfrac{13}{12} = \dfrac{3}{4} + \dfrac{1}{3}$, therefore $\dfrac{13\pi}{12} = \dfrac{3\pi}{4} + \dfrac{\pi}{3}$; use item (xii) and (xiii) of Table 2:
$$\cos\dfrac{13\pi}{12} = \left(\cos\dfrac{3\pi}{4}\right)\left(\cos\dfrac{\pi}{3}\right) - \left(\sin\dfrac{3\pi}{4}\right)\left(\sin\dfrac{\pi}{3}\right) = (\dfrac{-1}{\sqrt{2}})\left(\dfrac{1}{2}\right) - (\dfrac{1}{\sqrt{2}})(\dfrac{\sqrt{3}}{2}) = \dfrac{-1-\sqrt{3}}{2\sqrt{2}},$$
$$\sin\dfrac{13\pi}{12} = \left(\sin\dfrac{3\pi}{4}\right)\left(\cos\dfrac{\pi}{3}\right) + \left(\cos\dfrac{3\pi}{4}\right)\left(\sin\dfrac{\pi}{3}\right) = (\dfrac{1}{\sqrt{2}})\left(\dfrac{1}{2}\right) + (\dfrac{-1}{\sqrt{2}})(\dfrac{\sqrt{3}}{2}) = \dfrac{1-\sqrt{3}}{2\sqrt{2}}.$$

A1.2 — The trigonometric functions and basic identities

REMARK: Use items (iv), (xviii), (v), (xix) of Table 2; also note that the plus sign is used in (xviii) if $|x/2| \le \pi/2$ and the plus sign is used in (xix) if $0 \le x/2 \le \pi$.

$$\cos\frac{13\pi}{12} = \cos\left(\frac{\pi}{12} + \pi\right) = -\cos\frac{\pi}{12} = -\cos\frac{(\pi/6)}{2} = -\sqrt{\frac{1-\cos(\pi/6)}{2}} = -\sqrt{\frac{1-\sqrt{3}/2}{2}} = -\frac{\sqrt{2-\sqrt{3}}}{2}$$

$$\sin\frac{13\pi}{12} = \sin\left(\frac{\pi}{12} + \pi\right) = -\sin\frac{\pi}{12} = -\sin\frac{(\pi/6)}{2} = -\sqrt{\frac{1+\cos(\pi/6)}{2}} = -\sqrt{\frac{1+\sqrt{3}/2}{2}} = -\frac{\sqrt{2+\sqrt{3}}}{2}$$

9. $\frac{-1}{12} = \frac{1}{4} - \frac{1}{3}$, therefore $\frac{-\pi}{12} = \frac{\pi}{4} - \frac{\pi}{3}$; use item (xiv) and (xv) of Table 2:

$$\cos\frac{-\pi}{12} = \left(\cos\frac{\pi}{4}\right)\left(\cos\frac{\pi}{3}\right) + \left(\sin\frac{\pi}{4}\right)\left(\sin\frac{\pi}{3}\right) = \left(\frac{1}{\sqrt{2}}\right)\left(\frac{1}{2}\right) + \left(\frac{1}{\sqrt{2}}\right)\left(\frac{\sqrt{3}}{2}\right) = \frac{1+\sqrt{3}}{2\sqrt{2}},$$

$$\sin\frac{-\pi}{12} = \left(\sin\frac{\pi}{4}\right)\left(\cos\frac{\pi}{3}\right) - \left(\cos\frac{\pi}{4}\right)\left(\sin\frac{\pi}{3}\right) = \left(\frac{1}{\sqrt{2}}\right)\left(\frac{1}{2}\right) - \left(\frac{1}{\sqrt{2}}\right)\left(\frac{\sqrt{3}}{2}\right) = \frac{1-\sqrt{3}}{2\sqrt{2}}.$$

REMARK: We could use $\cos(-\pi/12) = \cos(\pi/12)$, $\sin(-\pi/12) = -\sin(\pi/12)$, and the preceding Remark.

11. Item (xviii) of Table 2 implies $\cos\dfrac{\pi}{8} = \sqrt{\dfrac{1+\cos(\pi/4)}{2}} = \sqrt{\dfrac{1+\sqrt{2}/2}{2}} = \dfrac{\sqrt{2+\sqrt{2}}}{2}$. Therefore

$$\cos\frac{\pi}{16} = \sqrt{\frac{1+\cos(\pi/8)}{2}} = \sqrt{\frac{1+\sqrt{2+\sqrt{2}}\big/2}{2}} = \frac{\sqrt{2+\sqrt{2+\sqrt{2}}}}{2}, \qquad \text{(xviii)}$$

$$\sin\frac{\pi}{16} = \sqrt{\frac{1-\cos(\pi/8)}{2}} = \sqrt{\frac{1-\sqrt{2+\sqrt{2}}\big/2}{2}} = \frac{\sqrt{2-\sqrt{2+\sqrt{2}}}}{2}. \qquad \text{(xix)}$$

13. $67\frac{1}{2}^{\circ} = \left(\dfrac{67.5}{180}\right)\pi = \dfrac{3}{8}\pi$ radians. Therefore

$$\cos\frac{3\pi}{8} = \cos\frac{(3\pi/4)}{2} = \sqrt{\frac{1+\cos(3\pi/4)}{2}} = \sqrt{\frac{1-\cos(\pi/4)}{2}} = \sqrt{\frac{1-\sqrt{2}/2}{2}} = \frac{\sqrt{2-\sqrt{2}}}{2} \qquad \text{(xviii, iv)}$$

$$\sin\frac{3\pi}{8} = \sin\frac{(3\pi/4)}{2} = \sqrt{\frac{1-\cos(3\pi/4)}{2}} = \sqrt{\frac{1+\cos(\pi/4)}{2}} = \sqrt{\frac{1+\sqrt{2}/2}{2}} = \frac{\sqrt{2+\sqrt{2}}}{2} \qquad \text{(xix, iv)}$$

REMARK: If you've worked Problem 10, you could use items (x) and (xi) together with the observation that $67.5^{\circ} = \dfrac{3\pi}{8} = \dfrac{\pi}{2} - \dfrac{\pi}{8}$ radians. Or, you might use items (xii) and (xiii) together with $\dfrac{3\pi}{8} = \dfrac{\pi}{4} + \dfrac{\pi}{8}$.

15. $-7\frac{1}{2}^{\circ} = \left(\dfrac{-7.5}{180}\right)\pi = \dfrac{-\pi}{24}$ radians. Therefore, using the answers to Problem 9, we compute

$$\cos\left(\frac{-\pi}{24}\right) = \cos\frac{(-\pi/12)}{2} = \sqrt{\frac{1+\cos(-\pi/12)}{2}} = \sqrt{\frac{1+(1+\sqrt{3})\big/(2\sqrt{2})}{2}} = \frac{1}{2}\sqrt{\frac{2\sqrt{2}+(1+\sqrt{3})}{2}} \qquad \text{(xviii)}$$

$$\sin\left(\frac{-\pi}{24}\right) = \sin\frac{(-\pi/12)}{2} = \sqrt{\frac{1-\cos(-\pi/12)}{2}} = \sqrt{\frac{1-(1+\sqrt{3})\big/(2\sqrt{2})}{2}} = \frac{1}{2}\sqrt{\frac{2\sqrt{2}-(1+\sqrt{3})}{2}} \qquad \text{(xix)}$$

REMARK: The alternate form for the answers to Problem 9 leads to

$$\cos\left(\frac{-\pi}{24}\right) = \frac{\sqrt{2+\sqrt{2+\sqrt{3}}}}{2} \qquad \text{and} \qquad \sin\left(\frac{-\pi}{24}\right) = -\frac{\sqrt{2-\sqrt{2+\sqrt{3}}}}{2}.$$

As another alternative, observe that $-7.5^{\circ} = 60^{\circ} - 67.5^{\circ}$ and use the answers to Problem 13:

$$\cos\left(\frac{-\pi}{24}\right) = \frac{\sqrt{2-\sqrt{2}}+\sqrt{3}\sqrt{2+\sqrt{2}}}{4} \qquad \text{and} \qquad \sin\left(\frac{-\pi}{24}\right) = \frac{\sqrt{3}\sqrt{2-\sqrt{2}}-\sqrt{2+\sqrt{2}}}{4}.$$

17.   Theorem 1 implies $\cos\left(\dfrac{\pi}{2} + x\right) = \left(\cos\dfrac{\pi}{2}\right)(\cos x) - \left(\sin\dfrac{\pi}{2}\right)(\sin x) = 0 \cdot \cos x - 1 \cdot \sin x = -\sin x.$ ∎

19.
$$\sin(x+y) = -\cos\left[\frac{\pi}{2} + (x+y)\right] = -\cos\left[\left(\frac{\pi}{2} + x\right) + y\right]$$
$$= -\left[\cos(\frac{\pi}{2} + x)\cos y - \sin(\frac{\pi}{2} + x)\sin y\right]$$
$$= -\left[(-\sin x)\cos y - \cos x \sin y\right] = \sin x \cos y + \cos x \sin y.$$ ∎

21. vi. $\cos(\pi - x) = \cos\left(\pi + (-x)\right) = \left(\cos\pi\right)\left(\cos(-x)\right) - \left(\sin\pi\right)\left(\sin(-x)\right) = (-1)(\cos x) - (0)(-\sin x) = -\cos x;$ Theorem 1 was used first, followed by formulas (5) and (6). ∎

    vii. $\sin(\pi - x) = \left(\sin\pi\right)\left(\cos(-x)\right) + \left(\cos\pi\right)\left(\sin(-x)\right) = (0)(\cos x) + (-1)(-\sin x) = \sin x;$ Problem 19 was used first, then formulas (5) and (6). ∎

    x. $\cos\left(\dfrac{\pi}{2} - x\right) = \left(\cos\dfrac{\pi}{2}\right)(\cos x) + \left(\sin\dfrac{\pi}{2}\right)(\sin x) = (0)(\cos x) + (1)(\sin x) = \sin x,$ using item (xiv). ∎

    xi. $\sin\left(\dfrac{\pi}{2} - x\right) = \left(\sin\dfrac{\pi}{2}\right)(\cos x) - \left(\cos\dfrac{\pi}{2}\right)(\sin x) = (1)(\cos x) - (0)(\sin x) = \cos x,$ using item (xv). ∎

REMARK: Items (x) and (xi) say the points $(\cos\theta, \sin\theta)$ and $\left(\cos\left(\dfrac{\pi}{2} - \theta\right), \sin\left(\dfrac{\pi}{2} - \theta\right)\right)$ are symmetric with respect to the line $y = x$. This is geometrically reasonable since the first point is reached by moving counterclockwise a distance θ from $(1,0)$ and the second is reached by moving clockwise a distance θ from $(0,1)$.

23.   Use Theorem 1 followed by formula (2):
$$\cos x = \cos\left(\frac{x}{2} + \frac{x}{2}\right) = \left(\cos\frac{x}{2}\right)\left(\cos\frac{x}{2}\right) - \left(\sin\frac{x}{2}\right)\left(\sin\frac{x}{2}\right)$$
$$= \left(\cos\frac{x}{2}\right)^2 - \left(\sin\frac{x}{2}\right)^2 = \left(\cos\frac{x}{2}\right)^2 - \left(1 - \left(\cos\frac{x}{2}\right)^2\right) = 2\left(\cos\frac{x}{2}\right)^2 - 1,$$

Therefore $1 + \cos x = 2\left(\cos\dfrac{x}{2}\right)^2$ which implies $\left(\cos\dfrac{x}{2}\right)^2 = \dfrac{1 + \cos x}{2}$ and $\cos\dfrac{x}{2} = \pm\sqrt{\dfrac{1 + \cos x}{2}}.$ ∎

25. xx.   Begin with item (xvi): $\cos 2x = 2\cos^2 x - 1 \iff 2\cos^2 x = 1 + \cos 2x \iff \cos^2 x = \dfrac{1 + \cos 2x}{2}.$ ∎

    xxi.   Again start with (xvi): $\cos 2x = 1 - 2\sin^2 x \iff 2\sin^2 x = 1 - \cos 2x \iff \sin^2 x = \dfrac{1 - \cos 2x}{2}.$ ∎

27.   Let $u = \dfrac{x+y}{2}$ and $v = \dfrac{x-y}{2}$. Then $u + v = x$, $u - v = y$, and
$$\sin x = \sin(u + v) = \sin u \cos v + \cos u \sin v,$$
$$\sin y = \sin(u - v) = \sin u \cos v - \cos u \sin v,$$
$$\sin x - \sin y = 2\cos u \sin v = 2\sin v \cos u = 2\sin\left(\frac{x-y}{2}\right)\cos\left(\frac{x+y}{2}\right).$$ ∎

REMARK: Items (xv), (xii), (xvii) yield
$$\sin(A - B)\cos(A + B) = (\sin A \cos B - \cos A \sin B)(\cos A \cos B - \sin A \sin B)$$
$$= \sin A \cos A \cos^2 B - \sin^2 A \sin B \cos B - \cos^2 A \sin B \cos B + \sin A \cos A \sin^2 B$$
$$= (\sin A \cos A)\left(\cos^2 B + \sin^2 B\right) - \left(\sin^2 A + \cos^2 A\right)(\sin B \cos B)$$
$$= \sin A \cos A - \sin B \cos B = \frac{\sin 2A - \sin 2B}{2}.$$

    Finish the proof by letting $A = x/2$ and $B = y/2$. ∎

29.   The periodic function $-2\cos x$ has minimum value -2 at $x = 0$ and has maximum value 2 at $x = \pi$; its amplitude is half the distance between those extremes, i.e., $\dfrac{2 - (-2)}{2} = 2.$

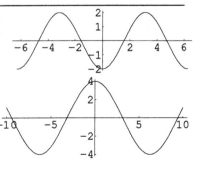

31.   Let $f(x) = 4\cos(x/2)$. Its minimum is $f(2\pi) = 4\cos\pi = -4$ and its maximum is $f(0) = 4\cos 0 = 4$, hence its amplitude is $\dfrac{4 - (-4)}{2} = 4$. We observe that $f(x + 4\pi) = 4\cos\left(\dfrac{x + 4\pi}{2}\right) = 4\cos\left(\dfrac{x}{2} + 2\pi\right) = 4\cos\left(\dfrac{x}{2}\right) = f(x)$ for all x; hence f is a periodic function.

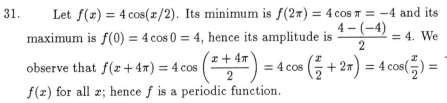

A1.2 — The trigonometric functions and basic identities

Since this argument is easily adapted to show f is periodic by proving $f(x + 8\pi) = f(x)$ for all x, it would be a mistake to jump to the conclusion that we have already shown the period of f is 4π; the following argument does that job.

Suppose α is positive and has the property that $f(x + \alpha) = f(x)$ for all x. Then, letting $x = 0$, we find α must satisfy $f(\alpha) = f(0)$, i.e., $4\cos\dfrac{\alpha}{2} = 4\cos\dfrac{0}{2} = 4\cos 0 = 4$. This implies $\cos\dfrac{\alpha}{2} = 1$, hence $\dfrac{\alpha}{2}$ is a multiple of 2π and α must be a multiple of 4π. Combining this with the earlier discussion, we see that $\alpha = 4\pi$ is the smallest positive number such that $f(x + \alpha) = f(x)$ for all x, i.e., f has period 4π. ∎

33.   Let $g(x) = \sin(x - 1)$ Because $\sin(u + 2\pi) = \sin u$ for all u, we find
$g(x + 2\pi) = \sin\big((x + 2\pi) - 1\big) = \sin\big((x - 1) + 2\pi\big) = \sin(x - 1) = g(x)$.
In fact, g has period 2π: $g(x + \alpha) = g(x)$ for all x implies $\sin(x + \alpha) = \sin(x + 1 + \alpha - 1) = g(x + 1 + \alpha) = g(x + 1) = \sin(x + 1 - 1) = \sin x$ for all x and this, in turn, implies α is a multiple of the known period, 2π, of the sine function.

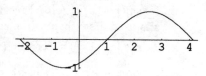

35.
Let $f(x) = 3\cos(3x - \dfrac{1}{2})$. Suppose α is positive and has the property
that $f(x + \alpha) = f(x)$ for all x. Then, letting $x = \dfrac{1}{6}$, we find

$$3\cos(3\alpha) = 3\cos\left(3\left(\frac{1}{6} + \alpha\right) - \frac{1}{2}\right) = 3\cos\left(\frac{3}{6} - \frac{1}{2}\right) = 3\cos 0 = 3.$$

Therefore $\cos(3\alpha) = 1$ and 3α must be a multiple of 2π. Because

$$f\left(x + \frac{2\pi}{3}\right) = 3\cos\left[3\left(x + \frac{2\pi}{3}\right) - \frac{1}{2}\right]$$
$$= 3\cos\left[\left(3x - \frac{1}{2}\right) + 2\pi\right] = 3\cos\left[3x - \frac{1}{2}\right] = f(x)$$

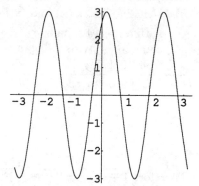

for all x, we see that $\alpha = 2\pi/3$ is the smallest positive choice of α such that $f(x + \alpha) = f(x)$ for all x. Hence $2\pi/3$ is the period of f. The minimum value of f is $-3 = f(1/6 - \pi/3)$ and the maximum value is $3 = f(1/6)$, therefore the amplitude of f is $\big(3 - (-3)\big)/2 = 3$.

Solutions A1.3   *Other trigonometric functions*   (page A9)

1. $\cos 6\pi = 1$ and $\sin 6\pi = 0$. Therefore $\tan 6\pi = \dfrac{\sin 6\pi}{\cos 6\pi} = \dfrac{0}{1} = 0$ and $\sec 6\pi = \dfrac{1}{\cos 6\pi} = \dfrac{1}{1} = 1$, but $\cot x = \dfrac{\cos x}{\sin x}$ and $\csc x = \dfrac{1}{\sin x}$ are undefined at $x = 6\pi$ because $\sin 6\pi = 0$.

3.
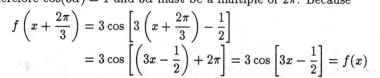

$$\sin\frac{7\pi}{6} = -\sin\left(\frac{7\pi}{6} - \pi\right) = -\sin\frac{\pi}{6} = -\frac{1}{2} \qquad \cos\frac{7\pi}{6} = -\cos\left(\frac{7\pi}{6} - \pi\right) = -\cos\frac{\pi}{6} = -\frac{\sqrt{3}}{2}$$

$$\tan\frac{7\pi}{6} = \frac{\sin(7\pi/6)}{\cos(7\pi/6)} = \frac{-1/2}{-\sqrt{3}/2} = \frac{1}{\sqrt{3}} \qquad \cot\frac{7\pi}{6} = \cot\frac{\pi}{6} = \frac{\sqrt{3}/2}{1/2} = \sqrt{3}$$

$$\csc\frac{7\pi}{6} = \frac{1}{\sin(7\pi/6)} = \frac{1}{-1/2} = -2 \qquad \sec\frac{7\pi}{6} = \frac{1}{\cos(7\pi/6)} = \frac{1}{-\sqrt{3}/2} = \frac{-2}{\sqrt{3}}$$

5.   Solution A1.2.5 supplies the entries for $\sin 75°$ and $\cos 75°$ shown in the following table.

$$\sin 75° = \frac{\sqrt{3}+1}{2\sqrt{2}} \qquad\qquad\qquad \cos 75° = \frac{\sqrt{3}-1}{2\sqrt{2}}$$

$$\tan 75° = \frac{(\sqrt{3}+1)/(2\sqrt{2})}{(\sqrt{3}-1)/(2\sqrt{2})} = \frac{\sqrt{3}+1}{\sqrt{3}-1} = 2+\sqrt{3} \qquad \cot 75° = \frac{1}{\tan 75°} = \frac{\sqrt{3}-1}{\sqrt{3}+1} = \frac{(\sqrt{3}-1)^2}{(\sqrt{3}+1)(\sqrt{3}-1)} = 2-\sqrt{3}$$

$$\csc 75° = \frac{1}{\sin 75°} = \frac{2\sqrt{2}}{\sqrt{3}+1} = \sqrt{2}(\sqrt{3}-1) \qquad \sec 75° = \frac{1}{\cos 75°} = \frac{2\sqrt{2}}{\sqrt{3}-1} = \sqrt{2}(\sqrt{3}+1)$$

7. Solution A1.2.7 supplies the entries for $\sin\dfrac{13\pi}{12}$ and $\cos\dfrac{13\pi}{12}$ shown in the following table.

$$\sin\frac{13\pi}{12} = \frac{1-\sqrt{3}}{2\sqrt{2}}$$

$$\cos\frac{13\pi}{12} = \frac{-1-\sqrt{3}}{2\sqrt{2}}$$

$$\tan\frac{13\pi}{12} = \frac{(1-\sqrt{3})/(2\sqrt{2})}{(-1-\sqrt{3})/(2\sqrt{2})} = \frac{1-\sqrt{3}}{-1-\sqrt{3}} = 2-\sqrt{3}$$

$$\cot\frac{13\pi}{12} = \frac{1}{\tan(13\pi/12)} = \frac{-1-\sqrt{3}}{1-\sqrt{3}} = 2+\sqrt{3}$$

$$\csc\frac{13\pi}{12} = \frac{1}{\sin(13\pi/12)} = \frac{2\sqrt{2}}{1-\sqrt{3}} = \sqrt{2}(-1-\sqrt{3})$$

$$\sec\frac{13\pi}{12} = \frac{2\sqrt{2}}{-1-\sqrt{3}} = \sqrt{2}(1-\sqrt{3})$$

9. Solution A1.2.9 supplies the entries for $\sin\dfrac{-\pi}{12}$ and $\cos\dfrac{-\pi}{12}$ shown in the following table.

$$\sin\frac{-\pi}{12} = \frac{1-\sqrt{3}}{2\sqrt{2}}$$

$$\cos\frac{-\pi}{12} = \frac{1+\sqrt{3}}{2\sqrt{2}}$$

$$\tan\frac{-\pi}{12} = \frac{(1-\sqrt{3})/(2\sqrt{2})}{(1+\sqrt{3})/(2\sqrt{2})} = \frac{1-\sqrt{3}}{1+\sqrt{3}} = -2+\sqrt{3}$$

$$\cot\frac{-\pi}{12} = \frac{1}{\tan(-\pi/12)} = \frac{1+\sqrt{3}}{1-\sqrt{3}} = -2-\sqrt{3}$$

$$\csc\frac{-\pi}{12} = \frac{1}{\sin(-\pi/12)} = \frac{2\sqrt{2}}{1-\sqrt{3}} = -\sqrt{2}(1+\sqrt{3})$$

$$\sec\frac{-\pi}{12} = \frac{1}{\cos(-\pi/12)} = \frac{2\sqrt{2}}{1+\sqrt{3}} = -\sqrt{2}(1-\sqrt{3})$$

11. Solution A1.2.11 supplies the entries for $\sin\dfrac{\pi}{16}$ and $\cos\dfrac{\pi}{16}$ shown in the following table.

$$\sin\frac{\pi}{16} = \frac{\sqrt{2-\sqrt{2+\sqrt{2}}}}{2}$$

$$\cos\frac{\pi}{16} = \frac{\sqrt{2+\sqrt{2+\sqrt{2}}}}{2}$$

$$\tan\frac{\pi}{16} = \frac{\sqrt{2-\sqrt{2+\sqrt{2}}}\Big/2}{\sqrt{2+\sqrt{2+\sqrt{2}}}\Big/2} = \frac{\sqrt{2-\sqrt{2+\sqrt{2}}}}{\sqrt{2+\sqrt{2+\sqrt{2}}}}$$

$$\cot\frac{\pi}{16} = \sqrt{\frac{2+\sqrt{2+\sqrt{2}}}{2-\sqrt{2+\sqrt{2}}}}$$

$$\csc\frac{\pi}{16} = \frac{1}{\sin(\pi/16)} = \frac{2}{\sqrt{2-\sqrt{2+\sqrt{2}}}}$$

$$\sec\frac{\pi}{16} = \frac{1}{\cos(\pi/16)} = \frac{2}{\sqrt{2+\sqrt{2+\sqrt{2}}}}$$

13. Solution A1.2.13 supplies the entries for $\sin 67.5°$ and $\cos 67.5°$ shown in the following table.

$$\sin 67.5° = \frac{\sqrt{2+\sqrt{2}}}{2}$$

$$\cos 67.5° = \frac{\sqrt{2-\sqrt{2}}}{2}$$

$$\tan 67.5° = \frac{\sqrt{2+\sqrt{2}}\Big/2}{\sqrt{2-\sqrt{2}}\Big/2} = \frac{\sqrt{2+\sqrt{2}}}{\sqrt{2-\sqrt{2}}} = \sqrt{2}+1$$

$$\cot 67.5° = \sqrt{\frac{2-\sqrt{2}}{2+\sqrt{2}}} = \sqrt{2}-1$$

$$\csc 67.5° = \frac{2}{\sqrt{2+\sqrt{2}}} = \sqrt{2}\,\sqrt{2-\sqrt{2}}$$

$$\sec 67.5° = \frac{2}{\sqrt{2-\sqrt{2}}} = \sqrt{2}\,\sqrt{2+\sqrt{2}}$$

15. Solution A1.2.15 supplies the entries for $\sin(-7.5°)$ and $\cos(-7.5°)$ shown in the following table.

$$\sin(-7.5°) = \frac{-\sqrt{2-\sqrt{2+\sqrt{3}}}}{2}$$

$$\cos(-7.5°) = \frac{\sqrt{2+\sqrt{2+\sqrt{3}}}}{2}$$

$$\tan(-7.5°) = \frac{-\sqrt{2-\sqrt{2+\sqrt{3}}}\Big/2}{\sqrt{2+\sqrt{2+\sqrt{3}}}\Big/2} = \frac{-\sqrt{2-\sqrt{2+\sqrt{3}}}}{\sqrt{2+\sqrt{2+\sqrt{3}}}}$$

$$\cot(-7.5°) = -\sqrt{\frac{2+\sqrt{2+\sqrt{3}}}{2-\sqrt{2+\sqrt{3}}}}$$

$$\csc(-7.5°) = \frac{2}{-\sqrt{2-\sqrt{2+\sqrt{3}}}} = \frac{-2}{\sqrt{2-\sqrt{2+\sqrt{3}}}}$$

$$\sec(-7.5°) = \frac{2}{\sqrt{2+\sqrt{2+\sqrt{3}}}}$$

17.

$$\cos \frac{x}{3} = \cos\left(\frac{1}{3} \cdot x\right) \text{ has period } \frac{2\pi}{1/3} = 6\pi,$$

therefore $\sec \dfrac{x}{3} = \dfrac{1}{\cos \frac{x}{3}}$ and $3 \sec \dfrac{x}{3}$ also have period 6π.

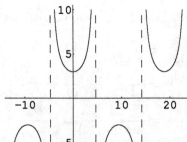

19.

$$\sin 6x \text{ has period } \frac{2\pi}{6} = \frac{\pi}{3},$$

therefore $\csc 6x = \dfrac{1}{\sin 6x}$ and $2 \csc 6x$ also have period $\dfrac{\pi}{3}$.

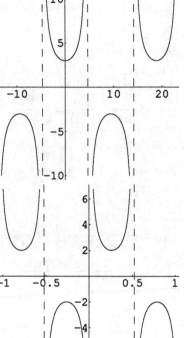

21.

$$\sin 5x \text{ and } \sin 5(x+1) \text{ both have period } \frac{2\pi}{5},$$

hence $\sec(5x+5) = \sec 5(x+1) = \dfrac{1}{\cos 5(x+1)}$ and $8 \sec(5x+5)$

also have period $\dfrac{2\pi}{5}$.

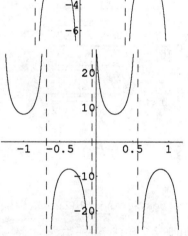

23.

$$\tan(x+y) = \frac{\sin(x+y)}{\cos(x+y)} = \frac{\sin x \cos y + \cos x \sin y}{\cos x \cos y - \sin x \sin y} = \frac{(\sin x \cos y + \cos x \sin y)/(\cos x \cos y)}{(\cos x \cos y - \sin x \sin y)/(\cos x \cos y)}$$

$$= \frac{\frac{\sin x}{\cos x} + \frac{\sin y}{\cos y}}{1 - \frac{\sin x}{\cos x} \cdot \frac{\sin y}{\cos y}} = \frac{\tan x + \tan y}{1 - \tan x \cdot \tan y}. \quad \blacksquare$$

Solutions A1.4   *Triangles*   (pages A10-1)

1.   Sketch a right triangle with hypotenuse of length $h = 11$ and side of length $a = 5$ adjacent to the angle of size θ; for this triangle we have $\cos\theta = \dfrac{5}{11}$. The side opposite θ has length op $= \sqrt{11^2 - 5^2} = 4\sqrt{6}$.

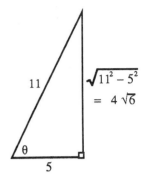

$0 \le \theta \le \dfrac{\pi}{2}$,	$\cos\theta$	$\sin\theta$	$\tan\theta$	$\cot\theta$	$\csc\theta$	$\sec\theta$
	$\dfrac{5}{11}$	$\dfrac{4\sqrt{6}}{11}$	$\dfrac{4\sqrt{6}}{5}$	$\dfrac{5}{4\sqrt{6}}$	$\dfrac{11}{4\sqrt{6}}$	$\dfrac{11}{5}$

3.   Sketch a right triangle with hypotenuse of length $h = 2$ and side of length $a = 1$ adjacent to the angle of size θ; for this triangle we have $\sec\theta = \dfrac{2}{1} = 2$ and $\cos\theta = \dfrac{1}{\sec\theta} = \dfrac{1}{2}$. The side opposite θ has length op $= \sqrt{2^2 - 1^2} = \sqrt{3}$; since θ finishes in the fourth quadrant, we have $0 > \sin\theta = -\sqrt{1 - \cos^2\theta} = -\sqrt{0.75}$.

$\dfrac{-\pi}{2} \le \theta \le 0$,	$\sec\theta$	$\cos\theta$	$\sin\theta$	$\tan\theta$	$\cot\theta$	$\csc\theta$
	2	$\dfrac{1}{2}$	$\dfrac{-\sqrt{3}}{2}$	$-\sqrt{3}$	$\dfrac{-1}{\sqrt{3}}$	$\dfrac{-2}{\sqrt{3}}$

5.   If $\csc\theta = 5$, then $\sin\theta = \dfrac{1}{\csc\theta} = \dfrac{1}{5}$. Since θ finishes in the second quadrant, we have $0 > \cos\theta = -\sqrt{1 - \sin^2\theta} = -\sqrt{0.96}$.

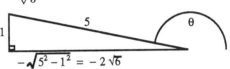

$\dfrac{\pi}{2} \le \theta \le \pi$,	$\csc\theta$	$\sin\theta$	$\cos\theta$	$\tan\theta$	$\cot\theta$	$\sec\theta$
	5	$\dfrac{1}{5}$	$\dfrac{-2\sqrt{6}}{5}$	$\dfrac{-1}{2\sqrt{6}}$	$-2\sqrt{6}$	$\dfrac{-5}{2\sqrt{6}}$

7.   If $\sin\theta = \dfrac{-2}{3}$ and θ finishes in the second quadrant, then $0 < \cos\theta = \sqrt{1 - \sin^2\theta} = \sqrt{5/9}$.

$\dfrac{-\pi}{2} \le \theta \le 0$,	$\sin\theta$	$\cos\theta$	$\tan\theta$	$\cot\theta$	$\sec\theta$	$\csc\theta$
	$\dfrac{-2}{3}$	$\dfrac{\sqrt{5}}{3}$	$\dfrac{-2}{\sqrt{5}}$	$\dfrac{-\sqrt{5}}{2}$	$\dfrac{3}{\sqrt{5}}$	$\dfrac{-3}{2}$

9.   If $\tan\theta = 10$ and θ finishes in the third quadrant, then

$$0 > \sec\theta = -\sqrt{1 + \tan^2\theta} = -\sqrt{101},$$
$$\cos\theta = \dfrac{1}{\sec\theta} = \dfrac{-1}{\sqrt{101}},$$
$$0 > \sin\theta = -\sqrt{1 - \cos^2\theta} = -\sqrt{\dfrac{100}{101}}.$$

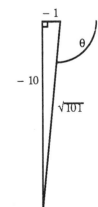

$-\pi \le \theta \le \dfrac{-\pi}{2}$,	$\tan\theta$	$\sec\theta$	$\cos\theta$	$\sin\theta$	$\csc\theta$	$\cot\theta$
	10	$-\sqrt{101}$	$\dfrac{-1}{\sqrt{101}}$	$\dfrac{-10}{\sqrt{101}}$	$\dfrac{-\sqrt{101}}{10}$	$\dfrac{1}{10}$

A1.4 — Triangles

11.   Place the triangle on a coordinate system with vertex C at the origin and vertex B on the positive x-axis at the point $(a, 0)$; let γ denote the radian measure of the angle at vertex C. The ray from C to the vertex A meets the unit circle at the point $(\cos\gamma, \sin\gamma)$. Because the distance from C to A is b, we infer A is at the point with coordinates $(b\cos\gamma, b\sin\gamma)$. The distance between vertices A and B is the distance between the points $(b\cos\gamma, b\sin\gamma)$ and $(a, 0)$, therefore (using $\cos^2\gamma + \sin^2\gamma = 1$)

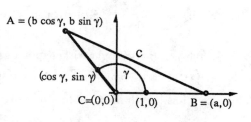

$$c = \sqrt{(b\cos\gamma - a)^2 + (b\sin\gamma - 0)^2} = \sqrt{b^2\cos^2\gamma - 2ab\cos\gamma + a^2 + b^2\sin^2\gamma} = \sqrt{b^2 - 2ab\cos\gamma + a^2}.$$

We conclude $c^2 = b^2 - 2ab\cos\gamma + a^2 = a^2 + b^2 - 2ab\cos\gamma$. ∎

13.   The Law of Sines will follow from computing the area of the triangle in three similar ways. Firstly, use the line segment $\overline{BC}$ as base. Drop a perpendicular, of length h, from vertex A to the line through B and C. If it meets the half-line from C through B, then $\dfrac{h}{b} = \sin\gamma$, otherwise $\dfrac{h}{b} = \sin(\pi - \gamma)$.

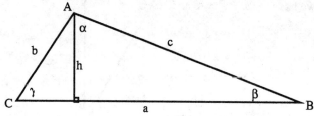 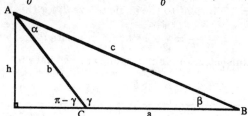

Because $\sin(\pi - \theta) = \sin\theta$ for all θ, we find $h = b\sin\gamma$ in either case and the area of the triangle is $\dfrac{1}{2}\cdot a\cdot h = \dfrac{1}{2}\cdot a\cdot b\sin\gamma$. Similarly, using the line segment $\overline{AC}$ as base, we compute the area to be $\dfrac{1}{2}\cdot b\cdot c\sin\alpha$; using the line segment $\overline{AB}$ as base, the area is seen to be $\dfrac{1}{2}\cdot c\cdot a\sin\beta$. These expressions for the triangle's area must be equal, i.e., $\dfrac{1}{2}\cdot a\cdot b\sin\gamma = \dfrac{1}{2}\cdot b\cdot c\sin\alpha = \dfrac{1}{2}\cdot c\cdot a\sin\beta$. Multiply by $\dfrac{2}{abc}$ to obtain $\dfrac{\sin\gamma}{c} = \dfrac{\sin\alpha}{a} = \dfrac{\sin\beta}{b}$ which is equivalent to $\dfrac{a}{\sin\alpha} = \dfrac{b}{\sin\beta} = \dfrac{c}{\sin\gamma}$. ∎

A 2
Mathematical Induction

1.   Let $P(n)$ represent the assertion: $1^3 + 2^3 + 3^3 + \cdots + (n-1)^3 + n^3 = \dfrac{n^2(n+1)^2}{4}$.

$P(1)$ is obviously true because $\dfrac{1^2(1+1)^2}{4} = \dfrac{1 \cdot 4}{4} = 1 = 1^3$. Suppose $P(k)$ were true.

$$
\begin{aligned}
1^3 + 2^3 + 3^3 + \cdots + (k-1)^3 + k^3 + (k+1)^3 &= (1^3 + 2^3 + 3^3 + \cdots + (k-1)^3 + k^3) + (k+1)^3 \\
&= \frac{k^2(k+1)^2}{4} + (k+1)^3 \qquad\qquad\qquad P(k) \\
&= (k+1)^2 \cdot \left(\frac{k^2}{4} + (k+1) \right) \\
&= (k+1)^2 \cdot \left(\frac{k^2 + 4k + 4}{4} \right) = (k+1)^2 \cdot \frac{(k+2)^2}{4} \\
&= \frac{(k+1)^2 \left((k+1)+1 \right)^2}{4}. \qquad\qquad\qquad P(k+1)
\end{aligned}
$$

$P(1)$ is true and $P(k)$ implies $P(k+1)$, therefore $P(n)$ is true for each positive integer n. ∎

REMARK: Combining this result with that of Example 2 proves that
$$
1^3 + 2^3 + 3^3 + \cdots + (n-1)^3 + n^3 = \left(1 + 2 + 3 + \cdots + (n-1) + n \right)^2.
$$

---

3.   Suppose $P_0(x) = 1$, $P_1(x) = x + a_0$, ..., $P_n(x) = x^n + a_{n-1}x^{n-1} + \cdots + a_1 a + a_0$. We are asked to prove that $\dfrac{d^n}{dx^n} P_n(x) = n!$. This problem is an instance where it is easier to prove a more general result. Let $P(n)$ represent the assertion: the nth derivative of $Q_n(x) = a_n x^n + a_{n-1}x^{n-1} + \cdots + a_1 x + a_0$ is equal to $n! \cdot a_n$.

$P(0)$.   If $n = 0$, then $\dfrac{d^n}{dx^n} f(x) = f(x)$ for any function f; in particular, $\dfrac{d^0}{dx^0} Q_0(x) = Q_0(x) = a_0 = 0! \cdot a_0$ because $0! = 1$.

$P(1)$. $\dfrac{d^1}{dx^1} Q_1(x) = \dfrac{d}{dx} (a_1 x + a_0) = a_1 = 1 \cdot a_1 = 1! \cdot a_1$.

$P(k+1)$.   Suppose the general result is true for $n = k$, i.e., $\dfrac{d^k}{dx^k} \left(b_k x^k + b_{k-1}x^{k-1} + \cdots + b_1 x + b_0 \right) = k! \cdot b_k$

for **any** polynomial of degree at most k. Let $R_k(x) = \dfrac{Q_{k+1}(x) - a_0}{x} = a_{k+1}x^k + a_k x^{k-1} + \cdots + a_2 x + a_1$.

$$
\begin{aligned}
\frac{d^{k+1}}{dx^{k+1}} Q_{k+1}(x) &= \frac{d^k}{dx^k} \left(\frac{d}{dx} Q_{k+1}(x) \right) = \frac{d^k}{dx^k} \left(\frac{d}{dx} \left(x \cdot R_k(x) + a_0 \right) \right) \\
&= \frac{d^k}{dx^k} \left(R_k(x) + x \cdot R_k'(x) \right) \\
&= \frac{d^k}{dx^k} R_k(x) + \frac{d^k}{dx^k} \left(x \cdot R_k'(x) \right) \\
&= \frac{d^k}{dx^k} \left(a_{k+1}x^k + \cdots \right) + \frac{d^k}{dx^k} \left(k a_{k+1}x^k + \cdots \right) \\
&= k! \cdot a_{k+1} + k! \cdot (k a_{k+1}) \qquad\qquad\qquad\qquad P(k) \\
&= k! \cdot (1 + k) \cdot a_{k+1} = (k+1)! \cdot a_{k+1}. \qquad\quad P(k+1)
\end{aligned}
$$

$P(0)$ and $P(1)$ are true and $P(k)$ implies $P(k+1)$, therefore $P(n)$ is true for each nonnegative integer n. ∎

A 2 — Mathematical Induction

REMARK: Problems 2.7.29-30 and Example 5 suggest an alternative way to organize this proof.

5.   Let $\mathcal{P}(n)$ represent the assertion: A set of n elements has exactly 2^n subsets.

$\mathcal{P}(0)$.   The only set with 0 elements is $\emptyset$, the empty set. $\emptyset$ is the only subset of $\emptyset$; hence $\mathcal{P}(0)$ is true since $1 = 2^0$.

$\mathcal{P}(k+1)$.   Suppose $\mathcal{P}(k)$ were true. Let A be an arbitrary set with $k+1$ elements. Select any one element x from A and let $B = A - \{x\}$, thus B is a set with k elements. The subsets of A can be classified into two non-overlapping collections: those excluding x and those including x. Those which exclude x are subsets of B, there are 2^k of these. Those which include x are of the form $\{x\} \cup C$ where C is a subset of B, there are 2^k of these. Therefore A has $2^k + 2^k = 2^{k+1}$ subsets.

$\mathcal{P}(0)$ is true and $\mathcal{P}(k)$ implies $\mathcal{P}(k+1)$, therefore $\mathcal{P}(n)$ is true for each nonnegative integer n. ∎

7.   Let $\mathcal{P}(n)$ represent the assertion: $\mathbf{u} \cdot (\mathbf{v}_1 + \mathbf{v}_2 + \cdots + \mathbf{v}_n) = \mathbf{u} \cdot \mathbf{v}_1 + \mathbf{u} \cdot \mathbf{v}_2 + \cdots + \mathbf{u} \cdot \mathbf{v}_n$.

$\mathcal{P}(2)$.   Theorem 11.2.1, parts (ii) and (i), implies $\mathcal{P}(2)$ is true.

$\mathcal{P}(k+1)$.   Suppose $\mathcal{P}(k)$ were true.

$$\begin{aligned}
\mathbf{u} \cdot (\mathbf{v}_1 + \mathbf{v}_2 + \cdots + \mathbf{v}_n + \mathbf{v}_{k+1}) &= \mathbf{u} \cdot \big((\mathbf{v}_1 + \mathbf{v}_2 + \cdots + \mathbf{v}_n) + \mathbf{v}_{k+1}\big) & \\
&= \mathbf{u} \cdot (\mathbf{v}_1 + \mathbf{v}_2 + \cdots + \mathbf{v}_n) + \mathbf{u} \cdot \mathbf{v}_{k+1} & \mathcal{P}(2) \\
&= (\mathbf{u} \cdot \mathbf{v}_1 + \mathbf{u} \cdot \mathbf{v}_2 + \cdots + \mathbf{u} \cdot \mathbf{v}_k) + \mathbf{u} \cdot \mathbf{v}_{k+1} & \mathcal{P}(k) \\
&= \mathbf{u} \cdot \mathbf{v}_1 + \mathbf{u} \cdot \mathbf{v}_2 + \cdots + \mathbf{u} \cdot \mathbf{v}_k + \mathbf{u} \cdot \mathbf{v}_{k+1}. & \mathcal{P}(k+1)
\end{aligned}$$

$\mathcal{P}(2)$ is true and $\mathcal{P}(k)$ implies $\mathcal{P}(k+1)$, therefore $\mathcal{P}(n)$ is true for each integer $n \geq 2$. ∎

A 3

The proofs of some theorems
on Limits, Continuity, and Differentiation

Solutions A 3   *Proofs of some limit theorems*   (page A23)

1.   Suppose c is a fixed real number and $f(x) = c$ for all real x. Use Definition 1.8.1 (page 124 of the text) and let $\delta = 1$ (independent of x_0 and ϵ). Then for every $\epsilon > 0$ and every x_0, we see that

$$0 < |x - x_0| < 1 \implies |f(x) - c| = |c - c| = 0 < \epsilon,$$

therefore $\lim_{x \to x_0} f(x) = c$ for every x_0. $\blacksquare$

3.   Suppose $g(x) = x$ for every real x. For each x_0 and for any ϵ, let $\delta = \epsilon$. Then

$$0 < |x - x_0| < \delta \implies |g(x) - x_0| = |x - x_0| < \epsilon,$$

therefore $\lim_{x \to x_0} x = \lim_{x \to x_0} g(x) = x_0$ for every x_0. $\blacksquare$

5.   The following proof is an adaptation of the text's proof of Theorem 1.

Suppose $\lim_{x \to \infty} f(x)$ exists and equals L. Then (see the text's Definition at the bottom of page 91) for every positive ϵ, there is a positive $N = N(\epsilon)$ such that $x > N$ implies $|f(x) - L| < \epsilon$.

$\boxed{c = 0}$ If $c = 0$, then $c \cdot f(x) = 0$ for all x and $\lim_{x \to \infty} c \cdot f(x) = 0$ (apply Problem 2). Since $0 \cdot L = 0$, this means $\lim_{x \to \infty} c \cdot f(x) = 0 = 0 \cdot \lim_{x \to \infty} f(x)$.

$\boxed{c \neq 0}$ If $c \neq 0$, then for each positive ϵ, let $M = N\left(\dfrac{\epsilon}{|c|}\right)$, i.e., $x > M \implies |f(x) - L| < \dfrac{\epsilon}{|c|}$. Therefore

$$x > M \implies |c \cdot f(x) - c \cdot L| = |c| \cdot |f(x) - L| < |c| \cdot \frac{\epsilon}{|c|} = \epsilon.$$

Therefore $\lim_{x \to \infty} c \cdot f(x) = c \cdot \lim_{x \to \infty} f(x)$. $\blacksquare$

7.   Let $f(x) = x + 3$ and $g(x) = 1 - x$. Then $\lim_{x \to \infty} f(x) = \infty$, $\lim_{x \to \infty} g(x) = -\infty$, $\lim_{x \to \infty} \big(f(x) + g(x)\big) = 4$, but $\left(\lim_{x \to \infty} f(x)\right) + \left(\lim_{x \to \infty} g(x)\right)$ is undefined.

9.   Let $f(x) = x^2$ and $g(x) = \dfrac{-5}{x^2}$. Then $\lim_{x \to \infty} f(x) = \infty$, $\lim_{x \to \infty} g(x) = 0$, $\lim_{x \to \infty} \big(f(x) \cdot g(x)\big) = -5$, but $\left(\lim_{x \to \infty} f(x)\right) \cdot \left(\lim_{x \to \infty} g(x)\right)$ is undefined.

11.   Let $f(x) = 2x$ and $g(x) = 3x^2$. Then $\lim_{x \to \infty} f(x) = \infty = \lim_{x \to \infty} g(x)$, $\lim_{x \to \infty} \dfrac{f(x)}{g(x)} = 0$, but $\dfrac{\lim_{x \to \infty} f(x)}{\lim_{x \to \infty} g(x)}$ is undefined.

A 4
Determinants

Solutions A 4   **Determinants**   (page A30)

1. The system $\begin{matrix} 2x_1 & + & 4x_2 & = & 6 \\ x_1 & + & x_2 & = & 3 \end{matrix}$ has determinant $\begin{vmatrix} 2 & 4 \\ 1 & 1 \end{vmatrix} = 2 \cdot 1 - 4 \cdot 1 = 2 - 4 = -2 \neq 0.$

$\begin{matrix} 2x_1 & + & 4x_2 & = & 6 \\ x_1 & + & x_2 & = & 3 \end{matrix} \iff \begin{matrix} x_1 & + & 2x_2 & = & 3 \\ x_1 & + & x_2 & = & 3 \end{matrix} \iff \begin{matrix} & & x_2 & = & 0 \\ x_1 & + & x_2 & = & 3 \end{matrix} \iff \begin{matrix} x_2 & = & 0 \\ x_1 & = & 3 \end{matrix}$

3. The system $\begin{matrix} 2x_1 & + & 4x_2 & = & 6 \\ x_1 & + & 2x_2 & = & 3 \end{matrix}$ has determinant $\begin{vmatrix} 2 & 4 \\ 1 & 2 \end{vmatrix} = 2 \cdot 2 - 4 \cdot 1 = 4 - 4 = 0.$

Observe that the first equation is simply 2 times the second equation; therefore, if x_2 is chosen arbitrarily, then $x_1 = 3 - 2x_2$, and there are an infinite number of solutions.

5. The system $\begin{matrix} 6x_1 & - & 3x_2 & = & 3 \\ -2x_1 & + & x_2 & = & 1 \end{matrix}$ has determinant $\begin{vmatrix} 6 & -3 \\ -2 & 1 \end{vmatrix} = 6 \cdot 1 - (-3) \cdot (-2) = 6 - 6 = 0.$

If we add 3 times the second equation to the first equation, the result is $0 = 6$ which is impossible; the system has no solutions.

7. The system $\begin{matrix} 2x_1 & + & 5x_2 & = & 0 \\ 3x_1 & - & 7x_2 & = & 0 \end{matrix}$ has determinant $\begin{vmatrix} 2 & 5 \\ 3 & -7 \end{vmatrix} = 2 \cdot (-7) - 5 \cdot 3 = -14 - 15 = -29 \neq 0.$

Because the determinant is not equal to zero, the obvious solution, $x_1 = 0 = x_2$, is the only solution.

9. $\begin{vmatrix} 1 & 2 & 3 \\ 6 & -1 & 4 \\ 2 & 0 & 6 \end{vmatrix} = 1 \cdot \begin{vmatrix} -1 & 4 \\ 0 & 6 \end{vmatrix} - 2 \cdot \begin{vmatrix} 6 & 4 \\ 2 & 6 \end{vmatrix} + 3 \cdot \begin{vmatrix} 6 & -1 \\ 2 & 0 \end{vmatrix} = 1 \cdot (-6 - 0) - 2 \cdot (36 - 8) + 3 \cdot (0 + 2) = -6 - 56 + 6 = -56.$

11. It is convenient to use Theorem 4 and to expand this determinant by its first column.

$\begin{vmatrix} 7 & 2 & 3 \\ 0 & 4 & 1 \\ 0 & 0 & 5 \end{vmatrix} = (-1)^{1+1} \cdot 7 \cdot \begin{vmatrix} 4 & 1 \\ 0 & 5 \end{vmatrix} + (-1)^{2+1} \cdot 0 \cdot \begin{vmatrix} 2 & 3 \\ 0 & 5 \end{vmatrix} + (-1)^{3+1} \cdot 0 \cdot \begin{vmatrix} 2 & 3 \\ 4 & 1 \end{vmatrix} = 1 \cdot 7 \cdot (20 - 0) + 0 + 0 = 140.$

13. $\begin{vmatrix} 4 & 2 & 7 \\ 1 & 5 & 3 \\ -1 & 1 & 4 \end{vmatrix} = 4 \cdot \begin{vmatrix} 5 & 3 \\ 1 & 4 \end{vmatrix} - 2 \cdot \begin{vmatrix} 1 & 3 \\ -1 & 4 \end{vmatrix} + 7 \cdot \begin{vmatrix} 1 & 5 \\ -1 & 1 \end{vmatrix} = 4 \cdot (20 - 3) - 2 \cdot (4 + 3) + 7 \cdot (1 + 5) = 68 - 14 + 42 = 96.$

15. Because of the many zeros, it is convenient to expand this determinant by its first column.

$\begin{vmatrix} 1 & 4 & 7 & 2 \\ 0 & 5 & 8 & 1 \\ 0 & 0 & -3 & 4 \\ 0 & 0 & 0 & 8 \end{vmatrix} = 1 \cdot \begin{vmatrix} 5 & 8 & 1 \\ 0 & -3 & 4 \\ 0 & 0 & 8 \end{vmatrix} - 0 + 0 - 0 = 1 \cdot 5 \cdot \begin{vmatrix} -3 & 4 \\ 0 & 8 \end{vmatrix} - 0 + 0 = 1 \cdot 5 \cdot (-3) \cdot 8 = -120.$

REMARK: All entries below the main diagonal are zero and the determinant is the product of the entries on that main diagonal.

17. Add 2 times the first row to the second row, then subtract the first row from the fourth row to get

$\begin{vmatrix} 2 & 1 & 3 & 4 \\ 3 & -2 & 5 & 1 \\ 4 & 0 & 4 & 5 \\ 2 & 1 & 7 & -4 \end{vmatrix} = \begin{vmatrix} 2 & 1 & 3 & 4 \\ 7 & 0 & 11 & 9 \\ 4 & 0 & 4 & 5 \\ 2 & 1 & 7 & -4 \end{vmatrix} = \begin{vmatrix} 2 & 1 & 3 & 4 \\ 7 & 0 & 11 & 9 \\ 4 & 0 & 4 & 5 \\ 0 & 0 & 4 & -8 \end{vmatrix}$

Now expand by the second column, then expand the resulting 3×3 determinant by its first column:

$\begin{vmatrix} 2 & 1 & 3 & 4 \\ 7 & 0 & 11 & 9 \\ 4 & 0 & 4 & 5 \\ 0 & 0 & 4 & -8 \end{vmatrix} = (-1)^{1+2} \cdot 1 \cdot \begin{vmatrix} 7 & 11 & 9 \\ 4 & 4 & 5 \\ 0 & 4 & -8 \end{vmatrix} = -1 \cdot \left(7 \cdot \begin{vmatrix} 4 & 5 \\ 4 & -8 \end{vmatrix} - 4 \cdot \begin{vmatrix} 11 & 9 \\ 4 & -8 \end{vmatrix} \right)$

$= -1 \cdot \left(7 \cdot (-32 - 20) - 4 \cdot (-88 - 36) \right) = -132.$

452

19. Since there is a 1 at the top of the third column, let's subtract that column twice from the first column, three times from the second column, and four times from the fourth column to obtain a determinant that expands simply in its first row.

$$\begin{vmatrix} 2 & 3 & 1 & 4 \\ 2 & 2 & 4 & 6 \\ 3 & -1 & -2 & 4 \\ 4 & 2 & -3 & -5 \end{vmatrix} = \begin{vmatrix} 0 & 0 & 1 & 0 \\ -6 & -10 & 4 & -10 \\ 7 & 5 & -2 & 12 \\ 10 & 11 & -3 & 7 \end{vmatrix} = 0 - 0 + 1 \cdot \begin{vmatrix} -6 & -10 & -10 \\ 7 & 5 & 12 \\ 10 & 11 & 7 \end{vmatrix} - 0$$

$$= -6 \cdot \begin{vmatrix} 5 & 12 \\ 11 & 7 \end{vmatrix} - (-10) \cdot \begin{vmatrix} 7 & 12 \\ 10 & 7 \end{vmatrix} + (-10) \cdot \begin{vmatrix} 7 & 5 \\ 10 & 11 \end{vmatrix}$$

$$= -6 \cdot (-97) + 10 \cdot (-71) + (-10) \cdot 27 = -398.$$

REMARK:

$$\begin{vmatrix} 2 & 3 & 1 & 4 \\ 2 & 2 & 4 & 6 \\ 3 & -1 & -2 & 4 \\ 4 & 2 & -3 & -5 \end{vmatrix} = 2 \cdot \begin{vmatrix} 2 & 3 & 1 & 4 \\ 1 & 1 & 2 & 3 \\ 3 & -1 & -2 & 4 \\ 4 & 2 & -3 & -5 \end{vmatrix} = 2 \cdot \begin{vmatrix} 0 & 1 & -3 & -2 \\ 1 & 2 & 2 & 3 \\ 0 & -4 & -8 & -5 \\ 0 & -2 & -11 & -17 \end{vmatrix} = 2 \cdot \begin{vmatrix} 0 & 1 & -3 & -2 \\ 1 & 0 & 5 & 5 \\ 0 & 0 & -20 & -13 \\ 0 & 0 & -17 & -21 \end{vmatrix}$$

$$= 2 \cdot (-1)^{2+1} \cdot (-1)^{1+1} \cdot \begin{vmatrix} -20 & -13 \\ -17 & -21 \end{vmatrix}.$$

21. The system $\begin{aligned} a_{11}x_1 &+ a_{12}x_2 &= b_1 \\ a_{21}x_1 &+ a_{22}x_2 &= b_2 \end{aligned}$ has determinant $D = \begin{vmatrix} a_{11} & a_{12} \\ a_{21} & a_{22} \end{vmatrix} = a_{11} \cdot a_{22} - a_{12} \cdot a_{21}$.

If we suppose a_{12} and a_{22} are both nonzero, then

$$D = 0 \iff a_{11} \cdot a_{22} - a_{12} \cdot a_{21} = 0 \iff -a_{12} \cdot a_{21} = -a_{11} \cdot a_{22} \iff \frac{-a_{21}}{a_{22}} = \frac{-a_{11}}{a_{12}};$$

$\dfrac{-a_{21}}{a_{22}}$ is the slope of the line determined by $a_{11}x_1 + a_{12}x_2 = b_1$ and $\dfrac{-a_{11}}{a_{12}}$ is the slope of the line determined by $a_{21}x_1 + a_{22}x_2 = b_2$.

If $a_{12} = 0$, then for $a_{11}x_1 + a_{12}x_2 = b_1$ to determine a line we must have $a_{11} \neq 0$. In this case,

$$D = 0 \iff a_{11}a_{22} = a_{11}a_{22} - a_{12}a_{21} = 0 \iff a_{22} = \frac{0}{a_{11}} = 0;$$

both lines are vertical. Similarly, if $a_{22} = 0$, then $D = 0 \iff a_{12} = 0 \iff$ both lines are vertical.

Therefore, the lines determined by the system are parallel (or are the same line) if and only if the determinant of the system is zero. ∎

Solutions A 5 *Complex numbers* (page A36)

1. $(2-3i)+(7-4i)=(2+7)+(-3-4)i=9-7i.$

3. $(1+i)(1-i)=1^2-i^2=1-(-1)=2.$

5. $(-3+2i)(7+3i)=((-3)(7)-(2)(3))+((-3)(3)+(2)(7))\,i=-27+5i.$

7. If $z=5+5i$, then $|z|=\sqrt{5^2+5^2}=5\sqrt{2}$ and $\arg(z)=\tan^{-1}(5/5)=\pi/4$. Therefore $5+5i=5\sqrt{2}e^{i\pi/4}$.

9. If $z=3-3i$, then $|z|=\sqrt{3^2+(-3)^2}=3\sqrt{2}$ and $\arg(z)=\tan^{-1}(-3/3)=-\pi/4$. Therefore $3-3i=3\sqrt{2}e^{-i\pi/4}$.

11. If $z=3\sqrt{3}+3i$, then $|z|=\sqrt{27+9}=6$ and $\arg(z)=\tan^{-1}\left(3/3\sqrt{3}\right)=\pi/6$. Therefore $3\sqrt{3}+3i=6e^{i\pi/6}$.

13. If $z=4\sqrt{3}-4i$, then $|z|=\sqrt{48+16}=8$ and $\arg(z)=\tan^{-1}\left(-4/4\sqrt{3}\right)=-\pi/6$. Therefore $4\sqrt{3}-4i=8e^{-i\pi/6}$.

15. If $z=-1-\sqrt{3}i$, then $|z|=\sqrt{1+3}=2$ and $\arg(z)=\tan^{-1}\left(-\sqrt{3}/(-1)\right)-\pi=\pi/3-\pi=-2\pi/3$. (Note that the graphical representation of $-1-\sqrt{3}i$ falls into the third quadrant.) Hence $-1-\sqrt{3}i=2e^{-2i\pi/3}$.

17. $2e^{-7\pi i}=2\left(\cos(-7\pi)+i\sin(-7\pi)\right)=2\left((-1)+i(0)\right)=-2.$

19. $(1/2)e^{(-3\pi/4)i}=(1/2)\left(\cos(-3\pi/4)+i\sin(-3\pi/4)\right)=(1/2)\left(-\sqrt{1/2}-i\sqrt{1/2}\right)=-(1+i)/\left(2\sqrt{2}\right).$

21. $4e^{(5\pi/6)i}=4\left(\cos(5\pi/6)+i\sin(5\pi/6)\right)=4\left(-\sqrt{3}/2+i(1/2)\right)=-2\sqrt{3}+2i.$

23. $3e^{(-2\pi/3)i}=3\left(\cos(-2\pi/3)+i\sin(-2\pi/3)\right)=3\left((-1/2)+i(-\sqrt{3}/2)\right)=-3/2-(3\sqrt{3}/2)i.$

25. $e^i=(1)(\cos 1+i\sin 1)=\cos 1+i\sin 1.$

27. $\overline{4+6i}=4-6i.$

29. $\overline{-7i}=-(-7)i=7i.$

31. $\overline{2e^{\pi i/7}}=2e^{-\pi i/7}$ (use formula 16).

33. $\overline{3e^{-4\pi i/11}}=3e^{4\pi i/11}.$

35. Suppose $z=\alpha+i\beta$. If z is real, then $\beta=0$ and $\overline{z}=\overline{\alpha+i(0)}=\alpha-i(0)=\alpha=z$. On the other hand, $z=\overline{z}\iff\alpha+i\beta=\alpha-i\beta\iff i\beta=-i\beta\iff\beta=-\beta\iff2\beta=0\iff\beta=0$. Since $\beta=0$ if and only if z is real, this means that $z=\overline{z}$ if and only if z is real. $\blacksquare$

37. Suppose $z=\alpha+i\beta$ where α and β are real. Then $z\overline{z}=(\alpha+i\beta)(\alpha-i\beta)=\alpha^2-(i\beta)^2=\alpha^2+\beta^2=|z|^2.$ $\blacksquare$

39. $|w|$ is the distance from w to the origin; this implies $|z-z_0|$ is the distance from z to z_0. If $a>0$, then $\{z:|z-z_0|=a\}$ is the circle in the complex plane centered at z_0 with radius a. If $a=0$, the set is $\{z_0\}$; while $a<0$ yields the empty set.

41. First we prove a three-part lemma (for complex w, z and integral n):

$$\text{(i)}\qquad \overline{w+z}=\overline{w}+\overline{z},$$
$$\text{(ii)}\qquad \overline{w\,z}=\overline{w}\,\overline{z},$$
$$\text{(iii)}\qquad \overline{z^n}=\overline{z}^n.$$

$\boxed{\text{Proof of lemma}}$ Suppose $w=a+ib$ and $z=\alpha+i\beta$.

i. $\overline{w+z}=\overline{(a+\alpha)+i(b+\beta)}=(a+\alpha)-i(b+\beta)=(a-ib)+(\alpha-i\beta)=\overline{w}+\overline{z}.$

ii. $\overline{w\,z}=\overline{(a\alpha-b\beta)+i(a\beta+b\alpha)}=(a\alpha-b\beta)-i(a\beta+b\alpha)=(a\alpha-(-b)(-\beta))+i\left(a(-\beta)+(-b)\alpha\right)=(a-ib)(\alpha-i\beta)=\overline{w}\,\overline{z}.$

iii. Part (ii) implies $\overline{z^2}=\overline{z\,z}=\overline{z}\,\overline{z}=\overline{z}^2$. Suppose $\overline{z^k}=\overline{z}^k$ for some integer $k\geq 2$, then $\overline{z^{k+1}}=\overline{z^k\,z}=\overline{z^k}\,\overline{z}=\overline{z}^k\,\overline{z}=\overline{z}^{k+1}$. Therefore, by Mathematical Induction, $\overline{z^n}=\overline{z}^n$ for every integer $n\geq 2$. $\blacksquare$

Suppose $a_0, a_1, \ldots, a_{n-1}, a_n$ are real numbers and $P(\lambda) = a_0 + a_1\lambda + a_2\lambda^2 + \cdots + a_{n-1}\lambda^{n-1} + a_n\lambda^n$

$$\overline{P(z)} = \overline{a_0 + a_1 z + a_2 z^2 + \cdots + a_{n-1} z^{n-1} + a_n z^n}$$
$$= \overline{a_0} + \overline{a_1 z} + \overline{a_2 z^2} + \cdots + \overline{a_{n-1} z^{n-1}} + \overline{a_n z^n}$$
$$= \overline{a_0} + \overline{a_1}\,\overline{z} + \overline{a_2}\,\overline{z}^2 + \cdots + \overline{a_{n-1}}\,\overline{z}^{n-1} + \overline{a_n}\,\overline{z}^n$$
$$= a_0 + a_1\overline{z} + a_2\overline{z}^2 + \cdots + a_{n-1}\overline{z}^{n-1} + a_n\overline{z}^n$$
$$= P\left(\overline{z}\right).$$

The hypothesis that all coefficients of P are real is used in the form (see Problem 15) that $\overline{a_k} = a_k$ for $0 \le k \le n$. If $P(w) = 0$, then $P\left(\overline{w}\right) = \overline{P(w)} = \overline{0} = 0$. Hence, the complex roots of a polynomial with real coefficients occur in conjugate pairs. ∎

43.   DeMoivre's formula says, for all integers n and all real θ, $(\cos\theta + i\sin\theta)^n = \cos(n\theta) + i\sin(n\theta)$.

i. $(\cos\theta + i\sin\theta)^0 = 1 = \cos 0 + i\sin 0$, the result is true if $n = 0$.

ii.   Suppose the formula is true for $n = k$. Then

$$(\cos\theta + i\sin\theta)^{k+1} = (\cos\theta + i\sin\theta)(\cos\theta + i\sin\theta)^k$$
$$= (\cos\theta + i\sin\theta)\left(\cos(k\theta) + i\sin(k\theta)\right)$$
$$= \left(\cos\theta\cos(k\theta) - \sin\theta\cos(k\theta)\right) + i\left(\sin\theta\cos(k\theta) + \cos\theta\sin(k\theta)\right)$$
$$= \cos(\theta + k\theta) + i\sin(\theta + k\theta)$$
$$= \cos\left((k+1)\theta\right) + \sin\left((k+1)\theta\right).$$

Hence, by Mathematical Induction, DeMoivre's formula is true for all integers $n \ge 0$.

iii.
$$(\cos\theta + i\sin\theta)^{-1} = \frac{1}{\cos\theta + i\sin\theta} = \left(\frac{1}{\cos\theta + i\sin\theta}\right)\left(\frac{\cos\theta - i\sin\theta}{\cos\theta - i\sin\theta}\right)$$
$$= \frac{\cos\theta - i\sin\theta}{\cos^2\theta + \sin^2\theta} = \cos\theta - i\sin\theta = \cos(-\theta) + i\sin(-\theta).$$

Therefore, if $n \ge 0$, then

$$(\cos\theta + i\sin\theta)^{-n} = \left((\cos\theta + i\sin\theta)^{-1}\right)^n = (\cos(-\theta) + i\sin(-\theta))^n$$
$$= \cos\left(n(-\theta)\right) + i\sin\left(n(-\theta)\right) = \cos\left((-n)\theta\right) + i\sin\left((-n)\theta\right).$$

Hence, using the result of part (ii), we find DeMoivre's formula is true for each negative integer. Putting parts (ii) and (iii) together, we conclude that DeMoivre's formula is true for all integers n. ∎

REMARK: In doing part (i) and the beginning of part (iii), you might want to note that $\cos\theta + i\sin\theta \ne 0$ for all choices of θ.

A 6
Graphing Using A Calculator

Solutions A 6   *Graphing using a calculator*   (pages A62-3)

1. $y = x^2 + 2$

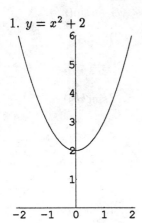

3. $y = x^3 - x$

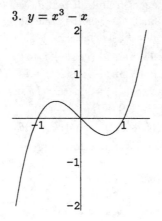

5. $y = -x^4 + 2x^2 - x + 3$

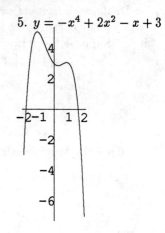

7. $y = x/(x+1)$

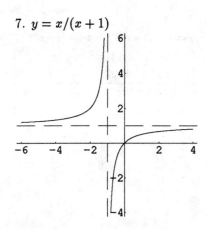

9. $y = (x+2)/(x-4)$

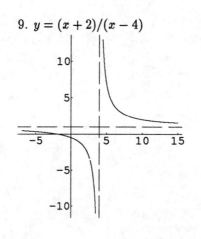

11. $y = (x-2)/(x^2-1)$

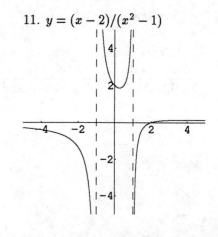

13. $y = \ln(x^2 - 4)$

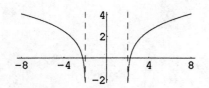

15. $y = e^{1/x}/(x+2)$

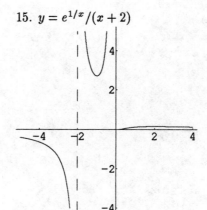

17. $y = e^{x-\ln x}$

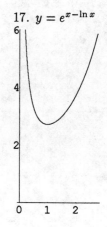

19. $y = 2\sin x - 5\cos 3x$

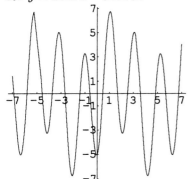

21. $y = x\cos x$

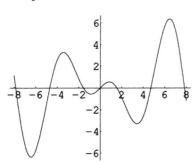

23. $y = e^{2x}\sin 3x$

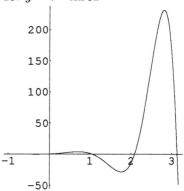

25. $y = 3\sec 5x$

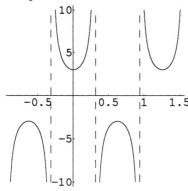

27. $y = \csc(x - 3)$

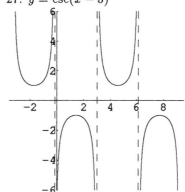

29. $y = x^2 - 3x - 10 = (2x+7)(3x-8)$

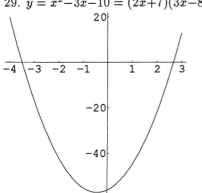

29. (neighborhood of $-7/2$)

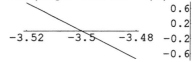

29. (neighborhood of $8/3$)

31. $y = x^3 - x^2 - 1$

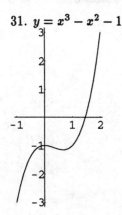

31. (neighborhood of 1.46557)

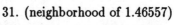
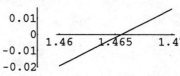

33. $y = x^7 - x^2 - 12$

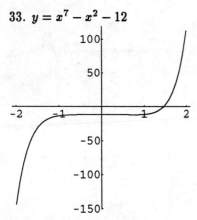

33. (neighborhood of 1.45986)

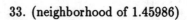
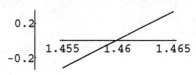

35. $y = x^3 - 6x^2 - 15x + 4$

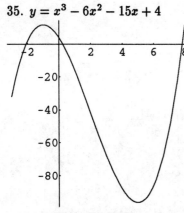

35. (near -2.09052)

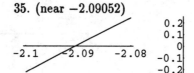

35. (near 7.84667)

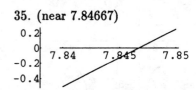

35. (near .243848)

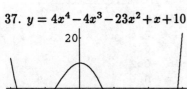

37. $y = 4x^4 - 4x^3 - 23x^2 + x + 10$

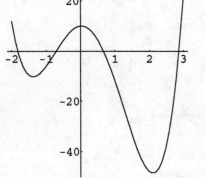

37. (near -1.820057)

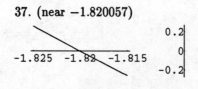

37. (near .668297)

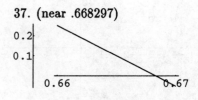

37. (near $-.716567$)

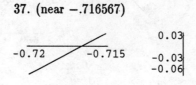

37. (near 2.868327)

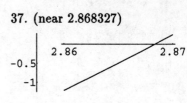

39. $y = e^{-x}$ & $y = \ln x$

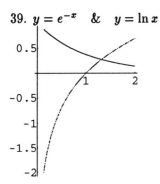

39. (neighborhood of 1.3098)

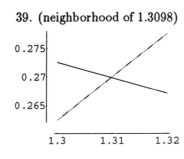

41. $y = x^2 - 3x + 5$ & $y = \ln(x^4 + x + 5)$

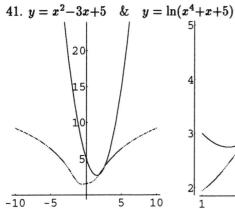

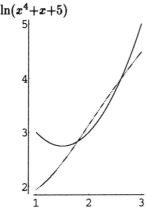

41. (near 1.7871)

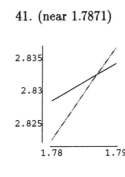

41. (near 2.6180)

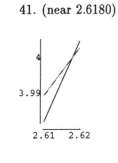

43. $y = 1 + 3\cos x$ & $y = \sqrt{x+5}$

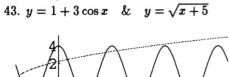

43. (near −4.59175)

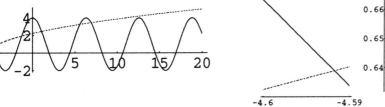

43. (near −1.25366)

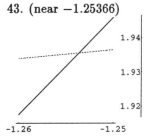

43. (near 1.06174)

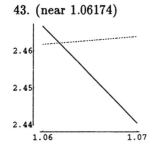

43. (near 5.56028)

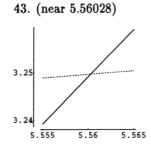

43. (near 6.89869)

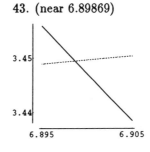

43. (no intersections)

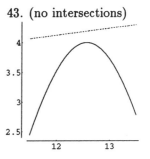

A 6 — Graphing using a calculator

45. $y = 4 - |3x - 5|$

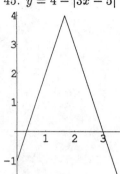

47. $y = (x^2 - 2) - (x + 5)$

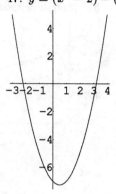

49. $y = 1/(2 - x) - x/(x^2 - 3)$

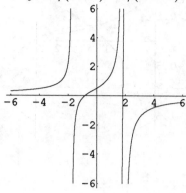

45. $|3x - 5| < 4 \iff 1/3 < x < 3.$

47. $x^2 - 2 > x + 5 \iff x < \left(1 - \sqrt{29}\right)/2 \approx -2.1926 \text{ or } x > \left(1 + \sqrt{29}\right)/2 \approx 3.1926.$

49. $x/(x^2 - 3) < 1/(2 - x) \iff x < -\sqrt{3} \approx 1.7321 \text{ or } -0.8229 \approx \left(1 - \sqrt{7}\right)/2 < x < \sqrt{3} \approx 1.7321 \text{ or } 1.8229 \approx \left(1 + \sqrt{7}\right)/2 < x < 2.$

51. $x^3 - 2x^2 + 7x - 5 > 2x^2 - x + 4 \iff 2.2225 < x.$

51. $y = (x^3 - 2x^2 + 7x - 5) - (2x^2 - x + 4)$

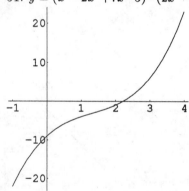

53. $x^2 + y^2 = 25$

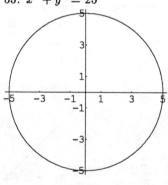

55. $x^2 + 2x + y^2 - 4y = 20$

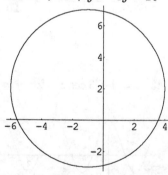

57. $7x^2 - 11y^2 = 47$

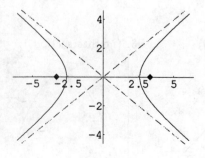

59. $x^2/37 + y^2/121 = 1$

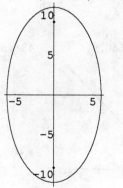

61. $3y^2 + 12x = 6$

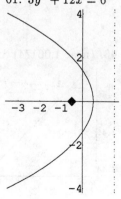

460

63. $2x^2 + 7x + 5y^2 - 2y = 8$

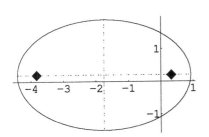

65. $2y^2 - 8y - 12x^2 + 12x = 16$

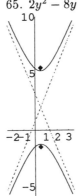

67. $\frac{x^2}{2} + 4x + \frac{y^2}{3} - 3y = 5$

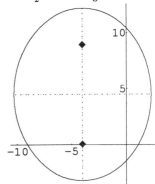

69. $r = 3\sin\theta$

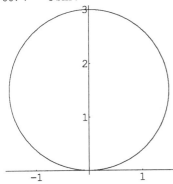

71. $r = 1 - 4\sin\theta$

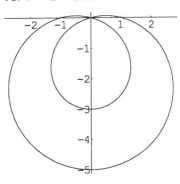

73. $r = 4 + 3\sin\theta$

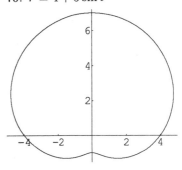

75. $r = 2\sin 3\theta$

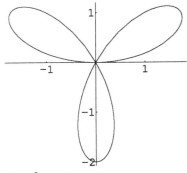

77. $r = \theta/3$

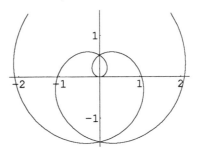

79. $r^2 = -9\sin 2\theta$

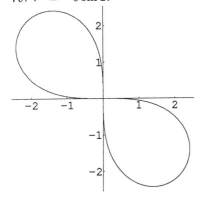

81. $r = 1 + 2\sec\theta$

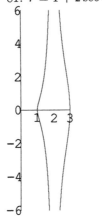